Plant Growth Regulators for Climate-Smart Agriculture

Footprints of Climate Variability on Plant Diversity

Series Editor

Shah Fahad

Climate Change and Plants
Biodiversity, Growth and Interactions
Shah Fahad, Osman Sönmez, Shah Saud, Depeng Wang, Chao Wu, Muhammad Adnan and Veysel Turan

Developing Climate Resilient Crops
Improving Global Food Security and Safety
Shah Fahad, Osman Sönmez, Shah Saud, Depeng Wang, Chao Wu, Muhammad Adnan and Veysel Turan

Sustainable Soil and Land Management and Climate Change
Shah Fahad, Osman Sönmez, Veysel Turan, Muhammad Adnan, Shah Saud, Chao Wu and Depeng Wang

Plant Growth Regulators for Climate-Smart Agriculture
Shah Fahad, Osman Sönmez, Veysel Turan, Muhammad Adnan, Shah Saud, Chao Wu and Depeng Wang

Plant Growth Regulators for Climate-Smart Agriculture

Edited By
Shah Fahad
Osman Sönmez
Shah Saud
Depeng Wang
Chao Wu
Muhammad Adnan
Veysel Turan

CRC Press
Taylor & Francis Group
Boca Raton London New York

CRC Press is an imprint of the
Taylor & Francis Group, an **informa** business

First edition published 2021
by CRC Press
6000 Broken Sound Parkway NW, Suite 300, Boca Raton, FL 33487-2742

and by CRC Press
2 Park Square, Milton Park, Abingdon, Oxon, OX14 4RN

CRC Press is an imprint of Taylor & Francis Group, LLC

Library of Congress Cataloging-in-Publication Data

Names: Fahad, Shah (Assistant professor in agriculture), editor. | Sönmez, Osman, editor. | Saud, Shah, editor. | Wang, Depeng (Professor in agriculture), editor. | Wu, Chao (Associate research fellow in agriculture), editor. | Adnan, Muhammad (Lecturer in agriculture), editor. | Turan, Veysel, editor.
Title: Plant growth regulators for climate-smart agriculture / edited by Shah Fahad, Osman Sönmez, Shah Saud, Depeng Wang, Chao Wu, Muhammad Adnan, Veysel Turan.
Other titles: Footprints of climate variability on plant diversity.
Description: First edition. | Boca Raton, FL : CRC Press, 2021. | Series: Footprints of climate variability on plant diversity | Includes bibliographical references and index.
Identifiers: LCCN 2021005843 | ISBN 9780367623432 (hardback) | ISBN 9780367623456 (paperback) | ISBN 9781003109013 (ebook)
Subjects: LCSH: Plant regulators. | Plants--Effect of temperature on. | Plant-atmosphere relationships.
Classification: LCC SB128 .P553 2021 | DDC 631.8/9--dc23
LC record available at https://lccn.loc.gov/2021005843

ISBN: 978-0-367-62343-2 (hbk)
ISBN: 978-0-367-62345-6 (pbk)
ISBN: 978-1-003-10901-3 (ebk)

Typeset in Times
by Deanta Global Publishing Services, Chennai, India

Contents

Acknowledgements

Words are bound and knowledge is limited to praise ALLAH, the Instant and Sustaining Source of all Mercy and Kindness, and the Sustainer of the Worlds. My greatest and ultimate gratitude is due to ALLAH (Subhanahu wa Taqadus). I thank ALLAH with all my humility, for everything that I can think of. His generous blessing and exaltation succeeded my thoughts and thrived my ambition to have the cherished fruit of my modest efforts in the form of this piece of literature from the blooming spring of blossoming knowledge. May ALLAH forgive my failings and weaknesses, strengthen and enliven my faith in HIM, and endow me with knowledge and wisdom. All praises and respects are for **Holy Prophet Muhammad** *Salle Allah Alleh Wassalam*, the greatest educator, the everlasting source of guidance and knowledge for humanity. He taught the principles of morality and eternal values and enabled us to recognise our Creator. I have a deep sense of obligation to my parents, my brothers, sisters, and son. Their unconditional love, care, and confidence in my abilities helped me achieve this milestone in my life. For this and much more, I am forever in their debt. It is to them that I dedicate this book. In this arduous time, I also appreciate the patience and serenity of my wife, who brought joy to my life in so many different ways. It is indeed on account of her affections and prayers that I was able to achieve something in my life.

Shah Fahad

Editors

Shah Fahad

Dr Shah Fahad is an Assistant Professor in the Department of Agronomy, University of Haripur, Khyber Pakhtunkhwa, Pakistan. He earned his PhD in Agronomy from Huazhong Agriculture University, China, in 2015. After doing his postdoctoral research in Agronomy at the Huazhong Agriculture University (2015–17), he accepted the position of Assistant Professor at the University of Haripur. He has published over 260 peer-reviewed papers (Impact factor 723.45), with more than 230 research and 30 review articles, on important aspects of climate change, plant physiology and breeding, plant nutrition, plant stress responses and tolerance mechanisms, and exogenous chemical priming induced abiotic stress tolerance. He has also contributed 50 book chapters to various book editions published by Springer, Wiley-Blackwell, and Elsevier. He has edited fifteen book volumes, including this one, published by CRC Press, Springer, and Intech Open. He won the Young Rice International Scientist award and Distinguish Scholar Award in 2014 and 2015, respectively. He won 13 projects from international and national donor agencies. Dr Shah Fahad's name has figured among the top two percent of scientists in a global list compiled by Stanford University, California. He has worked and is presently continuing on a wide range of topics, including climate change, greenhouse emission gasses, abiotic stresses tolerance, roles of phytohormones and their interactions in abiotic stress responses, heavy metals, and regulation of nutrient transport processes.

Osman Sönmez

Dr Osman Sönmez is a Professor in the Department of Soil Science, Faculty of Agriculture, Erciyes University, Kayseri, Turkey. He earned his MS and PhD in Agronomy from Kansas State University, Manhattan, Kansas in the years 1996–2004. In 2014, he accepted the position of Associate Professor at the University of Erciyes. Since 2014, he has worked in the Department of Soil Science, Faculty of Agriculture at Erciyes University. He has published over 90 peer-reviewed papers, research and review articles on soil pollution, plant physiology, and plant nutrition.

Veysel Turan

Dr Veysel Turan is an Assistant Professor in the Department of Soil Science and Plant Nutrition, Bingöl University, Turkey. He earned his PhD in Soil Science and Plant Nutrition from Atatürk University, Turkey, in 2016. Since completing his postdoctoral research in the Department of Microbiology, University of Innsbruck, Austria (2017–18), he has been working in Bingöl University. He has been and is continuing to work on a wide range of topics, such as soil-plant interaction, heavy metal accumulation, and bioremediation of soil by plant and soil amendment.

Muhammad Adnan

Dr Muhammad Adnan is a Lecturer in the Department of Agriculture at the University of Swabi (UOS), Pakistan. He earned his PhD (soil fertility and microbiology) from the Department of Soil and Environmental Sciences (SES), University of Agriculture Peshawar, Pakistan, and Department of Plant, Soil and Microbial Sciences, Michigan State University, Michigan. He earned his MSc and BSc (Hons) in Soil and Environmental Sciences, from Department of SES the University of Agriculture, Peshawar, Pakistan.

Shah Saud

Dr Shah Saud earned his PhD in turf grasses (Horticulture) from Northeast Agricultural University, Harbin, China. He is currently working as a Post Doctorate Researcher in the Department of Horticulture, Northeast Agricultural University. Dr Saud has published over 125 research publications in peer-reviewed journals. He has edited 3 books and written 25 book chapters on important aspects of plant physiology, plant stress responses, and environmental problems in relation to agricultural plants. According to Scopus®, Dr Shah Saud's publications have received roughly 2500 citations with an h-index of 24.

Chao Wu

Dr Chao Wu engages in the field crop cultivation and physiology, and plant phenomics. He earned his PhD during 2013–16 from Huazhong Agricultural University, Wuhan, China, and completed his post PhD during 2017–19 with Nanjing Agricultural University, Nanjing, China. He is currently an Associate Research Fellow in Guangxi Institute of Botany, Guangxi Zhuang Autonomous Region and the Chinese Academy of Sciences, Guilin, China. He chairs the Natural Science Foundation of Jiangsu Province and two Postdoctoral Science Foundation researches. Dr Wu's research mainly focuses on physiological mechanisms of abiotic stress tolerance (heat, drought) in crops and medicinal plants.

Depeng Wang

Dr Depeng Wang earned his PhD in 2016 in the field of Agronomy and Crop Physiology from Huazhong Agriculture University, Wuhan, China. Presently he is serving as a Professor in the College of Life Science, Linyi University, Linyi, China. He is the principal investigator of Crop Genetic Improvement at the Physiology & Ecology Center in Linyi University. His current research focus is on crop ecology, physiology, and agronomy. His main areas of effort are associated with high-yielding crops, the effects of temperature on crop grain yields, solar radiation utilisation, morphological plasticity to agronomic manipulation in leaf dispersion and orientation, and optimal integrated crop management practices for maximising crop grain yields. Dr Depeng Wang has published over 36 papers in reputed journals.

List of Contributors

Asad Abbas
College of Horticulture
Anhui Agricultural University
Hefei, China

Awatif M. Abdulmajeed
Biology Department
University of Tabuk
Umluj, Saudi Arabia

Abdelghafar M. Abu-Elsaoud
Botany Department
Suez Canal University
Ismailia, Egypt

Manzoor Ahmad
Department of Agriculture
Bacha khan University
Charsadda, Pakistan

Nisar Ahmad
Center for Biotechnology and Microbiology
University of Swat
Mingora, Pakistan

Fazal Akbar
Center for Biotechnology and Microbiology
University of Swat
Mingora, Pakistan

Mohammad Akmal
Department of Agronomy
The University of Agriculture Peshawar
Peshawar, Pakistan

Madeeha Alamzeb
Department of Agronomy
The University of Agriculture Peshawar
Peshawar, Pakistan

Amjed Ali
College of Agriculture
University of Sargodha
Sargodha, Pakistan

Mohammad Ali
Center for Biotechnology and Microbiology
University of Swat
Mingora, Pakistan

Hafiza Iqra Almas
Department of Botany
University of Agriculture
Faisalabad, Pakistan

Shazma Anwar
Department of Agronomy
The University of Agriculture
Peshawar, Pakistan

Muhammad Arif
Department of Agronomy
The University of Agriculture
Peshawar, Pakistan

Umair Ashraf
Department of Botany
University of Education
Lahore, Pakistan

Habib-ur-Rehman Athar
Institute of Pure and Applied Biology
Bahauddin Zakariya University
Multan, Pakistan

Mohammad Safdar Baloch
Department of Agronomy
Gomal University
Dera Ismail Khan, Pakistan

Pradipta Banerjee
Department of Biochemistry and Plant Physiology
Centurion University of Technology and Management
Odisha, India

Preetha Bhadra
Department of Biotechnology
Centurion University of Technology and Management
Odisha, India

Urjashi Bhattacharya
Bidhan Chandra Krishi Viswavidyalaya
Nadia, India

Muhammad Adnan Bukhari
Department of Agronomy
University College of Agriculture and Environmental Sciences
Bahawalpur, Pakistan

Sherien Bukhat
Institute of Molecular Biology and Biotechnology
Bahauddin Zakariya University
Multan, Pakistan

Qiang Dong
State Key Laboratory of Cotton Biology
Cotton Research Institute of CAAS
Anyang, China

Xiongming Du
State Key Laboratory of Cotton Biology
Cotton Research Institute of CAAS
Anyang, China

Debjani Dutta
Department of Plant Physiology
Bidhan Chandra Krishi Viswavidyalaya
Nadia, India

Shah Fahad
Department of Agronomy
University of Haripur
Khyber Pakhtunkhwa, Pakistan

Sourav Garai
Department of Agronomy
Bidhan Chandra Krishi Viswavidyalaya
Nadia, India

Huiping Gui
State Key Laboratory of Cotton Biology
Cotton Research Institute of CAAS
Anyang, China

Shoupu He
State Key Laboratory of Cotton Biology
Zhengzhou University
Zhengzhou, China

Akbar Hossain
Bangladesh Wheat and Maize Research Institute
Dinajpur, Bangladesh

Muhammad Ikram
Statistical Genome Laboratory
Huazhong Agricultural University
Wuhan, China

Inamullah
Department of Agronomy
The University of Agriculture
Peshawar, Pakistan

Inam-u-llah
Department of Food Science & Technology
The University of Haripur
Haripur, Pakistan

Asif Iqbal
State Key Laboratory of Cotton Biology
Cotton Research Institute of CAAS
Anyang, China

Javaid Iqbal
Department of Agronomy
Ghazi University
Dera Ghazi Khan, Pakistan

Mazhar Iqbal
Department of Botany
Shaheed Benazir Bhutto University Sheringal Dir (U)
Sheringal, Pakistan

Muhammad Israr
Department of Forensic Sciences
University of Swat
Mingora, Pakistan

Muhammad Jahangir
Department of Food Science & Technology
University of Haripur
Haripur, Pakistan

Yinhua Jia
State Key Laboratory of Cotton Biology
Anyang, China

Snehashis Karmakar
Department of Plant Physiology
Bidhan Chandra Krishi Viswavidyalaya
Nadia, India

Nosheen Khalid
Department of Food Science & Technology
University of Haripur
Haripur, Pakistan

Asad Ali Khan
Department of Agronomy
The University of Agriculture
Peshawar, Pakistan

Imran Khan
Department of Food Science & Technology
University of Haripur
Haripur, Pakistan

Hongge Li
State Key Laboratory of Cotton Biology
Cotton Research Institute of CAAS
Anyang, China

Muhammad Liaquat
Department of Food Science & Technology
University of Haripur
Haripur, Pakistan

Sammina Mahmood
Department of Botany
University of Education
Lahore, Pakistan

Sagar Maitra
Department of Agronomy
Centurion University of Technology and Management
Odisha, India

Hamid Manzoor
Institute of Molecular Biology and Biotechnology
Bahauddin Zakariya University
Multan, Pakistan

Sajid Mehmood
Guangdong Provincial Key Laboratory for Radionuclides Pollution Control and Resources
SGuangzhou University
Guanzhou, China

Udit Nandan Mishra
Department of Biochemistry and Plant Physiology
Centurion University of Technology and Management
Odisha, India

Mousumi Mondal
Research Scholar, Department of Agronomy
Bidhan Chandra Krishi Viswavidyalaya
Nadia, India

Jagamohan Nayak
Research Scholar, Department of Agronomy
Bidhan Chandra Krishi Viswavidyalaya
Nadia, India

Zhaoe Pan
State Key Laboratory of Cotton Biology
Cotton Research Institute of CAAS
Anyang, China

Zhen Peng
State Key Laboratory of Cotton Biology
Cotton Research Institute of CAAS
Anyang, China

Subhashisa Praharaj
Department of Agronomy
Centurion University of Technology and Management
Odisha, India

Abdul Qayyum
Department of Plant Breeding and Genetics
Bahauddin Zakariya University
Multan, Pakistan

Guangyong Qin
State Key Laboratory of Cotton Biology
Cotton Research Institute of CAAS
Anyang, China

Atta Ur Rahman
Center for Biotechnology and Microbiology
University of Swat
Mingora, Pakistan

Akhtar Rasool
Centre for Animal Sciences and Fisheries
University of Swat
Mingora, Pakistan

Sumaira Rasul
Institute of Molecular Biology and Biotechnology
Bahauddin Zakariya University
Multan, Pakistan

Abdul Rehman
State Key Laboratory of Cotton Biology
Zhengzhou University
Zhengzhou, China

and

Department of Plant Breeding and Genetics
College of Agriculture Bahauddin Zakariya
Multan, Pakistan

Zia ur Rehman
Institute of Soil and Environmental Sciences
University of Agriculture
Faisalabad, Pakistan

Muhammad Rizwan
Center for Biotechnology and Microbiology
University of Swat
Mingora, Pakistan

Anirban Roy
Department of Genetics and Plant breeding
Bidhan Chandra Krishi Viswavidyalaya
Nadia, India

Fozia Saeed
Institute of Molecular Biology and Biotechnology
Bahauddin Zakariya University
Multan, Pakistan

Muhammad Saeed
Department of Agronomy
The University of Agriculture
Peshawar, Pakistan

Chandrasekhar Sahu
Department of Biochemistry and Plant Physiology
Centurion University of Technology and Management
Odisha, India

Shehla Sammi
Department of Food Science & Technology
University of Haripur
Haripur, Pakistan

Muzafar Shah
Centre for Animal Sciences and Fisheries
University of Swat
Mingora, Pakistan

Shaheen Shah
Department of Agronomy
University of Agriculture Peshawar
Peshawar, Pakistan

Tariq Shah
Department of Agronomy
University of Agriculture Peshawar
Peshawar, Pakistan

Adnan Noor Shah
School of Agronomy
Anhui Agricultural University
Hefei, China

Tanmoy Shankar
Department of Agronomy
Centurion University of Technology and Management
Odisha, India

Mona H. Soliman
Botany and Microbiology Department
Cairo University
Giza, Egypt

Meizhen Song
State Key Laboratory of Cotton Biology
Cotton Research Institute of CAAS
Anyang, China

Muhammad Suleman
Center for Biotechnology and Microbiology
University of Swat
Mingora, Pakistan

Xiangru Wang
State Key Laboratory of Cotton Biology
Cotton Research Institute of CAAS
Anyang, China

Du Xiongming
State Key Laboratory of Cotton Biology
Cotton Research Institute of CAAS
Anyang, China

Hengheng Zhang
State Key Laboratory of Cotton Biology
Cotton Research Institute of CAAS
Anyang, China

Xiling Zhang
State Key Laboratory of Cotton Biology
Cotton Research Institute of CAAS
Anyang, China

1

Role of Gibberellins in Response to Stress Adaptation in Plants

Mousumi Mondal, Sourav Garai, Jagamohan Nayak, Anirban Roy, Debjani Dutta, Snehashis Karmakar, Shah Fahad, and Akbar Hossain

CONTENTS

1.1 Introduction

Several hormones, also known as plant growth regulators (PGRs) control plant growth and development and influence physiological and biochemical pathways in response to environmental stimuli for plant survival in critical situations (Davies 2010). These hormones play crucial roles when plants are exposed to abiotic stresses, such as drought, flood, or temperature stress and enable plants to check their growth, thereby conserving resources in order to survive (Bailey-Serres and Voesenek 2010). Such stresses significantly influence the biosynthesis, transportation, and signal transduction of specific stress hormones which may facilitate a protective environment. For example, stress hormones like abscisic acid (ABA) influence leaf stomata closing during dry conditions to protect the plant cell from dehydration

(Wilkinson and Davies 2002). This stress hormone acts as a signalling agent to communicate the stress between roots and leaves; however, the pace of response depends upon the inter-organ stimulus which leads to hormone biosynthesis in plant leaves (Christmann et al. 2007). Gibberellins (GAs) are well-studied PGRs that play a significant role in seed germination, vegetative and reproductive growth, and fruit and seed development. Recently, they have become a crucial factor in abiotic stress tolerance (Sun 2010). Research on GAs was first conducted in Japan in the 19th century and motivated by rice disease caused by the fungus *Gibberella fujikuroi* (Hedden and Sponsel 2015). Later, the growth stimulatory ability of *G. fujikuroi* was identified, and it was concluded that this effect on plants was mediated by a toxin secreted by the fungal strain – a mixture of GAs A and GAs B (Kurosawa 1926). More recent studies have uncovered biosynthesis of GAs in plant and fungus at the molecular level in terms of pathways, enzymes, and gene regulation (Hedden and Sponsel 2015; Salazar-Cerezo et al. 2018). GAs are diterpenoid phytohormones, synthesized by plants with the help of monooxygenase, dioxygenase, and cyclase enzymes. Application of chemical growth retardants is a common agronomic practice during the early occurrence of stress to check the plant stature as a stress tolerance strategy. It has been well reported that the primary mode of action of growth retardants is to inhibit the GA biosynthesis and signalling that helps to protect the plant from exposure to stresses (Colebrook et al. 2014). This chapter includes a comprehensive discussion of GA biosynthesis, its metabolism, and signalling cascades in response to various abiotic stresses.

1.2 Gibberellins Biosynthesis

Gibberellin biosynthesis and pathway details have been studied using gas chromatography and mass spectrometry for identification of chemical nature, gene identification, and assigning them to the pathway of the entire component. GA biosynthesis comprises three sub-parts involving ent-kaurene production, conversion, and production of GA_{20} and GA_{19} (Hedden and Phillips 2000). GAs have resulted in a product of diterpenoid pathway and C20 precursor compound, where cyclisation initiates the process of GA biosynthesis (Hedden and Proebsting 1999). The basic component that starts GA biosynthesis involves a preliminary component of 5-Carbon compound, i.e., isopentenyl pyrophosphate (IPP), which is also a component of terpenoid compounds (Sponsel and Hedden 2010). Plants produce IPP following two approaches, one of which involves mevalonic acid and another is methyl erythritol in cytoplasm and plastids, respectively. In the initial step, ent-kaurene is produced using a soluble enzyme in proplastid. In general, the GA precursor, which is produced from ent-kaurene and GA_{12} aldehyde, is catalysed at endoplasmic reticulum by Cytochrome P-450 monooxygenase. In the final stage, 2-oxoglutarate-dependent dioxygenases are the catalysing agents (Sun 2008). The first stage of GA synthesis follows an intermediate cyclisation process starting from GGDP via ent-copalyl diphosphate (CPP). Ent-kaurenoic acid is produced by stepwise oxidation specifically of ent-kaurene involving C19-based oxidation of ent-kaurenol and ent-kaurenal. During further oxidation, ent-kaurenoic acid is converted to ent-7α-hydroxykaurenoic acid followed by another oxidation event. In the G_{12} branch position of the synthesis chain, a C13-hydroxylation event converts GA_{12} to GA_{53}. Both of these GAs, i.e., GA_{12} and GA_{53}, initiating the formation of 13-hydroxylation Gas, while a parallel for non-13 hydroxylation pathway from GA_{12} results in GA_4 formation, respectively. GA_9 and GA_{20} are produced following another oxidation event at C-20 position (Hedden and Thomas). GA_9 and GA_{20} are converted to GA_4 and GA_1, respectively, involving 3-β hydroxylation, which are the ultimate steps in bioactive GA formation (Bomke and Tudzynski 2009).

1.2.1 Candidate Genes in Gibberellins Biosynthesis in Plants

Various molecular genetic studies and mutant characterisations have revealed various genes involved in the overall pathway for GA synthesis. The Arabidopsis Genome Initiative, which yielded a wealth of information during 2000, revealed the entire slate of probable candidate genes of gibberellin biosynthesis, paving the way for characterisation of all these genes (Bömke and Tudzynski 2009). Further gibberellin 13-hydroxylase cloned from rice helped to identify all respective genetic loci-contributing enzymes involved in gibberellin biosynthesis. Various GA-deficient mutants have been developed *viz.*, ga1, ga2, ga3, ga4, and ga5 (Koornneef and van der Veen 1980). A further detailed gene characterisation study of *Arabidopsis thaliana* shows that *GA1* encodes ent-copalyl-diphosphate synthase (Sun and Kamiya 1994), *GA2* encodes ent-kaurene synthase (Yamaguchi et al. 1998), *GA3* encodes ent-kaurene oxidase (Helliwell et al. 1998), *GA4* encodes *GA3* oxidase (Chiang et al. 1995), *GA5* encodes *GA20* oxidase (Afzal et al. 2017; Phillips et al. 1995). Similarly, two genes which are equivalent to *GA1* and *GA2* in Arabidopsis, i.e., OsCPS1 and OsKS1, were also identified in rice (Prisic et al. 2004). A total of seven enzymes are required for GA synthesis starting with GGDP. Enzymes responsible in the early steps of biosynthesis are monogenic, and, at a latter stage, gene families are responsible for enzyme synthesis (Hedden 2003). In pumpkin, one gene for KS and double copies for the same gene in *Stevia rebaudiana* have been reported (Sponsel and Hedden 2010) and describes stringent regulation of GA biosynthesis. GA_{12} is the precursor of all the GAs and conversion from this compound follows two paths, one of which involves early-13 hydroxylation which is encoded by CYP714B1 and CYP714B2 genetic loci in *Oryza sp.* (Magome et al. 2013). Genes responsible at latter stage, *viz.* of the C19-GA2ox and C20-GA2ox gene family, have been well characterised in Arabidopsis and rice. A total of five C19-GA2ox, two C20-GA2ox, seven C19-GA2ox, and three C20-GA2ox have been identified in *Arabidopsis* (Rieu et al. 2008) and rice, respectively. Homologues of these genes are also identified in *Zea mays*, *Pinus* sp, and *Phaseolus* sp. (Salazar-Cerezo et al. 2018).

1.3 Roles of Gibberellins in Stress Responses

Gibberellins (GAs) are growth-promoting phytohormones of which GA_1 and GA_4 are the predominant bioactive forms (Sponsel and Hedden 2004) of more than 130 types. These perform essential roles in several developmental processes of plants, including germination of seeds, elongation of the stem, expansion of leaves, development of trichomes, maturation of pollen, and the initiation of flowering (Achard and Genschik 2009). The enzymes, *viz.* monooxygenases, dioxygenases, and cyclases, act as catalysts for the synthesis of GAs in plants. The degradation of DELLA proteins improves the impacts of GAs on plant growth and development (Griffiths et al. 2006) which enable the modification of plant response to stress through the cumulative response of phytohormones to stress (Miransari 2012).

As natural growth hormones, the GAs are important targets for stress-induced growth modulation, and there is increasing evidence for the involvement of GA signalling in either growth suppression or promotion, depending on the response to specific abiotic stress (Colebrook et al. 2014) or biotic stress. The phytohormone GA is found to be involved with the adaptive response to varied abiotic stresses like cold, salinity, heat, flooding, and drought (Ahmad et al. 2017; Achard et al. 2008; Colebrook et al. 2014; Khan et al. 2015).

1.3.1 Gibberellins in Stress Response to Abiotic Stress

All types of abiotic stresses are lethal for crop production and effect the performance of crop production (Adnan et al. 2018, 2019, 2020; Ahmad et al. 2019; Akbar et al. 2020; Akram et al. 2018a, b; Amanullah et al. 2020; Amir et al. 2020; Amjad et al. 2020; Arif et al. 2020; Ayman et al. 2020; Aziz et al. 2017a, b; Baseer et al. 2019; Bayram et al. 2020; Depeng et al. 2018; Fahad and Bano 2012; Fahad et al. 2013, 2014a, b, 2015a, b, 2016a, b, c, d, 2017, 2018, 2019a, b; Farah et al. 2020; Farhana 2020; Fazli et al. 2020; Frahat et al. 2020; Gopakumar et al. 2020; Habib et al. 2017; Hafiz et al. 2016, 2018, 2019, 2020a, b; Tariq et al. 2018; Hesham and Fahad 2020; Hussain et al. 2020; Ibrar et al. 2020; Ilyas et al. 2020; Iqra et al. 2020; Jan et al. 2020; Kamaran et al. 2017; Mahar et al. 2020; Md Jakirand Allah 2020; Md. Enamul et al. 2020; Mohammad I. Al-Wabel et al. 2020a, b; Mubeen et al. 2020; Muhammad Tahir et al. 2020; Muhmmad et al. 2019; Noor et al. 2020; Qamar et al. 2017; Rashid et al. 2020; Rehman 2020; Sadam et al. 2020; Sajid et al. 2019, 2020; Saleem et al. 2020a, b, c; Saman et al. 2020; Saud et al. 2013, 2014, 2016, 2017, 2020; Senol 2020; Shafi et al. 2020; Shah et al. 2013; Subhan et al. 2020; Unsar et al. 2020; Wahid et al. 2020; Wajid et al. 2017; Wu et al. 2019, 2020; Yang et al. 2017; Zafar-ul-Hye et al. 2020a, b;

Zahida et al. 2017; Zia-ur-Rehman 2020). The role of the GA class of hormones has become increasingly obvious in response to several abiotic stresses. Upon exposure to stresses including cold, salt, osmotic stress, etc., restriction of plant growth has been found to be the result of the reduction of GA levels and signalling (Colebrook et al. 2014). The discovery of DELLA proteins that lead to a reduction in growth in response to abiotic stresses has proved to be a significant achievement in understanding the role of GA in plant growth regulation under stress (Achard et al. 2006, 2008; Magome et al. 2008).

1.3.1.1 Response to Temperature Stress

A. thaliana seedlings trigger a reduction in bioactive GA, stimulate the accumulation of DELLA proteins, and ensure DELLA-mediated growth restriction when exposed to cold stress (Achard et al. 2008a). The DELLA accumulation contributes to stress tolerance, whereas DELLA mutants show reduced survival under freezing temperature. In both cases, the observed reduction of bioactive GA is the result of overexpression of specific *GA2ox* genes by dehydration-responsive element-binding protein (DREB1)/C-repeat binding factor (CBF) (DREB1/CBF) family transcription factors (Achard et al. 2008a; Magome et al. 2008).

1.3.1.2 Response to Salt Stress

Salt stress is one of the major environmental stresses that limit plant growth and productivity. Plant adaptation to salt stress involves the adjustment of GA levels at different growth stages. GA binds to the pocket receptor GIBBERELLIN INSENSITIVE DWARF1 (GID1), initiating a conformational change in GID1 and recruiting the growth repressor DELLA proteins for the formation of a GA-GID1-DELLA complex (Yu et al. 2020). The interaction of E3 ubiquitin ligase F-box protein SLEEPY1 (SLY1) with DELLA leads to degradation of DELLAs by the 26S proteasome (Fazlullah et al. 2018; Bao et al. 2020). In some plants, the survival mechanism under salt stress is growth inhibition by DELLA protein SLR1 that inhibits GA signalling (Achard et al. 2006). Several genes associated with GA metabolism, like AtGA2ox7 (Magome et al. 2008), OsGA2ox5 (Shan et al. 2014), and OsMYB91 (Zhu et al. 2015) are also reported to improve tolerance of plants to salt stress through retarding growth.

1.3.1.3 Response to Submergence

Rapid internode elongation triggered by submergence is an escape strategy by the cultivars adapted to lowland areas where long-lived floods are common. This phenomenon helps the shoot to outgrow the flood-waters. The increased expression of the ethylene response factor (ERF) domain proteins *SNORKEL1* and *SNORKEL2* because ethylene accumulation triggers internode elongation (Hattori et al. 2009) directly or indirectly increases the levels of bioactive GA. However, rice varieties adapted to short-lived, deep floods adopt the quiescence strategy controlled by the *Sub1* locus (Xu et al. 2006). Upon submergence, in *Sub1A* gene-carrying rice plants, such a mechanism does not get activated (Fukao and Bailey-Serres 2008a). Instead, restriction in shoot elongation and conservation of carbohydrates to be utilized in re-growth after the flood are observed (Fukao et al. 2006; Xu et al. 2006). This limitation of shoot-stretching is associated with expanded degrees of the rice DELLA protein Slender Rice-1 (SLR1) and the negative controller of GA flagging SLR1 LIKE-1 (SLRL1) both of which decline in light of submergence in cultivars without *Sub1A* (Bailey-Serres and Voesenek 2010; Fukao and Bailey-Serres 2008b). The presence of *Sub1A* is moreover identified with a sensational expansion in resistance to submergence, with both leaf feasibility and recovery of leaf production substantially improved in *Sub1A* lines (Fukao et al. 2006).

1.3.1.4 Response to Shade

To avoid shade, plants have an escape strategy – that is, increasing growth during which plants alter their morphology due to the presence of close neighbours and changes in the light spectrum and intensity (Smith 1982). The responses include increased growth of the hypocotyl and stem, also leaf hyponasty and petiole extension involving the action of multiple hormones, including auxin, ethylene, and brassinosteroids along with GAs (Keuskamp et al. 2010b; Stamm and Kumar 2010).

1.3.1.5 Response to Mild Osmotic Stress

A 50% reduction in final leaf size is noticed in *Arabidopsis thaliana* seedlings when exposed to a low concentration of the solute mannitol due to influences on both cell proliferation and cell expansion (Skirycz et al. 2010, 2011). The sampling of leaves at points containing proliferating, expanding, or mature cells were possible by a particular analysis of cellular growth dynamics (Skirycz et al. 2010). This apparent developmental separation of hormone-mediated responses to osmotic stress suggests roles of ethylene and GAs in the regulation of cell proliferation and expansion and the role of ABA in the regulation of responses in mature tissues (Skirycz et al. 2010). The results indicate that regulation of the cell cycle and endoreduplication in proliferating *A. thaliana* exposed to mild osmotic stress are the impacts of ethylene and GA (Claeys et al. 2012; Skirycz et al. 2011). The action of the ERF transcription factor ERF6 is likely to be linked with the ethylene and GA-mediated responses (Dubois et al. 2013).

1.3.1.6 Response to Soil Drying

Soil drying exposes plants to varied abiotic stresses, the relative effects of which depend upon the characteristics of the soil and the degree of dryness (Chapman et al. 2011; Mittler 2006). The role of GAs during drought stress adaptation is still unexplored. During drought conditions, a decrease in levels of GAs has been reported in maize (Zea mays L.), wheat (*Triticum aestivum* L.) and ramie (*Boehmeria nivea* (L.) Gaud) (Wang et al. 2008; Coelho Filho et al. 2013). Moreover, a GA application could recover plant growth under stress conditions as it promotes greater growth and maintenance of photosynthesis as well as oxidative stress reduction (Kaya et al. 2006; Akter et al. 2009).

1.3.2 Gibberellins in Stress Response to Biotic Stress

The role of GAs and their signalling components in plant responses to pathogen attack is a recent field of study. Zhu et al. (2005) reported decreased GA levels in rice plants infected with rice dwarf virus because of interaction between the outer capsid protein P2 of the virus and plant ent-kaurene oxidases. High levels of SA-dependent resistance have been noticed in *Arabidopsis* mutants lacking four of the five DELLAs when infected with the hemibiotrophic pathogen *P. syringae* (Navarro et al. 2008). However, quadruple DELLA mutants demonstrated the reduced influence of the JA-reporter gene *PDF1.2*, which was associated with upgraded vulnerability against the necrotrophic fungal pathogen *Alternaria brassicola* (Navarro et al. 2008).

In rice, strikingly different outcomes have been obtained where exogenously regulated GA was found to increase susceptibility to the hemibiotrophic rice microbes *Magnaporthe oryzae* (*Mo*) and *Xanthomonas oryzae* pv. *oryzae* (*Xoo*) (Yang et al. 2008; Qin et al. 2013). However, GA can likewise act emphatically on rice immunity as seen for the necrotrophic rice root rot pathogen *P. graminicola* (De Vleesschauwer et al. 2012). Consequently, conversely with the circumstance in *Arabidopsis*, barley, and wheat, rice GA flagging seems to elevate protection from necrotrophs and proneness to (hemi) biotrophs.

It is clear from these investigations that GAs assume equivocal parts in the plant intrinsic resistant flagging organization. Now and again, they add to the advancement of infection indications making plants powerless; though, in different cases, they are needed for defense reactions and induction of plant opposition (De Bruyne et al. 2014).

1.4 Regulation of Gibberellins in Response to Stress Protection

A central role for GAs in response to various abiotic stresses is becoming increasingly evident. Reduced GA levels and signalling have been reported to have an important role in growth restriction of plants that are exposed to various abiotic stresses, such as cold stress, salinity stress, and osmotic stress (Khan et al. 2017; Colebrook et al. 2014; Iqbal et al. 2011). On the other hand, increased gibberellin biosynthesis and its signalling promote escape responses to submergence and shading of plants. In several cases, GA signalling has also been linked to abiotic stress tolerance. Here, we will discuss the pieces of evidence for the regulation of the gibberellin signalling pathway on the exposure of plants to various abiotic stresses and how reduced GA signalling results in enhancement of various stress tolerance (Colebrook et al. 2014).

1.4.1 Gibberellin Biosynthesis and Signal Transduction

GAs comprise a large group of tetracyclic diterpenoid carboxylic acids. Very few components of this large group function as growth hormones in higher plants. The most predominant bioactive forms of GAs are GA_1 and GA_4 (Sponsel and Hedden 2004). The GA signalling pathway consists of the biosynthesis of these active hormones, perception, signal transduction, and deactivation; each of which is regulated by environmental signals, including various abiotic stresses (Colebrook et al. 2014). GAs are biosynthesized from trans-geranylgeranyl diphosphate (Kasahara et al. 2002). They are formed via methylerythritol phosphate pathway in plastids through the action of two terpene cyclases, followed by oxidation by cytochrome P450 monooxygenases and thereafter sequentially by three 2-oxoglutarate-dependent dioxygenases (Hedden and Thomas 2012). The dioxygenases consist of small families of isozymes – GA_{20}-oxidase (GA_{20}ox) and GA_3-oxidase (GA_3ox). The third class of dioxygenases, GA_2-oxidases, produces inactive products and its function is to start GA turnover. Most studies and reports point to the genes which encode the dioxygenases as the key sites of regulation of gibberellin biosynthetic pathway by environmental signals, the GA_2-oxidase genes being especially responsive to various abiotic stresses. Expression of the genes encoding GA_{20}-oxidase, GA_3-oxidase, and GA_2-oxidase is regulated by gibberellin action. Expression of the biosynthetic genes (i.e., *GA_{20}-oxidase* and *GA_3-oxidase*) has been down-regulated, while *GA_2-oxidases* gene expression has been up-regulated and provides a mechanism for gibberellin homeostasis (O'Neill et al. 2010). Sometimes GA action is degraded by a group of transcriptional regulator proteins known as DELLA proteins. DELLA proteins have a conserved domain within the N-terminus that is unique and it is important for gibberellin-induced degradation. DELLA proteins perform as growth repressors, and they may activate or suppress some gene expressions. Some evidence shows that various DELLA-interacting proteins are the components of other plant hormone signalling pathways, providing a mechanism for GA signalling to interact with all of these pathways (Gallego-Bartolomé et al. 2012).

1.4.2 Regulation of Gibberellin Metabolism and Signalling Cascades in Response to Abiotic Stresses

1.4.2.1 Interaction between Signalling Pathways of Gibberellin and Other Plant Hormones

The main role of GA signalling in response to various abiotic stresses is to gather information from the signalling pathways of other plant hormones. The two classic stress hormones, i.e., ethylene and ABA, are closely integrated with GA signalling in many systems. In the case of ABA-treated *Arabidopsis thaliana*, root growth inhibition in seedlings was connected with DELLA protein accumulation (Achard et al. 2006). This accumulation of DELLA proteins could not be reproduced in a high amount of GA biosynthesis mutant ga1-3. To reduce DELLA protein levels, treatment with a lower gibberellin concentration suggested that DELLA protein accumulation was related to a reduced level of bioactive gibberellin instead of a direct influence on DELLA protein stability (Zentella et al. 2007). DELLA protein accumulation in ABA-treated plants was not noticed in the mutant abi1-1, an ABA receptor-insensitive mutant (Achard et al. 2006). These results propose that ABA-mediated growth restriction is partially DELLA-dependent, and it requires ABI1 signalling. In the case of submerged rice, the build-up of ethylene in the flooded tissues is considered the primary signal triggering the gibberellin-mediated responses to growth (Jackson 2008). The ethylene accumulation also promotes ABA catabolism. This occurs solely of Sub1A, but on submergence, strong upregulation of Sub1A may be required (Fukao and Bailey-Serres 2008b). In addition to promoting tolerance in case of submergence, Sub1A also encourages tolerance of dehydration and drought. In case of submerged dehydration and after withholding water, Sub1A line plants maintained a higher relative leaf water content, showed reduced reactive oxygen species accumulation, and recovered their growth easily on re-watering (Fukao et al. 2011). This enhanced stress-tolerant nature was associated with enhanced sensitivity to ABA and magnified expression of both ABA-independent and ABA-dependent drought-responsive transcripts (Jung et al. 2010).

1.4.2.2 Regulation of GA Metabolism and Its Signalling during Abiotic Stresses

Modulation of GA metabolism associated gene expression could be controlled by the regulation of activity of various transcription factors (TFs) through one mechanism which helps in the integration of GAs signalling into the varied stress response network. Though the precise mechanism is unclear, it was assumed that the ABA-independent expression of several DREB1/CBF TFs is responsible for the induced transcription of genes related to GA metabolism and signalling (Mizoi et al. 2012). A transcription factor, *A. thaliana* EIN3 protein, controls ethylene-dependent responses, regulates negatively the expression of and binds directly to the DREB1/CBF TFs, and specifies a probable mechanism for integration (Shi et al. 2012). In response to salt and cold stress in *A. thaliana*, two TFs belonging to the DREB1/CBF family have been observed to regulate GA inactivation (Magome et al. 2004, 2008; Achard et al. 2008a). Under cold stress, DREB1B/CBF1 was observed to control the expression of $AtGA_2ox6$ and $AtGA_2ox3$ (Achard et al. 2008a). Such types of regulation include a reduction in bioactive gibberellin and subsequent inhibition of DELLA-mediated root growth along with freezing tolerance when exposed to cold stress (Achard et al. 2008a). Also, the ERF6 transcription factor, belonging to the ERF family of TFs, has been intricated to regulate GA_2ox6 expression under slight osmotic stress (Dubois et al. 2013). A few other GA_2ox genes in such cases were also observed to be expressed differentially in response to several abiotic stress conditions. (Achard et al. 2008a; Magome et al. 2008). These suggest that in response to abiotic stresses, the decrease in bioactive gibberellin levels was regulated by GA deactivation and that different TFs of these gene families are regulated differentially, depending on the stress faced, the developmental stage, and the plant organs affected.

1.4.2.3 Gibberellins Signalling Integrates Various Developmental and Environmental Signals

The quickly growing evidence on the mechanism of DELLA signalling is delivering an essential understanding of how DELLA protein plays a main role as an integrator of signals of hormonal cross-talk in response to various stresses (Yang et al. 2012; Zhu et al. 2011). For instance, via interaction with DELLA signalling, jasmonic acid (JA) prompts both growth inhibition and pathogenic resistance (Yang et al. 2012). On direct interaction with the DELLA repressor, JAZ (JA ZIM-domain) proteins, the repressor proteins of jasmonates signalling, provides rivalry for binding to the MYC2 TF that regulates JA-dependent signalling (Boter et al. 2004; Hou et al. 2010). In addition, JAZ proteins compete with growth-promoting PIF TFs for binding to DELLA. JA signalling is also assumed to prevent growth via direct effects on the levels of DELLA proteins (Yang et al. 2012). A similar kind of interaction of repressor proteins, JAZ with EIN3/EIL1 TFs, regulates ethylene transcriptional response, suggesting that this mechanism is conserved among other hormonal signalling pathways (Zhu et al. 2011). In such types of interactions, RGL3, a DELLA repressor, has been shown to be induced in its expression in response to various abiotic stresses, for example, drought, cold, and high-salinity stress (Achard et al. 2006; Magome et al. 2008). Also, as a response to abiotic stresses, GA signalling has been associated with the regulation of ABA biosynthesis via XERICO, an early DELLA target gene, which in turn is up-regulated by DELLA repressors and is assumed to suppress a regulator which negatively regulates ABA biosynthesis (Ko et al. 2006; Zentellaet al. 2007). Altogether, such information recommends a few mechanisms by which GA signalling could integrate signals from various hormonal signalling pathways to coordinate several abiotic stress responses in plants.

1.5 Conclusion

The aforesaid discussions present evidence that the inhibition of GA signalling is a primary response to abiotic stress along with transcriptional up regulation of *GA2ox* genes, encoding GA-inactivating enzymes, and in *Arabidopsis*, the DELLA gene RGL3 encodes growth inhibitor, has been demonstrated in stress-related studies. Somewhere, the stress-induced *AP2/ERF*-type transcription factors directly target the genes, but understanding the signalling networks that link between stresses with the expression of genes remains in infancy. Additionally, the relationship between GA signalling and stress tolerance is not fully understood. Generally, stresses are studied in isolation such as the low water content in soil solution results in soil drying, higher mechanical resistance to root development, and restricts the access of water and nutrients. These individual stresses hamper physiological processes separately, but it is necessary to determine the component stress effect and predict the outcome from their combined effect. The identification of 'key' genes is necessary to increase the stress tolerance in abiotic stress research. In future, further studies are needed for deeper understanding of GA signalling responses to improve the crop production and stress tolerance under adverse environmental situations.

REFERENCES

Achard P, Cheng H, De Grauwe L, Decat J, Schoutteten H, Moritz T, Van Der Straeten D, Peng J, Harberd NP (2006) Integration of plant responses to environmentally activated phytohormonal signals. *Science* 311:91–94.

Achard P, Genschik P (2009) Releasing the brakes of plant growth: how GAs shutdown DELLA proteins. *J Exp Bot* 60:1085–1092.

Achard P, Gong F, Cheminant S, Alioua M, Hedden P, Genschik P (2008) The cold-inducible CBF1 factor-dependent signaling pathway modulates the accumulation of the growth-repressing DELLA proteins via its effect on gibberellin metabolism. *Plant Cell* 20:2117–2129.

Adnan M, Fahad S, Khan IA, Saeed M, Ihsan MZ, Saud S, Riaz M, Wang D, Wu C (2019) Integration of poultry manure and phosphate solubilizing bacteria improved availability of Ca bound P in calcareous soils. *Biotech* 9(10):368.

Adnan M, Fahad S, Muhammad Z, Shahen S, Ishaq AM, Subhan D, Zafar-ul-Hye M, Martin LB, Raja MMN, Beena S, Saud S, Imran A, Zhen Y, Martin B, Jiri H, Rahul D (2020) Coupling phosphate-solubilizing bacteria with phosphorus supplements improve maize phosphorus acquisition and growth under lime induced salinity stress. *Plan Theory* 9(900). https://doi.org/10.3390/plants9070900.

Adnan M, Shah Z, Sharif M, Rahman H (2018) Liming induces carbon dioxide (CO_2) emission in PSB inoculated alkaline soil supplemented with different phosphorus sources. *Environ Sci Pollut Res* 25(10):9501–9509.

Adnan M, Zahir S, Fahad S, Arif M, Mukhtar A, Imtiaz AK, Ishaq AM, Abdul B, Hidayat U, Muhammad A, Inayat-Ur R, Saud S, Muhammad ZI, Yousaf J, Amanullah HMH, Wajid N (2018) Phosphate-solubilizing bacteria nullify the antagonistic effect of soil calcification on bioavailability of phosphorus in alkaline soils. *Sci Rep* 8:4339. https://doi.org/10.1038/s41598-018-22653-7.

Afzal F, Adnan M, Rahman IU, Noor M, Khan A, Iqbal S, Nawaz S (2017) Growth response of olive cultivars to air layering. *Pure App Biol* 6(4):1403–1409.

Ahmad I, Jadoon SA, Said A, Adnan M, Mohammad F, Munsif F (2017) Response of sunflower varieties to NPK fertilization. *Pure App Biol* 6(1):272.

Ahmad S, Kamran M, Ding R, Meng X, Wang H, Ahmad I, Fahad S, Han Q (2019) Exogenous melatonin confers drought stress by promoting plant growth, photosynthetic capacity and antioxidant defense system of maize seedlings. *PeerJ* 7:e7793. https://doi.org/10.7717/peerj.7793.

Akbar H, Timothy JK, Jagadish TM, Golam M, Apurbo KC, Muhammad F, Rajan B, Fahad S, Hasanuzzaman M (2020) Agricultural land degradation: processes and problems undermining future food security. In: Fahad S, Hasanuzzaman M, Alam M, Ullah H, Saeed M, Khan AK, Adnan M (eds) *Environment, climate, plant and vegetation growth*. Springer Publ Ltd, Springer Nature Switzerland AG. Part of Springer Nature, pp 17–62. https://doi.org/10.1007/978-3-030-49732-3.

Akram R, Turan V, Hammad HM, Ahmad S, Hussain S, Hasnain A, Maqbool MM, Rehmani MIA, Rasool A, Masood N, Mahmood F, Mubeen M, Sultana SR, Fahad S, Amanet K, Saleem M, Abbas Y, Akhtar HM, Waseem F, Murtaza R, Amin A, Zahoor SA, ul Din MS, Nasim W (2018a) Fate of organic and inorganic pollutants in paddy soils. In: Hashmi MZ, Varma A (eds) *Environmental pollution of paddy soils, soil biology*. Springer International Publishing AG, Cham, Switzerland, pp 197–214.

Akram R, Turan V, Wahid A, Ijaz M, Shahid MA, Kaleem S, Hafeez A, Maqbool MM, Chaudhary HJ, Munis MFH, Mubeen M, Sadiq N, Murtaza R, Kazmi DH, Ali S, Khan N, Sultana SR, Fahad S, Amin A, Nasim W (2018b) Paddy land pollutants and their role in climate change. In: Hashmi MZ, Varma A (eds) *Environmental pollution of paddy soils, soil biology*. Springer International Publishing AG, Cham, Switzerland, pp 113–124.

Akter R, Hasan SMR, Hossain MM, Ahmed T, Majumder MM, Akter M (2009) In vitro evaluation of antioxidant potential of leaves of Opuntia dillenii Haw. *SJ Pharma Sci* 2(1):22–26.

Al-Wabel MI, Abdelazeem S, Munir A, Khalid E, Adel RAU (2020b) Extent of climate change in Saudi Arabia and its impacts on agriculture: a case study from Qassim Region. In: Fahad S, Hasanuzzaman M, Alam M, Ullah H, Saeed M, Khan AK, Adnan M (eds) *Environment, climate, plant and vegetation growth*. Springer Publ Ltd, Springer Nature Switzerland AG. Part of Springer Nature, pp 635–658. https://doi.org/10.1007/978-3-030-49732-3.

Al-Wabel MI, Ahmad M, Usman ARA, Akanji M, Rafique MI (2020a) Advances in pyrolytic technologies with improved carbon capture and storage to combat climate change. In: Fahad S, Hasanuzzaman M, Alam M, Ullah H, Saeed M, Khan AK, Adnan M (eds) *Environment, climate, plant and vegetation growth*. Springer Publ Ltd, Springer Nature Switzerland AG. Part of Springer Nature, pp 535–576. https://doi.org/10.1007/978-3-030-49732-3.

Amanullah SK, Imran HAK, Muhammad A, Abdel RA, Muhammad A, Fahad S, Azizullah S, Brajendra P (2020) Effects of climate change on irrigation water quality. In: Fahad S, Hasanuzzaman M, Alam M, Ullah H, Saeed M, Khan AK, Adnan M (eds) *Environment, climate, plant and vegetation growth*. Springer Publ Ltd, Switzerland. Part of Springer Nature, pp 123–132. https://doi.org/10.1007/978-3-030-49732-3.

Amir M, Muhammad A, Allah B, Sevgi Ç, Haroon ZK, Muhammad A, Emre A (2020) Bio fortification under climate change: the fight between quality and quantity. In: Fahad S, Hasanuzzaman M, Alam M, Ullah H, Saeed M, Khan AK, Adnan M (eds) *Environment, climate, plant and vegetation growth*. Springer Publ Ltd, Springer Nature Switzerland AG. Part of Springer Nature, pp 173–228. https://doi.org/10.1007/978-3-030-49732-3.

Amjad I, Muhammad H, Farooq S, Anwar H (2020) Role of plant bioactives in sustainable agriculture. In: Fahad S, Hasanuzzaman M, Alam M, Ullah H, Saeed M, Khan AK, Adnan M (eds) *Environment, climate, plant and vegetation growth*. Springer Publ Ltd, Springer Nature Switzerland AG. Part of Springer Nature, pp 591–606. https://doi.org/10.1007/978-3-030-49732-3.

Arif M, Talha J, Muhammad R, Fahad S, Muhammad A, Amanullah KA, Ishaq AM, Bushra K, Fahd R (2020) Biochar; a remedy for climate change. In: Fahad S, Hasanuzzaman M, Alam M, Ullah H, Saeed M, Khan AK, Adnan M (eds) *Environment, climate, plant and vegetation growth*. Springer Publ Ltd, Springer Nature Switzerland AG. Part of Springer Nature, pp 151–172. https://doi.org/10.1007/978-3-030-49732-3.

Ayman EL Sabagh, Hossain A, Barutçular C, Iqbal MA, Sohidul Islam M, Fahad S, Sytar O, Çig F, Meena RS, Erman M (2020) Consequences of salinity stress on the quality of crops and its mitigation strategies for sustainable crop production: an outlook of arid and semi- arid regions. In: Fahad S, Hasanuzzaman M, Alam M, Ullah H, Saeed M, Khan AK, Adnan M (eds) *Environment, climate, plant and vegetation growth*. Springer Publ Ltd, Springer Nature Switzerland AG. Part of Springer Nature, pp 503–534. https://doi.org/10.1007/978-3-030-49732-3.

Aziz K, Daniel KYT, Fazal M, Muhammad ZA, Farooq S, Fan W, Fahad S, Ruiyang Z (2017a) Nitrogen nutrition in cotton and control strategies for greenhouse gas emissions: a review. *Environ Sci Pollut Res* 24:23471–23487. https://doi.org/10.1007/s11356-017-0131-y.

Aziz K, Daniel KYT, Muhammad ZA, Honghai L, Shahbaz AT, Mir A, Fahad S (2017b) Nitrogen fertility and abiotic stresses management in cotton crop: a review. *Environ Sci Pollut Res* 24:14551–14566. https://doi.org/10.1007/s11356-017-8920-x.

Bailey-Serres J, Voesenek LACJ (2010) Life in the balance: a signalling network controlling survival of flooding. *Curr Opin Plant Biol* 13:489–494.

Bao S, Hua C, Shen L, Yu H (2020) New insights into gibberellin signaling in regulating flowering in Arabidopsis. *J Integr Plant Biol* 62:118–131.

Baseer M, Adnan M, Fazal M, Fahad S, Muhammad S, Fazli W, Muhammad A, Amanullah J, Depeng W, Saud S, Muhammad N, Muhammad Z, Fazli S, Beena S, Mian AR, Ishaq AM (2019) Substituting urea by organic wastes for improving maize yield in alkaline soil. *J Plant Nutr.* doi.org/10.1080/01904167.2019.1659344.

Bayram AY, Seher Ö, Nazlican A (2020) Climate Change Forecasting and Modeling for the Year of 2050. In: Fahad S, Hasanuzzaman M, Alam M, Ullah H, Saeed M, Khan AK, Adnan M (eds) *Environment, climate, plant and vegetation growth.* Springer Publ Ltd, Springer Nature Switzerland AG. Part of Springer Nature, pp 109–122. https://doi.org/10.1007/978-3-030-49732-3.

Bömke C, Tudzynski B (2009) Diversity, regulation, and evolution of the gibberellins biosynthetic pathway in fungi compared to plants and bacteria. *Phytochemistry* 70:1876–1893.

Boter M, Ruíz-Rivero O, Abdeen A, Prat S (2004) Conserved MYC transcription factors play a key role in jasmonatesignaling both in tomato and Arabidopsis. *Genes Dev* 18:1577–1591.

Chapman N, Whalley WR, Lindsey K, Miller AJ (2011) Water supply and not nitrate concentration determines primary root growth in Arabidopsis. *Plant Cell Environ* 34:1630–1638.

Chiang H-H, Hwang I, Goodman HM (1995) Isolation of the Arabidopsis GA4 locus. *Plant Cell* 7:195–201.

Christmann A, Weiler EW, Steudle E, Grill E (2007) A hydraulic signal in root-to-shoot signalling of water shortage. *Plant J* 52:167–174.

Claeys H, Skirycz A, Maleux K, Inzé D (2012) DELLA signaling mediates stress-induced cell differentiation in Arabidopsis leaves through modulation of anaphase-promoting complex/cyclosome activity. *Plant Physiol* 159:739–747.

Coelho Filho MA, Colebrook EH, Lloyd DPA, Webster CP, Mooney SJ, Phillips AL, Hedden P, Whalley WR (2013) The involvement of gibberellin signalling in the effect of soil resistance to root penetration on leaf elongation and tiller number in wheat. *Plant Soil* 371:81–94.

Colebrook EH, Thomas SG, Phillips AL, Hedden P (2014) The role of gibberellin signalling in plant responses to abiotic stress. *J Exp Biol* 217:67–75. https://doi.org/10.1242/jeb.089938.

Davies PJ (2010) The plant hormones: their nature, occurrence, and function. In: Davies PJ (ed) *Plant hormones: biosynthesis, signal transduction, action!* Springer, Dordrecht, pp 1–15.

De Bruyne L, Hofte M, De Vleesschauwer D (2014) Connecting growth and defense: the emerging roles of brassinosteroids and gibberellins in plant innate immunity. *Mol Plant* 7:943–959.

De Vleesschauwer D, Van Buyten E, Satoh K, Balidion J, Mauleon R, Choi IR, Vera-Cruz C, Kikuchi S, Höfte M (2012) Brassinosteroids antagonize gibberellin- and salicylatemediated root immunity in rice. *Plant Physiol* 158:1833–1846.

Depeng W, Fahad S, Saud S, Muhammad K, Aziz K, Mohammad NK, Hafiz MH, Wajid N (2018) Morphological acclimation to agronomic manipulation in leaf dispersion and orientation to promote "Ideotype" breeding: evidence from 3D visual modeling of "super" rice (Oryza sativa L.). *Plant Physiol Biochem.* https://doi.org/10.1016/j.plaphy.2018.11.010.

Dubois M, Skirycz A, Claeys H, Maleux K, Dhondt S, De Bodt S, Vanden Bossche R, De Milde L, Yoshizumi T, Matsui M, Dirk I (2013) The ethylene response factor 6 acts as central regulator of leaf growth under water limiting conditions in Arabidopsis thaliana. *Plant Physiol* 162:319–332.

Fahad S, Adnan M, Hassan S, Saud S, Hussain S, Wu C, Wang D, Hakeem KR, Alharby HF, Turan V, Khan MA, Huang J (2019b) Rice responses and tolerance to high temperature. In: Hasanuzzaman M, Fujita M, Nahar K, Biswas JK (eds) *Advances in rice research for abiotic stress tolerance.* Woodhead Publ Ltd, Abington Hall Abington, Cambridge, England, pp 201–224.

Fahad S, Bajwa AA, Nazir U, Anjum SA, Farooq A, Zohaib A, Sadia S, Nasim W, Adkins S, Saud S, Ihsan MZ, Alharby H, Wu C, Wang D, Huang J (2017, 1147) Crop production under drought and heat stress: plant responses and management options. *Front Plant Sci*:8. https://doi.org/10.3389/fpls.2017.01147.

Fahad S, Bano A (2012) Effect of salicylic acid on physiological and biochemical characterization of maize grown in saline area. *Pak J Bot* 44:1433–1438.

Fahad S, Chen Y, Saud S, Wang K, Xiong D, Chen C, Wu C, Shah F, Nie L, Huang J (2013) Ultraviolet radiation effect on photosynthetic pigments, biochemical attributes, antioxidant enzyme activity and hormonal contents of wheat. *J Food Agri Environ* 11(3&4):1635–1641.

Fahad S, Hussain S, Bano A, Saud S, Hassan S, Shan D, Khan FA, Khan F, Chen Y, Wu C, Tabassum MA, Chun MX, Afzal M, Jan A, Jan MT, Huang J (2014a) Potential role of phytohormones and plant growth-promoting rhizobacteria in abiotic stresses: consequences for changing environment. *Environ Sci Pollut Res* 22(7):4907–4921. https://doi.org/10.1007/s11356-014-3754-2.

Fahad S, Hussain S, Matloob A, Khan FA, Khaliq A, Saud S, Hassan S, Shan D, Khan F, Ullah N, Faiq M, Khan MR, Tareen AK, Khan A, Ullah A, Ullah N, Huang J (2014b) Phytohormones and plant responses to salinity stress: a review. *Plant Growth Regul* 75(2):391–404. https://doi.org/10.1007/s10725-014-0013-y.

Fahad S, Hussain S, Saud S, Hassan S, Chauhan BS, Khan F et al (2016a) Responses of rapid viscoanalyzer profile and other rice grain qualities to exogenously applied plant growth regulators under high day and high night temperatures. *PLoS One* 11(7):e0159590. https://doi.org/10.1371/journal.pone.0159590.

Fahad S, Hussain S, Saud S, Hassan S, Ihsan Z, Shah AN, Wu C, Yousaf M, Nasim W, Alharby H, Alghabari F, Huang J (2016c) Exogenously applied plant growth regulators enhance the morphophysiological growth and yield of rice under high temperature. *Front Plant Sci* 7:1250. https://doi.org/10.3389/fpls.2016.01250.

Fahad S, Hussain S, Saud S, Hassan S, Tanveer M, Ihsan MZ, Shah AN, Ullah A, Nasrullah KF, Ullah S, NW AH, Wu C, Huang J (2016d) A combined application of biochar and phosphorus alleviates heat-induced adversities on physiological, agronomical and quality attributes of rice. *Plant Physiol Biochem* 103:191–198.

Fahad S, Hussain S, Saud S, Khan F, Hassan S, Jr A, Nasim W, Arif M, Wang F, Huang J (2016b) Exogenously applied plant growth regulators affect heat-stressed rice pollens. *J Agron Crop Sci* 202:139–150.

Fahad S, Hussain S, Saud S, Tanveer M, Bajwa AA, Hassan S, Shah AN, Ullah A, Wu C, Khan FA, Shah F, Ullah S, Chen Y, Huang J (2015a) A biochar application protects rice pollen from high-temperature stress. *Plant Physiol Biochem* 96:281–287.

Fahad S, Muhammad ZI, Abdul K, Ihsanullah D, Saud S, Saleh A, Wajid N, Muhammad A, Imtiaz AK, Chao W, Depeng W, Jianliang H (2018) Consequences of high temperature under changing climate optima for rice pollen characteristics-concepts and perspectives. *Archives Agron Soil Sci.* https://doi.org/10.1080/03650340.2018.1443213.

Fahad S, Nie L, Chen Y, Wu C, Xiong D, Saud S, Hongyan L, Cui K, Huang J (2015b) Crop plant hormones and environmental stress. *Sustain Agric Rev* 15:371–400.

Fahad S, Rehman A, Shahzad B, Tanveer M, Saud S, Kamran M, Ihtisham M, Khan SU, Turan V, Rahman MHU (2019a) Rice responses and tolerance to metal/metalloid toxicity. In: Hasanuzzaman M, Fujita M, Nahar K, Biswas JK (eds) *Advances in rice research for abiotic stress tolerance.* Woodhead Publ Ltd, Abington Hall Abington, Cambridge, England, pp 299–312.

Farah R, Muhammad R, Muhammad SA, Tahira Y, Muhammad AA, Maryam A, Shafaqat A, Rashid M, Muhammad R, Qaiser H, Afia Z, Muhammad AA, Muhammad A, Fahad S (2020) Alternative and non-conventional soil and crop management strategies for increasing water use efficiency. In: Fahad S, Hasanuzzaman M, Alam M, Ullah H, Saeed M, Khan AK, Adnan M (eds) *Environment, climate, plant and vegetation growth.* Springer Publ Ltd, Springer Nature Switzerland AG. Part of Springer Nature, pp 323–338. https://doi.org/10.1007/978-3-030-49732-3.

Farhana G, Ishfaq A, Muhammad A, Dawood J, Fahad S, Xiuling L, Depeng W, Muhammad F, Muhammad F, Syed AS (2020) Use of crop growth model to simulate the impact of climate change on yield of various wheat cultivars under different agro-environmental conditions in Khyber Pakhtunkhwa, Pakistan. *Arab J Geosci* 13:112. https://doi.org/10.1007/s12517-020-5118-1.

Farhat A, Hafiz MH, Wajid I, Aitazaz AF, Hafiz FB, Zahida Z, Fahad S, Wajid F, Artemi C (2020) A review of soil carbon dynamics resulting from agricultural practices. *J Environ Manage* 268:110319.

Fazli W, Muhmmad S, Amjad A, Fahad S, Muhammad A, Muhammad N, Ishaq AM, Imtiaz AK, Mukhtar A, Muhammad S, Muhammad I, Rafi U, Haroon I, Muhammad A (2020) Plant-microbes interactions and functions in changing climate. In: Fahad S, Hasanuzzaman M, Alam M, Ullah H, Saeed M, Khan AK, Adnan M (eds) *Environment, climate, plant and vegetation growth.* Springer Publ Ltd, Springer Nature Switzerland AG. Part of Springer Nature, pp 397–420. https://doi.org/10.1007/978-3-030-49732-3.

Fazlullah AM, Fahad S, Iqbal S, Arshad M, Muhammad D, Wahid F, Noor MN (2018) Integrated application of phosphorus (P) and phosphate solubilizing bacteria (PSB) improve maize yield. *Pure Appl Biol* 7(2):494–499.

Fukao T, Bailey-Serres J (2008a) Ethylene—a key regulator of submergence responses in rice. *Plant Sci* 175:43–51.

Fukao T, Bailey-Serres J (2008b) Submergence tolerance conferred by Sub1A is mediated by SLR1 and SLRL1 restriction of gibberellin responses in rice. *Proc Natl Acad Sci U S A* 105:16814–16819.
Fukao T, Xu K, Ronald PC, Bailey-Serres J (2006) A variable cluster of ethylene response factor-like genes regulates metabolic and developmental acclimation responses to submergence in rice. *Plant Cell* 18:2021–2034.
Fukao T, Yeung E, Bailey-Serres J (2011) The submergence tolerance regulator SUB1A mediates crosstalk between submergence and drought tolerance in rice. *Plant Cell* 23:412–427.
Gallego-Bartolomé J, Minguet EG, Grau-Enguix F, Abbas M, Locascio A, Thomas SG, Alabadí D, Blázquez MA (2012) Molecular mechanism for the interaction between gibberellin and brassinosteroidsignaling pathways in Arabidopsis. *Proc Natl Acad Sci U S A* 109:13446–13451.
Gopakumar L, Bernard NO, Donato V (2020) Soil Microarthropods and Nutrient Cycling. In: Fahad S, Hasanuzzaman M, Alam M, Ullah H, Saeed M, Khan AK, Adnan M (eds) *Environment, climate, plant and vegetation growth.* Springer Publ Ltd, Springer Nature Switzerland AG. Part of Springer Nature, pp 453–472. https://doi.org/10.1007/978-3-030-49732-3.
Griffiths J, Murase K, Rieu I, Zentella R, Zhang ZL, Powers SJ, Gong F, Phillips AL, Hedden P, Sun TP et al (2006) Genetic characterization and functional analysis of the GID1 gibberellin receptors in Arabidopsis. *Plant Cell* 18:3399–3414.
Habib ur R, Ashfaq A, Aftab W, Manzoor H, Fahd R, Wajid I, Md. Aminul I, Vakhtang S, Muhammad A, Asmat U, Abdul W, Syeda RS, Shah S, Shahbaz K, Fahad S, Manzoor H, Saddam H, Wajid N (2017) Application of CSM-CROPGRO-Cotton model for cultivars and optimum planting dates: evaluation in changing semi-arid climate. *Field Crops Res.* https://doi.org/10.1016/j.fcr.2017.07.007.
Hafiz MH, Abdul K, Farhat A, Wajid F, Fahad S, Muhammad A, Ghulam MS, Wajid N, Muhammad M, Hafiz FB (2020) Comparative effects of organic and inorganic fertilizers on soil organic carbon and wheat productivity under arid region. *Commun Soil Sci Plant Anal.* https://doi.org/10.1080/00103624.2020.1763385.
Hafiz MH, Farhat A, Ashfaq A, Hafiz FB, Wajid F, Carol Jo W, Fahad S, Gerrit H (2020) Predicting Kernel growth of maize under controlled water and nitrogen applications. *Int J Plant Prod.* https://doi.org/10.1007/s42106-020-00110-8.
Hafiz MH, Farhat A, Shafqat S, Fahad S, Artemi C, Wajid F, Chaves CB, Wajid N, Muhammad M, Hafiz FB (2018) Offsetting land degradation through nitrogen and water management during maize cultivation under arid conditions. *Land Degrad Dev*:1–10. https://doi.org/10.1002/ldr.2933.
Hafiz MH, Muhammad A, Farhat A, Hafiz FB, Saeed AQ, Muhammad M, Fahad S, Muhammad A (2019) Environmental factors affecting the frequency of road traffic accidents: a case study of sub-urban area of Pakistan. *Environ Sci Pollut Res.* https://doi.org/10.1007/s11356-019-04752-8.
Hafiz MH, Wajid F, Farhat A, Fahad S, Shafqat S, Wajid N, Hafiz FB (2016) Maize plant nitrogen uptake dynamics at limited irrigation water and nitrogen. *Environ Sci Pollut Res* 24(3):2549–2557. https://doi.org/10.1007/s11356-016-8031-0.
Hattori Y, Nagai K, Furukawa S, Song XJ, Kawano R, Sakakibara H, Wu J, Matsumoto T, Yoshimura A, Kitano H et al (2009) The ethylene response factors SNORKEL1 and SNORKEL2 allow rice to adapt to deep water. *Nature* 460:1026–1030.
Hedden P, Phillips AL (2000) Gibberellin metabolism: new insights revealed by the genes. *Trends Plant Sci* 5:523–530.
Hedden P (2003) The genes of the green revolution. *Trends Genet* 19:5–9.
Hedden P, Proebsting WM (1999) Genetic analysis of gibberellin biosynthesis. *Plant Physiol* 119:365–370. https://doi.org/10.1104/pp.119.2.365.
Hedden P, Sponsel V (2015) A century of gibberellin research. *J Plant Growth Regul* 34:740–760. https://doi.org/10.1007/s00344-015-9546-1.
Hedden P, Thomas SG (2012) Gibberellin biosynthesis and its regulation. *Biochem J* 444:11–25.
Helliwell CA, Sheldon CC, Olive MR, Walker AR, Zeevaart JAD, Peacock WJ, Dennis ES (1998) Cloning of the Arabidopsis ent-kaurene oxidase gene GA3. *Proc Natl Acad Sci U S A* 95:9019–9024.
Hesham FA, Fahad S (2020) Melatonin application enhances biochar efficiency for drought tolerance in maize varieties: modifications in physio-biochemical machinery. *Agron J* 112(4):1–22.
Hou X, Lee LYC, Xia K, Yan Y, Yu H (2010) DELLAs modulate jasmonate signaling via competitive binding to JAZs. *Dev Cell* 19:884–894.

Hussain MA, Fahad S, Rahat S, Muhammad FJ, Muhammad M, Qasid A, Ali A, Husain A, Nooral A, Babatope SA, Changbao S, Liya G, Ibrar A, Zhanmei J, Juncai H (2020) Multifunctional role of brassinosteroid and its analogues in plants. *Plant Growth Regul.* https://doi.org/10.1007/s10725-020-00647-8.

Ibrar K, Aneela R, Khola Z, Urooba N, Sana B, Rabia S, Ishtiaq H, Mujaddad Ur Rehman, Salvatore M (2020) Microbes and environment: global warming reverting the frozen zombies. In: Fahad S, Hasanuzzaman M, Alam M, Ullah H, Saeed M, Khan AK, Adnan M (eds) *Environment, climate, plant and vegetation growth.* Springer Publ Ltd, Springer Nature Switzerland AG. Part of Springer Nature, pp 607–634. https://doi.org/10.1007/978-3-030-49732-3.

Ilyas M, Mohammad N, Nadeem K, Ali H, Aamir HK, Kashif H, Fahad S, Aziz K, Abid U (2020) Drought tolerance strategies in plants: a mechanistic approach. *J Plant Growth Regul.* https://doi.org/10.1007/s00344-020-10174-5.

Iqbal N, Nazar R, Iqbal M, Khan R, Masood A, Khan NA (2011) Role of gibberellins in regulation of source–sink relations under optimal and limiting environmental conditions. *Curr Sci* 100:998–1007.

Iqra M, Amna B, Shakeel I, Fatima K, Sehrish L, Hamza A, Fahad S (2020) Carbon Cycle in Response to Global Warming. In: Fahad S, Hasanuzzaman M, Alam M, Ullah H, Saeed M, Khan AK, Adnan M (eds) *Environment, climate, plant and vegetation growth.* Springer Publ Ltd, Springer Nature Switzerland AG. Part of Springer Nature, pp 1–16. https://doi.org/10.1007/978-3-030-49732-3.

Jackson MB (2008) Ethylene-promoted elongation: an adaptation to submergence stress. *Ann Bot* 101:229–248.

Jan M, Anwar-ul-Haq M, Adnan NS, Muhammad Y, Javaid I, Xiuling L, Depeng W, Fahad S (2019) Modulation in growth, gas exchange, and antioxidant activities of salt-stressed rice (Oryza sativa L.) genotypes by zinc fertilization. *Arab J Geosci* 12:775. https://doi.org/10.1007/s12517-019-4939-2.

Jung KH, Seo YS, Walia H, Cao P, Fukao T, Canlas PE, Amonpant F, Bailey-Serres J, Ronald PC (2010) The submergence tolerance regulator Sub1A mediates stress-responsive expression of AP2/ERF transcription factors. *Plant Physiol* 152:1674–1692.

Kamarn M, Wenwen C, Irshad A, Xiangping M, Xudong Z, Wennan S, Junzhi C, Shakeel A, Fahad S, Qingfang H, Tiening L (2017) Effect of paclobutrazol, a potential growth regulator on stalk mechanical strength, lignin accumulation and its relation with lodging resistance of maize. *Plant Growth Regul* 84:317–332. https://doi.org/10.1007/s10725-017-0342-8.

Kasahara H, Hanada A, Kuzuyama T, Takagi M, Kamiya Y, Yamaguchi S (2002) Contribution of the mevalonate and methylerythritol phosphate pathways to the biosynthesis of gibberellins in Arabidopsis. *J Biol Chem* 277:45188–45194.

Kaya MD, Okcu G, Atak M, Cıkıhı Y, Kolsarıc O (2006) Seed treatments to overcome salt and drought stress during germination in sunflower (Helianthus annuus L.). *Eur J Agron* 24(4):291–295.

Keuskamp DH, Sasidharan R, Pierik R (2010) Physiological regulation and functional significance of shade avoidance responses to neighbors. *Plant Signal Behav* 5:655–662.

Khan AF, Gul Y, Jamal J, Ali M, Adnan MA, Khan IN, Bacha Z (2017) Weeds management in maize hybrids with night plowing. *Pak J Weed Sci Res* 22(2):179–191.

Khan MIR, Asgher M, Fatma M, Per TS, Khan NA (2015) Drought stress vis a vis plant functions in the era of climate change. *Clim Change Environ Sust* 3:13–25.

Ko JH, Yang SH, Han KH (2006) Upregulation of an Arabidopsis RINGH2 gene, XERICO, confers drought tolerance through increased abscisic acid biosynthesis. *Plant J* 47:343–355.

Kurosawa E (1926) Experimental studies on the nature of the substance secreted by the "bakanae" fungus. *Trans Nat Hist Soc Formosa* 16:213–227.

Magome H, Nomura T, Hanada A, Takeda-Kamiya N, Ohnishi T, Shinma Y, Katsumata T, Kawaide H, Kamiya Y, Yamaguchi S (2013) CYP714B1 and CYP714B2 encode gibberellin 13-oxidases that reduce gibberellin activity in rice. *Proc Natl Acad Sci U S A* 110(5):1947–1952. https://doi.org/10.1073/pnas.1215788110.

Magome H, Yamaguchi S, Hanada A, Kamiya Y, Oda K (2004) Dwarf and delayed-flowering 1, a novel Arabidopsis mutant deficient in gibberellin biosynthesis because of overexpression of a putative AP2 transcription factor. *Plant J* 37:720–729.

Magome H, Yamaguchi S, Hanada A, Kamiya Y, Oda K (2008) The DDF1 transcriptional activator upregulates expression of a gibberellin-deactivating gene, GA2ox7, under high-salinity stress in Arabidopsis. *Plant J* 56:613–626.

Mahar A, Amjad A, Altaf HL, Fazli W, Ronghua L, Muhammad A, Fahad S, Muhammad A, Rafiullah IAK, Zengqiang Z (2020) Promising technologies for Cd-contaminated soils: drawbacks and possibilities. In: Fahad S, Hasanuzzaman M, Alam M, Ullah H, Saeed M, Khan AK, Adnan M (eds) *Environment, climate, plant and vegetation growth.* Springer Publ Ltd, Springer Nature Switzerland AG. Part of Springer Nature, pp 63–92. https://doi.org/10.1007/978-3-030-49732-3.

Md Enamul H, Shoeb AZM, Mallik AH, Fahad S, Kamruzzaman MM, Akib J, Nayyer S, Mehedi A KM, Swati AS, Md Yeamin A, Most SS (2020) Measuring vulnerability to environmental hazards: qualitative to quantitative. In: Fahad S, Hasanuzzaman M, Alam M, Ullah H, Saeed M, Khan AK, Adnan M (eds) *Environment, climate, plant and vegetation growth.* Springer Publ Ltd, Springer Nature Switzerland AG. Part of Springer Nature, pp 421–452. https://doi.org/10.1007/978-3-030-49732-3.

Md Jakir H, Allah B (2020) Development and applications of transplastomic plants; a way towards eco-friendly agriculture. In: Fahad S, Hasanuzzaman M, Alam M, Ullah H, Saeed M, Khan AK, Adnan M (eds) *Environment, climate, plant and vegetation growth.* Springer Publ Ltd, Springer Nature Switzerland AG. Part of Springer Nature, pp 285–322. https://doi.org/10.1007/978-3-030-49732-3.

Miransari M (2012) Role of phytohormone signaling during stress. In: Ahmad P, Prasad MNV (eds) *Environmental adaptations and stress tolerance of plants in the era of climate change.* Springer, New York, pp 381–393.

Mittler R (2006) Abiotic stress, the field environment and stress combination. *Trends Plant Sci* 11:15–19.

Mizoi J, Shinozaki K, Yamaguchi-Shinozaki K (2012) AP2/ERF family transcription factors in plant abiotic stress responses. *Biochim Biophys Acta* 1819:86–96.

Mubeen M, Ashfaq A, Hafiz MH, Muhammad A, Hafiz UF, Mazhar S, Din MS u, Asad A, Amjed A, Fahad S, Wajid N (2020) Evaluating the climate change impact on water use efficiency of cotton-wheat in semi-arid conditions using DSSAT model. *J Water Climate Change.* https://doi.org/10.2166/wcc.2019.179/622035/jwc2019179.pdf.

Muhammad Tahir ul Qamar, Amna F, Amna B, Barira Z, Xitong Z, Ling-Ling C (2020) Effectiveness of conventional crop improvement strategies vs. Omics. In: Fahad S, Hasanuzzaman M, Alam M, Ullah H, Saeed M, Khan AK, Adnan M (eds) *Environment, climate, plant and vegetation growth.* Springer Publ Ltd, Springer Nature Switzerland AG. Part of Springer Nature, pp 253–284. https://doi.org/10.1007/978-3-030-49732-3.

Muhammad Z, Abdul MK, Abdul MS, Kenneth BM, Muhammad S, Shahen S, Ibadullah J, Fahad S (2019) Performance of Aeluropus lagopoides (mangrove grass) ecotypes, a potential turfgrass, under high saline conditions. *Environ Sci Pollut Res.* https://doi.org/10.1007/s11356-019-04838-3.

Navarro L, Bari R, Achard P, Lison P, Nemri A, Harberd NP, Jones JDG (2008) DELLAs control plant immune responses by modulating the balance and salicylic acid signaling. *Curr Biol* 18:650–655.

Noor M, Naveed ur R, Ajmal J, Fahad S, Muhammad A, Fazli W, Saud S, Hassan S (2020) Climate change and costal plant lives. In: Fahad S, Hasanuzzaman M, Alam M, Ullah H, Saeed M, Khan AK, Adnan M (eds) *Environment, climate, plant and vegetation growth.* Springer Publ Ltd, Springer Nature Switzerland AG. Part of Springer Nature, pp 93–108. https://doi.org/10.1007/978-3-030-49732-3.

O'Neill DP, Davidson SE, Clarke VC, Yamauchi Y, Yamaguchi S, Kamiya Y, Reid JB, Ross JJ (2010) Regulation of the gibberellin pathway by auxin and DELLA proteins. *Planta* 232:1141–1149.

Phillips AL, Ward DA, Uknes S, Appleford NEJ, Lange T, Huttly A, Gaskin P, Graebe JE, Hedden P (1995) Isolation and expression of three gibberellin 20-oxidase cDNA clones from Arabidopsis. *Plant Physiol* 108:1049–1057.

Prisic S, Xu M, Wilderman PR, Peters RJ (2004) Rice contains two disparate entcopalyl diphosphate synthases with distinct metabolic functions. *Plant Physiol* 136:4228–4236.

Qamar-uz Z, Zubair A, Muhammad Y, Muhammad ZI, Abdul K, Fahad S, Safder B, Ramzani PMA, Muhammad N (2017) *Zinc biofortification in rice: leveraging agriculture to moderate hidden hunger in developing countries. Arch Agron Soil Sci* 64:147–161. https://doi.org/10.1080/03650340.2017.1338343.

Qin X, Liu JH, Zhao WS, Chen XJ, Guo ZJ, Peng YL (2013) Gibberellin 20-oxidase gene OsGA20ox3 regulates plant stature and disease development in rice. *Mol Plant-Microbe Interact* 26:227–239.

Rashid M, Qaiser H, Khalid SK, Mohammad I. Al-Wabel, Zhang A, Muhammad A, Shahzada SI, Rukhsanda A, Ghulam AS, Shahzada MM, Sarosh A, Muhammad FQ (2020) Prospects of biochar in alkaline soils to mitigate climate change. In: Fahad S, Hasanuzzaman M, Alam M, Ullah H,Saeed M, Khan AK, Adnan M (eds) *Environment, climate, plant and vegetation growth.* Springer Publ Ltd, Springer Nature Switzerland AG. Part of Springer Nature, pp 133–150. https://doi.org/10.1007/978-3-030-49732-3.

Rehman M, Fahad S, Saleem MH, Hafeez M, Rahman MH u, Liu F, Deng G (2020) Red light optimized physiological traits and enhanced the growth of ramie (Boehmeria nivea L.). *Photosynthetica* 58(4):922–931.

Rieu I, Eriksson S, Powers SJ, Gong F, Griffiths J, Woolley L, Benlloch R, Nilsson O, Thomas SG, Hedden P, Phillips AL (2008) Genetic analysis reveals that C19-GA 2-oxidation is a major gibberellin inactivation pathway in Arabidopsis. *Plant Cell* 20(9):2420–2436. https://doi.org/10.1105/tpc.108.058818.

Sadam M, Muhammad Tahir ul Qamar, Ghulam M, Muhammad SK, Faiz AJ (2020) Role of biotechnology in climate resilient agriculture. In: Fahad S, Hasanuzzaman M, Alam M, Ullah H, Saeed M, Khan AK, Adnan M (eds) *Environment, climate, plant and vegetation growth.* Springer Publ Ltd, Springer Nature Switzerland AG. Part of Springer Nature, pp 339–366. https://doi.org/10.1007/978-3-030-49732-3.

Sajid H, Jie H, Jing H, Shakeel A, Satyabrata N, Sumera A, Awais S, Chunquan Z, Lianfeng Z, Xiaochuang C, Qianyu J, Junhua Z (2020) Rice production under climate change: adaptations and mitigating strategies. In: Fahad S, Hasanuzzaman M, Alam M, Ullah H, Saeed M, Khan AK, Adnan M (eds) *Environment, climate, plant and vegetation growth.* Springer Publ Ltd, Springer Nature Switzerland AG. Part of Springer Nature, pp 659–686. https://doi.org/10.1007/978-3-030-49732-3.

Sajjad H, Muhammad M, Ashfaq A, Waseem A, Hafiz MH, Mazhar A, Nasir M, Asad A, Hafiz UF, Syeda RS, Fahad S, Depeng W, Wajid N (2019) Using GIS tools to detect the land use/land cover changes during forty years in Lodhran district of Pakistan. *Environ Sci Pollut Res.* https://doi.org/10.1007/s11356-019-06072-3.

Salazar-Cerezo S, Martínez-Montiel N, García-Sánchez J, Pérez-y-Terrón R, Martínez-Contreras RD (2018) Gibberellin biosynthesis and metabolism: a convergent route for plants, fungi and bacteria. *Microbiol Res* 208:85–98. https://doi.org/10.1016/j.micres.2018.01.010.

Saleem MH, Fahad S, Adnan M, Mohsin A, Muhammad SR, Muhammad K, Qurban A, Inas AH, Parashuram B, Mubassir A, Reem MH (2020a) Foliar application of gibberellic acid endorsed phytoextraction of copper and alleviates oxidative stress in jute (Corchorus capsularis L.) plant grown in highly copper-contaminated soil of China. *Environ Sci Poll Res.* https://doi.org/10.1007/s11356-020-09764-3.

Saleem MH, Fahad S, Shahid UK, Mairaj D, Abid U, Ayman ELS, Akbar H, Analía L, Lijun L (2020c) Copper-induced oxidative stress, initiation of antioxidants and phytoremediation potential of flax (Linum usitatissimum L.) seedlings grown under the mixing of two different soils of China. *Environ Sci Poll Res.* https://doi.org/10.1007/s11356-019-07264-7.

Saleem MH, Rehman M, Fahad S, Tung SA, Iqbal N, Hassan A, Ayub A, Wahid MA, Shaukat S, Liu L, Deng G (2020b) Leaf gas exchange, oxidative stress, and physiological attributes of rapeseed (Brassica napus L.) grown under different light-emitting diodes. *Photosynthetica* 58(3):836–845.

Saman S, Amna B, Bani A, Muhammad Tahir ul Qamar, Rana MA, Muhammad SK (2020) QTL mapping for abiotic stresses in cereals. In: Fahad S, Hasanuzzaman M, Alam M, Ullah H, Saeed M, Khan AK, Adnan M (eds) *Environment, climate, plant and vegetation growth.* Springer Publ Ltd, Springer Nature Switzerland AG. Part of Springer Nature, pp 229–252. https://doi.org/10.1007/978-3-030-49732-3.

Saud S, Chen Y, Fahad S, Hussain S, Na L, Xin L, Alhussien SA (2016) Silicate application increases the photosynthesis and its associated metabolic activities in Kentucky bluegrass under drought stress and post-drought recovery. *Environ Sci Pollut Res* 23(17):17647– 17655. https://doi.org/10.1007/s11356-016-6957-x.

Saud S, Chen Y, Long B, Fahad S, Sadiq A (2013) The different impact on the growth of cool season turf grass under the various conditions on salinity and drought stress. *Int J Agric Sci Res* 3:77–84.

Saud S, Fahad S, Cui G, Chen Y, Anwar S (2020) Determining nitrogen isotopes discrimination under drought stress on enzymatic activities, nitrogen isotope abundance and water contents of Kentucky bluegrass. *Sci Rep* 10:6415. https://doi.org/10.1038/s41598-020-63548-w.

Saud S, Fahad S, Yajun C, Ihsan MZ, Hammad HM, Nasim W, Jr A, Arif M, Alharby H (2017) Effects of nitrogen supply on water stress and recovery mechanisms in Kentucky bluegrass plants. *Front Plant Sci* 8:983. https://doi.org/10.3389/fpls.2017.00983.

Saud S, Li X, Chen Y, Zhang L, Fahad S, Hussain S, Sadiq A, Chen Y (2014) Silicon application increases drought tolerance of Kentucky bluegrass by improving plant water relations and morph physiological functions. *Sci World J* 2014:1–10. https://doi.org/10.1155/2014/368694.

Senol C (2020) The effects of climate change on human behaviors. In: Fahad S, Hasanuzzaman M, Alam M, Ullah H, Saeed M, Khan AK, Adnan M (eds) *Environment, climate, plant and vegetation growth.* Springer Publ Ltd, Springer Nature Switzerland AG. Part of Springer Nature, pp 577–590. https://doi.org/10.1007/978-3-030-49732-3.

Shafi MI, Adnan M, Fahad S, Fazli W, Ahsan K, Zhen Y, Subhan D, Zafar-ul-Hye M, Martin B, Rahul D (2020) Application of single superphosphate with humic acid improves the growth, yield and phosphorus uptake of wheat (Triticum aestivum L.) in calcareous soil. *Agronomie* 10:1224. https://doi.org/10.3390/agronomy10091224.

Shah F, Lixiao N, Kehui C, Tariq S, Wei W, Chang C, Liyang Z, Farhan A, Fahad S, Huang J (2013) Rice grain yield and component responses to near 2°C of warming. *Field Crop Res* 157:98–110.

Shan C, Mei Z, Duan J, Chen H, Feng H, Cai W (2014) OsGA2ox5, a gibberellin metabolism enzyme, is involved in plant growth, the root gravity response and salt stress. *PLoS One* e87110:9.

Shi Y, Tian S, Hou L, Huang X, Zhang X, Guo H, Yang S (2012) Ethylene signaling negatively regulates freezing tolerance by repressing expression of CBF and type-A ARR genes in Arabidopsis. *Plant Cell* 24:2578–2595.

Skirycz A, Claeys H, De Bodt S, Oikawa A, Shinoda S, Andriankaja M, Maleux K, Eloy NB, Coppens F, Yoo SD et al (2011) Pause-and-stop: the effects of osmotic stress on cell proliferation during early leaf development in Arabidopsis and a role for ethylene signaling in cell cycle arrest. *Plant Cell* 23:1876–1888.

Skirycz A, De Bodt S, Obata T, De Clercq I, Claeys H, De Rycke R, Andriankaja M, Van Aken O, Van Breusegem F, Fernie AR et al (2010) Developmental stage specificity and the role of mitochondrial metabolism in the response of Arabidopsis leaves to prolonged mild osmotic stress. *Plant Physiol* 152:226–244.

Smith H, quality L (1982) *Photoperception, and plant strategy. Annu. Rev Plant Physiol* 33:481–518.

Sponsel VM, Hedden P (2004) Gibberellin, biosynthesis and inactivation. In: Davies PJ (ed) *Plant hormones biosynthesis, signal transduction, action!* Springer, Dordrecht, pp 63–94.

Sponsel VM, Hedden P (2010) Gibberellin biosynthesis and inactivation: *plant hormones*. Springer, Netherlands. pp 63–94.

Stamm P, Kumar PP (2010) The phytohormone signal network regulating elongation growth during shade avoidance. *J Exp Bot* 61:2889–2903.

Subhan D, Zafar-ul-Hye M, Fahad S, Saud S, Martin B, Tereza H, Rahul D (2020) Drought Stress Alleviation by ACC Deaminase Producing Achromobacter xylosoxidans and Enterobacter cloacae, with and without timber waste biochar in maize. *Sustain* 12(6286):doi:10.3390/su12156286.

Sun TP (2008) Gibberellin metabolism, perception and signaling pathways in Arabidopsis. *The Arabidopsis Book*. p e0103.

Sun TP (2010) Gibberellin-GID1-DELLA: a pivotal regulatory module for plant growth and development. *Plant Physiol* 154:567–570.

Sun TP, Kamiya Y (1994) The Arabidopsis GA1 locus encodes the cyclase ent-kaurene synthetase A of gibberellin biosynthesis. *Plant Cell* 6:1509–1518.

Tariq M, Ahmad S, Fahad S, Abbas G, Hussain S, Fatima Z, Nasim W, Mubeen M, ur Rehman MH, Khan MA, Adnan M (2018) The impact of climate warming and crop management on phenology of sunflower-based cropping systems in Punjab, Pakistan. *Agri and Forest Met* 256:270–282.

Unsar Naeem U, Muhammad R, Syed HMB, Asad S, Mirza AQ, Naeem I, Muhammad H ur R, Fahad S, Shafqat S (2020) Insect pests of cotton crop and management under climate change scenarios. In: Fahad S, Hasanuzzaman M, Alam M, Ullah H, Saeed M, Khan AK, Adnan M (eds) *Environment, climate, plant and vegetation growth*. Springer Publ Ltd, Springer Nature Switzerland AG. Part of Springer Nature, pp 367–396. https://doi.org/10.1007/978-3-030-49732-3.

Wahid F, Fahad S, Subhan D, Adnan M, Zhen Y, Saud S, Manzer HS, Martin B, Tereza H, Rahul D (2020) Sustainable management with mycorrhizae and phosphate solubilizing bacteria for enhanced phosphorus uptake in calcareous soils. *Agri* 10(334). https://doi.org/10.3390/agriculture10080334.

Wajid N, Ashfaq A, Asad A, Muhammad T, Muhammad A, Muhammad S, Khawar J, Ghulam MS, Syeda RS, Hafiz MH, IAR M, Muhammad ZH, Muhammad Habib ur R, Veysel T, Fahad S, Suad S, Aziz K, Shahzad A (2017) Radiation efficiency and nitrogen fertilizer impacts on sunflower crop in contrasting environments of Punjab. *Pakistan Environ Sci Pollut Res* 25:1822–1836. https://doi.org/10.1007/s11356-017-0592-z.

Wang C, Yang A, Yin H, Zhang J (2008) Influence of water stress on endogenous hormone contents and cell damage of maize seedlings. *J Integr Plant Biol* 50:427–434.

Wilkinson S, Davies WJ (2002) ABA-based chemical signalling: the coordination of responses to stress in plants. *Plant Cell Environ* 25:195–210.

Wu C, Kehui C, She T, Ganghua L, Shaohua W, Fahad S, Lixiao N, Jianliang H, Shaobing P, Yanfeng D (2020) Intensified pollination and fertilization ameliorate heat injury in rice (Oryza sativa L.) during the flowering stage. *Field Crop Res* 252:107795.

Wu C, Tang S, Li G, Wang S, Fahad S, Ding Y (2019) Roles of phytohormone changes in the grain yield of rice plants exposed to heat: a review. *PeerJ* 7:e7792. https://doi.org/10.7717/peerj.7792.

Xu K, Xu X, Fukao T, Canlas P, Maghirang-Rodriguez R, Heuer S, Ismail AM, Bailey-Serres J, Ronald PC, Mackill DJ (2006) Sub1A is an ethylene-response-factor-like gene that confers submergence tolerance to rice. *Nature* 442:705–708.

Yamaguchi S, Sun T-p, Kawaide H, Kamiya Y (1998) The GA2 locus of Arabidopsis thaliana encodes ent-kaurene synthase of gibberellin biosynthesis. *Plant Physiol* 116:1271–1278.

Yang Z, Zhang Z, Zhang T, Fahad S, Cui K, Nie L, Peng S, Huang J (2017) The effect of season-long temperature increases on rice cultivars grown in the central and southern regions of China. *Front Plant Sci* 8:1908. https://doi.org/10.3389/fpls.2017.01908.

Yang DL, Li Q, Deng YW, Lou YG, Wang MY, Zhou GX, Zhang YY, He ZH (2008) Altered disease development in the Eui mutants and Eui overexpressors indicates that gibberellins negatively regulate rice basal disease resistance. *Mol Plant* 1:528–537.

Yang DL, Yao J, Mei CS, Tong XH, Zeng LJ, Li Q, Xiao LT, Sun TP, Li J, Deng XW et al (2012) Plant hormone jasmonate prioritizes defense over growth by interfering with gibberellin signaling cascade. *Proc Natl Acad Sci U S A* 109:E1192–E1200.

Yu Z, Duan X, Luo L, Dai S, Ding Z, Xia G (2020) How Plant Hormones Mediate Salt Stress Responses. *Trends Plant Sci* 25(11):1117–1130.

Zafar-ul-Hye M, Muhammad T ahzeebulHassan, Muhammad A, Fahad S, Martin B, Tereza D, Rahul D, Subhan D 2020b Potential role of compost mixed biochar with rhizobacteria in mitigating lead toxicity in spinach. *Sci Rep* 10:12159. https://doi.org/10.1038/s41598-020-69183-9.

Zafar-ul-Hye M, Muhammad N, Subhan D, Fahad S, Rahul D, Mazhar A, Ashfaq AR, Martin B, Jiˇrí H, Zahid HT, Muhammad N (2020a) Alleviation of Cadmium Adverse Effects by Improving Nutrients Uptake in Bitter Gourd through Cadmium Tolerant Rhizobacteria. *Environment* 7(54). https://doi.org/10.3390/environments7080054.

Zahida Z, Hafiz FB, Zulfiqar AS, Ghulam MS, Fahad S, Muhammad RA, Hafiz MH, Wajid N, Muhammad S (2017) Effect of water management and silicon on germination, growth, phosphorus and arsenic uptake in rice. *Ecotoxicol Environ Saf* 144:11–18.

Zentella R, Zhang ZL, Park M, Thomas SG, Endo A, Murase K, Fleet CM, Jikumaru Y, Nambara E, Kamiya Y et al (2007) Global analysis of DELLA direct targets in early gibberellin signaling in Arabidopsis. *Plant Cell* 19:3037–3057.

Zhu N, Cheng S, Liu X, Du H, Dai M, Zhou D, Yang W, Zhao Y (2015) The R2R3-type MYB gene OsMYB91 has a function in coordinating plant growth and salt stress tolerance in rice. *Plant Sci* 236:146–156.

Zhu S, Gao F, Cao X, Chen M, Ye G, Wei C, Li Y (2005) The rice dwarf virus P2 protein interacts with ent-kaurene oxidases in vivo, leading to reduced biosynthesis of gibberellins and rice dwarf symptoms. *Plant Physiol* 139:1935–1945.

Zhu Z, An F, Feng Y, Li P, Xue LAM, Jiang Z, Kim JM, To TK, Li W et al (2011) Derepression of ethylene-stabilized transcription factors (EIN3/EIL1) mediates jasmonate and ethylene signalingsynergy in Arabidopsis. *Proc Natl Acad Sci U S A* 108:12539–12544.

Zia-ur-Rehman M (2020) Environment, climate change and biodiversity. In: Fahad S, Hasanuzzaman M, Alam M, Ullah H, Saeed M, Khan AK, Adnan M (eds) *Environment, climate, plant and vegetation growth*. Springer Publ Ltd, Springer Nature Switzerland AG. Part of Springer Nature, pp 473–502. https://doi.org/10.1007/978-3-030-49732-3.

2

Abscisic Acid and Abiotic Stress Tolerance in Crops

Abdul Rehman, Hafiza Iqra Almas, Abdul Qayyum, Hongge Li, Zhen Peng, Guangyong Qin, Yinhua Jia, Zhaoe Pan, Shoupu He, and Xiongming Du

CONTENTS

2.1 Introduction

In plants, the signalling of abscisic acid is a complex mechanism involved in significant processes at the genetic level. It is a plant hormone that contains a small structure of lipophilic sesquiterpenoid (C15) (Seo and Koshiba 2002), and its basic function is the adjustment of stress as well as work in different biological mechanisms including the germination of seeds and the dormancy of buds (Wang et al. 2016). In the 1960s, this hormone was called dormin (abscisin II), but over time, innovative research found that it accumulates in mature cotton balls that surrendered to stimulate the abscission of ethylene and over-wintering buds (Addicott et al. 1968). Later it was revealed that, in different situations and phases of development, plants face many severe conditions of drought (Willmer et al. 1978). Hence, abscisic acid is a misnomer (Addicott and Lyon 1979), whereas it plays a vital function in the dormancy of seeds and senescence of leaf potentially by osmotic stress (Carrera et al. 2008). Studies suggest that the vegetative part of plants facing drought stress accumulate about 40-times higher than normal rates of ABA under osmotic stress. Therefore, ABA is a permanent stress indicator between roots and shoots (Ko and Helariutta 2017). As a result, the investigation of spatiotemporal gene expression which regulates the metabolism of ABA is crucial to our understanding of how plants survive under stress conditions. On the other hand, ABA is also involved in adjustment functions under abiotic stress (Adnan et al. 2016; Sakthivel et al. 2016; Stec et al. 2016); hence, understanding the synthesis of ABA and its ability to act as both a necrotroph and a biotroph is recommended (Siewers et al. 2006).

Gene products that interact with the cell wall, cell membrane, and cytoskeleton are considered to be the major contributors of initial stress. For example, gated aquaporins, i.e., PIP_2 and the channels of ion/Osmo at the interaction of the plasma membrane or cell wall may be involved in the perception

TABLE 2.1

Abscisic Acid and Its Associated Genes with Abiotic Stress Tolerance

Sr. #	Gene Name	Crop	Reference
1	ZmCBL9	*Zea mays*	Zhang et al. (2016)
2	BjABR1	*Brassica juncea*	Xiang et al. (2018)
3	ABI5	*Hordeum vulgare*	Collin et al. (2020)
4	IBI1	*Arabidopsis thaliana*	Schwarzenbacher et al. (2020)
5	AtMYBL-O	*Arabidopsis thaliana*	Jeong et al. (2018)
6	MEKK	*Capsicum annuum*	Lim et al. (2020)
7	GmNAC109	*Glycine max*	Yang et al. (2019)
8	AAI3	*Syntrichia caninervis*	Zhang et al. (2018)
9	GmSUMO2	*Glycine max*	Guo et al. (2020)
10	OsDT11	*Oryza sativa*	Zhao et al. (2020)
11	TtASR1	*Triticum aestivum*	Hamdi et al. (2020)
12	SiASR4	*Setaria italica*	Li et al. (2017)
13	GmWRKY54	*Glycine max*	Wei et al. (2019)
14	IbLCYB2	*Ipomoea batatas*	Kang et al. (2018)
15	GhPYL9-11A	*Gossypium hirsutum*	Liang et al. (2017)
16	GhWRKY6	*Gossypium hirsutum*	Ullah et al. (2018)
17	NDR1/HIN1	*Arabidopsis thaliana*	Bao et al. (2016)
18	ZmbZIP4	*Zea mays*	Ma et al. (2018)

of upstream (Cui et al. 2007). The ABA receptor remained unidentified until 2009. Since then, numerous receptors of ABA have been described (Guo et al. 2011). This protein family contains 14 members, all known as regulatory ABA receptors, RCAR1 to RCAR14 (Nishimura et als. 2010) or pyrabactin resistance 1 (PYR1) and PYR1-like (PYLs) 1–13 (Park et al. 2009). The finding of PYLs components arranges the basis for straightening out the pathways of ABA signalling comprehensively. That research gave way to the advancements of signalling of ABA pathways and was properly identified as scientific advances of the year (Adler 2010). Numerous studies about structure have explained the interfaces at the genetic level, such as a cascade of signalling comprising the ABA receptors of PYL, the essential signalling mechanisms of ABA are type 2C protein phosphatases (PP2Cs), and Snf1-related protein kinases 2 (SnRK2s). The binding of ABA prompts the interface of protein PYL at the active site of PP2C and stops the activity of phosphatase by hindering the catalytic site of PP2C by the substrate of SnRK2 (Ng et al. 2011; Soon et al. 2012). Such results highlight the signalling transduction mechanisms of ABA that could unravel the resistance during abiotic stress in addition to several processes of development in plants.

Comprehensively, many researchers have discovered the specific features of responses of ABA, as well as the interaction between responses of abiotic stress and signalling pathways of ABA, CA^{2+}, ubiquitination, and MAPK. Furthermore, tolerance under abiotic stress has been studied in detail, while less importance has been placed on the formation of lateral roots and germination and development of seed (de Zelicourt et al. 2016; Edel and Kudla 2016; Liu 2012; Nakashima and Yamaguchi-Shinozaki 2013; Vishwakarma et al. 2017; Yoshida et al. 2014; Yu et al. 2016). Signalling of ABA is a complex system that works with the interaction of other pathways. Here, we combined the pathways of ABA signalling, which is a complex system and active in various processes of development in plants under abiotic stress responses. As well as we merged the recent research on the signalling mechanisms of ABA in plants, we also provide a list of abscisic acids and their associated genes with abiotic stress tolerance in Table 2.1.

2.2 Ubiquitination in ABA Signalling

Ubiquitination of post-translational modification in protein has been studied in multiple features of responses under stress and growth and development of plants (Smalle and Vierstra 2004). Meanwhile,

ABA is a chief plant hormone that has important functions in the growth of plants and responses under stress. The regulation of this signalling mechanism must be exposed by ubiquitination. Reviews of ABA in E3 ligase-mediated ubiquitination concerning signalling components have emerged (Yu et al. 2016). In this chapter, the ubiquitin E3 ligases receptors of ABA, like RCAR, PYL, and PRY are controlled by water deficit conditions through the 26S ubiquitin proteasome network. The signalling mechanism of ABA, RSL1, works as the adverse regulator by controlling the receptors of the ABA via ubiquitination. ABI1 is interlinked with PUB12 and PUB13 that are the family members of U-box protein. Mutants PUB12 and PUB13 are persuaded by ABI1 than wild type regardless of the presence of ABA; though, it can ubiquitinate ABI1 only in the presence of both ABA and PYR1, indicating that the interference between PYR1 and ABI stimulates the dehydration of ABI1 via PUB12 and PUB13 (Kong et al. 2015). In Arabidopsis, Ubiquitin E3 ligase PUB10 modulates signalling of ABA. PUB10-OX is phenocopied *myc2* of plants, while the PUB10 phenocopied *MYC2-OX* plants under the response of ABA, representing the regulation of MYC2 by PUB10 (Seo et al. 2019). E3 ligase KEG with a trimmed domain RING also works as a bait for the interaction of the CIPK26, since it works as a negative regulator during signalling of ABA (Lyzenga et al. 2013). By self ubiquitinates, ABA also persuades the KEG degradation, resulting in the collection of ABI5 (Liu and Stone 2010). ABF1 and ABf3 are degraded by KEG because it is also ubiquitinated (Cheng et al. 2013). The above-mentioned studies recommended that ABF5, ABF3, and ABI1 were substrates for KEG.

DWA2, ABD1, and DWA1 are related to Cul4-based ubiquitin E3 ligases and regulate the signalling of ABA via ABI5 degradation through the ubiquitin regulation in the nucleus by 26S ubiquitin-proteasome network (Lee et al. 2010). Single mutants, DWA2, ABD1, and DWA1 and a double mutant, DWA2 and DWA1 show phenotypes of hypersensitive of ABA in the germination of seedlings and seeds (Seo et al. 2014), that represents ABI5 works as a target for DWA2, ABD1, and DWA1 that are related with Cul4-based ubiquitin E3 ligases that control the signalling of ABA of negative regulation in the nucleus. ABI3, an interacting protein, is a functional E3 RING-type ligase, which connects with an unstable protein and is degraded by the 26S ubiquitin proteasome network (Kurup et al. 2000). Various kinds of E3 ligases with dual functions have been investigated that are involved in the regulation of signalling ABA; but the understanding about their substrates and findings associated with signalling of ABA is still in the experimental phase.

2.3 ABA Signalling under Stress

2.3.1 Ca^{2+} Signalling with ABA Interaction and Regulation of Stomata

The accumulation of ABA and ROS in plants is stimulated by abiotic stress including hydrogen peroxide (Kumar 2013; Kumar and Kesawat 2018; Kumar et al. 2014, 2017). The concentration of cytosolic Ca^{2+} is activated by the maximum concentration of hydrogen peroxide by nitric oxide (Munemasa et al. 2013). Many studies have shed light on the association between signalling of Ca^{2+} and ABA networks at various stages. For that association of signalling of Ca^{2+} and ABA, ABI1 is a major regulator (Mitula et al. 2015). In favorable growth conditions, the activity of calcium and SnRK2, SnRK 3, SnRK 6, SnRK 7, SnRK 8, and CDPK are repressed through ABI1 or PP2C. This repression inhibits signalling of downstream (Brandt et al. 2015). Under developmental and stress conditions, the presence of ABA inhibits the activity of ABI1/PP2C, which stimulates RCARs and raises the concentration of H_2O_2, and sequentially the exchange of signals of ABA into proper cellular responses, where all SnRKs, like 2, 3, 6, 7, 8/CDPK phosphorylate the targets of downstream. These are typical signalling mechanisms of ABA, though, recent research has suggested that it is a complex mechanism. Possibly, it is combined with numerous pathways of signalling, like the pathway of Ca^{2+}. In eukaryotes, calcium, in addition to ABA, represents the most important secondary messenger and participates in critical features of signalling (Edel and Kudla 2015). Signals of stress that prompt ABA at cellular levels which can also increase the important cytosolic signals of Ca^{2+} in plants, are observed downstream by CBLs/ CBL/CIPKs (Batistič and Kudla 2012). The sensor of Ca^{2+}, such as CBL/CIPK, controls various kinds of downstream targets, like regulation of ion channels and stomata (Steinhorst and Kudla 2014). CDPKs and CBL/CIPKs have the same role in the signalling of ABA (Boudsocq and Sheen 2013; Maierhofer et al. 2014). OST1 has been described as a critical regulator in signalling of ABA (Jin et al. 2013). Mutant OST1 shows a closure of defective stomata during water deficit conditions. *OST1* and SnRK2.6 have similar positional cloning (Mustilli et al. 2002). The double mutant *OST1* and SnRK2.6 disturbs the closure of stomata during stress induced by signalling of ABA and in normal conditions. The mechanism of signalling of ABA is controlled via the direct association of SnRK2.6/*OST1* and PP2CA/ABI1 (Yoshida et al. 2006).

The specific proteins of calcium-independent and -dependent signalling of the ABA network have also targeted the pathways of other signalling networks. RBOHs (Reactive burst oxidases) are phosphorylated through CBL1-CIPK26, SnRK2.6-OST1, and CPK5 (Drerup et al. 2013). SnRK2.6 is dormant at the normal concentration of ABA, but (ABI1) PP2CA stops the SLAC1 and its homologous SLAH3 (Lee et al. 2013). Moreover, SnRK2.6 cannot prevent the activity of the K^+ channel (KAT1) due to its turgor pressure and effects on stomata opening (Sato et al. 2009). Plants could maintain closed stomata to avoid the loss of dehydration under stress conditions. Signalling of ABA would tend to close the stomata. During stress, the level of ABA is increased, then it stops the activity of PP2CA and activity of phosphorylation of different calcium-dependent and -independent pathways like SnRKs, CIPK, CDPK, and CBL, respectively that follows the SLAC1 or SLAH3 phosphorylation via CPK3/6/21/23, SnRK2.6/OST1, and CBL1/9-CIPK23 (Demir et al. 2013). The activity of KAT1 is stopped by SnRK2.6/OST1 (Kulik et al. 2011) and facilitates the influx of potassium and efflux of anions and reduces the turgor pressure eventually closing the stomata.

2.3.2 ABA Signalling Pathway Integration with Abiotic Stress

ABA signalling carries significance for plants in their response to stress generated by various environmental factors, especially during various developmental processes. Different abiotic stress factors for plants include drought, cold stress, and salt stress (Fujita et al. 2011). ABA is, additionally, used by plants to analyse the impact of stress. Depending on the environmental and physiological cues, ABA signalling may fluctuate to allow for delaying of processes such as development, germination, and formation of lateral roots. ABA pathway is known to upregulate various genes when exposed to stress conditions. The presence of many different *cis*-elements was found upon the analysis of the promoter regions of ABA-inducible genes, e.g., ABREs (PyACGTGG/TC). Analysis of dehydration-inducing promoters has revealed the presence of conserved ABREs *cis*-elements (Maruyama et al. 2012).

Environmental stresses, such as increased salinity and drought induce the expression of bZIP subfamily members such as AREB1/ABF2, AREB2/ABF4, and ABF3. These factors are observed to be overexpressed in transgenic plants thus imparting drought tolerance in the transformants (Fujita et al. 2005). A triple mutant *arebl/areb2/abf3* was developed by Yoshida et al. (2010) that helped regulate the activity of AREB/ABF in vegetative tissues as a response to stress. Other stress-responsive genes lying downstream of AREB/ABF factors include group A PP2Cs and LEA proteins, and those with impaired expressions were revealed through microarray analysis with the promoters consisting of ABRE sequences. In comparison to different single and double mutants, the triple mutant *arebl/areb2/abf3* was observed to have increased sensitivity to drought while resisting primary root growth. This suggests that ABA signalling under stress was heavily dependent upon ABF3, AREB1, and AREB2 TFs. HD-ZIP TF, or the HAT1, which carries significance in BR signalling, is seen to interact with SnRK2s. It is further known to regulate ABA in response to drought, thus, hinting at the interaction of ABA with BR signalling during abiotic stress tolerance (Tan et al. 2018).

ABA signalling is induced due to abiotic stresses and is also regulated by mitogen-activated protein kinases (MAPKs), which was confirmed through studies on MAPK inhibitors. For instance, ABA-induced MAPK activation was inhibited by phenyl arsine in barley (Danquah et al. 2014). ABA activated MAPKs include MPK3, MPK4, MPK6, MPK12, and C-clade MAPKs, such as MPK1/2/7/14. Upstream activity of MAP3K13 and MAP3K18 was observed in *Arabidopsis* as a response to ABA signalling, to allow for the activation of MKK3 and MAP2K and, ultimately, the C-clade MAPKs, i.e., MPK1/2/7/14 (Danquah et al. 2015). Interactions amongst various kinases in *Nicotiana benthamiana* were analysed using the BiFC and yeast 2-hybrid techniques, wherein a reduction in the activity of MPK7 by ABA was observed (Matsuoka et al. 2015). Regulation of MAP3K17/18 activated MAP3K, which allows an important signalling module to further the MAP3K17/18-MKK3-MPK1/2/7/14 activity in ABA signalling (Boudsocq et al. 2015). A direct interaction of PP2C phosphatase ABI1 with MAP3K18 is also indicated. Additionally, expression of abiotic stress-responsive and ABA genes *RD29B* and *RAB18* is also controlled by MAP3K18, suggesting the downstream role of ABRE genes to the MAPK cascade for ABA-signalling.

2.4 ABA Signalling in Plant Development

2.4.1 ABA Signalling in Lateral Root Formation and Seed Germination

During the seed development and germination, accumulation of ABA occurs in the seeds which, upon maturity, helps to tolerate desiccation through the LEA (late embryogenesis abundant) proteins synthesis. Moreover, germination is inhibited, allowing for dormancy in mature seeds due to the presence of ABA. Sensitivity of seeds and embryonic gene expression is controlled by *ABI3* and *ABI4* (Finkelstein et al. 2002). A reduction in dormancy and viviparity is displayed by the seeds of *abi3* mutant. Seed maturation is regulated through the binding of ABI3/VP1 with Sph/RY promoters. TFs that carry structural relevance to VP1/ABI3 are encoded by the *LEAFY COTYLEDON 2* (*LEC2*) and *FUSCA3* (*FUS3*) genes, which, in turn, interact with ABI5 (Brocard-Gifford et al. 2003), though instances of direct involvement of VP1/ABI3 in ABA signalling are also reported. Screening of ABA insensitive germination identified ABA-Insensitive 5 (ABI5), a bZIP protein (Finkelstein et al. 2002). The developing seed nuclei express EEL, AREB3, and AtbZIP67/AtDPBF2 (AREB/ABF type bZIP proteins) in addition to ABI5, all of which aid in the germination of the seed. Stress situations may induce early germination and maturation of seed wherein the expression of *LEA* genes (*AtEm1* and *AtEm6*) is directly regulated by ABI5 (Ali et al. 2016; Arafat et al. 2016; Bensmihen et al. 2005). Delay of germination 1 (*DOG1*) gene expression is critical for seed dormancy which is known to interact with ABI3 to influence ABI5 expression during the development of seed in *Arabidopsis* (Dekkers et al. 2016). PGIP1/2 is also linked to germination of seeds both of which are directly impacted by ABI5 (Skubacz et al. 2016). Thus, the aforementioned examples demonstrate the significance of ABI5 for ABA signalling pathway during seed development. Expression of *ABI5* is activated via the MYB96 which negatively regulates the lateral root formation and, additionally, affects plant response to drought and salinity. Expression of *ABI5* in seeds is also negatively regulated by MYB7. Hence, these studies show inhibition in the growth of lateral roots following stress conditions due to the role played by ABI5 in the ABA signalling pathway (Kim et al. 2015).

2.4.2 ABA and Light-Signalling Convergence

For seed germination and development, endogenous hormonal cues are provided by ABA while light provides the external environmental cues. The integration of both the internal and external stimulants is critical for the plants. However, the mechanisms involved in the interaction of ABA and light signalling need to be studied in detail for clarity. Extensive studies are present on the role of HY5 TF to promote photomorphogenesis of seedling, root development, and early growth of seedling. It is reported to play a role in the germination of seeds by directly binding and regulating the expression of *ABI5*. *ABI5* expression is positively induced for ABA signalling by two important TFs of the phytochrome A pathway, i.e., the FAR-RED IMPAIRED RESPONSE 1 (FAR1) and the FAR-RED ELONGATED HYPOCOTYL3 (FHY3). Further, *ABI5* is targeted by PIL5 (or PIF1), which is a phytochrome-interacting bHLH TF (Tang et al. 2013). On the contrary, *ABI5* expression is negatively regulated by BBX21 – a TF that regulates seedling photomorphogenesis – which hinders the binding of HY5 to *ABI5* promoter. Although BBX21 inhibits the activation of ABI5 through regulation of both ABI5 and HY5 binding to ABI5 promoter, ABI5 can itself regulate its expression. Hence, various TFs control ABI5 expression in the ABA signalling pathway (Xu et al. 2014).

2.4.3 ABA Signalling and Control of Flowering Time

ABA signalling also plays a role in controlling the functioning of meristem and flowering time. An ABA-deficient mutant demonstrated an inhibitory effect on floral transition brought about by the modulation of the DELLA protein thus acting as a floral repressor. A huge effect on flowering is observed by the FLOWERING LOCUS C (FLC) which, along with the *APETALA1*, *SOC1*, and *FT*, controls germination of seeds thus suggesting it to be critical in temperature during germination of seeds (Chiang et al. 2009). ABFs make up a class of bZIP TFs that play a role in ABA signalling during the germination of seeds. FD is another bZIP protein that controls FT signalling at the shoot apex (Abe et al. 2005). Up regulation of FLC expression and delay in the initiation of flowering is induced by the overexpression of *ABI5*. A direct impact on the floral transition is brought about by the ABI5/SnRK2 phosphorylation during ABA signalling. In the absence of phosphorylation, the ABI5 prompted inhibition of floral transition is also not observed. The direct binding of ABI5 to the FLC promoter region can allow for the transactivation of the FLC expression. Transactivation of FLC expression could occur by direct binding of ABI5 to *FLC* promoter regions (Wang et al. 2013). Pre-mRNA splicing of ABI5 and FLC is carried out by AtU2AF5b, which is a putative U2AF5 spliceosome, thus playing a role in ABA-mediated flowering indicating a positive regulation by ABI5 on the FLC activity (Xiong et al. 2019).

2.5 Other Aspects of ABA Signalling

ABA is transported from its synthesis site to the various sites of action, which is aided by ABA transporters, thus making them an important part of ABA signalling. For instance, in *Arabidopsis*, AtABCG25 exports ABA from the vasculature, AtABCG40 helps in stomata closure as a response to drought by playing its role as a plasma-membrane ABA transporter, while the AtABCG30 regulates the ABA uptake by embryo, and the AtABCG31 allows endosperm secretion of ABA. All the mentioned transporters are ATP-binding cassette transporter G subfamily members (Kang et al. 2015). According to a report on peanuts (*Arachis hypogaea* L.), the cognate protein AhATL1 of *ABA transporter-like 1* (*AhATL1*) gene is a plasma membrane transporter of the ATP-binding cassette transporter G subfamily (Kang et al. 2015). Exogenous ABA treatment and water stress unregulated both the AhATL1 transcript and its cognate protein. In *Medicago truncatula*, root morphology and germination of seeds were influenced by MtABCG20 that plays the role of an ABA exporter (Pawela et al. 2019). Thus, the reports suggest tolerance to drought and growth regulation to be lined with the ABA transport system.

Other hormones related to stress response and plant growth also undergo a cross-talk with the ABA signalling such as the cytokinin, strigolactone, and karrikin. Inhibition of shoot-branching is observed following the activity of strigolactone (SL), and its biosynthesis may be regulated through ABA signalling (Kumar et al. 2019b). In Arabidopsis, a response to drought stress is seen following an antagonistic activity amongst the ABA and cytokinin signalling (Huang et al. 2018). Alterations in gene expressions related to ABA are mediated by Karrikin, and this pathway is reported to be located upstream of the ABA signalling pathway (Wang et al. 2018). Germination of seeds is significantly impacted by DELLA proteins through the DELLA/ABI3/ABI5 complex, which is seen to interact with ABA (Kumar et al. 2019a).

2.6 Conclusions

The importance of ABA as a signalling compound is quite evident. ABA signalling is greatly influenced by SnRK protein kinases for the developmental and stress responses. Other signalling components including light, SA, JA, Ca^{2+}, MAP kinase, and ET signalling are integrated by ABA signalling following environmental stresses, cues, and developmental activities, all of which are important for plant development. But more study is required to be done on the extent of the frequency of these integrations. Moreover, genome-wide studies need to be incorporated for an in-depth analysis of the complex signalling procedures, while more efficient scientific tools need to be developed to engineer plants that are tolerant of various stresses. It would greatly help if all the genes involved in ABA signalling are determined and a thorough understanding of their function is known. This would help to unravel the signalling patterns amongst various genes in combined stress conditions in association with plant developmental processes running parallel to the ABA signalling pathway.

REFERENCES

Abe M, Kobayashi Y, Yamamoto S, Daimon Y, Yamaguchi A, Ikeda Y, Ichinoki H, Notaguchi M, Goto K, Araki T (2005) FD, a bZIP protein mediating signals from the floral pathway integrator FT at the shoot apex. *Science* 309:1052–1056.

Addicott F, Lyon J (1979) Citation classic-physiology of abscisic-acid and related substances. *Curr Cont Agr Biol Envir Sci*:12–12.

Addicott F, Lyon J, Ohkuma K, et al. (1968) Abscisic acid: a new name for abscisin II (dormin). *Science* 159:1493–1493.

Adler EM (2010). 2009: signaling breakthroughs of the year. *Am Assoc Adv Sci* 3:1. https://.org/10.1126/scisignal.3103eg1.

Adnan M (2016). Evaluation of soil for important properties and chromium concentration in the basin of chromite hills in lower Malakand. *Adv Environ Biol* 10(7):141–147.

Ali S, Alam M, Basir A, Adnan M, Malik M FA, Shah AS, Ibrahim M. (2016). Effect of seed priming on germination performance and yield of okra (Abelmoschus esculentus L.). *Pak J Agric Res* 29(3):253–262.

ArafatY, Shafi M, Khan MA, Adnan M, Basir A, Arshad M, Shah JA (2016). Yield response of wheat cultivars to zinc application rates and methods. *Pure Appl Biol* 5(4):1.

Bao Y, Song W-M, Pan J, et al. (2016) Overexpression of the NDR1/HIN1-like gene NHL6 modifies seed germination in response to abscisic acid and abiotic stresses in Arabidopsis. *PLoS One* 11: e0148572. https://doi.org/10.1371/journal.pone.0148572.

Batistič O, Kudla J (2012) Analysis of calcium signaling pathways in plants. *Biochim Biophys Acta (BBA) Gen Subj* 1820:1283–1293. https://doi.org/10.1016/j.bbagen.2011.10.012.

Bensmihen S, Giraudat J, Parcy F (2005) Characterization of three homologous basic leucine zipper transcription factors (bZIP) of the ABI5 family during Arabidopsis thaliana embryo maturation. *J Exp Bot* 56:597–603.

Boudsocq M, Danquah A, de Zélicourt A, et al. (2015) Plant MAPK cascades: Just rapid signaling modules? *Plant Signal Behav* 10:e1062197. https://doi.org/10.1080/15592324.2015.1062197.

Boudsocq M, Sheen J (2013) CDPKs in immune and stress signaling. *Trends Plant Sci* 18:30–40. https://doi.org/10.1016/j.tplants.2012.08.008.

Brandt B, Munemasa S, Wang C, et al. (2015) Calcium specificity signaling mechanisms in abscisic acid signal transductionin Arabidopsis guard cells. *elife* 4:e03599. https://doi.org/10.7554/eLife.03599.002.

Brocard-Gifford IM, Lynch TJ, Finkelstein RR (2003) Regulatory networks in seeds integrating developmental, abscisic acid, sugar, and light signaling. *Plant Physiol* 131:78–92.

Carrera E, Holman T, Medhurst A, et al. (2008) Seed after-ripening is a discrete developmental pathway associated with specific gene networks in Arabidopsis. *Plant J* 53:214–224.

Cheng M-C, Liao P-M, Kuo W-W, et al. (2013) The Arabidopsis ETHYLENE RESPONSE FACTOR1 regulates abiotic stress-responsive gene expression by binding to different cis-acting elements in response to different stress signals. *Plant Physiol* 162:1566–1582. https://doi.org/10.1104/pp.113.221911.

Chiang GC, Barua D, Kramer EM, et al. (2009) Major flowering time gene, FLOWERING LOCUS C, regulates seed germination in Arabidopsis thaliana. *Proc Natl Acad Sci U S A* 106:11661–11666.

Collin A, Daszkowska-Golec A, Kurowska M, et al. (2020) Barley ABI5 (Abscisic Acid Insensitive 5) Is Involved in Abscisic Acid-Dependent Drought Response. *Front Plant Sci* 11:1138. https://doi.org/10.3389/fpls.2020.01138.

Cui J, Li P, Li G, et al. (2007) AtPID: Arabidopsis thaliana protein interactome database—an integrative platform for plant systems biology. *Nucleic Acids Res* 36:D999–D1008.

Danquah A, de Zélicourt A, Boudsocq M, et al. (2015) Identification and characterization of an ABA-activated MAP kinase cascade in Arabidopsis thaliana. *Plant J* 82:232–244. https://doi.org/10.1111/tpj.12808.

Danquah A, de Zelicourt A, Colcombet J, et al. (2014) The role of ABA and MAPK signaling pathways in plant abiotic stress responses. *Biotechnol Adv* 32:40–52. https://doi.org/10.1016/j.biotechadv.2013.09.006.

de Zelicourt A, Colcombet J, Hirt H (2016) The role of MAPK modules and ABA during abiotic stress signaling. *Trends Plant Sci* 21:677–685. https://doi.org/10.1016/j.tplants.2016.04.004.

Dekkers BJ, He H, Hanson J, et al. (2016) The Arabidopsis delay of germination 1 gene affects abscisic acid insensitive 5 (ABI 5) expression and genetically interacts with ABI 3 during Arabidopsis seed development. *Plant J* 85:451–465. https://doi.org/10.1111/tpj.13118.

Demir F, Horntrich C, Blachutzik JO, et al. (2013) Arabidopsis nanodomain-delimited ABA signaling pathway regulates the anion channel SLAH3. *Proc Natl Acad Sci U S A* 110:8296–8301. https://doi.org/10.1073/pnas.1211667110.

Drerup MM, Schlücking K, Hashimoto K, et al. (2013) The calcineurin B-like calcium sensors CBL1 and CBL9 together with their interacting protein kinase CIPK26 regulate the arabidopsis NADPH oxidase RBOHF. *Mol Plant* 6:559–569. https://doi.org/10.1093/mp/sst009.

Edel KH, Kudla J (2015) Increasing complexity and versatility: how the calcium signaling toolkit was shaped during plant land colonization. *Cell Calcium* 57:231–246. https://doi.org/10.1016/j.ceca.2014.10.013.

Edel KH, Kudla J (2016) Integration of calcium and ABA signaling. *Curr Opin Plant Biol* 33:83–91. https://doi.org/10.1016/j.pbi.2016.06.010.

Finkelstein RR, Gampala SS, Rock CD (2002) Abscisic acid signaling in seeds and seedlings. *Plant Cell* 14(Supplement):S15–S45.

Fujita Y, Fujita M, Satoh R, et al. (2005) AREB1 is a transcription activator of novel ABRE-dependent ABA signaling that enhances drought stress tolerance in Arabidopsis. *Plant Cell* 17:3470–3488.

Fujita Y, Fujita M, Shinozaki K, et al. (2011) ABA-mediated transcriptional regulation in response to osmotic stress in plants. *J Plant Res* 124:509–525. https://doi.org/10.1007/s10265-011-0412-3.

Guo J, Wang S, Wang G, et al. (2020) Overexpression of GmSUMO2 gene confers increased abscisic acid sensitivity in transgenic soybean hairy roots. *Mol Biol Rep*:1–10. https://doi.org/10.1038/s41598-020-58403-x.

Guo J, Yang X, Weston DJ, et al. (2011) Abscisic acid receptors: Past, present and future F. *J Integr Plant Biol* 53:469–479. https://doi.org/10.1111/j.1744-7909.2011.01044.x.

Hamdi K, Brini F, Kharrat N, et al. (2020) Abscisic Acid, Stress, and Ripening (TtASR1) gene as a functional marker for salt tolerance in durum wheat. *BioMed Res Int.* https://doi.org/10.1155/2020/7876357.

Huang X, Hou L, Meng J, et al. (2018) The antagonistic action of abscisic acid and cytokinin signaling mediates drought stress response in Arabidopsis. *Mol Plant* 11:970–982. https://doi.org/10.1016/j.molp.2018.05.001.

Jeong CY, Lee SA, Kang C-S, et al. (2018) Overexpression of abiotic stress-induced AtMYBL-O results in negative modulation of abscisic acid signaling through the downregulation of abscisic acid-responsive genes in Arabidopsis thaliana. *Plant Growth Regul* 84:25–36. https://doi.org/10.1007/s10725-017-0318-8.

Jin X, Wang R-S, Zhu M, et al. (2013) Abscisic acid–responsive guard cell metabolomes of Arabidopsis wild-type and gpa1 G-protein mutants. *Plant Cell* 25:4789–4811. https://doi.org/10.1105/tpc.113.119800.

Kang C, Zhai H, Xue L, et al. (2018) A lycopene β-cyclase gene, IbLCYB2, enhances carotenoid contents and abiotic stress tolerance in transgenic sweetpotato. *Plant Sci* 272:243–254. https://doi.org/10.1016/j.plantsci.2018.05.005.

Kang J, Yim S, Choi H, et al. (2015) Abscisic acid transporters cooperate to control seed germination. *Nat Commun* 6:1–10. https://doi.org/10.1038/ncomms9113.

Kim JH, Hyun WY, Nguyen HN, et al. (2015) AtM yb7, a subgroup 4 R 2 R 3 M yb, negatively regulates ABA-induced inhibition of seed germination by blocking the expression of the bZIP transcription factor ABI 5. *Plant Cell Environ* 38:559–571. https://doi.org/10.1111/pce.12415.

Ko D, Helariutta Y (2017) Shoot–root communication in flowering plants. *Curr Biol* 27:R973–R978. https://doi.org/10.1016/j.cub.2017.06.054.

Kong L, Cheng J, Zhu Y, et al. (2015) Degradation of the ABA co-receptor ABI1 by PUB12/13 U-box E3 ligases. *Nat Commun* 6:1–13. https://doi.org/10.1038/ncomms9630.

Kulik A, Wawer I, Krzywińska E, et al. (2011) SnRK2 protein kinases—key regulators of plant response to abiotic stresses. *OMICS J Integr Biol* 15:859–872. https://doi.org/10.1089/omi.2011.0091.

Kumar M (2013) Crop plants and abiotic stresses. *J Biomol Res Ther* 3. http://dx.doi.org/10.4172/2167-7956.1000e125.

Kumar M, Gho Y-S, Jung K-H, et al. (2017) Genome-wide identification and analysis of genes, conserved between japonica and indica rice cultivars, that respond to low-temperature stress at the vegetative growth stage. *Front Plant Sci* 8:1120. https://doi.org/10.3389/fpls.2017.01120.

Kumar M, Kesawat MS (2018) Mechanism of salt stress tolerance and pathways in crop plants. *Int J Genom* 2014:27–44.

Kumar M, Kim I, Kim Y-K, et al. (2019b) Strigolactone signaling genes showing differential expression patterns in arabidopsis max mutants. *Plants* 8:352. https://doi.org/10.3390/plants8090352.

Kumar M, Le DT, Hwang S, et al. (2019a) Role of the indeterminate domain Genes in Plants. *Int J Mol Sci* 20:2286. https://doi.org/10.3390/ijms20092286.

Kumar M, Lee S-C, Kim J-Y, et al. (2014) Over-expression of dehydrin gene, OsDhn1, improves drought and salt stress tolerance through scavenging of reactive oxygen species in rice (*Oryza sativa* L.). *J Plant Biol* 57:383–393.

Kurup S, Jones HD, Holdsworth MJ (2000) Interactions of the developmental regulator ABI3 with proteins identified from developing Arabidopsis seeds. *Plant J* 21:143–155.

Lee J-H, Yoon H-J, Terzaghi W, et al. (2010) DWA1 and DWA2, two Arabidopsis DWD protein components of CUL4-based E3 ligases, act together as negative regulators in ABA signal transduction. *Plant Cell* 22:1716–1732. https://doi.org/10.1105/tpc.109.073783.

Lee SC, Lim CW, Lan W, et al. (2013) ABA signaling in guard cells entails a dynamic protein–protein interaction relay from the PYL-RCAR family receptors to ion channels. *Mol Plant* 6:528–538. https://doi.org/10.1093/mp/sss078.

Li J, Dong Y, Li C, et al. (2017) SiASR4, the target gene of SiARDP from Setaria italica, improves abiotic stress adaption in plants. *Front Plant Sci* 7:2053. https://doi.org/10.3389/fpls.2016.02053.

Liang C, Liu Y, Li Y, et al. (2017) Activation of ABA receptors gene GhPYL9-11A is positively correlated with cotton drought tolerance in transgenic Arabidopsis. *Front Plant Sci* 8:1453. https://doi.org/10.3389/fpls.2017.01453.

Lim CW, Jeong S, Lee SC (2020) Differential expression of MEKK subfamily genes in Capsicum annuum L. in response to abscisic acid and drought stress. *Plant Signal Behav*:1822019. https://doi.org/10.1080/15592324.2020.1822019.

Liu H, Stone SL (2010) Abscisic acid increases Arabidopsis ABI5 transcription factor levels by promoting KEG E3 ligase self-ubiquitination and proteasomal degradation. *Plant Cell* 22:2630–2641. https://doi.org/10.1105/tpc.110.076075.

Liu Y (2012) Roles of mitogen-activated protein kinase cascades in ABA signaling. *Plant Cell Rep* 31:1–12. https://doi.org/10.1007/s00299-011-1130-y.

Lyzenga WJ, Liu H, Schofield A, et al. (2013) Arabidopsis CIPK26 interacts with KEG, components of the ABA signalling network and is degraded by the ubiquitin–proteasome system. *J Exp Bot* 64:2779–2791. https://doi.org/10.1093/jxb/ert123.

Ma H, Liu C, Li Z, et al. (2018) ZmbZIP4 contributes to stress resistance in maize by regulating ABA synthesis and root development. *Plant Physiol* 178:753–770. https://doi.org/10.1104/pp.18.00436.

Maierhofer T, Diekmann M, Offenborn JN, et al. (2014) Site-and kinase-specific phosphorylation-mediated activation of SLAC1, a guard cell anion channel stimulated by abscisic acid. *Sci Signal* 7: ra86–ra86. https://doi.org/10.1126/scisignal.2005703.

Maruyama K, Todaka D, Mizoi J, et al. (2012) Identification of cis-acting promoter elements in cold-and dehydration-induced transcriptional pathways in Arabidopsis, rice, and soybean. *DNA Res* 19:37–49. https://doi.org/10.1093/dnares/dsr040.

Matsuoka D, Yasufuku T, Furuya T, et al. (2015) An abscisic acid inducible Arabidopsis MAPKKK, MAPKKK18 regulates leaf senescence via its kinase activity. *Plant Mol Biol* 87:565–575. https://doi.org/10.1007/s11103-015-0295-0.

Mitula F, Tajdel M, Cieśla A, et al. (2015) Arabidopsis ABA-activated kinase MAPKKK18 is regulated by protein phosphatase 2C ABI1 and the ubiquitin–proteasome pathway. *Plant Cell Physiol* 56:2351–2367. https://doi.org/10.1093/pcp/pcv146.

Munemasa S, Muroyama D, Nagahashi H, et al. (2013) Regulation of reactive oxygen species-mediated abscisic acid signaling in guard cells and drought tolerance by glutathione. *Front Plant Sci* 4:472. https://doi.org/10.3389/fpls.2013.00472.

Mustilli A-C, Merlot S, Vavasseur A, et al. (2002) Arabidopsis OST1 protein kinase mediates the regulation of stomatal aperture by abscisic acid and acts upstream of reactive oxygen species production. *Plant Cell* 14:3089–3099. https://doi.org/10.1105/tpc.007906.

Nakashima K, Yamaguchi-Shinozaki K (2013) ABA signaling in stress-response and seed development. *Plant Cell Rep* 32:959–970. https://doi.org/10.1007/s00299-013-1418-1.

Ng L-M, Soon F-F, Zhou XE, et al. (2011) Structural basis for basal activity and autoactivation of abscisic acid (ABA) signaling SnRK2 kinases. *Proc Natl Acad Sci U S A* 108:21259–21264. https://doi.org/10.1073/pnas.1118651109.

Nishimura N, Sarkeshik A, Nito K, et al. (2010) PYR/PYL/RCAR family members are major in-vivo ABI1 protein phosphatase 2C-interacting proteins in Arabidopsis. *Plant J* 61:290–299. https://doi.org/10.1111/j.1365-313X.2009.04054.x.

Park S-Y, Fung P, Nishimura N, et al. (2009) Abscisic acid inhibits type 2C protein phosphatases via the PYR/PYL family of START proteins. *Science* 324:1068–1071.

Pawela A, Banasiak J, Biała W, et al. (2019) MtABCG20 is an ABA exporter influencing root morphology and seed germination of *Medicago truncatula*. *Plant J* 98:511–523. https://doi.org/10.1111/tpj.14234.

Sakthivel P, Sharma N, Klahn P, et al. (2016) Abscisic acid: a phytohormone and mammalian cytokine as novel pharmacon with potential for future development into clinical applications. *Curr Med Chem* 23:1549–1570.

Sato A, Sato Y, Fukao Y, et al. (2009) Threonine at position 306 of the KAT1 potassium channel is essential for channel activity and is a target site for ABA-activated SnRK2/OST1/SnRK2.6 protein kinase. *Biochem J* 424:439–448. https://doi.org/10.1042/bj20091221.

Schwarzenbacher RE, Wardell G, Stassen J, et al. (2020) The IBI1 receptor of β-aminobutyric acid interacts with VOZ transcription factors to regulate abscisic acid signaling and callose-associated defense. *Mol Plant* 13:1455–1469. https://doi.org/10.1016/j.molp.2020.07.010.

Seo JS, Zhao P, Jung C, et al. (2019) PLANT U-BOX PROTEIN 10 negatively regulates abscisic acid response in Arabidopsis. *Appl Biol Chem* 62:39. https://doi.org/10.1186/s13765-019-0446-0.

Seo K-I, Lee J-H, Nezames CD, et al. (2014) ABD1 is an Arabidopsis DCAF substrate receptor for CUL4-DDB1–based E3 ligases that acts as a negative regulator of abscisic acid signaling. *Plant Cell* 26:695–711. https://doi.org/10.1105/tpc.113.119974.

Seo M, Koshiba T (2002) Complex regulation of ABA biosynthesis in plants. *Trends Plant Sci* 7:41–48.

Siewers V, Kokkelink L, Smedsgaard J, et al. (2006) Identification of an abscisic acid gene cluster in the grey mold Botrytis cinerea. *Appl Environ Microbiol* 72:4619–4626.

Skubacz A, Daszkowska-Golec A, Szarejko I (2016) The role and regulation of ABI5 (ABA-Insensitive 5) in plant development, abiotic stress responses and phytohormone crosstalk. *Front Plant Sci* 7:1884. https://doi.org/10.3389/fpls.2016.01884.

Smalle J, Vierstra RD (2004) The ubiquitin 26S proteasome proteolytic pathway. *Annu Rev Plant Biol* 55:555–590.

Soon F-F, Ng L-M, Zhou XE, et al. (2012) Molecular mimicry regulates ABA signaling by SnRK2 kinases and PP2C phosphatases. *Science* 335:85–88. https://doi.org/10.1126/science.1215106.

Stec N, Banasiak J, Jasiński M (2016) Abscisic acid-an overlooked player in plant-microbe symbioses formation? *Acta Biochim Pol* 63:53–58. https://doi.org/10.18388/abp.2015_1210.

Steinhorst L, Kudla J (2014) Signaling in cells and organisms—calcium holds the line. *Curr Opin Plant Biol* 22:14–21. https://doi.org/10.1016/j.pbi.2014.08.003.

Tan W, Zhang D, Zhou H, et al. (2018) Transcription factor HAT1 is a substrate of SnRK2. 3 kinase and negatively regulates ABA synthesis and signaling in Arabidopsis responding to drought. *PLoS Genet* 14: e1007336. https://doi.org/10.1371/journal.pgen.1007336.

Tang W, Ji Q, Huang Y, et al. (2013) FAR-RED ELONGATED HYPOCOTYL3 and FAR-RED IMPAIRED RESPONSE1 transcription factors integrate light and abscisic acid signaling in Arabidopsis. *Plant Physiol* 163:857–866. https://doi.org/10.1104/pp.113.224386.

Ullah A, Sun H, Yang X, et al. (2018) A novel cotton WRKY gene, GhWRKY6-like, improves salt tolerance by activating the ABA signaling pathway and scavenging of reactive oxygen species. *Physiol Plant* 162:439–454. https://doi.org/10.1111/ppl.12651.

Vishwakarma K, Upadhyay N, Kumar N, et al. (2017) Abscisic acid signaling and abiotic stress tolerance in plants: a review on current knowledge and future prospects. *Front Plant Sci* 8:161. https://doi.org/10.3389/fpls.2017.00161.

Wang D, Gao Z, Du P, et al. (2016) Expression of ABA metabolism-related genes suggests similarities and differences between seed dormancy and bud dormancy of peach (*Prunus persica*). *Front Plant Sci* 6:1248. https://doi.org/10.3389/fpls.2015.01248.

Wang L, Waters MT, Smith SM (2018) Karrikin-KAI2 signaling provides Arabidopsis seeds with tolerance to abiotic stress and inhibits germination under conditions unfavorable to seedling establishment. *New Phytol* 219:605–618. https://doi.org/10.1111/nph.15192.

Wang Y, Li L, Ye T, et al. (2013) The inhibitory effect of ABA on floral transition is mediated by ABI5 in Arabidopsis. *J Exp Bot* 64:675–684. https://doi.org/10.1093/jxb/ers361.

Wei W, Liang DW, Bian XH, et al. (2019) GmWRKY54 improves drought tolerance through activating genes in abscisic acid and Ca2+ signaling pathways in transgenic soybean. *Plant J* 100:384–398. https://doi.org/10.1111/tpj.14449.

Willmer C, Don R, Parker W (1978) Levels of short-chain fatty acids and of abscisic acid in water-stressed and non-stressed leaves and their effects on stomata in epidermal strips and excised leaves. *Planta* 139:281–287.

Xiang L, Liu C, Luo J, et al. (2018) A tuber mustard AP2/ERF transcription factor gene, BjABR1, functioning in abscisic acid and abiotic stress responses, and evolutionary trajectory of the ABR1 homologous genes in Brassica species. *PeerJ* 6: e6071. https://doi.org/10.7717/peerj.6071.

Xiong F, Ren JJ, Yu Q, et al. (2019) AtU2 AF 65b functions in abscisic acid mediated flowering via regulating the precursor messenger RNA splicing of ABI 5 and FLC in Arabidopsis. *New Phytol* 223:277–292. https://doi.org/10.1111/nph.15756.

Xu D, Li J, Gangappa SN, et al. (2014) Convergence of light and ABA signaling on the ABI5 promoter. *PLoS Genet* 10: e1004197. https://doi.org/10.1371/journal.pgen.1004197.

Yang X, Kim MY, Ha J, et al. (2019) Overexpression of the soybean NAC gene GmNAC109 increases lateral root formation and abiotic stress tolerance in transgenic Arabidopsis plants. *Front Plant Sci* 10:1036. https://doi.org/10.3389/fpls.2019.01036.

Yoshida R, Umezawa T, Mizoguchi T, et al. (2006) The regulatory domain of SRK2E/OST1/SnRK2. 6 interacts with ABI1 and integrates abscisic acid (ABA) and osmotic stress signals controlling stomatal closure in Arabidopsis. *J Biol Chem* 281:5310–5318.

Yoshida T, Fujita Y, Sayama H, et al. (2010) AREB1, AREB2, and ABF3 are master transcription factors that cooperatively regulate ABRE-dependent ABA signaling involved in drought stress tolerance and require ABA for full activation. *Plant J* 61:672–685. https://doi.org/10.1111/j.1365-313X.2009.04092.x.

Yoshida T, Mogami J, Yamaguchi-Shinozaki K (2014) ABA-dependent and ABA-independent signaling in response to osmotic stress in plants. *Curr Opin Plant Biol* 21:133–139. https://doi.org/10.1016/j.pbi.2014.07.009.

Yu F, Wu Y, Xie Q (2016) Ubiquitin–proteasome system in ABA signaling: from perception to action. *Mol Plant* 9:21–33. https://doi.org/10.1016/j.molp.2015.09.015.

Zhang F, Li L, Jiao Z, et al. (2016) Characterization of the calcineurin B-Like (CBL) gene family in maize and functional analysis of ZmCBL9 under abscisic acid and abiotic stress treatments. *Plant Sci* 253:118–129. https://doi.org/10.1016/j.plantsci.2016.09.011.

Zhang Y, Liu X, Zhang K, et al. (2018) An ABSCISIC ACID INSENSITIVE3-like gene from the desert moss Syntrichia caninervis confers abiotic stress tolerance and reduces ABA sensitivity. *Plant Cell Tissue Organ Cult* 133:417–435. https://doi.org/10.1007/s11240-018-1394-9.

Zhao M, Ju Y, Zhao B, et al. (2020) Transcriptional profiling analysis of OsDT11-mediated ABA-dependent signal pathway for drought tolerance in rice. *Plant Biotechnol Rep* 14:613–626. https://doi.org/10.1007/s11816-020-00637-2.

3

Plant Growth Regulators' Role in Developing Cereal Crops Resilient to Climate Change

Adnan Noor Shah, Asad Abbas, Mohammad Safdar Baloch, Javaid Iqbal, Amjed Ali, Shah Fahad, and Muhammad Adnan Bukhari

CONTENTS

3.1 Introduction

Climate change threatens crop production worldwide. There is an urgent need for new strategies which equip crops with the ability to adapt to the changing environment. Yield-impacting developmental and physiological processes are significantly influenced by elevated temperatures driven by climate change (Adnan et al. 2018a,b, 2019, 2020; Ahmad et al. 2019; Akbar et al. 2020; Akram et al. 2018a,b; Amanullah et al. 2020; Amir et al. 2020; Amjad et al. 2020; Arif et al. 2020; Ayman et al. 2020; Aziz et al. 2017a,b; Baseer et al. 2019; Bayram et al. 2020; Calleja-Cabrera et al. 2020; Depeng et al. 2018; Fahad and Bano 2012; Fahad et al. 2013, 2014a,b, 2015a,b, 2016a,b,c,d, 2017, 2018, 2019a,b; Farah et al. 2020; Farhana 2020; Fazli et al. 2020; Frahat et al. 2020; Gopakumar et al. 2020; Habib et al. 2017; Hafiz et al. 2016, 2018, 2019, 2020a,b; Tariq et al. 2018; Hesham and Fahad 2020; Iqra et al. 2020; Hussain et al. 2020; Ibrar et al. 2020; Ilyas et al. 2020; Jan et al. 2019; Kamaran et al. 2017; Mahar et al. 2020; Md Jakirand Allah 2020; Md. Enamul et al. 2020; Mohammad I. Al-Wabel et al. 2020a,b; Mubeen et al. 2020; Muhammad Tahir et al. 2020; Muhmmad et al. 2019; Noor et al. 2020; Qamar et al. 2017; Rashid et al. 2020; Rehman 2020; Sadam et al. 2020; Sajid et al. 2020; Sajjad et al. 2019; Saleem et al. 2020a,b,c; Saman et al. 2020; Saud et al. 2013, 2014, 2016, 2017, 2020; Senol 2020; Shafi et al. 2020; Shah et al. 2013; Subhan et al. 2020; Unsar et al. 2020; Wahid et al. 2020; Wajid et al. 2017; Wu et al. 2019, 2020; Yang et al. 2017; Zafar-ul-Hye et al. 2020a,b; Zahida et al. 2017; Zia-ur-Rehman 2020). Porter et al. (2014) and Reynolds et al. (2016) stated that climate change affects the production of cereals primarily through heat and water stress. Under climate change circumstances, yields of wheat (*Triticum aestivum* L.), rice (*Oryza sativa* L.), and maize (*Zea mays* L.) are expected to decrease in both tropical and temperate regions (Challinor et al. 2014; Reynolds et al. 2016; Zhao et al. 2017). Garrity and O'Toole (1994) reported that drought stress greatly disturbs the process of fertilisation and anthesis in rice crops, while the harvest index is decreased 60% due to water deficit. However, in many countries, the production of wheat is seriously affected by rising temperatures, decreasing by 6% for each increase in degree Celsius (Asseng et al. 2015).

Change in climate is altering rainfall patterns. More rainfall causing flooding, drought season, and off-season precipitation are expected to occur. In different forecast models, off-season rainfalls also cause yield reductions as they effect critical developmental stages (Lobell and Burke 2008). Urban et al. (2015) stated that higher precipitation in spring increases early damage to young maize plants. Another risk related to severe precipitation is the intensification of flooding. Floods threaten food security by damaging crop areas and cause delays in planting time because of high soil moisture, while the scarcity/long gaps between rainfalls lead to drought stress (Iizumi and Ramankutty 2015; Xu et al. 2013). Calleja-Cabrera et al. (2020) demonstrated that periods of drought cause the plants to minimise water consumption, leading to stomatal closure and lower CO_2 intake. CO_2 levels, due to environmental changes in plants, retard respiration and increase temperature. Rahman et al. (2016) reported that when temperatures rise to 15–40 °C, respiration rates of the plants were elevated, disturbing the morphological features of crop plants. Thus, water and heat stress are the key determinants of cereal yields. So, plant growth regulators (PGRs) mediate environmental stress. It has been determined that the direct application of PGRs to plants increases resistance to biotic and abiotic stress, activating dormant seeds and improving water use efficiency, drought tolerance, temperature tolerance, and nitrogen use efficiency. PGR application also promotes shoot elongation, increases shoot and root mass, stimulates root growth, and promotes photosynthesis (Small et al. 2019).

3.2 Plant Growth Regulators

PGRs can be classified into natural/synthetic compounds that influence high plant production and metabolic processes, often at low dosages. They have no nutritional value and are not usually phytotoxic (Rademacher 2015). Rademacher (2015) reported that PGRs are used in agriculture for particular goals, such as improving tolerance to biotic and abiotic stress and morphology, encouraging early harvesting, quantitative and qualitative increases in yield, and alterations of plant constituents.

When herbicides are used to cause a particular beneficial improvement, they are considered to be PGRS (Nickell 1979). Compounds produced inside the plants are called plant hormones. The word hormone is derived from a Greek word that means 'to activate or improve an activity' (Avery Jr et al. 1937). Farooq et al. (2011) stated that plant hormones play a role in the induction of plant stress responses. Abiotic stresses can cause higher levels of phytohormones and reductions in plant development (Colebrook et al. 2014) particularly; drought and heat stresses have a significant influence on crop plants (Pereira 2016). Alfonso and Brüggemann (2012) found that C3 (, i.e., wheat and rice) and C4 (i.e., maize and sorghum) plants vary in their response to drought and heat stress. Another study found that C4 plants are more susceptible to drought stress because of closures of stomata and reductions in photosynthetic enzymes (Ghannoum 2008). Nevertheless, Salvucci and Crafts-Brandner (2004) reported photosynthetic capacity is stronger in C3 under high temperature. In maize, high leaf temperature (i.e., more than 38°C) inhibited the net photosynthesis (Crafts-Brandner and Salvucci 2002). We can mitigate or minimise harmful effects of drought and heat stress through the exogenous application of PGRs. In this chapter, we will discuss PGRs in developing crop species that are resilient to these stresses.

3.3 Crop Growth and Yield Responses under Stress Conditions

3.3.1 Drought Stress

While scarcity of water at any plant growth stage is important, it is most impactful during reproductive and grain-filling stages (Ahmad et al. 2020). Weak germination and disrupted seedling establishment are the initial impact of drought on the plants. Kaya et al. (2006) reported that drought stress negatively

TABLE 3.1

Drought Stress Caused Yield Reduction in Cereals

Crop plants	Stress	Yield reduction (%)	Reference
Maize	Drought	87	Menkir et al. (2003)
Wheat	Drought	57	Balla et al. (2011)
Rice	Drought	92	Lafitte et al. (2007)

TABLE 3.2

Heat Stress Caused Yield Reduction in Cereals

Crop plants	Stress	Yield reduction (%)	Reference
Maize	Heat	43	Badu-Apraku et al. (1983)
Wheat	Heat	30	Balla and Rakszegi et al. (2011)
Rice	Heat	51	Li et al. (2010)

impacts germination rate, and later, seedling vigor and growth (Farooq et al. 2009). Drought stress decreases the photosynthetic rate, effects assimilate partitioning, and weakens flag leaf development (Farooq et al. 2009; Flexas et al. 2004; Rucker et al. 1995), ultimately causing a reduction in yield. Anjum et al. (2011) found that drought stress at the tussling stage of maize causes drastic reductions in yields (Table 3.1).

3.3.2 Heat Stress

The concept of heat stress relies on the optimum tolerance range of any living organism. For crop plants, it is characterised as a situation where temperature is sufficiently high for a sufficient time which may cause irreversible injury to plant function and development. High temperature reduces the performance of plant growth compared to optimum temperature. Maestri et al. (2002) reported that heat stress negatively affected the quality and final product in cereals and also considerably decreased the starch and protein contents. In another study, Fahad et al. (2016b) demonstrated that in wheat crops high temperatures reduced the grain numbers and weight. In rice, the tillering stage was sensitive to high night temperature, which decreased individual grain weight ultimately resulting in a reduction in grain production (Fahad et al. 2016b). Similarly, Wahid and Close (2007) reported that maize growth and net assimilation rates declined under high temperature. Drastic yield reduction and losses in cereal crops are the result of heat stress (Table 3.2).

3.4 Role of PGRs in Developing Crop Species Resilient to Climate Change

Foliar and exogenous application of PGRs is an effective way to develop tolerance against biotic stress. There are five key categories of PGRS, including abscisic acid, auxin, ethylene, gibberellins, and cytokinins. Newly identified PGRs, jasmonates and brassinosteroids, have been identified for controlling phytohormonal function in crop plants (Rademacher 2015). Tao et al. (2018) stated that agronomic applications of these PGRs focus on enhancing tolerance to abiotic stress in plants. In this chapter, we will discuss a number of PGRs that help in developing crop species resilient to climate change—particularly drought and heat stress.

3.4.1 Abscisic Acid (ABA)

Abscisic acid (ABA) is known for its role in stress signal transduction during changing physiological and environmental stimuli (Tuteja 2007). It is biosynthesised when the plants are subjected to abiotic stress stimuli like salinity, drought, and change in temperature (Hamayun et al. 2010). ABA development causes plant acclimatisation and stress tolerance (O'Brien and Benková 2013). Tuteja (2007) demonstrated that ABA also influences water use efficiency during drought by regulating stomatal conductance, transpiration, and intercellular water loss (Rademacher 2015). Gong et al. (1998) found ABA-inducing thermo tolerance in maize when pre-treatment of maize is done with 0.3mM/L ABA at 46°C. Another study reported that ABA exogenous application improves heat tolerance in some crops (Calleja-Cabrera et al. 2020). Hiron and Wright (1973) reported that increased ABA levels in the leaves of wheat plants enhanced leaf resistance under high air temperature. Some other roles that ABA controls are leaf abscission and senescence, regulation of protein coding genes, morphogenesis tissue growth, and osmotic stress (O'Brien and Benková 2013; Strydhorst et al. 2018; Tuteja 2007). Thus, ABA plays an important role in developing crop species resilient to climate change.

3.4.2 Gibberellins

Gibberellins are one of the vital PGRs that regulate physiology such as organ development either by cell enhancement, cell elongation, or cell division. Gibberellins are biosynthesised during plant development and in response to changing environmental factors (Iqbal and Ashraf 2013). Gupta and Chakrabarty (2013) named gibberellins after the sequel of their discovery, i.e., GA1–GA7. Gibberellins are usually produced and used in the form of gibberellic acid, i.e., GA3. In another study, it was found that foliar application of gibberellic acid reduced the drought stress effect (Lazar 2003). Akter et al. (2010) and Yang et al. (2013) studied exogenous foliar application of gibberellins to conquer abiotic stresses during the germination and seedling establishment stages of maize and wheat. Nevertheless, at present, giberrellins' influence on crop plants' tolerance for high temperature is unclear.

3.4.3 Brassinosteroids

Brassinosteroids (BRs) belong to the class of polyhydroxy steroids. This is the only steroid-based hormone in plants. Brassinosteroids are involved in various plant processes, i.e., cell expansion and elongation and vascular differentiation (Caño-Delgado et al. 2004; Clouse and Sasse 1998). Brassinosteroids provide protection to plants under drought and chilling stress (Clouse and Sasse 1998; Rao et al. 2002). Clouse and Sasse (1998) found there are no clear links between Brassinosteroids and cell division or cell regeneration. Other studies (Sharma and Bhardwaj 2007) determined that Brassinosteroids have a negative impact on biotic and abiotic stresses in plants. Grove et al. (1979) were the first to extract brassinosteroid in the form of Brassinolide from extracts of rapeseed (*Brassica napus* L.) pollen. Brassinolides are the major class of PGRs that play an important role in the metabolism of the plant (Kim et al. 2009; (Mori and Yokota 2017). Rao et al. (2002) and Vidya Vardhini and Seeta Ram Rao (2003) demonstrated Brassinolides significantly improved germination and seedling growth of sorghum during drought. Drought tolerance can be enhanced significantly in maize by the application of jasmonates along with Brassinolides due to the defensive action of antioxidants (Li et al. 1998). In another study, Farooq et al. (2009) reported that brassinolides improved rice and maize performance under water deficit conditions, enhanced the CO2 assimilation, and improved performance under water deficit conditions by enhancing the water relation and antioxidant defense (Anjum et al. 2011). Therefore, Brassinosteroids are vital growth regulators that help to develop crop species that are resilient to climate change. A summary of PGRs that help to develop crop species resilient to climate change – particularly in cereals – is presented in Figure 3.1.

3.5 Conclusion

Field crop production is significantly affected by abiotic stresses. Plants show a variety of responses to drought and heat stresses, which are mostly described physiologically by a variety of apparent changes in the growth and development of plants. Studies show that PGRs play significant roles in protecting against drought and heat and improving yields. While heat and drought stresses can cause harm to plant growth and development, the most affected stage, which will impact yields, is reproductive growth. Just moderate

FIGURE 3.1 Different plant growth regulators and their role in developing crop species resilient to climate change.

stress at anthesis and/or the grain filling phase can significantly reduce the crop yield. Other effects of these stresses are photosynthesis and oxidative damages. There is relatively less known about the effects of drought and heat stress. Future research is required to reveal the biochemical and molecular players behind all the above growth regulators influencing processes in plants under drought and heat stresses.

REFERENCES

Adnan M, Fahad S, Khan IA, Saeed M, Ihsan MZ, Saud S, Riaz M, Wang D, Wu C. (2019). Integration of poultry manure and phosphate solubilizing bacteria improved availability of Ca bound P in calcareous soils. *Biotech* 9(10):368.

Adnan M, Fahad S, Muhammad Z, Shahen S, Ishaq AM, Subhan D, Zafar-ul-Hye M, Martin LB, Raja MMN, Beena S, Saud S, Imran A, Zhen Y, Martin B, Jiri H, Rahul D (2020) Coupling phosphate-solubilizing bacteria with phosphorus supplements improve maize phosphorus acquisition and growth under lime induced salinity stress. *Plants* 9(900); doi: 10.3390/plants9070900.

Adnan M, Shah Z, Sharif M, Rahman H (2018) Liming induces carbon dioxide (CO_2) emission in PSB inoculated alkaline soil supplemented with different phosphorus sources. *Environ Sci Pollut Res* 25(10):9501–9509.

Adnan M, Zahir S, Fahad S, Arif M, Mukhtar A, Imtiaz AK, Ishaq AM, Abdul B, Hidayat U, Muhammad A, Inayat-Ur R, Saud S, Muhammad ZI, Yousaf J, Amanullah Hafiz MH, Wajid N (2018) Phosphate-solubilizing bacteria nullify the antagonistic effect of soil calcification on bioavailability of phosphorus in alkaline soils. *Sci Rep* 8:4339. https://doi.org/10.1038/s41598-018-22653-7.

Ahmad MI, Shah AN, Sun J, Song Y (2020) Comparative study on leaf gas exchange, growth, grain yield, and water use efficiency under irrigation regimes for two maize hybrids. *Agriculture* 10:369.

Ahmad S, Kamran M, Ding R, Meng X, Wang H, Ahmad I, Fahad S, Han Q (2019) Exogenous melatonin confers drought stress by promoting plant growth, photosynthetic capacity and antioxidant defense system of maize seedlings. *PeerJ* 7:e7793. http://doi.org/10.7717/peerj.7793.

Akbar H, Timothy JK, Jagadish T, Golam M, Apurbo KC, Muhammad F, Rajan B, Fahad S, Hasanuzzaman M (2020) Agricultural land degradation: processes and problems undermining future food security. In: Fahad S, Hasanuzzaman M, Alam M, Ullah H,Saeed M, Khan AK, Adnan M (eds) *Environment, climate, plant and vegetation growth.* Springer Publ Ltd, Springer Nature Switzerland AG. Part of Springer Nature, pp 17–62. https://doi.org/10.1007/978-3-030-49732-3.

Akram R, Turan V, Hammad HM, Ahmad S, Hussain S, Hasnain A, Maqbool MM, Rehmani MIA, Rasool A, Masood N, Mahmood F, Mubeen M, Sultana SR, Fahad S, Amanet K, Saleem M, Abbas Y, Akhtar HM, Waseem F, Murtaza R, Amin A, Zahoor SA, ul Din MS, Nasim W (2018a) Fate of organic and inorganic pollutants in paddy soils. In: Hashmi MZ, Varma A (eds) *Environmental pollution of paddy soils, soil biology*. Springer International Publishing AG, Cham, Switzerland, pp. 197–214.

Akram R, Turan V, Wahid A, Ijaz M, Shahid MA, Kaleem S, Hafeez A, Maqbool MM, Chaudhary HJ, Munis, MFH, Mubeen M, Sadiq N, Murtaza R, Kazmi DH, Ali S, Khan N, Sultana SR, Fahad S, Amin A, Nasim W (2018b) Paddy land pollutants and their role in climate change. In: Hashmi MZ, Varma A (eds) *Environmental pollution of paddy soils, soil biology*. Springer International Publishing AG, Cham, Switzerland, pp 113–124.

Akter R, Hasan R, Ahmed T, Majumder MM, Akter M (2010) In vitro evaluation of antioxidant potential of leaves of Opuntia dillenii Haw. *SJ Pharm Sci* 2(1): 22–26.

Alfonso S, Brüggemann W (2012) Photosynthetic responses of a C3 and three C4 species of the genus Panicum (s.l.) with different metabolic subtypes to drought stress. *Photosynth Res* 112:175–191.

Al-Wabel Mohammad I, Abdelazeem S, Munir A, Khalid E, Adel RAU (2020b) Extent of climate change in Saudi Arabia and its impacts on agriculture: a case study from Qassim region. In: Fahad S, Hasanuzzaman M, Alam M, Ullah H, Saeed M, Khan AK, Adnan M (eds) *Environment, climate, plant and vegetation growth*. Springer Publ Ltd, Springer Nature Switzerland AG. Part of Springer Nature, pp 635–658. https://doi.org/10.1007/978-3-030-49732-3.

Al-Wabel Mohammad I, Ahmad Munir, Usman Adel RA., Akanji Mutair, Rafique Muhammad Imran (2020a) Advances in pyrolytic technologies with improved carbon capture and storage to combat climate change. In: Fahad S, Hasanuzzaman M, Alam M, Ullah H, Saeed M, Khan AK, Adnan M (eds) *Environment, climate, plant and vegetation growth*. Springer Publ Ltd, Springer Nature Switzerland AG. Part of Springer Nature, pp 535–576. https://doi.org/10.1007/978-3-030-49732-3.

Amanullah Shah K, Imran Hamdan AK, Muhammad A, Abdel RA, Muhammad A, Fahad S, Azizullah S, Brajendra P (2020) Effects of climate change on irrigation water quality. In: Fahad S, Hasanuzzaman M, Alam M, Ullah H,Saeed M, Khan AK, Adnan M (eds) *Environment, climate, plant and vegetation growth*. Springer Publ Ltd, Springer Nature Switzerland AG. Part of Springer Nature, pp 123–132. https://doi.org/10.1007/978-3-030-49732-3.

Amir M, Muhammad A, Allah B, Sevgi Ç, Haroon ZK, Muhammad A, Emre A (2020) Bio fortification under climate change: the fight between quality and quantity. In: Fahad S, Hasanuzzaman M, Alam M, Ullah H, Saeed M, Khan AK, Adnan M (eds) *Environment, climate, plant and vegetation growth*. Springer Publ Ltd, Springer Nature Switzerland AG. Part of Springer Nature, pp 173–228. https://doi.org/10.1007/978-3-030-49732-3.

Amjad I, Muhammad H, Farooq S, Anwar H (2020) Role of plant bioactives in sustainable agriculture. In: Fahad S, Hasanuzzaman M, Alam M, Ullah H, Saeed M, Khan AK, Adnan M (eds) *Environment, climate, plant and vegetation growth*. Springer Publ Ltd, Springer Nature Switzerland AG. Part of Springer Nature, pp 591–606. https://doi.org/10.1007/978-3-030-49732-3.

Anjum SA, Wang LC, Farooq M, Hussain M, Xue LL, Zou CM (2011) Brassinolide application improves the drought tolerance in maize through modulation of enzymatic antioxidants and leaf gas exchange. *J Agron Crop Sci* 197:177–185.

Arif M, Talha J, Muhammad R, Fahad S, Muhammad A, Amanullah Kawsar A, Ishaq AM, Bushra K, Fahd R (2020) Biochar; a remedy for climate change. In: Fahad S, Hasanuzzaman M, Alam M, Ullah H, Saeed M, Khan AK, Adnan M (eds) *Environment, climate, plant and vegetation growth*. Springer Publ Ltd, Springer Nature Switzerland AG. Part of Springer Nature, pp 151–172. https://doi.org/10.1007/978-3-030-49732-3.

Asseng S et al. (2015) Rising temperatures reduce global wheat production. *Nat Clim Change* 5:143–147.

Avery, Jr GS, Burkholder PR, Creighton HB (1937) Nutrient deficiencies and growth hormone concentration in helianthus and nicotiana. *Am J Bot* 24:553–557.

Aziz K, Daniel KYT, Fazal M,Muhammad ZA, Farooq S, FanW, Fahad S, Ruiyang Z (2017a) Nitrogen nutrition in cotton and control strategies for greenhouse gas emissions: a review. *Environ Sci Pollut Res* 24:23471–23487. https://doi.org/10.1007/s11356-017-0131-y.

Aziz K, Daniel KYT, Muhammad ZA, Honghai L, Shahbaz AT, Mir A, Fahad S (2017b) Nitrogen fertility and abiotic stresses management in cotton crop: a review. *Environ Sci Pollut Res* 24:14551–14566. https://doi.org/10.1007/s11356-017-8920-x.

Badu-Apraku B, Hunter R, Tollenaar M (1983) Effect of temperature during grain filling on whole plant and grain yield in Maize (Zea mays L.). *Can J Plant Sci* 63:357–363.

Balla K, Rakszegi M, Li Z, Bekes F, Bencze S, Veisz O (2011) Quality of winter wheat in relation to heat and drought shock after anthesis. *Czech J Food Sci* 29:117–128.

Baseer M, Adnan M, Fazal M, Fahad S, Muhammad S, Fazli W, Muhammad A, Jr. Amanullah Depeng W, Saud S, Muhammad N, Muhammad Z, Fazli S, Beena S, Mian AR, Ishaq AM (2019) Substituting urea by organic wastes for improving maize yield in alkaline soil. *J Plant Nutr.* doi.org/10.1080/01904167.2019.1659344.

Bayram AY, Seher Ö, Nazlican A (2020) Climate change forecasting and modeling for the year of 2050. In: Fahad S, Hasanuzzaman M, Alam M, Ullah H,Saeed M, Khan AK, Adnan M (eds) *Environment, climate, plant and vegetation growth.* Springer Publ Ltd, Springer Nature Switzerland AG. Part of Springer Nature, pp 109–122. https://doi.org/10.1007/978-3-030-49732-3.

Calleja-Cabrera J, Boter M, Oñate-Sánchez L, Pernas M (2020) Root growth adaptation to climate change in crops. *Front Plant Sci* 11: 544.

Caño-Delgado A, Yin Y, Yu C, Vafeados D, Mora-García S, Cheng JC, Nam KH, Li J, Chory J (2004) BRL1 and BRL3 are novel brassinosteroid receptors that function in vascular differentiation in Arabidopsis. *Development (Cambridge England)* 131:5341–5351.

Challinor AJ, Watson J, Lobell DB, Howden SM, Smith DR, Chhetri N (2014) A meta-analysis of crop yield under climate change and adaptation. *Nat Clim Change* 4:287–291.

Clouse SD, Sasse JM (1998) Brassinosteroids: essential regulators of plant growth and development. *Annu Rev Plant Physiol Plant Mol Biol* 49:427–451.

Colebrook EH, Thomas SG, Phillips AL, Hedden P (2014) The role of gibberellin signalling in plant responses to abiotic stress. *J Exp Biol* 217:67–75.

Crafts-Brandner SJ, Salvucci ME (2002) Sensitivity of photosynthesis in a C4 plant, maize, to heat stress. *Plant Physiol* 129:1773–1780.

Depeng W, Fahad S, Saud S, Muhammad K, Aziz K, Mohammad NK, Hafiz MH, Wajid N (2018) Morphological acclimation to agronomic manipulation in leaf dispersion and orientation to promote "Ideotype" breeding: evidence from 3D visual modeling of "super" rice (*Oryza sativa* L.).*Plant Physiol Biochem* https://doi.org/10.1016/j.plaphy.2018.11.010.

Fahad S, Adnan M, Hassan S, Saud S, Hussain S, Wu C, Wang D, Hakeem KR, Alharby HF, Turan V, Khan MA, Huang J (2019b) Rice responses and tolerance to high temperature. In: Hasanuzzaman M, Fujita M, Nahar K, Biswas JK (eds) *Advances in rice research for abiotic stress tolerance.* Woodhead Publ Ltd, Cambridge, England, pp 201–224.

Fahad S, Bajwa AA, Nazir U, Anjum SA, Farooq A, Zohaib A, Sadia S, NasimW, Adkins S, Saud S, Ihsan MZ, Alharby H,Wu C,Wang D, Huang J (2017) Crop production under drought and heat stress: plant responses and management options. *Front Plant Sci* 8:1147. https://doi.org/10.3389/fpls.2017.01147.

Fahad S, Bano A (2012) Effect of salicylic acid on physiological and biochemical characterization of maize grown in saline area. *Pak J Bot* 44:1433–1438.

Fahad S, Chen Y, Saud S, Wang K, Xiong D, Chen C,Wu C, Shah F, Nie L, Huang J (2013) Ultraviolet radiation effect on photosynthetic pigments, biochemical attributes, antioxidant enzyme activity and hormonal contents of wheat. *J Food, Agri Environ* 11(3&4):1635–1641.

Fahad S, Hussain S, Bano A, Saud S, Hassan S, Shan D, Khan FA, Khan F, Chen Y, Wu C, Tabassum MA, Chun MX, Afzal M, Jan A, Jan MT, Huang J (2014a) Potential role of phytohormones and plant growth-promoting rhizobacteria in abiotic stresses: consequences for changing environment. *Environ Sci Pollut Res* 22(7):4907–4921. https://doi.org/10.1007/s11356-014-3754-2.

Fahad S, Hussain S, Matloob A, Khan FA, Khaliq A, Saud S, Hassan S, Shan D, Khan F, Ullah N, Faiq M, Khan MR, Tareen AK, Khan A, Ullah A, Ullah N, Huang J (2014b) Phytohormones and plant responses to salinity stress: a review. *Plant Growth Regul* 75(2):391–404. https://doi.org/10.1007/s10725-014-0013-y.

Fahad S, Hussain S, Saud S, Hassan S, Chauhan BS, Khan F et al (2016a) Responses of rapid viscoanalyzer profile and other rice grain qualities to exogenously applied plant growth regulators under high day and high night temperatures. *PLoS One* 11(7):e0159590. https://org/10.1371/journal.pone.0159590.

Fahad S, Hussain S, Saud S, Hassan S, Ihsan Z, Shah AN,Wu C, Yousaf M, Nasim W, Alharby H, Alghabari F, Huang J (2016c) Exogenously applied plant growth regulators enhance the morphophysiological growth and yield of rice under high temperature. *Front Plant Sci* 7:1250. https://doi.org/10.3389/fpls.2016.01250.

Fahad S, Hussain S, Saud S, Hassan S, Tanveer M, Ihsan MZ, Shah AN, Ullah A, Nasrullah KF, Ullah S, Alharby HNW, Wu C, Huang J (2016d) A combined application of biochar and phosphorus alleviates heat-induced adversities on physiological, agronomical and quality attributes of rice. *Plant Physiol Biochem* 103:191–198.

Fahad S, Hussain S, Saud S, Khan F, Hassan S, Jr A, Nasim W, Arif M, Wang F, Huang J (2016b) Exogenously applied plant growth regulators affect heat-stressed rice pollens. *J Agron Crop Sci* 202:139–150.

Fahad S, Hussain S, Saud S, Tanveer M, Bajwa AA, Hassan S, Shah AN, Ullah A,Wu C, Khan FA, Shah F, Ullah S, Chen Y, Huang J (2015a) A biochar application protects rice pollen from high-temperature stress. *Plant Physiol Biochem* 96:281–287.

Fahad S, Muhammad ZI, Abdul K, Ihsanullah D, Saud S, Saleh A, Wajid N, Muhammad A, Imtiaz AK, Chao W, Depeng W, Jianliang H (2018): Consequences of high temperature under changing climate optima for rice pollen characteristics-concepts and perspectives, *Arch Agron Soil Sci* DOI: 10.1080/03650340.2018.1443213.

Fahad S, Nie L, Chen Y, Wu C, Xiong D, Saud S, Hongyan L, Cui K, Huang J (2015b) Crop plant hormones and environmental stress. *Sustain Agric Res* 15:371–400.

Fahad S, Rehman A, Shahzad B, Tanveer M, Saud S, Kamran M, Ihtisham M, Khan SU, Turan V, Rahman MHU (2019a). Rice responses and tolerance to metal/metalloid toxicity. In: Hasanuzzaman M, Fujita M, Nahar K, Biswas JK (eds) *Advances in rice research for abiotic stress tolerance.* Woodhead Publ Ltd, Cambridge, England, pp 299–312.

Farah R, Muhammad R, Muhammad SA, Tahira Y, Muhammad AA, Maryam A, Shafaqat A, Rashid M, Muhammad R, Qaiser H, Afia Z, Muhammad AA, Muhammad A, Fahad S (2020) Alternative and non-conventional soil and crop management strategies for increasing water use efficiency. In: Fahad S, Hasanuzzaman M, Alam M, Ullah H, Saeed M, Khan AK, Adnan M (eds) *Environment, climate, plant and vegetation growth.* Springer Publ Ltd, Springer Nature Switzerland AG. Part of Springer Nature, pp. 323–338. https://doi.org/10.1007/978-3-030-49732-3.

Farhana G, Ishfaq A, Muhammad A, Dawood J, Fahad S, Xiuling L, Depeng W, Muhammad F, Muhammad F, Syed AS (2020) Use of crop growth model to simulate the impact of climate change on yield of various wheat cultivars under different agro-environmental conditions in Khyber Pakhtunkhwa, Pakistan. *Arabian J Geosci* 13:112 https://doi.org/10.1007/s12517-020-5118-1.

Farhat A, Hafiz MH, Wajid I, Aitazaz AF, Hafiz FB, Zahida Z, Fahad S, Wajid F, Artemi C (2020) A review of soil carbon dynamics resulting from agricultural practices. *J Environ Manage* 268 (2020) 110319.

Farooq M, Bramley H, Palta JA, Siddique KHM (2011) Heat stress in wheat during reproductive and grain-filling phases. *Crit Rev Plant Sci* 30:491–507.

Farooq M, Wahid A, Kobayashi N, Fujita D, Basra SMA (2009) Plant drought stress: effects, mechanisms and management. *Agron Sustain Dev* 29:185–212.

Fazli W, Muhmmad S, Amjad A, Fahad S, Muhammad A, Muhammad N, Ishaq AM, Imtiaz AK, Mukhtar A, Muhammad S, Muhammad I, Rafi U, Haroon I, Muhammad A (2020) Plant-microbes interactions and functions in changing climate. In: Fahad S, Hasanuzzaman M, Alam M, Ullah H, Saeed M, Khan AK, Adnan M (eds) *Environment, climate, plant and vegetation growth.* Springer Publ Ltd, Springer Nature Switzerland AG. Part of Springer Nature, pp. 397–420. https://doi.org/10.1007/978-3-030-49732-3.

Flexas J, Bota J, Loreto F, Cornic G, Sharkey TD (2004) Diffusive and metabolic limitations to photosynthesis under drought and salinity in C3 plants. *Plant Biol* 6:269–279.

Garrity DP, O'Toole JC (1994) Screening rice for drought resistance at the reproductive phase. *Field Crops Res* 39:99–110.

Ghannoum O (2008) C4 photosynthesis and water stress. *Ann Bot* 103:635–644.

Gong M, Li YJ, Chen SZ (1998) Abscisic acid-induced thermotolerance in maize seedlings is mediated by calcium and associated with antioxidant systems. *J Plant Physiol* 153: 488–496.

Gopakumar L, Bernard NO, Donato V (2020) Soil microarthropods and nutrient cycling. In: Fahad S, Hasanuzzaman M, Alam M, Ullah H, Saeed M, Khan AK, Adnan M (eds) *Environment, climate, plant and vegetation growth.* Springer Publ Ltd, Springer Nature Switzerland AG. Part of Springer Nature, pp 453–472. https://doi.org/10.1007/978-3-030-49732-3.

Grove MD, Spencer GF, Rohwedder W, Mandava N, Worley J, Warthen J, Steffens GL, Flippen-Anderson J, Cook J (1979) Brassinolide, a plant growth-promoting steroid isolated from Brassica napus pollen. *Nature* 281:216–217.

Gupta R, Chakrabarty SK (2013) Gibberellic acid in plant. *Plant Signal Behav* 8:e25504.

Habib ur R, Ashfaq A, Aftab W, Manzoor H, Fahd R, Wajid I, Md Aminul I, Vakhtang S, Muhammad A, Asmat U, Abdul W, Syeda RS, Shah S, Shahbaz K, Fahad S, Manzoor H, Saddam H, Wajid N (2017) Application of CSM-CROPGRO-Cotton model for cultivars and optimum planting dates: evaluation in changing semi-arid climate. *Field Crops Res* http://dx.doi.org/10.1016/j.fcr.2017.07.007.

Hafiz MH, Abdul K, Farhat A, Wajid F, Fahad S, Muhammad A, Ghulam MS, Wajid N, Muhammad M, Hafiz FB (2020b) Comparative effects of organic and inorganic fertilizers on soil organic carbon and wheat productivity under arid region. *Commun Soil Sci Plant Anal*. Doi: 10.1080/00103624.2020.1763385.

Hafiz MH, Farhat A, Ashfaq A, Hafiz FB, Wajid F, Carol Jo W, Fahad S, Gerrit H (2020a) Predicting Kernel growth of maize under controlled water and nitrogen applications. *Int J Plant Prod* https://doi.org/10.1007/s42106-020-00110-8.

Hafiz MH, Farhat A, Shafqat S, Fahad S, Artemi C, Wajid F, Chaves CB, Wajid N, Muhammad M, Hafiz FB (2018) Offsetting land degradation through nitrogen and water management during maize cultivation under arid conditions. *Land Degrad Dev* 1–10. DOI: 10.1002/ldr.2933.

Hafiz MH, Muhammad A, Farhat A, Hafiz FB, Saeed AQ, Muhammad M, Fahad S, Muhammad A (2019) Environmental factors affecting the frequency of road traffic accidents: a case study of sub-urban area of Pakistan. *Environ Sci Pollut Res* https://doi.org/10.1007/s11356-019-04752-8.

Hafiz MH, Wajid F, Farhat A, Fahad S, Shafqat S, Wajid N, Hafiz FB (2016) Maize plant nitrogen uptake dynamics at limited irrigation water and nitrogen. *Environ Sci Pollut Res* 24(3):2549–2557. https://doi.org/10.1007/s11356-016-8031-0.

Hamayun M, Khan SA, Khan AL, Shin JH, Ahmad B, Shin DH, Lee IJ (2010) Exogenous gibberellic acid reprograms soybean to higher growth and salt stress tolerance. *J Agric Food Chem* 58:7226–7232.

Hesham FA and Fahad S (2020) Melatonin application enhances biochar efficiency for drought tolerance in maize varieties: Modifications in physio-biochemical machinery. *Agron J* 112(4):1–22.

Hiron RWP, Wright STC (1973) the role of endogenous abscisic acid in the response of plants to stress. *J Exp Bot* 24: 769–780.

Hussain MA, Fahad S, Rahat S, Muhammad FJ, Muhammad M, Qasid A, Ali A, Husain A, Nooral A, Babatope SA, Changbao S, Liya G, Ibrar A, Zhanmei J, Juncai H (2020) Multifunctional role of brassinosteroid and its analogues in plants. *Plant Growth Regul* https://doi.org/10.1007/s10725-020-00647-8.

Ibrar K, Aneela R, Khola Z, Urooba N, Sana B, Rabia S, Ishtiaq H, Mujaddad Ur Rehman, Salvatore M (2020) Microbes and environment: global warming reverting the frozen zombies. In: Fahad S, Hasanuzzaman M, Alam M, Ullah H, Saeed M, Khan AK, Adnan M (eds) *Environment, climate, plant and vegetation growth*. Springer Publ Ltd, Springer Nature Switzerland AG. Part of Springer Nature, pp. 607–634. https://doi.org/10.1007/978-3-030-49732-3.

Iizumi T, Ramankutty N (2015) How do weather and climate influence cropping area and intensity? *Glob Food Sec* 4:46–50.

Ilyas M, Mohammad N, Nadeem K, Ali H, Aamir HK, Kashif H, Fahad S, Aziz K, Abid U, (2020) Drought tolerance strategies in plants: a mechanistic approach. *J Plant Growth Regul* https://doi.org/10.1007/s00344-020-10174-5.

Iqbal M, Ashraf M (2013) Gibberellic acid mediated induction of salt tolerance in wheat plants: growth, ionic partitioning, photosynthesis, yield and hormonal homeostasis. *Environ Exp Bot* 86:76–85.

Iqra M, Amna B, Shakeel I, Fatima K,Sehrish L, Hamza A, Fahad S (2020) Carbon cycle in response to global warming. In: Fahad S, Hasanuzzaman M, Alam M, Ullah H,Saeed M, Khan AK, Adnan M (eds) *Environment, climate, plant and vegetation growth*. Springer Publ Ltd, Springer Nature Switzerland AG. Part of Springer Nature, pp 1–16. https://doi.org/10.1007/978-3-030-49732-3.

Jan M, Muhammad Anwar-ul-Haq, Adnan NS, Muhammad Y, Javaid I, Xiuling L, Depeng W, Fahad S, (2019) Modulation in growth, gas exchange, and antioxidant activities of salt-stressed rice (Oryza sativa L.) genotypes by zinc fertilization. *Arabian J Geosci* 12:775 https://doi.org/10.1007/s12517-019-4939-2.

Kamarn M, Wenwen C, Irshad A, Xiangping M, Xudong Z, Wennan S, Junzhi C, Shakeel A, Fahad S, Qingfang H, Tiening L (2017) Effect of paclobutrazol, a potential growth regulator on stalk mechanical strength, lignin accumulation and its relation with lodging resistance of maize. *Plant Growth Regul* 84:317–332. https://doi.org/10.1007/ s10725-017-0342-8.

Kaya MD, Okcu G, Atak M, Cikili Y, Kolsarici O (2006) Seed treatments to overcome salt and drought stress during germination in sunflower (Helianthus annuus L.). European journal of agronomy. *J Eur Soc Agron* 24:291–295.

Kim TW, Guan S, Sun Y, Deng Z, Tang W, Shang JX, Sun Y, Burlingame AL, Wang ZY (2009) Brassinosteroid signal transduction from cell-surface receptor kinases to nuclear transcription factors. *Nat Cell Biol* 11:1254–1260.

Lafitte HR, Yongsheng G, Yan S, Li ZK (2007) Whole plant responses, key processes, and adaptation to drought stress: the case of rice. *J Exp Bot* 58:169–175.

Lazar T (2003) Taiz, L. and Zeiger, E. Plant physiology. *Ann Bot* 91:750–751.

Li L, Staden J, Jäger AK (1998) Effects of plant growth regulators on the antioxidant system in seedlings of two maize cultivars subjected to water stress. *Plant Growth Regul* 25:81–87.

Li Z, Peng T, Xie Q, Han S, Tian J (2010) Mapping of QTL for tiller number at different stages of growth in wheat using double haploid and immortalized F 2 populations. *J Genet* 89:409–415.

Lobell D, Burke M (2008) Why are agricultural impacts of climate change so uncertain? the importance of temperature relative to precipitation. *Environ Res Lett* 3: 034007.

Maestri E, Klueva N, Perrotta C, Gulli M, Nguyen HT, Marmiroli N (2002) Molecular genetics of heat tolerance and heat shock proteins in cereals. *Plant Mol Biol* 48:667–681.

Mahar A, Amjad A, Altaf HL, Fazli W, Ronghua L, Muhammad A, Fahad S, Muhammad A, Rafiullah Imtiaz AK, Zengqiang Z (2020) Promising technologies for cd-contaminated soils: drawbacks and possibilities. In: Fahad S, Hasanuzzaman M, Alam M, Ullah H,Saeed M, Khan AK, Adnan M (eds) *Environment, climate, plant and vegetation growth.* Springer Publ Ltd, Springer Nature Switzerland AG. Part of Springer Nature, pp. 63–92. https://doi.org/10.1007/978-3-030-49732-3.

Menkir A, Badu-Apraku B, Ibikunle O (2003) The influence of drought stress on growth, yield and yield components of selected maize genotypes. *J Agric Sci* 141:43–50.

Md Jakir H, Allah B (2020) Development and applications of transplastomic plants; a way towards eco-friendly agriculture. In: Fahad S, Hasanuzzaman M, Alam M, Ullah H, Saeed M, Khan AK, Adnan M (eds) *Environment, climate, plant and vegetation growth.* Springer Publ Ltd, Springer Nature Switzerland AG. Part of Springer Nature, pp 285–322. https://doi.org/10.1007/978-3-030-49732-3.

Md Enamul H, Shoeb AZM, Mallik AH, Fahad S, Kamruzzaman MM, Akib J, Nayyer S, Mehedi Adhan KM, Swati AS, Md Yeamin A, Most SS (2020) Measuring vulnerability to environmental hazards: qualitative to quantitative. In: Fahad S, Hasanuzzaman M, Alam M, Ullah H, Saeed M, Khan AK, Adnan M (eds) *Environment, climate, plant and vegetation growth.* Springer Publ Ltd, Springer Nature Switzerland AG. Part of Springer Nature, pp 421–452. https://doi.org/10.1007/978-3-030-49732-3.

Mori K, Yokota T (2017) Molecular structure and biological activity of brassinolide and related brassinosteroids. In: Martin Bohl, William L. Duax (eds) *Molecular structure and biological activity of steroids*, CRC Press, New York, pp. 317–340.

Mubeen M, Ashfaq A, Hafiz MH, Muhammad A, Hafiz UF, Mazhar S, Muhammad Sami ul Din, Asad A, Amjed A, Fahad S, Wajid N (2020) Evaluating the climate change impact on water use efficiency of cotton-wheat in semi-arid conditions using DSSAT model. *J Water Clim Change.* doi:10.2166/wcc.2019.179/622035/jwc2019179.pdf.

Muhammad Z, Abdul MK, Abdul MS, Kenneth BM, Muhammad S, Shahen S, Ibadullah J, Fahad S (2019) Performance of Aeluropus lagopoides (mangrove grass) ecotypes, a potential turfgrass, under high saline conditions. *Environ Sci Pollut Res* https://doi.org/10.1007/s11356-019-04838-3.

Nickell LG (1979) *Plant growth substances.* American Chemical Society, USA.

Noor M, Naveed ur R, Ajmal J, Fahad S, Muhammad A, Fazli W, Saud S, Hassan S (2020) Climate change and costal plant lives. In: Fahad S, Hasanuzzaman M, Alam M, Ullah H,Saeed M, Khan AK, Adnan M (eds) *Environment, climate, plant and vegetation growth.* Springer Publ Ltd, Springer Nature Switzerland AG. Part of Springer Nature, pp 93–108. https://doi.org/10.1007/978-3-030-49732-3.

O'Brien JA, Benková E (2013) Cytokinin cross-talking during biotic and abiotic stress responses. *Front Plant Sci* 4:451–451.

Pereira A (2016) Plant abiotic stress challenges from the changing environment. *Front Plant Sci* 7:1123–1123.

Porter JR, Xie L, Challinor AJ,Cochrane K, Howden SM, Iqbal MM, Lobell DB, Travasso MI, NetraChhetri NC, Garrett K, Ingram J, Lipper L, McCarthy N, McGrath J, Smith D, Thornton P, Watson J, Ziska L (2014) *Foodsecurity and food production systems.* Cambridge University Press, Cambridge, United Kingdom and New York, NY, USA pp. 485–533.

Qamar-uz Z, Zubair A, Muhammad Y, Muhammad ZI, Abdul K, Fahad S, Safder B, Ramzani PMA, Muhammad N (2017) Zinc biofortification in rice: leveraging agriculture to moderate hidden hunger in developing countries. *Arch Agron Soil Sci* 64:147–161. https://doi.org/10.1080/03650340.2017.1338343.

Rademacher W (2015) Plant growth regulators: backgrounds and uses in plant production. *J Plant Growth Regul* 34:845–872.

Rahman I, Ali S, Rahman I, Adnan M, Ullah H, Basir A, Malik FA, Shah AS, Ibrahim M, Arshad M (2016). Effect of pre-storage seed priming on biochemical changes in okra seed. *Pure Appl Biol* 5(1): 165–171.

Rao SSR, Vardhini BV, Sujatha E, Anuradha S (2002) Brassinosteroids—a new class of phytohormones. *Curr Sci* 82:1239–1245.

Rashid M, Qaiser H, Khalid SK, Mohammad I. Al-Wabel, Zhang A, Muhammad A, Shahzada SI, Rukhsanda A, Ghulam AS, Shahzada MM, Sarosh A, Muhammad FQ (2020) Prospects of biochar in alkaline soils to mitigate climate change. In: Fahad S, Hasanuzzaman M, Alam M, Ullah H,Saeed M, Khan AK, Adnan M (eds) *Environment, climate, plant and vegetation growth.* Springer Publ Ltd, Springer Nature Switzerland AG. Part of Springer Nature, pp 133–150. https://doi.org/10.1007/978-3-030-49732-3.

Rehman M, Fahad S, Saleem MH, Hafeez M, ur Rahman Muhammad Habib, Liu F, Deng G (2020) Red light optimized physiological traits and enhanced the growth of ramie (Boehmeria nivea L.). *Photosynthetica* 58(4):922–931.

Reynolds MP et al. (2016) An integrated approach to maintaining cereal productivity under climate change. *Glob Food Sec* 8:9–18.

Rucker KS, Kvien CK, Holbrook CC, Hook JE (1995) Identification of peanut genotypes with improved drought avoidance traits. *Peanut Sci* 22:14–18.

Sabagh Ayman EL, Hossain Akbar, Barutçular Celaleddin, Iqbal Muhammad Aamir, Sohidul Islam M, Fahad Shah, Sytar Oksana, Çig Fatih, Meena Ram Swaroop, Erman Murat (2020) Consequences of salinity stress on the quality of crops and its mitigation strategies for sustainable crop production: an outlook of arid and semi- arid regions. In: Fahad S, Hasanuzzaman M, Alam M, Ullah H, Saeed M, Khan AK, Adnan M (eds) *Environment, climate, plant and vegetation growth.* Springer Publ Ltd, Springer Nature Switzerland AG. Part of Springer Nature, pp 503–534. https://doi.org/10.1007/978-3-030-49732-3.

Sadam M, ul Qamar Muhammad Tahir, Ghulam M, Muhammad SK, Faiz AJ (2020) Role of biotechnology in climate resilient agriculture. In: Fahad S, Hasanuzzaman M, Alam M, Ullah H, Saeed M, Khan AK, Adnan M (eds) *Environment, climate, plant and vegetation growth.* Springer Publ Ltd, Springer Nature Switzerland AG. Part of Springer Nature, pp 339–366. https://doi.org/10.1007/978-3-030-49732-3.

Sajid H, Jie H, Jing H, Shakeel A, Satyabrata N, Sumera A, Awais S, Chunquan Z, Lianfeng Z, Xiaochuang C, Qianyu J, Junhua Z (2020) Rice production under climate change: adaptations and mitigating strategies. In: Fahad S, Hasanuzzaman M, Alam M, Ullah H, Saeed M, Khan AK, Adnan M (eds) *Environment, climate, plant and vegetation growth.* Springer Publ Ltd, Springer Nature Switzerland AG. Part of Springer Nature, pp 659–686. https://doi.org/10.1007/978-3-030-49732-3.

Sajjad H, Muhammad M, Ashfaq A, Waseem A, Hafiz MH, Mazhar A, Nasir M, Asad A, Hafiz UF, Syeda RS, Fahad S, Depeng W, Wajid N (2019) Using GIS tools to detect the land use/land cover changes during forty years in Lodhran district of Pakistan. *Environ Sci Pollut Res* https://doi.org/10.1007/s11356-019-06072-3.

Saleem MH, Fahad S, Adnan M, Mohsin A, Muhammad SR, Muhammad K, Qurban A, Inas AH, Parashuram B, Mubassir A, Reem MH (2020a) Foliar application of gibberellic acid endorsed phytoextraction of copper and alleviates oxidative stress in jute (Corchorus capsularis L.) plant grown in highly copper-contaminated soil of China. *Environ Sci Pollut Res* https://doi.org/10.1007/s11356-020-09764-3.

Saleem MH, Fahad S, Shahid UK, Mairaj D, Abid U, Ayman ELS, Akbar H, Analía L, Lijun L (2020c) Copper-induced oxidative stress, initiation of antioxidants and phytoremediation potential of flax (Linum usitatissimum L.) seedlings grown under the mixing of two different soils of China. *Environ Sci Pollut Res* https://doi.org/10.1007/s11356-019-07264-7.

Saleem MH, Rehman M, Fahad S, Tung SA, Iqbal N, Hassan A, Ayub A, Wahid MA, Shaukat S, Liu L, Deng G (2020b) Leaf gas exchange, oxidative stress, and physiological attributes of rapeseed (Brassica napus L.) grown under different light-emitting diodes. *Photosynthetica* 58 (3): 836–845.

Salvucci ME, Crafts-Brandner SJ (2004) Relationship between the heat tolerance of photosynthesis and the thermal stability of rubisco activase in plants from contrasting thermal environments. *Plant Physiol* 134:1460–1470.

Saman S, Amna B, Bani A, ul Qamar Muhammad Tahir, Rana MA, Muhammad SK (2020) QTL mapping for abiotic stresses in cereals. in: Fahad S, Hasanuzzaman M, Alam M, Ullah H, Saeed M, Khan AK, Adnan M (eds) *Environment, climate, plant and vegetation growth.* Springer Publ Ltd, Springer Nature Switzerland AG. Part of Springer Nature, pp 229–252. https://doi.org/10.1007/978-3-030-49732-3.

Saud S, Chen Y, Fahad S, Hussain S, Na L, Xin L, Alhussien SA (2016) Silicate application increases the photosynthesis and its associated metabolic activities in Kentucky bluegrass under drought stress and post-drought recovery. *Environ Sci Pollut Res* 23(17):17647–17655. https://doi.org/10.1007/s11356-016-6957-x.

Saud S, Chen Y, Long B, Fahad S, Sadiq A (2013) The different impact on the growth of cool season turf grass under the various conditions on salinity and drought stress. *Int J Agric Sci Res* 3:77–84.

Saud S, Fahad S, Cui G, Chen Y, Anwar S (2020) Determining nitrogen isotopes discrimination under drought stress on enzymatic activities, nitrogen isotope abundance and water contents of Kentucky bluegrass. *Sci Rep* 10:6415 | https://doi.org/10.1038/s41598-020-63548-w.

Saud S, Fahad S, Yajun C, Ihsan MZ, Hammad HM, Nasim W, Amanullah Jr Arif M and Alharby H (2017) Effects of nitrogen supply on water stress and recovery mechanisms in Kentucky bluegrass plants. *Front Plant Sci* 8:983. doi: 10.3389/fpls.2017.00983.

Saud S, Li X, Chen Y, Zhang L, Fahad S, Hussain S, Sadiq A, Chen Y (2014) Silicon application increases drought tolerance of Kentucky bluegrass by improving plant water relations and morph physiological functions. *Sci World J* 2014:1–10. https://doi.org/10.1155/2014/ 368694.

Senol C (2020) The effects of climate change on human behaviors. In: Fahad S, Hasanuzzaman M, Alam M, Ullah H, Saeed M, Khan AK, Adnan M (eds) *Environment, climate, plant and vegetation growth.* Springer Publ Ltd, Springer Nature Switzerland AG. Part of Springer Nature, pp 577–590. https://doi.org/10.1007/978-3-030-49732-3.

Shafi MI, Adnan M, Fahad S, Fazli W, Ahsan K, Zhen Y, Subhan D, Zafar-ul-Hye M, Martin B, Rahul D (2020) Application of single superphosphate with humic acid improves the growth, yield and phosphorus uptake of wheat (Triticum aestivum L.) in Calcareous Soil. *Agron* 10:1224. doi:10.3390/agronomy10091224.

Shah F, Lixiao N, Kehui C, Tariq S, Wei W, Chang C, Liyang Z, Farhan A, Fahad S, Huang J (2013) Rice grain yield and component responses to near 2°C of warming. *Field Crops Res* 157:98–110.

Sharma P, Bhardwaj R (2007) Effects of 24-epibrassinolide on growth and metal uptake in Brassica juncea L. under copper metal stress. *Acta Physiol Plant* 29:259–263.

Small CC, Degenhardt D, McDonald T (2019) Plant growth regulators for enhancing Alberta native grass and forb seed germination. *Ecol Eng* X(1):100003.

Strydhorst S, Hall L, Perrot L (2018). Plant growth regulators: what agronomists need to know. *Crops Soils* 51: 22–26.

Subhan D, Zafar-ul-Hye M, Fahad S, Saud S, Martin B, Tereza H, Rahul D (2020) Drought stress alleviation by ACC deaminase producing *achromobacter xylosoxidans* and *Enterobacter cloacae*, with and without timber waste biochar in maize. *Sustain* 12(6286):doi:10.3390/su12156286.

Tao Q, Zhou Y, Guo Q, Liu Y, Yu S, Yu C, Zhang M, Li Z, Duan L (2018) A novel plant growth regulator alleviates high-temperature stress in maize. *Agron J* 110:2350–2359.

Tariq M, Ahmad S, Fahad S, Abbas G, Hussain S, Fatima Z, Nasim W, Mubeen M, ur Rehman MH, Khan MA, Adnan M. (2018). The impact of climate warming and crop management on phenology of sunflower-based cropping systems in Punjab, Pakistan. *Agri Forest Meteorol* 15;256:270–282.

Tuteja N (2007) Abscisic acid and abiotic stress signaling. *Plant Signal Behav* 2:135–138.

ul Qamar Muhammad Tahir, Amna F, Amna B, Barira Z, Xitong Z, Ling-Ling C (2020) Effectiveness of conventional crop improvement strategies vs. omics. In: Fahad S, Hasanuzzaman M, Alam M, Ullah H, Saeed M, Khan AK, Adnan M (eds) *Environment, climate, plant and vegetation growth.* Springer Publ Ltd, Springer Nature Switzerland AG. Part of Springer Nature, pp. 253–284. https://doi.org/10.1007/978-3-030-49732-3.

Unsar Naeem-U, Muhammad R, Syed HMB, Asad S, Mirza AQ, Naeem I, Muhammad H ur R, Fahad S, Shafqat S (2020) Insect pests of cotton crop and management under climate change scenarios. In: Fahad S, Hasanuzzaman M, Alam M, Ullah H, Saeed M, Khan AK, Adnan M (eds) *Environment, climate, plant and vegetation growth.* Springer Publ Ltd, Springer Nature Switzerland AG. Part of Springer Nature, pp 367–396. https://doi.org/10.1007/978-3-030-49732-3.

Vidya Vardhini B, Seeta Ram Rao S (2003) Amelioration of osmotic stress by brassinosteroids on seed germination and seedling growth of three varieties of sorghum. *Plant Growth Regul* 41:25–31.

Wahid A, Close TJ (2007) Expression of dehydrins under heat stress and their relationship with water relations of sugarcane leaves. *Biol Plant* 51:104–109.

Wahid F, Fahad S, Subhan D, Adnan M, Zhen Y, Saud S, Manzer HS, Martin B, Tereza H, Rahul D (2020) Sustainable management with mycorrhizae and phosphate solubilizing bacteria for enhanced phosphorus uptake in calcareous soils. *Agri* 10(334). doi:10.3390/agriculture10080334.

Wajid N, Ashfaq A, Asad A, Muhammad T, Muhammad A,Muhammad S, Khawar J, Ghulam MS, Syeda RS, Hafiz MH, Muhammad IAR, Muhammad ZH, Muhammad Habib ur R, Veysel T, Fahad S, Suad S, Aziz K, Shahzad A (2017) Radiation efficiency and nitrogen fertilizer impacts on sunflower crop in contrasting environments of Punjab, Pakistan. *Environ Sci Pollut Res* 25:1822–1836. https://org/10.1007/s11356-017-0592-z.

Wu C, Kehui C, She T, Ganghua L, Shaohua W, Fahad S, Lixiao N, Jianliang H, Shaobing P, Yanfeng D (2020) Intensified pollination and fertilization ameliorate heat injury in rice (Oryza sativa L.) during the flowering stage. *Field Crops Res* 252: 107795.

Wu C, Tang S, Li G, Wang S, Fahad S, Ding Y (2019) Roles of phytohormone changes in the grain yield of rice plants exposed to heat: a review. *PeerJ* 7:e7792. doi: 10.7717/peerj.7792.

Xu L, Zhang Q, Zhou A-l, Huo R (2013) Assessment of flood catastrophe risk for grain production at the provincial scale in china based on the BMM method. *J Integr Agric* 12:2310–2320.

Yang W, Duan L, Chen G, Xiong L, Liu Q (2013) Plant phenomics and high-throughput phenotyping: accelerating rice functional genomics using multidisciplinary technologies. *Curr Opin Plant Biol* 16:180–187.

Yang Z, Zhang Z, Zhang T, Fahad S, Cui K, Nie L, Peng S, Huang J (2017) The effect of season-long temperature increases on rice cultivars grown in the central and southern regions of China. *Front Plant Sci* 8:1908. https://doi.org/10.3389/fpls.2017.01908.

Zafar-ul-Hye M, Muhammad T ahzeeb-ul-Hassan, Muhammad A, Fahad S, Martin B, Tereza D, Rahul D, Subhan D (2020b) Potential role of compost mixed biochar with rhizobacteria in mitigating lead toxicity in spinach. *Sci Rep* 10:12159. https://doi.org/10.1038/s41598-020-69183-9.

Zafar-ul-Hye M, Muhammad N, Subhan D, Fahad S, Rahul D, Mazhar A, Ashfaq AR, Martin B, Jiˇrí H, Zahid HT, Muhammad N (2020a) Alleviation of cadmium adverse effects by improving nutrients uptake in bitter gourd through cadmium tolerant rhizobacteria. *Environ* 7(54). doi:10.3390/environments7080054.

Zahida Z, Hafiz FB, Zulfiqar AS, Ghulam MS, Fahad S, Muhammad RA, Hafiz MH,Wajid N, Muhammad S (2017) Effect of water management and silicon on germination, growth, phosphorus and arsenic uptake in rice. *Ecotoxicol Environ Saf* 144:11–18.

Zhao C et al. (2017) Temperature increase reduces global yields of major crops in four independent estimates. *Proc Natl Acad Sci U. S. A.* 114:9326–9331.

Zia-ur-Rehman M (2020) Environment, climate change and biodiversity. In: Fahad S, Hasanuzzaman M, Alam M, Ullah H, Saeed M, Khan AK, Adnan M (eds) *Environment, climate, plant and vegetation growth.* Springer Publ Ltd, Springer Nature Switzerland AG. Part of Springer Nature, pp 473–502. https://doi.org/10.1007/978-3-030-49732-3.

4

Jasmonates: Debatable Role in Temperature Stress Tolerance

Sherien Bukhat, Habib-ur-Rehman Athar, Tariq Shah, Hamid Manzoor, Sumaira Rasul, and Fozia Saeed

CONTENTS

4.1 Introduction

As sedentary organisms, plants constantly face diverse environmental stresses, and these stresses cause impairment of their natural physiological mechanisms (Adnan et al. 2018, 2019, 2020; Ahmad et al. 2019; Akram et al. 2018a,b; Aziz et al. 2017a,b; Baseer et al. 2019; Depeng et al. 2018; Farhana 2020; Frahat et al. 2020; Habib et al. 2017; Hafiz et al. 2016, 2018, 2019, 2020a,b; Hussain et al. 2020; Ilyas et al. 2020; Jan et al. 2019; Kamaran et al. 2017; Mubeen et al. 2020; Muhmmad et al. 2019; Qamar et al. 2017; Rehman 2020; Sajjad et al. 2019; Saleem et al 2020a,b,c; Saud et al. 2013, 2014, 2016, 2017, 2020; Shafi et al. 2020; Shah et al. 2013; Subhan et al. 2020; Tariq et al. 2018; Wahid et al. 2020; Wajid et al. 2017; Wu et al. 2019, 2020; Yang et al. 2017; Zafar-ul-Hye et al. 2020a,b; Zahida et al. 2017). An ever-increasing concern worldwide is crop loss by numerous biotic and abiotic stresses. Because of a rapidly increasing world population that exerts tremendous pressure to provide more food to human populations, avoiding such losses are especially critical. Biotic stresses in plants primarily include bacterial, viral, and fungal infections that hinder the growth and productivity of crops, thereby ultimately impacting the economy of a country and human health (Wani et al. 2016). Major abiotic stresses comprising drought, high salinity, flooding/waterlogging, toxic metal/metalloids, nutrient deficiency, extreme temperature (heat, cold, and frost), and mechanical stress, *etc.* occur due to extreme climate change (Pereira 2016; Shah et al. 2019). Increased abiotic stresses reduce the overall yield of major crops by greater than 50% (Akbar et al. 2020; Amanullah et al. 2020; Amir et al. 2020; Amjad et al. 2020; Arif et al. 2020; Ayman et al. 2020; Bayram et al. 2020; Fahad and Bano 2012; Fahad et al. 2013, 2014a,b, 2015, 2016a,b,c,d, 2017, 2018, 2019a,b; Farah et al. 2020; Fazli et al. 2020; Gopakumar et al. 2020; Hesham and Fahad 2020; Ibrar et al. 2020; Iqra et al. 2020; Mahar et al. 2020; Md Jakirand Allah 2020; Md. Enamul et al. 2020; Mohammad I. Al-Wabel et al. 2020a,b; Muhammad Tahir et al. 2020; Noor et al. 2020;

Pereira 2016; Rashid et al. 2020; Sadam et al. 2020; Sajid et al. 2020; Saman et al. 2020; Senol 2020; Shah et al. 2020; Unsar et al. 2020; Wang et al. 2003; Zia-ur-Rehman 2020). Therefore, it is crucial to prevent crop loss due to abiotic stress for resilient cultivation. Plants cannot escape from stresses by moving to a more favourable environment. Therefore, plants defend themselves to combat these adverse stressful conditions by stimulating numerous metabolic, physiological, and molecular defense responses through cellular modifications (Harfouche et al. 2019). Plants have multiple mechanisms for inducing tolerance against several environmental stresses. These responses include genetic and epigenetic control in gene-expression level, biosynthesis of secondary metabolites and unique proteins, redox change, and modifications in anti-oxidative activities (Isah 2019; Khan et al. 2019; Zandalinas et al. 2020).

Plants also produce various volatile and non-volatile endogenous chemicals known as phytohormones, or plant growth regulators (PGRs), which are involved in defense signalling to resist environmental stress. These PGRs are key players and act as the chemical messengers in increasing acclimatisation of plants to the changing environmental conditions by assisting growth, productivity, and nutrient distribution (Ku et al. 2018; Wang et al. 2020). These PGRs include jasmonic acid (JA), brassinosteroids (BRs), auxin, cytokinins (CKs), salicylic acid (SA), abscisic acid (ABA), gibberellins (GAs), strigolactones (SLs), and ethylene (ET), which perform a major role in the development and production of plants (Lymperopoulos et al. 2018; Li et al. 2019; Khan et al. 2020). The non-traditional PGRs, known as oxylipins, comprise a highly diverse class of oxidised lipids. These oxylipins perform diverse roles, including developmental approaches in various plants (Andersson et al. 2006). Jasmonates (JAs) are the major best-characterised oxylipins. They are fatty acid derivatives and include various compounds, for example, JA, methyl jasmonate (MeJA), and jasmonate isoleucine conjugate (JA-Ile) (Wasternack and Strnad 2018). JAs regulate various developmental mechanisms in plants such as cell cycle regulation, vegetative growth, trichome and stamen development, anthocyanin biosynthesis, senescence, fruit ripening, stomatal opening, rubisco biosynthesis, glucose transport, and uptake of phosphorus and nitrogen (Koda 1997; Reinbothe et al. 2009; Wasternack and Hause 2013; Campos et al. 2014). As a signalling molecule, JA controls the expression of multiple genes in response to complex abiotic stresses and encourages specific defense mechanisms (Li et al. 2018). In this chapter, we target the multifunctional role of jasmonates (JA, JA-Ile, and MeJA) due to their high bioactivity in the alleviation of temperature stress.

4.2 Multifunctional Roles of Jasmonates in Mitigating Temperature Stress: A Mechanistic Approach

4.2.1 JA-Regulated Stress Signalling

JAs are diverse signalling molecules regulating a wide range of developmental processes and signalling cascades that facilitate plant acclimation to different stresses. Jasmonic acid is the linolenic acid-derived cyclopentanone compound produced from polyunsaturated fatty acids, including linoleic and linolenic acids, following the lipoxygenase (LOX) pathway (Wang et al. 2019). Over the decades, numerous genes and transcription factors (TFs) that are involved in JA synthesis and signalling pathways have been found, including various activators and inhibitors that take part in stress signalling (Howe et al. 2018; Howe and Yoshida 2019). There are different reviews on JA biosynthesis (Ruan et al. 2019; Ali and Baek 2020). This chapter avoids the repetition of JA synthesis; however, it focuses on JA's role in defense-signalling pathways against abiotic stresses, especially temperature stress. However, a brief overview of JA biosynthesis is in Figure 4.1. JA is a signalling molecule that regulates numerous physiological and molecular mechanisms to adapt and acquire tolerance to various environmental stresses. Physiological mechanisms include the activation of the enzymatic antioxidant system (peroxidase, catalase, superoxide anion radical, NADPH-oxidase), accumulation of soluble sugars and amino acids, and regulation of the closing or opening of stomata (Sadiq et al. 2020). Molecular processes often include various JA genes (*LOX2*, *AOC*, *AOS1*, *JAZ*, and *COI1*). After stress stimulation, JA's epimerisation undergoes the formation of JA-Ile, which starts to accumulate in the cytoplasm of stressed plant leaves. In Arabidopsis, jasmonic acid transfer protein 1 (*AtJAT1*) regulates the export of JA or JA-Ile in the nucleus and cytoplasm.

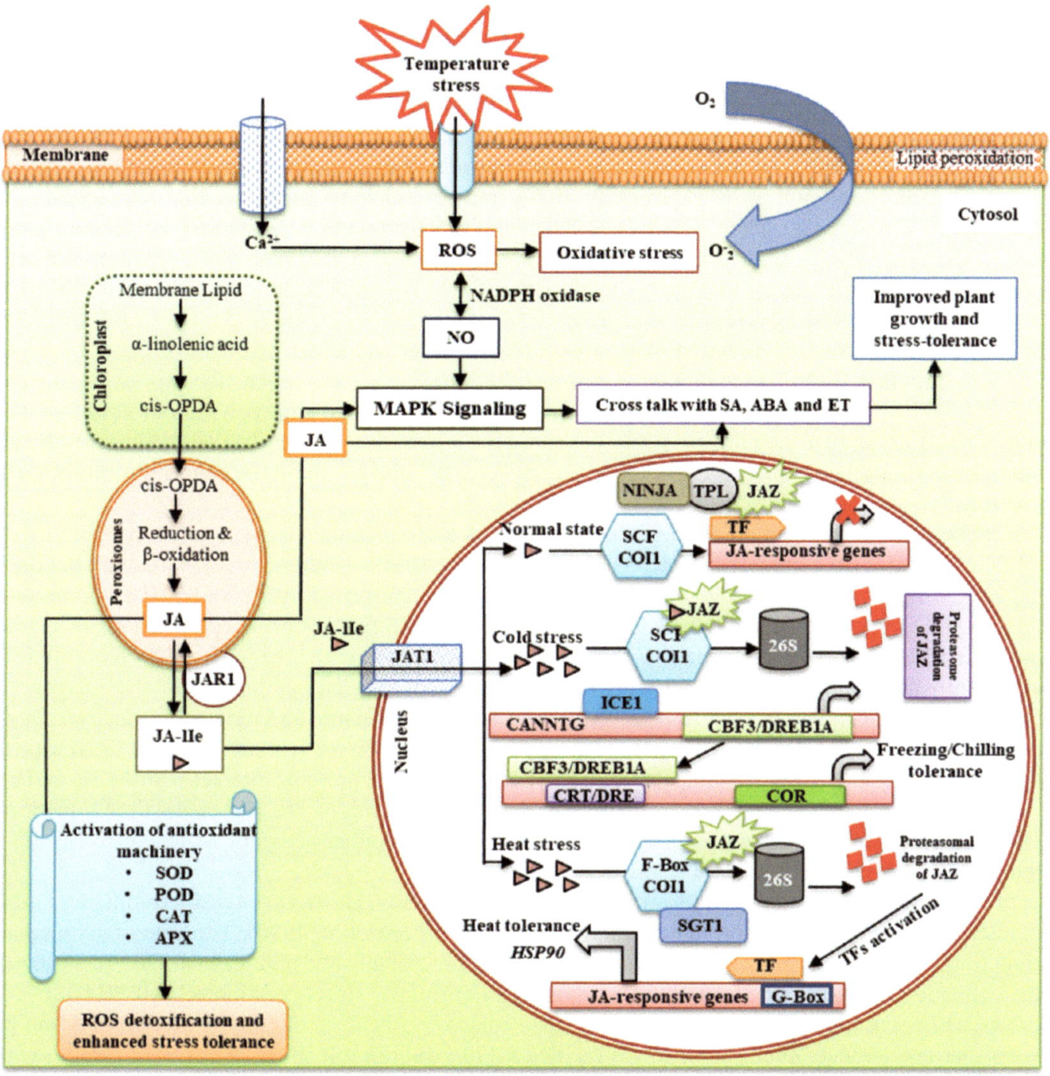

FIGURE 4.1 Model of JA signal transduction pathway during temperature stress.

In the apoplast, jasmonic acid further activates the signalling pathways in other cells. JA transmits long-distance signals through vascular bundles (Heil and Ton 2008).

In contrast to JA, MeJA can simply diffuse to faraway leaves due to its high volatility and ability to penetrate the cell membrane. Mostly, the Jasmonate-VQ-Motif GENE1 (*JAV1*) and Jasmonate-ZIM domain (JAZ) and families are involved in JA-mediated defense signalling. The JA-mediated degradation of JAZ proteins promotes the transcriptional reprogramming of enormous genes regulated by TFs, including MYC2 and MYB, which induce the activation of jasmonate-induced defense responses and growth. Moreover, JAZ also serves as the negative regulator of the JA-signalling pathway. The JAZ corepressors recruit the adaptor protein novel interactor of JAZ (NINJA) and the protein TOPLESS (TPL) to form a transcriptional repression complex (Figure 4.1). This complex inhibits jasmonate-responsive gene expression by transforming the open complex into a closed one. This conversion then recruits histone demethylases and deacetylases that induce chromatin modification (Howe et al. 2018; Pauwels et al. 2010). These epigenetic effects strongly repress the activation of numerous jasmonate responses. Under stress conditions, JA-Ile assists the proteolytic degradation of the JAZ proteins repressor with COI1 in the SCF complex, with the help of inositol pentakisphosphate, which efficiently converts the

JAZ-repressor to a transcriptional-activator complex (Sheard et al. 2010; Mosblech et al. 2011). JA-Ile-mediated degradation of JAZ corepressor modules (e.g., TPL-NINJA) subsequently converts the MYCs from the repressors to the transcriptional co-activators that activate the JA-signalling response (Howe et al. 2018). Studies have exposed that various other TFs, such as WRKY, GL1, AP2/ERF, and NAC, are also involved in JA-responsive signalling (Ali and Baek 2020).

In addition to various TFs, JA signalling further activates the mitogen-activated protein kinases, calcium channels (Kenton et al. 1999; Li et al. 2017), and several other processes that interact with SA, ABA, and ET to regulate plant growth and development during abiotic stresses (Santner and Estelle 2009). JA positively increases the C-repeat binding factor (ICE–CBF) pathway to induce cold stress tolerance and it works as an upstream signal of the ICE–CBF cold-responsive pathway to effectively regulate the expression of cold tolerance genes in Arabidopsis (Hu et al. 2013). JA is also an Inducer of *CBF EXPRESSION1* (ICE1), the MYC-type TF that serves as a chief regulator of the *CBF/DREB1* pathway (Figure 4.1). One study reports that three *CBF/DREB1* factors regulate the expression of *COLD REGULATED* (COR) genes and induce cold stress resistance in Arabidopsis (Gilmour et al. 2004). Li et al. (2018) found that the chilling stress increases the expression of dehydration-responsive element-binding (DREB1), late embryogenesis abundant (LEA), CBF, and JA and ABA concentrations in *Zoysia japonica*.

Receptors in plasma membranes detect temperature stress, which then triggers calcium signalling and produces reactive oxygen radicals, leading to oxidative stress. The synthesis of JA starts with the linolenic acid, which is released by the chloroplast membrane and converts it into cis-OPDA. In peroxisome, this cis-OPDA undergoes reduction and β-oxidation reactions, eventually producing JA. This JA is transported in the cytoplasm with the help of JAR1 and starts cross-talk with other defense hormone and activates antioxidant machinery for improving temperature tolerance. Temperature stress also increases JA biosynthesis and epimerises it into JA-Ile. JAT1 transporter helps to move JA-Ile into the nucleus. During normal conditions, JAZ proteins repress transcription factors and prevent activation of JA-responsive defense genes, while in cold and heat stress conditions, these JAZ proteins interact with JA-Ile and bind with SCF-COI1 complex or F-box protein-COI1 complex, leading to 26s proteasomal-degradation of JAZ. This helps in the activation of transcription factors which either activates JA-responsive genes for inducing cold stress (*COR*) or heat stress (*HSP90*).

In high-temperature stress, plants produce heat shock proteins (HSPs) that assist refolding and inhibit the denaturation of affected proteins (Khan et al. 2016; Boston et al. 1996). The role of JA-mediated signalling in thermo-tolerance of Arabidopsis has been long established (Clarke et al. 2009). JA induces the activation of HSPs and transcription factors (HSFs) in heat stress conditions. SUPPRESSOR OF G2 ALLELE OF SKP1 (SGT1) protein works as a cofactor of heat shock protein 90 (HSP90) and provides thermotolerance. Zhang et al. (2015) verified SGT1 protein's role in the JA-signalling pathway that requires F-box proteins along with ubiquitin ligases, like *COI1* of JA-signalling. *COI1* acts as a client protein of SGT1b–HSP70–HSP90 chaperone complexes and enhances this chaperone complex's functional capability to control JA responses (Sharma and Laxmi 2016).

4.2.2 Roles of Jasmonates in Cold Stress Tolerance

Among various environmental challenges, cold stress is the main abiotic stress that reduces crop production by influencing crop quality and post-harvest life. In a temperate environment, plants face two kinds of cold stress: chilling stress, which occurred at low temperatures (0–15°C), and freezing stress, occurring at temperatures below zero. Cold stress causes damage in tropical and subtropical plants by inducing cellular dehydration, the formation of cell-extracellular ice crystal, necrosis, chlorosis, membrane damage, changes in enzyme activities, cytoplasm viscosity, and eventually death (Ruelland and Zachowski 2010). These biochemical and physiological modifications elicit a series of events that help plants develop chilling and freezing tolerance through acclimation.

JA signalling plays a leading role in the acclimation of plants at low temperature or cold stress. JA maintains cold-treated plants' water status by inhibiting the stomata opening and increasing hydrolytic conductivity (Acharya and Assmann 2009). Researchers have identified the contribution of MeJA in chilling injury by producing cryo-protective agents, polyamines, ABA, proteinase inhibitors, lower activity of LOXs, and enzymatic antioxidants (González-Aguilar et al. 2000; Cao et al. 2009; Siboza et al. 2017; Wang et al. 2019). Du et al. (2013) described the increase in the endogenous JA upon cold stress treatment. Microarray analysis of cold-stressed rice seedlings showed up-regulation of JA biosynthesis genes such as *OsAOS1*, *OsAOS2*, *OsDAD1*, *OsAOC*, *OsLOX2*, *OsOPR1*, and *OsOPR7* and JA-signalling genes like *OsJAR1*, *OsCOI1a*, and *OsbHLH148*. Moreover, Hu et al. (2013) documented jasmonates' positive role in increasing cold acclimation-induced freezing resistance in Arabidopsis. The effective results were observed with 40 μM GABA, 50 μM MeJA, and 100 μM MeSA that enhanced the catalytic activity of enzymatic antioxidants like CAT, SOD, APX, and phenylalanine ammonia-lyase (PAL), while reducing the activity of POD and polyphenol oxidase (Habibi et al. 2020). MeJA and NO similarly enhance the CAT activity and gene expression and inhibit H_2O_2 generation from mitigating the chilling injury (Liu et al. 2016). Notably, MeJA and hot-air treatment was found effective in lowering the CI in peach fruit. Hot-air treatment was given to peach fruit for three days at 37°C, and 10 μmol L^{-1} vapor of MeJA was given for 24 h before storage under 5°C. These treatments resulted in a higher sucrose level than the control and improved sucrose phosphate synthase (*SPS*) expression and its activity while lowering invertase (*AI*) activity. These findings suggest that increased sucrose related to high *SPS* and low *AI* enhances tolerance to chilling stress witnessed in MeJA- and hot-air-treated peach fruits (Yu et al. 2016).

Exogenous JA application also promotes leaf senescence and regulates the expression of genes related to the leaf senescence, thus inducing freezing resistance in Arabidopsis (Hu et al. 2017). The expression of the MYC TFs and several cold-responsive genes, such as *MaCBF1*, *MaCBF2*, *MaKIN2*, *MaCOR1*, *MaRD2*, *MaRD5*, etc. was up-regulated during the cold storage of bananas (Zhao et al. 2013). MeJA could alleviate the cold stress in the tomato, loquat (Jin et al. 2014), guava (González-Aguilar et al. 2004), cowpea (Fan et al. 2016), and peach plants (Jin et al. 2009) by activating the antioxidants and synthesis of some defense compounds (e.g., heat shock proteins and phenolic compounds). These results suggest that JAs can ameliorate cold injury by stimulating active defense-signalling compounds and the antioxidant system. Table 4.1 shows the mitigating role of JAs in chilling stress.

4.2.3 Roles of Jasmonates in Heat Stress Tolerance

Heat stress is one of the most damaging environmental stresses that induce major crop losses worldwide. Temperatures that are too high for optimal growth may have a lethal impact on plant developmental processes as well as cell homeostasis by damaging the cellular machinery (Ding et al. 2019). Heat stress negatively influences the seed germination ability and photosynthetic efficiency that ultimately decreases plants' yields (Liu et al. 2019). Heat stress causes excessive accumulation of reactive oxygen species (ROS), which causes injury to plant nucleic acids, lipids, and proteins. Plants have a natural ability to alleviate the harmful effects of heat stress by synthesising the HSPs that inhibit denaturation and encourage re-folding damaged proteins (Saidi et al. 2011).

The application of JA is considered vital to enhancing heat stress (HS) tolerance (Table 4.2). Clarke et al. (2009) described that the low level of MeJA increased the accumulation of various jasmonates compounds such as OPDA, JA, JA-Ile, and MeJA. Further, JA's role in conferring thermo-tolerance was proved by the analysis of JA- and SA-signalling mutants, including *jar1-1cpr5-1*, *coi1-1*, and *opr3* that were witnessed to be sensitive under heat stress. It demonstrates that both JA and SA induce basal thermo-tolerance in plants. Pea plants pretreated with different MeJA concentrations, i.e., 50, 100, and 200 μM, were exposed to HS at 40°C showed up-regulation of the JA-signalling pathway and down-regulation of SA and ABA (Shahzad et al. 2015). Exogenous spray of low concentrations of jasmonic acid maintained the cell integrity in heat-stressed plants, validated through the leaf electrolyte leakage assays (Kazan 2015).

Moreover, Xu et al. (2016) studied that JA's foliar treatment and its methyl ester up-regulated the JA-signalling network in *Aquilaria sinensis*. The results validated that the MeJA treatment showed substantial positive effects on sesquiterpene's biosynthesis in contrast to SA and H_2O_2. These results propose that JA is a major signal-transducer in a stream of intracellular signals produced by HS, and, finally, JA shows an important part in sesquiterpenes compound build-up (Xu et al. 2016). The Arabidopsis response in high light (HL) and HS has been explored, and the findings reveal that the combined effect of both these stresses increases the accumulation of JA-IIe and JA. The JA biosynthesis mutants have

TABLE 4.1

List of Some Jasmonate-Mediated Chilling/Freezing Alleviation in Various Plant Species

Plant	JA type and dose employed	Defensive role	References
Punica granatum	0.01 and 0.1 mM MeJA	Enhanced the enzymes activity of peroxidase (POD) and catalase (CAT) and fruit total antioxidant contents	Sayyari et al. (2011)
Solanum lycopersicum	50 mM MeJA	Increased the arginine concentration, gene Expression, and enzymatic activity related to arginine and free polyamine levels	Zhang et al. (2012)
Prunus salicina	1120 and 2240 mg L^{-1} MeJA	Enhanced the chlorogenic acid contents and fruit firmness value	Karaman et al. (2013)
Eriobotrya japonica	16μM MeJA	Enhanced the ascorbate peroxidase (APX), superoxide dismutase (SOD), and CAT activities and inhibited lignin deposition and support pectin metabolism	Jin et al. (2014)
Fragaria ananassa	8 and 16 mol L^{-1} MeJA	Enhanced the enzymatic activity of CAT and POD	Asghari and Hasanlooe (2015)
Vaccinium corymbosum	0.05 mM L^{-1} MeJA	MeJA enhanced the anthocyanins content, total monomeric anthocyanins, and phenolic content and sustained fruit quality and yield	Huang et al. (2015)
Vigna sinensis	1 mM MeJA	Decreased the relative conductivity, weight loss, chilling injury index, malondialdehyde (MDA) contents; minimized the reduction in chlorophyll, ascorbic acid, and total soluble solids	Fan et al. (2016)
Averrhoa carambola	0.01, 0.1, 0.2, and 0.5 mM MeJA	Enhanced the ascorbic acid levels and POD activity and maintain the quality of fruit	Mustafa et al. (2016)
Citrus limon	10 μM MeJA	Enhanced the antioxidant activity of (CAT, APX, POD, GR, SOD) and decreased the ROS accumulation	Siboza et al. (2017)
Citrus sinensis	50 μM MeJA	Increased the activity of enzymatic antioxidant (SOD, CAT, POD, APX), proline content and PAL/PPO activity	Habibi et al. (2019)
Capsicum annuum	10 μmol L^{-1} MeJA	Decreased the MDA content, chlorophyll content, and chilling injury index and increased the activity of CAT, SOD, APX, and POD enzymes	Wang et al. (2019)
Punica granatum	0.1 mmol/l MeJA	Increased the contents of soluble proteins and maintain the integrity of epidermal structure	Chen et al. (2020)
Capsicum annuum	1 μmol L^{-1} MeJA	Decreased the lipid peroxidation and increased the proline content and maintain cell integrity	Ma et al. (2020)

demonstrated increased susceptibility to HL and HS when used at the same time. These results show that JA plays a major role in plant adaptations to heat stress (Balfagón et al. 2019).

Moreover, the JA application maintains the water potential of the plant cell, consequently preserving water contents in heat-stress conditions. JA also increases the synthesis of osmoregulators (involving soluble carbohydrates and proline) and enhances the water potential in plant cells (Bandurska et al. 2003). Further research is required to extend our knowledge of the mechanical action of JAs in alleviating the destructive effects of HS on plants.

4.3 Engineering Jasmonate Genes for Producing Temperature Tolerant Crops

Phytohormone engineering could be an ideal platform to strengthen the endogenous defense mechanism to improve the crop's developmental and stress-responsive pathways. Hormone-signalling pathways and

TABLE 4.2

List of Some Jasmonate Mediated Heat Stress Alleviation in Various Plant Species

Plant	JA type and dose employed	Defensive role	References
Capsicum annuum	100μM MeJA	Regulate the CaWRKY2 orthologs and improve heat stress tolerance	Dang et al. (2013)
Oryza japonica	MeJA	Enhanced enzymatic action of superoxide dismutase (SOD) and peroxidase (POD) and enhanced membrane stability	Liu et al. (2016)
Pisum sativum	50, 100, and 200 μM MeJA	Improved plant defense system by up-regulated the JA pathway and down-regulated the ABA and SA and in heat stress	Shahzad et al. (2015)
Aquilaria sinesisis	100 ng of dihydro-jasmonic acid (DHJA)	Stimulate the formation of sesquiterpene compounds and increased crop production	Xu et al. (2016)
Oryza sativa	MeJA	Enhanced synthesis of metabolites and improved germination and pollen retention	Fahad et al. (2016)

their interrelations in plant active-immunity specified the stage with increased resistance to stress and maintaining the overall fitness of plants (Wani et al. 2016). Genetic engineering of defense-signalling pathway genes enhances transgenic plants' tolerance and adaptive ability in diverse environmental conditions. Transgenic wheat plants over-expressing the *AtOPR3* gene showed up-regulated expression of ALLENE OXIDE SYNTHASE (*AOS*) with enhanced jasmonates' biosynthesis and improved freezing tolerance (Pigolev et al. 2018). *ZjICE1* over-expressed in transgenic Arabidopsis revealed enhanced cold tolerance with increased proline contents, improved enzymatic activity of SOD and POD, and decreased MDA contents. Moreover, it induced the transcription of cold tolerance genes including *COR47A*, *RD29A*, *KIN1*, *CBF1*, *CBF2*, and *CBF3* upon chilling stress stimulus (Zuo et al. 2019). Hu et al. (2013) indicated that over-expression of *TomloxD* (JA biosynthesis gene) induces the activation of octadecanoid defense-signalling genes, like LePR1, *LePR6*, *LeZAT*, and *LeHSP90* and enhances the synthesis of JA under high-temperature stress and *C. fulvum* infection.

Until now, much study has been devoted to describing JA-related gene transcription using the key TF known as MYC2 (Song et al. 2014). There is limited literature, however, regarding terminating JAs-regulated transcription along with their operating system. Liu et al. (2019) studied that MYC2 encoding *JIN* gene regulates a group of proteins known as MYC2-targeted bHLH (MTB). These MTB proteins negatively regulate the transcriptional activation and stop the JA signalling (Zhai and Li 2019). Numerous other genes have been found to involve JA receptors, although their specific functions have yet to be examined. Next-generation genome-editing techniques, including CRISPR/Cas9, have unlocked novel routes for targeting MTB-responsive genes for engineering resistance against various abiotic stresses. JA metabolic and signalling pathways are the most preferable positions for genetic engineering to acquire resistance in important plants. Engineering JA synthesis pathways without any negative effects will be an important challenge. Besides, the JA-signalling cascade could be analysed by comparing JA deficient/enriched in controlled and stressed environments.

4.4 Cross-Talk between Jasmonates and Other Growth Regulators

Jasmonates alter plant growth and development not through simple signalling pathways but by diverse interconnections among multifarious signalling pathways. Therefore, it is necessary to find out the hormonal interplay in stress conditions, plant growth, as well as development. A hormonal cross-talk includes both negative and positive feedback that may influence the synthesis and signalling of hormones (Adnan et al. 2016; Ku et al. 2018). In addition, this synergistic or antagonistic connection among JAs and different PGRs promotes plants' resistance under harsh conditions (Wang et al. 2020). Recently, Shi et al. (2012) described that ethylene negatively regulates cold tolerance. Another study revealed that

EIN3 attach to *CBF* promoters and negatively controls cold-related genes. EIN3 directly targets the JAZ repressors that control the JA-mediated signalling response.

Consequently, to regulate the CBF cold resistance via EIN3, JA signalling seems to be integrated with the ethylene pathway (Shi et al. 2012). JAs and ABA both positively regulate cold stress response, and JAs work downstream of ABA and regulate the CBF cold tolerance pathway. Consistent with these results, the ABA application also enhanced the expression of JA-synthesis genes which ultimately boost the accumulation of JA (Wang et al. 2016). Ding et al. (2015) stated that ABA-mediated *OST1* combines with *ICE1* and regulates its transcriptional function in chilling stress condition. Considering these facts, JA also promotes the transcriptional action of *ICE1*. Both JA- and ABA-activated *OST1* work synergistically and modulate the cold tolerance ICE–CBF pathway in plants (Hu et al. 2013).

Gibberellin (GA) also plays an active role in low-temperature stress responses. GA-insensitive or GA-deficient mutants reveal alterations in rice and Arabidopsis (Achard et al. 2008; Richter et al. 2013). For instance, DELLA genes knockout mutant, which represses the GA-pathway, showed a hyper-sensitive response to freezing stress. In contrast, the GA-deficit mutant exhibits greater up-regulation of *COR15A*, *CBF1*, and *CBF2* cold-related genes than wild type plants (Richter et al. 2013). Hou et al. (2010) presented that JAZ repressors work together with DELLA proteins and integrate GA as well as JA-signalling networks in the Arabidopsis. Recent study showed that GA promotes the DELLA proteins' degradation, assisting the JAZ1 binding to MYC2 and suppressing the JA-signalling events. These findings suggest that the interaction of JAZ proteins with DELLA possibly partially regulate GA signalling in response to cold tolerance by JAZ as well as JA-related pathways.

Conversely, SA and JA are two basic pathways that triggered a biochemical response in environmental stressors. Furthermore, various findings have revealed antagonistic functions of SA- and JA-signalling networks. A combined treatment of SA and MeJA boost chilling resistance in citrus. Moreover, the expression of CBF and ROS avoidance machinery by SA- and JA-induced signalling cascades draws attention towards the promising cross-talk among SA- and JA-signalling pathways in the alleviation of chilling stress (Sharma and Laxmi 2016).

4.5 Conclusion and Future Outlook

Jasmonates have substantial participation in the development and defense response to different environmental stresses. However, a remarkable body of studies indicates the central role of JAs in plant response against abiotic stresses. Various genes (*JAZ*, *LOX2*, *AOS1*, *COI1*, and *AOC*) and TFs (bHLH and MYC2) are the key mediators (activators or repressors) in JA-signalling pathway to regulate the responses and alleviate the stresses. Recent studies suggest that the interaction of JA-signalling components and transcriptional activators *ICE1* and *ICE2* trigger the regulation of cold-related genes (Hu et al. 2013) and indicate the involvement of similar signalling elements in the association of the two pathways. Additionally, multiple data sources suggest the contribution of JAs in conferring thermo-tolerance. Jasmonates further modify the biosynthesis of various other PGRs, like abscisic acid and ethylene, and show the synergistic/antagonistic interactions with these PGRs in harsh environments. Therefore, developments in omics-based strategies facilitate the study of JA-related gene–protein interactive networks and metabolites for the production of climate resilient plants. Moreover, the genetic engineering of JA-associated metabolic pathways by using genome-editing systems, such as CRISPR/Cas9, can open new insights to thoroughly understand the JA-mediated signalling network and enhance plant adaptability to temperature stresses.

REFERENCES

Achard P, Gong F, Cheminant S, Alioua M, Hedden P, Genschik P (2008) The cold-inducible CBF1 factor–dependent signaling pathway modulates the accumulation of the growth-repressing DELLA proteins via its effect on gibberellin metabolism. *Plant Cell* 20:2117–2129.

Acharya BR, Assmann SM (2009) Hormone interactions in stomatal function. *Plant Mol Biol* 69:451–462.

Adnan M, Fahad S, Khan IA, Saeed M, Ihsan MZ, Saud S, Riaz M, Wang D, Wu C. (2019). Integration of poultry manure and phosphate solubilizing bacteria improved availability of Ca bound P in calcareous soils. *Biotech* 9(10):368.

Adnan M, Fahad S, Muhammad Z, Shahen S, Ishaq AM, Subhan D, Zafar-ul-Hye M, Martin LB, Raja MMN, Beena S, Saud S, Imran A, Zhen Y, Martin B, Jiri H, Rahul D (2020) Coupling phosphate-solubilizing bacteria with phosphorus supplements improve maize phosphorus acquisition and growth under lime induced salinity stress. *Plants* 9(900). https://doi.org/10.3390/plants9070900.

Adnan M, Shah Z, Arif M, Khan M J, Mian I A, Sharif M, Saleem, N (2016). Impact of rhizobial inoculum and inorganic fertilizers on nutrients (NPK) availability and uptake in wheat crop. *Can J Soil Sci* 96(2):169–176.

Adnan M, Shah Z, Sharif M, Rahman H. (2018). Liming induces carbon dioxide (CO_2) emission in PSB inoculated alkaline soil supplemented with different phosphorus sources. *Environ Sci Pollut Res* 25(10):9501–9509.

Adnan M, Zahir S, Fahad S, Arif M, Mukhtar A, Imtiaz AK, Ishaq AM, Abdul B, Hidayat U, Muhammad A, Inayat-Ur R, Saud S, Muhammad ZI, Yousaf J, Amanullah Hafiz MH, Wajid N (2018) Phosphate-solubilizing bacteria nullify the antagonistic effect of soil calcification on bioavailability of phosphorus in alkaline soils. *Sci Rep* 8:4339. https://doi.org/10.1038/s41598-018-22653-7.

Ahmad S, Kamran M, Ding R, Meng X, Wang H, Ahmad I, Fahad S, Han Q (2019) Exogenous melatonin confers drought stress by promoting plant growth, photosynthetic capacity and antioxidant defense system of maize seedlings. *PeerJ* 7:e7793. http://doi.org/10.7717/peerj.7793.

Akbar H, Timothy JK, Jagadish T, Golam M, Apurbo KC, Muhammad F, Rajan B, Fahad S, Hasanuzzaman M (2020) Agricultural land degradation: processes and problems undermining future food security. In: Fahad S, Hasanuzzaman M, Alam M, Ullah H, Saeed M, Khan AK, Adnan M (eds) *Environment, climate, plant and vegetation growth.* Springer Publ Ltd, Springer Nature Switzerland AG. Part of Springer Nature, pp 17–62. https://doi.org/10.1007/978-3-030-49732-3.

Akram R, Turan V, Hammad HM, Ahmad S, Hussain S, Hasnain A, Maqbool MM, Rehmani MIA, Rasool A, Masood N, Mahmood F, Mubeen M, Sultana SR, Fahad S, Amanet K, Saleem M, Abbas Y, Akhtar HM, Waseem F, Murtaza R, Amin A, Zahoor SA, ul Din MS, Nasim W (2018a) Fate of organic and inorganic pollutants in paddy soils. In: Hashmi MZ, Varma A (eds) *Environmental pollution of paddy soils, soil biology.* Springer International Publishing AG, Cham, Switzerland, pp 197–214.

Akram R, Turan V, Wahid A, Ijaz M, Shahid MA, Kaleem S, Hafeez A, Maqbool MM, Chaudhary HJ, Munis MFH, Mubeen M, Sadiq N, Murtaza R, Kazmi DH, Ali S, Khan N, Sultana SR, Fahad S, Amin A, Nasim W (2018b) Paddy land pollutants and their role in climate change. In: Hashmi MZ and Varma A (eds) *Environmental pollution of paddy soils, soil biology.* Springer International Publishing AG, Cham, Switzerland, pp 113–124.

Ali M, Baek K-H (2020) Jasmonic acid signaling pathway in response to abiotic stresses in plants. *Int J Mol Sci* 21:621.

Al-Wabel Mohammad I, Abdelazeem S, Munir A, Khalid E, Adel RAU (2020b) Extent of climate change in Saudi Arabia and its impacts on agriculture: a case study from qassim region. In: Fahad S, Hasanuzzaman M, Alam M, Ullah H, Saeed M, Khan AK, Adnan M (eds) *Environment, climate, plant and vegetation growth.* Springer Publ Ltd, Springer Nature Switzerland AG. Part of Springer Nature, pp 635–658. https://doi.org/10.1007/978-3-030-49732-3.

Al-Wabel Mohammad I, Ahmad Munir, Usman Adel R. A., Akanji Mutair, Rafique Muhammad Imran (2020a) Advances in pyrolytic technologies with improved carbon capture and storage to combat climate change. In: Fahad S, Hasanuzzaman M, Alam M, Ullah H, Saeed M, Khan AK, Adnan M (eds) *Environment, climate, plant and vegetation growth.* Springer Publ Ltd, Springer Nature Switzerland AG. Part of Springer Nature, pp 535–576. https://doi.org/10.1007/978-3-030-49732-3.

Amanullah, Shah K, Imran Hamdan AK, Muhammad A, Abdel RA, Muhammad A, Fahad S, Azizullah S, Brajendra P (2020) Effects of climate change on irrigation water quality. In: Fahad S, Hasanuzzaman M, Alam M, Ullah H, Saeed M, Khan AK, Adnan M (eds) *Environment, climate, plant and vegetation growth.* Springer Publ Ltd, Springer Nature Switzerland AG. Part of Springer Nature, pp 123–132. https://doi.org/10.1007/978-3-030-49732-3.

Amir M, Muhammad A, Allah B, Sevgi Ç, Haroon ZK, Muhammad A, Emre A (2020) Bio fortification under climate change: the fight between quality and quantity. In: Fahad S, Hasanuzzaman M, Alam M, Ullah H, Saeed M, Khan AK, Adnan M (eds) *Environment, climate, plant and vegetation growth*. Springer Publ Ltd, Springer Nature Switzerland AG. Part of Springer Nature, pp 173–228. https://doi.org/10.1007/978-3-030-49732-3.

Amjad I, Muhammad H, Farooq S, Anwar H (2020) Role of plant bioactives in sustainable agriculture. In: Fahad S, Hasanuzzaman M, Alam M, Ullah H, Saeed M, Khan AK, Adnan M (eds) *Environment, climate, plant and vegetation growth*. Springer Publ Ltd, Springer Nature Switzerland AG. Part of Springer Nature, pp 591–606. https://doi.org/10.1007/978-3-030-49732-3.

Andersson MX, Hamberg M, Kourtchenko O, Brunnström Å, McPhail KL, Gerwick WH, Göbel C, Feussner I, Ellerström M (2006) Oxylipin profiling of the hypersensitive response in Arabidopsis thaliana formation of a novel oxo-phytodienoic acid-containing galactolipid, arabidopside E. *J Biol Chem* 281:31528–31537.

Arif M, Talha J, Muhammad R, Fahad S, Muhammad A, Amanullah Kawsar A, Ishaq AM, Bushra K, Fahd R (2020) Biochar; a remedy for climate change. In: Fahad S, Hasanuzzaman M, Alam M, Ullah H, Saeed M, Khan AK, Adnan M (eds) *Environment, climate, plant and vegetation growth*. Springer Publ Ltd, Springer Nature Switzerland AG. Part of Springer Nature, pp 151–172. https://doi.org/10.1007/978-3-030-49732-3.

Asghari M, Hasanlooe AR (2015) Interaction effects of salicylic acid and methyl jasmonate on total antioxidant content, catalase and peroxidase enzymes activity in “Sabrosa” strawberry fruit during storage. *Sci Hortic* 197:490–495.

Ayman ELS, Hossain Akbar, Barutçular Celaleddin, Iqbal Muhammad Aamir, Sohidul Islam M, Fahad Shah, Sytar Oksana, Çig Fatih, Meena Ram Swaroop, Erman Murat (2020) Consequences of salinity stress on the quality of crops and its mitigation strategies for sustainable crop production: an outlook of arid and semi- arid regions. In: Fahad S, Hasanuzzaman M, Alam M, Ullah H, Saeed M, Khan AK, Adnan M (eds) *Environment, climate, plant and vegetation growth*. Springer Publ Ltd, Springer Nature Switzerland AG. Part of Springer Nature, pp 503–534. https://doi.org/10.1007/978-3-030-49732-3.

Aziz K, Daniel KYT, Fazal M,Muhammad ZA, Farooq S, FanW, Fahad S, Ruiyang Z (2017a) Nitrogen nutrition in cotton and control strategies for greenhouse gas emissions: a review. *Environ Sci Pollut Res* 24:23471–23487. https://doi.org/10.1007/s11356-017-0131-y.

Aziz K, Daniel KYT, Muhammad ZA, Honghai L, Shahbaz AT, Mir A, Fahad S (2017b) Nitrogen fertility and abiotic stresses management in cotton crop: a review. *Environ Sci Pollut Res* 24:14551–14566. https://doi.org/10.1007/s11356-017-8920-x.

Balfagón D, Sengupta S, Gómez-Cadenas A, Fritschi FB, Azad RK, Mittler R, Zandalinas SI (2019) Jasmonic acid is required for plant acclimation to a combination of high light and heat stress. *Plant Physiol* 181:1668–1682.

Bandurska H, Stroiński A, Kubiś J (2003) The effect of jasmonic acid on the accumulation of ABA, proline and spermidine and its influence on membrane injury under water deficit in two barley genotypes. *Acta Physiol Plant* 25:279–285.

Baseer M, Adnan M, Fazal M, Fahad S, Muhammad S, Fazli W, Muhammad A Jr. Amanullah Depeng W, Saud S, Muhammad N, Muhammad Z, Fazli S, Beena S, Mian AR, Ishaq AM (2019) Substituting urea by organic wastes for improving maize yield in alkaline soil. *J Plant Nutr*. doi.org/10.1080/01904167.2019.1659344.

Bayram AY, Seher Ö, Nazlican A (2020) Climate change forecasting and modeling for the year of 2050. In: Fahad S, Hasanuzzaman M, Alam M, Ullah H,Saeed M, Khan AK, Adnan M (eds) *Environment, climate, plant and vegetation growth*. Springer Publ Ltd, Springer Nature Switzerland AG. Part of Springer Nature, pp 109–122. https://doi.org/10.1007/978-3-030-49732-3.

Boston RS, Viitanen PV, Vierling E (1996) Molecular chaperones and protein folding in plants. *Plant Mol Biol* 32:191–222.

Campos ML, Kang J-H, Howe GA (2014) Jasmonate-triggered plant immunity. *J Chem Ecol* 40:657–675.

Cao S, Zheng Y, Wang K, Jin P, Rui H (2009) Methyl jasmonate reduces chilling injury and enhances antioxidant enzyme activity in postharvest loquat fruit. *Food Chem* 115:1458–1463.

Chen L, Pan Y, Li H, Jia X, Guo Y, Luo J, Li X (2020) Methyl jasmonate alleviates chilling injury and keeps intact pericarp structure of pomegranate during low temperature storage. *Food Sci Technol Int*. Doi:1082013220921597.

Clarke SM, Cristescu SM, Miersch O, Harren FJ, Wasternack C, Mur LA (2009) Jasmonates act with salicylic acid to confer basal thermotolerance in Arabidopsis thaliana. *New Phytol* 182:175–187.

Dang FF, Wang YN, Yu L, Eulgem T, Lai Y, Liu ZQ, Wang X, Qiu AL, Zhang TX, Lin J (2013) CaWRKY40, a WRKY protein of pepper, plays an important role in the regulation of tolerance to heat stress and resistance to Ralstonia solanacearum infection. *Plant, Cell Environ* 36:757–774.

Depeng W, Fahad S, Saud S, Muhammad K, Aziz K, Mohammad NK, Hafiz MH, Wajid N (2018) Morphological acclimation to agronomic manipulation in leaf dispersion and orientation to promote "Ideotype" breeding: evidence from 3D visual modeling of "super" rice (*Oryza sativa* L.). *Plant Physiol Biochem*. https://doi.org/10.1016/j.plaphy.2018.11.010.

Ding Y, Li H, Zhang X, Xie Q, Gong Z, Yang S (2015) OST1 kinase modulates freezing tolerance by enhancing ICE1 stability in Arabidopsis. *Dev Cell* 32:278–289.

Ding Y, Shi Y, Yang S (2019) Advances and challenges in uncovering cold tolerance regulatory mechanisms in plants. *New Phytol* 222:1690–1704.

Du H, Liu H, Xiong L (2013) Endogenous auxin and jasmonic acid levels are differentially modulated by abiotic stresses in rice. *Front Plant Sci* 4:397.

Fahad S, Adnan M, Hassan S, Saud S, Hussain S, Wu C, Wang D, Hakeem KR, Alharby HF, Turan V, Khan MA, Huang J (2019b). Rice responses and tolerance to high temperature. In: Hasanuzzaman M, Fujita M, Nahar K, Biswas JK (eds) *Advances in rice research for abiotic stress tolerance*. Woodhead Publ Ltd, Cambridge, England, pp 201–224.

Fahad S, Bajwa AA, Nazir U, Anjum SA, Farooq A, Zohaib A, Sadia S, NasimW, Adkins S, Saud S, Ihsan MZ, Alharby H,Wu C,Wang D, Huang J (2017) Crop production under drought and heat stress: plant responses and management options. *Front Plant Sci* 8:1147. https://doi.org/10.3389/fpls.2017.01147.

Fahad S, Bano A (2012) Effect of salicylic acid on physiological and biochemical characterization of maize grown in saline area. *Pak J Bot* 44:1433–1438.

Fahad S, Chen Y, Saud S,Wang K, Xiong D, Chen C,Wu C, Shah F, Nie L, Huang J (2013) Ultraviolet radiation effect on photosynthetic pigments, biochemical attributes, antioxidant enzyme activity and hormonal contents of wheat. *J Food Agri Environ* 11(3&4):1635–1641.

Fahad S, Hussain S, Bano A, Saud S, Hassan S, Shan D, Khan FA, Khan F, Chen Y, Wu C, Tabassum MA, Chun MX, Afzal M, Jan A, Jan MT, Huang J (2014a) Potential role of phytohormones and plant growth-promoting rhizobacteria in abiotic stresses: consequences for changing environment. *Environ Sci Pollut Res* 22(7):4907–4921. https://doi.org/10.1007/s11356-014-3754-2.

Fahad S, Hussain S, Matloob A, Khan FA, Khaliq A, Saud S, Hassan S, Shan D, Khan F, Ullah N, Faiq M, Khan MR, Tareen AK, Khan A, Ullah A, Ullah N, Huang J (2014b) Phytohormones and plant responses to salinity stress: a review. *Plant Growth Regul* 75(2):391–404. https://doi.org/10.1007/s10725-014-0013-y.

Fahad S, Hussain S, Saud S, Hassan S, Chauhan BS, Khan F et al (2016a) Responses of rapid viscoanalyzer profile and other rice grain qualities to exogenously applied plant growth regulators under high day and high night temperatures. *PLoS One* 11(7):e0159590. https://org/10.1371/journal.pone.0159590.

Fahad S, Hussain S, Saud S, Hassan S, Ihsan Z, Shah AN,Wu C, Yousaf M, Nasim W, Alharby H, Alghabari F, Huang J (2016c) Exogenously applied plant growth regulators enhance the morphophysiological growth and yield of rice under high temperature. *Front Plant Sci* 7:1250. https://doi.org/10.3389/fpls.2016. 01250.

Fahad S, Hussain S, Saud S, Hassan S, Tanveer M, Ihsan MZ, Shah AN, Ullah A, Nasrullah KF, Ullah S, Alharby HNW, Wu C, Huang J (2016d) A combined application of biochar and phosphorus alleviates heat-induced adversities on physiological, agronomical and quality attributes of rice. *Plant Physiol Biochem* 103:191–198.

Fahad S, Hussain S, Saud S, Khan F, Hassan S Jr A, Nasim W, Arif M, Wang F, Huang J (2016b) Exogenously applied plant growth regulators affect heat-stressed rice pollens. *J Agron Crop Sci* 202:139–150.

Fahad S, Hussain S, Saud S, Khan F, Hassan S, Nasim W, Arif M, Wang F, Huang J (2016) Exogenously applied plant growth regulators affect heat-stressed rice pollens. *J Agron Crop Sci* 202:139–150.

Fahad S, Hussain S, Saud S, Tanveer M, Bajwa AA, Hassan S, Shah AN, Ullah A,Wu C, Khan FA, Shah F, Ullah S, Chen Y, Huang J (2015) A biochar application protects rice pollen from high-temperature stress. *Plant Physiol Biochem* 96:281–287.

Fahad S, Muhammad ZI, Abdul K, Ihsanullah D, Saud S, Saleh A, Wajid N, Muhammad A, Imtiaz AK, Chao W, Depeng W, Jianliang H (2018) Consequences of high temperature under changing climate optima for rice pollen characteristics-concepts and perspectives. *Arch Agron Soil Sci.* https://doi.org/10.1080/03650340.2018.1443213.

Fahad S, Nie L, Chen Y, Wu C, Xiong D, Saud S, Hongyan L, Cui K, Huang J (2015) Crop plant hormones and environmental stress. *Sustain Agric Res* 15:371–400.

Fahad S, Rehman A, Shahzad B, Tanveer M, Saud S, Kamran M, Ihtisham M, Khan SU, Turan V, Rahman MHU (2019a). Rice responses and tolerance to metal/metalloid toxicity. In: Hasanuzzaman M, Fujita M, Nahar K, Biswas JK (eds) *Advances in rice research for abiotic stress tolerance.* Woodhead Publ Ltd, Cambridge, England, pp 299–312.

Fan L, Wang Q, Lv J, Gao L, Zuo J, Shi J (2016) Amelioration of postharvest chilling injury in cowpea (Vigna sinensis) by methyl jasmonate (MeJA) treatments. *Sci Hortic* 203:95–101.

Farah R, Muhammad R, Muhammad SA, Tahira Y, Muhammad AA, Maryam A, Shafaqat A, Rashid M, Muhammad R, Qaiser H, Afia Z, Muhammad AA, Muhammad A, Fahad S (2020) Alternative and non-conventional soil and crop management strategies for increasing water use efficiency. In: Fahad S, Hasanuzzaman M, Alam M, Ullah H, Saeed M, Khan AK, Adnan M (eds) *Environment, climate, plant and vegetation growth.* Springer Publ Ltd, Springer Nature Switzerland AG. Part of Springer Nature, pp 323–338. https://doi.org/10.1007/978-3-030-49732-3.

Farhana G, Ishfaq A, Muhammad A, Dawood J, Fahad S, Xiuling L, Depeng W, Muhammad F, Muhammad F, Syed AS (2020) Use of crop growth model to simulate the impact of climate change on yield of various wheat cultivars under different agro-environmental conditions in Khyber Pakhtunkhwa, Pakistan. *Arabian J Geosci* 13:112 https://doi.org/10.1007/s12517-020-5118-1.

Farhat A, Hafiz MH, Wajid I, Aitazaz AF, Hafiz FB, Zahida Z, Fahad S, Wajid F, Artemi C (2020) A review of soil carbon dynamics resulting from agricultural practices. *J Environ Manage* 268(2020):110319.

Fazli W, Muhmmad S, Amjad A, Fahad S, Muhammad A, Muhammad N, Ishaq AM, Imtiaz AK, Mukhtar A, Muhammad S, Muhammad I, Rafi U, Haroon I, Muhammad A (2020) Plant-microbes interactions and functions in changing climate. In: Fahad S, Hasanuzzaman M, Alam M, Ullah H, Saeed M, Khan AK, Adnan M (eds) *Environment, climate, plant and vegetation growth.* Springer Publ Ltd, Springer Nature Switzerland AG. Part of Springer Nature, pp 397–420. https://doi.org/10.1007/978-3-030-49732-3.

Gilmour SJ, Fowler SG, Thomashow MF (2004) Arabidopsis transcriptional activators CBF1, CBF2, and CBF3 have matching functional activities. *Plant Mol Biol* 54:767–781.

González-Aguilar G, Fortiz J, Cruz R, Baez R, Wang C (2000) Methyl jasmonate reduces chilling injury and maintains postharvest quality of mango fruit. *J Agric Food Chem* 48:515–519.

González-Aguilar G, Tiznado-Hernandez M, Zavaleta-Gatica R, Martınez-Téllez M (2004) Methyl jasmonate treatments reduce chilling injury and activate the defense response of guava fruits. *Biochem Biophys Res Commun* 313:694–701.

Gopakumar L, Bernard NO, Donato V (2020) Soil microarthropods and nutrient cycling. In: Fahad S, Hasanuzzaman M, Alam M, Ullah H, Saeed M, Khan AK, Adnan M (eds) *Environment, climate, plant and vegetation growth.* Springer Publ Ltd, Springer Nature Switzerland AG. Part of Springer Nature, pp 453–472. https://doi.org/10.1007/978-3-030-49732-3.

Habib ur R, Ashfaq A, Aftab W, Manzoor H, Fahd R, Wajid I, Md. Aminul I, Vakhtang S, Muhammad A, Asmat U, Abdul W, Syeda RS, Shah S, Shahbaz K, Fahad S, Manzoor H, Saddam H, Wajid N (2017) Application of CSM-CROPGRO-Cotton model for cultivars and optimum planting dates: evaluation in changing semi-arid climate. *Field Crops Res* http://dx.doi.org/10.1016/j.fcr.2017.07.007.

Habibi F, Ramezanian A, Guillén F, Serrano M, Valero D (2020) Blood oranges maintain bioactive compounds and nutritional quality by postharvest treatments with γ-aminobutyric acid, methyl jasmonate or methyl salicylate during cold storage. *Food Chem* 306:125634.

Habibi F, Ramezanian A, Rahemi M, Eshghi S, Guillén F, Serrano M, Valero D (2019) Postharvest treatments with γ-aminobutyric acid, methyl jasmonate, or methyl salicylate enhance chilling tolerance of blood orange fruit at prolonged cold storage. *J Sci Food Agric* 99:6408–6417.

Hafiz MH, Abdul K, Farhat A, Wajid F, Fahad S, Muhammad A, Ghulam MS, Wajid N, Muhammad M, Hafiz FB (2020b) Comparative effects of organic and inorganic fertilizers on soil organic carbon and wheat productivity under arid region. *Commun Soil Sci Plant Anal.* https://doi.org/10.1080/00103624.2020.1763385.

Hafiz MH, Farhat A, Ashfaq A, Hafiz FB, Wajid F, Carol Jo W, Fahad S, Gerrit H (2020a) Predicting Kernel growth of maize under controlled water and nitrogen applications. *Int J Plant Prod* https://doi.org/10.1007/s42106-020-00110-8.

Hafiz MH, Farhat A, Shafqat S, Fahad S, Artemi C, Wajid F, Chaves CB, Wajid N, Muhammad M, Hafiz FB (2018) Offsetting land degradation through nitrogen and water management during maize cultivation under arid conditions. *Land Degrad Dev* 1–10. https://doi.org/10.1002/ldr.2933.

Hafiz MH, Muhammad A, Farhat A, Hafiz FB, Saeed AQ, Muhammad M, Fahad S, Muhammad A (2019) Environmental factors affecting the frequency of road traffic accidents: a case study of sub-urban area of Pakistan. *Environ Sci Pollut Res* https://doi.org/10.1007/s11356-019-04752-8.

Hafiz MH, Wajid F, Farhat A, Fahad S, Shafqat S, Wajid N, Hafiz FB (2016) Maize plant nitrogen uptake dynamics at limited irrigation water and nitrogen. *Environ Sci Pollut Res* 24(3):2549–2557. https://doi.org/10.1007/s11356-016-8031-0.

Harfouche AL, Jacobson DA, Kainer D, Romero JC, Harfouche AH, Mugnozza GS, Moshelion M, Tuskan GA, Keurentjes JJ, Altman A (2019) Accelerating climate resilient plant breeding by applying next-generation artificial intelligence. *Trends Biotechnol* 37:1217–1235.

Heil M, Ton J (2008) Long-distance signaling in plant defence. *Trends Plant Sci* 13:264–272.

Hesham FA, Fahad S (2020) Melatonin application enhances biochar efficiency for drought tolerance in maize varieties: modifications in physio-biochemical machinery. *Agron J* 112(4):1–22.

Hou X, Lee LYC, Xia K, Yan Y, Yu H (2010) DELLAs modulate jasmonate signaling via competitive binding to JAZs. *Dev Cell* 19:884–894.

Howe GA, Major IT, Koo AJ (2018) Modularity in jasmonate signaling for multistress resilience. *Annu Rev Plant Biol* 69:387–415.

Howe GA, Yoshida Y (2019) Evolutionary origin of JAZ proteins and jasmonate signaling. *Mol Plant* 12:153–155.

Hu Y, Jiang Y, Han X, Wang H, Pan J, Yu D (2017) Jasmonate regulates leaf senescence and tolerance to cold stress: crosstalk with other phytohormones. *J Exp Bot* 68:1361–1369.

Huang X, Li J, Shang H, Meng X (2015) Effect of methyl jasmonate on the anthocyanin content and antioxidant activity of blueberries during cold storage. *J Sci Food Agric* 95:337–343.

Hussain MA, Fahad S, Rahat S, Muhammad FJ, Muhammad M, Qasid A, Ali A, Husain A, Nooral A, Babatope SA, Changbao S, Liya G, Ibrar A, Zhanmei J, Juncai H (2020) Multifunctional role of brassinosteroid and its analogues in plants. *Plant Growth Regul* https://doi.org/10.1007/s10725-020-00647-8.

Ibrar K, Aneela R, Khola Z, Urooba N, Sana B, Rabia S, Ishtiaq H, Mujaddad Ur Rehman, Salvatore M (2020) Microbes and environment: global warming reverting the frozen zombies. In: Fahad S, Hasanuzzaman M, Alam M, Ullah H, Saeed M, Khan AK, Adnan M (eds) *Environment, climate, plant and vegetation growth.* Springer Publ Ltd, Springer Nature Switzerland AG. Part of Springer Nature, pp 607–634. https://doi.org/10.1007/978-3-030-49732-3.

Ilyas M, Mohammad N, Nadeem K, Ali H, Aamir HK, Kashif H, Fahad S, Aziz K, Abid U, (2020) Drought tolerance strategies in plants: a mechanistic approach. *J Plant Growth Regul* https://doi.org/10.1007/s00344-020-10174-5.

Iqra M, Amna B, Shakeel I, Fatima K,Sehrish L, Hamza A, Fahad S (2020) Carbon cycle in response to global warming. In: Fahad S, Hasanuzzaman M, Alam M, Ullah H,Saeed M, Khan AK, Adnan M (eds) *Environment, climate, plant and vegetation growth.* Springer Publ Ltd, Springer Nature Switzerland AG. Part of Springer Nature, pp 1–16. https://doi.org/10.1007/978-3-030-49732-3.

Isah T (2019) Stress and defense responses in plant secondary metabolites production. *Biol Res* 52:39.

Jan M, Anwar-ul-Haq Muhammad, Adnan NS, Muhammad Y, Javaid I, Xiuling L, Depeng W, Fahad S, (2019) Modulation in growth, gas exchange, and antioxidant activities of salt-stressed rice (Oryza sativa L.) genotypes by zinc fertilization. *Arabian J Geosci* 12:775 https://doi.org/10.1007/s12517-019-4939-2.

Jin P, Duan Y, Wang L, Wang J, Zheng Y (2014) Reducing chilling injury of loquat fruit by combined treatment with hot air and methyl jasmonate. *Food Bioproc Tech* 7:2259–2266.

Jin P, Zheng Y, Tang S, Rui H, Wang CY (2009) A combination of hot air and methyl jasmonate vapor treatment alleviates chilling injury of peach fruit. *Postharvest Biol Technol* 52:24–29.

Kamarn M, Wenwen C, Irshad A, Xiangping M, Xudong Z, Wennan S, Junzhi C, Shakeel A, Fahad S, Qingfang H, Tiening L (2017) Effect of paclobutrazol, a potential growth regulator on stalk mechanical strength, lignin accumulation and its relation with lodging resistance of maize. *Plant Growth Regul* 84:317–332. https://doi.org/10.1007/ s10725-017-0342-8.

Karaman S, Ozturk B, Genc N, Celik S (2013) Effect of preharvest application of methyl jasmonate on fruit quality of plum (P runus Salicina L indell cv."Fortune") at harvest and during cold storage. *J Food Process Preserv* 37:1049–1059.

Kazan K (2015) Diverse roles of jasmonates and ethylene in abiotic stress tolerance. *Trends Plant Sci* 20:219–229.

Kenton P, Mur LA, Draper J (1999) A requirement for calcium and protein phosphatase in the jasmonate-induced increase in tobacco leaf acid phosphatase specific activity. *J Exp Bot* 50:1331–1341.

Khan MS, Akther T, Ali DM, Hemalatha S (2019) An investigation on the role of salicylic acid alleviate the saline stress in rice crop (Oryza sativa (L)). *Biocatal Agric Biotechnol* 18:101027.

Khan N, Bano A, Ali S, Babar MA (2020) Crosstalk amongst phytohormones from planta and PGPR under biotic and abiotic stresses. *Plant Growth Regul*:1–15.

Khan N, Shah Z, Adnan M, Saeed M, Fahad S, Ali M, Zahoor M, Ali A, Khan WA, Arshad M, Alam A, Khan WA (2016). Assessment of plants for chromium concentration in the basin of chromite hills in lower Malakand-Pakistan. *Am Eurasian J Agric Environ Sci* 16(6):1172–1175.

Koda Y (1997) Possible involvement of jasmonates in various morphogenic events. *Physiol Plant* 100:639–646.

Ku Y-S, Sintaha M, Cheung M-Y, Lam H-M (2018) Plant hormone signaling crosstalks between biotic and abiotic stress responses. *Int J Mol Sci* 19:3206.

Li X, Zhang L, Ahammed GJ, Li Y-T, Wei J-P, Yan P, Zhang L-P, Han X, Han W-Y (2019) Salicylic acid acts upstream of nitric oxide in elevated carbon dioxide-induced flavonoid biosynthesis in tea plant (Camellia sinensis L.). *Environ Exp Bot* 161:367–374.

Li Y, Qin L, Zhao J, Muhammad T, Cao H, Li H, Zhang Y, Liang Y (2017) SlMAPK3 enhances tolerance to tomato yellow leaf curl virus (TYLCV) by regulating salicylic acid and jasmonic acid signaling in tomato (Solanum lycopersicum). *PLoS One* 12: e0172466.

Liu X, Tang S, Dou Z, Li G, Liu Z, Wang S, Ding C, Ding Y (2016) Effects of MeJA on the physiological characteristics of japonica rice wuyunjing 24 and ningjing 3 during early grain filling stage under heat stress. *Chin J Rice Sci* 30:291–303.

Lymperopoulos P, Msanne J, Rabara R (2018) Phytochrome and phytohormones: working in tandem for plant growth and development. *Front Plant Sci* 9:1037.

Ma M, Zhu Z, Cheng S, Zhou Q, Zhou X, Kong X, Hu M, Yin X, Wei B, Ji S (2020) Methyl jasmonate alleviates chilling injury by regulating membrane lipid composition in green bell pepper. *Sci Hortic* 266:109308.

Mahar A, Amjad A, Altaf HL, Fazli W, Ronghua L, Muhammad A, Fahad S, Muhammad A, Rafiullah, Imtiaz AK, Zengqiang Z (2020) Promising technologies for cd-contaminated soils: drawbacks and possibilities. In: Fahad S, Hasanuzzaman M, Alam M, Ullah H,Saeed M, Khan AK, Adnan M (eds) *Environment, climate, plant and vegetation growth.* Springer Publ Ltd, Springer Nature Switzerland AG. Part of Springer Nature, pp 63–92. https://doi.org/10.1007/978-3-030-49732-3.

Md Enamul H, Shoeb AZM, Mallik AH, Fahad S, Kamruzzaman MM, Akib J, Nayyer S, Mehedi AKM, Swati AS, Md Yeamin A, Most SS (2020) Measuring vulnerability to environmental hazards: qualitative to quantitative. In: Fahad S, Hasanuzzaman M, Alam M, Ullah H, Saeed M, Khan AK, Adnan M (eds) *Environment, climate, plant and vegetation growth.* Springer Publ Ltd, Springer Nature Switzerland AG. Part of Springer Nature, pp 421–452. https://doi.org/10.1007/978-3-030-49732-3.

Md Jakir H, Allah B (2020) Development and applications of transplastomic plants; a way towards eco-friendly agriculture. In: Fahad S, Hasanuzzaman M, Alam M, Ullah H, Saeed M, Khan AK, Adnan M (eds) *Environment, climate, plant and vegetation growth.* Springer Publ Ltd, Springer Nature Switzerland AG. Part of Springer Nature, pp 285–322. https://doi.org/10.1007/978-3-030-49732-3.

Mosblech A, Thurow C, Gatz C, Feussner I, Heilmann I (2011) Jasmonic acid perception by COI1 involves inositol polyphosphates in Arabidopsis thaliana. *Plant J* 65:949–957.

Mubeen M, Ashfaq A, Hafiz MH, Muhammad A, Hafiz UF, Mazhar S, Muhammad Sami ul Din, Asad A, Amjed A, Fahad S, Wajid N (2020) Evaluating the climate change impact on water use efficiency of cotton-wheat in semi-arid conditions using DSSAT model. *J Water Clim Change.* https://doi.org/10.2166/wcc.2019.179/622035/jwc2019179.pdf.

Muhammad Z, Abdul MK, Abdul MS, Kenneth BM, Muhammad S, Shahen S, Ibadullah J, Fahad S (2019) Performance of Aeluropus lagopoides (mangrove grass) ecotypes, a potential turfgrass, under high saline conditions. *Environ Sci Pollut Res* https://doi.org/10.1007/s11356-019-04838-3.

Mustafa MA, Ali A, Seymour G, Tucker G (2016) Enhancing the antioxidant content of carambola (Averrhoa carambola) during cold storage and methyl jasmonate treatments. *Postharvest Biol Technol* 118:79–86.

Noor M, Naveed ur R, Ajmal J, Fahad S, Muhammad A, Fazli W, Saud S, Hassan S (2020) Climate change and costal plant lives. In: Fahad S, Hasanuzzaman M, Alam M, Ullah H,Saeed M, Khan AK, Adnan M (eds) *Environment, climate, plant and vegetation growth.* Springer Publ Ltd, Springer Nature Switzerland AG. Part of Springer Nature, pp 93–108. https://doi.org/10.1007/978-3-030-49732-3.

Pauwels L, Barbero GF, Geerinck J, Tilleman S, Grunewald W, Pérez AC, Chico JM, Bossche RV, Sewell J, Gil E (2010) NINJA connects the co-repressor TOPLESS to jasmonate signaling. *Nature* 464:788–791.

Pereira A (2016) Plant abiotic stress challenges from the changing environment. *Front Plant Sci* 7:1123.

Pigolev AV, Miroshnichenko DN, Pushin AS, Terentyev VV, Boutanayev AM, Dolgov SV, Savchenko TV (2018) Overexpression of Arabidopsis OPR3 in hexaploid wheat (Triticum aestivum L.) alters plant development and freezing tolerance. *Int J Mol Sci* 19:3989.

Qamar-uz Z, Zubair A, Muhammad Y, Muhammad ZI, Abdul K, Fahad S, Safder B, Ramzani PMA, Muhammad N (2017) Zinc biofortification in rice: leveraging agriculture to moderate hidden hunger in developing countries. *Arch Agron Soil Sci* 64:147–161. https://doi.org/10.1080/03650340.2017.1338343.

Rashid M, Qaiser H, Khalid SK, Mohammad I. Al-Wabel Zhang A, Muhammad A, Shahzada SI, Rukhsanda A, Ghulam AS, Shahzada MM, Sarosh A, Muhammad FQ (2020) Prospects of biochar in alkaline soils to mitigate climate change. In: Fahad S, Hasanuzzaman M, Alam M, Ullah H,Saeed M, Khan AK, Adnan M (eds) *Environment, climate, plant and vegetation growth.* Springer Publ Ltd, Springer Nature Switzerland AG. Part of Springer Nature, pp 133–150. https://doi.org/10.1007/978-3-030-49732-3.

Rehman M, Fahad S, Saleem MH, Hafeez M, Habib ur Rahman Muhammad, Liu F, Deng G (2020) Red light optimized physiological traits and enhanced the growth of ramie (Boehmeria nivea L.). *Photosynthetica* 58(4):922–931.

Reinbothe C, Springer A, Samol I, Reinbothe S (2009) Plant oxylipins: role of jasmonic acid during programmed cell death, defence and leaf senescence. *The FEBS journal* 276:4666–4681.

Richter R, Bastakis E, Schwechheimer C (2013) Cross-repressive interactions between SOC1 and the GATAs GNC and GNL/CGA1 in the control of greening, cold tolerance, and flowering time in Arabidopsis. *Plant Physiol* 162:1992–2004.

Ruan J, Zhou Y, Zhou M, Yan J, Khurshid M, Weng W, Cheng J, Zhang K (2019) Jasmonic acid signaling pathway in plants. *Int J Mol Sci* 20:2479.

Ruelland E, Zachowski A (2010) How plants sense temperature. *Environ Exp Bot* 69:225–232.

Sadam M, ul Qamar Muhammad Tahir, Ghulam M, Muhammad SK, Faiz AJ (2020) Role of biotechnology in climate resilient agriculture. In: Fahad S, Hasanuzzaman M, Alam M, Ullah H, Saeed M, Khan AK, Adnan M (eds) *Environment, climate, plant and vegetation growth.* Springer Publ Ltd, Springer Nature Switzerland AG. Part of Springer Nature, pp 339–366. https://doi.org/10.1007/978-3-030-49732-3.

Sadiq Y, Zaid A, Khan MMA (2020) Adaptive physiological responses of plants under abiotic stresses: role of phytohormones. In: Hasanuzzaman M (ed) *Plant ecophysiology and adaptation under climate change: mechanisms and perspectives I.* Springer, Singapore, pp 797–824.

Saidi Y, Finka A, Goloubinoff P (2011) Heat perception and signaling in plants: a tortuous path to thermotolerance. *New Phytol* 190:556–565.

Sajid H, Jie H, Jing H, Shakeel A, Satyabrata N, Sumera A, Awais S, Chunquan Z, Lianfeng Z, Xiaochuang C, Qianyu J, Junhua Z (2020) Rice production under climate change: adaptations and mitigating strategies. In: Fahad S, Hasanuzzaman M, Alam M, Ullah H, Saeed M, Khan AK, Adnan M (eds) *Environment, climate, plant and vegetation growth.* Springer Publ Ltd, Springer Nature Switzerland AG. Part of Springer Nature, pp 659–686. https://doi.org/10.1007/978-3-030-49732-3.

Sajjad H, Muhammad M, Ashfaq A, Waseem A, Hafiz MH, Mazhar A, Nasir M, Asad A, Hafiz UF, Syeda RS, Fahad S, Depeng W, Wajid N (2019) Using GIS tools to detect the land use/land cover changes during forty years in Lodhran district of Pakistan. *Environ Sci Pollut Res* https://doi.org/10.1007/s11356-019-06072-3.

Saleem MH, Fahad S, Adnan M, Mohsin A, Muhammad SR, Muhammad K, Qurban A, Inas AH, Parashuram B, Mubassir A, Reem MH (2020a) Foliar application of gibberellic acid endorsed phytoextraction of copper and alleviates oxidative stress in jute (Corchorus capsularis L.) plant grown in highly copper-contaminated soil of China. *Environ Sci Pollut Res* https://doi.org/10.1007/s11356-020-09764-3.

Saleem MH, Fahad S, Shahid UK, Mairaj D, Abid U, Ayman ELS, Akbar H, Analía L, Lijun L (2020c) Copper-induced oxidative stress, initiation of antioxidants and phytoremediation potential of flax (Linum usitatissimum L.) seedlings grown under the mixing of two different soils of China. *Environ Sci Pollut Res* https://doi.org/10.1007/s11356-019-07264-7.

Saleem MH, Rehman M, Fahad S, Tung SA, Iqbal N, Hassan A, Ayub A, Wahid MA, Shaukat S, Liu L, Deng G (2020b) Leaf gas exchange, oxidative stress, and physiological attributes of rapeseed (Brassica napus L.) grown under different light-emitting diodes. *Photosynthetica* 58(3):836–845.

Saman S, Amna B, Bani A, ul Qamar Muhammad Tahir, Rana MA, Muhammad SK (2020) QTL mapping for abiotic stresses in cereals. In: Fahad S, Hasanuzzaman M, Alam M, Ullah H, Saeed M, Khan AK, Adnan M (eds) *Environment, climate, plant and vegetation growth.* Springer Publ Ltd, Springer Nature Switzerland AG. Part of Springer Nature, pp 229–252. https://doi.org/10.1007/978-3-030-49732-3.

Santner A, Estelle M (2009) Recent advances and emerging trends in plant hormone signaling. *Nature* 459:1071–1078.

Saud S, Chen Y, Fahad S, Hussain S, Na L, Xin L, Alhussien SA (2016) Silicate application increases the photosynthesis and its associated metabolic activities in Kentucky bluegrass under drought stress and post-drought recovery. *Environ Sci Pollut Res* 23(17):17647–17655. https://doi.org/10.1007/s11356-016-6957-x.

Saud S, Chen Y, Long B, Fahad S, Sadiq A (2013) The different impact on the growth of cool season turf grass under the various conditions on salinity and drought stress. *Int J Agric Sci Res* 3:77–84.

Saud S, Fahad S, Cui G, Chen Y, Anwar S (2020) Determining nitrogen isotopes discrimination under drought stress on enzymatic activities, nitrogen isotope abundance and water contents of Kentucky bluegrass. *Sci Rep* 10:6415 | https://doi.org/10.1038/s41598-020-63548-w.

Saud S, Fahad S, Yajun C, Ihsan MZ, Hammad HM, Nasim W, Amanullah, Jr, Arif M and Alharby H (2017) Effects of nitrogen supply on water stress and recovery mechanisms in Kentucky bluegrass plants. *Front Plant Sci* 8:983. https://doi.org/10.3389/fpls.2017.00983.

Saud S, Li X, Chen Y, Zhang L, Fahad S, Hussain S, Sadiq A, Chen Y (2014) Silicon application increases drought tolerance of Kentucky bluegrass by improving plant water relations and morph physiological functions. *Sci World J* 2014:1–10. https://doi.org/10.1155/2014/368694.

Sayyari M, Babalar M, Kalantari S, Martínez-Romero D, Guillén F, Serrano M, Valero D (2011) Vapour treatments with methyl salicylate or methyl jasmonate alleviated chilling injury and enhanced antioxidant potential during postharvest storage of pomegranates. *Food Chem* 124:964–970.

Senol C (2020) The effects of climate change on human behaviors. In: Fahad S, Hasanuzzaman M, Alam M, Ullah H, Saeed M, Khan AK, Adnan M (eds) *Environment, climate, plant and vegetation growth.* Springer Publ Ltd, Springer Nature Switzerland AG. Part of Springer Nature, pp 577–590. https://doi.org/10.1007/978-3-030-49732-3.

Shafi MI, Adnan M, Fahad S, Fazli W, Ahsan K, Zhen Y, Subhan D, Zafar-ul-Hye M, Martin B, Rahul D (2020) Application of single superphosphate with humic acid improves the growth, yield and phosphorus uptake of wheat (Triticum aestivum L.) in calcareous soil. *Agron* 10:1224. https://doi.org/10.3390/agronomy10091224.

Shah F, Lixiao N, Kehui C, Tariq S, Wei W, Chang C, Liyang Z, Farhan A, Fahad S, Huang J (2013) Rice grain yield and component responses to near 2°C of warming. *Field Crops Res* 157:98–110.

Shah T, Latif S, Khan H, Munsif F, Nie L(2019) Ascorbic acid priming enhances seed germination and seedling growth of winter wheat under low temperature due to late sowing in Pakistan. *Agron* 9, 757.

Shah T, Munsif F, D'amato R, Nie L(2020) Lead toxicity induced phytotoxic impacts on rapeseed and clover can be lowered by biofilm forming lead tolerant bacteria. *Chemosphere.* 246:125766.

Shahzad R, Waqas M, Khan AL, Hamayun M, Kang S-M, Lee I-J (2015) Foliar application of methyl jasmonate induced physio-hormonal changes in Pisum sativum under diverse temperature regimes. *Plant Physiol Biochem* 96:406–416.

Sharma M, Laxmi A (2016) Jasmonates: emerging players in controlling temperature stress tolerance. *Front Plant Sci* 6:1129.

Sheard LB, Tan X, Mao H, Withers J, Ben-Nissan G, Hinds TR, Kobayashi Y, Hsu F-F, Sharon M, Browse J (2010) Jasmonate perception by inositol-phosphate-potentiated COI1–JAZ co-receptor. *Nature* 468:400–405.

Shi Y, Tian S, Hou L, Huang X, Zhang X, Guo H, Yang S (2012) Ethylene signaling negatively regulates freezing tolerance by repressing expression of CBF and type-A ARR genes in Arabidopsis. *Plant Cell* 24:2578–2595.

Siboza XI, Bertling I, Odindo AO (2017) Enzymatic antioxidants in response to methyl jasmonate and salicylic acid and their effect on chilling tolerance in lemon fruit [Citrus limon (L.) Burm. F.]. *Sci Hortic* 225:659–667.

Song S, Huang H, Gao H, Wang J, Wu D, Liu X, Yang S, Zhai Q, Li C, Qi T (2014) Interaction between MYC2 and ETHYLENE INSENSITIVE3 modulates antagonism between jasmonate and ethylene signaling in Arabidopsis. *Plant Cell* 26:263–279.

Subhan D, Zafar-ul-Hye M, Fahad S, Saud S, Martin B, Tereza H, Rahul D (2020) Drought stress alleviation by ACC deaminase producing *achromobacter xylosoxidans* and *Enterobacter cloacae*, with and without timber waste biochar in Maize. *Sustain* 12(6286). https://doi.org/10.3390/su12156286.

Tariq M, Ahmad S, Fahad S, Abbas G, Hussain S, Fatima Z, Nasim W, Mubeen M, ur Rehman MH, Khan MA, Adnan M. (2018). The impact of climate warming and crop management on phenology of sunflower-based cropping systems in Punjab, Pakistan. *Agri Forest Meteorol* 15;256:270–282.

ul Qamar Muhammad Tahir, Amna F, Amna B, Barira Z, Xitong Z, Ling-Ling C (2020) Effectiveness of conventional crop improvement strategies vs. omics. In: Fahad S, Hasanuzzaman M, Alam M, Ullah H, Saeed M, Khan AK, Adnan M (eds) *Environment, climate, plant and vegetation growth.* Springer Publ Ltd, Springer Nature Switzerland AG. Part of Springer Nature, pp 253–284. https://doi.org/10.1007/978-3-030-49732-3.

Unsar Naeem-U, Muhammad R, Syed HMB, Asad S, Mirza AQ, Naeem I, Muhammad H ur R, Fahad S, Shafqat S (2020) Insect pests of cotton crop and management under climate change scenarios. In: Fahad S, Hasanuzzaman M, Alam M, Ullah H, Saeed M, Khan AK, Adnan M (eds) *Environment, climate, plant and vegetation growth.* Springer Publ Ltd, Springer Nature Switzerland AG. Part of Springer Nature, pp 367–396. https://doi.org/10.1007/978-3-030-49732-3.

Wahid F, Fahad S, Subhan D, Adnan M, Zhen Y, Saud S, Manzer HS, Martin B, Tereza H, Rahul D (2020) Sustainable management with mycorrhizae and phosphate solubilizing bacteria for enhanced phosphorus uptake in calcareous soils. *Agri* 10(334). https://doi.org/10.3390/agriculture10080334.

Wajid N, Ashfaq A, Asad A, Muhammad T, Muhammad A,Muhammad S, Khawar J, Ghulam MS, Syeda RS, Hafiz MH, Muhammad IAR, Muhammad ZH, Habib ur R Muhammad, Veysel T, Fahad S, Suad S, Aziz K, Shahzad A (2017) Radiation efficiency and nitrogen fertilizer impacts on sunflower crop in contrasting environments of Punjab, Pakistan. *Environ Sci Pollut Res* 25:1822–1836. https://org/10.1007/s11356-017-0592-z.

Wang F, Guo Z, Li H, Wang M, Onac E, Zhou J, Xia X, Shi K, Yu J, Zhou Y (2016) Phytochrome A and B function antagonistically to regulate cold tolerance via abscisic acid-dependent jasmonate signaling. *Plant Physiol* 170:459–471.

Wang J, Song L, Gong X, Xu J, Li M (2020) Functions of jasmonic acid in plant regulation and response to abiotic stress. *Int J Mol Sci* 21:1446.

Wang W, Vinocur B, Altman A (2003) Plant responses to drought, salinity and extreme temperatures: towards genetic engineering for stress tolerance. *Planta* 218:1–14.

Wang Y, Gao L, Wang Q, Zuo J (2019) Low temperature conditioning combined with methyl jasmonate can reduce chilling injury in bell pepper. *Sci Hortic* 243:434–439.

Wani SH, Kumar V, Shriram V, Sah SK (2016) Phytohormones and their metabolic engineering for abiotic stress tolerance in crop plants. *Crop J* 4:162–176.

Wasternack C, Hause B (2013) Jasmonates: biosynthesis, perception, signal transduction and action in plant stress response, growth and development. An update to the 2007 review in Annals of Botany. *Ann Bot* 111:1021–1058.

Wasternack C, Strnad M (2018) Jasmonates: news on occurrence, biosynthesis, metabolism and action of an ancient group of signaling compounds. *Int J Mol Sci* 19:2539.

Wu C, Kehui C, She T, Ganghua L, Shaohua W, Fahad S, Lixiao N, Jianliang H, Shaobing P, Yanfeng D (2020) Intensified pollination and fertilization ameliorate heat injury in rice (Oryza sativa L.) during the flowering stage. *Field Crops Res* 252:107795.

Wu C, Tang S, Li G, Wang S, Fahad S, Ding Y (2019) Roles of phytohormone changes in the grain yield of rice plants exposed to heat: a review. *PeerJ* 7:e7792. https://doi.org/10.7717/peerj.7792.

Xu Y-H, Liao Y-C, Zhang Z, Liu J, Sun P-W, Gao Z-H, Sui C, Wei J-H (2016) Jasmonic acid is a crucial signal transducer in heat shock induced sesquiterpene formation in Aquilaria sinensis. *Sci Rep* 6:21843.

Yang Z, Zhang Z, Zhang T, Fahad S, Cui K, Nie L, Peng S, Huang J (2017) The effect of season-long temperature increases on rice cultivars grown in the central and southern regions of China. *Front Plant Sci* 8:1908. https://doi.org/10.3389/fpls.2017.01908.

Yu L, Liu H, Shao X, Yu F, Wei Y, Ni Z, Xu F, Wang H (2016) Effects of hot air and methyl jasmonate treatment on the metabolism of soluble sugars in peach fruit during cold storage. *Postharvest Biol Technol* 113:8–16.

Zafar-ul-Hye M, ahzeeb-ul-Hassan Muhammad T, Muhammad A, Fahad S, Martin B, Tereza D, Rahul D, Subhan D (2020b) Potential role of compost mixed biochar with rhizobacteria in mitigating lead toxicity in spinach. *Sci Rep* 10:12159. https://doi.org/10.1038/s41598-020-69183-9.

Zafar-ul-Hye M, Muhammad N, Subhan D, Fahad S, Rahul D, Mazhar A, Ashfaq AR, Martin B, Ji˘rí H, Zahid HT, Muhammad N (2020a) Alleviation of cadmium adverse effects by improving nutrients uptake in bitter gourd through cadmium tolerant rhizobacteria. *Environ* 7(54). https://doi.org/10.3390/environments7080054.

Zahida Z, Hafiz FB, Zulfiqar AS, Ghulam MS, Fahad S, Muhammad RA, Hafiz MH,Wajid N, Muhammad S (2017) Effect of water management and silicon on germination, growth, phosphorus and arsenic uptake in rice. *Ecotoxicol Environ Saf* 144:11–18.

Zandalinas SI, Fritschi FB, Mittler R (2020) Signal transduction networks during stress combination. *J Exp Bot* 71:1734–1741.

Zhai Q, Li C (2019) The plant mediator complex and its role in jasmonate signaling. *J Exp Bot* 70:3415–3424.

Zhang X, Sheng J, Li F, Meng D, Shen L (2012) Methyl jasmonate alters arginine catabolism and improves postharvest chilling tolerance in cherry tomato fruit. *Postharvest Biol Technol* 64:160–167.

Zhang X-C, Millet YA, Cheng Z, Bush J, Ausubel FM (2015) Jasmonate signaling in Arabidopsis involves SGT1b–HSP70–HSP90 chaperone complexes. *Nature Plants* 1:1–8.

Zhao ML, Wang JN, Shan W, Fan JG, Kuang JF, Wu KQ, Li XP, Chen WX, He FY, Chen JY (2013) Induction of jasmonate signaling regulators MaMYC2s and their physical interactions with MaICE1 in methyl jasmonate-induced chilling tolerance in banana fruit. *Plant Cell Environ* 36:30–51.

Zia-ur-Rehman M (2020) Environment, climate change and biodiversity. In: Fahad S, Hasanuzzaman M, Alam M, Ullah H, Saeed M, Khan AK, Adnan M (eds) *Environment, climate, plant and vegetation growth*. Springer Publ Ltd, Springer Nature Switzerland AG. Part of Springer Nature, pp 473–502. https://doi.org/10.1007/978-3-030-49732-3.

Zuo Z-F, Kang H-G, Park M-Y, Jeong H, Sun H-J, Song P-S, Lee H-Y (2019) Zoysia japonica MYC type transcription factor ZjICE1 regulates cold tolerance in transgenic Arabidopsis. *Plant Sci* 289:110254.

5

The Role of Gibberellin against Abiotic Stress Tolerance in Plants

Sagar Maitra, Akbar Hossain, Chandrasekhar Sahu, Udit Nandan Mishra, Pradipta Banerjee, Preetha Bhadra, Subhashisa Praharaj, Tanmoy Shankar, and Urjashi Bhattacharya

CONTENTS

5.1 Introduction

Global climate change has made agriculture vulnerable to various abiotic stresses. The negative effect of climate change is observed as frequent extreme weather events. Such events are expected to affect food and nutritional security. Considering the fact that the global population is expected to reach 9 billion by 2050, addressing the issue of food security in the face of climate change is very important. Abiotic stresses are expected to rise; hence understanding their impact and developing a coping strategy needs attention (Adnan et al. 2018, 2019, 2020; Ahmad et al. 2019; Akram et al. 2018a,b; Aziz et al. 2017a,b; Baseer et al. 2019; Depeng et al. 2018; Farhana 2020; Frahat et al. 2020; Habib et al. 2017; Hafiz et al. 2016, 2018, 2019, 2020a,b; Hussain et al. 2020; Ilyas et al. 2020; Jan et al. 2019; Kamaran et al. 2017; Mubeen et al. 2020; Muhmmad et al. 2019; Qamar et al. 2017; Rehman 2020; Sajjad et al. 2019; Saleem et al 2020a,b,c; Saud et al. 2013, 2014, 2016, 2017, 2020; Shafi et al. 2020; Shah et al. 2013; Subhan et al. 2020; Tariq et al. 2018; Wahid et al. 2020; Wajid et al. 2017; Wu et al. 2019, 2020; Yang et al. 2017; Zafar-ul-Hye et al. 2020a,b; Zahida et al. 2017).

Plants face various kinds of abiotic stresses that may retard the growth and productivity of crops. Those include high or low temperature, high salinity, drought, waterlogging, submergence, and so on.

Plants respond to these stresses and make a suitable cellular response (Akbar et al. 2020; Amanullah et al. 2020; Amir et al. 2020; Amjad et al. 2020; Arif et al. 2020; Ayman et al. 2020; Bayram et al. 2020; Fahad and Bano 2012; Fahad et al. 2013, 2014a,b, 2015a,b, 2016a,b,c,d, 2017, 2018, 2019a,b; Farah et al. 2020; Fazli et al. 2020; Gopakumar et al.2020; Hesham and Fahad 2020; Ibrar et al. 2020; Iqra et al. 2020; Mahar et al. 2020; Md Jakirand Allah 2020; Md. Enamul et al. 2020; Mohammad I. Al-Wabel et al. 2020a,b; Muhammad Tahir et al. 2020; Noor et al. 2020; Rashid et al. 2020; Sadam et al. 2020; Sajid et al. 2020; Saman et al. 2020; Senol 2020; Unsar et al. 2020; Zia-ur-Rehman 2020). The morphological, physiological, and biochemical responses of a plant to a stress event often decide the extent of the stress tolerance. The impact of different abiotic stresses on the crop at the morphological, physiological and biochemical level, and adaptation mechanisms by crops to such stresses, needs proper understanding. Plants respond to stress by altering the production, distribution, and/or signal transduction of hormones; they will even modify their physiology and biochemistry depending on the extent of the abiotic stress (Colebrook et al. 2014). Hence, hormones play an important role in the stress tolerance of a plant. Gibberellic acid is an important plant hormone in this regard. Increased GA biosynthesis enhances the growth of plants to shading and submergence, while the reduction in GA biosynthesis and signalling has been found in other stresses (Colebrook et al. 2014).

This chapter focuses on the importance of GA as a plant growth regulator (PGR) and the regulation of the signal transduction pathway of GA when exposed to abiotic stress. The GA signalling mechanisms for the regulation of stress tolerance are also discussed.

5.2 Gibberellin Biosynthesis and Signal Transduction

Gibberellic acid was discovered while pathologists in Japan were investigating fungus-infected rice plants showing increased shoot height. It was found that a chemical secreted by the fungus *Gibberella fujikoroi* was responsible for this (Kurosawa 1926). GA_3 (Gibberellin A_3) was the first GA to be isolated from the fungus *Gibberella fujikoroi*. Presently, there are more than 100 GAs. GAs act throughout the plant life in multiple ways like, cell division, cell elongation, developmental phase transitions, and so on (Colebrook 2014).

GA is synthesised from geranylgeranyl diphosphate (GGDP) through various steps like the formation of GA_{12} and production of bioactive GAs such as GA_1, GA_3, GA_4, and GA_7. GGDP is primarily converted into ent-kaurene in the plastid, which is subsequently transported to the endoplasmic reticulum for GA_{12} biosynthesis. GA_{12} is then transformed into bioactive GAs (Sharan et al. 2017). The action of GA results in the degradation of DELLA proteins, a group of transcriptional regulators. DELLA proteins have the N-terminus domain, which is required for GA-induced degradation. The binding of GA to its soluble nuclear receptor, GID1, causes a conformational change in the protein that promotes its linking of N-terminal end of the DELLA protein with an SCF ubiquitin ligase. DELLA proteins act as growth repressors as well as the activator or suppressor of gene expression. There is no direct evidence that they bind to promoter regions but act in close association with transcription factors (Hirano et al. 2012) or as an inhibitor (Feng et al. 2008). Several DELLA-interacting proteins are also the components of other hormone-signalling pathways (Gallego-Bartolomé et al. 2012).

5.3 Gibberellins in Abiotic Stress Responses

During the growth and development period plant faces several abiotic stresses such as drought, submergence, salinity, and osmotic stress due to changing environmental conditions (Bohnert and Sheveleva 1998). Under these conditions, plants activate certain stress-responsive mechanisms through perception and transduction by enabling plant stress-tolerance or resistantance. Irreversible changes may take place causing decay of the cell if the plants fail to do so. PGRs play a vital role in abiotic stress response of plant because they are able to modify the biochemistry and physiology of plants in accordance with the changing environment (Ali et al. 2016; Franklin 2008). Among the known PGRs, GA plays an important role in abiotic stress tolerance (Sharan et al. 2017).

In specific concentrations, GA can be beneficial for plants under stress, as it regulates various metabolic processes through antioxidative enzymes and sugar signalling (Iqbal et al. 2011). Seed germination, stem elongation, leaf expansion, and fruit development are greatly influenced by the activity of GA (Yamaguchi 2008). As photosynthetic enzymes are also influenced by GA, it is known to improve photosynthetic efficiency, enhance nutrient efficiency, and also performs a vital role in diverse process modulation throughout the development of plants (Khan et al. 2007). GA increases the source potential and reallocation of assimilates produced and thus enhances sink strength (Khan et al. 2007). Further, the morphological and stress-protective effects of triazoles (TR) are inverted by GA3 (Gilley and Flecher 2007), thereby demonstrating a relationship between GA3 and plant stress protection. Reduced levels of bioactive GAs in salt-treated *Arabidopsis* plants suggest that retardation of plant growth due to salinity stress was the result of variations of the GA metabolism pathway (Achard et al. 2006).

5.3.1 Heat and Cold Stress

Increases in temperatures due to global warming have great impact on the growth and development of plants. The C-repeat/drought-responsive element-binding factor (CBF1/DREB1b) gene family plays a crucial role in this context. It was found that overexpressing the DREB1 gene of *Gossypium hirsutum* in tobacco increases tolerance to cold. Shan et al. (2007) demonstrated that the biological content of GA in transgenic plants is only half of the amount present in wild-type species. Exogenous use of GA3 represses GhDREB1 expression. This data indicates that the DREB1 and GA-signalling pathway cross-talk via a not well-understood mechanism (Shan et al. 2007). In 2008, Achard et al. (2008a) reported that the constitutive expression of CBF1/DREB1b gene in *Arabidopsis* increases lenience against stress due to low temperature. Overexpression of CBF1 results in RGA (DELLA) accumulation, which in turn results in stunted growth. This is accomplished by GA inactivation as CBF1-overexpressed plants showed enhanced expression of the GA 2-oxidase gene. The distinct function of DELLAs in responding to stress due to salinity is recognised (Achard et al. 2006).

Overexpression of another CBF1/DREB1 gene helps in the development of tolerance for cold stress (Kang et al. 2011). FREEZING TOLERANCE 1-1D (ftl1-1d) phenotype looks like ddf1, exhibiting dwarf stature with dark green leaves and deferred flowering. A DELLA gene, namely RGL3, exhibits up-regulation of gene expression. The information specifies that plants can tolerate abiotic stresses because of the relationship with the introduction of stress-responsive genes and gathering of DELLA protein. Treatment with exogenous GA to the FTL1/DDF1-overexpressing plants showed reduced tolerance to low temperature and impaired phenotypes were restored, indicating that reduction in GA level is key a player in plant responses to abiotic stresses (Kang et al. 2011). Chronic treatment at low temperature inhibited elongation of roots in GsGASA1-overexpressing transgenic *Arabidopsis* lines than the wild-type. These transgenic lines were shown to exhibit increased accumulation of two *Arabidopsis* DELLA proteins, viz., RGL2 and RGL3. Excessive heat adversely affects the growth of plants. Zhang and Wang (2011) showed the involvement of GASA5 in overcoming heat stress.

The transgenic plants which overexpress GASA5 exhibit less expression of genes that encode heat-shock proteins (HSPs), enhancing build-up of hydrogen peroxide and the DELLA protein GAI, and it has a role in accelerating cotyledon-yellowing along with slowing down the hypocotyl elongation phenotypes in the plants that overexpress GASA5. GA applied artificially saves the expression of HSP genes, but until recently, the interaction between HSP and GA was unrecognised. It is reported that heat induces the expression of GASA4, which is correlated with an increased expression of a known target of HSP, namely, Binding Protein (BiP). These findings may be considered direct evidence of the interrelationship between HSP and GASA4.

5.3.2 Response to Drought

Drought stress results in huge economic loss, especially in areas with poor irrigation facilities. Drought stress negatively affects the morphology and physiology of the crop resulting in poor crop yields. Understanding the role and mechanism of plant hormones in providing drought tolerance holds great significance in managing drought stress. Manipulation of the internal level of bioactive gibberellic acid can communicate tolerance to stress-sensitive plants (Sharan et al. 2017). Transgenic tomato with reduced bioactive GA levels by overexpression of the *Arabidopsis thaliana* GA METHYL TRANSFERASE 1 (AtGAMT1) gene showed tolerance to drought due to smaller stomata and reduced pore size, resulting in reduced whole-plant transpiration (Nir et al. 2014). It has been reported that GA deficiency confers both lodging and drought tolerance in small cereals (Plaza-Wüthrich et al. 2016).

5.3.3 Response to Submergence

Submergence stress is one of the most important oxygen-limiting factors that hinder normal plant growth and development. Because of submergence, lowland areas, in particular, face reduced crop production (Tamang et al. 2015). A vital role is played by PGRs, particularly GA, in all aspects and various signalling mechanisms for plant survival under this detrimental condition (Phukan et al. 2015). Expression of GA induced directly or indirectly by ethylene helps plants to carry out escape or quiescence strategies. These strategies help the plant in carbohydrate storage and internode or petiole elongation, both of which are crucial for the survival of a plant under submergence (Xiang et al. 2017). By both DELLA (N-terminal D-E-L-L-A amino acid sequence)-dependent and independent pathways, GAs are directly involved in submergence escape (Colebrook et al. 2014). The stomatal opening is checked by GA in submergence condition (Bashar et al. 2019). Various research has shown that ABA degradation is triggered during submergence and leads to an increase of sensitivity to GA-promoting elongation growth in deepwater rice (Fukao and Bailley-Serres 2008). GA-regulated processes negatively impact tolerance to submergence for a longer period by the promotion of elongation growth and compromises the survival of rice when GA3 treatment is carried out (Fukao and Bailley-Serres 2008). On the other hand, when rice seeds were treated with paclobutrazol (GA antagonist), biosynthesis restriction in elongation was observed under submergence. The GA-mediated growth delays the exhaustion of carbohydrates that ultimately compromises cell viability due to a deficiency in ATP.

5.3.4 Response to Salinity

Among the various stresses, salt stress is very detrimental to crop production. Out of the total agriculture area, 20% is affected by salinity. It is estimated that by 2050, salinity will affect around 50% of arable land (Machado and Serralheiro 2017). GA imparts stress tolerance of salinity (Hoque and Haque 2002). When the plant is exposed to abiotic stress, GA is accumulated and is reported to alleviate the adverse effects of salinity in plants (Yamaguchi 2008). Decreased stomatal resistance and improved plant water efficiency are noted in tomato by the application of GA at low salinity levels (Maggio et al. 2010). Under saline conditions, crop growth and yields can be increased by the application of GA (Iqbal et al. 2011). In soybean, growth is improved by GA by regulation of other PGRs. (Hamayun et al. 2010). This was observed due to increased GA and decreased ABA. In Brassica, salinity stress was alleviated by the application of GA with nitrogen (Siddiqui et al. 2008). GA has a positive effect under salinity because of enhanced water levels and maintenance of protein and RNA levels. Under saline conditions, the application of GA increases the growth of wheat (Kumar and Singh 1996). GA signalling is needed for plants to cope with salinity, as it helps to maintain the source-sink relationship because the reduction in the activity of sink enzyme is observed under salinity stress (Iqbal et al. 2011). It is observed that nitrogen and magnesium content also increase in the plant through the application of GA under salinity stress. In total, GA helps plants to maintain their normal growth and development under salinity stress conditions.

5.3.5 Shade Avoidance

Plants use various responses to avoid the effects of shading (hyponasty, shoot elongation, early flowering, etc.) by neighbouring plants (Zahoor et al. 2016a; Pierik et al. 2003). The ratio of red light (R) to far-red light (FR) is the major signal which triggers shade avoidance in plants. Research has found that GA is involved in the control of R:FR. For example, the elongation response occurs due to low R:FR and is mediated by an increased level of GA (Beall et al. 1996). The increased GA loosens the cell wall which then facilitates increased cell elongation. Because of the enhanced GA action plants respond to shade avoidance. Another study reported that ethylene also promotes hyponasty and stem elongation to promote shade avoidance, but it needs GA to carry out the function (Pierik et al. 2004). As per the investigation of Pierik et al. (2004), low R:FR induces shade avoidance and, for this purpose, GA plays a crucial role. GA is absolutely necessary for petiole and stems elongation response as well; however, GA is not necessary for low R:FR-induced hyponasty.

5.3.6 Response to Osmotic Stress

Various studies of *Arabidopsis thaliana* have proven that GA plays an important role in osmotic stress (Claeys et al. 2012; Skirycz et al. 2011). According to Skirycz et al. (2011), GA affects both cell proliferation and enlargement under osmotic stress conditions, which leads to a reduction in the final leaf size. GA generally affects young tissue during osmotic stress. Moreover, the cellular response to mild osmotic stress has led to speculation that phytohormone signals are introduced from other tissues rather than synthesised de novo (Verelst et al. 2010).

5.4 Stress Tolerance in the Context of GA Signalling

Nature's evolution process has bestowed upon the plants for their survival among various biotic and abiotic factors through signal perception and transduction mechanisms. The cascade of events can be triggered through phytohormones to battle against environmentally harsh conditions. Initial sensing of a threat causes the downstream signalling or pathways that are mediated through different phytohormones (Dolferus 2014). Needless to say, hormonal response to a stressful environment is one of many survival strategies for which there is limited scientific understanding so far. For successful crop establishment under a stressful environment, two events must be addressed: First, maintaining the successful growth and vigour; and second, maintaining a balance between the factors responsible for normal development and the factors responsible for mitigating the stress. These two underlying principles of stress mitigation occur through hormonal cross-talk (synergistic and/or antagonistic) which ultimately provide a basis for revolutionary signalling networks (Munné-Bosch and Müller 2013; Golldack et al. 2013; Nguyen et al. 2016; Zahoor et al. 2016b). In the signal transduction process, the underlying phytohormones crosstalk with other signalling modules (*viz.* mitogen-activated protein kinase [MAPK]) as the transduction mechanism involves several biological transducer molecules other than hormones alone to encounter stress (Roychoudhury and Banerjee 2017). This interactive networking is evident in the case of GA, which has stress-mitigating effects on plants either alone or through cross-talk with other hormones and secondary messengers (Alonso-Ramírez et al. 2009; Verma et al. 2016). Plant architecture in terms of its physiology and anatomy is significantly influenced by individual phenes (phenotypes under genetic control) when exposed to stress conditions, which is the consequence of co-regulatory interplay among different phytohormones out of which bioactive GAs (GA_1, GA_3, GA_4, and GA_7) play a pivotal role throughout the ups and downs of a plant's lifetime (Hedden and Thomas 2012; Daviere and Achard 2013).

5.5 Regulation of GA Metabolism and Signalling under Abiotic Stress

GA involvement in various physiological processes of plant growth and development makes it a key component in signalling. Regulation of this intricate signalling occurs at the molecular level through receptors and their associated proteins (activator and/or inhibitor). GA Insensitive Dwarf1 (GID1) receptor provides the active binding site for GA. The activated complex (GID1-GA) interacts with DELLA motif (aspartate-glutamate-leucine-leucine-alanine or D-E-L-L-A) present on DELLA protein followed by recruitment of F-box protein GA Insensitive Dwarf2/Sleepy1 (GID2/SLY1) (negative regulators of GA) leading to its ubiquitination, if required, (Ariizumi et al. 2011) as they are the key mediators involved in normal growth, development, and stress adaptation (Hou et al. 2010). Growing big and yielding more are not always optimal for a plant and its survival, especially under stress conditions where the metabolites have to be actively partitioned to form secondary metabolites by shifting the normal metabolism regime from primary (required for growth and development) to secondary (escape/quiescence strategy). Low cellular GA levels are maintained through stress-activated transcription factors, mediated DELLA activation (*viz.* cold-responsive transcription factor CBF1), and GA biosynthetic pathway enzyme (viz. GA2-oxidase) (Achard et al. 2008b). The regulatory role of DELLA proteins is, by virtue of its two unique domains (N-terminal and C-terminal), where DELLA-mediated repression of GA occurs through transcription factors binding at C-terminal, and receptor-mediated DELLA regulation for active GA availability occurs through N-terminal domain interaction (Davière and Achard 2013).

Nuclear specific location with a subset of plant-specific GAI-RGA-and-SCR (GRAS) family of proteins DELLA serves as a major transcription regulator of hormone biosynthesis (Bolle 2004). Conserved motifs and promoter-specific regulations make DELLA a suitable tissue-specific genetic regulatory element, as it offers the flexibility of protein–protein interaction including a wide array of regulatory proteins (Locascio et al. 2013). Moreover, targeted gene expression can be escalated to many folds through bridging transcription factors (TFs) between DELLA and DNA (Yoshida et al. 2014), whereas, on the other hand, there is an added effect of chromatin remodelling factor (SWI/SNF) in the DELLA transcripts by virtue of its core subunit interaction (Sarnowska et al. 2013). These distinguishing features of DELLA have made DELLA-mediated GA response and cross-talk more prominent. Differentially expressed genes (DEGs) analysis reveals up-regulation of GA upon multiple abiotic stress exposure/treatments through up-regulation of genes involved in jasmonic acid (JA), salicylic acid (SA), brassinosteroids (BRs) synthesis and/or genes encoding for GA degradation (Liu et al. 2017). Interactions between ABA and GA signalling is clear from the T-DNA mutant and overexpressing experiments by analyzing the conserved motif in GA signalling (Casein Kinase II; CKII), which provide evidence about the parallel influence of GA and ABA signalling through this evolutionarily conserved motif (Yuan et al. 2017). Further, the pleiotropic action of GA signalling in plant development is observed. Geranylgeranyl diphosphate (GGDP) is transformed into bioactive GAs through the conversion of GGDP to ent-kaurene inside plastid followed by transportation of ent-kaurene into the endoplasmic reticulum for GA12 biosynthesis, which is the precursor for bioactive GAs. DELLA can only be inhibited by the binding of GID1 (GA receptor) to DELLA, which allows further plant growth. The absolute requirement for GID1 binding is the bioactive GA which is obtained from the previously mentioned biosynthetic pathway (Ueguchi-Tanaka and Matsuoka 2010; Schwechheimer 2011).

5.5.1 Interactive Cross-Talk Networking between GA and Other Hormone-Signalling Pathways

Growth and stress responses are the two sides of a coin that are essential for the successful field establishment of any crop. Upon exposure to stress conditions, the plant exhibits phytohormonal regulation through the signal transduction mechanism, including antagonistic and synergistic hormones either to provide a balance between metabolite partitioning from primary growth to secondary stress response or to magnify the mitigation mechanism in order to survive under the stress while keeping the growth rate at a steady level. At this juncture, avoidance and escape, the mechanism is well evident in the case of submergence stress where the plant avoids stress by internodal elongation through GA accumulation. The GA accumulatory signal comes from ethylene response factor domain proteins (Hattori et al. 2009). On the contrary, the escape or quiescence mechanism can be seen by restricting shoot elongation thereby conserving carbohydrates (which are to be utilised post-submergence recovery). This process is facilitated by an increased level of DELLA proteins (a negative regulator of GA) (Xu et al. 2006; Fukao and Bailey-Serres 2008). GA signalling is antagonistically associated with ethylene signalling since elevated ethylene causes decline in bioactive GA levels through DELLA gathering.

Abscisic acid (ABA) is considered an inhibitory phytohormone due to its immediate growth inhibition effect upon stress exposure (Weiner et al. 2010). Like other hormones, ABA signalling controls different plant physiological processes driven by receptor and transcription factor-mediated ABA-responsive genes (Park et al. 2009). GA and ABA signalling is linked through DELLA protein involvement. Under stress condition, DELLA targets transcriptional downstream RING-H2 zinc finger factor (XERICO) which otherwise acts as an important regulator of ABA biosynthetic pathways (Ko et al. 2006; Zentella et al. 2007).

Biosynthesis of GA and salicylic acid (SA) are the two translational downstream consequences of heat stress-induced genes (Heckman et al. 2002; Larkindale and Vierling 2008). Among the different GA-induced genes, GASA is more prominent in imparting tolerance of stress in plants caused by abiotic factors. Transgenic work in *A. thaliana* shows additional supplements of GA to the GASA-transformed plants had enhanced expression of genes encoding SA biosynthetic pathway enzymes (Alonso-Ramírez et al. 2009). This experiment suggests the involvement of GA signalling in overexpression of the endogenous SA level with additive effect against heat stress.

In light of GA signals, the DELLA repressors, including REPRESSOR OF GA1-3 (RGA), GIBBERELLIC ACID INSENSITIVE (GAI), RGA-LIKE 1 (RGL1), RGL2, and RGL3, are engaged to the SLEEPY1 (SLY1)-based SCF complex for 26S proteasome-mediated ubiquitination, bringing about enactment of GA responses (Daviere and Achard 2013). GA and JA are antagonistic hormones to each other in managing seedling development and stress protection through interaction among JAZs and DELLAs. JAZs associate with and quell DELLAs to initiate the bHLH factor PHYTOCHROME INTERACTING FACTOR 3 (PIF3) in the GA pathway, while JA signals incite JAZs degradation and deliver DELLAs to stifle PIF3 and restrain GA-improved hypocotyl elongation (Yang et al. 2012). Alternately, DELLAs collaborate with and subdue JAZs to discharge MYC2 that decidedly controls JA-mediated root developmental aspects (Fernandez-Calvo et al. 2011; Wild et al. 2014; Hou et al. 2010). GA-actuated degradation of DELLA liberates JAZs to weaken MYC2, prompting diminished affectability of JA inhibitory root development (Hou et al. 2010). JA and GA likewise synergistically control trichome initiation, stamen development, and sesquiterpene biosynthesis. Both JAZs and DELLAs associate with similar downstream transcription factors, including WD-repeat/bHLH/MYB and MYC2, to tweak the JA and GA cooperativity in controlling trichome development and sesquiterpene biosynthesis. Furthermore, GA was found to direct JA biosynthesis by means of DELLAs which stifle articulation of DELAYED ANTHER DEHISCENCE1 (DAD1) and LIPOXYGENASE (LOX). Subsequently, MYB21, MYB24, and MYB57 are initiated for stamen development (Cheng et al. 2009). It remains to be clarified whether DELLAs additionally partner with MYB21, MYB24, and MYB57 for suppression of stamen development.

5.5.2 Unification of Plant Developmental and Environmental Signals

DELLA can act as transcriptional activators alone and/or in complexes with other transcription factors (TFs) (Hirano et al. 2012). On the other hand, they can act as repressors by inhibiting the gene-activating TFs (Feng et al. 2008). This opens up possible concepts of GA-signalling cross-talk with different pathways. The growing scientific evidence on DELLA signalling continues to provide significant understanding into how DELLA may work as an integrator of signals from different phytohormone pathways under stress conditions. For instance, the plant hormone jasmonic acid (JA) triggers both protection from biotic stress-causing variables and development hindrance, by means of communication with DELLA signalling (Yang et al. 2012). Ubiquitination of JA-signalling repressors (viz. DELLA and JAZ proteins) prevent the competitive binding of these repressors to the MYC2 transcription factor, which, under normal conditions, controls the JA-dependent transcriptional responses (Hou et al. 2010). Beyond this, JAZ proteins contend with PIF transcription factors for the binding site in DELLA. Thus, the DELLA-JAZ binding regulates both JA- and DELLA-assisted responses. JA signaling is likewise proved to repress development by means of impacts on DELLA levels (Yang et al. 2012). JA signaling up-regulates to DELLA gene RGL3, whose promoter was found to be an immediate target of the MYC2 transcription factor (Wild et al. 2012).

The level of DELLA is dependent on the levels of bioactive Gas, thus the level of JA responses are manipulated by DELLA-mediated GA signalling (Figure 5.1). This underlying repression mechanism mediated through JAZ repressors and other TFs suggests it may be conserved among other hormones also. DELLAs interact with the molecular components of other phytohormonal cascades (Bai et al. 2012). Investigating the putative relationship of DELLAs with fundamental leucine zipper (bZIP) or WRKY TFs can be embraced as future viewpoints (Banerjee and Roychoudhury 2015). Together, this

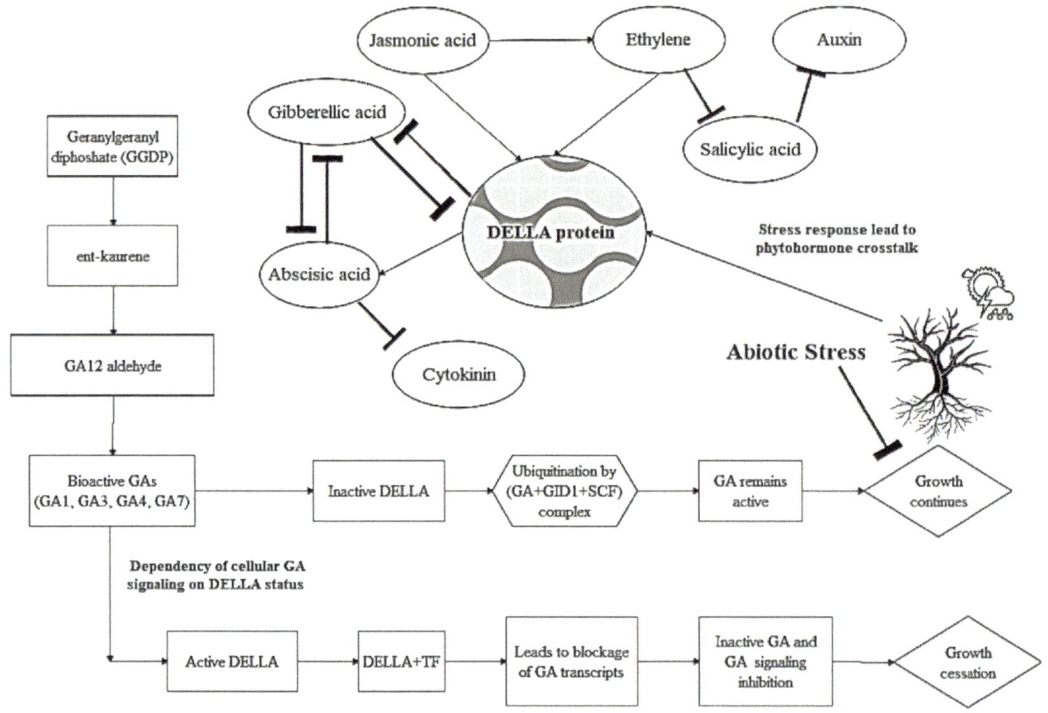

FIGURE 5.1 GA biosynthesis and DELLA mediated crosstalk of GA signalling with other phytohormones under abiotic stress.

information provides various components, and GA signalling could incorporate signals from different phytohormone-signalling pathways to respond to different abiotic stresses.

Under stress conditions, DELLA proteins are accumulated, and they build up cross-talks with different phytohormones. Basic TFs, like MYCs, DREBs, CBFs, JAZ-domain proteins, and PIFs partake in the GA-signalling pathway in a stressful environment. Such fast yet perplexing networking among phytohormones entangles definitive GA-intervened physiology under stress. Epigenetic changes and regulations of phytohormone-mediated abiotic stress response through phytohormonal cross-talking are not yet fully understood in the context of molecular physiology (Banerjee et al. 2017). In this manner, distinguishing such an epigenomic landscape under stress can be a novel undertaking. Further, genome-wide investigations must be led to recognise a greater amount of GA catabolic loci, which can be adequately planned for differential gene expression screening of greater yields and stress tolerance.

5.6 Conclusion

Abiotic stresses affect crop growth negatively, thus reducing crop productivity. As abiotic stresses are expected to be more frequent due to climate change, emphasis must be given to ameliorating the negative impacts of abiotic stress to ensure food and nutritional security. Plants respond to adverse conditions, inclusive of different stresses, at physiological and biochemical levels. Gibberellic acid is a key plant hormone, which plays a predominant role in abiotic stress tolerance in plants. To combat for their survival under stress conditions, plants adapt different mechanisms of stress responses. There can be numerous targets that control endogenous GA levels. GA 2-oxidase enzymes help in the deactivation of GA. Abiotic stress regulates the expression of GA2ox genes by DDF1 or DREB. A reduction in GA exhibits an increase in DELLA proteins. The DELLA-independent mechanism of stress tolerance in plants is mediated by the TSN protein via GA20ox gene and thus it is clear that GA-signalling cross-talk involves mediators of other phytohormones.

REFERENCES

Achard P, Cheng H, De Grauwe L, Decat J, Schoutteten H, Moritz T, Van Der Straeten D, Peng J, Harberd NP (2006) Integration of plant responses to environmentally activated phytohormonal signals. *Science* 311:91–94.

Achard P, Gong F, Cheminant S, Alioua M, Hedden P, Genschik P (2008a) The cold-inducible CBF1 factor-dependent signaling pathway modulates the accumulation of the growth-repressing DELLA proteins via its effect on GA metabolism. *Plant Cell* 20:2117–2129.

Achard P, Renou JP, Berthomé R, Harberd N, Genschik P (2008b) Plant DELLAs restrain growth and promote survival of adversity by reducing the levels of reactive oxygen species. *Curr Biol* 18:656–660.

Adnan M, Fahad S, Muhammad Z, Shahen S, Ishaq AM, Subhan D, Zafar-ul-HyeM, Martin LB, Raja MMN, Beena S, Saud S, Imran A, Zhen Y, Martin B, Jiri H, Rahul D (2020) Coupling phosphate-solubilizing bacteria with phosphorus supplements improve maize phosphorus acquisition and growth under lime induced salinity stress. *Plants* 9(900). https://doi.org/10.3390/plants9070900.

Adnan M, Fahad S, Khan IA, Saeed M, Ihsan MZ, Saud S, Riaz M, Wang D, Wu C (2019) Integration of poultry manure and phosphate solubilizing bacteria improved availability of Ca bound P in calcareous soils. *Biotech* 9(10):368.

Adnan M, Shah Z, Sharif M, Rahman H (2018) Liming induces carbon dioxide (CO_2) emission in PSB inoculated alkaline soil supplemented with different phosphorus sources. *Environ Sci Pollut Res* 25(10):9501–9509.

Adnan M, Zahir S, Fahad S, Arif M, Mukhtar A, Imtiaz AK, Ishaq AM, Abdul B, Hidayat U, Muhammad A, Inayat-Ur R, Saud S, Muhammad ZI, Yousaf J, Amanullah Hafiz MH, Wajid N (2018) Phosphate-solubilizing bacteria nullify the antagonistic effect of soil calcification on bioavailability of phosphorus in alkaline soils. *Sci Rep* 8:4339. https://doi.org/10.1038/s41598-018-22653-7.

Ahmad S, Kamran M, Ding R, Meng X, Wang H, Ahmad I, Fahad S, Han Q (2019) Exogenous melatonin confers drought stress by promoting plant growth, photosynthetic capacity and antioxidant defense system of maize seedlings. *PeerJ* 7:e7793. http://doi.org/10.7717/peerj.7793.

Akbar H, Timothy JK, Jagadish T, Golam M, Apurbo KC, Muhammad F, Rajan B, Fahad S, Hasanuzzaman M (2020) Agricultural land degradation: processes and problems undermining future food security. In: Fahad S, Hasanuzzaman M, Alam M, Ullah H, Saeed M, Khan AK, Adnan M (eds) *Environment, climate, plant and vegetation growth.* Springer Publ Ltd, Springer Nature Switzerland AG. Part of Springer Nature, pp 17–62. https://doi.org/10.1007/978-3-030-49732-3.

Akram R, Turan V, Hammad HM, Ahmad S, Hussain S, Hasnain A, Maqbool MM, Rehmani MIA, Rasool A, Masood N, Mahmood F, Mubeen M, Sultana SR, Fahad S, Amanet K, Saleem M, Abbas Y, Akhtar HM, Waseem F, Murtaza R, Amin A, Zahoor SA, ul Din MS, Nasim W (2018a) Fate of organic and inorganic pollutants in paddy soils. In: Hashmi MZ, Varma A (eds) *Environmental pollution of paddy soils, soil biology.* Springer International Publishing AG, Cham, Switzerland, pp 197–214.

Akram R, Turan V, Wahid A, Ijaz M, Shahid MA, Kaleem S, Hafeez A, Maqbool MM, Chaudhary HJ, Munis, MFH, Mubeen M, Sadiq N, Murtaza R, Kazmi DH, Ali S, Khan N, Sultana SR, Fahad S, Amin A, Nasim W (2018b) Paddy land pollutants and their role in climate change. In: Hashmi MZ, Varma A (eds) *Environmental pollution of paddy soils, soil biology.* Springer International Publishing AG, Cham, Switzerland, pp 113–124.

Ali M, Ullah Z, Mian IA, Khan N, Khan N, Adnan M, Saeed, M (2016) Response of maize to nitrogen levels and seed priming. *Pure Appl Biol* 5(3):578–587.

Alonso-Ramírez A, Rodríguez D, Reyes D, Jiménez JA, Nicolás G, López-Climent M, Gómez- Cadenas A, Nicolás C (2009) Developmental stage specifi city and the role of mitochondrial metabolism in the response of Arabidopsis leaves to prolonged mild osmotic stress. *Plant Physiol* 150:1335–1344.

Al-Wabel Mohammad I, Abdelazeem S, Munir A, Khalid E, Adel RAU (2020b) Extent of climate change in Saudi Arabia and its impacts on agriculture: a case study from qassim region. In: Fahad S, Hasanuzzaman M, Alam M, Ullah H, Saeed M, Khan AK, Adnan M (eds) *Environment, climate, plant and vegetation growth.* Springer Publ Ltd, Springer Nature Switzerland AG. Part of Springer Nature, pp 635–658. https://doi.org/10.1007/978-3-030-49732-3.

Al-Wabel Mohammad I, Ahmad Munir, Usman ARA., Akanji Mutair, Rafique Muhammad Imran (2020a) Advances in pyrolytic technologies with improved carbon capture and storage to combat climate change. In: Fahad S, Hasanuzzaman M, Alam M, Ullah H, Saeed M, Khan AK, Adnan M (eds) *Environment, climate, plant and vegetation growth.* Springer Publ Ltd, Springer Nature Switzerland AG. Part of Springer Nature, pp 535–576. https://doi.org/10.1007/978-3-030-49732-3.

Amanullah, Shah K, Imran Hamdan AK, Muhammad A, Abdel RA, Muhammad A, Fahad S, Azizullah S, Brajendra P (2020) Effects of climate change on irrigation water quality. In: Fahad S, Hasanuzzaman M, Alam M, Ullah H, Saeed M, Khan AK, Adnan M (eds) *Environment, climate, plant and vegetation growth.* Springer Publ Ltd, Springer Nature Switzerland AG. Part of Springer Nature, pp 123–132. https://doi.org/10.1007/978-3-030-49732-3.

Amir M, Muhammad A, Allah B, Sevgi Ç, Haroon ZK, Muhammad A, Emre A (2020) Bio fortification under climate change: the fight between quality and quantity. In: Fahad S, Hasanuzzaman M, Alam M, Ullah H, Saeed M, Khan AK, Adnan M (eds) *Environment, climate, plant and vegetation growth.* Springer Publ Ltd, Springer Nature Switzerland AG. Part of Springer Nature, pp 173–228. https://doi.org/10.1007/978-3-030-49732-3.

Amjad I, Muhammad H, Farooq S, Anwar H (2020) Role of plant bioactives in sustainable agriculture. In: Fahad S, Hasanuzzaman M, Alam M, Ullah H, Saeed M, Khan AK, Adnan M (eds) *Environment, climate, plant and vegetation growth.* Springer Publ Ltd, Springer Nature Switzerland AG. Part of Springer Nature, pp 591–606. https://doi.org/10.1007/978-3-030-49732-3.

Arif M, Talha J, Muhammad R, Fahad S, Muhammad A, Amanullah Kawsar A, Ishaq AM, Bushra K, Fahd R (2020) Biochar; a remedy for climate change. In: Fahad S, Hasanuzzaman M, Alam M, Ullah H, Saeed M, Khan AK, Adnan M (eds) *Environment, climate, plant and vegetation growth.* Springer Publ Ltd, Springer Nature Switzerland AG. Part of Springer Nature, pp 151–172. https://doi.org/10.1007/978-3-030-49732-3.

Ariizumi T, Lawrence PK, Steber CM (2011) The role of two f-box proteins, SLEEPY1 and SNEEZY, in Arabidopsis gibberellin signaling. *Plant Physiol* 155:765–775.

Ayman ELS, Hossain Akbar, Barutçular Celaleddin, Iqbal Muhammad Aamir, Sohidul Islam M, Fahad Shah, Sytar Oksana, Çig Fatih, Meena Ram Swaroop, Erman Murat (2020) Consequences of salinity stress on the quality of crops and its mitigation strategies for sustainable crop production: an outlook of arid and semi- arid regions. In: Fahad S, Hasanuzzaman M, Alam M, Ullah H, Saeed M, Khan AK, Adnan M (eds) *Environment, climate, plant and vegetation growth.* Springer Publ Ltd, Springer Nature Switzerland AG. Part of Springer Nature, pp 503–534. https://doi.org/10.1007/978-3-030-49732-3.

Aziz K, Daniel KYT, Fazal M, Muhammad ZA, Farooq S, FanW, Fahad S, Ruiyang Z (2017a) Nitrogen nutrition in cotton and control strategies for greenhouse gas emissions: a review. *Environ Sci Pollut Res* 24:23471–23487. https://doi.org/10.1007/s11356-017-0131-y.

Aziz K, Daniel KYT, Muhammad ZA, Honghai L, Shahbaz AT, Mir A, Fahad S (2017b) Nitrogen fertility and abiotic stresses management in cotton crop: a review. *Environ Sci Pollut Res* 24:14551–14566. https://doi.org/10.1007/s11356-017-8920-x.

Bai MY, Shang JX, Oh E, Fan M, Bai Y, Zentella R, Sun TP, Wang ZY (2012) Brassinosteroid, gibberellin and phytochrome impinge on a common transcription module in Arabidopsis. *Nat Cell Biol* 14(8):810–817.

Banerjee A, Roychoudhury A (2015) WRKY proteins: signalling and regulation of expression during abiotic stress responses. *Sci World J*:807560. https://doi.org/10.1155/2015/807560.

Banerjee A, Wani SH, Roychoudhury A (2017) Epigenetic control of plant cold responses. *Front Plant Sci* 8:1643. https://doi.org/10.3389/fpls.2017.01643.

Baseer M, AdnanM, Fazal M, Fahad S, Muhammad S, Fazli W, Muhammad A Jr Amanullah Depeng W, Saud S, Muhammad N, Muhammad Z, Fazli S, Beena S, Mian AR, Ishaq AM (2019) Substituting urea by organic wastes for improving maize yield in alkaline soil. *J Plant Nutr.* https://doi.org/10.1080/01904167.2019.1659344.

Bashar KK, Tareq M, Amin M, Honi U, Tahjib-Ul-Arif M, Sadat M, Hossen QM (2019) Phytohormone-mediated stomatal response, escape and quiescence strategies in plants under flooding stress. *Agronomy* 9(2):43.

Bayram AY, Seher Ö, Nazlican A (2020) Climate change forecasting and modeling for the year of 2050. In: Fahad S, Hasanuzzaman M, Alam M, Ullah H, Saeed M, Khan AK, Adnan M (eds) *Environment, climate, plant and vegetation growth.* Springer Publ Ltd, Springer Nature Switzerland AG. Part of Springer Nature, pp 109–122. https://doi.org/10.1007/978-3-030-49732-3.

Beall FD, Yeung EC, Pharis RP (1996) Far-red light stimulates internode elongation, cell division, cell elongation, and gibberellin levels in bean. *Can J Bot* 74:743–752.

Bohnert HJ, Sheveleva E (1998) Plant stress adaptations—making metabolism move. *Curr Opin Plant Biol* 1:267–274.

Bolle C (2004) The role of GRAS proteins in plant signal transduction and development. *Planta* 218:683–692.

Cheng H, Song S, Xiao L, Soo HM, Cheng Z, Xie D, Peng J (2009) Gibberellin acts through jasmonate to control the expression of MYB21, MYB24, and MYB57 to promote stamen filament growth in *Arabidopsis. PLoS Genet* 5(3):e1000440.

Claeys H, Skirycz A, Maleux K, Inzé D (2012) DELLA signaling mediates stress-induced cell differentiation in Arabidopsis leaves through modulation of anaphase-promoting complex/cyclosome activity. *Plant Physiol* 159:739–747.

Colebrook EH, Thomas SG, Phillips AL, Hedden P (2014) The role of gibberellin signaling in plant responses to abiotic stress. *J Exp Biol* 217:67–75.

Daviere JM, Achard P (2013) Gibberellin signaling in plants. *Development* 140:1147–1151.

Depeng W, Fahad S, Saud S, Muhammad K, Aziz K, Mohammad NK, Hafiz MH, Wajid N (2018) Morphological acclimation to agronomic manipulation in leaf dispersion and orientation to promote "Ideotype" breeding: evidence from 3D visual modeling of "super" rice (*Oryza sativa* L.). *Plant Physiol Biochem*. https://doi.org/10.1016/j.plaphy.2018.11.010.

Dolferus R (2014) To grow or not to grow: a stressful decision for plants. *Plant Sci.* 229:247–261.

Fahad S, Adnan M, Hassan S, Saud S, Hussain S, Wu C, Wang D, Hakeem KR, Alharby HF, Turan V, Khan MA, Huang J (2019b) Rice responses and tolerance to high temperature. In: Hasanuzzaman M, Fujita M, Nahar K, Biswas JK (eds) *Advances in rice research for abiotic stress tolerance*. Woodhead Publ Ltd, Cambridge, England, pp 201–224.

Fahad S, Bajwa AA, Nazir U, Anjum SA, Farooq A, Zohaib A, Sadia S, NasimW, Adkins S, Saud S, Ihsan MZ, Alharby H, Wu C, Wang D, Huang J (2017) Crop production under drought and heat stress: plant responses and management options. *Front Plant Sci* 8:1147. https://doi.org/10.3389/fpls.2017.01147.

Fahad S, Bano A (2012) Effect of salicylic acid on physiological and biochemical characterization of maize grown in saline area. *Pak J Bot* 44:1433–1438.

Fahad S, Chen Y, Saud S, Wang K, Xiong D, Chen C, Wu C, Shah F, Nie L, Huang J (2013) Ultraviolet radiation effect on photosynthetic pigments, biochemical attributes, antioxidant enzyme activity and hormonal contents of wheat. *J Food Agri Environ* 11(3&4):1635–1641.

Fahad S, Hussain S, Bano A, Saud S, Hassan S, Shan D, Khan FA, Khan F, Chen Y, Wu C, Tabassum MA, Chun MX, Afzal M, Jan A, Jan MT, Huang J (2014a) Potential role of phytohormones and plant growth-promoting rhizobacteria in abiotic stresses: consequences for changing environment. *Environ Sci Pollut Res* 22(7):4907–4921. https://doi.org/10.1007/s11356-014-3754-2.

Fahad S, Hussain S, Matloob A, Khan FA, Khaliq A, Saud S, Hassan S, Shan D, Khan F, Ullah N, Faiq M, Khan MR, Tareen AK, Khan A, Ullah A, Ullah N, Huang J (2014b) Phytohormones and plant responses to salinity stress: a review. *Plant Growth Regul* 75(2):391–404. https://doi.org/10.1007/s10725-014-0013-y.

Fahad S, Hussain S, Saud S, Hassan S, Chauhan BS, Khan F et al (2016a) Responses of rapid viscoanalyzer profile and other rice grain qualities to exogenously applied plant growth regulators under high day and high night temperatures. *PLoS One* 11(7):e0159590. https://org/10.1371/journal.pone.0159590.

Fahad S, Hussain S, Saud S, Khan F, Hassan S, Jr A, Nasim W, Arif M, Wang F, Huang J (2016b) Exogenously applied plant growth regulators affect heat-stressed rice pollens. *J Agron Crop Sci* 202:139–150.

Fahad S, Hussain S, Saud S, Hassan S, Ihsan Z, Shah AN, Wu C, Yousaf M, Nasim W, Alharby H, Alghabari F, Huang J (2016c) Exogenously applied plant growth regulators enhance the morphophysiological growth and yield of rice under high temperature. *Front Plant Sci* 7:1250. https://doi.org/10.3389/fpls.2016. 01250.

Fahad S, Hussain S, Saud S, Hassan S, Tanveer M, Ihsan MZ, Shah AN, Ullah A, Nasrullah KF, Ullah S, AlharbyH NW, Wu C, Huang J (2016d) A combined application of biochar and phosphorus alleviates heat-induced adversities on physiological, agronomical and quality attributes of rice. *Plant Physiol Biochem* 103:191–198.

Fahad S, Hussain S, Saud S, Tanveer M, Bajwa AA, Hassan S, Shah AN, Ullah A, Wu C, Khan FA, Shah F, Ullah S, Chen Y, Huang J (2015a) A biochar application protects rice pollen from high-temperature stress. *Plant Physiol Biochem* 96:281–287.

Fahad S, Muhammad ZI, Abdul K, Ihsanullah D, Saud S, Saleh A, Wajid N, Muhammad A, Imtiaz AK, Chao W, Depeng W, Jianliang H (2018) Consequences of high temperature under changing climate optima for rice pollen characteristics-concepts and perspectives. *Arch Agron Soil Sci.* https://doi.org/10.1080/03650340.2018.1443213.

Fahad S, Nie L, Chen Y, Wu C, Xiong D, Saud S, Hongyan L, Cui K, Huang J (2015b) Crop plant hormones and environmental stress. *Sustain Agric Res* 15:371–400.

Fahad S, Rehman A, Shahzad B, Tanveer M, Saud S, Kamran M, Ihtisham M, Khan SU, Turan V, Rahman MHU (2019a) Rice responses and tolerance to metal/metalloid toxicity. In: Hasanuzzaman, M and Fujita, M and Nahar, K and Biswas, JK (eds) *Advances in rice research for abiotic stress tolerance.* Woodhead Publ Ltd, Cambridge, England, pp 299–312.

Farah R, Muhammad R, Muhammad SA, Tahira Y, Muhammad AA, Maryam A, Shafaqat A, Rashid M, Muhammad R, Qaiser H, Afia Z, Muhammad AA, Muhammad A, Fahad S (2020) Alternative and non-conventional soil and crop management strategies for increasing water use efficiency. In: Fahad S, Hasanuzzaman M, Alam M, Ullah H, Saeed M, Khan AK, Adnan M (eds) *Environment, climate, plant and vegetation growth.* Springer Publ Ltd, Springer Nature Switzerland AG. Part of Springer Nature, pp 323–338. https://doi.org/10.1007/978-3-030-49732-3.

Farhana G, Ishfaq A, Muhammad A, Dawood J, Fahad S, Xiuling L, Depeng W, Muhammad F, Muhammad F, Syed AS (2020) Use of crop growth model to simulate the impact of climate change on yield of various wheat cultivars under different agro-environmental conditions in Khyber Pakhtunkhwa, Pakistan. *Arabian J Geosci* 13:112 https://doi.org/10.1007/s12517-020-5118-1.

Farhat A, Hafiz MH, Wajid I, Aitazaz AF, Hafiz FB, Zahida Z, Fahad S, Wajid F, Artemi C (2020) Areview of soil carbon dynamics resulting from agricultural practices. *J Environ Manage* 268(2020):110319.

Fazli W, Muhmmad S, Amjad A, Fahad S, Muhammad A, Muhammad N, Ishaq AM, Imtiaz AK, Mukhtar A, Muhammad S, Muhammad I, Rafi U, Haroon I, Muhammad A (2020) Plant-microbes interactions and functions in changing climate. In: Fahad S, Hasanuzzaman M, Alam M, Ullah H, Saeed M, Khan AK, Adnan M (eds) *Environment, climate, plant and vegetation growth.* Springer Publ Ltd, Springer Nature Switzerland AG. Part of Springer Nature, pp 397–420. https://doi.org/10.1007/978-3-030-49732-3.

Feng S H, Martinez C, Gusmaroli G, Wang Y, Zhou JL, Wang F, Chen LY, Yu L, Iglesias-Pedraz JM, Kircher S et al (2008) Coordinated regulation of Arabidopsis thaliana development by light and gibberellins. *Nature* 451:475–479.

Fernández-Calvo P, Chini A, Fernández-Barbero G, Chico JM, Gimenez-Ibanez S, Geerinck J, Eeckhout D, Schweizer F, Godoy M, Franco-Zorrilla JM, Pauwels L (2011) The Arabidopsis bHLH transcription factors MYC3 and MYC4 are targets of JAZ repressors and act additively with MYC2 in the activation of jasmonate responses. *Plant Cell* 23(2):701–715.

Franklin KA (2008) Shade avoidance. *New Phytol* 179:930–944.

Fukao T, Julia BS (2008) Submergence tolerance conferred by Sub1A is mediated by SLR1 and SLRL1 restriction of gibberellin responses in rice. *Proc Nat Acad Sci U S A* 105(43):16814–16819.

Gallego-Bartolomé J, Minguet EG, Grau-Enguix F, Abbas M, Locascio A, Thomas SG, Alabadí D, Blázquez MA (2012) Molecular mechanism for the interaction between gibberellin and brassinosteroid signaling pathways in Arabidopsis. *Proc Natl. Acad Sci U S A* 109:13446–13451.

Gilley A, Flecher RA (2007) Gibberellin antagonizes paclobutrazole induced stress protection in wheat seedlings. *J Plant Physiol* 103:200–207.

Golldack D, Li C, Mohan H, Probst N (2013) Gibberellins and abscisic acid signal crosstalk: living and developing under unfavorable conditions. *Plant Cell Rep* 32:1007–1016.

Gopakumar L, Bernard NO, Donato V (2020) Soil microarthropods and nutrient cycling. In: Fahad S, Hasanuzzaman M, Alam M, Ullah H, Saeed M, Khan AK, Adnan M (eds) *Environment, climate, plant and vegetation growth.* Springer Publ Ltd, Springer Nature Switzerland AG. Part of Springer Nature, pp 453–472. https://doi.org/10.1007/978-3-030-49732-3.

Habib ur R, Ashfaq A, Aftab W, Manzoor H, Fahd R, Wajid I, Md. Aminul I, Vakhtang S, Muhammad A, Asmat U, Abdul W, Syeda RS, Shah S, Shahbaz K, Fahad S, Manzoor H, Saddam H, Wajid N (2017) Application of CSM-CROPGRO-Cotton model for cultivars and optimum planting dates: evaluation in changing semi-arid climate. *Field Crops Res* http://dx.doi.org/10.1016/j.fcr.2017.07.007.

Hafiz MH, Abdul K, Farhat A, Wajid F, Fahad S, Muhammad A, Ghulam MS, Wajid N, Muhammad M, Hafiz FB (2020b) Comparative effects of organic and inorganic fertilizers on soil organic carbon and wheat productivity under arid region. *Commun Soil Sci Plant Anal.* https://doi.org/10.1080/00103624.2020.1763385.

Hafiz MH, Farhat A, Ashfaq A, Hafiz FB, Wajid F, Carol Jo W, Fahad S, Gerrit H (2020a) Predicting Kernel growth of maize under controlled water and nitrogen applications. *Int J Plant Prod.* https://doi.org/10.1007/s42106-020-00110-8.

Hafiz MH, Farhat A, Shafqat S, Fahad S, Artemi C, Wajid F, Chaves CB, Wajid N, Muhammad M, Hafiz FB (2018) Offsetting land degradation through nitrogen and water management during maize cultivation under arid conditions. *Land Degrad Dev* 1–10. https://doi.org/10.1002/ldr.2933.

Hafiz MH, Muhammad A, Farhat A, Hafiz FB, Saeed AQ, Muhammad M, Fahad S, Muhammad A (2019) Environmental factors affecting the frequency of road traffic accidents: a case study of sub-urban area of Pakistan. *Environ Sci Pollut Res.* https://doi.org/10.1007/s11356-019-04752-8.

Hafiz MH, Wajid F, Farhat A, Fahad S, Shafqat S, Wajid N, Hafiz FB (2016) Maize plant nitrogen uptake dynamics at limited irrigation water and nitrogen. *Environ Sci Pollut Res* 24(3):2549–2557. https://doi.org/10.1007/s11356-016-8031-0.

Hamayun M, Khan SA, Khan AL, Shin JH, Ahmad B, Shin DH, Lee IJ (2010) Exogenous gibberellic acid reprograms soybean to higher growth and salt stress tolerance. *J Agric Food Chem* 58:7226–7232.

Hattori Y, Nagai K, Furukawa S, Song X-J, Kawano R, Sakakibara H, Wu J, Matsumoto T, Yoshimura A, Kitano H, Matusoka M (2009)The ethylene response factors SNORKEL1 and SNORKEL2 allow rice to adapt to deep water. *Nature* 46:1026–1030.

Heckman NL, Horst GL, Gaussoin RE, Tavener BT (2002) Trinexapacethyl influence on cell membrane thermostability of Kentucky bluegrass leaf tissue. *Sci Hortic* 92:183–186.

Hedden P, Thomas SG (2012) Gibberellin biosynthesis and its regulation. *Biochem J* 444:11–25.

Hesham FA and Fahad S (2020) Melatonin application enhances biochar efficiency for drought tolerance in maize varieties: Modifications in physio-biochemical machinery. *Agron J* 112(4):1–22.

Hirano K, Kouketu E, Katoh H, Aya K, Ueguchi-Tanaka M, Matsuoka M (2012) The suppressive function of the rice DELLA protein SLR1 is dependent on its transcriptional activation activity. *Plant J* 71:443–453.

Hoque M, Haque S (2002) Effects of GA_3 and its mode of application on morphology and yield parameters of mungbean (*Vigna radiata* L.). *Pak J Biol Sci* 5:281–283.

Hou X, Lee LY, Xia K, Yan Y, Yu H (2010) DELLAs modulate jasmonate signaling via competitive binding to JAZs. *Dev Cell* 19:884–894.

Hussain MA, Fahad S, Rahat S, Muhammad FJ, Muhammad M, Qasid A, Ali A, Husain A, Nooral A, Babatope SA, Changbao S, Liya G, Ibrar A, Zhanmei J, Juncai H (2020) Multifunctional role of brassinosteroid and its analogues in plants. *Plant Growth Regul* https://doi.org/10.1007/s10725-020-00647-8.

Ibrar K, Aneela R, Khola Z, Urooba N, Sana B, Rabia S, Ishtiaq H, Mujaddad Ur Rehman, Salvatore M (2020) Microbes and environment: global warming reverting the frozen zombies. In: Fahad S, Hasanuzzaman M, Alam M, Ullah H, Saeed M, Khan AK, Adnan M (eds) *Environment, climate, plant and vegetation growth.* Springer Publ Ltd, Springer Nature Switzerland AG. Part of Springer Nature, pp 607–634. https://doi.org/10.1007/978-3-030-49732-3.

Ilyas M, Mohammad N, Nadeem K, Ali H, Aamir HK, Kashif H, Fahad S, Aziz K, Abid U (2020) Drought tolerance strategies in plants: a mechanistic approach. *J Plant Growth Regul.* https://doi.org/10.1007/s00344-020-10174-5.

Iqbal N, Nazar R, Iqbal MRK, Masood A, Nafees AK (2011) Role of gibberellins in regulation of source sink relations under optimal and limiting environmental conditions. *Curr Sci* 100:998–1007.

Iqra M, Amna B, Shakeel I, Fatima K, Sehrish L, Hamza A, Fahad S (2020) Carbon cycle in response to global warming. In: Fahad S, Hasanuzzaman M, Alam M, Ullah H, Saeed M, Khan AK, Adnan M (eds) *Environment, climate, plant and vegetation growth.* Springer Publ Ltd, Springer Nature Switzerland AG. Part of Springer Nature, pp 1–16. https://doi.org/10.1007/978-3-030-49732-3.

Jan M, Muhammad Anwar-ul-Haq, Adnan NS, Muhammad Y, Javaid I, Xiuling L, Depeng W, Fahad S, (2019) Modulation in growth, gas exchange, and antioxidant activities of salt-stressed rice (Oryza sativa L.) genotypes by zinc fertilization. *Arabian J Geosci* 12:775 https://doi.org/10.1007/s12517-019-4939-2.

Kamarn M, Wenwen C, Irshad A, Xiangping M, Xudong Z, Wennan S, Junzhi C, Shakeel A, Fahad S, Qingfang H, Tiening L (2017) Effect of paclobutrazol, a potential growth regulator on stalk mechanical strength, lignin accumulation and its relation with lodging resistance of maize. *Plant Growth Regul* 84:317–332. https://doi.org/10.1007/ s10725-017-0342-8.

Kang HG, Kim J, Kim B, Jeong H, Choi SH, Kim EK et al (2011) Overexpression of FTL1/DDF1, an AP2 transcription factor, enhances tolerance to cold, drought, and heat stresses in *Arabidopsis thaliana. Plant Sci* 180:634–641.

Khan NA, Singh S, Nazar R, Lone PM (2007) The source–sink relationship in mustard. *Asian Aust J Plant Sci Biotechnol* 1:10–18.

Ko J-H, Yang SH, Han K-H (2006) Upregulation of an *Arabidopsis* RING-H2 gene, XERICO, confers drought tolerance through increased abscisic acid biosynthesis. *Plant J* 47:343–355.

Kumar B, Singh B (1996) Effect of plant hormones on growth and yield of wheat irrigated with saline water. *Ann Agric Res*17:209–212.

Kurosawa E (1926) Experimental studies on the nature of the substance secreted by the "bakanae fungus". *Nat Hist Soc Formosa* 16:213–227.

Larkindale J, Vierling E (2008) Core genome responses involved in acclimation to high temperature. *Plant Physiol* 146:748–761.

Liu J, Cheng X, Liu P, Sun J (2017) miR156-targeted SBP-box transcription factors interact with DWARF53 to regulate TEOSINTE BRANCHED1 and BARREN STALK1 expression in bread wheat. *Plant Physiol* 174(3):1931–1948.

Locascio A, Bla´zquez MA, Alabadı´ D (2013) Genomic analysis of DELLA protein activity. *Plant Cell Physiol* 54:1229–1237.

Machado RMA, Serralheiro RP (2017) Soil salinity: effect on vegetable crop growth management practices to prevent and mitigate soil salinization. *Horticulturae* 3:30.

Maggio A, Barbieri G, Raimondi G, de Pascale S (2010) Contrasting effects of GA3 treatments on tomato plants exposed to increasing salinity. *J Plant Growth Regul* 29:63–72.

Mahar A, Amjad A, Altaf HL, Fazli W, Ronghua L, Muhammad A, Fahad S, Muhammad A, Rafiullah Imtiaz AK, Zengqiang Z (2020)Promising technologies for cd-contaminated soils: drawbacks and possibilities. In: Fahad S, Hasanuzzaman M, Alam M, Ullah H, Saeed M, Khan AK, Adnan M (eds) *Environment, climate, plant and vegetation growth*. Springer Publ Ltd, Springer Nature Switzerland AG. Part of Springer Nature, pp 63–92. https://doi.org/10.1007/978-3-030-49732-3.

Md Enamul H, Shoeb AZM, Mallik AH, Fahad S, Kamruzzaman MM, Akib J, Nayyer S, Mehedi AKM, Swati AS, Md Yeamin A, Most SS (2020) Measuring vulnerability to environmental hazards: qualitative to quantitative. In: Fahad S, Hasanuzzaman M, Alam M, Ullah H, Saeed M, Khan AK, Adnan M (eds) *Environment, climate, plant and vegetation growth*. Springer Publ Ltd, Springer Nature Switzerland AG. Part of Springer Nature, pp 421–452. https://doi.org/10.1007/978-3-030-49732-3.

Md Jakir H, Allah B (2020) Development and applications of transplastomic plants; a way towards eco-friendly agriculture. In: Fahad S, Hasanuzzaman M, Alam M, Ullah H, Saeed M, Khan AK, Adnan M (eds) *Environment, climate, plant and vegetation growth*. Springer Publ Ltd, Springer Nature Switzerland AG. Part of Springer Nature, pp 285–322. https://doi.org/10.1007/978-3-030-49732-3.

Mubeen M, Ashfaq A, Hafiz MH, Muhammad A, Hafiz UF, Mazhar S, ul Din Muhammad Sami, Asad A, Amjed A, Fahad S, Wajid N (2020) Evaluating the climate change impact on water use efficiency of cotton-wheat in semi-arid conditions using DSSAT model. *J Water Clim Change*. https://doi.org/10.2166/wcc.2019.179/622035/jwc2019179.pdf.

Muhammad Z, Abdul MK, Abdul MS, Kenneth BM, Muhammad S, Shahen S, Ibadullah J, Fahad S (2019) Performance of Aeluropus lagopoides (mangrove grass) ecotypes, a potential turfgrass, under high saline conditions. *Environ Sci Pollut Res* https://doi.org/10.1007/s11356-019-04838-3.

Munné-Bosch S, Müller M (2013) Hormonal cross-talk in plant development and stress responses. *Front Plant Sci* 4:529.

Nguyen D, Rieu I, Mariani C, van Dam NM (2016) How plants handle multiple stresses: hormonal interactions underlying responses to abiotic stress and insect herbivory. *Plant Mol Biol* 91(6):727–740.

Nir I, Moshelion M, Weiss D (2014) The Arabidopsis GIBBERELLIN METHYL TRANSFERASE 1 suppresses gibberellin activity, reduces whole-plant transpiration and promotes drought tolerance in transgenic tomato. *Plant Cell Environ* 37:113–123.

Noor M, Naveed ur R, Ajmal J, Fahad S, Muhammad A, Fazli W, Saud S, Hassan S (2020) Climate change and costal plant lives. In: Fahad S, Hasanuzzaman M, Alam M, Ullah H, Saeed M, Khan AK, Adnan M (eds) *Environment, climate, plant and vegetation growth*. Springer Publ Ltd, Springer Nature Switzerland AG. Part of Springer Nature, pp 93–108. https://doi.org/10.1007/978-3-030-49732-3.

Park SY, Fung P, Nishimura N, Jensen DR, Fujii H, Zhao Y, Lumba S, Santiago J, Rodrigues A, Chow TF, Alfred SE, Bonetta D, Finkelstein R, Provart NJ, Desveaux D, Rodriguez PL, McCourt P, Zhu JK, Schroeder JI, Volkman BF, Cutler SR (2009) Abscisic acid inhibits type 2 C protein phosphatases via the PYR/PYL family of START proteins. *Science* 324:1068–1071.

Phukan UJ, Mishra S, Shukla RK (2015) Waterlogging and submergence stress: affects and acclimation. *Crit Rev Biotechnol* 36:956–966.

Pierik R, Cuppens ML, Voesenek LA, Visser EJ (2004) Interactions between ethylene and gibberellins in phytochrome-mediated shade avoidance responses in tobacco. *Plant Physiol* 136(2):2928–2936.

Pierik R, Visser EJW, de Kroon H, Voesenek LACJ (2003)Ethylene is required in tobacco to succesfully compete with proximate neighbours. *Plant Cell Environ* 26:1229–1234.

Plaza-Wüthrich S, Blösch R, Rindisbacher A, Cannarozzi G, Tadele Z (2016) Gibberellin deficiency confers both lodging and drought tolerance in small cereals. *Front Plant Sci* 7:643.

Qamar-uz Z, Zubair A, Muhammad Y, Muhammad ZI, Abdul K, Fahad S, Safder B, Ramzani PMA, Muhammad N (2017) Zinc biofortification in rice: leveraging agriculture to moderate hidden hunger in developing countries. *Arch Agron Soil Sci* 64:147–161. https://doi.org/10.1080/03650340.2017.1338343.

Rashid M, Qaiser H, Khalid SK, Mohammad I. Al-Wabel Zhang A, Muhammad A, Shahzada SI, Rukhsanda A, Ghulam AS, Shahzada MM, Sarosh A, Muhammad FQ (2020) Prospects of biochar in alkaline soils to mitigate climate change. In: Fahad S, Hasanuzzaman M, Alam M, Ullah H, Saeed M, Khan AK, Adnan M (eds) *Environment, climate, plant and vegetation growth*. Springer Publ Ltd, Springer Nature Switzerland AG. Part of Springer Nature, pp 133–150. https://doi.org/10.1007/978-3-030-49732-3.

Rehman M, Fahad S, Saleem MH, Hafeez M, Habib ur Rahman Muhammad, Liu F, Deng G (2020) Red light optimized physiological traits and enhanced the growth of ramie (Boehmeria nivea L.). *Photosynthetica* 58(4):922–931.

Roychoudhury A, Banerjee A (2017) Abscisic acid signaling and involvement of mitogen activated protein kinases and calcium-dependent protein kinases during plant abiotic stress. In: Pandey GK (ed) *Wiley 2017; Mechanism of plant hormone signaling under stress*, Vol 1, pp 197–241. https://doi.org/10.1002/9781118889022.ch9.

Sadam M, ul Qamar Muhammad Tahir, Ghulam M, Muhammad SK, Faiz AJ (2020) Role of biotechnology in climate resilient agriculture. In: Fahad S, Hasanuzzaman M, Alam M, Ullah H, Saeed M, Khan AK, Adnan M (eds) *Environment, climate, plant and vegetation growth*. Springer Publ Ltd, Springer Nature Switzerland AG. Part of Springer Nature, pp 339–366. https://doi.org/10.1007/978-3-030-49732-3.

Sajid H, Jie H, Jing H, Shakeel A, Satyabrata N, Sumera A, Awais S, Chunquan Z, Lianfeng Z, Xiaochuang C, Qianyu J, Junhua Z (2020)Rice production under climate change: adaptations and mitigating strategies. In: Fahad S, Hasanuzzaman M, Alam M, Ullah H, Saeed M, Khan AK, Adnan M (eds) *Environment, climate, plant and vegetation growth*. Springer Publ Ltd, Springer Nature Switzerland AG. Part of Springer Nature, pp 659–686. https://doi.org/10.1007/978-3-030-49732-3.

Sajjad H, Muhammad M, Ashfaq A, Waseem A, Hafiz MH, Mazhar A, Nasir M, Asad A, Hafiz UF, Syeda RS, Fahad S, Depeng W, Wajid N (2019) Using GIS tools to detect the land use/land cover changes during forty years in Lodhran district of Pakistan. *Environ Sci Pollut Res*. https://doi.org/10.1007/s11356-019-06072-3.

Saleem MH, Fahad S, Adnan M, Mohsin A, Muhammad SR, Muhammad K, Qurban A, Inas AH, Parashuram B, Mubassir A, Reem MH (2020a) Foliar application of gibberellic acid endorsed phytoextraction of copper and alleviates oxidative stress in jute (Corchorus capsularis L.) plant grown in highly copper-contaminated soil of China. *Environ Sci Pollut Res*. https://doi.org/10.1007/s11356-020-09764-3.

Saleem MH, Fahad S, Shahid UK, Mairaj D, Abid U, Ayman ELS, Akbar H, Analía L, Lijun L (2020c) Copper-induced oxidative stress, initiation of antioxidants and phytoremediation potential of flax (Linum usitatissimum L.) seedlings grown under the mixing of two different soils of China. *Environ Sci Pollut Res*. https://doi.org/10.1007/s11356-019-07264-7.

Saleem MH, Rehman M, Fahad S, Tung SA, Iqbal N, Hassan A, Ayub A, Wahid MA, Shaukat S, Liu L, DengG (2020b) Leaf gas exchange, oxidative stress, and physiological attributes of rapeseed (Brassica napus L.) grown under different light-emitting diodes. *Photosynthetica* 58(3):836–845.

Saman S, Amna B, Bani A, ul Qamar Muhammad Tahir, Rana MA, Muhammad SK (2020) QTL mapping for abiotic stresses in cereals. In: Fahad S, Hasanuzzaman M, Alam M, Ullah H, Saeed M, Khan AK, Adnan M (eds) *Environment, climate, plant and vegetation growth*. Springer Publ Ltd, Springer Nature Switzerland AG. Part of Springer Nature, pp 229–252. https://doi.org/10.1007/978-3-030-49732-3.

Sarnowska EA, Rolicka AT, Bucior E, Cwiek P, Tohge T, Fernie AR, Jikumaru Y, Kamiya Y, Franzen R, Schmelzer E, Porri A (2013) DELLA-interacting SWI3C core subunit of switch/sucrose nonfermenting chromatin remodeling complex modulates gibberellin responses and hormonal cross talk in Arabidopsis. *Plant Physiol* 163(1):305–317.

Saud S, Chen Y, Fahad S, Hussain S, Na L, Xin L, Alhussien SA (2016) Silicate application increases the photosynthesis and its associated metabolic activities in Kentucky bluegrass under drought stress and post-drought recovery. *Environ Sci Pollut Res* 23(17):17647–17655. https://doi.org/10.1007/s11356-016-6957-x.

Saud S, Chen Y, Long B, Fahad S, Sadiq A (2013) The different impact on the growth of cool season turf grass under the various conditions on salinity and drought stress. *Int J Agric Sci Res* 3:77–84.

Saud S, Fahad S, Cui G, Chen Y, Anwar S (2020) Determining nitrogen isotopes discrimination under drought stress on enzymatic activities, nitrogen isotope abundance and water contents of Kentucky bluegrass. *Sci Rep* 10:6415 | https://doi.org/10.1038/s41598-020-63548-w.

Saud S, Fahad S, Yajun C, Ihsan MZ, Hammad HM, Nasim W, Amanullah Jr Arif M, Alharby H (2017) Effects of nitrogen supply on water stress and recovery mechanisms in Kentucky bluegrass plants. *Front Plant Sci* 8:983. https://doi.org/10.3389/fpls.2017.00983.

Saud S, Li X, Chen Y, Zhang L, Fahad S, Hussain S, Sadiq A, Chen Y (2014) Silicon application increases drought tolerance of Kentucky bluegrass by improving plant water relations and morph physiological functions. *Sci World J* 2014:1–10. https://doi.org/10.1155/2014/ 368694.

Schwechheimer C (2011) Gibberellin regulates PIN-FORMED abundance and is required for auxin transport-dependent growth and development in *Arabidopsis thaliana. Plant Cell* 23:2184–2195.

Senol C (2020) The effects of climate change on human behaviors. In: Fahad S, Hasanuzzaman M, Alam M, Ullah H, Saeed M, Khan AK, Adnan M (eds) *Environment, climate, plant and vegetation growth.* Springer Publ Ltd, Springer Nature Switzerland AG. Part of Springer Nature, pp 577–590. https://doi.org/10.1007/978-3-030-49732-3.

Shafi MI, Adnan M, Fahad S, Fazli W, Ahsan K, Zhen Y, Subhan D, Zafar-ul-Hye M, Martin B, Rahul D (2020) Application of single superphosphate with humic acid improves the growth, yield and phosphorus uptake of wheat (Triticum aestivum L.) in calcareous soil. *Agron* 10:1224. https://doi.org/10.3390/agronomy10091224.

Shah F, Lixiao N, Kehui C, Tariq S, Wei W, Chang C, Liyang Z, Farhan A, Fahad S, Huang J (2013) Rice grain yield and component responses to near 2°C of warming. *Field Crops Res* 157:98–110.

Shan DP, Huang JG, Yang YT, Guo YH, Wu CA, Yang GD et al (2007) Cotton GhDREB1 increases plant tolerance to low temperature and is negatively regulated by gibberellic acid. *New Phytol* 176:70–81.

Siddiqui MH, Khan MN, Mohammad F, Khan MMA (2008) Role of nitrogen and gibberellic acid (GA_3) in the regulation of enzyme activities and in osmoprotectant accumulation in Brassica juncea L. under salt stress. *J Agron Crop Sci* 194:214–224.

Skirycz A, Claeys H, De Bodt S, Oikawa A, Shinoda S, Andriankaja M, Maleux K, Eloy NB, Coppens F, Yoo SD, Saito K (2011) Pause-and-stop: the effects of osmotic stress on cell proliferation during early leaf development in Arabidopsis and a role for ethylene signaling in cell cycle arrest. *Plant Cell* 23(5):1876–1888.

Subhan D, Zafar-ul-HyeM, Fahad S, Saud S, Martin B, Tereza H, Rahul D (2020) Drought stress alleviation by ACC deaminase producing *achromobacter xylosoxidans* and *enterobacter cloacae*, with and without timber waste biochar in maize. *Sustain* 12(6286). https://doi.org/10.3390/su12156286.

Tamang BG, Fukao T (2015) Plant adaptation to multiple stresses during submergence and following desubmergence. *Int J Mol Sci* 16:30164–30180.

Tariq M, Ahmad S, Fahad S, Abbas G, Hussain S, Fatima Z, Nasim W, Mubeen M, ur Rehman MH, Khan MA, Adnan M (2018) The impact of climate warming and crop management on phenology of sunflower-based cropping systems in Punjab, Pakistan. *Agri Forest Meteorol* 15(256):270–282.

Ueguchi-Tanaka M, Matsuoka M (2010) The perception of gibberellins: clues from receptor structure. *Curr Opin Plant Biol* 13:503–508.

ul Qamar Muhammad Tahir, Amna F, Amna B, Barira Z, Xitong Z, Ling-Ling C (2020) Effectiveness of conventional crop improvement strategies vs. omics. In: Fahad S, Hasanuzzaman M, Alam M, Ullah H, Saeed M, Khan AK, Adnan M (eds) *Environment, climate, plant and vegetation growth.* Springer Publ Ltd, Springer Nature Switzerland AG. Part of Springer Nature, pp 253–284. https://doi.org/10.1007/978-3-030-49732-3.

Unsar Naeem-U, Muhammad R, Syed HMB, Asad S, Mirza AQ, Naeem I, Muhammad H ur R, Fahad S, Shafqat S (2020) Insect pests of cotton crop and management under climate change scenarios. In: Fahad S, Hasanuzzaman M, Alam M, Ullah H, Saeed M, Khan AK, Adnan M (eds) *Environment, climate, plant and vegetation growth.* Springer Publ Ltd, Springer Nature Switzerland AG. Part of Springer Nature, pp 367–396. https://doi.org/10.1007/978-3-030-49732-3.

Verelst W, Skirycz A, Inzé D (2010) Abscisic acid, ethylene and gibberellic acid act at different developmental stages to instruct the adaptation of young leaves to stress. *Plant Signal Behav* 5:473–475.

Verma V, Ravindran P, Kumar PP (2016) Plant hormone-mediated regulation of stress responses. *BMC Plant Biol* 16:86.

Wahid F, Fahad S, Subhan D, Adnan M, Zhen Y, Saud S, Manzer HS, Martin B, Tereza H, Rahul D (2020) Sustainable management with mycorrhizae and phosphate solubilizing bacteria for enhanced phosphorus uptake in calcareous soils. *Agri* 10(334). https://doi.org/10.3390/agriculture10080334.

Wajid N, Ashfaq A, Asad A, Muhammad T, Muhammad A, Muhammad S, Khawar J, Ghulam MS, Syeda RS, Hafiz MH, Muhammad IAR, Muhammad ZH, Habib ur R Muhammad, Veysel T, Fahad S, Suad S, Aziz K, Shahzad A (2017) Radiation efficiency and nitrogen fertilizer impacts on sunflower crop in contrasting environments of Punjab. *Pakistan Environ Sci Pollut Res* 25:1822–1836. https://org/10.1007/s11356-017-0592-z.

Weiner JJ, Peterson FC, Volkman BF, Cutler SR (2010) Structural and functional in sights in to core ABA signaling. *Curr Opin Plant Biol* 13:495–502.

Wild M, Davière JM, Cheminant S, Regnault T, Baumberger N, Heintz D, Baltz R, Genschik P, Achard P (2012) The Arabidopsis DELLA RGA-LIKE3 is a direct target of MYC2 and modulates jasmonate signaling responses. *The Plant Cell* 24(8):3307–3319.

Wild M, Daviere JM, Cheminant S, Regnault T, Baumberger N, Heintz D, Baltz R, Genschik P, Achard P (2014) The arabidopsis DELLA 118 cell signalling and gene regulation. *Curr Opin Plant Biol* 21:112–119.

Wu C, Kehui C, She T, Ganghua L, Shaohua W, Fahad S, Lixiao N, Jianliang H, Shaobing P, Yanfeng D (2020) Intensified pollination and fertilization ameliorate heat injury in rice (Oryza sativa L.) during the flowering stage. *Field Crops Res* 252:107795.

Wu C, Tang S, Li G, Wang S, Fahad S, Ding Y (2019) Roles of phytohormone changes in the grain yield of rice plants exposed to heat: a review. *PeerJ* 7:e7792. https://doi.org/10.7717/peerj.7792.

Xiang J, Wu H, Zhang Y, Zhang Y, Wang Y, Li Z, Lin H, Chen H, Zhang J, Zhu D (2017) Transcriptomic analysis of gibberellin- and paclobutrazol-treated rice seedlings under submergence. *Int J Mol Sci* 18:2225.

Xu X, Chen C, Fan B, Chen Z (2006) Physical and functional interactions between pathogeninduced Arabidopsis WRKY18, WRKY40, and WRKY60 transcription factors. *Plant Cell* 18:1310–1326.

Yamaguchi S (2008) Gibberellin metabolism and its regulation. *Annu Rev Plant Physiol* 59:225–251.

Yang DL, Yao J, Mei CS, Tong XH, Zeng LJ, Li Q, Xiao LT, Sun TP, Li J, Deng XW, Lee CM (2012) Plant hormone jasmonate prioritizes defense over growth by interfering with gibberellin signaling cascade. *Proc Natl Acad Sci U S A* 109(19), pp.E1192–E1200.

Yang Z, Zhang Z, Zhang T, Fahad S, Cui K, Nie L, Peng S, Huang J (2017) The effect of season-long temperature increases on rice cultivars grown in the central and southern regions of China. *Front Plant Sci* 8:1908. https://doi.org/10.3389/fpls.2017.01908.

Yoshida H, Hirano K, Sato T, Mitsuda N, Nomoto M, Maeo K, Koketsu E, Mitani R, Kawamura M, Ishiguro S, Tada Y (2014) DELLA protein functions as a transcriptional activator through the DNA binding of the indeterminate domain family proteins. *Proc Nat Acad Sci U S A* 111:7861–7866.

Yuan C, Ai J, Chang H, Xiao W, Liu L, Zhang C, He Z, Huang J, Li J, Guo X (2017) CKB1 is involved in abscisic acid and gibberellic acid signaling to regulate stress responses in Arabidopsis thaliana. *J Plant Res* 130(3):587–598.

Zafar-ul-Hye M, ahzeeb-ul-Hassan Muhammad T, Muhammad A, Fahad S, Martin B, Tereza D, Rahul D, Subhan D (2020b) Potential role of compost mixed biochar with rhizobacteria in mitigating lead toxicity in spinach. *Sci Rep* 10:12159. https://doi.org/10.1038/s41598-020-69183-9.

Zafar-ul-Hye M, Muhammad N, Subhan D, Fahad S, Rahul D, Mazhar A, Ashfaq AR, Martin B, Jiˇrí H, Zahid HT, Muhammad N (2020a) Alleviation of cadmium adverse effects by improving nutrients uptake in bitter gourd through cadmium tolerant rhizobacteria. *Environ* 7(54). https://doi.org/10.3390/environments7080054.

Zahida Z, Hafiz FB, Zulfiqar AS, Ghulam MS, Fahad S, Muhammad RA, Hafiz MH, Wajid N, Muhammad S (2017) Effect of water management and silicon on germination, growth, phosphorus and arsenic uptake in rice. *Ecotoxicol Environ Saf* 144:11–18.

Zahoor M, Afzal M, Ali M, Mohammad W, Khan N, Adnan M, Saeed M (2016a) Effect of organic waste and NPK fertilizer on potato yield and soil fertility. *Pure Appl Biol* 5(3):439–445.

Zahoor M, Khan N, Ali M, Saeed M, Ullah Z, Adnan M, Ahmad B (2016b) Integrated effect of organic waste and NPK fertilizers on nutrients uptake in potato crop and soil fertility. *Pure Appl Biol* 5(3):601–607.

Zentella R, Zhang ZL, Park M, Thomas SG, Endo A, Murase K, Fleet CM, Jikumaru Y, Nambara E, Kamiya Y, Sun TP (2007) Global analysis of DELLA direct targets in early gibberellin signaling in *Arabidopsis*. *Plant Cell* 19:3037–3057. https://doi.org/10.1105/tpc.107.054999.

Zhang S, Wang X (2011) Overexpression of GASA5 increases the sensitivity of Arabidopsis to heat stress. *J Plant Physiol* 168:2093–2101.

Zia-ur-Rehman M (2020) Environment, climate change and biodiversity. In: Fahad S, Hasanuzzaman M, Alam M, Ullah H, Saeed M, Khan AK, Adnan M (eds) *Environment, climate, plant and vegetation growth*. Springer Publ Ltd, Springer Nature Switzerland AG. Part of Springer Nature, pp 473–502. https://doi.org/10.1007/978-3-030-49732-3.

6

Role of Phytohormones in Drought Stress

Abdul Rehman, Hafiza Iqra Almas, Abdul Qayyum, Hongge Li, Zhen Peng, Guangyong Qin, Yinhua Jia, Zhaoe Pan, Fazal Akbar, Shoupu He, and Xiongming Du

CONTENTS

6.1 Introduction

The human population is increasing day by day and consequently, food demand around the globe is also increasing. But many key factors like abiotic and biotic stresses reduce agriculture production (Wani and Sah 2014). To fulfill the demand for food, agriculture production must be increased by almost 70% because it is supposed that the rate of population growth will lead to an additional 2.3 billion people by 2050 (Tilman et al. 2011). The basic and fundamental mechanisms against environmental stress and tolerance in plants are distinct and multifarious as compared to animals (Qin et al. 2011). It is a serious challenge for agricultural biotechnologists to identify the mechanisms that assist against environmental stresses. Different abiotic stresses like salt stress, water deficit, and high temperature are more significant and common worldwide (Adnan et al. 2018a,b, 2019, 2020; Ahmad et al. 2019; Akbar et al. 2020; Akram et al. 2018a,b; Amanullah et al. 2020; Amir et al. 2020; Amjad et al. 2020; Arif et al. 2020; Ayman et al. 2020; Aziz et al. 2017a,b; Baseer et al. 2019; Bayram et al. 2020; Depeng et al. 2018; Fahad and Bano 2012; Fahad et al. 2013, 2014a,b, 2015a,b, 2016a,b,c,d, 2017, 2018, 2019a,b; Farah et al. 2020; Farhana 2020; Fazli et al. 2020; Frahat et al. 2020; Gopakumar et al. 2020; Habib et al. 2017; Hafiz et al. 2016, 2018, 2019, 2020a,b; Tariq et al. 2018; Hesham and Fahad 2020; Hussain et al. 2020; Hussain Wani et al. 2013; Ibrar et al. 2020; Ilyas et al. 2020; Iqra et al. 2020; Jan et al. 2019; Kamaran et al. 2017; Mahar et al. 2020; Md Jakirand Allah 2020; Md. Enamul et al. 2020; Mohammad I. Al-Wabel et al. 2020a,b; Mubeen et al. 2020; Muhammad Tahir et al. 2020; Muhmmad et al. 2019; Noor et al. 2020; Qamar et al. 2017; Rashid et al. 2020; Rehman 2020; Sadam et al. 2020; Sajid et al. 2020.; Sajjad et al. 2019; Saleem et al 2020a,b,c; Saman et al. 2020; Saud et al. 2013, 2014, 2016, 2017, 2020; Senol 2020; Shafi et al. 2020;

Shah et al. 2013; Subhan et al. 2020; Unsar et al. 2020; Wahid et al. 2020; Wajid et al. 2017; Wu et al. 2019, Wu et al. 2020; Yang et al. 2017; Zafar-ul-Hye et al. 2020a,b; Zahida et al. 2017; Zia-ur-Rehman 2020).

Traditional breeding methods are not sufficient due to the complexity in the inheritance of traits against stresses. Efficient methods are urgently needed that maintain the increasing need for food supply over the entire world. In this chapter, unique and effective methods will be explained. The best method is the effect of phytohormones which develop environmental resilient crops with maximum yields.

Phytohormones are molecules that, despite being present in very small quantities, have the ability to control multifarious mechanisms in plants. They act like chemical messengers that interconnect with cellular actions in vascular plants (Voß et al. 2014). They are essential in coordinating different signal transduction processes against drought stress. Hence, they are internally controlled along with external stimuli (Kazan 2015). Various phytohormones, like ABA, have been recognised as stress hormones. In plant development, ABA has various functions, *i.e.*, fruit and germination inhibition, regulation of growth, closure of stomata, and preservation of seed dormancy as well as regulating biotic and abiotic stress (Li et al. 2010). Engineered phytohormones could be an important resource in the effort to increase agriculture production. In this chapter, plant hormones and their functions in plant development are elucidated comprehensively with recent achievements and future scenarios, as well as their role in growth and responses to drought stress. Furthermore, phytohormones improve the metabolic system of the plants and enhance the quality and quantity of food.

6.2 Phytohormones: Key Regulators of Plant Responses during Abiotic Stresses

The growth and development of plants can be disturbed in response to numerous external and internal stimuli (Wolters and Jürgens 2009). Phytohormones control these responses. Their critical function is to adjust the plants' responses to different environmental stresses by regulating development, source or sink, growth, and maintenance of nutrients. Though plant response during abiotic stress depends on several aspects, phytohormones are the most important endogenous molecule that plays a vital function in the response of stress, by adjusting the molecular, physiological, and biochemical activity – components that are critical to the survival of the plants (Fahad et al. 2015b). Phytohormones work as mobile and non-mobile; sometimes these are produced at the site of action and sometimes these are transported to where they are most needed (Peleg and Blumwald 2011). They are also involved in plant growth and development. Various phytohormones such as CKs (cytokinins), IAA (auxin), ET (ethylene), GAs (gibberellins), ABA (abscisic acid), JAs (jasmonates), BRs (brassinosteroids), and newly identified phytohormones SA (salicylic acid) and SL (strigolactone) are found to be involved in drought stress tolerance. Phytohormones and their genes associated with drought tolerance are listed in Table 6.1.

6.2.1 Abscisic Acid (ABA)

The role of abscisic acid is known by its name, as it is involved in leaf abscission. It also functions in the adjustment of different pathways of plants under abiotic stress conditions – that is why it is regarded as a stress hormone. ABA is an isoprenoid phytohormone that is produced through the plastidial 2-C methyl-D-erythritol-4-phosphate pathway. Its many functions in different stages of plant development as well as in physiological pathways include dormancy of seeds, opening/closing of stomata, morphogenesis of embryo, and formation and storage of lipids and proteins (Sreenivasulu et al. 2010). ABA is considered an important hormone in the adjustment of plants under abiotic stress, especially drought resistance. The levels of internal ABA increase rapidly stimulating signalling pathways that modify gene expression under drought stress conditions (O'Brien and Benková 2013). According to Nemhauser et al. (2006), transcriptionally ABA controls protein-encoding genes up to 10 percent. Similarly, it plays an important function as an internal stimulus that adjusts the mechanism of plants under stress conditions (Keskin et al. 2010).

Under drought stress, ABA enables the plants to send stress adjustment signals towards the shoots; ultimately, it acts as an antitranspirant by reducing leaf area and enabling closure of stomata (Wilkinson et al. 2012). During water deficit conditions, it enhances the development of root growth and other

TABLE 6.1

Phytohormones and Their Genes Associated with Drought Tolerance

Phytohormone	Genes	Crop	References
ABA	AtLOS5	*Gossypium hirsutum*	Yue et al. (2012)
	ABO3	*Arabidopsis thaliana*	Ren et al. (2010)
	OsSAPK2	*Oryza sativa*	Lou et al. (2017)
Cytokinin	IPT	*Gossypium hirsutum*	Kuppu et al. (2013)
	AtUGT76C2	*Arabidopsis thaliana*	Li et al. (2015)
	ARR1, ARR10, and ARR12	*Arabidopsis thaliana*	Nguyen et al. (2016)
Auxin	OsRGLP2	*Oryza sativa*	Ayub et al. (2019)
	OsIAA6	*Oryza sativa*	Jung et al. (2015)
	TaSAUR75	*Triticum aestivum*	Guo et al. (2018)
Ethylene	ARGOS	*Arabidopsis thaliana and Zea mays*	Shi et al. (2015)
	OsERF109	*Oryza sativa*	Yu et al. (2017)
	GmCYP82A3	*Glycine max*	Yan et al. (2016)
Brassinosteroids	WRKY46, WRKY54, and WRKY70	*Arabidopsis thaliana*	Chen et al. (2017)
	RD26	*Arabidopsis thaliana*	Ye et al. (2017)
	BdBRI1	*Brachypodium distachyon*	*Feng et al. (2015)*
Gibberellins	GhGA2ox1	*Gossypium hirsutum*	Shi et al. (2019)
	AtGA2ox1	*Zea mays*	Chen et al. (2019)
	OsCYP71D8L	*Oryza sativa*	Zhou et al. (2019)
Jasmonates	OsJAZ1	*Oryza sativa*	Fu et al. (2017)
	OsbHLH148	*Oryza sativa*	Seo et al. (2011)
Salicylic acid	LcSABP	*Nicotiana tabacum*	Li et al. (2019b)
	LcSAMT	*Nicotiana tabacum*	Wang et al. (2019)
	OsMYB–R1	*Oryza sativa*	Tiwari et al. (2020)

anatomical adjustments (Giuliani et al. 2005). It controls the numerous gene expressions that work under drought stress and is involved in the formation of different proteins like dehydrins, and protective LEA proteins (Sreenivasulu et al. 2012). ABA adjusts the turgor of cells and production of antioxidant assays and osmoprotectants that provide resistance during water deficit conditions (Chaves et al. 2003). Zhang et al. (2006) concluded that the concentration of ABA increases under saline conditions.

6.2.2 Cytokinins (CKs)

Cytokinins are important in numerous mechanisms of growth and development of plants and are considered master regulators (Kang et al. 2012). Alterations in internal plant CK levels under water stress indicate their involvement in stress tolerance (Kang et al. 2012). Transgenic tissues and mutants for cytokinin metabolic enzymes represented enhanced stress tolerance resulting in increased yields (Zalabák et al. 2013). Sometimes, the response of plants related to cytokinin have been estimated through their exogenous applications, but the internal level of CKs increases under stress conditions through uptake and improved biosynthesis (Pospíšilová 2003a). On the other hand, germination of seeds is inhibited by ABA, but seeds are released from dormancy via CKs (Fahad et al. 2015c). It's thought that CKs are antagonists to ABA (Pospíšilová 2003b). In water stress conditions, levels of CK reduce with an increase in ABA accumulation, hence, leading to the ABA/CK ratio. The decreased CK contents increase the apical dominance that combines with ABA which maintains stomatal aperture that helps in adjustment during drought conditions (O'Brien and Benková 2013).

6.2.3 Auxins (IAA)

Despite 100 years' worth of research on auxins, its many activities including transportation, signalling mechanism, and biosynthesis are not yet fully understood (Ke et al. 2015). Whereas in plants, several interrelating processes have been suggested until now for auxin biosynthesis, such as one Trp (Tryptophan)-independent and 4 dependent pathways (Mano and Nemoto 2012). IAA (indole-3-acetic acid) is a crucial multifunctional plant hormone and it is not only important for the growth and development of plants but also helpful in the maintenance and adjustment of various mechanisms and signals during stress conditions (Adnan et al. 2016; Kazan 2013). The existence of transportation, signalling, and biosynthesis of auxin in unicellular green algae is solid evidence of the evolutionary function performed by auxin in plants adjustment under various environmental stresses (De Smet et al. 2011). However, recently the role of auxin in growth and development has been further clarified: It also acts like a stress-responsive regulator (Kazan 2013).

In Arabidopsis, glucosinolate levels are regulated by aux/IAA (auxin sensitive) repressors. Glucosinolate levels are maintained by these proteins which act in a transcriptional network under drought conditions (Salehin et al. 2019). Likewise, the exogenous application of IAA ameliorates drought damage in white clover by mediating the endogenous hormone level and regulating the genes involved in drought tolerance (Arshad et al. 2016; Zhang et al. 2020). Remarkably, there is various advanced proof that auxin plays an essential role in the adjustment of plants under salinity (Iqbal et al. 2014). The growth of roots and shoots are upgraded under heavy metal and salt/drought stresses (Egamberdieva 2009). In maize, auxin decreases salinity stress, while salicylic acid enhances it (Fahad and Bano 2012), suggesting that the interaction and balance in hormones are most important for the perception of signals, mediation, and transduction under the response of stress. It regulates the several genes in transcription known as primary auxin response genes, and these genes have been recognised and categorised in many plant species such as soybean, rice, and Arabidopsis (Javid et al. 2011). IAA is considered a significant component of tolerance through regulation of a huge collection of genes and cross-talk mediation in response to biotic and abiotic stress (Fahad et al. 2015c). Auxin acts as an optimal solution in abiotic stress tolerance. In rice Zhang et al. (2012) introduced a putative auxin efflux transporter gene (OsPIN3t) working in the transmission of polar auxins that engaged under water deficit conditions. They also identified that, in the seedling stage, inactivation of the OsPIN3t gene results in crown root disorders, and drought tolerance is enhanced by its overexpression. GUS function grew massively under NAA treatment, in response to 20% polyethylene glycol stresses. The study revealed that OsPIN3t was used for the transportation of auxins under drought conditions. Eventually under stress, recognition of unique genes involved in the improvement of major crops during the response to abiotic stress.

6.2.4 Gibberellins (GAs)

GAs are a vast group of tetracyclic di-terpenoid carboxylic acids. Some work as growth regulators in vascular plants, major types are GA_1 and GA_4 (Sponsel and Hedden 2010). They show significant results on the elongation of stem, germination of seed, initiation of trichome and flowers, expansion of leaf, and development of fruits and flowers (Yamaguchi 2008). They have a significant role in the whole life cycle of plants in the stimulation of growth. Similarly, they stimulate all stages of development for transitions. Remarkably, there is rising indication for their critical functions under abiotic stress adjustment and response (Colebrook et al. 2014). In *Arabidopsis thaliana* seedlings, new research has been conducted to examine the function of gibberellins in osmoregulation (Claeys et al. 2012). In cotton, overexpression of gibberellins synthesis gene enhances drought tolerance (Shi et al. 2019). Similarly, GA plays a vital role in enhancing germination of rice seed by promoting the expressing genes, specifically the expression of alpha amylase (LI et al. 2019a). GAs are also related to all other plant hormones in many stimulus responsive and developmental mechanisms. The interface between GA and ethylene contains both positive and negative mutual adjustments to be determined by signalling pathways and tissue (Munteanu et al. 2014).

6.2.5 Brassinosteroids (BRs)

Comparatively, brassinosteroids include a new group of polyhydroxy steroidal phytohormones with strong stimulating potential for growth and development. This plant hormone was identified and separated from the pollen of *Brassica nupus*. More than 70 brassinosteroids have been separated from plants. Brassinolide has 2 major groups – one is 28 homobrassinolide while the other is 24 epibrassinolide – that organise the three highly bioactive BRs extensively used in research and physiological processes (Vardhini et al. 2006). These hormones are present in every part of a plant: in leaves, roots, pollen, shoots, vascular cambium, seeds, fruits, and flower buds. They perform significant functions in different growth and developmental mechanisms, including growth of roots and shoots, fruits and flower development, and floral initiation (Bajguz and Hayat 2009). Recent research suggest that BRs and different interlinked compounds act as alleviating hormones that reduce the effect of drought stress (Ahammed et al. 2013). In maize, it was also observed that BRs are highly linked with drought tolerance (Tůmová et al. 2018). Similarly, RD26 and WRKY transcription factors regulate the cross-talk among brassinosteroids' signalling and drought conditions (Chen and Yin 2017; Ye et al. 2017).

6.2.6 Ethylene (ET)

Among all plant hormones, ethylene is the only one that is gaseous. It is involved in different stages of growth and development of plants particularly ripening of fruits, senescence of flowers, abscission of petioles and leaves, as well as regulation of stress responses (Groen and Whiteman 2014). The precursor of its biosynthesised is methionine by ACC cyclic non protein amino acid and AdoMet (S-adenosyl-L-methionine). AdoMet converts into ACC by ACC synthase, but ACC is transferred into ethylene through ACC oxidase. In plants, the internal level of ethylene is changed under numerous abiotic stresses like drought. The maximum accumulation of ET was considered a high tolerance in plants (Shi et al. 2012). Similarly, under water stress, it plays a vital role as a defensive mechanism (Yu et al. 2017). Accumulation of ET enhances the rate of survival in plants under drought stress conditions with the regulation of gene expression (Hong et al. 2017). When ET combines with other plant hormones, such as SA and JA, then it works effectively. Because the mixture of all these phytohormones is considered a master hormone that supports the enhancement of defensive mechanisms against drought. In plant defensive processes, the cascade of signalling pathways is stimulated by the accumulation, biosynthesis, and transportation of these phytohormones (Matilla-Vazquez and Matilla 2014). Eventually, according to Liu et al. (2019), EIN3 (ethylene intensive 3 protein) is actively involved in drought tolerance and the expression of biosynthetic genes.

6.2.7 Jasmonates (JAs)

Cyclopentanone plant hormones derived from the membrane of fatty acid metabolisms, such as methyl jasmonate (MeJA), and its free acid, jasmonic acid (JA), are mutually known as jasmonates (JAs) and are extensively present in the plant kingdom. These phytohormones have participated significantly in different processes linked with development of plants and in stress survival comprising senescence, flowering, mechanisms of reproduction, fruiting, indirect response in defensive processes, and secondary metabolism (Fahad et al. 2015a). Besides functions of development in plants, JAs stimulate the defensive pathways of plants against the attack of pathogens and drought stress. Jasmonates (JAs) are crucial molecules of signalling prompted by numerous abiotic stresses like salt stress (Pauwels et al. 2009) and water deficit stress (Du et al. 2013). In oat, it is reported that endogenous JA significantly decreases the effect of drought stress with the help of fatty acids and lipids (Sánchez-Martín et al. 2018). MeJA could regulate drought stress tolerance in plants by expediting the antioxidant activities and osmotic adjustment substances (Xiong et al. 2020).

6.2.8 Salicylic acid (SA)

Naturally, SA occurs as a phenolic compound that performs a significant function in the regulation of proteins interlinked with pathogenesis. In addition to its defensive role, it plays a crucial function in different processes including plant growth and development, ripening, and response under abiotic stress conditions (Lahlali et al. 2014). Two pathways are involved in its synthesis: the PAL (phenylalanine ammonia lyase) and IC (iso-chorismate) pathways. The best pathway is iso-chorismate (IC) in tobacco (Uppalapati et al. 2007) and tomato (Catinot et al. 2008). Overall a remarkable or significant concept is that a minimum quantity of SA upgrades the machinery of antioxidants in plants, but a maximum quantity of SA is the reason for necrosis and exposure to abiotic stress. Ultimately, genes that respond effectively to the application of SA are linked with signalling mechanisms and stress that ultimately cause the death of cells. SAs comprise heat shock proteins, antioxidants, gene-encoding chaperones, and genes that work in the formation of secondary metabolites, including cytochrome P450, sinapyl, and cinnamyl alcohol dehydrogenase (Arshad et al. 2016; Jumali et al. 2011). In plants, SA plays a vital role against different abiotic stresses, including water deficit, salt stress (Khodary 2004), low temperature (Yang et al. 2012), and heat stress (Fayez and Bazaid 2014). In *Phillyrea angustifolia*, it is reported that five times the internal concentration of SA is enhanced by the drought stress, while in roots the concentration of SA increases two times (Bandurska 2005). Under drought stress, two types of SA genes are induced, PR1 and PR2 (pathogenesis-related genes) (Miura et al. 2013). Unfortunately, the advanced mechanisms of SAs are still a mystery.

6.2.9 Strigolactones (SLs)

SLs comprise a small class of carotenoid resultant compounds, first considered before forty-five years ago as germination of seeds stimulants in root-parasitic plants including Phelipanche, Striga, and Orobanche (Ruyter-Spira et al. 2013). Single species of plants can produce numerous kinds of SLs, while combinations of numerous kinds and concentration of strigolactones molecules are produced in intraspecific types (Yoneyama et al. 2013). While in roots, these are synthesised and released in low concentrations, but they are also present in other parts of the plant (Koltai and Beveridge 2013). In tomato, low levels of SL in roots predicts drought stress tolerance by systemic signalling (Visentin et al. 2016). Overexpression of D27 develops a link between ABA and SL which are collectively involved in drought tolerance (Haider et al. 2018). A similar interaction was also reported in barley in response to drought (Marzec et al. 2020). Research also revealed that MdIAA24 triggers the synthesis of SL and eventually leads to the drought stress-coping mechanism formation in apples which is called arbuscules (Huang et al. 2020). It seems that in its early evolution, SLs acted as a promoter against environmental stresses. In vascular plants, they are involved in the anatomy of both roots and shoots under different nutritional responses (Kapulnik and Koltai 2014). With the interaction of plant pathogens, SLs also work as signalling molecules. They increase nodulation in the interaction mechanisms of legume rhizobium (Foo and Davies 2011). Generally, it can be determined that SLs comprise the significant class of signalling molecules and have a main role in the adjustment of growth and development of plants under threat from natural stresses. They have many functions that are used in agriculture, such as the promoter of necrosis in germination in parasitic plants (Vurro and Yoneyama 2012).

6.3 Conclusion and Outlook

Overall, engineered phytohormones are the best source for producing tolerance against abiotic stress, hence, offering new opportunities to meet the demands under the current climate change scenario to preserve sustainable crop production. Due to the presence of several stress-reactive genes, these have a substantial contribution in plant stress resistance and adaptability to different stresses. In recent years, efficient research has been conducted to understand plant abiotic stress response, with the rapid advancement of genomic technology. Still, numerous challenges lie ahead in the effort to expose and recognise the hidden complexities of stress-induced signalling pathways. In order to acquire comprehensive understanding of plant interactions under drought stress in the future, maximum research should be carried out at the molecular level of the biosynthetic pathway of hormones, including auxin. Research reveals that engineering of plant hormones has been initiated. The roles of phytohormones have been demonstrated, and hormones perform a key role in regulating responses under stress. In this chapter, we summed up the functions of phytohormones and the significance of combined analysis of different stresses in plants for improving drought stress resistance in plants. Plant hormones are involved in a vast array of stresses, and it is evident that plant hormones take part in plant protection and plant environment interactions. In premise, while the engineering of phytohormones is encouraging for agri-biotechnologists, there is still a long way to go before technology can gain a competitive edge. The production of stable phytohormone-engineered crops that yield key staple foods such as rice, wheat, and maize is among the major hurdles that remain to be overcome. Future research must concentrate on a variety of stress responses in field environments, as increases in various stresses are predicted.

REFERENCES

Adnan M, Fahad S, Khan IA, Saeed M, Ihsan MZ, Saud S, Riaz M, Wang D, Wu C (2019) Integration of poultry manure and phosphate solubilizing bacteria improved availability of Ca bound P in calcareous soils. *Biotech* 9(10):368.

Adnan M, Fahad S, Muhammad Z, Shahen S, Ishaq AM, Subhan D, Zafar-ul-Hye M, Martin LB, Raja MMN, Beena S, Saud S, Imran A, Zhen Y, Martin B, Jiri H, Rahul D (2020) Coupling phosphate-solubilizing bacteria with phosphorus supplements improve maize phosphorus acquisition and growth under lime induced salinity stress. *Plants* 9(900). https://doi.org/10.3390/plants9070900.

Adnan M, Khan MA, Saleem N, Hussain Z, Arif M, Alam M, Ullah H (2016) Nitrogen depletion by weeds from organic and inorganic nitrogen sources in wheat crop. *Pak J Weed Sci Res* 22(1):103–110.

Adnan M, Shah Z, Sharif M, Rahman H (2018a) Liming induces carbon dioxide (CO_2) emission in PSB inoculated alkaline soil supplemented with different phosphorus sources. *Environ Sci Pollut Res* 25(10):9501–9509.

Adnan M, Zahir S, Fahad S, Arif M, Mukhtar A, Imtiaz AK, Ishaq AM, Abdul B, Hidayat U, Muhammad A, Inayat-Ur R, Saud S, Muhammad ZI, Yousaf J, Amanullah Hafiz MH, Wajid N (2018b) Phosphate-solubilizing bacteria nullify the antagonistic effect of soil calcification on bioavailability of phosphorus in alkaline soils. *Sci Rep* 8:4339. https://doi.org/10.1038/s41598-018-22653-7.

Ahammed GJ, Choudhary SP, Chen S, Xia X, Shi K, Zhou Y, Yu J (2013) Role of brassinosteroids in alleviation of phenanthrene–cadmium co-contamination-induced photosynthetic inhibition and oxidative stress in tomato. *J Exp Bot* 64:199–213. https://doi.org/10.1093/jxb/ers323.

Ahmad S, Kamran M, Ding R, Meng X, Wang H, Ahmad I, Fahad S, Han Q (2019) Exogenous melatonin confers drought stress by promoting plant growth, photosynthetic capacity and antioxidant defense system of maize seedlings. *PeerJ* 7:e7793. http://doi.org/10.7717/peerj.7793.

Akbar H, Timothy JK, Jagadish T, Golam M, Apurbo KC, Muhammad F, Rajan B, Fahad S, Hasanuzzaman M (2020) Agricultural land degradation: processes and problems undermining future food security. In: Fahad S, Hasanuzzaman M, Alam M, Ullah H, Saeed M, Khan AK, Adnan M (eds) *Environment, climate, plant and vegetation growth*. Springer Publ Ltd, Springer Nature Switzerland AG. Part of Springer Nature, pp 17–62. https://doi.org/10.1007/978-3-030-49732-3.

Akram R, Turan V, Hammad HM, Ahmad S, Hussain S, Hasnain A, Maqbool MM, Rehmani MIA, Rasool A, Masood N, Mahmood F, Mubeen M, Sultana SR, Fahad S, Amanet K, Saleem M, Abbas Y, Akhtar HM, Waseem F, Murtaza R, Amin A, Zahoor SA, ul Din MS, Nasim W (2018a) Fate of organic and inorganic pollutants in paddy soils. In: Hashmi MZ, Varma A (eds) *Environmental pollution of paddy soils, soil biology*. Springer International Publishing AG, Cham, Switzerland, pp 197–214.

Akram R, Turan V, Wahid A, Ijaz M, Shahid MA, Kaleem S, Hafeez A, Maqbool MM, Chaudhary HJ, Munis, MFH, Mubeen M, Sadiq N, Murtaza R, Kazmi DH, Ali S, Khan N, Sultana SR, Fahad S, Amin A, Nasim W (2018b) Paddy land pollutants and their role in climate change. In: Hashmi MZ, Varma A (eds) *Environmental pollution of paddy soils, soil biology*. Springer International Publishing AG, Cham, Switzerland, pp 113–124.

Al-Wabel Mohammad I, Ahmad Munir, Usman Adel R. A, Akanji Mutair, Rafique Muhammad Imran (2020) Advances in pyrolytic technologies with improved carbon capture and storage to combat climate change. In: Fahad S, Hasanuzzaman M, Alam M, Ullah H, Saeed M, Khan AK, Adnan M (eds) *Environment, climate, plant and vegetation growth*. Springer Publ Ltd, Springer Nature Switzerland AG. Part of Springer Nature, pp 535–576. https://doi.org/10.1007/978-3-030-49732-3.

Amanullah Shah K, Imran Hamdan AK, Muhammad A, Abdel RA, Muhammad A, Fahad S, Azizullah S, Brajendra P (2020) Effects of climate change on irrigation water quality. In: Fahad S, Hasanuzzaman M, Alam M, Ullah H, Saeed M, Khan AK, Adnan M (eds) *Environment, climate, plant and vegetation growth*. Springer Publ Ltd, Springer Nature Switzerland AG. Part of Springer Nature, pp 123–132. https://doi.org/10.1007/978-3-030-49732-3.

Amir M, Muhammad A, Allah B, Sevgi Ç, Haroon ZK, Muhammad A, Emre A (2020) Bio fortification under climate change: the fight between quality and quantity. In: Fahad S, Hasanuzzaman M, Alam M, Ullah H, Saeed M, Khan AK, Adnan M (eds) *Environment, climate, plant and vegetation growth.* Springer Publ Ltd, Springer Nature Switzerland AG. Part of Springer Nature, pp 173–228. https://doi.org/10.1007/978-3-030-49732-3.

Amjad I, Muhammad H, Farooq S, Anwar H (2020) Role of plant bioactives in sustainable agriculture. In: Fahad S, Hasanuzzaman M, Alam M, Ullah H, Saeed M, Khan AK, Adnan M (eds) *Environment, climate, plant and vegetation growth.* Springer Publ Ltd, Springer Nature Switzerland AG. Part of Springer Nature, pp 591–606. https://doi.org/10.1007/978-3-030-49732-3.

Arif M, Talha J, Muhammad R, Fahad S, Muhammad A, Amanullah Kawsar A, Ishaq AM, Bushra K, Fahd R (2020) Biochar; a remedy for climate change. In: Fahad S, Hasanuzzaman M, Alam M, Ullah H, Saeed M, Khan AK, Adnan M (eds) *Environment, climate, plant and vegetation growth.* Springer Publ Ltd, Springer Nature Switzerland AG. Part of Springer Nature, pp 151–172. https://doi.org/10.1007/978-3-030-49732-3.

Arshad M, Adnan M, Ahmed S, Khan A K, Ali I, Ali M, Khan M A (2016) Integrated effect of phosphorus and zinc on wheat crop. *Am Eurasian J Agric Environ Sci* 16(3):455–459.

Arshad M, Adnan M, AliA, Khan AK, Shahwar D (2016) Integrated effect of phosphorous and zinc on wheat quality and soil properties. *Adv Environ Bio* 10(2):40–45.

Ayman ELS, Hossain Akbar, Barutçular Celaleddin, Iqbal Muhammad Aamir, Sohidul Islam M, Fahad Shah, Sytar Oksana, Çig Fatih, Meena Ram Swaroop, Erman Murat (2020) Consequences of salinity stress on the quality of crops and its mitigation strategies for sustainable crop production: an outlook of arid and semi- arid regions. In: Fahad S, Hasanuzzaman M, Alam M, Ullah H, Saeed M, Khan AK, Adnan M (eds) *Environment, climate, plant and vegetation growth.* Springer Publ Ltd, Springer Nature Switzerland AG. Part of Springer Nature, pp 503–534. https://doi.org/10.1007/978-3-030-49732-3.

Ayub S, Hayat R, Zainab Z, Akhtar W, Mahmood T (2019) OsRGLP2 promoter derived GUS expression in transgenic tobacco in response to salicylic acid, H2O2, PEG, NaCl and auxins. *Plant Gene* 19:100190. https://doi.org/10.1016/j.plgene.2019.100190.

Aziz K, Daniel KYT, Fazal M, Muhammad ZA, Farooq S, FanW, Fahad S, Ruiyang Z (2017a) Nitrogen nutrition in cotton and control strategies for greenhouse gas emissions: a review. *Environ Sci Pollut Res* 24:23471–23487. https://doi.org/10.1007/s11356-017-0131-y.

Aziz K, Daniel KYT, Muhammad ZA, Honghai L, Shahbaz AT, Mir A, Fahad S (2017b) Nitrogen fertility and abiotic stresses management in cotton crop: a review. *Environ Sci Pollut Res* 24:14551–14566. https://doi.org/10.1007/s11356-017-8920-x.

Bajguz A, Hayat S (2009) Effects of brassinosteroids on the plant responses to environmental stresses. *Plant Physiol Biochem* 47:1–8. https://doi.org/10.1016/j.plaphy.2008.10.002.

Bandurska H (2005) The effect of salicylic acid on barley response to water deficit. *Acta Physiol Plant* 27:379–386.

Baseer M, Adnan M, Fazal M, Fahad S, Muhammad S, Fazli W, Muhammad A Jr. Amanullah Depeng W, Saud S, Muhammad N, Muhammad Z, Fazli S, Beena S, Mian AR, Ishaq AM (2019) Substituting urea by organic wastes for improving maize yield in alkaline soil. *J Plant Nutr.* doi.org/10.1080/01904167.2019.1659344.

Bayram AY, Seher Ö, Nazlican A (2020) Climate change forecasting and modeling for the year of 2050. In: Fahad S, Hasanuzzaman M, Alam M, Ullah H, Saeed M, Khan AK, Adnan M (eds) *Environment, climate, plant and vegetation growth.* Springer Publ Ltd, Springer Nature Switzerland AG. Part of Springer Nature, pp 109–122. https://doi.org/10.1007/978-3-030-49732-3.

Catinot J, Buchala A, Abou-Mansour E, Métraux J-P (2008) Salicylic acid production in response to biotic and abiotic stress depends on isochorismate in Nicotiana benthamiana. *FEBS Lett* 582:473–478. https://doi.org/10.1016/j.febslet.2007.12.039.

Chaves MM, Maroco JP, Pereira JS (2003) Understanding plant responses to drought—from genes to the whole plant. *Funct Plant Biol* 30:239–264. https://doi.org/10.1071/FP02076.

Chen J, Nolan TM, Ye H, Zhang M, Tong H, Xin P, Chu J, Chu C, Li Z, Yin Y (2017) Arabidopsis WRKY46, WRKY54, and WRKY70 transcription factors are involved in brassinosteroid-regulated plant growth and drought responses. *Plant Cell* 29:1425–1439. https://doi.org/10.1105/tpc.17.00364.

Chen J, Yin Y (2017) WRKY transcription factors are involved in brassinosteroid signaling and mediate the crosstalk between plant growth and drought tolerance. *Plant Signal Behav* 12:1365212. https://doi.org/10.1080/15592324.2017.1365212.

Chen Z, Liu Y, Yin Y, Liu Q, Li N, Li X, He W, Hao D, Liu X, Guo C (2019) Expression of AtGA2ox1 enhances drought tolerance in maize. *Plant Growth Regul* 89:203–215. https://doi.org/10.1007/s10725-019-00526-x.

Claeys H, Skirycz A, Maleux K, Inzé D (2012) DELLA signaling mediates stress-induced cell differentiation in Arabidopsis leaves through modulation of anaphase-promoting complex/cyclosome activity. *Plant Physiol* 159:739–747. https://doi.org/10.1104/pp.112.195032.

Colebrook EH, Thomas SG, Phillips AL, Hedden P (2014) The role of gibberellin signalling in plant responses to abiotic stress. *J Exp Biol* 217:67–75. https://doi.org/10.1242/jeb.089938.

De Smet I, Voß U, Lau S, Wilson M, Shao N, Timme RE, Swarup R, Kerr I, Hodgman C, Bock R (2011) Unraveling the evolution of auxin signaling. *Plant Physiol* 155:209–221. https://doi.org/10.1104/pp.110.168161.

Depeng W, Fahad S, Saud S, Muhammad K, Aziz K, Mohammad NK, Hafiz MH, Wajid N (2018) Morphological acclimation to agronomic manipulation in leaf dispersion and orientation to promote "Ideotype" breeding: evidence from 3D visual modeling of "super" rice (*Oryza sativa* L.). *Plant Physiol Biochem*. https://doi.org/10.1016/j.plaphy.2018.11.010.

Du H, Liu H, Xiong L (2013) Endogenous auxin and jasmonic acid levels are differentially modulated by abiotic stresses in rice. *Front Plant Sci* 4:397. https://doi.org/10.3389/fpls.2013.00397.

Egamberdieva D (2009) Alleviation of salt stress by plant growth regulators and IAA producing bacteria in wheat. *Acta Physiol Plant* 31:861–864. https://doi.org/10.1007/s11738-009-0297-0.

Fahad S, Adnan M, Hassan S, Saud S, Hussain S, Wu C, Wang D, Hakeem KR, Alharby HF, Turan V, Khan MA, Huang J (2019b) Rice responses and tolerance to high temperature. In: Hasanuzzaman M, Fujita M, Nahar K, Biswas JK (eds) *Advances in rice research for abiotic stress tolerance*. Woodhead Publ Ltd, Cambridge, England, pp 201–224.

Fahad S, Bajwa AA, Nazir U, Anjum SA, Farooq A, Zohaib A, Sadia S, NasimW, Adkins S, Saud S, Ihsan MZ, Alharby H, Wu C, Wang D, Huang J (2017) Crop production under drought and heat stress: plant responses and management options. *Front Plant Sci* 8:1147. https://doi.org/10.3389/fpls.2017.01147.

Fahad S, Bano A (2012) Effect of salicylic acid on physiological and biochemical characterization of maize grown in saline area. *Pak J Bot* 44:1433–1438.

Fahad S, Chen Y, Saud S, Wang K, Xiong D, Chen C, Wu C, Shah F, Nie L, Huang J (2013) Ultraviolet radiation effect on photosynthetic pigments, biochemical attributes, antioxidant enzyme activity and hormonal contents of wheat. *J Food Agri Environ* 11(3&4):1635–1641.

Fahad S, Hussain S, Bano A, Saud S, Hassan S, Shan D, Khan FA, Khan F, Chen Y, Wu C (2015b) Potential role of phytohormones and plant growth-promoting rhizobacteria in abiotic stresses: consequences for changing environment. *Environ Sci Pollut Res* 22:4907–4921. https://doi.org/10.1007/s11356-014-3754-2.

Fahad S, Hussain S, Bano A, Saud S, Hassan S, Shan D, Khan FA, Khan F, Chen Y, Wu C, Tabassum MA, Chun MX, Afzal M, Jan A, Jan MT, Huang J (2014a) Potential role of phytohormones and plant growth-promoting rhizobacteria in abiotic stresses: consequences for changing environment. *Environ Sci Pollut Res* 22(7):4907–4921. https://doi.org/10.1007/s11356-014-3754-2.

Fahad S, Hussain S, Matloob A, Khan FA, Khaliq A, Saud S, Hassan S, Shan D, Khan F, Ullah N (2015c) Phytohormones and plant responses to salinity stress: a review. *Plant Growth Regul* 75:391–404. https://doi.org/10.1007/s10725-014-0013-y.

Fahad S, Hussain S, Matloob A, Khan FA, Khaliq A, Saud S, Hassan S, Shan D, Khan F, Ullah N, Faiq M, Khan MR, Tareen AK, Khan A, Ullah A, Ullah N, Huang J (2014b) Phytohormones and plant responses to salinity stress: a review. *Plant Growth Regul* 75(2):391–404. https://doi.org/10.1007/s10725-014-0013-y.

Fahad S, Hussain S, Saud S, Hassan S, Chauhan BS, Khan F et al (2016a) Responses of rapid viscoanalyzer profile and other rice grain qualities to exogenously applied plant growth regulators under high day and high night temperatures. *PLoS One* 11(7):e0159590. https://org/10.1371/journal.pone.0159590.

Fahad S, Hussain S, Saud S, Hassan S, Ihsan Z, Shah AN, Wu C, Yousaf M, Nasim W, Alharby H, Alghabari F, Huang J (2016c) Exogenously applied plant growth regulators enhance the morphophysiological growth and yield of rice under high temperature. *Front Plant Sci* 7:1250. https://doi.org/10.3389/fpls.2016. 01250.

Fahad S, Hussain S, Saud S, Hassan S, Tanveer M, Ihsan MZ, Shah AN, Ullah A, Nasrullah KF, Ullah S, Alharby HNW, Wu C, Huang J (2016d) A combined application of biochar and phosphorus alleviates heat-induced adversities on physiological, agronomical and quality attributes of rice. *Plant Physiol Biochem* 103:191–198.

Fahad S, Hussain S, Saud S, Khan F, Hassan S Jr A, Nasim W, Arif M, Wang F, Huang J (2016b) Exogenously applied plant growth regulators affect heat-stressed rice pollens. *J Agron Crop Sci* 202:139–150.

Fahad S, Hussain S, Saud S, Tanveer M, Bajwa AA, Hassan S, Shah AN, Ullah A, Wu C, Khan FA, Shah F, Ullah S, Chen Y, Huang J (2015a) A biochar application protects rice pollen from high-temperature stress. *Plant Physiol Biochem* 96:281–287.

Fahad S, Muhammad ZI, Abdul K, Ihsanullah D, Saud S, Saleh A, Wajid N, Muhammad A, Imtiaz AK, Chao W, Depeng W, Jianliang H (2018) Consequences of high temperature under changing climate optima for rice pollen characteristics-concepts and perspectives, *Arch Agron Soil Sci*. https://doi.org/10.1080/03650340.2018.1443213.

Fahad S, Nie L, Chen Y, Wu C, Xiong D, Saud S, Hongyan L, Cui K, Huang J (2015b) Crop plant hormones and environmental stress. *Sustain Agric Res* 15:371–400.

Fahad S, Rehman A, Shahzad B, Tanveer M, Saud S, Kamran M, Ihtisham M, Khan SU, Turan V, Rahman MHU (2019a) Rice responses and tolerance to metal/metalloid toxicity. In: Hasanuzzaman M, Fujita M, Nahar K, Biswas JK (eds) *Advances in rice research for abiotic stress tolerance*. Woodhead Publ Ltd, Cambridge, England, pp 299–312.

Farah R, Muhammad R, Muhammad SA, Tahira Y, Muhammad AA, Maryam A, Shafaqat A, Rashid M, Muhammad R, Qaiser H, Afia Z, Muhammad AA, Muhammad A, Fahad S (2020) Alternative and non-conventional soil and crop management strategies for increasing water use efficiency. In: Fahad S, Hasanuzzaman M, Alam M, Ullah H, Saeed M, Khan AK, Adnan M (eds) *Environment, climate, plant and vegetation growth*. Springer Publ Ltd, Springer Nature Switzerland AG. Part of Springer Nature, pp 323–338. https://doi.org/10.1007/978-3-030-49732-3.

Farhana G, Ishfaq A, Muhammad A, Dawood J, Fahad S, Xiuling L, Depeng W, Muhammad F, Muhammad F, Syed AS (2020) Use of crop growth model to simulate the impact of climate change on yield of various wheat cultivars under different agro-environmental conditions in Khyber Pakhtunkhwa, Pakistan. *Arabian J Geosci* 13:112 https://doi.org/10.1007/s12517-020-5118-1.

Farhat A, Hafiz MH, Wajid I, Aitazaz AF, Hafiz FB, Zahida Z, Fahad S, Wajid F, Artemi C (2020) A review of soil carbon dynamics resulting from agricultural practices. *J Environ Manage* 268(2020):110319.

Fayez KA, Bazaid SA (2014) Improving drought and salinity tolerance in barley by application of salicylic acid and potassium nitrate. *J Saudi Soc* 13:45–55. https://doi.org/10.1016/j.jssas.2013.01.001.

Fazli W, Muhmmad S, Amjad A, Fahad S, Muhammad A, Muhammad N, Ishaq AM, Imtiaz AK, Mukhtar A, Muhammad S, Muhammad I, Rafi U, Haroon I, Muhammad A (2020) Plant-microbes interactions and functions in changing climate. In: Fahad S, Hasanuzzaman M, Alam M, Ullah H, Saeed M, Khan AK, Adnan M (eds) *Environment, climate, plant and vegetation growth*. Springer Publ Ltd, Springer Nature Switzerland AG. Part of Springer Nature, pp 397–420. https://doi.org/10.1007/978-3-030-49732-3.

Feng Y, Yin Y, Fei S (2015) Down-regulation of BdBRI1, a putative brassinosteroid receptor gene produces a dwarf phenotype with enhanced drought tolerance in Brachypodium distachyon. *Plant Sci* 234:163–173. https://doi.org/10.1016/j.plantsci.2015.02.015.

Foo E, Davies NW (2011) Strigolactones promote nodulation in pea. *Planta* 234:1073. https://doi.org/10.1007/s00425-011-1516-7.

Fu J, Wu H, Ma S, Xiang D, Liu R, Xiong L (2017) OsJAZ1 attenuates drought resistance by regulating JA and ABA signaling in rice. *Front Plant Sci* 8:2108. https://doi.org/10.3389/fpls.2017.02108.

Giuliani S, Sanguineti MC, Tuberosa R, Bellotti M, Salvi S, Landi P (2005) Root-ABA1, a major constitutive QTL, affects maize root architecture and leaf ABA concentration at different water regimes. *J Exp Bot* 56:3061–3070. https://doi.org/10.1093/jxb/eri303.

Gopakumar L, Bernard NO, Donato V (2020) Soil microarthropods and nutrient cycling. In: Fahad S, Hasanuzzaman M, Alam M, Ullah H, Saeed M, Khan AK, Adnan M (eds) *Environment, climate, plant and vegetation growth*. Springer Publ Ltd, Springer Nature Switzerland AG. Part of Springer Nature, pp 453–472. https://doi.org/10.1007/978-3-030-49732-3.

Groen SC, Whiteman NK (2014) The evolution of ethylene signaling in plant chemical ecology. *J Chem Ecol* 40:700–716. https://doi.org/10.1007/s10886-014-0474-5.

Guo Y, Jiang Q, Hu Z, Sun X, Fan S, Zhang H (2018) Function of the auxin-responsive gene TaSAUR75 under salt and drought stress. *Crop J* 6:181–190. https://doi.org/10.1007/s10886-014-0474-5.

Habib ur R, Ashfaq A, Aftab W, Manzoor H, Fahd R, Wajid I, Md. Aminul I, Vakhtang S, Muhammad A, Asmat U, Abdul W, Syeda RS, Shah S, Shahbaz K, Fahad S, Manzoor H, Saddam H, Wajid N (2017) Application of CSM-CROPGRO-Cotton model for cultivars and optimum planting dates: evaluation in changing semi-arid climate. *Field Crops Res*. http://dx.doi.org/10.1016/j.fcr.2017.07.007.

Hafiz MH, Abdul K, Farhat A, Wajid F, Fahad S, Muhammad A, Ghulam MS, Wajid N, Muhammad M, Hafiz FB (2020b) Comparative effects of organic and inorganic fertilizers on soil organic carbon and wheat productivity under arid region. *Communications in Soil Science and Plant Analysis*. https://doi.org/10.1080/00103624.2020.1763385.

Hafiz MH, Farhat A, Ashfaq A, Hafiz FB, Wajid F, Carol Jo W, Fahad S, Gerrit H (2020a) Predicting Kernel growth of maize under controlled water and nitrogen applications. *Int J Plant Prod*. https://doi.org/10.1007/s42106-020-00110-8.

Hafiz MH, Farhat A, Shafqat S, Fahad S, Artemi C, Wajid F, Chaves CB, Wajid N, Muhammad M, Hafiz FB (2018) Offsetting land degradation through nitrogen and water management during maize cultivation under arid conditions. *Land Degrad Dev* 1–10. https://doi.org/10.1002/ldr.2933.

Hafiz MH, Muhammad A, Farhat A, Hafiz FB, Saeed AQ, Muhammad M, Fahad S, Muhammad A (2019) Environmental factors affecting the frequency of road traffic accidents: a case study of sub-urban area of Pakistan. *Environ Sci Pollut Res* https://doi.org/10.1007/s11356-019-04752-8.

Hafiz MH, Wajid F, Farhat A, Fahad S, Shafqat S, Wajid N, Hafiz FB (2016) Maize plant nitrogen uptake dynamics at limited irrigation water and nitrogen. *Environ Sci Pollut Res* 24(3):2549–2557. https://doi.org/10.1007/s11356-016-8031-0.

Haider I, Andreo-Jimenez B, Bruno M, Bimbo A, Floková K, Abuauf H, Ntui VO, Guo X, Charnikhova T, Al-Babili S (2018) The interaction of strigolactones with abscisic acid during the drought response in rice. *J Exp Bot* 69:2403–2414. https://doi.org/10.1093/jxb/ery089.

Hesham FA, Fahad S (2020) Melatonin application enhances biochar efficiency for drought tolerance in maize varieties: Modifications in physio-biochemical machinery. *Agron J* 112(4):1–22.

Hong E, Lim CW, Han S-W, Lee SC (2017) Functional analysis of the pepper ethylene-responsive transcription factor, CaAIEF1, in enhanced ABA sensitivity and drought tolerance. *Front Plant Sci* 8:1407. https://doi.org/10.3389/fpls.2017.01407.

Huang D, Wang Q, Jing G, Ma M, Li C, Ma F (2020) Overexpression of MdIAA24 improves apple drought resistance by positively regulating strigolactone biosynthesis and mycorrhization. *Tree Physiol*. https://doi.org/10.1093/treephys/tpaa109.

Hussain MA, Fahad S, Rahat S, Muhammad FJ, Muhammad M, Qasid A, Ali A, Husain A, Nooral A, Babatope SA, Changbao S, Liya G, Ibrar A, Zhanmei J, Juncai H (2020) Multifunctional role of brassinosteroid and its analogues in plants. *Plant Growth Regul* https://doi.org/10.1007/s10725-020-00647-8.

Hussain Wani S, Brajendra Singh N, Haribhushan A, Iqbal Mir J (2013) Compatible solute engineering in plants for abiotic stress tolerance-role of glycine betaine. *Curr Genomics* 14:157–165.

Ibrar K, Aneela R, Khola Z, Urooba N, Sana B, Rabia S, Ishtiaq H, Mujaddad Ur Rehman, Salvatore M (2020) Microbes and environment: global warming reverting the frozen zombies. In: Fahad S, Hasanuzzaman M, Alam M, Ullah H, Saeed M, Khan AK, Adnan M (eds) *Environment, climate, plant and vegetation growth*. Springer Publ Ltd, Springer Nature Switzerland AG. Part of Springer Nature, pp 607–634. https://doi.org/10.1007/978-3-030-49732-3.

Ilyas M, Mohammad N, Nadeem K, Ali H, Aamir HK, Kashif H, Fahad S, Aziz K, Abid U, (2020) Drought tolerance strategies in plants: a mechanistic approach. *J Plant Growth Regul* https://doi.org/10.1007/s00344-020-10174-5.

Iqbal N, Umar S, Khan NA, Khan MIR (2014) A new perspective of phytohormones in salinity tolerance: regulation of proline metabolism. *Environ Exp Bot* 100:34–42. https://doi.org/10.1016/j.envexpbot.2013.12.006.

Iqra M, Amna B, Shakeel I, Fatima K, Sehrish L, Hamza A, Fahad S (2020) Carbon cycle in response to global warming. In: Fahad S, Hasanuzzaman M, Alam M, Ullah H, Saeed M, Khan AK, Adnan M (eds) *Environment, climate, plant and vegetation growth*. Springer Publ Ltd, Springer Nature Switzerland AG. Part of Springer Nature, pp 1–16. https://doi.org/10.1007/978-3-030-49732-3.

Jan M, Anwar-ul-Haq Muhammad, Adnan NS, Muhammad Y, Javaid I, Xiuling L, Depeng W, Fahad S, (2019) Modulation in growth, gas exchange, and antioxidant activities of salt-stressed rice (Oryza sativa L.) genotypes by zinc fertilization. *Arabian J Geosci* 12:775 https://doi.org/10.1007/s12517-019-4939-2.

Javid MG, Sorooshzadeh A, Moradi F, Modarres Sanavy SAM, Allahdadi I (2011) The role of phytohormones in alleviating salt stress in crop plants. *Aust J Crop Sci* 5:726.

Jumali SS, Said IM, Ismail I, Zainal Z (2011) Genes induced by high concentration of salicylic acid in'Mitragyna speciosa'. *Aust J Crop Sci* 5:296.

Jung H, Lee D-K, Do Choi Y, Kim J-K (2015) OsIAA6, a member of the rice Aux/IAA gene family, is involved in drought tolerance and tiller outgrowth. *Plant Sci* 236:304–312. https://doi.org/10.1016/j.plantsci.2015.04.018.

Kamarn M, Wenwen C, Irshad A, Xiangping M, Xudong Z, Wennan S, Junzhi C, Shakeel A, Fahad S, Qingfang H, Tiening L (2017) Effect of paclobutrazol, a potential growth regulator on stalk mechanical strength, lignin accumulation and its relation with lodging resistance of maize. *Plant Growth Regul* 84:317–332. https://doi.org/10.1007/ s10725-017-0342-8.

Kang NY, Cho C, Kim NY, Kim J (2012) Cytokinin receptor-dependent and receptor-independent pathways in the dehydration response of Arabidopsis thaliana. *J Plant Physiol* 169:1382–1391. https://doi.org/10.1016/j.jplph.2012.05.007.

Kapulnik Y, Koltai H (2014) Strigolactone involvement in root development, response to abiotic stress, and interactions with the biotic soil environment. *Plant Physiol* 166:560–569. https://doi.org/10.1104/pp.114.244939.

Kazan K (2013) Auxin and the integration of environmental signals into plant root development. *Ann Bot* 112:1655–1665. https://doi.org/10.1093/aob/mct229.

Kazan K (2015) Diverse roles of jasmonates and ethylene in abiotic stress tolerance. *Trends Plant Sci* 20:219–229. https://doi.org/10.1016/j.tplants.2015.02.001.

Ke Q, Wang Z, Ji CY, Jeong JC, Lee H-S, Li H, Xu B, Deng X, Kwak S-S (2015) Transgenic poplar expressing Arabidopsis YUCCA6 exhibits auxin-overproduction phenotypes and increased tolerance to abiotic stress. *Plant Physiol Biochem* 94:19–27. https://doi.org/10.1016/j.plaphy.2015.05.003.

Keskin BC, Yuksel B, Memon AR, Topal-Sarıkaya A (2010) Abscisic Acid Regulated Gene Expression in Bread Wheat ('Triticum aestivum'L.). *Aust J Crop Sci* 4:617.

Khodary S (2004) Effect of salicylic acid on the growth, photosynthesis and carbohydrate metabolism in salt stressed maize plants. *Int J Agric Biol* 6:5–8.

Koltai H, Beveridge CA (2013) Strigolactones and the coordinated development of shoot and root. In *Long-distance systemic signaling and communication in plants.* Springer, pp 189–204. https://doi.org/10.1007/978-3-642-36470-9_9.

Kuppu S, Mishra N, Hu R, Sun L, Zhu X, Shen G, Blumwald E, Payton P, Zhang H (2013) Water-deficit inducible expression of a cytokinin biosynthetic gene IPT improves drought tolerance in cotton. *PLoS One* 8:e64190. https://doi.org/10.1371/journal.pone.0064190.

Lahlali R, Jiang Y, Kumar S, Karunakaran C, Liu X, Borondics F, Hallin E, Bueckert R (2014) ATR–FTIR spectroscopy reveals involvement of lipids and proteins of intact pea pollen grains to heat stress tolerance. *Front Plant Sci* 5:747. https://doi.org/10.3389/fpls.2014.00747.

Li J, Li M, Han Y, Sun H, Du Y, Zhao Q (2019a) The crucial role of gibberellic acid on germination of drought-resistant upland rice. *Biol Plant* 63:529–535. https://doi.org/10.32615/bp.2019.049.

Li Q, Wang G, Guan C, Yang D, Wang Y, Zhang Y, Ji J, Jin C, An T (2019b) Overexpression of LcSABP, an orthologous gene for salicylic acid binding protein 2, enhances drought stress tolerance in transgenic tobacco. *Front Plant Sci* 10:200. https://doi.org/10.3389/fpls.2019.00200.

Li X-J, Yang M-F, Chen H, Qu L-Q, Chen F, Shen S-H (2010) Abscisic acid pretreatment enhances salt tolerance of rice seedlings: proteomic evidence. *Biochim Biophys Acta Proteins Proteom* 1804:929–940. https://doi.org/10.1016/j.bbapap.2010.01.004.

Li Y-j, Wang B, Dong R-r, Hou B-k (2015) AtUGT76C2, an Arabidopsis cytokinin glycosyltransferase is involved in drought stress adaptation. *Plant Sci* 236:157–167. https://doi.org/10.1016/j.plantsci.2015.04.002.

Liu C, Li J, Zhu P, Yu J, Hou J, Wang C, Long D, Yu M, Zhao A (2019) Mulberry EIL3 confers salt and drought tolerances and modulates ethylene biosynthetic gene expression. *PeerJ* 7:e6391. https://doi.org/10.7717/peerj.6391.

Lou D, Wang H, Liang G, Yu D (2017) OsSAPK2 confers abscisic acid sensitivity and tolerance to drought stress in rice. *Front Plant Sci* 8:993. https://doi.org/10.3389/fpls.2017.00993.

Mahar A, Amjad A, Altaf HL, Fazli W, Ronghua L, Muhammad A, Fahad S, Muhammad A, Rafiullah Imtiaz AK, Zengqiang Z (2020) Promising technologies for cd-contaminated soils: drawbacks and possibilities. In: Fahad S, Hasanuzzaman M, Alam M, Ullah H, Saeed M, Khan AK, Adnan M (eds) *Environment, climate, plant and vegetation growth.* Springer Publ Ltd, Springer Nature Switzerland AG. Part of Springer Nature, pp 63–92. https://doi.org/10.1007/978-3-030-49732-3.

Mano Y, Nemoto K (2012) The pathway of auxin biosynthesis in plants. *J Exp Bot* 63:2853–2872. https://doi.org/10.1093/jxb/ers091.

Marzec M, Daszkowska-Golec A, Collin A, Melzer M, Eggert K, Szarejko I (2020) Barley strigolactone signalling mutant hvd14. d reveals the role of strigolactones in abscisic acid-dependent response to drought. *Plant Cell Environ* 43:2239–2253. https://doi.org/10.1111/pce.13815.

Matilla-Vazquez M, Matilla A (2014) Ethylene: role in plants under environmental stress. In *Physiological mechanisms and adaptation strategies in plants under changing environment.* Springer, pp 189–222. https://doi.org/10.1007/978-1-4614-8600-8_7.

Md Enamul H, Shoeb AZM., Mallik AH, Fahad S, Kamruzzaman MM, Akib J, Nayyer S, Mehedi AKM, Swati AS, Md Yeamin A, Most SS (2020) Measuring vulnerability to environmental hazards: qualitative to quantitative. In: Fahad S, Hasanuzzaman M, Alam M, Ullah H, Saeed M, Khan AK, Adnan M (eds) *Environment, climate, plant and vegetation growth.* Springer Publ Ltd, Springer Nature Switzerland AG. Part of Springer Nature, pp 421–452. https://doi.org/10.1007/978-3-030-49732-3.

Md Jakir H, Allah B (2020) Development and applications of transplastomic plants; a way towards eco-friendly agriculture. In: Fahad S, Hasanuzzaman M, Alam M, Ullah H, Saeed M, Khan AK, Adnan M (eds) *Environment, climate, plant and vegetation growth.* Springer Publ Ltd, Springer Nature Switzerland AG. Part of Springer Nature, pp 285–322. https://doi.org/10.1007/978-3-030-49732-3.

Miura K, Okamoto H, Okuma E, Shiba H, Kamada H, Hasegawa PM, Murata Y (2013) SIZ1 deficiency causes reduced stomatal aperture and enhanced drought tolerance via controlling salicylic acid-induced accumulation of reactive oxygen species in A rabidopsis. *Plant J* 73:91–104. https://doi.org/10.1111/tpj.12014.

Mohammad I. Al-Wabel, Abdelazeem S, Munir A, Khalid E, Adel RAU (2020) Extent of climate change in Saudi Arabia and its impacts on agriculture: a case study from qassim region. In: Fahad S, Hasanuzzaman M, Alam M, Ullah H, Saeed M, Khan AK, Adnan M (eds) *Environment, climate, plant and vegetation growth.* Springer Publ Ltd, Springer Nature Switzerland AG. Part of Springer Nature, pp 635–658. https://doi.org/10.1007/978-3-030-49732-3.

Mubeen M, Ashfaq A, Hafiz MH, Muhammad A, Hafiz UF, Mazhar S, Muhammad Sami ul Din, Asad A, Amjed A, Fahad S, Wajid N (2020) Evaluating the climate change impact on water use efficiency of cotton-wheat in semi-arid conditions using DSSAT model. *J Water Clim Change.* https://doi.org/10.2166/wcc.2019.179/622035/jwc2019179.pdf.

Muhammad Z, Abdul MK, Abdul MS, Kenneth BM, Muhammad S, Shahen S, Ibadullah J, Fahad S (2019) Performance of Aeluropus lagopoides (mangrove grass) ecotypes, a potential turfgrass, under high saline conditions. *Environ Sci Pollut Res* https://doi.org/10.1007/s11356-019-04838-3.

Munteanu V, Gordeev V, Martea R, Duca M (2014) Effect of gibberellin cross talk with other phytohormones on cellular growth and mitosis to endoreduplication transition. *Int J Adv Res Biol Sci* 1:136–153.

Nemhauser JL, Hong F, Chory J (2006) Different plant hormones regulate similar processes through largely nonoverlapping transcriptional responses. *Cell* 126:467–475. https://doi.org/10.1016/j.cell.2006.05.050.

Nguyen KH, Van Ha C, Nishiyama R, Watanabe Y, Leyva-González MA, Fujita Y, Tran UT, Li W, Tanaka M, Seki M (2016) Arabidopsis type B cytokinin response regulators ARR1, ARR10, and ARR12 negatively regulate plant responses to drought. *Proc Natl Acad Sci U S A* 113:3090–3095. https://doi.org/10.1073/pnas.1600399113.

Noor M, Naveed ur R, Ajmal J, Fahad S, Muhammad A, Fazli W, Saud S, Hassan S (2020) Climate change and costal plant lives. In: Fahad S, Hasanuzzaman M, Alam M, Ullah H,Saeed M, Khan AK, Adnan M (eds) *Environment, climate, plant and vegetation growth.* Springer Publ Ltd, Springer Nature Switzerland AG. Part of Springer Nature, pp 93–108. https://doi.org/10.1007/978-3-030-49732-3.

O'Brien JA, Benková E (2013) Cytokinin cross-talking during biotic and abiotic stress responses. *Front Plant Sci* 4:451. https://doi.org/10.3389/fpls.2013.00451.

Pauwels L, Inzé D, Goossens A (2009) Jasmonate-inducible gene: what does it mean? *Trends Plant Sci* 14:87–91. https://doi.org/10.1016/j.tplants.2008.11.005.

Peleg Z, Blumwald E (2011) Hormone balance and abiotic stress tolerance in crop plants. *Curr Opin Plant Biol* 14:290–295. https://doi.org/10.1016/j.pbi.2011.02.001.

Pospíšilová J (2003a) Interaction of cytokinins and abscisic acid during regulation of stomatal opening in bean leaves. *Photosynthetica* 41:49–56. https://doi.org/10.1023/A:1025852210937.

Pospíšilová J (2003b) Participation of phytohormones in the stomatal regulation of gas exchange during water stress. *Biol Plant* 46:491–506. https://doi.org/10.1023/A:1024894923865.

Qamar-uz Z, Zubair A, Muhammad Y, Muhammad ZI, Abdul K, Fahad S, Safder B, Ramzani PMA, Muhammad N (2017) Zinc biofortification in rice: leveraging agriculture to moderate hidden hunger in developing countries. *Arch Agron Soil Sci* 64:147–161. https://doi.org/10.1080/03650340.2017.1338343.

Qin F, Shinozaki K, Yamaguchi-Shinozaki K (2011) Achievements and challenges in understanding plant abiotic stress responses and tolerance. *Plant Cell Physiol* 52:1569–1582. https://doi.org/10.1093/pcp/pcr106.

Rashid M, Qaiser H, Khalid SK, Mohammad I. Al-Wabel, Zhang A, Muhammad A, Shahzada SI, Rukhsanda A, Ghulam AS, Shahzada MM, Sarosh A, Muhammad FQ (2020) Prospects of biochar in alkaline soils to mitigate climate change. In: Fahad S, Hasanuzzaman M, Alam M, Ullah H, Saeed M, Khan AK, Adnan M (eds) *Environment, climate, plant and vegetation growth.* Springer Publ Ltd, Springer Nature Switzerland AG. Part of Springer Nature, pp 133–150. https://doi.org/10.1007/978-3-030-49732-3.

Rehman M, Fahad S, Saleem MH, Hafeez M, Muhammad Habib ur Rahman, Liu F, Deng G (2020) Red light optimized physiological traits and enhanced the growth of ramie (Boehmeria nivea L.). *Photosynthetica* 58(4):922–931.

Ren X, Chen Z, Liu Y, Zhang H, Zhang M, Liu Q, Hong X, Zhu JK, Gong Z (2010) ABO3, a WRKY transcription factor, mediates plant responses to abscisic acid and drought tolerance in Arabidopsis. *Plant J* 63:417–429. https://doi.org/10.1111/j.1365-313X.2010.04248.x.

Ruyter-Spira C, Al-Babili S, Van Der Krol S, Bouwmeester H (2013) The biology of strigolactones. *Trends Plant Sci* 18:72–83. https://doi.org/10.1016/j.tplants.2012.10.003.

Sadam M, ul Qamar Muhammad Tahir, Ghulam M, Muhammad SK, Faiz AJ (2020) Role of biotechnology in climate resilient agriculture. In: Fahad S, Hasanuzzaman M, Alam M, Ullah H, Saeed M, Khan AK, Adnan M (eds) *Environment, climate, plant and vegetation growth.* Springer Publ Ltd, Springer Nature Switzerland AG. Part of Springer Nature, pp 339–366. https://doi.org/10.1007/978-3-030-49732-3.

Sajid H, Jie H, Jing H, Shakeel A, Satyabrata N, Sumera A, Awais S, Chunquan Z, Lianfeng Z, Xiaochuang C, Qianyu J, Junhua Z (2020) Rice production under climate change: adaptations and mitigating strategies. In: Fahad S, Hasanuzzaman M, Alam M, Ullah H, Saeed M, Khan AK, Adnan M (eds) *Environment, climate, plant and vegetation growth.* Springer Publ Ltd, Springer Nature Switzerland AG. Part of Springer Nature, pp 659–686. https://doi.org/10.1007/978-3-030-49732-3.

Sajjad H, Muhammad M, Ashfaq A, Waseem A, Hafiz MH, Mazhar A, Nasir M, Asad A, Hafiz UF, Syeda RS, Fahad S, Depeng W, Wajid N (2019) Using GIS tools to detect the land use/land cover changes during forty years in Lodhran district of Pakistan. *Environ Sci Pollut Res* https://doi.org/10.1007/s11356-019-06072-3.

Saleem MH, Fahad S, Adnan M, Mohsin A, Muhammad SR, Muhammad K, Qurban A, Inas AH, Parashuram B, Mubassir A, Reem MH (2020a) Foliar application of gibberellic acid endorsed phytoextraction of copper and alleviates oxidative stress in jute (Corchorus capsularis L.) plant grown in highly copper-contaminated soil of China. *Environ Sci Pollut Res* https://doi.org/10.1007/s11356-020-09764-3.

Saleem MH, Fahad S, Shahid UK, Mairaj D, Abid U, Ayman ELS, Akbar H, Analía L, Lijun L (2020c) Copper-induced oxidative stress, initiation of antioxidants and phytoremediation potential of flax (Linum usitatissimum L.) seedlings grown under the mixing of two different soils of China. *Environ Sci Pollut Res* https://doi.org/10.1007/s11356-019-07264-7.

Saleem MH, Rehman M, Fahad S, Tung SA, Iqbal N, Hassan A, Ayub A, Wahid MA, Shaukat S, Liu L, Deng G (2020b) Leaf gas exchange, oxidative stress, and physiological attributes of rapeseed (Brassica napus L.) grown under different light-emitting diodes. *Photosynthetica* 58(3):836–845.

Saleh J, Maftoun M (2008) Interactive effect of NaCl levels and Zinc sources and levels on the growth and mineral composition of rice. *J Agric Sci Technol* 10:325–336.

Salehin M, Li B, Tang M, Katz E, Song L, Ecker JR, Kliebenstein DJ, Estelle M (2019) Auxin-sensitive Aux/IAA proteins mediate drought tolerance in Arabidopsis by regulating glucosinolate levels. *Nat Commun* 10:4021. https://dx.doi.org/10.1038%2Fs41467-019-12002-1.

Saman S, Amna B, Bani A, ul Qamar Muhammad Tahir, Rana MA, Muhammad SK (2020) QTL mapping for abiotic stresses in cereals. In: Fahad S, Hasanuzzaman M, Alam M, Ullah H, Saeed M, Khan AK, Adnan M (eds) *Environment, climate, plant and vegetation growth.* Springer Publ Ltd, Springer Nature Switzerland AG. Part of Springer Nature, pp 229–252. https://doi.org/10.1007/978-3-030-49732-3.

Sánchez-Martín J, Canales FJ, Tweed JK, Lee MR, Rubiales D, Gómez-Cadenas A, Arbona V, Mur LA, Prats E (2018) Fatty acid profile changes during gradual soil water depletion in oats suggests a role for jasmonates in coping with drought. *Front Plant Sci* 9:1077. https://doi.org/10.3389/fpls.2018.01077.

Saud S, Chen Y, Fahad S, Hussain S, Na L, Xin L, Alhussien SA (2016) Silicate application increases the photosynthesis and its associated metabolic activities in Kentucky bluegrass under drought stress and post-drought recovery. *Environ Sci Pollut Res* 23(17):17647–17655. https://doi.org/10.1007/s11356-016-6957-x.

Saud S, Chen Y, Long B, Fahad S, Sadiq A (2013) The different impact on the growth of cool season turf grass under the various conditions on salinity and drought stress. *Int J Agric Sci Res* 3:77–84.

Saud S, Fahad S, Cui G, Chen Y, Anwar S (2020) Determining nitrogen isotopes discrimination under drought stress on enzymatic activities, nitrogen isotope abundance and water contents of Kentucky bluegrass. *Sci Rep* 10:6415 | https://doi.org/10.1038/s41598-020-63548-w.

Saud S, Fahad S, Yajun C, Ihsan MZ, Hammad HM, Nasim W, Amanullah Jr Arif M and Alharby H (2017) Effects of nitrogen supply on water stress and recovery mechanisms in Kentucky bluegrass plants. *Front Plant Sci* 8:983. https://doi.org/10.3389/fpls.2017.00983.

Saud S, Li X, Chen Y, Zhang L, Fahad S, Hussain S, Sadiq A, Chen Y (2014) Silicon application increases drought tolerance of Kentucky bluegrass by improving plant water relations and morph physiological functions. *Sci World J* 2014:1–10. https://doi.org/10.1155/2014/ 368694.

Senol C (2020) The effects of climate change on human behaviors. In: Fahad S, Hasanuzzaman M, Alam M, Ullah H, Saeed M, Khan AK, Adnan M (eds) *Environment, climate, plant and vegetation growth.* Springer Publ Ltd, Springer Nature Switzerland AG. Part of Springer Nature, pp 577–590. https://doi.org/10.1007/978-3-030-49732-3.

Seo JS, Joo J, Kim MJ, Kim YK, Nahm BH, Song SI, Cheong JJ, Lee JS, Kim JK, Choi YD (2011) OsbHLH148, a basic helix-loop-helix protein, interacts with OsJAZ proteins in a jasmonate signaling pathway leading to drought tolerance in rice. *Plant J* 65:907–921. https://doi.org/10.1111/j.1365-313X.2010.04477.x.

Shafi MI, Adnan M, Fahad S, Fazli W, Ahsan K, Zhen Y, Subhan D, Zafar-ul-Hye M, Martin B, Rahul D (2020) Application of single superphosphate with humic acid improves the growth, yield and phosphorus uptake of wheat (Triticum aestivum L.) in calcareous soil. *Agron* 10:1224. https://doi.org/10.3390/agronomy10091224.

Shah F, Lixiao N, Kehui C, Tariq S, Wei W, Chang C, Liyang Z, Farhan A, Fahad S, Huang J (2013) Rice grain yield and component responses to near 2°C of warming. *Field Crops Res* 157:98–110.

Shi J, Habben JE, Archibald RL, Drummond BJ, Chamberlin MA, Williams RW, Lafitte HR, Weers BP (2015) Overexpression of ARGOS genes modifies plant sensitivity to ethylene, leading to improved drought tolerance in both Arabidopsis and maize. *Plant Physiol* 169:266–282. https://doi.org/10.1104/pp.15.00780.

Shi JB, Wang N, Zhou H, Xu QH, Yan GT (2019) The role of gibberellin synthase gene GhGA2ox1 in upland cotton (Gossypium hirsutum L.) responses to drought and salt stress. *Biotechnol Appl Biochem* 66:298–308. https://doi.org/10.1002/bab.1725.

Shi Y, Tian S, Hou L, Huang X, Zhang X, Guo H, Yang S (2012) Ethylene signaling negatively regulates freezing tolerance by repressing expression of CBF and type-A ARR genes in Arabidopsis. *Plant Cell* 24:2578–2595. https://doi.org/10.1105/tpc.112.098640.

Sponsel VM, Hedden P (2010) Gibberellin biosynthesis and inactivation. In *Plant Hormones.* Springer, pp 63–94. https://doi.org/10.1007/978-1-4020-2686-7_4.

Sreenivasulu N, Harshavardhan VT, Govind G, Seiler C, Kohli A (2012) Contrapuntal role of ABA: does it mediate stress tolerance or plant growth retardation under long-term drought stress? *Gene* 506:265–273. https://doi.org/10.1016/j.gene.2012.06.076.

Sreenivasulu N, Radchuk V, Alawady A, Borisjuk L, Weier D, Staroske N, Fuchs J, Miersch O, Strickert M, Usadel B (2010) De-regulation of abscisic acid contents causes abnormal endosperm development in the barley mutant seg8. *Plant J* 64:589–603. https://doi.org/10.1111/j.1365-313X.2010.04350.x.

Subhan D, Zafar-ul-Hye M, Fahad S, Saud S, Martin B, Tereza H, Rahul D (2020) Drought stress alleviation by ACC deaminase producing *achromobacter xylosoxidans* and *enterobacter cloacae*, with and without timber waste biochar in maize. *Sustain* 12(6286). https://doi.org/10.3390/su12156286.

Tariq M, Ahmad S, Fahad S, Abbas G, Hussain S, Fatima Z, Nasim W, Mubeen M, ur Rehman MH, Khan MA, Adnan M (2018) The impact of climate warming and crop management on phenology of sunflower-based cropping systems in Punjab, Pakistan. *Agri Forest Meteorol* 15;256:270–282.

Tilman D, Balzer C, Hill J, Befort BL (2011) Global food demand and the sustainable intensification of agriculture. *Proc Natl Acad Sci U S A* 108:20260–20264. https://doi.org/10.1073/pnas.1116437108.

Tiwari P, Indoliya Y, Chauhan AS, Singh P, Singh PK, Singh PC, Srivastava S, Pande V, Chakrabarty D (2020) Auxin-salicylic acid cross-talk ameliorates OsMYB–R1 mediated defense towards heavy metal, drought and fungal stress. *J Hazard Mater*:122811. https://doi.org/10.1016/j.jhazmat.2020.122811.

Tůmová L, Tarkowská D, Řehořová K, Marková H, Kočová M, Rothová O, Čečetka P, Holá D (2018) Drought-tolerant and drought-sensitive genotypes of maize (*Zea mays* L.) differ in contents of endogenous brassinosteroids and their drought-induced changes. *PloS one* 13:e0197870. https://doi.org/10.1371/journal.pone.0197870.

ul Qamar Muhammad Tahir, Amna F, Amna B, Barira Z, Xitong Z, Ling-Ling C (2020) Effectiveness of conventional crop improvement strategies vs. omics. In: Fahad S, Hasanuzzaman M, Alam M, Ullah H, Saeed M, Khan AK, Adnan M (eds) *Environment, climate, plant and vegetation growth.* Springer Publ Ltd, Springer Nature Switzerland AG. Part of Springer Nature, pp 253–284. https://doi.org/10.1007/978-3-030-49732-3.

Unsar Naeem-U, Muhammad R, Syed HMB, Asad S, Mirza AQ, Naeem I, Muhammad H ur R, Fahad S, Shafqat S (2020) Insect pests of cotton crop and management under climate change scenarios. In: Fahad S, Hasanuzzaman M, Alam M, Ullah H, Saeed M, Khan AK, Adnan M (eds) *Environment, climate, plant and vegetation growth.* Springer Publ Ltd, Springer Nature Switzerland AG. Part of Springer Nature, pp 367–396. https://doi.org/10.1007/978-3-030-49732-3.

Uppalapati SR, Ishiga Y, Wangdi T, Kunkel BN, Anand A, Mysore KS, Bender CL (2007) The phytotoxin coronatine contributes to pathogen fitness and is required for suppression of salicylic acid accumulation in tomato inoculated with Pseudomonas syringae pv. tomato DC3000. *Mol Plant Microbe Interact* 20:955–965. https://doi.org/10.1094/MPMI-20-8-0955.

Vardhini BV, Anuradha S, Rao S (2006) Brassinosteroids-New class of plant hormone with potential to improve crop productivity. *Indian J Plant Physiol* 11:1.

Visentin I, Vitali M, Ferrero M, Zhang Y, Ruyter-Spira C, Novák O, Strnad M, Lovisolo C, Schubert A, Cardinale F (2016) Low levels of strigolactones in roots as a component of the systemic signal of drought stress in tomato. *New Phytol* 212:954–963. https://doi.org/10.1111/nph.14190.

Voß U, Bishopp A, Farcot E, Bennett MJ (2014) Modelling hormonal response and development. *Trends Plant Sci* 19:311–319. https://doi.org/10.1016/j.tplants.2014.02.004.

Vurro M, Yoneyama K (2012) Strigolactones—intriguing biologically active compounds: perspectives for deciphering their biological role and for proposing practical application. *Pest Manage Sci* 68:664–668. https://doi.org/10.1002/ps.3257.

Wahid F, Fahad S, Subhan D, Adnan M, Zhen Y, Saud S, Manzer HS, Martin B, Tereza H, Rahul D (2020) Sustainable management with mycorrhizae and phosphate solubilizing bacteria for enhanced phosphorus uptake in calcareous soils. *Agri* 10(334). https://doi.org/10.3390/agriculture10080334.

Wajid N, Ashfaq A, Asad A, Muhammad T, Muhammad A,Muhammad S, Khawar J, Ghulam MS, Syeda RS, Hafiz MH, Muhammad IAR, Muhammad ZH, Habib ur R Muhammad, Veysel T, Fahad S, Suad S, Aziz K, Shahzad A (2017) Radiation efficiency and nitrogen fertilizer impacts on sunflower crop in contrasting environments of Punjab, Pakistan. *Environ Sci Pollut Res* 25:1822–1836. https://org/10.1007/s11356-017-0592-z.

Wang G, Li Q, Wang C, Jin C, Ji J, Guan C (2019) A salicylic acid carboxyl methyltransferase-like gene LcSAMT from Lycium chinense, negatively regulates the drought response in transgenic tobacco. *Environ Exp Bot* 167:103833. https://doi.org/10.1016/j.envexpbot.2019.103833.

Wani SH, Sah S (2014) Biotechnology and abiotic stress tolerance in rice. *J Rice Res* 2:1–2. https://doi.org/10.4172/jrr.1000e105.

Wilkinson S, Kudoyarova GR, Veselov DS, Arkhipova TN, Davies WJ (2012) Plant hormone interactions: innovative targets for crop breeding and management. *J Exp Bot* 63:3499–3509. https://doi.org/10.1093/jxb/ers148.

Wolters H, Jürgens G (2009) Survival of the flexible: hormonal growth control and adaptation in plant development. *Nat Rev Genet* 10:305–317. https://doi.org/10.1038/nrg2558.

Wu C, Kehui C, She T, Ganghua L, Shaohua W, Fahad S, Lixiao N, Jianliang H, Shaobing P, Yanfeng D (2020) Intensified pollination and fertilization ameliorate heat injury in rice (Oryza sativa L.) during the flowering stage. *Field Crops Res* 252:107795.

Wu C, Tang S, Li G, Wang S, Fahad S, Ding Y (2019) Roles of phytohormone changes in the grain yield of rice plants exposed to heat: a review. *PeerJ* 7:e7792 https://doi.org/10.7717/peerj.7792.

Xiong B, Wang Y, Zhang Y, Ma M, Gao Y, Zhou Z, Wang B, Wang T, Lv X, Wang X (2020) Alleviation of drought stress and the physiological mechanisms in citrus cultivar (Huangguogan) treated with methyl jasmonate. *Biosci Biotechnol Biochem* 84:1–8. https://doi.org/10.1080/09168451.2020.1771676.

Yamaguchi S (2008) Gibberellin metabolism and its regulation. *Annu Rev Plant Biol* 59:225–251. https://doi.org/10.1146/annurev.arplant.59.032607.092804.

Yan Q, Cui X, Lin S, Gan S, Xing H, Dou D (2016) GmCYP82A3, a soybean cytochrome P450 family gene involved in the jasmonic acid and ethylene signaling pathway, enhances plant resistance to biotic and abiotic stresses. *PloS one* 11:e0162253. https://doi.org/10.1371/journal.pone.0162253.

Yang Z, Cao S, Zheng Y, Jiang Y (2012) Combined salicyclic acid and ultrasound treatments for reducing the chilling injury on peach fruit. *J Agric Food Chem* 60:1209–1212. https://doi.org/10.1021/jf2041164.

Yang Z, Zhang Z, Zhang T, Fahad S, Cui K, Nie L, Peng S, Huang J (2017) The effect of season-long temperature increases on rice cultivars grown in the central and southern regions of China. *Front Plant Sci* 8:1908. https://doi.org/10.3389/fpls.2017.01908.

Ye H, Liu S, Tang B, Chen J, Xie Z, Nolan TM, Jiang H, Guo H, Lin H-Y, Li L (2017) RD26 mediates crosstalk between drought and brassinosteroid signalling pathways. *Nat Commun* 8:1–13. https://doi.org/10.1038/ncomms14573.

Yoneyama K, Kisugi T, Xie X, Yoneyama K (2013) Chemistry of strigolactones: why and how do plants produce so many strigolactones. *Mol Microb Eco Rhizo* 1:373–379.

Yu Y, Yang D, Zhou S, Gu J, Wang F, Dong J, Huang R (2017) The ethylene response factor OsERF109 negatively affects ethylene biosynthesis and drought tolerance in rice. *Protoplasma* 254:401–408. https://doi.org/10.1007/s00709-016-0960-4.

Yue Y, Zhang M, Zhang J, Tian X, Duan L, Li Z (2012) Overexpression of the AtLOS5 gene increased abscisic acid level and drought tolerance in transgenic cotton. *J Exp Bot* 63:3741–3748. https://doi.org/10.1093/jxb/ers069.

Zafar-ul-Hye M, ahzeeb-ul-Hassan Muhammad T, Muhammad A, Fahad S, Martin B, Tereza D, Rahul D, Subhan D (2020b) Potential role of compost mixed biochar with rhizobacteria in mitigating lead toxicity in spinach. *Sci Rep* 10:12159. https://doi.org/10.1038/s41598-020-69183-9.

Zafar-ul-Hye M, Muhammad N, Subhan D, Fahad S, Rahul D, Mazhar A, Ashfaq AR, Martin B, Ji˘rí H, Zahid HT, Muhammad N (2020a) Alleviation of cadmium adverse effects by improving nutrients uptake in bitter gourd through cadmium tolerant rhizobacteria. *Environ* 7(54). https://doi.org/10.3390/environments7080054.

Zahida Z, Hafiz FB, Zulfiqar AS, Ghulam MS, Fahad S, Muhammad RA, Hafiz MH, Wajid N, Muhammad S (2017) Effect of water management and silicon on germination, growth, phosphorus and arsenic uptake in rice. *Ecotoxicol Environ Saf* 144:11–18.

Zalabák D, Pospíšilová H, Šmehilová M, Mrízová K, Frébort I, Galuszka P (2013) Genetic engineering of cytokinin metabolism: prospective way to improve agricultural traits of crop plants. *Biotechnol Adv* 31:97–117. https://doi.org/10.1016/j.biotechadv.2011.12.003.

Zhang J, Jia W, Yang J, Ismail AM (2006) Role of ABA in integrating plant responses to drought and salt stresses. *Field Crops Res* 97:111–119. https://doi.org/10.1016/j.fcr.2005.08.018.

Zhang Q, Li J, Zhang W, Yan S, Wang R, Zhao J, Li Y, Qi Z, Sun Z, Zhu Z (2012) The putative auxin efflux carrier OsPIN3t is involved in the drought stress response and drought tolerance. *Plant J* 72:805–816. https://doi.org/10.1111/j.1365-313X.2012.05121.x.

Zhang Y, Li Y, Hassan MJ, Li Z, Peng Y (2020) Indole-3-acetic acid improves drought tolerance of white clover via activating auxin, abscisic acid and jasmonic acid related genes and inhibiting senescence genes. *BMC Plant Biol* 20:1–12. https://doi.org/10.1186/s12870-020-02354-y.

Zhou J, Li Z, Xiao G, Huang R, Zhang H (2019) OsCYP71D8L as a key regulator involved in growth and stress response by mediating gibberellins and cytokinins homeostasis in rice. *bioRxiv*:538785. https://doi.org/10.1101/538785.

Zia-ur-Rehman M (2020) Environment, climate change and biodiversity. In: Fahad S, Hasanuzzaman M, Alam M, Ullah H, Saeed M, Khan AK, Adnan M (eds) *Environment, climate, plant and vegetation growth*. Springer Publ Ltd, Springer Nature Switzerland AG. Part of Springer Nature, pp 473–502. https://doi.org/10.1007/978-3-030-49732-3.

7

Cross-Talk between Phytohormone-Signalling Pathways under Abiotic Stress Conditions

Asif Iqbal, Mazhar Iqbal, Madeeha Alamzeb, Shah Fahad, Mohammad Akmal, Shazma Anwar, Asad Ali Khan, Muhammad Arif, Inamullah, Shaheenshah, Muhammad Saeed, Manzoor Ahmad, Qiang Dong, Xiangru Wang, Huiping Gui, Hengheng Zhang, Xiling Zhang, Du Xiongming, and Meizhen Song

CONTENTS

7.1 Introduction

Naturally, plants are constantly exposed to various types of abiotic stresses like salinity, drought, cold, heat, heavy metals, and wounding (Singh et al. 2017; Suzuki et al. 2014; Tuteja and Sopory 2008; Adnan et al. 2018a,b; Adnan et al. 2019; Akram et al. 2018a,b; Aziz et al. 2017a,b; Habib et al. 2017; Hafiz et al. 2016; Hafiz et al. 2019; Kamaran et al. 2017; Muhmmad et al.2019; Sajjad et al. 2019; Saud et al. 2013; Saud et al. 2014; Saud et al. 2017; Saud et al. 2016; Shah et al. 2013; Saud et al. 2020; Qamar et al. 2017; Wajid et al. 2017; Yang et al. 2017; Zahida et al. 2017; Depeng et al. 2018; Hussain et al. 2020; Hafiz et al. 2020 a,b; Shafi et al. 2020; Wahid et al. 2020; Subhan et al. 2020; Zafar-ul-Hye et al. 2020a,b; Adnan et al. 2020; Ilyas et al. 2020; Saleem et al. 2020a,b,c; Rehman 2020; Frahat et al. 2020; Wu et al. 2020; Mubeen et al. 2020; Farhana 2020; Jan et al. 2019; Wu et al. 2019; Ahmad et al. 2019; Baseer et al. 2019; Hafiz et al. 2018;Tariq et al. 2018; Fahad and Bano 2012; Fahad et al. 2017; Fahad et al. 2013; Fahad et al. 2014a,b; Fahad et al. 2016a,b,c,d; Fahad et al. 2015a,b; Fahad et al. 2018; Fahad et al. 2019a,b; Hesham and Fahad 2020. Iqra et al. 2020; Akbar et al. 2020; Mahar et al. 2020; Noor et al. 2020; Bayram et al. 2020; Amanullah et al. 2020; Rashid et al. 2020; Arif et al. 2020; Amir et al. 2020; Saman et al. 2020; Muhammad Tahir et al. 2020; Md Jakirand Allah 2020; Farah et al. 2020; Sadam et al. 2020; Unsar et al. 2020; Fazli et al. 2020; Md. Enamul et al. 2020; Gopakumar et al. 2020; Zia-ur-Rehman 2020; Ayman et al. 2020; Mohammad I. Al-Wabel et al. 2020a,b; Senol 2020; Amjad et al. 2020; Ibrar et al. 2020; Sajid et al. 2020). These stresses lead to a significant reduction in plant growth, yield (Mahajan and Tuteja 2005), stomatal and non-stomatal limitations on photosynthesis, and changes in both hormonal balance and redox processes, leading to high lipid peroxidation, protein oxidation, and DNA damage (Munné-Bosch et al. 2013). The sensing of abiotic stresses in plants is controlled by molecular signalling transduction that interacts with downstream phytohormone-signalling pathways to respond or adapt to the onset of a particular abiotic stress (Dolferus 2014). As part of the signalling network, plant hormones act as a central integrator that can link and reprogram complex signal cascades that adapt to stress (Golldack et al. 2014). Studies have found that plant hormones, like ethylene (ET) (Cheng et al. 2013), abscisic acid (ABA) (Wu et al. 2007), auxin (He et al. 2005), gibberellins (Gas) (Magome et al. 2008, Shigenaga and Argueso 2016), cytokinins (CKs) (Wu et al. 2014), brassinosteroids (BRs) (Divi et al. 2010), jasmonates (JAs) (Cela et al. 2011), and salicylic acid (SA) (Jayakannan et al. 2015) play a great role in plant defense systems (Iqbal et al. 2020; Tuteja and Sopory 2008), with ABA playing the central role (Roychoudhury and Paul 2012). Together, hormonal interactions form a hormonal signalling network that mediates immunity, growth, and response to abiotic stresses (Pieterse et al. 2009). Hormone signal transduction pathways cross-talk and act conversely or antagonistically to defend plants from abiotic and biotic stresses (Nguyen et al. 2016). In the last thirty years, vast studies about molecular mechanisms of hormonal signal transduction and cross-talk in the defense process of plants have been acquired for only a limited number of species, such as the model plant Arabidopsis (Pieterse et al. 2009, Shigenaga and Argueso 2016). Cross-talk among the signalling networks of ABA, GA, ET, auxins, CTs, SA, JA, and ET further improves the tolerance of plants against multiple abiotic stresses (Golldack et al. 2014). There are some overlapping responses of different phytohormones which make plants tolerant to multiple stresses. Moreover, the phytohormone cross-talk also alters and regulates the expression of stress-responsive genes. In this chapter, we summarised the cross-talk between important phytohormones under an abiotic stress environment that strengthens the plant defense system. This chapter will further improve our understanding of phytohormone cross-talk during abiotic stress conditions, which will subsequently further widen our knowledge of improving tolerance in crop plants against abiotic stresses.

7.2 Phytohormone Cross-Talk Under Abiotic Stresses

It is assumed that the existence and conservation of hormonal cross-talk can bring fitness benefits to plants that are subjected to multiple stresses (Pieterse et al. 2009; Vos et al. 2015). There are various defensive phytohormone-signalling pathways in plants, such as SA, ET, and JA (Tuteja and Sopory 2008), and growth-regulating phytohormone pathways such as auxins, ABA, GA, and CKs. Under abiotic stress, these pathways also cross-talk with themselves (Knight and Knight 2001). To survive under abiotic stresses, plants have to maintain their health and growth as well as effectively minimise the harsh effects of stress (Mittler 2006). Furthermore, since activated immunity reduces the response to abiotic stress, the negative impact of ABA on JA and SA signals can enhance the response to abiotic stress, which may be helpful in increasing the survival rate of certain plant species under severe abiotic stress conditions (Mosher et al. 2010; Yasuda et al. 2008). Thus, the defense phytohormones need to cross-talk with the growth-regulating phytohormones to achieve these goals (Kohli et al. 2013). The response to particular abiotic stress is not the result of a single phytohormone, rather it is the result of cross-talk between two or more. Cross-talk among phytohormones may be positive or negative in the senses of the regulation and effect on each other (Pieterse et al. 2009). Apart from stress responses, hormonal cross-talk is also needed in the regulation of plant development and growth (Munné-Bosch and Müller 2013).

7.2.1 Abscisic Acid and Auxins

ABA and auxins work antagonistically for regulating developmental processes such as growth, differentiation, calcium level, and pH level of plant cells (Daminato et al. 2013; Sun and Li 2014). Molecular genetic analysis of the auxin and ABA response pathways provides evidence of auxin-ABA interaction (Brady et al. 2003). For instance, a SHATTERPROOF gene is regulated by ABA and auxin antagonistically for controlling fruit ripening (Daminato et al. 2013). ABSCISIC ACID INSENSITIVE3 (ABI3) is an auxin-regulated, ABRE-based TF that plays an important role in regulating seed dormancy

(Liu et al. 2013). Moreover, IBR5 supports the signalling pathways between ABA and auxin (Monroe-Augustus et al. 2003). During abiotic stress, the cross-talk between ABA and auxin assists the survival of seeds. During water stress, auxin transport is modulated by ABA for maintaining root growth in the root tip (Xu et al. 2013). Similarly, an R2R3-type MYB TF, MYB96, has been shown to modulate stress tolerance by integrating ABA and auxin signalling, which further results in the regulation of some auxin metabolism-related GH3 genes (Seo et al. 2009). Two Cys2/His2 zinc-finger proteins, AZF1 and AZF2, have been reported to suppress ABA-repressive and auxin-inducible genes during salinity stress, which indicates regulatory cross-talk (Kodaira et al. 2011). In a recent study, WRKY46 was shown to regulate lateral root development during salinity stress via regulating ABA signalling and auxin balance (Ding et al. 2015). Additionally, ASCORBATE PEROXIDASE6 (APX6) has been identified as a cross-talk-mediating enzyme between ABA, auxin, and reactive oxygen species for protecting plants from abiotic stress (Chen et al. 2014). The induced activity of ABA was found in the auxin-primed seed, which rescued plants under salt stress as a result of cross-talk between these hormones (Fahad et al. 2015). Thus, the cross-talk of ABA and auxin has a major role in the improvement of abiotic stress tolerance of germinating seed as well as developing plants.

7.2.2 Abscisic Acid and Gibberellins

Gibberellic acids (GAs) are considered primarily growth regulators, which are associated with various plant developmental processes and can also control certain physiological processes in stress conditions (Wang et al. 2017). ABA and GA antagonise and induce various plant developmental processes, such as seed germination, seed maturation and dormancy (Peng and Harberd 2002), stem elongation (Iwamoto et al. 2011), and primary root growth (Hyun et al. 2016) and control time of flowering (Luo et al. 2014, Shu et al. 2016). GAs cross-talk with other phytohormones to maintain growth and development under drought, salinity, and cold stress (Verma et al. 2016) especially via the involvement of cross-talk with ABA signalling (Skubacz et al. 2016). The production of ABA-INSENSITIVE5 (ABI5) that contains an abscisic acid-responsive element (ABRE) and is reportedly accumulated for growth, slows down under abiotic stress conditions. This growth inhibition is carried out by regulating stress adaptation genes such as LATE EMBRYOGENESIS ABUNDANT (LEA) genes under such conditions (Banerjee and Roychoudhury 2017). LEA proteins play an important role in abiotic stress tolerance such as cold, salt, drought, and heat. On the contrary, GAs work in the opposite manner. GA escalates seed germination when there are optimum conditions. GA suppresses the activity of ABA to enable successful germination. It is known that antagonistic cross-talk between GA and ABA is regulated by DELLA proteins (Jiang and Fu 2007). Some DELLA proteins are regulated by GAs which negatively affect ABA signalling and remove the growth-diminution effects of ABA. Some DELLA proteins are regulated by ABA for suppressing GA signalling to halt growth under an abiotic stress environment (Peng and Harberd 2002). Many key regulatory factors, including DELLAs and TFs that contain AP2 domains, have been widely studied (Liu et al. 2016). These factors have a significant role in the antagonism of ABA and GA, which affect the synthesis and/or signalling transduction of other plant hormones (Shu et al. 2018). This mechanism enables plants to escape harsh abiotic stress conditions by maintaining seed dormancy. As a survival strategy, this cross-talk results in delayed germination to protect seeds from the harmful effects of abiotic stresses. Additionally, ABA participates in a variety of stress responses, such as cascades associated with flood, drought, salt, and low temperature (Zhu et al. 2017). Therefore, ABA and GA cross-talk is a critical research point, and the detailed mechanisms that accurately mediate plant development against stress need to be further explored.

7.2.3 Abscisic Acid and Cytokinins

Individually, ABA and cytokinins (CKs) play important roles in stress tolerance, mainly during salinity stress. CKs are generally known as ABA antagonists in different plant biological processes (Pospíšilová 2003). It improves wheat tolerance against salt stress along with other phytohormones, particularly ABA and auxin (Iqbal et al. 2006). Additionally, there also exists cross-talk between ABA and CKs specifically under stress conditions (Verslues 2016). A reduction in CK content has been noted as an early response to salt stress, however, the influence of salinity on salt-sensitive genotypes is not induced by CKs, as the reduction in growth rate preceded any decline in CK level (Walker and Dumbroff 1981). The ARABIDOPSIS HISTIDINE KINASEs (AHKs) in Arabidopsis are known as CK-histidine kinase receptors. Under salt, drought, and cold stress, AHKs have proven to be important regulators (Jeon et al. 2010, Kumar and Verslues 2015). AHK1 is a positive regulator of ABA signalling, whereas AHK2 and AHK3 are negative regulators of ABA signalling (Tran et al. 2007). Functional analysis of CK-deficient plants has proven that CKs negatively modulate salt and drought stress signalling linked to induced ABA sensitivity. It was also observed that CK biosynthesis genes such as ISOPENTENYL-TRANSFERASEs and CYTOKININ OXIDASES/DEHYDROGENASES were suppressed by ABA (Nishiyama et al. 2011). In drought stress, ARABIDOPSIS-HISTIDINE PHOSPHOTRANSFER PROTEINs negatively regulate the expression of ABA-responsive genes. CK catabolism-related enzymes CYTOKININ OXIDASEs are also negatively regulated by ABA. During heat stress, CK and ABA balance has been observed to be an important factor for developing kernels in maize (Cheikh and Jones 1994). CK oxidase has also been reportedly affected by CK and ABA under abiotic stress (Brugière et al. 2003). Therefore, ABA cross-talks with CK under abiotic stress for maintaining CK homeostasis and stress tolerance (Li et al. 2016).

7.2.4 Abscisic Acid and Ethylene

The gaseous phytohormone ethylene (ET) cross-talks with ABA during abiotic stresses (Fujita et al. 2006). ET-mediated root growth inhibition needs ABA (Ma et al. 2014). Similarly, antagonistic cross-talk between these two hormones has been reported to regulate shoot growth when roots encounter hard soil (Hussain et al. 2000). ET inhibits ABA signalling (Harrison 2012) and ABA is the main internal signal that allows plants to survive in adverse environmental conditions. Salinity increases the ABA concentration of the plant in all compartments (Kefu et al. 1991); however, the role of ABA as a growth regulator is ambiguous (Dodd and Davies 1996), as it indirectly inhibits the growth by restricting ET (Sharp and LeNoble 2002). The exogenous application of ABA is reported to reduce ET and leaf abscission of citrus in saline conditions, which may be caused by reduced accumulation of toxic Cl ions in leaves (Gomez-Cadenas et al. 2002). Similarly, salinity-induced ABA inhibits leaf expansion and limits the foliar sodium and chloride ion (Cabot et al. 2009). ABA affects ET biosynthesis genes such as ethylene response factor 11 (ERF11), acyl-CoA synthetase 5 (ACS5), and 9-cis-epoxycarotenoid dioxygenase (NCED) (Li et al. 2011; Zhang et al. 2009). ABA affects the ET synthesis by regulating key ET biosynthesis genes. A mutant-enhanced response to ABA3 (era3) with increased sensitivity to ABA is related to ETHYLENE INSENSITIVE2 locus which confirmed that ET is a negative regulator of ABA, whereas ET signalling positively affects ABA action in the absence of ET (Ghassemian et al. 2000). At the molecular level, ABA-regulated DREB TFs which belong to the ethylene-responsive (ERF) family of TFs are known to be induced by ET (Lata and Prasad 2011). For breaking dormancy, ET intervenes with ABA signalling (Matilla and Matilla-Vázquez 2008) and for the regulation of stomatal closure, ABA also needs to cross-talk with ET signalling under drought and salt stress (Neill et al. 2008). Integration of ABA and ET signalling has shown to play a role in the improvement of salinity stress tolerance (Amjad et al. 2014). In a study, it was shown that ET receptors ETR1 and ETR2 have roles in stress signalling which are independent of ET signalling. Both ETR1 and ETR2 modulate ABA signalling for regulating germination under salinity stress (Wilson et al. 2014). Thus, the cross-talk between ABA and ET is also important in maintaining the hormonal level of each other for finalising decisions on growth, dormancy, fruit ripening, stomatal closing, etc. under abiotic stress conditions (Arc et al. 2013).

7.2.5 Abscisic Acid and Jasmonic Acid

Jasmonic acid (JA) biosynthesis is induced by stress conditions (Fahad et al. 2016a,b,c), and many signalling genes related to JA are associated with drought stress (Huang et al. 2008). The interaction of JA and ABA regulates stomatal closure by increasing Ca^{2+} flux, thereby stimulating the production of CDPK and the resulting signal cascade. The treatment of ABA or methyl JA (MeJA) leads to a reduction in stomatal opening within 10 minutes in excised Arabidopsis leaves (Munemasa et al. 2007). Inhibiting the biosynthesis of ABA by using chemical inhibitors or mutants lacking ABA can inhibit the Ca^{2+}

oscillations of guard cells induced by MeJA changes to stomatal closure (Hossain et al. 2011). Thus, stomatal closure regulation mediated by MeJA is in close interaction with ABA-induced Ca^{2+} signalling transduction pathways. In guard cells, the interaction of both has shown that they enhance the production of NO and ROS, as both were found in low concentration in plants that are insensitive to MeJA (Munemasa et al. 2007). It was noted that CPK6 plays a role downstream of ROS and NO signalling transduction and may be the target of NO activated Ca^{2+} into cytoplasm (Munemasa et al. 2011). CPK6 is also essential for ABA activation of cytoplasmic Ca^{2+} channels (Munemasa et al. 2011) and slow-type anion in guard cells (Munemasa et al. 2007). Thus, JA-mediated Ca^{2+} influx into the cytoplasm will activate CPK6, thereby activating the slow anion channel, which interacts with ABA-induced stomatal closure (Munemasa et al. 2011).

7.2.6 Gibberellins and Ethylene

ET and GA alleviate the harmful impacts of stresses by activating a series of defense responses or support normal plant growth. It was reported that GA increases ET synthesis, however, ET affects GA signal transduction, so this interaction opens the cross-talk between them. Individually, both ET and GA are involved in stress tolerance, but their interactive effect on stress tolerance has not been explored. In a study, it was found that exogenous application of ethephon and GA3 alleviated the inhibitory effect of salinity on *Amaranthus caudatus* seed germination (Bialecka and Kepczynski 2009). The stimulating effect of ethephon preceded GA3, and ethephon was found more effective than GA3 under salt stress (Mohammed 2007). The interaction between these two was found in the pea plant, where phytochromes negatively affect ET synthesis, thereby reducing the production of GA (Foo et al. 2006). GA-signalling components, DELLA proteins, are involved in ET signalling during salt tolerance. Plant mutants having disrupted quadruple-DELLA proteins have shown improved root growth even in salinity stress (Alvey and Boulton 2008). The results indicated that salt stress suppresses root growth by a DELLA protein-mediated mechanism. In other findings, ET signalling also affects the root growth via the DELLA mechanism. Therefore, it is clear that both GA and ET regulate root growth by crosstalking with each other (Achard and Genschik 2009, Golldack et al. 2014). DELLA and the CTR1-dependent ET response pathways cross-talk, downstream of the EIN3, for improving abiotic stress tolerance (Rahman et al., 2016, Achard et al. 2006). In other studies, expression of ET-responsive CBF1/DREB1B TF has been observed to show cold stress tolerance, via activating DELLA proteins (HUANG et al. 2009, Movahedi et al. 2012, Thomashow 2010). Therefore, the cross-talk between GA and ET is the result of the regulation of DELLA proteins which make the plants successful in tolerating cold, drought, and salt stresses.

7.2.7 Auxins and Ethylene

Auxins are one of the major hormones in plant development and have been shown to play an important role during stress survival. Along with ET, auxins control root development during abiotic stresses, such as salinity and drought (Muday et al. 2012). Both of the hormones strengthen the root system in such a way to enable plants to survive under abiotic stresses. In general, NAC2 TF acts downstream of ET and auxin-signalling pathways and plays a role in abiotic stress response and lateral root development for better survival in various plants (Nuruzzaman et al. 2013, Shan et al. 2014). Recently, ET-insensitive Never ripe (Nr) and auxin-insensitive diageotropica (dgt) tomato mutants were identified. In further studies, it was observed that the cross-talk between auxins and ET is important in maintaining Cd stress tolerance in tomato via communication between root and shoot parts (Alves et al. 2017). Under abiotic stress, auxin-triggered ET modulates ABA biosynthesis and growth inhibition, which leads to the rescue of plants from such conditions (Hansen and Grossmann 2000). Furthermore, in the single aux1-7 (AUXIN RESISTANT) and pin2 (PINFORMED2) mutants, which are insensitive to auxin, the aluminum inhibition effect on root elongation was less as compared to the wild type. These results specified that aluminum-induced ET biosynthesis may cause auxin redistribution by affecting auxin transport system via AUX1 and PIN1 (Sun et al. 2010), indicating the possible cross-talk between ET and auxin in plants against heavy metal (HM) stress. Therefore, it is clear from several studies that there is pivotal cross-talk between ET and auxins for improvement of abiotic stress tolerance.

7.3 Conclusions

Plants are constantly subjected to different stresses that significantly reduce plant growth and productivity. To overcome stressful conditions, plants have developed their defense mechanisms. In plants, the complex but specific cross-talk networks among phytohormones enable them to grow as well as survive under any kind of stressful environment. It seems that all phytohormones need to cross-talk among themselves to maintain their balance levels. Whether the levels of these hormones rise or fall, this is an adaptive mechanism that can resist stress. It was reported that ET increases, while GA decreases under stress conditions; however, both play an active role in stress tolerance. The application of exogenous ET and GA induced salt tolerance, however, it is still not confirmed whether this response is independent or both hormones depend on each other. It was reported that inhibition of GA biosynthesis affects the levels of auxins, cytokinins, ABA, and ET in rice. Therefore, there is a possibility of cross-talks at multiple levels between more than two phytohormones which may be complex, yet specific and precise. In a recent study, ABA-induced root elongation was found to be affected by ET and auxins. By using inhibitors of ET and auxin pathway-related genes, it was revealed that ABA-mediated growth occurs by cross-talk among ET and auxins. It was concluded that low ABA-mediated growth stimulation occurs by an ethylene-independent pathway which in turn cross-talks with auxin signalling and the PIN2/EIR1-mediated auxin-efflux-dependent pathway. On the other hand, high ABA-mediated growth inhibition follows ethylene-dependent pathway cross-talk with auxin signalling and the AUX1-mediated auxin-influx-dependent pathway. In Arabidopsis, mutant etrl-2 metabolic pathways of major phytohormones, such as ABA, auxin, cytokinin, and gibberellins are altered during moist chilling, seed dormancy, and germination. Under salt stress, the growth and development of plant roots are regulated by ABA signalling through cross-talk with other hormones such as auxin, CT, and ET. For a better understanding of cross-talk networking in plant cells, plant scientists are using robust and high-throughput techniques. In the coming years, the whole network of phytohormone signalling during multiple abiotic stresses will be revealed which will further help to improve the abiotic stress tolerance of crops for better yield and survival under such conditions.

REFERENCES

Achard P, Cheng H, De Grauwe L, Decat J, Schoutteten H, Moritz T, Van Der Straeten D, Peng J, Harberd NP (2006): Integration of plant responses to environmentally activated phytohormonal signals. *Science* 311(5757), 91–94.

Achard P, Genschik P (2009): Releasing the brakes of plant growth: How GAs shutdown DELLA proteins. *Journal of Experimental Botany* 60(4), 1085–1092.

Adnan M, Fahad S, Muhammad Z, Shahen S, Ishaq AM, Subhan D, Zafar-ul-Hye M, Martin LB, Raja MMN, Beena S, Saud S, Imran A, Zhen Y, Martin B, Jiri H, Rahul D (2020): Coupling phosphate-solubilizing bacteria with phosphorus supplements improve maize phosphorus acquisition and growth under lime induced salinity stress. *Plants* 9(900). doi: 10.3390/plants9070900.

Adnan M, Fahad S, Khan IA, Saeed M, Ihsan MZ, Saud S, Riaz M, Wang D, Wu C (2019): Integration of poultry manure and phosphate solubilizing bacteria improved availability of Ca bound P in calcareous soils. *3 Biotech* 1(10); 368.

Adnan M, Shah Z, Sharif M, Rahman H (2018a): Liming induces carbon dioxide (CO_2) emission in PSB inoculated alkaline soil supplemented with different phosphorus sources. *Environmental Science and Pollution Research International* 1(25(10)), 9501–9509.

Adnan M, Zahir S, Fahad S, Arif M, Mukhtar A, Imtiaz AK, Ishaq AM, Abdul B, Hidayat U, Muhammad A, Inayat-Ur R, Saud S, Muhammad ZI, Yousaf J, Amanullah Hafiz MH, Wajid N, Nasim W (2018b): Phosphate-solubilizing bacteria nullify the antagonistic effect of soil calcification on bioavailability of phosphorus in alkaline soils. *Scientific Reports* 8(1), 4339. doi: 10.1038/s41598-018-22653-7.

Ahmad S, Kamran M, Ding R, Meng X, Wang H, Ahmad I, Fahad S, Han Q (2019): Exogenous melatonin confers drought stress by promoting plant growth, photosynthetic capacity and antioxidant defense system of maize seedlings. *PeerJ* 7, e7793. doi: 10.7717/peerj.7793.

Akbar H, Timothy JK, Jagadish T, Golam M, Apurbo KC, Muhammad F, Rajan B, Fahad S, Hasanuzzaman M (2020): Agricultural land degradation: Processes and problems undermining future food security. In Fahad S, Hasanuzzaman M, Alam M, Ullah H, Saeed M, Khan AK, Adnan M (Eds.), *Environment, Climate, Plant and Vegetation Growth*, pp. 17–62. Springer Publ Ltd, Springer Nature. doi: 10.1007/978-3-030-49732-3.

Akram R, Turan V, Hammad HM, Ahmad S, Hussain S, Hasnain A, Maqbool MM, Rehmani MIA, Rasool A, Masood N, Mahmood F, Mubeen M, Sultana SR, Fahad S, Amanet K, Saleem M, Abbas Y, Akhtar HM, Waseem F, Murtaza R, Amin A, Zahoor SA, ul Din MS, Nasim W (2018a): Fate of organic and inorganic pollutants in paddy soils. In Hashmi MZ and Varma A (Eds.), *Environmental Pollution of Paddy Soils, Soil Biology*, pp. 197–214. Springer International Publishing AG, Cham, Switzerland.

Akram R, Turan V, Wahid A, Ijaz M, Shahid MA, Kaleem S, Hafeez A, Maqbool MM, Chaudhary HJ, Munis MFH, Mubeen M, Sadiq N, Murtaza R, Kazmi DH, Ali S, Khan N, Sultana SR, Fahad S, Amin A, Nasim W (2018b): Paddy land pollutants and their role in climate change. In Hashmi MZ and Varma A (Eds.), *Environmental Pollution of Paddy Soils, Soil Biology*, pp. 113–124. Springer International Publishing AG, Cham, Switzerland.

Alves LR, Monteiro CC, Carvalho RF, Ribeiro PC, Tezotto T, Azevedo RA, Gratão PL (2017): Cadmium stress related to root-to-shoot communication depends on ethylene and auxin in tomato plants. *Environmental and Experimental Botany* 134, 102–115.

Alvey L, Boulton MI (2008): DELLA proteins in signalling. *eLS*.

Amanullah Shah K, Imran Hamdan AK, Muhammad A, Abdel RA, Muhammad A, Fahad S, Azizullah S, Brajendra P (2020): Effects of climate change on irrigation water quality. In Fahad S, Hasanuzzaman M, Alam M, Ullah H, Saeed M, Khan AK, Adnan M (Eds.), *Environment, Climate, Plant and Vegetation Growth*, pp. 123–132. Springer Publ Ltd, Springer Nature, Switzerland. doi: 10.1007/978-3-030-49732-3.

Amir M, Muhammad A, Allah B, Sevgi Ç, Haroon ZK, Muhammad A, Emre A (2020): Bio fortification under climate change: The fight between quality and quantity. In Fahad S, Hasanuzzaman M, Alam M, Ullah H, Saeed M, Khan AK, Adnan M (Eds.), *Environment, Climate, Plant and Vegetation Growth*, pp. 173–228. Springer Publ Ltd, Springer Nature, Switzerland. doi: 10.1007/978-3-030-49732-3.

Amjad I, Muhammad H, Farooq S, Anwar H (2020): Role of plant bioactives in sustainable agriculture. In Fahad S, Hasanuzzaman M, Alam M, Ullah H, Saeed M, Khan AK, Adnan M (Eds.), *Environment, Climate, Plant and Vegetation Growth*, pp. 591–606. Springer Publ Ltd, Springer Nature, Switzerland. doi: 10.1007/978-3-030-49732-3.

Amjad M, Akhtar J, Anwar-ul-Haq M, Yang A, Akhtar SS, Jacobsen S-E (2014): Integrating role of ethylene and ABA in tomato plants adaptation to salt stress. *Scientia Horticulturae* 172, 109–116.

Arc E, Sechet J, Corbineau F, Rajjou L, Marion-Poll A (2013): ABA crosstalk with ethylene and nitric oxide in seed dormancy and germination. *Frontiers in Plant Science* 4, 63.

Arif M, Talha J, Muhammad R, Fahad S, Muhammad A, Amanullah Kawsar A, Ishaq AM, Bushra K, Fahd R (2020): Biochar; a remedy for climate change. In Fahad S, Hasanuzzaman M, Alam M, Ullah H, Saeed M, Khan AK, Adnan M (Ed.), *Environment, Climate, Plant and Vegetation Growth*, pp. 151–172. Springer Publ Ltd, Springer Nature, Switzerland. doi: 10.1007/978-3-030-49732-3.

Ayman EL Sabagh, Akbar Hossain, Celaleddin Barutçular, Muhammad Aamir Iqbal, M Sohidul Islam, Shah Fahad, Oksana Sytar, Fatih Çig, Ram Swaroop Meena, Murat Erman (2020): Consequences of salinity stress on the quality of crops and its mitigation strategies for sustainable crop production: An outlook of arid and semi- arid regions. In Fahad S, Hasanuzzaman M, Alam M, Ullah H, Saeed M, Khan AK, Adnan M (Eds.), *Environment, Climate, Plant and Vegetation Growth*, pp. 503–534. Springer Publ Ltd, Springer Nature, Switzerland. doi: 10.1007/978-3-030-49732-3.

Aziz K, Daniel KYT, Fazal M, Muhammad ZA, Farooq S, Fan W, Fahad S, Ruiyang Z (2017a): Nitrogen nutrition in cotton and control strategies for greenhouse gas emissions: A review. *Environmental Science and Pollution Research International* 24(30), 23471–23487. doi: 10.1007/s11356-017-0131-y.

Aziz K, Daniel KYT, Muhammad ZA, Honghai L, Shahbaz AT, Mir A, Fahad S (2017b): Nitrogen fertility and abiotic stresses management in cotton crop: A review. *Environmental Science and Pollution Research International* 24(17), 14551–14566. doi: 10.1007/s11356-017-8920-x.

Banerjee A, Roychoudhury A (2017): Abscisic-acid-dependent basic leucine zipper (bZIP) transcription factors in plant abiotic stress. *Protoplasma* 254(1), 3–16.

Baseer M, Adnan M, Fazal M, Fahad S, Muhammad S, Fazli W, Muhammad A, Jr, Amanullah Depeng W, Saud S, Muhammad N, Muhammad Z, Fazli S, Beena S, Mian AR, Ishaq AM, Mian IA (2019): Substituting urea by organic wastes for improving maize yield in alkaline soil. *Journal of Plant Nutrition*. doi: 10.1080/01904167.2019.1659344.

Bayram AY, Seher Ö, Nazlican A (2020): Climate change forecasting and modeling for the year of 2050. In Fahad S, Hasanuzzaman M, Alam M, Ullah H, Saeed M, Khan AK, Adnan M (Eds.), *Environment, Climate, Plant and Vegetation Growth*, pp. 109–122. Springer Publ Ltd, Springer Nature, Switzerland. doi: 10.1007/978-3-030-49732-3.

Bialecka B, Kepczynski J (2009): Effect of ethephon and gibberellin A3 on Amaranthus caudatus seed germination and alpha-and beta-amylase activity under salinity stress. *Acta Biologica Cracoviensia. Series Botanica* 2, 119–125.

Brady SM, Sarkar SF, Bonetta D, McCourt P (2003): The ABSCISIC ACID INSENSITIVE 3 (ABI3) gene is modulated by farnesylation and is involved in auxin signaling and lateral root development in Arabidopsis. *The Plant Journal* 34(1), 67–75.

Brugière N, Jiao S, Hantke S, Zinselmeier C, Roessler JA, Niu X, Jones RJ, Habben JE (2003): Cytokinin oxidase gene expression in maize is localized to the vasculature, and is induced by cytokinins, abscisic acid, and abiotic stress. *Plant Physiology* 132(3), 1228–1240.

Cabot C, Sibole JV, Barceló J, Poschenrieder C (2009): Abscisic acid decreases leaf Na+ exclusion in salt-treated Phaseolus vulgaris L. *Journal of Plant Growth Regulation* 28(2), 187–192.

Cela J, Chang C, Munné-Bosch S (2011): Accumulation of γ-rather than α-tocopherol alters ethylene signaling gene expression in the vte4 mutant of Arabidopsis thaliana. *Plant and Cell Physiology* 52(8), 1389–1400.

Cheikh N, Jones RJ (1994): Disruption of maize kernel growth and development by heat stress (role of cytokinin/abscisic acid balance). *Plant Physiology* 106(1), 45–51.

Chen C, Letnik I, Hacham Y, Dobrev P, Ben-Daniel B-H, Vanková R, Amir R, Miller G (2014): Ascorbate PEROXIDASE6 protects Arabidopsis desiccating and germinating seeds from stress and mediates cross talk between reactive oxygen species, abscisic acid, and auxin. *Plant Physiology* 166(1), 370–383.

Cheng M-C, Liao P-M, Kuo W-W, Lin T-P (2013): The Arabidopsis ethylene RESPONSE FACTOR1 regulates abiotic stress-responsive gene expression by binding to different cis-acting elements in response to different stress signals. *Plant Physiology* 162(3), 1566–1582.

Daminato M, Guzzo F, Casadoro G (2013): A SHATTERPROOF-like gene controls ripening in non-climacteric strawberries, and auxin and abscisic acid antagonistically affect its expression. *Journal of Experimental Botany* 64(12), 3775–3786.

Depeng W, Fahad S, Saud S, Muhammad K, Aziz K, Mohammad NK, Hafiz MH, Wajid N (2018): Morphological acclimation to agronomic manipulation in leaf dispersion and orientation to promote "Ideotype" breeding: Evidence from 3D visual modeling of "super" rice (*Oryza sativa* L.). *Plant Physiology and Biochemistry*. doi: 10.1016/j.plaphy.2018.11.010.

Ding ZJ, Yan JY, Li CX, Li GX, Wu YR, Zheng SJ (2015): Transcription factor WRKY 46 modulates the development of Arabidopsis lateral roots in osmotic/salt stress conditions via regulation of ABA signaling and auxin homeostasis. *The Plant Journal* 84(1), 56–69.

Divi UK, Rahman T, Krishna P (2010): Brassinosteroid-mediated stress tolerance in Arabidopsis shows interactions with abscisic acid, ethylene and salicylic acid pathways. *BMC Plant Biology* 10, 151.

Dodd I, Davies WJ (1996): The relationship between leaf growth and ABA accumulation in the grass leaf elongation zone. *Plant, Cell and Environment* 19(9), 1047–1056.

Dolferus R (2014): To grow or not to grow: A stressful decision for plants. *Plant Science: An International Journal of Experimental Plant Biology* 229, 247–261.

Fahad S, Adnan M, Hassan S, Saud S, Hussain S, Wu C, Wang D, Hakeem KR, Alharby HF, Turan V, Khan MA, Huang J (2019b): Rice responses and tolerance to high temperature. In Hasanuzzaman M and Fujita M, Nahar K, Biswas JK (Eds.), *Advances in Rice Research for Abiotic Stress Tolerance*, pp. 201–224. Woodhead PUBL LTD, Abington Hall.

Fahad S, Bajwa AA, Nazir U, Anjum SA, Farooq A, Zohaib A, Sadia S, Nasim W, Adkins S, Saud S, Ihsan MZ, Alharby H, Wu C, Wang D, Huang J (2017): Crop production under drought and heat stress: Plant responses and management options. *Frontiers in Plant Science* 8, 1147. doi: 10.3389/fpls.2017.01147.

Fahad S, Bano A (2012): Effect of salicylic acid on physiological and biochemical characterization of maize grown in saline area. *Pakistan Journal of Botany* 44, 1433–1438.

Fahad S, Chen Y, Saud S, Wang K, Xiong D, Chen C, Wu C, Shah F, Nie L, Huang J (2013): Ultraviolet radiation effect on photosynthetic pigments, biochemical attributes, antioxidant enzyme activity and hormonal contents of wheat. *Journal of Food, Agriculture and Environment* 11(3&4), 1635–1641.

Fahad S, Hussain S, Bano A, Saud S, Hassan S, Shan D, Khan FA, Khan F, Chen Y, Wu C, Tabassum MA, Chun MX, Afzal M, Jan A, Jan MT, Huang J (2014a): Potential role of phytohormones and plant growth-promoting rhizobacteria in abiotic stresses: Consequences for changing environment. *Environmental Science and Pollution Research International* 22(7), 4907–4921. doi: 10.1007/s11356-014-3754-2.

Fahad S, Hussain S, Matloob A, Khan FA, Khaliq A, Saud S, Hassan S, Shan D, Khan F, Ullah N, Faiq M, Khan MR, Tareen AK, Khan A, Ullah A, Ullah N, Huang J (2014b): Phytohormones and plant responses to salinity stress: A review. *Plant Growth Regulation* 75(2), 391–404. doi: 10.1007/s10725-014-0013-y.

Fahad S, Hussain S, Saud S, Hassan S, Chauhan BS, Khan F, Ihsan MZ, Ullah A, Wu C, Bajwa AA, Alharby H, Amanullah, Nasim W, Shahzad B, Tanveer M, Huang J (2016a): Responses of rapid viscoanalyzer profile and other rice grain qualities to exogenously applied plant growth regulators under high day and high night temperatures. *PLoS ONE* 11(7), e0159590. doi: 10.1371/journal.pone.0159590.

Fahad S, Hussain S, Saud S, Hassan S, Ihsan Z, Shah AN, Wu C, Yousaf M, Nasim W, Alharby H, Alghabari F, Huang J (2016c): Exogenously applied plant growth regulators enhance the morphophysiological growth and yield of rice under high temperature. *Frontiers in Plant Science* 7 01250, 1250. doi: 10.3389/fpls.2016.

Fahad S, Hussain S, Saud S, Hassan S, Tanveer M, Ihsan MZ, Shah AN, Ullah A, Nasrullah KF, Ullah S, Alharby H, NW, Wu C, Huang J, Huang J (2016d): A combined application of biochar and phosphorus alleviates heat-induced adversities on physiological, agronomical and quality attributes of rice. *Plant Physiology and Biochemistry* 103, 191–198.

Fahad S, Hussain S, Saud S, Khan F, Hassan S, Jr, A, Nasim W, Arif M, Wang F, Huang J (2016b): Exogenously applied plant growth regulators affect heat-stressed rice pollens. *Journal of Agronomy and Crop Science* 202(2), 139–150.

Fahad S, Hussain S, Saud S, Tanveer M, Bajwa AA, Hassan S, Shah AN, Ullah A, Wu C, Khan FA, Shah F, Ullah S, Chen Y, Huang J (2015a): A biochar application protects rice pollen from high-temperature stress. *Plant Physiology and Biochemistry* 96, 281–287.

Fahad S, Muhammad ZI, Abdul K, Ihsanullah D, Saud S, Saleh A, Wajid N, Muhammad A, Imtiaz AK, Chao W, Depeng W, Jianliang H (2018): Consequences of high temperature under changing climate optima for rice pollen characteristics-concepts and perspectives. *Archives of Agronomy and Soil Science.* doi: 10.1080/03650340.2018.1443213.

Fahad S, Nie L, Chen Y, Wu C, Xiong D, Saud S, Hongyan L, Cui K, Huang J (2015b): Crop plant hormones and environmental stress. *Sustainable Agriculture Research* 15, 371–400.

Fahad S, Rehman A, Shahzad B, Tanveer M, Saud S, Kamran M, Ihtisham M, Khan SU, Turan V, Rahman MHU (2019a): Rice responses and tolerance to metal/metalloid toxicity. In Hasanuzzaman M, Fujita M, Nahar K, Biswas JK (Eds.), *Advances in Rice Research for Abiotic Stress Tolerance*, pp. 299–312. Woodhead PUBL LTD, Abington Hall.

Farah R, Muhammad R, Muhammad SA, Tahira Y, Muhammad AA, Maryam A, Shafaqat A, Rashid M, Muhammad R, Qaiser H, Afia Z, Muhammad AA, Muhammad A, Fahad S (2020): Alternative and non-conventional soil and crop management strategies for increasing water use efficiency. In Fahad S, Hasanuzzaman M, Alam M, Ullah H, Saeed M, Khan AK, Adnan M (Eds.), *Environment, Climate, Plant and Vegetation Growth.* Springer Publ Ltd, Springer Nature, Switzerland, pp. 323–338. doi: 10.1007/978-3-030-49732-3.

Farhana G, Ishfaq A, Muhammad A, Dawood J, Fahad S, Xiuling L, Depeng W, Muhammad F, Muhammad F, Syed AS (2020): Use of crop growth model to simulate the impact of climate change on yield of various wheat cultivars under different agro-environmental conditions in Khyber Pakhtunkhwa, Pakistan. *Arabian Journal of Geosciences* 13(3), 112. doi: 10.1007/s12517-020-5118-1.

Farhat A, Hafiz MH, Wajid I, Aitazaz AF, Hafiz FB, Zahida Z, Fahad S, Wajid F, Artemi C (2020): A review of soil carbon dynamics resulting from agricultural practices. *Journal of Environment Management* 268, 110319.

Fazli W, Muhmmad S, Amjad A, Fahad S, Muhammad A, Muhammad N, Ishaq AM, Imtiaz AK, Mukhtar A, Muhammad S, Muhammad I, Rafi U, Haroon I, Muhammad A (2020): Plant-microbes interactions and functions in changing climate. In Fahad S, Hasanuzzaman M, Alam M, Ullah H, Saeed M, Khan AK, Adnan M (Eds.), *Environment, Climate, Plant and Vegetation Growth.* Springer Publ Ltd, Springer Nature, Switzerland, pp. 397–420. doi: 10.1007/978-3-030-49732-3.

Foo E, Ross JJ, Davies NW, Reid JB, Weller JL (2006): A role for ethylene in the phytochrome-mediated control of vegetative development. *The Plant Journal* 46(6), 911–921.

Fujita M, Fujita Y, Noutoshi Y, Takahashi F, Narusaka Y, Yamaguchi-Shinozaki K, Shinozaki K (2006): Crosstalk between abiotic and biotic stress responses: A current view from the points of convergence in the stress signaling networks. *Current Opinion in Plant Biology* 9(4), 436–442.

Ghassemian M, Nambara E, Cutler S, Kawaide H, Kamiya Y, McCourt P (2000): Regulation of abscisic acid signaling by the ethylene response pathway in Arabidopsis. *The Plant Cell* 12(7), 1117–1126.

Golldack D, Li C, Mohan H, Probst N (2014): Tolerance to drought and salt stress in plants: Unraveling the signaling networks. *Frontiers in Plant Science* 5, 151.

Gomez-Cadenas A, Arbona V, Jacas J, Primo-Millo E, Talon M (2002): Abscisic acid reduces leaf abscission and increases salt tolerance in citrus plants. *Journal of Plant Growth Regulation* 21(3), 234–240.

Gopakumar L, Bernard NO, Donato V (2020): Soil microarthropods and nutrient cycling. In Fahad S, Hasanuzzaman M, Alam M, Ullah H, Saeed M, Khan AK, Adnan M (Ed.), *Environment, Climate, Plant and Vegetation Growth*. Springer Publ Ltd, Springer Nature, Switzerland, pp. 453–472. doi: 10.1007/978-3-030-49732-3.

Habib ur R, Ashfaq A, Aftab W, Manzoor H, Fahd R, Wajid I, Md. Aminul I, Vakhtang S, Muhammad A, Asmat U, Abdul W, Syeda RS, Shah S, Shahbaz K, Fahad S, Manzoor H, Saddam H, Wajid N (2017): Application of CSM-CROPGRO-Cotton model for cultivars and optimum planting dates: Evaluation in changing semi-arid climate. *Field Crops Research*. doi: 10.1016/j.fcr.2017.07.007.

Hafiz MH, Abdul K, Farhat A, Wajid F, Fahad S, Muhammad A, Ghulam MS, Wajid N, Muhammad M, Hafiz FB (2020b): Comparative effects of organic and inorganic fertilizers on soil organic carbon and wheat productivity under arid region. *Communications in Soil Science and Plant Analysis*. doi: 10.1080/00103624.2020.1763385.

Hafiz MH, Farhat A, Ashfaq A, Hafiz FB, Wajid F, Carol Jo W, Fahad S, Gerrit H (2020a): Predicting kernel growth of maize under controlled water and nitrogen applications. *International Journal of Plant Production*. doi: 10.1007/s42106-020-00110-8.

Hafiz MH, Farhat A, Shafqat S, Fahad S, Artemi C, Wajid F, Chaves CB, Wajid N, Muhammad M, Hafiz FB (2018): Offsetting land degradation through nitrogen and water management during maize cultivation under arid conditions. *Land Degradation and Development*, 1–10. doi: 10.1002/ldr.2933.

Hafiz MH, Muhammad A, Farhat A, Hafiz FB, Saeed AQ, Muhammad M, Fahad S, Muhammad A (2019): Environmental factors affecting the frequency of road traffic accidents: A case study of sub-urban area of Pakistan. *Environmental Science and Pollution Research International*. doi: 10.1007/s11356-019-04752-8.

Hafiz MH, Wajid F, Farhat A, Fahad S, Shafqat S, Wajid N, Hafiz FB (2016): Maize plant nitrogen uptake dynamics at limited irrigation water and nitrogen. *Environmental Science and Pollution Research* 24(3), 2549–2557. doi: 10.1007/s11356-016-8031-0.

Hansen H, Grossmann K (2000): Auxin-induced ethylene triggers abscisic acid biosynthesis and growth inhibition. *Plant Physiology* 124(3), 1437–1448.

Harrison MA (2012): Cross-talk between phytohormone signaling pathways under both optimal and stressful environmental conditions. In Khan N., Nazar R., Iqbal N., Anjum N. (eds) *Phytohormones and Abiotic Stress Tolerance in Plants*, pp. 49–76. Springer.

He XJ, Mu RL, Cao WH, Zhang ZG, Zhang JS, Chen SY (2005): AtNAC2, a transcription factor downstream of ethylene and auxin signaling pathways, is involved in salt stress response and lateral root development. *The Plant Journal* 44(6), 903–916.

Hesham FA, Fahad S (2020): Melatonin application enhances biochar efficiency for drought tolerance in maize varieties: modifications in physio-biochemical machinery. *Agronomy Journal* 112(4), 1–22.

Hossain MA, Munemasa S, Uraji M, Nakamura Y, Mori IC, Murata Y (2011): Involvement of endogenous abscisic acid in methyl jasmonate-induced stomatal closure in Arabidopsis. *Plant Physiology* 156(1), 430–438.

Huang D, Wu W, Abrams SR, Cutler AJ (2008): The relationship of drought-related gene expression in Arabidopsis thaliana to hormonal and environmental factors. *Journal of Experimental Botany* 59(11), 2991–3007.

Huang JG, Yang M, Liu P, Yang GD, Wu CA, Zheng CC (2009): Gh: GhDREB1 enhances abiotic stress tolerance, delays GA-mediated development and represses cytokinin signalling in transgenic Arabidopsis. *Plant, Cell and Environment* 32(8), 1132–1145.

Hussain A, Black C, Taylor I, Roberts J (2000): Does an antagonistic relationship between ABA and ethylene mediate shoot growth when tomato (*Lycopersicon esculentum* Mill.) plants encounter compacted soil? *Plant, Cell and Environment* 23(11), 1217–1226.

Hussain MA, Fahad S, Rahat S, Muhammad FJ, Muhammad M, Qasid A, Ali A, Husain A, Nooral A, Babatope SA, Changbao S, Liya G, Ibrar A, Zhanmei J, Juncai H (2020): Multifunctional role of brassinosteroid and its analogues in plants. *Plant Growth Regulation*. doi: 10.1007/s10725-020-00647-8.

Hyun Y, Richter R, Vincent C, Martinez-Gallegos R, Porri A, Coupland G (2016): Multi-layered regulation of SPL15 and cooperation with SOC1 integrate endogenous flowering pathways at the Arabidopsis shoot meristem. *Developmental Cell* 37(3), 254–266.

Ibrar K, Aneela R, Khola Z, Urooba N, Sana B, Rabia S, Ishtiaq H, Mujaddad Ur Rehman, Salvatore M (2020): Microbes and environment: Global warming reverting the frozen zombies. In Fahad S, Hasanuzzaman M, Alam M, Ullah H, Saeed M, Khan AK, Adnan M (Eds.), *Environment, Climate, Plant and Vegetation Growth*, pp. 607–634. Springer Publ Ltd, Springer Nature, Switzerland. doi: 10.1007/978-3-030-49732-3.

Ilyas M, Mohammad N, Nadeem K, Ali H, Aamir HK, Kashif H, Fahad S, Aziz K, Abid U (2020): Drought tolerance strategies in plants: A mechanistic approach. *Journal of Plant Growth Regulation*. doi: 10.1007/s00344-020-10174-5.

Iqbal A, Fahad S, Iqbal M, Alamzeb M, Ahmad A, Anwar S, Khan AA, Arif M, Saeed M, Song M (2020): Special adaptive features of plant species in response to drought. In *Salt and Drought Stress Tolerance in Plants*, pp. 77–118. Springer.

Iqbal M, Ashraf M, Jamil A, Ur-Rehman S (2006): Does seed priming induce changes in the levels of some endogenous plant hormones in hexaploid wheat plants under salt stress? *Journal of Integrative Plant Biology* 48(2), 181–189.

Iqra M, Amna B, Shakeel I, Fatima K, Sehrish L, Hamza A, Fahad S (2020): Carbon cycle in response to global warming. In Fahad S, Hasanuzzaman M, Alam M, Ullah H, Saeed M, Khan AK, Adnan M (Eds.), *Environment, Climate, Plant and Vegetation Growth*, pp. 1–16. Springer Publ Ltd, Springer Nature, Switzerland. doi: 10.1007/978-3-030-49732-3.

Iwamoto M, Kiyota S, Hanada A, Yamaguchi S, Takano M (2011): The multiple contributions of phytochromes to the control of internode elongation in rice. *Plant Physiology* 157(3), 1187–1195.

Jan M, Muhammad Anwar-ul-Haq, Adnan NS, Muhammad Y, Javaid I, Xiuling L, Depeng W, Fahad S (2019): Modulation in growth, gas exchange, and antioxidant activities of salt-stressed rice (*Oryza sativa* L.) genotypes by zinc fertilization. *Arabian Journal of Geosciences* 12(24), 775. doi: 10.1007/s12517-019-4939-2.

Jayakannan M, Bose J, Babourina O, Shabala S, Massart A, Poschenrieder C, Rengel Z (2015): The NPR1-dependent salicylic acid signalling pathway is pivotal for enhanced salt and oxidative stress tolerance in Arabidopsis. *Journal of Experimental Botany* 66(7), 1865–1875.

Jeon J, Kim NY, Kim S, Kang NY, Novák O, Ku S-J, Cho C, Lee DJ, Lee E-J, Strnad M, Kim J (2010): A subset of cytokinin two-component signaling system plays a role in cold temperature stress response in Arabidopsis. *Journal of Biological Chemistry* 285(30), 23371–23386.

Jiang C, Fu X (2007): GA action: Turning on de-DELLA repressing signaling. *Current Opinion in Plant Biology* 10(5), 461–465.

Kamarn M, Wenwen C, Irshad A, Xiangping M, Xudong Z, Wennan S, Junzhi C, Shakeel A, Fahad S, Qingfang H, Tiening L (2017): Effect of paclobutrazol, a potential growth regulator on stalk mechanical strength, lignin accumulation and its relation with lodging resistance of maize. *Plant Growth Regulation* 84, 317–332. doi: 10.1007/ s10725-017-0342-8.

Kefu Z, Munns R, King R (1991): Abscisic acid levels in NaCl-treated barley, cotton and saltbush. *Functional Plant Biology* 18(1), 17–24.

Knight H, Knight MR (2001): Abiotic stress signalling pathways: Specificity and cross-talk. *Trends in Plant Science* 6(6), 262–267.

Kodaira K-S, Qin F, Tran L-SP, Maruyama K, Kidokoro S, Fujita Y, Shinozaki K, Yamaguchi-Shinozaki K (2011): Arabidopsis Cys2/His2 zinc-finger proteins AZF1 and AZF2 negatively regulate abscisic acid-repressive and auxin-inducible genes under abiotic stress conditions. *Plant Physiology* 157(2), 742–756.

Kohli A, Sreenivasulu N, Lakshmanan P, Kumar PP (2013): The phytohormone crosstalk paradigm takes center stage in understanding how plants respond to abiotic stresses. *Plant Cell Reports* 32(7), 945–957.

Kumar MN, Verslues PE (2015): Stress physiology functions of the Arabidopsis histidine kinase cytokinin receptors. *Physiologia Plantarum* 154(3), 369–380.

Lata C, Prasad M (2011): Role of DREBs in regulation of abiotic stress responses in plants. *Journal of Experimental Botany* 62(14), 4731–4748.

Li W, Herrera-Estrella L, Tran L-SP (2016): The Yin–Yang of cytokinin homeostasis and drought acclimation/adaptation. *Trends in Plant Science* 21(7), 548–550.

Li Z, Zhang L, Yu Y, Quan R, Zhang Z, Zhang H, Huang R (2011): The ethylene response factor AtERF11 that is transcriptionally modulated by the bZIP transcription factor HY5 is a crucial repressor for ethylene biosynthesis in Arabidopsis. *The Plant Journal* 68(1), 88–99.

Liu X, Zhang H, Zhao Y, Feng Z, Li Q, Yang H-Q, Luan S, Li J, He Z-H (2013): Auxin controls seed dormancy through stimulation of abscisic acid signaling by inducing ARF-mediated ABI3 activation in Arabidopsis. *Proceedings of the National Academy of Sciences of the United States of America* 110(38), 15485–15490.

Liu X, Hu P, Huang M, Tang Y, Li Y, Li L, Hou X (2016): The NF-YC–RGL2 module integrates GA and ABA signalling to regulate seed germination in Arabidopsis. *Nature Communications* 7, 1–14.

Luo X, Chen Z, Gao J, Gong Z (2014): Abscisic acid inhibits root growth in Arabidopsis through ethylene biosynthesis. *The Plant Journal* 79(1), 44–55.

Ma B, Yin C-C, He S-J, Lu X, Zhang W-K, Lu T-G, Chen S-Y, Zhang J-S (2014): Ethylene-induced inhibition of root growth requires abscisic acid function in rice (Oryza sativa L.) seedlings. *PLoS Genetics* 10(10), e1004701.

Magome H, Yamaguchi S, Hanada A, Kamiya Y, Oda K (2008): The DDF1 transcriptional activator upregulates expression of a gibberellin-deactivating gene, GA2ox7, under high-salinity stress in Arabidopsis. *The Plant Journal* 56(4), 613–626.

Mahajan S, Tuteja N (2005): Cold, salinity and drought stresses: An overview. *Archives of Biochemistry and Biophysics* 444(2), 139–158.

Mahar A, Amjad A, Altaf HL, Fazli W, Ronghua L, Muhammad A, Fahad S, Muhammad A, Rafiullah Imtiaz AK, Zengqiang Z (2020): Promising technologies for Cd-contaminated soils: Drawbacks and possibilities. In Fahad S, Hasanuzzaman M, Alam M, Ullah H, Saeed M, Khan AK, Adnan M (Eds.), *Environment, Climate, Plant and Vegetation Growth*, pp. 63–92. Springer Publ Ltd, Springer Nature, Switzerland. doi: 10.1007/978-3-030-49732-3.

Matilla A, Matilla-Vázquez M (2008): Involvement of ethylene in seed physiology. *Plant Science* 175(1–2), 87–97.

Md Jakir H, Allah B (2020): Development and applications of transplastomic plants; a way towards eco-friendly agriculture. In Fahad S, Hasanuzzaman M, Alam M, Ullah H, Saeed M, Khan AK, Adnan M (Eds.), *Environment, Climate, Plant and Vegetation Growth*, pp. 285–322. Springer Publ Ltd, Springer Nature, Switzerland. doi: 10.1007/978-3-030-49732-3.

Md. Enamul H, Shoeb AZM, Mallik AH, Fahad S, Kamruzzaman MM, Akib J, Nayyer S, Mehedi KM, Swati AS, Md Yeamin A, Most SS (2020): Measuring vulnerability to environmental hazards: Qualitative to quantitative. In Fahad S, Hasanuzzaman M, Alam M, Ullah H, Saeed M, Khan AK, Adnan M (Eds.), *Environment, Climate, Plant and Vegetation Growth*, pp. 421–452. Springer Publ Ltd, Springer Nature, Switzerland. doi: 10.1007/978-3-030-49732-3.

Mittler R (2006): Abiotic stress, the field environment and stress combination. *Trends in Plant Science* 11(1), 15–19.

Mohammed A (2007): Physiological aspects of mungbean plant (Vigna radiata L. Wilczek) in response to salt stress and gibberellic acid treatment. *Research Journal of Agriculture and Biological Sciences* 3, 200–213.

Mohammad I, Al-Wabel AS, Munir A, Adel RAU, Mutair A, and Muhammad Imran Rafique (2020a): Advances in pyrolytic technologies with improved carbon capture and storage to combat climate change. In Fahad S, Hasanuzzaman M, Alam M, Ullah H, Saeed M, Khan AK, Adnan M (Eds.), *Environment, Climate, Plant and Vegetation Growth*, pp. 535–576. Springer Publ Ltd, Springer Nature, Switzerland. doi: 10.1007/978-3-030-49732-3.

Mohammad I, Al-Wabel AS, Munir A, Khalid E, Adel RAU (2020b): Extent of climate change in Saudi Arabia and its impacts on agriculture: A case study from Qassim region. In Fahad S, Hasanuzzaman M, Alam M, Ullah H, Saeed M, Khan AK, Adnan M (Eds.), *Environment, Climate, Plant and Vegetation Growth*, pp. 635–658. Springer Publ Ltd, Springer Nature, Switzerland. doi: 10.1007/978-3-030-49732-3.

Monroe-Augustus M, Zolman BK, Bartel B (2003): IBR5, a dual-specificity phosphatase-like protein modulating auxin and abscisic acid responsiveness in Arabidopsis. *The Plant Cell* 15(12), 2979–2991.

Mosher S, Moeder W, Nishimura N, Jikumaru Y, Joo S-H, Urquhart W, Klessig DF, Kim S-K, Nambara E, Yoshioka K (2010): The lesion-mimic mutant cpr22 shows alterations in abscisic acid signaling and abscisic acid insensitivity in a salicylic acid-dependent manner. *Plant Physiology* 152(4), 1901–1913.

Movahedi S, Tabatabaei BS, Alizade H, Ghobadi C, Yamchi A, Khaksar G (2012): Constitutive expression of Arabidopsis DREB1B in transgenic potato enhances drought and freezing tolerance. *Biologia Plantarum* 56(1), 37–42.

Mubeen M, Ashfaq A, Hafiz MH, Muhammad A, Hafiz UF, Mazhar S, Muhammad Sami ul Din, Asad A, Amjed A, Fahad S, Wajid N (2020): Evaluating the climate change impact on water use efficiency of cotton-wheat in semi-arid conditions using DSSAT model. *Journal of Water and Climate Change.* doi: 10.2166/wcc.2019.179/622035/jwc2019179.pdf.

Muday GK, Rahman A, Binder BM (2012): Auxin and ethylene: Collaborators or competitors? *Trends in Plant Science* 17(4), 181–195.

Muhammad Z, Abdul MK, Abdul MS, Kenneth BM, Muhammad S, Shahen S, Ibadullah J, Fahad S (2019): Performance of *Aeluropus lagopoides* (mangrove grass) ecotypes, a potential turfgrass, under high saline conditions. *Environmental Science and Pollution Research International.* doi: 10.1007/s11356-019-04838-3.

Muhammad Tahir ul Q, Amna F, Amna B, Barira Z, Xitong Z, Ling-Ling C (2020): Effectiveness of conventional crop improvement strategies vs. omics. In Fahad S, Hasanuzzaman M, Alam M, Ullah H, Saeed M, Khan AK, Adnan M (Eds.), *Environment, Climate, Plant and Vegetation Growth*, pp. 253–284. Springer Publ Ltd, Springer Nature, Switzerland. doi: 10.1007/978-3-030-49732-3.

Munemasa S, Oda K, Watanabe-Sugimoto M, Nakamura Y, Shimoishi Y, Murata Y (2007): The coronatine-insensitive 1 mutation reveals the hormonal signaling interaction between abscisic acid and methyl jasmonate in Arabidopsis guard cells. Specific impairment of ion channel activation and second messenger production. *Plant Physiology* 143(3), 1398–1407.

Munemasa S, Hossain MA, Nakamura Y, Mori IC, Murata Y (2011): The Arabidopsis calcium-dependent protein kinase, CPK6, functions as a positive regulator of methyl jasmonate signaling in guard cells. *Plant Physiology* 155(1), 553–561.

Munné-Bosch S, Müller M (2013): Hormonal cross-talk in plant development and stress responses. *Frontiers in Plant Science* 4, 529.

Munné-Bosch S, Queval G, Foyer CH (2013): The impact of global change factors on redox signaling underpinning stress tolerance. *Plant Physiology* 161(1), 5–19.

Neill S, Barros R, Bright J, Desikan R, Hancock J, Harrison J, Morris P, Ribeiro D, Wilson I (2008): Nitric oxide, stomatal closure, and abiotic stress. *Journal of Experimental Botany* 59(2), 165–176.

Nguyen D, Rieu I, Mariani C, van Dam NM (2016): How plants handle multiple stresses: Hormonal interactions underlying responses to abiotic stress and insect herbivory. *Plant Molecular Biology* 91(6), 727–740.

Nishiyama R, Watanabe Y, Fujita Y, Le DT, Kojima M, Werner T, Vankova R, Yamaguchi-Shinozaki K, Shinozaki K, Kakimoto T, Sakakibara H, Schmülling T, Tran LS (2011): Analysis of cytokinin mutants and regulation of cytokinin metabolic genes reveals important regulatory roles of cytokinins in drought, salt and abscisic acid responses, and abscisic acid biosynthesis. *The Plant Cell* 23(6), 2169–2183.

Noor M, Naveed ur R, Ajmal J, Fahad S, Muhammad A, Fazli W, Saud S, Hassan S (2020): Climate change and costal plant lives. In Fahad S, Hasanuzzaman M, Alam M, Ullah H, Saeed M, Khan AK, Adnan M (Eds.), *Environment, Climate, Plant and Vegetation Growth*, pp. 93–108. Springer Publ Ltd, Springer Nature, Switzerland. doi: 10.1007/978-3-030-49732-3.

Nuruzzaman M, Sharoni AM, Kikuchi S (2013): Roles of NAC transcription factors in the regulation of biotic and abiotic stress responses in plants. *Frontiers in Microbiology* 4, 248.

Peng J, Harberd NP (2002): The role of GA-mediated signalling in the control of seed germination. *Current Opinion in Plant Biology* 5(5), 376–381.

Pieterse CM, Leon-Reyes A, Van der Ent S, Van Wees SC (2009): Networking by small-molecule hormones in plant immunity. *Nature Chemical Biology* 5(5), 308–316.

Pospíšilová J (2003): Interaction of cytokinins and abscisic acid during regulation of stomatal opening in bean leaves. *Photosynthetica* 41(1), 49–56.

Qamar-uz Z, Zubair A, Muhammad Y, Muhammad ZI, Abdul K, Fahad S, Safder B, Ramzani PMA, Muhammad N (2017) Zinc biofortification in rice: Leveraging agriculture to moderate hidden hunger in developing countries. *Archives of Agronomy and Soil Science* 64, 147–161. doi: 10.1080/03650340.2017.1338343.

Rahman IU, Ali S, Adnan M, Ibrahim M, Saleem N, Khan IA (2016): Effect of seed priming on growth parameters of okra (*Abelmoschus esculentus* L.). *Pure and Applied Biology* 5(1), 165–171.

Rashid M, Qaiser H, Khalid SK, Mohammad I, Al-Wabel Zhang A, Muhammad A, Shahzada SI, Rukhsanda A, Ghulam AS, Shahzada MM, Sarosh A, Muhammad FQ (2020): Prospects of biochar in alkaline soils to mitigate climate change. In Fahad S, Hasanuzzaman M, Alam M, Ullah H, Saeed M, Khan AK, Adnan M (Eds.), *Environment, Climate, Plant and Vegetation Growth*, pp. 133–150. Springer Publ Ltd, Springer Nature, Switzerland. doi: 10.1007/978-3-030-49732-3.

Rehman M, Fahad S, Saleem MH, Hafeez M, Muhammad Habib ur Rahman, Liu F, Deng G (2020): Red light optimized physiological traits and enhanced the growth of ramie (Boehmeria nivea L.). *Photosynthetica* 58(4), 922–931.

Roychoudhury A, Paul A (2012): Abscisic acid-inducible genes during salinity and drought stress. *Advances in Medicine and Biology* 51, 1–78.

Sadam M, Muhammad Tahir ul Qamar, Ghulam M, Muhammad SK, Faiz AJ (2020): Role of biotechnology in climate resilient agriculture. In Fahad S, Hasanuzzaman M, Alam M, Ullah H, Saeed M, Khan AK, Adnan M (Eds.), *Environment, Climate, Plant and Vegetation Growth*, pp. 339–366. Springer Publ Ltd, Springer Nature, Switzerland. doi: 10.1007/978-3-030-49732-3.

Sajid H, Jie H, Jing H, Shakeel A, Satyabrata N, Sumera A, Awais S, Chunquan Z, Lianfeng Z, Xiaochuang C, Qianyu J, Junhua Z (2020): Rice production under climate change: Adaptations and mitigating strategies. In Fahad S, Hasanuzzaman M, Alam M, Ullah H, Saeed M, Khan AK, Adnan M (Eds.), *Environment, Climate, Plant and Vegetation Growth*, pp. 659–686. Springer Publ Ltd, Springer Nature, Switzerland. doi: 10.1007/978-3-030-49732-3.

Sajjad H, Muhammad M, Ashfaq A, Waseem A, Hafiz MH, Mazhar A, Nasir M, Asad A, Hafiz UF, Syeda RS, Fahad S, Depeng W, Wajid N (2019): Using GIS tools to detect the land use/land cover changes during forty years in Lodhran district of Pakistan. *Environmental Science and Pollution Research.* doi: 10.1007/s11356-019-06072-3.

Saleem MH, Fahad S, Adnan M, Mohsin A, Muhammad SR, Muhammad K, Qurban A, Inas AH, Parashuram B, Mubassir A, Reem MH (2020a): Foliar application of gibberellic acid endorsed phytoextraction of copper and alleviates oxidative stress in jute (*Corchorus capsularis* L.) plant grown in highly copper-contaminated soil of China. *Environmental Science and Pollution Research International.* doi: 10.1007/s11356-020-09764-3.

Saleem MH, Fahad S, Shahid UK, Mairaj D, Abid U, Ayman ELS, Akbar H, Analía L, Lijun L (2020c): Copper-induced oxidative stress, initiation of antioxidants and phytoremediation potential of flax (*Linum usitatissimum* L.) seedlings grown under the mixing of two different soils of China. *Environmental Science and Pollution Research International.* doi: 10.1007/s11356-019-07264-7.

Saleem MH, Rehman M, Fahad S, Tung SA, Iqbal N, Hassan A, Ayub A, Wahid MA, Shaukat S, Liu L, Deng G (2020b): Leaf gas exchange, oxidative stress, and physiological attributes of rapeseed (Brassica napus L.) grown under different light-emitting diodes. *Photosynthetica* 58(3), 836–845.

Saman S, Amna B, Bani A, Muhammad Tahir ul Q, Rana MA, Muhammad SK (2020): QTL mapping for abiotic stresses in cereals. In Fahad S, Hasanuzzaman M, Alam M, Ullah H, Saeed M, Khan AK, Adnan M (Ed.), *Environment, Climate, Plant and Vegetation Growth*, pp. 229–252. Springer Publ Ltd, Springer Nature, Switzerland. doi: 10.1007/978-3-030-49732-3.

Saud S, Chen Y, Fahad S, Hussain S, Na L, Xin L, Alhussien SA (2016): Silicate application increases the photosynthesis and its associated metabolic activities in Kentucky bluegrass under drought stress and post-drought recovery. *Environmental Science and Pollution Research International* 23(17), 17647–17655. doi: 10.1007/s11356-016-6957-x.

Saud S, Chen Y, Long B, Fahad S, Sadiq A (2013): The different impact on the growth of cool season turf grass under the various conditions on salinity and drought stress. *International Journal of Agricultural Science and Research* 3, 77–84.

Saud S, Fahad S, Cui G, Chen Y, Anwar S (2020): Determining nitrogen isotopes discrimination under drought stress on enzymatic activities, nitrogen isotope abundance and water contents of Kentucky bluegrass. *Scientific Reports* 10(1), 6415. doi: 10.1038/s41598-020-63548-w.

Saud S, Fahad S, Yajun C, Ihsan MZ, Hammad HM, Nasim W, Amanullah Jr Arif M, Alharby H (2017): Effects of nitrogen supply on water stress and recovery mechanisms in Kentucky bluegrass plants. *Frontiers in Plant Science* 8, 983. doi: 10.3389/fpls.2017.00983.

Saud S, Li X, Chen Y, Zhang L, Fahad S, Hussain S, Sadiq A, Chen Y (2014): Silicon application increases drought tolerance of Kentucky bluegrass by improving plant water relations and morph physiological functions. *Scientific World Journal* 2014, 1–10. doi: 10.1155/2014/ 368694.

Senol C (2020): The effects of climate change on human behaviors. In Fahad S, Hasanuzzaman M, Alam M, Ullah H, Saeed M, Khan AK, Adnan M (Eds.), *Environment, Climate, Plant and Vegetation Growth*, pp. 577–590. Springer Publ Ltd, Springer Nature, Switzerland. doi: 10.1007/978-3-030-49732-3.

Seo PJ, Xiang F, Qiao M, Park J-Y, Lee YN, Kim S-G, Lee Y-H, Park WJ, Park C-M (2009): The MYB96 transcription factor mediates abscisic acid signaling during drought stress response in Arabidopsis. *Plant Physiology* 151(1), 275–289.

Shafi MI, Adnan M, Fahad S, Fazli W, Ahsan K, Zhen Y, Subhan D, Zafar-ul-Hye M, Martin B, Rahul D (2020): Application of single superphosphate with humic acid improves the growth, yield and phosphorus uptake of wheat (Triticum aestivum L.) in calcareous soil. *Agronomy* 10(9), 1224. doi: 10.3390/agronomy10091224.

Shah F, Lixiao N, Kehui C, Tariq S, Wei W, Chang C, Liyang Z, Farhan A, Fahad S, Huang J (2013): Rice grain yield and component responses to near 2°C of warming. *Field Crops Research* 157, 98–110.

Shan W, Kuang JF, Lu WJ, Chen JY (2014): Banana fruit NAC transcription factor MaNAC 1 is a direct target of MaICE 1 and involved in cold stress through interacting with MaCBF. *Plant, Cell and Environment* 37 1, 2116–2127.

Sharp RE, LeNoble ME (2002): ABA, ethylene and the control of shoot and root growth under water stress. *Journal of Experimental Botany* 53(366), 33–37.

Shigenaga AM, Argueso CT (2016): No hormone to rule them all: Interactions of plant hormones during the responses of plants to pathogens. *Seminars in Cell and Developmental Biology*. Elsevier, 174–189.

Shu K, Chen Q, Wu Y, Liu R, Zhang H, Wang P, Li Y, Wang S, Tang S, Liu C, Yang W, Cao X, Serino G, Xie Q (2016): ABI 4 mediates antagonistic effects of abscisic acid and gibberellins at transcript and protein levels. *The Plant Journal* 85(3), 348–361.

Shu K, Zhou W, Yang W (2018): APETALA 2-domain-containing transcription factors: Focusing on abscisic acid and gibberellins antagonism. *New Phytologist* 217(3), 977–983.

Singh S, Tripathi DK, Singh S, Sharma S, Dubey NK, Chauhan DK, Vaculík M (2017): Toxicity of aluminium on various levels of plant cells and organism: A review. *Environmental and Experimental Botany* 137, 177–193.

Skubacz A, Daszkowska-Golec A, Szarejko I (2016): The role and regulation of ABI5 (ABA-Insensitive 5) in plant development, abiotic stress responses and phytohormone crosstalk. *Frontiers in Plant Science* 7, 1884.

Subhan D, Zafar-ul-Hye M, Fahad S, Saud S, Martin B, Tereza H, Rahul D (2020): Drought stress alleviation by ACC deaminase producing *Achromobacter xylosoxidans* and *Enterobacter cloacae*, with and without timber waste biochar in maize. *Sustainability* 12(6286). doi: 10.3390/su12156286.

Sun J, Li C (2014): Cross talk of signaling pathways between ABA and other phytohormones. In *Abscisic Acid: Metabolism, Transport and Signaling*, pp. 243–253. Springer.

Sun P, Tian Q-Y, Chen J, Zhang W-H (2010): Aluminium-induced inhibition of root elongation in Arabidopsis is mediated by ethylene and auxin. *Journal of Experimental Botany* 61(2), 347–356.

Suzuki N, Rivero RM, Shulaev V, Blumwald E, Mittler R (2014): Abiotic and biotic stress combinations. *New Phytologist* 203(1), 32–43.

Tariq M, Ahmad S, Fahad S, Abbas G, Hussain S, Fatima Z, Nasim W, Mubeen M, ur Rehman MH, Khan MA, Adnan M (2018): The impact of climate warming and crop management on phenology of sunflower-based cropping systems in Punjab, Pakistan. *Agriculture and Forest Meteorology* 15(256), 270–282.

Thomashow MF (2010): Molecular basis of plant cold acclimation: Insights gained from studying the CBF cold response pathway. *Plant Physiology* 154(2), 571–577.

Tran L-SP, Urao T, Qin F, Maruyama K, Kakimoto T, Shinozaki K, Yamaguchi-Shinozaki K (2007): Functional analysis of AHK1/ATHK1 and cytokinin receptor histidine kinases in response to abscisic acid, drought, and salt stress in Arabidopsis. *Proceedings of the National Academy of Sciences of the United States of America* 104(51), 20623–20628.

Tuteja N, Sopory SK (2008): Chemical signaling under abiotic stress environment in plants. *Plant Signaling and Behavior* 3(8), 525–536.

Unsar Naeem U, Muhammad R, Syed HMB, Asad S, Mirza AQ, Naeem I, Muhammad H ur R, Fahad S, Shafqat S (2020): Insect pests of cotton crop and management under climate change scenarios. In Fahad S, Hasanuzzaman M, Alam M, Ullah H, Saeed M, Khan AK, Adnan M (Eds.), *Environment, Climate, Plant and Vegetation Growth*, pp. 367–396. Springer Publ Ltd, Springer Nature, Switzerland. doi: 10.1007/978-3-030-49732-3.

Verma V, Ravindran P, Kumar PP (2016): Plant hormone-mediated regulation of stress responses. *BMC Plant Biology* 16, 86.

Verslues PE (2016): ABA and cytokinins: Challenge and opportunity for plant stress research. *Plant Molecular Biology* 91(6), 629–640.

Vos IA, Moritz L, Pieterse CM, Van Wees S (2015): Impact of hormonal crosstalk on plant resistance and fitness under multi-attacker conditions. *Frontiers in Plant Science* 6, 639.

Wahid F, Fahad S, Subhan D, Adnan M, Zhen Y, Saud S, Manzer HS, Martin B, Tereza H, Rahul D (2020): Sustainable management with mycorrhizae and phosphate solubilizing bacteria for enhanced phosphorus uptake in calcareous soils. *Agriculturists* 10(334). doi: 10.3390/agriculture10080334.

Wajid N, Ashfaq A, Asad A, Muhammad T, Muhammad A, Muhammad S, Khawar J, Ghulam MS, Syeda RS, Hafiz MH, Muhammad IAR, Muhammad ZH, Muhammad Habib ur R, Veysel T, Fahad S, Suad S, Aziz K, Shahzad A (2017): Radiation efficiency and nitrogen fertilizer impacts on sunflower crop in contrasting environments of Punjab. *Pakistan Environmental Science and Pollution Research International* 25, 1822–1836. doi: 10.1007/s11356-017-0592-z.

Walker M, Dumbroff E (1981): Effects of salt stress on abscisic acid and cytokinin levels in tomato. *Zeitschrift für Pflanzenphysiologie* 101(5), 461–470.

Wang B, Wei H, Xue Z, Zhang WH (2017): The role of gibberellin in iron homeostasis in rice. *Annals of Botany* 119(6), 945–956.

Wilson RL, Kim H, Bakshi A, Binder BM (2014): The ethylene receptors ethylene RESPONSE1 and ethylene RESPONSE2 have contrasting roles in seed germination of Arabidopsis during salt stress. *Plant Physiology* 165(3), 1353–1366.

Wu C, Kehui C, She T, Ganghua L, Shaohua W, Fahad S, Lixiao N, Jianliang H, Shaobing P, Yanfeng D (2020): Intensified pollination and fertilization ameliorate heat injury in rice (Oryza sativa L.) during the flowering stage. *Field Crops Research* 252. http://www.ncbi.nlm.nih.gov/pubmed/107795.

Wu C, Tang S, Li G, Wang S, Fahad S, Ding Y (2019): Roles of phytohormone changes in the grain yield of rice plants exposed to heat: A review. *PeerJ* 7, e7792. doi: 10.7717/peerj.7792.

Wu L, Chen X, Ren H, Zhang Z, Zhang H, Wang J, Wang X-C, Huang R (2007): ERF protein JERF1 that transcriptionally modulates the expression of abscisic acid biosynthesis-related gene enhances the tolerance under salinity and cold in tobacco. *Planta* 226(4), 815–825.

Wu X, He J, Chen J, Yang S, Zha D (2014): Alleviation of exogenous 6-benzyladenine on two genotypes of eggplant (Solanum melongena Mill.) growth under salt stress. *Protoplasma* 251(1), 169–176.

Xu W, Jia L, Shi W, Liang J, Zhou F, Li Q, Zhang J (2013): Abscisic acid accumulation modulates auxin transport in the root tip to enhance proton secretion for maintaining root growth under moderate water stress. *New Phytologist* 197(1), 139–150.

Yang Z, Zhang Z, Zhang T, Fahad S, Cui K, Nie L, Peng S, Huang J (2017): The effect of season-long temperature increases on rice cultivars grown in the central and southern regions of China. *Frontiers in Plant Science* 8, 1908. doi: 10.3389/fpls.2017.01908.

Yasuda M, Ishikawa A, Jikumaru Y, Seki M, Umezawa T, Asami T, Maruyama-Nakashita A, Kudo T, Shinozaki K, Yoshida S, Nakashita H (2008): Antagonistic interaction between systemic acquired resistance and the abscisic acid–mediated abiotic stress response in Arabidopsis. *The Plant Cell* 20(6), 1678–1692.

Zafar-ul-Hye M, Muhammad N, Subhan D, Fahad S, Rahul D, Mazhar A, Ashfaq AR, Martin B, Jiˇrí H, Zahid HT, Muhammad N (2020a): Alleviation of cadmium adverse effects by improving nutrients uptake in bitter gourd through cadmium tolerant rhizobacteria. *Environmentalist* 7(54). doi: 10.3390/environments7080054.

Zafar-ul-Hye M, Muhammad T, Ahzeeb-ul-Hassan, Muhammad A, Fahad S, Martin B, Tereza D, Rahul D, Subhan D (2020b): Potential role of compost mixed biochar with rhizobacteria in mitigating lead toxicity in spinach. *Scientific Reports* 10, 12159; doi: 10.1038/s41598-020-69183-9.

Zahida Z, Hafiz FB, Zulfiqar AS, Ghulam MS, Fahad S, Muhammad RA, Hafiz MH, Wajid N, Muhammad S (2017): Effect of water management and silicon on germination, growth, phosphorus and arsenic uptake in rice. *Ecotoxicology and Environment Safety* 144, 11–18.

Zhang M, Yuan B, Leng P (2009): The role of ABA in triggering ethylene biosynthesis and ripening of tomato fruit. *Journal of Experimental Botany* 60(6), 1579–1588.

Zhu Y, Wang B, Tang K, Hsu C-C, Xie S, Du H, Yang Y, Tao WA, Zhu J-K (2017): An Arabidopsis Nucleoporin NUP85 modulates plant responses to ABA and salt stress. *PLoS Genetics* 13(12), e1007124.

Zia-ur-Rehman M (2020): Environment, climate change and biodiversity. In Fahad S, Hasanuzzaman M, Alam M, Ullah H, Saeed M, Khan AK, Adnan M (Eds.), *Environment, Climate, Plant and Vegetation Growth*, pp. 473–502. Springer Publ Ltd, Springer Nature, Switzerland. doi: 10.1007/978-3-030-49732-3.

8

Salicylic Acid: Its Role in Temperature Stress Tolerance

Nosheen Khalid, Imran Khan, Shehla Sammi, Inam-u-llah, Muhammad Liaquat, and Muhammad Jahangir

CONTENTS

8.1 Introduction

Salicylic acid (SA), also known as 2-hydroxy benzoic acid, belongs to the phenolic group of phytochemicals which acts as an endogenous growth regulator in plants. It is comprised of a hydroxyl group as its functional derivative, connected to an aromatic ring (Dempsey et al. 2011b). SA is commonly synthesised by plants. It was thought to be a nonessential compound for a long time (Hadacek et al. 2010); however, this concept was rejected, as phenolic compounds play a key role in various metabolic progressions in plants, such as allelopathy, disease resistance, and regulation of responses to abiotic stresses (Dempsey et al. 2011a; Malamy and Klessig 1992; Raskin 1992). Recently, SA has also gained attention due to its manipulation role for multiple modes of plant stress resistance (Ding and Wang 2003; Ding et al. 2001). Many factors are involved after exposure of plants to high temperature stress, such as high ambient temperature, effect of soil exposed to infrared radiation absorption from sun on germinating seeds, increased plant transpiration, lower transpiration ability of certain organs of plants, forest fires, natural gas blowout (Sharkey and Schrader 2006; Adnan et al. 2018; Adnan et al. 2019; Akram et al. 2018a,b; Aziz et al. 2017a,b; Habib et al. 2017; Hafiz et al. 2016; Hafiz et al. 2019; Kamaran et al. 2017; Muhammad et al. 2019; Sajjad et al. 2019; Saud et al. 2013; Saud et al. 2014; Saud et al. 2017; Saud et al. 2016; Shah et al. 2013; Saud et al. 2020; Qamar et al. 2017; Wajid et al. 2017; Yang et al. 2017; Zahida et al. 2017; Depeng et al. 2018; Hussain et al. 2020; Hafiz et al. 2020a,b; Shafi et al. 2020; Wahid et al. 2020; Subhan et al. 2020; Zafar-ul-Hye et al. 2020a,b); Adnan et al. 2020; Ilyas et al. 2020; Saleem et al. 2020a,b,c; Rehman 2020; Frahat et al. 2020; Wu et al. 2020; Mubeen et al. 2020; Farhana 2020; Jan et al. 2020; Wu et al. 2019; Ahmad et al. 2019; Baseer et al. 2019; Hafiz et al. 2018; Tariq et al. 2018; Fahad and Bano 2012; Fahad et al. 2017; Fahad et al. 2013; Fahad et al. 2014a,b; Fahad et al. 2016a,b,c,d; Fahad et al. 2015a,b; Fahad et al. 2018; Fahad et al. 2019a,b; Hesham and Fahad 2020. Iqra et al. 2020; Akbar et al. 2020; Mahar et al. 2020; Noor et al. 2020; Bayram et al. 2020; Amanullah et al. 2020; Rashid et al. 2020; Arif et al. 2020; Amir et al. 2020; Saman et al. 2020; Muhammad Tahir et al. 2020; Md Jakirand Allah 2020; Farah et al. 2020; Sadam et al. 2020; Unsar et al. 2020; Fazli et al. 2020; Md.

Enamul et al. 2020; Gopakumar et al. 2020; Zia-ur-Rehman 2020; Ayman et al. 2020; Mohammad I. Al-Wabel et al. 2020a,b; Senol 2020; Amjad et al. 2020; Ibrar et al. 2020; Sajid et al. 2020). Although SA is well studied and understood for its roles in plants under biotic stress, there are a number of reports that support the extensive role of SA for its modulating effects in plants against various abiotic stresses, such as drought, salinity, chilling stress, UV light, and heat shock (Ding and Wang 2003; Ding et al. 2001). It further regulates several aspects of plant defense, growth, development, and has a role in disease resistance and thermogenesis signalling (Anelia 2017; Vlot et al. 2009).

The role of SA as an intrinsic signalling compound is also supported by the fact that it is present in the phloem of flowering and has the ability to induce flowering in Lemna gibba G3 (Cleland and Ajami 1974). However, further research in the area of flowering induction produced conflicting results regarding SA's role, but there remains massive literature regarding the active role of SA in flowering induction (Ali et al. 2016; Adnan et al. 2015; Jin et al. 2008; Martínez et al. 2004). Further, it also regulates heat production (thermogenesis) in the flowers of certain angiosperms as well as the cycad's reproductive structures (Raskin 1992; Vlot et al. 2009). This compound induces alternative oxidase expression to stimulate thermogenesis that leads to an increase in the ability of the mitochondrial alternative respiratory pathway, which produces ATP, and the heat is released as remaining available energy. Furthermore, SA treatment regulates alternative respiratory oxidative expression in non-thermogenic plant species, such as *Arabidopsis* (Clifton et al. 2005) and tobacco (Saleem et al. 2015; Norman et al. 2004). Following the discovery of SA's regulation of thermogenesis, another key function was revealed which was its role in the activation of disease resistance (Tsuda et al. 2008).

For the first time it was revealed in tobacco plants that SA is responsible for disease resistance as an endogenous signalling compound, where SA application on tobacco leaves developed an increased resistance under infestation with Tobacco Mosaic Virus (TMV), which induces pathogenesis-related (PR) protein accretion (White 1979). More evidence came in support of SA's role as a disease resistance-signalling molecule from further analysis of tobacco and *Arabidopsis*. That accrued to contain either minute or no SA, due to production of SA-degrading salicylate hydroxylase as an expression of *nahG* bacterial gene or mutation/alteration of genes which biosynthesised SA. In such conditions, the application of SA on these plants exhibited disease resistance against their virulent and a-virulent pathogenesis (Dewdney et al. 2000; Gaffney et al. 1993; Hao et al. 2019). Interestingly, the salicylate is active in both of its forms, i.e., methyl salicylate (MeSA) and SA, as endogenous signal molecules and plays a key role in the modulation of stress responses and developmental processes in plants (Ding and Wang 2003; Ding et al. 2001).

The chorismate is the end product of the shikimate pathway and acts as a precursor for both isochorismate (IC) and phenylalanine ammonia-lyase (PAL) pathways. Both of these pathways are responsible for the SA biosynthesis in plants (Figure 8.1). The IC is the primary route for SA biosynthesis in *A. thaliana*, while the phenylalanine ammonia-lyase pathway has also been associated with SA biosynthesis in a number of species that play a direct or indirect role in SA production in *Arabidopsis*. However, to date, both biosynthesis pathways have not been fully elucidated (Dempsey et al. 2011b).

8.2 Plant Temperature Stress Response

All living organisms, including higher plants, have the highest rate of development under optimal conditions, including temperature, where at over a diurnal temperature range they become acclimatised (Fitter and Hay 2012). Continuous global environmental changes in the form of both high and low temperatures pose a high abiotic stress to crop plants. Such stresses not only affect plant physiology but also affect biochemical processes in order to cope with the variations and obtain optimal growth conditions (Khan et al. 2013a).

As the surrounding temperature of a plant deviates from the optimum, physiological, metabolic, biochemical, and molecular changes take place within the plants. These changes are important in order to maximise the developmental processes and growth and to sustain cellular homeostasis (Guy et al. 2008). When plants are exposed to continuous stressful conditions, they experience abnormal, reduced, or dysfunctional cellular processes until the cardinal temperatures (T_{min} and T_{max}, i.e., survival temperature

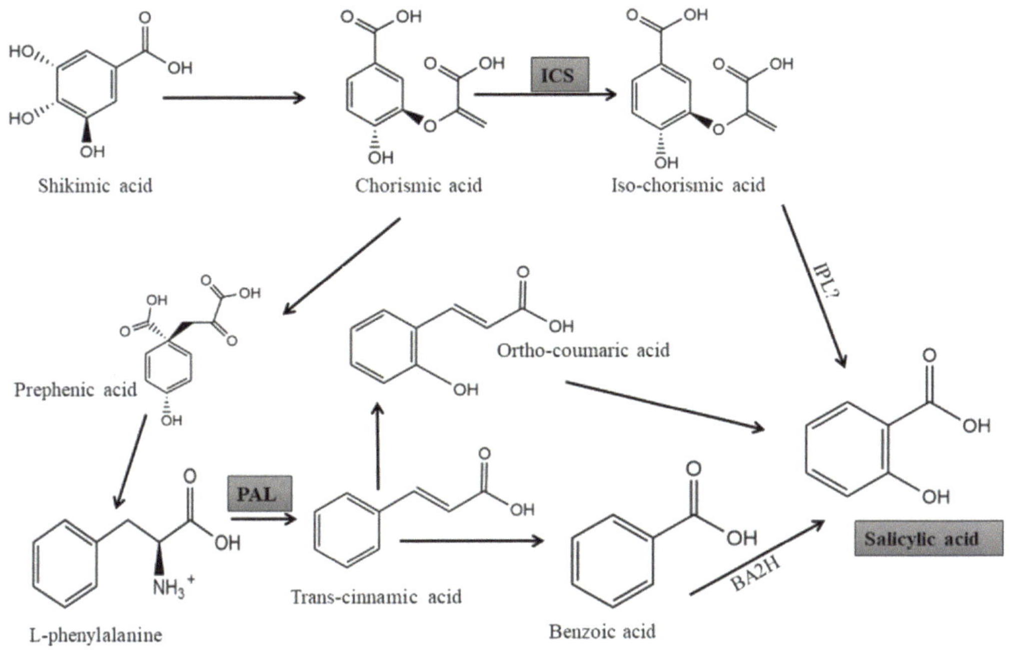

FIGURE 8.1 Biosynthetic pathways (ICS = Isochorismate synthase; PAL = Phenylalanine ammonia-lyase pathway; BA2H = benzoicacid-2-hydroxylase) of salicylic acid in plants.

between minimum and maximum, respectively) are reached (Fitter and Hay 2012). In addition, plant metabolism changes in two different ways which are necessary for survival. Firstly, plant cellular metabolic network must adapt to the new environment in which they experience low or higher temperature on metabolic fluxes affecting overall metabolism. Secondly, changes in metabolism in response to temperature would be those allied with boosted tolerance mechanisms. For example, many of plants' metabolites are believed to have important properties that could participate in triggering thermo-tolerance as an associated stress response (Guy 1990; Levitt 1972; Thomashow 1999).

The progression of high or low temperature stress also depends on its duration, rate of temperature changes, and plant developmental stage. Temperature stresses could be categorised into high, chilling, and freezing temperatures. Higher temperatures stresses could be long-term exposure (>27°C) at vegetative and reproductive stage and/or short-term exposure (37°C) at vegetative stage (Kai and Iba 2014). To cope with this issue, most plants have multiple interconnected metabolic mechanisms, such as antioxidant-including proteins, membrane lipids, or other metabolites, as regulatory factors (Balogh et al. 2013; Kai and Iba 2014). Therefore, it is an area of scientific interest and leads to explorations to investigate underlying molecular mechanisms adopted by plants against adverse temperature conditions. The understanding of these mechanisms will help to develop temperature-tolerant plant varieties for the agriculture sector.

8.3 High Temperature Stress and Salicylic Acid

All the cells of a plant react to high temperature stress. If a plant is exposed to a sudden rise in ambient temperature, usually 5 to 10 °C for a few minutes to a few hours above the normal growth temperature, it produces an elite class of proteins called heat shock proteins (HSPs). These proteins in normal conditions are absent in unstressed plants or present in very small amounts (Rao et al. 2006). The heat shock effects were studied in *Arabidopsis* plant in form of metabolism. It was revealed that carbohydrate and amino acid metabolisms are affected by heat shock and changed various gene expressions resulting in the increase of GABA, alanine, threonine, valine, leucine, isoleucine, and asparagine, which are derived from oxaloacetate and pyruvate (Guy et al. 2008).

In such extreme temperature conditions, it is the role of phytohormones to mediate heat stresses. Among these phytohormones SA is important, in addition to others, i.e., gibberellic acid (GA), abscisic acid (ABA), auxin, cytokinins (CKs), polyamines (PAs), ethylene (ET), brassinosteroids (BRs), jasmonic acid (JA), and nitric oxide (NO) regarding their active role to regulate plant growth-related biological processes under alleviated heat stress (Asensi-Fabado et al. 2013; Clarke et al. 2009; Clarke et al. 2004; Khan et al. 2013b; Lubovská et al. 2014; Song et al. 2013; Zhang and Wang 2011).

As aforementioned, the role of SA in the modulation of various developmental processes in plant under heat stress condition, resulting in cross-talk between chemical-signalling pathways is well-reported (Nazar et al. 2017). It happens in mustard after spraying with SA (Dat et al. 1998) where the endogenous levels of SA and its glucoside in plants increased after exposure to ozone, heat, and UV (Nazar et al. 2017). The SA further regulates the actions of antioxidant enzymes by scavenging reactive oxygen species (ROS) for plant cell protection against oxidative damage and to sustain the overall growth and photosynthetic performance of plant cells under stress (Nazar et al. 2015; Radwan 2012). As evidence, it is observed in field and pot experiments of wheat that the SA treatment as foliar application as well as on seed has improved the chlorophyll content, sugar accumulation, soluble protein, proline, and increased the net yield (Munir and Shabbir 2018). Although the hydroponic application of SA has shown opposite results (Khan et al. 2013b; Nazar et al. 2017), still, in the field, the high temperature stress severely impacted reproductive stage (fertility), grain fillings, and quality in cereals (Jagadish et al. 2010; Kai and Iba 2014).

SA helps in alleviating the adverse effects of heat stress on photosynthesis in wheat (*Triticum aestivum* L.) with a 0.5mM application. It copes with heat stress by an increase in proline formation through the increase in γ-glutamyl kinase (GK) and decrease in proline oxidase (PROX) activity. It results in an improved osmotic pressure for water potential as required for well-maintained and regulated photosynthetic activity. In addition to this, the ET production is also inhibited by SA application in heat-stressed plants to optimal range through the inhibitory effects of 1-aminocyclopropane carboxylic acid (ACC) synthase (ACS) activity (Khan et al. 2013b). Similarly, the foliar spray treatment of 1mM SA on cucumber plants (*Cucumis sativus* L.), induced heat tolerance by way of Cucumis sativus lowering H_2O_2 levels, electrolyte leakage parameter, high catalase activity, lipid peroxide levels, and higher Fv/Fm chlorophyll, a fluorescence value. The catalase activity plays a key role in the removal of H_2O_2 during heat stress, as it was found to be improved by the foliar application of SA. (Shi et al. 2006). Interestingly, the photosynthesis rate in grape leaves was also reported to be increased under heat stress subsequent to the FV application of SA (Wang et al. 2010) and reduced the heat-induced damage by up-regulating the antioxidant system in plants (Wang and Li 2006). In a recent study, the pretreatment (1 mM) of SA was also investigated in tomato (*Solanum lycopersicum* L.) and proved to suppress the adverse effects of heat stress and increased anti-oxidative enzymes and photosynthesis activity (Shah Jahan et al. 2019).

8.4 Chilling Stress and Salicylic Acid

Chilling stress leads to chilling injury with continuous low temperature stress due to the oxidative burst (Figure 8.2). This also commonly refers to low temperature stress (0–15°C) and is capable of affecting plants at all stages of life (Theocharis et al. 2012; Wang et al. 2016; Xia et al. 2018). Each plant has its own capacity to tolerate chilling stress. Plants in different regions behave differently to chilling stress, for example, many tropical and subtropical plants such as maize, tobacco, and rice cannot survive chilling stress while *Arabidopsis* and some overwintering cereals can tolerate chilling stress (Stitt and Hurry 2002; Wang et al. 2016). When plants are exposed to chilling stress, different physiological and cellular changes occur, such as membrane structure alterations, calcium signals, photosynthesis, and metabolism, in order to adapt to chilling stress (Liu et al. 2018; Theocharis et al. 2012). Therefore, tolerance of plants against chilling temperature stress is an important property.

It is well-proven that SA plays an important role in the regulation of plant responses against chilling stress. Regardless of its mode of action, SA treatment has already helped various plant species to build

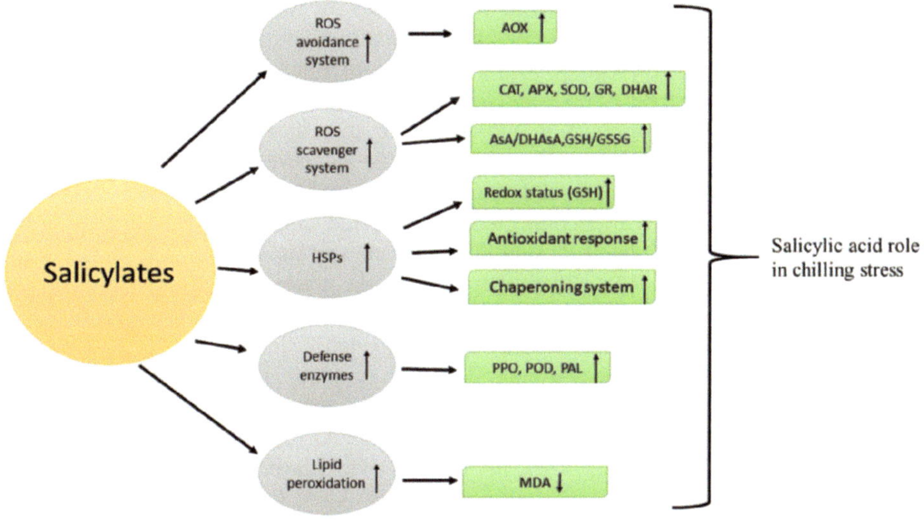

FIGURE 8.2 Role of salicylates during chilling stress environment in plants.

their chilling stress tolerance (Horváth et al. 2007). The SA and acetyl SA at 0.5 mM concentration have shown chilling tolerance capacity in young maize plants in hydroponic conditions (Janda et al. 1999). Although it is well-reported that the hydroponic application of SA increased the chilling tolerance of the shoots, this effect is not evident in the roots of maize, rice, and cucumber (Kang and Saltveit 2002). Plants, such as potato, banana seedlings, tomatoes, and peaches have also shown chilling tolerance after treatment with SA at low concentrations (Ding et al. 2002; Kang et al. 2003; Mora-Herrera et al. 2005; Wang et al. 2006; Arif et al. 2015).

Moreover, exogenous treatment of SA promotes plant chilling tolerance in cucumber by inducing the production of hydrogen sulfide (H_2S) endogenously, which in turn enhanced antioxidant activity while activating the expression of chilling-sensitive genes (Pan et al. 2020). This indicates that SA may act in combination with other plant substances to produce chilling tolerance in plants, such as the SA with H_2O_2 acts synergistically to promote signal transduction and hormone metabolism and also increased energy supply and antioxidant enzyme activities under such conditions. These mechanisms and plant responses are closely relevant to chilling injury alleviation and chilling-tolerance enhancement in maize seed as well (Li et al. 2017).

Under chilling stress, there are two lines (Figure 8.2) of defense action of plants against oxidative stress – avoiding the production of reactive oxygen species (ROS) through alternative oxidase (AOX) production and activating the ROS-scavenging genes for superoxide dismutase (SOD), catalase (CAT), the glutathione peroxidase system, the ascorbate/glutathione cycle, and thioredoxin system activation. SA has been shown to induce expression of AOX and increase the antioxidant capacity of the cells under stress environments (Asghari and Aghdam 2010). For example, SA induces the production of antioxidant enzymes and polyphenol oxidase (PPO), PAL, and β-1, 3-glucanase in sweet cherries (Qin et al. 2003). Application of SA dramatically decreases CAT activity followed by significant increase in peroxidase (POD) and ascorbate peroxidase (APX) activities relative to control plants in fresh leaf and root tissues of tomato under chilling stress (Orabi et al. 2015). An increase in AOX expression levels after treatment with SA and MeSA induced tolerance in fresh sweet green peppers against chilling stress (Fung et al. 2004).

8.5 Freezing Stress

Normal functions of plants are reduced when exposed to freezing temperatures. Plants in temperate and cool regions are most commonly exposed to freezing temperatures (Saleem et al. 2020). Freezing stress badly affects plant growth regulation resulting in significant adverse effects on plant yield (Guo et al. 2018). It causes nucleation of ice, loss of water, protein denaturation, alteration in metabolite levels, and may result in the death of the plant (Guy 1990; Rao et al. 2006). Interestingly, plants have shown varying capabilities to resist freezing stress (< 0°C) and develop tolerance (Saleem et al. 2020).

It is well-established that SA application enhances plant tolerance against freezing stress, however, its continuous use may also lower plant tolerance potentials. SA increased cold tolerance in a variety of plant species such as potato, rice, maize, and many others (Ahmad and Prasad 2011). It has been shown that SA and cold stress are highly interconnected and the effect could be influenced by SA application (concentration) and tissue involvement. A simple signal pathway showing SA involvement during cold stress is depicted in Figure 8.3. It is evident that SA aids the accumulation of abscisic acid (ABA) in plants (Szepesi et al. 2009), such as in *Cynodon dactylon* (L). Pers. ABA leads to cold tolerance when applied exogenously by cell membrane stability and maintenance, increasing carbon isotopic fractionation, and improving the process of photosystem II (Huang et al. 2017). As calcium signalling has shown to play a key role in cold tolerance in plants, interestingly the one mediating mechanism of SA against cold tolerance is through regulating Ca^{2+} homeostasis (Wang and Li 2006).

These studies prove that under cold stress SA acts as a key player in modulation and regulation of growth by transducing signals through various components (ABA, Ca, NO, ROS, cGMP, MAPKs, and phospholipid cascades) of signal transduction pathways. These activated signal components lead to the activation of different transcription factors responsible for the gene expression required for plant survival

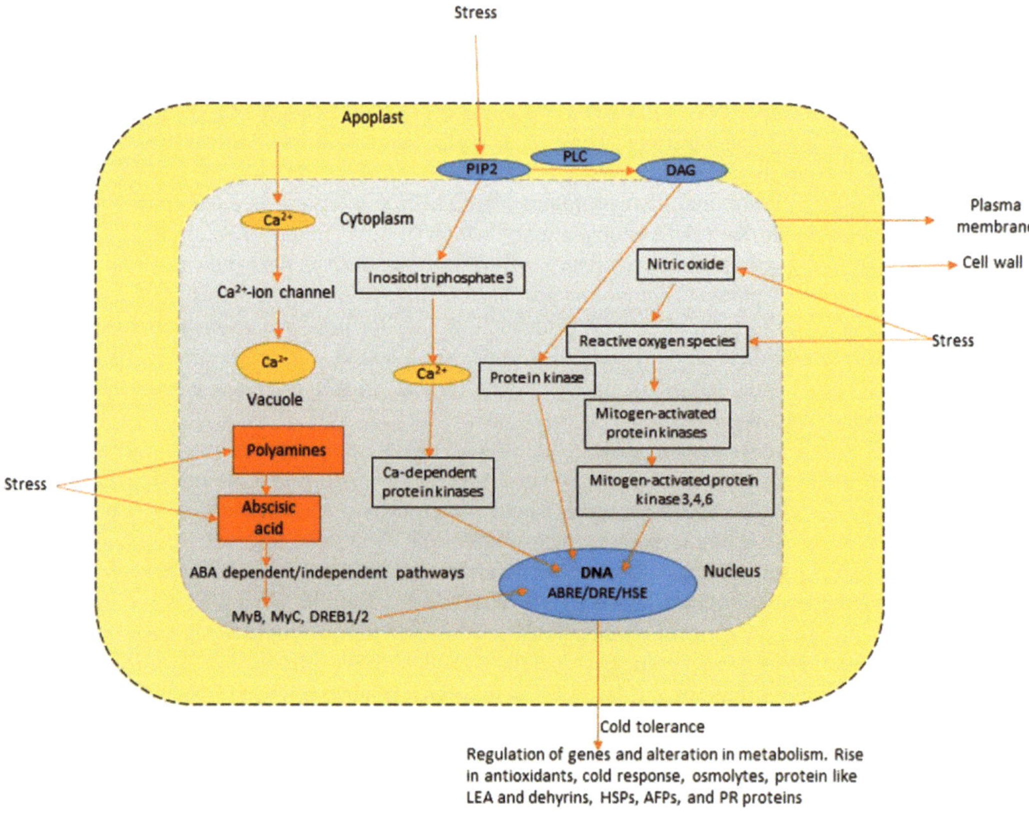

FIGURE 8.3 Signaling pathways of salicylic acid during cold stress. ABRE = ABA response element; AFPs = antifreezing proteins; Ca = calcium; DAG = diacylglycerol; DRE = dehydration response element; DREB = dehydration-responsive element-binding protein; HSE = heat shock element; HSPs = heat shock proteins; LEA = late embryogenesis abundant proteins; PIP2 = phosphotidyl inositol-2-phosphate; PLC = phospholipase C; PS = pathogenesis-related proteins; ROS/RNS = reactive oxygen/nitrogen species.

during cold stress. Moreover, SA-mediated signalling acts by enhancing antioxidant activity, and increasing accumulation of compatible solutes and other cold-responsive proteins, such as anti-freezing proteins (AFPs), heat shock proteins (HSPs), late embryogenesis abundant (LEA), and pathogenesis-related (PR) proteins.

8.6 Conclusion

Temperature stresses affect different morphological, physiological, cytological, and biochemical characteristics in plants. Alteration in these attributes of plants caused by temperature stress adversely affect the growth and development of plants. The SA is an important phytochemical, which is a mediator and an agent of cross-talk to various signalling molecules during temperature stress. It has multiple actions while acting as a key player in plant growth regulation, plant development, and growth during temperature stress. But the application of SA is a double-edged sword; at optimal concentration it enhances plants developmental processes, but on the other hand, at abnormal concentration it increases the burden of stress in plants. So, a care and an evidence-based approach must be used for application of SA on crops. However, there is still a huge gap which needs to be investigated, such as the molecular mechanisms of SA transport system, interaction with other pathways, receptors, mediators, and targets in plants.

REFERENCES

Adnan M, Basir A, Arif M, Shah SRA, Khan M, Jamal Y (2015) Impact of grazing on wheat yield and associated weeds. *Pak J Weed Sci Res* 21(3):351–358.

Adnan M, Fahad S, Khan IA, Saeed M, Ihsan MZ, Saud S, Riaz M, Wang D, Wu C (2019) Integration of poultry manure and phosphate solubilizing bacteria improved availability of Ca bound P in calcareous soils. *Biotech* 9(10):368.

Adnan M, Fahad S, Zamin M, Shah S, Mian IA, Danish S, Zafar-Ul-Hye M, Battaglia ML, Naz RMM, Saeed B, Saud S, Ahmad I, Yue Z, Brtnicky M, Holatko J, Datta R (2020) Coupling phosphate-solubilizing bacteria with phosphorus supplements improve maize phosphorus acquisition and growth under lime induced salinity stress. *Plants* 9(7). https://doi.org/10.3390/plants9070900.

Adnan M, Shah Z, Fahad S, Arif M, Alam M, Khan IA, Mian IA, Basir A, Ullah H, Arshad M, Rahman IU, Saud S, Ihsan MZ, Jamal Y, Amanullah, Hammad HM, Nasim W (2018a) Author correction: phosphate-solubilizing bacteria nullify the antagonistic effect of soil calcification on bioavailability of phosphorus in alkaline soils. *Sci Rep* 8(1):4339. https://doi.org/10.1038/s41598-018-22653-7.

Adnan M, Shah Z, Sharif M, Rahman H (2018b) Liming induces carbon dioxide (CO2) emission in PSB inoculated alkaline soil supplemented with different phosphorus sources. *Environ Sci Pollut Res Int* 25(10):9501–9509.

Ahmad P, Prasad MNV (2011) *Abiotic stress responses in plants: metabolism, productivity and sustainability*. Springer Science & Business Media, New York.

Ahmad S, Kamran M, Ding R, Meng X, Wang H, Ahmad I, Fahad S, Han Q (2019) Exogenous melatonin confers drought stress by promoting plant growth, photosynthetic capacity and antioxidant defense system of maize seedlings. *PeerJ* 7:e7793. http://doi.org/10.7717/peerj.7793.

Akbar H, Timothy JK, Jagadish T, Golam MM, Apurbo KC, Muhammad F, Rajan B, Fahad S, Hasanuzzaman M (2020) Agricultural land degradation: processes and problems undermining future food security. In: Fahad S, Hasanuzzaman M, Alam M, Ullah H, Saeed M, Khan AK, Adnan M (eds) *Environment, climate, plant and vegetation growth*. Springer Publ Ltd, Springer Nature Switzerland AG. Part of Springer Nature, pp 17–62. https://doi.org/10.1007/978-3-030-49732-3.

Akram R, Turan V, Hammad HM, Ahmad S, Hussain S, Hasnain A, Maqbool MM, Rehmani MIA, Rasool A, Masood N, Mahmood F, Mubeen M, Sultana SR, Fahad S, Amanet K, Saleem M, Abbas Y, Akhtar HM, Waseem F, Murtaza R, Amin A, Zahoor SA, ul Din MS, Nasim W (2018a) Fate of organic and inorganic pollutants in paddy soils. In: Hashmi MZ, Varma A (eds) *Environmental pollution of paddy soils, soil biology*. Springer International Publishing AG, Cham, Switzerland, pp 197–214.

Akram R, Turan V, Wahid A, Ijaz M, Shahid MA, Kaleem S, Hafeez A, Maqbool MM, Chaudhary HJ, Munis MFH, Mubeen M, Sadiq N, Murtaza R, Kazmi DH, Ali S, Khan N, Sultana SR, Fahad S, Amin A,

Nasim W (2018b) Paddy land pollutants and their role in climate change. In: Hashmi MZ, Varma A (eds) *Environmental pollution of paddy soils, soil biology*. Springer International Publishing AG, Cham, Switzerland, pp 113–124.

Ali A, Imtiyaz Adnan M, Arshad M, Inayat UR, Muhammad YJ, Saleem N, Rahman Z (2016) Effect of zinc activities on shoot, root biomass and phosphorus uptake in wheat gynotypes. *Am Eurasian J Agric Environ Sci* 16(1):204–208.

Al-Wabel Mohammad I, Abdelazeem S, Munir A, Khalid E, Adel RAU (2020b) Extent of climate change in Saudi Arabia and its impacts on agriculture: a case study from Qassim region. In: Fahad S, Hasanuzzaman M, Alam M, Ullah H, Saeed M, Khan AK, Adnan M (eds) *Environment, climate, plant and vegetation growth*. Springer Publ Ltd, Springer Nature Switzerland AG. Part of Springer Nature, pp 635–658. https://doi.org/10.1007/978-3-030-49732-3.

Al-Wabel Mohammad I, Ahmad Munir, Usman Adel RA, Akanji Mutair, Rafique Muhammad Imran (2020a) Advances in pyrolytic technologies with improved carbon capture and storage to combat climate change. In: Fahad S, Hasanuzzaman M, Alam M, Ullah H, Saeed M, Khan AK, Adnan M (eds) *Environment, climate, plant and vegetation growth*. Springer Publ Ltd, Springer Nature Switzerland AG. Part of Springer Nature, pp 535–576. https://doi.org/10.1007/978-3-030-49732-3.

Amanullah Shah K, Imran Hamdan AK, Muhammad A, Abdel RA, Muhammad A, Fahad S, Azizullah S, Brajendra P (2020) Effects of climate change on irrigation water quality. In: Fahad S, Hasanuzzaman M, Alam M, Ullah H, Saeed M, Khan AK, Adnan M (eds) *Environment, climate, plant and vegetation growth*. Springer Publ Ltd, Springer Nature Switzerland AG. Part of Springer Nature, pp 123–132. https://doi.org/10.1007/978-3-030-49732-3.

Amir M, Muhammad A, Allah B, Sevgi Ç, Haroon ZK, Muhammad A, Emre A (2020) Bio fortification under climate change: the fight between quality and quantity. In: Fahad S, Hasanuzzaman M, Alam M, Ullah H, Saeed M, Khan AK, Adnan M (eds) *Environment, climate, plant and vegetation growth*. Springer Publ Ltd, Springer Nature Switzerland AG. Part of Springer Nature, pp 173–228. https://doi.org/10.1007/978-3-030-49732-3.

Amjad I, Muhammad H, Farooq S, Anwar H (2020) Role of plant bioactives in sustainable agriculture. In: Fahad S, Hasanuzzaman M, Alam M, Ullah H, Saeed M, Khan AK, Adnan M (eds) *Environment, climate, plant and vegetation growth*. Springer Publ Ltd, Springer Nature Switzerland AG. Part of Springer Nature, pp 591–606. https://doi.org/10.1007/978-3-030-49732-3.

Anelia GD (2017) Signaling molecules in plants: exogenous application. *Acta Sci Agric* 1:38–41.

Arif M, Shah T, Ilyas M, Ahmad W, Mian AA, Jadoon MA, Adnan M (2015) Effect of organic manures and their levels on weeds density and maize yield. *Pak J Weed Sci Res* 21(4):517–522.

Arif M, Talha J, Muhammad R, Fahad S, Muhammad A, Amanullah Kawsar A, Ishaq AM, Bushra K, Fahd R (2020) Biochar; a remedy for climate change. In: Fahad S, Hasanuzzaman M, Alam M, Ullah H, Saeed M, Khan AK, Adnan M (eds) *Environment, climate, plant and vegetation growth*. Springer Publ Ltd, Springer Nature Switzerland AG. Part of Springer Nature, pp 151–172. https://doi.org/10.1007/978-3-030-49732-3.

Asensi-Fabado MA, Oliván A, Munné-Bosch S (2013) A comparative study of the hormonal response to high temperatures and stress reiteration in three Labiatae species. *Environ Exp Bot* 94:57–65.

Asghari M, Aghdam MS (2010) Impact of salicylic acid on post-harvest physiology of horticultural crops. *Trends Food Sci Technol* 21(10):502–509.

Ayman ELS, Hossain Akbar, Barutçular Celaleddin, Iqbal Muhammad Aamir, Sohidul Islam M, Fahad Shah, Sytar Oksana, Çig Fatih, Meena Ram Swaroop, Erman Murat (2020) Consequences of salinity stress on the quality of crops and its mitigation strategies for sustainable crop production: an outlook of arid and semi- arid regions. In: Fahad S, Hasanuzzaman M, Alam M, Ullah H, Saeed M, Khan AK, Adnan M (eds) *Environment, climate, plant and vegetation growth*. Springer Publ Ltd, Springer Nature Switzerland AG. Part of Springer Nature, pp 503–534. https://doi.org/10.1007/978-3-030-49732-3.

Balogh G, Péter M, Glatz A, Gombos I, Török Z, Horváth I, Harwood JL, Vígh L (2013) Key role of lipids in heat stress management. *FEBS Lett* 587(13):1970–1980. https://doi.org/10.1016/j.febslet.2013.05.016.

Baseer M, Adnan M, Fazal M, Fahad S, Muhammad S, Fazli W, Muhammad A, Jr. Amanullah DW, Saud S, Muhammad N, Muhammad Z, Fazli S, Beena S, Mian AR, Ishaq AM (2019) Substituting urea by organic wastes for improving maize yield in alkaline soil. *J Plant Nutr.*

Bayram AY, Seher Ö, Nazlican A (2020) Climate change forecasting and modeling for the year of 2050. In: Fahad S, Hasanuzzaman M, Alam M, Ullah H, Saeed M, Khan AK, Adnan M (eds) *Environment, climate, plant and vegetation growth*. Springer Publ Ltd, Springer Nature Switzerland AG. Part of Springer Nature, pp 109–122. https://doi.org/10.1007/978-3-030-49732-3.

Clarke SM, Cristescu SM, Miersch O, Harren FJ, Wasternack C, Mur LA (2009) Jasmonates act with salicylic acid to confer basal thermotolerance in Arabidopsis thaliana. *New Phytol* 182(1):175–187.

Clarke SM, Mur LA, Wood JE, Scott IM (2004) Salicylic acid dependent signaling promotes basal thermotolerance but is not essential for acquired thermotolerance in Arabidopsis thaliana. *Plant J* 38(3):432–447.

Cleland CF, Ajami A (1974) Identification of the flower-inducing factor isolated from aphid honeydew as being salicylic acid. *Plant Physiol* 54(6):904–906.

Clifton R, Lister R, Parker KL, Sappl PG, Elhafez D, Millar AH, Day DA, Whelan J (2005) Stress-induced co-expression of alternative respiratory chain components in Arabidopsis thaliana. *Plant Mol Biol* 58(2):193.

Danish S, Zafar-ul-Hye M, Fahad S, Saud S, Brtnicky M, Hammerschmiedt T, Datta R (2020) Drought stress alleviation by ACC deaminase producing *Achromobacter xylosoxidans* and *Enterobacter cloacae*, with and without timber waste biochar in maize. *Sustainability* 12(15). https://doi.org/10.3390/su12156286.

Dat JF, Foyer CH, Scott IM (1998) Changes in salicylic acid and antioxidants during induced thermotolerance in mustard seedlings. *Plant Physiol* 118(4):1455–1461.

Dempsey DA, Vlot AC, Wildermuth MC, Klessig DF (2011) Salicylic acid biosynthesis and metabolism. *Arabidopsis Book* 9:e0156. https://doi.org/10.1199/tab.0156.

Depeng W, Fahad S, Saud S, Muhammad K, Aziz K, Mohammad NK, Hafiz MH, Wajid N (2018) Morphological acclimation to agronomic manipulation in leaf dispersion and orientation to promote "Ideotype" breeding: evidence from 3D visual modeling of "super" rice (*Oryza sativa* L.). *Plant Physiol Biochem* 135:499–510.

Dewdney J, Reuber TL, Wildermuth MC, Devoto A, Cui J, Stutius LM, Drummond EP, Ausubel FM (2000) Three unique mutants of Arabidopsis identify eds loci required for limiting growth of a biotrophic fungal pathogen. *Plant J* 24(2):205–218.

Ding C-K, Wang CY (2003) The dual effects of methyl salicylate on ripening and expression of ethylene biosynthetic genes in tomato fruit. *Plant Sci* 164(4):589–596.

Ding C-K, Wang CY, Gross KC, Smith DL (2001) Reduction of chilling injury and transcript accumulation of heat shock proteins in tomato fruit by methyl jasmonate and methyl salicylate. *Plant Sci* 161(6):1153–1159.

Ding C-K, Wang C, Gross KC, Smith DL (2002) Jasmonate and salicylate induce the expression of pathogenesis-related-protein genes and increase resistance to chilling injury in tomato fruit. *Planta* 214(6):895–901.

Enamul Md H, Shoeb AZM, Mallik AH, Fahad S, Kamruzzaman MM, Akib J, Nayyer S, Mehedi KMA, Swati AS, Yeamin Md A, Most SS (2020) Measuring vulnerability to environmental hazards: qualitative to quantitative. In: Fahad S, Hasanuzzaman M, Alam M, Ullah H, Saeed M, Khan AK, Adnan M (eds) *Environment, climate, plant and vegetation growth*. Springer Publ Ltd, Springer Nature Switzerland AG. Part of Springer Nature, pp 421–452. https://doi.org/10.1007/978-3-030-49732-3.

Fahad S, Adnan M, Hassan S, Saud S, Hussain S, Wu C, Wang D, Hakeem KR, Alharby HF, Turan V, Khan MA, Huang J (2019b) Rice responses and tolerance to high temperature. In: Hasanuzzaman M, Fujita M, Nahar K, Biswas JK (eds) *Advances in rice research for abiotic stress tolerance*. Woodhead Publ Ltd, Cambridge, England, pp 201–224.

Fahad S, Bajwa AA, Nazir U, Anjum SA, Farooq A, Zohaib A, Sadia S, Nasim W, Adkins S, Saud S, Ihsan MZ, Alharby H, Wu C, Wang D, Huang J (2017) Crop production under drought and heat stress: plant responses and management options. *Front Plant Sci* 8:1147. https://doi.org/10.3389/fpls.2017.01147.

Fahad S, Bano A (2012) Effect of salicylic acid on physiological and biochemical characterization of maize grown in saline area. *Pak J Bot* 44:1433–1438.

Fahad S, Chen Y, Saud S,Wang K, Xiong D, Chen C,Wu C, Shah F, Nie L, Huang J (2013) Ultraviolet radiation effect on photosynthetic pigments, biochemical attributes, antioxidant enzyme activity and hormonal contents of wheat. *J Food, Agri Environ* 11(3&4):1635–1641.

Fahad S, Hussain S, Bano A, Saud S, Hassan S, Shan D, Khan FA, Khan F, Chen Y, Wu C, Tabassum MA, Chun MX, Afzal M, Jan A, Jan MT, Huang J (2015a) Potential role of phytohormones and plant growth-promoting rhizobacteria in abiotic stresses: consequences for changing environment. *Environ Sci Pollut Res* 22(7):4907–4921. https://doi.org/10.1007/s11356-014-3754-2.

Fahad S, Hussain S, Matloob A, Khan FA, Khaliq A, Saud S, Hassan S, Shan D, Khan F, Ullah N, Faiq M, Khan MR, Tareen AK, Khan A, Ullah A, Ullah N, Huang J (2015b) Phytohormones and plant responses to salinity stress: a review. *Plant Growth Regul* 75(2):391–404. https://doi.org/10.1007/s10725-014-0013-y.

Fahad S, Hussain S, Saud S, Hassan S, Chauhan BS, Khan F, Ihsan MZ, Ullah A, Wu C, Bajwa AA, Alharby H, Amanullah, Nasim W, Shahzad B, Tanveer M, Huang J (2016a) Responses of rapid viscoanalyzer profile and other rice grain qualities to exogenously applied plant growth regulators under high day and high night temperatures. *PLOS ONE* 11(7):e0159590. https://doi.org/10.1371/journal.pone.0159590.

Fahad S, Hussain S, Saud S, Hassan S, Ihsan Z, Shah AN, Wu C, Yousaf M, Nasim W, Alharby H, Alghabari F, Huang J (2016c) Exogenously applied plant growth regulators enhance the morphophysiological growth and yield of rice under high temperature. *Front Plant Sci* 7:1250. https://doi.org/10.3389/fpls.2016.

Fahad S, Hussain S, Saud S, Hassan S, Tanveer M, Ihsan MZ, Shah AN, Ullah A, Nasrullah, Khan F, Ullah S, Alharby H, Nasim W, Wu C, Huang J (2016d) A combined application of biochar and phosphorus alleviates heat-induced adversities on physiological, agronomical and quality attributes of rice. *Plant Physiol Biochem* 103:191–198.

Fahad S, Hussain S, Saud S, Khan F, Hassan S Jr A, Nasim W, Arif M, Wang F, Huang J (2016b) Exogenously applied plant growth regulators affect heat-stressed rice pollens. *J Agron Crop Sci* 202(2):139–150.

Fahad S, Hussain S, Saud S, Tanveer M, Bajwa AA, Hassan S, Shah AN, Ullah A, Wu C, Khan FA, Shah F, Ullah S, Chen Y, Huang J (2015c) A biochar application protects rice pollen from high-temperature stress. *Plant Physiol Biochem* 96:281–287.

Fahad S, Ihsan MZ, Khaliq A, Daur I, Saud S, Alzamanan S, Nasim W, Abdullah M, Khan IA, Wu C, Wang D, Huang J (2018) Consequences of high temperature under changing climate optima for rice pollen characteristics-concepts and perspectives. *Arch Agron Soil Sci*. https://doi.org/10.1080/03650340.2018.1443213.

Fahad S, Nie L, Chen Y, Wu C, Xiong D, Saud S, Hongyan L, Cui K, Huang J (2015d) Crop plant hormones and environmental stress. *Sustain Agric Rev* 15:371–400.

Fahad S, Rehman A, Shahzad B, Tanveer M, Saud S, Kamran M, Ihtisham M, Khan SU, Turan V, Rahman MHU (2019a) Rice responses and tolerance to metal/metalloid toxicity. In: Hasanuzzaman M, Fujita M, Nahar K, Biswas JK (eds) *Advances in rice research for abiotic stress tolerance*. Woodhead Publ Ltd, Cambridge, England, pp 299–312.

Farah R, Muhammad R, Muhammad SA, Tahira Y, Muhammad AA, Maryam A, Shafaqat A, Rashid M, Muhammad R, Qaiser H, Afia Z, Muhammad AA, Muhammad A, Fahad S (2020) Alternative and non-conventional soil and crop management strategies for increasing water use efficiency. In: Fahad S, Hasanuzzaman M, Alam M, Ullah H, Saeed M, Khan AK, Adnan M (eds) *Environment, climate, plant and vegetation growth*. Springer Publ Ltd, Springer Nature Switzerland AG. Part of Springer Nature, pp 323–338. https://doi.org/10.1007/978-3-030-49732-3.

Farhat A, Hafiz MH, Wajid I, Aitazaz AF, Hafiz FB, Zahida Z, Fahad S, Wajid F, Artemi C (2020) A review of soil carbon dynamics resulting from agricultural practices. *J Environ Manag* 268. http://www.ncbi.nlm.nih.gov/pubmed/110319.

Fazli W, Muhmmad S, Amjad A, Fahad S, Muhammad A, Muhammad N, Ishaq AM, Imtiaz AK, Mukhtar A, Muhammad S, Muhammad I, Rafi U, Haroon I, Muhammad A (2020) Plant-microbes interactions and functions in changing climate. In: Fahad S, Hasanuzzaman M, Alam M, Ullah H, Saeed M, Khan AK, Adnan M (eds) *Environment, climate, plant and vegetation growth*. Springer Publ Ltd, Springer Nature Switzerland AG. Part of Springer Nature, pp 397–420. https://doi.org/10.1007/978-3-030-49732-3.

Fitter AH, Hay RK (2012) *Environmental physiology of plants*. Academic Press, London.

Fung RW, Wang CY, Smith DL, Gross KC, Tian M (2004) MeSA and MeJA increase steady-state transcript levels of alternative oxidase and resistance against chilling injury in sweet peppers (Capsicum annuum L.). *Plant Sci* 166(3):711–719.

Gaffney T, Friedrich L, Vernooij B, Negrotto D, Nye G, Uknes S, Ward E, Kessmann H, Ryals J (1993) Requirement of salicylic acid for the induction of systemic acquired resistance. *Science* 261(5122):754–756.

Gopakumar L, Bernard NO, Donato V (2020) Soil microarthropods and nutrient cycling. In: Fahad S, Hasanuzzaman M, Alam M, Ullah H, Saeed M, Khan AK, Adnan M (eds) *Environment, climate, plant and vegetation growth*. Springer Publ Ltd, Springer Nature Switzerland AG. Part of Springer Nature, pp 453–472. https://doi.org/10.1007/978-3-030-49732-3.

Gul F, Ahmed I, Ashfaq M, Jan D, Fahad S, Li X, Wang D, Fahad M, Fayyaz M, Shah SA (2020) Use of crop growth model to simulate the impact of climate change on yield of various wheat cultivars under different agro-environmental conditions in Khyber Pakhtunkhwa, Pakistan. *Arab J Geosci* 13(3):112. https://doi.org/10.1007/s12517-020-5118-1.

Guo X, Liu D, Chong K (2018) Cold signaling in plants: insights into mechanisms and regulation. *J Integr Plant Biol* 60(9):745–756.

Guy C, Kaplan F, Kopka J, Selbig J, Hincha DK (2008) Metabolomics of temperature stress. *Physiol Plant* 132(2):220–235. https://doi.org/10.1111/j.1399-3054.2007.00999.x.

Guy CL (1990) Cold acclimation and freezing stress tolerance: role of protein metabolism. *Annu Rev Plant Physiol Plant Mol Biol* 41(1):187–223.

Habib ur R, Ashfaq A, Aftab W, Manzoor H, Fahd R, Wajid I, Md. Aminul I, Vakhtang S, Muhammad A, Asmat U, Abdul W, Syeda RS, Shah S, Shahbaz K, Fahad S, Manzoor H, Saddam H, Wajid N (2017) Application of CSM-CROPGRO-Cotton model for cultivars and optimum planting dates: evaluation in changing semi-arid climate. *Field Crops Res.*

Hadacek F, Bachmann G, Engelmeier D, Chobot V (2010) Hormesis and a chemical raison d'ětre for secondary plant metabolites. *Dose-Response* 9(1):79–116.

Hafiz MH, Farhat A, Shafqat S, Fahad S, Artemi C, Wajid F, Chaves CB, Wajid N, Muhammad M, Hafiz FB (2018) Offsetting land degradation through nitrogen and water management during maize cultivation under arid conditions. *Land Degrad Dev* 1–10.

Hafiz MH, Wajid F, Farhat A, Fahad S, Shafqat S, Wajid N, Hafiz FB (2016) Maize plant nitrogen uptake dynamics at limited irrigation water and nitrogen. *Environ Sci Pollut Res* 24(3):2549–2557.

Hafiz MH, Muhammad A, Farhat A, Hafiz FB, Saeed AQ, Muhammad M, Fahad S, Muhammad A (2019) Environmental factors affecting the frequency of road traffic accidents: a case study of sub-urban area of Pakistan. *Environ Sci Pollut Res.*

Hammad HM, Abbas F, Ahmad A, Bakhat HF, Farhad W, Wilkerson CJ, Fahad S, Hoogenboom G (2020a) Predicting kernel growth of maize under controlled water and nitrogen applications. *Int J Plant Prod.* https://doi.org/10.1007/s42106-020-00110-8.

Hammad HM, Abbas F, Saeed S, Fahad S, Cerdà A, Farhad W, Bernardo CC, Nasim W, Mubeen M, Bakhat HF (2018) Offsetting land degradation through nitrogen and water management during maize cultivation under arid conditions. *Land Degrad Dev* 29(5):1366–1375.

Hammad HM, Khaliq A, Abbas F, Farhad W, Fahad S, Aslam M, Shah GM, Nasim W, Mubeen M, Bakhat HF (2020b) Comparative effects of organic and inorganic fertilizers on soil organic carbon and wheat productivity under arid region. *Commun Soil Sci Plant Anal.* https://doi.org/10.1080/00103624.2020.1763385.

Hao G, Naumann TA, Vaughan MM, McCormick S, Usgaard T, Kelly A, Ward TJ (2018) Characterization of a Fusarium graminearum salicylate hydroxylase. *Front Microbiol* 9:3219–3219. https://doi.org/10.3389/fmicb.2018.03219.

Hesham FA, Fahad S (2020) Melatonin application enhances biochar efficiency for drought tolerance in maize varieties: modifications in physio-biochemical machinery. *Agron J* 112(4):1–22.

Horváth E, Szalai G, Janda T (2007) Induction of abiotic stress tolerance by salicylic acid signaling. *J Plant Growth Regul* 26(3):290–300.

Huang X, Shi H, Hu Z, Liu A, Amombo E, Chen L, Fu J (2017) ABA is involved in regulation of cold stress response in Bermudagrass. *Front Plant Sci* 8. https://doi.org/10.3389/fpls.2017.01613.

Hussain MA, Fahad S, Rahat S, Muhammad FJ, Muhammad M, Qasid A, Ali A, Husain A, Nooral A, Babatope SA, Changbao S, Liya G, Ibrar A, Zhanmei J, Juncai H (2020) Multifunctional role of brassinosteroid and its analogues in plants. *Plant Growth Regul.* https://doi.org/10.1007/s10725-020-00647-8.

Hussain S, Mubeen M, Ahmad A, Akram W, Hammad HM, Ali M, Masood N, Amin A, Farid HU, Sultana SR, Fahad S, Wang D, Nasim W (2020) Using GIS tools to detect the land use/land cover changes during forty years in Lodhran District of Pakistan. *Environ Sci Pollut Res Int* 27(32):39676–39692. https://doi.org/10.1007/s11356-019-06072-3.

Ibrar K, Aneela R, Khola Z, Urooba N, Sana B, Rabia S, Ishtiaq H, Mujaddad Ur Rehman, Salvatore M (2020) Microbes and environment: global warming reverting the frozen zombies. In: Fahad S, Hasanuzzaman M, Alam M, Ullah H, Saeed M, Khan AK, Adnan M (eds) *Environment, climate, plant and vegetation growth.* Springer Publ Ltd, Springer Nature Switzerland AG. Part of Springer Nature, pp 607–634. https://doi.org/10.1007/978-3-030-49732-3.

Ilyas M, Nisar M, Khan N, Hazrat A, Khan AH, Hayat K, Fahad S, Khan A, Ullah A (2020) Drought tolerance strategies in plants: a mechanistic approach. *J Plant Growth Regul*. https://doi.org/10.1007/s00344-020-10174-5.

Iqra M, Amna B, Shakeel I, Fatima K, Sehrish L, Hamza A, Fahad S (2020) Carbon cycle in response to global warming. In: Fahad S, Hasanuzzaman M, Alam M, Ullah H, Saeed M, Khan AK, Adnan M (eds) *Environment, climate, plant and vegetation growth*. Springer Publ Ltd, Springer Nature Switzerland AG. Part of Springer Nature, pp 1–16. https://doi.org/10.1007/978-3-030-49732-3.

Izhar Shafi M, Adnan M, Fahad S, Wahid F, Khan A, Yue Z, Danish S, Zafar-ul-Hye M, Brtnicky M, Datta R (2020) Application of single superphosphate with humic acid improves the growth, yield and phosphorus uptake of wheat (Triticum aestivum L.) in calcareous soil. *Agronomy* 10(9):1224. https://doi.org/10.3390/agronomy10091224.

Jagadish S, Muthurajan R, Oane R, Wheeler TR, Heuer S, Bennett J, Craufurd PQ (2010) Physiological and proteomic approaches to address heat tolerance during anthesis in rice (Oryza sativa L.). *J Exp Bot* 61(1):143–156.

Jakir Md H, Allah B (2020) Development and applications of transplastomic plants; a way towards eco-friendly agriculture. In: Fahad S, Hasanuzzaman M, Alam M, Ullah H, Saeed M, Khan AK, Adnan M (eds) *Environment, climate, plant and vegetation growth*. Springer Publ Ltd, Springer Nature Switzerland AG. Part of Springer Nature, pp 285–322. https://doi.org/10.1007/978-3-030-49732-3.

Jan M, Anwar-ul-Haq M, Shah AN, Yousaf M, Iqbal J, Li X, Wang D, Fahad S (2019) Modulation in growth, gas exchange, and antioxidant activities of salt-stressed rice (Oryza sativa L.) genotypes by zinc fertilization. *Arab J Geosci* 12(24):775. https://doi.org/10.1007/s12517-019-4939-2.

Janda T, Szalai G, Tari I, Páldi E (1999) Hydroponic treatment with salicylic acid decreases the effects of chilling injury in maize (Zea mays L.) plants. *Planta* 208(2):175–180.

Jin JB, Jin YH, Lee J, Miura K, Yoo CY, Kim WY, Van Oosten M, Hyun Y, Somers DE, Lee I, Yun DJ, Bressan RA, Hasegawa PM (2008) The SUMO E3 ligase, AtSIZ1, regulates flowering by controlling a salicylic acid-mediated floral promotion pathway and through affects on FLC chromatin structure. *Plant J* 53(3):530–540.

Kai H, Iba K (2014) *Temperature stress in plants*. eLS. John Wiley & Sons, Ltd, Chichester, England.

Kamaran M, Wenwen C, Irshad A, Xiangping M, Xudong Z, Wennan S, Junzhi C, Shakeel A, Fahad S, Qingfang H, Tiening L (2017) Effect of paclobutrazol, a potential growth regulator on stalk mechanical strength, lignin accumulation and its relation with lodging resistance of maize. *Plant Growth Regul* 84:317–332.

Kang G, Wang C, Sun G, Wang Z (2003) Salicylic acid changes activities of H2O2-metabolizing enzymes and increases the chilling tolerance of banana seedlings. *Environ Exp Bot* 50(1):9–15.

Kang HM, Saltveit ME (2002) Chilling tolerance of maize, cucumber and rice seedling leaves and roots are differentially affected by salicylic acid. *Physiol Plant* 115(4):571–576.

Khan A, Tan DKY, Afridi MZ, Luo H, Tung SA, Ajab M, Fahad S (2017b) Nitrogen fertility and abiotic stresses management in cotton crop: a review. *Environ Sci Pollut Res* 24(17):14551–14566. https://doi.org/10.1007/s11356-017-8920-x.

Khan A, Tan DKY, Munsif F, Afridi MZ, Shah F, Wei F, Fahad S, Zhou R (2017a) Nitrogen nutrition in cotton and control strategies for greenhouse gas emissions: a review. *Environ Sci Pollut Res* 24(30):23471–23487. https://doi.org/10.1007/s11356-017-0131-y.

Khan MIR, Asgher M, Khan NA (2013a) Rising temperature in the changing environment: a serious threat to plants. *Clim Chan and Envi Sust* 1(1):25–36. https://doi.org/10.5958/j.2320-6411.1.1.004.

Khan MIR, Iqbal N, Masood A, Per TS, Khan NA (2013b) Salicylic acid alleviates adverse effects of heat stress on photosynthesis through changes in proline production and ethylene formation. *Plant Signal Behav* 8(11):e26374.

Levitt J (1972) *Responses of plants to environmental stresses*. Academic Press, New York, p 697.

Li Z, Xu J, Gao Y, Wang C, Guo G, Luo Y, Huang Y, Hu W, Sheteiwy MS, Guan Y, Hu J (2017) The synergistic priming effect of exogenous salicylic acid and H2O2 on chilling tolerance enhancement during maize (Zea mays L.) seed germination. *Front Plant Sci* 8. https://doi.org/10.3389/fpls.2017.01153.

Liu X, Zhou Y, Xiao J, Bao F (2018) Effects of chilling on the structure, function and development of chloroplasts. *Front Plant Sci* 9. https://doi.org/10.3389/fpls.2018.01715.

Lubovská Z, Dobrá J, Štorchová H, Wilhelmová N, Vanková R (2014) Cytokinin oxidase/dehydrogenase overexpression modifies antioxidant defense against heat, drought and their combination in Nicotiana tabacum plants. *J Plant Physiol* 171(17):1625–1633.

Mahar A, Amjad A, Altaf HL, Fazli W, Ronghua L, Muhammad A, Fahad S, Muhammad A, Rafiullah Imtiaz AK, Zengqiang Z (2020) Promising technologies for Cd-contaminated soils: drawbacks and possibilities. In: Fahad S, Hasanuzzaman M, Alam M, Ullah H, Saeed M, Khan AK, Adnan M (eds) *Environment, climate, plant and vegetation growth.* Springer Publ Ltd, Springer Nature Switzerland AG. Part of Springer Nature, pp 63–92. https://doi.org/10.1007/978-3-030-49732-3.

Malamy J, Klessig DF (1992) Salicylic acid and plant disease resistance. *Plant J* 2(5):643–654.

Martínez C, Pons E, Prats G, León J (2004) Salicylic acid regulates flowering time and links defence responses and reproductive development. *Plant J* 37(2):209–217.

Mora-Herrera ME, López-Delgado H, Castillo-Morales A, Foyer CH (2005) Salicylic acid and H2O2 function by independent pathways in the induction of freezing tolerance in potato. *Physiol Plant* 125:430–440.

Mubeen M, Ahmad A, Hammad HM, Awais M, Farid HU, Saleem M, Din MSu, Amin A, Ali A, Fahad S, Nasim W (2020) Evaluating the climate change impact on water use efficiency of cotton-wheat in semi-arid conditions using DSSAT model. *J Water Clim Change* 11(4):1661–1675. https://doi.org/10.2166/wcc.2019.179/622035/jwc2019179.pdf.

Muhammad B, Adnan M, Munsif F, Fahad S, Saeed M, Wahid F, Arif M, Amanullah J, Wang D, Saud S, Noor M, Zamin M, Subhan F, Saeed B, Raza MA, Mian IA (2019) Substituting urea by organic wastes for improving maize yield in alkaline soil. *J Plant Nutr.* https://doi.org/10.1080/01904167.2019.1659344.

Munir M, Shabbir G (2018) Salicylic acid mediated heat stress tolerance in selected bread wheat genotypes of Pakistan. *Pak J Bot* 50:2141–2146.

Nazar R, Iqbal N, Umar S (2017) Heat stress tolerance in plants: action of salicylic acid. In: *Salicylic acid multifaceted horm*, Springer, Singapore, pp 145–161.

Nazar R, Umar S, Khan N, Sareer O (2015) Salicylic acid supplementation improves photosynthesis and growth in mustard through changes in proline accumulation and ethylene formation under drought stress. *S Afr J Bot* 98:84–94.

Noor M, ur Naveed R, Ajmal J, Fahad S, Muhammad A, Fazli W, Saud S, Hassan S (2020) Climate change and costal plant lives. In: Fahad S, Hasanuzzaman M, Alam M, Ullah H, Saeed M, Khan AK, Adnan M (eds) *Environment, climate, plant and vegetation growth.* Springer Publ Ltd, Springer Nature Switzerland AG. Part of Springer Nature, pp 93–108. https://doi.org/10.1007/978-3-030-49732-3.

Norman C, Howell KA, Millar AH, Whelan JM, Day DA (2004) Salicylic acid is an uncoupler and inhibitor of mitochondrial electron transport. *Plant Physiol* 134(1):492–501.

Orabi SA, Dawood MG, Salman S (2015) Comparative study between the physiological role of hydrogen peroxide and salicylic acid in alleviating the harmful effect of low temperature on tomato plants grown under sand-ponic culture. *Sci Agric* 9(1):49–59. https://doi.org/10.15192/PSCP.SA.2015.1.9.4959.

Pan D-Y, Fu X, Zhang X-W, Liu F-J, Bi H-G, Ai X-Z (2020) Hydrogen sulfide is required for salicylic acid–induced chilling tolerance of cucumber seedlings. *Protoplasma* https://doi.org/10.1007/s00709-020-01531-y.

Qamar Z, Zubair A, Muhammad Y, Muhammad ZI, Abdul K, Fahad S, Safder B, Ramzani PMA, Muhammad N (2017) Zinc biofortification in rice: leveraging agriculture to moderate hidden hunger in developing countries. *Arch Agron Soil Sci* 64:147–161.

Qin GZ, Tian SP, Xu Y, Wan YK (2003) Enhancement of biocontrol efficacy of antagonistic yeasts by salicylic acid in sweet cherry fruit. *Physiol Mol Plant Pathol* 62(3):147–154.

Radwan DEM (2012) Salicylic acid induced alleviation of oxidative stress caused by clethodim in maize (Zea mays L.) leaves. *Pestic Biochem Physiol* 102(2):182–188.

Rahman MHu, Ahmad A, Wajid A, Hussain M, Rasul F, Ishaque W, Islam MA, Shelia V, Awais M, Ullah A, Wahid A, Sultana SR, Saud S, Khan S, Fahad S, Hussain M, Hussain S, Nasim W (2019) Application of CSM-CROPGRO-Cotton model for cultivars and optimum planting dates: evaluation in changing semi-arid climate. *Field Crops Res* 238:139–152.

Rao K, Raghavendra A, Reddy K (2006) Introduction. In: Madhava Rao K, Raghavendra A, Janardhan Reddy K (eds) *Physiology and molecular biology of stress tolerance.* Springer, Dordrecht, Nertherlands.

Rashid M, Qaiser H, Khalid SK, Mohammad I, Al-Wabel Zhang A, Muhammad A, Shahzada SI, Rukhsanda A, Ghulam AS, Shahzada MM, Sarosh A, Muhammad FQ (2020) Prospects of biochar in alkaline soils to mitigate climate change. In: Fahad S, Hasanuzzaman M, Alam M, Ullah H, Saeed M, Khan AK, Adnan M (eds) *Environment, climate, plant and vegetation growth.* Springer Publ Ltd, Springer Nature Switzerland AG. Part of Springer Nature, pp 133–150. https://doi.org/10.1007/978-3-030-49732-3.

Raskin I (1992) Role of salicylic acid in plants. *Annu Rev Plant Physiol Plant Mol Biol* 43(1):439–463.

Rehman M, Fahad S, Saleem MH, Hafeez M, Rahman MHu, Liu F, Deng G (2020) Red light optimized physiological traits and enhanced the growth of ramie (Boehmeria nivea L.). *Photosynthetica* 58(4): 922–931.

Sadam M, ul Qamar Muhammad Tahir, Ghulam M, Muhammad SK, Faiz AJ (2020) Role of biotechnology in climate resilient agriculture. In: Fahad S, Hasanuzzaman M, Alam M, Ullah H, Saeed M, Khan AK, Adnan M (eds) *Environment, climate, plant and vegetation growth.* Springer Publ Ltd, Springer Nature Switzerland AG. Part of Springer Nature, pp 339–366. https://doi.org/10.1007/978-3-030-49732-3.

Sajid H, Jie H, Jing H, Shakeel A, Satyabrata N, Sumera A, Awais S, Chunquan Z, Lianfeng Z, Xiaochuang C, Qianyu J, Junhua Z (2020) Rice production under climate change: adaptations and mitigating strategies. In: Fahad S, Hasanuzzaman M, Alam M, Ullah H, Saeed M, Khan AK, Adnan M (eds) *Environment, climate, plant and vegetation growth.* Springer Publ Ltd, Springer Nature Switzerland AG. Part of Springer Nature, pp 659–686. https://doi.org/10.1007/978-3-030-49732-3.

Saleem M, Fariduddin Q, Janda T (2020) Multifaceted role of salicylic acid in combating cold stress in plants: a review. *J Plant Growth Regul.* https://doi.org/10.1007/s00344-020-10152-x.

Saleem MH, Fahad S, Adnan M, Ali M, Rana MS, Kamran M, Ali Q, Hashem IA, Bhantana P, Ali M, Hussain RM (2020a) Foliar application of gibberellic acid endorsed phytoextraction of copper and alleviates oxidative stress in jute (Corchorus Capsularis L.) plant grown in highly copper-contaminated soil of China. *Environ Sci Pollut Res.* https://doi.org/10.1007/s11356-020-09764-3.

Saleem MH, Fahad S, Khan SU, Din M, Ullah A, Sabagh AE, Hossain A, Llanes A, Liu L (2020c) Copper-induced oxidative stress, initiation of antioxidants and phytoremediation potential of flax (Linum usitatissimum L.) seedlings grown under the mixing of two different soils of China. *Environ Sci Pollut Res Int* 27(5):5211–5221. https://doi.org/10.1007/s11356-019-07264-7.

Saleem MH, Rehman M, Fahad S, Tung SA, Iqbal N, Hassan A, Ayub A, Wahid MA, Shaukat S, Liu L, Deng G (2020b) Leaf gas exchange, oxidative stress, and physiological attributes of rapeseed (Brassica napus L.) grown under different light-emitting diodes. *Photosynthetica* 58(3):836–845.

Saleem N, Adnan M, Khan NA, Zaheer S, Jalal F, Amin M, Jamal Y (2015) Dual purpose canola: grazing and grains options. *Pak J Weed Sci Res* 21(2):295–304.

Saleh J, Maftoun M (2008) Interactive effect of NaCl levels and zinc sources and levels on the growth and mineral composition of rice. *J Agric Sci Tech* 10:325–336.

Saman S, Amna B, Bani A, ul Qamar Muhammad Tahir, Rana MA, Muhammad SK (2020) QTL mapping for abiotic stresses in cereals. In: Fahad S, Hasanuzzaman M, Alam M, Ullah H, Saeed M, Khan AK, Adnan M (eds) *Environment, climate, plant and vegetation growth.* Springer Publ Ltd, Springer Nature Switzerland AG. Part of Springer Nature, pp 229–252. https://doi.org/10.1007/978-3-030-49732-3.

Saud S, Chen Y, Long B, Fahad S, Sadiq A (2013) The different impact on the growth of cool season turf grass under the various conditions on salinity and drought stress. *Int J Agric Sci Res* 3:77–84.

Saud S, Fahad S, Cui G, Yajun C, Anwar S (2020) Determining nitrogen isotopes discrimination under drought stress on enzymatic activities, nitrogen isotope abundance and water contents of Kentucky bluegrass. *Sci Rep* 10(1):6415. https://doi.org/10.1038/s41598-020-63548-w.

Saud S, Fahad S, Yajun C, Ihsan MZ, Hammad HM, Nasim W, Amanullah Jr, Arif M, Alharby H (2017) Effects of nitrogen supply on water stress and recovery mechanisms in Kentucky bluegrass plants. *Front Plant Sci* 8:983. https://doi.org/10.3389/fpls.2017.00983.

Saud S, Li X, Chen Y, Zhang L, Fahad S, Hussain S, Sadiq A, Chen Y (2014) Silicon application increases drought tolerance of kentucky bluegrass by improving plant water relations and morphophysiological functions. *Sci World J* 2014:368694. https://doi.org/10.1155/2014/368694.

Saud S, Yajun C, Fahad S, Hussain S, Na L, Xin L, Alhussien SA (2016) Silicate application increases the photosynthesis and its associated metabolic activities in Kentucky bluegrass under drought stress and post-drought recovery. *Environ Sci Pollut Res* 23(17):17647–17655. https://doi.org/10.1007/s11356-016-6957-x.

Senol C (2020) The effects of climate change on human behaviors. In: Fahad S, Hasanuzzaman M, Alam M, Ullah H, Saeed M, Khan AK, Adnan M (eds) *Environment, climate, plant and vegetation growth.* Springer Publ Ltd, Springer Nature Switzerland AG. Part of Springer Nature, pp 577–590. https://doi.org/10.1007/978-3-030-49732-3.

Shah F, Lixiao N, Kehui C, Tariq S, Wei W, Chang C, Liyang Z, Farhan A, Fahad S, Huang J (2013) Rice grain yield and component responses to near 2°C of warming. *Field Crops Res* 157:98–110.

Shah Jahan M, Wang Y, Shu S, Zhong M, Chen Z, Wu J, Sun J, Guo S (2019) Exogenous salicylic acid increases the heat tolerance in Tomato (Solanum Lycopersicum L) by enhancing photosynthesis efficiency and improving antioxidant defense system through scavenging of reactive oxygen species. *Sci Hortic* 247:421–429. https://doi.org/10.1016/j.scienta.2018.12.047.

Sharkey T, Schrader S (2006) High temperature stress. In: *Physiology and molecular biology of stress tolerance in plants*. Springer, Berlin, Germany.

Shi Q, Bao Z, Zhu Z, Ying Q, Qian Q (2006) Effects of different treatments of salicylic acid on heat tolerance, chlorophyll fluorescence, and antioxidant enzyme activity in seedlings of Cucumis sativa L. *Plant Growth Regul* 48(2):127–135.

Song L, Yue L, Zhao H, Hou M (2013) Protection effect of nitric oxide on photosynthesis in rice under heat stress. *Acta Physiol Plant* 35(12):3323–3333.

Stitt M, Hurry V (2002) A plant for all seasons: alterations in photosynthetic carbon metabolism during cold acclimation in Arabidopsis. *Curr Opin Plant Biol* 5(3):199–206.

Szepesi Á, Csiszár J, Gémes K, Horváth E, Horváth F, Simon ML, Tari I (2009) Salicylic acid improves acclimation to salt stress by stimulating abscisic aldehyde oxidase activity and abscisic acid accumulation, and increases Na+ content in leaves without toxicity symptoms in Solanum Lycopersicum L. *J Plant Physiol* 166(9):914–925. https://doi.org/10.1016/j.jplph.2008.11.012.

Tariq M, Ahmad S, Fahad S, Abbas G, Hussain S, Fatima Z, Nasim W, Mubeen M, Rehman MHu, Khan MA, Adnan M, Wilkerson CJ, Hoogenboom G (2018) The impact of climate warming and crop management on phenology of sunflower-based cropping systems in Punjab, Pakistan. *Agri Forest Met* 256–257:270–282.

Theocharis A, Clément C, Barka EA (2012) Physiological and molecular changes in plants grown at low temperatures. *Planta* 235(6):1091–1105.

Thomashow MF (1999) Plant cold acclimation: freezing tolerance genes and regulatory mechanisms. *Annu Rev Plant Physiol Plant Mol Biol* 50:571–599.

Tsuda K, Sato M, Glazebrook J, Cohen JD, Katagiri F (2008) Interplay between MAMP-triggered and SA-mediated defense responses. *Plant J* 53(5):763–775.

ul Qamar Muhammad Tahir, Amna F, Amna B, Barira Z, Xitong Z, Ling-Ling C (2020) Effectiveness of conventional crop improvement strategies vs. omics. In: Fahad S, Hasanuzzaman M, Alam M, Ullah H, Saeed M, Khan AK, Adnan M (eds) *Environment, climate, plant and vegetation growth*. Springer Publ Ltd, Springer Nature Switzerland AG. Part of Springer Nature, pp 253–284. https://doi.org/10.1007/978-3-030-49732-3.

Unsar Naeem-U, Muhammad R, Syed HMB, Asad S, Mirza AQ, Naeem I, Muhammad H ur R, Fahad S, Shafqat S (2020) Insect pests of cotton crop and management under climate change scenarios. In: Fahad S, Hasanuzzaman M, Alam M, Ullah H, Saeed M, Khan AK, Adnan M (eds) *Environment, climate, plant and vegetation growth*. Springer Publ Ltd, Springer Nature Switzerland AG. Part of Springer Nature, pp 367–396. https://doi.org/10.1007/978-3-030-49732-3.

Vlot AC, Dempsey DA, Klessig DF (2009) Salicylic acid, a multifaceted hormone to combat disease. *Annu Rev Phytopathol* 47:177–206.

Wahid F, Fahad S, Subhan D, Adnan M, Zhen Y, Saud S, Manzer HS, Martin B, Tereza H, Rahul D (2020) Sustainable management with mycorrhizae and phosphate solubilizing bacteria for enhanced phosphorus uptake in calcareous soils. *Agriculturists* 10(334). https://doi.org/10.3390/agriculture10080334.

Wajid N, Ashfaq A, Asad A, Muhammad T, Muhammad A,Muhammad S, Khawar J, Ghulam MS, Syeda RS, Hafiz MH, Muhammad IAR, Muhammad ZH, Muhammad Habib ur R, Veysel T, Fahad S, Suad S, Aziz K, Shahzad A (2017) Radiation efficiency and nitrogen fertilizer impacts on sunflower crop in contrasting environments of Punjab, Pakistan. *Environ Sci Pollut Res* 25:1822–1836.

Wang L, Chen S, Kong W, Li S, Archbold DD (2006) Salicylic acid pretreatment alleviates chilling injury and affects the antioxidant system and heat shock proteins of peaches during cold storage. *Postharvest Biol Technol* 41(3):244–251.

Wang L-J, Fan L, Loescher W, Duan W, Liu GJ, Cheng JS, Luo HB, Li SH (2010) Salicylic acid alleviates decreases in photosynthesis under heat stress and accelerates recovery in grapevine leaves. *BMC Plant Biol* 10:34.

Wang L-J, Li S-H (2006) Salicylic acid-induced heat or cold tolerance in relation to Ca2+ homeostasis and antioxidant systems in young grape plants. *Plant Sci* 170(4):685–694.

Wang S, Bai G, Wang S, Yang L, Yang F, Wang Y, Zhu JK, Hua J (2016) Chloroplast RNA-binding protein RBD1 promotes chilling tolerance through 23S rRNA processing in Arabidopsis. *PLOS Genet* 12(5):e1006027.

White R (1979) Acetylsalicylic acid (aspirin) induces resistance to tobacco mosaic virus in tobacco. *Virology* 99(2):410–412.

Wu C, Cui K, Tang S, Li G, Wang S, Fahad S, Nie L, Huang J, Peng S, Ding Y (2020) Intensified pollination and fertilization ameliorate heat injury in rice (Oryza sativa L.) during the flowering stage. *Field Crops Res* 252. http://www.ncbi.nlm.nih.gov/pubmed/107795.

Wu C, Tang S, Li G, Wang S, Fahad S, Ding Y (2019) Roles of phytohormone changes in the grain yield of rice plants exposed to heat: a review. *PeerJ* 7:e7792. https://doi.org/10.7717/peerj.7792.

Xia XJ, Fang PP, Guo X, Qian XJ, Zhou J, Shi K, Zhou YH, Yu JQ (2018) Brassinosteroid-mediated apoplastic H2O2-glutaredoxin 12/14 cascade regulates antioxidant capacity in response to chilling in tomato. *Plant Cell Environ* 41(5):1052–1064.

Yang Z, Zhang Z, Zhang T, Fahad S, Cui K, Nie L, Peng S, Huang J (2017) The effect of season-long temperature increases on rice cultivars grown in the central and southern regions of China. *Front Plant Sci* 8:1908. https://doi.org/10.3389/fpls.2017.01908.

Zahida Z, Hafiz FB, Zulfiqar AS, Ghulam MS, Fahad S, Muhammad RA, Hafiz MH,Wajid N, Muhammad S (2017) Effect of water management and silicon on germination, growth, phosphorus and arsenic uptake in rice. *Ecotoxicol Environ Saf* 144:11–18.

Zafar-ul-Hye M, Muhammad T ahzeeb-ul-Hassan, Muhammad A, Fahad S, Martin B, Tereza D, Rahul D, Subhan D (2020b) Potential role of compost mixed biochar with rhizobacteria in mitigating lead toxicity in spinach. *Sci Rep* 10:12159. https://doi.org/10.1038/s41598-020-69183-9.

Zafar-ul-Hye M, Naeem M, Danish S, Fahad S, Datta R, Abbas M, Rahi AA, Brtnicky M, Holátko J, Tarar ZH, Nasir M (2020a) Alleviation of cadmium adverse effects by improving nutrients uptake in bitter gourd through cadmium tolerant rhizobacteria. *Environments* 7(8). https://doi.org/10.3390/environments7080054.

Zhang S, Wang X (2011) Overexpression of GASA5 increases the sensitivity of Arabidopsis to heat stress. *J Plant Physiol* 168(17):2093–2101.

Zia-ur-Rehman M (2020) Environment, climate change and biodiversity. In: Fahad S, Hasanuzzaman M, Alam M, Ullah H, Saeed M, Khan AK, Adnan M (eds) *Environment, climate, plant and vegetation growth.* Springer Publ Ltd, Springer Nature Switzerland AG. Part of Springer Nature, pp 473–502. https://doi.org/10.1007/978-3-030-49732-3.

9

Ethylene: A Key Regulatory Molecule in Plant Appraisal of Abiotic Stress Tolerance

Mona H. Soliman, Awatif M. Abdulmajeed, and Abdelghafar M. Abu-Elsaoud

CONTENTS

9.1 Introduction

9.1.1 Regulatory Role of Ethylene as a Signalling Molecule for Plant Growth and Responses

Phytohormones including auxins, gibberellins, and cytokinins act as promoters, and abscisic acid and ethylene as inhibitors (Gray, 2004; Iqbal et al., 2020) are the major regulators for various physiological and biochemical pathways as well as for enhanced crop growth and performance. Ethylene (ET) is a true biogas phytohormone that regulates several functions, including seed viability and fruit ripening and mediates interactions between plants and environmental changes at the sub-cellular, molecular, and tissue levels (Bleecker and Kende, 2000; Khan and Khan, 2014; Lin et al., 2009; Pieterse et al., 2009; Schaller, 2012). One of the open questions discussed in this chapter is how ethylene functions as a signal in sustainable development. Ethylene is involved in controlling cell division and elongation in terms of plant growth and improves the development phase, including ripening process, abscission, and cellular damage. Ethylene also contributes to the regulation of cell division and elongation and promotes fruit reproduction, cell death, and leaf senescence (Schaller, 2012).

9.2 Ethylene Regulates Morphological and Structural Behaviour

9.2.1 Root Architecture as a Key Trait for Plant Morphology and Growth Responses

Optimum root growth that serves as an interface between plants and the soil is considered a key factor in counteracting abiotic stresses (Davies and Bacon, 2003; Wells and Eissenstat, 2002). Such attributes determine plants' ability to use water and mineral nutrients to avoid water loss resulting from climate change (Baker et al., 1992). The proliferated root system, including primary and complex branching architecture, help plants to provide deeper soil penetration for maximum water acquisition and adapt to unfavourable circumstances (Franco et al., 2011; Negi et al., 2008). Numerous experiments have shown that the gaseous phytohormone ethylene is a vital component of root growth for the promotion of advantageous root development as well as the mediation of auxin-signalling molecules via ethylene responsive factors as modules of the major cross-talk (Lewis et al., 2011; Mao et al., 2016; Miao et al., 2018; Swarup et al., 2007; Xu et al., 2020). The direct consequence of increased accumulation of intracellular ethylene or ACC as a major contributing factor is the rapid transfer of IAA and as an outcome, the lessening of adventitious rooting by inhibition of ERF required by lateral early root (Chilley et al., 2006; Hirota et al., 2007; Lewis et al., 2011). The input of ethylene as a signal molecule may contribute to root growth and penetration via bioavailability of soil nutrients (Lynch and Brown, 1997) since later root output is correlated with inadequate soil phosphate as reported by Borch et al. (1999).

9.2.2 Triple Response

During the nineteenth century, coal gas was used for street lighting. A significant phenomenon was observed: Plants in the vicinity of the lamps lost leaves to a greater degree than other trees growing in the area. Later on, it was found that coal gas affects the growth and development of plants. And the presence of ethylene was identified in the coal gas as a signalling molecule causing the symptoms. A Russian graduate student, namely Dimitry Neljubov, in 1901, at the Botanical Institute of St. Petersburg, noticed that pea seedlings under dark conditions in the laboratory exhibited triple response symptoms: Reduced elongation of stem, increased shoot thickness (swelling), and abnormal horizontal growth. But when plants were grown under normal air, they had normal growth without any symptoms.

9.2.3 Nodule Development and Autoregulation

Ethylene is potentially triggered in rhizobacteria and regulates nodule development, integrity, and senescence (Caba et al. 1998; Ligero et al. 1986) and seriously impacts the developmental cycle. The penetration of the infection thread is hindered by ethylene (Penmetsa and Cook, 1997) and restricts the radial location of the cortical cellular division (Heidstra et al. 1997; Penmetsa et al. 2003). As a consequence, a faster approach should be adopted to define that ethylene is a key nodulation molecule that particularly affects hormonal signal transduction pathways and shifts on/off to manage each level of developmental formation mechanisms (Abeles et al. 2012; Guinel, 2015; Wang et al. 2017). Due to the normal impact of a higher accumulation of ethylene under challenging circumstances, several developmental changes, including activation of the nodule, amounts, thread growth, and orientation have been altered (Glick, 2005; Guinel, 2015).

9.2.4 Ethylene and a Dual Outcome Interface

Triple growth activity in *Pisum sativum* seedlings is considered a valuable attribute in accordance with ethylene biological markers (Guzmán and Ecker, 1990). The overall reduction in the development of hypocotyl biomass is a significant drop in the length of hypocotyl, root elongation, and apical hook emergence (Guzmán and Ecker, 1990; Stepanova and Alonso, 2005). In addition, certain crops have automatically shown visceral attitudes to the growth of Poa leaves and reduction in the overall root elongation and penetration of *Rumex palustris* and *Cucumis sativus* due to massive ethylene promotion (Binder et al. 2004; Fiorani et al. 2002; Visser et al. 1997). A plethora of measures to evaluate the application of ethylene involved in the improvement of major crop species by rising adaptive tolerance mechanisms have been established (Rajala and Peltonen-Sainio, 2002; Zhang et al. 2020). As shown through rice seedlings or *Arabidopsis thaliana*, the encouraging or hindering ethylene interaction is strongly attributable to sensitive proteins (Fukao and Bailey-Serres, 2008; Yamauchi et al. 2016). Finally, previous studies indicate that the availability of ethylene, plant species, and environmental growth determines whether the behaviour pattern is promotive or inhibitor (Pierik et al. 2006a).

9.3 Ethylene as a Signal during Morphological and Anatomical Responses

9.3.1 Root System as a Close-Range Interface to Plant Morphology

The root systems of plants display a considerable plasticity in response to various endogenous and stress factors. Plant hormone ethylene controls the initiation of the formation of laterals in the pericycle and outgrowth and elongation of lateral root (Negi et al. 2008). Ivanchenko et al. (2008) studied the ethylene–auxin interactions in regulating the initiation and emergence of lateral root in *Arabidopsis thaliana*. The results suggest that the enhanced ethylene synthesis is triggered by application of ethylene precursor 1-aminocyclopropane-1-carboxylic acid (ACC) and promotes the development of primordial lateral root. The application of plants with high ACC was found to inhibit the ability of pericycle cells to initiate the formation of new lateral root primordial. When auxin and ACC was applied simultaneously, ACC is unable to prevent the auxin stimulation of lateral root. Van Hoewyk et al. (2008) by employing transcriptomic analyses revealed that selenium (Se) up-regulated the ethylene associated genes in *Arabidopsis thaliana*. In yet another study, Feigl et al. (2019) studied the interplay of ethylene and nitric oxide in lateral root emergence in *Arabidopsis thaliana*. The study aimed to investigate the effect of Se on ethylene and nitric oxide in wild-type and ethylene insensitive *etr1-1 Arabidopsis*. They suggested the antagonistic interplay of nitric oxide and ethylene ET and NO in regulating lateral root emergence in *Arabidopsis thaliana* under Se stress conditions. It has been reported that the ethylene-induced modulation of synthesis and transport of auxin develops auxin gradients which is necessary for the lateral root primordia induction in the pericycle (Fukao and Bailey-Serres, 2008).

9.3.2 Aerenchyma Formation and Adaptation

The mechanism of aerenchyma formation under ethylene and stress conditions is a well-studied topic. Various studies have shown that the deficiency of oxygen triggers the production of ethylene which results in the formation of aerenchyma in submerged plant tissues (Dauphinee et al. 2012; He et al. 1992; Kawase, 1981). Ethylene, along with ROS, acts as a signalling molecule to induce the formation of aerenchyma for facilitating the exchange of gases and the process of programmed cell death (PCD) was involved in aerenchyma formation in *Helianthus annuus* plants. Their results suggest that ethylene-mediated ROS signalling plays a crucial role in aerenchyma formation. In roots of gramineous plants (rice and maize), Yamauchi et al. (2016) studied the ethylene-dependent aerenchyma formation. They showed that ethylene-dependent formation of aerenchyma is regulated differently in rice and maize plants. The attribution of the difference in the formation of aerenchyma in rice and maize plants resulted from their different mechanisms for ethylene biosynthesis regulation. The results of Yamauchi et al. (2014) also suggest that in rice very long-chain fatty acids increase ethylene production by promoting the biosynthesis of 1-aminocyclopropane-1-carboxylic acid, which, in turn, induces lysigenous aerenchyma formation in the root cortex. In tomato plants, Vidoz et al. (2016) showed that the rapid adventitious root formation and partial ethylene sensitivity result is responsible for adaptation to flooding stress. In a recent study, Yamauchi et al. (2020) unravelled the role of auxin in ethylene-mediated aerenchyma formation in rice plants.

9.3.3 Shoot Systems as Agronomic Traits and Adaptation Mechanisms

Phytohormone ethylene regulates the growth of the shoot system under stress and optimal conditions. Van Dijck et al. (1988) studied the role of ethylene along with cytokinins in shoot growth in bromeliads. They observed that the stimulation of ethylene production is engaged with the outgrowth of the lateral buds in the tested plants. The ethylene rise mediated promotion of growth was attributed to (a) shoots' outgrowth being inhibited by preventing ethylene production, (b) ACC substituted for another hormone in the promotion of ethylene production and lateral bud outgrowth. Deep water rice is an *Oryza sativa* variety that is grown in flood plains and could remain covered by up to 5 feet of water. The primary effect in deep water rice plants under submergence is an increase in cell number per cell file which is resulted from a decrease in cell cycle time from 24 h to 7 h in the intercalary meristem (Bleecker et al. 1986).

9.3.4 Leaf Anatomy and Morphology

Plant hormone ethylene regulates both growth and senescence of plants. The promotion or inhibition of these processes depends on its concentration, plant species, and timing of application. In mustard plants, it has been reported that an ethylene-releasing compound 'ethephon' enhanced stress ethylene levels and increased leaf area at a lower concentration (Khan et al. 2008). The development of leaves, flowers, and fruits is also governed by ethylene. In *Arabidopsis*, it has been observed that ETHYLENE RESPONSE FACTOR5 (ERF5) and ERF6 improved the leaf growth in response to environmental conditions (Dubois et al. 2018). In two separate experiments, it was observed that treatment with ethephon increases the expansion of leaf area to a considerable extent (Khan, 2005; Khan et al. 2008). In young leaves, ethylene and the downstream ERFs played a key role in the regulation of leaf growth inhibition, thereby orchestrating the process of cell division and cell expansion (Dubois et al. 2018).

9.4 Interaction of Ethylene with Other Signalling Molecules under Abiotic Stress

9.4.1 The Potential Role of Ethylene in Waterlogging Mitigation

Ethylene initiates and controls many adaptive plant responses, which help the plants to avoid anaerobiosis at the molecular and morphological levels under waterlogged conditions. Visser et al. (1997) reported the formation of aerenchyma to be coincident with the enhancement of ethylene synthesis in roots under hypoxia. Similarly, in maize roots, external application of ethylene, even at very low rates of 0.1 cm^3/dm^3 stimulated aerenchyma development, which was repressed in the presence of an ethylene inhibitor, Ag+ ions (Drew et al. 1981). The formation of adventitious roots with large air spaces is another major response to ethylene accumulation under waterlogged conditions. The application of an ethylene inhibitor, L-α-(2-amino-ethoxyvinyl)-glycine or aminobutyric acid, led to a decreased formation of adventitious roots in waterlogged *Rumex palustris* (Visser et al. 1997; Hadi et al. 2014). Hypoxia in maize seedlings caused increased ethylene production as well as enhanced expression of enzymes capable of destroying cell walls in adventitious roots.

One mechanism that causes ethylene accumulation is its diffusion into water from roots at rates that are 10 times slower than its diffusion into the air (Stunzi and Kende, 1989). Ethylene is released from flooded soil into the internal aerenchyma channels from where it moves on to the root and, eventually, the shoot. Another mechanism behind the increase in ethylene concentrations under waterlogged conditions is its enhanced synthesis in hypoxic roots and aerobic shoots (Jackson, 1985). Concentrations of endogenous ethylene in organs of a flooded plant are rapidly elevated to physiologically saturating amounts within the first hour of flooding (Sasidharan and Voesenek, 2015). Synthesis of the immediate precursor of ethylene, ACC, occurs throughout the plant but largely takes place in the root through an oxygen-dependent pathway (Bradford and Yang, 1980). However, all preceding biochemical reactions are oxygen-independent; in fact, the activity of ACS is stimulated by hypoxic conditions (Cohen and Kende,

1987). Therefore, when anaerobic root cells are unable to generate ACC under hypoxic waterlogged conditions, the process of ethylene production is not stopped; instead, it is shifted toward the shoot cells, where oxygen is still available. The slow diffusion of ethylene into water from the roots further promotes its entrapment within the plant's mass, regardless of other environmental conditions. Therefore, ethylene is considered to be an excellent signal of waterlogging-induced stress in plants.

9.4.2 Role of Ethylene in Heat Stress

High temperatures alter the production of ethylene to different extents in different plant species. For instance, while ethylene content was enhanced in wheat grains under the influence of heat stress as described by Hays et al. (2007) and Khan et al. (2015), another study reported it to decrease (Arshad and Frankenberger, 2002). At high temperatures, ethylene production was enhanced in *Glycine max* that, in turn, caused ROS generation and premature leaf senescence (Firon et al. 2012). The application of 1-methylcyclopropene (1-MCP), an inhibitor of ethylene, inhibited its production and delayed leaf senescence by strengthening the antioxidant defence system, even under heat stress. On the contrary, ethylene confers thermo-tolerance to tomato pollens (Klay et al. 2018). Exposure to high temperatures induced the expression of ethylene-responsive (ER) genes, including *ER5*, *ER21*, *LeJERF1*, and *ER24*, along with the expression of the gene-encoding ACS in developing pollen grains (Frank et al. 2009).

9.4.3 Cross-Talk between Ethylene and Heavy-Metal Stress

The production of ethylene was enhanced in plants exposed to high concentrations of cadmium (Cd), copper (Cu), iron (Fe), nickel (Ni), and zinc (Zn) (Maksymiec, 2007). Cadmium and Cu have been reported to stimulate ethylene synthesis by increasing the expression of *ACS* (Khan et al. 2015). Increased production of ethylene under conditions of heavy-metal stress is attributed to enhanced expression of *ACS* and *ACO* (Schellingen et al. 2014). *Sesbania drummondii* is a Pb-hyperaccumulator that exhibits increased levels of these two enzymes, as well ethylene production, on exposure to Pb (Srivastava et al. 2007). Similarly, transcriptome analysis of the Arsenic (As)-tolerant Columbia ecotype of *A. thaliana* shows it to be enriched in gene-encoding components of ethylene biosynthesis or ethylene-signalling pathways on exposure to As, as compared to a less tolerant Wassilewskija ecotype (Fu et al. 2014). The increased generation of ethylene in plants under heavy-metal stress is often accompanied by a reduction in plant growth and development, suggesting that it is a negative regulator in the response of plants to heavy metals. In *A. thaliana*, exposure to Cd up-regulated the expression of *ERF2*, *ETR2*, and *ACO2*, and thus, resulted in enhanced production of ethylene, which caused a reduction in leaf biomass (Schellingen et al. 2014).

9.4.4 Ethylene as a Response Factor under Salinity Stress

Ethylene is one of the main response factors working against salinity in most plants. Central membrane proteins, including ethylene overproducer 2 (ETO2), ethylene response sensor 1 (ERS1), and EIN2 are responsible for ethylene signalling during plant growth. Application of ethylene or ACC to plants grown under high salinity improved plant tolerance (Cao et al. 2009), mainly via increased expression of ROS-scavenging proteins (Peng et al. 2014). Ethylene has been reported to mediate response to salinity in rice as well (Steffens, 2014). On the contrary, research has shown a negative role of ACC in the growth regulation of tomato seedlings grown under saline conditions (Albacete et al. 2009). In tomato leaves, ACC was observed to increase before Na+ accumulation under saline conditions; this accumulation was coincident with the onset of oxidative stress and leaf senescence (Freitas, 2018; Ghanem et al. 2008). Overproduction of ethylene by plants themselves or their external treatment with ethylene-producing compounds, such as ethephon or ethylene precursors, such as ACC, has shown to increase tolerance to salinity in various plants including *A. thaliana* (Yang et al. 2013) and maize (Freitas, 2018). Ethylene is known to be a positive mediator of tolerance against salinity in grapevines (Xu et al. 2019), maize (Freitas, 2018), and tomato (Gharbi et al. 2017).

9.4.5 Ethylene Production and Drought Stress

Ethylene aids plants in responding to drought by controlling the opening and closing of stomata (Husain et al. 2020). It was established that ethylene controls stomatal closure under stress in conjunction with flavanol produced in guard cells by an *EIN2*-mediated pathway. Under conditions of drought, plants synthesise ABA and ethylene. Although these two hormones cause stomatal closure independently, when working together, they cannot induce the closure of stomata (Beguerisse-Dıaz et al. 2012). A deficit of water has been observed to enhance ethylene production, along with an increase and subsequent decrease in ACC, suggesting that drought induces the expression of *ACS*, which encodes the rate-controlling enzyme in biosynthesis of ethylene. Increased production of ethylene under drought may reduce photosynthesis (Rajala and Peltonen-Sainio, 2002). Furthermore, under drought conditions, ethylene causes leaf abscission that further hampers water loss (Gomez-Cadenas et al. 1996).

9.4.6 Cross-Talk between Ethylene and UV-B Stress

UV-B stress has various deleterious and damaging effects on plants and plants' cell metabolism. Concerning interaction between ethylene hormone and UV-B stress-signalling, little data can be found (Vanhaelewyn et al. 2016). It was indicated that the lack of the positive regulator and central component of ethylene signalling, EIN2, enhances UV-B stress protection because of enhanced production of flavonoids (Sun et al. 2011). Results are in accordance with ethylene stimulating COP1 function that inhibits photomorphogenesis (Yu et al. 2013). Flavonoids accumulate in high levels in the epidermis under the effect of UV-B. Fascinatingly, ethylene mediates a specific epidermal stomatal closure response to UV-B in *Vicia faba* L., acting upstream of H_2O_2 and NO (He et al. 2011). Notably, most of the 1-aminocyclopropane-1-carboxylic acid (ACC) synthase genes of Arabidopsis, encoding the rate-limiting step in ethylene biosynthesis, are expressed in the epidermis but not in the mesophyll (Tsuchisaka and Theologis, 2004). UV-B radiation, at various – mainly high – intensities, stimulates ethylene production in many plant species including pear (Predieri et al. 1995), Arabidopsis (Mackerness et al. 1999; Rakitina et al. 2001), tobacco (McLeod et al. 2008; Nara and Takeuchi, 2002), maize (Wang et al. 2006), and canola (Qaderi et al. 2010). However, it is not clear whether this production is enzymatic or not. It has been suggested that UV-B-induced ethylene is derived from pectin in the leaves (McLeod et al. 2008), while production through the canonical ethylene biosynthesis pathway with ACC as intermediate has also been documented (Nara and Takeuchi, 2002; Wang et al. 2006). The accumulation of ACC was UV-B dose-dependent in pea, with more ACC accumulating at higher UV-B doses (Katerova et al. 2009). Part of the underlying molecular mechanism may lie in the transcriptional up-regulation of ACC synthases. In tomato, ACS genes are up-regulated in the presence of enhanced UV-B (An et al. 2006). However, this seems an exception. Many data point to stress signalling enhancing ethylene biosynthesis rather than the UVR8 pathway.

BIBLIOGRAPHY

Abeles, F.B., Morgan, P.W., Saltveit, Jr M.E., 2012. *Ethylene in plant biology.* Academic Press.

Adato, I., Gazit, S., 1974. Water-deficit stress, ethylene production, and ripening in avocado fruits. *Plant Physiol.* 53(1), 45–46. https://doi.org/10.1104/pp.53.1.45.

Albacete, A., Martínez-Andújar, C., Ghanem, M.E., Acosta, M., Sánchez-Bravo, J., Asins, M.J., Cuartero, J., Lutts, S., Dodd, I.C., Pérez-Alfocea, F., 2009. Rootstock-mediated changes in xylem ionic and hormonal status are correlated with delayed leaf senescence, and increased leaf area and crop productivity in salinized tomato. *Plant Cell Environ.* 32(7), 928–938. https://doi.org/10.1111/j.1365-3040.2009.01973.x.

An, L., Xu, X., Tang, H., Zhang, M., Hou, Z., Liu, Y., Zhao, Z., Feng, H., Xu, S., Wang, X., 2006. Ethylene production and 1-aminocyclopropane-1-carboxylate (ACC) synthase gene expression in tomato (Lycopsicon esculentum mill.) leaves under enhanced UV-B radiation. *J. Integr. Plant Biol.* 48(10), 1190–1196. https://doi.org/10.1111/j.1744-7909.2006.00331.x.

Arshad, M., Frankenberger, W.T., 2002. *Ethylene: agricultural sources and applications.* Kluwer Academic/ Plenum Publishers, New York.

Baker, J.M., Wraith, J.M., Dalton, F.N., 1992. Root function in water transport. In: Hatfield, J.L., Stewart, B.A. (eds) *Limitations to plant root growth: advances in soil science*, vol 19. Springer, New York, NY, pp 53–72. https://doi.org/10.1007/978-1-4612-2894-3_3.

Beguerisse-Dıaz, M., Hernández-Gómez, M.C., Lizzul, A.M., Barahona, M., Desikan, R., 2012. Compound stress response in stomatal closure: a mathematical model of ABA and ethylene interaction in guard cells. *BMC Syst. Biol.* 6, 146. https://doi.org/10.1186/1752-0509-6-146.

Binder, B.M., O'Malley, R.C., Wang, W., Moore, J.M., Parks, B.M., Spalding, E.P., Bleecker, A.B., 2004. Arabidopsis seedling growth response and recovery to ethylene: a kinetic analysis. *Plant Physiol.* 136(2), 2913–2920.

Bleecker, A.B., Kende, H., 2000. Ethylene: a gaseous signal molecule in plants. *Annu. Rev. Cell Dev. Biol.* 16, 1–18.

Bleecker, A.B., Schuette, J.L., Kende, H., 1986. Anatomical analysis of growth and developmental patterns in the internode of deepwater rice. *Planta* 169(4), 490–497.

Borch, K., Bouma, T.J., Lynch, J.P., Brown, K.M., 1999. Ethylene: a regulator of root architectural responses to soil phosphorus availability. *Plant Cell Environ.* 22(4), 425–431. https://doi.org/10.1046/j.1365-3040.1999.00405.x.

Bradford, K.J., Yang, S.F., 1980. Xylem transport of 1-aminocyclopropane-1-carboxylic acid, an ethylene precursor, in waterlogged tomato plants. *Plant Physiol.* 65(2), 322–326. https://doi.org/10.1104/pp.65.2.322.

Caba, J.M., Recalde, L., Ligero, F., 1998. Nitrate-induced ethylene biosynthesis and the control of nodulation in alfalfa. *Plant Cell Environ.* 21(1), 87–93. https://doi.org/10.1046/j.1365-3040.1998.00242.x.

Cao, S., Chen, Z., Liu, G., Jiang, L., Yuan, H., Ren, G., Bian, X., Jian, H., Ma, X. 2009. The Arabidopsis ethylene-Insensitive 2 gene is required for lead resistance. *Plant Physiol. Biochem.* 47(4), 308–312. https://doi.org/10.1016/j.plaphy.2008.12.013.

Cao, W., Dong, Y., Zhang, J., Chen, S., 2003. Characterization of an ethylene receptor homolog gene from rice. *Sci. China C Life Sci.* 46(4), 370–378.

Cao, W.-H., Liu, J., He, X.-J., Mu, R.-L., Zhou, H.-L., Chen, S.-Y., Zhang, J.-S., 2007. Modulation of ethylene responses affects plant salt-stress responses. *Plant Physiol.* 143(2), 707–719.

Chibi, F., Matilla, A.J., 1994. The involvement of ethylene in in vitro maturation of mid-binucleate pollen of nicotiana tabacum. *J. Exp. Bot.* 45(5), 529–532. https://doi.org/doi.org/10.1093/jxb/45.5.529.

Chilley, P.M., Casson, S.A., Tarkowski, P., Hawkins, N., Wang, K.L.-C., Hussey, P.J., Beale, M., Ecker, J.R., Sandberg, G.K., Lindsey, K., 2006. The POLARIS peptide of Arabidopsis regulates auxin transport and root growth via effects on ethylene signaling. *Plant Cell* 18(11), 3058–3072. https://doi.org/10.1105/tpc.106.040790.

Cohen, E., Kende, H., 1987. In vivo 1-aminocyclopropane-1-carboxylate synthase activity in internodes of deepwater rice: enhancement by submergence and low oxygen levels. *Plant Physiol.* 84(2), 282–286. https://doi.org/10.1104/pp.84.2.282.

Dauphinee, A.N., Wright, H., Rantong, G., Gunawardena, A.H.L.A.N., 2012. The involvement of ethylene in programmed cell death and climacteric-like behaviour during the remodelling of lace plant (*Aponogeton madagascariensis*) leaves. *Botany* 90(12), 1237–1244. https://doi.org/10.1139/b2012-093.

Davies, W.J., Bacon, M.A., 2003. Adaptation of roots to drought. In: de Kroon, H., Visser, E.J.W. (eds) *Root ecology. Ecological studies (analysis and synthesis)*, vol 168. Springer, Berlin, Heidelberg.

Drew, M.C., Jackson, M.B., Giffard, S.C., Campbell, R., 1981. Inhibition by silver ions of gas space (aerenchyma) formation in adventitious roots of Zea mays L. subjected to exogenous ethylene or to oxygen deficiency. *Planta* 153(3), 217–224. https://doi.org/10.1007/BF00383890.

Dubois, M., Van den Broeck, L., Inzé, D., 2018. The pivotal role of ethylene in plant growth. *Trends Plant Sci.* 23(4), 311–323. https://doi.org/10.1016/j.tplants.2018.01.003.

Dugardeyn, J., Van Der Straeten, D., 2008. Ethylene: fine-tuning plant growth and development by stimulation and inhibition of elongation. *Plant Sci.* 175, 59–70.

Feigl, G., Horváth, E., Molnár, Á., Oláh, D., Poór, P., Kolbert, Z., 2019. Ethylene-nitric oxide interplay during selenium-induced lateral root emergence in Arabidopsis. *J. Plant Growth Regul.* 38(4), 1481–1488.

Fiorani, F., Bögemann, G.M., Visser, E.J., Lambers, H., Voesenek, L.A., 2002. Ethylene emission and responsiveness to applied ethylene vary among Poa species that inherently differ in leaf elongation rates. *Plant Physiol.* 129(3), 1382–1390.

Firon, N., Pressman, E., Meir, S., Khoury, R., Altahan, L., 2012. Ethylene is involved in maintaining tomato (Solanum Lycopersicum) pollen quality under heat-stress conditions. *AoB Plants*. https://doi.org/10.1093/aobpla/pls024.

Franco, J.A., Bañón, S., Vicente, M.J., Miralles, J., Martínez-Sánchez, J.J., 2011. Review article: root development in horticultural plants grown under abiotic stress conditions—a review. *J. Hortic. Sci. Biotechnol.* 86(6), 543–556.

Frank, G., Pressman, E., Ophir, R., Althan, L., Shaked, R., Freedman, M., Shen, S., Firon, N., 2009. Transcriptional profiling of maturing tomato (Solanum Lycopersicum L.) microspores reveals the involvement of heat shock proteins, ROS scavengers, hormones, and sugars in the heat stress response. *J. Exp. Bot.* 60(13), 3891–3908. https://doi.org/10.1093/jxb/erp234.

Freitas, V.S., 2018. Ethylene triggers salt tolerance in maize genotypes by modulating polyamine catabolism enzymes associated with H2O2 production. *Environ. Exp. Bot.* 12: 75–86.

Fu, S.F., Chen, P.Y., Nguyen, Q.T., Huang, L.Y., Zeng, G.R., Huang, T.L., Lin, C.Y., Huang, H.J., 2014. Transcriptome profiling of genes and pathways associated with arsenic toxicity and tolerance in Arabidopsis. *BMC Plant Biol.* 94. https://doi.org/10.1186/1471-2229-14-94.

Fukao, T., Bailey-Serres, J., 2008. Ethylene—a key regulator of submergence responses in rice. *Plant Sci.* 175(1–2), 43–51. https://doi.org/10.1016/j.plantsci.2007.12.002.

Gallie, D.R., 2015. Ethylene receptors in plants—why so much complexity? *F1000Prime Rep.* 7: 1–12.

Ghanem, M.E., Albacete, A., Martínez-Andújar, C., Acosta, M., Romero-Aranda, R., Dodd, I.C., Lutts, S., Pérez-Alfocea, F., 2008. Hormonal changes during salinity-induced leaf senescence in tomato (Solanum Lycopersicum L.). *J. Exp. Bot.* 59(11), 3039–3050. https://doi.org/10.1093/jxb/ern153.

Gharbi, E., Martínez, J.-P., Benahmed, H., Lepoint, G., Vanpee, B., Quinet, M., Lutts, S., 2017. Inhibition of ethylene synthesis reduces salt-tolerance in tomato wild relative species Solanum chilense. *J. Plant Physiol.* 210, 24–37. https://doi.org/10.1016/j.jplph.2016.12.001.

Glick, B.R., 2005. Modulation of plant ethylene levels by the bacterial enzyme ACC deaminase. *FEMS Microbiol. Lett.* 251(1), 1–7. https://doi.org/10.1016/j.femsle.2005.07.030.

Gomez-Cadenas, A., Tadeo, F.R., Talon, M., Primo-Millo, E., 1996. Leaf abscission induced by ethylene in water-stressed intact seedlings of Cleopatra mandarin requires previous abscisic acid accumulation in roots. *Plant Physiol.* 112(1), 401–408. https://doi.org/10.1104/pp.112.1.401.

González Besteiro, M.A., Bartels, S., Albert, A., Ulm, R., 2011. Arabidopsis MAP kinase phosphatase 1 and its target MAP kinases 3 and 6 antagonistically determine UV-B stress tolerance, independent of the UVR8 photoreceptor pathway. *Plant J.* 68(4), 727–737. https://doi.org/10.1111/j.1365-313X.2011.04725.x.

Gray, W.M., 2004. Hormonal regulation of plant growth and development. *PLOS Biol.* 2(9), e311. https://doi.org/10.1371/journal.pbio.0020311.

Guinel, F.C., 2015. Ethylene, a hormone at the center-stage of nodulation. *Front. Plant Sci.* 6. https://doi.org/10.3389/fpls.2015.01121.

Guzmán, P., Ecker, J.R., 1990. Exploiting the triple response of Arabidopsis to identify ethylene-related mutants. *Plant Cell* 2(6), 513–523. https://doi.org/10.1105/tpc.2.6.513.

Hadi, F., Ibrar, M., Dastagir, G., Arif, M., Naveed, K., Adnan, M., 2014. Weed diversity in wheat and maize with special reference to their ethnomedicinal uses at rich valley, HinduKush range, Chitral, Pakistan. *Pak. J. Weed Sci. Res.* 20(3), 335–346.

Han, L., Li, G.-J., Yang, K.-Y., Mao, G., Wang, R., Liu, Y., Zhang, S., 2010. Mitogen-activated protein kinase 3 and 6 regulate Botrytis cinerea-induced ethylene production in Arabidopsis. *Plant J* 64(1), 114–127. https://doi.org/10.1111/j.1365-313X.2010.04318.x.

Hays, D.B., Do, J.H., Mason, R.E., Morgan, G., Finlayson, S.A., 2007. Heat stress induced ethylene production in developing wheat grains induces kernel abortion and increased maturation in a susceptible cultivar. *Plant Sci.* 172(6), 1113–1123. https://doi.org/10.1016/j.plantsci.2007.03.004.

He, C.-J., Morgan, P.W., Drew, M.C., 1992. Enhanced sensitivity to ethylene in nitrogen- or phosphate-starved roots of *Zea mays* L. during aerenchyma formation. *Plant Physiol.* 98(1), 137–142. https://doi.org/10.1104/pp.98.1.137.

He, J.-M., Zhang, Z., Wang, R.-B., Chen, Y.-P., 2011. UV-B-induced stomatal closure occurs via ethylene-dependent NO generation in Vicia faba. *Funct. Plant Biol.* 38(4), 293. https://doi.org/10.1071/FP10219.

Hectors, K., Prinsen, E., De Coen, W., Jansen, M.A.K., Guisez, Y., 2007. Arabidopsis thaliana plants acclimated to low dose rates of ultraviolet B radiation show specific changes in morphology and gene expression in the absence of stress symptoms. *New Phytol.* 175(2), 255–270. https://doi.org/10.1111/j.1469-8137.2007.02092.x.

Heidstra, R., Yang, W.C., Yalcin, Y., Peck, S., Emons, A., van Kammen A, Bisseling, T., 1997. Ethylene provides positional information on cortical cell division but is not involved in Nod factor-induced root hair tip growth in rhizobium-legume interaction. *Development* 124(9), 1781–1787.

Hirota, A., Kato, T., Fukaki, H., Aida, M., Tasaka, M., 2007. The auxin-regulated AP2/EREBP gene PUCHI is required for morphogenesis in the early lateral root primordium of Arabidopsis. *Plant Cell* 19(7), 2156–2168. https://doi.org/10.1105/tpc.107.050674.

Husain, T., Fatima, A., Suhel, M., Singh, S., Sharma, A., Prasad, S.M., Singh, V.P., 2020. A brief appraisal of ethylene signaling under abiotic stress in plants. *Plant Signal. Behav.* 15(9), 1782051. https://doi.org/10.1080/15592324.2020.1782051.

Iqbal, S., Hussain, S., Qayyaum, M.A., Ashraf, M., Saifullah, 2020. The response of maize physiology under salinity stress and its coping strategies. *Plant Stress Physiol.* https://doi.org/10.5772/intechopen.92213.

Ivanchenko, M.G., Muday, G.K., Dubrovsky, J.G., 2008. Ethylene–auxin interactions regulate lateral root initiation and emergence in Arabidopsis thaliana. *Plant J.* 55(2), 335–347.

Jackson, M.B., 1985. Ethylene and responses of plants to soil waterlogging and submergence. *Annu. Rev. Plant Physiol.* 36(1), 145–174.

Jackson, M.B., 2008. Ethylene-promoted elongation: an adaptation to submergence stress. *Ann. Bot.* 101(2), 229–248. https://doi.org/10.1093/aob/mcm237.

Katerova, Z., Ivanov, S., Prinsen, E., Onckelen, H., Alexieva, V., Azmi, A., 2009. Low doses of ultraviolet-B or ultraviolet-C radiation affect phytohormones in young pea plants. *Biologia Plant* 53(2), 365–368. https://doi.org/10.1007/s10535-009-0068-1.

Kawase, M., 1981. Effect of ethylene on aerenchyma development. *Am. J. Bot.* 68(5), 651–658. https://doi.org/10.1002/j.1537-2197.1981.tb12396.x.

Khan, M.I.R., Khan, N.A., 2014. Ethylene reverses photosynthetic inhibition by nickel and zinc in mustard through changes in PS II activity, photosynthetic nitrogen use efficiency, and antioxidant metabolism. *Protoplasma* 251(5), 1007–1019.

Khan, M.I.R., Nazir, F, Asgher, M., Per, T.S., Khan, N.A., 2015. Selenium and sulfur influence ethylene formation and alleviate cadmium-induced oxidative stress by improving proline and glutathione production in wheat. *J. Plant Physiol.* 173, 9–18. https://doi.org/10.1016/j.jplph.2014.09.011.

Khan, N.A., 2005. The influence of exogenous ethylene on growth and photosynthesis of mustard (Brassica juncea) following defoliation. *Scientia Horticulturae* 105(4), 499–505. https://doi.org/10.1016/j.scienta.2005.02.004.

Khan, N.A., Mir, M.R., Nazar, R., Singh, S., 2008. The application of ethephon (an ethylene releaser) increases growth, photosynthesis and nitrogen accumulation in mustard (Brassica juncea L.) under high nitrogen levels. *Plant Biol.* 10(5), 534–538.

Klay, I., Gouia, Sandra, Liu, M., Mila, I., Khoudi, H., Bernadac, A., Bouzayen, M., Pirrello, J., 2018. Ethylene Response Factors (ERF) are differentially regulated by different abiotic stress types in tomato plants. *Plant Sci.* 274, 137–145. https://doi.org/10.1016/j.plantsci.2018.05.023.

Larkindale, J., Knight, M.R., 2002. Protection against heat stress-induced oxidative damage in Arabidopsis involves calcium, abscisic acid, ethylene, and salicylic acid. *Plant Physiol.* 128(2), 682–695. https://doi.org/10.1104/pp.010320.

Lewis, D.R., Negi, S., Sukumar, P., Muday, G.K., 2011. Ethylene inhibits lateral root development, increases IAA transport and expression of PIN3 and PIN7 auxin efflux carriers. *Development* 138(16), 3485–3495. https://doi.org/10.1242/dev.065102.

Li, Z., Zhang, L., Yu, Y., Quan, R., Zhang, Z., Zhang, H., Huang, R., 2011. The ethylene response factor AtERF11 that is transcriptionally modulated by the bZIP transcription factor HY5 is a crucial repressor for ethylene biosynthesis in Arabidopsis. *Plant J.* 68(1), 88–99. https://doi.org/10.1111/j.1365-313X.2011.04670.x.

Ligero, F., Lluch, C., Olivares, J., 1986. Evolution of ethylene from roots of Medicago sativa Plants inoculated with Rhizobium meliloti. *J. Plant Physiol.* 125(3–4), 361–365. https://doi.org/10.1016/S0176-1617(86)80158-4.

Lin, Z., Zhong, S., Grierson, D., 2009. Recent advances in ethylene research. *J. Exp. Bot.* 60(12), 3311–3336.

Lynch, J., Brown, K.M., 1997. Ethylene and plant responses to nutritional stress. *Physiol. Plant.* 100, 613–619.

Mackerness, S.A.H., Surplus, S.L., Blake, P., John, C.F., Buchanan-Wollaston, V., Jordan, B.R., Thomas, B., 1999. Ultraviolet-B-induced stress and changes in gene expression in Arabidopsis thaliana: role of signalling pathways controlled by jasmonic acid, ethylene and reactive oxygen species. *Plant Cell Environ.* 22(11), 1413–1423. https://doi.org/10.1046/j.1365-3040.1999.00499.x.

Maksymiec, W., 2007. Signaling responses in plants to heavy metal stress. *Acta Physiol. Plant.* 29(3). https://doi.org/10.1007/s11738-007-0036-3.

Maksymiec, W., 2011. Effects of jasmonate and some other signalling factors on bean and onion growth during the initial phase of cadmium action. *Biologia Plant* 55(1), 112–118. https://doi.org/10.1007/s10535-011-0015-9.

Mao, J.-L., Miao, Z.-Q., Wang, Z., Yu, L.-H., Cai, X.-T., Xiang, C.-B., 2016. Arabidopsis ERF1 mediates cross-talk between ethylene and auxin biosynthesis during primary root elongation by regulating ASA1 expression. *PLOS Genet.* 12(1), e1005760. https://doi.org/10.1371/journal.pgen.1005760.

McLeod, A.R., Fry, S.C., Loake, G.J., Messenger, D.J., Reay, D.S., Smith, K.A., Yun, B.-W., 2008. Ultraviolet radiation drives methane emissions from terrestrial plant pectins. *New Phytol.* 180(1), 124–132. https://doi.org/10.1111/j.1469-8137.2008.02571.x.

Métraux, J.-P., Kende, H., 1983. The role of ethylene in the growth response of submerged deep water rice. *Plant Physiol.* 72(2), 441–446. https://doi.org/10.1104/pp.72.2.441.

Miao, Z.-Q., Zhao, P.-X., Mao, J.-L., Yu, L.-H., Yuan, Y., Tang, H., Liu, Z.-B., Xiang, C.-B., 2018. HOMEOBOX PROTEIN52 mediates the crosstalk between ethylene and auxin signaling during primary root elongation by modulating auxin transport-related gene expression. *Plant Cell* 30(11), 2761–2778. https://doi.org/10.1105/tpc.18.00584.

Musgrave, A., Jackson, M.B., Ling, E., 1972. Callitriche Stem Elongation is controlled by ethylene and gibberellin. *Nat. New Biol.* 238(81), 93–96. https://doi.org/10.1038/newbio238093a0.

Nara, A., Takeuchi, Y., 2002. Ethylene evolution from tobacco leaves irradiated with UV-B. *J. Plant Res.* 115(4), 247–253. https://doi.org/10.1007/s10265-002-0031-0.

Negi, S., Ivanchenko, M.G., Muday, G.K., 2008. Ethylene regulates lateral root formation and auxin transport in *Arabidopsis thaliana. Plant J.* 55(2), 175–187. https://doi.org/10.1111/j.1365-313X.2008.03495.x.

Ni, X.-L., Gui, M.-Y., Tan, L.-L., Zhu, Q., Liu, W.-Z., Li, C.-X., 2018. Programmed cell death and aerenchyma formation in water-logged sunflower stems and its promotion by ethylene and ROS. *Front. Plant Sci.* 9, 1928.

Pan, W.S., Zheng, L.P., Tian, H., Li, W.Y., Wang, J.W., 2014. Transcriptome responses involved in artemisinin production in Artemisia annua L. under UV-B radiation. *J. Photochem. Photobiol. B* 140, 292–300. https://doi.org/10.1016/j.jphotobiol.2014.08.013.

Peng, J., Li, Z., Wen, X., Li, W., Shi, H., Yang, L., Zhu, H., Guo, H., 2014. Salt-induced stabilization of EIN3/EIL1 confers salinity tolerance by deterring ROS accumulation in Arabidopsis. *PLOS Genet.* 10(10), e1004664. http://dx.doi.org.ezproxy.library.dal.ca/10.1371/journal.pgen.1004664.

Penmetsa, R.V., Cook, D.R., 1997. A legume ethylene-insensitive mutant hyperinfected by its rhizobial symbiont. *Science* 275(5299), 527–530.

Penmetsa, R.V., Frugoli, J.A., Smith, L.S., Long, S.R., Cook, D.R., 2003. Dual genetic pathways controlling nodule number inMedicago truncatula. *Plant Physiol.* 131(3), 998–1008.

Pierik, R., Tholen, D., Poorter, H., Visser, E.J.W., Voesenek, L.A.C.J., 2006. The Janus face of ethylene: growth inhibition and stimulation. *Trends Plant Sci.* 11(4), 176–183. https://doi.org/10.1016/j.tplants.2006.02.006.

Pieterse, C.M.J., Leon-Reyes, A., Van der Ent, S., Van Wees, S.C.M., 2009. Networking by small-molecule hormones in plant immunity. *Nat. Chem. Biol.* 5(5), 308.

Pontin, M.A., Piccoli, P.N., Francisco, R., Bottini, R., Martinez-Zapater, J.M., Lijavetzky, D., 2010. Transcriptome changes in grapevine (Vitis vinifera L.) cv. Malbec leaves induced by ultraviolet-B radiation. *BMC Plant Biol.* 10, 224. https://doi.org/10.1186/1471-2229-10-224.

Predieri, S., Norman, H.A., Krizek, D.T., Pillai, P., Mirecki, R.M., Zimmerman, R.H., 1995. Influence of UV-B radiation on membrane lipid composition and ethylene evolution in 'Doyenne d'hiver' pear shoots grown in vitro under different photosynthetic photon fluxes. *Environ. Exp. Bot.* 35(2), 151–160. https://doi.org/10.1016/0098-8472(95)00003-2.

Qaderi, M.M., Basraon, N.K., Chinnappa, C.C., Reid, D.M., 2010. Combined effects of temperature, ultraviolet-B radiation, and watering regime on growth and physiological processes in canola (Brassica napus) seedlings. *Int. J. Plant Sci.* 171(5), 466–481. https://doi.org/10.1086/652389.

Rajala, A., Peltonen-Sainio, P., 2002. Timing applications of growth regulators to alter spring cereal development at high latitudes. *Agric. Food Sci.* 11(3), 233–244. https://doi.org/10.23986/afsci.5721.

Rakitina, T.Ya., Vlasov, P.V., Rakitin, V.Yu., 2001. Hormonal aspects of different resistance of arabidopsis thalianamutants to ultraviolet radiation. *Russ. J. Plant Physiol.* 48(3), 353–358. https://doi.org/10.1023/A:1016618517615.

Raskin, I., Kende, H., 1984. Regulation of growth in stem sections of deep-water rice. *Planta* 160(1), 66–72.

Rébeillé, F., Jabrin, S., Bligny, R., Loizeau, K., Gambonnet, B., Van Wilder, V., Douce, R., Ravanel, S., 2006. Methionine catabolism in Arabidopsis cells is initiated by a gamma-cleavage process and leads to S-methylcysteine and isoleucine syntheses. *Proc Natl Acad Sci U S A* 103(42), 15687–15692. https://doi.org/10.1073/pnas.0606195103.

Sasidharan, R., Voesenek, L.A.C.J., 2015. Ethylene-mediated acclimations to flooding stress. *Plant Physiol.* 169(1), 3–12. https://doi.org/10.1104/pp.15.00387.

Schaller, G.E., 2012. Ethylene and the regulation of plant development. *BMC Biol.* 10, 9. https://doi.org/10.1186/1741-7007-10-9.

Schellingen, K., Van Der Straeten, D., Vandenbussche, F., Prinsen, E., Remans, T., Vangronsveld, J., Cuypers, A., 2014. Cadmium-induced ethylene production and responses in Arabidopsis thaliana rely on ACS2 and ACS6 gene expression. *BMC Plant Biol.* 14. https://doi.org/10.1186/s12870-014-0214-6.

Shibli, R.A., Kushad, M., Yousef, G.G., Lila, M.A., 2007. Physiological and biochemical responses of tomato microshoots to induced salinity stress with associated ethylene accumulation. *Plant Growth Regul.* 51(2), 159–169. https://doi.org/10.1007/s10725-006-9158-7.

Srivastava, A.K., Venkatachalam, P., Raghothama, K.G., Sahi, S.V., 2007. Identification of lead-regulated genes by suppression subtractive hybridization in the heavy metal accumulator Sesbania drummondii. *Planta* 225(6), 1353–1365. https://doi.org/10.1007/s00425-006-0445-3.

Steffens, B., 2014. The role of ethylene and ROS in salinity, heavy metal, and flooding responses in rice. *Front. Plant Sci.* 5. https://doi.org/10.3389/fpls.2014.00685.

Stepanova, A.N., Alonso, J.M., 2005. Ethylene signalling and response pathway: a unique signalling cascade with a multitude of inputs and outputs. *Physiol. Plant.* 123(2), 195–206.

Stünzi, J.T., Kende, H., 1989. Gas composition in the internal air spaces of deepwater rice in relation to growth induced by submergence. *Plant Cell Physiol.* 30(1), 49–56.

Sun, Z., Wen, C., Chen, Z., Zhang, Q., Fang, L., Li, J., Jiang, S., Cao, S., 2011. A role for ethylene-Insensitive 2 gene in the regulation of the ultraviolet-B response in Arabidopsis. *Acta Physiol. Plant.* 33(3), 1025–1030. https://doi.org/10.1007/s11738-010-0593-8.

Swarup, R., Perry, P., Hagenbeek, D., Van Der Straeten, D.V.D., Beemster, G.T.S., Sandberg, G., Bhalerao, R., Ljung, K., Bennett, M.J., 2007. Ethylene upregulates auxin biosynthesis in Arabidopsis seedlings to enhance inhibition of root cell elongation. *Plant Cell* 19(7), 2186–2196. https://doi.org/10.1105/tpc.107.052100.

Tsuchisaka, A., Theologis, A., 2004. Unique and overlapping expression patterns among the Arabidopsis 1-amino-cyclopropane-1-carboxylate synthase gene family members. *Plant Physiol.* 136(2), 2982–3000. https://doi.org/10.1104/pp.104.049999.

Ulm, R., Baumann, A., Oravecz, A., Máté, Z., Ádám, É., Oakeley, E.J., Schäfer, E., Nagy, F., 2004. Genome-wide analysis of gene expression reveals function of the bZIP transcription factor HY5 in the UV-B response of Arabidopsis. *Proc Natl Acad Sci U S A* 101(5), 1397–1402. https://doi.org/10.1073/pnas.0308044100.

Van Dijck, R., De Proft, M., De Greef, J., 1988. Role of ethylene and cytokinins in the initiation of lateral shoot growth in bromeliads. *Plant Physiol.* 86(3), 836–840.

Van Hoewyk, D., Takahashi, H., Inoue, E., Hess, A., Tamaoki, M., Pilon-Smits, E.A., 2008. Transcriptome analyses give insights into selenium-stress responses and selenium tolerance mechanisms in Arabidopsis. *Physiol. Plant.* 132(2), 236–253.

Vanhaelewyn, L., Prinsen, E., Van Der Straeten, D., Vandenbussche, F., 2016. Hormone-controlled UV-B responses in plants. *J. Exp. Bot.* 67(15), 4469–4482. https://doi.org/10.1093/jxb/erw261.

Vidoz, M.L., Mignolli, F., Aispuru, H.T., Mroginski, L.A., 2016. Rapid formation of adventitious roots and partial ethylene sensitivity result in faster adaptation to flooding in the aerial roots (aer) mutant of tomato. *Scientia Horticulturae* 201, 130–139.

Visser, E.J.W., Nabben, R.H.M., Blom, C.W.P., Voesenek, L.A.C.J., 1997. Elongation by primary lateral roots and adventitious roots during conditions of hypoxia and high ethylene concentrations. *Plant Cell Environ.* 20(5), 647–653. https://doi.org/10.1111/j.1365-3040.1997.00097.x.

Wang, Y., Feng, H., Qu, Y., Cheng, J., Zhao, Z., Zhang, M., Wang, X., An, L., 2006. The relationship between reactive oxygen species and nitric oxide in ultraviolet-B-induced ethylene production in leaves of maize seedlings. *Environ. Exp. Bot.* 57(1–2), 51–61. https://doi.org/10.1016/j.envexpbot.2005.04.009.

Wang, Y., Yuan, J., Yang, W., Zhu, L., Su, C., Wang, X., Wu, H., Sun, Z., Li, X., 2017. Genome wide identification and expression profiling of ethylene receptor genes during soybean nodulation. *Front. Plant Sci.* 8, 859. https://doi.org/10.3389/fpls.2017.00859.

Wells, C.E., Eissenstat, D.M., 2002. Beyond the roots of young seedlings: the influence of age and order on fine root physiology. *J. Plant Growth Regul.* 21(4), 324–334. https://doi.org/10.1007/s00344-003-0011-1.

Wilkinson, S., Davies, W.J., 2010. Drought, ozone, ABA and ethylene: new insights from cell to plant to community. *Plant Cell Environ.* 33(4), 510–525. https://doi.org/10.1111/j.1365-3040.2009.02052.x.

Xu, L., Xiang, G., Sun, Q., Ni, Y., Jin, Z., Gao, S., Yao, Y., 2019. Melatonin enhances salt tolerance by promoting MYB108A-mediated ethylene biosynthesis in grapevines. *Hortic Res* 6, 114. http://dx.doi.org.ezproxy.library.dal.ca/10.1038/s41438-019-0197-4.

Xu, P., Zhao, P.-X., Cai, X.-T., Mao, J.-L., Miao, Z.-Q., Xiang, C.-B., 2020. Integration of jasmonic acid and ethylene Into auxin signaling in root development. *Front. Plant Sci.* 11. https://doi.org/10.3389/fpls.2020.00271.

Yamauchi, T., Tanaka, A., Mori, H., Takamure, I., Kato, K., Nakazono, M., 2016. Ethylene-dependent aerenchyma formation in adventitious roots is regulated differently in rice and maize. *Plant Cell Environ.* 39(10), 2145–2157. https://doi.org/10.1111/pce.12766.

Yamauchi, T., Tanaka, A., Tsutsumi, N., Inukai, Y., Nakazono, M., 2020. A role for auxin in ethylene-dependent inducible aerenchyma formation in rice roots. *Plants* 9(5), 610.

Yamauchi, T., Watanabe, K., Fukazawa, A., Mori, H., Abe, F., Kawaguchi, K., Oyanagi, A., Nakazono, M., 2014. Ethylene and reactive oxygen species are involved in root aerenchyma formation and adaptation of wheat seedlings to oxygen-deficient conditions. *J. Exp. Bot.* 65(1), 261–273. https://doi.org/10.1093/jxb/ert371.

Yang, L., Zu, Y.-G., Tang, Z.-H., 2013. Ethylene improves Arabidopsis salt tolerance mainly via retaining K+ in shoots and roots rather than decreasing tissue Na+ content. *Environ. Exp. Bot.* 86, 60–69. https://doi.org/10.1016/j.envexpbot.2010.08.006.

Yu, Y., Wang, J., Zhang, Z., Quan, R., Zhang, H., Deng, X.W., Ma, L., Huang, R., 2013. Ethylene promotes hypocotyl growth and HY5 degradation by enhancing the movement of COP1 to the nucleus in the light. *PLOS Genet.* 9(12), e1004025. https://doi.org/10.1371/journal.pgen.1004025.

Yukiyoshi, K., Karahara, I., 2014. Role of ethylene signalling in the formation of constitutive aerenchyma in primary roots of rice. *AoB Plants* 6.

Zhang, Y., Wang, Y., Ye, D., Xing, J., Duan, L., Li, Z., Zhang, M., 2020. Ethephon-regulated maize internode elongation associated with modulating auxin and gibberellin signal to alter cell wall biosynthesis and modification. *Plant Sci.* 290, 110196. https://doi.org/10.1016/j.plantsci.2019.110196.

Zhu, C.Q., Hu, W.J., Cao, X.C., Zhu, L.F., Bai, Z.G., Huang, J., Liang, Q.D., Jin, Q.Y., Zhang, J.H., 2020. Role of salicylic acid in alleviating the inhibition of root elongation by suppressing ethylene emission in rice under Al toxicity conditions. *Plant Growth Regul.* 90(3), 475–487.

10

The Role of Phytohormones in Heat Stress Tolerance in Plants

Sagar Maitra, Akbar Hossain, Pradipta Banerjee, and Preetha Bhadra

CONTENTS

10.1 Introduction

During the nineteenth century, rice seedlings were found to be tainted with the parasite Gibberella fujikuroi in advanced development. This upgraded development gave its name to the ailment, 'bakanae' (absurd seedling) sickness of rice. Afterwards, the compound answerable for this phenotype was named after the microorganism, gibberellin (GA). Other plant sicknesses are regularly connected with either hindering or upgraded development and tumours, which drove scientists at the time to presume irritated hormone levels in these tissues. Biochemical methods permitted the confinement of development hormones, for example, auxin and cytokinin from both microscopic organisms and parasites. Regardless of proof of a tight connection between plant development hormones and infection, scarcely any reports illustrated the role of those hormones in plant-microbe cooperation.

Rather, protection-related flagging atoms, for example, SA, JA, and ET have gotten a lot of consideration, and their significance in plant destruction is currently very much recorded. Late subatomic and articulation profiling methods permit '- omics' approaches and have given new insights about potential roles of hormones in plant/microorganism collaborations. Here, we centre around late reports on the contribution of hormones during pathogenesis just as communications between hormone flagging and safeguard flagging. Microorganism diseases prompt hormone biosynthesis and reactions in which SA, JA, and ET figure unmistakably. SA is by and large connected with biotroph opposition, while JA and ethylene are for the most part connected with protection from necrotrophs. The collaboration between these two is frequently adversarial and the acceptance of one constricts the other. This opposition can be abused by a microorganism to advance an ailment. The customary perspective on the big three 'stress' hormones associated with plant-microbe collaboration is substantial, yet distorted; and recent investigations propose a more mind-boggling reality. In tobacco, it was shown that salicylate hydroxylase down-regulated SA without influencing JA during the extremely touchy reaction (HR). JA is, by all accounts, significant in the foundational signal. It was demonstrated that foundational incited helplessness towards it is subject to coronatine, a JA imitator delivered by the microbes. Strangely, this wonder isn't restricted to microorganisms and can likewise actuate vulnerability to an herbivore. In their examination, they exhibited that fast enlistment of JA biosynthetic and responsive qualities in fundamental tissue is essential for SAR to happen against PstDC3000. These ongoing investigations propose that a synergistic connection can likewise exist between the SA and JA pathways.

The plant hormones salicylic corrosive (SA), jasmonates (JAs), ethylene (ET), abscisic corrosive (ABA), auxin (AUX), cytokinin (CK), gibberellin (GA), strigolactone (SL), and brassinosteroid (BR) assume a significant function in intercellular and fundamental flagging frameworks, setting off articulation of different safeguard responsive qualities. The SA–JA–ET flagging frameworks are considered the foundation of the plant-flagging framework, while ABA, auxin, cytokinin, GA, and BR are associated with tweaking plant reactions by directing host safeguard reactions set off by the SA–JA–ET flagging frameworks.

SA flag is required for the appearance of foundational obtained opposition (SAR). Methyl salicylate, dehydroabietinal, pipecolic corrosive, azelaic corrosive, a lipid move protein (DIR1), a lipid-inferred atom (glycerol 3-phosphate), and a glycerol-3-phosphate-subordinate factor have been accounted for as portable flagging segments in SA-initiated SAR. Mediators, MED16 and MED15, are engaged with setting off SA-intervened SAR. SAR is related to the preparation of guard reactions, which brings about a quicker and more grounded enlistment of safeguard reactions after microorganism assault. Histone modifications are fundamentally set during a preparing occasion. The preparation can be acquired epigenetically, and relatives of prepared plants show cutting edge foundational procured opposition. DNA methylation assumes a significant part in the transgenerational SAR. JA flagging triggers a foundational invulnerability called 'initiated fundamental opposition (ISR).' JA-Ile might be the portable sign engaged with the acceptance of ISR. The ISR includes preparing of JA-subordinate reactions. MED25, MED16, and MED8, subunits of the Mediator complex, cooperate with a few record factors known to work in the control of JA-related quality articulation. Ethylene may go about as a two-faceted part in the plant invulnerable reaction organization, setting off obstruction or helplessness against various microorganisms. ABA signal observation and sign transduction pathways incorporate PYR/PYL/RCAR (an abscisic corrosive receptor), type 2C protein phosphatase (PP2C, a negative controller), and SNF1-related protein kinase (SnRK2, a positive controller). Auxin ties to TIR1/AFB atomic receptors, which are F-box subunits of SCF ubiquitin ligase complex. The auxin signal is then adjusted by the Aux/IAA repressors and the auxin reaction factor (ARF) record factors. Auxin flagging is additionally associated with setting off SAR. Auxin flagging builds SA levels, which trigger SAR. Cytokinins manage the host protection reactions either decidedly or adversely, relying upon the centralisations of cytokinins accessible at the contamination site. Key parts in the GA-flagging pathway incorporate the GA receptor GID1, the DELLA proteins, and the F-box proteins. Diverse flagging systems all include hormone-actuated focussing of transcriptional controllers for debasement. How the focussing is enacted, and the idea of the objectives themselves, fluctuate over these pathways and are point by point underneath. Notwithstanding, a couple of regular highlights develop. To start with, proteolytic focussing on is interceded by Skpl/Cullin/F-box (SCF)- type E3 ubiquitin ligase buildings, which comprise of an Skpl connector protein, a Cullin framework protein, an F-box protein, and a Ring-box (RBX) protein that ties the E2 ubiquitin-forming chemical (126). The F-box protein presents substrate explicitness to the complex. Upon relationship with a substrate protein, or focus by the F-box subunit, the SCF complex joins monoubiquitin or polyubiquitin chains to the substrate. The last alteration denotes the objective for fast corruption by the 26S proteasome. A subsequent regular component is that the proteolysed focusses of these hormone reaction pathways are transcriptional controllers that don't straightforwardly tie DNA. Rather, the objectives work through the relationship with DNA-restricting record factors. A third normal property is that the objectives of the auxin, JA and SL pathways, either have an ethylene-responsive component restricting element-related amphiphilic restraint (EAR) theme or partner legitimately with a protein that has an EAR theme. The EAR theme empowers authority to proteins with C-terminal LisH areas, most unmistakably the TOPLESS (TPL) and TOPLESS-RELATED (TPR) transcriptional corepressors. Accordingly, the actuation of transcriptional programs because of these three hormones is the result of the alleviation of the hindrance. The current chapter highlights the major role of phytohormones in heat stress tolerance in plants.

10.2 Sensing Thermal Stimuli by Plants

All most all plants respond to ambient changes in temperature changes. Studies on plant genetics and physiology have revealed innumerable thermal-sensitive pathways in plants. It was revealed by several studies that thermal changes transduce signal influence on gene expression. Heat and cold treatment can alter cell membrane fluidity and cause an elevation in intracellular free-calcium levels, which is the key event that controls plants' responses to various thermal stimuli. Calcium acts as a secondary messenger in signal transduction. Due to the influx of calcium from the cell wall, there is an increase of the cytoplasmic calcium level, which results in cooling (Knight et al. 1996; Ali et al. 2013). The influx of calcium occurs in milliseconds after stimulation, indicating proximity to the primary sensing event. Plasma membrane transient receptor potentials (TRPs) are responsible for the main thermal sensor (McKemy et al. 2002). Cold-menthol receptor 1 (CMR1) belongs to the TRP family that responds to cold and small molecules (e.g., Icilin). Cold acclimation introduces many biochemical changes in plants' physiology that include alterations of cell membrane composition, increase in total soluble-proteins, and increase in cryo-protectant levels, like proline and sugars. Cold-responsive genes (COR) play an important role in cold acclimatisation. The C-repeat-binding factors (CBFs) regulate expression of COR genes (Liu et al. 2019). Li et al. (2013) reported that a transcription factor, namely, C-repeat (CRT)/ dehydration responsive element (DRE) binding factor, plays a role in the cold response of plants. One of the CBF genes is activated by a calmodulin-binding transcription activator (CAMTA) that indicates a connection to the calcium signalling at the onset of cooling. It should be noted that the influx of calcium in cytosol also increases in response to heat. Saidi et al. (2009) demonstrated that, in the lower plant, the heat-shock response is dependent on the calcium channels of the cell membrane. This leads to the paradox of how different responses are caused by the same change in cytosolic calcium influx. This paradox led to a hypothesis that calcium ion transients are the primary events in temperature sensitivity. Cytoskeletal/ membrane dynamics may play a role, along with calcium ion transients, in differentiating the hot and cold temperature. Plasma membrane may also be one major site of ambient temperature perception in plants.

Macromolecules present in the plant cell, e.g., protein complexes, membranes, nucleic acids, can sense the heat stimuli from the surrounding environment and act as thermo-sensors through reversible physical changes like increase in membrane fluidity, partial melting of nucleic acid strands, proteins dissociation, etc. (Mittler et al. 2012; Ahmad et al. 2015). Heat-responsive macromolecules can perceive heat stimuli followed by differential response to temperature changes. They upregulate the expression of innumerable heat stress response (HSR) genes. HSR genes are highly conserved and play an important role in the acclimation response (Bahuguna and Jagadish 2015; Tariq et al. 2013). It can rightly be said that calcium fluxes, calmodulin, calcium-dependent protein kinases (CDPKs), mitogen-activated protein kinases (MAPKs), phosphatases, and transcriptional regulators, such as heat shock transcription factors (HSFs), multiprotein bridging factor 1c or MBF1c, WRKY, dehydration-responsive element-binding transcription factors or DREB, and basic leucine zippers or bZIPs transcription factors play a pivotal role in sensing thermal stimuli in plants.

10.3 The Concept of Basal and Acquired Thermo-Tolerance

Basal or intrinsic thermo-tolerance is defined as the inherent ability of plants to withstand temperatures above the optimum temperature required for growth, not preceded by acclimation to nonlethal temperature elevations. Whereas adaptive or acquired thermo-tolerance (ATT) is induced by pre-exposure to elevated but nonlethal temperatures that give the ability to withstand a series of severe heat stresses that would otherwise be lethal in the absence of the preconditioning heat treatment. ATT is short-lived in nature and enhances basal thermo-tolerance and heat endurance via a transition to 'efficient' cellular performance when acclimatory homeostasis is reached (Bokszczanin et al. 2013). Differences between acquired and basal thermo-tolerance has been studied in wheat, where Qin et al. (2008) demonstrated that 1314 transcripts are differentially expressed followed by heat treatments with or without pre-acclimation.

Some regulatory and acclimation proteins, namely, multi-protein bridging factorlc (MBF1c) (Suzuki et al. 2008) and ROS detoxifying enzyme catalase, play important roles in basal thermo-tolerance but are not at all required for acquired thermo-tolerance (Vanderauwera et al. 2011). However, some HSFs, and the disaggregating chaperone HSP101, are required for basal as well as acquired thermo-tolerance (Liu et al. 2011).

10.4 Hormonal Cross-Talk

Plants react to biotic and abiotic stresses in a highly coordinated and interactive ways (Adnan et al. 2018; Adnan et al. 2019; Akram et al. 2018a,b; Aziz et al. 2017a,b; Hafiz et al. 2016; Hafiz et al. 2019; Kamaran et al. 2017; Muhammad et al. 2019; Sajjad et al. 2019; Saud et al. 2013; Saud et al. 2014; Saud et al. 2017; Saud et al. 2016; Shah et al. 2014; Saud et al. 2020; Qamar et al. 2020; Wajid et al. 2017; Yang et al. 2017; Zahida et al. 2017; Depeng et al. 2018; Hussain et al. 2020; Shafi et al. 2020; Wahid et al. 2020; Subhan et al. 2020; Zafar-ul-Hye et al. 2020a,b; Adnan et al. 2020; Ilyas et al. 2020; Saleem et al 2020a,b,c; Rehman 2020; Frahat et al. 2020; Wu et al. 2020; Mubeen et al. 2020; Farhana 2020; Jan et al. 2019; Wu et al. 2019; Ahmad et al. 2019; Baseer et al. 2019; Tariq et al. 2018; Fahad and Bano 2012; Fahad et al. 2017; Fahad et al. 2013; Fahad et al. 2014a,b; Fahad et al. 2016a,b,c,d; Fahad et al. 2015a,b; Fahad et al. 2018; Fahad et al. 2019a,b; Hesham and Fahad 2020; Iqra et al. 2020; Akbar et al. 2020; Mahar et al. 2020; Noor et al. 2020; Bayram et al. 2020; Amanullah et al.2020; Rashid et al. 2020; Arif et al. 2020; Amir et al. 2020; Saman et al. 2020; Md Jakirand Allah 2020; Farah et al. 2020; Sadam et al. 2020; Unsar et al. 2020; Fazli et al. 2020; Md. Enamul et al. 2020; Gopakumar et al. 2020; Zia-ur-Rehman 2020; Ayman et al. 2020; Mohammad I. Al-Wabel et al. 2020a,b; Senol 2020; Amjad et al. 2020; Ibrar et al. 2020; Sajid et al. 2020). Although plants are stationary, they show plasticity in growth and adapt to various environmental conditions. This adaptation is facilitated by phytohormones. The cascade of signal transduction starts when the hormones are triggered by abiotic stresses (Harrison et al. 2012). The regulation of stress-responsive hormones alters the cellular dynamics important in regulating growth responses under different stress conditions (Kohli et al. 2013). The meeting points of phytohormone signal transduction cascades are known as cross-talk that constitute a signalling network (Harrison et al. 2012). Hormones interact either by activating a common second messenger, like cAMP, or by a phosphorylating cascade reaction. In recent years, insights into the biosynthesis and signalling pathways of major phytohormones, including abscisic acid (ABA), auxins (IAA), brassinosteroids (BRs), gibberellins (GAs), jasmonates (JAs), and ethylene (ET) have been revealed (Singh and Jwa 2013). The mechanism of stress tolerance remains mostly unclear due to a lack of proper information regarding cell plasticity and tissue-specific response (Kohli et al. 2013). Stomatal closure due to water deficiency and the role of hormonal cross-talk in the regulation of stomatal closure is one of the examples of signal transduction. A complex signal transduction pathway tightly regulates the opening and closing of stomata. ABA is primarily responsible for stomatal movements, and it regulates the gene expression of root growth-controlling genes, which in turn influence the water uptake from soil (Zhang et al. 2006). Again, ABA is found to cross-talk with jasmonates (JAs) and nitric oxide (NO) to stimulate the closure of stomata. The regulation of gene expression by ABA includes the genes responsible for ethylene, cytokinin, and auxin synthesis (Harrison et al. 2012). Desikan et al. (2006) showed that the closing of stomata by ethylene is controlled by a signal transduction pathway, which needs hydrogen peroxide synthesis. This again shows the involvement of JA and JA-associated genes. JA coordinates with the stomatal closure by increasing calcium ion influx that, in turn, stimulates calcium-dependent protein kinase (CDPK) production resulting in signal cascade. The long-term abiotic stress response is influenced by the ABA-mediated regulation of transcription factors that can bind to ABA-responsive elements. It is noted that phosphorylation may lead to changes in ABA-regulated transcription factors. For example, the ABA-responsive transcription factors namely, ABF1 and ABF4, are activated when they are phosphorylated by ABA-inducible kinases (Zhu et al. 2007; Harrison et al. 2012). Walia et al. (2006) studied the regulation of JA pathway genes under high saline soil conditions in barley plants. JA assumes a hormone, which signals cell death (Kohli et al. 2013). Nishiyama et al. (2011) showed that exogenous ABA treatment can downregulate a major cytokine biosynthetic enzyme, called isopentenyl transferase (IPT), but increases the expression of genes that encode cytokinin oxidases and dehydrogenases. Stamm and Kumar (2013) clearly demonstrated the cross-talk between auxins and brassinosteroids; Linkies and Leubner-Metzger (2012) showed that GA engages in cross-talk in response to environmental stress.

Environmentally activated phytohormones like ABA, ET, GA, and CK play an important role in determining stress responses in plants (Krouk et al. 2012; Iqbal et al. 2011). Capiati et al. (2006) reported that some phytohormones could combat adverse conditions. For instance, a signal peptide system coupled with JA was found to be associated in wound-induced salt-stress adaptation in tomato, pointing to cross-tolerance signalling of phytohormones.

10.5 Hormones Involved in the Response of Plants to Heat Stress

10.5.1 Abscisic Acid (ABA)

Plants experience different types of abiotic stresses, and, to cope with adverse environmental conditions, they have several mechanisms (Nishizawa et al. 2008). ABA can fight heat against stress by enhancing its antioxidant property and/or by modulating HSPs. Accumulation of ROS leads to oxidative damage and inhibits growth (Nishizawa et al. 2008; Bolouri-Moghaddam et al. 2010). Heat stress may be seen as an indirect effect of oxidative stress, rather than a direct effect of heat stress as observed in rice (Zhang et al. 2016). To downregulate excess ROS generation, ABA induces the antioxidant capacity of plants to combat heat stress. Due to heat stress, increase in endogenous ABA levels is observed that promotes water balance and regulates heat tolerance by modulating stomatal closure and increasing antioxidant ability (Hu et al. 2010a; Hsieh et al. 2013). The ABA-mediated heat stress response is done via antioxidant property as observed in cucumber and rice leaves (Tang and MA, 2005). The response to heat stress induced by ABA is dependent on the intervention of ROS. ROS is a signalling molecule that mediates the abiotic stress response in plants (Baxter et al. 2014). Due to the attack of free radicals, ROS-scavenging genes are activated and the enzymes responsible for detoxification such as catalase, ascorbate peroxidase, superoxide dismutase, and antioxidants like ascorbic acid and flavonoids (Mittler et al. 2004) come into action to maintain the balance between production and scavenging via heat stress response (Driedonks et al. 2015). Hydrogen peroxide (H_2O_2) mediates the role of ABA in response to heat tolerance. A significant increase in hydrogen peroxide is observed when plants are sprayed with ABA (Larkindale et al. 2005). Marino et al. (2012) found that plants which are deficient in ABA synthesis, or those that are treated with exogenously sprayed ABA synthesis inhibitors (e.g., fluridone) show defective heat tolerance response by reducing ABA and hydrogen peroxide levels when subjected to heat stress. NADPH oxidases, also known as respiratory burst oxidase homologues (RBOHs), are responsible for generating ROS. ABA can induce the expression of AtrbohD and AtrbohF in guard cells to generate ROS. It was reported that the double mutant of atrbohD/atrbohF has impaired ability for stomatal closure (Kwak et al. 2003). The expression of RBOH mRNA was found to be upregulated in grafted cucumber under heat stress when pretreated with ABA (Li et al. 2014). All these data suggest that ABA increases hydrogen peroxide in plants by regulating RBOH gene expression regardless of the heat conditions. However, it is not clear how ABA affects the expression of RBOH genes under heat stress.

10.5.2 Auxins

Auxins, or Indole-3-Acetic Acid (IAA), play a role in cell division and cell elongation as well as act as a key regulator of adaptive growth modulator in response to high-temperature conditions. Biosynthesis of auxin includes four tryptophan-dependent and one tryptophan-independent pathway (Mano and Nemoto 2012). There is evidence that IAA plays role in salinity and heavy metal stress. Auxin is an important defence element, involved in the regulation of numerous genes and cross-talk in response to high temperature (Gray et al. 1998).

The phytohormone auxin orchestrates many physiological and developmental processes including growth control, organ patterning, and root and shoot architecture. In Arabidopsis seedlings, moderately HT stimulates the elongation of hypocotyls by activation of auxin biosynthetic pathways with the tryptophan aminotransferase-encoding gene TAA1/TIR2 (Gray et al. 1998). This transcription is positively upregulated by increased temperature in hypocotyls, cotyledons, and roots. By contrast, endogenous auxin levels specifically decrease in the developing anthers of barley and Arabidopsis during the early response to HT conditions (Sakata et al. 2010a). Immunohistochemical analysis using anti-IAA antibodies reveals that endogenous auxin levels are specifically reduced by increasing temperatures in barley early developing anther cells (Sakata et al. 2010a). In an Arabidopsis DR5-GUS recombinant line that places the GUS gene under the control of a synthetic auxin response element, the strongest GUS activity appears in PMCs and tapetum cells at stage 10 of floral development when the microspores are visible in the anthers (Sakata et al. 2010a). In recombinants exposed to moderate HT (31 or 33°C) for 1 day, the GUS signals significantly decreased in an anther-specific manner (Sakata et al. 2010a,b). HT represses the expression of the YUCCA auxin biosynthesis genes in both barley and Arabidopsis (Sakata et al. 2010a). YUCCA flavin monooxygenases are implicated in the biosynthesis of indole-3-acetaldoxime (IAOx), which is an intermediate for one of several tryptophan-dependent auxin biosynthetic pathways. Expression of YUCCA genesis temporally and spatially controlled. In Arabidopsis, double or triple mutants that include yuc2 and yuc6 have a total loss of male fertility and form short stamens that lack pollen grains. Given that these mutant phenotypes are quite similar to the phenotypes of HT injury, biosynthesis of endogenous auxin with YUCCA in developing anthers is a major factor responsible for the HT-induced abortion of pollen and reduction of auxin. These phenotypes are highly conserved in monocots and dicots. Tissue- and organ-specific auxin biosynthesis independently and differentially controls in response to fluctuation of temperatures (Sakata et al. 2010a,b).

10.5.2.1 Application of Auxin Reverses HT Injury

Comprehensive transcriptome analyses using more than 600 microarray datasets of barley show high and positive pairwise correlations between the expression profiles of auxin-induced genes, DNA replication-related genes, and mitochondrial-related genes. These genes are negatively correlated with the expression profiles of auxin-repressed protein genes and photosynthesis- or chloroplast-related genes. In developing anthers, HT stress inhibits the expression of the former genes and upregulates the expression of the latter genes. Therefore, the application of exogenous auxin recovers the expression of DNA replication-related genes under HT conditions (Sakata et al. 2010a). Anther cell proliferation and development, and subsequent fertilization and seed set, are restored under HT conditions by application of exogenous auxin (Figure 10.1). Therefore, HT-tolerant plants might be achieved by controlling anther-specific auxin biosynthesis genes using recombinant DNA technology. The current data indicate that the appropriate application of auxin is useful for the promotion of plant fertility and maintenance of crop yields under the threat of global warming.

10.5.3 Cytokinins

Even though auxins and cytokinins (CKs) have been studied mainly in relation to their growth and developmental regulatory functions, their involvement in stress responses is recognized. In the case

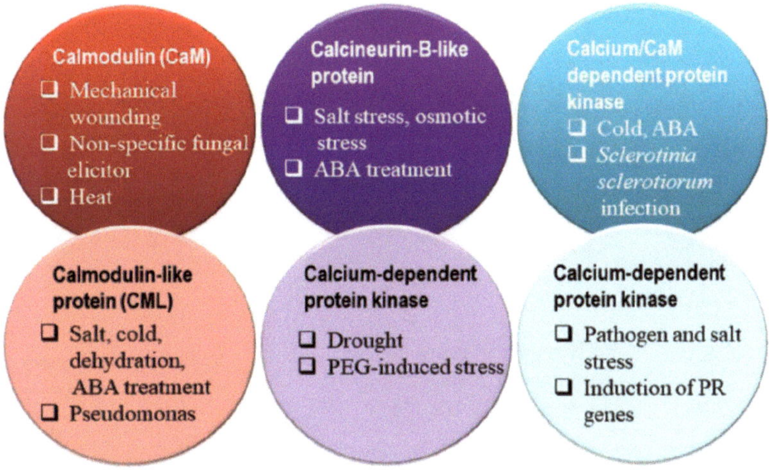

FIGURE 10.1 Hormone-enzyme crosstalk.

of heat stress, CK effect on stomata opening followed by stimulation of leaf transpiration was found to be crucial at the early phase of the stress response. Further, a large number of heat stress affected genes are CK-responsive (Cerny et al. 2014). CKs stimulate the antioxidant system upon heat stress. Their exogenous application can positively affect photosynthesis in heat stress. However, the precise mechanism of CK functions remains to be understood. This study was conducted to elucidate the role of CKs (in cross-talk with other hormones) in heat stress responses and to evaluate whether raising the level of CKs could increase plant heat stress tolerance. Modulation of endogenous levels of CKs can be achieved by their exogenous application. However, such disturbance of endogenous CK homeostasis usually stimulates degradation and deactivation mechanisms, especially in the absence of the protective moiety of the exogenous CK, e.g., tetrahydropyranyl or tetrahydrofuranyl. The other possibility is to insert the CK biosynthetic gene isopentenyl transferase under suitable promoter. The choice of the promoter is crucial due to the organ-specific CK dynamics in different phases of the stress response. Another approach is the suppression of CK degradation by inhibition of CK oxidase/dehydrogenase (CKX), which causes cleavage of the side chain from the adenine moiety. INCYDE (INhibitor of CYtokinin DEgradation) was found to be a potent CKX inhibitor. INCYDE has been successfully used for elevation of plant tolerance to abiotic and biotic stresses, e.g., to cadmium stress in Bulbine natalensis and *Rumex crispus*, to salt stress in tomato, and Verticillium longisporum in Brassicaceae (Figure 10.2).

10.5.4 Gibberellins

Gibberellins (GAs) are significant plant development hormones engaged with plant inborn insusceptibility. GA tweaks plant safeguard reactions, generally by controlling SA–JA–ET flagging frameworks. Key segments in the GA-flagging pathway incorporate the GA receptor GID1, the DELLA proteins, and the F-box proteins. Upon GA official, the GA-subordinate conformational change causes the GID1 N-terminal helical cover area to act like 'atomic paste' to frame the GA–GID1–DELLA complex. DELLAs quell GA-subordinate safeguard reactions, though GA calms their abusive movement. GA lifts the DELLA constraint by focussing on DELLA for annihilation through the ubiquitin-proteasome pathway. GA controls the plant's natural invulnerable reactions either decidedly or contrarily. It initiates helplessness or opposition against various parasitic and bacterial microbes. GA regulates plant ailment opposition or helplessness by inciting the debasement of DELLA proteins. GA manages plant invulnerable reactions by balancing JA- and SA-flagging frameworks, and it has been found to upgrade the SAR against microbes. GA facilitates SA collection and advances opposition against microorganisms by

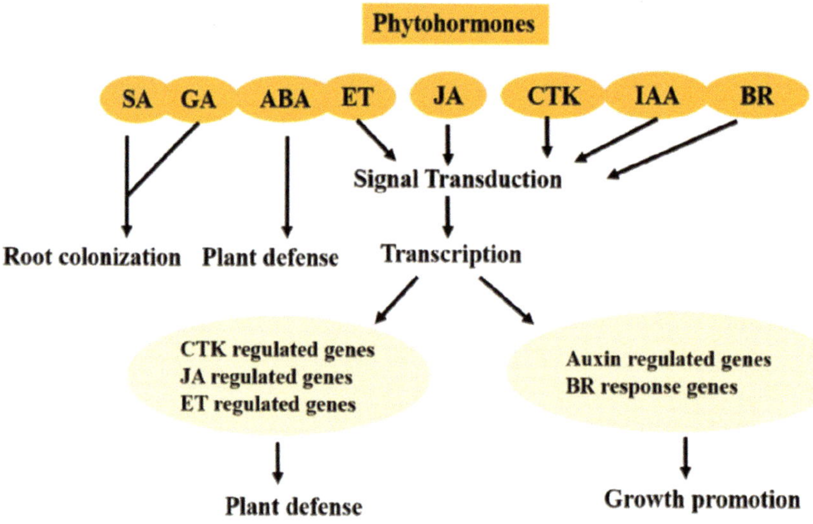

FIGURE 10.2 Hormones and biotic stress.

corrupting DELLA proteins. Misfortune of function freaks in DELLAs, the silencers of GA motioning, up-control the SA-intervened protection and down-manage JA/ET-interceded safeguards in Arabidopsis. GA interfaces inimically with JA flagging, while GA weakens the JA-incited articulation of various JA-responsive qualities. Both JA- and GA-flagging frameworks balance plant insusceptible reactions. JA meddles with gibberellin flagging. JA defers GA-interceded DELLA protein corruption. The JAZ protein JAZ9 restrains RGA cooperation with the record factor PIF3. JA flagging appears to actuate articulation of DELLA qualities associated with the GA-flagging pathway. JA-incited RGL3 articulation works using the COI1/MYC2-subordinate flagging pathway. JA-interceded enlistment of RGL3 articulation was annulled in the coi1-1 freak, demonstrating that RGL3 is downstream of COI1. BRs contrarily control intrinsic safe reactions prompted by GAs. BR and GA cause cross-inhibitory impacts on the complementary hormone biosynthesis pathways to associate in a commonly hostile way, and microorganisms trigger overexpression of DELLA proteins to smother GA-controlled protection reactions (Emamverdian et al. 2020; Khan et al. 2020; Banerjee and Roychoudhury 2019).

10.5.5 Salicylic Acid

Foundational obtained obstruction (SAR) is an elevated condition of protection against a wide range of microbes initiated all through a plant following nearby contamination. The incited obstruction is recognized in the immunized foundational tissue of a microbe-tainted plant. Limited assault by a necrotizing microbe actuates SAR to ensuing assault by an expansive scope of typically harmful microorganisms. Salicylic corrosive (SA) gathering is required for actuation of nearby safeguards at the underlying site of the assault, and in the far-off microorganism-free organs for the enlistment of SAR. SA aggregation and motioning in these removed microorganism-free organs are prepared to additionally increment to more elevated levels upon challenge with a microbe. SA studies have uncovered that few salicylate biosynthetic qualities are up-controlled in the midribs or entire leaves of grapevine tainted with the phytoplasma 'Ca. P. solani', e.g., VvICS, which encodes isochorismate synthase, and VvSamt, which encodes S-adenosyl-L-methionine: Salicylic corrosive carboxyl methyltransferase, which catalyses the development of the unpredictable ester methyl salicylate from SA. What's more, evaluation of SA and its forms in the primary leaf veins during contamination of grapevine with 'Ca. *P. solani*' indicated a 26-overlap increment in SA 2-O-glucopyranosyl, and essentially higher free and absolute SA, when contrasted with uninfected grapevine. Salicylic corrosive glucopyranoside and salicylate were likewise extraordinarily expanded in the grapevine primary leaf veins tainted with the flavescence dorée phytoplasma.

SA flagging is required for the appearance of SAR. Isochorismate synthase movement is required for SA combination. The Arabidopsis icsl freak, which is insufficient in isochorismate synthase 1 action, is SAR-inadequate. Transgenic plants communicating the SA-debasing protein salicylate hydroxylase encoded by the Pseudomonas putida nahG quality were discovered to be inadequate in communicating SAR. The FMO1 quality is required for the fundamental gathering of SA in far-off microorganism-free leaves, and the FMO1 quality is required for the enlistment of SAR. Methyl esterase 1 is required for SAR in potato, and MES is required for amassing of free SA in the distal foundational tissue, the tissue that does not get the essential (starting) contamination. NPR1 quality is an ace controller of the SA-interceded acceptance of protective qualities. PR1 straightforwardly ties SA enacts SA-flagging framework. NPR1 is associated with setting off SAR, and the nprl freak of Arabidopsis thaliana is inadequate in SAR (Bailey et al. 2009; Chen et al. 2013; Denancě et al. 2013).

10.5.6 Jasmonic Acid

As a pressure-related hormone, JA has additionally been associated with salt-prompted development restraint. JA levels were raised and JA flagging was initiated on salt pressure. The F-box protein Coronatine Insensitivel (COI1), otherwise called the JA receptor, frames the SCFCOI1 E3 ligase complex with SKP1 and CULLIN1, which intercedes Jasmonates ZIM area (JAZ) debasement by the 26S proteasome. When JAZ was being eliminated, the hindered record factor (e.g., MYC) actuates the statement of JA-responsive qualities. Enactment of JA motioning in light of salt pressure inevitably prompts the restraint of essential root development. Transformations of COI1 and MYC2/3/4 alongside the balanced-out JAZ freak jaz3-1 all gave more noteworthy cell prolongation under salt pressure. In addition, the JA-safe freak jasmonates insensitive3 (jai3) was found to recoup before and demonstrated an expanded development rate during salt pressure, further showing that JA intervenes in development suppression in the salt reaction. Exogenous use of JA lightened salt toxicity through keeping up ROS or particle homeostasis. In tomato, the high-JA-aggregation res freak showed more prominent salt resistance with expanded K+ collection. Reliably, decreased JA creation or collection prompted touchiness to saltiness in tomato and rice, though raised JA biosynthesis upgraded salt resilience in wheat and Arabidopsis. In synopsis, JA-interceded development, restraint may be a significant versatile technique in salt conditions. Salt-directed qualities elevated by JA would, in general, be instigated in the inward tissue layer of roots in light of saltiness, suggesting the spatial guideline of JA flagging. Reliably, two JA biosynthetic rice freaks, cpm2 and hebiba, were less vulnerable to salt because of lower Na+ collection in shoots; however, no distinction was identified in roots. Also, the maize JA biosynthesis freak, opr7opr8, showed a diminished salt reaction in shoots yet showed an overstated reaction in roots with a progression of antagonistic physiological records. These perceptions propose that JA intercedes in plant development reaction to salt stress in a tissue/organ-subordinate way. What's more, the coil freak is more vulnerable to salt stress at the underlying phase of seed germination yet shows no distinction at the late development stage after seed germination, suggesting JA-interceded plant salt reactions can be additionally managed in a transiently explicit way. Under salt pressure, JA treatment was joined by an expansion in endogenous ABA content and checked the NaCl-incited GA decrease in rice and chard plants, showing that other plant hormones are associated with the JA-interceded salt reaction. ABA content was lower in the JA biosynthesis freak aos and JA levels were diminished in the ABA biosynthesis freak aba. Moreover, MYC2, the ace controller working in JA flagging, transcriptionally initiates articulation of the ABA-inducible qualities RD22 and AtADH1 for plant adaption to osmotic pressure and, thus, PnJAZ1 was discovered to be actuated by ABA. In this manner, salt resilience is advanced in plants through mitigating exorbitant ABA motioning under salt pressure. In synopsis, JA–ABA exchange assumes a significant function in the plant salt reaction (Wang et al. 2020; Ali and Baek 2020).

10.5.7 Brassinosteroids

Brassinosteroids (BRs) are polyhydroxy steroidal plant hormones with solid development-advancing potential. They were first detached and described in the dust of the assault plant. In excess of 70 BRs have been disengaged from plants. Nevertheless, brassinolide, 28-homobrassinolide, and 24-epibrassinolide comprise the three most bioactive BRs and are generally utilised in physiological and trial studies. They are available in all aspects of plants including dust, blossom buds, organic products, seeds, vascular cambium, leaves, shoots, and roots. They assume a basic function in various formative cycles, e.g., stem and root development, botanical commencement, and improvement of blossoms and natural products. Notwithstanding, ongoing discoveries propose pressure sway moderating functions of BRs and related mixes in different plants exposed to different abiotic stresses. These abiotic stresses are high temperature, chilling, soil saltiness, light, dry season, flooding, metals/metalloids, and natural toxins. Recent research has demonstrated the enormous capability of BRs and related mixes in the regulation of parts of cell reinforcement protection framework because of and to check the abiotic stress-instigated oxidative burst, audited by Vardhini and Anjum. Nevertheless, there is the colossal extension for additional examination zeroed in on the destinations, pathways, and enzymology of their biosynthesis, source-sink connections, formative and stress physiology, their cooperation with microorganisms, parasites, and creatures, and the acknowledgement of their incredible applications (Ahammed et al. 2020; Kaur and Pati 2019).

10.5.8 Ethylene

Ethylene (ET) flagging framework is a significant segment in plant intrinsic safe framework. ET-flagging framework appears to actuate opposition against the necrotrophic parasitic microbe *Botrytis cinerea* in Arabidopsis. Ethylene-heartless ein2 freaks demonstrated upgraded weakness to *B. cinerea* in Arabidopsis, proposing that ethylene might assume a significant part in giving opposition against *B. cinerea*. ET-flagging pathway contributed to obstruction against the impact microbe *Magnaporthe oryzae* in rice. The spermidine-aggregating transgenic tomato lines overexpressing yeast spermidine synthase were more vulnerable to *B. cinerea* than the wild-type plants. Exogenous use of ET forerunners S-adenosyl-Met and 1-aminocyclopropane-1-carboxylic corrosive (ACC) switched the reaction of the transgenic plants to *B. cinerea*. The expanded defencelessness of the transgenic tomato plants to *B. cinerea* was related to down-guideline of quality records engaged with ethylene biosynthesis and flagging. These outcomes recommend that ET flagging is a significant part in the plant resistant framework setting off safeguard reactions against *B. cinerea*. ET prompted obstruction against the necrotrophic microorganism *Cercospora nicotianae* and the bacterium *Erwinia carotovora*. Conversely, the ET framework did not prompt opposition against the biotrophic parasite *Oidium neolycopersici* and initiated vulnerability against the oomycete *Peronospora tabacina and Tobacco mosaic* infection in tobacco. Ethylene initiated fractional obstruction in Medicago truncatula against the charcoal decay microbe Macrophomina haseolina. Antisense concealment of OsEIN2 presents ET harshness and builds powerlessness against the rice impact microorganism *M. oryzae*. ET biosynthesis has been demonstrated to be an essential segment of R gene-mediated impact opposition. Ethephon treatment brings down vulnerability in viable rice impact associations. These outcomes propose that ET is associated with setting off resistant reactions against *M. oryzae* in rice (Debbarma et al. 2019; Sharma et al. 2019; Kazan 2015).

10.5.9 Strigolactones

The system of strigolactones activity relies upon the unit of the D-ring by a catalyst shaping a hydroxyl butenolide structure, which is fit for actuating an adjustment in adaptation in a receptor pocket, which will start a biochemical course of responses, reflecting in a sign transduction pathway. The initial step of SLs movement is its collaboration with a protein receptor called strigolactone esterase or D14 protein. A hidden subatomic component of germination measure requires the inclusion of some portion of dynamic protein is called α/β-type hydrolase. This hydrolase contains an entrenched synergist chain of serine–histidine–aspartate having a major cone-formed interior depression that can exemplify a SL atom to achieve the unit of a hydroxy D-ring. It was accounted for that SLs change the phenotypic yield of PIN-FORMED (PIN) protein group of auxin carriers for aligning of advancement and developmental responses. PIN qualities are available in the genomes of multicellular plants solely and direct topsy-turvy auxin conveyance in numerous formative cycles that incorporate embryogenesis, organogenesis, tissue separation, and jungle reactions. Flagging pathway of SLs in rice is depicted in Figure 4. The flagging segments in rice incorporate a putative αβ hydrolase receptor (D14), F-box protein, a segment of SCF complex (D3), and a ClpATPase (D53). This complex is known to control the quality articulation by balancing the corruption of different record factors, which go about as either repressors or activators of record. Strigolactone circulation is directed by means of PDR1 carrier inside the plants and outside into the rhizosphere. PDR1 encodes Pdr1p (Pleiotropic Drug Resistance), a record consider included general medication reaction (Banerjee and Bhadra, 2020; De Saint Germain et al. 2013).

10.6 Biotechnological Applications

Hereditary designing has opened new roads for presenting abiotic stress resistance in various financially significant harvest species. The ongoing achievement has made hereditary designing methodologies one of the strongest solutions for improving harvest efficiency under testing conditions. The achievement of transgenic research, nevertheless, relies largely upon compelling plant change strategies for stable reconciliation and practical articulation of unfamiliar qualities in the plant genome. Since beginning reports in tobacco, quick improvements in change innovation have brought about the hereditary adjustment of an enormous number of plant species. Two general methodologies are accessible for moving gene(s) to plants: one vector-interceded, by a natural vector, e.g., Agrobacterium, and the other direct quality exchange approach, where DNA is brought into cells by a physical concoction or even electrical methods. The utilisation of biolistic or molecule assault is by a long shot the most generally utilised direct quality exchange technique. With its ever-expanding host extend, Agrobacterium tumefaction has become the away from of decision for the quality exchange to the entirety of the principal crop plants. This dirt bacterium has the normal capacity to convey a particular aspect of its plasmid DNA (move or T-DNA) into the atomic genome of its host plant. This condition has emerged additionally halfway because albeit direct quality exchange strategies are valuable for stable change just as transient articulation, there are still a few issues that permit recurrence of stable change, undesirable hereditary revisions because of the high duplication of qualities, and the extensive stretch needed to recover entire transgenic plants. In Agrobacterium, bacterial qualities can be effectively supplanted with gene(s) of interest, and curiously, this substitution does not influence the change cycle or recurrence. For productive change and ensuing recovery, Agrobacterium-intervened techniques are reliant on a few components. They are the decision of explant, the hormonal structure of the medium utilised, dietary enhancements, culture conditions previously and during vaccination, length of co-development, the destructiveness of the Agrobacterium strain, and focus and arrangement of the bacteriostatic specialist utilised. The length of choice and grouping of the anti-toxin determination marker are likewise significant. Other significant variables are the plant cultivar and different states of tissue culture, including a strong arrangement of plant recovery. Notwithstanding, through broad endeavours, wonderful advancement has been accomplished during the last thirty years in plant hereditary designing. Advanced conventions have been produced for the Agrobacterium interceded hereditary change of an expansive scope of yield plants including monocots, recently considered to lie outside the Agrobacterium. As of late, achievement has been made in creating in planta change techniques. Nonetheless, propelled by the overwhelming patent scene of Agrobacterium innovation and looking for an ideal (open source) stage for plant biotechnology, have distinguished three non-Agrobacterium species: Rhizobium sp. NGR234, Sinorhizobium meliloti, and Mesorhizobium loti as being fit for fruitful hereditary change of various plant species. An ongoing expansion to this rundown of natural vectors for hereditary change of plants is infection-based vectors.

10.7 Conclusion and Future Perspective

Some significant plant hormones, e.g., ABA and JA, are firmly organized with Ca^{2+} flagging. In this manner, it has been posited that ABA and JA are the focalized focusses between abiotic stresses and biotic burdens. This hypothesis is upheld by contemplates demonstrating the double functions of Ca2+ sensors as both the underlying pressure signal identifiers s furthermore the controllers of ABA and JA flagging. Simultaneously, G-proteins are additionally known to assume double functions in abiotic stress and biotic pressure reactions, and it has been accounted for that G-proteins are likewise associated with the flagging pathways of plant hormones, e.g., JA. Nonetheless, as G-proteins comprised of so wide a range of sub-gatherings, their various conduct could be an obstruction to completely understanding the parts of G-proteins in plant hormone flagging occasions. By and by, the assorted qualities of G-proteins likewise allude to the numerous prospects of how G-proteins can direct plant hormone motioning under biotic and abiotic stress.

The atomic components managing hormone combination, flagging, and activity have been explained during the previous few years, and the parts of plant hormones for reactions to changing conditions have been illustrated. These discoveries will encourage the alteration of hormone biosynthetic pathways for the age of transgenic plants with improved abiotic stress resilience. Controlling the hormone portion/reaction proportion remains to be tested, since the hormone levels achieved ought to be moderate to keep up a harmony between the constructive outcomes of plant hormones on stress resistance and the negative impacts on development and advancement. The utilisation of contingent advertisers driving quality articulation at explicit formative stages, in explicit tissues/organs or potentially in light of explicit natural prompts goes around this issue and will encourage the age of transgenic crops ready to develop under different abiotic stresses with negligible yield misfortunes.

REFERENCES

Adnan M, Fahad S, Khan IA, Saeed M, Ihsan MZ, Saud S, Riaz M, Wang D, Wu C (2019) Integration of poultry manure and phosphate solubilizing bacteria improved availability of Ca bound P in calcareous soils. *Biotech* 9(10):368.

Adnan M, Fahad S, Zamin M, Shah S, Mian IA, Danish S, Zafar-Ul-Hye M, Battaglia ML, Naz RMM, Saeed B, Saud S, Ahmad I, Yue Z, Brtnicky M, Holatko J, Datta R (2020) Coupling phosphate-solubilizing bacteria with phosphorus supplements improve maize phosphorus acquisition and growth under lime induced salinity stress. *Plants* 9(7). https://doi.org/10.3390/plants9070900.

Adnan M, Shah Z, Fahad S, Arif M, Alam M, Khan IA, Mian IA, Basir A, Ullah H, Arshad M, Rahman IU, Saud S, Ihsan MZ, Jamal Y, Amanullah, Hammad HM, Nasim W (2018a) Author correction: phosphate-solubilizing bacteria nullify the antagonistic effect of soil calcification on bioavailability of phosphorus in alkaline soils. *Sci Rep* 8(1):4339. https://doi.org/10.1038/s41598-018-22653-7.

Adnan M, Shah Z, Sharif M, Rahman H (2018) Liming induces carbon dioxide (CO2) emission in PSB inoculated alkaline soil supplemented with different phosphorus sources. *Environ Sci Pollut Res Int* 25(10):9501–9509.

Ahammed GJ, Li X, Liu A, Chen S (2020) Brassinosteroids in plant tolerance to abiotic stress. *J Plant Growth Regul*:1451–1464.

Ahammed GJ, Li X, Zhou J, Zhou Y-H, Yu J-Q (2016) Role of hormones in plant adaptation to heat stress. In: *Plant horm challenging environ factors*. Springer, pp 1–21.

Ahmad K, Adnan M, Khan MA, Hussain Z, Junaid K, Saleem N, Ali A (2015) Bioactive neem leaf powder enhances the shelf life of stored mungbean grains and extends protection from pulse beetle. *Pak J Weed Sci Res* 21(1):71–81.

Ahmad S, Kamran M, Ding R, Meng X, Wang H, Ahmad I, Fahad S, Han Q (2019) Exogenous melatonin confers drought stress by promoting plant growth, photosynthetic capacity and antioxidant defense system of maize seedlings. *PeerJ* 7:e7793. http://doi.org/10.7717/peerj.7793.

Akbar H, Timothy JK, Jagadish T, Golam MM, Apurbo KC, Muhammad F, Rajan B, Fahad S, Hasanuzzaman M (2020) Agricultural land degradation: processes and problems undermining future food security. In: Fahad S, Hasanuzzaman M, Alam M, Ullah H,Saeed M, Khan AK, Adnan M (eds) *Environment, climate, plant and vegetation growth*. Springer Publ Ltd, Springer Nature Switzerland AG. Part of Springer Nature, pp 17–62. https://doi.org/10.1007/978-3-030-49732-3.

Akram R, Turan V, Hammad HM, Ahmad S, Hussain S, Hasnain A, Maqbool MM, Rehmani MIA, Rasool A, Masood N, Mahmood F, Mubeen M, Sultana SR, Fahad S, Amanet K, Saleem M, Abbas Y, Akhtar HM, Waseem F, Murtaza R, Amin A, Zahoor SA, ul Din MS, Nasim W (2018a) Fate of organic and inorganic pollutants in paddy soils. In: Hashmi MZ, Varma A (eds) *Environmental pollution of paddy soils, soil biology*. Springer International Publishing AG, Cham, Switzerland, pp 197–214.

Akram R, Turan V, Wahid A, Ijaz M, Shahid MA, Kaleem S, Hafeez A, Maqbool MM, Chaudhary HJ, Munis MFH, Mubeen M, Sadiq N, Murtaza R, Kazmi DH, Ali S, Khan N, Sultana SR, Fahad S, Amin A, Nasim W (2018b) Paddy land pollutants and their role in climate change. In: Hashmi MZ, Varma A (eds) *Environmental pollution of paddy soils, soil biology*. Springer International Publishing AG, Cham, Switzerland, pp 113–124.

Ali K, Arif M, Khan Z, Tariq M, Waqas M, Gul B, ... Adnan M (2013) Effect of cutting on productivity and associated weeds of canola. *Pak J Weed Sci Res* 19(4):393–401.

Ali M, Baek KH (2020) Jasmonic acid signaling pathway in response to abiotic stresses in plants. *Int J Mol Sci* 21(2):621.

Al-Wabel Mohammad I, Abdelazeem S, Munir A, Khalid E, Adel RAU (2020b) Extent of climate change in Saudi Arabia and its impacts on agriculture: a case study from Qassim region. In: Fahad S, Hasanuzzaman M, Alam M, Ullah H, Saeed M, Khan AK, Adnan M (eds) *Environment, climate, plant and vegetation growth*. Springer Publ Ltd, Springer Nature Switzerland AG. Part of Springer Nature, pp 635–658. https://doi.org/10.1007/978-3-030-49732-3.

Al-Wabel Mohammad I, Ahmad Munir, Usman Adel RA, Akanji Mutair, Rafique Muhammad Imran (2020a) Advances in pyrolytic technologies with improved carbon capture and storage to combat climate change. In: Fahad S, Hasanuzzaman M, Alam M, Ullah H, Saeed M, Khan AK, Adnan M (eds) *Environment, climate, plant and vegetation growth*. Springer Publ Ltd, Springer Nature Switzerland AG. Part of Springer Nature, pp 535–576. https://doi.org/10.1007/978-3-030-49732-3.

Amanullah Shah K, Imran Hamdan AK, Muhammad A, Abdel RA, Muhammad A, Fahad S, Azizullah S, Brajendra P (2020) Effects of climate change on irrigation water quality. In: Fahad S, Hasanuzzaman M, Alam M, Ullah H,Saeed M, Khan AK, Adnan M (eds) *Environment, climate, plant and vegetation growth*. Springer Publ Ltd, Springer Nature Switzerland AG. Part of Springer Nature, pp 123–132. https://doi.org/10.1007/978-3-030-49732-3.

Amir M, Muhammad A, Allah B, Sevgi Ç, Haroon ZK, Muhammad A, Emre A (2020) Bio fortification under climate change: the fight between quality and quantity. In: Fahad S, Hasanuzzaman M, Alam M, Ullah H, Saeed M, Khan AK, Adnan M (eds) *Environment, climate, plant and vegetation growth*. Springer Publ Ltd, Springer Nature Switzerland AG. Part of Springer Nature, pp 173–228. https://doi.org/10.1007/978-3-030-49732-3.

Amjad I, Muhammad H, Farooq S, Anwar H (2020) Role of plant bioactives in sustainable agriculture. In: Fahad S, Hasanuzzaman M, Alam M, Ullah H, Saeed M, Khan AK, Adnan M (eds) *Environment, climate, plant and vegetation growth*. Springer Publ Ltd, Springer Nature Switzerland AG. Part of Springer Nature, pp 591–606. https://doi.org/10.1007/978-3-030-49732-3.

Arif M, Talha J, Muhammad R, Fahad S, Muhammad A, Amanullah Kawsar A, Ishaq AM, Bushra K, Fahd R (2020) Biochar; a remedy for climate change. In: Fahad S, Hasanuzzaman M, Alam M, Ullah H,Saeed M, Khan AK, Adnan M (eds) *Environment, climate, plant and vegetation growth*. Springer Publ Ltd, Springer Nature Switzerland AG. Part of Springer Nature, pp 151–172. https://doi.org/10.1007/978-3-030-49732-3.

Ayman ELS, Hossain Akbar, Barutçular Celaleddin, Iqbal Muhammad Aamir, Sohidul Islam M, Fahad Shah, Sytar Oksana, Çig Fatih, Meena Ram Swaroop, Erman Murat (2020) Consequences of salinity stress on the quality of crops and its mitigation strategies for sustainable crop production: an outlook of arid and semi- arid regions. In: Fahad S, Hasanuzzaman M, Alam M, Ullah H, Saeed M, Khan AK, Adnan M (eds) *Environment, climate, plant and vegetation growth*. Springer Publ Ltd, Springer Nature Switzerland AG. Part of Springer Nature, pp 503–534. https://doi.org/10.1007/978-3-030-49732-3.

Bahuguna RN, Jagadish KSV (2015) Temperature regulation of plant phenological development. *Environ Exp Bot* 111:83–90.

Bailey TA, Zhou X, Chen J, Yang Y (2009) Role of ethylene, abscisic acid and MAP kinase pathways in rice blast resistance. In: Wang GL, Valent B (eds) *Advances in genetics, genomics and control of rice blast disease*. Springer, Dordrecht, pp 185–190.

Banerjee A, Roychoudhury A (2019) The regulatory signaling of gibberellin metabolism and its crosstalk with phytohormones in response to plant abiotic stresses. In: *Plant signaling molecules*. Woodhead Publishing, pp 333–339.

Banerjee P, Bhadra P (2020) Mini review on strigolactones: newly discovered plant hormones. *Biosc Biotech Res Comm* 13(3):1121–1127. http://doi.org/10.21786/bbrc/13.3/20.

Baseer M, Adnan M, Fazal M, Fahad S, Muhammad S, Fazli W, Muhammad A, Jr. Amanullah DW, Saud S, Muhammad N, Muhammad Z, Fazli S, Beena S, Mian AR, Ishaq AM (2019) Substituting urea by organic wastes for improving maize yield in alkaline soil. *J Plant Nutr.*

Baxter A, Mittler R, Suzuki N (2014) ROS as key players in plant stress signalling. *J Exp Bot* 65(5):1229–1240.

Bayram AY, Seher Ö, Nazlican A (2020) Climate change forecasting and modeling for the year of 2050. In: Fahad S, Hasanuzzaman M, Alam M, Ullah H,Saeed M, Khan AK, Adnan M (eds) *Environment, climate, plant and vegetation growth.* Springer Publ Ltd, Springer Nature Switzerland AG. Part of Springer Nature, pp 109–122. https://doi.org/10.1007/978-3-030-49732-3.

Bokszczanin KL, Fragkostefanakis S, Bostan H, Bovy A, Chaturvedi P, Chiusano ML, Firon N, Iannacone R, Jegadeesan S, Klaczynskid K, Li H (2013) Perspectives on deciphering mechanisms underlying plant heat stress response and thermotolerance. *Front Plant Sci* 4:315. https://doi.org/10.3389/fpls.2013.00315.

Bolouri-Moghaddam MR, Le Roy K, Xiang L, Rolland F, Van den Ende W (2010) Sugar signalling and antioxidant network connections in plant cells. *FEBS J* 277(9):2022–2037.

Capiati DA, Paıs SM, Téllez-Iñón MT (2006) Wounding increases salt tolerance in tomato plants: evidence on the participation of calmodulin-like activities in cross-tolerance signalling. *J Exp Bot* 57(10):2391–2400.

Chen L, Zhang L, Li D, Wang F, Yu D (2013) WRKY8 transcription factor functions in the TMV-cg defense response by mediating both abscisic acid and ethylene signaling in Arabidopsis. *Proc Natl Acad Sci U S A* 110(21):E1963–E1971. https://doi.org/10.1073/pnas.1221347110.

de Saint Germain A, Ligerot Y, Dun EA, Pillot JP, Ross JJ, Beveridge CA, Rameau C (2013) Strigolactones stimulate internode elongation independently of gibberellins. *Plant Physiol* 163(2):1012–1025.

Debbarma J, Sarki YN, Saikia B, Boruah HP, Singha DL, Chikkaputtaiah C (2019) Ethylene response factor (ERF) family proteins in abiotic stresses and CRISPR–Cas9 genome editing of ERFs for multiple abiotic stress tolerance in crop plants: a review. *Rev. Mol Biotechnol* 61(2):153–172.

Denancĕ N, Sánchez-Vallet A, Goffner D, Molina A (2013) Disease resistance or growth: the role of plant hormones in balancing immune responses and fi tness costs. *Front Plant Sci* 4:1–12. https://doi.org/10.3389/fpls.2013.00155.

Depeng W, Fahad S, Saud S, Muhammad K, Aziz K, Mohammad NK, Hafiz MH, Wajid N (2018) Morphological acclimation to agronomic manipulation in leaf dispersion and orientation to promote "Ideotype" breeding: evidence from 3D visual modeling of "super" rice (*Oryza sativa* L.). *Plant Physiol Biochem* 135:499–510.

Desikan R, Last K, Harrett-Williams R, Tagliavia C, Harter K, Hooley R, Hancock JT, Neill SJ (2006) Ethylene-induced stomatal closure in Arabidopsis occurs via Atrboh F-mediated hydrogen peroxide synthesis. *Plant J* 47(6):907–916.

Driedonks N, Xu J, Peters JL, Park S, Rieu I (2015) Multi-level interactions between heat shock factors, heat shock proteins, and the redox system regulate acclimation to heat. *Front Plant Sci* 6:999. https://doi.org/10.3389/fpls.2015.00999.

Emamverdian A, Ding Y, Mokhberdoran F (2020) The role of salicylic acid and gibberellin signaling in plant responses to abiotic stress with an emphasis on heavy metals. *Plant Signal Behav*:1777372. https://doi.org/10.1080/15592324.2020.1777372.

Enamul Md H, Shoeb AZM, Mallik AH, Fahad S, Kamruzzaman MM, Akib J, Nayyer S, Mehedi KMA, Swati AS, Yeamin Md A, Most SS (2020) Measuring vulnerability to environmental hazards: qualitative to quantitative. In: Fahad S, Hasanuzzaman M, Alam M, Ullah H, Saeed M, Khan AK, Adnan M (eds) *Environment, climate, plant and vegetation growth.* Springer Publ Ltd, Springer Nature Switzerland AG. Part of Springer Nature, pp 421–452. https://doi.org/10.1007/978-3-030-49732-3.

Fahad S, Adnan M, Hassan S, Saud S, Hussain S, Wu C, Wang D, Hakeem KR, Alharby HF, Turan V, Khan MA, Huang J (2019b) Rice responses and tolerance to high temperature. In: Hasanuzzaman M, Fujita M, Nahar K, Biswas JK (eds) *Advances in rice research for abiotic stress tolerance.* Woodhead Publ Ltd, Cambridge, England, pp 201–224.

Fahad S, Bajwa AA, Nazir U, Anjum SA, Farooq A, Zohaib A, Sadia S, Nasim W, Adkins S, Saud S, Ihsan MZ, Alharby H,Wu C,Wang D, Huang J (2017) Crop production under drought and heat stress: plant responses and management options. *Front Plant Sci* 8:1147. https://doi.org/10.3389/fpls.2017.01147.

Fahad S, Bano A (2012) Effect of salicylic acid on physiological and biochemical characterization of maize grown in saline area. *Pak J Bot* 44:1433–1438.

Fahad S, Chen Y, Saud S,Wang K, Xiong D, Chen C,Wu C, Shah F, Nie L, Huang J (2013) Ultraviolet radiation effect on photosynthetic pigments, biochemical attributes, antioxidant enzyme activity and hormonal contents of wheat. *J Food Agri Environ* 11(3&4):1635–1641.

Fahad S, Hussain S, Bano A, Saud S, Hassan S, Shan D, Khan FA, Khan F, Chen Y, Wu C, Tabassum MA, Chun MX, Afzal M, Jan A, Jan MT, Huang J (2015a) Potential role of phytohormones and plant growth-promoting rhizobacteria in abiotic stresses: consequences for changing environment. *Environ Sci Pollut Res* 22(7):4907–4921. https://doi.org/10.1007/s11356-014-3754-2.

Fahad S, Hussain S, Matloob A, Khan FA, Khaliq A, Saud S, Hassan S, Shan D, Khan F, Ullah N, Faiq M, Khan MR, Tareen AK, Khan A, Ullah A, Ullah N, Huang J (2015b) Phytohormones and plant responses to salinity stress: a review. *Plant Growth Regul* 75(2):391–404. https://doi.org/10.1007/s10725-014-0013-y.

Fahad S, Hussain S, Saud S, Hassan S, Chauhan BS, Khan F, Ihsan MZ, Ullah A, Wu C, Bajwa AA, Alharby H, Amanullah, Nasim W, Shahzad B, Tanveer M, Huang J (2016a) Responses of rapid viscoanalyzer profile and other rice grain qualities to exogenously applied plant growth regulators under high day and high night temperatures. *PLOS ONE* 11(7):e0159590. https://doi.org/10.1371/journal.pone.0159590.

Fahad S, Hussain S, Saud S, Hassan S, Ihsan Z, Shah AN,Wu C, Yousaf M, Nasim W, Alharby H, Alghabari F, Huang J (2016c) Exogenously applied plant growth regulators enhance the morphophysiological growth and yield of rice under high temperature. *Front Plant Sci* 7:1250. https://doi.org/10.3389/fpls.2016.

Fahad S, Hussain S, Saud S, Hassan S, Tanveer M, Ihsan MZ, Shah AN, Ullah A, Nasrullah, Khan F, Ullah S, Alharby H, Nasim W, Wu C, Huang J (2016d) A combined application of biochar and phosphorus alleviates heat-induced adversities on physiological, agronomical and quality attributes of rice. *Plant Physiol Biochem* 103:191–198.

Fahad S, Hussain S, Saud S, Khan F, Hassan S Jr A, Nasim W, Arif M, Wang F, Huang J (2016b) Exogenously applied plant growth regulators affect heat-stressed rice pollens. *J Agron Crop Sci* 202(2):139–150.

Fahad S, Hussain S, Saud S, Tanveer M, Bajwa AA, Hassan S, Shah AN, Ullah A,Wu C, Khan FA, Shah F, Ullah S, Chen Y, Huang J (2015a) A biochar application protects rice pollen from high-temperature stress. *Plant Physiol Biochem* 96:281–287.

Fahad S, Ihsan MZ, Khaliq A, Daur I, Saud S, Alzamanan S, Nasim W, Abdullah M, Khan IA, Wu C, Wang D, Huang J (2018) Consequences of high temperature under changing climate optima for rice pollen characteristics-concepts and perspectives. *Arch Agron Soil Sci*. https://doi.org/10.1080/03650340.2018.1443213.

Fahad S, Nie L, Chen Y, Wu C, Xiong D, Saud S, Hongyan L, Cui K, Huang J (2015b) Crop plant hormones and environmental stress. *Sustain Agric Rev* 15:371–400.

Fahad S, Rehman A, Shahzad B, Tanveer M, Saud S, Kamran M, Ihtisham M, Khan SU, Turan V, Rahman MHU (2019a) Rice responses and tolerance to metal/metalloid toxicity. In: Hasanuzzaman M, Fujita M, Nahar K, Biswas JK (eds) *Advances in rice research for abiotic stress tolerance*. Woodhead Publ Ltd, Cambridge, England, pp 299–312.

Farah R, Muhammad R, Muhammad SA, Tahira Y, Muhammad AA, Maryam A, Shafaqat A, Rashid M, Muhammad R, Qaiser H, Afia Z, Muhammad AA, Muhammad A, Fahad S (2020) Alternative and non-conventional soil and crop management strategies for increasing water use efficiency. In: Fahad S, Hasanuzzaman M, Alam M, Ullah H, Saeed M, Khan AK, Adnan M (eds) *Environment, climate, plant and vegetation growth*. Springer Publ Ltd, Springer Nature Switzerland AG. Part of Springer Nature, pp 323–338. https://doi.org/10.1007/978-3-030-49732-3.

Farhana G, Ishfaq A, Muhammad A, Dawood J, Fahad S, Xiuling L, Depeng W, Muhammad F, Muhammad F, Syed AS (2020) Use of crop growth model to simulate the impact of climate change on yield of various wheat cultivars under different agro-environmental conditions in Khyber Pakhtunkhwa, Pakistan. *Arabian J Geosci* 13:112.

Farhat A, Hafiz MH, Wajid I, Aitazaz AF, Hafiz FB, Zahida Z, Fahad S, Wajid F, Artemi C (2020) A review of soil carbon dynamics resulting from agricultural practices. *J Environ Manag* 268. http://www.ncbi.nlm.nih.gov/pubmed/110319.

Fazli W, Muhmmad S, Amjad A, Fahad S, Muhammad A, Muhammad N, Ishaq AM, Imtiaz AK, Mukhtar A, Muhammad S, Muhammad I, Rafi U, Haroon I, Muhammad A (2020) Plant-microbes interactions and functions in changing climate. In: Fahad S, Hasanuzzaman M, Alam M, Ullah H, Saeed M, Khan AK, Adnan M (eds) *Environment, climate, plant and vegetation growth*. Springer Publ Ltd, Springer Nature Switzerland AG. Part of Springer Nature, pp 397–420. https://doi.org/10.1007/978-3-030-49732-3.

Gopakumar L, Bernard NO, Donato V (2020) Soil microarthropods and nutrient cycling. In: Fahad S, Hasanuzzaman M, Alam M, Ullah H, Saeed M, Khan AK, Adnan M (eds) *Environment, climate, plant and vegetation growth*. Springer Publ Ltd, Springer Nature Switzerland AG. Part of Springer Nature, pp 453–472. https://doi.org/10.1007/978-3-030-49732-3.

Gray WM, Ostin A, Sandberg G, Romano CP, Estelle M (1998) High temperature promotes auxin-mediated hypocotyl elongation in Arabidopsis. *Proc Natl Acad Sci USA* 95(12):7197–7202.

Hafiz MH, Muhammad A, Farhat A, Hafiz FB, Saeed AQ, Muhammad M, Fahad S, Muhammad A (2019) Environmental factors affecting the frequency of road traffic accidents: a case study of sub-urban area of Pakistan. *Environ Sci Pollut Res.*

Hafiz MH, Wajid F, Farhat A, Fahad S, Shafqat S, Wajid N, Hafiz FB (2017) Maize plant nitrogen uptake dynamics at limited irrigation water and nitrogen. *Environ Sci Pollut Res* 24(3):2549–2557.

Hammad HM, Abbas F, Ahmad A, Bakhat HF, Farhad W, Wilkerson CJ, Fahad S, Hoogenboom G (2020a) Predicting kernel growth of maize under controlled water and nitrogen applications. *Int J Plant Prod.* https://doi.org/10.1007/s42106-020-00110-8.

Hammad HM, Abbas F, Saeed S, Fahad S, Cerdà A, Farhad W, Bernardo CC, Nasim W, Mubeen M, Bakhat HF (2018) Offsetting land degradation through nitrogen and water management during maize cultivation under arid conditions. *Land Degrad Dev* 29(5):1366–1375. https://doi.org/10.1002/ldr.2933.

Hammad HM, Khaliq A, Abbas F, Farhad W, Fahad S, Aslam M, Shah GM, Nasim W, Mubeen M, Bakhat HF (2020b) Comparative effects of organic and inorganic fertilizers on soil organic carbon and wheat productivity under arid region. *Commun Soil Sci Plant Anal.* https://doi.org/10.1080/00103624.2020.1763385.

Harrison MA (2012) Cross-talk between phytohormone signalling pathways under both optimal and stressful environmental conditions. In: Khan NA, Nazar R, Iqbal N, Anjum NA (eds) *Phytohormones and abiotic stress tolerance in plants.* Springer-Verlag, Berlin Heidelberg, pp 49–76.

Hesham FA, Fahad S (2020) Melatonin application enhances biochar efficiency for drought tolerance in maize varieties: modifications in physio-biochemical machinery. *Agron J* 112(4):1–22.

Hsieh EJ, Cheng MC, Lin TP (2013) Functional characterization of an abiotic stress-inducible transcription factor AtERF53 in Arabidopsis thaliana. *Plant Mol Biol* 82(3):223–237.

Hu X, Liu R, Li Y, Wang W, Tai F, Xue R, Li C (2010) Heat shock protein 70 regulates the abscisic acid-induced antioxidant response of maize to combined drought and heat stress. *Plant Growth Regul* 60(3):225–235.

Hussain MA, Fahad S, Rahat S, Muhammad FJ, Muhammad M, Qasid A, Ali A, Husain A, Nooral A, Babatope SA, Changbao S, Liya G, Ibrar A, Zhanmei J, Juncai H (2020) Multifunctional role of brassinosteroid and its analogues in plants. *Plant Growth Regul.* https://doi.org/10.1007/s10725-020-00647-8.

Hussain S, Mubeen M, Ahmad A, Akram W, Hammad HM, Ali M, Masood N, Amin A, Farid HU, Sultana SR, Fahad S, Wang D, Nasim W (2020) Using GIS tools to detect the land use/land cover changes during forty years in Lodhran District of Pakistan. *Environ Sci Pollut Res Int* 27(32):39676–39692. https://doi.org/10.1007/s11356-019-06072-3.

Ibrar K, Aneela R, Khola Z, Urooba N, Sana B, Rabia S, Ishtiaq H, Mujaddad Ur Rehman, Salvatore M (2020) Microbes and environment: global warming reverting the frozen zombies. In: Fahad S, Hasanuzzaman M, Alam M, Ullah H, Saeed M, Khan AK, Adnan M (eds) *Environment, climate, plant and vegetation growth.* Springer Publ Ltd, Springer Nature Switzerland AG. Part of Springer Nature, pp 607–634. https://doi.org/10.1007/978-3-030-49732-3.

Ilyas M, Nisar M, Khan N, Hazrat A, Khan AH, Hayat K, Fahad S, Khan A, Ullah A (2020) Drought tolerance strategies in plants: a mechanistic approach. *J Plant Growth Regul.* https://doi.org/10.1007/s00344-020-10174-5.

Iqbal N, Nazar R, Khan MI, Masood A, Khan NA (2011) Role of gibberellins in regulation of source sink relations under optimal and limiting environmental conditions. *Curr Sci* 100:998–1007.

Iqra M, Amna B, Shakeel I, Fatima K,Sehrish L, Hamza A, Fahad S (2020) Carbon cycle in response to global warming. In: Fahad S, Hasanuzzaman M, Alam M, Ullah H,Saeed M, Khan AK, Adnan M (eds) *Environment, climate, plant and vegetation growth.* Springer Publ Ltd, Springer Nature Switzerland AG. Part of Springer Nature, pp 1–16. https://doi.org/10.1007/978-3-030-49732-3.

Izhar Shafi M, Adnan M, Fahad S, Wahid F, Khan A, Yue Z, Danish S, Zafar-ul-Hye M, Brtnicky M, Datta R (2020) Application of single superphosphate with humic acid improves the growth, yield and phosphorus uptake of wheat (Triticum aestivum L.) in calcareous soil. *Agronomy* 10(9):1224. https://doi.org/10.3390/agronomy10091224.

Jakir Md H, Allah B (2020) Development and applications of Transplastomic plants; A way Towards eco-friendly agriculture. In: Fahad S, Hasanuzzaman M, Alam M, Ullah H, Saeed M, Khan AK, Adnan M (eds) *Environment, climate, plant and vegetation growth.* Springer Publ Ltd, Springer Nature Switzerland AG. Part of Springer Nature, pp 285–322. https://doi.org/10.1007/978-3-030-49732-3.

Jan M, Anwar-ul-Haq M, Shah AN, Yousaf M, Iqbal J, Li X, Wang D, Fahad S (2019) Modulation in growth, gas exchange, and antioxidant activities of salt-stressed rice (Oryza sativa L.) genotypes by zinc fertilization. *Arab J Geosci* 12(24):775. https://doi.org/10.1007/s12517-019-4939-2.

Kamran M, Cui W, Ahmad I, Meng X, Zhang X, Su W, Chen J, Ahmad S, Fahad S, Han Q, Liu T (2018) Effect of paclobutrazol, a potential growth regulator on stalk mechanical strength, lignin accumulation and its relation with lodging resistance of maize. *Plant Growth Regul* 84(2):317–332. https://doi.org/10.1007/ s10725-017-0342-8.

Kamarn M, Wenwen C, Irshad A, Xiangping M, Xudong Z, Wennan S, Junzhi C, Shakeel A, Fahad S, Qingfang H, Tiening L (2017) Effect of paclobutrazol, a potential growth regulator on stalk mechanical strength, lignin accumulation and its relation with lodging resistance of maize. *Plant Growth Regul.*

Kaur N, Pati PK (2019) Harnessing the potential of brassinosteroids in abiotic stress tolerance in plants. In: *Brassinosteroids: plant growth and development*. Springer, Singapore, pp 407–423.

Kazan K (2015) Diverse roles of jasmonates and ethylene in abiotic stress tolerance. *Trends Plant Sci* 20(4):219–229.

Khan A, Bilal S, Khan AL, Imran M, Shahzad R, Al-Harrasi A, Al-Rawahi A, Al-Azhri M, Mohanta TK, Lee IJ (2020) Silicon and gibberellins: synergistic function in harnessing ABA signaling and heat stress tolerance in date palm (*Phoenix dactylifera* L.). *Plants* 9(5):620.

Khan A, Tan DKY, Afridi MZ, Luo H, Tung SA, Ajab M, Fahad S (2017b) Nitrogen fertility and abiotic stresses management in cotton crop: a review. *Environ Sci Pollut Res* 24(17):14551–14566. https://doi.org/10.1007/s11356-017-8920-x.

Khan A, Tan DKY, Munsif F, Afridi MZ, Shah F, Wei F, Fahad S, Zhou R (2017a) Nitrogen nutrition in cotton and control strategies for greenhouse gas emissions: a review. *Environ Sci Pollut Res* 24(30):23471–23487. https://doi.org/10.1007/s11356-017-0131-y.

Knight H, Trewavas AJ, Knight MR (1996) Cold calcium signaling in Arabidopsis involves two cellular pools and a change in calcium signature after acclimation. *Plant Cell* 8(3):489–503.

Kohli A, Sreenivasulu N, Lakshmanan P, Kumar PP (2013) The phytohormone crosstalk paradigm takes center stage in understanding how plants respond to abiotic stresses. *Plant Cell Rep* 32(7):945–957.

Krouk G, Ruffel S, Gutiérrez RA, Gojon A, Crawford NM, Coruzzi GM, Lacombe B (2011) A framework integrating plant growth with hormones and nutrients. *Trends Plant Sci* 16(4):178–182.

Kwak JM, Mori IC, Pei ZM, Leonhardt N, Torres MA, Dangl JL, Bloom RE, Bodde S, Jones JDG, Schroeder JI (2003) NADPH oxidase AtrbohD and AtrbohF genes function in ROS-dependent ABA signaling in Arabidopsis. *EMBO J* 22(11):2623–2633.

Larkindale J, Hall JD, Knight MR, Vierling E (2005) Heat stress phenotypes of Arabidopsis mutants implicate multiple signaling pathways in the acquisition of thermotolerance. *Plant Physiol* 138(2):882–897.

Li H, Liu SS, Yi CY, Wang F, Zhou J, Xia XJ, Shi K, Zhou YH, Yu JQ (2014) Hydrogen peroxide mediates abscisic acid-induced HSP70 accumulation and heat tolerance in grafted cucumber plants. *Plant Cell Environ* 37(12):2768–2780.

Li Z, Zhang L, Wang A, Xu X, Li J (2013) Ectopic overexpression of SlHsfA3, a heat stress transcription factor from tomato, confers increased thermotolerance and salt hypersensitivity in germination in transgenic Arabidopsis. *PLOS ONE* 8(1):e54880.

Linkies A, Leubner-Metzger G (2012) Beyond gibberellins and abscisic acid: how ethylene and jasmonates control seed germination. *Plant Cell Rep* 31(2):253–270.

Liu HC, Liao HT, Charng YY (2011) The role of class A1 heat shock factors (HSFA1s) in response to heat and other stresses in Arabidopsis. *Plant Cell Environ* 34(5):738–751.

Liu Y, Dang P, Liu L, He C (2019) Cold acclimation by the CBF-COR pathway in a changing climate: lessons from Arabidopsis thaliana. *Plant Cell Rep* 38(5):511–519.

Mahar A, Amjad A, Altaf HL, Fazli W, Ronghua L, Muhammad A, Fahad S, Muhammad A, Rafiullah Imtiaz AK, Zengqiang Z (2020) Promising technologies for Cd-contaminated soils: drawbacks and possibilities. In: Fahad S, Hasanuzzaman M, Alam M, Ullah H,Saeed M, Khan AK, Adnan M (eds) *Environment, climate, plant and vegetation growth*. Springer Publ Ltd, Springer Nature Switzerland AG. Part of Springer Nature, pp 63–92. https://doi.org/10.1007/978-3-030-49732-3.

Mano Y, Nemoto K (2012) The pathway of auxin biosynthesis in plants. *J Exp Bot* 63(8):2853–2872.

Marino D, Dunand C, Puppo A, Pauly N (2012) A burst of plant NADPH oxidases. *Trends Plant Sci* 17(1): 9–15.

McKemy DD, Neuhausser WM, Julius D (2002) Identification of a cold receptor reveals a general role for TRP channels in thermosensation. *Nature* 416(6876):52–58.

Mittler R, Finka A, Goloubinoff P (2012) How do plants feel the heat? *Trends Biochem Sci* 37(3):118–125.

Mittler R, Vanderauwera S, Gollery M, Van Breusegem F (2004) Reactive oxygen gene network of plants. *Trends Plant Sci* 9(10):490–498.

Mubeen M, Ahmad A, Hammad HM, Awais M, Farid HU, Saleem M, Din MSu, Amin A, Ali A, Fahad S, Nasim W (2020) Evaluating the climate change impact on water use efficiency of cotton-wheat in semi-arid conditions using DSSAT model. *J Water Clim Change* 11(4):1661–1675. https://doi.org/10.2166/wcc.2019.179/622035/jwc2019179.pdf.

Muhammad B, Adnan M, Munsif F, Fahad S, Saeed M, Wahid F, Arif M, Amanullah J, Wang D, Saud S, Noor M, Zamin M, Subhan F, Saeed B, Raza MA, Mian IA (2019) Substituting urea by organic wastes for improving maize yield in alkaline soil. *J Plant Nutr.* https://doi.org/10.1080/01904167.2019.1659344.

Nasim W, Ahmad A, Amin A, Tariq M, Awais M, Saqib M, Jabran K, Shah GM, Sultana SR, Hammad HM, Rehmani MIA, Hashmi MZ, Rahman MHU, Turan V, Fahad S, Suad S, Khan A, Ali S (2018) Radiation efficiency and nitrogen fertilizer impacts on sunflower crop in contrasting environments of Punjab, Pakistan. *Environ Sci Pollut Res Int* 25(2):1822–1836. https://doi.org/10.1007/s11356-017-0592-z.

Nishiyama R, Watanabe Y, Fujita Y, Le DT, Kojima M, Werner T, Vankova R, Yamaguchi-Shinozaki K, Shinozaki K, Kakimoto T, Sakakibara H, Schmülling T, Tran LS (2011) Analysis of cytokinin mutants and regulation of cytokinin metabolic genes reveals important regulatory roles of cytokinins in drought, salt and abscisic acid responses, and abscisic acid biosynthesis. *Plant Cell* 23(6):2169–2183.

Nishizawa A, Yabuta Y, Shigeoka S (2008) Galactinol and raffinose constitute a novel function to protect plants from oxidative damage. *Plant Physiol* 147(3):1251–1263.

Noor M, ur Naveed R, Ajmal J, Fahad S, Muhammad A, Fazli W, Saud S, Hassan S (2020) Climate change and costal plant lives. In: Fahad S, Hasanuzzaman M, Alam M, Ullah H,Saeed M, Khan AK, Adnan M (eds) *Environment, climate, plant and vegetation growth.* Springer Publ Ltd, Springer Nature Switzerland AG. Part of Springer Nature, pp 93–108. https://doi.org/10.1007/978-3-030-49732-3.

Qin D, Wu H, Peng H, Yao Y, Ni Z, Li Z, Zhou C, Sun Q (2008) Heat stress-responsive transcriptome analysis in heat susceptible and tolerant wheat (Triticum aestivum L.) by using Wheat Genome Array. *BMC Genomics* 9:432.

Rahman MHu, Ahmad A, Wajid A, Hussain M, Rasul F, Ishaque W, Islam MA, Shelia V, Awais M, Ullah A, Wahid A, Sultana SR, Saud S, Khan S, Fahad S, Hussain M, Hussain S, Nasim W (2019) Application of CSM-CROPGRO-Cotton model for cultivars and optimum planting dates: evaluation in changing semi-arid climate. *Field Crops Res* 238:139–152. http://doi.org/10.1016/j.fcr.2017.07.007.

Rashid M, Qaiser H, Khalid SK, Mohammad I, Al-Wabel Zhang A, Muhammad A, Shahzada SI, Rukhsanda A, Ghulam AS, Shahzada MM, Sarosh A, Muhammad FQ (2020) Prospects of biochar in alkaline soils to mitigate climate change. In: Fahad S, Hasanuzzaman M, Alam M, Ullah H,Saeed M, Khan AK, Adnan M (eds) *Environment, climate, plant and vegetation growth.* Springer Publ Ltd, Springer Nature Switzerland AG. Part of Springer Nature, pp 133–150. https://doi.org/10.1007/978-3-030-49732-3.

Rehman M, Fahad S, Saleem MH, Hafeez M, Rahman MHu, Liu F, Deng G (2020) Red light optimized physiological traits and enhanced the growth of ramie (Boehmeria nivea L.). *Photosynthetica* 58(4):922–931.

Sadam M, ul Qamar Muhammad Tahir, Ghulam M, Muhammad SK, Faiz AJ (2020) Role of biotechnology in climate resilient agriculture. In: Fahad S, Hasanuzzaman M, Alam M, Ullah H, Saeed M, Khan AK, Adnan M (eds) *Environment, climate, plant and vegetation growth.* Springer Publ Ltd, Springer Nature Switzerland AG. Part of Springer Nature, pp 339–366. https://doi.org/10.1007/978-3-030-49732-3.

Saidi Y, Finka A, Muriset M, Bromberg Z, Weiss YG, Maathuis FJ, Goloubinoff P (2009) The heat shock response in moss plants is regulated by specific calcium-permeable channels in the plasma membrane. *Plant Cell* 21(9):2829–2843.

Sajid H, Jie H, Jing H, Shakeel A, Satyabrata N, Sumera A, Awais S, Chunquan Z, Lianfeng Z, Xiaochuang C, Qianyu J, Junhua Z (2020) Rice production under climate change: adaptations and mitigating strategies. In: Fahad S, Hasanuzzaman M, Alam M, Ullah H, Saeed M, Khan AK, Adnan M (eds) *Environment, climate, plant and vegetation growth.* Springer Publ Ltd, Springer Nature Switzerland AG. Part of Springer Nature, pp 659–686. https://doi.org/10.1007/978-3-030-49732-3.

Sajjad H, Muhammad M, Ashfaq A, Waseem A, Hafiz MH, Mazhar A, Nasir M, Asad A, Hafiz UF, Syeda RS, Fahad S, Depeng W, Wajid N (2019) Using GIS tools to detect the land use/land cover changes during forty years in Lodhran district of Pakistan. *Environ Sci Pollut Res.*

Saleem MH, Fahad S, Adnan M, Ali M, Rana MS, Kamran M, Ali Q, Hashem IA, Bhantana P, Ali M, Hussain RM (2020a) Foliar application of gibberellic acid endorsed phytoextraction of copper and alleviates oxidative stress in jute (Corchorus Capsularis L.) plant grown in highly copper-contaminated soil of China. *Environ Sci Pollut Res.* https://doi.org/10.1007/s11356-020-09764-3.

Saleem MH, Fahad S, Khan SU, Din M, Ullah A, Sabagh AE, Hossain A, Llanes A, Liu L (2020) Copper-induced oxidative stress, initiation of antioxidants and phytoremediation potential of flax (Linum

usitatissimum L.) seedlings grown under the mixing of two different soils of China. *Environ Sci Pollut Res Int* 27(5):5211–5221. https://doi.org/10.1007/s11356-019-07264-7.
Saleem MH, Rehman M, Fahad S, Tung SA, Iqbal N, Hassan A, Ayub A, Wahid MA, Shaukat S, Liu L, Deng G (2020b) Leaf gas exchange, oxidative stress, and physiological attributes of rapeseed (Brassica napus L.) grown under different light-emitting diodes. *Photosynthetica* 58(3):836–845.
Saleh J, Maftoun M (2008) Interactive effect of NaCl levels and zinc sources and levels on the growth and mineral composition of rice. *J Agric Sci Tech* 10:325–336.
Saman S, Amna B, Bani A, ul Qamar Muhammad Tahir, Rana MA, Muhammad SK (2020) QTL mapping for abiotic stresses in cereals. In: Fahad S, Hasanuzzaman M, Alam M, Ullah H, Saeed M, Khan AK, Adnan M (eds) *Environment, climate, plant and vegetation growth.* Springer Publ Ltd, Springer Nature Switzerland AG. Part of Springer Nature, pp 229–252. https://doi.org/10.1007/978-3-030-49732-3.
Saud S, Fahad S, Cui G, Yajun C, Anwar S (2020) Determining nitrogen isotopes discrimination under drought stress on enzymatic activities, nitrogen isotope abundance and water contents of Kentucky bluegrass. *Sci Rep* 10(1):6415. https://doi.org/10.1038/s41598-020-63548-w.
Saud S, Fahad S, Yajun C, Ihsan MZ, Hammad HM, Nasim W, Amanullah Jr, Arif M, Alharby H (2017) Effects of nitrogen supply on water stress and recovery mechanisms in Kentucky bluegrass plants. *Front Plant Sci* 8:983. https://doi.org/10.3389/fpls.2017.00983.
Saud S, Li X, Chen Y, Zhang L, Fahad S, Hussain S, Sadiq A, Chen Y (2014) Silicon application increases drought tolerance of kentucky bluegrass by improving plant water relations and morphophysiological functions. *Sci World J* 2014:368694. https://doi.org/10.1155/2014/368694.
Saud S, Yajun C, Fahad S, Hussain S, Na L, Xin L, Alhussien SA (2016) Silicate application increases the photosynthesis and its associated metabolic activities in Kentucky bluegrass under drought stress and post-drought recovery. *Environ Sci Pollut Res* 23(17):17647–17655. https://doi.org/10.1007/s11356-016-6957-x.
Saud S, Yajun C, Long B, Fahad S, Sadiq A (2013) The different impact on the growth of cool season turf grass under the various conditions on salinity and drought stress. *Int J Agric Sci Res* 3:77–84.
Senol C (2020) The effects of climate change on human behaviors. In: Fahad S, Hasanuzzaman M, Alam M, Ullah H, Saeed M, Khan AK, Adnan M (eds) *Environment, climate, plant and vegetation growth.* Springer Publ Ltd, Springer Nature Switzerland AG. Part of Springer Nature, pp 577–590. https://doi.org/10.1007/978-3-030-49732-3.
Shah F, Nie L, Cui K, Shah T, Wu W, Chen C, Zhu L, Ali F, Fahad S, Huang J (2014) Rice grain yield and component responses to near 2°C of warming. *Field Crop Res* 157:98–110.
Sharma A, Shahzad B, Kumar V, Kohli SK, Sidhu GP, Bali AS, Handa N, Kapoor D, Bhardwaj R, Zheng B (2019) Phytohormones regulate accumulation of osmolytes under abiotic stress. *Biomolecules* 9(7):285. https://doi.org/10.3390/biom9070285.
Singh R, Jwa NS (2013) Understanding the responses of rice to environmental stress using proteomics. *J Proteome Res* 12(11):4652–4669.
Stamm P, Kumar PP (2013) Auxin and gibberellin responsive Arabidopsis SMALL AUXIN UP RNA36 regulates hypocotyl elongation in the light. *Plant Cell Rep* 32(6):759–769.
Subhan D, Zafar-ul-Hye M, Fahad S, Saud S, Martin B, Tereza H, Rahul D (2020) Drought stress alleviation by ACC deaminase producing *Achromobacter xylosoxidans* and *Enterobacter cloacae*, with and without timber waste biochar in maize. *Sustain* 12(6286).
Suzuki N, Bajad S, Shuman J, Shulaev V, Mittler R (2008) The transcriptional co-activator MBF1c is a key regulator of thermotolerance in Arabidopsis thaliana. *J Biol Chem* 283(14):9269–9275.
Tang HB, Ma BJ (2005) Study on synthesis technology and properties of hydroxypropyl potato starch. *Food Sci* 26:167–170.
Tariq M, Ahmad S, Fahad S, Abbas G, Hussain S, Fatima Z, Nasim W, Mubeen M, Rehman MHu, Khan MA, Adnan M, Wilkerson CJ, Hoogenboom G (2018) The impact of climate warming and crop management on phenology of sunflower-based cropping systems in Punjab, Pakistan. *Agri Forest Met* 256–257:270–282.
Tariq M, Khan Z, Arif M, Ali K, Waqas M, Naveed K, Adnan M (2013) Effect of nitrogen application timings on the seed yield of Brassica cultivars and associated weeds. *Pak J Weed Sci Res* 19(4):493–502.
ul Qamar Muhammad Tahir, Amna F, Amna B, Barira Z, Xitong Z, Ling-Ling C (2020) Effectiveness of conventional crop improvement strategies vs. omics. In: Fahad S, Hasanuzzaman M, Alam M, Ullah H, Saeed M, Khan AK, Adnan M (eds) *Environment, climate, plant and vegetation growth.* Springer Publ Ltd, Springer Nature Switzerland AG. Part of Springer Nature, pp 253–284. https://doi.org/10.1007/978-3-030-49732-3.
Unsar Naeem-U, Muhammad R, Syed HMB, Asad S, Mirza AQ, Naeem I, Muhammad H ur R, Fahad S, Shafqat S (2020) Insect pests of cotton crop and management under climate change scenarios. In: Fahad

S, Hasanuzzaman M, Alam M, Ullah H, Saeed M, Khan AK, Adnan M (eds) *Environment, climate, plant and vegetation growth*. Springer Publ Ltd, Springer Nature Switzerland AG. Part of Springer Nature, pp 367–396. https://doi.org/10.1007/978-3-030-49732-3.

Vanderauwera S, Suzuki N, Miller G, Van De Cotte B, Morsa S, Ravanat JL, Hegie A, Triantaphylidès C, Shulaev V, Van Montagu MC, Van Breusegem F, Mittler R (2011) Extranuclear protection of chromosomal DNA from oxidative stress. *Proc Natl Acad Sci USA* 108(4):1711–1716.

Wahid F, Fahad S, Danish S, Adnan M, Yue Z, Saud S, Siddiqui MH, Brtnicky M, Hammerschmiedt T, Datta R Sustainable management with mycorrhizae and phosphate solubilizing bacteria for enhanced phosphorus uptake in calcareous soils. *Agriculture* 10(8). https://doi.org/10.3390/agriculture10080334.

Walia H, Wilson C, Wahid A, Condamine P, Cui X, Close TJ (2006) Expression analysis of barley (Hordeum vulgare L.) during salinity stress. *Funct Integr Genom* 6(2):143–156.

Wang D, Fahad S, Saud S, Kamran M, Khan A, Khan MN, Hammad HM, Nasim W (2019) Morphological acclimation to agronomic manipulation in leaf dispersion and orientation to promote "Ideotype" breeding: evidence from 3D visual modeling of "super" rice (Oryza sativa L.). *Plant Physiol Biochem* 135:499–510. https://doi.org/10.1016/j.plaphy.2018.11.010.

Wang J, Song L, Gong X, Xu J, Li M (2020) Functions of jasmonic acid in plant regulation and response to abiotic stress. *Int J Mol Sci* 21(4):1446. https://doi.org/10.3390/ijms21041446.

Wu C, Cui K, Tang S, Li G, Wang S, Fahad S, Nie L, Huang J, Peng S, Ding Y (2020) Intensified pollination and fertilization ameliorate heat injury in rice (Oryza sativa L.) during the flowering stage. *Field Crops Res* 252. http://www.ncbi.nlm.nih.gov/pubmed/107795.

Wu C, Tang S, Li G, Wang S, Fahad S, Ding Y (2019) Roles of phytohormone changes in the grain yield of rice plants exposed to heat: a review. *PeerJ* 7:e7792. https://doi.org/10.7717/peerj.7792.

Yang Z, Zhang Z, Zhang T, Fahad S, Cui K, Nie L, Peng S, Huang J (2017) The effect of season-long temperature increases on rice cultivars grown in the central and southern regions of China. *Front Plant Sci* 8:1908. https://doi.org/10.3389/fpls.2017.01908.

Zafar-ul-Hye M, Muhammad T ahzeeb-ul-Hassan, Muhammad A, Fahad S, Martin B, Tereza D, Rahul D, Subhan D (2020b) Potential role of compost mixed biochar with rhizobacteria in mitigating lead toxicity in spinach. *Sci Rep* 10:12159. https://doi.org/10.1038/s41598-020-69183-9.

Zafar-ul-Hye M, Naeem M, Danish S, Fahad S, Datta R, Abbas M, Rahi AA, Brtnicky M, Holátko J, Tarar ZH, Nasir M Alleviation of cadmium adverse effects by improving nutrients uptake in bitter gourd through cadmium tolerant rhizobacteria. *Environments* 7(8). https://doi.org/10.3390/environments7080054.

Zahida Z, Hafiz FB, Zulfiqar AS, Ghulam MS, Fahad S, Muhammad RA, Hafiz MH,Wajid N, Muhammad S (2017) Effect of water management and silicon on germination, growth, phosphorus and arsenic uptake in rice. *Ecotoxicol Environ Saf* 144:11–18.

Zaman Qu, Aslam Z, Yaseen M, Ihsan MZ, Khaliq A, Fahad S, Bashir S, Ramzani PMA, Naeem M (2018) Zinc biofortification in rice: leveraging agriculture to moderate hidden hunger in developing countries. *Arch Agron Soil Sci* 64(2):147–161. https://doi.org/10.1080/03650340.2017.1338343.

Zamin M, Khattak AM, Salim AM, Marcum KB, Shakur M, Shah S, Jan I, Fahad S (2019) Performance of Aeluropus lagopoides (mangrove grass) ecotypes, a potential turfgrass, under high saline conditions. *Environ Sci Pollut Res*. https://doi.org/10.1007/s11356-019-04838-3.

Zhang CX, Fu GF, Yang XQ, Yang YJ, Zhao X, Chen TT, Zhang XF, Jin QY, Tao LX (2016) Heat stress effects are stronger on spikelets than on flag leaves in rice due to differences in dissipation capacity. *J Agron Crop Sci* 202(5):394–408.

Zhang J, Jia W, Yang J, Ismail AM (2006) Role of ABA in integrating plant responses to drought and salt stresses. *Field Crops Res* 97(1):111–119.

Zhu SY, Yu XC, Wang XJ, Zhao R, Li Y, Fan RC, Shang Y, Du SY, Wang XF, Wu FQ, Xu YH, Zhang XY, Zhang DP (2007) Two calcium-dependent protein kinases, CPK4 and CPK11, regulate abscisic acid signal transduction in Arabidopsis. *Plant Cell* 19(10):3019–3036.

Zia-ur-Rehman M (2020) Environment, climate change and biodiversity. In: Fahad S, Hasanuzzaman M, Alam M, Ullah H, Saeed M, Khan AK, Adnan M (eds) *Environment, climate, plant and vegetation growth*. Springer Publ Ltd, Springer Nature Switzerland AG. Part of Springer Nature, pp 473–502. https://doi.org/10.1007/978-3-030-49732-3.

11

Plant Resilience to Abiotic Stress Mitigated through Phytohormones' Production and Their Transcriptional Control

Sammina Mahmood, Umair Ashraf, Zia ur Rehman, Muhammad Ikram, and Sajid Mehmood

CONTENTS

11.1 Introduction

The global food supply is threatened due to several factors, like socio-economic instability, ever-increasing populations, urbanisation, desertification, and limited land amenable to agriculture. The situation is further accentuated by unpredictable patterns of climate change, specifically abiotic factors which limit crop productivity and sustainability (Adnan et al. 2018; Adnan et al. 2019; Akram et al. 2018a,b; Aziz et al. 2017a,b; Habib et al. 2017; Hafiz et al. 2016; Hafiz et al. 2019; Kamaran et al. 2017; Muhammad et al. 2019; Sajjad et al. 2019; Saud et al. 2013; Saud et al. 2014; Saud et al. 2017; Saud et al. 2016; Shah et al. 2013; Saud et al. 2020; Qamar et al. 2020; Wajid et al. 2017; Yang et al. 2017; Zahida et al. 2017; Depeng et al. 2018; Hussain et al. 2020; Hafiz et al. 2020 a,b; Shafi et al. 2020; Wahid et al. 2020; Subhan et al. 2020; Zafar-ul-Hye et al. 2020a,b; Adnan et al. 2020; Ilyas et al. 2020; Saleem et al 2020a,b,c; Rehman 2020; Frahat et al. 2020; Wu et al. 2020; Mubeen et al. 2020; Farhana 2020; Jan et al. 2019; Wu et al.

2019; Ahmad et al. 2019; Baseer et al. 2019; Hafiz et al. 2018;Tariq et al. 2018); Fahad and Bano 2012; Fahad et al. 2017; Fahad et al. 2013; Fahad et al. 2014a,b; Fahad et al. 2016a,b,c,d; Fahad et al. 2015a, b; Fahad et al. 2018; Fahad et al. 2019a,b; Hesham and Fahad 2020. Iqra et al. 2020; Akbar et al. 2020; Mahar et al. 2020; Noor et al. 2020; Bayram et al. 2020; Amanullah et al. 2020; Rashid et al. 2020; Arif et al. 2020; Amir et al. 2020; Saman et al. 2020; Muhammad Tahir et al. 2020; Md Jakirand Allah 2020; Farah et al. 2020; Sadam et al. 2020; Unsar et al. 2020; Fazli et al. 2020; Md. Enamul et al. 2020; Gopakumar et al. 2020; Zia-ur-Rehman 2020; Ayman et al. 2020; Mohammad I. Al-Wabel et al. 2020a,b; Senol 2020; Amjad et al. 2020; Ibrar et al. 2020; Sajid et al. 2020). The climatic abiotic factors include drought, heat, salinity, flooding, water logging, UV rays, pH alteration, nutrient imbalance, freezing, gaseous pollutants (NO, SO2, CO, ozone, etc.) (Tardieu and Tuberosa 2010). Plants endure the brutality of these natural factors by executing variations in morphological, physiological, biochemical, and molecular response alone or in combination, depending upon the severity of stress, to cope with the prevailing abiotic stress. Plants successfully coping with the stress conditions are called tolerant plant types, executing the ability of the system to come back to its natural state with minimal internal damages upon relief from the stress. This ability of plant is called 'plant resilience.' Once stress is perceived, plant resilience is activated through the regulation of an output channel – 'the transcriptional network' – and downstream signalling molecules, known as plant growth regulators (PGRs) or phytohormones (Khan et al. 2019). Phytohormones are nontoxic, environmentally friendly biomolecules, synthesised naturally under normal or stress condition. They regulate the plant abiotic stress response attributes at all levels (Passioura 2002). The functional importance of phytohormones to plant resilience to abiotic stress has shifted the focus of plant research into 'the regulatory role of phytohormones to control plant internal homeostasis and their transcriptional reprogramming during defense against abiotic stress' worldwide since the last decade. Intensive research of both the enhancement of endogenous levels and exogenous application has enlightened the role of phytohormones in abiotic stress mitigation in plants. Phytohormones have been divided into two broad categories, the traditional/classical phytohormones, and new era novel phytohormones. Traditional well-known phytohormones include five classical phytohormones, i.e., auxin, gibberellin (GA), cytokinin (CK), abscisic acid (ABA), and ethylene (ET). The new era novel hormones include brassinosteroids (BRs), jasmonate (JA), strigolactone (SL), salicylic acid (SA), and nitric oxide (NO), which have become the research hotspot and may have great potential for developing abiotic stress tolerant crops. A phytohormone does not play an independent regulatory role but works in a complex network with other phytohormones. Researchers and scientists have been engaged in investigating the multifunctional roles of phytohormones in stress mitigation and their transcriptional control under unfavorable environmental conditions. This chapter mainly outlines the role of phytohormones in response variations during defence against different abiotic stresses and transcriptional activation and control of phytohormones in different crop plants. Fundamental information of endogenous regulators and their transcriptional control is essential to build up strategies to improve abiotic stress tolerance in crop plants. Based on this knowledge, several success stories have been generated by introgression of tolerance traits from tolerant to susceptible plants via genetic transformation (Zhang et al. 2020).

11.2 Classical Phytohormones: Role in the Regulation of Plant Internal Homeostasis, Adaptive Variation in Plant Architecture and Their Transcriptional Reprogramming Under Abiotic Stress Conditions

11.2.1 Auxin

Auxin, or indol-acetic acid (IAA), belongs to the family of indol derivatives and is an important growth hormone. Auxin has a cardinal role in the life cycle of plant: Concentration of auxin within the cell, specifically in roots, acts as a developmental switch for transition from cell division to cell differentiation (Di-Mambro 2017). Stress-induced changes in concentration of phenolic (quercetin and kaempferol), flavonoids, pH of apoplast, and ionic fluctuations trigger the plant immune system that regulates the auxin homeostasis within a cell. These biomolecules act as auxin transport inhibitors and disturb the auxin efflux carriers (Potters et al. 2009). Heavy metal stress and nutrient deficiency, e.g., phosphorous and sulfur, also affect auxin metabolism (Potters et al. 2007). Puterbance in the auxin homeostasis, in turn, showed developmental defects, such as inhibition of root elongation, shoot growth, cotyledon and leaf epinasty, lateral growth of shoot and root (Di-Mambro 2017; Saini et al. 2017). Salt stress affects free and conjugated IAA levels in the developing xylem in poplar, resulting in negative effects on the expansion of xylem vessels (Junghans et al. 2006). H_2O_2 level is a co-factor for peroxidases to catalyze the oxidative degradation of IAA. Stress-induced variation in H_2O_2 concentration regulates the metabolism of IAA within the cell (Tognetti et al. 2012). Apoplastic reactive oxygen species (ROS) transiently decrease auxin signalling, modified auxin homeostasis, and altered leaf morphology in ozone (O_3) treated plants (Blomster et al. 2011).

Several genes are involved in the regulation of auxin transport, biosynthesis, conjugation, perception, and signalling pathways. Multiple abiotic stresses differentially effect and intersect auxin homeostasis at different levels. Auxin response factors (ARF) transcription factors and their associated genes are considered the main regulators of auxin homeostasis under abiotic stress conditions (Jain and Khurana 2009; Wang et al. 2010). SIARF1, SIARF8, SIARF19, and SIARF24 were induced under salt stress in tomato, while SIARF4, SIARF6A, SIARF6B, SIARF7A, and SIARF7B were upregulated by drought stress. OsARF11 and OsARF15 upregulated in rice under drought stress (Jain and Khurana 2009). The *bZIP11* transcription factor in Arabidopsis interacts with one adapted protein via an amino-terminal activation domain to align the histone acetylation system to specific auxin responsive genes. Drought stress induced the expressions of *WRKY2*, *WRKY56*, and *WRKY87* genes which is further upregulated by exogenous application of IAA in white clover (Zhang et al. 2020).

11.2.2 Gibberellin

The gibberellins (GAs) belong to the tetracyclic diterpenoid carboxylic acid group, famous for its large number of members. However, only a very few members of the group function as growth hormones in plants among which the predominant bioactive forms are GA1 and GA4 (Sponsel and Hedden 2004). They remain active throughout the life cycle of plants and control cell division, elongation, developmental phase transition (including seed dormancy to germination), juvenile to adult phase, and vegetative to reproductive phase. Non-availability or restrictions in synthesis slows the developmental processes, which are controlled by GA (Griffith et al. 2006; Ueguchi-Tanaka et al. 2005). The signalling pathway, which is regulated by developmental and environmental stimuli comprised of stress perception, signals transduction and inactivation of synthesis. Dioxygenases act as main regulatory sites for GA biosynthesis particularly the GA-2 oxidases (GA2ox) under abiotic stress conditions (Kasahara et al. 2002).

Abiotic stresses reduce the contents of bioactive GAs in plants. Reduction in GA content due to salinity (Magome et al. 2008) and cold stress in Arabidopsis is strongly linked with accumulation of DELLA proteins (Achard et al. 2006). In both cases, GA2ox genes are upregulated by DREB1/CBF transcription family members, which enhances the ability of plants to cope with stress conditions (Achard et al. 2006).

11.2.3 Cytokinins

Stress conditions in the root zone can trigger a change in concentration of CK not only in roots but also in leaves, shoot tips, and buds accompanied by increased ABA levels in the plant body (Schachtman and Goodger 2008; Kudoyarova et al. 2007). In response to drought stress, reduced CK levels trigger apical dominance and regulated stomata movement together with ABA. This negative regulatory function of CK has been confirmed by exogenous application in genetic studies, evidenced either by loss of function of biosynthesis gene isopentyl transferase (IPT) or overexpression of cytokinin oxidase (CKK) degradation genes. Reduced effect of IPT or overexpression of CKX1 gene increased the resistance to both salt and drought stress (Nishiyama et al. 2011; Wang et al. 2011). The study of mutant strategy in genetics paved the way to greater understanding of the function of genes. In studies, loss-of-function mutation Arabidopsis histidine kinase (AHK), acting as a stress sensor, was found more resistant to freezing and drought stress (Kang et al. 2012; Jeon et al. 2010). These mutants also exhibited hypersensitivity to ABA treatment (Tran et al. 2007). This response of CK receptor functionality is in agreement with increased abiotic stress at low CK concentration levels (Kang et al. 2012; Jeon et al. 2010; Tran et al. 2007). Three types of CK response regulators were found in Arabidopsis, type A-ARR (Arabidopsis response regulators), type B-ARR, and type C-ARR, in which type A-ARR belongs to CK early responsive genes. Arabidopsis phosphotransfer (AHP) proteins mediate stress signals by translocation into the nucleus and activate type B-ARR factors which in turn triggers the transcription of specific genes in response to CK (Hwang et al. 2012; To and Kieber 2008; Hutchison et al. 2006). Type A-ARR acts as a negative regulator and inhibits the activity of type B-ARR by a mechanism that is still unknown. ARR5, ARR7, ARR8, and ARR15 upregulated under cold stress, all are the type A-ARR (Jeon and Kim 2013; Jeon et al. 2010). Expression of ARR5, ARR7, ARR15, and type C-ARR22 increased under dehydration stress (Kang et al. 2012), ARR5 under salt stress (Mason et al. 2010), ARR5, ARR6, ARR7, and ARR15 under cold stress (Jeon and Kim 2013). The high redundancy of CK pathway components including 3 CK receptors, 6AHPs, 10 type of ARR, 11 type B-ARR, and several type C-ARRs, makes it difficult to specify the role of a specific component (Hwang et al. 2012). The signalling and cross-talk among CK components have been found to be controlled by cytokine response factor (CRF) transcription factors of the family APETALA2 (AP2) (Rashotte et al. 2006). CRF6 is responsive to several abiotic stress stimuli and characterised in leaf senescence control (Zwack et al. 2013).

11.2.4 Abscisic Acid

Abscisic acid (ABA) is one of the intensively studied endogenous tolerance-inducing phytohormones, known for its rapid and most conspicuous role in abiotic stress tolerance. ABA concentration increases rapidly in response to stress stimuli by modulating the gene expression level (Sah et al. 2016; Fujita et al. 2011). The rapid increase of ABA under drought and salinity stress modulates the gene expression which initiates adaptive physiological responses of plants, like ionic balance, stomatal closure (Sah et al. 2016), leaf senescence, CO2 assimilation, and controlling root architecture (Ng et al. 2014). ABA triggers stomatal regulation based on water potential status of guard cells and production of H_2O_2. Stomata get closed as the permeability of guard cells to water increases through various aquaporins (Verma et al. 2016). ABA can also influence the activity of aquaporins either by changing gene expression or modifying the post-transcriptional gene product (Sharipova et al. 2016).

ABA is reported to be regulated by a wide variety of environmental stimuli as a first line of defence. Hence, its large set of transcription factors play their role of ABA reprogramming in response to abiotic stresses, e.g., drought stress upregulates the expression of Zeaxanthin epoxidase (ZEP), Molybdenum cofactor sulfurase (MCSU), Aldehydeboxidase (AAO3), and 9-cisepoxycarotenoid dioxygenase (NCED3). All are controlled by MYC, ERF, NAC, MYB, DREB/CBF, AREB, and bZIP (Verma et al. 2016). Each transcription factor acts as a master regulator to control the expression of stress-inducible genes, hence coordinating and controlling the hormonal, developmental, and environmental stimuli in plant systems (Jaradat et al. 2013). AREB1/ABF2, AREB2/ABF4, ABF3, RD29A, RD29B, COR78, and LT178 genes were found responsive to salinity, drought, and cold. Whole sets of reported genes belong to AREBs/ABFs (ABA response element binding proteins) which is considered the major ABA-based transcription factor imparting abiotic stress tolerance to plants (Chen et al. 2014). Genes OsNAC3, 4, 5, 6, and SNAC1, belonging to the NAC transcription factor family, showed enhanced expression under salinity, drought, and low temperature stress conditions (Danquah et al. 2014; Fujita et al. 2011). ONAC022 upregulated under drought and salt stress (Hong et al. 2016). Other studies elaborating the role of NAC include (Hong et al. 2016), WRKY (Jiang et al. 2014), bZIP (Huang et al. 2016), MYB (Fang et al. 2017) and MYC (Le-Hir et al. 2017). The activity of all the above-mentioned genes is dependent on the concentration of ABA within the cell (Liang et al. 2014). Activation promotes the expression of genes linked to particular transcription factors, which ultimately leads towards physiological, metabolic, and chemical variations in cellular homeostasis (Fang et al. 2017; Huang et al. 2016).

11.2.5 Ethylene

Ethylene, a gaseous molecule with two carbon skeletons (C_2H_2), acts as a crucial natural phytohormone mediating several plant developmental processes under normal circumstances. The role of ethylene as a stress hormone is also very well-characterised, particularly in reference to abiotic stress tolerance

(Upreti et al. 2012; Colmer 2003). Ethylene is a simple, unsaturated hydrocarbon found in all plant parts, however, its production rate is dependent upon the type of the tissue, its growth and development stage, and intensity and duration of stress (Upreti et al. 2012). Leaf abscission, flower senescence, and fruit ripening elevate ethylene synthesis in plants. Plants can form ethylene both photochemically and enzymatically. Although it can be formed by all the organs of higher plants, its synthesis is more common in tissues undergoing senescence. Advances in research on ethylene production and function began after the introduction of gas chromatography. Ethylene modulates a broad array of responses in plants, like breakdown seed dormancy and initiate seed germination (Linkies and Leubner-Metzger 2012), expansion of cells, flowering, ripening of fruits (Barry and Giovannoni 2007), and leaf senescence (Koyama 2014). Abiotic stress conditions lead to the synthesis of ethylene by modulating the activities of ACC oxidase and ACC synthase (Upreti et al 2012) under cold (Zhao et al. 2009) and heat stress (Upreti et al. 2012). Low oxygen levels and the formation of free radicals under flooding conditions trigger the ACC biosynthesis and hence enhance ethylene production (Colmer 2003). Salinity stress positively regulates ethylene biosynthesis, which contributes to enhanced salt resistance in plants by regulating Na/K homeostasis (Archard et al. 2006). Further, ethylene controls the movement of stomata, permitting more inflow of CO2 for carboxylation and enhanced photosynthesis, which might be due to enhanced carboxylation as well as water use efficiency (Iqbal et al. 2011)

Ethylene is also known to regulate the gene expression of plants by regulating ethylene response factors (ERFs) transcriptional levels under abiotic stress (Hussain et al. 2011). ERFs possess the ability to bind with GCC box as well as dehydration response element (DRE)/C-repeat (CRT) motifs and cis-acting elements involved in the cellular response to cold/osmotic stress (Wang et al. 2004). SodERF3 is an ERF that plays a crucial role in enhancing resistance against salt and drought stress. Mild drought did not have a significant effect on ethylene or ACC levels; however, a swift change in the drought conditions may alter concentrations of ethylene and ACC (Trujillo et al. 2008). Overexpression of wheat transcription factor TaERF1 regulated stomatal apertures and provided adaptation towards drought and salt stress (Xu et al. 2007). AP2/EREBP family transcription factors also reduced stomatal density in Arabidopsis mutant shine (shn) in response to drought stress (Aharoni et al. 2004). Quan et al. (2010) demonstrated increased expression of photosynthesis-related genes in tomato, owing to the synthesis of ERFs in response to drought stress. ERF7 imparts salt tolerance by increasing chlorophyll and carbohydrate content and decreasing malondialdehyde content in transgenic tobacco plants (Zhai et al. 2013). ERFs bind to DREs to trigger transcription of stress-responsive genes on exposure to high temperature stress (Ahammed et al. 2016). Ethylene in combination with ABA and JA also induced the expression of CaWRKY6 gene in Capsicum annum, which plays a crucial role in high temperature stress (Cai et al. 2015). ERF proteins bind to DRE/CRT under cold stress. Expression of TERF2 enhanced the concentration of osmotic solutes and chlorophyll pigment accompanied by a reduction in the levels of ROS species and malondialdehyde content in rice seedlings grown in the presence of high level of cold (Tian et al. 2011). However, contradicting results were found during investigation where overexpression of EIN3 impedes ethylene production in *A. thaliana* and hence negatively affects cold tolerance (Shi et al. 2012). The overexpression of ERF genes increased expression of glutathione synthesis genes, leading to increased Cd tolerance (Guan et al. 2015). Additionally, EIN2 is crucial in providing metal tolerance against Pb by enhancing glutathione concentration (Cao et al. 2009). Ethylene's role in heavy metal stress tolerance has also been reported in several studies (Carrió-Seguí et al 2015).

11.3 New Era Phytohormones: Role in the Regulation of Plant Internal Homeostasis, Adaptive Variation in Plant Architecture, and Their Transcriptional Reprogramming under Abiotic Stress Conditions

11.3.1 Brassinosteroids

Brassinosteroid (BR), a polyhydroxylated plant steroid hormone, is involved in the regulation of various physiological and developmental processes in plants, the most important of which are seed germination, embryogenesis, cell division, elongation and differentiation, cell cycle, root cell elongation and root meristem growth (Khatoon et al. 2017; Gonzalez et al. 2011), photosynthesis, leaf senescence, leaf expansion, flower initiation, and ethylene biosynthesis (Zhiponova et al. 2013; Vardhini et al. 2008).

BR is an emerging plant hormone that is getting attention in ameliorating the effects of abiotic stress on plants. BRs evoke stress-specific defence responses in plants. Under chilling stress (4°C), BRs involved in the control of ion leakage and outflow and activation of the antioxidant defence system, i.e., CAT, APX, SOD, to ease the lethal effect of H_2O_2 in different plant species (Aghdam and Mohammadkhani 2014; Xi et al. 2013). Activation of the defence system in turn maintains the membrane integrity, enhances osmoprotectants, and reduces the effects of chilling injury. BRs mitigated the effect of drought stress by improving contents of soluble protein and proline and by upregulating stress responsive and inductive proteins in maize (El-Khallal 2002). Exogenous application of BRs application has been reported to reduce the toxic effect and impact of salinity on *Phaseolus vulgaris* L., and to mitigate the effect of cadmium metal stress in *Solanum lycopersicum* and nickel in *Brassica juncea* L. (Kanwar et al 2013). Although the relationship between BRs and stress response was first observed long ago, the underlying molecular mechanisms are far from being completely understood. The Brassinozol resistant (BZR) TFs family was found to be involved in the initiation of the BR-based response to sustained stress conditions. Brassinazole resistant 1 (BZR1) and Brassinosteroid insensitive 1-ethyl methanesulfonate suppressor 1(BES1) both belong to BZR family, regulating a huge number of genes by either activation or suppression of expression. As BRs are mainly associated with roots and cannot travel long distances, the BZR also serve as connecting points among other signalling pathways to impart influence over large surface areas (Saini et al 2015). Under nitrogen starvation, ZmBZR genes are upregulated in maize (Singh et al. 2014) in contrast to BRI1, a BR receptor-like kinase that acts as a negative regulator in BR-signalling pathways that is upregulated under the same stress conditions in other studies (Zhao et al. 2015). These conflicting reports indicate that BRs perform diverse functions by sharing signalling pathways with other phytohormones, proving the antagonistic interaction between ABA and BR synthesis (Zhang et al. 2009); this is also true of the GA and BR relationship, in which GA acts as a repressor of DELLA (Li et al. 2012).

11.3.2 Jasmonate (JA)

Jasmonate is a naturally occurring oxylipin phytohormone that acts as an endogenous signalling molecule in diverse developmental and stress-related processes. The important members of this family include jasmonic acid, methyl jasmonate (MeJA), and jasmonyl isoleucine (JA-Ile). As a crucial developmental component within the plant cell, it plays its role in seed development, leaf growth, and senescence under normal environmental conditions. It is also found involved in the natural defence systems of plants in response to wound injury, disease, and insect attack. Under abiotic stress conditions, jasmonate modulates the growth and developmental processes via upregulation of gene expression of traits it is controlling under normal conditions (Llanes et al. 2016). The role of JA in mitigating environmental stress by enhancing tolerance capacity of plants had been reported widely under several abiotic stress conditions such as salinity (Farhangi-Abriz and Ghassemi-Golezani 2018), water stress (Wasternack 2014), heavy metal stress (Dar et al. 2015), and heat stress (Sharma and Laxmi 2015). JA concentration drastically increases under stress conditions and struggles to maintain cell integrity, activates the antioxidant response system, and remobilises photoassimilates to vegetative sinks (Farhangi-Abriz and Ghassemi-Golezani 2018). It also enhances the antioxidative defence system and root growth under drought stress (Alavi-shamani et al 2015), enhances ABA contents and proline concentrations (De-Ollas et al. 2015), enhances nitric oxide and ascorbate glutathione (Shan et al. 2015), and is involved in antioxidative activities, chlorophyll content, and glyoxalase in *Brassica napus* (Alam et al. 2014) in order to promote maximum plant growth. Seed-priming with jasmonate also has a promising role in mitigating drought stress (Sheteiwy et al. 2018). However, it is still not clear how JA acts to modulate the antioxidant system. It is possible JA influences the enzymes and changes the transcription of responsive genes. Several studies on both endogenously enhanced production and exogenous application have been conducted on various crop plants to determine the role of JA in the mitigation of salt stress. JA is also responsible for enhanced ATPase activity and reduced sodium uptake in soybean (Farhangi-Abriz and Ghassemi-Golezani 2018) and the enhanced accumulation of osmoprotectants, flavonoids, proline content, and glycinebetaine synthesis (Talebi et al. 2018) and sugar accumulation in rapeseed (Harpeet et al. 2013). Jasmonate also helps to maintain cell integrity and reduced electrolyte leakage under heat stress in plants (Kazan 2015). Jasmonate maintains RWC under chilling stress by increasing hydrolytic conductivity and decreasing stomatal opening size. Physiological responses often include accumulation of amino acids (methionine and isoleucine) and soluble sugars, regulation of stomatal opening and closing (Wasternack 2014), and the enhancement of arsenic tolerance by increased production of ascorbate-glutathione in *Brassica napus* (Farooq et al. 2018).

At the molecular level JAZ, AOS1, AOC, LOX2, and COII genes are associated with JA metabolism (Robson et al. 2010), and their expression depends on their interaction with TFs MYC2 and bHLH148 (Zhao et al. 2013). OsJAZ1 plays a negative regulatory role in rice drought stress tolerance, which suggests that JA ZIM domain proteins (JAZ) act as regulators, typically repressors, in JA-signalling pathways (Fu et al. 2017). Meanwhile, OsbHLH148, a basic helix-loop-helix protein, also acts as a transcriptional regulator and upregulates the OsJAZ and OsDREB1, which are involved in JA-signalling and drought stress response, respectively (Seo et al. 2011). Reduced expression of COI1 (coronatine-insensitive 1), a central component of JA signalling, exhibited the low level of response to high irradiance far red (FR) light (Robson et al. 2010). Other reports indicate that phytochrome A and JA are both required for high intensity light sensation and presences of both molecules interact with JAZ1 domain for their regulation via COI1 (Chen et al. 2018; Ahmad et al. 2017). Circadian stress also activates JA signalling, which interacts with MYC2 and JAZ1 for response initiation (Nitschke et al. 2016).

11.3.3 Strigolactones

Strigolactone, a phytohormone that promotes seed germination (Abe et al. 2014), inhibits branching by which it determines the overall structure of the plant (Umehara et al. 2008). These are mainly identified in root exudates of plants, having ABC ring structure which share conserved tricyclic lactone components linked through the enol-ether bridge to the unsaturated furanone ring called a ‘D ring’ (Al-Babili and Bouwmeester, 2015). Carotenoid-cleavage dioxygenase (CCD) enzymes in sequential steps cleave the carotenoids into carlactone (CL) (Alder et al. 2012). CL is further converted into CLA (carlactonic acid) and then to MeCLA (Methyl carlactonoate); all these conversions convert the CL into functional SLs. All these CL conversions are executed by a gene, MAX1, which encodes a cytochrome P450 (Abe et al. 2014). There is a variation in the biosynthetic pathway based on target crop species (Saeed et al. 2017). The signal for SLs synthesis is initiated by proteasomal degradation (Arite et al. 2009). Structural orientation, biosynthesis, and regulation of Sla have been addressed and updated in recent reviews (Bythell-Douglas et al. 2017; Waters et al. 2017). SL regulation under stress conditions is largely associated with auxin concentration. The stress condition that disturbs the balance of auxin also initiates the SLs cycle in the cell which, in turn, inhibits the branching, auxiliary bud growth (Crawford et al. 2010), seminal root length, lateral root formation, and root hair elongation.

Nutritional deficiency or imbalance is reported to regulate SLs production. Different effects of various nutrients have been observed in different studies. For example, low P levels trigger SLs synthesis (Jamil et al. 2014), while in leguminous/non-leguminous species, both P and/or N deficiency regulates SLs. Most legume crops showed increased SL exudation under P starvation, while the same was true for non-legume crops under N stress; however exceptional cases exist in both groups. General mechanism to nutrient deficiency outlines as N starvation affects P level in shoot which in turn triggers SL induction. SL activation, however, depends upon plant type, type of depriving nutrient, degree of stress, and plant macro-nutrient uptake strategy (Yoneyama et al. 2013). However, the exact mechanism of SL exudation and its effect on nodulation modulation is still unclear (De-Cuyper et al. 2015). SL-mediated regulation of root architecture requires MAX2, MAX3, and MAX4 genes for initiation of response (Ito et al. 2016).

11.3.4 Salicylic Acid

Salicylic acid (SA) is a beta hydroxy acid phenolic compound, endogenously synthesised by plants. It is synthesised by following different metabolic pathways, however the main compounds of reaction are benzoic acid, cinnamic acid, and coumaric acid and the chief enzymes involved are benzoic acid 2-hydroxylase, cinnamate 4-hydroxylase, and phenylalanine ammonia-lyase (PAL). The details of SA synthesis and its metabolic pathways have been well described in several studies (Herrera-Vasquez et al. 2015; An and Mou 2011). Its synthesis and metabolism inside the plant are highly dependent on plant species, organ, and age (Herrera-Vasquez et al. 2015). SA emerged as a new candidate mediating the abiotic stress response of plant due to its involvement in the regulation of important physiological processes of plants such as proline metabolism, glycinebetaine synthesis, photosynthesis, nitrogen metabolism, plant–water relationship, antioxidant defence system (Khan et al. 2013; Miura and Tada 2014; Nazar et al. 2011), membrane protection from H_2O_2 damages (Chao et al. 2010), and inhibits Fe deficiency, which causes chlorosis in plants (Kong et al. 2014). SA-mediated oxidative stress control is mainly attributed to the uptake of plant-beneficial elements, such as Cu, Ca, Zn, P, Fe, and Mn (Wang et al. 2011) and photosynthetic system to nutrient content level of N, P, K, and Ca (Syeed et al. 2011). As the main regulatory component of physiological parameters, SA plays a role in plant stress tolerance against multiple stresses such as heavy metals (Zhang et al. 2015), drought (Fayaz and Bazaid 2014), salinity (Nazar et al. 2015), osmotic stress, and Ozone (Ogawa et al. 2005). OsWRKY45 gene expression is linked with SA content. A high content of SA prevents membrane damage by detoxifying the effect of H_2O_2 under Cd stress in rice (Chao et al. 2010).

11.3.5 Nitric Oxide

Nitric oxide (NO) is the only gaseous phytohormone and a highly reactive nitrogen species (RNS) produced within the cell under normal as well as stress conditions. Its synthesis mechanism is complicated and not fully understood, as it is the product of several oxidative and reductive pathways which do not produce NO (Crawford and Guo 2005). The source of origin and biosynthesis pathways are posited in several studies (Prochazkova et al. 2014; Salgado et al. 2013), but it is a challenging task and continues to be a subject of debate (Salgado et al. 2013). Few properties aid into dynamism of NO molecules like presence of free radical, small size redox molecule, easily diffusible across membranes and neutral nature, and make it a very significantly favorable molecule under stress conditions (Domingos et al. 2015). The physiological plant parameters controlled by NO are germination, development, flowering, programmed cell death, stomatal movement regulation, and senescence (Bright et al. 2006). Under stress conditions, the enhanced production of NO activates the antioxidant defence system that, in turn, acts as a detoxifying agent against ROS and minimises the detrimental effects within the plant (Hu et al. 2016). Increasing levels of NO also elevate the concentration of antioxidant enzymes SOD (superoxide dismutase), CAT (catalase), APX (ascorbate peroxidase), and DHAR (dehydroascorbate reductase). Several reports have described NO's role in multiple abiotic stress mitigations, e.g., drought (Farooq et al. 2009) and salt (Du et al 2015). However, an intensive study is still required to explain the molecular mechanism behind NO production and its role in abiotic stress mitigation.

11.4 Conclusion

Phytohormones play an indispensable role in the growth and developmental processes of plants. Under natural conditions, plants are exposed to a wide spectrum of stresses, both biotic and abiotic, which threaten crop sustainability. Credible research on the fluctuation of phytohormone levels of plants in response to abiotic stress, their association with developmental traits, and phytohormone-induced transcriptional regulation of genes, has validated that they play a critical role in plant defences against environmental stress stimuli. As phytohormones play a direct role in multiple and specific developmental processes, they modulate the same traits to ensure survival under stress conditions. However, it is difficult to isolate the effects of one phytohormone from the other, as plants are usually exposed to multiple stresses at a time, and, to cope with that, multiple hormones are activated. The situation is further accentuated by the cross-talk which is stress perception, hormone activation, transcriptional activation, gene expression, and developmental variations which occur in response to all the above factors. Overall, cross-talk among phytohormones at the cellular level, cross-talk at transcriptional level, and generation of same physiological or biochemical variation output in response to multiple stresses make this study domain highly complicated, and even after immense study remains elusive at several levels. All domains of stress biology are hot spots and huge gaps exist in the elaboration of the mechanism of stress tolerance; hence, each study targeting phytohormones under stress conditions will contribute greatly to this domain in the future.

REFERENCES

Abe S, Sado A, Tanaka K, Kisugi T, Asami K, Ota S, Kim HI, Yoneyama K, Xie X, Ohnishi T, Seto Y, Yamaguchi S, Akiyama K, Yoneyama K, Nomura T (2014). Carlactone is converted to carlactonoic acid by max1 in Arabidopsis and its methyl ester can directly interact with AtD14 in vitro. *Proc Natl Acad Sci U S A* 111(50):18084–18089.

Achard P, Cheng H, De Grauwe L, Decat J, Schoutteten H, Moritz T, Van Der Straeten D, Peng J, Harberd NP (2006). Integration of plant responses to environmentally activated phytohormonal signals. *Science* 311(5757):91–94.

Achard P, Gong F, Cheminant S, Alioua M, Hedden P, Genschik P (2008a). The cold-inducible CBF1 factor–dependent signaling pathway modulates the accumulation of the growth-repressing DELLA proteins via its effect on gibberellin metabolism. *Plant Cell* 20(8):2117–2129.

Achard P, Renou J-P, Berthomé R, Harberd NP, Genschik P (2008b). Plant DELLAs restrain growth and promote survival of adversity by reducing the levels of reactive oxygen species. *Curr Biol* 18(9):656–660.

Adnan M, Fahad S, Khan IA, Saeed M, Ihsan MZ, Saud S, Riaz M, Wang D, Wu C (2019). Integration of poultry manure and phosphate solubilizing bacteria improved availability of Ca bound P in calcareous soils. *Biotech* 9(10):368.

Adnan M, Fahad S, Zamin M, Shah S, Mian IA, Danish S, Zafar-Ul-Hye M, Battaglia ML, Naz RMM, Saeed B, Saud S, Ahmad I, Yue Z, Brtnicky M, Holatko J, Datta R (2020). Coupling phosphate-solubilizing bacteria with phosphorus supplements improve maize phosphorus acquisition and growth under lime induced salinity stress. *Plants* 9(7). https://doi.org/10.3390/plants9070900.

Adnan M, Shah Z, Fahad S, Arif M, Alam M, Khan IA, Mian IA, Basir A, Ullah H, Arshad M, Rahman IU, Saud S, Ihsan MZ, Jamal Y, Amanullah, Hammad HM, Nasim W (2018a). Author correction: phosphate-solubilizing bacteria nullify the antagonistic effect of soil calcification on bioavailability of phosphorus in alkaline soils. *Sci Rep* 8(1):4339. https://doi.org/10.1038/s41598-018-22653-7.

Adnan M, Shah Z, Sharif M, Rahman H (2018b). Liming induces carbon dioxide (CO_2) emission in PSB inoculated alkaline soil supplemented with different phosphorus sources. *Environ Sci Pollut Res Int* 25(10):9501–9509.

Aghdam MS, Mohammadkhani N (2014). Enhancement of chilling stress tolerance of tomato fruit by postharvest brassinolide treatment. *Food Bioproc Tech* 7(3):909–914.

Ahammed GJ, Li X, Zhou J, Zhou Y-H, Yu J-Q (2016). Role of hormones in plant adaptation to heat stress. In *Plant hormones under challenging environmental factors*. Springer, pp 1–21.

Aharoni A, Dixit S, Jetter R, Thoenes E, Van Arkel G, Pereira A (2004). The SHINE clade of AP2 domain transcription factors activates wax biosynthesis, alters cuticle properties, and confers drought tolerance when overexpressed in Arabidopsis. *Plant Cell* 16(9):2463–2480.

Ahmad P, Abdel Latef AA, Hashem A, Abd Allah EF, Gucel S, Tran LS (2016). Nitric oxide mitigates salt stress by regulating levels of osmolytes and antioxidant enzymes in chickpea, Abd_Allah EF, Gucel S, Tran L-SP. *Front Plant Sci* 7:347.

Ahmad P, Alyemeni MN, Wijaya L, Alam P, Ahanger MA, Alamri SA (2017). Jasmonic acid alleviates negative impacts of cadmium stress by modifying osmolytes and antioxidants in faba bean (Vicia faba L.). *Arch Agron Soil Sci* 63(13):1889–1899.

Ahmad S, Kamran M, Ding R, Meng X, Wang H, Ahmad I, Fahad S, Han Q (2019). Exogenous melatonin confers drought stress by promoting plant growth, photosynthetic capacity and antioxidant defense system of maize seedlings. *PeerJ* 7:e7793. http://doi.org/10.7717/peerj.7793.

Akbar H, Timothy JK, Jagadish T, Golam MM, Apurbo KC, Muhammad F, Rajan B, Fahad S, Hasanuzzaman M (2020). Agricultural land degradation: processes and problems undermining future food security. In: Fahad S, Hasanuzzaman M, Alam M, Ullah H, Saeed M, Khan AK, Adnan M (eds) *Environment, climate, plant and vegetation growth.* Springer Publ Ltd, Springer Nature Switzerland AG. Part of Springer Nature, pp 17–62. https://doi.org/10.1007/978-3-030-49732-3.

Akram R, Turan V, Hammad HM, Ahmad S, Hussain S, Hasnain A, Maqbool MM, Rehmani MIA, Rasool A, Masood N, Mahmood F, Mubeen M, Sultana SR, Fahad S, Amanet K, Saleem M, Abbas Y, Akhtar HM, Waseem F, Murtaza R, Amin A, Zahoor SA, ul Din MS, Nasim W (2018a). Fate of organic and inorganic pollutants in paddy soils. In: Hashmi MZ, Varma A (eds) *Environmental pollution of paddy soils, soil biology.* Springer International Publishing AG, Cham, Switzerland, pp 197–214.

Akram R, Turan V, Wahid A, Ijaz M, Shahid MA, Kaleem S, Hafeez A, Maqbool MM, Chaudhary HJ, Munis MFH, Mubeen M, Sadiq N, Murtaza R, Kazmi DH, Ali S, Khan N, Sultana SR, Fahad S, Amin A, Nasim W (2018b). Paddy land pollutants and their role in climate change. In: Hashmi MZ, Varma A (eds) *Environmental pollution of paddy soils, soil biology.* Springer International Publishing AG, Cham, Switzerland, pp 113–124.

Alam MM, Nahar K, Hasanuzzaman M, Fujita M (2014). Exogenous jasmonic acid modulates the physiology, antioxidant defense and glyoxalase systems in imparting drought stress tolerance in different brassica species. *Plant Biotechnol Rep* 8(3):279–293.

Alavi-Samani SM, Kachouei MA, Pirbalouti AG (2015). Growth, yield, chemical composition, and antioxidant activity of essential oils from two thyme species under foliar application of jasmonic acid and water deficit conditions. *Hortic Environ Biotechnol* 56(4):411–420.

Al-Babili S, Bouwmeester HJ (2015). Strigolactones, a novel carotenoid-derived plant hormone. *Annu Rev Plant Biol* 66:161–186.

Alder A, Jamil M, Marzorati M, Bruno M, Vermathen M, Bigler P, Ghisla S, Bouwmeester H, Beyer P, Al-Babili S (2012). The path from β-carotene to carlactone, a strigolactone-like plant hormone. *Science* 335(6074):1348–1351.

Al-Wabel Mohammad I, Abdelazeem S, Munir A, Khalid E, Adel RAU (2020b). Extent of climate change in Saudi Arabia and its impacts on agriculture: a case study from Qassim region. In: Fahad S, Hasanuzzaman M, Alam M, Ullah H, Saeed M, Khan AK, Adnan M (eds) *Environment, climate, plant and vegetation growth.* Springer Publ Ltd, Springer Nature Switzerland AG. Part of Springer Nature, pp 635–658. https://doi.org/10.1007/978-3-030-49732-3.

Al-Wabel Mohammad I, Ahmad Munir, Usman Adel RA, Akanji Mutair, Rafique Muhammad Imran (2020a). Advances in pyrolytic technologies with improved carbon capture and storage to combat climate change. In: Fahad S, Hasanuzzaman M, Alam M, Ullah H, Saeed M, Khan AK, Adnan M (eds) *Environment, climate, plant and vegetation growth.* Springer Publ Ltd, Springer Nature Switzerland AG. Part of Springer Nature, pp 535–576. https://doi.org/10.1007/978-3-030-49732-3.

Amanullah Shah K, Imran Hamdan AK, Muhammad A, Abdel RA, Muhammad A, Fahad S, Azizullah S, Brajendra P (2020). Effects of climate change on irrigation water quality. In: Fahad S, Hasanuzzaman M, Alam M, Ullah H, Saeed M, Khan AK, Adnan M (eds) *Environment, climate, plant and vegetation growth.* Springer Publ Ltd, Springer Nature Switzerland AG. Part of Springer Nature, pp 123–132. https://doi.org/10.1007/978-3-030-49732-3.

Amir M, Muhammad A, Allah B, Sevgi Ç, Haroon ZK, Muhammad A, Emre A (2020). Bio fortification under climate change: the fight between quality and quantity. In: Fahad S, Hasanuzzaman M, Alam M, Ullah H, Saeed M, Khan AK, Adnan M (eds) *Environment, climate, plant and vegetation growth.* Springer Publ Ltd, Springer Nature Switzerland AG. Part of Springer Nature, pp 173–228. https://doi.org/10.1007/978-3-030-49732-3.

Amjad I, Muhammad H, Farooq S, Anwar H (2020). Role of plant bioactives in sustainable agriculture. In: Fahad S, Hasanuzzaman M, Alam M, Ullah H, Saeed M, Khan AK, Adnan M (eds) *Environment, climate, plant and vegetation growth.* Springer Publ Ltd, Springer Nature Switzerland AG. Part of Springer Nature, pp 591–606. https://doi.org/10.1007/978-3-030-49732-3.

An C, Mou Z (2011). Salicylic acid and its function in plant immunity F. *J Integr Plant Biol* 53(6):412–428.

Arif M, Talha J, Muhammad R, Fahad S, Muhammad A, Amanullah Kawsar A, Ishaq AM, Bushra K, Fahd R (2020). Biochar; a remedy for climate change. In: Fahad S, Hasanuzzaman M, Alam M, Ullah H, Saeed M, Khan AK, Adnan M (eds) *Environment, climate, plant and vegetation growth.* Springer Publ Ltd, Springer Nature Switzerland AG. Part of Springer Nature, pp 151–172. https://doi.org/10.1007/978-3-030-49732-3.

Arite T, Umehara M, Ishikawa S, Hanada A, Maekawa M, Yamaguchi S, Kyozuka J (2009). d14, a strigolactone-insensitive mutant of rice, shows an accelerated outgrowth of tillers. *Plant Cell Physiol* 50(8):1416–1424.

Ayman ELS, Hossain Akbar, Barutçular Celaleddin, Iqbal Muhammad Aamir, Sohidul Islam M, Fahad Shah, Sytar Oksana, Çig Fatih, Meena Ram Swaroop, Erman Murat (2020). Consequences of salinity stress on the quality of crops and its mitigation strategies for sustainable crop production: an outlook of arid and semi- arid regions. In: Fahad S, Hasanuzzaman M, Alam M, Ullah H, Saeed M, Khan AK, Adnan M (eds) *Environment, climate, plant and vegetation growth.* Springer Publ Ltd, Springer Nature Switzerland AG. Part of Springer Nature, pp 503–534. https://doi.org/10.1007/978-3-030-49732-3.

Barry CS, Giovannoni JJ (2007). Ethylene and fruit ripening. *J Plant Growth Regul* 26(2):143.

Bayram AY, Seher Ö, Nazlican A (2020). Climate change forecasting and modeling for the year of 2050. In: Fahad S, Hasanuzzaman M, Alam M, Ullah H, Saeed M, Khan AK, Adnan M (eds) *Environment, climate, plant and vegetation growth.* Springer Publ Ltd, Springer Nature Switzerland AG. Part of Springer Nature, pp 109–122. https://doi.org/10.1007/978-3-030-49732-3.

Blomster T, Salojärvi J, Sipari N, Brosché M, Ahlfors R, Keinänen M, Overmyer K, Kangasjärvi J (2011). Apoplastic reactive oxygen species transiently decrease auxin signaling and cause stress-induced morphogenic response in Arabidopsis. *Plant Physiol* 157(4):1866–1883.

Bright J, Desikan R, Hancock JT, Weir IS, Neill SJ (2006). ABA-induced NO generation and stomatal closure in Arabidopsis are dependent on H2O2 synthesis. *Plant J Cell Mol Biol* 45(1):113–122.

Baseer M, Adnan M, Fazal M, Fahad S, Muhammad S, Fazli W, Muhammad A, Jr. Amanullah DW, Saud S, Muhammad N, Muhammad Z, Fazli S, Beena S, Mian AR, Ishaq AM (2019). Substituting urea by organic wastes for improving maize yield in alkaline soil. *J Plant Nutr.*

Bythell-Douglas R, Rothfels CJ, Stevenson DW, Graham SW, Wong GK-S, Nelson DC, Bennett T (2017). The complex origins of strigolactone signalling in land plants. *bioRxiv.* http://www.ncbi.nlm.nih.gov/pubmed/102715.

Cai H, Yang S, Yan Y, Xiao Z, Cheng J, Wu J, Qiu A, Lai Y, Mou S, Guan D, Huang R, He S (2015). CaWRKY6 transcriptionally activates CaWRKY40, regulates Ralstonia solanacearum resistance, and confers high-temperature and high-humidity tolerance in pepper. *J Exp Bot* 66(11):3163–3174.

Cao S, Chen Z, Liu G, Jiang L, Yuan H, Ren G, Bian X, Jian H, Ma X (2009). The Arabidopsis ethylene-Insensitive 2 gene is required for lead resistance. *Plant Physiol Biochem* 47(4):308–312.

Cao W-H, Liu J, He X-J, Mu R-L, Zhou H-L, Chen S-Y, Zhang J-S (2007). Modulation of ethylene responses affects plant salt-stress responses. *Plant Physiol* 143(2):707–719.

Carrió-Seguí A, Garcia-Molina A, Sanz A, Peñarrubia L (2015). Defective copper transport in the copt5 mutant affects cadmium tolerance. *Plant Cell Physiol* 56(3):442–454.

Chao Y-Y, Chen C-Y, Huang W-D, Kao CH (2010). Salicylic acid-mediated hydrogen peroxide accumulation and protection against Cd toxicity in rice leaves. *Plant Soil* 329(1–2):327–337.

Chen H-J, Fu T-Y, Yang S-L, Hsieh H-L (2018). FIN219/JAR1 and cryptochromel antagonize each other to modulate photomorphogenesis under blue light in Arabidopsis. *PLOS Genet* 14(3):e1007248.

Chen X, Wang Y, Lv B, Li J, Luo L, Lu S, Zhang X, Ma H, Ming F (2014). The NAC family transcription factor OsNAP confers abiotic stress response through the ABA pathway. *Plant Cell Physiol* 55(3):604–619.

Colmer T (2003). Long-distance transport of gases in plants: a perspective on internal aeration and radial oxygen loss from roots. *Plant Cell Environ* 26(1):17–36.

Crawford NM, Guo F-Q (2005). New insights into nitric oxide metabolism and regulatory functions. *Trends Plant Sci* 10(4):195–200.

Crawford S, Shinohara N, Sieberer T, Williamson L, George G, Hepworth J, Müller D, Domagalska MA, Leyser O (2010). Strigolactones enhance competition between shoot branches by dampening auxin transport. *Development* 137(17):2905–2913.

Danquah A, de Zelicourt A, Colcombet J, Hirt H (2014). The role of ABA and MAPK signaling pathways in plant abiotic stress responses. *Biotechnol Adv* 32(1):40–52.

Dar TA, Uddin M, Khan MMA, Hakeem KR, Jaleel H (2015). Jasmonates counter plant stress: a review. *Environ Exp Bot* 115:49–57.

De Cuyper C, Fromentin J, Yocgo RE, De Keyser A, Guillotin B, Kunert K, Boyer F-D, Goormachtig S (2015). From lateral root density to nodule number, the strigolactone analogue GR24 shapes the root architecture of Medicago truncatula. *J Exp Bot* 66(1):137–146.

De Ollas C, Arbona V, Gómez-Cadenas A (2015). Jasmonic acid interacts with abscisic acid to regulate plant responses to water stress conditions. *Plant Signal Behav* 10(12):e1078953.

Depeng W, Fahad S, Saud S, Muhammad K, Aziz K, Mohammad NK, Hafiz MH, Wajid N (2018). Morphological acclimation to agronomic manipulation in leaf dispersion and orientation to promote "Ideotype" breeding: evidence from 3D visual modeling of "super" rice (*Oryza sativa* L.). *Plant Physiol Biochem* 135:499–510.

Di Mambro R, De Ruvo M, Pacifici E, Salvi E, Sozzani R, Benfey PN, Busch W, Novak O, Ljung K, Di Paola L, Marée AFM, Costantino P, Grieneisen VA, Sabatini S (2017). Auxin minimum triggers the developmental switch from cell division to cell differentiation in the Arabidopsis root. *Proc Natl Acad Sci U S A* 114(36):E7641–E7649.

Domingos P, Prado AM, Wong A, Gehring C, Feijo JA (2015). Nitric oxide: a multitasked signaling gas in plants. *Mol Plant* 8(4):506–520.

Du S-T, Liu Y, Zhang P, Liu H-J, Zhang X-Q, Zhang R-R (2015). Atmospheric application of trace amounts of nitric oxide enhances tolerance to salt stress and improves nutritional quality in spinach (Spinacia oleracea L.). *Food Chem* 173:905–911.

El-Khallal S (2002). The influence of some phyto-growth regulators on the activity of antioxidant system in maize plant under water stressed conditions. *Bull Fac Sci.*

Enamul Md H, Shoeb AZM, Mallik AH, Fahad S, Kamruzzaman MM, Akib J, Nayyer S, Mehedi KM, A, Swati AS, Yeamin Md A, Most SS (2020). Measuring vulnerability to environmental hazards: qualitative to quantitative. In: Fahad S, Hasanuzzaman M, Alam M, Ullah H, Saeed M, Khan AK, Adnan M (eds) *Environment, climate, plant and vegetation growth.* Springer Publ Ltd, Springer Nature Switzerland AG. Part of Springer Nature, pp 421–452. https://doi.org/10.1007/978-3-030-49732-3.

Fahad S, Adnan M, Hassan S, Saud S, Hussain S, Wu C, Wang D, Hakeem KR, Alharby HF, Turan V, Khan MA, Huang J (2019b). Rice responses and tolerance to high temperature. In: Hasanuzzaman M and Fujita M, Nahar K, Biswas JK (eds) *Advances in rice research for abiotic stress tolerance.* Woodhead Publ Ltd, Cambridge, England, pp 201–224.

Fahad S, Bajwa AA, Nazir U, Anjum SA, Farooq A, Zohaib A, Sadia S, Nasim W, Adkins S, Saud S, Ihsan MZ, Alharby H, Wu C, Wang D, Huang J (2017). Crop production under drought and heat stress: plant responses and management options. *Front Plant Sci* 8:1147. https://doi.org/10.3389/fpls.2017.01147.

Fahad S, Bano A (2012). Effect of salicylic acid on physiological and biochemical characterization of maize grown in saline area. *Pak J Bot* 44:1433–1438.

Fahad S, Chen Y, Saud S, Wang K, Xiong D, Chen C, Wu C, Shah F, Nie L, Huang J (2013). Ultraviolet radiation effect on photosynthetic pigments, biochemical attributes, antioxidant enzyme activity and hormonal contents of wheat. *J Food Agric Environ* 11(3&4):1635–1641.

Fahad S, Hussain S, Bano A, Saud S, Hassan S, Shan D, Khan FA, Khan F, Chen Y, Wu C, Tabassum MA, Chun MX, Afzal M, Jan A, Jan MT, Huang J (2015a). Potential role of phytohormones and plant growth-promoting rhizobacteria in abiotic stresses: consequences for changing environment. *Environ Sci Pollut Res Int* 22(7):4907–4921. https://doi.org/10.1007/s11356-014-3754-2.

Fahad S, Hussain S, Matloob A, Khan FA, Khaliq A, Saud S, Hassan S, Shan D, Khan F, Ullah N, Faiq M, Khan MR, Tareen AK, Khan A, Ullah A, Ullah N, Huang J (2015b). Phytohormones and plant responses to salinity stress: a review. *Plant Growth Regul* 75(2):391–404. https://doi.org/10.1007/s10725-014-0013-y.

Fahad S, Hussain S, Saud S, Hassan S, Chauhan BS, Khan F, Ihsan MZ, Ullah A, Wu C, Bajwa AA, Alharby H, Amanullah, Nasim W, Shahzad B, Tanveer M, Huang J (2016a). Responses of rapid viscoanalyzer profile and other rice grain qualities to exogenously applied plant growth regulators under high day and high night temperatures. *PLOS ONE* 11(7):e0159590. https://doi. https://doi.org/10.1371/journal.pone.0159590.

Fahad S, Hussain S, Saud S, Hassan S, Ihsan Z, Shah AN, Wu C, Yousaf M, Nasim W, Alharby H, Alghabari F, Huang J (2016c). Exogenously applied plant growth regulators enhance the morphophysiological growth and yield of rice under high temperature. *Front Plant Sci* 7:1250. https://doi.org/10.3389/fpls.2016.

Fahad S, Hussain S, Saud S, Hassan S, Tanveer M, Ihsan MZ, Shah AN, Ullah A, Nasrullah, Khan F, Ullah S, Alharby H, Nasim W, Wu C, Huang J (2016d). A combined application of biochar and phosphorus alleviates heat-induced adversities on physiological, agronomical and quality attributes of rice. *Plant Physiol Biochem* 103:191–198.

Fahad S, Hussain S, Saud S, Khan F, Hassan S Jr A, Nasim W, Arif M, Wang F, Huang J (2016b). Exogenously applied plant growth regulators affect heat-stressed rice pollens. *J Agron Crop Sci* 202(2):139–150.

Fahad S, Hussain S, Saud S, Tanveer M, Bajwa AA, Hassan S, Shah AN, Ullah A, Wu C, Khan FA, Shah F, Ullah S, Chen Y, Huang J (2015a). A biochar application protects rice pollen from high-temperature stress. *Plant Physiol Biochem* 96:281–287.

Fahad S, Ihsan MZ, Khaliq A, Daur I, Saud S, Alzamanan S, Nasim W, Abdullah M, Khan IA, Wu C, Wang D, Huang J (2018). Consequences of high temperature under changing climate optima for rice pollen characteristics-concepts and perspectives. *Arch Agron Soil Sci.* https://doi.org/10.1080/03650340.2018.1443213.

Fahad S, Nie L, Chen Y, Wu C, Xiong D, Saud S, Hongyan L, Cui K, Huang J (2015b). Crop plant hormones and environmental stress. *Sustain Agric Rev* 15:371–400.

Fahad S, Rehman A, Shahzad B, Tanveer M, Saud S, Kamran M, Ihtisham M, Khan SU, Turan V, Rahman MHu (2019a). Rice responses and tolerance to metal/metalloid toxicity. In: Hasanuzzaman M, Fujita M, Nahar K, Biswas JK (eds) *Advances in rice research for abiotic stress tolerance.* Woodhead Publ Ltd, Cambridge, England, pp 299–312.

Fang Q, Jiang T, Xu L, Liu H, Mao H, Wang X, Jiao B, Duan Y, Wang Q, Dong Q, Yang L, Tian G, Zhang C, Zhou Y, Liu X, Wang H, Fan D, Wang B, Luo K (2017). A salt-stress-regulator from the Poplar R2R3 MYB family integrates the regulation of lateral root emergence and ABA signaling to mediate salt stress tolerance in Arabidopsis. *Plant Physiol Biochem* 114:100–110.

Farah R, Muhammad R, Muhammad SA, Tahira Y, Muhammad AA, Maryam A, Shafaqat A, Rashid M, Muhammad R, Qaiser H, Afia Z, Muhammad AA, Muhammad A, Fahad S (2020). Alternative and non-conventional soil and crop management strategies for increasing water use efficiency. In: Fahad S, Hasanuzzaman M, Alam M, Ullah H, Saeed M, Khan AK, Adnan M (eds) *Environment, climate, plant and vegetation growth.* Springer Publ Ltd, Springer Nature Switzerland AG. Part of Springer Nature, pp 323–338. https://doi.org/10.1007/978-3-030-49732-3.

Farhangi-Abriz S, Ghassemi-Golezani K (2018). How can salicylic acid and jasmonic acid mitigate salt toxicity in soybean plants? *Ecotoxicol Environ Saf* 147:1010–1016.

Farhat A, Hafiz MH, Wajid I, Aitazaz AF, Hafiz FB, Zahida Z, Fahad S, Wajid F, Artemi C (2020). A review of soil carbon dynamics resulting from agricultural practices. *J Environ Manag* 268. http://www.ncbi.nlm.nih.gov/pubmed/110319.

Farooq M, Basra S, Wahid A, Rehman H (2009). Exogenously applied nitric oxide enhances the drought tolerance in fine grain aromatic rice (Oryza sativa L.). *J Agron Crop Sci* 195(4):254–261.

Farooq MA, Islam F, Yang C, Nawaz A, Athar H, Gill RA, Ali B, Song W, Zhou W (2018). Methyl jasmonate alleviates arsenic-induced oxidative damage and modulates the ascorbate–glutathione cycle in oilseed rape roots. *Plant Growth Regul* 84(1):135–148.

Fayez KA, Bazaid SA (2014). Improving drought and salinity tolerance in barley by application of salicylic acid and potassium nitrate. *J Saudi Soc Agric Sci* 13(1):45–55.

Fazli W, Muhmmad S, Amjad A, Fahad S, Muhammad A, Muhammad N, Ishaq AM, Imtiaz AK, Mukhtar A, Muhammad S, Muhammad I, Rafi U, Haroon I, Muhammad A (2020). Plant-microbes interactions and functions in changing climate. In: Fahad S, Hasanuzzaman M, Alam M, Ullah H, Saeed M, Khan AK, Adnan M (eds) *Environment, climate, plant and vegetation growth.* Springer Publ Ltd, Springer Nature Switzerland AG. Part of Springer Nature, pp 397–420. https://doi.org/10.1007/978-3-030-49732-3.

Fu J, Wu H, Ma S, Xiang D, Liu R, Xiong L (2017). OsJAZ1 attenuates drought resistance by regulating JA and ABA signaling in rice. *Front Plant Sci* 8:2108.

Fujita Y, Fujita M, Shinozaki K, Yamaguchi-Shinozaki K (2011). ABA-mediated transcriptional regulation in response to osmotic stress in plants. *J Plant Res* 124(4):509–525.

González-García M-P, Vilarrasa-Blasi J, Zhiponova M, Divol F, Mora-García S, Russinova E, Caño-Delgado AI (2011). Brassinosteroids control meristem size by promoting cell cycle progression in Arabidopsis roots. *Development* 138(5):849–859.

Gopakumar L, Bernard NO, Donato V (2020). Soil microarthropods and nutrient cycling. In: Fahad S, Hasanuzzaman M, Alam M, Ullah H, Saeed M, Khan AK, Adnan M (eds) *Environment, climate, plant and vegetation growth.* Springer Publ Ltd, Springer Nature Switzerland AG. Part of Springer Nature, pp 453–472. https://doi.org/10.1007/978-3-030-49732-3.

Griffiths J, Murase K, Rieu I, Zentella R, Zhang Z-L, Powers SJ, Gong F, Phillips AL, Hedden P, Sun T-p, Thomas SG (2006). Genetic characterization and functional analysis of the GID1 gibberellin receptors in Arabidopsis. *Plant Cell* 18(12):3399–3414.

Guan C, Ji J, Wu D, Li X, Jin C, Guan W, Wang G (2015). The glutathione synthesis may be regulated by cadmium-induced endogenous ethylene in Lycium chinense, and overexpression of an ethylene responsive transcription factor gene enhances tolerance to cadmium stress in tobacco. *Mol Breed* 35(5):123.

Gul F, Ahmed I, Ashfaq M, Jan D, Fahad S, Li X, Wang D, Fahad M, Fayyaz M, Shah SA (2020). Use of crop growth model to simulate the impact of climate change on yield of various wheat cultivars under different agro-environmental conditions in Khyber Pakhtunkhwa, Pakistan. *Arab J Geosci* 13(3):112. https://doi.org/10.1007/s12517-020-5118-1.

Habib ur R, Ashfaq A, Aftab W, Manzoor H, Fahd R, Wajid I, Md. Aminul I, Vakhtang S, Muhammad A, Asmat U, Abdul W, Syeda RS, Shah S, Shahbaz K, Fahad S, Manzoor H, Saddam H, Wajid N (2017). Application of CSM-CROPGRO-Cotton model for cultivars and optimum planting dates: Evaluation in changing semi-arid climate. *Field Crops Res.*

Hafiz MH, Farhat A, Shafqat S, Fahad S, Artemi C, Wajid F, Chaves CB, Wajid N, Muhammad M, Hafiz FB (2018). Offsetting land degradation through nitrogen and water management during maize cultivation under arid conditions. *Land Degrad Dev* 1–10.

Hafiz MH, Muhammad A, Farhat A, Hafiz FB, Saeed AQ, Muhammad M, Fahad S, Muhammad A (2019). Environmental factors affecting the frequency of road traffic accidents: a case study of sub-urban area of Pakistan. *Environ Sci Pollut Res.*

Hafiz MH, Wajid F, Farhat A, Fahad S, Shafqat S, Wajid N, Hafiz FB (2016). Maize plant nitrogen uptake dynamics at limited irrigation water and nitrogen. *Environ Sci Pollut Res* 24(3):2549–2557.

Hammad HM, Abbas F, Ahmad A, Bakhat HF, Farhad W, Wilkerson CJ, Fahad S, Hoogenboom G (2020a). Predicting kernel growth of maize under controlled water and nitrogen applications. *Int J Plant Prod.* https://doi.org/10.1007/s42106-020-00110-8.

Hammad HM, Khaliq A, Abbas F, Farhad W, Fahad S, Aslam M, Shah GM, Nasim W, Mubeen M, Bakhat HF (2020b). Comparative effects of organic and inorganic fertilizers on soil organic carbon and wheat productivity under arid region. *Commun Soil Sci Plant Anal.* https://doi.org/10.1080/00103624.2020.1763385.

Harpreet K, Poonam S, Geetika S (2013). Sugar accumulation and its regulation by jasmonic acid in Brassica napus L. under salt stress. *J Stress Physiol Biochem* 9:53–64.

Herrera-Vásquez A, Salinas P, Holuigue L (2015). Salicylic acid and reactive oxygen species interplay in the transcriptional control of defense genes expression. *Front Plant Sci* 6:171.

Hesham FA, Fahad S (2020). Melatonin application enhances biochar efficiency for drought tolerance in maize varieties: modifications in physio-biochemical machinery. *Agron J* 112(4):1–22.

Hong Y, Zhang H, Huang L, Li D, Song F (2016) Overexpression of a stress-responsive NAC transcription factor gene ONAC022 improves drought and salt tolerance in rice. *Front Plant Sci* 7:4.

Huang C, Zhou J, Jie Y, Xing H, Zhong Y, She W, Wei G, Yu W, Ma Y (2016). A ramie (Boehmeria nivea) bZIP transcription factor BnbZIP3 positively regulates drought, salinity and heavy metal tolerance. *Mol Breed* 36(8):120–129.

Hussain MA, Fahad S, Rahat S, Muhammad FJ, Muhammad M, Qasid A, Ali A, Husain A, Nooral A, Babatope SA, Changbao S, Liya G, Ibrar A, Zhanmei J, Juncai H (2020). Multifunctional role of brassinosteroid and its analogues in plants. *Plant Growth Regul.* https://doi.org/10.1007/s10725-020-00647-8.

Hussain SS, Kayani MA, Amjad M (2011). Transcription factors as tools to engineer enhanced drought stress tolerance in plants. *Biotechnol Progr* 27(2):297–306.

Hutchison CE, Li J, Argueso C, Gonzalez M, Lee E, Lewis MW, Maxwell BB, Perdue TD, Schaller GE, Alonso JM, Ecker JR, Kieber JJ (2006). The Arabidopsis histidine phosphotransfer proteins are redundant positive regulators of cytokinin signaling. *Plant Cell* 18(11):3073–3087.

Hwang I, Sheen J, Müller B (2012). Cytokinin signaling networks. *Annu Rev Plant Biol* 63:353–380.

Ibrar K, Aneela R, Khola Z, Urooba N, Sana B, Rabia S, Ishtiaq H, Mujaddad Ur Rehman, Salvatore M (2020). Microbes and environment: global warming reverting the frozen zombies. In: Fahad S, Hasanuzzaman M, Alam M, Ullah H, Saeed M, Khan AK, Adnan M (eds) *Environment, climate, plant and vegetation growth.* Springer Publ Ltd, Springer Nature Switzerland AG. Part of Springer Nature, pp 607–634. https://doi.org/10.1007/978-3-030-49732-3.

Ilyas M, Nisar M, Khan N, Hazrat A, Khan AH, Hayat K, Fahad S, Khan A, Ullah A (2020). Drought tolerance strategies in plants: a mechanistic approach. *J Plant Growth Regul.* https://doi.org/10.1007/s00344-020-10174-5.

Iqbal N, Nazar R, Syeed S, Masood A, Khan NA (2011). Exogenously-sourced ethylene increases stomatal conductance, photosynthesis, and growth under optimal and deficient nitrogen fertilization in mustard. *J Exp Bot* 62(14):4955–4963.

Iqra M, Amna B, Shakeel I, Fatima K, Sehrish L, Hamza A, Fahad S (2020). Carbon cycle in Response to Global Warming. In: Fahad S, Hasanuzzaman M, Alam M, Ullah H, Saeed M, Khan AK, Adnan M (eds) *Environment, climate, plant and vegetation growth.* Springer Publ Ltd, Springer Nature Switzerland AG. Part of Springer Nature, pp 1–16. https://doi.org/10.1007/978-3-030-49732-3.

Ito S, Ito K, Abeta N, Takahashi R, Sasaki Y, Yajima S (2016). Effects of strigolactone signaling on Arabidopsis growth under nitrogen deficient stress condition. *Plant Signal Behav* 11(1):e1126031.

Jakir Md H, Allah B (2020). Development and applications of Transplastomic plants; A way Towards eco-friendly agriculture. In: Fahad S, Hasanuzzaman M, Alam M, Ullah H, Saeed M, Khan AK, Adnan M (eds) *Environment, climate, plant and vegetation growth.* Springer Publ Ltd, Springer Nature Switzerland AG. Part of Springer Nature, pp 285–322. https://doi.org/10.1007/978-3-030-49732-3.

Jain M, Khurana JP (2009). Transcript profiling reveals diverse roles of auxin-responsive genes during reproductive development and abiotic stress in rice. *FEBS J* 276(11):3148–3162.

Jamil M, Charnikhova T, Verstappen F, Ali Z, Wainwright H, Bouwmeester H (2014). Effect of phosphate-based seed priming on strigolactone production and S triga hermonthica infection in cereals. *Weed Res* 54(3):307–313.

Jan M, Anwar-ul-Haq M, Shah AN, Yousaf M, Iqbal J, Li X, Wang D, Fahad S (2019). Modulation in growth, gas exchange, and antioxidant activities of salt-stressed rice (Oryza sativa L.) genotypes by zinc fertilization. *Arab J Geosci* 12(24):775. https://doi.org/10.1007/s12517-019-4939-2.

Jaradat MR, Feurtado JA, Huang D, Lu Y, Cutler AJ (2013) Multiple roles of the transcription factor AtMYBR1/AtMYB44 in ABA signaling, stress responses, and leaf senescence. *BMC Plant Biol* 13:192.

Jeon J, Kim J (2013). Arabidopsis response Regulator1 and Arabidopsis histidine phosphotransfer Protein2 (AHP2), AHP3, and AHP5 function in cold signaling. *Plant Physiol* 161(1):408–424.

Jeon J, Kim NY, Kim S, Kang NY, Novák O, Ku S-J, Cho C, Lee DJ, Lee E-J, Strnad M, Kim J (2010). A subset of cytokinin two-component signaling system plays a role in cold temperature stress response in Arabidopsis. *J Biol Chem* 285(30):23371–23386.

Jiang X, Zhang C, Lü P, Jiang G, Liu X, Dai F, Gao J (2014) RhNAC3, a stress-associated NAC transcription factor, has a role in dehydration tolerance through regulating osmotic stress-related genes in rose petals. *Plant Biotechnol J* 12(1):38–48.

Junghans U, Polle A, Düchting P, Weiler E, Kuhlman B, Gruber F, Teichmann T (2006). Adaptation to high salinity in poplar involves changes in xylem anatomy and auxin physiology. *Plant Cell Environ* 29(8):1519–1531.

Kamaran M, Wenwen C, Irshad A, Xiangping M, Xudong Z, Wennan S, Junzhi C, Shakeel A, Fahad S, Qingfang H, Tiening L (2018). Effect of paclobutrazol, a potential growth regulator on stalk mechanical strength, lignin accumulation and its relation with lodging resistance of maize. *Plant Growth Regul* 84:317–332.

Kang G, Li G, Guo T (2014). Molecular mechanism of salicylic acid-induced abiotic stress tolerance in higher plants. *Acta Physiol Plant* 36(9):2287–2297.

Kang NY, Cho C, Kim NY, Kim J (2012). Cytokinin receptor-dependent and receptor-independent pathways in the dehydration response of Arabidopsis thaliana. *J Plant Physiol* 169(14):1382–1391.

Kanwar MK, Bhardwaj R, Chowdhary SP, Arora P, Sharma P, Kumar S (2013). Isolation and characterization of 24-Epibrassinolide from Brassica juncea L. and its effects on growth, Ni ion uptake, antioxidant defense of Brassica plants and in vitro cytotoxicity. *Acta Physiol Plant* 35(4):1351–1362.

Karpets YV, Kolupaev YE, Lugovaya A, Oboznyi A (2014). Effect of jasmonic acid on the pro-/antioxidant system of wheat coleoptiles as related to hyperthermia tolerance. *Russ J Plant Physiol* 61(3):339–346.

Kasahara H, Hanada A, Kuzuyama T, Takagi M, Kamiya Y, Yamaguchi S (2002). Contribution of the mevalonate and methylerythritol phosphate pathways to the biosynthesis of gibberellins inArabidopsis. *J Biol Chem* 277(47):45188–45194.

Kazan K (2015). Diverse roles of jasmonates and ethylene in abiotic stress tolerance. *Trends Plant Sci* 20(4):219–229.

Khan A, Tan DKY, Afridi MZ, Luo H, Tung SA, Ajab M, Fahad S (2017b). Nitrogen fertility and abiotic stresses management in cotton crop: a review. *Environ Sci Pollut Res Int* 24(17):14551–14566. https://doi.org/10.1007/s11356-017-8920-x.

Khan A, Tan DKY, Munsif F, Afridi MZ, Shah F, Wei F, Fahad S, Zhou R (2017a). Nitrogen nutrition in cotton and control strategies for greenhouse gas emissions: a review. *Environ Sci Pollut Res Int* 24(30):23471–23487. https://doi.org/10.1007/s11356-017-0131-y.

Khan MIR, Iqbal N, Masood A, Per TS, Khan NA (2013). Salicylic acid alleviates adverse effects of heat stress on photosynthesis through changes in proline production and ethylene formation. *Plant Signal Behav* 8(11):e26374.

Khan MS, Akther T, Mubarak Ali DM, Hemalatha S (2019). An investigation on the role of salicylic acid alleviate the saline stress in rice crop (Oryza sativa (L)). *Biocatal Agric Biotechnol* 18. http://www.ncbi.nlm.nih.gov/pubmed/101027.

Khatoon H, Singh A, Ahmad F, Kama A (2017). Biosciences and plant biology. *Int J Curr Res Biosci Plant Biol* 4(7):88–96.

Koyama T (2014). The roles of ethylene and transcription factors in the regulation of onset of leaf senescence. *Front Plant Sci* 5:650.

Kudoyarova GR, Vysotskaya LB, Cherkozyanova A, Dodd IC (2007). Effect of partial rootzone drying on the concentration of zeatin-type cytokinins in tomato (Solanum Lycopersicum L.) xylem sap and leaves. *J Exp Bot* 58(2):161–168.

Le Hir R, Castelain M, Chakraborti D, Moritz T, Dinant S, Bellini C (2017). AtbHLH68 transcription factor contributes to the regulation of ABA homeostasis and drought stress tolerance in Arabidopsis thaliana. *Physiol Plant* 160(3):312–327.

Li S, Yang Y, Zhang Q, Liu N, Xu Q, Hu L (2018). Differential physiological and metabolic response to low temperature in two zoysiagrass genotypes native to high and low latitude. *PLOS ONE* 13(6): e0198885.

Li QF, Wang C, Jiang L, Li S, Sun SS, He JX (2012) An interaction between BZR1 and DELLAs mediates direct signaling crosstalk between brassinosteroids and gibberellins in Arabidopsis. *Sci Signal* 5(244):ra72.

Liang C, Wang Y, Zhu Y, Tang J, Hu B, Liu L, Ou S, Wu H, Sun X, Chu J, Chu C (2014):OsNAP connects abscisic acid and leaf senescence by fine-tuning abscisic acid biosynthesis and directly targeting senescence-associated genes in rice. *Proc Natl Acad Sci U S A* 111(27):10013–10018.

Linkies A, Leubner-Metzger G (2012). Beyond gibberellins and abscisic acid: how ethylene and jasmonates control seed germination. *Plant Cell Rep* 31(2):253–270.

Liu J, Lovisolo C, Schubert A, Cardinale F (2013). Signaling role of strigolactones at the interface between plants,(micro) organisms, and a changing environment. *J Plant Interact* 8(1):17–33.

Llanes A, Andrade A, Alemano S, Luna V (2016). Alterations of endogenous hormonal levels in plants under drought and salinity. *Am J Plant Sci* 7(9):1357.

Magome H, Yamaguchi S, Hanada A, Kamiya Y, Oda K (2008). The DDF1 transcriptional activator upregulates expression of a gibberellin-deactivating gene, GA2ox7, under high-salinity stress in Arabidopsis. *Plant J Cell Mol Biol* 56(4):613–626.

Mahar A, Amjad A, Altaf HL, Fazli W, Ronghua L, Muhammad A, Fahad S, Muhammad A, Rafiullah Imtiaz AK, Zengqiang Z (2020). Promising technologies for Cd-contaminated soils: drawbacks and possibilities. In: Fahad S, Hasanuzzaman M, Alam M, Ullah H, Saeed M, Khan AK, Adnan M (eds) *Environment, climate, plant and vegetation growth.* Springer Publ Ltd, Springer Nature Switzerland AG. Part of Springer Nature, pp 63–92. https://doi.org/10.1007/978-3-030-49732-3.

Mason MG, Jha D, Salt DE, Tester M, Hill K, Kieber JJ, Schaller G (2010). Type-B response regulators ARR1 and ARR12 regulate expression of AtHKT1; 1 and accumulation of sodium in Arabidopsis shoots. *Plant J Cell Mol Biol* 64(5):753–763.

Miura K, Tada Y (2014). Regulation of water, salinity, and cold stress responses by salicylic acid. *Front Plant Sci* 5:4.

Muhammad B, Adnan M, Munsif F, Fahad S, Saeed M, Wahid F, Arif M, Amanullah J, Wang D, Saud S, Noor M, Zamin M, Subhan F, Saeed B, Raza MA, Mian IA (2019). Substituting urea by organic wastes for improving maize yield in alkaline soil. *J Plant Nutr.* https://doi.org/10.1080/01904167.2019.1659344.

Mühlenbock P, Szechyńska-Hebda M, Płaszczyca M, Baudo M, Mateo A, Mullineaux PM, Parker JE, Karpińska B, Karpiński S (2008). Chloroplast signaling and lesion simulating disease1 regulate crosstalk between light acclimation and immunity in Arabidopsis. *Plant Cell* 20(9):2339–2356.

Mubeen M, Ahmad A, Hammad HM, Awais M, Farid HU, Saleem M, Din MSu, Amin A, Ali A, Fahad S, Nasim W (2020). Evaluating the climate change impact on water use efficiency of cotton-wheat in semi-arid conditions using DSSAT model. *J Water Clim Change* 11(4):1661–1675. https://doi.org/10.2166/wcc.2019.179/622035/jwc2019179.pdf.

Nazar R, Iqbal N, Syeed S, Khan NA (2011). Salicylic acid alleviates decreases in photosynthesis under salt stress by enhancing nitrogen and sulfur assimilation and antioxidant metabolism differentially in two mungbean cultivars. *J Plant Physiol* 168(8):807–815.

Nazar R, Umar S, Khan NA (2015). Exogenous salicylic acid improves photosynthesis and growth through increase in ascorbate-glutathione metabolism and S assimilation in mustard under salt stress. *Plant Signal Behav* 10(3):e1003751.

Ng LM, Melcher K, Teh BT, Xu HE (2014). Abscisic acid perception and signaling: structural mechanisms and applications. *Acta Pharmacol Sin* 35(5):567–584.

Nishiyama R, Watanabe Y, Fujita Y, Le DT, Kojima M, Werner T, Vankova R, Yamaguchi-Shinozaki K, Shinozaki K, Kakimoto T, Sakakibara H, Schmülling T, Tran LS (2011). Analysis of cytokinin mutants and regulation of cytokinin metabolic genes reveals important regulatory roles of cytokinins in drought, salt and abscisic acid responses, and abscisic acid biosynthesis. *Plant Cell* 23(6):2169–2183.

Nitschke S, Cortleven A, Iven T, Feussner I, Havaux M, Riefler M, Schmülling T (2016). Circadian stress regimes affect the circadian clock and cause jasmonic acid-dependent cell death in cytokinin-deficient Arabidopsis plants. *Plant Cell* 28(7):1616–1639.

Noor M, ur Naveed, R, Ajmal J, Fahad S, Muhammad A, Fazli W, Saud S, Hassan S (2020). Climate change and costal plant lives. In: Fahad S, Hasanuzzaman M, Alam M, Ullah H, Saeed M, Khan AK, Adnan M (eds) *Environment, climate, plant and vegetation growth.* Springer Publ Ltd, Springer Nature Switzerland AG. Part of Springer Nature, pp 93–108. https://doi.org/10.1007/978-3-030-49732-3.

Ogawa D, Nakajima N, Sano T, Tamaoki M, Aono M, Kubo A, Kanna M, Ioki M, Kamada H, Saji H (2005). Salicylic acid accumulation under O3 exposure is regulated by ethylene in tobacco plants. *Plant Cell Physiol* 46(7):1062–1072.

Passioura JB (2002). Review: environmental biology and crop improvement. *Funct Plant Biol* 29(5):537–546.

Potters G, Pasternak TP, Guisez Y, Jansen MA (2009). Different stresses, similar morphogenic responses: integrating a plethora of pathways. *Plant Cell Environ* 32(2):158–169.

Potters G, Pasternak TP, Guisez Y, Palme KJ, Jansen MA (2007). Stress-induced morphogenic responses: growing out of trouble? *Trends Plant Sci* 12(3):98–105.

Procházková D, Haisel D, Pavlíková D (2014). Nitric oxide biosynthesis in plants—the short overview. *Plant Soil Environ* 60(3):129–134.

Quan R, Hu S, Zhang Z, Zhang H, Zhang Z, Huang R (2010). Overexpression of an ERF transcription factor TSRF1 improves rice drought tolerance. *Plant Biotechnol J* 8(4):476–488.

Rashid M, Qaiser H, Khalid SK, Mohammad I, Al-Wabel Zhang A, Muhammad A, Shahzada SI, Rukhsanda A, Ghulam AS, Shahzada MM, Sarosh A, Muhammad FQ (2020). Prospects of biochar in alkaline soils to mitigate climate change. In: Fahad S, Hasanuzzaman M, Alam M, Ullah H, Saeed M, Khan AK, Adnan M (eds) *Environment, climate, plant and vegetation growth.* Springer Publ Ltd, Springer Nature Switzerland AG. Part of Springer Nature, pp 133–150. https://doi.org/10.1007/978-3-030-49732-3.

Rashotte AM, Mason MG, Hutchison CE, Ferreira FJ, Schaller GE, Kieber JJ (2006). A subset of Arabidopsis AP2 transcription factors mediates cytokinin responses in concert with a two-component pathway. *Proc Natl Acad Sci U S A* 103(29):11081–11085.

Rehman M, Fahad S, Saleem MH, Hafeez M, Rahman MHu, Liu F, Deng G (2020). Red light optimized physiological traits and enhanced the growth of ramie (Boehmeria nivea L.). *Photosynthetica* 58(4):922–931.

Robson F, Okamoto H, Patrick E, Harris S-R, Wasternack C, Brearley C, Turner JG (2010). Jasmonate and phytochrome A signaling in Arabidopsis wound and shade responses are integrated through JAZ1 stability. *Plant Cell* 22(4):1143–1160.

Saeed W, Naseem S, Ali Z (2017). Strigolactones biosynthesis and their role in abiotic stress resilience in plants: a critical review. *Front Plant Sci* 8:1487.

Sadam M, ul Qamar Muhammad Tahir, Ghulam M, Muhammad SK, Faiz AJ (2020). Role of biotechnology in climate resilient agriculture. In: Fahad S, Hasanuzzaman M, Alam M, Ullah H, Saeed M, Khan AK, Adnan M (eds) *Environment, climate, plant and vegetation growth.* Springer Publ Ltd, Springer Nature Switzerland AG. Part of Springer Nature, pp 339–366. https://doi.org/10.1007/978-3-030-49732-3.

Sah SK, Reddy KR, Li J (2016). Abscisic acid and abiotic stress tolerance in crop plants. *Front Plant Sci* 7:571.

Saini K, Markakis MN, Zdanio M, Balcerowicz DM, Beeckman T, De Veylder L, Prinsen E, Beemster GT, Vissenberg K (2017). Alteration in auxin homeostasis and signaling by overexpression of PINOID kinase causes leaf growth defects in Arabidopsis thaliana. *Front Plant Sci* 8:1009.

Saini S, Sharma I, Pati PK (2015). Versatile roles of brassinosteroid in plants in the context of its homoeostasis, signaling and crosstalks. *Front Plant Sci* 6:950.

Sajjad H, Muhammad M, Ashfaq A, Waseem A, Hafiz MH, Mazhar A, Nasir M, Asad A, Hafiz UF, Syeda RS, Fahad S, Depeng W, Wajid N (2019). Using GIS tools to detect the land use/land cover changes during forty years in Lodhran district of Pakistan. *Environ Sci Pollut Res.*

Sajid H, Jie H, Jing H, Shakeel A, Satyabrata N, Sumera A, Awais S, Chunquan Z, Lianfeng Z, Xiaochuang C, Qianyu J, Junhua Z (2020). Rice production under climate change: adaptations and mitigating strategies. In: Fahad S, Hasanuzzaman M, Alam M, Ullah H, Saeed M, Khan AK, Adnan M (eds) *Environment, climate, plant and vegetation growth.* Springer Publ Ltd, Springer Nature Switzerland AG. Part of Springer Nature, pp 659–686. https://doi.org/10.1007/978-3-030-49732-3.

Saleem MH, Fahad S, Adnan M, Ali M, Rana MS, Kamran M, Ali Q, Hashem IA, Bhantana P, Ali M, Hussain RM (2020a). Foliar application of gibberellic acid endorsed phytoextraction of copper and alleviates oxidative stress in jute (Corchorus Capsularis L.) plant grown in highly copper-contaminated soil of China. *Environ Sci Pollut Res Int.* https://doi.org/10.1007/s11356-020-09764-3.

Saleem MH, Fahad S, Khan SU, Din M, Ullah A, Sabagh AE, Hossain A, Llanes A, Liu L (2020c). Copper-induced oxidative stress, initiation of antioxidants and phytoremediation potential of flax (Linum usitatissimum L.) seedlings grown under the mixing of two different soils of China. *Environ Sci Pollut Res Int.* https://doi.org/10.1007/s11356-019-07264-7.

Saleem MH, Rehman M, Fahad S, Tung SA, Iqbal N, Hassan A, Ayub A, Wahid MA, Shaukat S, Liu L, Deng G (2020b). Leaf gas exchange, oxidative stress, and physiological attributes of rapeseed (Brassica napus L.) grown under different light-emitting diodes. *Photosynthetica* 58(3):836–845.

Saleh J, Maftoun M (2008). Interactive effect of NaCl levels and zinc sources and levels on the growth and mineral composition of rice. *J Agric Sci Technol* 10:325–336.

Salgado I, Carmen Martínez M, Oliveira HC, Frungillo L (2013). Nitric oxide signaling and homeostasis in plants: a focus on nitrate reductase and S-nitrosoglutathione reductase in stress-related responses. *Braz J Bot* 36(2):89–98.

Saman S, Amna B, Bani A, ul Qamar Muhammad Tahir, Rana MA, Muhammad SK (2020). QTL mapping for abiotic stresses in cereals. In: Fahad S, Hasanuzzaman M, Alam M, Ullah H, Saeed M, Khan AK, Adnan M (eds) *Environment, climate, plant and vegetation growth.* Springer Publ Ltd, Springer Nature Switzerland AG. Part of Springer Nature, pp 229–252. https://doi.org/10.1007/978-3-030-49732-3.

Saud S, Fahad S, Cui G, Yajun C, Anwar S (2020). Determining nitrogen isotopes discrimination under drought stress on enzymatic activities, nitrogen isotope abundance and water contents of Kentucky bluegrass. *Sci Rep* 10(1):6415. https://doi.org/10.1038/s41598-020-63548-w.

Saud S, Fahad S, Yajun C, Ihsan MZ, Hammad HM, Nasim W, Amanullah Jr Arif M, Arif M, Alharby H (2017). Effects of nitrogen supply on water stress and recovery mechanisms in Kentucky bluegrass plants. *Front Plant Sci* 8:983. https://doi.org/10.3389/fpls.2017.00983.

Saud S, Li X, Chen Y, Zhang L, Fahad S, Hussain S, Sadiq A, Chen Y (2014). Silicon application increases drought tolerance of Kentucky bluegrass by improving plant water relations and morphophysiological functions. *Sci World J* 2014:368694. https://doi.org/10.1155/2014/368694.

Saud S, Yajun C, Fahad S, Hussain S, Na L, Xin L, Alhussien SA (2016). Silicate application increases the photosynthesis and its associated metabolic activities in Kentucky bluegrass under drought stress and post-drought recovery. *Environ Sci Pollut Res Int* 23(17):17647–17655. https://doi.org/10.1007/s11356-016-6957-x.

Saud S, Yajun C, Long B, Fahad S, Sadiq A (2013). The different impact on the growth of cool season turf grass under the various conditions on salinity and drought stress. *Int J Agric Sci Res* 3:77–84.

Schachtman DP, Goodger JQ (2008). Chemical root to shoot signaling under drought. *Trends Plant Sci* 13(6):281–287.

Senol C (2020). The effects of climate change on human behaviors. In: Fahad S, Hasanuzzaman M, Alam M, Ullah H, Saeed M, Khan AK, Adnan M (eds) *Environment, climate, plant and vegetation growth.* Springer Publ Ltd, Springer Nature Switzerland AG. Part of Springer Nature, pp 577–590. https://doi.org/10.1007/978-3-030-49732-3.

Seo JS, Joo J, Kim MJ, Kim YK, Nahm BH, Song SI, Cheong JJ, Lee JS, Kim JK, Choi YD (2011). OsbHLH148, a basic helix-loop-helix protein, interacts with OsJAZ proteins in a jasmonate signaling pathway leading to drought tolerance in rice. *Plant J Cell Mol Biol* 65(6):907–921.

Shah F, Lixiao N, Kehui C, Tariq S, Wei W, Chang C, Liyang Z, Farhan A, Fahad S, Huang J (2014). Rice grain yield and component responses to near 2°C of warming. *Field Crops Res* 157:98–110.

Shafi MI, Adnan M, Fahad S, Fazli W, Ahsan K, Zhen Y, Subhan D, Zafar-ul-Hye M, Martin B, Rahul D (2020). Application of single superphosphate with humic acid improves the growth, yield and phosphorus uptake of wheat (*Triticum aestivum* L.) in calcareous soil. *Agron* 10:1224.

Shan C, Zhou Y, Liu M (2015). Nitric oxide participates in the regulation of the ascorbate-glutathione cycle by exogenous jasmonic acid in the leaves of wheat seedlings under drought stress. *Protoplasma* 252(5):1397–1405.

Sharipova G, Veselov D, Kudoyarova G, Fricke W, Dodd IC, Katsuhara M, Furuichi T, Ivanov I, Veselov S (2016). Exogenous application of abscisic acid (ABA) increases root and cell hydraulic conductivity and abundance of some aquaporin isoforms in the ABA-deficient barley mutant Az34. *Ann Bot* 118(4):777–785.

Sharma M, Laxmi A (2015). Jasmonates: emerging players in controlling temperature stress tolerance. *Front Plant Sci* 6:1129.

Sheteiwy MS, Gong D, Gao Y, Pan R, Hu J, Guan Y (2018). Priming with methyl jasmonate alleviates polyethylene glycol-induced osmotic stress in rice seeds by regulating the seed metabolic profile. *Environ Exp Bot* 153:236–248.

Shi Y, Tian S, Hou L, Huang X, Zhang X, Guo H, Yang S (2012). Ethylene signaling negatively regulates freezing tolerance by repressing expression of CBF and type-A ARR genes in Arabidopsis. *Plant Cell* 24(6):2578–2595.

Singh AP, Fridman Y, Friedlander-Shani L, Tarkowska D, Strnad M, Savaldi-Goldstein S (2014). Activity of the brassinosteroid transcription factors brassinazole resistant1 and brassinosteroid insensitive1-ethyl methanesulfonate-suppressor1/brassinazole resistant2 blocks developmental reprogramming in response to low phosphate availability. *Plant Physiol* 166(2):678–688.

Sponsel V, Hedden P (2004). Gibberellin biosynthesis and inactivation. In: Davies PJ (ed) *Plant hormones: biosynthesis, signal transduction, action. Cap. 2.* Kluwer Acad Pub.

Subhan D, Zafar-ul-Hye M, Fahad S, Saud S, Martin B, Tereza H, Rahul D (2020). Drought stress alleviation by ACC deaminase producing *Achromobacter xylosoxidans* and *Enterobacter cloacae*, with and without timber waste biochar in maize. *Sustain* 12(6286).

Syeed S, Anjum NA, Nazar R, Iqbal N, Masood A, Khan NA (2011). Salicylic acid-mediated changes in photosynthesis, nutrients content and antioxidant metabolism in two mustard (Brassica juncea L.) cultivars differing in salt tolerance. *Acta Physiol Plant* 33(3):877–886.

Talebi M, Moghaddam M, Pirbalouti AG (2018). Methyl jasmonate effects on volatile oil compounds and antioxidant activity of leaf extract of two basil cultivars under salinity stress. *Acta Physiol Plant* 40(2):34.

Tardieu F, Tuberosa R (2010). Dissection and modelling of abiotic stress tolerance in plants. *Curr Opin Plant Biol* 13(2):206–212.

Tariq M, Ahmad S, Fahad S, Abbas G, Hussain S, Fatima Z, Nasim W, Mubeen M, Rehman MHu, Khan MA, Adnan M, Wilkerson CJ, Hoogenboom G (2018). The impact of climate warming and crop management on phenology of sunflower-based cropping systems in Punjab, Pakistan. *Agric Forest Meteorol* 256–257:270–282.

Tian Y, Zhang H, Pan X, Chen X, Zhang Z, Lu X, Huang R (2011). Overexpression of ethylene response factor TERF2 confers cold tolerance in rice seedlings. *Transgenic Res* 20(4):857–866.

To JP, Kieber JJ (2008). Cytokinin signaling: two-components and more. *Trends Plant Sci* 13(2):85–92.

Tognetti VB, Mühlenbock P, Van Breusegem F (2012). Stress homeostasis–the redox and auxin perspective. *Plant Cell Environ* 35(2):321–333.

Tran LS, Urao T, Qin F, Maruyama K, Kakimoto T, Shinozaki K, Yamaguchi-Shinozaki K (2007). Functional analysis of AHK1/ATHK1 and cytokinin receptor histidine kinases in response to abscisic acid, drought, and salt stress in Arabidopsis. *Proc Natl Acad Sci U S A* 104(51):20623–20628.

Trujillo L, Sotolongo M, Menéndez C, Ochogavía M, Coll Y, Hernández I, Borrás-Hidalgo O, Thomma B, Vera P, Hernández L (2008). SodERF3, a novel sugarcane ethylene responsive factor (ERF), enhances salt and drought tolerance when overexpressed in tobacco plants. *Plant Cell Physiol* 49(4): 512–525.

Ueguchi-Tanaka M, Ashikari M, Nakajima M, Itoh H, Katoh E, Kobayashi M, Chow T-y, Hsing YI, Kitano H, Yamaguchi I, Matsuoka M (2005). Gibberellin INSENSITIVE DWARF1 encodes a soluble receptor for gibberellin. *Nature* 437(7059):693–698.

ul Qamar Muhammad Tahir, Amna F, Amna B, Barira Z, Xitong Z, Ling-Ling C (2020). Effectiveness of conventional crop improvement strategies vs. omics. In: Fahad S, Hasanuzzaman M, Alam M, Ullah H, Saeed M, Khan AK, Adnan M (eds) *Environment, climate, plant and vegetation growth*. Springer Publ Ltd, Springer Nature Switzerland AG. Part of Springer Nature, pp 253–284. https://doi.org/10.1007/978-3-030-49732-3.

Umehara M, Hanada A, Yoshida S, Akiyama K, Arite T, Takeda-Kamiya N, Magome H, Kamiya Y, Shirasu K, Yoneyama K, Kyozuka J, Yamaguchi S (2008). Inhibition of shoot branching by new terpenoid plant hormones. *Nature* 455(7210):195–200.

Unsar Naeem-U, Muhammad R, Syed HMB, Asad S, Mirza AQ, Naeem I, Muhammad H ur R, Fahad S, Shafqat S (2020) Insect pests of cotton crop and management under climate change scenarios. In: Fahad S, Hasanuzzaman M, Alam M, Ullah H, Saeed M, Khan AK, Adnan M (eds) *Environment, climate, plant and vegetation growth*. Springer Publ Ltd, Springer Nature Switzerland AG. Part of Springer Nature, pp 367–396. https://doi.org/10.1007/978-3-030-49732-3.

Upreti K, Srinivasa Rao N, Jayaram H (2012). Floral abscission in capsicum under high temperature: role of endogenous hormones and polyamines. *Ind J Plant Physiol* 17:207–214.

Vardhini BV, Rao S, Rao K (2008). Effect of brassinolide on growth, yield, metabolite content and enzyme activities of tomato (Lycopersicon esculentum) mill. In: Ashwani Kumar, Sopory Sudhir K (eds) *Recent advances in plant biotechnology and its applications*. IK International Publishing House Ltd., New Delhi, India, pp 133–139.

Verma V, Ravindran P, Kumar PP (2016). Plant hormone-mediated regulation of stress responses. *BMC Plant Biol* 16:86.

Wahid F, Fahad S, Subhan D, Adnan M,, Zhen Y, Saud S, Manzer HS, Martin B, Tereza H, Rahul D (2020). Sustainable management with mycorrhizae and phosphate solubilizing bacteria for enhanced phosphorus uptake in calcareous soils. *Agri* 10(334).

Wajid N, Ashfaq A, Asad A, Muhammad T, Muhammad A,Muhammad S, Khawar J, Ghulam MS, Syeda RS, Hafiz MH, Muhammad IAR, Muhammad ZH, Muhammad Habib ur R, Veysel T, Fahad S, Suad S, Aziz K, Shahzad A (2017). Radiation efficiency and nitrogen fertilizer impacts on sunflower crop in contrasting environments of Punjab, Pakistan. *Environ Sci Pollut Res* 25:1822–1836.

Wang C, Zhang S, Wang P, Hou J, Qian J, Ao Y, Lu J, Li L (2011). Salicylic acid involved in the regulation of nutrient elements uptake and oxidative stress in Vallisneria natans (Lour.) Hara under Pb stress. *Chemosphere* 84(1):136–142.

Wang H, Huang Z, Chen Q, Zhang Z, Zhang H, Wu Y, Huang D, Huang R (2004). Ectopic overexpression of tomato JERF3 in tobacco activates downstream gene expression and enhances salt tolerance. *Plant Mol Biol* 55(2):183–192.

Wang S, Bai Y, Shen C, Wu Y, Zhang S, Jiang D, Guilfoyle TJ, Chen M, Qi Y (2010). Auxin-related gene families in abiotic stress response in Sorghum bicolor. *Funct Integr Genom* 10(4):533–546.

Wasternack C (2014). Action of jasmonates in plant stress responses and development—applied aspects. *Biotechnol Adv* 32(1):31–39.

Waters MT, Gutjahr C, Bennett T, Nelson DC (2017). Strigolactone signaling and evolution. *Annu Rev Plant Biol* 68.

Wu C, Cui K, Tang S, Li G, Wang S, Fahad S, Nie L, Huang J, Peng S, Ding Y (2020). Intensified pollination and fertilization ameliorate heat injury in rice (Oryza sativa L.) during the flowering stage. *Field Crop Res* 252. http://www.ncbi.nlm.nih.gov/pubmed/107795.

Wu C, Tang S, Li G, Wang S, Fahad S, Ding Y (2019). Roles of phytohormone changes in the grain yield of rice plants exposed to heat: a review. *PeerJ* 7:e7792. https://doi.org/10.7717/peerj.7792.

Xi Z, Wang Z, Fang Y, Hu Z, Hu Y, Deng M, Zhang Z (2013). Effects of 24-epibrassinolide on antioxidation defense and osmoregulation systems of young grapevines (V. vinifera L.) under chilling stress. *Plant Growth Regul* 71(1):57–65.

Xu Z-S, Xia L-Q, Chen M, Cheng X-G, Zhang R-Y, Li L-C, Zhao Y-X, Lu Y, Ni Z-Y, Liu L, Qiu ZG, Ma YZ (2007). Isolation and molecular characterization of the Triticum aestivum L. ethylene-responsive factor 1 (TaERF1) that increases multiple stress tolerance. *Plant Molecul Biol* 65(6):719–732.

Yang Z, Zhang Z, Zhang T, Fahad S, Cui K, Nie L, Peng S, Huang J (2017). The effect of season-long temperature increases on rice cultivars grown in the central and southern regions of China. *Front Plant Sci* 8:1908. https://doi.org/10.3389/fpls.2017.01908.

Yoneyama K, Xie X, Kim HI, Kisugi T, Nomura T, Sekimoto H, Yokota T, Yoneyama K (2012). How do nitrogen and phosphorus deficiencies affect strigolactone production and exudation? *Planta* 235(6):1197–1207.

Yoneyama K, Xie X, Kisugi T, Nomura T, Yoneyama K (2013). Nitrogen and phosphorus fertilization negatively affects strigolactone production and exudation in sorghum. *Planta* 238(5):885–894.

Zafar-ul-Hye M, Muhammad T ahzeeb-ul-Hassan, Muhammad A, Fahad S, Martin B, Tereza D, Rahul D, Subhan D (2020b) Potential role of compost mixed biochar with rhizobacteria in mitigating lead toxicity in spinach. *Sci Rep* 10:12159. https://doi.org/10.1038/s41598-020-69183-9.

Zafar-ul-Hye M, Naeem M, Danish S, Fahad S, Datta R, Abbas M, Rahi AA, Brtnicky M, Holátko J, Tarar ZH, Nasir M (2020a). Alleviation of cadmium adverse effects by improving nutrients uptake in bitter gourd through cadmium tolerant rhizobacteria. *Environments* 7(8). https://doi.org/10.3390/environments7080054.

Zahida Z, Hafiz FB, Zulfiqar AS, Ghulam MS, Fahad S, Muhammad RA, Hafiz MH, Wajid N, Muhammad S (2017). Effect of water management and silicon on germination, growth, phosphorus and arsenic uptake in rice. *Ecotoxicol Environ Saf* 144:11–18.

Zaman Qu, Aslam Z, Yaseen M, Ihsan MZ, Khaliq A, Fahad S, Bashir S, Ramzani PMA, Naeem M (2018). Zinc biofortification in rice: leveraging agriculture to moderate hidden hunger in developing countries. *Arch Agron Soil Sci* 64(2):147–161. https://doi.org/10.1080/03650340.2017.1338343.

Zamin M, Khattak AM, Salim AM, Marcum KB, Shakur M, Shah S, Jan I, Fahad S (2019). Performance of Aeluropus lagopoides (mangrove grass) ecotypes, a potential turfgrass, under high saline conditions. *Environ Sci Pollut Res Int*. https://doi.org/10.1007/s11356-019-04838-3.

Zhai Y, Wang Y, Li Y, Lei T, Yan F, Su L, Li X, Zhao Y, Sun X, Li J, Wang Q (2013). Isolation and molecular characterization of GmERF7, a soybean ethylene-response factor that increases salt stress tolerance in tobacco. *Gene* 513(1):174–183.

Zhang S, Cai Z, Wang X (2009). The primary signaling outputs of brassinosteroids are regulated by abscisic acid signaling. *Proc Natl Acad Sci U S A* 106(11):4543–4548.

Zhang Y, Li Y, Hassan MJ, Li Z, Peng Y (2020). Indole-3-acetic acid improves drought tolerance of white clover via activating auxin, abscisic acid and jasmonic acid related genes and inhibiting senescence genes. *BMC Plant Biol* 20(1):1–12.

Zhang Y, Xu S, Yang S, Chen Y (2015). Salicylic acid alleviates cadmium-induced inhibition of growth and photosynthesis through upregulating antioxidant defense system in two melon cultivars (Cucumis melo L.). *Protoplasma* 252(3):911–924.

Zhao D, Shen L, Fan B, Yu M, Zheng Y, Lv S, Sheng J (2009). Ethylene and cold participate in the regulation of LeCBF1 gene expression in postharvest tomato fruits. *FEBS Lett* 583(20):3329–3334.

Zhao ML, Wang JN, Shan W, Fan JG, Kuang JF, Wu KQ, Li XP, Chen WX, He FY, Chen JY, Lu WJ (2013). Induction of jasmonate signalling regulators MaMYC2s and their physical interactions with MaICE1 in methyl jasmonate-induced chilling tolerance in banana fruit. *Plant Cell Environ* 36(1):30–51.

Zhao W, Yang X, Yu H, Jiang W, Sun N, Liu X, Liu X, Zhang X, Wang Y, Gu X (2015). RNA-Seq-based transcriptome profiling of early nitrogen deficiency response in cucumber seedlings provides new insight into the putative nitrogen regulatory network. *Plant Cell Physiol* 56(3):455–467.

Zhiponova MK, Vanhoutte I, Boudolf V, Betti C, Dhondt S, Coppens F, Mylle E, Maes S, González-García MP, Caño-Delgado AI, Inzé D, Beemster GT, De Veylder L, Russinova E (2013). Brassinosteroid production and signaling differentially control cell division and expansion in the leaf. *New Phytol* 197(2):490–502.

Zia-ur-Rehman M (2020). Environment, climate change and biodiversity. In: Fahad S, Hasanuzzaman M, Alam M, Ullah H, Saeed M, Khan AK, Adnan M (eds) *Environment, climate, plant and vegetation growth*. Springer Publ Ltd, Springer Nature Switzerland AG. Part of Springer Nature, pp 473–502. https://doi.org/10.1007/978-3-030-49732-3.

Zwack PJ, Robinson BR, Risley MG, Rashotte AM (2013). Cytokinin response factor 6 negatively regulates leaf senescence and is induced in response to cytokinin and numerous abiotic stresses. *Plant Cell Physiol* 54(6):971–981.

Zwiewka M, Nodzyński T, Robert S, Vanneste S, Friml J (2015). Osmotic stress modulates the balance between exocytosis and clathrin-mediated endocytosis in Arabidopsis thaliana. *Mol Plant* 8(8):1175–1187.

12

The Role of Phytohormones in Combating Biotic Stress

Fazal Akbar, Atta Ur Rahman, Abdul Rehman, Nisar Ahmad, Mohammad Ali, Akhtar Rasool, Muzafar Shah, Muhammad Israr, Muhammad Suleman, and Muhammad Rizwan

CONTENTS

12.1 Introduction

Plants, being immobile organisms, are frequently exposed to an extensive range of biotic and abiotic stresses. Abiotic stresses include salinity, cold, heat, and drought, while biotic stresses mainly include plant attack by viruses, fungi, bacteria, nematodes, and insects (Adnan et al. 2018; Adnan et al. 2019; Akram et al. 2018a,b; Aziz et al. 2017a,b; Habib et al. 2017; Hafiz et al. 2016; Hafiz et al. 2019; Kamaran et al. 2018; Muhammad et al. 2019; Sajjad et al. 2019; Saud et al. 2013; Saud et al. 2014; Saud et al. 2017; Saud et al. 2016; Shah et al. 2013; Saud et al. 2020; Qamar et al. 2017; Wajid et al. 2017; Yang et al. 2017; Zahida et al. 2017; Depeng et al. 2018; Hussain et al. 2020; Hafiz et al. 2020a,b; Shafi et al. 2020; Wahid et al. 2020; Subhan et al. 2020; Zafar-ul-Hye et al. 2020a,b; Adnan et al. 2020; Ilyas et al. 2020; Saleem et al 2020a,b,c; Rehman 2020; Frahat et al. 2020; Wu et al. 2020; Mubeen et al. 2020; Farhana 2020; Jan et al. 2019; Wu et al. 2019; Ahmad et al. 2019; Baseer et al. 2019; Hafiz et al. 2018;Tariq et al. 2018; Fahad and Bano 2012; Fahad et al. 2017; Fahad et al. 2013; Fahad et al. 2014a,b; Fahad et al. 2016a,b,c,d; Fahad et al. 2015a,b; Fahad et al.2018; Fahad et al. 2019a,b; Hesham and Fahad 2020). Iqra et al. 2020; Akbar et al. 2020; Mahar et al. 2020; Noor et al. 2020; Bayram et al.

2020; Amanullah et al. 2020; Rashid et al. 2020; Arif et al. 2020; Amir et al. 2020; Saman et al. 2020; Muhammad Tahir et al. 2020; Farah et al. 2020; Sadam et al. 2020; Unsar et al. 2020; Fazli et al. 2020; Md. Enamul et al. 2020; Gopakumar et al. 2020; Zia-ur-Rehman 2020; Ayman et al. 2020; Mohammad I. Al-Wabel et al. 2020a,b; Senol 2020; Amjad et al. 2020; Ibrar et al. 2020; Sajid et al. 2020). To cope with these adverse environmental situations, plants have adopted well-built mechanisms that recognise the stress signals and enable optimum growth response. Plant hormones play critical roles in supporting the plants to adjust to adverse environmental conditions (Verma et al. 2016).

Plant hormones (also generally known as phytohormones) are signalling compounds within the plant and occur in very small concentrations. They were initially termed as transported regulators required for plant growth, development, and other physiological processes. Like animal hormones, plant hormones have been identified more recently, they function not only in endocrine but also in paracrine and autocrine signalling to facilitate both local and also systemic reactions to environmental signals. The main hormones produced by plants are abscisic acid (ABA), auxins (Aux), brassinosteroids (BR), cytokinins (CK), ethylene (ET), gibberellins (GA), jasmonates (JA), salicylic acid (SA), and strigolactones. Out of these, JA, SA, ET, and ABA are well-known to play key roles in mediating plant protection responses against pathogens and abiotic stresses (Bari and Jones, 2009; Nakashima and Yamaguchi-Shinozaki, 2013). Generally, ABA is accountable for plant defence against abiotic stresses because environmental conditions, such as salinity, drought, heat and cold stress, and wounding are recognised to stimulate increases in ABA levels (Lata and Prasad, 2011; Zhang et al. 2006). Contrastingly, JA, ET, and SA play a main role in reply to biotic stress conditions, as their quantities rise with pathogen infection (Bari and Jones, 2009). In this chapter, we will discuss various plant hormones and their role in biotic stress.

12.2 Regulation of Biotic Stress Responses by Plant Hormones

Pathogens are generally divided into necrotrophs and biotrophs. Biotrophs feed on living host tissue, while necrotrophs destroy host tissues and feed on their remains. In the case of biotrophs, it is easy to conceive that resistance mediated by R gene and SA signalling can result in resistance. The hypersensitive response (HR) will dispossess such pathogens of a diet source. Conversely, in the case of necrotrophs, it looks that programmed cell death (PCD) in the host will only make life easier for the pathogen (Glazebrook, 2005). SA, JA, and ET are generally known to play substantial roles in controlling plant defence responses against different pests and pathogens (Bari and Jones, 2009). SA is mostly involved in the stimulation of resistance response against biotrophic and hemi-biotrophic pathogens (Loake and Grant, 2007), while JA and ET are accountable for protection against herbivorous insects and necrotrophic pathogens (Gamalero and Glick, 2012; Wasternack and Hause, 2013). Figure 12.1 describes the summary of the main components involved in hormone signalling after biotic stress in plants. The involvement of various plant hormones in both positive and negative regulation of plant resistance to different pathogens is shown in Figure 12.2.

12.3 Salicylic Acid

Plants synthesise SA in response to a challenge by various phytopathogens – this is essential for establishing both local and systemic acquired resistance (SAR). The application of SA induces pathogenesis-related protein accumulation. Mutations that lead to either a low level of SA production or reduced SA perception increase vulnerability to both virulent and avirulent pathogens (Loake and Grant, 2007).

12.3.1 Biosynthesis of SA and Its Conjugates in Plant Defence

Pathogen-derived SA is synthesised from chorismate bisochorismate synthase. Chorismate resulting from the shikimate pathway whose components are intensely transcriptionally up-regulated after pathogen attack (Truman et al. 2006). Recombinant ICS1 binds high affinity to chorismate and is active with

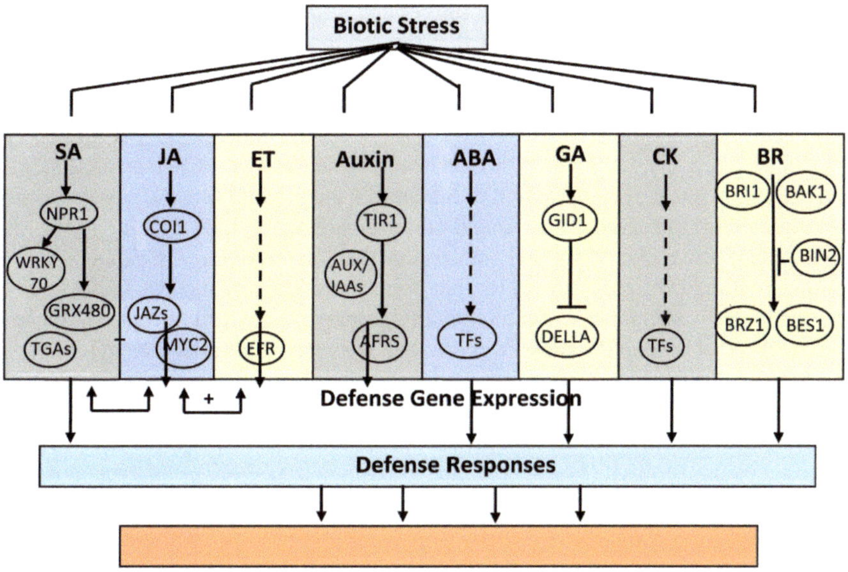

FIGURE 12.1 A summary of the major components involved in different hormones signaling after biotic stress in plants. Biotic stress results in alterations in the levels of different plant hormones. Changes in the levels of plant hormone results in the alterations in the expression of defense-related genes and triggering of defense-responses. Major components involved in different hormone perception and signaling are shown. The minus (–) sign indicates negative interaction while a plus (+) sign indicates positive interaction. It is suggested to see the text for complete details and abbreviations. (Abbreviations: Aux for Auxin, IAA for Indole-3 acetic acid; ABA for abscisic acid; ARFs for Auxin response factors; BR for brassinosteroid; BRI1 for BR insensitive 1; BAK1 for BRI1-associated kinase1; BIN2 for BR insensitive 2; BRZ1 for brassinazole resistant 1; BES1 for BRI1 ems suppressor 1; CK for cytokinin; ERF for ethylene response factor; ET for ethylene; GA for gibberellins; GID1 for gibberellin insensitive dwarf1; JA for jasmonates; SA for salicylic acid.)

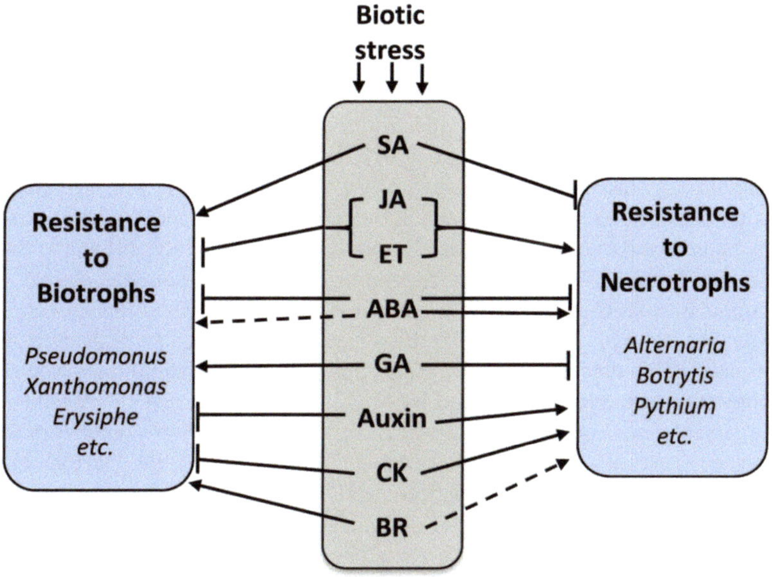

FIGURE 12.2 A simple model showing the participation of different plant hormones in the regulation (positive or negative) of plant resistance to different pathogens. The arrows indicate positive interaction/ activation while the blocked lines indicate negative interaction /repression. It is suggested to see the text for complete details and abbreviations.

a variety of physiological parameters and is introduced and immuno-localised to chloroplast stroma (Strawn et al. 2007). Current studies suggest that the modifications of SA by amino acid conjugation, methylation, and glucosylation form bioactive SA conjugates that may provide biological specificity in plant defence responses (Dean et al. 2005). A volatile ester Methyl salicylate (MeSA), is generally lacking in plants but is intensely induced when pathogen infects the plant (Huang et al. 2003; Seskar et al. 1998). MeSA is produced by SA carboxyl methyltransferase (SAMT) (Chen et al. 2003; Zubieta et al. 2003). It has been shown that overexpression of SA SAMT reduced PR1, SAG, and SA accumulation that augmented vulnerability to bacterial and fungal pathogens (Koo et al. (2007). It has also been observed that JA could induce SA carboxyl methyl transferase that may provide a mechanistic link for JA/SA antagonism in signalling over-reduction of the SA pool. The transformation of MeSA to SA seems to be catalysed by the high SA affinity cytosolic SA methyl esterase salicylic acid-binding protein 2 (SABP2) (Forouhar et al. 2005). The SABP2-silenced plants had reduced local resistance to tobacco mosaic virus (Kumar and Klessig, 2003). MeSA rises in both local infected and systemic tissue after TMV inoculation, which is reduced by the overexpression of a SABP2 mutant with unregulated MeSA esterase activity in the local infected tissue. By contrast, current studies implicate that lipid signals, such as JA-derived molecules, are necessary for SAR (Nandi et al. 2004; Truman et al. 2007), and a recognised lipid transfer protein is required in tested tissue to start a mobile signal (Maldonado et al. 2002). Nevertheless, it is equally possible that both JA and SA derivatives support temporally and spatially distinct modes of action to activate SAR. The SA amino acid conjugates are also evolving as bioactive inducers of defence responses (Nobuta et al. 2007).

12.3.2 SA-Dependent Signalling

Substantial developments have been made in connecting defence responses and SA signalling. Plants attack when certain pathogens or the HR stimulates activation of SA-dependent signalling. This increases SA levels that activates various presumed defence effector genes (including *PR-1* gene). *PAD4* and *EDS1* genes are necessary for triggering SA accumulation in response to specific, but not all, SA-inducing stimuli (Falk et al. 1999; Zhou et al. 1998). The proteins encoded by these genes are identical to triacylglycerol lipases that interact with each other, but there is no solid proof that lipase activity is essential for the function of these proteins (Falk et al. 1999; Feys et al. 2001). *PAD4* appears to trigger defence responses in addition to SA signalling. Expression profiling discovered that there are several genes induced by pathogens that need *PAD4* for complete expression but do not need *EDS5* and *SID2* genes vital for SA biosynthesis (Glazebrook et al. 2003).

The isochorismate synthase is encoded by *sid2*, and SA production is severely decreased in *sid2* plants. This indicates that the majority of SA is made from isochorismate pathway instead of phenylalanine pathway (Coquoz et al. 1998; Wildermuth et al. 2001). But, there is some SA existing in *sid2* mutants that may be produced through the phenylalanine pathway (Wildermuth et al. 2001). *EDS5* is also necessary for the production of SA in response to an attack by pathogens. It encodes a transporter of the MATE family that may be involved in transportation of intermediates for the biosynthesis of SA (Nawrath et al. 2002; Nawrath and Métraux, 1999). The expression-profiling experiment has shown that the effect of *eds5* is very similar to *sid2*; this is consistent with the idea that *EDS5* may be essential for the biosynthesis of SA (Glazebrook et al. 2003). The expression of *EDS5* induced by pathogens needs both *EDS1* and *PAD4* that place *EDS5* downstream from *EDS1* and *PAD4* in SA signalling (Nawrath et al. 2002).

Bacterial salicylate hydroxylase is encoded by *NahG* that abolishes SA by transforming it into catechol. The expression of *NahG* in transgenic plants has confirmed that SA is obligatory for the expression of defence effector genes, such as *PR-1*, and for SAR (Delaney et al. 1994; Gaffney et al. 1993). *NPR1* acts downstream from SA. When SA level is lower, NPR1 is present in the cytoplasm in an oligomeric form. When the SA level increases, the NPR1 oligomers detach into monomers due to the reduction of disulfide linkages that hold the monomers together. These monomers then enter the nucleus, where they interact with TGA-type transcription factors (TFs) (Després et al. 2003; Després et al. 2000; Fan and Dong, 2002; Johnson et al. 2003). TGAs 2, 5, and 6 are required for the activation of PR-1 expression by SA. The TF WRKY70 is also required for complete expression of PR-1 in response to pathogen infection (Li et al. 2004; Zhang et al. 2003).

Certain SA-dependent defence responses are not dependent on NPR1, so there must be other branches of the SA-signalling pathway (Clarke et al. 2000; Uquillas et al. 2004). The TF AtWhyl is stimulated by SA treatment and pathogen infection, and this induced expression is independent of *NPR1*. TF AtWhyl is requisite for SA induction of PR-1. These findings suggest that AtWhyl may contain portions of the NPR1-independent and SA-dependent signalling pathways, and that both of the NPR1-independent and NPR1-dependent branches of the pathway contribute to SA-induced PR-1 expression (Desveaux et al. 2004).

12.3.3 SA/ABA Antagonism

ABA is well-known as a plant hormone controlling abiotic stress responses in plants. Distinct organisms have developed dissimilar biosynthesis and signal transduction pathways for ABA. In spite of this, there are also fascinating mutual focusses where ABA frequently downturns host immune responses and is used by pathogens as an effector molecule. The ABA also appears to play a crucial role in compatible mutualistic interactions, such as mycorrhiza and rhizospheric bacteria with plants and probably the microbiome of animal gut.

Ectopic expression of the effector, AvrPtoB, mimicked ABA induction and suppression of defence genes. Some pathogens use ABA indirectly as an effector molecule by tempering the endogenous ABA biosynthesis and response pathways in the plant host (de Torres-Zabala et al. 2007), which, in reply, indirectly controls mutually exclusive pathogen responses, like the antagonistic association between callose-dependent responses and SA (Oide et al. 2013). Other pathogens, most remarkably numerous plant pathogenic fungi such as *B. cinerea*, have their own ABA biosynthesis (Ding et al. 2016). *M. oryzae* (rice blast fungus), which relies on the same ABA biosynthesis pathway as *B. cinerea*, also depends on ABA for virulence and responds to ABA in order to make appressoria for infection (Spence et al. 2015). *Toxoplasma gondii* (a protozoan/animal parasite) also uses ABA as a virulence factor to defeat animal host immunity via inducing autophagy (Wang et al. 2009).

12.4 Jasmonates

Jasmonate (JA) and its derivatives methyl ester methyl jasmonate are phytohormones. These are cyclopentanone which is a class of oxidized lipids that are derived from α-linolenic acids. JA was isolated from the culture filtrate of a plant fungal pathogen, the *Botryodiplodia* and was recognised as a plant growth inhibitor. Methyl jasmonate was initially isolated from the petal extract of jasmine (Sembdner and Parthier, 1993). Several environmental and developmental elements are accountable for the synthesis of JA in membranes and trigger the expression of stress tolerant genes. JA is universally present in the kingdom *Plantae* and has effects on the expression of genes at the levels of transcription and post-transcription (Schaller, 2011). A vital role of jasmonate (when applied exogenously in small doses) is to enhance plant resistance against pathogens.

12.4.1 Jasmonate Biosynthesis

The biosynthetic pathway of jasmonate was first recognised in 1984 by Vick and Zimmerman who illustrated the biosynthetic pathway in a simple method and showed that linolenic acid can be transformed by lipoxygenase enzyme into cyclopentanone 12-oxo-phytodienoic acid (Vick and Zimmerman, 1984). The octadecanoid pathway of jasmonates completes in cell peroxisomes and chloroplasts (these two cellular organelles are considered the primary sites for this pathway) (Cheong and Do Choi, 2003; Sembdner and Parthier, 1993). The main enzymes invloved in the biosynthesis of jasmonates are lipoxygenase, allene oxide synthase, phospholipases, and allene oxide cyclase.

12.4.2 Signal Perception and Transduction Pathways of Jasmonates

Numerous transcription factors, repressors, up-regulation or down-regulation of some genes, and participants of ubiquitin-proteasome complexes are playing their role in the signal perception and transduction pathway of JA. Various independent studies discovered that the only natural and direct JA-signalling ligand in plants is (+)-7-iso-JA-L-Ile (Fonseca et al. 2009; Glauser et al. 2009).

The Skp1, Cullin, and F-box collectively make a proteinous complex known as SCF ubiquitinproteasome complex. The *coronatine insensitive 1 (COI1)* encodes an F-box protein in jasmonate signal transduction pathway that links with its other complements, Cullin, SKP1, and Rbx proteins, to make an E3 ubiquitin ligase (Xie et al. 1998). Thus *COI1* is considered the jasmonate receptor in the signal transduction pathway. Scientists have also identified a novel group of proteins called jasmonate zim-domain (JAZ), which are supposed to be an important negative regulator of JA signalling and also acts as substrate for other complexes (Chini et al. 2007; Thines et al. 2007). Furthermore, the interaction of COI1 with the other proteins/complexes make the co-receptor complex which is considered to be the true jasmonates receptors (Sheard et al. 2010; Yan et al. 2009). Numerous co-repressors also play their part concurrently in the signal transduction pathway. Co-repressors in transcriptional regulators prevent the initiation of transcription by causing histone de-acetylation and de-methylation that result in chromatin modification and, ultimately, mediate repression of genes (Macrae and Long, 2011; Pauwels et al. 2010).

12.4.3 The Role of Jasmonates in Biotic Stress Tolerance

Plants show two types of responses (local response and systemic response) to tissue injury. Jasmonates play a vital role in these responses via signalling in plants. During tissue damage in a local response, numerous damage-associated plant-derived signals or attacker-derived signals are formed. These signals are either physical or chemical in nature and are identified by specialized receptors present on the cell surface known as pattern recognition receptors. This recognition episode triggers *de novo synthesis* of JA-Ile and JA by an unidentified pathway. JA-Ile activates SCF^{COI} 1/26 proteasome that results in JAZ protein degradation. These proteins are responsible for suppression of TFs that are involved in the expression of defence-related characters.

Systemic defence responses are mediated by two different pathways that involve JA. One pathway is known as a cell-autonomous pathway which is a fast pathway. The second pathway is known as a cell-nonautonomous pathway which is a slow pathway. JA is made upon leaf damage in the cell-nonautonomous pathway that translocates to undamaged leaf where it activates JA responses in the target cells. In the cell-autonomous pathway, a mobile signal (other than JA) is produecd upon wound-induction that activates JA/JA-Ile synthesis and results in responses in distal tissues. Both of these pathways may work synergistically to enhance the responses of temporal and spatial expression (Koo et al. 2007).

12.4.4 JA/SA Antagonism

Using the signalling mutants of jasmonate, the genetic evidence for JA antagonism of SA-signalling pathways is recognised (Glazebrook, 2005). However, new evidence proposes a more complex signalling network inducing both negative and positive regulatory interactions. Concurrent application of JA/SA in small amounts results in synergistic expression of classical JA defence markers and the *PR1* gene. In contrast, higher phytohormone concentrations have antagonistic effects that trigger cell death and the production of apoplastic reactive oxygen (Mur et al. 2006). Likewise, JA, or its derivatives, are essential for complete development of SAR (Truman et al. 2007). It is noteworthy that SA and allene oxide synthase (AOS), respectively, share a chloroplastic site for synthesis and activity. AOS is the first enzymatic phase in the biosynthesis of JA. SA has been described to either stimulate (Laudert and Weiler, 1998) or hinder (Norton et al. 2007) the activity of AOS. The levels of SA in chloroplast may be significantly higher than those measured in whole leaf, and are thus possibly capable to mediate local JA antagonism by inhibiting AOS.

The production of phytotoxin coronatine, a JA-Ile mimic, by the conjugation of coronafacic acid to coronamic acid (Brooks et al. 2004) is providing a convincing model of how a pathogen has used negative interactions to overcome plant defence responses. The COR is thought to trigger or moderate JA signalling to downturn SA defences (Laurie-Berry et al. 2006). Consistent with these observations, mutants impaired in jasmonate signalling showed higher resistance to *P. syringae* (Laurie-Berry et al. 2006).

Remarkably, the bacterial necrotroph *Erwinia caratovora* also seems to have horizontally obtained a pathogenicity island containing genes for the synthesis of CFA (Toth et al. 2006). CFA may be combined with a range of amino acids, providing another potential approach for controlling host defence responses.

12.5 Ethylene

Ethylene (also known as ethene) is the simplest olefin that occurs in the gaseous state under ambient conditions. It controls several features of the plant life cycle that include seed germination, sex determination, root initiation, flower development, fruit ripening, root hair development, senescence, etc. In addition, ethylene also plays an important role in response to abiotic stresses, such as chilling, hypoxia, and wounding and biotic stresses, such as pathogen attack (Khan et al. 2014; Lin et al. 2009; Pierik et al. 2006; Schaller, 2012).

12.5.1 Ethylene Biosynthesis

Due to ethylene's simple structure, numerous compounds, such as methionine, propanal, b-alanine, and linolenic acid, have been suggested as its precursors. Mapson and Lieberman revealed that methionine was the precursor of ethylene in a chemical model system and indicated that ethylene was derived from C4 and C3 of methionine (Yang and Hoffman, 1984). Three main enzymatic reactions convert methionine into ethylene: (i) the enzyme S-AdoMet synthetase first converts methionine into to S-AdoMet; (ii) S-AdoMet is then converted to ACC by ACC synthase (ACS); and (iii) ACC is degraded by ACC oxidase (ACO) to release ethylene. The transformation of ACC to ethylene catalysed by ACO is dependent on oxygen. The formation of ethylene is entirely suppressed under anaerobic conditions. In this reaction, ascorbate is used as a co-substrate and Fe^{2+} is used as a cofactor. In plants, multigene families encode ACS isoforms and these genes are controlled differentially at transcription level via environmental and developmental signals in response to both external and internal stimuli.

12.5.2 Ethylene Signalling

The triple response to ethylene is very simple making it perfect for rapidly recognising ethylene signalling mutants. The initial discovery occurred when a dominant ethylene-insensitive mutant, *ethylene response 1*, was isolated. This encodes histidine kinase-like receptor (a two-component receptor) that fails to bind ethylene in the mutated form (Bleecker et al. 1988; Chang et al. 1993). Consequently, the first tomato ethylene receptor, NR, was identified (Wilkinson et al. 1995). The second element of the pathway was also recognised from a recessive mutant *constitutive triple response 1* that exhibited constitutive triggering of ethylene responses in exogenous ethylene absence. The studies of epistasis have placed *CTR1* downstream of the ethylene receptors. *CTR1* ciphers a Raf-like threonine/serine kinase that can interact with the receptors by physical interaction (Clark et al. 1998; Kieber et al. 1993). *Ethylene insensitive 2* (*EIN2*) is a positive regulator of the signalling pathway and is located downstream of the receptors–CTR1 complex (Alonso et al. 1999). The EIN2 N-terminus has sequence similarity with NRAMP ion transporters. The EIN2 C-terminal domain interacts with two (ETP1 and ETP2) F-box proteins. These are EIN2 targeting proteins that target it for degradation. This complex interaction between ethylene, ETP1/ ETP2, and EIN2 degradation is proposed to activate ethylene responses (Qiao et al. 2009). Inside the nucleus, ethylene responses are augmented by a cascade of TFs that includes EIN3, ethylene response factor 1, and their homologues (Solano et al. 1998). The ethylene responses are negatively regulated by the ethylene receptors through CTR1. In brief, the receptors trigger CTR1, which signals downstream to suppress the ethylene response in ethylene absence. The binding of ethylene leads to receptor inactivation and the inhibitory signal from CTR1 is then turned off. When CTR1 is inactivated, EIN2 triggers the ethylene response via downstream TFs such as EIN3 and other EIN3-like proteins, which consequently control other ethylene-responsive genes in the transcription cascade (Tieman et al. 2001).

12.5.3 The Role of Ethylene in Biotic Stress

The role of ethylene in plants exposed to various biotic stresses has been studied in distinct plants and numerous diseases. Wang et al. (2017) have revealed that ethylene shows a crucial role in the sensitivity to *A. alternata* in *Pyrus pyrifolia*. The biosynthesis of high levels of ethylene stimulates fungus development, while low ethylene progression was linked to plant resistance (Wang et al. 2017). A relationship was found between ethylene production and detoxifying enzyme activities. Particularly, a close connection was found between ethylene and catalase activity. In susceptible cultivars, it was revealed that an increase in ethylene production is linked with high levels of hydrogen peroxidase, and small CAT activity was the suitable conditions for the growth of *A. alternata* and induction of PCD.

Analogously, it has been revealed that ethylene has a crucial role in the development of endophytic fungi as detected in *A. lancea*. The endophytic fungus *Gilmaniella sp.* stimulated ethylene in *A. lancea* and consequently the accumulation of sesquiterpenoids. The ethylene appears to upregulate the biosynthesis of sesquiterpenes and also interacts with other phytohormones, such as JA and SA (Yuan et al. 2016).

Wang et al. (2017) revealed that ethylene has a role in the vulnerability of maize to *A. flavus*. Unusually, both conidiation and colonization of *A. flavus* were decreased in kernels treated with ethylene biosynthesis inhibitor. Kernels of acs2 and acs6, in case of the two ethylene biosynthetic mutants, showed greater conidiation and seed colonization but did not increase the quantity of aflatoxin after infection. These findings suggest that ethylene released by infected seeds supported colonization by *A. flavus* but without any aflatoxin biosynthesis. Numerous research studies show that JA- and ET-signalling frequently function synergistically to stimulate the expression of certain defence-related genes after inoculation by pathogen (Glazebrook, 2005; Thomma et al. 2001). It has been shown that TF and ERF1 act as positive regulators of ET and JA signalling (Lorenzo et al. 2003).

12.6 Abscisic Acid

Abscisic acid (ABA) is a plant hormone mainly involved in regulating abiotic stress responses in plants. Diverse organisms have developed divergent biosynthesis and signal transduction pathways for ABA. ABA has a role in the regulation of several aspects of plant development and growth, such as embryo maturation, seed germination, stomatal aperture, leaf senescence, and adaptation to environmental stresses (Wasilewska et al. 2008). There are also fascinating mutual focusses where ABA frequently downturns host immune responses and is used by pathogens as an effector molecule. ABA also appears to play a crucial role in compatible mutualistic interactions, such as mycorrhiza and rhizospheric bacteria with plants, and probably the microbiome of animal gut.

12.6.1 The Role of Abscisic Acid in Biotic Stress

Numerous current studies have asserted that ABA plays a vital role in plant defence responses (Adie et al. 2007; de Torres-Zabala et al. 2007; Mauch-Mani and Mauch, 2005; Mohr and Cahill, 2007). But the role of ABA in plant defence seems to be more complex, differing among diverse kinds of plant-pathogen interactions. ABA can positively control plant resistance at the initial phases of infection by mediating the closure of stomata against attackers or inducing deposition of callose if the pathogen escapes the primary defence response. But, if triggered in late phases, ABA can inhibit ROS stimulation and JA- or SA-signalling transduction, thus opposing defences regulated by these two pathways (Asselbergh, De Vleesschauwer et al. 2008; Ton et al. 2009).

Generally, it has been suggested that ABA is negatively regulatiing plant defence against numerous necrotrophic and biotrophic pathogens. For instance, the ABA-deficient tomato (*sitiens* mutant) displayed more resistance to *Oidium neolycopersici* (Achuo et al. 2006), *B. cinerea* (Audenaert et al. 2002), *Pst* (Thaler and Bostock, 2004), and *Erwinia chrysanthemi* (Asselbergh, Achuo, et al. 2008) in comparison to wild type plants. Likewise, the ABA-deficient *Arabidopsis* (*aba2-1* mutant) exhibited more resistance to *Fusarium oxysporum* (Anderson et al. 2004) and the *aba1-1* mutant exhibited less vulnerability to *Hyaloperonospora arabidopsidis* (Mohr and Cahill, 2003) in comparison to wild type plants. Similarly, applying ABA exogenously increases vulnerability of numerous plant species to fungi and bacteria. For instance, ABA application improves the vulnerability of soybean to *Phytophthora sojae* (Mohr and Cahill, 2001), rice to *Magnaporthe grisea* (Koga et al. 2004), and *Arabidopsis* to *Pst* (de Torres-Zabala et al. 2007).

Nevertheless, the role of ABA in the positive regulation of a plant defence system has also been documented (Mauch-Mani and Mauch, 2005). ABA triggers the closure of stomata that acts as an obstacle against bacterial pathogens (Melotto et al. 2006). Consequently, ABA-deficient mutants display more vulnerability to *Pst*. Treatment of plants with ABA guards plants against *P. cucumerina* and *A. brassicicola*. This indicates that ABA acts as a positive signal for defence against certain necrotrophs (Ton and Mauch-Mani, 2004). In comparison, mutants deficient in ABA are more sensitive to fungal pathogen infections by *L. maculans* (Kaliff et al. 2007), and *A. brassicicola,* and *P. irregulare* (Adie et al. 2007). These outcomes reveal that ABA is not always a positive regulator of plant defence against all kinds of necrotrophs, and its role depends on specific interactions with each plant pathogen. The precise molecular mechanism of the role of ABA on plant defences against different pathogens still remains uncertain. Subsequently, ABA has a role in both abiotic and biotic stresses, although the cross-talk between these signalling pathways and the molecular mechanisms involved remain unclear. Identifying the main components involved in ABA mediated cross-talk between abiotic and biotic stress signalling needs future comprehensive studies.

12.7 Conclusion

SA, JA, and ET are the three main phytohormones that play important roles in controlling plant defence responses against numerous pathogens and pests. Other phytohormones, such as ABA, auxin, cytokinins, gibberellic acid, brassinosteroids, and peptides may also be involved in plant defence-signalling pathways, however, their role in plant defence needs further investigation.

REFERENCES

Achuo E, Prinsen E, Höfte M (2006). Influence of drought, salt stress and abscisic acid on the resistance of tomato to Botrytis cinerea and Oidium neolycopersici. *Plant Pathol* 55(2):178–186.

Adie BA, Pérez-Pérez J, Pérez-Pérez MM, Godoy M, Sánchez-Serrano J-J, Schmelz EA, Solano R (2007). ABA is an essential signal for plant resistance to pathogens affecting JA biosynthesis and the activation of defenses in Arabidopsis. *Plant Cell* 19(5):1665–1681.

Adnan M, Fahad S, Khan IA, Saeed M, Ihsan MZ, Saud S, Riaz M, Wang D, Wu C (2019). Integration of poultry manure and phosphate solubilizing bacteria improved availability of Ca bound P in calcareous soils. *Biotech* 9(10):368.

Adnan M, Fahad S, Zamin M, Shah S, Mian IA, Danish S, Zafar-Ul-Hye M, Battaglia ML, Naz RMM, Saeed B, Saud S, Ahmad I, Yue Z, Brtnicky M, Holatko J, Datta R (2020). Coupling phosphate-solubilizing bacteria with phosphorus supplements improve maize phosphorus acquisition and growth under lime induced salinity stress. *Plants* 9(7). https://doi.org/10.3390/plants9070900.

Adnan M, Shah Z, Fahad S, Arif M, Alam M, Khan IA, Mian IA, Basir A, Ullah H, Arshad M, Rahman IU, Saud S, Ihsan MZ, Jamal Y, Amanullah, Hammad HM, Nasim W (2018a). Author correction: phosphate-solubilizing bacteria nullify the antagonistic effect of soil calcification on bioavailability of phosphorus in alkaline soils. *Sci Rep* 8(1):4339. https://doi.org/10.1038/s41598-018-22653-7.

Adnan M, Shah Z, Sharif M, Rahman H (2018b). Liming induces carbon dioxide (CO_2) emission in PSB inoculated alkaline soil supplemented with different phosphorus sources. *Environ Sci Pollut Res Int* 25(10):9501–9509.

Ahmad S, Kamran M, Ding R, Meng X, Wang H, Ahmad I, Fahad S, Han Q (2019). Exogenous melatonin confers drought stress by promoting plant growth, photosynthetic capacity and antioxidant defense system of maize seedlings. *PeerJ* 7:e7793. http://doi.org/10.7717/peerj.7793.

Akbar H, Timothy JK, Jagadish T, Golam MM, Apurbo KC, Muhammad F, Rajan B, Fahad S, Hasanuzzaman M (2020). Agricultural land degradation: processes and problems undermining future food security. In: Fahad S, Hasanuzzaman M, Alam M, Ullah H, Saeed M, Khan AK, Adnan M (eds) *Environment, climate, plant and vegetation growth*. Springer Publ Ltd, Springer Nature Switzerland AG. Part of Springer Nature, pp 17–62. https://doi.org/10.1007/978-3-030-49732-3.

Akram R, Turan V, Hammad HM, Ahmad S, Hussain S, Hasnain A, Maqbool MM, Rehmani MIA, Rasool A, Masood N, Mahmood F, Mubeen M, Sultana SR, Fahad S, Amanet K, Saleem M, Abbas Y, Akhtar HM, Waseem F, Murtaza R, Amin A, Zahoor SA, ul Din MS, Nasim W (2018a). Fate of organic and inorganic pollutants in paddy soils. In: Hashmi MZ, Varma A (eds) *Environmental pollution of paddy soils, soil biology*. Springer International Publishing AG, Cham, Switzerland, pp 197–214.

Akram R, Turan V, Wahid A, Ijaz M, Shahid MA, Kaleem S, Hafeez A, Maqbool MM, Chaudhary HJ, Munis MFH, Mubeen M, Sadiq N, Murtaza R, Kazmi DH, Ali S, Khan N, Sultana SR, Fahad S, Amin A, Nasim W (2018b). Paddy land pollutants and their role in climate change. In: Hashmi MZ, Varma A (eds) *Environmental pollution of paddy soils, soil biology*. Springer International Publishing AG, Cham, Switzerland, pp 113–124.

Alonso JM, Hirayama T, Roman G, Nourizadeh S, Ecker JR (1999). EIN2, a bifunctional transducer of ethylene and stress responses in Arabidopsis. *Science* 284(5423):2148–2152.

Al-Wabel Mohammad I, Abdelazeem S, Munir A, Khalid E, Adel RAU (2020b). Extent of climate change in Saudi Arabia and its impacts on agriculture: a case study from Qassim region. In: Fahad S, Hasanuzzaman M, Alam M, Ullah H, Saeed M, Khan AK, Adnan M (eds) *Environment, climate, plant and vegetation growth*. Springer Publ Ltd, Springer Nature Switzerland AG. Part of Springer Nature, pp 635–658. https://doi.org/10.1007/978-3-030-49732-3.

Al-Wabel Mohammad I, Ahmad Munir, Usman Adel RA, Akanji Mutair, Rafique Muhammad Imran (2020a). Advances in pyrolytic technologies with improved carbon capture and storage to combat climate change. In: Fahad S, Hasanuzzaman M, Alam M, Ullah H, Saeed M, Khan AK, Adnan M (eds) *Environment, climate, plant and vegetation growth*. Springer Publ Ltd, Springer Nature Switzerland AG. Part of Springer Nature, pp 535–576. https://doi.org/10.1007/978-3-030-49732-3.

Amanullah Shah K, Imran Hamdan AK, Muhammad A, Abdel RA, Muhammad A, Fahad S, Azizullah S, Brajendra P (2020). Effects of climate change on irrigation water quality. In: Fahad S, Hasanuzzaman M, Alam M, Ullah H,Saeed M, Khan AK, Adnan M (eds) *Environment, climate, plant and vegetation growth*. Springer Publ Ltd, Springer Nature Switzerland AG. Part of Springer Nature, pp 123–132. https://doi.org/10.1007/978-3-030-49732-3.

Amir M, Muhammad A, Allah B, Sevgi Ç, Haroon ZK, Muhammad A, Emre A (2020). Bio fortification under climate change: the fight between quality and quantity. In: Fahad S, Hasanuzzaman M, Alam M, Ullah H, Saeed M, Khan AK, Adnan M (eds) *Environment, climate, plant and vegetation growth*. Springer Publ Ltd, Springer Nature Switzerland AG. Part of Springer Nature, pp 173–228. https://doi.org/10.1007/978-3-030-49732-3.

Amjad I, Muhammad H, Farooq S, Anwar H (2020). Role of plant bioactives in sustainable agriculture. In: Fahad S, Hasanuzzaman M, Alam M, Ullah H, Saeed M, Khan AK, Adnan M (eds) *Environment, climate, plant and vegetation growth*. Springer Publ Ltd, Springer Nature Switzerland AG. Part of Springer Nature, pp 591–606. https://doi.org/10.1007/978-3-030-49732-3.

Anderson JP, Badruzsaufari E, Schenk PM, Manners JM, Desmond OJ, Ehlert C, Maclean DJ, Ebert PR, Kazan K (2004). Antagonistic interaction between abscisic acid and jasmonate-ethylene signaling pathways modulates defense gene expression and disease resistance in Arabidopsis. *Plant Cell* 16(12):3460–3479.

Arif M, Talha J, Muhammad R, Fahad S, Muhammad A, Amanullah Kawsar A, Ishaq AM, Bushra K, Fahd R (2020). Biochar; a remedy for climate change. In: Fahad S, Hasanuzzaman M, Alam M, Ullah H, Saeed M, Khan AK, Adnan M (eds) *Environment, climate, plant and vegetation growth*. Springer Publ Ltd, Springer Nature Switzerland AG. Part of Springer Nature, pp 151–172. https://doi.org/10.1007/978-3-030-49732-3.

Asselbergh B, Achuo AE, Höfte M, Van Gijsegem F (2008). Abscisic acid deficiency leads to rapid activation of tomato defence responses upon infection with Erwinia chrysanthemi. *Mol Plant Pathol* 9(1):11–24.

Asselbergh B, De Vleesschauwer D, Höfte M (2008). Global switches and fine-tuning-ABA modulates plant pathogen defense. *Mol Plant Microbe Interact* 21(6):709–719.

Audenaert K, De Meyer GB, Höfte MM (2002). Abscisic acid determines basal susceptibility of tomato toBotrytis cinerea and suppresses salicylic acid-dependent signaling mechanisms. *Plant Physiol* 128(2):491–501.

Ayman ELS, Hossain Akbar, Barutçular Celaleddin, Iqbal Muhammad Aamir, Sohidul Islam M, Fahad Shah, Sytar Oksana, Çig Fatih, Meena Ram Swaroop, Erman Murat (2020). Consequences of salinity stress on the quality of crops and its mitigation strategies for sustainable crop production: an outlook of arid and semi- arid regions. In: Fahad S, Hasanuzzaman M, Alam M, Ullah H, Saeed M, Khan AK, Adnan M (eds) *Environment, climate, plant and vegetation growth*. Springer Publ Ltd, Springer Nature Switzerland AG. Part of Springer Nature, pp 503–534. https://doi.org/10.1007/978-3-030-49732-3.

Bari R, Jones JD (2009). Role of plant hormones in plant defence responses. *Plant Mol Biol* 69(4):473–488.

Baseer M, Adnan M, Fazal M, Fahad S, Muhammad S, Fazli W, Muhammad A, Jr. Amanullah DW, Saud S, Muhammad N, Muhammad Z, Fazli S, Beena S, Mian AR, Ishaq AM (2019). Substituting urea by organic wastes for improving maize yield in alkaline soil. *J Plant Nutr.*

Bayram AY, Seher Ö, Nazlican A (2020). Climate change forecasting and modeling for the year of 2050. In: Fahad S, Hasanuzzaman M, Alam M, Ullah H,Saeed M, Khan AK, Adnan M (eds) *Environment, climate, plant and vegetation growth*. Springer Publ Ltd, Springer Nature Switzerland AG. Part of Springer Nature, pp 109–122. https://doi.org/10.1007/978-3-030-49732-3.

Bleecker AB, Estelle MA, Somerville C, Kende H (1988). Insensitivity to ethylene conferred by a dominant mutation in Arabidopsis thaliana. *Science* 241(4869):1086–1089.

Brooks DM, Hernández-Guzmán G, Kloek AP, Alarcón-Chaidez F, Sreedharan A, Rangaswamy V, Peñaloza-Vázquez A, Bender CL, Kunkel BN (2004). Identification and characterization of a well-defined series of coronatine biosynthetic mutants of Pseudomonas syringae pv. tomato DC3000. *Mol Plant Microbe Interact* 17(2):162–174.

Chang C, Kwok SF, Bleecker AB, Meyerowitz EM (1993). Arabidopsis ethylene-response gene ETR1: similarity of product to two-component regulators. *Science* 262(5133):539–544.

Chen F, D'Auria JC, Tholl D, Ross JR, Gershenzon J, Noel JP, Pichersky E (2003). An Arabidopsis thaliana gene for methylsalicylate biosynthesis, identified by a biochemical genomics approach, has a role in defense. *Plant J Cell Mol Biol* 36(5):577–588.

Cheong J-J, Do Choi Y (2003). Methyl jasmonate as a vital substance in plants. *Trends Genet* 19(7):409–413.

Chini A, Fonseca S, Fernández G, Adie B, Chico J, Lorenzo O, García-Casado G, López-Vidriero I, Lozano F, Ponce M, Micol JL, Solano R (2007). The JAZ family of repressors is the missing link in jasmonate signalling. *Nature* 448(7154):666–671.

Clark KL, Larsen PB, Wang X, Chang C (1998). Association of the Arabidopsis CTR1 Raf-like kinase with the ETR1 and ERS ethylene receptors. *Proc Natl Acad Sci U S A* 95(9):5401–5406.

Clarke JD, Volko SM, Ledford H, Ausubel FM, Dong X (2000). Roles of salicylic acid, jasmonic acid, and ethylene in cpr-induced resistance in Arabidopsis. *Plant Cell* 12(11):2175–2190.

Coquoz J-L, Buchala A, Métraux J-P (1998). The biosynthesis of salicylic acid in potato plants. *Plant Physiol* 117(3):1095–1101.

Danish S, Zafar-ul-Hye M, Fahad S, Saud S, Brtnicky M, Hammerschmiedt T, Datta R (2020). Drought stress alleviation by ACC deaminase producing Achromobacter xylosoxidans and Enterobacter cloacae, with and without timber waste biochar in maize. *Sustainability* 12(15). https://doi.org/10.3390/su12156286.

de Torres-Zabala M, Truman W, Bennett MH, Lafforgue G, Mansfield JW, Rodriguez Egea P, Bögre L, Grant M (2007). Pseudomonas syringae pv. tomato hijacks the Arabidopsis abscisic acid signalling pathway to cause disease. *EMBO J* 26(5):1434–1443.

Dean JV, Mohammed LA, Fitzpatrick T (2005). The formation, vacuolar localization, and tonoplast transport of salicylic acid glucose conjugates in tobacco cell suspension cultures. *Planta* 221(2):287–296.

Delaney TP, Uknes S, Vernooij B, Friedrich L, Weymann K, Negrotto D, Gaffney T, Gut-Rella M, Kessmann H, Ward E, Ryals J (1994). A central role of salicylic acid in plant disease resistance. *Science* 266(5188):1247–1250.

Depeng W, Fahad S, Saud S, Muhammad K, Aziz K, Mohammad NK, Hafiz MH, Wajid N (2019). Morphological acclimation to agronomic manipulation in leaf dispersion and orientation to promote "Ideotype" breeding: Evidence from 3D visual modeling of "super" rice (*Oryza sativa* L.). *Plant Physiol Biochem* 135:499–510.

Després C, Chubak C, Rochon A, Clark R, Bethune T, Desveaux D, Fobert PR (2003). The Arabidopsis NPR1 disease resistance protein is a novel cofactor that confers redox regulation of DNA binding activity to the basic domain/leucine zipper transcription factor TGA1. *Plant Cell* 15(9):2181–2191.

Després C, DeLong C, Glaze S, Liu E, Fobert PR (2000). The Arabidopsis NPR1/NIM1 protein enhances the DNA binding activity of a subgroup of the TGA family of bZIP transcription factors. *Plant Cell* 12(2):279–290.

Desveaux D, Subramaniam R, Després C, Mess J-N, Lévesque C, Fobert PR, Dangl JL, Brisson N (2004). A "Whirly" transcription factor is required for salicylic acid-dependent disease resistance in Arabidopsis. *Dev Cell* 6(2):229–240.

Ding Z, Zhang Z, Zhong J, Luo D, Zhou J, Yang J, Xiao L, Shu D, Tan H (2016). Comparative transcriptome analysis between an evolved abscisic acid-overproducing mutant Botrytis cinerea TBC-A and its ancestral strain Botrytis cinerea TBC-6. *Sci Rep* 6:37487.

Enamul Md H, Shoeb AZM, Mallik AH, Fahad S, Kamruzzaman MM, Akib J, Nayyer S, Mehedi KMA, Swati AS, Yeamin Md, A, Most SS (2020). Measuring vulnerability to environmental hazards: qualitative to quantitative. In: Fahad S, Hasanuzzaman M, Alam M, Ullah H, Saeed M, Khan AK, Adnan M (eds) *Environment, climate, plant and vegetation growth*. Springer Publ Ltd, Springer Nature Switzerland AG. Part of Springer Nature, pp 421–452. https://doi.org/10.1007/978-3-030-49732-3.

Fahad S, Adnan M, Hassan S, Saud S, Hussain S, Wu C, Wang D, Hakeem KR, Alharby HF, Turan V, Khan MA, Huang J (2019b). Rice responses and tolerance to high temperature. In: Hasanuzzaman M and Fujita M, Nahar K, Biswas JK (eds) *Advances in rice research for abiotic stress tolerance*. Woodhead Publ Ltd, Cambridge, England, pp 201–224.

Fahad S, Bajwa AA, Nazir U, Anjum SA, Farooq A, Zohaib A, Sadia S, Nasim W, Adkins S, Saud S, Ihsan MZ, Alharby H,Wu C,Wang D, Huang J (2017). Crop production under drought and heat stress: plant responses and management options. *Front Plant Sci* 8:1147. https://doi.org/10.3389/fpls.2017.01147.

Fahad S, Bano A (2012). Effect of salicylic acid on physiological and biochemical characterization of maize grown in saline area. *Pak J Bot* 44:1433–1438.

Fahad S, Chen Y, Saud S,Wang K, Xiong D, Chen C,Wu C, Shah F, Nie L, Huang J (2013). Ultraviolet radiation effect on photosynthetic pigments, biochemical attributes, antioxidant enzyme activity and hormonal contents of wheat. *J Food Agric Environ* 11(3&4):1635–1641.

Fahad S, Hussain S, Bano A, Saud S, Hassan S, Shan D, Khan FA, Khan F, Chen Y, Wu C, Tabassum MA, Chun MX, Afzal M, Jan A, Jan MT, Huang J (2015a). Potential role of phytohormones and plant growth-promoting rhizobacteria in abiotic stresses: consequences for changing environment. *Environ Sci Pollut Res Int* 22(7):4907–4921. https://doi.org/10.1007/s11356-014-3754-2.

Fahad S, Hussain S, Matloob A, Khan FA, Khaliq A, Saud S, Hassan S, Shan D, Khan F, Ullah N, Faiq M, Khan MR, Tareen AK, Khan A, Ullah A, Ullah N, Huang J (2015b). Phytohormones and plant responses to salinity stress: a review. *Plant Growth Regul* 75(2):391–404. https://doi.org/10.1007/s10725-014-0013-y.

Fahad S, Hussain S, Saud S, Hassan S, Chauhan BS, Khan F, Ihsan MZ, Ullah A, Wu C, Bajwa AA, Alharby H, Amanullah, Nasim W, Shahzad B, Tanveer M, Huang J (2016a). Responses of rapid viscoanalyzer profile and other rice grain qualities to exogenously applied plant growth regulators under high day and high night temperatures. *PLOS ONE* 11(7):e0159590. https://doi.org/10.1371/journal.pone.0159590.

Fahad S, Hussain S, Saud S, Hassan S, Ihsan Z, Shah AN,Wu C, Yousaf M, Nasim W, Alharby H, Alghabari F, Huang J (2016c). Exogenously applied plant growth regulators enhance the morphophysiological growth and yield of rice under high temperature. *Front Plant Sci* 7:1250. https://doi.org/10.3389/fpls.2016.

Fahad S, Hussain S, Saud S, Hassan S, Tanveer M, Ihsan MZ, Shah AN, Ullah A, Nasrullah, Khan F, Ullah S, Alharby H, Nasim W, Wu C, Huang J (2016d). A combined application of biochar and phosphorus alleviates heat-induced adversities on physiological, agronomical and quality attributes of rice. *Plant Physiol Biochem* 103:191–198.

Fahad S, Hussain S, Saud S, Khan F, Hassan S Jr A, Nasim W, Arif M, Wang F, Huang J (2016b). Exogenously applied plant growth regulators affect heat-stressed rice pollens. *J Agron Crop Sci* 202(2):139–150.

Fahad S, Hussain S, Saud S, Tanveer M, Bajwa AA, Hassan S, Shah AN, Ullah A,Wu C, Khan FA, Shah F, Ullah S, Chen Y, Huang J (2015a). A biochar application protects rice pollen from high-temperature stress. *Plant Physiol Biochem* 96:281–287.

Fahad S, Ihsan MZ, Khaliq A, Daur I, Saud S, Alzamanan S, Nasim W, Abdullah M, Khan IA, Wu C, Wang D, Huang J (2018). Consequences of high temperature under changing climate optima for rice pollen characteristics-concepts and perspectives. *Arch Agron Soil Sci.* https://doi.org/10.1080/03650340.2018.1443213.

Fahad S, Nie L, Chen Y, Wu C, Xiong D, Saud S, Hongyan L, Cui K, Huang J (2015b). Crop plant hormones and environmental stress. *Sustain Agric Rev* 15:371–400.

Fahad S, Rehman A, Shahzad B, Tanveer M, Saud S, Kamran M, Ihtisham M, Khan SU, Turan V, Rahman MHU (2019a). Rice responses and tolerance to metal/metalloid toxicity. In: Hasanuzzaman M and Fujita M, Nahar K, Biswas JK (eds) *Advances in rice research for abiotic stress tolerance.* Woodhead Publ Ltd, Cambridge, England, pp 299–312.

Falk A, Feys BJ, Frost LN, Jones JD, Daniels MJ, Parker JE (1999). EDS1, an essential component of R gene-mediated disease resistance in Arabidopsis has homology to eukaryotic lipases. *Proc Natl Acad Sci U S A* 96(6):3292–3297.

Fan W, Dong X (2002). In vivo interaction between NPR1 and transcription factor TGA2 leads to salicylic acid–mediated gene activation in Arabidopsis. *Plant Cell* 14(6):1377–1389.

Farah R, Muhammad R, Muhammad SA, Tahira Y, Muhammad AA, Maryam A, Shafaqat A, Rashid M, Muhammad R, Qaiser H, Afia Z, Muhammad AA, Muhammad A, Fahad S (2020). Alternative and non-conventional soil and crop management strategies for increasing water use efficiency. In: Fahad S, Hasanuzzaman M, Alam M, Ullah H, Saeed M, Khan AK, Adnan M (eds) *Environment, climate, plant and vegetation growth.* Springer Publ Ltd, Springer Nature Switzerland AG. Part of Springer Nature, pp 323–338. https://doi.org/10.1007/978-3-030-49732-3.

Farhat A, Hafiz MH, Wajid I, Aitazaz AF, Hafiz FB, Zahida Z, Fahad S, Wajid F, Artemi C (2020). A review of soil carbon dynamics resulting from agricultural practices. *J Environ Manag*:268. http://www.ncbi.nlm.nih.gov/pubmed/110319.

Fazli W, Muhmmad S, Amjad A, Fahad S, Muhammad A, Muhammad N, Ishaq AM, Imtiaz AK, Mukhtar A, Muhammad S, Muhammad I, Rafi U, Haroon I, Muhammad A (2020). Plant-microbes interactions and functions in changing climate. In: Fahad S, Hasanuzzaman M, Alam M, Ullah H, Saeed M, Khan AK, Adnan M (eds) *Environment, climate, plant and vegetation growth.* Springer Publ Ltd, Springer Nature Switzerland AG. Part of Springer Nature, pp 397–420. https://doi.org/10.1007/978-3-030-49732-3.

Feys BJ, Moisan LJ, Newman MA, Parker JE (2001). Direct interaction between the Arabidopsis disease resistance signaling proteins, EDS1 and PAD4. *EMBO J* 20(19):5400–5411.

Fonseca S, Chini A, Hamberg M, Adie B, Porzel A, Kramell R, Miersch O, Wasternack C, Solano R (2009). (+)-7-iso-Jasmonoyl-L-isoleucine is the endogenous bioactive jasmonate. *Nat Chem Biol* 5(5):344–350.

Forouhar F, Yang Y, Kumar D, Chen Y, Fridman E, Park SW, Chiang Y, Acton TB, Montelione GT, Pichersky E, Klessig DF, Tong L (2005). Structural and biochemical studies identify tobacco SABP2 as a methyl salicylate esterase and implicate it in plant innate immunity. *Proc Natl Acad Sci U S A* 102(5):1773–1778.

Gaffney T, Friedrich L, Vernooij B, Negrotto D, Nye G, Uknes S, Ward E, Kessmann H, Ryals J (1993). Requirement of salicylic acid for the induction of systemic acquired resistance. *Science* 261(5122):754–756.

Gamalero E, Glick BR (2012). Ethylene and abiotic stress tolerance in plants. In: *Environmental adaptations and stress tolerance of plants in the era of climate change.* Springer, pp 395–412.

Glauser G, Dubugnon L, Mousavi SA, Rudaz S, Wolfender J-L, Farmer EE (2009). Velocity estimates for signal propagation leading to systemic jasmonic acid accumulation in wounded Arabidopsis. *J Biol Chem* 284(50):34506–34513.

Glazebrook J (2005). Contrasting mechanisms of defense against biotrophic and necrotrophic pathogens. *Annu Rev Phytopathol* 43:205–227.

Glazebrook J, Chen W, Estes B, Chang HS, Nawrath C, Métraux JP, Zhu T, Katagiri F (2003). Topology of the network integrating salicylate and jasmonate signal transduction derived from global expression phenotyping. *Plant J Cell Mol Biol* 34(2):217–228.

Gopakumar L, Bernard NO, Donato V (2020). Soil microarthropods and nutrient cycling. In: Fahad S, Hasanuzzaman M, Alam M, Ullah H, Saeed M, Khan AK, Adnan M (eds) *Environment, climate, plant and vegetation growth.* Springer Publ Ltd, Springer Nature Switzerland AG. Part of Springer Nature, pp 453–472. https://doi.org/10.1007/978-3-030-49732-3.

Gul F, Ahmed I, Ashfaq M, Jan D, Fahad S, Li X, Wang D, Fahad M, Fayyaz M, Shah SA (2020). Use of crop growth model to simulate the impact of climate change on yield of various wheat cultivars under different agro-environmental conditions in Khyber Pakhtunkhwa, Pakistan. *Arab J Geosci* 13(3):112. https://doi.org/10.1007/s12517-020-5118-1.

Habib ur R, Ashfaq A, Aftab W, Manzoor H, Fahd R, Wajid I, Md. Aminul I, Vakhtang S, Muhammad A, Asmat U, Abdul W, Syeda RS, Shah S, Shahbaz K, Fahad S, Manzoor H, Saddam H, Wajid N (2019). Application of CSM-CROPGRO-Cotton model for cultivars and optimum planting dates: evaluation in changing semi-arid climate. *Field Crops Res.*

Hafiz MH, Farhat A, Shafqat S, Fahad S, Artemi C, Wajid F, Chaves CB, Wajid N, Muhammad M, Hafiz FB (2018). Offsetting land degradation through nitrogen and water management during maize cultivation under arid conditions. *Land Degrad Dev* 1–10.

Hafiz MH, Muhammad A, Farhat A, Hafiz FB, Saeed AQ, Muhammad M, Fahad S, Muhammad A (2019). Environmental factors affecting the frequency of road traffic accidents: a case study of sub-urban area of Pakistan. *Environ Sci Pollut Res.*

Hafiz MH, Wajid F, Farhat A, Fahad S, Shafqat S, Wajid N, Hafiz FB (2017). Maize plant nitrogen uptake dynamics at limited irrigation water and nitrogen. *Environ Sci Pollut Res* 24(3):2549–2557.

Hammad HM, Abbas F, Ahmad A, Bakhat HF, Farhad W, Wilkerson CJ, Fahad S, Hoogenboom G (2020a). Predicting kernel growth of maize under controlled water and nitrogen applications. *Int J Plant Prod.* https://doi.org/10.1007/s42106-020-00110-8.

Hammad HM, Khaliq A, Abbas F, Farhad W, Fahad S, Aslam M, Shah GM, Nasim W, Mubeen M, Bakhat HF (2020b). Comparative effects of organic and inorganic fertilizers on soil organic carbon and wheat productivity under arid region. *Commun Soil Sci Plant Anal.* https://doi.org/10.1080/00103624.2020.1763385.

Hesham FA, Fahad S (2020). Melatonin application enhances biochar efficiency for drought tolerance in maize varieties: modifications in physio-biochemical machinery. *Agron J* 112(4):1–22.

Huang J, Cardoza YJ, Schmelz EA, Raina R, Engelberth J, Tumlinson JH (2003). Differential volatile emissions and salicylic acid levels from tobacco plants in response to different strains of Pseudomonas syringae. *Planta* 217(5):767–775.

Hussain MA, Fahad S, Rahat S, Muhammad FJ, Muhammad M, Qasid A, Ali A, Husain A, Nooral A, Babatope SA, Changbao S, Liya G, Ibrar A, Zhanmei J, Juncai H (2020). Multifunctional role of brassinosteroid and its analogues in plants. *Plant Growth Regul.* https://doi.org/10.1007/s10725-020-00647-8.

Ibrar K, Aneela R, Khola Z, Urooba N, Sana B, Rabia S, Ishtiaq H, Mujaddad Ur Rehman, Salvatore M (2020). Microbes and environment: global warming reverting the frozen zombies. In: Fahad S, Hasanuzzaman M, Alam M, Ullah H, Saeed M, Khan AK, Adnan M (eds) *Environment, climate, plant and vegetation growth.* Springer Publ Ltd, Springer Nature Switzerland AG. Part of Springer Nature, pp 607–634. https://doi.org/10.1007/978-3-030-49732-3.

Ilyas M, Nisar M, Khan N, Hazrat A, Khan AH, Hayat K, Fahad S, Khan A, Ullah A (2020). Drought tolerance strategies in plants: a mechanistic approach. *J Plant Growth Regul.* https://doi.org/10.1007/s00344-020-10174-5.

Iqra M, Amna B, Shakeel I, Fatima K,Sehrish L, Hamza A, Fahad S (2020). Carbon cycle in Response to Global Warming. In: Fahad S, Hasanuzzaman M, Alam M, Ullah H,Saeed M, Khan AK, Adnan M (eds) *Environment, climate, plant and vegetation growth.* Springer Publ Ltd, Springer Nature Switzerland AG. Part of Springer Nature, pp 1–16. https://doi.org/10.1007/978-3-030-49732-3.

Izhar Shafi M, Adnan M, Fahad S, Wahid F, Khan A, Yue Z, Danish S, Zafar-ul-Hye M, Brtnicky M, Datta R (2020). Application of single superphosphate with humic acid improves the growth, yield and phosphorus uptake of wheat (Triticum aestivum L.) in calcareous soil. *Agronomy* 10(9):1224. https://doi.org/10.3390/agronomy10091224.

Jakir Md H, Allah B (2020). Development and applications of transplastomic plants; a way towards eco-friendly agriculture. In: Fahad S, Hasanuzzaman M, Alam M, Ullah H, Saeed M, Khan AK, Adnan M (eds) *Environment, climate, plant and vegetation growth.* Springer Publ Ltd, Springer Nature Switzerland AG. Part of Springer Nature, pp 285–322. https://doi.org/10.1007/978-3-030-49732-3.

Jan M, Anwar-ul-Haq M, Shah AN, Yousaf M, Iqbal J, Li X, Wang D, Fahad S (2019). Modulation in growth, gas exchange, and antioxidant activities of salt-stressed rice (Oryza sativa L.) genotypes by zinc fertilization. *Arab J Geosci* 12(24):775. https://doi.org/10.1007/s12517-019-4939-2.

Johnson C, Boden E, Arias J (2003). Salicylic acid and NPR1 induce the recruitment of trans-activating TGA factors to a defense gene promoter in Arabidopsis. *Plant Cell* 15(8):1846–1858.

Kaliff M, Staal J, Myrenås M, Dixelius C (2007). ABA is required for Leptosphaeria maculans resistance via ABI1- and ABI4-dependent signaling. *Mol Plant Microbe Interact* 20(4):335–345.

Kamran M, Cui W, Ahmad I, Meng X, Zhang X, Su W, Chen J, Ahmad S, Fahad S, Han Q, Liu T (2018). Effect of paclobutrazol, a potential growth regulator on stalk mechanical strength, lignin accumulation and its relation with lodging resistance of maize. *Plant Growth Regul* 84(2):317–332. https://doi.org/10.1007/ s10725-017-0342-8.

Khan A, Tan DKY, Afridi MZ, Luo H, Tung SA, Ajab M, Fahad S (2017b). Nitrogen fertility and abiotic stresses management in cotton crop: a review. *Environ Sci Pollut Res Int* 24(17):14551–14566. https://doi.org/10.1007/s11356-017-8920-x.

Khan A, Tan DKY, Munsif F, Afridi MZ, Shah F, Wei F, Fahad S, Zhou R (2017a). Nitrogen nutrition in cotton and control strategies for greenhouse gas emissions: a review. *Environ Sci Pollut Res Int* 24(30):23471–23487. https://doi.org/10.1007/s11356-017-0131-y.

Khan NA, Khan MIR, Asgher M, Fatma M, Masood A, Syeed S (2014). Salinity tolerance in plants: revisiting the role of sulfur metabolites. *J Plant Biochem Physiol* 2(120):2.

Kieber JJ, Rothenberg M, Roman G, Feldmann KA, Ecker JR (1993). CTR1, a negative regulator of the ethylene response pathway in Arabidopsis, encodes a member of the raf family of protein kinases. *Cell* 72(3):427–441.

Koga H, Dohi K, Mori M (2004). Abscisic acid and low temperatures suppress the whole plant-specific resistance reaction of rice plants to the infection of Magnaporthe grisea. *Physiol Mol Plant Pathol* 65(1):3–9.

Koo YJ, Kim MA, Kim EH, Song JT, Jung C, Moon J-K, Kim J-H, Seo HS, Song SI, Kim J-K, Lee JS, Cheong JJ, Choi YD (2007). Overexpression of salicylic acid carboxyl methyltransferase reduces salicylic acid-mediated pathogen resistance in Arabidopsis thaliana. *Plant Mol Biol* 64(1–2):1–15.

Kumar D, Klessig DF (2003). High-affinity salicylic acid-binding protein 2 is required for plant innate immunity and has salicylic acid-stimulated lipase activity. *Proc Natl Acad Sci U S A* 100(26):16101–16106.

Lata C, Prasad M (2011). Role of DREBs in regulation of abiotic stress responses in plants. *J Exp Bot* 62(14):4731–4748.

Laudert D, Weiler EW (1998). Allene oxide synthase: a major control point in Arabidopsis thaliana octadecanoid signalling. *Plant J Cell Mol Biol* 15(5):675–684.

Laurie-Berry N, Joardar V, Street IH, Kunkel BN (2006). The Arabidopsis thaliana jasmonate INSENSITIVE 1 gene is required for suppression of salicylic acid-dependent defenses during infection by Pseudomonas syringae. *Mol Plant Microbe Interact* 19(7):789–800.

Li J, Brader G, Palva ET (2004). The WRKY70 transcription factor: a node of convergence for jasmonate-mediated and salicylate-mediated signals in plant defense. *Plant Cell* 16(2):319–331.

Lin Z, Zhong S, Grierson D (2009). Recent advances in ethylene research. *J Exp Bot* 60(12):3311–3336.

Loake G, Grant M (2007). Salicylic acid in plant defence—The players and protagonists. *Curr Opin Plant Biol* 10(5):466–472.

Lorenzo O, Piqueras R, Sánchez-Serrano JJ, Solano R (2003). Ethylene RESPONSE FACTOR1 integrates signals from ethylene and jasmonate pathways in plant defense. *Plant Cell* 15(1):165–178.

Macrae R, Long J (2011). Interactions between TOPLESS and histone-modifying enzymes. 22nd International Conference on Arabidopsis Research. The North American Arabidopsis Steering Committee, Madison, WI.

Mahar A, Amjad A, Altaf HL, Fazli W, Ronghua L, Muhammad A, Fahad S, Muhammad A, Rafiullah Imtiaz AK, Zengqiang Z (2020). Promising technologies for Cd-contaminated soils: drawbacks and possibilities. In: Fahad S, Hasanuzzaman M, Alam M, Ullah H,Saeed M, Khan AK, Adnan M (eds) *Environment, climate, plant and vegetation growth.* Springer Publ Ltd, Springer Nature Switzerland AG. Part of Springer Nature, pp 63–92. https://doi.org/10.1007/978-3-030-49732-3.

Maldonado AM, Doerner P, Dixon RA, Lamb CJ, Cameron RK (2002). A putative lipid transfer protein involved in systemic resistance signalling in Arabidopsis. *Nature* 419(6905):399–403.

Mauch-Mani B, Mauch F (2005). The role of abscisic acid in plant–pathogen interactions. *Curr Opin Plant Biol* 8(4):409–414.

Melotto M, Underwood W, Koczan J, Nomura K, He SY (2006). Plant stomata function in innate immunity against bacterial invasion. *Cell* 126(5):969–980.

Mohr PG, Cahill DM (2001). Relative roles of glyceollin, lignin and the hypersensitive response and the influence of ABA in compatible and incompatible interactions of soybeans with Phytophthora sojae. *Physiol Mol Plant Pathol* 58(1):31–41.

Mohr PG, Cahill DM (2003). Abscisic acid influences the susceptibility of Arabidopsis thaliana to Pseudomonas syringae pv. tomato and Peronospora parasitica. *Funct Plant Biol* 30(4):461–469.

Shah F, Lixiao N, Kehui C, Tariq S, Wei W, Chang C, Liyang Z, Farhan A, Fahad S, Huang J (2014). Rice grain yield and component responses to near 2°C of warming. *Field Crops Res* 157:98–110.

Sheard LB, Tan X, Mao H, Withers J, Ben-Nissan G, Hinds TR, Kobayashi Y, Hsu F-F, Sharon M, Browse J, He SY, Rizo J, Howe GA, Zheng N (2010). Jasmonate perception by inositol-phosphate-potentiated COI1–JAZ co-receptor. *Nature* 468(7322):400–405.

Solano R, Stepanova A, Chao Q, Ecker JR (1998). Nuclear events in ethylene signaling: a transcriptional cascade mediated by ethylene-INSENSITIVE3 and ethylene-RESPONSE-FACTOR1. *Genes Dev* 12(23):3703–3714.

Spence CA, Lakshmanan V, Donofrio N, Bais HP (2015). Crucial roles of abscisic acid biogenesis in virulence of rice blast fungus Magnaporthe oryzae. *Front Plant Sci* 6:1082.

Strawn MA, Marr SK, Inoue K, Inada N, Zubieta C, Wildermuth MC (2007). Arabidopsis isochorismate synthase functional in pathogen-induced salicylate biosynthesis exhibits properties consistent with a role in diverse stress responses. *J Biol Chem* 282(8):5919–5933.

Tariq M, Ahmad S, Fahad S, Abbas G, Hussain S, Fatima Z, Nasim W, Mubeen M, Rehman MHu, Khan MA, Adnan M, Wilkerson CJ, Hoogenboom G (2018). The impact of climate warming and crop management on phenology of sunflower-based cropping systems in Punjab, Pakistan. *Agric Forest Meteorol* 256–257:270–282.

Thaler JS, Bostock RM (2004). Interactions between abscisic-acid-mediated responses and plant resistance to pathogens and insects. *Ecology* 85(1):48–58.

Thines B, Katsir L, Melotto M, Niu Y, Mandaokar A, Liu G, Nomura K, He S, Howe G, Browse J (2007). JAZ repressor proteins are targets of the SCF (COI1) complex during jasmonate signalling. *Nature* 448(7154):661–665.

Thomma BP, Penninckx IA, Broekaert WF, Cammue BP (2001). The complexity of disease signaling in Arabidopsis. *Curr Opin Immunol* 13(1):63–68.

Tieman DM, Ciardi JA, Taylor MG, Klee HJ (2001). Members of the tomato LeEIL (EIN3-like) gene family are functionally redundant and regulate ethylene responses throughout plant development. *Plant J Cell Mol Biol* 26(1):47–58.

Ton J, Flors V, Mauch-Mani B (2009). The multifaceted role of ABA in disease resistance. *Trends Plant Sci* 14(6):310–317.

Ton J, Mauch-Mani B (2004). β-amino-butyric acid-induced resistance against necrotrophic pathogens is based on ABA-dependent priming for callose. *Plant J Cell Mol Biol* 38(1):119–130.

Toth IK, Pritchard L, Birch PR (2006). Comparative genomics reveals what makes an enterobacterial plant pathogen. *Annu Rev Phytopathol* 44:305–336.

Truman W, Bennett MH, Kubigsteltig I, Turnbull C, Grant M (2007). Arabidopsis systemic immunity uses conserved defense signaling pathways and is mediated by jasmonates. *Proc Natl Acad Sci U S A* 104(3):1075–1080.

Truman W, de Zabala MT, Grant M (2006). Type III effectors orchestrate a complex interplay between transcriptional networks to modify basal defence responses during pathogenesis and resistance. *Plant J Cell Mol Biol* 46(1):14–33.

ul Qamar Muhammad Tahir, Amna F, Amna B, Barira Z, Xitong Z, Ling-Ling C (2020). Effectiveness of conventional crop improvement strategies vs. omics. In: Fahad S, Hasanuzzaman M, Alam M, Ullah H, Saeed M, Khan AK, Adnan M (eds) *Environment, climate, plant and vegetation growth.* Springer Publ Ltd, Springer Nature Switzerland AG. Part of Springer Nature, pp 253–284. https://doi.org/10.1007/978-3-030-49732-3.

Unsar Naeem-U, Muhammad R, Syed HMB, Asad S, Mirza AQ, Naeem I, Muhammad H ur R, Fahad S, Shafqat S (2020) Insect pests of cotton crop and management under climate change scenarios. In: Fahad S, Hasanuzzaman M, Alam M, Ullah H, Saeed M, Khan AK, Adnan M (eds) *Environment, climate, plant and vegetation growth.* Springer Publ Ltd, Springer Nature Switzerland AG. Part of Springer Nature, pp 367–396. https://doi.org/10.1007/978-3-030-49732-3.

Uquillas C, Letelier I, Blanco F, Jordana X, Holuigue L (2004). NPR1-independent activation of immediate early salicylic acid-responsive genes in Arabidopsis. *Mol Plant Microbe Interact* 17(1):34–42.

Verma V, Ravindran P, Kumar PP (2016). Plant hormone-mediated regulation of stress responses. *BMC Plant Biol* 16(1):86.

Vick BA, Zimmerman DC (1984). Biosynthesis of jasmonic acid by several plant species. *Plant Physiol* 75(2):458–461.

Wahid F, Fahad S, Danish S, Adnan M, Yue Z, Saud S, Siddiqui MH, Brtnicky M, Hammerschmiedt T, Datta R. Sustainable management with mycorrhizae and phosphate solubilizing bacteria for enhanced phosphorus uptake in calcareous soils. *Agriculture* 10(8). https://doi.org/10.3390/agriculture10080334.

Wajid N, Ashfaq A, Asad A, Muhammad T, Muhammad A,Muhammad S, Khawar J, Ghulam MS, Syeda RS, Hafiz MH, Muhammad IAR, Muhammad ZH, Muhammad Habib ur R, Veysel T, Fahad S, Suad S, Aziz K, Shahzad A (2017). Radiation efficiency and nitrogen fertilizer impacts on sunflower crop in contrasting environments of Punjab, Pakistan. *Environ Sci Pollut Res* 25:1822–1836.

Wang H, Lin J, Chang Y, Jiang C-Z (2017). Comparative transcriptomic analysis reveals that ethylene/H2O2-mediated hypersensitive response and programmed cell death determine the compatible interaction of sand pear and Alternaria alternata. *Front Plant Sci* 8:195.

Wang S, Park Y-S, Yang Y, Borrego EJ, Isakeit T, Gao X, Kolomiets MV (2017). Seed-derived ethylene facilitates colonization but not aflatoxin production by Aspergillus flavus in maize. *Front Plant Sci* 8:415.

Wang Y, Weiss LM, Orlofsky A (2009). Host cell autophagy is induced by Toxoplasma gondii and contributes to parasite growth. *J Biol Chem* 284(3):1694–1701.

Wasilewska A, Vlad F, Sirichandra C, Redko Y, Jammes F, Valon C, Frei dit Frey NF, Leung J (2008). An update on abscisic acid signaling in plants and more.... *Mol Plant* 1(2):198–217.

Wasternack C, Hause B (2013). Jasmonates: biosynthesis, perception, signal transduction and action in plant stress response, growth and development. An update to the 2007 review in Annals of Botany. *Ann Bot* 111(6):1021–1058.

Wildermuth MC, Dewdney J, Wu G, Ausubel FM (2001). Isochorismate synthase is required to synthesize salicylic acid for plant defence. *Nature* 414(6863):562–565.

Wilkinson JQ, Lanahan MB, Yen H-C, Giovannoni JJ, Klee HJ (1995). An ethylene-inducible component of signal transduction encoded by Never-ripe. *Science* 270(5243):1807–1809.

Wu C, Cui K, Tang S, Li G, Wang S, Fahad S, Nie L, Huang J, Peng S, Ding Y (2020). Intensified pollination and fertilization ameliorate heat injury in rice (Oryza sativa L.) during the flowering stage. *Field Crops Res* 252. http://www.ncbi.nlm.nih.gov/pubmed/107795.

Wu C, Tang S, Li G, Wang S, Fahad S, Ding Y (2019). Roles of phytohormone changes in the grain yield of rice plants exposed to heat: a review. *PeerJ* 7:e7792. https://doi.org/10.7717/peerj.7792.

Xie D-X, Feys BF, James S, Nieto-Rostro M, Turner JG (1998). COI1: an Arabidopsis gene required for jasmonate-regulated defense and fertility. *Science* 280(5366):1091–1094.

Yan J, Zhang C, Gu M, Bai Z, Zhang W, Qi T, Cheng Z, Peng W, Luo H, Nan F, Wang Z, Xie D (2009). The Arabidopsis coronatine INSENSITIVE1 protein is a jasmonate receptor. *Plant Cell* 21(8):2220–2236.

Yang SF, Hoffman NE (1984). Ethylene biosynthesis and its regulation in higher plants. *Annu Rev Plant Physiol* 35(1):155–189.

Yang Z, Zhang Z, Zhang T, Fahad S, Cui K, Nie L, Peng S, Huang J (2017). The effect of season-long temperature increases on rice cultivars grown in the central and southern regions of China. *Front Plant Sci* 8:1908. https://doi.org/10.3389/fpls.2017.01908.

Yuan J, Sun K, Deng-Wang M-Y, Dai C-C (2016). The mechanism of ethylene signaling induced by endophytic fungus Gilmaniella sp. AL12 mediating sesquiterpenoids biosynthesis in Atractylodes Lancea. *Front Plant Sci* 7:361.

Zafar-ul-Hye M, Muhammad T ahzeebulHassan, Muhammad A, Fahad S, Martin B, Tereza D, Rahul D, Subhan D (2020b) Potential role of compost mixed biochar with rhizobacteria in mitigating lead toxicity in spinach. *Sci Rep* 10:12159. https://doi.org/10.1038/s41598-020-69183-9.

Zafar-ul-Hye M, Naeem M, Danish S, Fahad S, Datta R, Abbas M, Rahi AA, Brtnicky M, Holátko J, Tarar ZH, Nasir M. Alleviation of cadmium adverse effects by improving nutrients uptake in bitter gourd through cadmium tolerant rhizobacteria. *Environments* 7(8). https://doi.org/10.3390/environments 7080054.

Zahida Z, Hafiz FB, Zulfiqar AS, Ghulam MS, Fahad S, Muhammad RA, Hafiz MH,Wajid N, Muhammad S (2017). Effect of water management and silicon on germination, growth, phosphorus and arsenic uptake in rice. *Ecotoxicol Environ Saf* 144:11–18.

Zamin M, Khattak AM, Salim AM, Marcum KB, Shakur M, Shah S, Jan I, Fahad S (2019). Performance of Aeluropus lagopoides (mangrove grass) ecotypes, a potential turfgrass, under high saline conditions. *Environ Sci Pollut Res Int*. https://doi.org/10.1007/s11356-019-04838-3.

Zhang J, Jia W, Yang J, Ismail AM (2006). Role of ABA in integrating plant responses to drought and salt stresses. *Field Crops Res* 97(1):111–119.

Biodiversity

The term "biodiversity" or "biological diversity" describes the variety of living beings on Earth encompassing microorganisms, plants, animals, and ecosystems, such as coral reefs, forests, and deserts. In fact, it also represents a wealth of biological resources available to us.

Today, instead of exploring and preserving its unmapped biodiversity, mankind has gone away from mother nature, conquering the domain through relentless exploitation. This has resulted in an ecological imbalance and thereby has posed serious threat to biodiversity.

Declining biodiversity is, therefore, a concern for countless reasons. Biodiversity's wealth of genetic information can be conserved by storing biological material from endangered species. Cell lines, gametes, and embryos are preserved so that in case a species becomes extinct or there is a need to increase the population of the species, then cloning and artificial reproductive techniques can be used to revive and reproduce the said species.

The present volume discusses the biodiversity of specific animal groups, such as cnidarians and fish, as well as potential risks to and methods for conserving some of them.

The following important aspects are addressed:

1. The importance of biodiversity in maintaining ecosystem balance for sustainability
2. The emerging role of biodiversity as a source of important materials, pharmaceuticals, food, and so on as such or through genetic manipulation to meet the contemporary and future challenges
3. The impact of environment degradation on biodiversity and conservation in a changing environment
4. Conventional and emerging biotechnological techniques to conserve biodiversity

The book is intended for academics, scientists, and naturalists working in the fields of biology, biochemistry, biophysics, and biomedical sciences.

Translating Animal Science Research

Series Editor

RC Sobti, Department of Biotechnology, Panjab University, Chandigarh, India

Biodiversity: Threats and Conservation
RC Sobti

Environmental Studies and Climate Change
RC Sobti, S.K. Malhotra, Kamal Jaiswal, and Sanjeev Puri

Genomics, Proteomics and Biotechnology
RC Sobti, Manishi Mukesh, and Aastha Sobti

Biomedical Research, Medicine and Disease
RC Sobti and Aastha Sobti

For more information, visit webpage: www.routledge.com/Translating-Animal-Science-Research/book-series/TASR

Biodiversity

Threats and Conservation

Edited by RC Sobti

CRC Press
Taylor & Francis Group
Boca Raton London New York

CRC Press is an imprint of the
Taylor & Francis Group, an **informa** business

First edition published 2023
by CRC Press
6000 Broken Sound Parkway NW, Suite 300, Boca Raton, FL 33487–2742

and by CRC Press
4 Park Square, Milton Park, Abingdon, Oxon, OX14 4RN

CRC Press is an imprint of Taylor & Francis Group, LLC

Library of Congress Cataloging-in-Publication Data
Names: Sobti, R. C., editor.
Title: Biodiversity : threats and conservation / edited by R.C. Sobti.
Description: First edition. | Boca Raton : CRC Press, 2023. | Series: Translating animal science research | Includes bibliographical references and index.
Identifiers: LCCN 2022023844 | ISBN 9781032115443 (hardback) | ISBN 9781032115474 (paperback) | ISBN 9781003220398 (ebook)
Subjects: LCSH: Biodiversity.
Classification: LCC QH541.15.B56 B563 2023 | DDC 333.95—dc23/eng/20220831
LC record available at https://lccn.loc.gov/2022023844

ISBN: 978-1-032-11544-3 (hbk)
ISBN: 978-1-032-11547-4 (pbk)
ISBN: 978-1-003-22039-8 (ebk)

DOI: 10.1201/9781003220398

Typeset in Times
by Apex CoVantage, LLC

Contents

Editor Bio

Ranbir Chander Sobti, the former vice chancellor of Panjab University, Chandigarh, and Babasaheb Bhimrao Ambedkar University (Central University), Lucknow, is a scientist, an able administrator, and dynamic institution builder.

He has published more than 300 papers in the journals of national and international repute, such as Mutation Research, Carcinogenesis, Archives of Toxicology, Cancer Genetics and Cytogenetics, Molecular Cell Biochemistry, and PLOS ONE, to name a few. He has also published more than 50 books by international publishers, such as Springers, Academic Press, CRC, Elsevier, and others.

He is a fellow of the Third World Academy of Sciences, National Academy of Sciences India, Indian National Science Academy, National Academy of Medical Sciences, National Academy of Agricultural Sciences Canadian Academy of Cardiovascular Diseases, and a few others. He was the general president of Indian Science Congress for the 102nd session held at University of Jammu in 2013. Dr. Sobti is the recipient of many prestigious awards, such as the INSA Young Scientist Medal, UGC Career Award, Punjab Rattan Award, JC Bose Oration and Sriram Oration Awards, and Lifetime Achievement Awards of the Punjab Academy of Sciences, Zoological Society of India, and the Environment Academy of India, besides many other medals and awards at the national and international levels.

Contributors

S. M. A. Abidi
Section of Parasitology, Department of Zoology
Aligarh Muslim University
Aligarh—202002, India.

A. Ali
Department of Zoology
Panjab University
Sector 14, Chandigarh, India 160014

D. Antil
Department of Botany
Panjab University
Chandigarh

Anudeep
Dr. Y. S. Parmar University of Horticulture and Forestry
Solan, HP

M. K. Arya
Insect Biodiversity Laboratory, Department of Zoology, DBS Campus
Kumaon University
Uttrakhand, India

M. Bala
Department of Zoology and Environmental Sciences
Punjabi University
Patiala, Punjab, India

M. H. Balkhi
Division of Aquatic Environmental Management, Faculty of Fisheries
Rangil, Ganderbal, Srinagar, Kashmir

T. Banday
SKAUST Srinagar, J&K

F. A. Bhat
Division of Fisheries Resource Management, Faculty of Fisheries
Rangil, Ganderbal, Srinagar, Kashmir

S. S. Bisht
Department of Zoology
Kumaun University
Nainital-263002, Uttarakhand, India

S. Bisht
Insect Biodiversity Laboratory, Department of Zoology, DBS Campus
Kumaon University
Uttrakhand, India

S. Bora
Department of Zoology
Kumaun University
Nainital-263002, Uttarakhand, India

M. Budhwar
Department of Zoology
Panjab University
Chandigarh

S. Burman
Mosquito, Microbiology, and Nanotechnology Research Units
Parasitology Laboratory, Department of Zoology
University of Burdwan
PurbaBardhaman, West Bengal, India

G. Chandra
Mosquito, Microbiology, and Nanotechnology Research Units
Parasitology Laboratory, Department of Zoology
University of Burdwan
PurbaBardhaman, West Bengal, India

H. Chandra
Insect Biodiversity Laboratory, Department of Zoology, DBS Campus
Kumaon University
Uttrakhand, India

A. Chauhan
Department of Zoology
Panjab University
Sector 14, Chandigarh, India 160014

R. S. Chauhan
College of Fisheries
G.B. Pant University of Agriculture & Technology
Pantnagar-263145, U.S. Nagar, Uttarakhand, India

M. Chopra
Department of Zoology
Panjab University
Chandigarh

S. B. Dandin
Department of Zoology
Kumaun University
Nainital, Uttarakhand, India

S. Dewali
Department of Zoology
Kumaun University
Nainital-263002, Uttarakhand, India

M. S. Dhanya
Department of Environment Sciences and Technology
Central University of Punjab
Bathinda, Punjab

P. Dolma
Department of Zoology
Panjab University
Sector 14, Chandigarh, India 160014

M. R. Gora
College of Temperate Sericulture
Mirgund, SKUAST, J&K

A. Gupta
School of Zoology, Gerge S. Wise, Faculty of Life sciences
Tel Aviv University, Israel

A. K. Gupta
Department of Zoology, School of Life Sciences
BabasahebBhimraoAmbedkar University
Lucknow

H. P. Gurushankara
Department of Zoology
Central University of Kerala
TejaswiniHills, Periya—671 320, Kasaragod, India

S. Hussain
Department of Botany
Panjab University
Chandigarh

M. F. Jagnesh
Department of Environment Sciences and Technology
Central University of Punjab
Bathinda, Punjab

N. Jaiswal
Department of Zoology, School of Life Sciences
BabasahebBhimraoAmbedkar University
Lucknow

S. Jasrotia
Department of Zoology
Jammu University
Jammu

B. D. Joshi
Department of Zoology and Environmental Science
GurukulaKangri University
Haridwar

N. Joshi
Department of Environment Science
GurukulaKangri University, KanyaGurukul Campus
Haridwar, India

P. C. Joshi
Department of Zoology and Environmental Science
GurukulaKangri University
Haridwar

A. Kadwalia
Department of Zoology
Panjab University
Sector 14, Chandigarh, India 160014

N. Kapoor
Department of Zoology
Allahabad University
Prayagraj

H. Kaur
Department of Zoology
Panjab University
Chandigarh 160014

K. Kaur
Department of Anthropology
Panjab University
Sector-14, Chandigarh, India

P. Kaur
Department of Zoology
Punjabi University, Patiala

T. Kaur
Department of Zoology
DAV University
Jalandhar

A. Kaushal
Union Christian College
Aluva, Kerala

I. L. Khan
Department of Zoology

K. Krishan
Department of Anthropology
Panjab University, Chandigarh

A. Kumar
Department of Botany
Panjab University
Chandigarh

A. Kumar
Department of Zoology
Kumaun University
Nainital, Uttarakhand, India

P. Kumar
Department of Botany
Panjab University
Chandigarh

R. Kumar
Dr. Y. S. Parmar University of Horticulture and Forestry
Solan, HP

S. K. Malhotra
Department of Zoology
Allahabad University,
Pryagraj

S. Malviya
Department of Zoology
Allahabad University
Prayagraj

E. V. Mathew
Department of Zoology
Union Christian College
Aluva, Kerala

S. Mehraj
College of Temperate Sericulture
Mirgund, SKUAST, J&K

R. Miglani
Department of Zoology
Kumaun University
Nainital, Uttarakhand, India

A. P. Minhas
Biotechnology Branch
University Institute of Engineering and Technology, UIET
Panjab University, Chandigarh

A. Mishra
College of Fisheries
G.B. Pant University of Agriculture & Technology
Pantnagar-263145, U.S. Nagar
Uttarakhand, India

I. Mishra
School of Smart Agricultural Science
University of Engineering and Technology
Roorkee, Uttarakhand, India

R. Mishra
Department of Botany
Polosora Science College
Ganjam, Odisha, India

S. Mishra
Department of Zoology
Babasaheb Bhimrao Ambedkar University
Lucknow, India

Mubashir
SKAUST Srinagar, J&K

P. Nath
School of Smart Agricultural Science
University of Engineering and Technology
Roorkee, Uttarakhand, India
Department of Agriculture Sciences
College of Engineering Roorkee, COER-SM
Roorkee, Uttarakhand, India

N. Nazim
College of Temperate Sericulture
Mirgund, SKUAST, J&K

K. N. Niladevi
Department of Zoology
Union Christian College, Aluva, Kerala

A. K. Panda
Department of Biotechnology
Sant Gahira Guru University
Ambikapur, Chhattisgarh, India

M. Patil
Department of Botany
Panjab University
Chandigarh

N. Prabhu
SKAUST Srinagar, J&K

S. Puri
Department of Biotechnology, UIET
Panjab University, Chandigarh

J. Rai
Forensic Biological Sciences
IFSC, Panjab University
Chandigarh

A. Rehman
Section of Parasitology
Department of Zoology
Aligarh Muslim University
Aligarh—202002, India.

Reshu
Department of Zoology
School of Life Sciences
Babasaheb Bhimrao Ambedkar University
Lucknow

K. A. Sana
Department of Zoology
Union Christian College
Aluva, Kerala

K. M. Sandeep
Department of Zoology
Allahabad University
Prayagraj

Shahana
SKAUST Srinagar, J&K

S. Shanaz
SKAUST Srinagar, J&K

Anika Sharma
Department of Zoology and Environmental Sciences
Punjabi University
Patiala, Punjab, India

Anuradha Sharma
Department of Zoology
Panjab University
Chandigarh

I. Sharma
Department of Zoology
Panjab University
Chandigarh

R. K. Sharma
Department of Zoology
Jammu University
Jammu

M. Sharma
Department of Zoology
Panjab University
Chandigarh

R. Sharma
University School of Pharmaceutical Sciences
Rayat-Bahra University
SAS Nagar (Mohali), Punjab, India

S. Sharma
Department of Botany
Panjab University
Chandigarh

K. V. Singh
National of Genetic Resources
Karnal

A. N. Singh
Department of Botany
Panjab University
Chandigarh

G. Singh
University School of Pharmaceutical Sciences
Rayat-Bahra University
SAS Nagar (Mohali), Punjab, India

P. Singh
Rayat-Bahra University
SAS Nagar (Mohali), Punjab, India

Shivani Singh
Department of Zoology
School of Life Sciences
Babasaheb Bhimra oAmbedkar University
Lucknow

Siril Singh
Department of Botany
Panjab University
Chandigarh

R. C. Sobti
Department of Biotechnology
Panjab University
Chandigarh

R. Srivastava
Department of Zoology
School of Life Sciences
Babasaheb Bhimrao Ambedkar University
Lucknow

T. Kanchan
Department of Forensic Medicine and Toxicology
All India Institute of Medical Sciences
Jodhpur, India.

M. Thakur
Department of Microbiology
DAV University
Jalandhar

D. Tokas
Department of Botany
Panjab University
Chandigarh

R. Tramboo
SKAUST Srinagar, J&K

V. Vasudev
Former Director and Professor of Bioscience
University of Mysore
P. G. Center, Hemagangothri

A. Verma
Department of Zoology
Kumaun University
Nainital-263002, Uttarakhand, India

R. Yadav
Department of Botany
Panjab University
Chandigarh

A. V. Yadav
Department of Zoology
Allahabad University
Prayagraj

1

Biodiversity: Threats and Conservation Strategies

Ranbir Chander Sobti, Manish Thakur, Tejinder Kaur, and Suman Mishra

CONTENTS

DOI: 10.1201/9781003220398-1

1.1 Introduction

Living organisms (such as animals, plants, and microbes), the environment, and the sum of numerous ecological processes associated to them, such as ecosystem diversity, species diversity, and genetic diversity, make up the ecological complex known as biodiversity (Adom et al., 2019). The term "biological diversity" was first coined by Thomas Lovejoy in 1980 (Lovejoy, 1980). It refers to the sum of all the distinct species of animals, plants, fungi, and microorganisms living on Earth, as well as the variety of habitats in which they reside. Scientists believe that there are approximately 8.7 million plant and animal species in existence. But only 1.2 million species have been detected and described, and millions of different organisms still remain mysterious to scientists (National Geographic, 2021). Natural and human-engineered ecosystems, as well as the ecosystem services that nature delivers free of charge to human society, rely on biological variety to function.

1.2 Definition of Biodiversity

Despite the fact that "biodiversity" is a relatively new compound word, the term "biological diversity" has been widely used for a long time to describe the number of species. This definition lacks specificity and context. A lot of people have claimed that species richness is a more accurate indicator of biodiversity than species count (Fiedler and Jain, 1992).

DeLong (1996) provides a more comprehensive definition:

> Biodiversity is an attribute of an area and specifically refers to the variety within and among living organisms, assemblages of living organisms, biotic communities, and biotic processes, whether naturally occurring or modified by humans. Biodiversity can be measured in terms of genetic diversity and the identity and number of different types of species, assemblages of species, biotic communities, and biotic processes, and the amount (e.g., abundance, biomass, cover, rate) and structure of each. It can be observed and measured at any spatial scale ranging from microsites and habitat patches to the entire biosphere.

The United Nations Earth Summit in 1992 defined "biological diversity" as "variability among living organisms from all sources, including, inter alia, terrestrial, marine, and other aquatic ecosystems, and the ecological complexes of which they are a part: this includes diversity within species, between species, and of ecosystems" (Hawksworth, 1996).

The definition given by Gaston and Spicer in their book Biodiversity: An Introduction is "variation of life at all levels of biological organisation" (Gaston and Spicer, 2004).

While ecological processes are typically viewed as being vital to sustaining biodiversity, other studies claim that this justifies their inclusion in the definition of biodiversity (Swingland, 2013).

The Convention on Biological Diversity defines biodiversity as "the variability among living organisms from all sources including, inter alia, terrestrial, marine and other aquatic ecosystems and the ecological complexes of which they are a part; this includes diversity within species, between species, and of ecosystems."

In totality, "biodiversity" is a term that cannot be defined, but its invention has had a positive impact in that it has spurred many research projects on ecology and systematics, as well as attracted many scientists to contribute to the research on it because biodiversity is useful and necessary for ecosystems to function properly.

1.3 Types of Biodiversity

One method to express biodiversity is to divide it into distinct levels of biological organization. Diversity within species (genetic diversity) is one level of biodiversity, followed by species diversity and ecosystem diversity. Interaction and mutual influence exist among the three levels of biodiversity. They also fluctuate over time and location-by-location.

1.3.1 Genetic Diversity

The members of each species have unique gene makeup. In individuals, mutations in genes and chromosomes lead to new genetic variations, and in sexually reproducing organisms, recombination is responsible for spreading them throughout the population. A species' progeny has new combinations of genes as a result of interbreeding between individuals of the species. Genetic diversity refers to the heritable variety within and between populations of organisms.

Genetic variety is necessary for the survival of healthy populations in natural communities; nevertheless, environmental changes and interbreeding can cause it to change or be affected. When a community's environment changes over time, certain individuals will have qualities that are more suited to the new environment, making them more likely to survive and generate offspring who are also suited to the new environment (NBII, 2021). Genetic diversity information can be used to better characterize and construct core collections so that the available resources can be used more effectively (Ramanatha and Hodgkin, 2002). Population and community dynamics are impacted by human-mediated environmental changes, resulting in population genetic diversity loss, species extinctions, changes in ecosystem functioning, and loss of ecosystem services (Kort et al., 2021).

1.3.2 Species Diversity

In biology, species are described as "a group or population of similar organisms that reproduce via interbreeding within the group, and do not generally reproduce with members of any other species." Members of a certain species share features that set them apart from other species, and these qualities remain constant no matter where they live. In this way, species are used to measure biological diversity, and species diversity refers to the number and variety of species found in a certain habitat or geographic location. It can refer to species richness, taxonomic diversity or phylogenetic diversity, or even species evenness. Species richness is just a basic count of the number of species in a given area or habitat. Species evenness is a measure of how evenly distributed the species are in terms of abundance (Tuomisto, 2010). Around 8 to 8.7 million species are estimated by scientists (Borenstein, 2019). Only about 14% of these had been described by 2011, though (Camilo et al., 2011).

1.3.3 Ecosystem Diversity

Ecological diversity refers to the complex web of species that make up a local ecosystem and the dynamic interaction between those species and their surroundings. An ecosystem is made up of both living (organisms, such as microbes, plants, animals, and fungi) and non-living (climate and chemicals) components that are linked by the flow of energy, nutrients, and matter. Each ecosystem contains a variety of habitats that are home to a wide range of species. Although different species serve different functional roles in ecosystems, they interact not only with one another but also with their non-living surroundings. Each ecosystem contains a wide range in species and genetic diversity on the biggest scale of biodiversity, which is ecological diversity (Elizabeth, 2012). When it comes to species composition, ecosystems might experience a number of changes before they become stable. Among the most biologically and ecologically varied ecosystems are the rainforests and coral reef systems. Any evolutionary interaction, no matter how minor, can have a significant impact on the diversity of ecosystems around the planet.

1.4 Keystone Species

Robert T. Paine, a zoologist, coined the term "keystone species" back in 1969 (Mills et al., 1993). He described a keystone species as a species that has a disproportionately high impact on its environment compared to its abundance (Paine, 1995). Some species in biological communities may have a significant impact on the ability of many other species to survive and these species are known as keystone species.

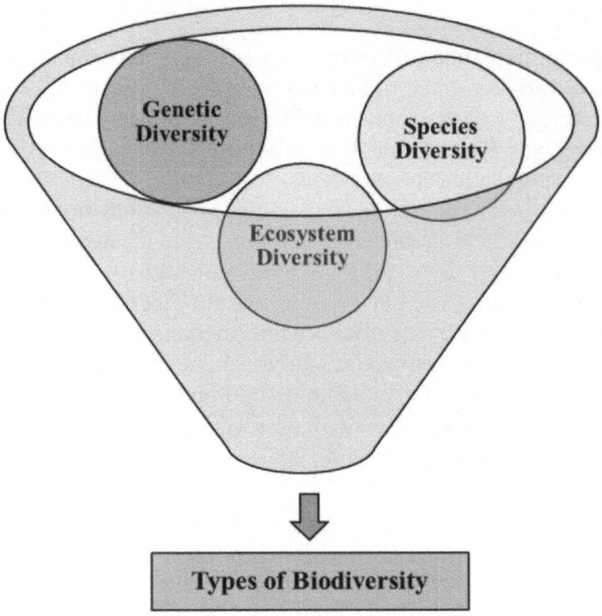

FIGURE 1.1 Types of biodiversity.

1.4.1 Significance of Keystone Species

- The disappearance of a keystone species in a population can lead to the extinction of many other species, such as pteropid bats, or flying foxes.
- In order to protect a monkey species, it may be required to protect keystone species such as fig trees and dung beetles that it depends on.
- If the keystone species of a community can be identified, they can be carefully protected if the area is being impacted by human activity.

1.5 Measurement of Biodiversity

There are numerous measures of biodiversity, just as there are numerous ways to define biodiversity. Most measures count the number of traits, individuals, or species in a given area while taking their degree of dissimilarity into account. Some assess biodiversity on a genetic level, while others assess biodiversity within a single habitat or across ecosystems. Measuring biodiversity at the ecosystem level rather than at the species level is thought to be a better way of looking at the overall shape of the system.

Whittaker (1972) introduced three concepts for measuring biodiversity over spatial scales in a seminal work on the measurement of diversity: alpha (diversity values for single sites), beta (changes in diversity between sites at local scales), and gamma diversity (changes in diversity between sites at geographical scales).

1.5.1 Alpha Diversity

Diversification within a community or ecosystem is known as alpha diversity and is measured by the number of species present (Sahney and Benton, 2008).

1.5.2 Beta Diversity

It is used to compare the number of taxa that are unique to each habitat, as well as the variations in species diversity between ecosystems.

1.5.3 Gamma Diversity

This is a measure of total biodiversity of several ecosystems within an entire region. According to Whittaker (1972), gamma diversity refers to the richness in species that may be found in a range of environments (e.g., a landscape, an island).

1.6 Biodiversity Hotspots

According to British environmental specialist Norman Myers, this concept was first proposed in 1998. Just 1.4% of the land surface of the Earth is covered by a total of 25 hotspots, where 44% of all vascular plant species and 35% of all four vertebrate species can be present at the same time (Myers et al., 2000). Afterward, ten more hotspots were added to the network (Pimm et al., 2014). The Himalayas, the Western Ghats, the Indo-Burma region, and Andaman and Nicobar Islands are among India's biodiversity hotspots (ENVIS, 2021).

1.6.1 The Himalayas

Eastern and Western Himalayan ranges are home to a variety of ecosystems that are home to a wide variety of plant life and wildlife. These mountains are home to large mammals, such as the tiger and the elephant, as well as other species, such as the snow leopard, the Himalayan monal, and the western tragopan. Other species include the blue sheep, the black bear, and the cheer pheasant. Thirty-five endemic reptile species may be found in the Eastern Himalayan region, which is particularly diverse and endemism-rich. Eastern Himalaya is home to at least 68 species of Indian amphibians, out of a total of 341 species recorded so far. A total of 20 amphibian species inhabit the area (Conservation International, 2012).

1.6.2 The Western Ghats

This location is in the world's top ten biodiversity hotspots. As a result, there are more than 7,402 types of flowering plants and 1,814 types of non-flowering plants. There are also 139 mammal species, 508 bird species, 179 amphibian species, and 290 freshwater fish species. The Western Ghats are presumably home to a large number of species that have yet to be discovered. It is estimated that the Western Ghats are home to 325 species that are internationally threatened (Nayar et al., 2014).

According to the Western Ghats Ecology Expert Panel (WGEEP), the entire Western Ghats is an ecologically sensitive area (Gadgil, 2012).

1.6.3 The Indo-Burma Region

Topography in this hotspot is marked by 1,277 bird species (74 are endemic), 430 mammal species (71 are endemic), 519 non-marine reptile species (189 are endemic), and 323 amphibian species (139 are endemic). It is likely that Indo-Burma has the world's highest diversity of freshwater turtles. With 1,262 confirmed freshwater fish species, the hotspot has roughly 10% of the world's total freshwater fish species, including 566 endemic species (Tordoff et al., 2012).

1.6.4 Andaman and Nicobar Islands

Some of the world's most diverse ecosystems may be found on the Andaman and Nicobar Islands in the Indian Ocean. Over 110 types of indigenous orchids can be found on the Andaman and Nicobar Islands alone. Like other tropical regions, the Andaman and Nicobar Islands feature a diverse insect fauna; however, many are still undiscovered. According to India's zoological survey, between 1,500 and 2,000 bug species have been discovered on these islands (Maruthamuthu et al., 2008). Also, 8,425 species of fauna have been identified and 846 are endemic, and marine habitats account for more than 60% of biodiversity (Shivaperuman, 2014).

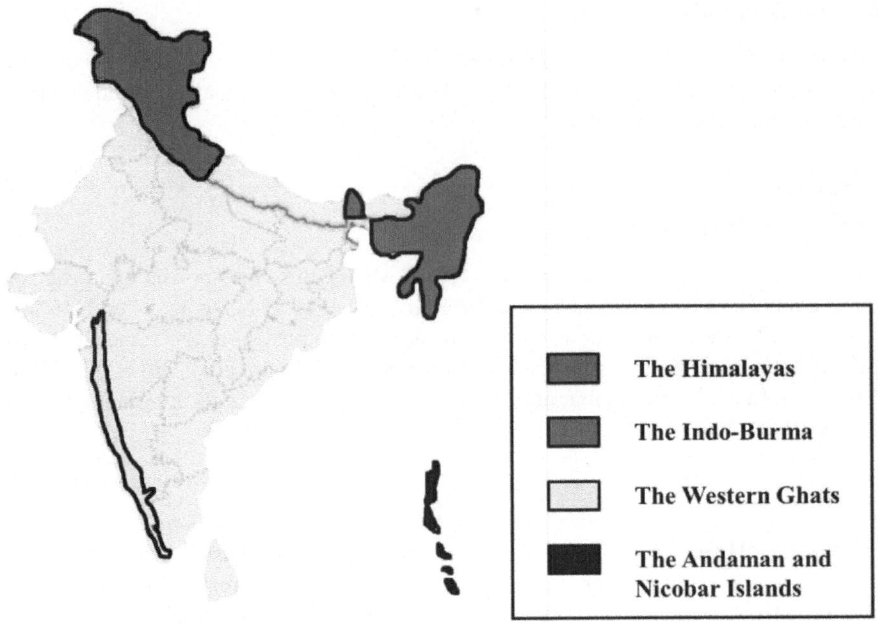

FIGURE 1.2 A map showing the biodiversity hotspots in India.

1.7 Threats to Indian Biodiversity

In 2012, the IUCN (International Union for Conservation of Nature) released a Red List of 63,837 species, of which 19,817 were found to be threatened (Ayre, 2012). A total of 3,947 of these species are classified as critically endangered and 5,766 as endangered, while more than 10,000 species are classified as vulnerable (Giangaspero et al., 2021). All this is the result of the following anthropogenic activities.

1.7.1 Urbanization

Most countries are urbanizing, with more than half of the world's population now living in cities. Disaster risks are on the rise as cities become more densely populated. People still build on unsuitable land and urbanize earthquake-prone areas. Large populations concentrated in a small area (such as through urbanization) are more vulnerable to high winds, heavy rains, and heatwaves (Pagett, 2021). Land-use changes due to urbanization are one of the main reasons for global insect collapse (Fenoglio, 2021). Urbanization leads to habitat loss and habitat fragmentation for many species. As a result, there are significant losses in biodiversity and animal ecosystems (Salomão et al., 2020).

1.7.2 Farming Activities

Ecosystems are being threatened by agricultural intensification. These semi-natural, biodiversity-rich grasslands are either intensified or replaced by more remunerative crops in the last few decades (Brambilla et al., 2021). To make way for the green revolution, thousands of native crop types were replaced by a handful of crossbred varieties. This resulted in the destruction of our genetic resource of crop plants, especially wheat and rice. Because of the disappearance of the plants, the associated microorganisms and flora were also extinct. The indiscriminate use of fertilizers and other chemicals has also led to a decrease in the diversity of microorganisms. Populations of livestock have already been homogenized, and their genetic diversity has been eradicated.

1.7.3 Mining Activities

In addition to the contamination of soil and water sources, mineral exploitation threatens wildlife because of vegetation suppression and changes in landscape configuration. To meet their ecological needs, large and medium-sized forest-dependent mammals require large areas of land and are frequently impacted by these landscape modifications. As a result, forest suppression around mining zones is the most common threat assessed and is reported as the primary impact on these mammals (Martins-Oliveira, 2021).

1.7.4 Air Pollution

Animals' respiratory systems are affected by air pollution, which has a negative impact on their well-being, including their ability to lay eggs and their behavior (Sanderfoot and Holloway, 2017). According to research, air pollution has a negative impact on animals' reproductive abilities and their breeding success (Mohallem et al., 2005).

1.7.5 Water Pollution

To protect food production and meet global food demand, pesticides are widely used. However, pesticides are also ubiquitous environmental pollutants, causing adverse effects on the water quality, biodiversity, and human health (Tang et al., 2021). Heavy metals, such as arsenic, cadmium, and mercury, are introduced into water bodies as a result of anthropogenic activity, such as the production of cement, automobiles, and mining (Singh and Verma, 2018).

1.7.6 Soil Pollution

Agriculture's excessive use of fertilizers, pesticides, and antibiotics is also tremendously harmful to biodiversity. Agricultural contaminants, such as nitrogen from fertilizers, change the soil's pH and nutrient levels. The increased presence of nutrients in the soil causes grass species to grow rapidly, stifling the growth of wildflowers, which are necessary for bees and other pollinating insects (Klein et al., 2007).

1.7.7 Natural Calamities

Biodiversity is severely affected by natural disasters, such as volcano eruptions or wildfires, and other natural calamities, such as storms, droughts, diseases, and tsunamis (Rondeau et al., 2020). Consequently, animals and plants suffer, and biodiversity is affected.

1.7.8 Invasive Species

The introduction of invasive species poses the greatest threat to averting a global biodiversity crisis. Invasive species are those that are not native to the ecosystem they live in and are mostly brought by humans into the new ecosystem and begin to affect its balance. Because they affect the ecosystem in a way that no other species can, these species are harmful and negatively impact the ecology (Linders et al., 2019).

1.7.9 Climate Change

Climate change and biodiversity are closely linked. Rapid climate change, resulting in biodiversity loss, puts human interests and security at risk, as it makes it impossible to access clean water, air, medicines, and other natural resources we rely on due to the reduction or disappearance of the flora and fauna from which they are obtained (Abrahms et al., 2017).

1.7.10 Overexploitation

Ongoing overexploitation occurs when native populations are overburdened with species from their natural habitat. About one-third of the planet's vertebrates are in danger of extinction owing to overexploitation. Overhunting of passenger pigeons resulted in their extinction (Hammerschlag and Gallagher, 2017) The overexploitation of medicinal plants resulted in their disappearance from their native environments (Bodeker et al., 2014). Other examples include overhunting and extreme fishing.

1.8 Conservation Methods

The most sustainable value for present and future generations comes from the control of human use of biodiversity. Therefore, biodiversity conservation encompasses the protection, maintenance, sustainable use, restoration, and enhancement of biodiversity (Kasso and Balakrishnan, 2013). The conservation procedures can be classified into two categories—*in situ* and *ex situ*. To maintain the genetic integrity of a specimen, *ex situ* conservation involves storing genetic material outside of the "normal" environment, whereas *in situ* conservation allows biological resources to change and evolve through time in their natural environment through natural or human-driven methods.

1.8.1 *In Situ* Conservation

In situ conservation refers to the conservation of different species in their natural habitat or natural ecosystem. Natural surroundings or ecosystems are saved and maintained as a result of this process, which benefits all species (even the ones which are undiscovered). The causes that endanger the existence of the species are eradicated by means of appropriate mechanisms. There are a number of distinct types of protected areas, such as national parks, wildlife sanctuaries, and biosphere reserves that fall under the category of *in situ* conservation

1.8.1.1 National Parks

Natural items and wildlife are protected in a national park, which is a place dedicated to preserving the scenery (or ecosystem). For example, grazing of domestic animals is not allowed within the boundaries of a national park because private rights do not exist there. Public access is permitted, however, for the purposes of observation and research. There are 104 existing national parks in India to date (Wildlife Institute of India, 2021).

1.8.1.2 Wildlife Sanctuaries

A wildlife sanctuary focuses on the conservation of a particular species only. The killing of birds or mammals is also prohibited in the sanctuary. As long as forestry and other uses do not adversely affect wildlife, private ownership may be permitted in a sanctuary. Chapter IV of the Wildlife Protection Act, 1972, describes the activities that can be conducted in the sanctuary area. To date, India has 566 existing wildlife sanctuaries (Wildlife Institute of India, 2021).

1.8.1.3 Biosphere Reserves

UNESCO's Man and the Biosphere (MAB) Programme recognizes biosphere reserves, which are designated by governments to encourage sustainable development based on local community initiatives and sound scientific research. UNESCO launched the biosphere reserve program in 1971. Creating biosphere reserves has the goal of preserving *in situ* all forms of life, along with their support systems, in the whole, so that they may be used as reference systems for monitoring and evaluating changes in natural ecosystems in the near future. As of 1979, there were 631 biosphere reserves in 119 nations, making it the world's largest network of biosphere reserves (Wildlife Institute of India, 2021).

1.8.2 *Ex Situ* Conservation

Ex situ conservation refers to the conservation of organisms outside their natural environment. When a plant or animal's native habitat is not suitable for its protection and care, it is placed in a man-made environment that mimics the animal's natural habitat and cared for by humans. Zoos, captive breeding, aquariums, botanical gardens, and gene banks are some of the strategies used to conserve biological variety outside of its native home.

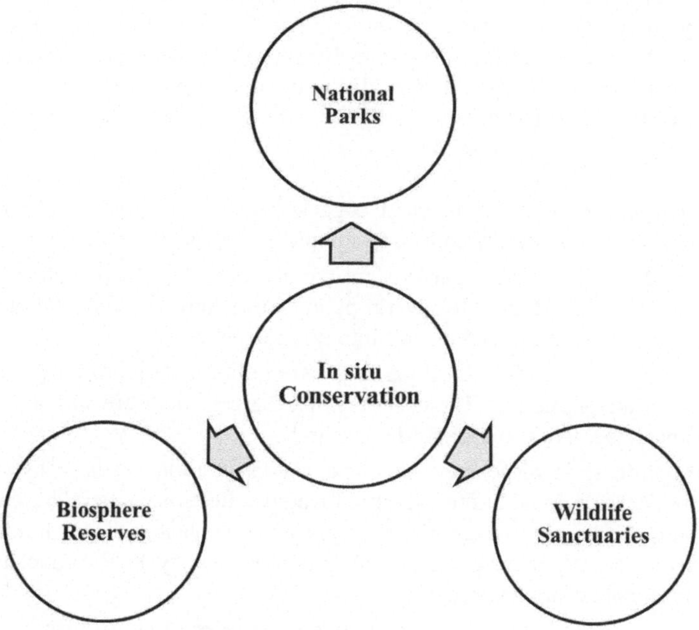

FIGURE 1.3 *In situ* conservation methods to protect biodiversity loss.

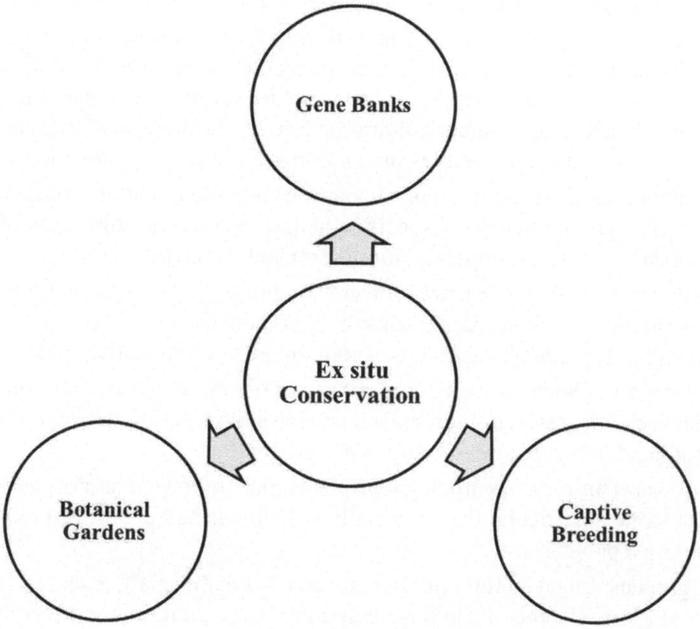

FIGURE 1.4 *Ex situ* conservation methods to protect biodiversity loss.

1.8.2.1 Botanic Gardens

The IUCN Botanic Gardens Conservation Secretariat (BGCS) was established in 1987. The year 1989 saw the creation of the BGCS, and the following year, the BGCS became independent from the IUCN and was renamed Botanic Gardens Conservation International (BGCI). An incredibly significant rationale and framework for having botanic gardens worldwide was created with the release of the BGCS (1989). Many conservation programs were developed, and new botanic gardens were created or older ones were renovated, all over the world as a result (Wyse and Sutherland, 2000). By the Convention on Biological Diversity, in 2002, the Global Strategy for Plant Conservation (GSPC) was adopted, which was then amended in 2010 (CBD, 2010). More than 105,634 plant species are managed in botanic gardens, which represents 30% of all plant species diversity, and over 41% of all vulnerable species are conserved (Mounce et al., 2017). The botanic gardens can be of the following different types (Wyse and Sutherland, 2000):

1. Classic multipurpose gardens: As the name suggests, such areas are meant for a wide range of activities like education, research, collecting samples for herbaria, and laboratory experiments.
2. Ornamental gardens: Ornamental gardens are the places with attractive plant collections that are well-documented. In addition to several public parks, other ornamental gardens privately owned by individuals or organizations fall into this category.
3. Historical gardens: Some of the oldest gardens were built for religious purposes, while others were built to educate physicians. The upshot is that these gardens are still actively involved in medicinal plant conservation and scientific research.
4. Conservation gardens: In response to local plant conservation needs, conservation gardens have lately been created. In addition to their cultivated collections, several have wild vegetation on their grounds or adjacent to them. For example, a native plant garden is limited to the plants indigenous to the area where the garden is located or the country. Public education is a primary function of most conservation gardens.
5. University gardens: Many universities have botanic gardens that are used for teaching and research purposes. A large number of them are accessible to the general public.
6. Combined botanical and zoological gardens: This includes developing plant collections that offer habitats for the displayed fauna and educating the general population on how these habitats work.
7. Agro-botanical and germplasm gardens: Plants having economic value or potential are collected in these gardens as an *ex situ* collection for the sake of conservation or study, plant breeding, and agriculture. They also serve as a source of information for scientists and researchers. Most experimental stations are affiliated with agricultural or forestry institutes and include accompanying laboratories, plant breeding, and seed-testing facilities; however, most are not open to the public.
8. Alpine or mountain gardens: Some tropical countries and the mountains of Europe have many alpine or mountain gardens. They are specifically designed for the cultivation of mountain and alpine flora, or subtropical or temperate flora in tropical countries.
9. Natural or wild gardens: Plants that are protected and managed in their natural or semi-natural state are called natural or wild gardens. Almost all of them have been created for the purpose of conservation and public education, and they contain areas where native species are cultivated.
10. Horticultural gardens: Society-owned and managed botanical gardens are often open to the public for viewing. As a result of their existence, horticulturists are trained, plant varieties are registered, and garden plant varieties are conserved.
11. Thematic gardens: This type of garden grows plants that are related or morphologically similar in order to illustrate a particular theme, usually to help educate people or to save the environment or to put on a public show.
12. Community gardens: Local communities create and maintain small, resource-constrained gardens for a variety of reasons, including recreation, education, and conservation. They also produce medicinal and other economic plants.

1.8.2.2 Captive Breeding in Zoos and Aquariums

Anthropogenic and natural factors pose a serious hazard to people, populations, and species. To prevent the extinction of species, subspecies, or populations, captive breeding is a critical component of the overall conservation action plan. In recent years, zoos and aquariums (hereinafter, "zoos") are becoming more widely acknowledged as crucial partners in the conservation of rare and endangered species (Che-Castaldo, 2018). The rationale is that animals kept in these facilities are ambassadors for their species and, through their reproduction, can serve as an insurance measure against the species' extinction in the wild (McGregor Reid and Zippel, 2008). The World Association of Zoos and Aquaria (WAZA) has officially pledged to align its efforts with the goal of "enhancing the state of biodiversity by protecting habitats, species, and genetic diversity" (Barongi et al., 2015). Numerous endangered species have been saved thanks to such captive breeding.

1.8.2.3 Gene Banks

Biodiversity conservation also relies on the use of genome resource banking. Depending on the type of materials conserved, different types of gene banks have been formed for the storage of biodiversity so that it may be used in the future, either directly or as material in breeding programs, in anticipation of any changes in environmental conditions or social demands (Diez et al., 2018).

1. **Seed banks**: Because seeds can be collected and stored in a viable state for long periods of time, they are preferred for gene bank conservation. Temperature, seed moisture content, and relative humidity are the most important factors affecting seed longevity in seed banking techniques, which store seeds at low temperatures (Kasso and Balakrishnan, 2013). Community seed banks (CSBs) are collective-action entities that are locally governed and administered, with the primary goal of preserving seeds for local usage. Plant genetic resources for food and agriculture are to be conserved and revitalized, strengthened, and improved, with a particular focus on local types of crops, but not exclusively. Major and small crops, as well as neglected and underutilized species, are all handled by CSBs and their staff. They are helping farmers and communities all across the world become back in charge of their seeds (Vernooy et al., 2020).
2. **Field gene banks**: Live plants are traditionally used in *ex situ* field gene banks for plants with very long life cycle and plants that cannot be preserved as seeds due to their recalcitrant or because of asexual mode of reproduction (Kasso and Balakrishnan, 2013).
3. ***In vitro* conservation**: *In vitro* conservation refers to a particular form of gene bank known as the slow-growth conservation approach. Growing diverse plant parts in pathogen-free sterile culture in synthetic media with growth retardants has been found to be a suitable technique to supplement and provide backup to field collection efforts (Kasso and Balakrishnan, 2013).
4. **Cryopreservation**: This technique is used to conserve living tissues by freezing them at very low temperatures (–196°C) in liquid nitrogen. This will stop all the metabolic activities and arrest cell cycle (Kasso and Balakrishnan, 2013).
5. **De-extinction by gene editing**: De-extinction was proposed as a way to "correct" historic extinctions by reintroducing extinct species into their old habitats (Novak, 2018). The first experiment to revive an extinct species was done by Folch et al. in 2009. The scientists tried to clone the already extinct goat, commonly known as Bucardo, by using the cryopreserved fibroblasts of the last living female of the subspecies *Capra pyrenaica pyrenaica* (Bucardo), which died in the year 2000. The team was able to produce one normal female out of 280 embryos. Though the newborn survived only for a minute, the experiment paved a way for the revival of many extinct and endangered species. With the aim to bring back a long extinct animal, the wooly mammoth, scientists at Harvard University started a Wooly Mammoth Revival project in association with an NGO called Revive and Restore in 2015. They plan to do this by editing the genome of Asian elephant, the closest living relative of the wooly mammoth, by using

CRISPR-cas9. The species created by this technique will be a hybrid of the Asian elephant and the wooly mammoth (Devlin, 2017). To date, the scientists have not achieved success in this experiment, but the whole team is confident in getting a positive outcome in the next few years. Novak (2018) mentions in his article that currently seven projects to revive different species are active globally. However, the ethics of conducting such experiments are questionable. Keeping this in view, the International Union for Conservation of Nature (IUCN) launched a De-extinction Task Force under the umbrella of the Species Survival Commission (SSC) in April 2014. For conservation purposes, the Task Force published in 2016 some guidelines on constructing proxies of extinct species (IUCN, 2016). These guidelines define de-extinction as any attempt to create some proxy of an extinct species or subspecies (hereinafter, "species") through any technique, including methods such as selective back breeding, somatic cell nuclear transfer (cloning), and genome engineering. The term "proxy" is used to clarify that the newborn produced is not an exact copy of the extinct species but is similar to the one in some aspects.

The *ex situ* conservation methods provide a good option for conserving and reviving endangered species in a short span of time, but because species must be kept in artificial settings, genetic diversity is lost, inbreeding depression occurs, adaptations to confinement develop, and detrimental alleles accumulate. Also, the technique of animal cloning has ethical constraints. All these limits the use of this method for conserving biodiversity.

1.9 Concluding Remarks

Biodiversity is essential to keeping the natural world healthy and functioning. It is crucial that we gather and share essential knowledge that may empower people, companies, and governments and help them live in harmony with nature in order to ensure that we protect and nourish biodiversity.

REFERENCES

Abrahms, B., Dipietro, D., Graffis, A., Hollander, A. (2017). Managing biodiversity under climate change: challenges, frameworks, and tools for adaptation. *Biodivers. Conserv.*, 26: 2277–2293.

Adom, D., Umachandran, K., Ziarati, P., Sawicka, B., Sekyere, P. (2019). The concept of biodiversity and its relevance to mankind: a short review. *J. Agric. Sustain.*, 12(2): 219–231.

Ayre, J. (2012). The Red List of Threatened Species, Annual Report Released. *Planetsave.com*. Archived from the original on 10 November 2013. Retrieved on 14 September 2021.

Barongi, R., Fisken, F.A., Parker, M., Gusset, M. (2015). *Committing to Conservation*. World Association of Zoos and Aquariums (WAZA) Executive Office: 69.

Bodeker, G., van't Klooster, C., Weisbord, E. (2014). *Prunus africana* (Hook.f.) Kalkman: the overexploitation of a medicinal plant species and its legal context. *J. Altern. Complement. Med.*, 810–822.

Borenstein, S. (2019). *UN Report: Humans Accelerating Extinction of Other Species*. New York: Associated Press.

Brambilla, M., Gubert, F., Pedrinia, P. (2021). The effects of farming intensification on an iconic grassland bird species, or why mountain refuges no longer work for farmland biodiversity. *Agric. Ecosyst. Environ.*, 319.

Camilo, M., Tittensor, D.P., Sina, A., Simpson, Alastair, G.B., Boris, W. (2011). How many species are there on earth and in the ocean? *PloS. Biol.*, 9(8): e1001127.

CBD. (2010). *Conference of the Parties 10 Decision X/17. Consolidated Update of the Global Strategy for Plant Conservation 2011–2020*. Secretariat of the Convention on Biological Diversity (2010). www.cbd.int/doc/decisions/cop-10/cop-10-dec-17-en.pdf.

Che-Castaldo, J.P., Grow, S.A., Faust, L.J. (2018). Evaluating the Contribution of North American Zoos and Aquariums to Endangered Species Recovery. *Sci. Rep.*, 8: 9789.

Conservation International. (2012). www.conservation.org/. Retrieved on 14 September 2021.

De Kort, H., Prunier, J.G., Ducatez, S., et al. (2021). Life history, climate and biogeography interactively affect worldwide genetic diversity of plant and animal populations. *Nat. Commun.*, 12: 516.

DeLong, D.C. (1996). Defining biodiversity. *Wildl. Soc. Bull.*, 24: 738–749.

Devlin, H. (2017). Woolly mammoth on verge of resurrection, scientists reveal. *The Guardian*. www.theguardian.com/science/2017/feb/16/woolly-mammoth-resurrection-scientists.

Diez, M.J., De la Rosa, L., Martin, I., Guasch, L., et al. (2018). Plant Genebanks: Present situation and proposals for their improvement. The case of the Spanish network. *Front. Plant Sci.* 9: 1794.

Elizabeth, P. (2012). Ecosystems. In S. George Philander (ed.), *Encyclopedia of Global Warming & Climate Change* (Vol. 1, 2nd ed.). Thousand Oaks, CA: SAGE Reference, pp. 485–487.

ENVIS Centre on Floral Diversity. (2016). *Global biodiversity hotspots with special emphasis on Indian hotspots*. Botanical Survey of India, Kolkata, 30 June 2016. http://bsienvis.nic.in/files/biodiversity%20hotspots%20in%20india.pdf. Retrieved on 14 September 2021.

Fenoglio, M.S., Calviño, A., González, E., Salvo, A., Videla, M. (2021). Urbanisation drivers and underlying mechanisms of terrestrial insect diversity loss in cities. *Ecol. Entomol.*, 46(4): 757–771.

Fiedler, P.L., Jain, S.K. (eds.). (1992). *Conservation Biology: The Theory and Practice of Nature Conservation, Preservation and Management*. New York: Chapman and Hall.

Folch, J., Cocero, M.J., Chesné, P., Alabart, J.L., Domínguez, V., Cognié, Y., Roche, A., Fernández-Arias, A., Martí, J.I., Sánchez, P., Echegoyen, E., Beckers, J.F., Bonastre, A.S., Vignon, X. (2009). First birth of an animal from an extinct subspecies (Capra pyrenaica pyrenaica) by cloning. *Theriogenology*, 71(6): 1026–1034.

Gadgil, M. (2012). Report of the western ghats ecology expert panel. *Westernghatindia.org*. Ministry of Environment and Forests, Government of India. Part 1: Summary XIX. https://www.cppr.in/wp-content/uploads/2013/03/Gadgil-report.pdf. Retrieved on 14 September 2021.

Gaston, K.J., Spicer, J.I. (2004). *Biodiversity: An Introduction* (2nd ed.). Hoboken: Wiley-Blackwell. Retrieved on 14 September 2021.

Giangaspero, M., Al-Ghafri, M. K. S. (2014). Vulnerable Wild Animal Species, Conservation and Genetic Investigations. *Trop. Med. Surg.*, 2: e116

Hammerschlag, N., Gallagher, J.A. (2017). Extinction risk and conservation of the earth's national animal symbols. *Bioscience*, 67: 744–749.

Hawksworth, D.L. (1996). Biodiversity: measurement and estimation. Philosophical transactions of the royal society of London. Series B, *Biol. Sci.*, 345: 6.

IUCN. (2016). *IUCN SSC guiding principles on creating proxies of extinct species for conservation benefit*. https://portals.iucn.org/library/sites/library/files/documents/Rep-2016-009.pdf

Kasso, M., Balakrishnan, M. (2013). Ex situ conservation of biodiversity with particular emphasis to ethiopia. *Int. Sch. Res. Notices*, 985037.

Klein, A.M., Vaissière, B.E., Cane, J.H., Steffan-Dewenter, I., Cunningham, S.A., et al. (2007). Importance of pollinators in changing landscapes for world crops. *Proc. Biol. Sci.*, 274: 303–313.

Linders, T.E.W., Schaffner, U., Eschen, R., Abebe, A., Choge, S.K., et al. (2019). Direct and indirect effects of invasive species: biodiversity loss is a major mechanism by which an invasive tree affects ecosystem functioning. *J. Ecol.*, 107: 2660–2672.

Lovejoy, T.E. (1980). The global 2000 report to the president. In G.O. Barney (ed.), *The Technical Report* (Vol. 2). New York: Penguin, pp. 327–332.

Martins-Oliveira, A.T., Zanin, M., Canale, G.R., da Costa, C.A., Eisenlohr, P.V., Silveira Alves de Melo, F.C., de Melo, F.R. (2021). A global review of the threats of mining on mid-sized and large mammals. *J. Nat. Conserv.*, 62.

Maruthamuthu, B., Srivastava, R., Pokhriyal, M. (2008). Biodiversity of Andaman and Nicobar Islands, 9: 9–12.

McGregor Reid, G., Zippel, K.C. (2008). Can zoos and aquariums ensure the survival of amphibians in the 21st century? *Int. Zoo Yb.*, 42: 1–6.

Mills, L.S., Soule, M.E., Doak, D.F. (1993). The keystone-species concept in ecology and conservation. *Bioscience*, 43(4): 219.

Mohallem, S.V., de Araújo Lobo, D.J., Pesquero, C.R., Assunção, J.V., de Andre, P.A., et al. (2005). Decreased fertility in mice exposed to environmental air pollution in the city of Sao Paulo. *Environ. Res.*, 98: 196–202.

Mounce, R., Smith, P., Brockington, S. (2017). Ex situ conservation of plant diversity in the world's botanic gardens. *Nature Plants*, 3: 795–802.

Myers, N., Mittermeier, R.A., Mittermeier, C.G., da Fonseca, G.A.B., Jennifer, K. (2000). Biodiversity hotspots for conservation priorities. *Nature*, 403(6772): 853–858.

National Biological Information Infrastructure. (2011). *Introduction to Genetic Diversity*. U.S. Geological Survey. Reston, Virginia, United States. Retrieved on 14 September 2021.

National Geographic. (2021). www.nationalgeographic.org/encyclopedia/biodiversity/. Retrieved on 14 September 2021.

Nayar, T.S., Rasiya Beegam, A., Sibi, M. (2014). *Flowering Plants of the Western Ghats, India* (2 Vol.). Thiruvananthapuram, India: Jawaharlal Nehru Tropical Botanic Garden and Research Institute, p. 1700.

Novak, B.J. (2018). De-extinction. *Genes*, 9(11): 548.

Pagett, R. (2021). Principles regarding urbanisation, disaster risks and resilience. In S. Eslamian, F. Eslamian (eds.), *Handbook of Disaster Risk Reduction for Resilience*. Cham: Springer.

Paine, R.T. (1995). A conversation on refining the concept of keystone species. *Conserv. Biol.*, 9(4): 962–964.

Pimm, S.L., Jenkins, C.N., Abell, R., Brooks, T.M., Gittleman, J.L., Joppa, L.N., Raven, P.H., Roberts, C.M., Sexton, J.O. (2014). The biodiversity of species and their rates of extinction, distribution, and protection. *Science*, 344(6187).

Ramanatha Rao, V., Hodgkin, T. (2002). Genetic diversity and conservation and utilization of plant genetic resources. *Plant Cell Tissue Organ Cult.*, 68: 1–19.

Rondeau, D., Perry, B., Grimard, F. (2020). The consequences of COVID-19 and other disasters for wildlife and biodiversity. *Environ. Resource. Econ.*, 76: 945–961.

Sahney, S., Benton, M.J. (2008). Recovery from the most profound mass extinction of all time. Proceedings of the Royal Society B: Biol. *Sci.*, 275(1636): 759–765.

Salomão, R.P., Alvarado, F., Baena-Díaz, F., Favila, M.E., Iannuzzi, L., Liberal, C.N., Santos, B.A., Villegas-Guzmán, G.A., González-Tokman, D. (2020). Negative effects of urbanisation on the physical condition of an endemic dung beetle from a neotropical hotspot. *Ecol. Entomol.*, 45(4): 886–895.

Sanderfoot, O.V., Holloway, T. (2017). Air pollution impacts on avian species via inhalation exposure and associated outcomes. *Environ. Res. Lett.*, 12: 083002.

Shivaperuman, C. (2014). Faunal resources in the Ritechie's Archipleago, Andaman and Nicobar Islands. *Rec. Zool. Surv. India, Dee.*, 360: 1–76+16.

Singh, V., Verma, K. (2018). Metals from cell to environment: connecting metallomics with other omics. *Open J. Plant Sci.*, 3: 001–014.

Swingland, I.R. (2013). Biodiversity Definition of. In Simon Levin (ed.). *Encyclopedia of Biodiversity* (2nd ed.). Amsterdam: Academic Press.

Tang, F.H.M., Lenzen, M., McBratney, A., et al. (2021). Risk of pesticide pollution at the global scale. *Nat. Geosci.*, 14: 206–210.

Tordoff, A.W., Baltzer, M.C., Fellowes, J.R., Pilgrim, J.D., Langhammer, P.F. (2012). Key biodiversity areas in the Indo-Burma hotspot: Process, progress and future directions. *J. Threat. Taxa.*, 4(8): 2779–2787.

Tuomisto, H. (2010). A diversity of beta diversities: straightening up a concept gone awry. Part 1. Defining beta diversity as a function of alpha and gamma diversity. *Ecography*, 33: 2–22.

Vernooy, R., Mulesa, T.H., Gupta, A., Johny, J.A., Koffi, K.E., Mbozi, H., et al. (2020). The role of community seed banks in achieving farmers' rights. *Dev. Pract.*, 30(5): 561–574.

Whittaker, R.H. (1972). Evolution and measurement of species diversity. *Taxon*, 21: 213–251.

Wildlife Institute of India. (2021). https://wii.gov.in/nwdc_conservation_reserves. Retrieved on 14 September 2021.

Wyse J., P.S., Sutherland, L.A. (2000). International agenda for botanic gardens in conservation. *Botanic Gardens Conservation International*, U.K. www.bgci.org/files/All/Key_Publications/interagendaeng2580.pdf.

2

Resilience Strategies for Biodiversity Conservation

Anu Priya Minhas, Sanjeev Puri, and Ranbir Chander Sobti

CONTENTS

2.1 Introduction

The effect of overall climate change on the intensity, timing, and frequency of extreme events is often more damaging than the effect of mean climate (Turner et al., 2010). Novel conservation and restoration approaches are the key scientific solutions to and/or societal resilience to climate change toward ecosystems and biodiversity management. A long-term understanding of community response to change is required before working on resilience strategies toward ecological communities. Iglesias and Whitlock (2020) analyze pollen and charcoal records from temperate forests in the Northern and Southern Hemispheres to understand the role of fire in changing forest tree composition. Soil, climate, or historical legacies or biophysical feedbacks under extreme conditions are the deciding factors for vulnerability and resilience of any forest species composition to changing fire regimes. Synergetic interaction between the extreme climatic conditions, local anthropogenic disturbances, and mean climate trends need to be addressed for effective post disturbance ecosystem recovery in high biodiversity tropical ecosystems (França et al., 2020). Multitrophic animal-mediated processes assist ecosystem recovery in tropical forests and coral reefs. Seed dispersal by dung beetles and grazing by parrotfish are the examples where local actions can increase resilience to extreme events toward ecosystem protection or restoration.

The majority of the literature on nature-based approaches to climate mitigation and adaptation has focused purely on terrestrial ecosystems (e.g., forests and peatlands) or terrestrial-coastal systems (e.g., mangroves and salt marshes). Solan et al. (2020) examine the climate mitigation and adaptation potential of marine benthic soft-sediment ecosystems. Soft-sediment ecosystems in terms of biodiversity of benthic fauna and flora are the most enriched and expensive habitats on Earth. These play a key role in regulating climate-active gases, biogeochemical cycling, ocean chemistry, and long-term removal of carbon from the ocean-atmosphere system. Sediment-dwelling invertebrates can significantly intensify, neutralize, or alleviate the effects of acidification, warming, deoxygenation and rise in sea level. Later effects are mediated through bioturbation and ventilation behavior of these invertebrates. Interest in climate change adaptation is driving interest in benthic habitat restoration, but the science is in its infancy.

DOI: 10.1201/9781003220398-2

Adopting strategic protection plans in the key areas in a network is required for wider and larger-scale resilience and ecosystem function.

Protection strategies generally strengthen the capacity of ecosystems to retain carbon. These also work to impound additional carbon and so enhance ecological resilience to climate change. However, much of the connection between ecosystem intactness and carbon sequestration stems from terrestrial ecosystems. Marine ecosystems with conserved fish and marine mammal populations might help magnify the ocean nutrient cycle and associated sequestration rates. However, marine ecosystems have been less appreciated so far. Roberts et al. (2020) emphasize the potential synergies of marine biodiversity protection with the mitigation of and adaptation to climate change. Recently, marine megafauna's role in enhancing vertical nutrient transfer, modifying ocean fertility, and carbon sequestration at large scales have been highlighted. However, to achieve this target, the expansion of marine protected areas from the current 10% of sea area in the Aichi biodiversity to 30% of sea area need emphasis.

For conterminous United States (US), taking into account both current and projected distributions of biodiversity, Lawler et al. (2020) suggest protected area network as a cost-effective strategy to configure a terrestrial ecosystem under climate change. The cost of planning protected area network can increase modestly. However, this is incomparable to high cost involved in the expanding the reserve network without considering climate change. Additionally, protecting some kinds of climate refugia may be an economical conservation option.

2.2 Practical Solutions for Ecosystem Management and Biodiversity Conservation

The final focus should be on the opportunities and challenges corresponding to the practical management, restoration and protection of biodiversity, and the solutions for ecosystem management. These interventions have been described further as follows.

2.2.1 The Changing Goals of Conservation

From the century, nature conservation implies preserving the original state of nature, a state before the introduction of humans. According to this approach, humans are outsiders and are harmful to nature. This approach ignores sustainable practices and social justice to indigenous peoples affected by conservation (IPBES, 2019). Overall, conservation efforts are insufficient compared to the level of threats to biodiversity loss. Only less than 3% of marine and 25% of terrestrial areas have been protected under this approach so far.

Of the 16,495 protected areas, this approach occupy approximately 25,033,869 km^2 of area, of which 7% correspond to seas (Marine Conservation Institute, 2021). Conservation policies and practices presently focus on the territory and resource co-management toward people's benefit. These also manage biodiversity per the demand of indigenous peoples and scientific communities (Grorud-Colvert et al., 2021).

The UNESCO biosphere reserves and world heritage cultural landscapes range from small sites to thousands of square kilometers. This landscape focuses on cultural and social needs, regulated sustainable development and strict no-entry zones for biodiversity conservation. Small community-managed marine conservation zones have been successfully scaled up to regional networks (Gilchrist et al., 2020). Although this method has implementation challenges, it is still emerging as an appropriate conservation model for climate mitigation and adaptation (Baird et al., 2018). Species, ecosystems, and protected areas must be analyzed for susceptibility to climate change. The method of conservation must be socially flexible. Due to the fresh implementation and the unavailability of adequate data, its effectiveness of this technique in the real world has not been assessed. Conservation is inevitably linked to human and societal adaptation; therefore, conservation practitioners should adapt biodiversity changes toward better adaptation (Whitney et al., 2017).

2.2.2 Specific Area-Based Conservation

Traditional area-based conservation approaches focus on hotspots of overlapping biodiversity richness and species vulnerability without aligning to climate change (Myers et al., 2000; Hindell et al., 2020). Species vulnerable to climate change but not to non-climatic factors have never been prioritized in the "hotspot" category (Foden et al., 2013). However, the impact of variables over time and space emphasizes considering climate change a pragmatic threat to such protected areas (Elsen et al., 2020). Hannah et al. (2020) analyze the adequacy of existing protected area networks toward climate change-based species range shifts. Protected areas are considered carbon-rich ecosystems and play a role in preventing emissions resulting from deforestation and degradation. Information about climate velocity trajectories is an important parameter to be considered in the establishment and planning of climate-smart conservation area networks (Arafeh-Dalmau et al., 2021). Therefore, deliberately establishing protecting area networks across underrepresented climate spaces, hotspots, elevation gaps, and climate refugia significantly enhance biodiversity conservation at a global level.

The scattered climate velocity trajectories across small islands and limited areas may lead to poorly protected areas. Instead, a specific, flexible, smart way out for conservation focusing both protected and anthrome landscapes is required (Carlson et al., 2019). Areas with low climate velocity, refugia, and climate analogues should be given priority while selecting and connecting conservation locations (Keeley et al., 2018). New connections may have varied effects on diverse species. These may favor generalized, more flexible, invasive, and/or pathogenic interactions, which later may need redress for balance. Furthermore, sudden traumas such as the recent COVID-19 pandemic can impact the operational and financial aspects of conservation and therefore need to be appraised (Northrop et al., 2020).

2.2.3 Focus on Dynamic Species

For successful dispersal and institution of a species in new niches, a strong association between land and water habitats is necessary. Species relocation and assisted migration are assured conservation tools to assist in achieving rapid migration and enhanced habitat fragmentation (Costanza et al., 2020). In the case of animal species, reintroduction guidelines should carefully follow the choice of species based on their functional relevance and species distribution models (Pörtner et al., 2021).

For land ecosystems, tree plantations have proven to be a potential strategy capable of shifting the original ecosystem to a suitable future bioclimatic space. The assisted migration of tree species may result in genetic overload, promote pest and disease transfer, and increase the gap between predicted and realized climate. However, the overall associated risks of assisted migration are insignificantly compared with the major consequences of successful invasion in colonized communities (Mueller & Hellmann, 2008). Therefore, legal and policy frameworks and strength and weakness analysis along with local relevance should be performed before selecting a multi-tool approach for restoring species (Rinkevich, 2019; Sansilvestri et al., 2015).

2.2.4 Bringing Down Non-Climate Stressors

Reducing the detrimental effects of non-climatic stresses should be the major focus of biodiversity conservation. The implementation of conservation practices is dependent on the efficacy of treatment (Sutherland et al., 2016). For better adaptation and survival of wild creatures and ecosystems to climate change, there is a need to minimize non-climatic stressors. This includes illness, habitat loss, overexploitation, and invasive species (Wanger et al., 2020). Climate change can deteriorate ecosystems, increase disease susceptibility, and change migration patterns. Wild animals and ecosystems display a negative correlation between non-climatic stresses, adaptability, and resistance to climate change (Räsänen et al., 2016). Climate change has multiple effects on non-climatic stressors. Therefore, non-climatic stressor mitigation strategies must be scaled up and integrated into climate change-focused conservation policies with a focus on maintaining multi-stressor solutions (Pörtner et al., 2021).

2.2.5 Genetic Diversity Preservation

Protecting genetic diversity is crucial to achieve species and ecosystem variance for sustainable benefits to humanity (Des Roches et al., 2021). Genetic diversity can assist populations withstanding climate change. It helps in shifting and rebounding depleted populations by recreating genetic heterogeneity in a species (Crow et al., 2021). In agriculture, strategies employed to preserve diversity include *in situ* and *ex situ* conservation, less monoculture cropping, and use of heritage landraces in plant breeding to preserve ancestral traits (Ebert & Waqainabete, 2018; Mastretta-Yanes et al., 2018). In aquaculture and livestock, introducing ancestral breeds with higher genetic diversity and better adaptability or enhanced productivity is employed for preserving genetic resources (Gicquel et al., 2020). Keeping in mind the balance between yield and resilience, focus should be on a more balanced, economical biodiversity conservation model. Furthermore, with advances in genomic research globally, data need to be easily accessed and utilized for capacity building and innovation (Blasiak et al., 2018).

2.2.6 Multifunctional Scape Approach

Generally, biodiversity protection efforts are implemented across large wild areas to completely transformed spaces. Large wild areas refer to wild, intact, or slightly altered ecosystems. Completely transformed spaces refer to cities and farms dominated by human infrastructure and fulfilling human requirements (Locke et al., 2019). Natural ecosystems span approximately 55% of the land and approximately >95% of the ocean, referred to as "shared lands managed ecosystems" or "working landscapes" (Locke et al., 2019; Díaz et al., 2020; Garibaldi et al., 2021). In general, people exploit nature in all states from intact to shared spaces for diverse reasons except in anthromes with a limited set of benefits. Nature, in its intact environment, is hardly present. A multifunctional scape approach (scape is shorthand for landscapes, freshwater-scapes, and seascapes) involves conservation at different levels of biodiversity from wild, shared to seascapes with natural and anthrome pocket. The multifunctional scape paradigm can link conservation between broader spatial features of land and sea dynamics. Research in the multifunctional scape approach focuses on "strength, uses and dependency on local factors, self-supporting biodiversity, suitable climate and quality life for all." Shift in mixed savannas and grasslands in southern and eastern Africa to forestland due to increased CO_2 levels is an example of multifunctional scape approach Midgley and Bond (2015). Entire herbivore populations, sufficient cattle grazing, and controlled wood cutting modulate this shift and therefore benefit native biodiversity and humans in the form of livestock and meat production, wood use, and tourism (Pörtner et al., 2021).

2.2.7 Nature-Based Solutions (NBSs)

NBSs address the socio-environmental challenges adequately and attentively toward human well-being for sustainable management, protection, and repair of modified ecosystems. Under the broad umbrella of NBS, natural climate solutions (NCSs) have gained increasing drag for mitigating climate change, protecting and restoring ecosystems (Griscom et., 2017). NBSs can account for carbon sequestration under climate change and disaster risk management under natural calamities and can also create green facilities in cities (Dorst et al., 2019). NBSs can partially manage global warming with potential support for biodiversity and ecosystem services. NBSs help to integrate the ecosystem with socio-economic systems and therefore strengthen the adaptive capacity of both ecosystems and societies (Seddon et al., 2020). This study emphasizes nature's role in reducing social-ecological sensitivity to climate change. However, the undesirable effects on NBSs designed for specific objectives is the key concern (Seddon et al., 2020). Planting trees could be effective at sequestering carbon but can also affect biodiversity. The structure and function of natural savannas and peatland ecosystems deteriorate with afforestation leading to loss of unique species (Abreu et al., 2017; Samways et al., 2020). Keeping in mind biodiversity-climate connections, NBSs should be positive toward climate mitigation or at least neutral toward biodiversity benefits. If applied correctly, NBSs action may result in more gifts from nature to humanity. Given its dual responsibilities in producing beneficial effects for biodiversity and people's well-being, NBSs have become a significant tactic in multifunctional scapes (Garibaldi et al., 2021)

2.2.8 Rehabilitation

In terms of climate change, the term "rehabilitation" stands for upgraded and restored state or functions suitable for new climate conditions (Kleypas et al., 2021). In damaged natural systems with no chance of recovery, restoration of species and/or natural habitats is critical or too slow (IPCC, 2000; (Munasinghe & Swart, 2005). Adaptation can be incorporated into the restoration process. Restoration, such as NBS, may help to fulfill people, biodiversity, and climate-associated objectives (Martin, 2017). Under climate change, it is critical that restoration targets be in line with future requirements (Harris et al., 2006).

2.3 Conclusion

Rapid alterations in location, operation, and relation of organisms with ecosystems due to progressive climate change are posing serious threats to biodiversity. Inflexible and vulnerable species are more susceptible to climate change. Adopting sustainable practices in all economic sectors is emphasized to halt and reverse the impacts of climate change on biodiversity. This is considered a crucial step to cease or undo the footprints of climate change on biodiversity. This chapter emphasizes various practical interventions for managing, protecting, and restoring biodiversity and ecosystems. Some of the interventions include redefining conservation goals to a broader level, co-managing area and resources, reduction of no-climate factors, introduction of dynamic species, protection of natural and anthropogenic pockets, and adaptation of nature-based solutions with species/habitat restoration. Further, successful implementation of all these interventions can enhance the involvement and relationship of people with nature and biodiversity.

BIBLIOGRAPHY

Abreu, R. C. R., Hoffmann, W. A., Vasconcelos, H. L., Pilon, N. A., Rossatto, D. R., & Durigan, G. (2017). The biodiversity cost of carbon sequestration in tropical savanna. *Science Advances*, *3*(8). https://doi.org/10.1126/SCIADV.1701284/SUPPL_FILE/1701284_SM.PDF

Arafeh-Dalmau, N., Brito-Morales, I., Schoeman, D. S., Possingham, H. P., Klein, C. J., & Richardson, A. J. (2021). Incorporating climate velocity into the design of climate-smart networks of marine protected areas. *Methods in Ecology and Evolution*, *12*(10), 1969–1983. https://doi.org/10.1111/2041-210X.13675

Baird, J., Plummer, R., Schultz, L., Armitage, D., & Bodin, O. (2018). Integrating conservation and sustainable development through adaptive co-management in UNESCO biosphere reserves. *Conservation and Society*, *16*(4), 409. https://doi.org/10.4103/CS.CS_17_58

Blasiak, R., Jouffray, J. B., Wabnitz, C. C. C., Sundström, E., & Osterblom, H. (2018). Corporate control and global governance of marine genetic resources. *Science Advances*, *4*(6). https://doi.org/10.1126/SCIADV.AAR5237/SUPPL_FILE/AAR5237_SM.PDF

Carlson, R. R., Foo, S. A., & Asner, G. P. (2019). Land use impacts on coral reef health: A ridge-to-reef perspective. *Frontiers in Marine Science*, *6*, 562. https://doi.org/10.3389/FMARS.2019.00562/BIBTEX

Costanza, J. K., Watling, J., Sutherland, R., Belyea, C., Dilkina, B., Cayton, H., Bucklin, D., Romañach, S. S., & Haddad, N. M. (2020). Preserving connectivity under climate and land-use change: No one-size-fits-all approach for focal species in similar habitats. *Biological Conservation*, *248*, 108678. https://doi.org/10.1016/J.BIOCON.2020.108678

Crow, T. M., Buerkle, C. A., Runcie, D. E., & Hufford, K. M. (2021). Implications of genetic heterogeneity for plant translocation during ecological restoration. *Ecology and Evolution*, *11*(3), 1100–1110. https://doi.org/10.1002/ECE3.6978

Des Roches, S., Pendleton, L. H., Shapiro, B., & Palkovacs, E. P. (2021). Conserving intraspecific variation for nature's contributions to people. *Nature Ecology & Evolution*, *5*(5), 574–582. https://doi.org/10.1038/s41559-021-01403-5

Díaz, S., Zafra-Calvo, N., Purvis, A., Verburg, P. H., Obura, D., Leadley, P., Chaplin-Kramer, R., De Meester, L., Dulloo, E., Martín-López, B., Shaw, M. R., Visconti, P., Broadgate, W., Bruford, M. W., Burgess, N. D., Cavender-Bares, J., DeClerck, F., Fernández-Palacios, J. M., Garibaldi, L. A., . . . Zanne, A. E. (2020). Set ambitious goals for biodiversity and sustainability. *Science*, *370*(6515), 411–413. https://doi.org/10.1126/SCIENCE.ABE1530

Dorst, H., van der Jagt, A., Raven, R., & Runhaar, H. (2019). Urban greening through nature-based solutions—Key characteristics of an emerging concept. *Sustainable Cities and Society*, *49*, 101620. https://doi.org/10.1016/J.SCS.2019.101620

Ebert, A. W., & Waqainabete, L. M. (2018). Conserving and sharing taro genetic resources for the benefit of global taro cultivation: A core contribution of the centre for pacific crops and trees. *Biopreservation and Biobanking*, *16*(5), 361–367. https://doi.org/10.1089/BIO.2018.0017

Elsen, P. R., Monahan, W. B., Dougherty, E. R., & Merenlender, A. M. (2020). Keeping pace with climate change in global terrestrial protected areas. *Science Advances*, *6*(25). https://doi.org/10.1126/SCIADV.AAY0814/SUPPL_FILE/AAY0814_SM.PDF

Foden, W. B., Butchart, S. H. M., Stuart, S. N., Vié, J. C., Akçakaya, H. R., Angulo, A., DeVantier, L. M., Gutsche, A., Turak, E., Cao, L., Donner, S. D., Katariya, V., Bernard, R., Holland, R. A., Hughes, A. F., O'Hanlon, S. E., Garnett, S. T., Şekercioğlu, Ç. H., & Mace, G. M. (2013). Identifying the world's most climate change vulnerable species: A systematic trait-based assessment of all birds, amphibians and corals. *PloS ONE*, *8*(6), e65427. https://doi.org/10.1371/JOURNAL.PONE.0065427

França, F. M., Benkwitt, C. E., Peralta, G., Robinson, J. P. W., Graham, N. A. J., Tylianakis, J. M., Berenguer, E., Lees, A. C., Ferreira, J., Louzada, J., & Barlow, J. (2020). Climatic and local stressor interactions threaten tropical forests and coral reefs. *Philosophical Transactions of the Royal Society B*, *375*(1794). https://doi.org/10.1098/RSTB.2019.0116

Garibaldi, L. A., Oddi, F. J., Miguez, F. E., Bartomeus, I., Orr, M. C., Jobbágy, E. G., Kremen, C., Schulte, L. A., Hughes, A. C., Bagnato, C., Abramson, G., Bridgewater, P., Carella, D. G., Díaz, S., Dicks, L. V., Ellis, E. C., Goldenberg, M., Huaylla, C. A., Kuperman, M., . . . Zhu, C. D. (2021). Working landscapes need at least 20% native habitat. *Conservation Letters*, *14*(2), e12773. https://doi.org/10.1111/CONL.12773

Gicquel, E., Boettcher, P., Besbes, B., Furre, S., Fernández, J., Danchin-Burge, C., Berger, B., Baumung, R., Feijóo, J. R. J., & Leroy, G. (2020). Impact of conservation measures on demography and genetic variability of livestock breeds. *Animal*, *14*(4), 670–680. https://doi.org/10.1017/S1751731119002672

Gilchrist, H., Rocliffe, S., Anderson, L. G., & Gough, C. L. A. (2020). Reef fish biomass recovery within community-managed no take zones. *Ocean and Coastal Management*, *192*. https://doi.org/10.1016/J.OCECOAMAN.2020.105210

Griscom, B. W., et al. (2017). Natural climate solutions. *Proceedings of the National Academy of Sciences of the United States of America*, 114, 11645–11650. http://doi.org/10.1073/pnas.1710465114.

Grorud-Colvert, K., Sullivan-Stack, J., Roberts, C., Constant, V., Horta E Costa, B., Pike, E. P., Kingston, N., Laffoley, D., Sala, E., Claudet, J., Friedlander, A. M., Gill, D. A., Lester, S. E., Day, J. C., Gonçalves, E. J., Ahmadia, G. N., Rand, M., Villagomez, A., Ban, N. C., . . . Lubchenco, J. (2021). The MPA guide: A framework to achieve global goals for the ocean. *Science*, *373*(6560). https://doi.org/10.1126/SCIENCE.ABF0861

Hannah, L., Roehrdanz, P. R., Marquet, P. A., Enquist, B. J., Midgley, G., Foden, W., Lovett, J. C., Corlett, R. T., Corcoran, D., Butchart, S. H. M., Boyle, B., Feng, X., Maitner, B., Fajardo, J., McGill, B. J., Merow, C., Morueta-Holme, N., Newman, E. A., Park, D. S., . . . Svenning, J. C. (2020). 30% land conservation and climate action reduces tropical extinction risk by more than 50%. *Ecography*, *43*(7), 943–953. https://doi.org/10.1111/ECOG.05166

Harris, J. A., Hobbs, R. J., Higgs, E., & Aronson, J. (2006). Ecological restoration and global climate change. *Restoration Ecology*, *14*(2), 170–176. https://doi.org/10.1111/J.1526-100X.2006.00136.X

Hindell, M. A., Reisinger, R. R., Ropert-Coudert, Y., Hückstädt, L. A., Trathan, P. N., Bornemann, H., Charrassin, J. B., Chown, S. L., Costa, D. P., Danis, B., Lea, M. A., Thompson, D., Torres, L. G., Van de Putte, A. P., Alderman, R., Andrews-Goff, V., Arthur, B., Ballard, G., Bengtson, J., . . . Raymond, B. (2020). Tracking of marine predators to protect Southern Ocean ecosystems. *Nature*, *580*(7801), 87–92. https://doi.org/10.1038/s41586-020-2126-y

Iglesias, V., & Whitlock, C. (2020). If the trees burn, is the forest lost? Past dynamics in temperate forests help inform management strategies. *Philosophical Transactions of the Royal Society B*, *375*(1794). https://doi.org/10.1098/RSTB.2019.0115

IPBES. (2019). *Global Assessment Report on Biodiversity and Ecosystem Services of the Intergovernmental Science-Policy Platform on Biodiversity and Ecosystem Services*. E. S. Brondizio, J. Settele, S. Díaz, and H. T. Ngo (editors). IPBES Secretariat, Bonn, Germany, 1148 pages. https://doi.org/10.5281/zenodo.3831673

IPCC. (2000). *Land Use, Land-Use Change, and Forestry*. R. T. Watson, I. R. Noble, B. Bolin, N. H. Ravindranath, D. J. Verardo, and D. J. Dokken (editors.). Cambridge University Press. www.ipcc.ch/report/land-use-land-use-change-and forestry/

Keeley, A. T. H., Basson, G., Cameron, D. R., Heller, N. E., Huber, P. R., Schloss, C. A., Thorne, J. H., & Merenlender, A. M. (2018). Making habitat connectivity a reality. *Conservation Biology: The Journal of the Society for Conservation Biology*, *32*(6), 1221–1232. https://doi.org/10.1111/COBI.13158

Kleypas, J., Allemand, D., Anthony, K., Baker, A. C., Beck, M. W., Hale, L. Z., Hilmi, N., Hoegh-Guldberg, O., Hughes, T., Kaufman, L., Kayanne, H., Magnan, A. K., Mcleod, E., Mumby, P., Palumbi, S., Richmond, R. H., Rinkevich, B., Steneck, R. S., Voolstra, C. R., . . . Gattuso, J. P. (2021). Designing a blueprint for coral reef survival. *Biological Conservation*, *257*, 109107. https://doi.org/10.1016/J.BIOCON.2021.109107

Lawler, J. J., Rinnan, D. S., Michalak, J. L., Withey, J. C., Randels, C. R., & Possingham, H. P. (2020). Planning for climate change through additions to a national protected area network: Implications for cost and configuration. *Philosophical Transactions of the Royal Society B*, *375*(1794). https://doi.org/10.1098/RSTB.2019.0117

Locke, H., Ellis, E. C., Venter, O., Schuster, R., Ma, K., Shen, X., Woodley, S., Kingston, N., Bhola, N., Strassburg, B. B. N., Paulsch, A., Williams, B., & Watson, J. E. M. (2019). Three global conditions for biodiversity conservation and sustainable use: An implementation framework. *National Science Review*, *6*(6), 1080–1082. https://doi.org/10.1093/NSR/NWZ136

Marine Conservation Institute. (2021). *SeaStates 2021: Marine Protection in America's Ocean Areas*. http://marine-conservation.org/seastates

Martin, D. M. (2017). Ecological restoration should be redefined for the twenty-first century. *Restoration Ecology*, *25*(5), 668–673. https://doi.org/10.1111/REC.12554

Mastretta-Yanes, A., Acevedo Gasman, F., Burgeff, C., Cano Ramírez, M., Piñero, D., & Sarukhán, J. (2018). An initiative for the study and use of genetic diversity of domesticated plants and their wild relatives. *Frontiers in Plant Science*, *9*, 209. https://doi.org/10.3389/FPLS.2018.00209/BIBTEX

Midgley, G. F., & Bond, W. J. (2015). Future of African terrestrial biodiversity and ecosystems under anthropogenic climate change. *Nature Climate Change*, *5*(9), 823–829. https://doi.org/10.1038/nclimate2753

Munasinghe, M., & Swart, R. J. (2005). *Primer on Climate Change and Sustainable Development: Facts, Policy Analysis and Applications*. 445 pp. ISBN: 9780521810661. DOI:10.1017/CBO9780511622984

Mueller, J. M., & Hellmann, J. J. (2008). An assessment of invasion risk from assisted migration. *Conservation Biology: The Journal of the Society for Conservation Biology*, *22*(3), 562–567. https://doi.org/10.1111/J.1523-1739.2008.00952.X

Myers, N., Mittermeler, R. A., Mittermeler, C. G., Da Fonseca, G. A. B., & Kent, J. (2000). Biodiversity hotspots for conservation priorities. *Nature*, *403*(6772), 853–858. https://doi.org/10.1038/35002501

Northrop, E., Konar, M., Frost, N., & Hollaway, E. (2020). *A Sustainable and Equitable Blue Recovery to the COVID-19 Crisis* (p. 10). www.oceanpanel.org/bluerecovery.

Pörtner, H.O., Scholes, R.J., Agard, J., Archer, E., Arneth, A., Bai, X., Barnes, D., Burrows, M., Chan, L., Cheung, W.L., Diamond, S., Donatti, C., Duarte, C., Eisenhauer, N., Foden, W., Gasalla, M. A., Handa, C., Hickler, T., Hoegh-Guldberg, O., Ichii, K., Jacob, U., Insarov, G., Kiessling, W., Leadley, P., Leemans, R., Levin, L., Lim, M., Maharaj, S., Managi, S., Marquet, P. A., McElwee, P., Midgley, G., Oberdorff, T., Obura, D., Osman, E., Pandit, R., Pascual, U., Pires, A. P. F., Popp, A., Reyes-García, V., Sankaran, M., Settele, J., Shin, Y. J., Sintayehu, D. W., Smith, P., Steiner, N., Strassburg, B., Sukumar, R., Trisos, C., Val, A.L., Wu, J., Aldrian, E., Parmesan, C., Pichs-Madruga, R., Roberts, D.C., Rogers, A.D., Díaz, S., Fischer, M., Hashimoto, S., Lavorel, S., Wu, N., Ngo, H.T. 2021. *IPBES-IPCC co-sponsored workshop report on biodiversity and climate change; IPBES and IPCC,* DOI:10.5281/zenodo.4782538

Räsänen, A., Juhola, S., Nygren, A., *et al.* (2016). Climate change, multiple stressors and human vulnerability: a systematic review. *Regional Environmental Change*, *16*, 2291–2302. https://doi.org/10.1007/s10113-016-0974-7

Rinkevich, B. (2019). The active reef restoration toolbox is a vehicle for coral resilience and adaptation in a changing world. *Journal of Marine Science and Engineering*, *7*(7), 201. https://doi.org/10.3390/JMSE7070201

Roberts, C. M., O'Leary, B. C., & Hawkins, J. P. (2020). Climate change mitigation and nature conservation both require higher protected area targets. *Philosophical Transactions of the Royal Society B*, *375*(1794). https://doi.org/10.1098/RSTB.2019.0121

Samways, M. J., Barton, P. S., Birkhofer, K., Chichorro, F., Deacon, C., Fartmann, T., Fukushima, C. S., Gaigher, R., Habel, J. C., Hallmann, C. A., Hill, M. J., Hochkirch, A., Kaila, L., Kwak, M. L., Maes, D., Mammola, S., Noriega, J. A., Orfinger, A. B., Pedraza, F., . . . Cardoso, P. (2020). Solutions for humanity on how to conserve insects. *Biological Conservation*, *242*. https://doi.org/10.1016/J.BIOCON.2020.108427

Sansilvestri, R., Frascaria-Lacoste, N., & Fernández-Manjarrés, J. F. (2015). Reconstructing a deconstructed concept: Policy tools for implementing assisted migration for species and ecosystem management. *Environmental Science & Policy*, *51*, 192–201. https://doi.org/10.1016/J.ENVSCI.2015.04.005

Seddon, N., Chausson, A., Berry, P., Girardin, C. A. J., Smith, A., & Turner, B. (2020). Understanding the value and limits of nature-based solutions to climate change and other global challenges. *Philosophical Transactions of the Royal Society B*, *375*(1794). https://doi.org/10.1098/RSTB.2019.0120

Solan, M., Bennett, E. M., Mumby, P. J., Leyland, J., & Godbold, J. A. (2020). Benthic-based contributions to climate change mitigation and adaptation. *Philosophical Transactions of the Royal Society B*, *375*(1794). https://doi.org/10.1098/RSTB.2019.0107

Sutherland, W. J., Broad, S., Caine, J., Clout, M., Dicks, L. V., Doran, H., Entwistle, A. C., Fleishman, E., Gibbons, D. W., Keim, B., LeAnstey, B., Lickorish, F. A., Markillie, P., Monk, K. A., Mortimer, D., Ockendon, N., Pearce-Higgins, J. W., Peck, L. S., Pretty, J., . . . Wright, K. E. (2016). A horizon scan of global conservation issues for 2016. *Trends in Ecology & Evolution*, *31*(1), 44–53. https://doi.org/10.1016/J.TREE.2015.11.007

Turner, W. R., Bradley, B. A., Estes, L. D., Hole, D. G., Oppenheimer, M., & Wilcove, D. S. (2010). Climate change: Helping nature survive the human response. *Conservation Letters*, *3*(5), 304–312. https://doi.org/10.1111/J.1755-263X.2010.00128.X

Wanger, T. C., Ainun, N., Brook, B. W., Friess, D. A., Oh, R. R. Y., Rusdin, A., Smithers, S., & Tjoa, A. (2020). Ecosystem-based tsunami mitigation for tropical biodiversity hotspots. *Trends in Ecology & Evolution*, *35*(2), 96–100. https://doi.org/10.1016/J.TREE.2019.10.008

Whitney, C. K., Bennett, N. J., Ban, N. C., Allison, E. H., Armitage, D., Blythe, J. L., Burt, J. M., Cheung, W., Finkbeiner, E. M., Kaplan-Hallam, M., Perry, I., Turner, N. J., Yumagulova, L., Whitney, C. K., Bennett, N. J., Ban, N. C., Allison, E. H., Armitage, D., Blythe, J. L., . . . Yumagulova, L. (2017). Adaptive capacity: from assessment to action in coastal social-ecological systems. *Ecology and Society*, *22*(2). https://doi.org/10.5751/ES-09325-220222

3

Biodiversity Resources and Ecosystem Services in the Context of the Industrialized World

Pardeep Kumar, Abhishek Kumar, Meenu Patil, Sabir Hussain, Sheenu Sharma, Siril Singh, Rajni Yadav, Diksha Tokas, and Anand Narain Singh

CONTENTS

3.1 Introduction

The convention on biological diversity (CBD) in 1992 defines biodiversity as the variability among living organisms from all sources, including terrestrial, marine, and other aquatic ecosystems and the ecological complexes of which they are part: this includes diversity within species, between species, and of ecosystems. This simple definition does not merely restrict to the species level; however, it embodies the diversity of genes, landscapes, and biomes. Biodiversity provides valuable ecosystem services, including provisioning services such as food and water; regulating services in terms of flood control, disease control, and maintenance of nutrient cycling; and cultural services, such as recreational, spiritual, and aesthetic importance (Elmqvist and Maltby 2010). Besides, biodiversity also supplies all kinds of medicines, raw materials for industries, and agricultural products to sustain rural livelihood. Therefore, life subsistence on Earth is inherently linked to biodiversity, and any loss of biological resources will intimately influence our ecosystem services.

Unfortunately, biological resources are being exploited at an unprecedented rate that has posed various threats to biodiversity. Biodiversity loss primarily occurs due to commercial exploitation of resources (illegal trading for medicines and cosmetics), clearing forests for agricultural land, and mining activities. In addition to this, global climate change and associated environmental perturbations, pollution, and

DOI: 10.1201/9781003220398-3

biological invasion augment this substantially. With the rapidly growing population, modernization, and changed lifestyle, technological advancement is predominately achieved through the industrial revolution; therefore, the industry is partially linked with biodiversity loss. Industrialization has resulted in atmospheric pollution, such as the deposition of atmospheric nitrogen, which has detrimental impacts on marine life (Bustamante et al. 2018; Mader and Scheyven 2019). Discharge of industrial effluents into water bodies, mainly rivers, leach from wastelands, dumping grounds, and mine spoils, deteriorates the water quality and affects aquatic life, mainly fish. Conservation efforts to mitigate biodiversity loss are being implemented worldwide, and international communities have come together to tackle the issue and introduced a convention on biodiversity (CBD) in 1992.

Several non-government organizations (NGOs), such as the International Union for Conservation of Nature and Natural Resources (IUCN), the World Wide Fund for Nature (WWF), and Conservation International, work at the global level to conserve biodiversity. Despite these efforts, biological resources are being degraded and exploited unharmoniously at unprecedented rates. The strategic plan for biodiversity (2011–2020) provides an overarching framework on biodiversity for the biodiversity-related conventions, the entire United Nations system and all other partners engaged in biodiversity management and policy development. In Aichi targets, strategic goal B, "*reduce the direct pressures on biodiversity and promote sustainable use*," describes the management of and the sustainable and legal harvesting of fish, invertebrate stocks, and aquatic plants, aiming to reduce the pollution level and entry of invasive alien species and to manage anthropogenic pressure on coral reefs to maintain their ecology and integrity. Unfortunately, the efforts to achieve a significant reduction in biodiversity loss have not been met yet. Biodiversity provides essential resources and raw materials for industrial innovation, while the industry is a driver of biodiversity loss and innovation that may lead to biodiversity conservation (Mader and Scheyvens 2019). Such a mutual interdependence of biodiversity and industry and their risks and benefits need to be given due attention for a sustainable future.

3.2 Global Status of Biodiversity

The variety of living organisms, including plants, animals, and microbes, that exist on the Earth is quite remarkable, and it is difficult to determine the exact number of species that inhabit the Earth. According to a UNEP estimate, approximately 9–52 million species exist on Earth. However, recent global biodiversity estimates predict that there are around 8.7 million eukaryotic species perhaps existing on Earth, of which 2.2 million are marine (Mora et al. 2011). Nevertheless, approximately 90% of the existing species on Earth are still unexplored. India is concerned with the many species that inhabit India due to its variability in climatic conditions, including snow-laden mountains, coasts, floodplains, and deserts. India is one of the 17 megadiverse countries globally and harbors 7–8% of the world's recorded species. To date, 91,200 species of animals and 45,500 species of plants have been described across India's ten biogeographic regions (NBA 2018).

3.3 Values of Bioresources

Biodiversity or bioresources are an indispensable part of human life. Biodiversity serves humanity both directly and indirectly. In an immediate sense, people use biodiversity (plants, animals, and microorganisms) for food, fuel, clothing, medicines, shelter, and so on. Major staple crops, such as rice, wheat, and maize, together feed more than 90% of the world's population, and several plant species that now attained the status of weeds have been consumed by our ancestors for millennia. Wild relatives of these crops provide great insight into the development of disease-resistant and high-yield varieties. Biodiversity is also valuable to us indirectly in various forms, such as plant-animal-microbial interaction. It maintains the ecosystem food chain and regulates nutrient cycling; plants sequester atmospheric CO_2 and produce oxygen, prevent soil erosion, and maintain the hydrological cycle and local climatic phenomena.

Biodiversity also plays a vital role in traditional practices of medicine to cure various health ailments. According to the WHO, 80% of the world's population still depends on natural medicines (Hamilton 2004). Several modern therapeutic drugs, such as quinine (from *Cinchona*), digitalin (from *Digitalis*), ergotamine (from ergot fungus), paclitaxel (from *Taxus*), and aspirin (from *Salix*), are extracted from plants that were used once traditionally to cure disease (Kumar 2006; Sen and Samanta 2014). Thus, the vast genetic diversity available in plants, animals, and microorganisms offers a bulk of opportunities to develop several therapeutic drugs of interest (Sen and Samanta 2014). Biodiversity is a source of aesthetic beauty and recreation, especially in the urban environment where the cities' hustle and bustle deprived one of enjoying nature's spiritual and serene value. Studies have shown that increased psychological benefits are associated with green spaces in urban areas and demonstrate that biodiversity in urban areas enhances positivity and overall human well-being (Fuller et al. 2007; Lee et al. 2009). For example, ornamental plants such as orchids, highly praised for their intriguing appearance, are a billion-dollar industry. Besides, countries rich in biodiversity, especially island nations or those possessing abundant marine life, are major tourist attractions worldwide and draw a major proportion of their revenue from the tourism industry. Detailed information about essential ecosystem services and provisions of biodiversity is depicted in Figure 3.1.

3.4 Threats to Biodiversity

Biodiversity loss and climate change are major concerns today for the world. Among the several drivers causing biodiversity loss are land-use change, habitat fragmentation and degradation, overutilization of natural resources, pollution, climate change, and invasive alien species (Armenteras and Finlayson 2012). It is speculated that these factors mentioned will further exacerbate biodiversity loss, due to which global biodiversity is projected to decrease about 63% by 2050 (Slingenberg et al. 2009). According to a Millennium Ecosystem Assessment (MEA 2005) report, habitat change, overexploitation, and N and P pollution have caused tremendous biodiversity loss over the past 100 years. However, the relative

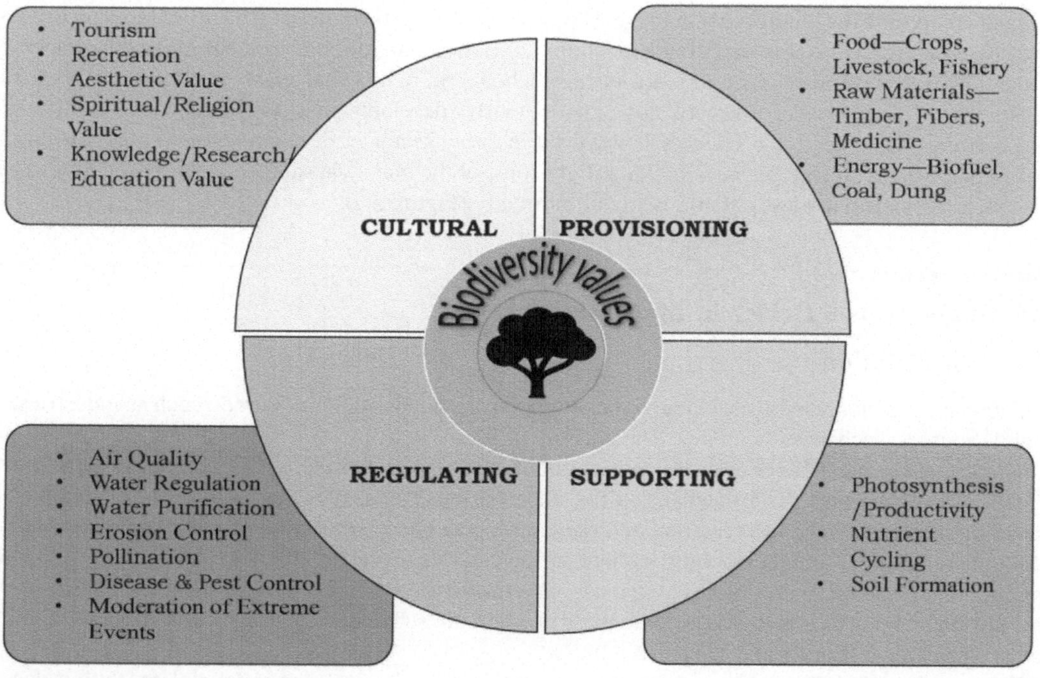

FIGURE 3.1 Types of ecosystem services via biodiversity resources.

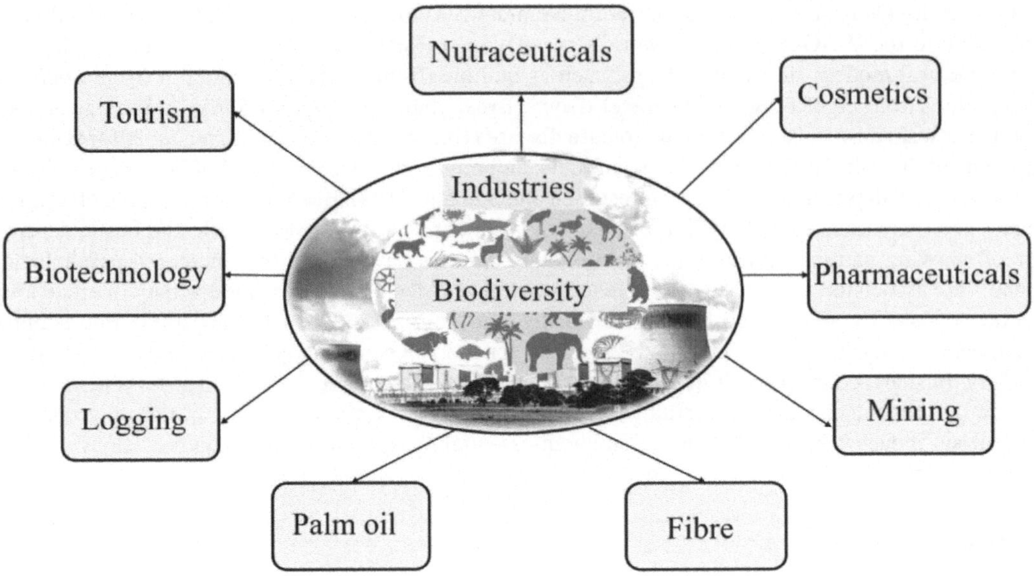

FIGURE 3.2 A pictorial representation of biodiversity-based industries.

significance of these factors varies region-wise. It has been projected that in South Asia, by the year 2050, infrastructure development and agricultural expansion will be the primary causal factors for biodiversity loss. Population explosion, poverty, or underdevelopment of some sections of human society are often linked with biodiversity loss.

The prevailing lack of basic amenities and poverty have led people to encroach on wildlife areas and subsequent clearing of forests and land degradation. However, development cannot be ignored since it is important to alleviate the standard of life and eradicate poverty, especially in developing countries, without compromising biodiversity damage. Human economic growth lies at the cost of biodiversity loss until potential measures and strict implementation of policies are not undertaken. As briefly discussed earlier, there are several causes for biodiversity loss; however, in this chapter, we restrict ourselves to the causes which are directly or indirectly linked with industry development as a result of which biodiversity loss is occurring. Indeed, the industry is involved in every sector of development and thus also in the loss of biodiversity; therefore, we aim to highlight some of the major causal factors where the industry is complicit in biodiversity loss either directly or indirectly (Figure 3.2).

3.5 Industry as a Driver of Biodiversity Loss

3.5.1 Air Pollution

Pollution affects biodiversity on a large scale. Air pollution can have point sources, such as industries, or mobile sources, such as automobiles. The burning of fossil fuels, thermal power plants, pulp mills, ore smelters, and chemical factories emits various pollutants, such as NO_2, SO_2, CO, and particulate matter. These compounds, such as nitrogen and sulfur oxides, after secondary reactions, form acid rain and can cause significant damage to vegetation and regional biota. The rising level of particulate matter in the atmosphere reduces sunlight reaching the leaf surface and hence retards the photosynthesis rate, resulting in stunted plant growth. Unlike habitat loss and climate change, air pollution impacts on biodiversity loss are challenging to investigate because these changes are subtle and elusive and may take a long time to become evident (Barker and Tingey 1992).

The concentration of reactive nitrogen has risen significantly since the industrial revolution due to fertilizers and fossil fuels burning. After biodiversity loss and climate change, nitrogen pollution is considered the third major threat to Earth (Giles 2005). Atmospheric deposition of reactive nitrogen can enrich the nitrogen content of soils and their acidification; some species may find it difficult to survive in acidic environments, ultimately leading to biodiversity loss. Several studies (Stevens et al. 2004; Jones et al. 2004; Maskell et al. 2010; Stevens et al. 2010) conducted in temperate (acidic) grassland ecosystems in Europe have revealed a linear decline in species richness in response to chronic nitrogen deposition, thus presenting a global threat to the biodiversity of these unique ecosystems. However, further studies are needed to elucidate its impacts on a broader scale. However, birds are very prone to harmful air pollutants. For example, a significant decline has been observed in the population of Indian sparrows (*Passer domesticus*) in several cities across India (Dandapat et al. 2010). Among the several factors affecting the house sparrow population are habitat loss, urbanization, and invasive species; the low quality of air due to industrial pollution is also responsible for the declining population (Sharma and Binner 2020).

3.5.2 Water Pollution

Water pollution is mainly due to the discharge of industrial effluents and harmful chemicals from textile industries, hospitals, thermal power plants, and fertilizer industries. Water pollution is detrimental to aquatic life, mainly fish, and can cause long-term alterations. One such phenomenon is eutrophication, where the excess release of nitrogen and phosphorus in the aquatic bodies causes excessive growth of certain algae. These algae form a layer over the water surface and block the oxygen supply to lower organisms, subsequently leading to the death of aquatic organisms. The most prominent example of the impact of industries on water pollution is the Ganges River. The primary cause of water pollution in the holy Ganges River is the release of effluents from the leather-tanning industry around the city of Kanpur (MOEFCC 2017; Chaudhary and Walker 2019). The tannery effluents, ranging between 5.8 and 8.8 MLD, carry a high concentration of toxic metals, such as chromium and mercury. These pollutants render the water unusable and pose a severe threat to human and aquatic life (Chaudhary et al. 2017).

The Ganges River flows through 30 major cities and 70 towns and thousands of villages. Therefore, several other industries, such as cotton, sugar mills, fertilizers, and pulp, release their effluents into the river (Behera et al. 2011). Apart from this, more than six million tonnes of chemical fertilizers and nine thousand tonnes of pesticides are used annually in the Ganges basin (Behera et al. 2011; Wong et al. 2007). The severely degraded water quality of the Ganges has affected the aquatic life most such as freshwater turtles, reptiles, and fish. The Ganges basin is reported to harbor 140 fish species, the richest freshwater fish fauna in India (Jones et al. 2003), which provide sustenance for a significant part of the Ganges River basin. The greatest cause of concern is the declining population of endangered Ganges River dolphins, a top predator and better indicator of river ecology and its ecosystem health. If the prevailing situation is not reversed, the Ganges River dolphin might face extinction in this century (Behera et al. 2011).

Due to accidental oil spills in oceans, water pollution is another cause of concern affecting marine biodiversity and the sea ecosystem. Studies have estimated that 3.2 million tonnes of oil per year are being released into the environment, most of which come from shipping tankers and industries (ITOPF 1990; Saadoun 2015). Oil spill causes significant damage to aquatic life, mainly fish and coral reefs. Recently (on August 6, 2020), a major oil spill incident happened in Mauritius when the cargo ship MV *Wakashio* ran aground on a coral reef on the southeast tip of the country, which is regarded as a hotspot and Ramsar site and estimated to release about 1,000 tonnes of diesel and fuel oil into the ocean (Lewis 2020). The ingestion of oil spillage by marine organisms can cause mortality and reproductive and developmental abnormalities in zooplankton, invertebrates, and other benthic organisms (Jiang et al. 2010). Apart from this, sea birds are one of the groups of animals most vulnerable to oil spills, and ingestion of a small quantity of oil can be lethal to sea birds and affect the waterproofing of their plumage, leading to loss of insulation and buoyancy and causing rapid death by hypothermia, starvation, or drowning (Piatt et al. 1990; Leighton 1993).

3.5.3 Mining

The mining industry plays a significant role in powering the economics of many countries worldwide. India's mining sector contributes 6.8% of coal and lignite production globally and a significant producer of metallic material, industrial material, and metal alloys (Datar et al. 2011). Mineral resources are mainly confined to biodiversity-rich areas (Murguía et al. 2016). Mining for coal, quarrying for minerals, and other processes are prerequisites for the industrialized world to meet the ever-increasing energy demand. Therefore, tensions between conservation priorities for biodiversity loss and mining are expected to intensify in the future (Ali et al. 2017; Sonter et al. 2018). Environmental trade-offs are associated with the mining industry, such as air pollution, water pollution, soil acidification, and soil erosion. Most of the mine reservoirs, particularly coal mines, are deposited under thick age-old forests, formed by the deposition and fossilization of forests biomass over thousands of years. In the process to recover coal reserves, a substantial proportion of the Earth's surface is removed through open cast mining, thereby causing inevitable biodiversity loss, such as destruction of vegetation, loss of fertile soil, and alteration of microbial communities (Singh et al. 2002; Kumar et al. 2020).

The impact of mining on biodiversity depends on different stages of the project, such as exploration, development, production, transportation, and post-closure; therefore, each stage is needed to be assessed cautiously. Therefore, mainstreaming biodiversity into the energy and mining sector is a top priority by the conference of parties to the convention on biodiversity (UNEP-WCMC 2017). Sonter et al. (2020) stress that a switch to renewable energy would drive an increase in mining activities for the procurement of metals required for technology and infrastructure development. The excessive deposition of materials at the time of mining operations poses a severe threat to the surroundings, especially aquifers, through the leaching of toxic elements (Jain and Das 2017). The water discharged through mined areas carries harmful chemical pollutants and sediments deposited in nearby water bodies, increasing water turbidity, reducing light penetration to lower depths, and diminishing algal and planktonic biomass (Niyogi et al. 2002). Elevated chemical concentration, varied pH, and temperature conditions also negatively impact fish (Wong et al. 1999).

Apart from coal mining, sand mining is also a significant threat to biodiversity loss. According to the Ministry of Mines, sand is the fourth most crucial minor element in production. In India, excessive and illegal sand extraction from rivers poses a severe threat to biodiversity and river ecology. The typical example is of Chambal River (the Chambal Valley traverses between the states of Rajasthan and Madhya Pradesh), where excessive sand mining over the years had forced the government to declare a major part of the river as a sanctuary for the protection of critically endangered species, such as gharial, Ganges River dolphins, Indian otters, tortoises, and other IUCN red-listed species (Kukreti 2017). The Chambal River is also a breeding place for the Indian gharial, and the unmanaged sand mining destroys the nesting sites of gharial, posing a severe threat to this critically endangered reptile. Despite the declaration of the sanctuary, mining laws are being openly flouted, as sand mafias still practice illegal sand extraction (Koehnken and Rintoul 2018). The chief factors responsible for illicit sand extraction and the subsequent deteriorating state of rivers and wildlife are poor management, weak mining policies, and rampant corruption, which has left several Indian rivers in a dire state. Several political support incidents for the sand mafia are anticipated in UP, MP, and Punjab states. Since sand is the main element in concrete and cement making and its demand will further surge in the future, it emphasizes the need to enforce strict laws and policies for sustainable use and protect river hydrology.

Coral blasting for mining limestone and other construction materials has degraded a significant population of coral reefs in South and Southeast Asia (Rajasuriya et al. 2004). Limestone quarrying has caused a significant decline in biodiversity in Southeast Asia's karst ecosystem (Malaysia, Vietnam, Thailand, and Indonesia). Limestone karsts are deposits of calcium carbonate formed by the secretion of marine organisms (corals and brachiopods) over thousands of years and lifted above sea level due to tectonic movements (Clements et al. 2006). The karst areas are generally unique ecosystems due to varied climates and habitats, such as cliffs, caves, and rocks, which harbor various endemic species, such as orchids, balsams, begonias, and bryophytes (calcicolous bryophytes), and large trees, such as dipterocarps (Clements et al. 2006). Several species of butterflies, land snails, birds, and monkeys are endemic to this region. The mining of limestone and elements such as antimony, gold, and iron has caused large-scale degradation of karsts ecosystems, especially in Southeast Asia. Therefore, mining and quarrying activities in such biodiversity-rich hotspots must be restrained or be persuaded sustainably without compromising the monetary benefits to local people and biodiversity loss.

3.5.4 Overexploitation of Biodiversity by the Pharmaceutical, Cosmetic, and Food Industries

Plants and animal bioresources have been used as a source of medicine by humans since time immemorial. The majority of the drugs used in clinical applications are primarily derived from higher plants (Farnsworth and Morris 1976), followed by microbial, animal, and mineral products (Alves and Rosa 2007). Plant-based drugs have been used in the treatment of various disorders for several millennia. It has been mentioned in the world's major pharmacopeia, including Ayurveda, Unani, and Siddha systems of medicine. The practice of collecting medicinal plants to cure ailments is still prevalent today in many developing countries as more than 80% of the world population relies on traditional medicine (Sen and Samanta 2014). Approximately 50,000 to 80,000 flowering plants are used for medicinal purposes worldwide (IUCN 2007), of which 15,000 are facing extinction due to overexploitation and habitat destruction (Bentley 2010; Chen et al. 2016). Medicinal and aromatic plants provide raw materials (flavonoids, alkaloids, saponins) for the pharmaceutical and cosmetic industries.

According to the World Health Organization (WHO), the current demand for medicinal plants is around USD 14 billion per year, and with the rise in the current population, it is expected to be more than USD 5 trillion by the year 2050 (Sharma and Thokchom 2014). In recent years, commercial exploitation of medicinal plants for foreign export has surged exponentially. According to Kumar (2006), India's export of medicinal plants has risen exponentially during the last three decades. Probable reasons for this could be the heavy dependence of people on plant-based drugs because of associated side effects of the indiscriminate use of allopathic medicine, which prompts people to opt for plant-based remedies with minimum side effects, the needs for foreign exchange, and the rapidly growing global bioprospecting industry and cosmetics and herbal manufacturers (Kumar 2006). This has led to the imprudent harvesting of several medicinal plants' wild populations to such an extent that they face extinction threats. For example, *Dactylorhiza hatagirea*, an orchid species commonly found in the Indian Himalayan region, has immense medicinal significance and has been collected excessively for its crude drug extract (salep).

The plant has high demand in the pharmaceutical sector; around 5,000 tons per annum are harvested (Kala 2004), and its global trade is worth around USD 71,583 (Dhiman et al. 2019; Wani et al. 2020). Therefore, this plant's overexploitation has made it critically endangered and put in the CITES Appendix II (Uniyal et al. 2002). The prevailing situation is critical for several other Himalayan species as well, such as *Taxus baccata*, *Aconitum heterophyllum*, *Nardostachys grandiflora*, *Gloriosa superba*, *Arnebia benthami*, *Picrorhiza kurroa*, and *Podophyllum hexandrum*, which need due concern and better policymaking for collection and pharmaceutical trade. Some of the overexploited medicinal plants used for therapeutic purposes, cosmetic products, and food facing severe threats are listed in Table 3.1. Besides, land-use change, deforestation, climate change, slow growth, and poor seed germination of species in natural environments further reduce the wild populations. In recognition of this, several medicinal plants have been brought under cultivation to meet the rising demands. Although the cultivation of medicinal plants can reduce the existing pressure on their wild populations, at the same time, it may reduce the genetic diversity or the number of natural compounds with medicinal importance and remove the incentive to conserve wild populations (Anon 2002).

Medicinal properties in plants are attributed to the prevalence of specific secondary metabolites and bioactive molecules synthesized under specific environmental conditions, including stress and competition, which might not be available in artificial conditions (Schippmann et al. 2002). Also, certain active substances in medicinal plants can be enhanced deliberately under cultivation, which might be the reason for the qualitative inferiority compared to plants gathered from natural populations (Uniyal et al. 2000). Therefore, the collection of medicinal plants for drug development and pharmacological evaluation should be done sustainably and adequately to ensure that ecosystem, and natural balance is not harmed.

Apart from plants, overutilization and harvesting of animal organs, such as skin, bones, fins, and horns for medicinal purposes, have threatened several animal species. International regulations and strict national laws against hunting and poaching do not discourage poachers and illegal traders from flourishing due to the high market price for medicines made from animal parts (Alves and Rosa 2007). For example, the use of tiger parts in treating several diseases has been mentioned in various medicine

TABLE 3.1

A List of Illegally Traded Medicinal Plants Used in Traditional Medicines and Therapeutic Purposes

Species name	Family	Medicinal uses	Threat status	Source
Dactylorhiza hatagirea (salam panja)	Orchidaceae	Chronic diarrhea, nerve tonic, aphrodisiac, neurasthenia, urinary and respiratory disorders	High demand in the pharmaceutical industry and commercial trade has led to overexploitation of this species' wild populations.	Wani et al. 2020
Podophyllum hexandrum (ban kakri)	Berberidaceae	Lung cancer, leukemia, and solid tumors	Critically endangered (CR). Due to the abundance of compound podophyllotoxin used in treating lung cancer and other tumors, pharmaceutical companies have recklessly exploited the plant.	Pandey et al. 2007
Nardostachys jatamansi (jatamansi)	Caprifoliaceae	Circulatory, respiratory, nervous, and reproductive disorders	CR; illegal and unsustainable harvesting. The plant has high demand in the international market, especially in European and American cosmetic industries, to prepare an essential oil, spikenard, which costs around USD 70 per kilogram. In India, the plant's rhizome is in high demand (500–1,000 metric tons) in the herbal industry and its oil costs around 12,000–13,000 per liter.	Jhunjhunwalla 2010; Dhiman and Bhattacharya 2020
Taxus wallichiana (Himalayan yew)	Taxaceae	Arrests cell division in ovarian, breast, and lung cancerous cells	Destructive harvesting of bark to extract the anticancerous compound paclitaxel. The drug is sold under the name Taxol. Worldwide demand for Taxol is 80–1,000 kg.	Pant and Samant 2008; Nimasow et al. 2015
Swertia chirayita (chirayata)	Gentianaceae	Possesses antifungal, anti-inflammatory, antibacterial, and antidiabetic properties	CR; overexploitation for national and international trade due to its high medicinal bitter roots. Listed as one of the 32 highly medicinal plants of India.	Kumar and Van staden 2016
Picrorhiza Kurroa (kutki)	Plantaginaceae	Possesses anticancer, anti-inflammatory, antibacterial, and antiviral properties; also used to treat cardiac problems	Overexploitation, illicit trade. The annual production of kutki is 375 tons, while the global annual demand is 500 tons per year. India alone contributes to 75% of global demand. Harvesting of around 300–400 plants will yield 1 kg of roots.	Thani et al. 2018; Bhattacharjee et al. 2013; Debnath et al. 2020
Gentiana kurroo (chireta, kadu)	Gentianaceae	Used to cure digestive problems, asthma, and liver disorders	Unregulated extraction of commercially important roots and rhizome; mining and stone quarrying.	Behera and Raina 2012
Meconopsis aculeata (Himalayan blue poppy)	Papaveraceae	Roots possess narcotic properties and leaves used in treatment of headache.	Overexploitation and collection by herbalists and chemists due to the prevalence of therapeutic substances.	Rana and Samant 2011
Saussurea costus (kuth, kushta)	Asteraceae	Rheumatism, dysentery, cholera, bronchitis, jaundice, skin diseases, and stomach disorders	CR; excessive demand in the herbal industry. The essential oil from the roots is used in the perfume industry and is of high international value; because of this, the plant has been overexploited and now at the verge of extinction. Listed in Appendix I of CITES.	Nadda et al. 2020; Rathore et al. 2021

systems, such as traditional Chinese medicine (TCM), Indian materia medica (including Ayurveda, Unani, and home remedies), and Japanese and Korean traditional medicine. However, TCM led to the overexploitation of tiger parts for today's commercial market and left the species on the brink of extinction. China had a population of 4,000 South China tigers in the 1950s, but due to the exponential rise in the Chinese population and the industrial revolution in the late 1980s, the demand for tiger parts surged for traditional medicine, leaving only 1,000 tigers in China in 1993. Though the Chinese government has banned tiger-based medicine, now illegal selling of tiger-based medicines is practiced (Nowell and Ling 2007). Due to its proximity to China, India also suffered a major loss in tiger population due to illegal poaching and the transport of tiger parts from the Indian side.

Every part of the tiger, such as its tail, nose, skin, whiskers, bone, and flesh, is believed to possess powers and healing properties to cure various disorders, including liver and kidney diseases, baldness, epilepsy, joint pains, and arthritis (CITES 2018). Tiger bone is used in rice wine to prepare a highly prized elixir known to give great strength (IFAW 2007). A recent study by World Wildlife Crime states that India and Thailand together constitute 82% of the illegal trade of tiger parts (UNODC 2020). Apart from tigers, other endangered species, such as the great Indian single-horned rhinos, pangolins, elephants, monkeys, and snakes, are being targeted for illegal hunting of their body parts to be used in medicines, fetching hefty prices in the international black market. The African elephant has lost 90% of its population within the last century (TRAFFIC 2019), which is the greatest example of overexploitation and illegal trade of wildlife. The actual number of wild animals commercialized in traditional medicine remains unknown (Alves and Alves 2011). At its extreme, overhunting and exploitation can cause extinction, as happened with the passenger pigeon and the Dodo bird of Mauritius (Freedman 2018). Nests built by swiftlets (*Collocalia fuciphagus* and *C. maximus*) are highly prized delicacies in the market, especially in China. The Gomantong karsts of Sabah, Malaysia, alone produce five tonnes of the nest, which cost around USD 2.5 million per annum (Lim and Cranbrook 2002; Clements et al. 2006). So there is a strong need to prepare baseline data of commercial trade in wildlife for medicinal purposes and to ensure their wild populations are not harmed beyond their replenishing capacity. This section has highlighted that medical applications of plants, animals, and marine organisms constitute an integral component of biodiversity, and the loss of any such components will degrade its value.

The recent surge in commercial trade of medicinal bioresources (plants and animals) is also driven by poverty and unemployment, and therefore, to generate income, local people or forest dwellers often recklessly harvest bioresources without recognizing the habitat loss and extinction threats to the species. Local people sell their products at a low price to the pharmaceutical, cosmetics, and herbal industries without realizing the commercial importance and international market price.

Conservation efforts are mainly restricted to check habitat loss because biodiversity is conserved as a byproduct (Alves et al. 2013). The commercial trade of bioresources, especially the harvesting of animal parts for traditional medicine, has gained the least attention from conservationists (Whiting et al. 2011; Alves et al. 2013).

So there is an urgent need to (1) increase our knowledge on trade and commercial exploitation of those species, (2) enforce strict conservation measures either to impose a complete ban or their utilization sustainably, (3) make sure that laws or norms are being appropriately implemented, and (4) develop a socio-economic relationship between forest-dwelling communities and industries for efficient resource management and phase out illegal trade of medicinal resources. Furthermore, conservation biologists and policymakers' main focus often tend toward megafauna or charismatic species threatened by illegal trade, while species that are uncharismatic but important to poor people, such as pollinators, soil microbes, crop varieties, and fiber-yielding and medicinally important plants, are less prioritized (Roe et al. 2019).

3.5.4.1 Overexploitation of Marine Resources

In terms of threat and overexploitation, even marine resources are not left intact by anthropogenic activities. The primary cause of concern to marine life is the threat to coral reefs. Approximately 500 million people depend directly or indirectly on coral reefs for their livelihoods and food (Wilkinson 2004), and coral reefs regulate ecosystem balance by protecting coastal beaches, controlling flooding, and being an essential part of land accretion. They are home to 25% of marine fish and form an interdependent community of photosynthetic organisms (algae and corals), regulating the sea layer's top-down effect and maintaining nutrient cycling (Miththapala 2008). Also, coral reefs harbor a great variety of species and act as a breeding place for shrimp and small fish (Perkol-Finkel and Benayahu 2007), and they sequester atmospheric CO_2 and release oxygen and, therefore, regulate the ecological state and the climatic phenomenon of the world.

Overfishing has been considered the main driver of biodiversity loss in the sea, as more than 80% of the world's fish stocks are fully overexploited (FAO 2012). One-fifth of the world's population relies on fish as the primary protein source, providing economic and monetary incentives to developing countries. Certain species of fish, due to overfishing, have been exploited to near extinction. For example, the population of Atlantic cod, a major predator of the seafood chain, has declined severely that it lost its control over its prey, such as shrimp and snow crabs, which proliferate extensively in the absence of top predators, disturbing the entire food web and ecosystem balance (Frank et al. 2005). The rising industrial demand for shark fins as meat and medicine has significantly reduced the shark population in the Southeastern United States (Kumar 2006).

A study conducted by Myers et al. (2007) states that the loss of a top predator has resulted in community restructuring in the Southeastern United States, where in the absence of the shark, its prey, the cownose ray, has completely scrapped the scallop fish (commercially important for fishing activities) and ultimately shut down the fishing activities, exemplifying how the removal of single species by overfishing can have consequences for the entire ecosystem. With the rise in the human population, fish consumption has doubled in the past 50 years, and the demand will rise in the future again to keep pace with the rising population. Several marine organisms, such as marine turtles, seahorses, sea cucumbers, sharks, and blue whales, are being harvested for medicinal purposes, especially in TCM, resulting in a significant decline in their population. Widespread overharvesting of cone mollusks for medicinal purposes, therapeutic research, and sale to tourists has led many species to the verge of extinction (Lead et al. 2005).

Therefore, there is an urgent need to pay due attention to overfishing. Since fishery is an industry and inherently linked to history, civilization, and food sources in developing countries, conservation efforts should be taken to ensure the long-term sustainable utilization of fisheries.

3.5.5 Timber Industry and Biodiversity Loss

Wood is the most versatile raw material provided by forests and is predominately used for energy generation and in the construction sector to make building materials, houses, furniture, pulp and paper, and musical instruments. Wood harvesting from forests is an ancient and age-old practice of our ancestors to sustain their livelihood. However, the demand for wood and wood products has increased significantly today, owing to the growing human population, industrialization, and declining rate of forest cover. Illegal logging occurs either due to the demand for valuable timber species or land clearance for agricultural purposes.

As a result, the habitat of some primates and large keystone mammals, such as tigers and elephants, is degraded. Elephants play a crucial role in shaping forest ecology. They regulate tree regeneration and canopy formation, disperse seeds of rare species, and maintain the ecosystem, especially in the African tropics. The loss of keystone species, such as elephants, leads to the loss of several other species as well.

Another key example of illegally traded species is the Madagascar rosewood (*Dalbergia* sp.), which has been exported to China recklessly, where it is used in traditional Chinese furniture *hongmu*. Despite being listed in CITES Appendix II, rosewood is the most trafficked form of flora or fauna globally and is traded far more than elephant ivory, rhino horn, and pangolin scales (Ong and Carver 2019; FAO 2020). The tall rosewood trees act as a key nesting site for endemic species of ruffed lemurs. Therefore, the excessive logging of rosewood trees has dire consequences for the lemurs and other associated species (Vasey et al. 2018). In Asia, countries such as China, India, Japan, and South Korea account for ~92% of hardwood timber import and consumption, mainly from countries with inadequate forest management plans. Therefore, high illegal logging risks prevail there (Hoare 2015), causing ecosystem degradation and biodiversity loss. India ranks third after China and Vietnam in the illegal import of timber in the world (Kleinschmit et al. 2016) and accounts for approximately 10% of the global wood trade. Therefore, the importing countries should strengthen the enforcement of laws against the import of illegal wood.

3.5.6 Fiber Industry

The natural fiber is obtained from wild plants, such as wood pulp and animal skin or furs, which can cause significant biodiversity loss if overexploited. Fibers obtained from cultivated plants, such as cotton, can have negative consequences on biodiversity because cotton constitutes 22.5% of the world's insecticide use and 10% of pesticide use (Aiama et al. 2016) and requires a high water supply for irrigation. The Aral Sea, a saline lake in Central Asia, is reported to have lost approximately 60% of its area between 1960 and 2000 due to the diversion of its two main feeding rivers for cotton cultivation in the downstream states of Uzbekistan (Shi et al. 2014; Loodin 2020). The water shortage in the Aral Sea due to irrigation has resulted in water salinization, pollution, and food insecurity. Nonetheless, cotton is a major cash crop and consumed worldwide, and its impact on biodiversity loss needs further attention.

3.5.7 Oil Palm Industry

In vegetable oil consumption, palm oil (*Elaeis guineensis*) ranks first globally and constitutes about 21% of edible oil production, only second after soybean oil. Oil is a crucial ingredient in the production of food, cosmetics, soap, and margarine across many countries (Nicholas et al. 2018). Southeast Asia, mainly Indonesia and Malaysia, fulfill 89% of global oil palm requirements. A large-scale palm oil plantation is grown across Malaysia, Indonesia, and Latin America after clearing age-old thick tropical rain forests.

The commercial farming of oil palm at the cost of tropical forest degradation is a primary cause of concern because these forests have great biodiversity, serve as storehouses for atmospheric CO2, and regulate climatic conditions. Industry-oriented and monoculture farming significantly reduce biodiversity through deforestation and affect smallholder farming and monetary incentives to poor communities.

3.5.8 Tourism Industry

Tourism is a major source of revenue, and foreign exchange, especially in developing countries, employs various sectors of society. International tourism is a USD 1 trillion industry in global exports and can drive the world economy (Sajjad et al. 2014). The tourism industry has a significant impact on the national economy, but it has negative consequences as well, in particular to the environment (Habibullah et al. 2016). The rise in tourism and tourism-related activities is thought to exert extra pressure on biodiversity and threatened species.

Tourism and tourism-related activities affect biodiversity in several ways, such as (1) degradation of habitats due to landscape transformation to construct hotels and lodging facilities in biodiversity-rich areas; (2) pollution of water bodies and marine areas due to the release of sewage effluents from nearby hotels; (3) excessive generation of waste material, mainly plastic; (4) mass (excessive) tourism, leading to rising incidents of human-wildlife conflicts and forced wild animals to change their habitats regularly; (5) recreation activities like scuba diving, often leading to damaged coral reefs and their pristine environment (UNEP 2013); and (6) the introduction of invasive alien species, either deliberately or accidentally. The rising number of trekking activities to remote and untouched places in the mountains often destroys rare plant species habitats.

Accepting all these facts cited, it would not be impertinent to say that tourism development is linked to major drivers of biodiversity loss, such as land-use change, pollution, and climate change (Hall 2010). A modeling study conducted by Habibullah et al. (2016) shows a significant decline in biodiversity with the increase in tourism-related activities. They anticipate a 10% rise in international tourists would lead to biodiversity loss of between 2 and 4%. Since tourism and tourism-related activities depend on biodiversity, and loss of biodiversity will ultimately lead to the cessation of the tourism industry and the loss of income for the poor communities who draw their sustenance from the tourism industry (Habibullah et al. 2016). The strong inter-relationships between tourism and biodiversity suggest an urgent need to develop steady-state approaches to tourism that may focus on conserving natural capital (Hall 2010). Therefore, biodiversity conservation should be the top priority while making tourism policies to ensure that tourism activities are carried out sustainably without compromising environmental degradation (UNEP 2013).

3.6 New Emerging Threats to Biodiversity

In recent decades several new threats have emerged with potential impacts on biodiversity. The advancement of biotechnological applications has led to the development of genetically modified (GM) crops with increased yield, better disease and pest resistance, and improved nutritional status. It has been speculated that the introduction of GM crops will reduce the overall genetic diversity in crops because the ultimate goal of breeding programs is to focus on the small number of genetically uniform, high-yielding cultivars instead of multiple landraces and local crop varieties (Ammann 2005; Roe et al. 2019). In developed nations, the ongoing scientific innovations and technological advancements further pose a threat to biodiversity in radioactive waste, dangerous chemicals, and nanoparticles (Slingenberg et al. 2009). Notwithstanding this, biotechnology is an innovation that can reduce biodiversity loss and environmental issues that deteriorate biodiversity values. However, it is not sure whether technological innovations and development are beneficial or not on a longer scale.

3.7 Conclusions

The planet Earth has faced several mass extinctions since its origin; however, the cause of concern is that the current species extinction is 1,000 times faster than the previous ones. The industry is partially or wholly involved in biodiversity loss, either directly or indirectly. The degradation of biodiversity and overexploitation of resources, such as plants, animals, and marine organisms, for traditional medicines, food, commercial use in cosmetics, and therapeutic drug development have made several species become extinct and left many species severely threatened. Land-use change and degradation due to mining, logging, and commercial plantations affect biodiversity. Air pollution from the burning of fossil fuels in industries and thermal power plants has degraded terrestrial biodiversity. The release of industrial effluents and agricultural runoff into the water bodies poses a threat to marine life. With the increase in global population and subsequent increase in demand for natural products, a substantial loss of biodiversity is projected in the future as well. Overharvesting of industrial consumption resources is detrimental to biodiversity and affects the local communities that sustain their livelihoods. However, the industry itself depends on biodiversity, and a decline in biodiversity will directly influence the industry. Therefore, it should be recognized that biodiversity is an inevitable part of the industry; therefore, the conservation of biodiversity and its sustainable use should become a top priority in the industrial and commercial sectors.

Furthermore, there is a strong need to develop dialogues between policymakers, stakeholders, and local communities to achieve sustainability in biodiversity use. In addition to this, industry regulation and management approaches should be targeted in high biodiversity-rich areas or regions with high species endemism. Apart from technological innovations and conservation organizations' efforts, some socio-economic issues, such as poverty and unemployment, need to be addressed to halt biodiversity loss. Alternate sources of income should be created where people solely depend on natural resources such as fisheries. However, developing countries are at greater risk of biodiversity loss because of the heavy dependence of the population on natural resources, weak management, poor laws, and inadequate enforcement, causing greater biodiversity loss. To avoid overharvesting resources while developing conservation programs and designs, local and marginalized people should be consulted to alleviate poverty and unemployment and be better tools for biodiversity conservation.

REFERENCES

Aiama, D., G. Carbone, D. Cator, and D. Challender. 2016. *Biodiversity risks and opportunities in the apparel sector*. International Union for Conservation of Nature: Gland, Switzerland. https://portals.iucn.org/library/efiles/documents/Rep-2016-001.pdf.

Ali, S. H., D. Giurco, N. Arndt, et al. 2017. Mineral supply for sustainable development requires resource governance. *Nature* 543, no. 7645 (March): 367–372.

Alves, R. R. N., and H. N. Alves. 2011. The faunal drugstore: Animal-based remedies used in traditional medicines in Latin America. *Journal of Ethnobiology and Ethnomedicine* 7, no. 1 (March): 1–43.

Alves, R. R. N., and I. M. L. Rosa. 2007. Biodiversity, traditional medicine and public health: Where do they meet? *Journal of Ethnobiology and Ethnomedicine* 3, no. 1 (March): 1–9.

Alves, R. R. N., I. L. Rosa, U. P. Albuquerque, and A. B. Cunningham. 2013. Medicine from the Wild: An overview of the use and trade of animal products in traditional medicines. In *Animals in Traditional Folk Medicine*, ed. R. R. N, Alves and I. L. Rosa, 25–42. Springer, Heidelberg, New York, Dordrecht, London.

Ammann, K. 2005. Effects of biotechnology on biodiversity: Herbicide-tolerant and insect-resistant gm crops. *Trends in Biotechnology* 23, no. 8 (August): 388–394.

Anon. 2002. *Conservation Impacts of Commercial Captive Breeding Workshop*. Jacksonville Cambridge, IUCN/SSC Wildlife Trade Programme.

Armenteras, D., and M. Finlayson. 2012. Biodiversity. In *UNEP, Global Environmental Outlook 5. Environment for the Future We Want*. Progress Press Ltd, Malta.

Barker, J. R., and D. T. Tingey. 1992. *Air Pollution Effects on Biodiversity*. Springer Science and Business Media, New York.

Behera, M. C., and R. Raina. 2012. Gentiana Kurroo Royle—A critically endangered bitter herb. *International Journal of Medicinal and Aromatic Plants* 2, no. 1 (March): 2249–4340.

Behera, S. G., P. G. Areendran, and V. Sagar. 2011. *For a Living Ganga—Working with People and Aquatic Species*, WWF-India, New Delhi, 84.

Bentley, R. 2010. *Medicinal Plants*. Domville-Fife Press, London, 23–46.

Bhattacharjee, S., S. Bhattacharya, S. Jana, and D. S. Bhagel. 2013. A review on medicinally important species of Picrorhiza. *International Journal of Pharmaceutical Research and Bio-Science* 2, no. 4: 1–6.

Bustamante, M., E. H. Helmer, S. Schill, et al. 2018. Direct and indirect drivers of change in biodiversity and nature's contributions to people. In *IPBES Regional Assessment Report on Biodiversity and Ecosystem Services for The Americas*, eds. J. Rice et al. 335–509. Secretariat of the Intergovernmental Science-Policy Platform on Biodiversity and Ecosystem Services, Bonn, Germany. Available at https://www.ipbes.net/assessment-reports/americas.

Chaudhary, M., S. Mishra, and A. Kumar. 2017. Estimation of water pollution and probability of health risk due to imbalanced nutrients in River Ganga, India. *International Journal of River Basin Management* 15, no. 1 (January): 53–60. https://doi.org/10.1080/15715124.2016.1205078.

Chaudhary, M., and T. R. Walker. 2019. River Ganga pollution: Causes and failed management plans (Correspondence on Dwivedi et al. 2018. Ganga Water pollution: A potential health threat to inhabitants of Ganga basin. *Environment International* 126 (February): 202–206. https://doi.org/10.1016/j.envint.2019.02.033.

Chen, S. L., H. Yu, H. M. Luo, Q. Wu, C. F. Li, and A. Steinmetz. 2016. Conservation and sustainable use of medicinal plants: Problems, progress, and prospects. *Chinese Medicine* 11, no. 1 (July): 1–10. https://doi.org/10.1186/s13020-016-0108-7.

CITES. 2018. *Seventieth Meeting of the Standing Committee, Species Specific Matters: Asian Big Cats (Felidae spp.): Report of the Secretariat*, SC70 Doc. 51. Retrieved from https://cites.org/eng/com/sc/70/index.php.

Clements, R., N. S. Sodhi, M. Schilthuizen, and P. K. L. Ng. 2006. Limestone karsts of Southeast Asia: Imperiled arks of biodiversity. *BioScience* 56, no. 9 (September): 733–742. https://doi.org/10.1641/0006-3568(2006)56[733:LKOSAI]2.0.CO;2.

Convention on Biological Diversity. 1992. *Convention on Biological Diversity*. Secretariat of the Convention on Biological Diversity, Montreal, Canada.

Dandapat, A., D. Banerjee, and D. Chakraborty. 2010. The case of the disappearing house sparrow (Passer Domesticus Indicus). *Veterinary World* 3, no. 2 (February): 97–100.

Datar, A., P. Audet, and D. Mulligan. 2011. *Post-Mined land rehabilitation in india: Cataloguing plant species used in land revegetation*. Centre for Mined Land Rehabilitation, Sustainable Minerals Institute, The University of Queensland, St. Lucia Campus, Brisbane, 4072 QLD, Australia.

Debnath, P., S. Rathore, and S. Walia. 2020. Picrorhiza Kurroa: A promising traditional therapeutic herb from higher altitude of Western Himalayas. *Journal of Herbal Medicine* 23 (October): 100358. https://doi.org/10.1016/j.hermed.2020.100358.

Dhiman, N., N. K. Sharma, P. Thapa, et al. 2019. De novo transcriptome provides insights into the growth behaviour and resveratrol and trans-stilbenes biosynthesis in Dactylorhiza Hatagirea—an endangered alpine terrestrial orchid of western himalaya. *Scientific Reports* 9, no. 1 (September): 1–13. https://doi.org/10.1038/s41598-019-49446-w.

Dhiman, N., and A. Bhattacharya. 2020. *Nardostachys jatamansi* (D.Don) DC.-Challenges and opportunities of harnessing the untapped medicinal plant from the Himalayas. *Journal of Ethnopharmacology* 246 (January): 112211. https://doi.org/10.1016/j.jep.2019.112211.

Elmqvist, T., and E. Maltby. 2010. Biodiversity, ecosystems and ecosystem services. In *The Economics of Ecosystems and Biodiversity: Ecological and Economic Foundations*, ed. P. Kumar, 41–111. Earthscan, Oxford.

FAO. 2012. *The State of World Fisheries and Aquaculture—2012*. Rome, 209 pp. Retrieved from https://www.fao.org/publications/card/en/c/81ea54da-0f65-56c6-91d2-8257b8ac4136/.

FAO and UNEP. 2020. The State of the World's Forests. *Forests, Biodiversity and People*. Rome. https://doi.org/10.4060/ca8642en.

Farnsworth, N. R., and R. W. Morris. 1976. Higher plants: The sleeping giant for drug development. *American Journal of Pharmacology* 148, no. 2: 46–52.

Frank, K. T., B. Petrie, J. S. Choi, and W. C. Leggett. 2005. Ecology: Trophic cascades in a formerly cod-dominated ecosystem. *Science* 308, no. 5728 (June): 1621–1623. https://doi.org/10.1126/science.1113075.

Freedman, B. 2018. The biodiversity crisis. In *Environmental Science: A Canadian Perspective*, 6th edition, Dalhousie University Libraries, Nova Scotia, Canada.

Fuller, R. A., K. N. Irvine, P. Devine-Wright, P. H. Warren, and K. J. Gaston. 2007. Psychological benefits of greenspace increase with biodiversity. *Biology Letters* 3, no. 4 (August 22): 390–394. https://royalsocietypublishing.org/doi/abs/10.1098/rsbl.2007.0149.

Giles, J. 2005. Nitrogen study fertilizes fears of pollution. *Nature* 433, no. 7028 (February 24): 791.

Habibullah, M. S., B. H. Din, C. W. Chong, and A. Radam. 2016. Tourism and biodiversity loss: Implications for business sustainability. *Procedia Economics and Finance* 35 (January 1): 166–172.

Hall, C. M. 2010. Changing paradigms and global change: From sustainable to steady-state tourism. *Tourism Recreation Research* 35, no. 2: 131–143. www.tandfonline.com/doi/abs/10.1080/02508281.2010.11081629.

Hamilton, A. C. 2004. Medicinal plants, conservation and livelihoods. *Biodiversity and Conservation* 13, no. 8 (July): 1477–1517.

Hoare, A. 2015. Tackling illegal logging and the related trade: What progress and where next? *Chatham House Report*: 79. www.chathamhouse.org/%0Ahttps://forestgovernance.chathamhouse.org/countries/republic-of-the-congo.

IFAW. 2007. *Made in China—Farming Tigers to Extinction*. International Fund for Animal Welfare, Yarmouth.

International Tanker Owners Pollution Federation (ITOPF). 1990. *Response to Marine Oil Spills*. Witherby, London.

IUCN Species Survival Commission. Medicinal Plant Specialist Group. 2007. *Why Conserve and Manage Medicinal Plants*. https://www.google.com/search?q=IUCN+Species+Survival+Commission.+Medicinal+Plant+Specialist+Group.+2007.+Why+Conserve+and+Manage+Medicinal+Plants.

Jain, M. K., and A. Das. 2017. Impact of mine waste leachates on aquatic environment: A review. *Current Pollution Reports* 3, no. 1 (March 1): 31–37. https://link.springer.com/article/10.1007/s40726-017-0050-z.

Jhunjhunwalla, A. 2010. Local market report; natural essential oil of Indian origin. *Indian Perfum* 54, no. 4: 13.

Jiang, Z., Y. Huang, X. Xu, Y. Liao, L. Shou, J. Liu, Q. Chen, and J. Zeng. 2010. Advance in the toxic effects of petroleum water accommodated fraction on marine plankton. *Acta Ecologica Sinica* 30, no. 1 (February 1): 8–15.

Jones, M. L. M., H. L. Wallace, D. Norris, et al. 2004. Changes in vegetation and soil characteristics in coastal sand dunes along a gradient of atmospheric nitrogen deposition. *Plant Biology* 6, no. 5 (September 1): 598–605. https://onlinelibrary.wiley.com/doi/full/10.1055/s-2004-821004.

Jones, T., B. Phillips, C. E. Williams, and J. Pittock. 2003. *Managing Rivers Wisely: Lessons from WWF's Work for Integrated River Basin Management*. WWF International, Gland, Switzerland.

Kala, C. P. 2004. Assessment of species rarity. *Current Science* 86: 1058–1059.

Kleinschmit, D., S. Mansourian, C. Wildburger, and A. Purret. 2016. Illegal logging and related timber trade—dimensions, drivers, impacts and responses. A global scientific rapid response assessment report. *IUFRO World Series Volume 35*. Vienna, p. 148. https://dspace.stir.ac.uk/bitstream/1893/24818/1/Ch2fromWorldSeries35-FullReport(low-resolution).pdf.

Koehnken, L., and M. Rintoul. 2018. *Impacts of Sand Mining on Ecosystem Structure, Process and Biodiversity in Rivers*. WWF, Gland, Switzerland.

Kukreti, I. 2017. How will India address illegal sand mining without any data? *Down to Earth*. www.downtoearth.org.in/news/mining/flouted-with-impunity-58736 (accessed April 8, 2021).

Kumar, P. 2006. The Economics of medicinal plants: Are high commercial values enough to ensure biodiversity conservation? In *Conserving Medicinal Species: Securing a Healthy Future*, ed. S. Miththapala, 184. IUCN: Ecosystems and Livelihoods Group, Colombo, Sri Lanka.

Kumar, P., A. Kumar, M. Patil, N. K. Sharma, and A. N. Singh. 2020. Plant diversity development under planted woody species on coal mine spoil in a dry tropical environment, India: A case study. *Journal of Ecological Engineering* 21, no. 2 (January): 228–243. https://doi.org/10.12911/22998993/116330.

Kumar, V., and J. Van Staden. 2016. A review of Swertia Chirayita (Gentianaceae) as a traditional medicinal plant. *Frontiers in Pharmacology* 6 (January): 308.

Lead, C., A. J. Beattie, W. Barthlott, et al. 2005. New products and industries from biodiversity. *Ecosystems and Human Well-Being: Current State and Trends* 1 (December): 271.

Lee, J., B. J. Park, Y. Tsunetsugu, T. Kagawa, and Y. Miyazaki. 2009. Restorative effects of viewing real forest landscapes, based on a comparison with urban landscapes. *Scandinavian Journal of Forest Research* 24, no. 3 (June): 227–234.

Leighton, F. A. 1993. The toxicity of petroleum oils to birds. *Environtal Reviews* 1, no. 2: 92–103. https://cdnsciencepub.com/doi/abs/10.1139/a93-008.

Lewis, D. 2020. Cleaning up after Mauritius oil spill. *Nature* 585, no.7824 (September): 172.

Lim, C. K., G. Cranbrook. 2002. Swiftlets of Borneo: builders of edible nests. *Natural History Publications Borneo*: 171. http://myagric.upm.edu.my/id/eprint/10324/.

Loodin, N. 2020. Aral Sea: an environmental disaster in twentieth century in Central Asia. *Modeling Earth Systems and Environment* 6, no. 4: 2495–2503. https://doi.org/10.1007/s40808-020-00837-3.

Mader, A., and H. Scheyvens. 2019. *Biodiversity and Industry*. GEO-6 for Industry in Asia-Pacific, pp. 46–54.

Maskell, L. C., S. M. Smart, J. M. Bullock, K. Thompson, and C. J. Stevens. 2010. Nitrogen deposition causes widespread loss of species richness in british habitats. *Global Change Biology* 16, no. 2 (February 1): 671–679.

Millennium Ecosystem Assessment (MEA). 2005. *Ecosystems and Human Well-being: Synthesis*. Island Press, Washington, DC. https://www.millenniumassessment.org/en/index.html.

Miththapala, S. 2008. *Coral Reefs. Coastal Ecosystems Series* (Vol. 1). Ecosystems and Livelihoods Group Asia, IUCN, Colombo, Sri Lanka.

MoEFCC (Ministry of Environment, Forest and Climate Change). 2017. http://envfor.nic.in/legis/ucp/ucpsch8.html (accessed April 3, 2021).

Mora, C., D.P. Tittensor, S. Adl, A. G. B. Simpson, and B. Worm. 2011. How many species are there on Earth and in the Ocean? *PloS Biology* 9, no. 8 (August): 1001127. www.plosbiology.org.

Murguía, D. I., S. Bringezu, and R. Schaldach. 2016. Global direct pressures on biodiversity by large-scale metal mining: Spatial distribution and implications for conservation. *Journal of Environmental Management* 180 (September): 409–420.

Myers, R. A., J. K. Baum, T. D. Shepherd, S. P. Powers, and C. H. Peterson. 2007. Cascading effects of the loss of apex predatory sharks from a coastal ocean. *Science* 315, no. 5820 (March 30): 1846–1850.

Nadda, R. K., A. Ali, R. C. Goyal, P. K. Khosla, and R. Goyal. 2020. Aucklandia costus (Syn. Saussurea Costus): ethnopharmacology of an endangered medicinal plant of the Himalayan region. *Journal of Ethnopharmacology* 263 (December): 113199. https://doi.org/10.1016/j.jep.2020.113199.

National Biodiversity Authority (NBA). 2018. *Annual Report 2017–18*. http://nbaindia.org/uploaded/Annual_report_2017-18_english.pdf (accessed April 8, 2021).

Nicholas, K., J. Fanzo, and K. MacManus. 2018. Palm oil in Myanmar: A spatiotemporal analysis of the effects of industrial farming on biodiversity loss. *Global Health Science and Practice* 6, no. 1 (March 1): 210–222. www.ghspjournal.org.

Nimasow, G., O. D. Nimasow, J. S. Rawat, and L. Norbu. 2015. Conservation efforts of an important medicinal plant (Taxus baccata Linn.) in West Kameng district of Arunachal Pradesh (India). *Earth* 4, no. 3–1 (March): 1–10.

Niyogi, D. K., W. M. Lewis, and D. M. McKnight. 2002. Effects of stress from mine drainage on diversity, biomass, and function of primary producers in mountain streams. *Ecosystems* 5, no. 6 (September): 554–567. https://link.springer.com/article/10.1007/s10021-002-0182-9.

Nowell, K., and X. Ling. 2007. *Taming the Tiger Trade*. Hong Kong: TRAFFIC East Asia.

Ong, S., and E. Carver. 2019. The rosewood trade: An illicit trail from forest to furniture. In *Yale Environment* 360 [online]. New Haven, CT, USA. https://e360.yale.edu/features/the-rosewood-trade-theillicit-trail-from-forest-to-furniture.

Pandey, H., S. K. Nandi, A. Kumar, U. T. Palni, and L. M. S. Palni. 2007. Podophyllotoxin content in Podophyllum hexandrum Royle plants of known age of seed origin and grown at a lower altitude. *Acta Physiologiae Plantarum* 29, no. 2 (January): 121–126.

Pant, S., and S. S. Samant. 2008. Population ecology of the endangered Himalayan Yew in Khokhan Wildlife Sanctuary of North Western Himalaya for conservation management. *Journal of Mountain Science* 5, no. 3 (June): 257–264.

Perkol-Finkel, S., and Y. Benayahu. 2007. Differential recruitment of benthic communities on neighboring artificial and natural reefs. *Journal of Experimental Marine Biology and Ecology* 340, no. 1 (January): 25–39.

Piatt, J. F., C. J. Lensink, W. Butler, M. Kendziorek, and D. R. Nysewander. 1990. Immediate impact of the'Exxon Valdez'oil spill on marine birds. *The Auk* 107, no. 2 (April): 387–397.

Rajasuriya, A., H. Zahir, K. Venkataraman, Z. Islam, and J. Tamelander. 2004. Status of coral reefs in South Asia: Bangladesh, Chagos, India, Maldives and Sri Lanka. In *Coral Reefdegradation in the Indian Ocean Status Report*, ed. D. Souter and O. Linden, 213–233. CORDIO, Sweden.

Rana, M. S., and S. S. Samant. 2011. Diversity, indigenous uses and conservation status of medicinal plants in Manali wildlife sanctuary, North western Himalaya. *Indian Journal of Traditional Knowledge* 10, no. 3 (July): 439–459.

Rathore, S., P. Debnath, and R. Kumar. 2021. Kuth {Saussurea costus (Falc.) Lipsch.}: A critically endangered medicinal plant from Himalaya. *Journal of Applied Research on Medicinal and Aromatic Plants* 20 (February): 100277.

Roe, D., N. Seddon, and J. Elliott. 2019. Biodiversity loss is a development issue: A rapid review of evidence. *IIED Issue Paper* no. 798 (April): 678–683.

Saadoun, I. M. K. 2015. Impact of oil spills on marine life, emerging pollutants in the environment—current and further implications, Marcelo L. *Larramendy and Sonia Soloneski*, IntechOpen.

Sajjad, F., U. Noreen, and K. Zaman. 2014. Climate change and air pollution jointly creating nightmare for tourism industry. *Environmental Science and Pollution Research* 21, no. 21 (June): 12403–12418.

Schippmann, U., D. J. Leaman, and A. B. Cunningham. 2002. Impact of cultivation and gathering of medicinal plants on biodiversity: global trends and issues. FAO (2002) Biodiversity and the ecosystem approach in agriculture. Forestry and Fisheries, Rome, 1–21.

Sen, T., and S.K. Samanta. 2014. Medicinal plants, human health and biodiversity: A broad review. *Biotechnological Applications of Biodiversity*: 59–110.

Sharma, P., and M. Binner. 2020. The decline of population of house sparrow in India. *International Journal of Agricultural Science* 5 (January): 1–4.

Sharma, S., and R. Thokchom. 2014. A review on endangered medicinal plants of India and their conservation. *Journal of Crop and Weed* 10, no. 2: 205–218.

Shi, W., M. Wang, and W. Guo. 2014. Long-term hydrological changes of the Aral Sea observed by satellites. *Journal of Geophysical Research: Oceans* 119, no. 6 (May): 3313–3326.

Singh, A. N., A. S. Raghubanshi, and J. S. Singh. 2002. Plantations as a tool for mine spoil restoration. *Current Science* 82, no. 12 (June): 1436–1441.

Slingenberg, A., L. Braat, H. V. D. Windt, K. Rademaekers, L. Eichler, and K. Turner. 2009. Study on understanding the causes of biodiversity loss and the policy assessment framework. *European Commission Directorate-General for Environment*. http://ec.europa.eu/environment/enveco/biodiversity/pdf/causes_biodiv_loss. Pdf.

Sonter, L. J., S. H. Ali, and J. E. M. Watson. 2018. Mining and biodiversity: Key issues and research needs in conservation science. *Proceedings of the Royal Society B* 285, no. 1892 (December): 20181926.

Sonter, L. J., M. C. Dade, J. E. Watson, and R. K. Valenta. 2020. Renewable energy production will exacerbate mining threats to biodiversity. *Nature Communications* 11, no. 1 (September): 1–6.

Stevens, C. J., N. B. Dise, J. O. Mountford, and D. J. Gowing. 2004. Impact of nitrogen deposition on the species richness of grasslands. *Science* 303, no. 5665 (March): 1876–1879.

Stevens, C. J., C. Duprè, E. Dorland, et al. 2010. Nitrogen deposition threatens species richness of grasslands across Europe. *Environmental Pollution* 158, no. 9 (September): 2940–2945.

Thani, P. R. 2018. Phytochemical studies on Indian market. *Research Journal of Agriculture and Forest* 6, no. 2 (February): 1–5.

TRAFFIC. 2019. *African Elephants. Elephant Conservation and the Global Trade in Ivory.* https://blog.iufro.org/tag/illegal-logging/ (accessed April 7, 2021).

UNEP. 2013. *Green Economy and Trade—Trends, Challenges and Opportunities.* www.unep.org/greeneconomy/GreenEconomyandTrade (accessed April 5, 2021).

UNEP-WCMC. 2017. *Mainstreaming of Biodiversity into the Energy and Mining Sectors: An Information Document for the 21st Meeting of the Subsidiary Body on Scientific, Technical and Technological Advice (SBSTTA-21).* UNEP-WCMC, Cambridge.

Uniyal, R. C., M. R. Uniyal, and P. Jain. 2000. *Cultivation of Medicinal Plants in India: A Reference Book.* TRAFFIC-India, New Delhi, India.

Uniyal, S. K., A. Awasthi, and G. S. Rawat. 2002. Current status and distribution of commercially exploited medicinal and aromatic plants in upper Gori valley, Kumaon Himalaya, Uttaranchal. *Current Science* 82, no. 10 (May): 1246–1252.

UNODC. 2020. *World Wildlife Crime Report: Trafficking in Protected Species.* www.unodc.org/documents/data-and-analysis/wildlife/2020/World_Wildlife_Report_2020_9July.pdf.

Vasey, N., M. Mogilewsky, and G. E. Schatz. 2018. Infant nest and stash sites of variegated lemurs (Varecia rubra): The extended phenotype. *American Journal of Primatology* 80, no. 9 (September): e22911.

Wani, I. A., V. Kumar, S. Verma, A. T. Jan, and I. A. Rather. 2020. Dactylorhiza hatagirea (D. Don) Soo: A critically endangered perennial orchid from the North-West Himalayas. *Plants* 9, no. 12 (November): 1644.

Whiting, M. J., L. W. Vivienne, and J. H. Toby. 2011. Animals traded for traditional medicine at the Faraday market in South Africa: Species diversity and conservation implications. In *Animals in Traditional Folk Medicine*, 421–473. Springer, Berlin, Heidelberg.

Wilkinson, C. 2004. *Status of Coral Reefs of the World, 2004* (Vol. 1). Australian Institute of Marine Science, Townsville, Australia, Xiv, 301.

Wong, C. M., C. E. Williams, J. Pittock, U. Collier, and P. Schelle. 2007. *World's Top 10 Rivers at Risk.* WWF International, Gland, Switzerland.

Wong, H. K. T., A. Gauthier, and J. O. Nriagu. 1999. Dispersion and toxicity of metals from abandoned gold mine tailings at Goldenville, Nova Scotia, Canada. *Science of the Total Environment* 228, no. 1 (March): 35–47.

4

Extremophilic Diversity and Climate Change

Amrita Kumari Panda, Rojita Mishra, Rashi Miglani, Seeta Dewali, Ankit Kumar, Shikha Bora, and Satpal Singh Bisht

CONTENTS

4.1 Introduction to Extremophiles: Life at the Extremes

Microorganisms can grow and thrive in each possible condition, including broader ranges of pH, temperature, radiation, pressure, and nutrient-depriving events on Earth. They are adapted to survive in severe space conditions with life-threatening radiation, pressure, highly flexible temperature, and microgravity (Yamagishi et al., 2018). Extremophiles are highly adapted organisms and remain metabolically active in extreme conditions across the planet. Extremophilic microorganisms have been identified from various environments, wherever water is available, including extreme hot and cold environments and acidic and alkaline conditions. Extremophiles have been isolated from the Earth's crust, deep-sea hydrothermal vents, permafrost, alkaline lakes, acid mine drainages, radioactive sites, and many more. Though extremophiles include archaea, bacteria, and eukarya, archaea constitute the core group to survive in extreme conditions. *Methanopyrus kandleri* strain 116 grows at 122°C (highest documented temperature), whereas *Picrophilus torridus* is able to grow at a pH of 0.06, the most acidophilic microorganism identified (Rampelotto et al., 2013).

It has long been in science's desire to investigate unexplained phenomena and create scientific hypotheses. In the fascinating world of microbes, extremophiles are the most intriguing group of life on Earth (Rothschild and Mancinelli, 2001) and perhaps also on other planets (Navarro-Gonzalez et al., 2003).

DOI: 10.1201/9781003220398-4

Nature, of course, provides a plethora of chances for living forms that can consume or create enough energy to survive. Normal survival may not be feasible in severe settings (e.g., temperature, pressure, pH, salinity, geological scale and barriers, radiation, chemical extremes, lack of nourishment, osmotic barriers, or polyextremophilia). Extremophiles can live in such severe environments due to exceptional characteristics (mainly bacteria and archaea, with a few eukarya). It would be beneficial for human civilization to learn from extremophiles; they have the potential to help us deal with developing illnesses due to their clever adaptations and metabolic methods for survival under extreme environmental circumstances.

The ocean, which comprises 75% of the Earth, provides a varied habitat for life. Rasmussen (2000) discovered deep-sea microfossils of threadlike bacteria in 3,235-million-year-old volcanogenic sulfide deposits, the earliest fossil evidence for microbial life, in a Precambrian underwater thermal spring system. Other studies have given the facts of a suitable environment for all life-forms due to the one significant factor of life, water, which astrobiologists are investigating on other worlds. Following the discovery of hydrothermal vents in 1979, a distinct ecology with a range of prokaryotic and eukaryotic microorganisms that had adapted to the harsh climate and lack of energy from sunlight was seen. Researchers were inspired to explore microbial life under extreme conditions of temperature, pressure, oxygen, pH, and so on due to a lack of information and technology. Cold habitats, including the ocean, comprise the bulk of the biosphere on Earth and other planets. Lakes, deserts, caverns, and the high atmosphere (upper troposphere and lower stratosphere) are also being evaluated as permanent cold habitats for living species. Another severe situation in which bacteria may survive is anoxia. Anoxic environments (deep underground, sedimented bottoms of water bodies, deep sea, higher altitudes, and industrial effluent sites) and animal gut microbial flora exhibit a diverse range of anaerobic bacteria with extensive histories in chemical and fuel generation (Zeikus, 1980).

4.2 Extremophiles: Source of Novel Metabolites

Extremolytes are the results of extremophilic microbial metabolisms: They have proved their relevance to biotechnology in the form of enzymes, proteins, and primary and secondary metabolic products. On an industrial basis, there has been some success in manufacturing a variety of extremolytes. Uncultivable microbes can be discovered in severe environmental niches (Deppe et al., 2005). These bacteria obtain energy from unknown sources and contemporary science has yet to identify a suitable growing medium. On the other hand, advances in metagenomics may aid in the investigation of the unique characteristics of such uncultivable microbes (Singh, 2010; Singh and MacDonald, 2010). If suitable sources for growing uncultivable microorganisms can be identified, it may open new doors to the fascinating microbial world and its distinct features. Food is essential for organisms to sustain the appropriate energy levels for survival. Regardless of the abundance of food on Earth, human civilization is constantly concerned about food safety and security. Modern and sophisticated genetic engineering (GE) technologies can provide food security, but food safety remains a concern (Singh et al., 2006). Food restrictions enforced by government bodies rely on data given by food farmers (Singh, 2010). Most organisms employed in food processing are mesophiles; however, severe conditions are necessary for specific applications. Microorganisms that thrive in hostile conditions to other organisms provide a source of new bioproducts (extremozymes), which are byproducts of primary and secondary metabolites. To achieve the ever-increasing demand for energy, humanity may turn to nature, which provides vast renewable resources to replace fossil fuels. However, several concerns, including economics and technological preparedness, must be addressed. Taylor et al., (2010) discuss that extremophiles are utilized in biofuel research, while thermophiles are used in second-generation bioethanol production as cellulosic ethanol and biodiesel are the most prominent and urgent target fuels.

Microbial metabolic products with distinct properties, such as exopolysaccharides, provide a diverse spectrum of chemical structures with diverse uses in the food, pharmaceutical, and other industrial areas. Radiation in the form of particles or electromagnetic waves (for example, UV radiation, gamma rays, x-rays, radio waves, and so on) causes severe oxidative damage to essential biomolecules, such as proteins and nucleic acids. UV radiation and other radioactive chemicals have historically been associated

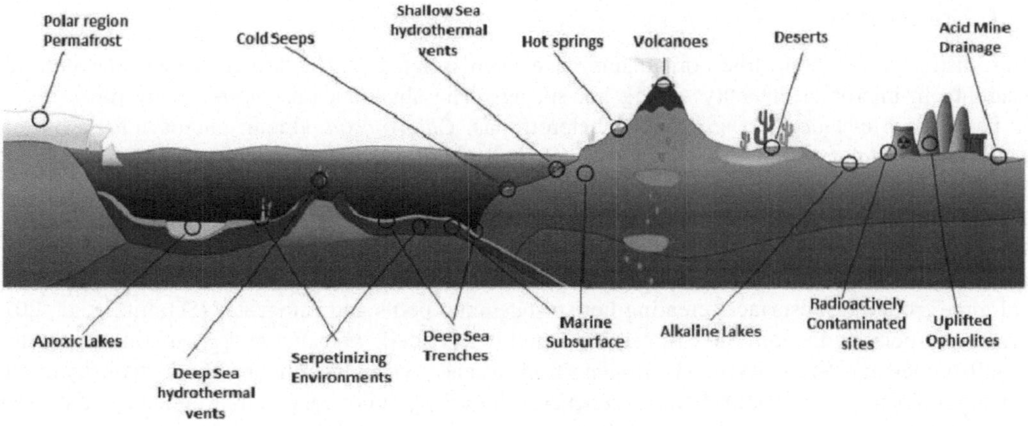

FIGURE 4.1 Earth's crust showing various extreme habitats.

(*Source:* Adapted from Merino et al., 2019, with permission)

with various negative consequences, including immunological suppression, dermatitis, premature aging, neurotoxicity, and skin cancer. Extremolytes are distinct organic molecules that are not directly engaged in the normal growth, development, or reproduction of organisms; nonetheless, their absence affects the organism's survival, fecundity, or aesthetics in the long run. These microbial reserves have been extensively researched for industrial applications; however, their therapeutic implications have yet to be examined. Exploration of strategic therapeutic uses of extremophiles in military and homeland security holds real promise. The potential for radioprotective medication development utilizing radioresistant extremophiles has yet to be identified. A quick overview of unknown applications will allow industry experts to look beyond the box by investing in research and technology development.

Due to rising industrial demand, significant effort has been expended in the quest for biocatalysts, enzymes and metabolites that can withstand industrial process conditions. Extremophiles have a high potential for future breakthroughs in biotechnology, medicines, and the elimination of certain hazardous chemicals from the environment due to their capacity to live in severe environments that would kill other creatures practically quickly. Extremozymes, such as thermostable amylase, are integrated into biochemical processes that occur at high temperatures in water-based solutions and could be replaced by high-cost reactants to reduce end-product costs. Besides that, the current nuclear arms race, environmental instability caused by ozone depletion, and solar flares reaching the Earth's surface ("M8.7 Solar Flare and Earth Directed CME," available at www.nasa.gov/mission pages/sunearth/news/News012312-M8.7.html) make normal life susceptible to both natural and man-made radiation. Radiation-resistant microorganisms include chemicals that might be used to develop radioprotective medicines, which could be beneficial in space projects to minimize undesired radiation exposure. Extremophile exploitation will undoubtedly advance in the next years to find treatments for radiation-mediated cancer illnesses and fulfill other industrial needs.

4.3 Abundant Taxa in Various Extreme Environments

Extreme habitats harbor distinctive microbial diversity, and many researchers are attracted to study the microbial community of these ecosystems due to their varied industrial applications. Extreme ecosystems have been proved as eukaryogenesis sites with the isolation of archaea *Candidatus Prometheoarchaeum syntrophicum* strain MK-D from deep marine sediment and provide new insights into the evolution of eukaryotes from such habitats (Imachi et al., 2020). Extremophiles are also striking models for astrobiology due to their connection with ancestral lineages of life on Earth, the potential to survive in various planetary bodies (DasSarma et al., 2020).

4.3.1 Hot Springs

Many distinctive endemic microorganisms have been isolated and identified despite the unlimited variability in microbial diversity among hot springs. The abundant taxa found many times in various hot springs include *Thermus* among heterotrophs, *Chloroflexus* among photosynthetic bacteria, *Synechococcus* among cyanobacteria, *Sulfolobus* and *Acidithiobacillus* among chemolithotrophs, and so on (Brock, 1978). Phylum Aquificae is also found in various geothermal ecosystems (Ferrera et al., 2007). The genus *Thermus* is widely exploited and bioprospecting as a source of many thermozymes among all hot spring microbes. *Aspergillus* and *Penicillium* are the dominant fungal taxa in hot springs (Odilia et al., 2018). Acidic volcanos are formed where mud, gas, and water are ejected due to high fluid force from the subsurface, creating hot springs, mud pools and fumaroles (Schmitz et al., 2021). Verrucomicrobial methanotrophs are usually found in hot acidic volcanic and geothermal ecosystems and utilize methane, hydrogen, and hydrogen sulfide as energy sources. The metabolic diversity of these methanotrophs plays a crucial role in biogeochemical cycling in hot acidic environments and mitigates greenhouse gas emissions from geothermal environments.

4.3.2 Deep-Sea and Hydrothermal Vents

Deep-sea hydrothermal vents harbor varied microbial communities due to high atmospheric pressure and environmental ascents. Many distinctive microbial lineages in these niches are reported to be less abundant. Epsilonproteobacteria and Gammaproteobacteria among bacteria and Archaeoglobi and Halobacteria among archaea occur abundantly in hydrothermal vent samples of the Hulk vent on the Juan de Fuca Ridge (Anderson et al., 2015). They also reported the high abundance of Thermococcales in high-temperature vent samples and Thermoplasmata in diffuse flow samples. Proteobacteria (54.74%), Bacteroidetes (24.36%) and Firmicutes (9.48%) were abundant phyla in deep-sea sediment samples of the southern Colombian Caribbean Sea (Franco et al., 2021). They also reported the declined bacterial alpha diversity, whereas the dominance of several genera in deeper marine sediments. Organic matter plays a crucial role in determining the microbial diversity in all deep-sea samples. Ojos del Salado, the highest volcano on Earth and its fumarolic creek sediments, contains acidophilic iron oxidizers (*Ferrithrix, Gallionella*) and iron reducers (*Acidiphilium*). The pH of this sediment was reported to be 4.48 (Aszalos et al., 2020).

4.3.3 Cold Environment

The dry valleys of Antarctica are a unique, extreme, low-complex niche to study microbial diversity as there are no higher plants, plant-soil interactions and very narrow soil invertebrate (Treonis et al., 1999) and protozoan (Fell et al., 2006) diversity. Van Horn et al., 2014 studied the bacterial community composition in McMurdo Dry Valleys, Antarctica, and found Acidobacteria (33%), followed by Actinobacteria (17%), Firmicutes (15%), unknown bacteria (15%), Proteobacteria (7%), Deinococcus-Thermus (3%), Verrucomicrobia (3%), Bacteroidetes (2%), and Spirochaetes (2%). The pH, soil organic carbon (SOC), microbial biomass carbon (MBC), and conductivity are the four variables that regulate the community structure in Antarctic soil (Lee et al., 2012; Zeglin et al., 2011). The soil organic matter (SOM) in the dry valley is lower and the ratio of MBC to SOM is higher due to the absence of higher plants compared to other ecosystems. The negative relationship between MBC and SO_4 indicates rises in salt-related variables that negatively impact soil microbes (Van Horn et al., 2014).

4.3.4 Acidic and Alkaline Environments

A meta-analysis study reported that the dominance of chemolithotrophs and *Hydrogenobaculum* was abundant in acidic springs (Massello et al., 2020). The microbiome study of two soda-saline lakes in Inner Mongolia of China reported 48 metagenome-assembled genomes from class Halobacteria among Euryarchaeota (Zhao et al., 2020). They reported Proteobacteria to be the abundant phyla, followed by Firmicutes, Bacteroidetes, and Actinobacteria. *Acidiella polonica*, a fungus isolated from highly acidic (pH 2.7–3.8) technogenic soil in Poland (Kolarik et al., 2021). Few acidophiles, *Acidithiobacillus*, *Ferroplasma*, and *Leptospirillum* have been discovered from the Copahue Volcano–Río Agrio system (Lopez et al., 2020). Lake Paitu of China has *Ferrovum*, the major iron-oxidizing bacterium in most seasons (pH 3.01–3.16). Lake Tafang of China has *Leptospirillum and Acidithiobacillus* (pH 2.43–2.75) as abundant taxa. Similarly, *Methylacidiphilum* and *Methylacidimicrobium* are the abundant taxa from extremely acidic geothermal systems (Kucera et al., 2020). The Euryarchaeotal community and *Halomonas*- and *Nitrincola*-dominated bacterial community are found in the production water of the alkaline oil reservoir (Chen et al., 2020). Leptolyngbyaceae, Oscillatoriales, and Pseudanabaenaceae are the abundant cyanobacteria found in the Alkaline Springs of the Voltri Massif, Italy (Kamran et al., 2021). Bacteroidetes, Alphaproteobacteria, Gammaproteobacteria, and Euryarchaeota were identified as abundant taxa in the brines of alkaline soda lakes. (Vavourakis et al., 2016). Phototrophic Chloroflexi are the abundant taxa in alkaline hot springs (Bennett et al., 2020).

4.4 Climate Change Affecting the Extremophilic Community

Extremophiles harbor novel adaptation mechanisms to exist in the era of climate change and it is expected to lose many extremophiles, especially Psychrophiles, as their habitats such as sea ice, permafrost vanished due to climate change.

4.4.1 Cold Environment

The Arctic includes the largest cold environments on Earth and is heating at a very high rate with the highest surface temperature reported in the year 2020 per the Arctic Report (https://arctic.noaa.gov/Report-Card/Report-Card-2020). The Greenland ice sheet faced greater ice loss, whereas glaciers outside of Greenland have sustained ice loss, mainly from Alaska and Arctic Canada (www.arctic.noaa.gov/Report-Card/Report-Card-2020). The sea ice amount has declined approximately 50% from 1979 to 2019 during the melting month of September (Fetterer et al., 2017; Serreze and Meier, 2019). Arctic sea ice loss occurs intensely in the 21st century due to Arctic amplification and global warming due to greenhouse gas emissions. A significant amount of sea ice was found in the Arctic Ocean 30 years ago, but only small patches of the oldest ice persisted (Figure 4.2) on the Greenland coast.

There are reports that expansion of higher plant populations occurred due to global warming (Amesbury et al., 2017), and the rate of photosynthesis in mosses rises with increasing temperatures (Perera-Castro et al., 2020). The community structure and soil microbial activity either surge or decline with the rise of temperature, a subject to water and nutrient accessibility (Benhua et al., 2014). Yergeau et al. (2009) reported a decrease in bacterial and archaeal communities with increasing latitude in Antarctic soils. The Antarctic soils are estimated to have more water and nutrient accessibility due to permafrost thawing (Fountain et al., 2014). Buelow et al. (2016) reported a loss of taxonomic diversity in Antarctic soils due to a rise in organic matter and soil moisture due to climate change. They observed taxonomic loss within the bacteria and many transcripts of specific pathways lost by bacteria. Proteobacteria, Actinobacteria, and Firmicutes were dominant in soil with increased water and many oligotrophic endemic taxa lost, diversity declined due to climate change (Van Horn et al., 2014).

The role of microorganisms in polar ecosystems is crucial in the biogeochemical cycle and carbon sequestration (Cavicchioli, 2015), especially the phytoplankton fixing carbon and driving the nutrient cycling. The microbial diversity associated with sea ice is unique, and when sea ice melts, the microbial processes regulate the nutrient delivery to surrounding ocean water (Boetius et al., 2015). Recent studies reported large diversity of cyanobacterial nifH gene has been found in sea-ice brine. A five-year field study in the fell fields of southern maritime Antarctic reported that the metabolism of decomposer fungus *Pseudogymnoascus roseus* in Antarctic soil was reduced not only by hyphal extension rate but also the extracellular enzyme activities reduced by 54% (Misiak et al., 2021). They observed climate warming inhibits the growth of cold-adapted microbes, negatively impacting soil microbial activities as the Earth's climate continues to warm. Substantial summer warming in Arctic regions considerably reduced litter mass loss by 32% and litter moisture content in the dry sites of Greenland (Christiansen et al., 2017).

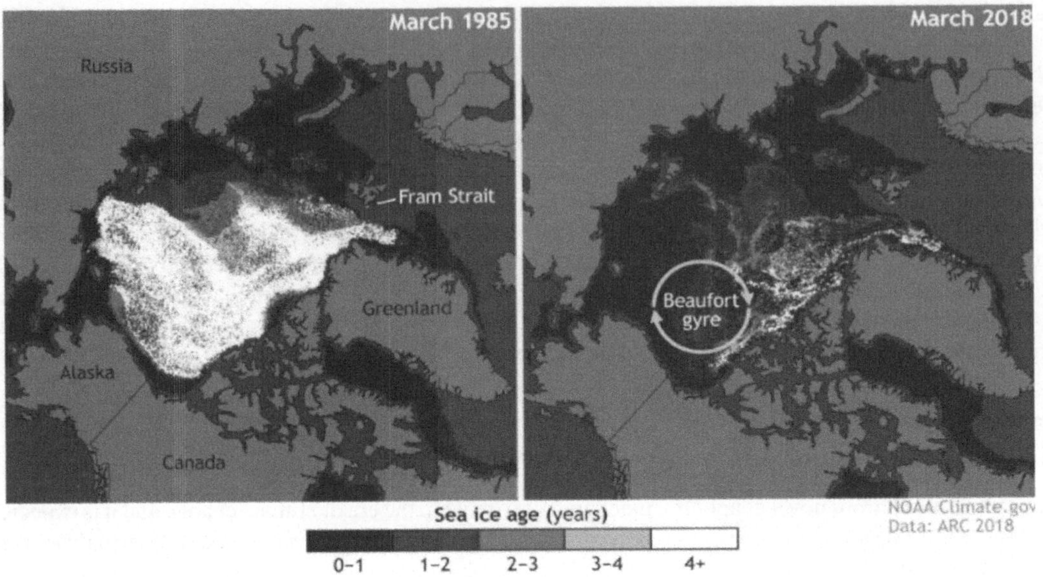

FIGURE 4.2 Loss of old ice over the years.

(*Source:* NOAA Climate.gov based on data from Climate Change: Arctic sea ice by Lindsey and Scott 2020 with permission.)

4.4.2 Hot Environment

It was reported that global mean surface temperature has ascended from 0.5 to 1°C from the year 1980 to 2018 (Hoegh-Guldberg et al., 2019). California, Arizona, Colorado, and Minnesota wildfires of 2021 are examples of the disastrous impact of climate change. Panda et al., 2016 reported the shift in the bacterial community in two alkaline hot springs of Northeastern India with different temperatures; at 46°C Firmicutes, Chloroflexi, and Thermi were dominant bacterial phyla in Jakrem, Meghalaya, India, whereas at 39°C, Proteobacteria were found to be dominant in the Yumthang hot spring, Sikkim, India. Future research is needed to explore extremophiles and exploit them in the production of secondary metabolites, combating multiple stress conditions. For example a novel strain, *Micromonospora* YIM 78104, from hot springs in Tengchong, China, proved to have a broad antibacterial properties (Liu et al., 2016), the metabolites from the particular strain can be further prospected to develop new antibiotics. Similarly, Gholami et al. (2015) identified a novel strain, *Kocuria rosea* MG2, from the radioactive Ab-e-Siah spring, Iranand, were able to withstand the high levels of UV radiation and desiccation. They reported the occurrence of carotenoids that helps with the photoprotective mechanism in this polyextremophile. Such studies support that identifying and exploring these extremophiles from hot environments could be the potential source of therapeutic agents, such as antioxidants and antibiotics.

4.4.3 Deep-Sea Sediment

The deep sea is an extreme environment where there is no light and very high atmospheric pressure exists. Few extremophiles still adapted and established unlimited biosynthetic capacity in these niches (Dalmaso et al., 2015). The microbiome of these ecosystems is less explored due to sampling difficulties, but the recent advances in next-generation sequencing technologies identified novel species. Extracts of *Streptomyces cyaneofuscatus* M-157 and M-192 exhibited the maximum cytotoxic activity against the cancer cells. *Streptomyces xiamenensis* M186 and *S. cyaneofuscatus* M190 produce β-elemene, which is used in brain and breast cancer therapy (Hui et al., 2021). Lai et al. (2020) reported that the organic carbon content and glacial-interglacial cycles shaped the archaeal community and observed a probable connection between climate change and the deep-sea microbial community.

4.5 Adaption of Extremophiles to Changing Climate Conditions

Extremophiles have evolved many mechanisms for their survival. In archaea, sugar glycerates like mannosyglycerates and glucosylglycerates are produced in thermal stress and halophilic and hyperthermophilic conditions (Singh et al., 2019). Life at high temperatures is possible if microbes learn to protect their proteins and nucleic acids in such high temperatures. One such adaptation is high genomic GC content, which helps in the stability of nucleic acids. Another aspect of stable DNA in thermophiles is the presence of enzyme reverse gyrase, mainly in hyperthermophiles (Schroder et al., 2020). This enzyme helps in the stabilization of DNA by modifying the DNA topology. Some have intracellular high salt concentration to manage depurination and cytosine deamination (Wang et al., 2006). Heat shock proteins are also present in extremophiles, which prevent protein denaturation. They have ether lipids in their membrane, mainly tetraether structures of the membrane help them to survive in high-temperature conditions (Rampelotto, 2013). Psychophiles have enzymes having few charged surface amino acids, while thermophiles have enzymes with more hydrophobic core (Schroder et al., 2020). Various types of adaptation mechanisms are present in different forms of extremophiles as described in the following sections.

4.5.1 Adaptations of Psychrophiles

Psychrophiles survive in a temperature from 0–20°C. They have unsaturated fatty acids in the cell membrane, cold shock proteins and cryoprotectants, and a low transcription and translation level. They have compatible solutes inside the cell, like glycine betaine, mannitol, and sucrose. Psychrophiles induce exopolysaccharides to protect them from cold stress. These microbes developed altered mechanisms for nutrient transport (D'Amico et al., 2006; Margesin et al., 2007; Pulicherla et al., 2011; De Maayer et al., 2014). The adaptive strategies of Psychrophiles are summarized in Table 4.1.

4.5.2 Adaptations of Thermophiles

Thermophiles grow in high-temperature habitats as they have ether lipids in their cell membrane. They can control nutrient flow from inside and outside of the cell. They have high GC content in their genome and more charged amino acids on the surface of the protein and intramolecular salt bridges. These adaptations parameters help the thermophiles to adapt more because of environmental changes (Mehta et al., 2016).

4.5.3 Adaptations of Alkaliphiles

The organisms that can grow in a high-pH environment are known as alkaliphiles. These organisms regulate the intake of hydrogen with the help of secondary antiporters. They have acidic amino acids, teichuronic acids outside the cell. They have cytochrome carbon that regulates the transfer of protons on the membrane. They synthesize phosphoserine aminotransferase, which helps them to survive in high alkaline conditions (Dubnovitsky et al., 2005; Coker, 2019)

4.5.4 Adaptations of Acidophiles

The organisms found in high-acidic conditions like acid sulfur springs, marine volcanic vents, and acid mine drainage are called acidophiles. They require intracellular neutral pH as they have membrane impermeability due to the presence of tetra-ether lipids. They adapt to the changing climate by reducing the membrane pore size. They possess proton efflux systems like H1 ATPases, antiporters, and sympoters. Inside their cells, arginine, histidine, and lysine are present in a higher quantity, which

TABLE 4.1
Adaptive Strategies by Psychrophiles in Cold Environments

Strategy	Habitat	Extremophilic organism	Adaptation mechanism	Reference
Pigment	Glaciers	*Flavobacterium bomense* sp. nov.	Adaptation to cold and harsh environments by the production of flexirubin-type pigments	Liu et al., 2019
Enhanced unsaturated fatty acids	Antarctic sea ice	*Glaciozyma antarctica* PI12	Enhanced level of PUFA to fuel the membrane fluidity, particularly at the freezing temperature of −12°C	Yusof et al., 2021
Antifreeze proteins	Antarctic sea ice	*Glaciozyma antarctica* PI12	Produces antifreeze proteins to inhibit the growth of ice crystals	Hashim et al., 2013 Muñoz et al., 2017
Stress-related proteins	Glaciers	*Cryobacterium* sp.	Genomes with a significantly higher number of genes involved in stress response Presence of expansin-like proteins, which target sea ice loosening for efficient substrate hydrolysis	Liu et al., 2019 Yusof et al., 2021
RNA-binding proteins	Antarctic sea ice	*Glaciozyma antarctica*	Protein stability and flexibility	Charles et al., 2020
Accumulation of cryoprotectants	Permafrost on the Qinghai-arles Tibet Plateau	*Mrakia psychrophila*	Accumulating glycerol as a cryoprotectant	Su et al., 2016
Oxidative stress response	Arctic permafrost	*Nesterenkonia* sp. PF2B19	Combats reactive oxygen species damage	Singh et al., 2021

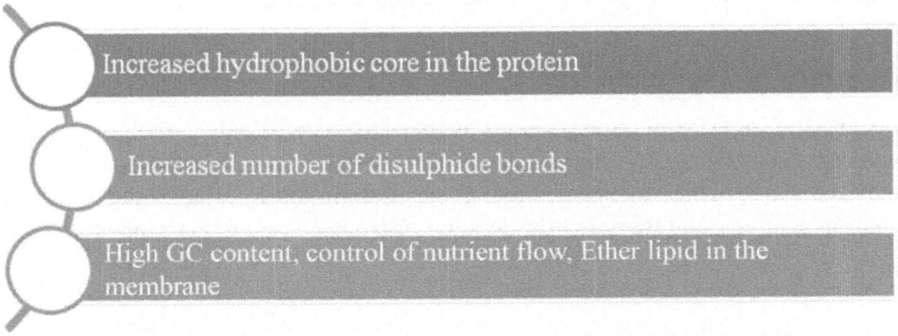

FIGURE 4.3 Various strategies adapted by thermophiles.

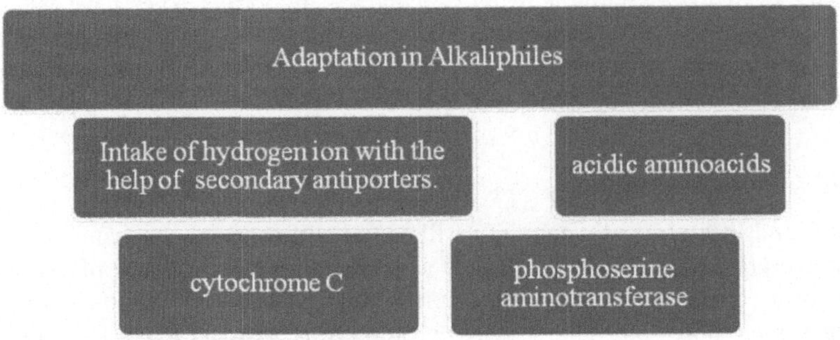

FIGURE 4.4 Various strategies adapted by alkaliphiles.

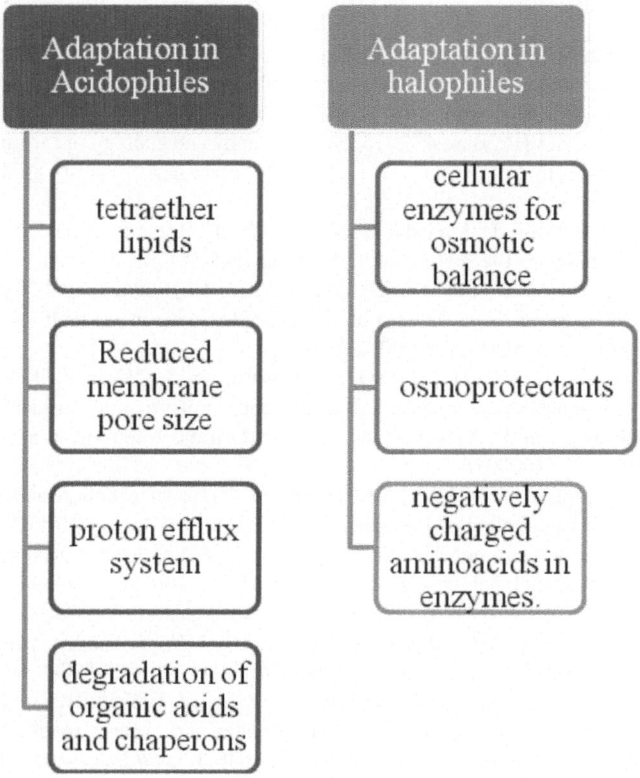

FIGURE 4.5 Various strategies by acidophiles and halophiles.

helps in proton sequestration. Synthesis of chaperons and degradation of organic acids are the known strategies to adjust in the acidic environment (Tyson et al., 2004; Baker-Austin and Dopson, 2007; Dopson, 2012).

4.5.5 Adaptations of Halophiles

Halophiles survive in high salt concentration as they have cellular enzymes specialized for osmotic balance. Their cell produces various osmoprotectants. They have enzymes that contain more amounts of negatively charged amino acids (Coker, 2019).

4.6 Conclusion

Extremophiles are organisms that live in extreme environmental conditions. They have adapted themselves by using various mechanisms. Recent advances in multi-omics technologies give insight into the potentiality of these microorganisms for various commercial applications and climate restoration. The survival of human beings and the entire ecosystem balance is at risk due to climate change. These microbes in the extreme environment are providing some hope for survival. Studies related to diversity after climate change in extreme environments and their adaptations give abundant knowledge and information to address various problems associated with human sustainability. Previous studies on methanotrophic bacteria reported that they utilize methane and reduces the methane load by 15% (Masood et al., 2021). This can be further investigated to use methanotrophs in moderating greenhouse gas emissions and mitigating the issues of climate change.

REFERENCES

Amesbury, M.J., T.P. Roland, J. Royles, D.A. Hodgson, P. Convey, H. Griffiths, and D.J. Charman. 2017. Widespread biological response to rapid warming on the Antarctic Peninsula. *Curr. Biol.* 27(11): 1616–1622. https://doi.org/10.1016/j.cub.2017.04.034

Anderson, R.E., M.L. Sogin, and J.A. Baross. 2015. Biogeography and ecology of the rare and abundant microbial lineages in deep-sea hydrothermal vents. *FEMS Microbio. Eco.* 91(1): 1–11. https://doi.org/10.1093/femsec/fiu016

Aszalos, J.M., A. Szabo, T. Felfoldi, L. Jurecska, B. Nagy, and A.K. Borsodi. 2020. Effects of active volcanism on bacterial communities in the highest-altitude crater lake of Ojos del Salado (Dry Andes, Altiplano-Atacama Region). *Astrobiol.* 20(6): 741–753. http://doi.org/10.1089/ast.2018.2011

Baker-Austin, C., and M. Dopson. 2007. Life in acid: pH homeostasis in acidophiles. *Trends Microbiol.* 15: 165–171. https://doi.org/10.1016/j.tim.2007.02.005

Benhua, S., P.G. Dennis, V.A. Laudicina, V.J. Ord, S.P. Rushton, A.G. O'donnell, K.K. Newsham, and D.W. Hopkins. 2014. Biogeochemical responses to nutrient, moisture and temperature manipulations of soil from Signy Island, South Orkney Islands in the Maritime Antarctic. *Antarct. Sci.* 1–8. http://doi.org/10.1017/S0954102014000030

Bennett, A.C., S.K. Murugapiran, E.D. Kees, T.L. Hamilton. 2020. Distribution and functional potential of photoautotrophic bacteria in alkaline hot springs. *bioRxiv*. https://doi.org/10.1101/2020.12.16.423123

Boetius, A., A.M. Anesio, J.W. Deming, J.W. Mikucki, and J.Z. Rapp. 2015. Microbial ecology of the cryosphere: Sea ice and glacial habitats. *Nat. Rev. Microbiol.* 13(11): 677–690. http://doi.org/10.1038/nrmicro3522

Brock, T.D. 1978. *Thermophilic Microorganisms and Life at High Temperatures*, Springer Series in Microbiology. Berlin and Heidelberg: Springer. https://doi.org/10.1007/978-1-4612-6284-8_2

Buelow, H.N., A.S. Winter, D.J. Van Horn, J.E. Barrett, M.N. Gooseff, E. Schwartz, and C.D. Takacs-Vesbach. 2016. Microbial community responses to increased water and organic matter in the arid soils of the McMurdo Dry Valleys, Antarctica. *Fron. Microbio.* 7: 1040. https://doi.org/10.3389/fmicb.2016.01040

Cavicchioli, R. 2015. Microbial ecology of Antarctic aquatic systems. *Nat Rev Microbiol.* 13: 691–706. https://doi.org/10.1038/nrmicro3549

Charles, J., M. Masnoddin, F. Nazaie, et al. 2020. Structure and function of a novel cold regulated cold shock domain containing protein from an obligate psychrophilic yeast, *Glaciozyma antarctica*. *Adv. Polar. Sci.* 31(2): 137–145. http://doi.org/10.13679/j.advps.2019.0030

Chen, Y., J. Pan, Y. Yun, B. Zhi, G. Li, M. Li, and T. Ma. 2020. Halomonas plays a central role in the syntrophic community of an alkaline oil reservoir with alkali-surfactant-polymer (ASP) flooding. *Sci. Total Env.* 141333. http://doi.org/10.1016/j.scitotenv.2020.1413

Christiansen, C.T., M.S. Haugwitz, A. Prieme, C.S. Nielsen, B. Elberling, A. Michelsen, P. Grogan, and D. Blok. 2017. Enhanced summer warming reduces fungal decomposer diversity and litter mass loss more strongly in dry than in wet tundra. *Glo. Chan. Biol.* 23: 406–420. https://doi.org/10.1111/gcb.13362

Coker, J.A. 2019. Recent advances in understanding extremophiles. *F1000Research*. 8(F1000 Faculty Rev): 1917. http://doi.org/10.12688/f1000research.20765.1

Dalmaso, G.Z.L., D. Ferreira, and A.B. Vermelho. 2015. Marine extremophiles: a source of hydolase for biotechnological application. *Mar. Drugs.* 13: 1925–1965. http://doi.org/10.3390/md13041925

D'Amico, S., T. Collins, and C. Gerday. 2006. Psychrophilic microorganisms: challenges for life. *EMBO Rep.* 7(4): 385–389. https://doi.org/10.1038/sj.embor.7400662

DasSarma, S., P. DasSarma, V.J. Laye, and E.W. Schwieterman. 2020. Extremophilic models for astrobiology: haloarchaeal survival strategies and pigments for remote sensing. *Extremo.* 24(1): 31–41. https://doi.org/10.1007/s00792-019-01126-3

De Maayer, P., D. Anderson, C. Cary, and D.A. Cowan. 2014. Some like it cold: understanding the survival strategies of psychrophiles. *EMBO Rep.* 15(5): 508–517.

Deppe, U., H.H. Richnow, W. Michaelis, and G. Antranikian. 2005. Degradation of crude oil by an arctic microbial consortium. *Extremo.* 9: 461–470. https://doi.org/10.1002/embr.201338170

Dopson, M. 2012. Physiological adaptations and biotechnological applications of acidophiles. In: Anitori RP, editor. *Extremophiles: Microbiology and Biotechnology*. Norfolk, UK: Caister Academic Press; pp. 265–294. https://doi.org/10.21775/9781910190838

Dubnovitsky, A.P., E.G. Kapetaniou, and A.C. Papageorgiou. 2005. Enzyme adaptation to alkaline pH: atomic resolution (1.08 A) structure of phosphoserine aminotransferase from *Bacillus alcalophilus*. *Protein. Sci.* 14: 97–110. https://doi.org/10.1110/ps.041029805

Fell, J.W., G. Scorzetti, L. Connell, and S. Craig. 2006. Biodiversity of micro-eukaryotes in Antarctic Dry Valley soils with 5% soil moisture. *Soil Bio. Biochem.* 38: 3107–3119. https://doi.org/10.1016/j.soilbio.2006.01.014

Ferrera, I., S. Longhorn, A.B. Banta, Y. Liu, D. Preston, and A.L. Reysenbach. 2007. Diversity of 16S rRNA gene, ITS region and aclB gene of the Aquificales. *Extremo.* 11: 57–64. https://doi.org/10.1007/s00792-006-0009-2

Fetterer, F., K. Knowles, W.N. Meier, M. Savoie, and A.K. Windnagel. 2017. *Sea Ice Index, Version 3: Regional Daily Data*. National Snow and Ice Data Center (NSIDC), Boulder, CO. https://doi.org/10.7265/N5K072F8

Fountain, A.G., J.S. Levy, M.N. Gooseff, and D. Van Horn. 2014. The McMurdo Dry Valleys: a landscape on the threshold of change. *Geomo.* 225: 25–35. http://doi.org/10.1016/j.geomorph.2014.03.044

Franco, N.R., M.A. Giraldo, D. Lopez-Alvarez, J.J. Gallo-Franco, L.F. Dueñas, V. Puentes, and A. Castillo. 2021. Bacterial composition and diversity in deep-sea sediments from the Southern Colombian Caribbean Sea. *Diver.* 13(1): 10. https://doi.org/10.3390/d13010010

Gholami, M., Z. Etemadifar, and M. Bouzari. 2015. Isolation a new strain of *Kocuria rosea* capable of tolerating extreme conditions. *J. Environ. Radioact.* 144: 113–119. https://doi.org/10.1016/j.jenvrad.2015.03.010

Hashim, N.H.F., I. Bharudin, D.L.S. Nguong, S. Higa, F.D.A. Bakar, S. Nathan, and A.M.A. Murad. 2013. Characterization of Afp1, an antifreeze protein from the psychrophilic yeast *Glaciozyma antarctica* PI12. *Extremophiles* 17(1): 63–73. https://doi.org/10.1007/s00792-012-0494-4

Hoegh-Guldberg, O., E. Northrop, and J. Lubchenco. 2019. The ocean is key to achieving climate and societal goals. *Sci.* 365(6460): 1372–1374. http://doi.org/10.1126/science.aaz4390

Hui, M.L.Y., L.T.H. Tan, V. Letchumanan, Y.W. He, C.M. Fang, K.G. Chan, and L.H. Lee. 2021. The extremophilic actinobacteria: From microbes to medicine. *Antibio.* 10(6): 682. https://doi.org/10.3390/antibiotics10060682

Imachi, H., M.K. Nobu, N. Nakahara, Y. Morono, M. Ogawara, Y. Takaki, Y. Takano, K. Uematsu, T. Ikuta, M. Ito, Y. Matsui, M. Miyazaki, K. Murata, Y. Saito, S. Sakai, C. Song, E. Tasumi, Y. Yamanaka, T. Yamaguchi, Y. Kamagata, H. Tamaki, and K. Takai. 2020. Isolation of an archaeon at the prokaryote—eukaryote interface. *Nature*. 577: 519–525. http://doi.org/10.1038/s41586-019-1916-6

Kamran, A., K. Sauter, A. Reimer, T. Wacker, J. Reitner, and M. Hoppert. 2021. Cyanobacterial mats in calcite-precipitating serpentinite-hosted alkaline springs of the Voltri Massif, Italy. *Microorg.* 9: 62. https://doi.org/10.3390/microorganisms9010062

Kolarik, M., H. Stępniewska, and R. Jankowiak. 2021. Taxonomic revision of the acidophilic genus Acidiella (*Dothideomycetes, Capnodiales*) with a description of new species from Poland. *Plant. Syst. Evol.* 307: 38. https://doi.org/10.1007/s00606-021-01753-4

Kucera, J., J. Lochman, P. Bouchal, E. Pakostova, K. Mikulasek, S. Hedrich, and D.B. Johnson. 2020. A model of aerobic and anaerobic metabolism of hydrogen in the extremophile *Acidithiobacillus ferrooxidans*. *Front. Microbiol.* 11. http://doi.org/10.3389/fmicb.2020.610836

Lai, D., B.P. Hedlund, W. Xie, J. Liu, T.J. Phelps, C. Zhang, and P. Wang. 2020. Impact of terrestrial input on deep-sea benthic archaeal community structure in South China Sea sediments. *Front. Microbiol.* 11: 572017. http://doi.org/10.3389/fmicb.2020.572017

Lee, C.K., B.A. Barbier, E.M. Bottos, I.R. McDonald, and S.C. Cary. 2012. The inter-valley soil comparative survey: the ecology of Dry Valley edaphic microbial communities. *The ISME J.* 6: 1046–1057. https://doi.org/10.1038/ismej.2011.170

Lindsey, R., and M. Scott. 2020. *Climate Change: Arctic Sea Ice*. NOAA Climate.gov. www.climate.gov/news-features/understanding-climate/climate-change-minimum-arctic-sea-ice-extent

Liu, L., N. Salam, J.Y. Jiao, H.C. Jiang, E.M. Zhou, Y.R. Yin, and W.J. Li. 2016. Diversity of culturable thermophilic actinobacteria in hot springs in Tengchong, China and studies of their biosynthetic gene profiles. *Microb. Ecol.* 72: 150–162. https://doi.org/10.1007/s00248-016-0756-2

Liu, Q., H.C. Liu, Y.G. Zhou, and Y.H. Xin, 2019. Microevolution and adaptive strategy of psychrophilic species *Flavobacterium bomense* sp. nov. isolated from glaciers. *Front. Microbiol.* 10: 1069. https://doi.org/10.3389/fmicb.2019.01069

Lopez, B.G., F.L. Massello, A. Giaveno, E.R. Donati, and M.S. Urbieta. 2020. A deeper look into the biodiversity of the extremely acidic copahue volcano-rioagrio system in Neuquen, Argentina. *Microorg.* 8: 58. https://doi.org/10.3390/microorganisms8010058

Margesin, R., G. Neuner, and K.B. Storey. 2007. Cold-loving microbes, plants, and animals-fundamental and applied aspects. *Naturwissenschaften* 94: 77–99. http://doi.org/10.1007/s00114-006-0162-6

Masood, F., S. Ahmad, and A. Malik. 2021. Role of methanotrophs in mitigating global warming. In *Microbiomes and the Global Climate Change*. Singapore: Springer; pp. 43–60. https://doi.org/10.1007/978-981-33-4508-9_4

Massello, F.L., C.S. Chan, K.-G. Chan, K.M. Goh, E. Donati, and M.S. Urbieta. 2020. Meta-analysis of microbial communities in hot springs: Recurrent taxa and complex shaping factors beyond pH and temperature. *Microorganisms* 8: 906. https://doi.org/10.3390/microorganisms 8060906

Mehta, R., P. Singh, H. Singh, D. Damle, and A.K. Sharma. 2016. Insight into thermophiles and their wide-spectrum applications. *3 Biotech.* 6: 81. http://doi.org/10.1007/s13205-016-0368-z

Merino N., H.S. Aronso, D.P. Bojanov, J. Feyhl-Buska, M.L. Wong, S. Zhang, and D. Giovannell. 2019. Living at the extremes: Extremophiles and the limits of life in a planetary context. *Front. Microbial.* 10: 780. https://doi.org/10.3389/fmicb.2019.00780

Misiak, M., W.P. Goodall-Copestake, T.H. Sparks, M.R. Worland, L. Boddy, N. Magan, and K.K. Newsham. 2021. Inhibitory effects of climate change on the growth and extracellular enzyme activities of a widespread Antarctic soil fungus. *Glob. Chan. Bio.* 27(5): 1111–1125. https://doi.org/10.1111/gcb.15456

Muñoz, P.A., S.L. Marquez, F.D. Gonzalez-Nilo, V. Marquez-Miranda, and J.M. Blamey. 2017. Structure and application of antifreeze proteins from Antarctic bacteria. *Micro. Cell Fac.* 16(1): 1–13. https://doi.org/10.1186/s12934-017-0737-2

Navarro-Gonzalez, L., F.A. Rainey, P. Molina, D.R. Bagaley, B.J. Hollen, J.D.A. Rosa, A.M. Small, R.C. Quinn, F.K.J. Grunthaner, L. Cáceres, B. Gomez-Silva, and C.P. McKay. 2003. Mars-like soils in the atacama desert, chile, and the dry limit of microbial life. *Sci.* 302(5647): 1018–1021. http://doi.org/10.1126/science.1089143

Odilia, A.S., M.M. Huxley, W.K. Remmy, and I.B. Hamadi. 2018. Isolation and characterization of fungi from a hot-spring on the shores of Lake Bogoria, Kenya. *J. Yeast. Fungal. Res.* 9: 1–13. https://doi.org/10.5897/JYFR2018.0182

Panda, A.K., S.S. Bisht, S.D. Mandal, and N.S. Kumar. 2016. Bacterial and archeal community composition in hot springs from Indo-Burma region, North-east India. *AMB Express.* 6(1): 111. http://doi.org/10.1186/s13568-016-0284-y

Perera-Castro, A.V., M.J. Waterman, J.D. Turnbull, M.B. Ashcroft, E. McKinley, J.R. Watling, and S.A. Robinson. 2020. It is hot in the sun: Antarctic mosses have high temperature optima for photosynthesis despite cold climate. *Front. Plant Sci.* 11: 1178. https://doi.org/10.3389/fpls.2020.01178

Pulicherla, K.K., M. Ghosh, P.S. Kumar, and K.R.S.S. Rao. 2011. Psychrozymes the next generation industrial enzymes. *J. Mar. Sci.: Res. Dev.* 1: 102. http://doi.org/10.4172/2155-9910.1000102

Rampelotto, P.H. 2013. Extremophiles and extreme environments. *Life* 3: 482–485. https://doi.org/10.3390/life3030482

Rasmussen, B. 2000. Filamentous microfossils in a 3,235-million-year-old volcanogenic massive sulphide deposit. *Nature* 405(6787): 676–679. http://doi.org/10.1038/35015063

Rothschild, L.J., and R.L. Mancinelli. 2001. Life in extreme environments. *Nature* 409: 1092–1101. http://doi.org/10.1038/35059215

Schmitz, R.A., S.H. Peeters, W. Versantvoort, N. Picone, A. Pol, M.S. Jetten, and H.J. Op den Camp. 2021. Verrucomicrobial methanotrophs: ecophysiology of metabolically versatile acidophiles. *FEMS Microbiol. Rev.* 45(5): fuab007. https://doi.org/10.1093/femsre/fuab007

Schroder, C., C. Burkhardt, and G. Antranikian. 2020. What we learn from extremophiles. *Chem. Texts.* 6(1): 1–6. https://doi.org/10.1007/s40828-020-0103-6

Serreze, M.C., and W.N. Meier. 2019. The Arctic's sea ice cover: trends, variability, predictability, and comparisons to the Antarctic. *Ann. New York Aca. Sci.* 1436(1): 36–53. https://doi.org/10.1111/nyas.13856

Singh, B.K. 2010. Exploring microbial diversity for biotechnology: the way forward. *Trends Biotechnol.* 28: 111–116. https://doi.org/10.1016/j.tibtech.2009.11.006

Singh, B.K., and C.A. Macdonald. 2010. Drug discovery from uncultivable microorganisms. *Drug Discov. Today* 15(17–18): 792–799. https://doi.org/10.1016/j.drudis.2010.07.002

Singh, O.V., S. Ghai, D. Paul, and R.K. Jain. 2006. Genetically modified crops: success, safety assessment, and public concern. *Appl. Microbiol. Biotechnol.* 71: 598–607. https://doi. org/10.1007/s00253-006-0449-8

Singh, P., K. Jain, C. Desai, O. Tiwari, and D. Madamwar. 2019. Microbial community dynamics of extremophiles/extreme environment. In: Das S, Dash HR, editors. *Microbial Diversity in the Genomic Era.* London, UK: Academic Press, pp. 323–332. https://doi.org/10.1016/B978-0-12-814849-5.00018-6

Singh, P., N. Kapse, V. Gowdaman, M. Tsuji, S.M. Singh, and P.K. Dhakephalkar. 2021. Comparative genomic analysis of arctic permafrost bacterium *Nesterenkonia* sp. PF2B19 to gain insights into its cold adaptation tactic and diverse biotechnological potential. *Sustainability.* 13(8): 4590. https://doi.org/10.3390/su13084590

Su, Y., X. Jiang, W. Wu, M. Wang, M.I. Hamid, M. Xiang, and X. Liu. 2016. Genomic, transcriptomic, and proteomic analysis provide insights into the cold adaptation mechanism of the obligate psychrophilic fungus *Mrakia psychrophila. G3: Genes, Genomes, Genetics.* 6(11): 3603–3613. http://doi.org/10.1534/g3.116.033308

Taylor, M.P., R. Bauer, S. Mackay, M. Tuffin, and D.A. Cowan. 2010. Extremophiles and their application to biofuel research. *Environ. Technol.* 1–40. http://doi.org/10.1080/09593331003710236

Treonis, A.M., D.H. Wall, and R.A. Virginia. 1999. Invertebrate biodiversity in Antarctic Dry Valley soils and sediments. *Ecosystems.* 2: 482–492.

Tyson, G.W., J. Chapman, P. Hugenholtz, E.E. Allen, R.J. Ram, P.M. Richardson, V.V. Solovyev, E.M. Rubin, D.S. Rokhsar, and J.F. Banfield. 2004. Community structure and metabolism through reconstruction of microbial genomes. *Nature* 428: 37–43. https://doi.org/10.1038/nature02340.

Van Horn, D.J., J.G. Okie, H.N. Buelow, M.N. Gooseff, J.E. Barrett, and C.D. Takacs-Vesbach. (2014). Soil microbial responses to increased moisture and organic resources along a salinity gradient in a polar desert. *Appl. Environ. Microbiol.* 80: 3034–3043. http://doi.org/10.1128/AEM.03414-13

Vavourakis, C.D., R. Ghai, F. Rodriguez-Valera, D.Y. Sorokin, S.G. Tringe, P. Hugenholtz, and G. Muyzer. 2016. Metagenomic insights into the uncultured diversity and physiology of microbes in four hypersaline soda lake brines. *Front. Microbiol.* 7: 211. http://doi.org/10.3389/Fmicb.2016.00211

Wang, H.C., E. Susko, and A.J. Roger. 2006. On the correlation between genomic G+C content and optimal growth temperature in prokaryotes: data quality and confounding factors. *Biochem. Biophys. Res. Commun.* 342(3): 681–684. https://doi.org/10.1016/j.bbrc.2006.02.037

Yamagishi, A., Y. Kawaguchi, H. Hashimoto, H. Yano, E. Imai, S. Kodaira, Y. Uchihori, and K. Nakagawa. 2018. Environmental data and survival data of *Deinococcus aetherius* from the exposure facility of the Japan experimental module of the international space station obtained by the tanpopo mission. *Astrobiol.* 18: 1369–1374. http://doi.org/10.1089/ast.2017.1751

Yergeau, E., S.A. Schoondermark-Stolk, E.L. Brodie, S. Déjean, T.Z. DeSantis, O. Gonçalves, and G.A. Kowalchuk. 2009. Environmental microarray analyses of Antarctic soil microbial communities. *The ISME J.* 3(3): 340–351. http://doi.org/10.1038/ismej.2008.111

Yusof, N.A., N.H.F. Hashim, and I. Bharudin. 2021. Cold adaptation strategies and the potential of psychrophilic enzymes from the antarctic yeast, *Glaciozyma antarctica* PI12. *J. Fungi.* 7(7): 528. https://doi.org/10.3390/jof7070528

Zeglin, L., C. Dahm, J. Barrett, M. Gooseff, S. Fitpatrick, and C.D. Takacs-Vesbach. 2011. Bacterial community structure along moisture gradients in the parafluvial sediments of two ephemeral desert streams. *Microb. Ecol.* 61: 543–556. www.jstor.org/stable/41489077

Zeikus, J.G. 1980. Chemical and fuel production by anaerobic bacteria. *Ann. Rev. Microbiol.* 34: 423–464. https://doi.org/10.1146/annurev.mi.34.100180.002231

Zhao, D., S. Zhang, Q. Xue, J. Chen, J. Zhou, F. Cheng, and H. Xiang. 2020. Abundant taxa and favorable pathways in the microbiome of soda-saline lakes in Inner Mongolia. *Front. Microbiol.* 11: 1740. https://doi.org/10.3389/fmicb.2020.01740

5

Coral Reef Biodiversity—Threats and Management

Dhanya M. S. and Jignesh M. E.

CONTENTS

5.1 Introduction

The coral reef ecosystem is commonly known as tropical rainforest of oceans because of its productive, diverse, and rich nature (Knowlton and Jackson, 2001). The coral reef ecosystem is one of the most sensitive and most threatened ecosystems (Burke et al., 2011; Gattuso et al., 2015; Hughes et al., 2018; Wagner, 2020). The symbionts of zooxanthellae with polyps are a vulnerable group to climate change. The world's largest coral reef is in Indonesia, and the Great Barrier Reef of the Queensland

DOI: 10.1201/9781003220398-5

coast of Australia has more assemblies of coral reefs. There is a need to understand the importance of the coral reef ecosystem and to protect the diverse, rich coral community by following necessary conservation steps.

5.2 Corals and Their Classification

The corals are marine invertebrates belong to Class Anthozoa in Phylum Cnidaria (also called Coelenterata). Scleractinian corals are the major coral-reef-forming organisms, commonly known as ecosystem engineers (Jones et al., 1994). Symbiosis occurs between the coral host and zooxanthellae (Allemand and Furla, 2018). The density is about one million zooxanthellae per square centimeter of coral tissue (Stambler, 2011). The coral reefs are distributed mainly in tropical oceans receiving high sunlight (Goreau et al., 1979). A few species are even distributed to the Arctic, the Antarctic, and the deep sea. Darwin (1859) explained the formation of fringing reefs to barrier reefs and atolls by subsidence of volcanic islands. Dana (1853) first published the global distribution of coral reefs. The reefs are formed by the deposition of calcium carbonate and are limited to geographic locations with a minimum temperature of 20°C (Veron, 1986).

Coral reefs provide habitats to over a million marine species (Knowlton et al., 2010; Pratchett et al., 2014). The coral reefs are steno type with its distribution over a narrow range of environmental conditions (Kleypas et al., 1999). The largest coral reef ecosystem, Great Amazon Reef, with an area of 56,000 km^2 and depth of up to 220 m, has diverse habitats, such as reef platforms, reef walls, rhodolith beds, and sponge bottoms (Francini-Filho et al., 2018). Dana (1853) explained the optimum temperature of 21°C and the requirement of light for reef growth.

The coral reefs are diverse, rich ecosystems with a variety of invertebrates, vertebrates, and algae (Richmond, 1993; Reaka-Kudla, 1997). They cover only 0.2% area of total seafloor but contributes near about 35% of the total marine biodiversity (Reaka-Kudla, 2005; Knowlton et al., 2010). Around 845 coral species are known to be reef builders in the photic zone (Carpenter et al., 2008). The Anthozoa and Porifera groups are prominent groups of reef-building corals (Neigel et al., 2007). The coral species dominant in Caribbean is *Porites branneri,* and a rare species found in some parts of Cuba was *Dendrogyra cylindrus* (Galford et al., 2018).

The coral reefs are classified based on type, geomorphology and depth (Hoegh-Guldberg et al., 2017; Mallik, 2017). The major type of coral reefs found in India are fringing reefs, atolls, and patches. Fringing reefs are reefs found close to shorelines or coasts and are shallow in nature. The linear offshore reefs, parallel to coastline, form barrier reefs, which are distant from coast. The patch reefs are seen as isolated and discontinuous patches distant from the shoreline and in the middle of the ocean. The atoll is circular or horseshoe shaped reef enclosing a central lagoon. The fringing reefs are present in the Gulf of Mannar, Gulf of Kutch and Andaman, and Nicobar. The coral reefs in Lakshadweep are atoll types. The patch reefs are found in the Ratnagiri, Malvan, and Gaveshana banks of Mangalore and Quilon to Enayam in the Kerala coast (Saravanan et al., 2017).

5.3 Importance/Value of Coral Reefs

The coral reefs have ecological and economic importance in fulfilling essential ecosystem services for the global population (Moberg and Folke, 1999).

5.3.1 Habitat Provision

The coral reefs act as habitats for fish, sea anemones, starfish, and other marine species, covering one-third of marine species in the 0.2% of the total ocean area (Pandolfi et al., 2011; Wild et al., 2011). The reefs act as feeding, spawning, and nursery areas for fish and many other marine organisms. The *Lophelia pertusa* reefs in the Northeast Atlantic reported to be habitat for around 900 species (Rogers, 1999). Wagner et al. (2020) compiled reef-building corals had habitats in escarpments, seamounts, and submarine ridges, contributing more than 74% of the records in international waters and hence reported the necessity of protection of hidden hotspots of coral reefs of high seas.

5.3.2 Productive Functions

The coral reefs provide food to a large population (Harvey et al., 2018). Fishermen depends on coral reefs for their livelihood (Teh et al., 2013). The productive function should be protected for the resilience of both coral reefs and the communities that depend on them (Cinner et al., 2012; Pratchett et al., 2016; Anthony et al., 2017).

5.3.3 Protective Functions

The coral reef ecosystems provide ecosystem services (Cinner et al., 2012). The coral reef controls the erosion of the coastal ecosystem by acting as a barrier to waves. It provides protection to beaches from erosion, storms, tsunamis, and floods and maintain the stability of the marine ecosystem. The reef also protects coastal infrastructure. The coral reefs also protect other breeding sites of marine animals, such as mangroves and seagrass beds.

5.3.4 Aesthetic Value

The coral reef ecosystem has recreation and tourism opportunities, such as scuba diving and snorkeling, and contributes billions of dollars to the local economy.

5.4 Threats to the Coral Reef Ecosystem

The coral reefs have already been considered as one of the most endangered ecosystems. The reasons for coral damage are natural factors and human-made activities (Hoegh-Guldberg et al., 2007; Burke et al., 2011; Gattuso et al., 2015; Hughes et al., 2017; França et al., 2020; Wagner, 2020). The human impacts on coral reefs are huge and irreversible (Halpern et al., 2008). The changes in the environmental parameters are major factors for coral loss (Hughes et al., 2017; Hughes et al., 2018). Chatragadda et al. (2019) identified stress factors causing threat to 21 islands of the Gulf of Mannar Marine Biosphere Reserve (GOMMBRE). They found two coral species, *Porites* sp. and *Acropora* sp., that were mostly affected and pink line disease all over the islands. Benkwitt et al. (2020) reported human-induced stressors like extreme heatwaves increased coral bleaching, and invasive rats caused imbalance in the ecosystem functioning. The various threats to the coral reef ecosystem are given in Figure 5.1.

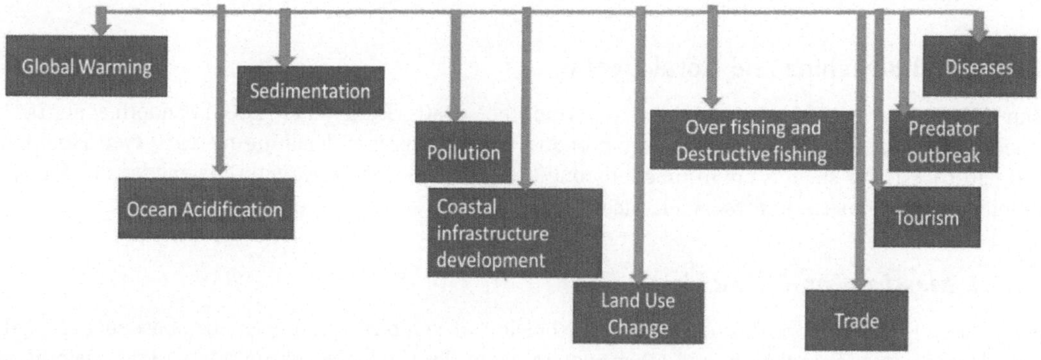

FIGURE 5.1 Abiotic and biotic stressors and threats to the coral reef ecosystem.

Source: Compiled from Wilson et al., 2005; Crain et al., 2009; Praveena et al., 2012; Weeks, et al., 2013; Hughes et al., 2017; Hughes et al., 2018; Magris et al., 2018.

5.4.1 Impacts of Climate Change

Climate change is the major threat to coral reef ecosystem (Pandolfi et al., 2011; Heron et al., 2016; Hughes et al., 2018).

The biodiversity in coral reefs responds to climate change in different ways. The response mechanisms followed by the coral reef community are adaptation, acclimatization, or dispersal (Hoegh-Guldberg et al., 2014). Climate change affects individual organisms and the coral reef ecosystem as a whole. The immediate effect is the reduction in survival, recruitment, growth, and reproduction (Ateweberhan et al., 2013). The direct effects are damage to the reefs, alteration in their structure, and increased susceptibility to bleaching and diseases (Burke et al., 2011; Bijma et al., 2013). According to IPCC (2018), the projections of coral reef loss at 1.5°C was 70–90%, which can be turned to irreversible loss at 2°C.

Climate change and human activities resulting coral loss are well reported (Hoegh-Guldberg et al., 2007; McClanahan et al., 2014; IPCC, 2014). The rise of sea surface temperature by 1–2°C causes coral bleaching (Jones et al., 2004; IPCC, 2007). The rise in sea surface temperature has the worst impact on symbiotic relationship in corals (Hoegh-Guldberg et al., 2007; Wooldridge, 2013). The ocean warming reduces the growth of corals like *Porites* spp. and *Diploastrea heliopore* in the Great Barrier Reef and the Red Sea, respectively (Cooper et al., 2008; Cantin et al., 2010). The Great Barrier Reef has a high acceleration rate of decline in coral cover (Hughes et al., 2017). Hard corals are also destroyed by tropical cyclones (Elsner et al., 2008; Cheal et al., 2017). The massive removal of endosymbionts was reported in the Jamaican reef after the hurricane (Goreau, 1964). Global warming increases the impact of cyclones, typhoons, and hurricanes (Osborne et al., 2011). The Keppel Bay fringing reefs were highly disturbed by cyclones and floods (GBRMPA, 2008).

5.4.2 Ocean Acidification

Ocean acidification is the decrease in pH of the ocean water over a long period caused by CO_2 uptake. It shifts the acid-base balance of the carbonate system in oceanic water. This global phenomenon is caused mainly by anthropogenic activities that had occurred over decades or even longer. The CO_2 in atmosphere increased and resulted in the increase in absorption of CO_2 by seawater and reduction in pH of the water, thereby causing ocean acidification and change in ocean chemistry (Hoegh-Guldberg et al., 2007). The oceanic CO_2 absorption results in the decreased pH of oceanic water, causing acidification and affecting $CaCO_3$ formation in shell-forming organisms (Bijma et al., 2013). The average decrease in ocean pH is estimated to be 0.1, which increases the acidity of the ocean by 26% from the era of the industrial revolution (IPCC, 2014; Jewett and Romanou, 2017). The projections for the decline of aragonite saturation values (Ωar) in Western Central Atlantic and Western and Central Pacific due to CO_2 absorption was below 3. This results in a net erosion of coral reefs. Yeakel et al. (2015) reported interannual acidification in the Bermuda coral reef resulting from high calcification and respiration due to shift in reef biogeochemical processes.

5.4.3 Coral Bleaching and Coral Deaths

Many bleaching events of coral reefs were reported since 1980s (Baker et al., 2008; Pandolfi et al., 2011). Ninety percent of coral reefs is definitely prone to annual massive bleaching by 2050 (van Hooidonk et al., 2016). Factors such as environmental conditions, reef morphology, and the genetics of hosts and symbionts caused micro- to meso-scale bleaching of corals.

5.4.3.1 Massive Coral Bleaching

The whitening of the coral tissues with the expulsion of symbionts is the major consequence of the hyperthermia, intense radiation, and other contaminants. The coral bleaching refers to the visibility of aragonite skeleton of cnidarian host tissues due to the removal of symbionts (zooxanthellae/algae). The bleaching of corals can be of physiological bleaching, algal stress bleaching, or animal stress bleaching (Figure 5.2).

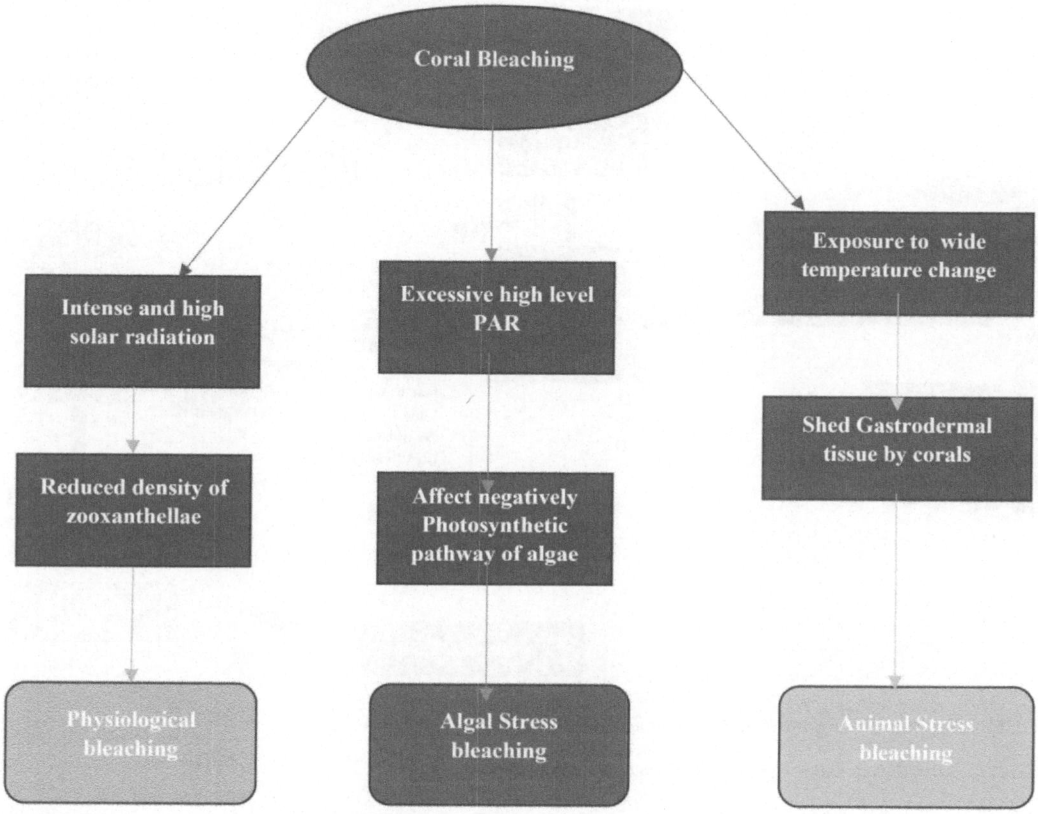

FIGURE 5.2 Major kinds of coral bleaching.

Source: McAuley, 1994; Osmond, 1994; Fitt and Warner, 1995; Fitt et al., 2001.

The mass bleaching is mainly connected with the increase in seawater temperature. The rise in temperature associated with the El Nino event in 1982–1983 resulted in large-scale deaths of corals in the eastern Pacific (Glynn, 1990). The massive bleaching of coral reefs and their death was also reported due to the El Nino Southern Oscillation (ENSO) in 1998 (Baker et al., 2008). National Oceanic and Atmospheric Administration (2017) reported that the 2014–2017 period witnessed the most destructive coral bleaching because of the combined effect of El Nino, La Nina, and ocean warming.

5.4.3.2 Mechanism of Coral Bleaching

The abiotic and biotic stressors induce the production of radical oxygen species, which causes oxidative stress, resulting in disturbing the coral-algae symbiosis. The radical oxygen species (ROS) mediated mechanism (Figure 5.3) of coral death was explained by Goreau (1964), Weis (2008), and Baird et al. (2009). Nielsen et al. (2018) explained the excessive production of ROS-induced physiological stress in *Symbiodinium* sp., an endosymbiont in coral hosts.

The thermal stress and UV radiation resulted in photoinhibition and oxidative stress in corals, leading to coral bleaching (Downs et al., 2013; Nielsen et al., 2018). The reduction in photosynthetic electron transport damaging PS-II decreases the photosynthetic yield in comparison to ambient conditions. The overproduction of ROS leads to oxidative stress, resulting in further damage to cells, DNA, lipids, and proteins. The breakdown in symbiotic relation between cnidarian hosts and endosymbionts are by *in situ* degradation of symbionts, apoptosis, necrosis, and exocytosis (Figure 5.3).

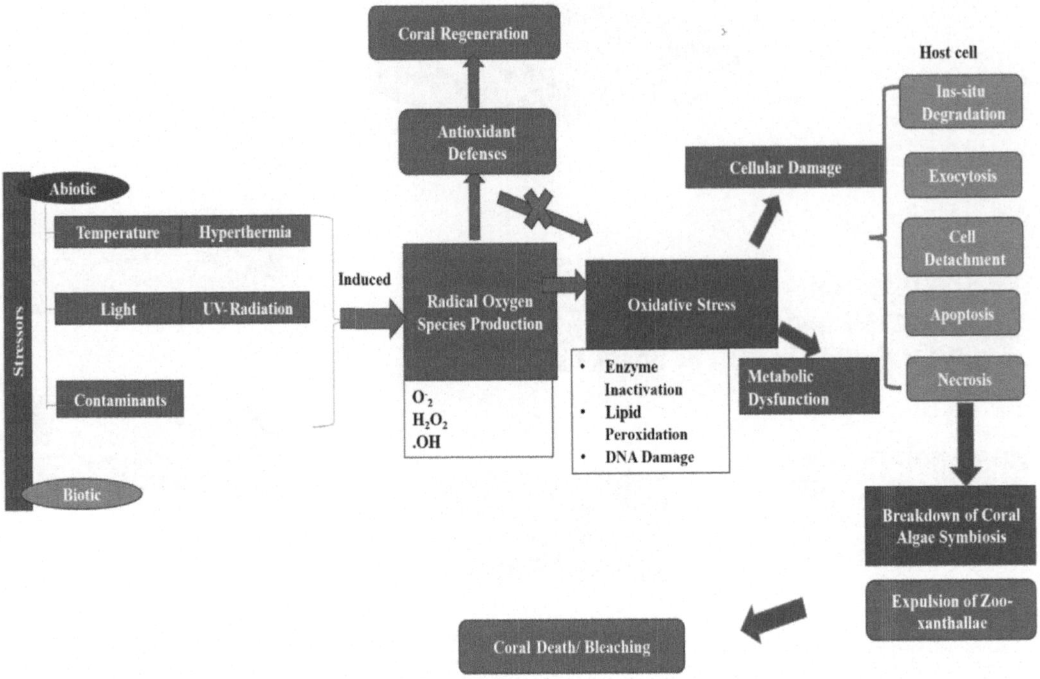

FIGURE 5.3 Mechanisms involved in coral bleaching.

Source: Compiled from Gates et al., 1992; Weis et al., 2008; Baird et al., 2009; Nielsen et al., 2018.

Coral bleaching initially reduces the growth of corals, which on long duration leads to mass mortality (Veron, Hoegh-Guldberg et al., 2009). Bleaching affects not only corals but also gorgonians, soft corals, and anemones (Jones et al., 2004; Hallock et al., 2006). Altieria et al. (2017) explained the importance of hypoxia and the associated dead zones in the massive death of coral reefs. Bahiá Almirante in the Bocas del Toro region of Panama was studied for its hypoxic stress and for its benthic microbial activity. The coral loss is directly linked with the fishing community in the coral reef ecosystem (Bellwood et al., 2019).

5.4.4 Destructive Fishing

The destructive fishing is another cause exerting pressure on coral reefs (Burke et al., 2011; Praveena et al., 2012; The et al., 2013; Norström et al., 2016; Hughes et al., 2017; Magris et al., 2018). The destructive fishing practices of using dynamite for blast fishing, cyanide fishing, and using fishing gear like gill nets (Jones et al., 1999; Fox and Caldwell, 2006). The coral mortality and bleaching were also enhanced by cyanide (Jones and Hoegh-Guldberg, 1999).

The impact on corals by fishing pressure was also studied by Cinner et al. (2012) and Maynard et al. (2015). The fish are the major herbivores in food chain of coral ecosystem, so overfishing results in the increase of algal growth, which in turn decline space for coral growth (Jackson et al., 2001; Rogers et al., 2014; Robinson et al., 2019). Riniwati et al. (2019) reported the environmental conditions and distance from fishing area are the most vulnerable variables for the loss of coral reef susceptibility.

5.4.5 Diseases

Black band disease (BBD) is the most infectious and contagious disease of coral reef caused by the cyanophyte *Phormidium corallyticum*. The disease affects individual corals, ecosystems, and communities.

The other coral diseases are white band disease, yellow band disease, white patch disease, aspergillosis, and dark spot disease. Algal blooms increase the disease occurrence in corals by increasing microbial growth (Sandin et al., 2008).

5.4.6 Pollution

Most of the untreated wastes are dumped into the ocean. The fertilizers and pesticides from the agricultural runoff reach coral reefs. The excessive nutrient loss from agriculture also adds to eutrophication (Burke et al., 2011). The land use change and soil pollution also cause damage to coral reefs Fredston-Hermann et al., 2016). The nutrient enrichment increases the vulnerability to coral bleaching by reducing coral calcification (Wiedenmann et al., 2013). The nutrient enrichment is directly related with increase in diseases, turbidity, COTS outbreaks, and bleaching (Veron, Hoegh-Guldberg et al., 2009; Pollock et al., 2014; Vega Thurber et al., 2014; D'Angelo and Wiedenmann, 2014).

Rocha et al. (2018) reported the mesophotic corals in shallow and deep reefs are biologically different. Furthermore, deep (or mesophotic) reefs are also suffering from human impacts.

The negative impact of tourism on coral reef was also reported by Jobbins (2006) and Spalding et al. (2017). The oil spill in the Gulf of Mexico, which resulted from the explosion of the BP-operated Deepwater Horizon offshore oil rig, along with the use of a dispersant, Corexit R, increased the negative impact on benthic communities and coral larvae (Goodbody-Gringley et al., 2013). The threats to coral reefs from oil exploration were also reported by Francini-Filho et al. (2018).

5.4.7 Sedimentation

One of the reasons for sedimentation is the coastal development (Burke et al., 2011). The deforestation, especially in catchment areas of water sheds, adds to the sedimentation in the coral reef ecosystem (Richmond et al., 2007). The turbidity in water and reduction in photosynthesis affect the coral growth (Fabricius et al., 2014). De'ath and Fabricius (2010) reported the sedimentation is directly linked with disease outbreak. The survival of larvae is affected by sedimentation, which in turn affects coral recruitment (Perez et al., 2014; Pollock et al., 2014).

The sedimentation in coral reef resulted from deforestation was reported by Wilson et al. (2005) and Hamilton et al. (2017). Lamb et al. (2018) estimated entanglement of 11.1 billion plastic wastes on coral reefs from the survey conducted at 159 coral reefs across the Asia-Pacific and reported its impact on the rise in diseases.

5.4.8 Predator Outbreak

The outbreak of the coral-eating starfish (COTS), *Acanthaster planci*, affects corals. The frequency and intensity of COTS outbreak increases with the anthropogenic stressors (Babcock et al., 2016). The larval growth of COTS is linked with nutrient limitation (Fabricius et al., 2010). Haszprunar et al. (2017) studied the diversity of COTS by DNA barcoding and reported four species from the Indo-Pacific region. The other corallivores are coral-eating snails *Drupella cornus* and *Coralliophila abbreviate*, the fireworm (*Hermodice carunculata*), the stoplight parrotfish (*Sparisoma viride*), and the three-spot damselfish (*Stegastes planifrons*).

5.4.9 Coral Mining

The mining of corals is mainly meant for construction material, and it exploits dense calcium carbonate skeletons of corals such as *Platygyra, Porites*, and *Favia* (Sarjana, 1992; Dulvy et al., 1995; Caras and Pasternak, 2009). Coral is converted into limestone. Asir et al. (2020) reported that the long-term impact of coral mining results in the reduction in coral cover in islands of the Gulf of Mannar of India. High-value coral species were overharvested in large scale for aquarium trade and jewellery (EPA, 2021). The coral reefs in Maldives, Sri Lanka, India, Indonesia, Philippines, Saba, and Pacific atolls are highly exploited for coral mining (Brown and Dunne, 1988; Wells, 1984; Asir et al., 2020).

5.5 A Need for the Conservation of Coral Reefs

The loss of coral reefs in the world accounts to 30% from 1980s (Hoegh-Guldberg, 2005). The lack of conservation measures has the potential of losing 90% of coral reefs in the near future (Hoegh-Guldberg et al., 2018; Hughes et al., 2018). Climate change and ocean acidification have an impact on the abundance of marine organisms, largely affecting fish (IPCC, 2014). The pressure exerted by local stressors prevents coral reefs from coping with climate change factors and recovering from bleaching (Anthony, 2016). The coral biodiversity is more susceptible to heat stress and photoinhibition (Fitt and Warne, 1995).

The diverse, rich marine organisms, ecosystem services and the threats made coral reef conservation and management as the world's one of the most important conservation area (Klein et al., 2010). The coral community had high species richness and endemism (Roberts et al., 2002). The high richness in species among various coral reefs are present in the Coral Triangle (Veron, DeVantier et al., 2009), which confirm the necessity for conservation priority. Roberts et al. (2002) estimated that 7.2 to 53.6% of reef fish, corals, snails, and lobsters, with the reef's narrow range are vulnerable to extinction. Global warming and other factors cause the decline in abundance of coral reefs by 70–90% (Hoegh-Guldberg et al., 2018).

5.6 Strategies in the Conservation and Management of the Coral Reef Ecosystem

Conservation and management strategies like expansion of marine protected areas (MPAs), decreasing local stressors, fishing control, and alternative restriction, restoration, promotion, adaptation, and mitigation efforts in relation to climate change are needed for the positive response of coral reef ecosystems (Baums, 2008; Anthony et al., 2017; Walsworth et al., 2019). The reduction in carbon dioxide emissions and preventing the rise in temperature beyond 2°C is vital for the protection from climate change impacts on coral reefs (Hoegh-Guldberg et al., 2017). The geographic prioritization is very much required for the conservation of coral reefs for increased action (Klein et al., 2010). DNA barcoding is an effective tool for the conservation of coral reefs (Neigel et al., 2007).

5.6.1 Marine Protected Areas (MPAs)

The most commonly followed method of coral reef conservation and management is creating marine protected areas or no-take marine reserves (McLeod et al., 2021). MPAs aim to reduce the local factors damaging coral reefs by prioritizing conservation measures. This method protects biodiversity, especially in fisheries, by increasing the resilience to the various factors resulting in coral loss (Rinkevich, 2015; Mellin et al., 2016; Hughes et al., 2017; Roberts et al., 2017). The maintenance of the stability in reef fish biomass was observed in 310 MPAs (Cinner et al., 2016). Edgar et al. (2014) reported no-take, enforced, old, large, and isolated MPAs must be effective with the reduction in carbon dioxide emissions. The reefs of Keppel Bay in the southern Great Barrier Reef used differentiation in the Great Barrier Reef Marine Park, zoning with no-take marine reserves and general-use zones (GBRMPA, 2008). The maintenance of biodiversity improves the growth and survival of corals and also suppresses macroalgae (Clements and Hay, 2019).

MPAs have several limitations. The inability to conserve most threatened reefs and restoring ecosystem services in changing climatic conditions is one of the most important challenges (Burke et al., 2011; Sale et al., 2014; Norström et al., 2016). The benefits from MPA had uneven distribution (Cinner et al., 2014). MPA had no safeguard to sedimentation from land runoff (Fabricius et al., 2014). The integrated approach of combining MPAs with participatory approaches and other management strategies are required for effective coral reef management (Rogers et al., 2015).

5.6.2 Reduction in Human-Induced Factors on Coral Reef Damage

The increased recovery of coral reefs from bleaching by decreasing local stressors was prominent from a 20-year study in the communities near the Great Barrier Reef (Mellin et al., 2016). The improvement in water quality by sedimentation reduction and nutrient addition helped retain the ecosystem services of coral reefs (Fabricius et al., 2010; Pollock et al., 2014). Watershed management, restoration of riparian reserves, and control of wastewater discharge decreased sedimentation and coral bleaching (Wagner et al., 2010). The avoidance of fishing in the protected areas also improved resilience of reefs by means of herbivory, decline in predation, structural maintenance, and increased diversity, ultimately promoting coral recovery from bleaching (Robinson et al., 2019; Benkwitt et al., 2020). Restrictions in fishing by controlling the species to be caught, breeding site closures according the season, and using alternative fishing gear helped in restoring fish communities and their ecological functions (Burke et al., 2011; McClanahan et al., 2015). The measures to control different types of pollution, eutrophication and siltation are needed to reduce the negative impacts on coral reefs (He and Silliman, 2019).

5.6.3 Coral Reef Restoration

Coral reef restoration and reef resilience reduce the damage and restores the reef to its original state (Rogers et al., 2015; Abelson et al., 2016; Anthony, 2016). The recovery processes consisting of coral recruitment, growth, or dispersal by overcoming local stressors improve the reestablishment of the coral reef ecosystem. The strategies for coral restoration are *in situ* and *ex situ* nursing of coral stock (Amar and Rinkevich, 2007. The marine spatial planning by involving MPAs and zoning is a good management strategy (Sale et al., 2014).

There is still facing challenges in recovery of degraded coral reefs by simply following the reduction in human-based stressors. Active management measures are to be undertaken for the efficient recovery of coral reefs despite climate change impacts (Rogers et al., 2015; Abelson et al., 2016).

5.6.4 Coral Gardening

Coral gardening is an important option for active coral reef management (Rinkevich, 2015). The success of artificial coral transplanting was reported in the Gulf of Aqaba, Red Sea (Rogers et al., 2015). The additional benefit of coral gardening is the reduction in tourism pressure on natural coral reefs. The steps in coral gardening are given in Figure 5.4. (Rinkevich, 2015).

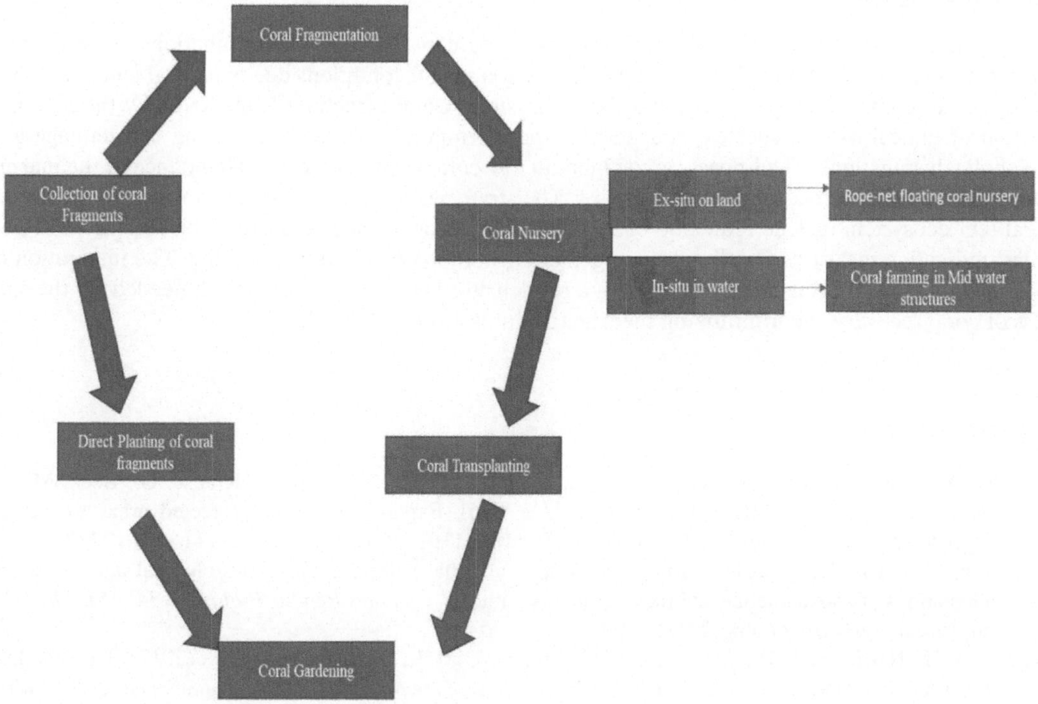

FIGURE 5.4 Schematic representation of coral gardening.

The fragmentation or recruitment for coral growth is the first step in coral gardening followed by transplanting (Rinkevich, 2014). Shafir et al. (2006) demonstrated rope-net floating coral nursery with coral fragments in plastic pin further connected to nets. The fragments were prepared from ten branching coral colonies of *Stylophora pistillata*, *Pocillopora damicornis*, *Acropora pharaonic*, *A. eurystoma*, and *A. valida*. The gardening of denuded reef areas is an effective restoration technique (Rinkevich, 2015). The transplanting of genetically modified stress-tolerant corals onto reefs is an emerging coral reef management strategy that helps in improving resistance to various stressors and fast recovery (Anthony et al., 2017).

5.7 Global Initiatives in Coral Reef Management

The major goal of the Paris Climate Agreement is to restrain rise in temperature below 2°C to reduce the severe impacts on coral (Hoegh-Guldberg et al., 2018). Various global initiatives, such as Global Coral Reef Monitoring Network (GCRMN), Global Coral Reef Alliance (GCRA), and International Coral Reef Initiative (ICRI), play significant roles in raising public awareness for the conservation of coral reefs. Zoccola et al. (2020) proposed for a biological repository for corals that taps into the network of the world's public aquaria and coral reef scientists. The Aichi Biodiversity Target 10 of the Convention on Biological Diversity (CBD) aims to minimize the anthropogenic pressures on coral reefs and other vulnerable ecosystems impacted by climate change or ocean acidification by maintaining integrity and functioning. Aichi Biodiversity Target 6 and Aichi Biodiversity Target 11 also focus on the protection of coral reef ecosystems (Obura et al., 2019). The management, conservation, and development of the marine ecosystem was also the focus of the United Nations Sustainable Development Goals (SDGs) of Agenda 2030. The vulnerability of global coral reefs was assessed by the Intergovernmental Platform on Biodiversity and Ecosystem Services, the Intergovernmental Panel on Climate Change, and the UN World Ocean Assessment (WOA, 2016; IPCC, 2018; IPBES, 2019).

5.8 Conclusion

The coral ecosystem, with its productive, protective, regulative, and aesthetic functions, is an integral part of the marine ecosystem. But the change in environmental conditions due to natural and man-made stressors cause huge damage to the coral reef ecosystem. Coral bleaching is mediated by the overproduction of radical oxygen species induced by hyperthermia and irradiance, causing cell damage and metabolic dysfunctions. The impacts are evident in low coral cover, decreased abundance of the marine community, and disturbed ecosystem services. The strategies for conservation and management of the coral reef ecosystem include reduction of local stressors, global-scale measures with marine protected areas, mitigation measures for climate change, and restoration by coral engineering. The integration of local and global management measures with a participatory approach is very much needed for the success of coral recovery and minimizing the negative impacts on corals.

REFERENCES

Abelson, A., Nelson, P. A., Edgar, G. J., Shashar, N., Reed, D. C., Belmaker, J., Krause, G., Beck, M. W., Brokovich, E., France, R. and Gaines, S. D. (2016). Expanding marine-protected areas to include degraded coral reefs. *Conservation Biology*. 30: 1182–1191. https://doi.org/10.1111/cobi.12722.

Allemand, D. and Furla, P. (2018). How does an animal behave like a plant? Physiological and molecular adaptations of zooxanthellae and their hosts to symbiosis. *Comptes Rendus Biologies*. 341 (5): 276–280. https://doi.org/10.1016/j.crvi.2018.03.007.

Altieria, A. H., Harrisona, S. B., Seemanna, J., Collina, R., Diaz, R. J. and Knowlton, N. (2017). Tropical dead zones and mass mortalities on coral reefs. *Proceedings of the National Academy of Sciences of the United States of America*. 114 (4): 3660–3665.

Amar, K. O. and Rinkevich, B. (2007). A floating mid-water coral nursery as larval dispersion hub: Testing an idea. *Marine Biology*. 151: 713–718. https://doi.org/10.1007/s00227-006-0512-0

Anthony, K. R. N. (2016). Coral reefs under climate change and ocean acidification: challenges and opportunities for management and policy. *Annual Review of Environment and Resources*. 41: 59–81. https://doi.org/10.1146/annurev-environ-110615-085610

Anthony, K. R. N., Bay, L. K., Costanza, R., Firn, J., Gunn, J., Harrison, P., Heyward, A., Lundgren, P., Mead, D., Moore, T., Mumby, P. J., Madeleine J. H. van Oppen, Robertson, J., Runge, M. C., Suggett, D. J., Schafelke, B., Wachenfeld, D. and Walshe, T. (2017). New interventions are needed to save coral reefs. *Nature Ecology & Evolution*. 1: 1420–1422.

Asir, N. G. G., Kumar, P. D., Arasamuthu, A., Mathews, G., Diraviya Raj, K., Ashok Kumar, T. K., Bilgi, D. S. and Patterson Edward, J. K. (2020). Eroding islands of Gulf of Mannar, Southeast India: a consequence of long-term impact of coral mining and climate change. *Natural Hazards*. 103: 103–119. https://doi.org/10.1007/s11069-020-03961-6

Ateweberhan, M., Feary, D. A., Keshavmurthy, S., Chen, A., Schleyer, M. H. and Sheppard, C. R. C. (2013). Climate change impacts on coral reefs: synergies with local effects, possibilities for acclimation, and management implications. *Marine Pollution Bulletin*. 74: 526–539. https://doi.org/10.1016/j.marpolbul.2013.06.011

Babcock, R. C., Dambacher, J. M., Morello, E. B., Plagányi, É. E., Hayes, K. R., Sweatman, H. P. A. and Pratchett, M. S. (2016). Assessing different causes of crown-of-thorns starfish outbreaks and appropriate responses for management on the great barrier reef. *PLoS ONE*. 11 (12): e0169048. https://doi.org/10.1371/journal.pone.0169048.

Baird, A. H., Bhagooli, R., Ralph, P. J. and Takahashi, S. (2009). Coral bleaching: the role of the host. *Trends in Ecology & Evolution*. 24 (1): 16–20. https://doi.org/10.1016/j.tree.2008.09.005

Baker, A. C., Glynn, P. W. and Riegl, B. (2008). Climate change and coral reef bleaching: an ecological assessment of long-term impacts, recovery trends and future outlook. *Estuarine, Coastal and Shelf Science*. 80: 435–471. https://doi.org/10.1016/j.ecss.2008.09.003

Baums, L. B. (2008). A restoration genetics guide for coral reef conservation. *Molecular Ecology*. 17: 2796–2811. http://doi.org/10.1111/j.1365-294X.2008.03787.x

Bellwood, D. R., Streit, R. S., Brandl, S. J. and Tebbett, S. B. (2019). The meaning of the term "function" in ecology: A coral reef perspective. *Functional Ecology*. 33: 948–961. http://doi.org/10.1111/1365-2435.13265

Benkwitt, C. E., Wilson, S. K. and Graham, N. A. J. (2020). Biodiversity increases ecosystem functions despite multiple stressors on coral reefs. *Nature Ecology & Evolution*. 4: 919–926.

Bijma, J., Portner, H. O., Yesson, C. and Rogers, A. D. (2013). Climate change and the oceans—what does the future hold? *Marine Pollution Bulletin*. 74: 495–505. https://doi.org/10.1016/j.marpolbul.2013.07.022

Brown, B. E. and Dunne, R. P. (1988). The environmental impact of coral mining on coral reefs in the Maldives. *Environmental Conservation*. 15: 159–165.

Burke, L., Reytar, K., Spalding, M. and Perry, A. (2011). *Reefs at Risk—Revisited*. World Resources Institute, Washington, DC.

Cantin, N. E., Cohen, A. L., Karnauskas, K. B., Tarrant, A. M. and McCorkle, D. C. (2010). Ocean warming slows coral growth in the Central Red Sea. *Science*. 329: 322–325.

Caras, T. and Pasternak, Z. (2009). Long-term environmental impact of coral mining at the Wakatobi marine park, Indonesia. *Ocean & Coastal Management*. 52: 539–544. http://doi.org/10.1016/j.ocecoaman.2009.08.006

Carpenter, K. E., Abrar, M., Aeby, G., Aronson, R. B., Banks, S., Bruckner, A., Chiriboga, A., Cortés, J., Delbeek, J. C., Devantier, L., Edgar, G. J., Edwards, A. J., Fenner, D., Guzmán, H. M., Hoeksema, B. W., Hodgson, G., Johan, O., Licuanan, W. Y., Livingstone, S. R., Lovell, E. R., Moore, J. A., Obura, D. O., Ochavillo, D., Polidoro, B. A., Precht, W. F., Quibilan, M. C., Reboton, C., Richards, Z. T., Rogers, A. D., Sanciangco, J., Sheppard, A., Sheppard, C., Smith, J., Stuart, S., Turak, E. N., Wallace, C., Weil, E. and Wood, E. (2008). One-third of reef-building corals face elevated extinction risk from climate change and local impacts. *Science*. 321: 560–563. http://doi.org/10.1126/science.1159196

Chatragadda, R., Koushik, S., Shunmugaraj, T. and Murthy, M. V. R. (2019). Baseline survey on coral diseases, stress factors and new threats in coral reefs of Gulf of Mannar Marine Biosphere, India. *Journal of Marine Science* [S.l.]. 1 (1). https://doi.org/10.30564/jms.v1i1.835

Cheal, A., Macneil, A., Emslie, M. and Sweatman, H. (2017). The threat to coral reefs from more intense cyclones under climate change. *Global Change Biology*. 23: 1511–1524. http://doi.org/10.1111/gcb.13593

Cinner, J. E., Daw, T., Huchery, C., Thoya, P., Wamukota, A., Cedras, M. and Abunge, C. (2014). Winners and losers in marine conservation: Fishers' displacement and livelihood benefits from Marine Reserves. *Society & Natural Resources: An International Journal*. 27 (9): 994–1005.

Cinner, J. E., Huchery, C., MacNeil, M. A., Graham, N. A. J., McClanahan, T. R., Maina, J., Maire, E., Kittinger, J. N., Hicks, C. C., Mora, C., Allison, E. H., D'Agata, S., Hoey, A., Feary, D. A., Crowder, L., Williams, I. D., Kulbicki, M., Vigliola, L., Wantiez, L., Edgar, G., Stuart-Smith, R. D., Sandin, S. A., Green, A. L., Hardt, M. J., Beger, M., Friedlander, A., Campbell, S. J., Holmes, K. E., Wilson, S. K., Brokovich, E., Brooks, A. J., Cruz-Motta, J. J., Booth, D. J., Chabanet, P., Gough, C., Tupper, M., Ferse, S. C. A., Sumaila, U. R. and Mouillot, D. (2016). Bright spots among the world's coral reefs. *Nature*. 535: 416–419. https://doi.org/10.1038/nature18607

Cinner, J. E., McClanahan, T. R., Graham, N. A. J., Daw, T. M., Maina, J., Stead, S. M., Wamukota, A., Brown, K. and Bodin, O. (2012). Vulnerability of coastal communities to key impacts of climate change on coral reef fisheries. *Global Environmental Change*. 22 (1): 12–20. https://doi.org/10.1016/j.gloenvcha.2011.09.018

Clements, C. S. and Hay, M. E. (2019). Biodiversity enhances coral growth, tissue survivorship and suppression of macroalgae. *Nature Ecology & Evolution*. 3: 178–182.

Cooper, T. F., De'Ath, G., Fabricius, K. E. and Lough, J. M. (2008). Declining coral calcification in massive Porites in two nearshore regions of the northern Great Barrier Reef. *Global Change Biology*. 14: 529–538.

Crain, C. M., Halpern, B. S., Beck, M. W. and Kappel, C. V. (2009). Understanding and managing human threats to the coastal marine environment. *Annals of the New York Academy of Sciences*. 1162: 39–62.

Dana, J. D. (1853). On an isothermal oceanic chart illustrating the geographical distribution of marine animals. *American Journal of Science*. 16: 314–327.

D'Angelo, C. and Wiedenmann, J. (2014). Impacts of nutrient enrichment on coral reefs: New perspectives and implications for coastal management and reef survival. *Current Opinion in Environmental Sustainability*. 7: 82–93. https://doi.org/10.1016/j.cosust.2013.11.029

Darwin, C. (1859). *The Origin of Species by Means of Natural Selection, or the Preservation of Favoured Races in the Struggle for Life*. John Murray, London.

De'ath, G. and Fabricius, K. E. (2010). Water quality as a regional driver of coral biodiversity and macroalgae on the Great Barrier Reef. *Ecological Applications*. 20: 840–850. https://doi.org/10.1890/08-2023.1

Downs, C. A., McDougall, K. E., Woodley, C. M., Fauth, J. E., Richmond, R. H., Kushmaro, A., Gibb, S. W., Loya, Y., Ostrander, G. K. and Kramarsky-Winter, E. (2013). Heat-stress and light-stress induce different cellular pathologies in the symbiotic dinoflagellate during coral bleaching. *PLoS ONE* 8 (12): e77173. https://doi.org/10.1371/journal.pone.0077173

Dulvy, N. K., Stanwell-Smith, D., Darwall, W. R. T. and Horrill, C. J. (1995). Coral mining at Mafia Island, Tanzania: A management dilemma. *Ambio*. 24 (6): 358–365.

Edgar, G. J., Stuart-Smith, R. D., Willis, T. J., Kininmonth, S., Baker, S. C., Banks, S., Barrett, N. S., Becerro, M. A., Bernard, A. T. F., Berkhout, J., Buxton, C. D., Campbell, S. J., Cooper, A. T., Davey, M., Edgar, S. C., Försterra, G., Galván, D. E., Irigoyen, A. J., Kushner, D. J., Moura, R., Parnell, P. Ed., Shears, N. T., Soler, G., Strain, E. M. A. and Thomson, R. J. (2014). Global conservation outcomes depend on marine protected areas with five key features. *Nature*. 506: 216–220. https://doi.org/10.1038/nature13022

Elsner, J. B., Kossin, J. P. and Jagger, T. H. (2008). The increasing intensity of the strongest tropical cyclones. *Nature*. 455: 92–95.

EPA. (2021). *Threats to Coral Reefs*. www.epa.gov/coral-reefs/threats-coral-reefs. Accessed on 10 September 2021.

Fabricius, K. E., Logan, M., Weeks, S. and. Brodie, J. (2014). The effects of river run-off on water clarity across the central Great Barrier Reef. *Marine Pollution Bulletin*. 84: 191–200. https://doi.org/10.1016/j.marpolbul.2014.05.012

Fabricius, K. E., Okaji, K. and De'ath, G. (2010). Three lines of evidence to link outbreaks of the crown-of-thorns seastar Acanthasterplanci to the release of larval food limitation. *Coral Reefs*. 29: 593–605.

Fitt, K. W., Brown, B., Warner, M. and Dunne, R. (2001). Coral bleaching: Interpretation of thermal tolerance limits and thermal thresholds in tropical corals. *Coral Reefs*. 20: 51–65. http://doi.org/10.1007/s003380100146

Fitt, K. W. and Warner, M. E. (1995). Bleaching patterns of four species of Caribbean Reef Corals. *Biological Bulletin*. 189 (3): 298–307.

Fox, H. E. and Caldwell, R. L. (2006). Recovery of blast fishing on coral reefs: a tale of two scales. *Ecological Applications*. 16: 1631–1635.

França, F. M., Benkwitt, B. C., Peralta, G., Robinson, J. P. W., Graham, N. A. J., Tylianakis, J. M., Berenguer, E., Lees, A. C., Ferreira, J., Louzada, J. and Barlow, J. (2020). Climatic and local stressor interactions threaten tropical forests and coral reefs. *Philosophical Transactions of the Royal Society B*. 375: 20190116. http://doi.org/10.1098/rstb.2019.0116

Francini-Filho, R. B., Asp, N. E., Siegle, E., Hocevar, J., Lowyck, K., D'Avila, N., Vasconcelos, A. A., Baitelo, R., Rezende, C. E., Omachi, C. Y., Thompson, C. C. and Thompson, F. L. (2018). Perspectives on the Great Amazon Reef: Extension, biodiversity, and threats. *Frontiers in Marine Science*. 5: 142. http://doi.org/10.3389/fmars.2018.00142

Fredston-Hermann, A., Brown, C. J., Albert, S., Klein, C. J., Mangubhai, S., Nelson, J. L., Teneva, L., Wenger, A., Gaines, S. D. and Halpern, B. S. (2016). Where does river runoff matter for coastal marine conservation? *Frontiers in Marine Science*. 3: 273.

Galford, G. L., Fernandez, M., Roman, J., Monasterolo, I., Ahamed, S., Fiske, G., González-Díaz, P. and Kaufman, L. (2018). Cuban land use and conservation, from rainforests to coral reefs. *Bulletin of Marine Science*. 94 (2): 171–191. https://doi.org/10.5343/bms.2017.1026

Gates, R. D., Baghdasarian, G. and Muscatine, L. (1992). Temperature stress causes host cell detachment in symbiotic cnidarians: implications for coral bleaching. *Biology Bulletin*. 182: 324–332.

Gattuso, J. P., Magnan, A., Billé, R., Cheung, W. W. L., Howes, E. L., Joos, F., Allemand, D., Bopp, L., Cooley, S. R., Eakin, C. M., Hoegh-Guldberg, O., Kelly, R. P., Pörtner, H. O., Rogers, A. D., Baxter, J., Laffoley, D., Osborn, D., Rankovic, A., Rochette, J., Sumaila, U. R., Treyer, S. and Turley, C. (2015). Contrasting futures for ocean and society from different anthropogenic CO2 emissions scenarios. *Science*. 349 (6243): aac4722.

GBRMPA. (2008). *Bio-physical Assessment of the Reefs of Keppel Bay: A Baseline Study*. Great Barrier Reef Marine Park Authority, Townsville.

Glynn, P. W. and D'Croz, L. (1990). Experimental evidence for high temperature stress as the cause of El Nino-coincident coral mortality. *Coral Reefs*. 8: 181–191.

Goodbody-Gringley, G., Wetzel, D. L., Gillon, D., Pulster, E., Miller, A. and Ritchie, K. B. (2013). Toxicity of deepwater horizon source oil and the chemical dispersant, Corexit R 9500, to coral larvae. *PLoS ONE*. 8: e45574. http://doi.org/10.1371/journal.pone.0045574

Goreau, T. F. 1964. Mass expulsion of zooxanthellae from Jamaican reef communities after Hurricane Flora. *Science*. 145: 383–386.

Goreau, T. F., Goreau, N. I. and Goreau, T. J. (1979). Corals and coral reefs. *Scientific American*. 241 (2): 124–137.

Hallock, P., Williams, D. E., Fisher, E. M. and Toler, S. K. (2006). Bleaching in Foraminifera with algal symbionts: Implications for reed monitoring and risk assessment. *Marine Science Faculty Publications*. 1249.

Halpern, B. S., Walbridge, S., Selkoe, K. A., Kappel, C. V., Micheli, F., D'Agrosa, C., Bruno, J. F., Casey, K. S., Ebert, C., Fox, H. E., Fujita, R., Heinemann, D., Lenihan, H., Madin, E., Perry, M., Selig, E., Spalding, M., Steneck, R. and Watson, R. (2008). A global map of human impact on marine ecosystems. *Science*. 319: 948–952. http://doi.org/10.1126/science.1149345.2008

Hamilton, R. J., Almany, G. R., Brown, C., Pita, J., Peterson, N. A. and Choat, J. H. (2017). Logging degrades nursery habitat for an iconic coral reef fish. *Biological Conservation*. 210: 273–280. https://doi.Org/10.1016/j.biocon.2017.04.024

Harvey, B. J., Nash, K. L., Blanchard, J. L. and Edwards, D. P. (2018). Ecosystem-based management of coral reefs under climate change. *Ecology and Evolution*. 8: 6354–6368. http://doi.org/10.1002/ece3.4146

Haszprunar, G., Vogler, C. and Wörheide, G. (2017). Persistent gaps of knowledge for naming and distinguishing multiple species of Crown-of-Thorns-Seastar in the Acanthasterplanci species complex. *Diversity*. 9 (2): 22. https://doi.org/10.3390/d9020022

He, Q. and Silliman, B. R. (2019). Climate change, human impacts, and coastal ecosystems in the Anthropocene. *Current Biology*. 29 (19): R1021–R1035. https://doi.org/10.1016/j.cub.2019.08.042

Heron, S. F., Maynard, J. A., Van Hooidonk, R. and Eakin, C. M. (2016). Warming trends and bleaching stress of the world's coral reefs 1985–2012. *Science Reporter*. 6: 38402.

Hoegh-Guldberg, O. (2005). Low coral cover in a high-CO2 world. *Journal of Geophysical Research C: Oceans*. 110: C09S06. https://doi.org/10.1029/2004JC002528

Hoegh-Guldberg, O. (2014). Coral reef sustainability through adaptation: Glimmer of hope or persistent mirage? *Current Opinion in Environmental Sustainability*. 7: 127–133. https://doi.org/10.1016/j.cosust.2014.01.005.

Hoegh-Guldberg, O., Kennedy, E. V., Beyer, H. L., McClennen, C. and Possingham, H. P. (2018). Securing a long-term future for coral reefs. *Trends in Ecology & Evolution*. 33: 12. https://doi.org/10.1016/j.tree.2018.09.00

Hoegh-Guldberg, O., Mumby, P., Hooten, A. J., Steneck, R. S., Greenfield, P., Gomez, E., Harvell, C., Sale, P., Edwards, A., Caldeira, K., Knowlton, N., Eakin, C. M., Iglesias-Prieto, R., Muthiga, N., Bradbury, R., Dubi, A. and Hatziolos, M. (2007). Coral reefs under rapid climate change and ocean acidification. *Science*. 318: 1737–1742. http://doi.org/10.1126/science.1152509

Hoegh-Guldberg, O., Poloczanska, E. S., Skirving, W. and Dove, S. (2017). Coral Reef ecosystems under climate change and ocean acidification. *Frontiers in Marine Science*. 4: 158. https://doi.org/10.3389/fmars.2017.00158

Hughes, T. P., Kerry, J. T., Alvarez-Noriega, M., Alvarez-Romero, J. G., Anderson, K. D., Baird, A. H. and Bridge, T. C. (2017). Global warming and recurrent mass bleaching of corals. *Nature*. 543: 373–377.

Hughes, T. P., Kerry, J. T., Baird, A. H., Connolly, S. R., Dietzel, A., Eakin, C. M., Heron, S. F., Hoey, A. S., Hoogenboom, M. O., Liu, G., McWilliam, M. J., Pears, R. J., Pratchett, M. S., Skirving, W. J., Stella, J. S. and Torda, G. (2018). Global warming transforms coral reef assemblages. *Nature*. 556 (7702): 492–496. http://doi.org/10.1038/s41586-018-0041-2

IPBES. (2019). Summary for policymakers of the global assessment report on biodiversity and ecosystem services. In: *Intergovernmental Science-policy Platform on Biodiversity and Ecosystem Services*. S. Díaz, J. Settele, E. S. Brondizio, H. T. Ngo, M. Guéze, J. Agard, A. Arneth, P. Balvanera, K. A. Brauman, S. H. M. Butchart, K. M. A. Chan, L. A. Garibaldi, K. Ichii, J. Liu, S. M. Subramanian, G. F. Midgley, P. Miloslavich, Z. Molnár, D. Obura, A. Pfaff, S. Polasky, A. Purvis, J. Razzaque, B. Reyers, R. Roy Chowdhury, Y. J. Shin, I. J. Visseren-Hamakers, K. J. Willis, and C. N. Zayas (Eds). IPBES Secretariat, Bonn, 39.

IPCC. (2007). Climate change 2007: The physical science basis. In: *Contribution of Working Group I to the Fourth Assessment Report of the Intergovernmental Panel on Climate Change*. S. Solomon (Ed). Cambridge University Press, Cambridge.

IPCC. (2014). Summary for policymakers. In: *Climate Change 2014: Impacts, Adaptation, and Vulnerability. Part A: Global and Sectoral Aspects. Contribution of Working Group II to the Fifth Assessment Report of the Intergovernmental Panel on Climate Change*. C. B. Field, V. R. Barros, D. J. Dokken, K. J. Mach, M. D. Mastrandrea, T. E. Bilir, M. Chatterjee, K. L. Ebi, Y. O. Estrada, R. C. Genova, B. Girma, E. S. Kissel, A. N. Levy, S. MacCracken, P. R. Mastrandrea, and L. L. White (Eds). Cambridge University Press, Cambridge and New York, 1–32.

IPCC. (2018). Summary for policymakers. In: *The Context of Strengthening the Global Response to the Threat of Climate Change, Sustainable Development, and Efforts to Eradicate Poverty*. P. Zhai, H. O. Portner, D. Roberts, J. Skea, P. R. Shukla, A. Pirani, W. Moufouma-Okia, C. Péan, R. Pidcock, S. Connors, J. B. R. Matthews, Y. Chen, X. Zhou, M. I. Gomis, E. Lonnoy, T. Maycock, M. Tignor, and T. Waterfield (Eds). V. Masson-Delmotte. World Meteorological Organization, Geneva, Switzerland, 32.

Jackson, J. B. C., Kirby, M. X., Berger, W. H., Bjorndal, K. A., Botsford, L. W., Bourque, B. J., Bradbury, R. H., Cooke, R., Erlandson, J., Estes, J. A., Hughes, T. P., Kidwell, S., Lange, C. B., Lenihan, H. S., Pandolfi, J. M., Peterson, C. H., Steneck, R. S., Tegner, M. J. and Warner, R. R. (2001). Historical overfishing and the recent collapse of coastal ecosystems. *Science*. 293: 629–638.

Jewett, L. and Romanou, A. (2017). Ocean acidification and other ocean changes. In: *Climate Science Special Report: Fourth National Climate Assessment, Volume I*. D. J. Wuebbles, D. W. Fahey, K. A. Hibbard, D. J. Dokken, B. C. Stewart, and T. K. Maycock (Eds). U.S. Global Change Research Program, Washington, DC, USA, pp.364–392. http://doi.org/10.7930/J0QV3JQB

Jobbins, G. (2006). Tourism and coral-reef based conservation: can they coexist? In: *Coral Reef Conservation*. I. M. Cote and J. D. Reynolds (Eds). Cambridge University Press, Cambridge, 237–263.

Jones, C. G., Lawton, J. H. and Shachak, M. (1994). Organisms as ecosystem engineers. *Oikos*. 69: 373–386.

Jones, G. P., McCormick, M. I., Srinivasan, M. and Eagle, J. V. (2004). Coral decline threatens fish biodiversity in marine reserves. *Proceedings of the National Academy of Sciences of the United States of America*. 101: 8251–8253.

Jones, R. J. and Hoegh-Guldberg, O. (1999). Effects of cyanide on coral photosynthesis: Implications for identifying the cause of coral bleaching and for assessing the environmental effects of cyanide fishing. *Marine Ecology Progress Series*. 177: 83–91. https://doi.org/10.3354/meps177083

Jones, R. J., Kildea, T. and Hoegh-Guldberg, O. (1999). PAM chlorophyll fluorometry: a new in situ technique for satres Assessment in Scleractinianceorals, used to examine the effects of cyanide from cyanide fishing. *Marine Pollution Bulletin*. 38 (10): 864–874.

Klein, C. J., Ban, N. C., Halpern, B. S., Beger, M., Game, E. T., Grantham, H. S., Green, A., Klein, T. J., Kininmonth, S., Treml, E., Wilson, K. and Possingham, H. P. (2010). Prioritizing land and sea conservation investments to protect coral reefs. *PLoS ONE*. 5 (8): e12431. https://doi.org/10.1371/journal.pone.0012431

Kleypas, J. A., Buddemeier, R. W., Archer, D., Gattuso, J. P., Langdon, C. and Opdyke, B. N. (1999). Geochemical consequences of increased atmospheric carbon dioxide on coral reefs. *Science*. 284: 118–120.

Knowlton, N., Brainard, R. E., Fischer, R., Moews, M., Plaisance, L. and Caley, M. J. (2010). Coral reef biodiversity. In: *Life in the World's Oceans: Diversity, Distribution, and Abundance*. A. D. McIntyre (Ed). Wiley-Blackwell, Oxford, 65–74.

Knowlton, N. and Jackson, J. B. C. (2001). The ecology of coral reefs. In: *Marine Community Ecology*. M. D. Bertness, S. D. Gaines, and M. Hay (Eds), Sinauer, Sunderland, MA, 395–422.

Lamb, J. B., Willis, B. L., Fiorenza, E. A., Couch, C. S., Howard, R., Rader, D. N., True, J. D., Kelly, L. A., Ahmad, A., Jompa, J. and Harvell, C. D. (2018). cPlastic waste associated with disease on coral reefs. *Sience*. 359 (6374): 460–462. http://doi.org/10.1126/science.aar3320

Magris, R., Grech, A. and Pressey, R. (2018). Cumulative human impacts on coral reefs: Assessing risk and management implications for Brazilian coral reefs. *Diversity*. 10: 26.

Mallik, T. K. (2017). Coral atolls of Lakshadweep, Arabian Sea, Indian Ocean. *MOJ Ecology and Environmental Sciences*. 2 (2): 68–83. http://doi.org/10.15406/mojes.2017.02.00021

Maynard, J. A., McKagan, S., Raymundo, L., Johnson, S., Ahmadia, G. N., Johnston, L., Houk, P., Williams, G. J., Kendall, M., Heron, S. F., van Hooidonk, R., Mcleod, E., Tracey, D. and Planes, S. (2015). Assessing relative resilience potential of coral reefs to inform management. *Biological Conservation*. 192: 109–119. https://doi.org/10.1016/j.biocon.2015.09.001

McAuley, P. J. (1994) Interactions between host and symbionts in algal invertebrate intracellular symbioses. *Botanical Journal of Scotland*. 47 (1): 97–112.

McClanahan, T. R., Graham, N. A. J. and Darling, E. S. (2014). Coral reefs in a crystal ball: Predicting the future from the vulnerability of corals and reef fishes to multiple stressors. *Current Opinion in Environmental Sustainability*. 7: 59–64. https://doi.org/10.1016/j.cosust.2013.11.028

McClanahan, T. R., Graham, N. A. J., Macneil, M. A. and Cinner, J. E. (2015). Biomass-based targets and the management of multispecies coral reef fisheries. *Conservation Biology*. 29: 409–417. https://doi.Org/10.1111/cobi.12430

McLeod, E., Shaver, E. C., Beger, M., Koss, J. and Grimsditch, G. (2021). Using resilience assessments to inform the management and conservation of coral reef ecosystems. *Journal of Environmental Management*. 277: 111384. https://doi.org/10.1016/j.jenvman.2020.111384.

Mellin, C., MacNeil, M. A., Cheal, A. J., Emslie, M. J. and Caley, M. J. (2016). Marine Protected Areas increase resilience among coral reef communities. *Ecology Letters*. 19: 629–637.

Moberg, F. and Folke, C. (1999). Ecological goods and services of coral reef ecosystems. *Ecological Economics*. 29: 215–233. http://doi.org/10.1016/s0921-8009(99)00009-9

National Oceanic and Atmospheric Administration (NOAA). (2017). *Global Coral Bleaching Event Likelyending*, June 19. www.noaa.gov/media-release/global-coral-ble aching-event-likely-ending

Neigel, J., Domingo, A. and Stake, J. (2007). DNA barcoding as a tool for coral reef conservation. *Coral Reefs*. 26: 487–499. http://doi.org/10.1007/s00338-007-0248-4

Nielsen, D. A., Petrou, K. and Gates, R. D. (2018). Coral bleaching from a single cell perspective. *ISME of Journal*. 12: 1558–1567. https://doi.org/10.1038/s41396-018-0080-6

Norström, A. V., Nyström, M., Jouffray, J. B., Folke, C., Graham, N. A. J., Moberg, F., Olsson, P. and Williams, G. J. (2016). Guiding coral reef futures in the Anthropocene. *Frontiers in Ecology and the Environment*. 14: 490–498. https://doi.org/10.1002/fee.1427

Obura, D. O., Greta, A., Natchanon, A., Ward, A., Nicholas, B., Joe, B., Brainard, R. E., Samuel, C., Pamela, F., Gordon, T. A. C., Lew, G., Mishal, G., John, H., Hendee, J., Hodgson, G., Huang, D., Jankulak, M., Jones, A., Kimura, T., Levy, J., Miloslavich, P., Chou, L. M., Muller-Karger, F., Osuka, K., Samoilys, M., Simpson, S. D., Tun, K. and Wongbusarakum, S. (2019). Coral reef monitoring, reef assessment technologies, and ecosystem-based management. *Frontiers in Marine Science*. 6: 580. http://doi.org/10.3389/fmars.2019.00580

Osborne, K., Dolman, A. M., Burgess, S. C. and Johns, K. A. (2011). Disturbance and the dynamics of coral cover on the Great Barrier Reef (1995–2009). *PLoS ONE*. 6: e17516. http://doi.org/10.1371/journal.pone.0017516

Osmond, C. B. (1994). What is photoinhibition? Some insights from comparisons of shade and sun plants. In: *Photoinhibition of Photosynthesis: From Molecular Mechanisms to the Field*. N. R. Baker and J. R. Bowyer. BIOS Scientific Publishers, Abingdon, pp. 1–24.

Pandolfi, J. M., Connolly, S. R., Marshall, D. J. and Cohen, A. L. (2011). Projecting coral reef futures under global warming and ocean acidification. *Far-Field Optical Nanoscopy, Science*. 333: 418–422.

Perez, K. III, Rodgers, K. S., Jokiel, P. L., Lager, C. V. and Lager, D. J. (2014). Effects of terrigenous sediment on settlement and survival of the reef coral Pocilloporadamicornis. *PeerJ*. 2: e387. https://doi.org/10.7717/peerj.387

Pollock, F. J., Lamb, J. B., Field, S. N., Heron, S. F., Schaffelke, B., Shedrawi, G., Bourne, D. G. and Willis, B. L. (2014). Sediment and turbidity associated with offshore dredging increase coral disease prevalence on nearby reefs. *PLoS ONE*. 9: e102498. https://doi.org/10.1371/journal.pone.0102498

Pratchett, M. S., Hoey, A. S. and Wilson, S. K. (2014). Reef degradation and the loss of critical ecosystem goods and services provided by coral reef fishes. *Current Opinion in Environmental Sustainability*. 7: 37–43.

Pratchett, M. S., Hoey A. S. and Wilson, S. K. (2016). Habitat-use and specialisation among coral reef Damselfishes. In: *Biology of Damselfishes* (1st Edition). B. Frédérich and E. Parmentier (Eds). CRC Press, Boca Raton, 38.

Praveena, S. M., Siraj, S. S. and Aris, A. Z. (2012). Coral reefs studies and threats in Malaysia: A mini review. *Reviews in Environmental Science and Bio/Technology*. 11: 27–39.

Reaka-Kudla, M. L. (1997). The global biodiversity of coral reefs: a comparison with rain forests. In: *Biodiversity II: Understanding and Protecting Our Biological Resources*. M. L. Reaka-Kudla, D. E. Wilson and E. O. Wilson (Eds). Joseph Henry Press, Washington, DC, 83–108.

Reaka-Kudla, M. L. (2005). Biodiversity of Caribbean coral reefs. In: *Caribbean Marine Biodiversity*. P. Miloslavich and E. Klein (Eds). Des Tech Publishers, Lancaster, PA, 259–276.

Richmond, R. H. (1993). Coral Reefs: Present problems and future concerns resulting from anthropogenic disturbance. *American Zoologist*. 33: 524–536.

Richmond, R. H., Rongo, T., Golbuu, Y., Victor, S., Idechong, N., Davis, G., Kostka, W., Neth, L., Hamnett, M. and Wolanski, E. (2007). Watersheds and coral reefs: conservation science, policy and implementation. *BioScience*. 57: 598–607. https://doi.org/10.1641/B570710

Riniwati, H., Harahab, N. and Abidin, Z. (2019). A vulnerability analysis of coral reefs in coastal ecotourism areas for conservation management. *Diversity*. 11: 107. http://doi.org/10.3390/d11070107

Rinkevich, B. (2014). Rebuilding coral reefs: Does active reef restoration lead to sustainable reefs? *Current Opinion in Environmental Sustainability*. 7: 28–36. https://doi.org/10.1016/j.cosust.2013.11.018

Rinkevich, B. (2015). Climate change and active reef restoration—ways of constructing the "reefs of tomorrow." *Journal of Marine Science and Engineering*. 3: 111–127. https://doi.org/10.3390/jmse3010111

Roberts, C. M., McClean, C. J., Veron, J. E., Hawkins, J. P., Allen, G. R., McAllister, D. E., Mittermeier, C. G., Schueler, F. W., Spalding, M., Wells, F., Vynne, C. and Werner, T. B. (2002). Marine biodiversity hotspots and conservation priorities for tropical reefs. *Science*. 295 (5558): 1280–1284. http://doi.org/10.1126/science.1067728

Roberts, C. M., O'Leary, B. C., McCauley, D. J., Cury, P. M., Duarte, C. M., Lubchenco, J., Pauly, D., Sáenz-Arroyo, A., Sumaila, U. R., Wilson, R. W., Worm, B. and Castilla, J. C. (2017). Marine reserves can mitigate and promote adaptation to climate change. *Proceedings of the National Academy of Sciences of the United States of America*. 114 (24): 6167–6175. https://doi.org/10.1073/pnas.1701262114

Robinson, J. P. W., McDevitt-Irwin, J. M., Dajka, J. C., Hadj-Hammou, J., Howlett, S., Graba-Landry, A., Hoey, A. S., Nash, K. L., Wilson, S. K. and Graham, N. A. J. (2019). Habitat and fishing control grazing potential on coral reefs. *Functional Ecology*. 34 (1): 240–251. https://doi.org/10.1111/1365-2435.13457

Rocha, L. A., Pinheiro, H. T., Shepherd, B., Papastamatiou, Y. P., Luiz, O. J., Pyle, R. L. and Bongaerts, P. (2018). Mesophotic coral ecosystems are threatened and ecologically distinct from shallow water reefs. *Science*. 361 (6399): 281–284. http://doi.org/10.1126/science.aaq1614

Rogers, A. D. (1999). The biology of Lopheliapertusa (Linnaeus 1758) and other deep-water reef-forming corals and impacts from human activities. *Hydrobiologia*. 84: 315–406. http://doi.org/10.1002/iroh.199900032

Rogers, A. D., Blanchard, J. L. and Mumby, P. J. (2014). Vulnerability of coral reef fisheries to a loss of structural complexity. *Current Biology*. 24: 1000–1005.

Rogers, A. D., Harborne, A. R., Brown, C. J., Bozec, Y. M., Castro, C., Chollett, I., Hock, K., Knowland, C. A., Marshell, A., Ortiz, J. C., Razak, T., Roff, G., Samper-Villarreal, J., Saunders, M. I., Wolff, N. H. and Mumby, P. J. (2015). Anticipative management for coral reef ecosystem services in the 21st century. *Global Change Biology*. 21: 504–514. https://doi.org/10.1111/gcb.12725

Sale, P. F., Agardy, T., Ainsworth, C. H., Feist, B. E., Bell, J. D., Christie, P., Hoegh-Guldberg, O., Mumby, P. J., Feary, D. A., Saunders, M. I., Daw, T. M., Foale, S. J., Levin, P. S., Lindeman, K. C., Lorenzen, K., Pomeroy, R. S., Allison, E. H., Bradbury, R., Corrin, J., Edwards, A. J., Obura, D. O., Sadovy De M. Y. J., Samoilys, M. A. and Sheppard, C. R. C. (2014). Transforming management of tropical coastal seas to cope with the challenges of the 21st century. *Marine Pollution Bulletin*. 85: 8–23. https://doi.org/10.1016/j.marpolbul.2014.06.005

Sandin, S. A., Smith, J. E., DeMartini, E. E., Dinsdale, E. A., Donner, S. D., Friedlander, A. M. and Sala, O. (2008). Baselines and degradation of coral reefs in the Northern Line Islands. *PLoS ONE*. 3: e1548. https://doi.org/10.1371/journal.pone.0001548

Saravanan, R., Lakshmanan, R., Jasmine, S. and Joshi, K. K. (2017). Coral bleaching: causes, consequences and mitigation. *Marine Fisheries Information Service, Technical and Extension Series*. No. 231: 3–9.

Sarjana, P. K. (1992). *The Impact of Coral Mining on Coral Reef Condition in the East and South Coast of Bali, Indonesia*. M.Sc. Thesis, University of Newcastle-UponTyne (UK).

Shafir, S., Van Rijn, J. and Rinkevich, B. (2006). Steps in the construction of underwater coral nursery, an essential component in reef restoration acts. *Marine Biology*. 149: 679–687. https://doi.org/10.1007/s00227-005-0236-6

Spalding, M., Burke, L., Wood, S. A., Ashpole, J., Hutchison, J. and Ermgassene, P. Z. (2017). Mapping the global value and distribution of coral reef tourism. *Marine Policy*. 82: 104–113. http://doi.org/10.1016/j.marpol.2017.05.014

Stambler, N. (2011). Zooxanthellae: The Yellow Symbionts inside a nimals. In: *Coral Reefs: An Ecosystem in Transition*. Z. Dubinsky and N. Stambler (Eds). Springer, Dordrecht. https://doi.org/10.1007/978-94-007-0114-4_7

Teh, L. S. L., Teh, L. C. L. and Sumaila, U. R. (2013). A global estimate of the number of coral reef fishers. *PLoS ONE*. 8: e65397. https://doi.org/10.1371/journal.pone.0065397

van Hooidonk, R., Maynard, J., Tamelander, J., Gove, J., Ahmadia, G., Raymundo, L., Williams, G., Heron, S. F. and Planes, S. 2016. Local-scale projections of coral reef futures and implications of the Paris Agreement. *Scientific Reports*. 6: 39666.

Vega Thurber, R. L., Burkepile, D. E., Fucs, C., Shantz, A. A., Mcminds, R. and Zaneveld, J. R. (2014). Chronic nutrient enrichment increases prevalence and severity of coral disease and bleaching. *Global Change Biology*. 20: 544–554. https://doi.org/10.1111/gcb.12450.

Veron, J. E. N. (1986). *Corals of Australia and the Indo-Pacific*. University of Hawai Press, Angus & Robertson, Australia.

Veron, J. E. N., DeVantier, L. M., Turak, E., Green, A. L., Kininmonth, S., Stafford-Smith, M. and Peterson, N. (2009). Delineating the Coral Triangle. *Galaxea, Journal of Coral Reef Studies*. 11: 91–100. http://doi.org/10.3755/galaxea.11.91

Veron, J. E. N., Hoegh-Guldberg, O., Lenton, T. M., Lough, J. M., Obura, D. O., Pearce-Kelly, P., Sheppard, C. R., Spalding, M., Stafford-Smith, M. G. and Rogers, A. D. (2009). The coral reef crisis: the critical importance of<350 ppm CO2. *Marine Pollution Bulletin*. 58 (10): 1428–1436. http://doi.org/10.1016/j.marpolbul.2009.09.009

Wagner, D. E., Friedlander, A. M., Pyle, R. L., Brooks, C. M., Gjerde, K. M. and Wilhelm, T. A. (2020). Coral Reefs of the High Seas: Hidden Biodiversity Hotspots in need of protection. *Frontiers in Marine Science*. 7: 567428. http://doi.org/10.3389/fmars.2020.567428

Wagner, D. E., Kramer, P. and Van Woesik, R. (2010). Species composition, habitat, and water quality influence coral bleaching in southern Florida. *Marine Ecology Progress Series*. 408: 65–78. https://doi.org/10.3354/meps08584

Walsworth, Schindler, D. E., Colton, M. A., Webster, M. S., Palumbi, S. R., Mumby, P. J., Essington, T. and Pinsky, M. L. (2019). Management for network diversity speeds evolutionary adaptation to climate change. *Nature Climate Change*. 9 (8): 632–636. https://doi.org/10.7282/t3-zs9g-s663

Weeks, E. S., Walker, S., Overton, J. M. and Clarkson, B. (2013). The value of validated vulnerability data for conservation planning in rapidly changing landscapes. *Environmental Management*. 51: 1055–1066.

Weis, V. M. (2008). Cellular mechanisms of Cnidarian bleaching: Stress causes the collapse of symbiosis. *The Journal of Experimental Biology*. 211 (pt 19): 3059–3066. http://doi.org/10.1242/jeb.009597.

Weis, V. M., Davy, S. K., Hoegh-Guldberg, O., Rodriguez-Lanetty, M. and Pringle, J. R. (2008). Cell biology in model systems as the key to understanding corals. *Trends in Ecology & Evolution*. 23 (7): 369–376. https://doi.org/10.1016/j.tree.2008.03.004.

Wells, S. M. (1984). What's happening to coral reefs?—Part 1. *NAGA, ICLARM News's- Letter*. 2 (4): 5–7.

Wiedenmann, J., D'Angelo, C., Smith, E. G., Hunt, A. N., Legiret, F. E., Postle, A. D. and Achterberg, E. P. (2013). Nutrient enrichment can increase the susceptibility of reef corals to bleaching. *Nature Climate Change*. 3: 160–164. https://doi.org/10.1038/nclimate1661

Wild, C., Hoegh-Guldberg, O., Naumann, M. S., Colombo-Pallotta, M. F., Ateweberhan, M., Fitt, W. K., Iglesias-Prieto, R., Palmer, C., Bythell, J. C., Ortiz, J., Loya, Y. and van Woesik, R. (2011). Climate change impedes scleractinian corals as primary reef ecosystem engineers. *Marine and Freshwater Research*. 62: 205–215.

Wilson, K., Pressey, R. L., Newton, A., Burgman, M., Possingham, H. and Weston, C. (2005). Measuring and incorporating vulnerability into conservation planning. *Environmental Management*. 35: 527–543.

WOA. (2016). *The First Global Integrated Marine Assessment World Ocean Assessment I; by the Group of Experts of the Regular Process*. United Nations General Assembly and its Regular Process for Global Reporting and Assessment of the State of the Marine Environment, Including Socioeconomic Aspects.

Wooldridge, S. A. (2013). Breakdown of the coral-algae symbiosis: towards formalising a linkage between warm-water bleaching thresholds and the growth rate of the intracellular zooxanthellae. *Biogeosciences*. 10: 1647–1658.

Yeakel, K. J., Andersson, A. J., Bates, N. R., Noyes, T. J., Collins, A. and Garley, R. (2015). Shifts in coral reef biogeochemistry and resulting acidification linked to offshore productivity. *Proceedings of the National Academy of Sciences of the United States of America*. 112 (47): 14512–14517. https://doi.org/10.1073/pnas.1507021112

Zoccola, D., Ounais, N., Barthelemy, D., Calcagno, R., Gaill, F., Henard, S., Hoegh-Guldberg, O., Janse, M., Jaubert, J., Putnam, H., Salvat, B., Voolstra, C. R. and Allemand, D. (2020) The World Coral Conservatory (WCC): A Noah's ark for corals to support survival of reef ecosystems. *PLoS Biology*. 18 (9): e3000823. https://doi.org/10.1371/journal.pbio.3000823

6

Brachyuran Crab Diversity of Mangroves of Kochi, Kerala: An Overlooked and Threatened Face of Urban Biodiversity

Elizabeth V. Mathew, Sana K. A., and Niladevi K. N.

CONTENTS

6.1 Introduction

Mangroves are one of the most productive, salt-tolerant, and unique ecosystems found mostly in the tropical and subtropical regions of the world. The uniqueness of mangrove ecosystems could be attributed to the extreme abiotic conditions prevailing in these ecosystems, including high salinity, heavy tidal actions, anaerobic soils, and abundant nutrient supply (Kathiresan and Bingham, 2001; Sandilyan, 2010). The mangroves provide innumerable ecosystem services to mankind. Being located at the interface between sea and land, mangroves serve as natural buffers to protect the coastal lines by minimizing the effect of strong waves and cyclones (Kathiresan and Bingham, 2001; Kathiresan and Rajendran, 2005; Das and Vincent, 2009). Nutrient recycling and carbon sequestration have been recognized as other important ecosystem functions of mangroves (Krauss et al., 2008; Spalding et al., 2010), especially in the current scenario of global warming and similar environmental issues. A rich assemblage of organisms, including microorganisms, phytoplankton, zooplankton, crabs, shrimp, lobsters, fish, and other vertebrates, inhabits the mangroves (Selvam et al., 2002; Lee, 2008). Mangroves are the feeding and breeding sites for large number of organisms, including fish (Mumby et al., 2004; Koenig et al., 2007; Kumar and Khan, 2013; Pawar, 2012). The major invertebrate taxa represented in mangroves include mollusks, crustaceans, polychaetes, oligochaetes, and amphipods (Kumar and Khan, 2013; Sasikala et al., 2017; Parvez Al-Usmani, 2018). The mangrove crustacean fauna is mostly dominated by brachyuran crabs (Lee, 2008), and their diversity in various mangrove ecosystems of the Indian subcontinent has been well accounted for (Fredrick and Ravichandran, 2013; Roy and Das, 2000; Ravichandran et al., 2007; Pawar, 2012). Crabs are one of the most important groups of organisms in a mangrove ecosystem, owing to their roles in nutrient cycling, formation of detritus, carbon fluxes, and also aerating the soils of mangroves (Schories et al., 2003; Pradnya et al., 2011), and they are often considered as keystone species of mangrove ecosystems (Cannicci et al., 2008). Even though the brachyuran crab diversity of Kochi backwaters has been studied extensively (Devi, 2015), studies on their diversity in the mangrove patches of Kochi are

DOI: 10.1201/9781003220398-6

scanty except for a few reports (Apreshgi and Abraham, 2019; Abraham and Apreshgi, 2020). The available literature shows that there is a lacuna in assessing the brachyuran crab diversity of various isolated patches of Kochi, irrespective of their sizes. Despite the endless ecological roles, the mangroves are facing great threats all over the world, and it is estimated that the rate of disappearance of global mangroves is 1–2% per year (Spalding et al., 2010), and the complete destruction of mangroves is predicted to occur by 2100 (Duke et al., 2007).

The development of major cities has always been linked to areas of high productivity, such as coastlines, riverbanks, and estuaries (Faeth et al., 2011). This intimate link between aquatic ecosystems and cities has also led to the massive destruction of aquatic ecosystems closer to the urban landscapes as they are more frequently treated as the dumping site for all sorts of refuse from cities. It is noteworthy that urbanization is occurring at a rapid pace worldwide, and it is estimated that more than two-thirds of the world's population will be living in major cities by 2050 (Brears, 2017). This tremendous increase in urbanization is not without its impact on the associated ecosystems. The negative impacts on urban ecosystems due to human interventions are multifaceted. Urbanization and industrialization have already turned the majority of the urban ecosystems into pollution cesspools, and urban mangroves are not an exemption. Conversion of natural habitats for urban development and land usage patterns pertaining to urban ecosystems significantly contribute to the loss of biodiversity (McKinney, 2002; Sukopp, 2004; Whitford et al., 2001). However, urban environments offer ample opportunities for the conservation of biodiversity as the creation and implementation of government policies could be enacted more successfully in cities (Puppim de Oliveira et al., 2012).

India, being one of the fastest-growing economies, also experiences the highest rate of urbanization on a global scale (Nijman, 2012). The improved socio-economic growth of the nation was achieved certainly by compromising a great deal of biodiversity. Indiscriminate expansion of urban landscapes generally results in considerable destruction of the various ecosystems, including mangroves that border the cities (Penha-Lopes et al., 2011; Wang et al., 2020). Mangroves in India expand over an area of 4,66,156 ha, of which only less than 1% is in the state of Kerala (Hema and Indira Devi, 2014). The mangroves of Kerala have undergone considerable degradation in the past few decades, and per the reports in 2013, mangroves in the state shrank to around 600 ha from a great mangrove cover of 70,000 ha reported in 1975 (FSI Report, 2013). Mangroves in Kerala occur as isolated patches, and the majority of them are located in the districts of Kannur and Ernakulam (Apreshgi and Abraham, 2019; Hema and Indira Devi, 2014). Kochi (Ernakulam District), the commercial, industrial, and financial hub of Kerala is blessed with many small patches of mangroves, most of which grow along the Kochi backwaters, which are part of Vembanad Lake, a renowned Ramsar site. The mangrove belts of Kochi are of great significance, owing to their location within the city and also their biodiversity. The mangrove patches of Kochi include Mangalavanam, Puthuvype, Vallarpadam, Nettoor, Aroor, and Malippuram, of which the first two are more important due to their size and composition (Jayson, 2001; Rani et al., 2016; Sahadevan et al., 2017). The mangroves of Mangalavanam are well known as a bird sanctuary, situated in the heart of Kochi, while Puthuvype is a suburb on the western part of the city and is also an upcoming industrial area (Sahadevan et al., 2017). The biodiversity of these two major mangroves have been recorded (Jayson, 2001; Sahadevan et al., 2017; Apreshgi and Abraham, 2019; Vinod et al., 2020), and much of the studies have focused on the avian fauna of Mangalavanam (Jayson, 2001; Azeez and Bhupathy, 2006; Karunakaran et al., 2015).

The mangrove stretches of Kochi, especially the Puthuvype region, are considered as the fast-depleting wetlands in Kerala, and many large-scale projects have already been implemented in this eco-sensitive region (Sahadevan, 2016), thereby threatening the biodiversity of these urban ecosystems (Rani et al., 2016). Hence, a thorough investigation on the impact of such developmental activities on the biodiversity of these urban ecosystems has become a need of the hour. Urban ecosystems, such as the isolated mangrove patches of Kochi, are deteriorating at a faster rate as compared to other natural ecosystems as they are the immediate receptors of the negative effects imposed by anthropogenic activities; this is well exemplified by the pollution load of all the three mangrove regions under study. It is high time that the scientific community pays due attention to the conservation of environments like urban ecosystems and their biodiversity, in the absence of which the ever-increasing rates of urbanization and industrialization would deteriorate them in an irreparable manner. The present work is a step in this direction.

6.2 Methodology

The study sites were selected with a view to encompass mangrove patches of Kochi with widely differing conservation strategies. The study area 1 (Mangalavanam) is primarily a bird sanctuary that is protected and maintained by the state government. Study area 2 (Puthuvype) includes large mangrove belts and is a site for several developmental projects, including LNG terminal, bunkering terminal, and a proposed site for mega oceanarium project. Study area 3 (Nettoor) is a mangrove stretch that has caught least attention from the authorities, environmentalists, and scientific community and remains mostly overlooked.

6.2.1 Study Area 1: Mangalavanam

Mangalavanam is located in the heart of the Kochi metro city, covering an area of 2.74 ha of wetlands with mangrove vegetation. It houses a shallow tidal lake that is connected to the Kochi backwaters by a canal. The associated mangroves provide habitat to diverse invertebrate fauna including brachyuran crab species.

6.2.2 Study Area 2: Puthuvype

Puthuvype is the major mangrove patch of Kochi. The mangrove patches of this region remain interspersed with settlements and are endowed with numerous canals, extending several kilometers in length. The mangrove cover includes 206.80 ha of mangrove forest spread over three survey numbers in Puthuvype village.

6.2.3 Study Area 3: Nettoor

Nettoor, is located on Panangad Island and is a part of Maradu Municipality. The area is traversed by many small canals and is bordered by huge residential complexes and the stretch of mangrove nearer to human habitation is polluted to a great extent.

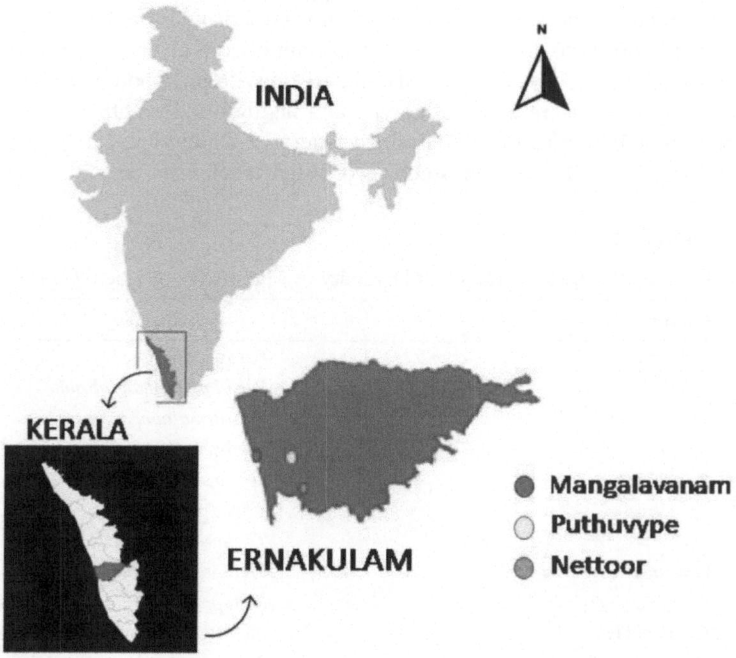

FIGURE 6.1 Map of the study areas (Mangalavanam, Puthuvype, and Nettoor, Ernakulum District, Kerala, India).

6.2.4 Sampling and Identification

Brachyuran crab samples were collected on a monthly basis for four consecutive months from September to December 2019. Specimens were mostly hand-picked. To collect the specimens from their burrows, long sticks and twines were used. Help from local men was sought to bring the crabs out of their burrows and collect them alive. On an average, each session of sampling took 15 minutes. Photographs of the representative specimens were taken at the collection site itself. The specimens were preserved in 5% formalin and brought to the lab for identification. Identification followed (Jayabaskaran et al., 1999; Ng et al., 2008; and Ajmal Khan and Ravichandran, 2009). The World Register of Marine Species (WoRMS, 2021; www.marinespecies.org) was used to cross-check the names of the brachyuran crabs.

6.3 Result and Discussion

A total of 12 species belonging to nine genera and four families (Portunidae, Grapsidae, Sesarmidae, and Ocypodidae) were recorded during the period of sampling (Table 6.1). Of these, Sesarmidae is the dominant family, represented by four species, closely followed by Portunidae and Ocypodidae, with three species each, and Grapsidae with two species. Ten species were recorded from Puthuvype, followed by five species from Nettoor and four species from Mangalavanam. The pollution status of the surveyed mangrove patches was found to be appalling. Sewage, sewerage, and domestic refuse stagnated vast areas in mangrove patches closer to human habitation. Plastic waste is dumped indiscriminately and is often burned by residents (Plate 4). The luxuriant mangrove forests of Nettoor, though undisturbed in many stretches, present such sights toward its boundaries with human dwellings. The Mangalavanam bird sanctuary, significant for its location in the heart of the city and for its status as a protected area in the city (ICLEI South Asia, 2020a), has a sewage canal running through it. Sediments from Vembanad Lake and sewage inflow from the city take a toll on the pristineness of the sanctuary (Karunakaran et al., 2015). The increasing number of huge apartment complexes around this eco-sensitive area contributes to pollution. The shallow tidal lake in the middle of the sanctuary is stagnant due to heavy siltation. Presence of hydrocarbons, including oil, has been reported in samples collected from Mangalavanam (Karunakaran et al., 2015); this can be detrimental to aquatic life. Diversity profile of most classes of animals in Kochi is limited to reports from Mangalavanam (Jayson and Easa, 1999; Madhusudhanan and Jayesh, 2011), yet it is important to note that a proper account of crab diversity of this area is missing. A report on the diversity of mangroves of Kerala in comparison with pre-flood data (Bhat and Sreekanth, 2018) documents the absence of crabs in Puthuvype and negligible diversity in Mangalavanam. This could have been the immediate effect of floods since otherwise Puthuvype has been a haven for crabs (Apreshgi and Abraham, 2019). The present study records ten species of crabs from this site.

TABLE 6.1

Checklist of Brachyuran Crab Diversity of Mangroves of Kochi City

Family	Species
Sesarmidae	*Parasesarma* sp.
	Pseudosesarma glabrum
	Neosarmatium sp.
	Nanosesarma andersoni
Portunidae	*Scylla serrata*
	Scylla olivacea
	Thalamita crenata
Grapsidae	*Metopograpsus latifrons*
	Metopograpsus messor
Ocypodidae	*Austruca annulipes*
	Austruca perplexa
	Macrophthalmus depressus

There have been a few studies on the crab diversity of Cochin backwaters and Puthuvype, yet a consolidated checklist of mangrove crab diversity of the city and its suburbs is missing. Devi et al. (2015) documented 24 crab species from Cochin backwaters. Sahadevan (2016) reported crabs of four families from mangrove patches of Ernakulam District. Ng et al. (2017) discovered a new species of *Pseudosesarma* (Serene & Soh, 1970) from the Njarakkal estuary in the suburbs of Kochi. Crab diversity of the mangrove belt of Puthuvype was documented by Apreshgi and Abraham (2019); 12 species of true mangrove crabs belonging to four families were reported. The checklist of brachyuran crabs of mangroves of Kerala published by Abraham and Apreshgi (2020) documents 18 species belonging to four families.

The current study views brachyuran crab diversity as an integral component of urban biodiversity and shows way in documentation of lesser-known organisms. Since all the sampling sites lie along Kochi backwaters—at the core of the city and its immediate suburbs—crab diversity in these patches contributes to urban biodiversity of the area. Kochi, despite being the densely populated industrial and

Plates: 1 – Mangroves of Mangalavanam Bird sanctuary; 2 – Mangroves of Puthuvype; 3 –Mangroves of Nettoor; 4 – Solid and plastic waste pollution in Nettoor mangrove belt bordering human dwelling; *5 – Scylla serrata; 6 – Scylla olivacea; 7 – Austruca annulipes; 8 – Parasesarma* sp.; *9 – Neosarmatium* sp. *10 – Metopograpsus latifrons; 11 – Thalamita crenata*

FIGURE 6.2 Study sites and some of the crabs spotted therein.

commercial hub of Kerala, is endowed with many natural areas (ICLEI South Asia, 2020a), and it is the second city in India to have a biodiversity index. Yet a comprehensive account of biodiversity of these green patches is lacking. The People's Biodiversity Register (PBR) reports only six crustacean species in corporation area (ICLEI South Asia, 2020a). Two species of crabs reported by Asha et al. (2014) from Vembanad Lake have been documented in a report by ICLEI South Asia (2020b). The Environmental Impact Assessment (EIA) for Multi-User Liquid Terminal Project (MULT) at Puthuvypeen (WAPCOS, 2015) reports four crab species in the study site. These data point to a gross understatement of brachyuran crab diversity of Kochi when viewed in the light of the current study and previous studies (Devi et al., 2015; Apreshgi and Abraham, 2019). Further, a checklist of crabs is missing in technical reports, citing lack of scientific data (ICLEI South Asia, 2020a). This underscores the need for scientific documentation of brachyuran crab diversity of the city. The role of crabs as crucial components of biodiversity of the urban landscape should not be overlooked, considering the vast extent of mangroves and backwaters in the region. It is important that proper checklists be prepared and published for crab diversity to be actively discussed as part of conservation goals and reclamation efforts. Not any different is the case of other lesser-known invertebrates.

It is equally important to study the effect of pollution and habitat fragmentation on the diversity of crabs. Considering the rate at which urbanization and associated conversion of natural habitats is progressing in Kochi, it is imperative to document and conserve the diversity of these green zones, which would otherwise be left ecologically bereft. The alarming rates of pollution can alter the soil and water profiles of natural habitats, leading to loss of diversity and shrinkage of natural populations (Bhat and Sreekanth, 2018).

This study is a preliminary attempt toward creating an inventory of crabs of Kochi, and it highlights the need for effective documentation and preservation of crab diversity of Kochi in specific and urban landscapes in general. Only a scientific profile of urban biodiversity can flag this resource for further research and conservation. A comprehensive biodiversity profile of Kochi should be a priority in urban planning, and the same applies to all cities. Close monitoring of biodiversity through consistent research will pave way for sustainable development in cities and their suburbs.

REFERENCES

Abraham, K.M. and Apreshgi, K.P. 2020. Checklist of brachyuran mangrove crabs of Kerala, India. *Journal of Threatened Taxa* 12:17153–17160.

Ajmal Khan, S. and Ravichandran, S. 2009. *Brachyuran Crabs, ENVIS Publication, CAS in Marine Biology.* Annamalai University, Parangipettai, 321–338.

Apreshgi, K.P. and Abraham, K.M. 2019. Brachyuran crab diversity in an isolated mangrove patch of the Cochin backwaters, Central Kerala, India. *Journal of Aquatic Biology & Fisheries* 7:8–14.

Asha, C.V., Suson, P.S., Cleetus, R. and Sivasankaran, B.N. 2014. Decline in diversity and production of exploited fishery resources in Vembanad wetland system: Strategies for better management and conservation. *Open Journal of Marine Science* 4:344–357.

Azeez, P.A. and Bhupathy, S. 2006. *The Mangalavanam Bird Sanctuary/Mangrove Area, Ernakulam Kerala.* Report submitted to Kerala Forest & Wildlife Department, Thiruvananthapuram.

Bhat, S.G. and Sreekanth, P.M. 2018. *Riverine Biodiversity Monitoring with Reference to True Mangroves in Comparison with Preflood Data.* Final Report to Kerala State Biodiversity Board, 3371/A8/2018/KSBB. Dept. of Biotechnology, CUSAT, Kochi, Kerala, 66pp.

Brears, R. 2017. *Urban Water Security.* John Wiley and Sons, Ltd. https://doi.org/10.1002/9781119131755

Cannicci, S.D., et al. 2008. Faunal impact on vegetation structure and ecosystem function in mangrove forests: A review. *Aquatic Botany* 89:186–200.

Das, S. and Vincent, J.R. 2009. Mangroves protected villages and reduced death toll during Indian super cyclone. *Proceedings of the National Academy of Sciences of the United States of America* 106:7357–7360.

Devi, P.L. 2015. *Brachyuran Crabs of Cochin Backwaters. Diversity, Habitat Ecology and Systematics, with Special Emphasis to Genus Scylla.* Ph.D. thesis, Faculty of science, Cochin University of Science and Technology, Kochi, India.

Duke, N.C., Meynecke, J.O., Dittmann, S., et al. 2007. A world without mangroves. *Science* 317:41–42.

Faeth, S.H., Bang, C. and Saari, S. 2011. Urban biodiversity: Patterns and mechanisms. *Annals of the New York Academy of Sciences* 1223:69–81.

Fredrick, W.S. and Ravichandran, S. 2013. Diversity of brachyuran crabs in the mangrove environment of Tamil Nadu. *World Journal of Fisheries and Marine Sciences* 5:441–444.

FSI. 2013. *The State of Forest Report, Forest Survey of India*. Ministry of Environment and Forests, Dehradun.

Hema, M. and Indira Devi, P. 2014. Factors of mangrove destruction and management of mangrove ecosystem of Kerala, India. *Journal of Aquatic Biology and Fisheries* 2:184–196.

ICLEI South Asia. 2020a. *City Biodiversity Index—Kochi*. Prepared under INTERACT-Bio, Implemented by ICLEI South Asia and c-hed. ICLEI South Asia.

ICLEI South Asia. 2020b. *Local Biodiversity Strategy and Action Plan for Kochi Municipal Corporation*. Prepared under the BMU Supported INTERACT-Bio Project. ICLEI South Asia.

Jayabaskaran, R., Ajmal Khan, S. and Ramaiyan, V. 1999. *Brachyuran crabs of Gulf of Mannar. CAS in Marine Biology Publication*, Annamalai University, India. 99.

Jayson, E.A. 2001. Structure, composition and conservation of birds in Mangalavanam mangroves, Cochin, Kerala. *Zoos' Print Journal* 16:471–478.

Jayson, E.A. and Easa, P.S. 1999. Documentation of vertebrate fauna in Mangalavanam mangrove area. *KFRI Research Report* 183:42.

Karunakaran, P.V., Jaypal, R., Anjan Kumar Prusty, B. and Kumar, M. 2015. *Habitat Assessment of Mangalavanam Bird Sanctuary, Kerala*. Report Submitted to Submitted to Kerala Forest & Wildlife Department, Thiruvananthapuram.

Kathiresan, K. and Bingham, B.L. 2001. Biology of mangroves and mangrove ecosystems. *Advances in Marine Biology* 40:81–251.

Kathiresan, K. and Rajendran, N. 2005. Coastal mangrove forests mitigated tsunami. *Estuarine Coastal and Shelf Science* 65:601–606.

Koenig, C.C., Coleman, F.C., Eklund, A., Schull, J. and Ueland, J. 2007. Mangroves as essential nursery habitat for Goliath Grouper (EpinephulusItajara). *Bulletin of Marine Science* 80(3):567–585.

Krauss, K.W., Lovelock, C.E., McKee, K.L., Lopez-Hoffman, L., Ewe, S.M.L. and Sousa, W.P. 2008. Environmental drivers in mangrove establishment and early development: A review. *Aquatic Botany* 89:105–127.

Kumar, P.S. and Khan, A.B. 2013. The distribution and diversity of benthic macroinvertebrate fauna in Pondicherry mangroves, India. *Aquatic Biosystems* 9:15.

Lee, S.Y. 2008. Mangrove macrobenthos: Assemblages, services, and linkages. *Journal of Sea Research* 59:16–29.

Madhusudhanan, K. and Jayesh, R. 2011. Physico-chemical and floristic studies of mangalavanam mangrove ecosystem in Ernakulam district, Kerala, South India. *Nature, Environment and Pollution Technology* 10(1):15–20.

McKinney, M.L. 2002. Urbanization, biodiversity, and conservation. *BioScience* 52:883–890.

Mumby, P.J., Edwards, A.J. and Arisa-Gonza'Lez, J.E. 2004. Mangroves enhance the biomass of coral reef fish communities in the Caribbean. *Nature* 427:533–536.

Ng, P.K.L., Guinot, D. and Davie, P.J.F. 2008. SystemaBrachyurorum: Part 1. An annotated list of extant brachyuran crabs of the world. *The Raffles Bulletin of Zoology* 17:1–286.

Ng, P.K.L., Rani, V. and Nandan, S.B. 2017. A new species of *Pseudosesarma* Serène&Soh, 1970 (Crustacea: Brachyura: Sesarmidae) from Cochin in southwestern India. *Zootaxa* 4311(2):263–270.

Nijman, J. 2012. India's urban challenge. *Eurasian Geography and Economics* 53:7–20.

Parvez Al-Usmani, S.M. 2018. Habitat heterogeneity and spatio-temporal distribution of macrobenthos in in a tropical estuarine mangrove ecosystem. *Journal of Ecophysiology and Occupational Health* 18:80–85.

Pawar, P.R. 2012. Diversity of decapods fauna from mangrove ecosystem of Uran (Raigad), Navi Mumbai, Maharashtra, West coast of India. *Indian Journal of Scientific Research* 3:87–90.

Penha-Lopes, G., Torres, P., Cannicci, S., Narciso, L. and Paula, J. 2011. Monitoring anthropogenic sewage pollution on mangrove creeks in southern mozambique: A test of *Palaemonconcinnusdana*, 1852 (Palaemonidae) as a biological indicator. *Environmental Pollution* 159:636–645.

Pradnya, D.B., Kusuma, N. and Kakati, V.S. 2011. Biodiversity of crabs in Karwar mangrove environment West coast of India. *Recent Research in Science and Technology* 3:1–5.

Puppim de Oliveira, J.A., Suwa, A., Balaban, O., et al. 2012. Good governance in cities for promoting a greener economy. In Jose Antonio Puppim de Oliveira (ed.), *Green Economy and Good Governance for Sustainable Development: Opportunities, Promises and Concerns*. United Nations University Press, Tokyo.

Rani, V., Sreelekshmi, S., Asha, C.V. and BijoyNandan, S. 2016. Forest structure and community composition of cochin mangroves, South-West coast of India. *Proceedings of the National Academy of Sciences, India—Section B: Biological Sciences* 88.

Ravichandran, S., Anthonisami, A., Kannupandi, T. and Balasubramanian, T. 2007. Habitat preference of crabs in Pichavaram mangrove environment. *Journal of Environmental Sciences* 2:47–55.

Roy, M.K.D. and Das, A. K. 2000. Taxonomy, ecology and distribution pattern of brachyuran crabs of mangrove ecosystem in Andaman Islands. *Records of the Zoological Survey of India* 185:1–211.

Sahadevan, P. 2016. Diversity of fishes, crustaceans and molluscs of puthuvypeen of ernakulam district, Kerala, South India. *International Journal of Fisheries and Aquatic Studies* 4:101–107.

Sahadevan, P., Dinesan, C. and Suma, C.C. 2017. On mangroves and mangrove associates of Puthuvypin of Ernakulam district of Kerala (South India). *International Journal of Multidisciplinary Research and Development* 4:134–139.

Sandilyan, S. 2010. Climate change and mangrove wetlands. *Emerging Science* 2:18–19.

Sasikala, V., Saravanakumar, A. and Balasubramnaian, T. 2017. Spatial and temporal distribution of macrobenthos in Point Calimere of Southeast coast of India. *Indian journal of Geo Marine Sciences* 46:377–384.

Schories, D.A., Barletta, B., Barletta, M., Krumme, U., Mehling, U. and Rademaker, V. 2003. The keystone role of leaf removing crabs in mangrove forests of Brazil. *Wetlands Ecology and Management* 11:243–255.

Serène, R. and Soh, S.L. 1970. New Indo-Pacific genera allied to *Sesarma* Say 1817 (Brachyura, Decapoda, Crustacea). *Treubia* 27(4): 387–416.

Selvam, V., Gnanappazham, V.L. and Navamuniyammal, M. 2002. *Atlas of Mangrove Wetlands of India. Part-1*. M.S. Swaminathan Research Foundation, Chennai.

Spalding, M., Kainuma, M. and Collins, L. 2010. *World Atlas of Mangroves*. Earthscan, London and Washington, DC, 319.

Sukopp, H. 2004. Human-caused impact on preserved vegetation. *Landscape and Urban Planning* 68:347–355.

Vinod, K.N., Jasna, V. and Parvathi, A. 2020. Comparison of bacterial diversity from two mangrove ecosystems from India through metagenomic sequencing. *Regional Studies in Marine Science* 35.

Wang, W., Wu, T., Li, Y., et al. 2020. Urbanization impacts on natural habitat and ecosystem services in the Guangdong Hong Kong-Macao "Megacity." *Sustainability* 12:6675.

WAPCOS Limited. 2015. Environmental Impact Assessment Study for Multi—User Liquid Terminal Project (MULT) at Puthuvypeen, Cochin Port, Vol: 1. Cochin Port Trust.

Whitford, V., Ennos, A.R. and Handley J.F. 2001. City form and natural process—indicators for the ecological performance of urban areas and their application to Merseyside, U K. *Landscape and Urban Planning* 57:91–103.

WoRMS Editorial Board. 2021. *World Register of Marine Species*. Checklist Dataset. https://doi.org/10.14284/170

7

Glimpses of the Freshwater Zooplankton Biodiversity and Conservation in the Wetlands of Kerala: A Review

Anju Kaushal

CONTENTS

7.1 Introduction

"God's own country" is a phrase synonymous with the state of Kerala of India. The justification that this region is especially favored by God lies in its super abundant tropical greenery, spectacular countryside beauty, rich biodiversity of Western Ghats, and munificent rainfall all year around. This wettest region is studded and nurtured by numerous water bodies and inland and coastal wetlands. These include rivers, lakes, lagoons, estuaries, unique wetland ecosystems of kol (or kole), polders, *Myristica* swamps, and kayals (or backwaters) along with their waterways traversing throughout the state. Other natural resources include ponds (or kulam), wells, springs, and surangams (horizontal wells). Rivers, lakes, kol, and ponds hold fresh water, while estuaries, backwaters, lagoons, mangroves enclose both fresh water and saline water. Ponds, lakes and reservoirs and wetlands are examples of lentic whereas rivers, springs, streams are types of lotic ecosystems.

DOI: 10.1201/9781003220398-7

Location-wise, Kerala is a south western state toward Malabar Coast of India (Menon 2007). It shares its border with Karnataka to the north, Tamil Nadu to the south and east, and Lakshadweep Sea covering on the west side (IHO 1953). Latitude and longitude coordinates of Kerala are 10.850516 and 76.271080 (www.latlong.net). The wetlands of the state are well recognized for their expanse and distinctiveness. They are classified based on their geomorphological features and diverse terrain of high land, midland, and low land. There are 217 identified wetlands in Kerala (Kokkal et al. 2008). Three of them—namely, Ashtamudi, Sasthamkotta Lake, and Vembanad Kol—are designated Ramsar sites (Ramsar 2020). These three Ramsar sites, along with Kadalundi and Kottuli, are also identified under the National Wetland Conservation Programme (KSESTE 2007).

Wetlands all around the world are currently more and more threatened due to fast-approaching eutrophication processes and callous anthropogenic activities, like industrial use, spills, transport, and so forth (Xu et al. 2019). They are the most productive ecosystems; they support life and are important in many ways. Apart from meeting the water necessities of the population, these water resources provide essential ecosystem services in the form of fish as food and pharmaceutical products. Fish, both adults and larvae, are dependent on free-floating and microscopic phytoplankton and zooplankton for food. Zooplankton serve as the first link in the food chain, feed on phytoplankton, and are termed as primary consumers (Madin et al. 2001). They play a crucial role as secondary producers in energy transfer to fish and organisms feeding on them (Edmondson 1974; Deivanai et al. 2004). They also manipulate cycling of matter, which is another functional aspect of aquatic ecosystems (Park and Shin 2007). They not only act as pollution indicators but are also helpful for mitigating the effects of pollution (Mukhopadhyay et al. 2007). Thus, they occupy a very crucial position in the ecosystem; therefore, reviewed and discussed here are biodiversity and conservation issues of their habitats in the wetlands of Kerala.

7.2 Literature Survey of Zooplankton in the Wetlands of Kerala

The majority of studies discussed at the forefront pertain mainly to finding the biodiversity of zooplankton, especially mesoplankton of Rotifera, Cladocera, Copepoda, and Ostracoda within the 0.2–20 millimeters of size range. Other parameters include counting zooplankton for abundance and measuring associated atmospheric and physico-chemical conditions, such as temperature, turbidity, dissolved oxygen, dissolved carbon dioxide, pH, salinity, dissolved solids, nitrates, phosphates, conductivity, and so on. It has become an accepted protocol among researchers to observe them typically during pre-monsoon (or summer), monsoon, and post-monsoon seasons for seasonal fluctuations. The information thus obtained has helped them in exploring the composition, the dominant species, and the contribution of zooplankton in differentiating between masses of water and changes brought about in them. Some discussion is allocated to their dispersal to new localities, colonizing habits, and the effects of altered habitat on them. Here are the glimpses of the fascinating and unique wetlands of Kerala and equally remarkable zooplankton held by them.

7.2.1 Riverine Wetlands

A study was undertaken from July 2009 to June 2010 to observe the seasonal fluctuations in the diversity of zooplankton of the Achancovil River, flowing through Kollam, Alappuzha, and Pathanamthitta Districts of Kerala. The Achancovil River is composed of 28 species of cladocerans (eleven species), copepods (nine species), and rotifers (eight species). Each group has its own peak periods of density, depending on prevailing ecological conditions. Cladocerans peaked at monsoon, whereas copepods and rotifers peaked at post-monsoon and summer only. Overall, cladocerans were richer and more abundant than copepods and rotifers (Jose and Sanalkumar 2012).

The zooplankton diversity of the Bharathapuzha (Nile) River was examined for two years (June 2011–May 2013) in different seasons. Groups taken up for studies included protozoans, rotifers, cladocerans, and copepods. Protozoans were represented by ten species, rotifers by seven species, and cladocerans and copepods by four species individually. Pre-monsoon or summer season maximally favors the species diversity in comparison to other seasons (Divya and Manonmani 2014).

7.2.2 Lake Wetlands

Sasthamkotta Lake is the largest freshwater lake in Kollam District of Kerala. The lake water body area is 3.4 km^2 in measurement. This lake is named after the ancient Shasta temple situated on its bank. The lake water is potable and thus used by a population of some five-lakh people. The water of this majestic and grand lake is devoid of any salt, metal, minerals, and aquatic vegetation, and only fish abound. A study established that plankton were the dominant species during the monsoon season. As many as 83 species of phytoplankton were identified. Major zooplankton reported were nauplius, copepods, rotifers, and cladocerans. They have not been described by name although they are reported to be low in abundance and richness (Alexander et al. 2014).

The opinion regarding the scanty population of zooplankton in the lake stems from activities of the larval *Chaoborus*, commonly called phantom midges, which devour large amounts of plankton and bacteria and are thus incidentally responsible for cleaning the lake also. The gut content analysis of *Chaoborus* larvae revealed that the diet is composed of cladocerans, *Daphnia*, *Keratella*, *Nauplius*, and *Diaptomus* sp. The findings also revealed that the abundant diet in *Chaoborus* is cladocerans (Sheeba 2015). Further, this larva is found to be most abundant in February, which can also be related to maximum abundance of zooplankton in the pre-monsoon season.

A recent study on lake plankton was carried out at Vattakayal Lake, located in the Sakthikulangara Panchayath, under Kollam Corporation. A total of 23 species of zooplankton belonging to five classes of rotifers (nine species), copepods (four species), protozoans (six species), cladocerans (three species), and a single species of Ostracoda were recorded. Observably, rotifers were most abundant and rich in species diversity, followed by cladocerans, protozoans, copepods, and ostracods, in that order. Results revealed that in the pre-monsoon season, rotifers were the dominant group. Pre-monsoon and monsoon seasons had the abundance of cladocerans, while post-monsoon season was dominated by copepod (Sherly Williams et al. 2020).

7.2.3 Kole Wetlands

Kol wetlands are freshwater in nature that runs parallel to sea. They lie 0.5 to 1 meter below the sea level. They are the most fertile as floodwater from rivers brings nutrient rich alluvium, and hence, the paddy yield is four to five times as compared to other places. These lands are thus known as "kol lands" signifying bumper crop in Malayalam. Karuvannur and Keecheri Rivers drain these lands and ultimately confluence with Arabian Sea. The Muriyad Kayal is situated little away from the town of Irinjalakuda in Thrissur District. A rich biodiversity of 64 species of zooplankton were identified from the Muriyad wetlands. Eight species of cladocerans, eleven species of copepods, and five species of rotifers were identified in this new study. One cladocera species called *Disparlona* sp. was newly identified. *Allodiaptomus mirabilibus* is reported for the first time from India (Thomas et al. 2003).

The composition, distribution, abundance, and biodiversity of zooplankton of unique rain-fed kole wetlands with an area of 500 acres at Maranchery in Malappuram District was counted in a project report submitted to Kerala State Diversity Board. This wetland is mainly occupied by waterfowl. Field sampling was done monthly for two years at eight selected stations. Stations 1–5 vary in land use pattern, but stations 6–8 were inundated with rainwater throughout the study period. There was fair enough zooplankton biodiversity consisting of 18 genera belonging to Rotifera (six families), Cladocera (six families), Copepoda (two families), and Ostracoda (one family). Copepods dominated the zooplankton populations during all seasons, followed by cladocerans and rotifers. Ostracods were lowest in diversity count. Nauplii were found in large numbers during the pre-monsoon period. Large *Daphnia* were observed in pre-monsoon periods at Stations 2–3, which appears to be associated with clear water and healthy sport fish population (Nandan 2011).

Thrissur and Malappuram contain major Kole freshwater wetlands. They are the most productive yet threatened wetlands in Kerala. Rotifers constitute major portion of the freshwater zooplankton and serve as an important source of food for many larger aquatic organisms, thereby forming an integral part of the aquatic food web. Rotifers were examined for the biodiversity and abundance from November 2017 to October 2019 from different stations of Thrissur Kole wetlands. The highest diversity and density of rotifers was recorded during the pre-monsoon season followed by post-monsoon and lowest during monsoon season. A total number of 40 species of rotifers were recorded. Dominant species of rotifers were *Keratella cochlearis* and *Brachionus falcatus* during all seasons. Rotifers constituted an important group in this study area and 50% of them constituted novel species. Richness and abundance of rotifer species was related to favorable aquatic conditions (Fathibi et al. 2020).

7.2.4 Pond Wetlands

Anicadu Chira (pond) is situated in Avoly Panchayat, Ernakulam District. A single year study from April 2014 to March 2015 was related to physico-chemical parameters and zooplankton biodiversity. A total of 28 species of zooplankton comprising five orders—Rotifera (twelve species), Cladocera (seven species), Copepoda (six species), Ostracoda (three species), and Protozoa (one species)—were identified in this perennial pond. *Brachionus falcatus*, *Daphnia carinata*, *Diaptamus*, and *Cypris* were dominant species among rotifers, cladocerans, copepods, and ostracods, respectively. *Vorticella* was the only representative of Protozoa. Summer favored the density of zooplankton. Rotifers were the most abundant, followed by cladocerans and copepods (Jose and Senthilkumaar 2015).

Plankton biodiversity and hydrological parameters for seasonal variations were studied for the old village pond at Dhanuvachapuram, Trivandrum, from August 2014 to March 2015. This pond is almost oval in shape with an extent of 23 m^2. The pond also receives water from the neighboring Neyyar dam, apart from seasonal rain. In this study, 32 species of zooplankton were identified from different classes. Rotifers showed maximum richness, followed by copepods and cladocerans (Nath et al. 2015).

Venganoor Panchayat has plentiful water bodies, like natural ponds, temple ponds, lakes, and pools. The study area was two ponds, Valiya Kulam, Azhakulam, and Sarkar Kulam, Muttacaud, situated in Thiruvananthapuram District of Kerala. During the study period, a total number of 32 zooplankton species—thirteen rotifers, five copepods and nauplii, six cladocerans, four protozoans, two malacostracans, and a single species of ostracod—were encountered. The diversity of zooplankton was maximum during the pre-monsoon or summer season and minimum in monsoon season. These reports testified the importance of studying even smaller ponds (Thampiraj and Prakash 2019).

Zooplankton biodiversity and physico-chemical characteristics of Nelanattukulam (Sreemoolanagaram) or Kaladi Pond (150 m^2) in Ernakulam District were enumerated for a year (2017–2018). It is located two kilometers away from the Sreemoolanagaram. This mud-colored pond is mostly filled with plastic garbage and some other forms of waste materials but has rich algal diversity. There were seven species of rhizopods, eight species of rotifers, fourteen species of cladocerans and copepods each, and a single species of ostracod. Another Chathamkulam (Allapra) or Perumbavoor pond (approximately 45 m^2) was found to contain seven species of rhizopods, twenty-two species of rotifers, thirteen species of cladocerans, and four species of ostracods. This pond, in contrast to the Kaladi Pond is clean, highly transparent, and pollution-free. Statistically, there was adequate abundance, diversity, evenness, and richness both pond-wise and season-wise (Thomas and Lakhumna 2019). A small man-made pond at UC College, Aluva, Ernakulam, was found to contain a total of 61 species with fourteen rhizopods and ciliata, twenty-four rotifers, six cladocerans, five copepods, and twelve ostracods (Kaushal et al. 2020a). In another study, a rainfed Irumbakulam Pond, Aluva, Ernakulam (240 m^2) contained 65 species of zooplankton, including fifteen rhizopods, thirty-three rotifers, eight cladocerans, and eight copepods in a study conducted for a year where the pond has a natural pond color with good algal and fish diversity (Kaushal et al. 2020b).

7.2.5 Estuary Wetlands

Faunal diversity of the Ashtamudi wetlands is described in a survey report based on several studies. Ashtamudi is the second largest brackish estuary containing a set of eight lakes in the Kollam District of Kerala. It has an area spread of 4,335 ha. The source of fresh water in the estuary originates from the Kallada River at Western Ghats, and it drains into Arabian Sea at Neendakara near Kollam.

The estuary contains the Neendakara Marine Zone, Ashtamudi Brackish Water Zone I, Kanjirakode Brackish Water Zone II, and Kadapuzha Tidal River Zone, which is the lower reach of the Kallada River.

Though highly productive, the estuary is currently facing pollution, encroachments and siltation from various sources situated near its banks.

Zooplankton of Ashtamudi contained all the three marine, brackish, and freshwater forms. Freshwater copepods—*Mesocyclops* sp., Paradiaptomus sp., and *Pseudodiaptomus* sp.—represented the post-monsoon survey sample. On the other hand, marine copepods represented the pre-monsoon survey. During the monsoon period, few cladocerans were also encountered (Raghunathan 2007).

Mesozooplankton belonging to the Vembanad Estuary, the largest estuary of Kerala spread in three districts, from Alappuzha (south) to Cochin (north), have been studied. This lake has a surface area of 36,500 ha, a length of 96 km, and a depth of 1 to 9 m. Owing to the construction of the Thanneermukkom Barrage (TMB), the southern part has turned into a freshwater zone and the northern part into a saline zone. The study was undertaken at ten stations—Stations 1–6 toward the south side and Stations 7–10 on the northern region. The information regarding spatiotemporal distribution, abundance, and community structure of mesozooplankton with special reference to calanoid copepods and rotifers was collected and analyzed. Altogether, ten species of calanoid copepods and twenty-two species of rotifers were present (Cleetus et al. 2016). Another study found the Panangad-Kumbalam region of Vembanad Lake containing all major types of zooplankton (Jyotsna Rajeswari et al. 2018).

An analysis of zooplankton and their monthly variation was carried out during July 2016 to June 2017 at three different stations of the Kadalundi River estuary, situated at borderline of Malappuram and Kozhikode of Kerala. Seventeen zooplankton species were reported from the three selected stations belonging to Crustacea, Maxillopoda, Nematoda, Protozoa, Rhizopoda, and Rotifera. Out of these 17 species, six are rotiferans, five are protozoans, and three are maxillopods and crustaceans. Rhizopods and nematodes contributed a single species each (Ali Akshad et al. 2018).

7.2.6 Backwater Wetlands

An investigation was carried out during August 2000 to July 2002 on different species of *Brachionus* in nine selected stations along the Cochin backwater system in the southwest coast of India. It is reported that a total of 13 species of *Brachionus* were recorded for the first time from this area. The water is brackish in nature and contains the rotifers found in fresh water, meaning that rotifers can tolerate a wide range of saline conditions (Varghese 2006).

7.2.7 Mangrove Wetlands

Ayiramthengu mangroves are an important part of the Kayamkulam estuary, and it is situated about six kilometers west of Ochira town and the bank of the Kayamkulam estuary. Mangroves area specialized ecosystem that develops along estuarine seacoasts and river mouths in tropical and subtropical regions of the world. The mangrove ecosystem and its biological components are under the influence of both freshwater and marine conditions and have developed a set of physiological adaptations to overcome problems of anoxia, salinity, and frequent tidal inundations. A study was aimed to analyze zooplankton diversity in this mangrove region. During the study period (September 2015–August 16), a total 50 marine or freshwater zooplanktonic fauna were encountered. Out of 50 species, freshwater zooplankton diversity consisted of four out of total twelve species of protozoans, eight species of rotifers, eight species of cladocerans, and fourteen out of fifteen species of copepods (Ranjana and Amina 2018).

7.2.8 Ephemeral Wetlands

Ephemeral wetlands represent distinct habitats that develop and survive during rains only. They are seasonal and temporary and hence support a relatively small predictable assemblage of animal species with high adaptability and survival strategies. They are different from the species of permanent water.

A first of its kind survey of ephemeral wetlands consisting of 22 pools from eight study sites along a 50 km coastal belt of Thiruvananthapuram District was conducted. Physical and chemical parameters of the habitat were recorded. Fish were absent, so these habitats were predator-free, and hence, invertebrates and amphibian tadpoles flourished there. Seventy-four genera of invertebrates were identified.

There were eleven species of cladocerans, two copepods and three ostracods, followed by insects (four), protozoans (eighteen), and rotifers (twelve). The Brachiopods, the quintessential animal of temporary water bodies, were represented by one species of conchostracans. Though only 22 pools from eighr sites were surveyed, three sites were found to be undisturbed, harboring the typical inhabitants of temporary pools. Anthropogenic activities have led to acute habitat loss as the survey area lies in a densely populated southern coastal zone. This suggests that many undescribed pool species might have already become extinct. This survey is an important step toward documenting the largely unknown inhabitants of such habitats (Balaraman 2008).

7.2.9 Reservoir Wetlands

The Idukki Reservoir was studied for a period of three years, from February 2007 to January 2010, for seasonal (pre-monsoon, monsoon, and post-monsoon) variations in physico-chemical parameters and plankton and fish community structure. Nineteen species and *Nauplius* belonging to four groups of zooplankton—Rotifera, Copepoda, Cladocera, and Ostracoda—were recorded in the reservoir. Numerically, Copepoda was the dominant group, and *Nauplius* were the most abundant larvae (Sebastian 2012).

7.3 Biodiversity of Zooplankton

Zooplankton are generally classified into rotifers, cladocerans, copepods, and ostracods for studying biodiversity and comparison purposes. Rotifers are most rich, followed by cladocerans, copepods, and ostracods. Occasionally, protozoans are also studied along with these groups. Biodiversity of zooplankton among different wetlands of Kerala as referred to earlier is described in this section of the chapter. Figure 7.1 presents various commonly encountered species for biodiversity studies of zooplankton groups.

7.3.1 Rotifera

Rotifers are the most easily identifiable group of zooplankton. They are generally transparent or yellowish in the intestinal region, except *Brachionus sessilis* and *Ptygura pilula*, which are brightly yellow in color. They are mostly identified on the basis of foot structure, corona, and lorica. Most diverse rotifer assemblages can be found in soft, slightly acidic, and oligo- to mesotrophic waters.

In the Achencovil River, eight species of rotifers—*Brachionus angularis*, *B. calyciflorus*, *B. caudatus*, *Keratella tropica*, *Asplanchna priodonta*, *Trichocera rattus*, *Lecane bulla*, and *Testudinella patina* are found. The Bharathapuzha River holds rotifers, such as *Brachionus falcatus*, *B. rubens*, *Keratella vulga*, *K. cochlearis*, *Lecane similis*, *Monostyla* sp., *Monostyla bulla*, and *Lepadella ovalis*.

Vattakkayal Lake has rotifer species—*Keratella* sp., *Monostyla* sp., *Brachionus quadridentatus*, *B. angularis*, *B. calyciflorus*, *Notholca* sp., *Plationus* sp., *Lecane* sp., and *Polyarthra vulgaris*.

In the Kol wetlands, 40 species are reported from the Maranchery and Trissur Kol wetlands. They include *Anuraeopsis fissa*, *A. navicula*, *Brachionus angularis*, *B. calyciflorus*, *B.dichotomous*, *B. falcatus*, *B. forficula*, *B. sessilis*, *Platyias patulus*, *P. quadricornis*, *Keratella cochlearis*, *K. tropica*, *Asplanchna brightwelli*, *A. herricki*, *A. priodonta*, *Mytilina* sp., *Filinia opoliensis*, *Lecane bulla*, *L. cornuta*, *L. aculeata*, *L. furcata*, *L. monostyla*, *L. luna*, *L. lateralis*, *L. inopinata*, *L. pyriformis*, *Lecane* sp. 1, *L.* sp. 2, *Polyarthra vulgaris*, *P.* sp., *Testudinella patina*, *Trichocerca cylindrica*, *T. kostei*, *T. longiseta*, *T. similis*, *T. rattus*, *Synchaeta* sp., *Conchilus unicornis*, *C. hippocrepis*, and *Horaella brehmi*.

The Muriyad wetlands rotifers reported are *Brachionus calyciflorus*, *B. caudatus*, *B. falcatus*, *B. qudridentatus*, *Keratella cochlearis*, *Keratella tropica*, *Plationus platulus*, *Platyias quadricornis*, *Euchlanis dilatata*, *Macrochetus collinsi*, *Lepadella patella*, *Lecane bulla*, *L. inopinata*, *L. lateralis*, *L. ludwigii*, *L. papuana*, *L. quadridentata*, *L. ungulata*, *Filinia opoliensis*, and *Testudinella patina*.

The Cochin backwater rotifers include *Brachionus plicatilis*, *B. rotundiformis*, *B. angularis*, *B. urceolaris*, *B. rubens*, *B. calyciflorus*, *B.caudatus*, *B. falcatus*, *B. forficula*, *B. quadridentatus*, *B. patulus*, *B. bidentata*, and *B. mirabilis*.

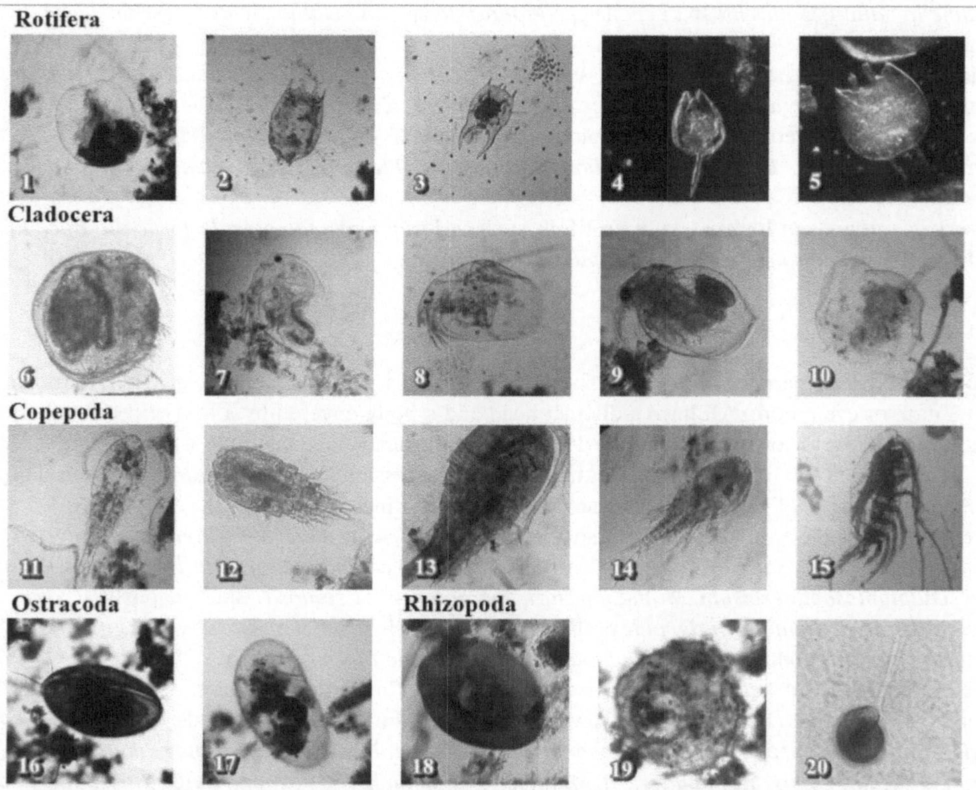

Rotifera: 1. Asplanchna priodonta 2. Brachionus calyciflorus 3. B. forficula 4. Lecane bulla 5. Platyias quadricornis
Cladocera: 6. Chydorus sphaericus 7. Macrothrix rosea 8. Alona rectangula 9. Ceriodaphnia sp. 10. Bosmina longirostris
Copepoda: 1 1. Microcyclops varicans 12. Paracyclops sp. 13. Mesocyclops hyalinus 14. Tropocyclops prasinus 15. Heliodiaptomus viduus
Ostracoda: 16. Physocypria pustulosa 17. Strandesia elongata :: **Rhizopoda:** 18. Arcella vulgaris 19. Centropyxis aculeata 20. Lesquereusia spiralis

FIGURE 7.1 Various representative species of zooplankton groups of Kerala (not to the scale).

Zooplankton investigation in ponds for rotifers is reported mainly from the Ernakulam and Thiruvananthapuram Districts of Kerala. The 75 species listed here are *Anuraeopsis fissa*, *A. navicula*, *A. psudonavicula*, *Brachionus angularis*, *B. calyciflorus*, *B. falcatus*, *B. forficula*, *B. sessilis*, *B. quadridentatus*, *B. bidentatus*, *Platyias quadricornis*, *Horaella brehmi*, *Keratella cochlearis*, *K. tropica*, *Monostyla quadridentatus*, *Notholca squamula*, *N. Lebes*, *Notholca* sp., *Trichocerca rattus*, *T. porcellus*, *Colurella uncinata*, *C. sulcata*, *C. colurus*, *Conochilus unicornis*, *Depleuchlaris propatus*, *Euchlanis dilatata*, *E. incisa*, *Filinia longiseta*, *Hexathra* sp., *Mytilina ventralis*, *Trichotria truncata*, *T. buchneri*, *Wolga spinifera*, *Lepadella ovalis*, *L. aspida*, *L. patella*, *Macrochaetus danneli*, *Lecane bulla*, *L. lunaris*, *L papuana*, *L. closterocerca*, *L. inopinata*, *L. decipiens*, *L. sinuata*, *L. leontina*, *L. aculeata*, *L. haliclystis*, *L. hamata*, *L. obtusa*, *L. curvicornis*, *L. hastata*, *L. quadridentata*, *Monostyla crenata*, *Cephalodella catellina*, *C. auriculata*, *Notommata aurita*, *Monommata longiseta*, *M. maculata*, *Scaridium longicaudum*, *Asplanchna brightwelli*, *A.* sp., *A. priodonta*, *A. herricki*, *Asplanchnopus multiceps*, *Testudinella elliptica*, *T. patina*, *Ptygura pilula*, *Squatinella geleii*, *Plationus patulus*, *Proales sordida*, *Rotaria tardigrada*, *R. neptunia*, *Trochosphaera aequatorialis*, and *Ascomorphella volvocicola*.

Estuary wetlands contains rotifers *Asplanchna herricki*, *A. priodonta*, *Brachionus plicatilis*, *B. urceolaris*, *B. angularis*, *B. calyciflorus*, *B. diversicornis*, *B. falcatus*, *B. forficula*, *B. quadridentatus*, *B. caudatus*, *B. rotundiformis*, *B. havanaensis*, *Platyias quadricornis*, *Keratella cochlearis*, *K. tropica*, *Polyarthra vulgaris*, *Trichocerca cylindrica*, *Hexarthra intermedia*, *Filinia terminalis*, *F. longiseta*,

Testudinella patina, Colurella sp., *Lecane* sp., *Monostyla* sp., *and Notholca* sp. These are some 26 species listed here.

Mangrove wetlands has a few rotifers, such as *Brachionus angularis, B. falcatus, B. caudatus, B. plicatilis, B. quadridentatus, B. calyciflorus, K. tropica*, and *K. vulga.*

Even ephemeral wetlands contain some rotifers, such as *Philodina, Trichocerca, Keratella, Cephalodella, Lecane, Lepadella, Euchlanis, Monostyla, Platyias, Epiphanes, Dicronophorus*, and *Conochilis.*

Reservoir wetlands of Idukki contain rotifers, such as *Brachionus falcatus, B. forficula, B. caudatus, F. longiseta, Polyarthra* sp., *Keratella tropica*, and *Lecane* sp.

7.3.2 Cladocera

Cladocerans are transparent and bigger in size. The vast majority of cladocerans range in size from about 0.2 to 3 mm, or even more. All have a distinct head and a body covered by a fold of the cuticle, which extends backward and downward from the dorsal side of the head and constitutes a bivalve carapace. The junction of the head and body is sometimes marked by a depression, the *cervical sinus* or *notch*. They are identified on the basis of carapace, legs, and post abdomen spines.

The Achancovil River holds cladocera species like *Chydorus ventricosus, Alona quadrangularis, A. rectangula, A. verrucosa, Ceriodaphnia cornuta, Daphnia carinata, Bosmina longirostris, Bosmina fatalis, Diaphanosoma excisum, Moina micrura*, and *Macrothrix spinosa.* Bharathapuzha River carries species of *Daphnia lumphaltzi, D. pulex, Diphanosoma brachyurum*, and *Bosmina longirostris.*

Vattakkayal Lake contains *Daphnia, Simocephalus* sp., and *Moina* sp. only.

The Maranchery Kol wetland contains typical species of *Bosmina, Chydorus, Alona, Ceriodaphnia, Daphnia, Moina* and *Diaphanosoma* sp. Cladocera of Muriyad kol wetlands include *Pseudosida zalayi, Latonopsis australis, Diaphanosoma sarsi, Scapholebris kingii, Simocephalus serrulatus, Simocephalus latirostris, Moina micrura, Moino daphnia macleayi, Bosmina longirostris, Bosminopsis deitersi, Macrothrix spinosa, Macrothrix triserialis, Macrothrix odiosa, Gurnella raphalis, Ilyocryptus spinifera, Disparlona* sp., *Alonella clathratula, Chydorus spharicus, Chydorus parvus, Chydorus ventricosus, Chydorus reticulatus, Ephimeroporus barrosi, Dadaya macropus, Alona davidi, Alona monacantha tridentata, Alona pulchella, Alona verrucosa, Alona affinis, Gratolebris testudinaria, Oxyurella singalensis, Kurzia longirostris, Euryalona orientalis*, and *Notalona globulosa.*

Pond cladocera species of Anicadu Chira are *Bosmina longirostris, Daphnia carinata, Daphnia similis, Diaphanosoma* sp., *Leydigia* sp, *Moina brachiata, Moina daphnia, Alona* sp.and *Bosmina cornuta.*

Ernakulam ponds contains *Polyphemus pediculus, Diaphanosoma sarsi, Scapholeberis kingi, Moinodaphnia macleayii, Ilyocryptus spinifera, Macrothrix rosea, M. goeldii, M. laticornis, M. hirsuticornis, Macrothrix* sp, *Chydorus sphaericus, C. bicornutus, C. ciliata, C. ovalis, Chydorus* sp., *C. barroisi, C. parvus, Dunhevedia crassa, Alona intermedia, A. rectangula, A. monacantha, A. karau, Oxyurella longicaudis, Pleuroxus denticulatus, Kurzia latissima, Ceriodaphnia dubia, C.* sp., *C. rigaudi, Alona pulchella, Alona monacantha, A. karau*, and *Evdne spinifera.*

The Ashtamudi estuary zooplankton are freshwater life-forms like *Moina micrura, Diaphanosoma sarsi, Penilia avirostris, Evadne tergestina, Ceriodaphnia cornuta, Bosminopsis deitersi, and Chydorus barroisi.*

The Ayiramthengu mangrove wetlands carry the species of *Daphnia* sp., *Daphnia pulex, Bosmina longirostris, Penilia avirostris, Diaphanosoma sarsi, Moina* sp., *Podon* sp., and *Evadne tergestina.*

Ephimeral cladocera species are different than usually found. Apart from typical cladocera species, *Simonocephalous, Eurycercus, Alonella, Pleuroxis, Macrothrix, Illocryptus* and *Latonopsis* are also seen.

The Idukki Reservoir contains *Daphnia carinata, Daphnia pulex, Ceriodaphnia cornuta, Moina brachiata, Moina micrura*, and *Alona pulchella.*

7.3.3 Copepoda

Copepods are the most eye-catching in the whole assemblage. Mostly freshwater copepods have an average size range of 0.5 to 15 mm. It is delightful to watch tiger-striped (yellow-and-white) *Eucyclops agilis* or the ornamental-striped carapace of *Paracyclops*. The number of first antenna segments, structure of the fifth leg, caudal rami, and so on help in identifying the copepods.

The Achencovil River carries copepods, such as *Mesocyclops leukartii*, *M. hyalinus*, *Eucyclops serrulatus*, *Trophocyclops prasinus*, *Mysis* spp., *Pseudodiaptomus nostradamus*, *Microcyclops varicans*, *Mesocyclops varicans*, and *Filipinodiaptomus insulanus*.

The Bharathapuzha River is inhabited by *Cylops leuckarti*, *Mesocyclops hyalinus*, *M. leuckarti*, and *Paracyclops affinis*. *Paracyclops*, *Thermocyclops*, *Mesocyclops* sp., and *Cyclopoid nauplii* are present in Vattakkayal Lake.

The Maranchery Kol wetlands has *Mesocyclops*, *Microcyclops*, *Paracylops*, and *Heliodiptomus* sp. The Muriyad wetlands copepods include *Ectocyclops rubescens*, *Eucyclops serrulatus*, *Tropocyclops prasinus*, *Mesocyclops* sp., *Mesocyclops agunus*, *Mesocyclops splendidus*, *Microcyclops varicans*, *Thermocyclops decipiens*, *Allodiaptomus mirabalipus*, *Tropodiaptomus informis*, and *Heliodiaptomus cincatus*.

Pond copepods include *Heliodiaptomus viduus*, *Microcyclops varicans*, *Mesocyclops hyalinus*, *Tropocyclops*, *Thermocyclops* sp., *Thermocyclops crassus*, *Cyclops scutifer*, *Cyclops viridis*, *Paracyclops* sp., *Eucyclops* sp., *E. macrurus*, *E. agilis*, *E. speratus*, *Macrobrachium*, *Tropocyclops prasinus*, *Ectocyclops phleratus*, *Diaptomus*, *Neodiaptomas*, and *Gammarus pulex*.

Estuary copepods are more variable with species of *Diaptomus*, *Harpacticoid* sp., *Nauplius* sp., *Mesocyclops*, *Cyclops*, *Oithona* sp., *Paradiaptomus*, *Pseudodiaptomus*, *Undinula vulgaris*, *Heliodiaptomus cinctus*, *Allodiaptomus mirabilis*, *Pseudodiaptomus serricaudatus*, and *P. annandeli*.

Mangroves copepods are *Mesocyclops leukartii*, *M. hyalinus*, *M. aspericornis*, *Thermocyclops* sp., *Thermocyclops crassus*, *Eucyclops agilis*, *Acarita major*, *Diaptomus* sp., *Paraclanus parvus*, *Acrocalanus longicornis*, *A. gibber*, *Pontella danae*, *Microstella rosea*, and *Longipedia weberi*.

Ephemeral copepods are *Diaptomus* and *Cyclops* spp.

Copepods of reservoir contain usual species of *Mesocyclops leukartii*, *Tropocyclops prasinus*, *Mesocyclops* sp., *Heliodiaptomus viduus*, and *Diaptomid calanoid*.

7.3.4 Ostracoda

Ostracods are mostly 1.0–30 mm long. They can be identified properly with shell open and details on the valves of the shell-like hairs, spines, pore canals, puncta, tubercles, and so on.

Very few ostracods have been identified. Vattakayal Lake is represented by a single species of *Cypris*. The Kol wetlands contain the ostracod *Stentor coeruleus* only.

There are pond species, such as *Cypris* sp., *Stenocypris malcolmsonii*, *Eucypris bispinosa*, *Cyprinotus glaucus*, *Cyprinotus unispinifera*, *Cyprinotus salinus*, *Physocypria pustulosa*, *Cypridopsis viduella*, *Cyprideis torosa*, *Strandesia elongata*, *S. obtusata*, *Cypris subglobosa*, *Cypricercus tuberculatus*, and *Cypricercus reticulatus*.

Ephemeral wetland species of ostracod include *Cypris*, *Cypricercus*, and *Cyprois*, and the reservoir contains *Stenocypris* sp. only.

7.3.5 Rhizopoda

Rhizopods are mainly faint grayish or reddish in colour. They are testate so they are identified on the basis of the structure of the shell and its mouth. The size of rotifers ranges between 0.1 and 0.5 mm (although their size can range from 50 μm to over 2 mm) and are common in freshwater environments throughout the world with a few saltwater species.

The Bharathapuzha River holds protozoa *Arcella discoides*, *A. vulgaris*, *Peridinium* sp., *Phacus* sp., *Nebela* sp., *Euglena* sp., *Paramecium* sp., *Vorticella globosa*, *Amoeba radiosa*, and *Didinium* sp.

Vattakayal Lake contains protozoans, such as *Centropyxis spinosa*, *Vorticella* sp., *Arcella discoides*, *Euglypha acanthophora*, *Difflugia* sp., *Assulina muscorum*. Ponds in Trivandrum contain *Amoeba*, *Paramoecium*, *Vorticella*, and *Euglena*. Ponds in Ernakulam contain *Vorticella microstoma*, *Rugipes bilzi*, *Arcella mitrata spectabilis*, *A. conica*, *A. discoides*, *A. vulgaris*, *A. gibbosa*, *Centropyxis arcelloides*, *C. aculeata*, *C. constricta*, *C. haemispherica*, *Difflugia acuminata*, *D. corona*, *D. lebes*, *D. bacillifera*,

D. arcula, D. oblonga, Lequereusia modesta, L. spiralis, and Euglypha ciliata, Pelomyxa palustris, Nebella collaris, Euglypha alveolata, Euplotes eurystomus, and *Uroleptus piscis.*

The Kadalundi River estuary contains protozoa species like *Difflugia* sp., *Loxophyllum* sp., *Paramoecium* sp., *Pelomyxa* sp., and *Vorticella* sp., along with Rhizopoda of Foraminifera.

7.4 Dispersal of Zooplankton

Dispersal is predicted to play a major role in the origin and maintenance of species diversity (Lengyel et al. 2009). Dispersal is a regional process and, along with the local environment, regulates the distribution of zooplankton. Evidence for different mechanisms of dispersal among lakes suggests that zooplankton move via human or animal vectors, flowing surface water and wind, though the relative importance of different modes of transport is poorly understood. Lakes separated over short spatial scales (<10 km) allows rapid dispersal with less local interactions. Broader scales (tens to thousands of kilometers) impose dispersal limitations with larger probability of influencing community structure (Havel and Shurin 2004). In an experimental setup, even if given the identical conditions, the zooplankton community composition differed throughout the year due to differences both in vagility—that is, ability to move and viability (Jenkins 1995). In systems of interconnected ponds or lakes, the dispersal of zooplankton through unidirectional dispersal pathways such as rivulets and overflows are observed. It is found that the number of dispersing zooplankton sampled in connecting elements was on average high (almost 7,000 ind h^{-1}). However, the size of dispersing individuals to the total population size in the target ponds was restricted (<1% 24 h^{-1}; Michels et al. 2001). It is also noted in experimental setup that the dispersal magnitudes in excess of ~1% results in convergence or homogenization of zooplankton communities (Vogt and Beisner 2011).

Dispersal ability determines species potential for further migration and colonization and therefore has ecological and evolutionary significance. There is not much information about dispersal abilities on spatial and temporal scales in many taxonomic units. Zooplankton inhabit lakes and ponds that functionally are "aquatic islands" or patches on the land, and their dispersal abilities affect both community composition and richness, along with colonizing abilities post-dispersal. Among other means of allochory, zoochory, or more specifically ornithochory (dispersal by birds), may have a role in shaping diversity and community composition, by direct dispersal, and also via affecting post-dispersal species selection by changing water quality and productivity (Hessen 2019). Dispersal of zooplankton through wind or anemochory and rain is relatively low and infrequent and probably limited to a few zooplankton species. It is also found to be depending on wind direction and temperature (Moreno et al. 2019). Diapausing propagules of zooplankton, especially rotifers and cladocerans, can be geographically dispersed by abiotic vectors (wind, rain and, water flow) or organisms (mainly by insects, fish, and water birds) with a fair amount of hatching success (Moreno et al. 2016). Habitat fragmentation is posing restrictions for dispersal from one aquatic system to another. In a bid to quantify dispersal, it is seen that significant movement is possible if the aquatic habitats are plentiful in the vicinity of the parent body. Wetland destruction may inhibit dispersal and promote inbreeding, endangering the evolutionary process (Allen 2007).

The dispersal and settlement of zooplankton across the wetlands of Kerala state can be explained in the light of their diversity and dispersal information obtained from reports presented here. The emerging point of view directs the conclusion somewhat toward homogenization of species. A few characteristic species of particular wetland and more common species are present in all systems. Kerala, being home to plenty of water bodies in close vicinity to each other, allows easy intermingling. The diaspora theory is fermenting up to the broader role of rains in Kerala, along with birds or humans, in facilitating zooplankton dispersal and approach in the next water body. Viable conditions allow them to settle and colonize as a community. The colonization process is complicated and depends on local conditions and individual capacities along with competition among species.

7.5 Conservation

There are no threatened, endangered, or red-listed species of zooplankton. However, they are facing multiple threats due to environmental factors, such as climatic changes, nutrients imbalances, eutrophication, predators, and stressors. Anthropogenic activities, like oil spills and pollution, are detrimental to biodiversity. Drastic procedures, such as degradation, destruction, encroachment, unscientific land use, mining, extracting sand, siltation of rivers, deforestation, and reclamation for development and urbanization, have turned habitats into graveyards. Thus, wetlands are losing their environments because of natural and anthropogenic reasons. These activities are rampant as there are no notified boundaries or demarcations of wetlands, and they are shrinking in each passing day. There is very sparse information regarding threats to zooplankton and more inputs are required to highlight the urgency of the situation in order to avoid hunger in future. Both wetlands and biodiversity need to be rescued from disappearance and should be comprehensively managed with accountability as many critical ecological services are contributed by them.

7.5.1 Threats

An experiment demonstrates that higher temperature and shortened growing season reduced the survival rate of zooplankton propagules, thereby resulting in reduced population and extinction risk projections in future (Pinceel et al. 2018). In Kerala, 11 wetlands—Anchuthengu, Kadinamkulam, Azhikode, Kadalundi, Chettuva, Ashtamudi, Ponnani, Kadalundi–Beypore, Mahe, Valapattanam, and Neeleswaram—spread from Kasargod to Thiruvananthapuram have high diversity in plankton but reduced abundance during pre-monsoon period (1994–2005). This was due to the coconut-husk-retting process at its peak in the pre-monsoon season and related industries that drained wastewater into these wetlands. This process imparts acidity to water and subsequently high amounts of sulfide, coupled with high carbon dioxide (Nandan 2008). Chavara-titanium and Neendakara were the two sites that were studied for the impact of heavy metals (Hg, Pd, Cd, and Cr) on plankton population. Both sites were environmentally disturbed and received heavy metal influx from industries and fish-landing-related activities, respectively. Only metal-pollution-tolerant species of *Navicula, Trochophore* larvae, and *Discorbis* sp. were found at Chavara-titanium. Neendakara had a better condition in terms of abundance and diversity of plankton (Yahiya and Miranda 2016).

Kottakayal is a wetland in South Kerala that was previously a paddy field. This wetland is subjected to mining for sand and clay, causing water logging and ecological threats. Zooplankton studies scored only 38 species from November 2015 to October 2017. Organic pollution is indicated due to presence of rotifer species of *Keratella*, *Brachionus*, *Monostyla*, *Lecane*, and Rhizopoa species of *Arcella* and *Difflugia* (Roshan and Sreejai 2020).

7.5.2 Habitat Destruction

Sasthamcotta is a freshwater lake in the Kollam District of Kerala. This lake is heading toward big loss following unstoppable human interference. There is lowering of the water table as water also gets removed during the process of sand mining. Construction of embankment on the south side is creating a water deficit, thus threatening the very existence of the lake (Sreekumari et al. 2015). Soil erosion, siltation, exposure of the lakebed to dryness at some places during extreme summer, sewage, and agrochemicals are polluting the lake. Washing clothes, growth of weeds is causing anoxia endangering the life of plankton, benthos, fish, and so on. With negligible corrective measures appearing on ground, it is feared the water will not remain potable at all (Chackacherry and Jayakumar 2011).

The Ashtamudi estuary degradation is one glaring example of excessive human interference. It is a designated Ramsar site, and it supports endangered species of mangrove trees and birds, fish, and many unique copepod species along with visiting 20,000 waterfowl. The lake has been subjected to deforestation of wetland catchment areas, leading to soil erosion, reduction in nutrient load, summer flows, and increase of flood peaks. There is disturbance to the ecosystem by construction of godowns, acacia plants, construction of hotels, tourist resorts, and parks. Pollution through effluents from the paper, starch, and ceramic factories situated on the bank and siltation make it look like a wasteland. Reclamation for agriculture, aquaculture, harbor development, urbanization, and industrialization, along with construction of dams, bunds, barriers, dredging, and reclamation, result in hostile physico-chemical characteristics. This habitat shows visible signs of shrinkage, pollution, and overexploitation. There are conservation plans in place with funding from the Ministry of Environment and Forests, which needs to be implemented wisely at ground level after assessing the situation comprehensively (Jayakumar 2011).

The coastal wetlands of Vembanad and Kuttanad, measuring thousands of hectares, have undergone conversion in Kerala, known as backwater reclamation. This is not only an irreversible ecological process but also contains an element of uncertainty. The reclamation of Vembanad Lake for activities like paddy cultivation, prawn farming, husk retting, agriculture activities, industrial purposes, and harbor and urban development have significantly reduced the area of wetlands (Jeena 2000). The maximum reclamation occurred from 2001 to 2012 as the population boomed (Krishnadas 2014). The Vembanad Kol wetlands now stand totally environmentally degraded. Constructions of regulators, spillways, overexploitation of resources, and uncontrolled urbanization have pushed the ecology of the backwater system into big trouble (Remani et al. 2010). Elevated levels of heavy metals due to local industrialization are affecting biota in the northern part of the backwater (Anu et al. 2014). The Muriyad wetland system, which forms part of the Vembanad Kol, is freshwater in nature. Clay and sand mining, reclamation of paddy land for housing, coconut plantations, and pollution are some of the major threats upsetting the well-being of the wetlands. This is just a reflection of the situation of wetlands elsewhere in the state. Conditions of other freshwater wetlands—Vellyani, Sasthamkotta, Pookot, Enamakkal, Manakkodi, Eravikulam, Devikulam, and Elephant Lake—can be guessed as not different from these wetlands (Thomas et al. 2003).

In this paper, literature on heavy metal sources, levels, and spatio-temporal distribution in water, sediments, and organisms were reviewed for a comprehensive understanding of the status of heavy metal pollution in the Cochin backwater. Increased levels of the heavy metals encountered were linked to local industrialization. The northern part of the backwater is more polluted by heavy metals compared to southern and middle parts of the backwater. High levels of heavy metals were also found in biota.

The Kavvayi backwaters form the biggest wetland in Kannur District, North Kerala. Also called Malabar's own backwaters, it is touted as the next Ramsar site in Kerala. Reclamation, sand and laterite mining, tourism, overexploitation of minerals and resources, aquatic weeds, destruction of mangroves, and construction of bunds are some major threats to this wetland of brackish nature. The biodiversity is under pressure because of the need for fuel, food, and fodder. Other threats include dumping of wastes, coconut husk retting, sand and shell removal, aquaculture, and agriculture. The pollutants causing havoc with water quality are fertilizers, human sewage, animal waste, pesticides, and heavy metals (Harikumar 2016). The Canadian Council of Ministers of the Environment (CCME) water quality index and biological water quality criteria (BWQC) indicated poor quality and moderate level of pollution (Shiji et al. 2016).

The Kottuli wetland is one among the 94 wetlands of national importance identified by the government of India for conservation action under the National Wetland Conservation Programme (MoEF, 2006–2007). Located at the middle of Kozhikode (Calicut) City, this perennial wetland is linked to the brackish "Connolly" canal, built in 1848 by the then district collector. This wetland is encountering reclamation of low-lying areas, dwindling mangroves, open defecation by laborers and the homeless, solid waste and weed infestation, municipal sewage, chemical discharge, vehicular noise, and sound pollution. An ecotourism Sarovaram Project is proposed and projected to revive the wetland at some initial environmental cost but with commitment for preserving the wetland in the long run (Azeez et al. 2008).

7.5.3 Management

Kerala has a total geographical area of 3,886,287 ha, and this includes a total wetland area of 160,590 ha. Kerala has the largest proportion of wetland area in comparison to all the states of India. If compared to other states, the wetlands in Kerala are at risk because of developmental activities (Abraham 2015).

Many management programs for wetlands have been started during the last five years for evolving and executing sustainable management plans. Management action plans have been prepared by the Kerala State Council for Science, Technology, and Environment (KSCSTE) for all the three Ramsar sites. The Centre for Water Resources Development and Management (CWRDM), in collaboration with the Ministry of Environment and Forests (MoEF), has initiated a management action plan for three Ramsar sites and the Kottuli wetlands. Vembanad-Kole and few other wetlands are being monitored as part of the Asian Waterfowl Census. The need for protection of the lake, river, and even large pond wetlands has gained much attention due to activities of many community-based organizations and protection groups in the state. Authorized agencies looking after exclusively for wetlands should be constituted so that problems being faced by the wetland ecosystems in Kerala are removed (Kokkal et al. 2008). Strictly following the Coastal Regulation Zone (CRZ) 2011 regulation for Kerala and wetland rule 2010, the relocation of industries away from wetlands can bring reprieve to all the wetland inhabitants (Krishnadas 2014).

7.6 Conclusions

The literature available on the diversity of zooplankton in scanty and patchy and reflects many lacunas and gaps. For example, there is no information on plankton species of *Myristica* swamps and Thuruthu backwaters. There are 44 rivers in Kerala, but almost nil data is available as far as zooplankton are concerned. As noticeable many a times the identification has been done up to Genus's level only, probably due to the lack of experts available for identification. There are two monsoons in a year that Kerala experiences, but workers have barely related to the second monsoon season. It is only referred to as the post-monsoon season. Nonetheless, the literature available gives insight to the rich diversity of zooplankton assemblages. Rotifers exhibit a great level of adaptability to varied environmental and internal conditions. They have shown high diversity in kol and pond wetlands and appear to thrive better in static water.

Cladocerans dominate in lotic freshwater wetlands, like rivers. They are also present in estuary freshwater zones during monsoon whereas copepods dominate in lentic systems like kol wetlands. Nauplii, the juvenile stages of copepods abound in the pre monsoon period. Ostracoda are few and need substratum like aquatic vegetation etc. to attach and feed themselves. Rhizopods have made known their preference to the static rich environment. Overall zooplankton show abundance and diversity during pre and post monsoon seasons as these seasons are environmentally settled with no turbulence.

To sum up, many more comprehensive studies are required to unearth the micro-faunal biodiversity of the unexplored, unique wetlands of Kerala. A more controlled and cautious approach is needed to manage and preserve the unique backwater ecosystems of Kerala. Wetland and faunal biodiversity should not be allowed to bear the brunt because of human greed and ignorance. Wetland (Conservation and Management) Rules 2017 has already been notified by MoEFCC to replace the Wetland Rules 2010. The current version has at least civil society representation for good outcomes. Wetlands, if protected in a stringent manner like forests, could save many, if not all. To conclude, the discussion converges to the reality that the biodiversity loss is unidirectional though habitat loss and can be reversed to some extent for existing species. Hence, a holistic approach of what and how much is allowed, regulated, or prohibited within wetlands and their zone of influence helps strengthen the conservation strategies and wise use of wetlands.

REFERENCES

Abraham, S. 2015. "The Relevance of Wetland Conservation in Kerala." *International Journal of Fauna and Biological Studies* 2, no. 3: 1–5.

Alexander, T., M.P. Nayar, and L. Thushara. 2014. "Status and Seasonal Distribution of Plankton Diversity in Sasthamkotta Lake—The Largest Fresh Water Lake of Kerala." *Indian Journal of Environmental Health* 56, no. 4: 423–430.

Ali Akshad, M., O. Sathick, and V. Shaheer Ansari. 2018. "Analysis of Zooplankton and Their Monthly Variation of Kadalundi River Estuary, Kerala." *International Journal of Emerging Technologies and Innovative Research* 5, no. 12: 583–591.

Allen, M.R. 2007. "Measuring and Modeling Dispersal of Adult Zooplankton." *Oecologia* 153, no. 1: 135–143.

Anu, P.R., P.R. Jayachandran, P.K. Sreekumar, and Sivasankaran Bijoy Nandan. 2014. "A Review on Heavy Metal Pollution in Cochin Backwaters, Southwest Coast of India." *International Journal of Marine Science* 4, no. 10: 92–98.

Azeez, P.A., S. Bhupathy, Nikhil Raj, and R. Chandra. 2008. "Conservation of Kottuli Wetlands, Calicut, Kerala." Report submitted to Tourist Resorts (Kerala) Limited.

Balaraman, U. 2008. "The Unique Fauna of Ephemeral Wetlands of Southern Kerala." In: *Proceedings of Taal 2007: The 12th World Lake Conference*, Jaipur 28 October–2 November, 2007, eds. M. Senupta and R. Dalwani, International Lake Environment Committee, Kusatsu, 401–409.

Chackacherry, G., and K.V. Jayakumar. 2011. "Sasthamcotta Freshwater Lake, Kerala: Management Issues." In: *Proceedings: National Workshop on Ramsar Designated Wetlands of India*, Kolkotta.

Cleetus, R.I., C.V. Asha, P.S. Suson, and S. Bijoy Nandan. 2016. "Meso Zooplankton Abundance and Community Structure in Vembanad Kol Wetland Ecosystem, Kerala, India." *Indian Journal of Geo-Marine Sciences* 45, no. 4: 533–545.

Deivanai, K., S. Arunprasath, M.K. Rajan, and S. Baskaran. 2004. "Biodiversity of Phyto and Zooplankton in Relation to Water Quality Parameters in a Sewage Polluted Pond at Ellayirampannai, Virudhunagar District." In: *The Proceedings of National Symposium on Biodiversity Resources Management and Sustainable Use, Organized by the Center for Biodiversity and Forest Studies*, Madurai Kamaraj University, Madurai.

Divya, K.R., and K. Manonmani. 2014. "Impact of Domestic Sewage Pollution on Zooplankton Diversity in Bharathapuzha River." *Journal of Pharmaceutical Biology* 4, no. 1: 51–53.

Edmondson, W.T. 1974. "Secondary Production." *Mitt. Internt. Verein. Limnology* 20: 241–264.

Fathibi, K., Ambalaparambil V. Sudhikumar, and Embalil M. Aneesh. 2020. "Species Composition and Abundance of Rotifers (Rotifera: Eurotatoria) in Thrissur Kole Wetland, Kerala, India." *Egyptian Journal of Aquatic Biology and Fisheries* 24, no. 6: 439–451.

Harikumar, P.S. 2016. "Wetlands of Kerala: Degradation, Restoration and Future Management-a Case Study of Kavvayi Wetland-a Coastal Wetland in the Northern Kerala." In: *Proceedings Lake 2016: Conference on Conservation and Sustainable Management of Ecologically Sensitive Regions in Western Ghats*, Moodbidri Karnataka, 28th–31st December 2016, 250–253. http://ces.iisc.ernet.in/energy

Havel, John E., and Jonathan B. Shurin. 2004. "Mechanisms, Effects, and Scales of Dispersal in Freshwater Zooplankton." *Limnology and Oceanography* 49, no. 4: Part 2. http://doi.org/10.4319/lo.2004.49.4_part_2.1229

Hessen, Dag O., Thomas C. Jensen, and Bjorn Walseng. 2019. "Zooplankton Diversity and Dispersal by Birds; Insights from Different Geographical Scales." *Frontiers in Ecology and Evolution* 7: 74. https://doi.org/10.3389/fevo.2019.00074

International Hydrographic Organization. Limits of Oceans and Seas (1953) (Special Publication No. 28). 3rd edition (PDF) 1–45. https://doi.pangaea.de/10013/epic.37175.d001

Jayakumar, K.V. 2011. "Ashtamudi Wetland, Kerala: Values and Threats." In: *Conference: National Workshop on Ramsar Designated Wetlands of India*, Kolkota.

Jeena, T. S. 2000. "Wetland Conversion: The Case of Backwater Reclamation in Kerala." *Working Papers 74*, Institute for Social and Economic Change, Bangalore.

Jenkins, D.G. 1995. "Dispersal-limited Zooplankton Distribution and Community Composition in New Ponds." *Hydrobiologia* 313/314: 15–20.

Jose, D., and P. Senthilkumaar. 2015. "Physicochemical Parameters and Zooplankton Diversity in Anicadu Chira, Kerala, India." *International Research Journal of Biological Sciences* 4, no. 11: 55–60.

Jose, R., and M.G. Sanalkumar. 2012. "Seasonal Variations in the Zooplankton Diversity of River Achencovil." *International Journal of Scientific and Research Publications* 2, no. 11: 1–5.

Jyotsna Rajeswari K., G. Suresh, P. Srikanth, and N.N. Raman. 2018. "Study of Temporal Variations in Physico-Chemical Parameters of Vembanad Lake at Panangad-Kumbalam Region, Kochi, Kerala, India." *International Journal of Current Microbiology and Applied Sciences* 7, no. 6: 2905–2917.

Kaushal, A., P. Thomas, K.C. Pradeep, and S.E. Varghese. 2020a. "Annual Physico-chemical Characteristics and Zooplankton Diversity of Botany Pond at U. C. College, Aluva, Kerala, India." In: *Proceedings of*

18th K.V Thomas Endowment National Seminar. New Frontiers in Material and Environmental Sciences, eds. S. Ramakrishnan and Ammu Rosin Jose, Thevara, Kochi 28–29 January 2020. Research and Postgraduate Department of Chemistry, Sacred Heart College (Autonomous) Thevara, Kochi-682013, 95–100.

Kaushal, A., P. Thomas, G. Prasad, and K.C. Pradeep. 2020b. "Role of Zooplanktons for Ecological Status of Irumbakulam Pond, Aluva, Kerala, India." In: *Proceedings of the Virtual International Conference on Energy, Environment and Health. GRENZE Scientific Society Trivandrum, Kerala, India*, eds. K.C. Prakash, S. Suresh, N. Madhavan, and R.T. Ananth Kumar. Abs. 31. http://thegrenze.com

Kokkal, K., P. Harinarayanan, and K.K. Sabu. 2008. "Wetlands of Kerala." In: *Proceedings of Taal 2007: The 12th World Lake Conference. Jaipur. 28 Oct–2 Nov, 2007*, eds. M. Sengupta and R. Dalwani. International Lake Environment Committee, Kusatsu, 1889–1893.

Krishnadas, H. 2014. "Assessment of Land Reclamation of Wetlands: A Case of Vembanad Wetland, Kerala." In: *International Conference on Urban and Regional Planning, 2014*. Johor, Malaysia. http://doi.org/10.13140/2.1.3925.0242

KSESTE. 2007. "Land Environment, Wetlands of Kerala and Environmental Health." *State of Environment Report Kerala. Kerala State Council for Science, Technology and Environment*, Sasthra Bhavan, Pattom, Thiruvananthapuram—695004 Volume (1), 270.

Lengyel, Szabolcs, Aaron D. Gove, Andrew M. Latimer, Jonathan D. Majer, and Robert R. Dunn. 2009. "Ants Sow the Seeds of Global Diversification in Flowering Plants." *PloS ONE* 4, no. 5: e5480. https://doi.org/10.1371/journal.pone.0005480

Madin, L.P., Erich F. Horgan, and Deborah K. Steinberg. 2001. "Zooplankton at the Bermuda Atlantic Time-series Study (BATS) Station: Diel, Seasonal and Interannual Variation in Biomass 1994–1998." *Deep Sea Research II* 48, no. 8–9: 2063–2082.

Menon, A. Sreedhara. 2007. *Kerala Charitram* (2007 ed.). DC Books, Kottayam, 1–441.

Michels, E., Karl Cottenie, Liesbeth Neys, and K. Luc De Meester. 2001. "Zooplankton on the Move: First Results on the Quantification of Dispersal of Zooplankton in a Set of Interconnected Ponds." *Hydrobiologia* 442: 117–126.

MoEF (2006–2007). Annual Report. Annexure VI A & VI B. pp. 273–276.

Moreno, E., C. Pérez-Martínez, and J. M. Conde-Porcuna. 2016. "Dispersal of Zooplankton Dormant Propagules by Wind and Rain in Two Aquatic Systems." *Limnetica* 35, no. 2: 323–336.

Moreno, E., C. Pérez-Martínez, and J. M. Conde-Porcuna. 2019. "Dispersal of Rotifers and Cladocerans by Waterbirds: Seasonal Changes and Hatching Success." *Hydrobiologia* 834: 145–162.

Mukhopadhyay, Subhra Kumar, Buddhadeb Chattopadhyay, Abhishek Roy Goswami, and Asitava Chatterjee. 2007. "Spatial Variations in Zooplankton Diversity in Waters Contaminated with Composite Effluents." *Journal of Limnology* 66, no. 2: 97–106.

Nandan, S. Bijoy. 2008. "Current Status and Biodiversity Modification in the Coastal Wetland Ecosystems of India with Objectives for its Sustainable Management." In: *Proceedings of the Conserv-Vision Conference, University of Waikato, 2–4 July 2007*. The University of Waikato, Private Bag 3105, Hamilton, New Zealand. www.waikato.ac.n/wtass/conserve-vision

Nandan, S. Bijoy. 2011. "Final Project Report: Diversity and Abundance of Plankton and Benthic Fauna of Maranchery Wetland." Submitted to Kerala State Biodiversity Board, Thiruvananthapuram, Government of Kerala. 1–136.

Nath, A., R.V. Neethu, and J.S. Revathy. 2015. "Diversity of Plankton in a Village Pond at Dhanuvachapuram, Trivandrum." *International Journal of Innovative Research in Science, Engineering and Technology* 4, no. 9: 9189–9193.

Park, Kwan Soon, and Hyun Woung Shin. 2007."Studies on Phyto-and-zooplankton Composition and its Relation to Fish Productivity in a West Coast Fish Pond Ecosystem." *Journal of Environmental Biology* 28, no. 2: 415–422.

Pinceel, Tom, Falko Buschke, Margo Weckx, Luc Brendonck, and Bram Vanschoenwinkel. 2018. "Climate Change Jeopardizes the Persistence of Freshwater Zooplankton by Reducing Both Habitat Suitability and Demographic Resilience." *BMC Ecology* 18, no. 2. https://doi.org/10.1186/s12898-018-0158-z

Raghunathan, M.B. 2007. "Faunal Diversity of Ashtamudi Wetlands, Kerala, India." *Records of the Zoological Survey of India*, Occ. Paper No. 276, Published by the Director Zoological Survey of India Kolkata, 1–38.

Ramsar. 2020. "The List of Wetlands of International Importance." Published 25 February 2020. www.ramsar.org/sites/default/files/documents/library/sitelist.pdf

Ranjana, V.S., and S. Amina. 2018. "Zooplankton Diversity of Ayiramthengu Mangrove Region Kollam (Dist.), Kerala." *Indian Journal of Scientific Research* 20, no. 1: 91–95.

Remani, K.N., P. Jayakumar, and T. K. Jalaja. 2010. "Environmental Problems and Management Aspects of Vembanad Kol Wetlands in South West Coast of India." *Nature Environment and Pollution Technology* 9, no. 2: 247–254.

Roshan, F. J., and R. Sreejai. 2020. "Zooplankton Diversity of Kottakayal—a Wetland in South Kerala That is Under the Threat of Sand and Clay Mining." *Uttar Pradesh Journal of Zoology* 41, no. 8: 52–58. https://mbimph.com/index.php/UPJOZ/article/view/1551

Sebastian, S. 2012. "Studies on Physico Chemical Characteristics, Plankton Diversity and Ichthyofauna of Idukki Reservoir, Kerala, India." PhD diss., Mahatma Gandhi University, Kerala.

Sheeba, S. 2015. "Abundance and Feeding Habits of the Chaoborus Larvae—A Case Study in Sasthamkotta Lake System, Kerala." Project Report submitted to Kerala State Biodiversity Board. Order No 2281 A1 12012/KSBB.

Sherly Williams, E., P. Nisha Thomas, and S. Vineetha. 2020. "A First Report on the Plankton Status of Vattakkayal Lake, Kerala." *International Journal of Scientific and Technology Research* 9, no. 3: 260–263.

Shiji, M., A.R. Sabitha, Kavya Prabhakar, and P.S. Harikumar. 2016. "Water Quality Assessment of Kavvayi Lake of Northern Kerala, India Using CCME Water Quality Index and Biological Water Quality Criteria." *Journal of Environmental Biology* 37, no. 6: 1265–1272.

Sreekumari, Vishnu Mohan, Shiekha Elizabeth John, Rajimol Thengumparambil Rajan, Maya Kesavan, Sajan Kurian, and Padmalal Damodaran. 2015. "Human Interventions and Consequent Environmental Degradation of a Protected Freshwater Lake in Kerala, SW India." *Geosciences Journal* 20: 391–402. https://doi.org/10.1007/s12303-015-0049-7

Thampiraj, Reju J., and J. W. Prakash. 2019. "Seasonal Variations of Zooplanktons in the Selected Ponds of Venganoor Grama Panchayath, Southern Western Ghats." *Plant Archives* 19, no. 1: 1534–1540.

Thomas, K. John, S. Sreekumar, and Jaya Cheriyan. 2003. "Muriyad Wetlands: Ecological Changes and Human Consequences." *Project Report submitted to Kerala Research Programme on Local Development*, Centre for Developmental Studies, Thiruvananthapuram. 1–83.

Thomas, P., and Anju Lakhumna. 2019. "Zooplanktons as Indicators in the Seasonal Ecology of Three Ponds of Ernakulam District of Kerala." In: *Our Environment—Our Future: Science and Technology for Rebuilding Kerala 31st Kerala Science Congress, Kollam, Kerala*. 02–03 February, 2019 Eds. R. Prakash Kumar, S. Shaheem, R. Raj Vikraman, P. S. Jothish, Arun Chandran, Ebin Sam, S. Suresh, S. R. Bobby, R. J. Sanjai. Kerala State Council for Science, Technology and Environment, Sasthra Bhavan, Pattom, Thiruvananthapuram—695004 Abs, 124.

Varghese, M., L. Krishnan, and V.J. Kuttyamma. 2006. "Systematic Account on Rotifers of the Genus *Brachionus* from Cochin Backwaters." *Journal of the Marine Biological Association of India* 48, no. 2: 147–155.

Vogt, Richard J., and Beatrix E. Beisner. 2011. "Assessing the Impact of Dispersal on Zooplankton Community Structure." *Journal of Plankton Research* 33, no. 11: 1757–1761.

Xu, T., Baisha Weng, Denghua Yan, et al. 2019. "Wetlands of International Importance: Status, Threats, and Future Protection." *International Journal of Environmental Research and Public Health* 16, no. 10: 1818.

Yahiya, M., and Mary Teresa P. Miranda. 2016. "Planktons as Indicators of Heavy Metal Pollution Along South West Coast of India." *International Journal of Science and Research* 5, no. 12: 977–982.

8

New Innovative Technologies for Sustainable Aqua Production

Rikhi S. Chauhan and Ashutosh Mishra

CONTENTS

8.1 Introduction

Aquaculture pertains to farming of aquatic organisms, including fish, mollusks, crustaceans, and aquatic plants. Farming implies some form of intervention in the rearing process to enhance production, such as regular stocking, feeding, and protection from predators. Thus, aquaculture is under water agriculture, or aquafarming. Aquaculture is almost synonymous with fish culture, or pisciculture, as most of the cultivated aquatic species belong to the finfish group. Pisciculture is an age-old practice. It developed as hunting was replaced by farming. The Egyptians were probably the first to culture fish (keeping fish) as far back as 2500 BC, about 4,500 years ago. Fish culture was being practiced in China in 2500 BC, but no information is available about techniques used for fish culture until 475 BC. In India, *Arthashastra*, written by Kautilya (321–300 BC), mentions raising fish in poisonous conditions in reservoirs during times of war (Jhingran, 1991). Until about 19th century, the collection and transport of carp spawn from rivers, stocking them in ponds, and adopting empirically developed traditional methods of pond management were mainly confined to Bengal, Bihar, and Orissa. The beginning of 20th century marked the introduction of several exotic species in Indian waters for various purpose, including food production. Successful induced breeding of Indian major carp in 1957 marked an indelible milestone on the food fish cultivation map.

Today, Indian aquaculture comprises about 40 species belonging to varied groups of finfish, crustaceans, mollusks, and aquatic plants and contributing significantly to the food basket of the country. Commercial fish hatching operations commenced by mid-19th century. In fact, this outcome was the result of tireless efforts by pioneering biologists to understand and manipulate the reproductive

DOI: 10.1201/9781003220398-8

physiology of the targeted aquatic species and with this, commenced the inquiry into all associated fields of husbandry, nutrition, health, harvest, post-harvest, genetics, and others that form the subject matter of formal education in fisheries today.

8.2 Fish in Nutrition

Fish is one of the promising and cheaper sources of animal protein. It is a source of high-quality, balanced, and easily digestible proteins, vitamins, and polyunsaturated fatty acids. Fish provides 17% of protein consumed by low-income families in different countries (LIFDCs), while it accounts for 7% of intake of all proteins consumed by global population (FAO, 2020a). Eating fish is a great alternative to eating red meat. A number of health benefits accrue by consuming fish as it is loaded with many nutrients that most people are lacking. This includes high-quality protein, iodine, and various vitamins and minerals. Fatty species like salmon, trout, sardine, tuna, and mackerel contain higher amount of fat-based nutrients like vitamin D, a fat-soluble nutrient that many people are lacking. Fish also has omega-3 fatty acids, which are crucial for optimal body and brain function and strongly linked to a reduced risk of many diseases. The omega-3 docosahexaenoic acid (DHA) is especially important for brain and eye development. Fish is considered as one of the most heart-healthy foods one can eat (Bennett et al., 2018). Unsurprisingly, many large observational studies show that people who eat fish regularly have a low risk of heart attacks, strokes, and death from heart disease (Mohan et al., 2021). Eating fish one or two times in a week is considered sufficient to reap its benefits. Shark liver oil contains squalene (polysaturated triterpene), which carries oxygen independent of the hemoglobin. It has been reported to have an inhibitory effect on cancer promotion and a high anti-tumor activity, thus, it is primarily used as a supportive therapy in a variety of cancers (Gunes, 2013).

8.3 National and Global Fish Production Scenario

Aquaculture is one of the fastest-growing food production sectors in India, engaging over 16 million people at the primary level and many more along the value chain. India ranks second in global aquaculture production, after China. The strategies for increasing fish production of the country to meet its ever-growing demand and overcoming the issue of malnutrition through assured supply of nutrient-rich fish have been the center point of discussions among the researchers, planners, and policymakers today. The impressive growth in aquaculture production registered during last three decades has been instrumental in achieving fish production of 13.75 million tonnes from 2018 to 2019 (DoF, 2020). Inland aquaculture contributes more than 70% of the inland fish production of the county. Considering the fact that the marine capture fish production is almost stagnating and as such no substantial increase from this sector may be possible, the bulk of 16 million tonnes of the domestic demand of fish by 2025 is expected to come from freshwater aquaculture. The export earnings from fisheries sector from 2018 to 2019 registered the value of Rs. 46,598.37 crore (USD 6.73 billion). India shares 7.58% to the global production. It contributes 1.24% to India's gross value added (GVA) and 7.28% (2018–2019) to the agricultural GVA (DoF, 2020).

The global fish production peaked at about 179 million tonnes in 2018 with estimated sale value of about USD 401 billion. Aqua-animal production contributed 46.0% (82.1 million tonnes with value of USD 200 billion) to the total fish production. In per capita terms, food fish consumption grew from 9 kg in 1961 to 20.5 kg in 2018, at an average rate of about 1.5% per year (FAO, 2020a). During 2018, 31.8 million tonnes of aquatic plants and 37,900 tonnes of non-food aquatic products were also produced. Asia contributes 89.39% to the world aquaculture production. A high proportion of aquaculture in Asian countries continues to be from traditional pond culture of freshwater fish, especially carp and other cyprinids.

After adoption of FAO Code of Conduct for Responsible Fisheries in 1995, the importance of utilizing fisheries and aquaculture resources responsibly is now widely recognized and prioritized. The United Nations' 2030 Agenda for Sustainable Development includes 17 Sustainable Development Goals (SDGs) (UN, 2015), many of those are relevant to fisheries, particularly SDG 14, "Life below water."

TABLE 8.1

Aquaculture Production of the Main Groups of Species by Continent ('000T)

	Continent						
Category	**Africa**	**America**	**Asia**	**Europe**	**Oceania**	**World**	**Share (%)**
Finfish	1,972	1,978	47,722	2,332	87	54,300	67.6
Crustaceans	5	795	7,055	0	7	9.4	9.8
Mollusks	6	574	15,835	613	112	17.7	21.4
Other aquatic animals	0	1	933	0	5	939	1.2
Total	1,983	3,348	71,546	2,945	210	80,031	100
Share (%)	2.47	4.18	89.39	3.67	0.26		

8.4 Main Cultivated Aquatic Species

Fish have inhabited virtually all aquatic habitats, and more than 34,700 species have been described (Fish Base, 2021). Domestication efforts for number of aquatic species are rising rapidly, to fulfill the national food security. The main groups being cultivated in varied aquatic systems are finfish, crustaceans, mollusks, and other aquatic animals (Table 8.1). The aquatic plant production is dominated by farmed seaweeds representing 97.1% by volume of the total of 32.4 million tonnes of wild—collected and cultivated aquatic algae combined. The main cultivated seaweeds are species of *Laminaria*, *Eucheuma*, *Gracilaria*, *Undaria*, *Porphyra*, and *Kappaphycus* (FAO, 2020b).

The global production has been recorded for a total of 598 species items ever farmed in the world. A species item refers to a single species, a group of species (where identification of the species level is not possible), or an interspecific hybrid (FAO, 2020a). Species items recorded so far include 369 finfish, (including 5 hybrids), 109 mollusks, 64 crustaceans, 7 amphibians and reptiles (excluding alligators and crocodiles), 9 aquatic invertebrates, and 40 aquatic algae. Over 90% of the total finfish production in the world is dominated by 27 species. Top 30 cultivated aquatic animal species contribute 65.66% of the world aquaculture production of all aquatic animals. These species are *Ctenopharyngodon idella*, *Litopenaeus vannamei*, *Hypophthalmichthys molitrix*, *Oreochromis niloticus*, *Cyprinus carpio*, *Ruditapes phillippinarum*, *Hypophthalmichthys nobilis*, *Catla catla*, *Salmo salar*, *Pangasianodon hypophthalmus*, *Labeo rohita*, *Procambarus clarkii*, *Chanos chanos*, *Sinonovacula constricta*, *Oncorhynchus mykiss*, *Megalobrama amblycephala*, *Eriocheir sinensis*, *Penaeus monodon*, *Mylopharyngodon piceus*, *Crassostrea gigas*, *Pelteobagrus fulvidraco*, *Cirrhinus mrigala*, *Channa argus*, *Micropterus salmoides*, *Anadara granosa*, *Icatalurus punctatus*, *Barbonymus gonionotus*, *Silurus asotus*, *Mytilus chilensis*, and *Misgurnus angullicaudatus* (FAO, 2020b).

8.5 Innovative Aquafarming Technologies

The rapid growth of aquaculture during recent years has raised questions concerning its environmental sustainability. To minimize the potential threat for climate change, there is need to modernize the sector in a manner that is more responsible, sustainable, and transparent and generates enhanced public and private confidence in aquaculture. Therefore, future aquaculture should follow the principle of responsible fisheries as defined by the FAO (FAO, 2011) and be targeted to produce healthier and disease-free fish using less water in order to sustain the ecosystems and biodiversity. Major innovations in the aquaculture systems developed to meet the these objectives is discussed in the following sections.

- Recirculatory aquaculture system (RAS)
- Running-water fish culture
- Biofloc technology

- Aquaponics
- Cage culture

8.5.1 Recirculatory Aquaculture System (RAS)

The recirculatory aquaculture system (RAS) is a new water quality improvement technology, much of it developed for the wastewater treatment industry. It has revolutionized our ability to grow fish in tank culture. The water quality in RAS depends on various factors, most significantly the source, the level of recirculation, the species being cultured, and the wastewater treatment process inside the system (Sanni and Forsberg, 1996; Losordo et al., 1999). The physico-chemical parameters of water are important as these directly affect its quality and subsequently its suitability for the distribution and production of fish and other aquatic animals (Toma, 2012). These factors encourage the use of RAS, where a large amount of high-quality fish can be grown in a small volume of water.

The earliest scientific research on RAS conducted in Japan in 1950's focused on biofilter design for carp production driven by the need to use limited water resources for more productively (Martins et al., 2010). The interest of growing fish in RAS has been increasing especially in intensive finfish culture in the world. This is due to the perceived advantages that RAS greatly reduces land and water requirements and offers a high degree of control of the culture environment that allows year-round growth at optimal rates and that fish biomass can be determined more accurately than in ponds. A typical RAS consists of a water supply system, mechanical and biological filtration pumps to maintain water flows, aeration and oxygenation systems, and other water treatment components that deliver optimal water quality for fish growth within the system (Figures 8.1 and 8.2). RAS also offers other potential advantages, including

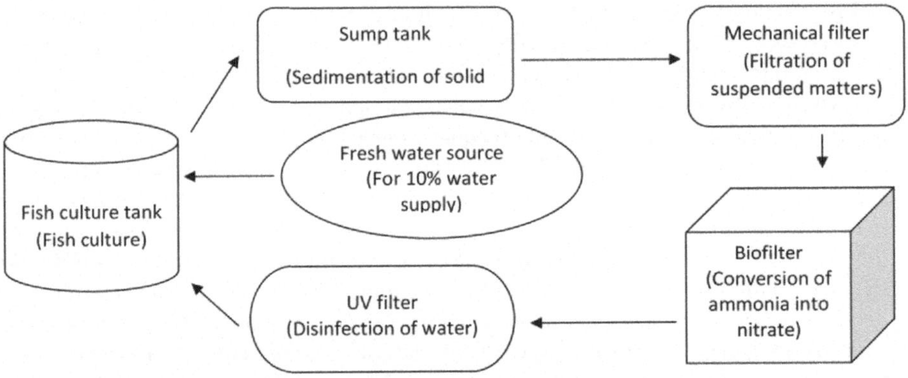

FIGURE 8.1 Schematic representation of a recirculatory aquaculture system.

FIGURE 8.2 Fish production in a recirculatory aquaculture system.

the ability to place the farm in locations where water resources are limited and near to the market to reduce product transport time and costs. With more stringent water pollution control, a recirculatory system provides greater environmental sustainability than traditional pond aquaculture in managing waste production. Depending on the location and situation, the systems can be designed to cater for different capacities and efficiencies. The development of such aquaculture technology appears to be feasible, and it can bring about major advances in the development of high-tech in aquafarming with maximum sustainable fish production.

Research and development in recirculatory systems have been going on for nearly four decades. There are many alternative machineries for each process and operation. The selection of a particular technology depends on the species being reared, site, management expertise, infrastructure, production, and other factors (Dunning et al., 1998). In RAS, more than 90% of the water can be recirculated by a series of biological and mechanical filtration systems so that only a part of water is expended. New water is added to the tanks only to balance for losses through splash-outs, evaporation, and loss from flushing out waste materials.

An RAS unit is designed for holding and growing a wide variety of aquatic species that recycle the water by passing it through filters for removing metabolic and other waste products (Kazmierczak and Caffey, 1995). Recirculating systems can control the aquatic environment to achieve better hygiene and disease management (Martins et al., 2010). RAS can be used as grow-out systems to produce food fish or as hatcheries to produce spawn, fry, and fingerlings for stocking and ornamental fish for home aquaria (Helfrich and Libey, 1991). In contrast to the traditional aquaculture practices, RAS offers more impartiality from the external environment, which affords a basis for enhanced risk management (Rawlinson, 2002).

Knowing appropriate stocking density is recognized as an essential aspect because it plays a big role in increasing the fish production to meet the continuous increase in fish demand and maintain the profitable and economic sustainable culture system for fish producers (Rafatnezhad et al., 2008). Stocking density is well known as the weight of fish per unit volume (Ellis, 2001) or the number of fish stocked at the beginning of the experiment (Ruane et al., 2002). In high stocking density, usually fish exhibit aggressive behavior, especially when food availability is limited. Therefore, food availability is very important when fish density is concerned (Holm et al., 1990). Other problems encountered in a recirculatory system are low water temperature, low dissolved oxygen, and high waste metabolite concentration (Twarowska et al., 1997).

However, keeping good water quality is of primary importance in RAS. Potential users of aquaculture systems need to know about the essential water treatment processes to control temperature, pH, pathogens, dissolved gases (oxygen, carbon dioxide, and nitrogen), and fish metabolites, such as dissolved nitrogen compounds (ammonia, nitrite, and nitrate) and solids (both dissolved and particulate) in the culture water; the components are accessible for each process and the technology behind each component (Losordo et al., 1999). The economics (essentially competing in price with wild-caught fish) is currently the limiting factor preventing the broader commercial development of aquaculture. At this time, most successful recirculation initiatives focus on niche markets of high-value fish, such as trout, eel, catfish, turbot, sea bass, and sole (Martins et al., 2010).

Recirculation can be carried out at different intensities, depending on how much water is recirculated or reused. The super-intensive farming systems developed in insulated building using as little as 300 liters of new water, and sometimes even less, per kilogram of fish produced per year, while intensive farming system requires 1,000 liters of water per kilogram per year (Martins et al., 2009). The limited use of water makes it much easier and cheaper to remove the nutrients excreted from fish. This new, innovative fish-farming technology has a great potential to enhance quality fish production from a limited area and play an important role in contributing a big share in country's GVA.

8.5.2 Running-Water Fish Culture

In order to have most efficient utilization of water resources, running water aquaculture may be practiced in streams having suitable water quality conditions for fish growth. It is one of the technologies of intensive fish culture, first developed in Indonesia during early '70s (Suprayitno, 1986). There are great advantages in cultivating fish in water, which is actually flowing (Figure 8.3). The water will supply abundant

FIGURE 8.3 Fish culture in a running-water system.

amount of dissolved oxygen and will flush away any byproduct of the fish or the uneaten artificial feed supplied in pelleted form (Sharma and Mishra, 2009). There are additional benefits in running-water fish culture, such as increase demand of yearlings of about 100 g as running water requires fish stock of a minimum 100 g weight, decrease in the utilization of wide land as running water system needs a small area, shortening of the culture period for fish to reach marketable size, and increasing the income of fish farmers. The fish species suitable for this culture system should be strong enough to live with fast water current, sociable in habit, and highly disease-resistant due to high stocking density and should have the ability to accept artificial pelleted feed and to transform the feed into flesh efficiently (Singh and Mishra, 2019). The quantity of water required for fish culture depends on the species being raised and the number of fish in the tank, as well as the method of fish culture. The water supply used in running-water fish culture is between 0.8 and 3.2 liters per second per cubic meter of tank volume with average value of 1.6 liters per second per cubic meter. According to available information, the water in running water tank/raceways must totally be changed within a period of ten minutes. The best area for running-water fish culture is a ground with slight slope (between 0.5 and 1% but maximum slope can be about 2.5%), free from occurrence of flood, landslide, and storm (Suprayatino, 1986). The site of the raceways must be near the source of water. The minimum height difference of 30 cm between stream surface water and the tank surface water is preferable in order to drain and fill the tank by gravity. The raceways should not be built on an area that is affected by agricultural wastes, such as pesticides and other wastes. Among all shapes, the oval-shaped raceways are most preferable. Stocking density commonly applied in running water raceways ranges between 5 and 20 kg of fish per cubic meter. Usually, fish are fed with artificial feed pellets of 28% protein with daily feeding rate ranges from 3 to 4% of body weight. It should be revised after every 10 or 14 days adjusted by weight of sampled fish. Usually, feeding frequencies vary between 5 to 12 times per day (Suprayitno, 1986).

There are large water resources in India with perennial water flow and availability of flat land along the banks of streams where ponds can be dug out or cemented raceways can be constructed and stream water can pass in required quantities for cultivating fast-growing fish, such as Indian major carp, exotic carp, and mahseer. Though most of the cultivable fish can be cultured in this system, most suitable fish for this culture system are trout, mahseer, and common carp.

8.5.3 Biofloc Technology

Biofloc technology is a technique of enhancing water quality in aquaculture through balancing carbon and nitrogen in the system. It is the retention of waste and its conversion to biofloc as a natural food within the culture system. Aquatic ecosystems recycle huge hoards of dissolved and particulate organic matter, playing one of the prime important roles in the food webs. In the biofloc setup, the bacteria surround the feces, molts, dead organisms, and surplus food to foster bacterial biomass, which is consumed by detritivores (Sunitha et al., 2018). This promotes nitrogen uptake by bacterial growth and decreases the ammonium concentration more rapidly than nitrification.

An Israeli brainchild, biofloc technology was first devised by Avnimelech (2009). In aquaculture, it is termed as the new "blue revolution" as nutrients can be continually used up and refurnished in the water system, benefitted by minimum or zero-water exchange. These flocs are enriched with proteins, which are retained in the water column all the time due to the complex reactions that occur among the organic matter, physical substrate and large range of microorganisms. As a result, the water is brimming with natural productivity, which manifolds the chances of recycling of the nutrients. It also serves as an aqua-feed ingredient, also called biofloc meal. Of late, it has been introduced in carp culture, catfish culture, and cachama culture.

Bioflocs are the aggregates (flocs) of algae, bacteria, protozoa, and other kinds of particulate organic matter, such as feces and uneaten feed (Figure 8.4). Each floc is held together in a loose matrix of mucus that is secreted by bacteria. Biofloc technology makes it possible to minimize water exchange and water usage in adequate water quality within the culture unit while producing low cost bioflocs rich in protein, which in turn can serve as a feed for aquatic organisms (Avnimelech et al., 1994). Compared to conventional water treatment technologies used in aquaculture, biofloc provides a more economical alternative (decrease of water treatment expenses in the order of 30%) and, additionally, a potential gain on feed expenses (the efficiency of protein utilization is twice as high in biofloc technology systems than conventional ponds), making it a low-cost sustainable constituent to modern aquaculture development. Also, biofloc technology is robust, economical, and easy in operation. The process of constant aeration maintains the high level of microbial floc in suspension and agitation of water column with the addition of carbon sources as organic substrate to allow aerobic decomposition.

The growth increment of different species in biofloc varies according to the carbon sources. The growth performance, digestive enzyme activity, immune response, and water quality in several aquatic species is reported to be positively affected by this technology. In this technology, fish and shrimp are reared (minimum of 300 g of biomass per square meter) with zero or little water exchange by maintaining carbon and nitrogen ratio in the range of 12–20:1, which naturally generate heterotrophic microbial community formation and stabilization that play three major roles: (1) maintenance of water quality, by the conversion of nitrogenous compounds into microbial protein; (2) nutrition, by improving culture feasibility with reduced feed conversion ratio (FCR); and (3) competition with pathogens (Pérez-Fuentes et al., 2016). High carbon-to-nitrogen ratio is prerequisite of optimum heterotrophic bacterial growth, which produces energy for maintenance (respiration, feeding, movement, digestion, etc.) but also for the proliferation of new bacterial cells. Under feasible ambience of temperature, carbon-to-nitrogen ratio, pH, and so on, the bacteria show a fast growth. Leonard et al. (2000) calculated that birth time for the free viable heterotrophic populations was around 2.5 h in laboratory conditions. Phytoplankton and heterotrophic and nitrifying bacteria have the overarching role in the nitrogen and organic reutilization.

FIGURE 8.4 Biofloc fish culture system.

Fungi, ciliates, protozoa, rotifers, copepods, and nematodes benefit the biofloc community, lending a hand in the recycling of organic load, forming complex food webs, which include the cultured species.

Brown et al. (1996) assessed the biochemical make up of seven strains of marine bacteria, the main component of biofloc and documented protein content (dry weight) of 29–49%; carbohydrates, 2.5–11.2%; and lipids, 4–6% and, additionally, the presence of all essential amino acids. Bacterial groups are involved in autotrophic process in two ways: (1) the ammonia-oxidizing bacteria, which acquire the energy by catabolizing unionized ammonia to nitrite, including the genera *Nitrosomonas*, *Nitrosococcus*, *Nitrosospira*, *Nitrosolobus*, and *Nitrosovibrio* and (2) the nitrite-oxidizing bacteria, which metabolize nitrite to nitrate, including the genera *Nitrobacter*, *Nitrococcus*, *Nitrospira*, and *Nitrospina*.

The main aim of this technology is to provide a sustainable approach, utilizing minimum space, offering maximum affordability. The eating up of biofloc by fish has proved to be beneficial for growth rate, lowering in feed conservation ratio and costs drawn out on feed (Haziq, 2020). Using up sugar, alcohol, and organic acids as energy source, heterotrophic bacteria exist in specialized species that are adept of decomposing cellulose, lignin, chitin, keratin, hydrocarbons, phenol, and other substances. Mostly spread in soil, freshwater, and saltwater, they are skilled to settle in a variety of diverse environment.

8.5.4 Aquaponics

The population rise, climate change, soil degradation, water scarcity, and food insecurity are among the important issues before the humanity. There is a need to develop a well-designed bio-integrated food production system to increase economic efficiency in limited resources with regular food production. Aquaponics may be an attractive opportunity of producing both plant and fish crops together. Aquaponics is a soilless culture technique, combining hydroponics and aquaculture in a system. In a well-designed and efficiently run aquaponics facility, the ability to profit is greater because of increased product output, efficient use of resources and regularity of production. Aquaponic system provides a simple solution to the main issues of these two systems normally face—management of nutrient-rich effluent water from aquaculture and water scarcity.

Although the production of fish and vegetables/fruits/herbs are the most visible output of aquaponic units, it is essential to understand that aquaponic system is the management of a complete ecosystem that consists of three major groups of organisms: fish, plants, and bacteria. The symbiotic relation among the different components of aquaponic unit encourages sustainable use of water and nutrients and recycling of nutrients and water both (Figure 8.5).

The water flowing from fish tank is nutrient-rich with a load of ammonia. The fecal matter and leftover fish feed are the main reason for total ammonia nitrogen (TAN), which needs to be converted into more

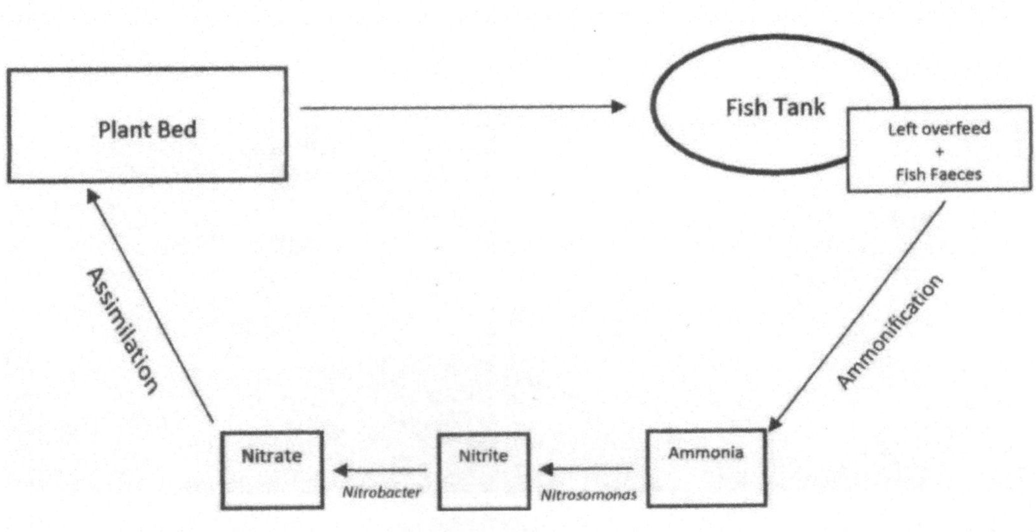

FIGURE 8.5 Schematic representation of an aquaponic system.

accessible or nontoxic form nitrate. The ammonia is metabolized by a specific group of bacteria called nitrifying bacteria. Nitrifying bacteria, which live in diverse environments, such as soil, water, and air, are an essential component of the nitrification process that converts plant and animal waste into accessible nutrients for plants. These bacteria first convert the ammonia (TAN) into nitrite compounds (NO^{2-}) and then finally into nitrate compounds (NO^{3-}) (Tyson et al., 2008). Plants are able to assimilate nitrates (NO^{3-}) by their roots to perform their growth processes. Two major classes of nitrifying bacteria are involved in this to perform the bi-chemical reaction.

1. Ammonia oxidized into nitrite form by *Nitrosomonas* bacteria
2. Nitrite further oxidized into nitrate by *Nitrobacter* or *Nitrospira* bacteria

The bacteria are completely dependent on the environment within the aquaponic unit to keep ammonia levels down. Adequate surface area and suitable water conditions are two key aspects to consider when ensuring a consistent bacterial colony. Fish constantly produce ammonia and excrete it into the water which is converted in nitrate in the biofilter (bedding material). The content of ammonia in a healthy system is always within nontoxic thresholds for the fish. Fish are also harmed by nitrite, which is produced during the initial step of ammonium conversion. As a result, both bacterial colonies must be completely operational (Chen et al., 2006). Adequate surface area and proper water conditions are two significant factors that influence bacteria proliferation and should be considered while maintaining a healthy biofilter.

The biological activity of nitrifying bacteria and their ability to convert ammonia and nitrite is influenced by the pH level, temperature, and DO level of the water. However, for aquaponics, a pH range of 7.2–7.8 and 7.2–8.2 is more favorable for *Nitrosomonas* spp. And *Nitrobacter* spp., respectively (Loveless and Painter, 1968; Antoniou et al., 1990). Temperatures of 17–34°C are good for bacterial growth and productivity (Zhu and Chen, 2002). Microbial productivity will decrease if the water temperature falls below 17°C (Malone and Pfeiffer, 2006). Nitrification is an oxidative reaction in which oxygen is utilized as a reagent; the reaction will cease if oxygen is not available in the system. The DO level of 4–8 mg/liter is ideal for bacterial growth. Nitrifying bacteria are photosensitive microorganisms. When a new aquaponic system is set up, light has the most impact on the initial bacteria colonies. UV radiation is not a significant issue once bacteria have colonized a surface (three to five days). As a result, cover the fish tank and filter components with UV protective material while ensuring that no water in the hydroponic component is exposed to the sun at least until the bacteria colonies are fully developed.

The plant-growing areas of the aquaponic unit are referred to as hydroponic beds. Hydroponics is a way of producing agricultural crops without the use of soil and within various media types, providing a nutrient-rich solution to the root zones. This solution provides all of the nutrients necessary for plant growth. Media beds, floating rafts, or deep-water culture (DWC) and nutrient film technology (NFT) are the three basic aquaponic techniques used around the world. The three designs are as follows:

- Media-based growing beds, also known as particulate beds, in which plants grow in a substrate. The water is commonly supplied in an ebb-and-flow pattern, ensuring sequential nutrition and aeration.
- Nutrient film technique (NFT) units, in which plants grow with their roots in wide pipes supplied with a trickle of culture water (Adler et al., 2000; Lennard and Leonard, 2006).
- Deep-water culture (DWC) units, also known as raft aquaponics or floating bed systems, in which plants are suspended above a tank of water using a floating raft (Rakocy et al., 2004).

Media-Based Growing Bed: Its designs are simple and space-efficient, have a cheap initial cost, and are excellent for novices. It is a solid media-filled bed system with more surface area for biofiltration/nitrification, mineralization, and thus, effective plant growth. It is filled with gravel, expanded clay, or pumice. Under low fish stocking densities, media bed techniques are frequently used to remove solids. The medium is used to support the roots of the plants in media bed units, and it also serves as a mechanical and biological filter. Flood and drain (also known as ebb and flow) is one of the methods of irrigating plants used in a media bed unit, which involves filling the hydroponic unit with nutrient-rich water and then draining it to allow air to enter the root zone. Periodic immersion in water followed by air exposure delivers oxygen to the roots, building a favorable root environment. Because of the good mechanical support given for the roots of the plants, the grow bed is ideal for fruity plants like tomatoes and cucumbers, but it may also be used for strawberries and leafy vegetables.

Deep-Water Culture (DWC) Bed: Another system that has great potential for commercial use is DWC, the deep-water culture system. Plants are suspended on polystyrene sheets with their roots hanging into the water in the DWC method. It consists of large troughs with perforated floating rafts where net plant pots are inserted. These plant pots are generally filled with media, such as rock wool, coco, or pumice, that support the roots, which are then continually submerge in the water tank. This is the most frequent design for large commercial aquaponic unit for the growth of any specific plant in bed, like lettuce or basil. Although the raft is normally made of polystyrene insulation and the plants are supported by plastic net pots, various novel food-grade materials have been developed. The rafts have a number of advantages, including usability, mobility, ease of cleaning, and a lower risk of plant mortality during power outages. Floating raft system is one of the most commercially adopted techniques under DWC. A major difference between the raft system, and the NFT and media-based systems is the amount of water used. The water level beneath the rafts is anywhere from 12 to 24 inches deep and, as a result the volume of water, is approximately four times greater than other systems.

Nutrient Film Technique: The nutrient film technique (NFT) consists of narrow channels of perforated squared pipes where the roots are partially immersed in a thin layer of streaming water. Plants are placed within holes in the top of the pipes and are able to use this thin film of nutrient-rich water. The shallow film of water at the bottom of each perforated pipe ensures that the roots receive large amounts of oxygen at the root zone along with moisture and nutrition. NFT systems are most suitable for smaller plants, such as herbs and plants used in salads, and may even be used for larger plants, including tomatoes and okra. The light-weight design of this system enables for vertical placement of channels over one another or atop a deep-water culture system (DWC). The water flow in each grow pipe should be not greater than 1–2 liters per minute. The flow rate is used to control from the Y-valve, with all excess water flow returning to the fish tank. The length of the grow pipe can range from 1 to 12 meters. Longer than 12 meters is not recommended because the initial plants have already consumed most of the nutrients, nutrient deficits can emerge in plants toward the end of pipes.

System Cycling

System cycling is described as the initial process of building a nitrifying bacteria colony in a new aquaponic system. This three-to-five-week process involves adding an ammonia source into the system (fish feed/ammonia-based fertilizer, up to a concentration in water of 1–2 mg/liter) in order to stimulate nitrifying bacteria growth. This should be done slowly and consistently. The progress is measured by monitoring the nitrogen levels. Ammonia, nitrite, and nitrate are monitored to determine the status of the biofilter (Seawright et al., 1998). Within five to seven days after the first addition of ammonia, the ammonia-oxidizing bacteria start forming a colony and begin to oxidize the ammonia into nitrite. The nitrite levels in the water will have started to rise after another five to seven days, attracting the nitrite-oxidizing bacteria (NOB). As the population of NOBs grows, the amount of nitrite in the water decreases as nitrite is oxidized to nitrate. When the nitrate level is increasing rapidly, the nitrite level is 0 mg/liter, and the ammonia level is less than 1 mg/liter, the cycling process is complete (Lekong and Kleppe, 2000; Chen et al., 2006). This takes roughly 25–40 days in ideal conditions, but it could take up to two months if the water is cold. Plants and fish should be added only after the cycle is complete. Only once the ammonia and nitrite levels are below 1 mg/liter, it is safe to initiate stocking the fish.

Production Management

To achieve and maintain system balance and ensure optimal crop yields, fish and plant output must be kept consistent, and environmental parameters must be managed. Air temperature, water temperature, macro- and micronutrient concentrations, DO and CO_2 level in air and water, pH, and light are the most important production attributes. These important parameters should be tuned to fit the needs of the fish and plant species being farmed. In terms of nutrition, maintaining to simple plant-to-fish ratios would make nutritional management easier, as long as limiting elements like calcium, phosphorus, and iron are consistently supplied which could be done through buffering. Based on a FCR feed conversion ratio of 1, the plant biomass should be seven to ten times that of the fish. Fish management requires the maintenance of optimal growth conditions for the species being cultured.

Major Fish in Aquaponics: The type of fish desirable for consumption by consumers and the type of fish feed available to the grower is the prerequisite consideration point during species selection. Fish species suitable for aquaponic farming include tilapia, common carp, silver carp, grass carp, barramundi, perch, catfish, trout, salmon, Murray cod, and sea bass. Tilapia is a eurythermal species and therefore a popular fish in aquaponic systems. The fish growth rate, measured as specific growth rate (SGR % per day), can be 0.91–5.1% per day in aquaponics (Seawright et al., 1998; Al-Hafedh et al., 2008). Performances on feed conversion ratio range within 1.1–1.8 (Tyson et al., 2008).

Plants in Aquaponics: Several crops have been produced at hydroponic units in aquaponic systems and many of them have performed very well. Plants need a number of micro- and macronutrients, also referred to as inorganic salts. These nutrients are required to facilitate photosynthesis, growth, and reproduction. In soilless plant production all of these essential nutrients come from the fish waste. There are two major categories of nutrients: macronutrients and micronutrients. There are six nutrients that plants need in relatively large amounts. These nutrients are nitrogen, phosphorous, potassium, calcium, magnesium, and sulfur. Leafy vegetables survive well in aquaponics. More than 150 different vegetables, herbs, flowers, and small trees have been grown successfully in aquaponic systems, including research, commercial, and domestic units. The main vegetable plant species grown so far in aquaponic systems are lettuce, beans, zucchini, broccoli, peppers, cucumbers, peas, and spinach. While some herbs, like basil, thyme, cilantro, sage, lemongrass, wheatgrass, oregano, and parsley have been grown in aquaponics. Lettuce cultivation, which has been grown under different densities from 16 to 44 plants per square meter mainly on floating raft system, provide variable yields ranging from 1.4 to 6.5 kg/m^2 per crop (Dediu et al., 2012). Basil is also a widely tested crop, with densities of 8 to 36 plants per square meter, which brings yields of 1.4 to 4.4 kg/m^2 (Rakocy et al., 2004; Pantanella et al., 2011).

There is no waste in aquaponics because all materials are important inputs in the production cycle. As a result, aquaponics aspires to mimic a natural system by making the best use of all nutrients, water, and energy. With almost no application of pesticides, herbicides, hormones, and drugs, we can consider it as an industrial chemical-free production system. Currently, aquaponics is gaining popularity around the world. Several ongoing research in semi-commercial aquaponics and research units are generating results that will help to enable successfully scaled aquaponic systems that can contribute to a new integrated and sustainable food production approach. The potential of these systems is, however, not fully understood. Further research needs to address the nutrient dynamics of different growing media and to optimize system design and nutritional requirement of vegetables in those water bodies with limited dissolved nutrients. The possibilities of this integrated system are quite high, and it can provide sensitive benefits to small holders and big aquaculture enterprises.

8.5.5 Cage Culture

In aquaculture, the popularity of fish culture is an artificially enclosed area of open water bodies is increasing worldwide. Cage and pen culture is being considered as an opportunity to utilize existing reservoirs and lakes with great production potential to enhance fish production and is being posed as an answer to increasing demand for animal protein in the country (Sugunan et al., 2013). The cage is enclosed by bamboo mat or other screen made up of nylon, plastic, or metal at the bottom and sides, sometimes top is also covered. Netting system allows free circulation of water that supplies fresh oxygen for the stocked fish. Cages can be used for both fish fingerling raising and production of table-size fish. This will further reduce the dependence on land-based nurseries and also improve per unit fish productivity and production from reservoirs. The success of cage farming in states like Jharkhand and Chhattisgarh has set a benchmark and proved the potential of cage culture technology in reservoirs of the country. Currently, cage culture is being adopted by more than 12 states in India and yielding promising results. The ICAR-CIFRI has demonstrated and facilitated the implementation of cage culture technology in the country and achieved a production level of 50 kg/m^3 of Sutchi catfish (*Pangasianodon sutchi*) against a moderate stocking of 60 fish per cubic meter (Jha et al., 2013).

Cage culture possibly first originated nearly over 200 years ago in Cambodia. It is traditional in Indonesia. Common carp culture in bamboo cages is practiced in West Java since early the 1940s. Cage culture in open water bodies has spread over 35 countries (Chattopadhyay, 2009). It has become a high-tech fish-farming method in the developed countries, such as Canada, Norway, and Scotland, which raise salmonids; Japan, which raises tuna and yellow fins; and Southern USA, which raises catfish. In India, cage culture was attempted for the first time in the case of air-breathing fish in swamps and for major carp in running water in the rivers Ganga and Yamuna from 1972 to 1976 at Allahabad and common carp, catla, silver carp, grass carp, rohu, murrels, and tilapia in stagnant water bodies in Karnataka (NFDB, 2016).

Today, cage culture is receiving more attention by both researchers and commercial producers. Factors such as increasing consumption of fish, declining stocks of wild fish, and poor farm economy have increased interest in fish production in cages. Cage aquaculture involves the growing of fish in existing water resources while being enclosed in a net cage, which allows free flow of water (Figure 8.6). It is an aquaculture production system made of a floating frame, net materials, and mooring system, such as rope, buoy, and anchor, with a round/square floating net to hold and culture a large number of fish, which can be installed in reservoir, river, lake, or sea. A catwalk and hand rail is built around a battery of floating cages (NFDB, 2016). Economically, cage aquaculture is a low-impact farming practice with high returns and least carbon emission. Cage farms are positioned in such a way to utilize natural currents, which provide the fish with oxygen and other appropriate natural conditions (Manna and Hassan, 2004).

In view of the high production attainable in cage culture system, it can play a significant role in increasing the overall fish production from aquatic resources of India. India has vast inland water resources in the form of ponds and tanks (2.381 m ha), reservoirs (3.153 m ha), floodplain oxbow lakes (0.202 m ha), upland lakes (0.72 m ha), irrigation canals (0.162 m ha), and rivers (29,000 km). These resources are mostly underused and suitable for cage culture of cultivable fish for enhancing fish production. Also, the inshore coastal waters of maritime states have a great potential to raise suitable fish in cages. As the investment is low and requires very little area, this farming system is ideal for small-scale fisherfolk as an alternative income source (Karnatak and Kumar, 2014).

Cage culture can be practiced at various management levels, such as extensive, semi- intensive, and intensive methods. Among them, semi-intensive and intensive culture can be practiced in cages in fairly productive water bodies. In intensive culture system, the fish species are reared exclusively at high stocking density (100–200 fingerlings per square meter) and quality supplemental feeds with >30% crude protein (Chattopadhyay., 2009).

FIGURE 8.6 Installation of a cage in open water.

8.6 Conclusion

These new fish culture technologies are very efficient in sustainable aqua production. Also, there is a complete utilization of nutrients and natural food produced there. In these systems the quality of water is maintained throughout the culture period; hence, the risk of disease spread is minimum. Though in some, the input cost is high, but all are very efficient in producing enormous quantity of aquatic animals in minimum water volume with low input cost.

REFERENCES

Adler PR, Harper JK, Wade EM (2000). Economic analysis of an aquaponic system for the integrated production of rainbow trout and plants. *Int J Rec Aqua* 1:15 –34.

Al-Hafedh YS, Alam A, Beltagi MS (2008). Food production and water conservation in a recirculating aquaponic system in Saudi Arabia at different ratios of fish feed to plants. *J World Aqu Soc* 39:510–520.

Antoniou P, Hamilton JB, Koopman R, Jain B, Holloway GL, Svoronos SA (1990). Effect of temperature and pH on the effective maximum specific growth rate of nitrifying bacteria. *Wat Res* 24(1):97–101.

Avnimelech Y (2009). *Biofloc Technology: A Practical Guide Book*. World Aquaculture Society, 182p.

Avnimelech Y, Kochva M, Diab S (1994). Development of controlled intensive aquaculture systems with a limited water exchange and adjusted carbon to nitrogen ratio. *Bamidgeh* 46:119–131.

Bennett A, Patil P, Kleisner K, Rader, D, Virdin J, Basurto X (2018). Contribution of fisheries to food and nutrition security: Current knowledge, policy and research. *NI Report 18–02*, Durham, NC, Duke University, 45p.

Brown MR, Barrett SM, Volkman JK, Nearhos SP, Nell JA, Allan GL (1996). Biochemical composition of new yeasts and bacteria evaluated as food for bivalve aquaculture. *Aquaculture* 143(3):341–360.

Chattopadhyay DN (2009). Carp seed raising in cages. In: *Recent Trends in Seed Production of Freshwater Finfishes and Shellfishes*, Central Institute of Freshwater Aquaculture, Bhubaneswar, pp. 59–67.

Chen S, Ling J, Blancheton JP (2006). Nitrification kinetics of biofilm as affected by water quality factors. *Aquac Eng* 34:179–197.

Dediu L, Cristea V, Xiaoshuang Z (2012). Waste production and valorization in an integrated aquaponic system with bester and lettuce. *Af J Biot* 11(9):2349–2358.

DeLong DP, Losordo TM, Rakocy JE (2009). *Tank Culture of Tilapia*. SRAC Publication 282, South Regional Aquaculture Center.

DoF (2020). *Handbook on Fisheries Statistics 2020*. Department of Fisheries, Ministry of Fisheries, Animal Husbandry & Dairying, Government of India, 196p.

Dunning RD, Losordo TM, Hobbs AO (1998). *The Economics of Recirculating Tank Systems: A Spreadsheet for Individual Analysis*. SRAC Publication No. 456, 8p.

Ellis T (2001). What is stocking density? *Trout News CEFAS* 32:35–37.

FAO (2011). *Code of Conduct of Responsible Fisheries*. FAO, Rome, 108p.

FAO (2020a). *The State of World Fisheries and Aquaculture: Sustainability in Action*. FAO, Rome, 224p.

FAO (2020b). *FAO Yearbook. Fishery and Aquaculture Statistics 2018*. FAO Rome, 110p.

Fish Base (2021). www.fishbase.org b06/2021.

Gunes FE (2013). Medical use of squalene as a natural antioxidant. *MOSBED* 3(4):220–228.

Haziq Qayoom Lone (2020). *Application of Biofloc Technology for Water Quality Improvement in Fish Production System*. Thesis submitted to GB Pant University of Agric & Tech, Pantnagar 53p.

Helfrich LA, Libey G (1991). *Fish Farming in Recirculating Aquaculture Systems*. Department of Fisheries and Wildlife Sciences, Virginia Tech, VA.

Holm JC, Refstie T, Bo S (1990). The effects of fish density on feeding regime on individual growth rate and mortality in rainbow trout (*Oncorhynchus mykiss*). *Aquaculture* 89:225–232.

Jha BC, Sharma AP, Das AK 2013. Status, scope and potential of enclosure (cage and pen) fish farming in reservoirs and floodplain wetlands. In: VV Sugunan, AP Sharma and BC Jha (eds), *Recent Advances in Culture Based Fisheries in India, Pillay Aquaculture Foundation*, CIFRI, Inland Fisheries Society of India, pp. 157–168.

Jhingran VG (1991). *Fish and Fisheries of India*. Hindustan Publishing Corporation, 727p.

Karnatak, G, Kumar V (2014). Potential of cage aquaculture in Indian reservoirs. *Int J Fish Aq St* 1(6):108–112.

Kazmierczak Jr RF, Caffey RH (1995). Management ability and the economics of recirculating aquaculture production systems. *Mar Res Econ* 10:187–209.

Lekong OI, Kleppe H (2000). Efficiency of nitrification in trickling filters using different filter media. *Aqua Engg* 21:181–199.

Lennard WA, Leonard BV (2006). A comparison of three different hydroponic sub-systems (gravel bed, floating and nutrient film technique) in an aquaponic test system. *Aqua Int* 14:539–550.

Leonard N, Blancheton JP, Guiraud JP (2000). Populations of heterotrophic bacteria in an experimental recirculating aquaculture system. *Aqua Engg* 22(1):109–120.

Losordo TM, Masser MP, Rakocy J (1999). Recirculating aquaculture tank production systems: A review of component options. *SRAC Publication No* 453:12.

Loveless JE, Painter HA (1968). The influence of metal ion concentration and pH value on the growth of a *Nitrosomonas* strain isolated from activated sludge. *J Gen Micro* 52:1–4.

Malone RF, Pfeiffer TJ (2006). Rating fixed film nitrifying biofilters used in recirculating aquaculture systems. *Aqua Engg* 34:389–402.

Manna RK, Hassan MA (2004). Cage culture of fish. *Bulletin No. 132*, Central Inland Fisheries Research Institute, Barrackpore, 26p.

Martins CIM, Eding EH, Verreth JAJ (2010). The effect of recirculating aquaculture systems on the concentrations of heavy metals in culture water and tissues of Nile tilapia *Oreochromis niloticus*. *Food Chem* 126:1001–1005.

Martins CIM, Ochola D, Ende SSW, Eding EH, Verreth JAJ (2009). Is growth retardation present in Nile tilapia *Oreochromis niloticus* cultured in low water exchange recirculating aquaculture systems? *Aquaculture* 298:43–50.

Mohan Deepa, Mente A, Dehghan M (2021). Associations of fish consumption with risk of cardiovascular disease and mortality among individuals with or without vascular disease from 58 countries. *Jama Intern Med* 181(5):631–649.

NFDB (2016). *Guidelines for Cage Culture in Inland Open Water Bodies of India*. National Fisheries Development Board, Hyderabad, 14p.

Pantanella E, Danaher JJ, Rakocy JE, Shultz RC, Bailey DS (2011). Alternative media types for seedling production of lettuce and basil. *Acta Hort (ISHS)* 891:257–264.

Pérez-Fuentes JA, Hernández-Vergara MP, Pérez-Rostro CI, Fogel I (2016). C: N ratios affect nitrogen removal and production of Nile tilapia Oreochromis niloticus raised in a biofloc system under high density cultivation. *Aquaculture* 452:247–251.

Rafatnezhad S, Falahatkar B, Gilani MHT (2008). Effects of stocking density on haematological parameters, growth and fin erosion of great sturgeon (*Huso huso*) juveniles. *Aqua Res* 39:1506–1513.

Rakocy JE, Shultz RC, Bailey DS, Thoman ES (2004). Aquaponic production of tilapia and basil: Comparing a batch and staggered cropping system. *Acta Hort* 648:63–69.

Rawlinson P (2002). The economic efficiencies of partial and intensive recirculation aquaculture systems for Murray cod. In: BA Ingram (ed), *Murray Cod Aquaculture: Now and Into the Future*. Victorian Institute of Animal Science, Attwood, VictoriaAustralia, pp. 17–18.

Ruane NM, Carballo EC, Komen J (2002). Increasing stocking density influences the acute physiological stress response of Common carp *Cyprinus carpio* (L). *Aqua Res* 33:777–784.

Sanni S, Forsberg OI (1996). Modelling pH and carbon dioxide in single-pass seawater aquaculture systems. *Aqua Engg* 15(2):91–110.

Seawright DE, Stickney RR, Walker RB (1998). Nutrient dynamics in integrated aquaculture-hydroponics systems. *Aquaculture* 160:215–237.

Sharma AP, Mishra A (2009). Fisheries in Uttarakhand: Status prospects and development. In: BS Bisht and DP Singh (eds), *Agriculture in Uttarakhand Hills* by Directorate of Experiment Station. GBPUA&T, Pantnagar, pp. 217–226.

Singh IJ, Mishra A (2019). Innovation in aquaculture production. *Indian Farmer's Digest*, April 2019, 4–7.

Sugunan VV, Sharma AP, Jha BC (2013). *Recent Advances in Culture Based Fisheries in India Pillay Aquaculture Foundation Central Inland Fisheries Research Institute Inland Fisheries Society of India*. Hindustan Publishing Corporation, 276p.

Sunitha K, Kavitha K, Darwin CH (2018). Biofloc technology an energy tool for sustainable aquaculture. *Int J Zoo St* 3(1):87–90.

Suprayitno Hartati (1986). *Mannual of Running Water Fish Culture.* ASEAN/SF/86/Manual No. 1, Manila Philippines.

Toma JJ (2012). Limnological study in Dokan lake kurdistan region of Iraq. *J Env Stud* 6:1–12.

Twarowska JG, Westerman PW, Losordo TM (1997). Water treatment and waste characterization evaluation of an intensive recirculating fish production system. *Aqua Engg* 16:133–147.

Tyson RV, Simonne EH, Treadwell DD, White JM, Simonne A (2008). Reconciling pH for ammonia biofiltration and cucumber yield in a recirculating aquaponic system with perlite biofilters. *Hort Sci* 43:719–724.

UN (2015). *Transforming Our World: The 2030 Agenda for Sustainable Development.* United Nations, 41p.

Zhu S, Chen S (2002). The impact of temperature on nitrification rate in fixed film biofilters. *Aqua Engg* 26:221–237.

9

Biodiversity and Conservation of Sheep Genetic Resources in Jammu and Kashmir

Syed Shanaz, Tufail Banday, Shahana Riyaz Tramboo, and Mubashir

CONTENTS

DOI: 10.1201/9781003220398-9

9.1 Introduction

India, being an agrarian country, is bestowed with rich wealth of livestock resources. These livestock resources are usually considered profitable, but they have both positive and negative effects on the natural resource base, public health, social equity, and economic growth (World Bank, 2009). Livestock plays an important role in Indian economy and provides livelihood to about 20.5 million people, contributing about 16% to the income of small farm households as against an average of 14% for all rural households and provides employment to about 8.8% of the population in India. The livestock sector contributes 4.11% of the GDP and 25.6% of the total agriculture GDP. For those farmers who maintain few head of animals, livestock is also a source of subsidiary income. In India, farmers usually follow mixed farming system involving a combination of crop and livestock where the output of one enterprise becomes the input of another enterprise. The landless and people with meager amount of land depend on livestock for utilizing their labor during lean agricultural season. The livestock provides food items such as milk, meat, and eggs for human consumption. India, being the highest milk producer in the world, is producing about 176.34 million tonnes of milk in a year (2017–2018) and 95.22 billion eggs and 7.70 million tonnes of meat in a year. Livestock also contributes in terms of wool, hair, hides, and pellets. India produces about 41.5 million kilograms of wool per annum (Singh, 2020).

Among livestock species, sheep and goats are considered an important livestock species in India, which contribute greatly to the agrarian economy, particularly in areas where crop and dairy farming are not economical, besides playing an important role in the livelihood of a large proportion of farmers and landless laborers (Acharya, 1982). The climatic conditions, rich alpine pastures and host of other natural endowments in the United Territory of Jammu and Kashmir (UT of J&K) create conditions suitable for profitable rearing of sheep and goats (Want *et al.*, 2020). Sheep and goat rearing plays a vital role in socio-economic upliftment of weaker sections of the society, particularly Gujjars, Bakharwals, Chopans, Gaddies, and Changpas (Rather *et al.*, 2020a). Sheep, in particular, form an important component of rural economy, particularly in the arid, semi-arid, and mountainous areas of the country owing to its multifaceted utility for wool, meat, milk, skins, and manure. Manure, a rich source of nitrogen, phosphorus, and potassium, is used to improve soil fertility. It constitutes important source of earning to the shepherds through sale of wool and meat. Sheep plays an important role in the livelihood of a large percentage of small and marginal farmers and landless laborers engaged in sheep rearing (Rather *et al.*, 2020a).

9.2 Sheep Population

Total sheep population in the country is 74.26 million, contributing about 13.8% of the total livestock population (Livestock Census, 2020). The UT of J&K has 3.2 million sheep and ranks sixth with regard to sheep population and second in wool production (Anonymous, 2019). The native breeds of sheep in J&K include Gaddi, Gurez, Bakharwal, Poonchi, Karnah, and Changthangi (Anonymous, 2017). In order to improve production of mutton and wool both in terms of quality and quantity, a vast exotic repository of sheep genetic resources have been imported from different countries time to time. Rambouillet and Dallen Merino were imported from USA in 1951, and after these breeds, Merino, Corriedale, Polwarth, Somalia, South Down, and Soviet Merino were also subsequently imported (Kaushish, 2019).

TABLE 9.1

Sheep Population of Major States (2012 and 2019)

S. no.	States/Uts	Population (in million) 2012	Population (in million) 2019	% Change
1	Telangana	12.8	19.1	48.51
2	Andhra Pradesh	13.6	17.6	30.00
3	Karnataka	9.6	11.1	15.31
4	Rajasthan	9.1	7.9	−12.95
5	Tamil Nadu	4.8	4.5	−5.98
6	Jammu and Kashmir	3.4	3.2	−4.19
7	Maharashtra	2.6	2.7	3.87
8	Gujarat	1.7	1.8	4.66
9	Odisha	1.6	1.3	−19.10
10	Uttar Pradesh	1.4	1.0	−27.25

(*Source:* Livestock Census, 2020)

TABLE 9.2

Performance Traits of Sheep in J&K

	Production	
Year	Mutton (in lakh kg)	Wool (in lakh kg)
1950	7.4	0.89
1990	53.99	15.98
2019	112.1	32.75

(*Source:* Directorate of Sheep Husbandry Kashmir)

9.3 Importance of Sheep in India

India is one among the largest producers of wool in the world, standing ninth in position. Despite 2% global production share, its wool consumption is expected to reach 260 million kg by 2019–2020 (Anonymous, 2021). The wool industries are located mainly in Punjab, Haryana, Rajasthan, Uttar Pradesh, Maharashtra, and Gujarat, among which about 35% of the wool production units are concentrated in Punjab, followed by Maharashtra and Rajasthan (Anonymous, 2021). Sheep are primarily reared for meat (Sharma *et al.*, 2020), and India produces 648,660 tonnes of meat (Department of Animal husbandry).

9.4 Present Status of Sheep and Its Importance in J&K

Among all the states of India, J&K ranks second in wool production, contributing about 18.9% and is also the largest producer of fine wool in the country (Directorate of Sheep Husbandry Kashmir). However, due to non-availability of wool processing facilities in J&K, almost the entire quantity of wool produced in J&K is exported to neighboring states, which is then imported back at enhanced rates (Anonymous, 2020b). The sheep industry provides employment to more than 2,00,000 people, and from them, the wool collected is sent for processing outside the state. The J&K government has taken several initiatives to upgrade the quality of livestock by enhancing wool production through incorporation of appropriate selection and breeding methods. J&K produces about 70–75 lakh kg of raw wool, and about 70% of wool is sold in raw form and large quantity of raw wool is exported to other parts of the country at meager rates. Also, the quality of raw wool gets degraded due to dust and dirt and along with transportation and handling costs reduces the cost of wool (Anonymous, 2020b). Demand for mutton is more in J&K due to larger sections of people consuming it. Sheep produces 112.1 and 32.75 lakh kg of mutton and wool in the state (Table 9.2).

9.5 Sheep Diversity in India

The sheep breeds in India have been classified into four groups, based on agro-ecological regions (Acharya, 1982): (1) north temperate region (NT), (2) northwestern arid and semi-arid region (NWASA), (3) southern peninsular region (SP), and (4) eastern region (ET) (Bhatia and Arora, 2005). The Himalaya mountains constitute the NT region. The NWASA, having an undulating landscape of the sandy desert, consists of scattered hills in vast alluvial plains is the primary region for carpet-wool production and has the second largest population of sheep. The SP region, being a hot and humid coastline and semi-arid central peninsular region, includes sandy tracts, mountain ranges, plateaus, deltas, plains, and even highlands and has the largest sheep population in the country which are primarily maintained for meat. The climate of ET region is primarily hot and humid, apart from a few subtemperate and humid areas of eastern states. Its landscape includes alluvial plains along with undulations, plateaus and tablelands, and Himalaya hill ranges, as well as valleys in the northeastern area. The sheep are primarily maintained for meat and produce extremely coarse and hairy fleece (Sharma *et al.*, 2020).

The adaptability of different Indian breeds to a wide range of agro-climatic zones offers the variability, which provides the opportunity to meet the future demands for food, and also the flexibility to respond to altered markets and needs (Sheriff and Alemaheyu, 2018). The characterization of sheep in India has largely been done using microsatellites, with limited studies utilizing mitochondrial DNA and are mainly restricted to a single population (Singh *et al.*, 2017), with very few studies targeting more number of breeds (Arora *et al.*, 2011a; Arora *et al.*, 2011b). Among the four agro-ecological regions of India, genetic variation and relationships among the majority of recognized sheep breeds have been carried out in NWASA region by employing microsatellite markers (Arora *et al.*, 2011c). For the other three regions, comprehensive knowledge about the sheep breeds is sought for designing suitable and sustainable sheep breeding programs and conservation strategies (Sharma *et al.*, 2020).

9.6 Sheep Genetic Resources of J&K

Sheep and goat rearing is considered as the core activity of rural masses of the J&K State, which is possible due to favorable agro-climatic conditions and availability of lush green alpine pastures. In spite of these advantages, economic returns from sheep and goat rearing are insignificant due to poor productivity of the available genetic material. No concrete sheep developmental program was undertaken except for crossbreeding experiments/trials till 1962. After Sheep Husbandry Department was separated from Animal Husbandry Department, it laid maximum emphasis on crossbreeding program resulting in substantial progress in production of wool and mutton. Knowledge of the indigenous breeds (types or strains of sheep) is essential in order to improve quantity and quality of these traits up to desired level. Initially, no survey of sheep was made in the state in this regard, but efforts made by some researchers are appreciable in this respect (Anonymous, 2020a).

Based on diverse geographical locations, the state has been divided into three distinct regions—the Kashmir region (temperate), Ladakh region (cold arid), and Jammu region (subtropical) (Baba *et al.*, 2011; Mathur, 2015). These agro-climatic zones inhabit a rich genetic resource of small ruminants. Among the 42 registered breeds of sheep in the country, Bakharwal, Changthangi, Gurez, Karnah, and Poonchi constitute the local genetic resource of the state (National Bureau of Animal Genetic Resources, Karnal). Many other breeds, such as Malluk and Purgi, and so on have not been registered yet (Table 9.5). The native sheep breeds are disease-resistant and have sturdiness and presence of fecundity gene. Gurez and Karnah breeds are found in Kashmir, while Bakharwal and Poonchi are occur particularly in Jammu. In the cold arid region of Leh and Kargil, Changthangi sheep are predominantly found there. Despite the harsh climatic conditions prevalent in high altitude arid regions of Ladakh, the Changthangi breed thrives well and produces good quality coarse and long wool which is used by carpet industry. (Kour *et al.*, 2018). Despite the bulky size of the Bhakarwal breed, it is still hardy and sturdy, proving to be the best climber on the mountains (Directorate of Sheep Husbandry, Kashmir). The Gurez breed, being the largest one, shows the better tendency of survival in the cold regions of the valley as

TABLE 9.3

List of Farm Animal Genetic Resources of Jammu and Kashmir

Species	List of registered breeds	Main uses/important traits
Sheep	Bakharwal	Carpet wool
	Gaddi	Adaptation traits
	Poonchi	Diseases resistant
	Gurezi	Easy to manage
	Karnah	
	Malluk	
	Purgi	
	Changthagi	
	Changluk	
	Kashmir Merino	Apparel wool

(*Source:* Rather *et al.*, 2020b)

compared to the exotic breeds. Since 1960s, native breeds were continuously manipulated by breeders due to introduction of Merino and Rambouillet breeds for upgrading the local sheep (Wani, 2014). This might have increased the demand of fine wool in the markets leading to a much higher population of crossbreeds, but the necessity to conserve the native breeds is also alarming. The Bhakarwal breed is found in the Rajouri and Poonch districts of Jammu, while Changthangi is a registered breed in cold arid regions, such as Leh and Kargil, and Malluk and Purgi are the unregistered breeds of Ladakh. Breeds like Malluk and Changluk produce better quality wool for cloth or carpet industry, but they require further characterization (Shergojry *et al.*, 2017). Kashmir Merino, a synthetic breed (not yet registered) was developed around 1960 at the sheep breeding and research farm in Reasi, Jammu, by crossing Gaddi, Poonchi, and Bakharwal ewes with Australian Merino rams, and the females of the F1 generation so produced were bred with Delain rams (imported from USA. Inter se mating of F2 was done till a breed (Kashmir Merino) with stable and uniform characteristics was evolved. The breed is found in every nook and corner of Kashmir Valley is used for the upgradation of local sheep. All native sheep genetic resources are constantly being crossed with Kashmir Merino and Rambouillet, respectively, for fine wool. Crossbreeding is being practiced without giving any regard to conservation of native breeds. While continuing this crossbreeding practice and along with negligence, we may lose the precious native germplasm.

9.7 Risk Status of Genetic Resources

Several factors, like extent of crossbreeding, management system, utility of breed, and rate of decline, under local agro-climatic conditions determine the breed to be at risk. The size of breeding stock, which can be articulated by the number of breeding females, the sex ratio, the number of breeding males, the proportion of females bred to males of the same breed, the trend in population size, and effective population size under a particular production system help to estimate the risk status of a breed. A working rule is that when population size approaches 5,000 breeding females (a total population of about 10,000 animals) the survival risk of the breed should be studied, and appropriate measures would be needed for its conservation. It is suggested that whenever the total number of animals falls below 10,000, one must start preserving semen and embryos (Bhatia and Arora, 2005). The condition may altogether be different where not even initial surveys have been conducted on breeds and the breeding policies are framed generally on population estimates based on census reports of species. Therefore, Nivsarkar (1994) and Acharya (1999) recommended an action plan for determination of risk status and conservation of breeds/strains under Indian conditions. The risk status under Indian conditions are presented in Table 9.4 (Nivsarkar, 1994; Acharya, 1999), and accordingly, the risk status of farm animal genetic resources found in J&K is presented in Table 9.5.

TABLE 9.4

Risk Status Under Indian Conditions

S. no	Population	Risk status
1	50,000 with no serious declining trend	Can be considered a normal population
2	30,000 to 50,000 animals with constant declining trend	Insecure category
3	15,000–30,000	Vulnerable
4	8,000–1,500	Endangered
5	<8,000	Critical and needs immediate conservation

(*Source:* Rather *et al.*, 2020b)

TABLE 9.5

Risk Status of Farm Animal Genetic Resources of Jammu and Kashmir, as well as Reasons and Possible Intervention

Species	N	Breeds	Population	Risk status	Reasons
Sheep	1	Bakharwal	24,575	Vulnerable	• Indiscriminate crossbreeding • Uncontrolled amalgamation with migratory • Flocks • Lack of breeding policies and strategies • Conservation not included in breeding policy • Shrinking of grazing areas
	2	Gaddi	92,194	Normal, decreasing	
	3	Poonchi	2,643	Critical	
	4	Gurezi	17,207	Vulnerable	
	5	Karnah	2,946	Critical	
	6	Malluk	26,328	Insecure	
	7	Purgi	NA	Not documented	
	8	Changthagi	65,115	Normal, but decreasing population	• Smaller flocks scattered over large areas of fragile ecology, which resulted in inbreeding depression and genetic drift • Inbreeding depression and genetic drift • Lack of organized efforts for selection and improvement • Availability of alternate livelihood activities
	9	Kashmir Merino	>11 lakhs		

(*Source:* Rather *et al.*, 2020b)

9.8 Importance of Animal Genetic Resources of J&K

The native farm animal genetic resources of J&K are the backbone of its agrarian economy. They provide milk, meat, eggs, fibers, pelts, hides, manure, draft power, genetic material for research, and germplasm for the development new breeds and strains and are a source of revenue for poor rural communities, insurance against future changes in production system and opportunities to meet future demands. Therefore, losing them will have deleterious effects on their fragile economy and livelihood of poor rural communities. The native animal genetic resources also help to withstand the pressure of change in production circumstances and are adapted to harsh environmental conditions and possess unique traits like fiber fineness in the Changthangi goat is regarded as best in the world. Due to their high adaptation to the agroclimatic conditions, they possess traits like resistance to diseases and to heat and cold. The farm animal genetic resources of J&K are easy to manage on low input methods in terms of feed/fodder, housing, and healthcare, and these convert low-quality feed/fodder in animal products (meat, milk, egg, etc.) efficiently. They have draft ability to work in challenging environmental conditions, as seen in yak, double-humped camel and Zanskari horse work in hypoxic conditions. They provide genetic material for crossbreeding and for the development of new and improved breeds and are important to preserve the cultural and historical value, sustain the inheritance value of livestock, and fulfill the rights of an existing genetic resource to continue to exist (Ganai *et al.*, 2016; Rather *et al.*, 2020b).

9.9 Reasons for the Decline of Farm Animal Genetic Resources of J&K

If concrete steps are not taken, the native animal genetic resources may be lost forever. The important reasons behind the loss or dilution of native breeds include use of non-adapted exotic germplasm through indiscriminate crossbreeding, which arose through improper recommendation and biased and misleading comparisons between the native-bred and the crossbred populations, as a result of which the Kashmir Valley breed has been lost completely and the other genetic resources are getting endangered too. Other reasons include uncontrolled intermixing with nondescript animals, such as sheep breeds and Zanskari horses. Conservation has not been included in the breeding policy until recently. Species hybridization (e.g., yak) is crossed with cattle. There is pasture degradation and shrinking of grazing areas, and there are other alternate livelihood activities on the rise. The availability of commercial animal products everywhere, like the Kashmir Favorolla, Kashmir Anz, Kashmir duck, have gone into eclipse due to the easy availability of broiler meat and eggs in the valley. The genetic potential of native animal genetic resources has not been harvested fully due to the lack of organized breeding programs and basic requirement for genetic improvement (e.g., date recording, which is either nonexistent or poorly recoded). The selection process in sheep breeding programs is not based on performance evaluation. Therefore, the Department of Sheep Husbandry should adopt modern techniques and algorithms for the selection of animals of high genetic merit without any available animal breeder and geneticist, which would serve the purpose. Non-availability or nonexistence of livestock database management systems for real-time decision-making and policy planning and evaluation is also the reason behind loss or dilution of native breeds (Rather *et al.*, 2020b).

9.10 Strategies Needed for Conservation of Farm Animal Genetic Resources

Efforts and strategies needed for conservation at ground level ensures that the farm animal diversity is being maintained for contribution to food production and agricultural productivity through strategic planning and policies for future purposes. The loss of animal genetic resources, particularly from native breeds, means the loss of unique genes/gene combinations associated with adaptability and utilization of low-quality feeds, as well as loss of genetic material for research purposes. Further crossbreeds also fail to exploit their genetic worth and production level goes down after F1 stage, so in order to produce F1, we need native breeds to harvest hybrids with vigor. Native breeds also supply germplasm for the improvement of crossbreeds.

Effective conservation of genetic resources is possible only if the animal genetic resources are recognized and documented adequately, and there is a need of full participation toward conservation efforts of communities keeping the animals, as well as scientists and authorities concerned, like development authorities, universities, and animal and sheep husbandry departments. Proper characterization (phenotypic and molecular) to assess within and between breed diversity, followed by documentation and proper management, is the first and most important step for efficient conservation policy for animal genetic resources. So all the native undocumented breeds should be characterized and studied for performance evaluation on priority basis. The native animal genetic resources can be made self-sustainable through selection and breeding, and modern breeding techniques (ranking of sires and BLUP) and biotechnological tools (MOET and embryo transfer technology) can be utilized to develop highly productive livestock populations, which can sustain the modern economic pressures. However, for modern breeding techniques, reliable and accurate data is needed, which should be uniform and digital, and the data management in the farms should be done with the help of ICT-based management information systems and modern breeding tools to warrant reliable and accurate data recording and analysis. The crossbreeding should be restricted to nondescript populations only. The pastures and grasslands need to be improved and protected and all the encroached grasslands, pastures, and meadows should be protected from illegal encroachments. All sheep breeds should be kept at government sheep-breeding farms in respective niche areas for genetic improvement and conservation under an open nucleus breeding system (ONBS) for bidirectional gene flow. Therefore, all sheep-breeding farms must have a dual mandate genetic improvement of crossbred germplasm and genetic improvement conservation of native germplasm. The

TABLE 9.6

Possible Interventions to Conserve Farm Animal Genetic Resources of J&K

Species	Breeds	Possible intervention
Sheep	Bakharwal Gaddi Poonchi Gurezi Karnah Malluk Purgi Changthangi	• Halt crossbreeding in respective breeding tracts • Establish breeding farms in the respective ecological niches for its genetic improvement • *In situ* conservation at government sheep-breeding farms in respective niches • Selection for pure breeding under ONBS • Molecular characterization for diversity
	Kashmir Merino	• Registration and documentation

(*Source:* Rather *et al.*, 2020b)

crossed animals need to be improved through inter se mating with stabilized exotic inheritance levels of 52–62.5% for semi-intensive and 62–75% for intensive production systems as increasing exotic blood to higher levels deteriorates the adaptability traits. The crossbreeding should be only practiced for nondescript populations and purebred native breeds should be preserved for future to harvest heterosis and for breed development of new breeds. Progressive farmers' flocks need to be linked with nucleus government breeding farms to allow the transfer of elite animals to and from the nucleus. The Department of Sheep Husbandry should boost farmers to adopt systematic feeding and breeding practices. Education, incentives, subsidies on feed and fodders should be provided to the farmers for rearing of native breeds. State agricultural universities should act as nodal agencies to conserve all animal genetic resources of J&K, and all local Kashmiri breeds should be characterized and studied for performance evaluation and selected for genetic improvement. Malra, Bakharwal, Purgi, Malluk, and so on need to be characterized on same lines. Some possible interventions for the conservation of animal genetic resources in J&K are given in Table 9.6 (Rather *et al.*, 2020b).

9.11 Various Methods of Conservation

9.11.1 *In Situ* Conservation

It involves the maintenance of live populations of animals in their adaptive environment, and these continue to evolve and develop for more sustainable use. The active *in situ* conservation is equivalent to breed development by conducting well designed animal breeding programs, while the passive *in situ* conservation is concerned with the maintenance of live animal breeds at risk of loss in their ecosystem. Therefore, a strict national policy should be brought into enforcement not only to preserve the live animals within their native environment but also to make them self-sustainable under their natural ecosystem. *In situ* conservation has several advantages as the breeds can gradually adapt to changing environmental conditions. Improvement through selective breeding can be carried out by genetic improvement and sustainable management, institutional flocks, farmers' flocks, pastoralists' flocks, and stack-holders' involvement.

a. *Genetic Improvement and Sustainable Management*

Improvement of the breed through selection and breeding can help in making it self-sustainable. There are a number of biotechnological tools (MOET and embryo transfer technology), which can help in development of a highly productive population of sheep genetic resources, capable of sustaining the economic pressures. Nevertheless, ram selection for weights and wool are the accepted and proven methodologies for achieving desired genetic gains (Taneja and Bhat, 1999).

b. Institutional Flocks

Institutional flocks managed under scientific lines can maintain a minimum breedable population required in recreation of breed for future and also supply superior rams for conservation of the breed. Smith (1984) provides estimates of the minimum size of a breeding unit and the replacement rate for sheep species if inbreeding levels are to be kept at about 0.2% per year (Table 10). A tolerable inbreeding level is 1% per generation for which a herd size of hundred animals is necessary (Hodges, 1991). Replacement is one of the most important factors in deciding the minimum number which must be regulated to preserve the gene structure of the given population as far as possible.

The rules for placement are as follows:

1. Every breeding animal must be replaced by his/her progeny.
2. The percentage of the known or visible characteristics must be kept approximately around the foundation frequency.
3. Possibly the immunogenetics characteristics and DNA levels polymorphism data should also be taken into consideration to the extent possible.
4. The generation interval must be kept for as long as possible according to the breed or species.
5. The number of males in consecutive years must be as many as possible.

In India, besides having a large network of infrastructural facilities in terms of research institutes/universities/central and state animal-breeding farms and artificial insemination (AI) centers, the only way of improvement was through crossbreeding. Therefore, these institutional herds need to be strengthened both in terms of superior germplasm and by modern tools and techniques (Bhatia and Arora, 2005).

c. Farmers' Flocks

Involvement of the farmers who have been keeping a particular breed for years is highly desirable. The hindrances against preservation are the cost and the modern society may be reluctant to fund project like conservation from which little economic or financial return can be expected in near future. Under Indian farming system a number of indigenous breeds are to stay in for long run due to zero input requirement and disease resistance. Progressive farmers need to be identified and encouraged to conserve these breeds and compensate them for the low returns, as a national obligation for maintenance of sheep biodiversity. Besides financial assistance, increased extension services, provision of improved breeding rams, public awareness, and mass movement, incentives and awards are suggested to meet the challenge (Khan, 2001).

d. Pastoralists' Flocks

Pastoralists keep animals on natural grazing areas and usually have a highly complex indigenous knowledge with regard to animal breeding. These breeding strategies then result in production of animals, which are not only able to survive and reproduce in hostile environments but also fairly productive due to presence of closed gene pools. Pastoral breeds are usually categorized as genetically superior by farmers (George, 1985). Since pastoralists keep animals under wild conditions and without much protection against the elements and climatic extremes, their breeds may carry fitness traits of potential interest for maintaining the vitality of high-performance breeds. Pastoralists play an important role in the conservation of indigenous livestock. For many pastoralists, these farm animal genetic resources are the basis of their cultural identity and they have a moral and social attachment to them (Bhatia and Arora, 2005).

e. *Stack-Holders' Involvement*

In India, sheep are raised by farmers and no data recording system is followed in field; thus, proper evaluation of indigenous animals for further improvement at present is not feasible. It is necessary to form the breed societies in India for the conservation of indigenous breeds. Moreover, the improvement of sheep breeds through selection of outstanding animals true to their breed types from the larger population maintained in the breeding tracts can be simultaneously taken up in collaboration with these breed societies (Acharya, 1999). Dairy cooperative network in India has initiated breed improvement programs in cattle and buffaloes for milk, besides marketing their products at reasonable costs. This model is being replicated for sheep and goats also (Taneja and Bhat, 1999).

9.11.2 *Ex Situ* Conservation

It involves conservation of genetic material out of the environment in which it developed (FAO, 2000).

a. *In Vivo Methods*

Ex situ conservation of live animals in the form of an organized herd maintained in a research institution, state-owned sheep farm, zoo, or breeding safari involves *in vivo* conservation. Keeping large flock of relatively less productive breeds is not feasible on economic grounds. In small populations, inbreeding and appearance of deleterious genetic defects are commonly noticed. It is very important to breed the population in such a manner so that the inbreeding rate is kept at minimal level and production performance at higher level to make the breed self-sustaining. The population size of breeding females and males generated by natural mating or through AI can be maintained either under scientifically managed farms of organized sectors or breeding tracts or under breeding safaris/parks. The biggest limitation of conservation of breeds outside their home tract is the ill effects due to inbreeding. Thus, population size is very important consideration, which depends on the ratio of males and females under different systems of selection (Bhatia and Arora, 2005).

Livestock Breed Safari

It has been globally recognized that there may be a place for the animal resources on the pattern of wild animals. This may help preserving the natural habitats, which include all species of plants, animals, and other organisms. If managed appropriately, this type of natural system can be an ideal example for conserving the precious germplasm of domestic breed of livestock, including sheep. Moreover, it should be self-sustaining unit. This approach, in the form of amusement parks, is gaining popularity in developed countries with a tourist industry. They give opportunity to urban people to get an impression of diverse population of distinct breeds of livestock species.

b. *In Vitro Methods*

Cryo-conservation of wide variety of living cells or tissues for long periods of time (sperms, oocytes, embryos, DNA, etc.) comprises *in vitro ex situ* conservation. The basic objectives of *in vitro* conservation are as follows:

1. Regeneration of endangered breeds
2. Development of new breed s
3. Supporting the *in vivo* populations
4. Research for determining the effect of single major gene
5. DNA studies and genome mapping (Bhatia and Arora, 2005)

Sperm and Ova

In sheep the deep-freezing of semen in various forms has been standardized. For the successful conservation of breed, a minimum of 30 rams, each contributing around 500 doses of semen, would be essential. In order to avoid any risk, the semen should be stored in frozen conditions at least in two locations. It has been documented in several studies that sheep oocytes, like oocytes of other farm animals, can be stored and fertilized *in vitro* after thawing as usual (Khan, 2001).

Embryos

This is an excellent tool of conservation as all the genetic information is stored in one diploid zygote, which can give rise to a new progeny. Embryos are usually frozen between 30- and 120-cell stage in 0.25 ml straws. These straws can be stored at −198°C for long-term storage and can be implanted successfully in the recipients. With limited resources in terms of facilities and trained manpower at some places this technique can be utilized for the *ex situ* conservation of only declining/endangered breeds of sheep. However, it is still relatively expensive to obtain embryos, and their cryopreservation technique needs further refinement for economical use (Bhatia and Arora, 2005).

Cloning

Currently, cloning has created revolutionary opportunities in both animal breeding and research by development of procedure for nuclear transfer in sheep, where unlimited number of cloned animals are bred by nuclear transfer, using somatic cells cultured *in vitro*. This technology shall hold for conservation of available genetic diversity of threatened genetic resources (Anon, 1998; Cunninghum, 1999). Cloning can be used to produce identical copies of elite stud rams or ewes, more rapidly than conventional breeding. Scientists are using the technique to improve the quality of wool and meat.

DNA

Conservation of breeds through storage of DNA has the advantage of disease-free transportation across the countries, but this method has disadvantages that genome maps are not yet available to identify the sequences of DNA that are responsible for specific traits in the highly performing animal, and the use of stored DNA to recreate an animal with specific traits is yet to be achieved because DNA reinsertion techniques with animal cells still produce random results. Other new methods of conservation include establishment of embryonic stem cell lines and conservation of somatic cells. It is presumed that when fully developed, these techniques will provide an alternative approach to the preservation of genetic material of endangered breeds in developing countries. The other important aspects of conservation of sheep genetic resources involve establishment/strengthening of a national animal data bank, a national gene bank, a national animal conservation board, a regional gene cum data bank, and a global data bank (Bhatia and Arora, 2005).

9.12 Agencies for Improvement and Conservation Programs

Several agencies, including private farmers, must be involved in facilitating the various steps of conservation and mobilization of individual enthusiasts.

1. National focal point
2. Central government
3. ICAR Institutes and Agriculture Universities
4. State government
5. Non-governmental organizations (NGOs)
6. Private companies

9.12.1 National Focal Point

National Bureau of Animal Genetic Resources (NBAGR) has now basic infrastructure to meet all the requirements of becoming the national focal point for the conservation programs on all domesticated livestock and poultry breeds and has recently developed and submitted country report on the guidelines of the Intergovernmental Technical Working Group on Animal Genetic Resources (ITWG on AnGR, FAO, 2000).

9.12.2 Central Government

The overall responsibility and control for the conservation of domestic livestock breeds is within the purview of the Department of Animal Husbandry and Dairying, the Ministry of Agriculture, the government of India (http://dahd.nic.in/sheep.htm). Other ministries, such as the Ministry of Environment and Natural Resources or the Ministry of Commerce and Industry, keep necessary liaisons and coordination with other related ministries and government bodies to realize effective policy, planning, and operations. The government influences conservation programs through budget allocation, mandate the state farms to keep the breeds at risk besides the moneymaking commercial flocks or herds in order to maintain national heritage. The population of various species of purebred animals has come down to the category of threatened breeds in the country. The farms or farmers' units, in their respective breeding tracts, should be provided with 100% central assistance along with active participation of state governments and NGOs for those breeds in which population declines to 10,000. A new centrally sponsored scheme for conservation of such threatened breeds has been started by the government of India during Tenth Five-Year Plan with a budget outlay of Rs. 1,500 lakhs and a sum of Rs. 295.50 lakhs has been released to the State of Gujarat, Karnataka, Punjab, Rajasthan, and Tripura in 2002–2003 for scheme implementation (Bhatia and Arora, 2005).

9.12.3 State Government

As the development activities are involved in maintenance of livestock in the state, implementation of breeding programs and providing health to all the animals are chiefly looked after by state governments, and these should develop and implement technically sound breeding programs for the improvement of indigenous breeds. The artificial insemination network of the Animal Husbandry Department should be effectively strengthened to provide extensive breeding services to cover the Desi breeds using semen from genetically superior rams for the improvement of animals under field conditions. Survey in the breeding tracts of indigenous breeds to know the present status, demographic and geographic distribution may also be entrusted with state animal husbandry departments. The NER Databank (http://databank), brought in association with Nation Informatics Centre, Assam, and North Eastern Council by North Eastern Development Finance Corporation Ltd., provides information on sheep population, facilities available, and production in entire the North Eastern state as a whole. (Bhatia and Arora, 2005). Such activities should be extended to J&K also.

9.12.4 ICAR Institute and Agriculture Universities

Universities and research institutes, as public institutions, could be entrusted with the conservation of endangered breeds and maintaining them as control populations for research, teaching/training, and development aspects.

9.12.5 Non-Governmental Organizations (NGOs)

Sheep-related activities are an important component of NGO activities in India and play a crucial role to reach indigenous communities for establish linkages between them and the national bodies responsible for domestic animal diversity. Voluntary or non-governmental organizations, like Neighbors, Dharwad, PETA, and Livestock Heritage of India assist farmers in keeping rare breeds, extending appropriate technologies, upgrading skills, improving the livelihood of marginalized farmers and shepherds by creating awareness in simple improved sheep husbandry practices, raising financial assistance by reducing high mortality among sheep, and also enforcing animal protection laws. One NGO-based initiative is the LIFE Network/Movement for Peoples' Conservation of Domestic Animal Diversity, which was approved the Sadri Declaration (2000) at a recent workshop in India. The League for Pastoral Peoples is an advocacy and support group for pastoralists who depend on common property resources and are engaged in researching and working with pastoral communities. Several breed associations, like Andhra Pradesh Sheep and Goat Rearers Association, Kurma Sheep Breeders, and Vembur Sheep Breeders have been launched by active involvement of NGOs for leading and motivating the farmers' movements (Alsipura Statement, 2002).

9.12.6 Private Companies

Commercial breeding companies, processing companies, and agriculture support services may have become more interested and are increasingly involved in conservation activities. The role of these organizations is considered more important in poultry, sheep, and goat keeping, which are related with commercial production of meat and meat products. They can provide the necessary research and development support, financial assistance, and marketing of products (Bhatia and Arora, 2005).

9.13 Evaluation, Conservation, and Improvement Programs in Sheep Genetic Resources

Various programs being carried out in India are as under the projects at NBAGR. During last 20 years, concerted efforts have been made to create awareness among all the developmental organizations in the country about the usage of indigenous animal genetic resources, including phenotypic and genetic characterization and strategies outlined for their conservation and sustainable management, as detailed in the following sections.

9.13.1 National Agricultural Technology Programs (NATPs)

Animal genetic resource biodiversity:

- To generate information on population status, breed utility, management practices, and farming system, through systematic survey
- To conserve breeds in their native tracts
- To preserve germplasm for posterity using cryopreservation
- To molecularly characterize breeds using microsatellite molecular markers
- To characterize commercially and economically important genes

Genetic characterization and conservation of important sheep and goat breeds of the arid zone:

- To generate information on population trends, management practices, socio-economic importance, performance levels, breed utility, and farming system from the breeding tract of the breed
- To identify the microsatellite markers specific for each species for genetic characterization
- To identify the unique characteristics of the breed
- To develop the complete breed descriptor of each breed
- To develop the technology for conservation of endangered/declining breeds of sheep and goats for posterity

Integrated national agricultural resources information system:

- To integrate the existing database on animal genetic resources in the country
- To design and develop a database on animal genetic resources in the country
- To develop a resource map on animal genetic resources in the country through GiaS

- To perform a molecular-marker-based pilot genetic test for fecundity genes in indigenous sheep and goat breeds
- To standardize/optimize the microsatellite/molecular-marker-based genetic test (PCR based) for fecundity genes
- To screen DNA markers linked to prolificacy genes in indigenous sheep and goat breeds (Bhatia and Arora, 2005)

9.13.2 ICAR's SRC Project at NBAGR

Molecular genetic characterization of indigenous breeds of farm animals/sheep species:

- To characterize and establish genetic relationships among various breeds of sheep using microsatellite markers

9.13.3 Network Project on Animal Genetic Resources at NBAGR

The project aims at genetic characterization and conservation of indigenous breeds of livestock and poultry. The project has been in operation since the VIII plan involving various agencies, such as state agriculture universities, state animal husbandry departments, and NGOs, with the coordinating unit at NBAGR. Two core laboratories stationed at TANUVAS, Chennai, and GAU, Anand, were identified to carry out the genetic characterization/distancing work based on microsatellite markers for the southern and western zones, respectively, while NBAGR was identified to cater the needs of the northern zone. The objectives of the project are as follows:

- To characterize the breeds in terms of both qualitative and quantitative traits
- To study the molecular genetic characterization of the breed
- To develop breed descriptors
- To conserve the germplasm of elite/unique animals/birds (Bhatia and Arora, 2005)

9.13.4 Network Project in Sheep Improvement at Central Sheep and Wool Research Institute (CSWRI), Avikanagar

All India Coordinated Research Project (AICRP) on sheep breeding was converted into the Network Project on Sheep Improvement (NWPSI) with effect from 1.4.90. The breeding policy for NWPSI was selective breeding in indigenous breeds. There are five farm-based cooperating units and two field based cooperating units of project in addition to project coordination cell situated at CSWRI, Avikanagar (Bhatia and Arora, 2005).

9.13.5 National Ram/Buck Production Program

The centrally sponsored scheme of National Ram/Buck Production Program provides scope to states of strengthening farms of indigenous sheep and goat breeds and for distribution of rams and bucks of such breeds to the farmers for breeding purposes. Breeding policy advocated by the government for sheep envisages selective breeding of indigenous breeds for their improvement (Botabyal, 1999).

9.14 Efforts Made So Far at NBAGR for the Evaluation and Conservation of Sheep Genetic Resources

Regarding database management, the data bank on census, literature, germplasm resources and breed characteristics has been developed in ovines.

9.15 Phenotypic Characterization

Twenty-eight breeds of sheep covered for phenotypic characterization include Bandur, Bellary, Bakharwal, Bonpala, Changthangi, Chokla, Coimbatore, Chottanagpuri, Deccani, Gaddi, Garole, Gurez, Jalauni, Jaisalmeri, Hassan, Karnah, Kenguri, Kheri, Magra, Mandya, Malpura, Marwari, Mecheri, Muzaffarnagari, Nali, Pugal, Rampur-Bushair, and Sonadi.

9.16 Molecular Genetic Characterization

Nineteen breeds covered under this program are Bandur, Bellary, Bakharwal, Chokla, Gaddi, Garole, Gurez, Hassan, Jaisalmeri, Kenguri, Karnah, Kheri, Magra, Malpura, Marwari, Muzaffarnagari, Nali, Pugal, and Sonadi.

9.17 Recommendations

The various recommendations made for the evaluation of sheep biodiversity and conservation of sheep genetic resources are as follows:

1. The breed improvement policy can be framed properly if information on the status of breed is available as livestock census provides information only on sex ratio, population of different age groups, and production status. The animal census, therefore, should be conducted on breed basis so that the population of sheep breeds showing declining trend can be monitored and conservation programs are undertaken accordingly.
2. The status of indigenous sheep and their performance should be brought out in the form of newsletters in the form of periodic publications. A web page about indigenous breeds may also help in popularizing these breeds.
3. For bringing out national watch list, efforts should be carried out and the national watch list should be updated periodically with additional and latest information sent by other related agencies, which will ensure better use of information on both genetic characterization and census data, in addition to restriction on repetitive work.
4. The biggest task of characterizing the indigenous sheep genetic resources using FAO recommended markers should be accomplished on priority basis in a time bound frame, which could prove to be of utmost importance in deciding priorities and further planning strategies for conservation.
5. Funding and scientific support for the conservation and sustainable utilization of sheep genetic resources should be provided to various breed societies.
6. Awareness and interest in rare breeds should be generated by conservationists and voluntary agencies.
7. In order to determine the basic norms for a breed registration, an expert panel should be appointed at national level.
8. Security of minimum population size of pure indigenous breed must be ensured by law while introducing exotic breeds.
9. The live animal gene-banks (*in situ* conservation) should be established in the native ecology of the breed.
10. Financial support and capacity-building must be extended to appropriately qualified rural development NGOs in order to assist indigenous communities in conservation/development and facilitate interaction between them and scientists, bureaucrats, and policymakers.

11. It is presumed that with the increase in human population, the consumption of meat and milk in developing countries will rise exponentially, and this demand will be then met by expansion of industrialized livestock production into these countries (IFPRI, 2000). This expansion by industrialized livestock companies in these developing countries must be stopped (Blair, 1994); otherwise, it will have negative consequences for all marginal livestock keepers and for domestic animal diversity. ICAR, the state Animal Husbandry Department, agricultural universities, and NGOs have initiated genetic improvement and conservation programs for the small ruminants, but strong linkages need to be established for achievement of conservation goals. Moreover, timely and infrastructural facilities should be made available to make the conservation a success (Bhatia and Arora, 2005).

9.18 Conservation and Its Rationale Approach to Conservation of Sheep Genetic Resources

Approach to conservation of sheep genetic resources involve combination of a number of integrally related components, such as the following:

- Monitoring and describing of existing animal genetic resources
- Breed characterization at the molecular level to assess between breed diversity in order to maximize cost-effectiveness of management
- Accessible documentation
- Informed use
- Appropriate conservation *in situ* and/or *ex situ*
- National watch list (Bhatia and Arora, 2005)

9.19 Conclusion

There is a rich biodiversity among the indigenous sheep breeds of India reflected by more than 40 listed breeds of sheep. On the basis of phenotypic/major products and geographical locations, four types/groups have been recognized. Molecular genetic characterization, currently underway, will help to characterize and clarify the relationships among the breeds of existing ovine biodiversity—a major prerequisite in deciding priorities for their conservation and improvement programs. J&K also possesses a rich animal biodiversity. These unique native breeds provide a livelihood to the locals and are also a part of the fragile ecosystem of the region. The future of sheep in J&K lies in the appropriate approaches to conservation combining a number of integrally related components and effective action programs approached holistically for successful conservation of sheep genetic resources at national level. Therefore, their conservation through programs and schemes is the need of the hour for sustainable development, meeting the demands in future, preservation of cultural heritage, and improving the productivity of the precious native germplasm for biodiversity purposes.

REFERENCES

Acharya, R. M. 1982. Sheep and goat breeds of India. *FAO Animal Health and Production Paper No. 30*, 2: 190.

Acharya, R. M. 1999. Management and conservation of livestock genetic resources. *Indian Journal of Animal Genetics and Breeding*, 21(1–2): 7–19.

Alsipura Statement. 2002. *Alsipura Declaration from India Herder Conference*. www.pastoralpeoples.org/alsipuradecl.htm

Anon. 1998. *The Implication of Reversible DNA Quiescence and Somatic Cloning for the Conservation of Animal Genetic Resources*. FAO Meeting Report. www.fao.org/dad-is

Anonymous. 2004. *Directorate of Sheep Husbandry Kashmir Division*, Government of Jammu and Kashmir.

Anonymous. 2017. *Livestock Census Report. Animal Husbandry Statistics Division, Department of Animal Husbandry, Dairying and Fisheries*, Government of India, New Delhi.

Anonymous. 2019. *Directorate of Sheep Husbandry Kashmir Division*, Government of Jammu and Kashmir.

Anonymous. 2020a. *Wool Industry in India*, Indian Economy.

Anonymous. 2020b. *Draft Wool Processing, Handloom*, Handicrafts Policy 2020 Jammu and Kashmir.

Anonymous. 2021. *Wool and Woolen Textile Exports from India*. Indian Brand Equity Foundation. www.ibef.org.

Arora, R., Bhatia, S., Mishra, B.P., Jain, A. and Prakash, B. 2011b. Diversity analysis of sheep breeds from Southern peninsular and Eastern regions of India. *Tropical Animal Health Production*, 43: 401–408.

Arora, R., Bhatia, S., Mishra, B.P. and Joshi, B.K. 2011a. Population structure in Indian sheep ascertained using microsatellite information. *Animal Genetics*, 42(3): 242–250.

Arora, R., Bhatia, S., Yadav, D.K. and Mishra, B.P. 2011c. Current genetic profile of sheep breeds/populations from North-western semi-arid zone of India. *Livestock Sciences*, 135: 193–198.

Baba, S.H., Wani, M.H. and Zargar, B.A. 2011. Dynamics and sustainability of livestock sector in Jammu & Kashmir. *Agricultural Economics Research Review*, 24: 119–132.

Bhatia, S. and Arora, R. 2005. Biodiversity and conservation of Indian sheep genetic resources-an overview. *Asian-Australas. Journal of Animal Sciences*, 18: 1387–1402.

Blair, H.T. 1994. Domestic livestock conservation: Is it achieving its goal. In: *Proceedings of the Third Global Conference on Conservation of Domestic Animal Genetic Resources*, 1–3 August 1994, Queen's University in Kingston, Ontario, Canada. Rare Breeds International.

Botabyal, A. 1999. Management of India's programme of management and conservation of livestock genetic resources. *The Indian Journal of Animal Genetics and Breeding*, 21(1): 27–43.

Cunninghum, E.P. 1999. Recent developments in biotechnology as they are related to animal genetic resources for food and agriculture. *Background Study Paper No. 10*. Commission Genetic Resources for Food and Agriculture.

FAO. 2000. *World Watch List for Domestic Animal Diversity*. 3rd edition (Ed. Beate D. Scherf). FAO of United Nations, Rome, Italy.

Ganai, N.A., Bhat, Z.A. and Shiekh, F.A. 2016. Policy planning for conservation and sustainable use of farm animal genetic resources in J&K. National Symposium on Policy Planning for Livelihood Security through Domestic Animal Biodiversity & XIII Annual Convention of Society for Conservation of Domestic Animal Biodiversity. *Compendium*, 2: 84–94.

George, S. 1985. Nomadic cattle breeders and dairy policy in India. *Nomadic Peoples*, 19: 1–19.

Hodges, J. 1991. Sustainable development of animal genetic resources. *World Animal Review*, 68(3). www.fao.org/ag/Aga/AGAP/FRG/FEEDback/War/u4900b/u4900b04.

IFPRI. 2000. (International Food Policy Research Institute). www.cgiar.org/ifpri

Kaushish, S.K. 2019. *Sheep Production in Tropics and Subtropics*. Scientific Publishers, India.

Khan, B.U. 2001. Conservation and management of sheep genetic resources of India. In: *Proceedings of Summer School on Recent advances in Characterization and Conservation of Animal Genetic Resources*, 16th July–6th August. NBAGR, Karnal, pp. 162–176.

Kour, G., Singh, A., Kumar, P. and Kumar, D. 2018. An overview of diversified animal genetic resources in the Indian State of Jammu and Kashmir. *International Journal of Current Microbiology and Applied Sciences*, 7(10): 3113–3121.

Livestock Census. 2020. FAO.

Mathur, B.K. 2015. Strategies of livestock feeding and health management in arid regions of India. *Annals of Arid Zone*, 54(3&4): 95–108.

Nivsarkar, A.E. 1994. Small ruminants and an action plan for conservation in India. In: *Domestic Animal Diversity* (Ed. R. Sahai and R.K. Vijh). SBI Publications, Karnal, India, pp. 49–57.

Rather, M.A., Bashir, I., Shah, M.M., Ahanger, S.A. and Hamadani, A. 2020a. A brief overview of sheep husbandry in Jammu and Kashmir. *International Journal of Current Microbiology and Applied Sciences*, 9(5): 1644–1651.

Rather, M.A., Shanaz, S., Ganai, N.A. and Hamadani, A. 2020b. Status of farm animal genetic resources of Jammu and Kashmir: A review. *International Journal of Livestock Research*, 10(4): Apr'2020. eISSN: 2277–1964.

Sadri Declaration. 2000. *Proceedings International Congress and Workshop on Local Livestock Breeds for Sustainable Rural Livelihoods*. November 2000. Udaipur and Sadri, India, pp. 1–4.

Sharma, R., Ahlawat, S., Sharma, H., Sharma, P., Panchal, P., Arora, R. and Tantia, M.S. 2020. Microsatellite and mitochondrial DNA analyses unveil the genetic structure of native sheep breeds from three major agro-ecological regions of India. *Scientific Reports*, 10: 20422. https://doi.org/10.1038/s41598-020-77480-6

Shergojry, S.A., Akhoon, Z.A., Mubarak, T. and Namgyal, D. 2017. Socio-economic impact of livestock in tribal areas of Leh. *Journal of Krishi Vigyan*, 6(1): 187–190.

Sheriff, O. and Alemayehu, K. 2018. Genetic diversity studies using microsatellite markers and their contribution in supporting sustainable sheep breeding programs: A review. *Cogent Food and Agriculture*. 4: 1459062.

Singh, R. 2020. *Importance of Livestock in Indian Economy*.

Singh, S., Mishra, A.K., Vohra, V., Raja, K.N., Singh, Y., Singh, K., Ganguly, I. and Arora, R. 2017. Microsatellite based genetic diversity estimation in Kajali sheep and its phylogenetic relationship with other indigenous sheep breeds. *Indian Journal of Animal Sciences*, 87: 1097–1101.

Smith, C. 1984. Genetic aspects of conservation in farm animals. *Livestock Production Science*, 11: 37–48.

Taneja, V.K. and Bhat, P.N. 1999. Livestock genetic resources in Asia-issues, constraints, strategies for conservation, management and cost framework. *The Indian Journal of Animal Genetics and Breeding*, 21(1): 1–6.

Wani, S.A., Shaheen, F.A., Baba, S.H., Naqash, F. and Manzoor, M. (2014). Value Chains for livestock products in Himalayan mountains: Studies from Jammu and Kashmir. *Indian Journal of Agricultural Economics*, 69(3): 280–289.

Want, Q.H., Banday, M.T. and Adil, S. 2020. A study on constraints perceived by Sheep farmers of District Srinagar, Jammu And Kashmir. *Animal Science Quarterly*, 1(1): 26–30.

World Bank. 2009. *Minding the Stock: Bringing Public Policy to Bear on Livestock Sector Development*. Washington, DC. Report no. 44010-GLB.

10

Fish Biodiversity of Jammu and Kashmir and Conservation Measures

Masood H. Balkhi and Farooz Ahmad Bhat

CONTENTS

10.1 Introduction

Jammu and Kashmir (J&K) is bestowed with plenty of aquatic ecosystems that harbor economically important bioresources, including fishery resources. The erstwhile state has three agro-climatic regions—temperate Kashmir, tropical Jammu, and cold arid Ladakh, differing in altitude, catchment, and watershed. In the Kashmir Valley, the Jhelum River, with its numerous feeder streams, lakes, and wetlands, forms an important fishery resource, while in Jammu and Ladakh regions, the role is played by the Chenab and its tributaries and the Indus and its tributaries respectively. The present fish production in the state is 21,350 tons, which includes 650 tons of trout (J&K Fisheries Dept., 2021). Fish is preferred over meat due to its health benefits in view of the presence of omega-3 fatty acids that helps in checking the cholesterol level and protecting the heart. In order to meet the demand from growing population of the state, considerable quantities of fish are also imported from the neighboring states. However, if the huge fishery resources are properly managed, the production of fish in J&K can be increased manifold to meet the requirement of the people.

J&K is rich in aquatic resources, ranging from roadside ditches, ponds, pools, water-flows, fountains, wetlands, springs, and rivers to the voluminous lakes in the plains and the high altitudes. These water bodies play a great role in the socio-economic status of the people of J&K and produce a variety of flora and fauna. The ichthyofauna of Kashmir is appreciably different from rest of the country and is mainly represented by the cold-water schizothoracine group. About 41 species of fish are reported from Kashmir, majority of which belong to the genus *Schizothorax*. The present research work believes presence of only 23 species from Kashmir. About 100 species of fish are recorded from Jammu waters. The validity of many of these species needs to be ascertained, as there exists a lot of confusion in the present status of the exact number of fish species in J&K. Therefore, fresh studies are warranted to update the present status of ichthyofauna in the state in general and Kashmir in particular. *Barilius* spp. And the golden mahseer (*Tor putitora*) dominate the fish catch composition in the Tawi River system and the

DOI: 10.1201/9781003220398-10

Chenab River system, respectively, in Jammu, while in Kashmir the catches are dominated by the common carp in the lacustrine waters and the snow trout in the riverine waters.

It is observed that some of the *Schizothorax* fish species have almost disappeared from the scenario. A few species have either been synonymized or renamed by some recent workers. The population of schizothoracine fish has drastically decreased over the years either due to the introduction of exotics or due to the anthropogenic pressures. *Schizothorax esocinus* that was once reported abundantly in Dal Lake is poorly represented in the lake. Similarly, almost all species of schizothoracine fish have been wiped off from Khushal Sar Lake and Brari Nambal Basin of Dal Lake due to organic pollution. Among other things, breeding grounds of the fish have been destroyed over the years, especially by filling large areas of waters. In order to save the fish biodiversity of the state especially the schizothoracine fish from further deterioration, immediate conservation measures need to be undertaken, which may include checking on pollution, abstinence from the destruction of breeding grounds of fish, and development of hatcheries for mass seed production of trout, snow trout, and golden mahseer for aquaculture and/or ranching. Detailed studies are also warranted on impact of introduction of exotics on the local fishery and its environment before their release in open waters. Among the inland fisheries research in the country, the cold-water fisheries research is only three decades old.

10.2 Topography of the Area

J&K (33.34°N–34.34°N latitude and 75.13°E–76.11°E longitude) is blessed with innumerable cold-water resources and has a tremendous potential for the development of fisheries. J&K exhibits a great diversity in its geo-climatic conditions. The Jammu region extends from the Shivalik plains to high altitudes, with climate ranging from temperate (Kishtwar, Bhaderwah, and Banihal) to the (sub)tropical conditions. The valley of Kashmir is situated in the midst of western Himalaya at an average height of 6,000 ft above sea level and exhibits temperate climate. The cold waters of Kashmir are generally located at >1,500 m above sea level. Most of the fish are small, showing a distribution pattern that depends on rate of flow of water, nature of substrata, and availability of food. Some species have developed special organs of attachment to enable them to live in fast turbulent streams, as is common with fish of such waters. The important river systems in the erstwhile state include the the Indus River system (Ladakh), the Jhelum River system (Kashmir), and the Chenab River system (including the Tawi, Ravi, and Rajouri Tawi systems in Jammu).

J&K can also be divided into three regions with respect to fish: (1) fish of the Kashmir Valley, with many fish of Central Asia and endemic species; (2) fish of Kashmir Province, consisting of outlying areas of the Kashmir region, including Gulmarg, Baramulla, Lolab Valley, Pahalgam, and Sonamarg up to Kishanganga-Sindh Valley and the Ladakh area within Central Asian Ichthyofauna; and (3) fish of Jammu Province, including Poonch, Kathua, Bhaderwah, and Kishtwar with fish mostly of Indian origin. The origin of Jammu Province fish is from the Indo-Gangetic region (Das & Subla, 1963).

10.3 Important Aquatic Resources

Among the lakes of Kashmir, the important ones are Dal (1,100 ha), Manasbal (280 ha), Wular (2,400 ha), Anchar (700 ha), Khushal Sar (70 ha), and Malpursar (1.6 ha), situated at 1,584 m. above sea level, while Nilnag (25 ha), Alapather (8 ha), Marsar (8.5 ha), Tarsar (5 ha), Krishansar (40 ha), Gangabal (175 ha), Kausar Nag (140 ha), and Seshanag (51 ha) are all high-altitude lakes, situated above 2,000 m. above sea level. Eutrophic lakes (most of the lower-altitude lakes) produce a variety of fish dominated by the common carp (*Cyprinus carpio*). The important wetlands in the valley include Hokersar (Srinagar District), Shalabugh (Ganderbal District), and Mirgund and Hygam (in the district of Baramulla), and these constitute ideal homes for the migratory birds. These wetlands are dominated by the common carp and may turn eventually dry up if measures to save them are not enforced well in time. Among the lotic habitats, Jhelum River occupies an important place in the valley followed by the trout streams. The important lakes in Jammu include Kailash, Sanasar, Malisar, Bhadnaz, Talva or Sharoth or Dhudhari Lake, Hadri-Kabansar, Mansar, and Surinsar.

10.4 Fish and Fisheries

The major fishery of the Jhelum River is constituted by the schizothoracids (an endemic group of *Schizothorax* spp.). Fish like *Glyptothorax* spp. *Glyptosternum* spp., *Nemacheilus* (=*Triplophysa*) spp., and salmonids (*Salmo* spp.) prefer the clean flowing waters of snow-fed and spring-fed streams, like Lidder, Bringhi, Erin, Boniyar, Kishanganga, and Sindh. The *Barilius* spp. And the golden mahseer (*Tor putitora*) dominate the fish catch composition in Tawi River system and Chenab River system, respectively, in Jammu, while in Kashmir, the catches are dominated by the common carp in the lacustrine waters and the snow trout in the riverine waters. Trout is the most prized fish of these waters and constitutes an important resource in Kashmir both from sport and aquaculture angle. J&K has more than 27 trout hatcheries located at different places. Each year, lakhs of trout fingerlings are produced in these hatcheries/farms and stocked in the hill streams. A hatchery was built at Achabal in 1908 from where eyed ova were distributed to all over Kashmir and were even sent to other parts of India like Kulu, Shimla, Shillong, and so on. Important trout hatcheries in Kashmir include Laribal, Achabal, and Kokarnag—the latter acting as *mother seed farm* for trout fry and fingerlings. Rainbow trout are best suited for culture practices because they are easily domesticated and adapt to the environmental conditions without much strain than the brown trout. Further, rainbow trout easily take to artificial feeding, can withstand higher temperature and reduced oxygen and are also resistant to certain diseases. Their incubation period is short, and development is rapid as compared to the brown trout. Thus, there is great need to study the breeding and feeding behavior of the brown trout with evolving artificial feed to bring it to the desired level of farming and subsequent stocking in the natural resources.

In case of the Jhelum, Sunder and Subla (1984) reported a fishing effort of 261 g–829 g/man-hour, while Yousuf et al. (2006) reported it to be 173 g–360 g/man-hour. During the present study, the fishing effort varied from 210 g in November to 793 g/man-hour in May at Anantnag (Batengoo) and Sopore, respectively. In case of Dal Lake CIFRI (1977) reported the fish catch per man-hour in the range of 156–978 g at Saida Kadal and 117–797 g in Hazratbal basins, while Sunder at al. (1978) reported the mean catch per man-hour as 369 g. The fishing effort varies with season and is dependent upon several factors, including the time of the day and the behavior of the fish, whether migratory or resident. Shafi et al. (2005) reported the overall mean effort as 382 g/man-hour in this lake. They reported the fishing effort rise from 122 g/man-hour (Hazratbal Basin) in December to 1,015 g/man-hour (Nishat Basin) in October. During the present study the mean fishing effort from the Dal Lake was recorded as 398 g/man-hour. As the fishing effort is an index of the abundance of the fish population/stocks, it is, therefore, clear that over the years the fish populations, rather their density, has decreased from the aquatic habitats of the Kashmir Valley.

10.5 Historical Background

The fish fauna of the valley is quite different from rest of the country and is dominated by the schizothoracine group. Although a number of researchers have worked on the taxonomy of this group of fish, still the validity of different species has not been resolved and a lot of confusion prevails about the identification of different species. Fish fauna studies in Kashmir were initiated by Heckel (1838). During this study, he reported 16 species of fish that were all new to science and published them in his book *Fishe aus caschmir.* Later on, McClelland (1839), Heckel (1844), Misra (1949), Malik and Khajuria (1955), Silas (1960), Sunder and Subla (1960), Das and Subla (1963–1964), Malik (1966), Yousuf (1989, 1996, 2000), Balkhi (1995), Wanganeo et al. (2003), and Bhat et al. (2010) also reported on the Ichthyofauna of Kashmir in different habitats.

10.6 Fish Biodiversity

About 41 species of fish are reported from Kashmir (Table 10.1), excluding the grass carp and the silver carp, which are reported to be maintained in the closed ecosystems only by state fisheries department. More recently, Kullander et al. (1999) revised fish of Kashmir from the Jhelum River and a few surrounding habitats. The authors have synonymized/renamed some species of *Schizothorax* and opened new challenges for updating the fish species of valley. They recorded 20 valid fish species and reported *Tor putitora* as extirpated from Kashmir. However, Balkhi (2004) recovered specimens of *Tor putitora* from the explored area between Parampila and Red Bridge in the Uri sector in Kashmir, thus negating reports of its extirpation. Bhat et al. (2010) recovered *Diptychus maculatus* from the Wanghat Stream at Naranag. Also, Bhat et al. (2020) reported the presence of grass carp in Dal Lake and Manasbal Lake. With this, the number of valid fish species in Kashmir rise to 23 (Table 10.1, Plates 1–12). Regarding

TABLE 10.1

The Ichthyofauna of Kashmir

Order: Cypriniformes	**Order: Siluriformes**
Family: Cyprinidae	**Family: Cobitidae**
Subfamily: Cyprininae	***Subfamily: Botinae***
1. *Cyprinus carpio* (Linnaeus), introduced#	26. *Botia birdi* (Chaudhuri)#
C. c. communis	
C. c. specularis	
**C. c. nudus*	
2. *Carassius carassius* (Linnaeus), introduced#	***Subfamily: Nemachilinae***
3. **Labeo dero* (Heckel)	27. **Nemachilus gracilis* (Day)
4. **L. diocheilus* (McClelland)	28. **N. vitatus* (Heckel)
5. *Bangana diplostoma* (Heckel) #	
6. *Puntius conchonius* (Hamilton), introduced#	29. *N. kashmirensis* (Hora)
	(=*Triplophysa kashmirensis* [Hora])#
7. ***Tor putitora* (Hamilton), migratory	30. *N. marmoratus* (Heckel)
	(=*Triplophysa marmorata* [Heckel])#
Subfamily: Schizothoracinae	31. **N. rupicola* (McClelland)
8. *Schizothorax curvifrons* (Heckel)#	32. **N. yasinensis* (Alcock)
9. *S. esocinus* (Heckel)#	33. **N. stoliczki* (Steind)
S. montanus, from Ladakh	34. *Schistura punjabensis* (Hora)#
10. *S. huegelii* (Heckel)	**Family: Sisoridae**
11. *S. lingipinnis* (Heckel)	35. *Glyptothorax kashmirensis* (Hora)#
12. *S. micropogon* (Heckel)	36. *G. pectinopterus* (McClelland & Griffith)#
13. *S. nasus* (Heckel)	37. *Exostoma stoliczki* (Day)
	(=*Glyptosternon reticulatum* [McCleland & Griffith])#
14. *S. niger* Heckel#	**Family: Siluridae**
15. *S. labiatus* (McClelland & Griffith)#	
(=*S. ladacensis*, from Ladakh)	
16. *S. planifrons* (Heckel)	38. **Ompok bimaculatus* (Bloch)
17. **S. progastus* (Heckel)	**Order: Cyprinidontiformes**
18. **S. punctatus* (Heckel)	**Family: Poecilidae**
19. *S. plagiostomus* (Heckel)#	39. *Gambusia affinis* (Baird & Girard), introduced
(=*O. plagiostomus* =*S. skardensis*, from Ladakh)	(=*G. holbrooki* [Girard])#
20. **O. sinuatus* (Heckel)	**Order: Salmoniformes**
21. **Ptychobarbus conirostris* (Steind)	**Family: Salmonidae**
22. ##*Diptychus maculatus* (Steind)	40. *Salmo trutta* (Linnaeus) introduced#
23. **Schizopygopsis stoliczkae* (Steind)	41. *Salmo gairdneri* (Richardson), introduced
	(=*Oncorhynchus mykiss* [Walbaum])#
Subfamily: Garrinae	
24. **Gara gotyla* (Gray)	
25. *Crossocheilus diplochilus* (Heckel)#	

*Uncertain validity

**Species recovered from Uri, Kashmir (Balkhi, 2004)

#Valid species from Kashmir (Kullander et al., 1999), including *T. putitora* (Balkhi, 2004)

##Species recovered from Wanghat Stream (Naranag) by Bhat et al. (2010)

Plate 1: ***Schizothorax niger***

Plate 2: ***Schizothorax esocinus***

Plate 3: ***Schizothorax labiatus***

Plate 4: ***Schizothorax curvifrons***

Plate 5: ***Schizothorax plagiostomus***

Plate 6: ***Crossocheilus diplochilus***

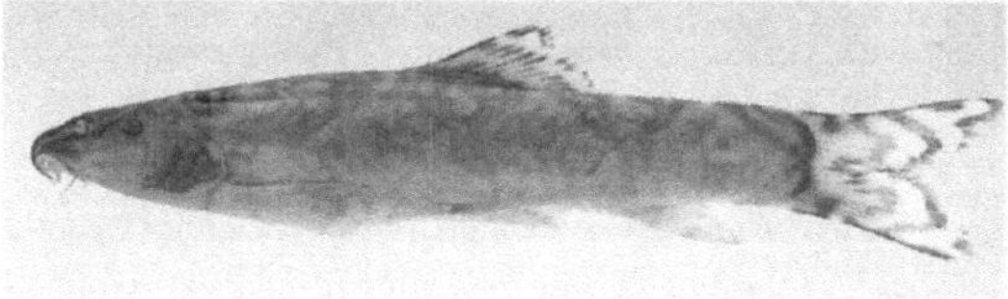

Plate 7: ***Botia birdi***

Plate 8: ***Tripophysa kashmirensis***

Plate 9: ***Glyptosternon reticulatum***

Plate 10: ***Glyptothorax kashmirensis***

Plate 11: ***Glyptothorax pectinopterus***

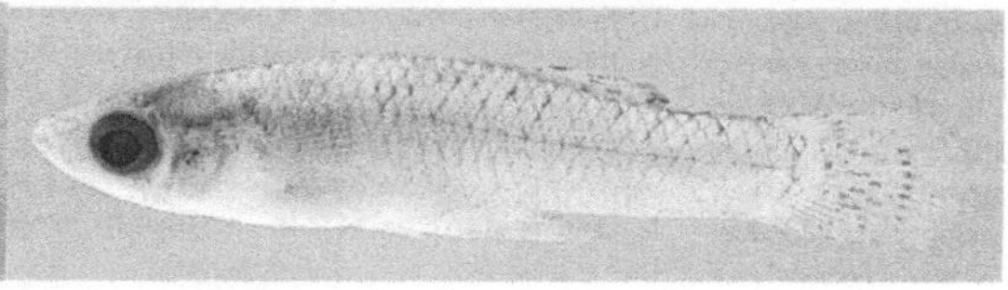

Plate 12: ***Gambusia holbrooki***

exact status of fish fauna, a lot of confusion exists and fresh detailed studies are warranted on the actual number of fish species that are also of great significance in fishery management.

Common carp and the schizothoracids (especially *S. esocinus*, *S. niger*, and *S. plagiostomus*) constitute the important fisheries of the valley. The common carp in the lacustrine waters constitutes more than 50% of the fish catch, while *Schizothorax* species form a major catch in the riverine waters. Among the endemic fish species, *Bangana diplostoma* (roput in Kashmiri), *Schizothorax esocinus* (churoo in Kashmiri), *Schizothorax niger* (ale gad), *S. curvifrons* (satter gad), and *Crossocheilus diplochilus* (tethur) have commercial importance and have a wider distribution in clean waters.

The ichthyofauna of Jammu has not been adequately investigated until Das and Nath (1965, 1966), who reported on the fish of Jammu, including Poonch. Subsequent investigations were conducted in Samba, Doda, Kathua, Udhampur, Rajouri, Poonch, and Chenab/Tawi Rivers by Das and Nath (1971), Malhotra et al. (1975), Malhotra and Dutta (1976), Dutta and Malhotra (1984), Dutta and Kour (1999), Dutta et al. (2001), and Dutta and Fayaz (2003). Das and Nath (1971) reported 23 species of fish from Jammu. Jyoti et al. (2004) reported 63 species of fish from Jammu, including the five exotic species. More recently, Bhat et al. (2020) reported about 100 fish species from the Jammu region (Table 10.2; Plates 13–20).

The validity of many species needs to be ascertained and fresh taxonomic studies are warranted to remove all confusion in the present status of the fish diversity. During a recent study of the Tawi and Chenab Rivers in Jammu (Balkhi, 2004), 12 valid species of fish were recorded with catch composition dominated by the *Barilius* spp. And the golden mahseer as compared to any other fish species. During the present study, the author did not record a single specimen of *Tor khudree*, and it was a common observation to catch small specimens of the golden mahseer weighing hardly up to 500 g, probably due to great pressure of fishing in the available waters in J&K. The river systems of Jehlum (Kashmir) and Chenab (including Tawi and Jammu) have remained subjected to intensive exploitation, making the capture fisheries of past years no longer existing.

TABLE 10.2

Ichthyofauna of Jammu

Order: Cypriniformes

Family: Ralitoridae

1. *Laliamsdaocoiritis botia* (Ham & Buch)

Family: Cobilidae

2. *Hada dayi* (Ham)
3. *Lepidocephalus guntea* (Ham & Buch)
4. *Nemacheilus botia* (Ham)
5. *N. prashadi* (Hora)
6. *N. punjabensis* (Hora)
7. *N. stoliczkae* (Steind)*

Family: Cyprinidae

8. *Amblypharyngodon gadigarhi*
9. *Aspidoparia morar* (Ham)
10. *Barbus hexagonolepis* (McClelland)*
11. *B. wynaadensis* (Day)*
12. *Barilius bendelesis* (Ham)+
13. *B. vagra* (Ham)+
14. *Calla catla* (Ham & Buch), exotic
15. *Chela atpar laubuca* (Ham)*
16. *C. cachius* (Ham & Buch)
17. *C. laubuca* Ham)
18. *Cirrhinus mrigala* (Ham & Buch), exotic
19. *Cirrhinus reba* (Ham)
20. *Crossocheilus latius diplochilus* (Heckel)
21. *Ctenopharyngodon idella* (Valenciennes), exotic
22. *Cyprinus carpio communis* (Linnaeus), exotic
    C. c. specularis (Linnaeus), exotic
23. *Danio devario* (Ham)
24. *D. rerio* (Ham)
25. *Esomus danricus* (Ham)
26. *Garra gotylas gotyla* (Gray)
27. *G. lamta* (Ham)
28. *G. mullya* (Sykes)*
29. *G. variabilis* (Heckel)*
30. *Hypophthahnichthys molitrix*
31. *Labeo bata* (Ham)
32. *L. boggut* (Sykes)*
33. *L. calcasu* (Ham & Buch)
34. *L. dero* (Ham)
35. *L. diplostomus*
36. *L. gonius* (Ham & Buch)

37. *L. microphthahnus* (Beavan)*
38. *L. pangusia* (Ham)
39. *L. rohita* (Ham & Buch)
40. *Oreinus plagiostomus* (Heckel)
41. *Osteobrama cotio* (Ham)
42. *Oxygaster bacaila* (Ham)
43. *Puntius chrysopterus* (McClelland)*
44. *P. conchonius* (Ham)
45. *P. chola* (Ham & Buch)
46. *P. sarana* (Ham)
47. *P. sophore* (Ham)
48. *P. stigma* (Ham)*
49. *P. ticto* (Ham)
50. *Rasbora daniconius* (Ham)*
51. *R. rasbora* (Ham)
52. *Salmostoma bacalia* (Ham & Buch)
53. *S. punjabensis* (Ham & Buch)
54. *Schizothorax esocinus* (Heckel)
55. *S. curvifrons* (Heckel)
56. *S. progastus* (McClelland)
57. *S. richardsoni* (Heckel)
58. *Tor khudree* (Sykes)*
59. *T. putitora***
60. *T. for*
61. *Triplophysa yasinensis* (Alcock)

Order: Perciformes

Family: Gobidae

62. *Glossogobius giuris* (Ham)

Family: Osphronemidae

63. *Trichogaster fasciatus* (Bloch & Schneider)

Family: Nandidae

64. *Badis badis* (Ham & Buch)
65. *Nandus nandus* (Ham & Buch)

Family: Channidae

66. *Channa gachua* (Ham)*
67. *C. marulius* (Ham & Buch)
68. *C. orientalis* (Bloch & Schneider)
69. *C. punctatus* (Bloch)
70. *C. striatus* (Bloch)

Family: Belontidae

Order: Beloniformes

Family: Belonidae

72. *Xenoniodon cancila* (Ham & Buch)

Order: Synbrachiformes

Family: Mastacembelidae

73. *Iacrognathus jammuensis*
74. *Mastocembelus armatus* (Lacepede)
75. *M. pancalus*

Order: Salmoniformes

Family: Salmonidae

76. *Salmo gairdneri gairdneri* (Richardson)
77. *S. trutta fario* (Linnaeus)

Order: Osteoglossiformes

Family: Notopteridae

78. *Notopterus notopterus* (Pallas)

Order: Siluriformes

Family: Sisoridae

79. *Amblyceps mangois* (Ham & Buch)
80. *Bagarius bagarius* (Ham & Buch)
81. *Gagata cenia* (Ham & Buch)
82. *Glyptothorax botium* (Ham & Buch)*
83. *G. cavia* (Ham & Buch)*
84. *G. garhwali* (Tilak)
85. *G. kashmerenses* (Hora)
86. *G. pectinopterus* (McClelland)
87. *G. punjabensis* (Mirza & Kashmiri)
88. *G. telchitta telchitta* (Ham. & Buch)
89. *Glyptosternon reticulatum* McClelland

Family: Bagridae

90. *Mystus bleekeri* (Day)
91. *M. cavassius* (Ham & Buch)
92. *M. seenghala* (Sykes)
93. *M. vittatus* (Bloch)
94. *Rita rita* (Ham & Buch)

Family: Schilbeidae

95. *Clupisoma garua* (Ham & Buch)
96. *Eutropiichthys vacha* (Ham & Buch)
97. *Pseeutropiusud atherinoides* (Bloch)
98. *Heteropneustes fossilis* (Bloch)

Family: Siluridae

99. *Ompok bimaculatus* (Bloch)
100. *Wallago attu* (Schneider)

---

*Validity uncertain
+Constituted dominant catches in River Tawi and its tributaries
**Constituted dominant catches in the Tawi and Chenab river systems

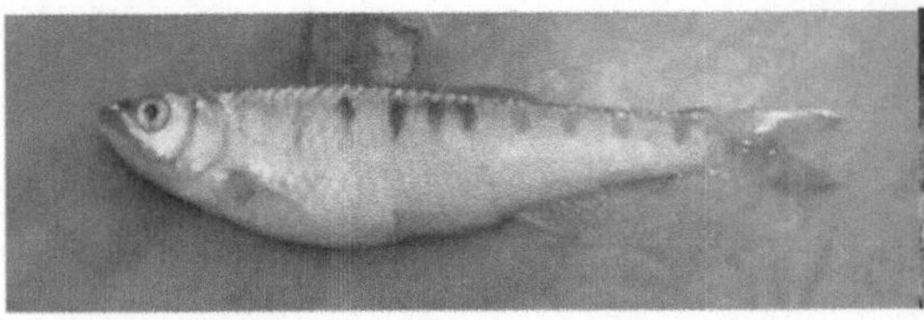

Plate 13: ***Barilius* sp.**

Plate 14: ***Nemacheilus* sp.**

Plate 15: ***Catla catla***

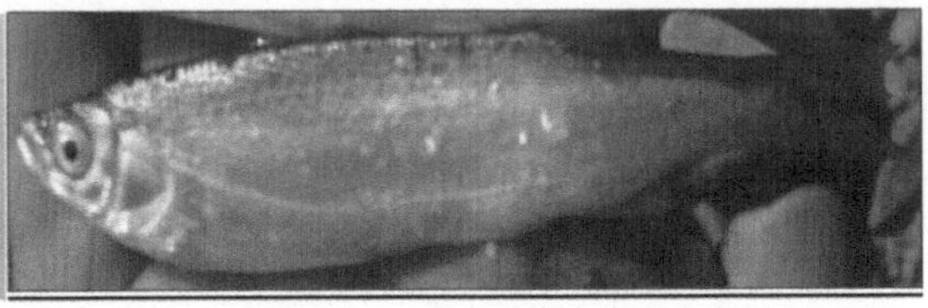

Plate 16: ***Chelalaubuca***

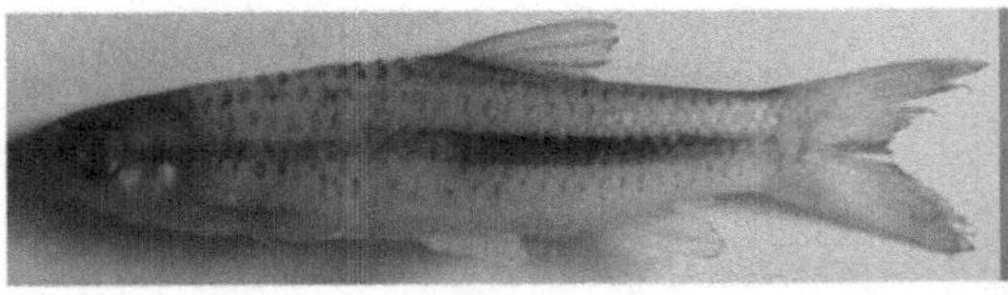

Plate 17: ***Esomus dendricus***

Plate 18: ***Garragotyla***

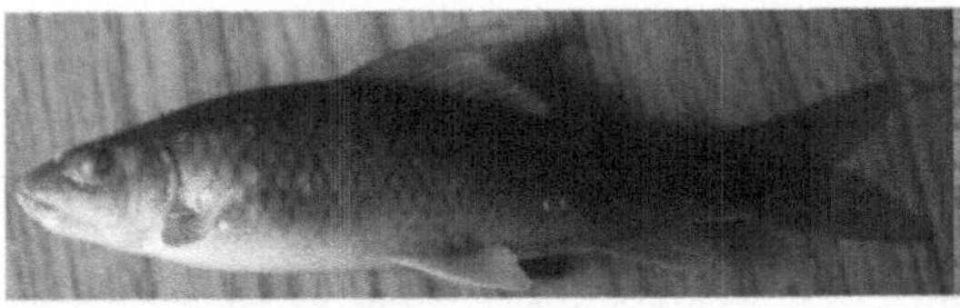

Plate 19: ***Labeorohita***

Plate 20: ***Puntius conchonius***

Plate 21: ***Puntius ticto***

Plate 22: ***Puntius sarana***

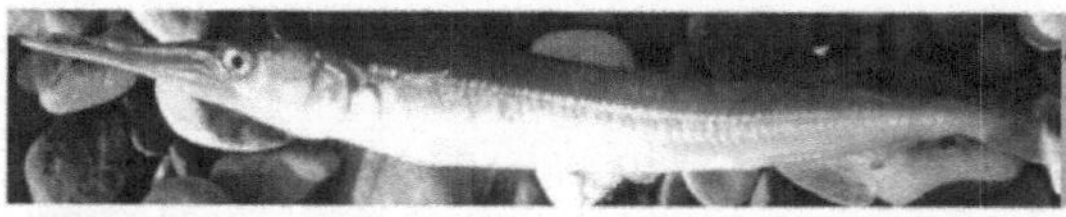

Plate 23: ***Xenontodon cancila***

Plate 24: ***Macrognathus aural***

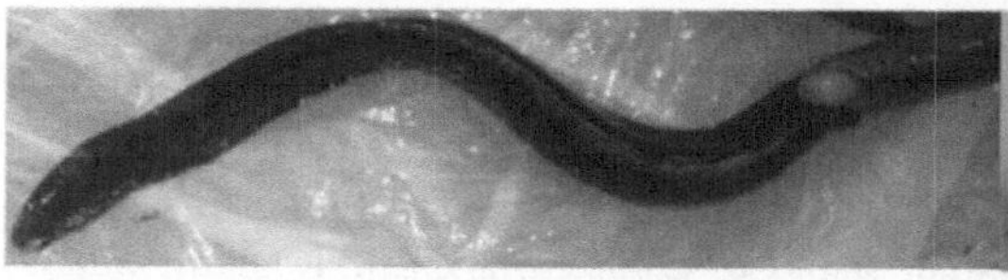

Plate 25: ***Monopterus cuchia***

Plate 26: ***Lepdiocephalus guntea***

Plate 27: ***Mystus cavasicus***

Plate 28: ***Ompok bimaculatus***

Plate 29: ***Channa punctatus***

Plate 30: ***Paraambassis ranga***

10.7 Concerns Vis-a-Vis Effect of Exotics on Local Fauna

As a consequence of dam construction and other obstructions in rivers, disruptions of the reproductive cycle of the migrant species (e.g., mahseer), and alterations in the breeding grounds of other resident species (schizothoracids) have been widely reported. The construction of the Salal hydroelectric power project on River Chenab in Jammu and the Mangla Dam in POK and Uri hydroelectric power projects in Kashmir have proved highly detrimental for the migratory mahseer and the snow trout, respectively. The mahseer fish cannot make it from Muzaffarabad in POK to come to up to the Jhelum in Kashmir due to absence of a proper fish pass (ladder) in the barrage. The specimens recovered during the recent study by the authors from Uri Jhelum could be the old stock that has learned to survive in the stretch of water between the line of control and Parampilla in Uri Kashmir. Similarly, the mahseer in Chenab cannot migrate beyond Reasi where the Salal project/barrage has come up without providing a fish pass (ladder) in it. The fish pass (ladder) provided in the Uri barrage is not efficient enough to allow all fish to migrate upstream. Therefore, in order to conserve the fishery of the rivers, it is first essential to provide proper fish passes (ladders) in the power projects so that the migratory route of the fish is not cut off, or else it has a detrimental impact on their breeding and subsequent population dynamics.

The introduction of exotic species has become a common practice. Although some introductions have been successful and have augmented fish production in the country, others have shown a strong negative impact on native population. The population of schizothoracine fish in the valley has considerably decreased over the years particularly after the introduction of common carp in 1956. The fish has not only flourished well in the aquatic habitats of the valley but also provides the maximum overall fish catch (about 70% of total fish catch). However, common carp has proved to be a continuous threat to the prized local species of *Schizothorax* whose population has considerably decreased during the past. This is, in particular due to the deterioration of the habitat caused by the feeding habitats of the common carp together with its high fecundity. *Gambusia affinis* is believed to have been introduced accidentally some 60 years back. The fish is not commercially important but has got well established in Kashmir waters. There is also no definite record about the introduction of *Carassius carassius* in Kashmir waters. However, the fish forms a good catch in certain valley lakes and has also gained commercial importance.

Grass carp and silver carp have been introduced in Kashmir under pond conditions about two decades back by the Fisheries Development Department. There has been much debate about the possibilities of biological control measures to cope up with the weed problem in the lakes. The overall ecological balance has to be kept in mind before any decision in this direction is taken. The ecology and behavior of the grass carp in relation to the local fauna and flora has to be studied in detail in some closed system. The closed system is very essential as otherwise the fish may be accidentally released into other water bodies, which can have far reaching consequences as almost all the major aquatic habitats of the valley are interconnected. However, presently grass carp has also invaded the open waters of Kashmir and have been found in the commercial catch of Dal and Manasbal lakes, which is a matter of concern for all of us and also quite embarrassing from scientific point of view. Its presence in natural waters without any scientific study seems to create further ecological problems in Kashmir waters as almost all the water bodies are interconnected.

The other principal factor that has possibly contributed to the decline in endemic fishery includes the rapidly changing ecological scenario in the aquatic habitats. All the water bodies situated in close vicinity of human settlements have been receiving large quantities of untreated sewage that has affected the water quality. This has greatly harmed the local fish fauna that are adapted to live in relatively clean waters. Fish of almost all kinds (schizothoracids) have been wiped out from Khushal Sar Lake and Brari Nambal areas of Dal Lake due to organic pollution. Measures need to be taken to ensure that no (untreated) domestic/industrial effluents are allowed to be released into the open waters.

Another factor responsible for the decline in the schizothoracine population is the encroachment of the shallow peripheral areas of the lakes. This has been particularly responsible for the decrease in the population of *Schizothorax niger* (locally called ale gaad). This fish breeds in the shallow areas of the lake particularly in sandy areas with springs oozing from bottom. Since large areas of lakes Dal, Anchar, Wular, and so on have been filled up, the breeding grounds of the fish have greatly shrank, and this has led to the decrease in population/production of fish. Therefore, in order to stop further devastation to the breeding grounds of the fish, it should be ensured that no encroachments are allowed in the peripheries of the water bodies.

Fluctuating water level in tributaries of some lakes and streams have also a great bearing on the fish populations. Most of the schizothoracine fish breed in the upper reaches of the connecting streams especially in areas with sandy and gravelly bottom. The breeding season of *Schizothorax* spp. In Kashmir starts in March and continues up to June in different species. During this period, the streams are generally full of water, and as such, the fish move as far upward as possible and spawn in shallow areas of the streams. However, soon after the eggs hatch out and the larval stage is reached, the quantity of water in the streams decreases, thus resulting into formation of side pools, wherein a large number of the fish fry get trapped and ultimately perish due to the drying up of these side pockets just before the fish reach the feeding ground in the lower reaches of the streams and into the lakes. This is especially true for Telbal and Sindh nallas where the larvae of *Scisizothorax esocinus* get trapped and perish.

10.8 Conservation Measures

The capture fishery of the prized mahseer and the snow trout in these rivers has led to a drastic decline in number and size as a result of indiscriminate catching of brood stock and juveniles. Therefore, strict measures need to be taken by the government functionaries to ensure fishing in proper seasons by using nets of desirable mesh size. Besides the fishing pressure, there are other natural and anthropogenic pressures that have contributed to decline in fish stocks. They are habitat destruction due to over abstraction of water, construction of dams and barrages, siltation, and pollution due to urban, industrial, and agricultural runoff. Government machinery needs to be geared up to rescue such trapped population for safe release into the open waters. The dams, which lack the fish ladders, need to be demolished and constructed again with designs of fish ladders. There is an urgent need for conservation of fish genetic resources of the state. Among other things, the establishment of a fish sanctuary at the Faculty of Fisheries, Rangil, Ganderbal, could serve a long way for conservation of fish germplasm.

10.9 Conclusion

On the basis of the present survey, coupled with earlier reports, it may be safely concluded that the aquatic habitats of J&K are inhabited by about 100 to 120 species of fish. The variations in altitude and topography have led to different sequence of successions of the water bodies in different areas of the state, which has ultimately resulted in the colonization of these waters by different sets of fish. In the valley of Kashmir, five valid species of *Schizothorax* (*S. plagiostomus*, *S. niger*, *S. esocinus*, *S. labiatus*, and *S. curvifrons*) and one species of *Diptychus* (*D. maculatus*) could be easily located, while the other previously reported schizothoracine species like *P. conirostris*, *S. stoliczki*, and *S. progastus* were untraceable. This is also true of some species of *Nemacheilus*. However, these are reported from the Ladakh region. Since many of the early workers didn't provide the details about the locations where from the species reported were collected (see, e.g., Silas, 1961; Das & Subla, 1964; Das, 1965), it is most probable that the lists provided were based on clubbing of fish of Kashmir and Ladakh together. The high species diversity in Jammu region is mainly due to presence of warm-water fish. J&K is the leading state in trout production in India and has a great potential for cold water fisheries. Aquatic resources are available in plenty for the development of the aquaculture of the native and exotic species throughout the state. If proper management, conservation, and regulations of the aquatic/fishery resources of the state are taken, it is expected that the fish production from the natural resources can be increased manifold. Propagation/culture of indigenous and fast-growing fish will also help in conservation and production of fish in the state.

11

Biodiversity and Conservation Status of Indigenous Livestock of India

Karan Veer Singh

CONTENTS

11.1 Introduction

Biodiversity broadly refers to the wealth of the Earth, or the variety of life-forms it has. In other words, it means variety and variability among living organisms, their genetic differences, and the ecosystem of which they are part. This diversity is a vital resource for sustainable existence on Earth, through judicious exploitation, utilization, and conservation we can meet our food and nutritional security. Biological diversity (often referred to as biodiversity) is defined in Article 2 of the Convention on Biological Diversity. The objectives of the CBD are the conservation of biological diversity, its sustainable use, and equitable sharing of the benefits that arise out of the use of genetic resources. The Commission on Genetic Resources for Food and Agriculture (CGRFA) of the Food and Agriculture Organisation (FAO) reviews all matters relating to policies, programs, and activities like their characterization, conservation, and sustainable use. Agricultural biodiversity has coevolved with economies, cultures, knowledge systems, and societies moreover it requires continuous active human management (FAO, 2009).

Animal biodiversity is the diversity of animals at genetic, species, and ecosystem levels managed by farmers and livestock keepers for hundreds of generations. Livestock genetic diversity is usually discussed in terms of the genetic components of breeds. Livestock breeds are recognized as important components of world biodiversity because the genes and gene combinations they carry may be useful to agriculture in the future. Adaptation in diverse habitats and selection strategies leads to evolution

DOI: 10.1201/9781003220398-11

of unique traits, it is very dynamic in nature new breeds has emerged and others have disappeared as environments and societies have changed. Animal genetic resources (AnGR) remain at risk and many drivers of loss are common with those of wild biodiversity, such as demand, intensification, degradation of natural resources, and climate change. In AnGR, in particular, the loss of within-breed diversity is difficult to assess.

The Domestic Animal Diversity Information System (DAD-IS), maintained and developed by the FAO, contains data from 182 countries on a total of more than 8,000 livestock breeds belonging to 38 species. From small gene pool only few thousand of livestock breeds have evolved over the time to suit particular environments. And out of these few thousand species, only 40 species were selected for domestication by different settlements. The five major livestock species (cattle, sheep, goats, pigs, and chickens) have spread globally.

However genetic diversity can help cope with the challenge of climate change. The ability of selective breeds to make use of poor-quality pasture and good grazing abilities can definitely help in replacing the animals in drought prone areas. The Rashtriya Gokul Mission (RGM) of government of India is meant for conserving indigenous breeds of cattle through selective breeding.

Early genetic improvement was mainly dependent on the selective breeding with superior phenotypes, the use of present-day molecular genetics techniques with conventional animal breeding tools has optimized the process. Genetic and genomic data set has been used to estimate genetic diversity, population structure, and demographic history. Genomics is playing a very crucial role in biodiversity conservation; it helps in identifying the segments of genome responsible for breed adaptation and ultimately for developing modern and fast monitoring tools for endangered biodiversity. High-throughput sequencing techniques have opened up the possibility of investigating and understanding the genetic differences between breeds. Whole transcriptome analysis by next-generation sequencing (NGS) technologies or RNA-Seq has been utilized to reveal the differentially expressed genes for various characteristics in various species. Several important genes and pathways have been identified in livestock species that can be used for improvement of the meat, milk, and egg quality. This chapter focuses on the present status of domestic livestock biodiversity of India.

11.2 Threats to Biodiversity

The main causes of biodiversity loss include land-use changes, environmental change, and biological invasion of non-required species or introduction of exotic germplasm. The causes of biodiversity loss are varied and often interrelated. The greatest threat to wild plant and animal species is due to habitat destruction or loss due to destruction or alteration of their habitat. This area is used by species for shelter and reproduction, and if it is destroyed or disrupted, species must adapt to the new changes or migrate to new area, or else they simply die. Consequently, it must compete with the locally adapted population of the same species or other species sharing the same food source. The selective destruction of species can produce equally unfortunate results. Global climate change and wildlife poaching are also equally significant contributing factors.

The introduction of non-native species into new areas where the species have few or no natural predators to keep their populations in check is considered the most important cause of native biodiversity loss. Few elite breeds of livestock are preferred by industries for production and supply. Crossbreeding and the use of selected breeds' semen for artificial insemination (AI) are also two of the major causes of genetic erosion. The increasing use of non-native breeds, decline of traditional livestock production systems, and neglect of breeds that are considered "not competitive enough" are some of major threat to livestock genetic diversity. Increasing global trade in breeding animals and livestock semen, often for crossbreeding purposes, if not checked and planned, can lead to the loss of valuable characteristics (such as tolerance to extreme temperature), poor-quality feed, and other challenging aspects.

According to FAO, more than 90% of livestock breeds are local breeds, and of these, 26% are at the risk of extinction. Further, more than 50% of the breeds are under threat as no recent population data is available (FAO, 2015). In India, the wild Bactrian camel population in Nubra Valley, Ladakh, are critically endangered per the IUCN Red List. The wild water buffalo has also been listed as endangered a small number still exist in Balpakram National Park, Meghalaya, while the Indian bison, wild yak, wild goat, and swan goose have been listed as vulnerable (Baillie et al., 2004).

11.3 Livestock Diversity of India

The advent of new era of national sovereignty over natural genetic resources under the Convention on Biological Diversity (CBD) requires a new approach to describe, evaluate, and catalog livestock breeds. It is an international legal framework for the conservation and sustainable use of biodiversity. The Nagoya Protocol ensures fair and equitable benefit sharing arising from the utilization of genetic resources while under the World Trade Organization (WTO) and intellectual property rights protection for the indigenous genetic diversity and recognition for the livestock keepers are provided (The Nagoya Protocol on Access and Benefit-sharing, 2014). The documentation and inventorization of indigenous breeds and other nondescript populations help in their sustainable use, improvement, and preservation for future generations.

The diversity of livestock is not accidental, nor it is purely natural; the range of livestock is the outcome of thousands of years of field selection by keepers and exposure to a range of natural conditions (Hall 2019). Indian farmers, for centuries, have continuously modified the genetic material available to them by following simple but meticulous breeding schemes from generation to generation. Apart from physical and biological adaptation forced by the environment, the economic, cultural, religious, and survival factors have also played a role in this diversification. India can be divided into ten major regions from the snow-covered regions to the Thar Desert, from Tarai and the plains of Ganga to Himalayan mountain range, from grasslands to the western and eastern coastal belts of more than 6,000 km. Different livestock breeds have adapted to diverse local conditions. For instance, India has sheep breeds that are adapted to the harsh summers of the West Indian Desert and others that can survive equally harsh winters of Himalayan tracts. There are cattle breeds that thrive in humid hills of the Western Ghats, while other breeds do well in the driest regions of Rajasthan and Gujarat.

FAO has defined the breed that is widely accepted as a group of animals selected by man to have a uniform appearance that distinguishes them visual from other similarly defined groups within same species, or a group that is geographical separated from phenotypically similar groups having separate identity. The evaluation and characterization of indigenous breeds provide guidance for decision-makers for successful conservation of livestock biodiversity. Breeds can be subject to breeding programs to improve their productivity or promote other desirable characteristics. Total number of indigenous breeds of livestock, poultry, and dog in the country is 200, which include 50 cattle, 17 buffalo, 34 goat, 44 sheep, 7 horse/pony, 9 camel, 10 pig, 3 donkey, 1 yak, 19 chicken, 2 duck, 1 goose, and 3 dog breeds (ICAR-NBAGR 2022). The total livestock population is 535.78 million in the country and total bovine population (cattle, buffalo, mithun, and yak) is 302.79 million in 2019 with a share of 35.94% (cattle), 27.80% (goat), 20.45% (buffalo), 13.87% (sheep), 1.69% (pigs), 0.23% (mithun, yaks, horses, ponies, mules, donkeys, and camels all together) of the total livestock (20th Livestock Census 2019).

11.3.1 Breeds of Cattle

It is generally accepted that the domestication of cattle followed the domestication of sheep, goats, pigs, and dogs. India has 192.49 million cattle which account for 35% of total livestock (20th Livestock Census 2019). About 70% of Indian cows are owned by economically poor strata of the society. Indigenous cattle of India belong to the species, *Bos indicus*. Many of the indigenous breed have been used in foreign countries for selection, as early as 1854, Gir, Kankrej, Nelore and Ongole cows were imported to Brazil, Argentine and US and it has been used to develop beef brands.

India has 50 registered indigenous cattle breeds, which includes world's top milch breeds like Sahiwal, Gir, Red Sindhi, Tharparkar, and Rathi. A few others, as dual-purpose breeds Kankrej, Ongole, and

FIGURE 11.1 Vechur cow, the smallest cattle breed in the world.

(*Source:* ICAR-NBAGR)

Hariana, that have both milch and drought qualities. The rest are drought breeds. India has indigenous mini cow breeds the Vechur and Kasaragod breed from Kerala and the Malnad Gidda breed of Karnataka state. According to the *Guinness Book of Records*, Vechur is the smallest cattle breed in the world.

There are several lesser-known indigenous breeds adapted to particular local environment, which needs evaluation and characterization. In the last few decades, India imported several exotic cow varieties to enhance its milk production. The phylogenetic analysis and genetic distance estimation from the whole mitochondrial DNA (mtDNA) of seven Indian cattle breeds showed the close genetic relationship among the Indian cattle breeds, whereas distinct genetic differences were observed between *Bos indicus* and *Bos taurus* cattle.

11.3.2 Breeds of Buffalo

Buffalo are elite dairy animals of India and hold the greatest promise and potential for milk production and meat. Buffalo has been rightly christened as a black gold of India due to the presence of world's best milch breeds in India: Murrah, Mehsana, Jaffarabadi, Banni, and Nili- Ravi. The riverine buffalo has been selected as a dairy animal with several well-recognized breeds, whereas the swamp buffalo are primarily used for draft purposes (Colli et al., 2018). India is fortunate in terms of largest buffalo population with 109.85 million, which includes 18 recognized riverine breeds and one swamp buffalo (luit) and eight to ten lesser-known defined population groups. Many undescribed indigenous population are now being characterized as swamp-like Bhangor buffalo from Tripura (Singh et al., 2022) and the Meghalaya buffalo population (Singh et al., 2019).

Buffalo breeds of India have been traditionally divided into five major groups: the Murrah group (Murrah and Nili-Ravi), the Gujarat group (Jaffarabadi, Surti, Mehsana, and Banni), the Uttar Pradesh breeds (Bhadawari and Tarai), the Central India group (Nagpuri, Pandharpuri, Kalahandi, Manda, Jerangi, Sambalpur, Chilika, and Marathwadi), and the South India group (Toda, South Kanara, and Bargur).

11.3.3 Breeds of Goat

Goats also known as "poor man's cow" have been an integral component of Indian farming system; it plays a significant role in providing supplementary income and livelihood to millions of resource-poor

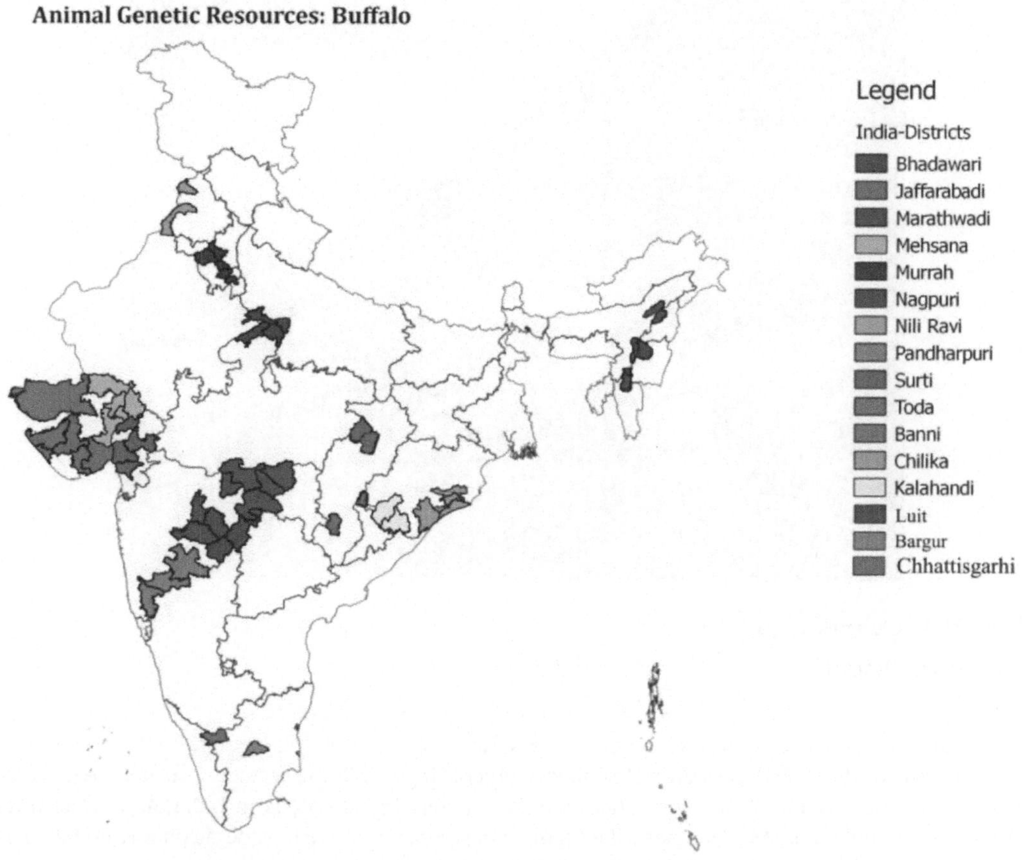

FIGURE 11.2 Geographical distribution of buffalo breeds of India.

FIGURE 11.3 Swamp buffalo from Meghalaya.

(*Source:* ICAR-NBAGR)

FIGURE 11.4 Changthangi (pashmina) goats.

(*Source:* ICAR-NBAGR)

farmers and supports a large population of landless farmers. Goats, being key livestock species, are an important part of the global agricultural economy. According to 20th Livestock Census, 2019, the goat population of India is 148.88 million. There are 34 registered goat breeds in India, as well as several lesser-known populations. Many undescribed indigenous population are now being characterized. These breeds vary notably in their morphology and reproductive traits.

The goat production system of the country is highly diverse, and these populations are well adapted to varied agro-climatic regions of the country, ranging from inhabitants like Chegu, Changthangi, and Gaddi from the cold desert of Ladakh in the Western Himalayan region to the dry northern regions covering the plains of Punjab, Haryana, Delhi, Rajasthan, Gujarat, Uttar Pradesh, Northern Madhya Pradesh, and Maharashtra, with valuable 11 goat breeds accounting to 33% of the goat population of the country. Semi-arid regions of MP, Maharashtra, AP, Tamil Nadu, Karnataka, and Kerala have the major mutton goat breeds like Osmanabadi, Sangamneri, Kanni-adu, Malabari, and Attappady.

The comparative transcriptome analysis of longissimus thoracis muscles has led to the identification of several genes that are differentially expressed in Barbari and Changthangi goats, which may be reflective of genetic selection and/or adaptation. The study highlights some important genes and pathways related to lipid metabolism in Barbari and muscle fiber contraction in Changthangi goats (Ashish et al., 2020). A recent study on gene networks and metabolic pathways that play significant role in pashmina production in Changthangi goats identified genes that can be to improve quality and quantity of the finest natural fiber, pashmina (Sonika et al., 2020).

11.3.4 Breeds of Sheep

Sheep were among the first animals domesticated. An archaeological site in Iran suggests that selection for woolly sheep begun to occur over 6,000 years ago. Sheep are the only species of livestock unable to return to a feral or wild state. With about 74.26 million sheep (Livestock Census, 2019), India ranks sixth in the world. The sheep breeds have been classified into wool and meat types. Out of the registered 44 sheep breeds, some of the important breeds are Balangir, Bellary, Bakharwal, Changthangi, Chokla, Chota Nagpuri, Coimbatore, Deccani, Gaddi, Ganjam, Garole, Jaisalmeri, Jalauni, Kenguri,

Magra, Marwari, Mecheri, Muzaffarnagari, Nali, Nelore, Nilgiri, Patanwadi, Poonchi, Ramnad White, Shahbadi, Sonadi, Tibetan, and Vembur.

11.3.5 Breeds of Camel

With the availability of draft camel genome as early as 2012, the high-quality genome sequences of Bactrian and dromedary camels and alpacas provided new insights into the camel adaptations to the harsh desert environment (Ramadan and Inoue-Murayama 2017). India has around 0.25 million camels, which are mostly concentrated in eastern parts of India. The major registered nine breeds of camel are Bikaneri, Jaisalmeri, Jalori, Kutchi, Malvi, Kharai, Marwari, Mewari, and Mewati. Kharai camels possess a significant trait of surviving In marshy and saline areas of Kutch region of Gujarat, and they are very good swimmer. Microsatellite markers (SSR) are constantly being used for identification of country's germplasm and its existing diversity for designing conservation plan (Rekha et al., 2020).

11.3.6 Breeds of Pig

The domestic pig/swine/hog (*Sus scrofa domesticus*, or only *Sus domesticus*) is an omnivorous domesticated animals. The present-day pig breeds have descended from the Eurasian wild boar (*Sus scrofa*), and the domestication of the pig started as early as 9,000 years ago. It is the first animal that indicates the start of settled farming communities since pigs are difficult to herd and move for long distances. About 2% of the total livestock in India is contributed by pigs. India has 7.16 million indigenous/nondescript breeds and about 2 million exotic animals imported from Europe. There are ten registered indigenous pig breeds, which include Ghungroo, Gurrah, Doom, Agonda Goan, Tenyi Vo, and Niang Megha, and still many undescribed.

FIGURE 11.5 Camels.

(*Source:* ICAR-NBAGR)

FIGURE 11.6 Traditional animal house for pigs, Jharkhand.

FIGURE 11.7 Arunachali yak.

(*Source:* ICAR-NBAGR)

11.3.7 Breeds of Yak

The yak (*Peophagus grunniens*) is a unique and multipurpose domestic animal in high mountainous areas of the Himalayas. It provides the indigenous people with meat, milk, butter, cheese, wool, fiber, leather, fuel, and travel and tourism requirements. Yaks have got adapted to very high-altitude regions where the oxygen content is very low due to anatomical and physiological advantages that help them, which includes large lungs and hearts, lack of hypoxic pulmonary vasoconstriction, strong environmental sense, and high energy metabolism.

Yaks are classified into Arunachali yak, Himachali yak, Ladakhi yak, and Sikkimi yak, based on the geographical area of inhabitation in India. Arunachali yak is the only registered breed in India. Per the 19th Livestock Census, the yak population in the country is 0.07 million.

The genome-wide SNPs are nucleotide variations in the DNA sequence of individuals of a population and are the most abundant molecular marker in the genome, with the double-digest restriction representation approach (ddRAD), genomic diversity; also, genetic candidate genes responsible for high-altitude adaptation in the Indian yak populations have been studied (Jayakumar et al., 2020).

11.3.8 Breeds of Chicken, Duck, and Geese

The chicken is a major source of protein in the world, with billions of birds used in meat and egg production. Since domestication about 5,400 years in Southeast Asia, selective breeding has made spectacular progress in both egg and meat production traits in the chicken (Burt, 2002). Chickens are widespread fowl and have been domesticated in India and Southeast Asia for various purposes, including decoration, religion, cockfighting, and food production. India harbors 19 native breeds of chicken, including the red jungle fowl, the wild ancestor of modern chicken breeds. These are distributed in various parts of the country and are well adapted to unfavorable changes in climate.

The whole mitochondrial sequence of six Indian native chickens—Aseel, Ghagus, Nicobari, Tellicherry, Kadaknath, and Haringhata Black—have been explored and analyzed. The chicken mitochondrial genome contains 37 genes encoding 13 proteins, 2 rRNAs, and 22 tRNAs encompassing 16,775 bp circular DNA.

Pati of Assam and Maithili from Bihar are the duck breeds registered. One goose breed registered is Kashmir Anz from Jammu and Kashmir. These are reared for meat, eggs, and feathers or as a hobby in areas located around the water bodies. Kashmir Anz geese are easy to maintain with minimum input and are hardy and disease-resistant.

FIGURE 11.8 Kadaknath chicken.

11.3.9 Breeds of Dog

Dogs are the most loyal, affectionate, and intelligent animals. Archaeological findings suggest that dogs first appeared 15,000–33,000 years ago in Europe and eastern Siberia, when humans were largely hunter-gatherers (Ovodov et al., 2011). Recent studies based on haplotype-sharing analyses show that contrary to previous mtDNA analyses, the nuclear genome of dogs derives primarily from Middle Eastern or European wolves, a result more consistent with the archaeological record. However, many populations—including those of Vietnam, India, and Egypt—show minimal evidence of European admixture. Dogs demonstrate extreme phenotypic diversity, as they are selected for phenotypic traits such as size, shape, coat color, conformation, and behavior.

Many dog breeds were developed in India but have been ignored. The Rajapalayam breed is an Indian guard dog developed in the Rajapalayam region in Tamil Nadu. The Chippiparai breed from Tamil Nadu, having milky-white coat with a pink nose, earlier bred by the Madurai royal family, is being used by the Indian army as guard dogs. The Karnataka Mudhol hound breed is very good companion for hunting and guarding. There are other breeds that still need recognition like the Rampur hound from Rampur, the Gaddi from the Himalayan region, the Kombai and Kanni from Tamil Nadu, the Indian mastiff from Punjab, and the Indian pariah dog.

11.3.10 Breeds of Horse, Pony, Mule, and Donkey

Horses were being used in central Asia for plowing the agricultural fields by humans around 4000 BC. After that, they were progressively used on sleds and later on mounted rollers or wheels. There are seven registered breeds of horse and pony, which are Bhutia, Kathiawari, Manipuri, Marwari, Spiti, Kachchhi-Sindhi, and Zanskari. Marwari horses got patronage from the erstwhile rulers of Marwar and are considered as the symbol of the warrior kings, while Manipuri horses are famous for their utility in polo. On the other hand, Spiti and Zanskari horses are found in high-altitude terrain areas.

FIGURE 11.9 Changkhi indigenous dog, Ladakh.

FIGURE 11.10 Mithun (*Bos frontalis*), Nagaland

The total population of horses, ponies, mules, and donkeys has drastically decreased by 51.9%, according to the 2012 census. Presently, the population of horses, ponies, mules, and donkeys in the country is 0.62 million. India has three registered breeds of donkey, which are Spiti of Himachal and Hallari and Kacchi of Gujrat.

11.3.11 Mithun

Mithun (*Bos frontalis*), also called gayal, is an endangered bovine species, has a limited geographical distribution restricted to the North-Eastern Hill (NEH) region of India. The first de novo draft genome assembly of the Indian mithun was published in 2018 (Mukherji et al., 2019).

11.4 Need for Conservation

The last few decades have witnessed serious erosion and even extinction of some indigenous livestock breeds in the country. Many existing breeds are facing various effects, which are pushing them toward extinction. In all the states, the crossbreeding of cattle is now occupying a dominant position in the production program, and in the process, the native cattle breeds that are well adapted have been neglected from the production system. Hence, there is an urgent need for preserving the biodiversity and conservation of breeds endangered of extinction.

Different conservation strategies have to be adopted based on the species or habitat. It can be done at *in situ* on-site-conservation and *ex situ* off-site-conservation conditions, based on needs and requirements. *In situ* conservation deals with maintaining species in their natural habitats and is the best way to maintain the biological diversity. This can be done in protected areas, national parks, and reserves, which provide ideal environments for the captive breeding programs. *Ex situ* conservation involves lab intervention for animals and helps in adaptation problems or managing loss of genetic variability due to inbreeding. Marker-assisted selection (MAS) is being used for selecting the best-suited animals, which could be helpful in increasing the frequency of desirable alleles/traits. Embryo storage, cryopreservation, seed banks, AI, and foster care are some examples. Judicious utilization and enhancement of the indigenous diversity is important for sustainability. Hence, there is an urgent need for preserving the biodiversity and conservation of breeds endangered of extinction.

11.5 Conclusion

Livestock diversity and agricultural ecosystems are developed and managed by humans. Human decisions are most important for their future fate. India is one of the mega-biodiverse hotspots, and it maintains a wide variety of livestock breeds that are the result of thousands of years of selection and environment effect. Different livestock breeds of the country are well adapted to diverse local conditions due to their inbuilt traits, like superior disease resistance, better tolerance to heat, and other characteristics suitable to particular agro-climate environments. The new era of omics technology provides us with genomic charts and genetic variations among individuals and groups that may prove beneficial. Identification and use of molecular markers for breed quality and production traits, disease resistance, and thermo-tolerance will ensure better productivity. Molecular markers for the improvement of temperament will ensure better management, production, and welfare of indigenous breeds. Apart from these, markers such as microsatellites for assessment of biodiversity will help in making conservation plans for populations.

REFERENCES

Baillie, J.E.M., C. Hilton-Taylor and S.N. Stuart (eds) (2004). *2004 IUCN Red List of Threatened Species. A Global Species Assessment.* IUCN, Gland, Switzerland.

Burt, D. (2002). Applications of biotechnology in the poultry industry. *World's Poultry Science Journal* 58: 5–14.

The Commission on Genetic Resources for Food and Agriculture. www.fao.org/cgrfa/en/

The Convention on Biological Diversity. www.cbd.int/

Colli, L., M. Milanesi, E. Vajana, D. Iamartino, L. Bomba, F. Puglisi, M. Del Corvo, E.L. Nicolazzi, S.S.E. Ahmed, J.R.V. Herrera, L. Cruz, S. Zhang, A. Liang, G. Hua, L. Yang, X. Hao, F. Zuo, S.-J. Lai, S. Wang, R. Liu, Y. Gong, M. Mokhber, Y. Mao, F. Guan, A. Vlaic, B. Vlaic, L. Ramunno, G. Cosenza, A. Ahmad, I. Soysal, E. Ö. Ünal, M. Ketudat-Cairns, J.F. Garcia, Y.T. Utsunomiya, P.S. Baruselli, M.E.J. Amaral, R. Parnpai, M.G. Drummond, P. Galbusera, J. Burton, E. Hoal, Y. Yusnizar, C. Sumantri, B. Moioli, A. Valentini, A. Stella, J.L. Williams and P. Ajmone-Marsan (2018). New insights on water buffalo genomic diversity and post-domestication migration routes from medium density SNP chip data. *Frontiers in Genetics*. 9: 53. doi: 10.3389/fgene.2018.00053

The Domestic Animal Diversity Information System. www.fao.org/dad-is/en/

FAO (2015). *The Second Report on the State of the World's Animal Genetic Resources for Food and Agriculture*, edited by B.D. Scherf and D. Pilling. FAO Commission on Genetic Resources for Food and Agriculture Assessments, Rome. www.fao.org/3/a-i4787e/index.html.

Food and Agriculture Organization of the United Nations (FAO) (2009). *The Roles of Small-Scale Livestock Keepers as Custodians of Animal Genetic Resources, CGRFA/WG-AnGR- 5/09/5.* FAO, Rome.

Hall, S.J.G. (2019). Livestock biodiversity as interface between people, landscapes and nature. *People and Nature* 1: 284–290. https://doi.org/10.1002/pan3.23

ICAR-NBAGR. (2022). National Bureau of Animal Genetic Resources. https://nbagr.icar.gov.in/en/home/

Jayakumar, S., M.R. Vineeth, T. Surya, K.V. Singh, S.P. Dixit, S.K. Niranjan, M.S. Tantia, I.D. Gupta, and D. Ravikumar (2020). Genomic divergence reveals unique populations among Indian Yaks. *Scientific Reports* 10: 3636. https://doi.org/10.1038/s41598-020-59887-3

Kumar, Ashish, Mandeep Kaur, Sonika Ahlawat, Upasna Sharma, Manoj Kumar Singh, Karan Veer Singh, Pooja Chhabra, Ramesh Kumar Vijh, Anita Yadav, and Reena Arora (2020). Transcriptomic diversity in longissimus thoracis muscles of Barbari and Changthangi goat breeds of India. *Genomics* 113(2021): 1639–1646.

Livestock Census. (2019). https://dahd.nic.in/documents/statistics/livestock-census

Mukherjee, S., Z. Cai, A. Mukherjee, et al. (2019). Whole genome sequence and de novo assembly revealed genomic architecture of Indian Mithun (Bos frontalis). *BMC Genomics* 20: 617. https://doi.org/10.1186/s12864-019-5980-y

The Nagoya Protocol on Access and Benefit-sharing. (2014). www.cbd.int/abs/

Ovodov, N.D., S.J. Crockford, Y.V. Kuzmin, T.F. Higham, G.W. Hodgins, et al. (2011). A 33,000-year-old incipient dog from the Altai Mountains of Siberia: evidence of the earliest domestication disrupted by the Last Glacial Maximum. *Plos ONE* 6: e22821.

Ramadan, Sherif and Miho Inoue-Murayama (2017). Advances in camel genomics and their applications: A review. *The Journal of Animal Genetics* 45(2): 49–58. http://doi.org/10.5924/abgri.45.49

The Rashtriya Gokul Mission (RGM). http://dahd.nic.in/rashtriya-gokul-mission

Sharma, Rekha, Sonika Ahlawat, Himani Sharma, Ved Prakash, Shilpa, Sunita Khatak, R.K. Sawal and M.S. Tantia (2020). Identification of a new Indian camel germplasm by microsatellite markers based genetic diversity and population structure of three camel populations. *Saudi Journal of Biological Sciences* 27(7): 1699–1709.

Singh, K.V., Uma Kant Verma, Manisha Arora, R.S Kataria and Saket Niranjan (2019). Cytogenetic analysis reveals existence of swamp buffalo population in Meghalaya state. *Indian Journal of Animal Sciences* 90(6): 868–870.

Singh, K.V., Ramendra Das, Saket Niranjan and R.S. Kataria (2022). Cytogenetic analysis reveals swamp status of the indigenous 'Bhangor' buffalo population from Tripura. *Indian Journal of Animal Sciences* 92(1): 65–67.

Sonika, A., A. Reena, Rekha Sharma, Upasna Sharma, Mandeep Kaur, Ashish Kumar, Karan Veer Singh, Manoj Kumar Singh and Ramesh Kumar Vijh (2020). Skin transcriptome profiling of Changthangi goats highlights the relevance of genes involved in Pashmina production. *Scientific Reports* 10: 6050. https://doi.org/10.1038/s41598-020-63023-6

12

Silkworm Biodiversity, Conservation, Threats, and Sustainable Utilization in Indian Subcontinent

Dandin S. B. and Nagesh Prabhu

CONTENTS

12.1 Introduction

Biodiversity of both plants and animals is becoming increasingly important for the survival of humankind on planet Earth. In the Sustainable Development Goals of the United Nations, the role of biodiversity is overwhelmingly emphasized because of its multiple ecosystem service and the associated benefits it offers. Human malnutrition is directly related to the diversity of foods. Mineral and vitamin deficiency

DOI: 10.1201/9781003220398-12

has become rampant among women and children and has resulted in large-scale nutrition-associated health disorders. Another important and highly visible effect of biodiversity loss is climate change and its effects in the ecosystem. Unless the process is not halted and reversed, the survival of life on Earth is very much jeopardized. Realizing the great threat being posed the biodiversity, several international treaties—IBPGR (FAO), TRIPS (WTO), GATT, UPOV, and so on—have been brought in to action. As a step forward, the Convention on Biological Diversity (CBD) was the first major legally bound approach signed by more than 170 countries, including the European Union, in Rio de Janeiro at the Earth Summit in 1992. The main objectives of the CBD are the conservation of biodiversity, the sustainable use of its components, and fair and equitable sharing of the benefits from of the use of genetic resources. The outcome of the CBD recognizes the sovereign rights of the nations over its biodiversity, environment, and ecosystems.

As part of animal biodiversity, insect diversity is the largest in terms of distribution and number. This is very much evident from the figure shown in this section. Insects are often referred to as pests that causes crop diseases and decreases crop yield. However, the group of economically important insects, comprising pollinators, silk-producing insects, lac insects, and so on, are well recognized for their economic products, besides the ecosystem services they provide for the maintenance of ecological balance. Of late, the role of insects as biocontrol agents is increasingly recognized due to residual toxicity of pesticides in food crops. Among all the beneficial insects, silk-producing insects stand out because of the silk processing is very significant and is recognized as a major farm activity that generates high income and employment all over the world. India is the only country where all the five commercially raised silkworms—*Bombyx mori*, *Antheraea myllitta*, *Antheraea royli*, *Antheraea perni*, *Antheraea assama*, and *Philosomia ricini*—are found. Of these, *Antheraea assama* is endemic to India and produces the much-valued golden silk. Rich food plant flora distributed in the majority of the Indian states support all types of silkworms year-round. In mori silkworm (*Bombyx mori*), there are three geographical races—univoltine, bivoltine, and multivoltine—whereas the tasar silkworm has three distinct species—*Antheraea mylitta*, *Antheraea perni*, and *Antheraea royli* which are popularly known as tropical tasar and have wild and domesticated populations. *Antheraea assama* is popularly known as the muga silkworm and is confined only to the Northeastern and Central Himalayan region. Similar to tasar, the *Antheraea assama* has both wild and domesticated populations. *Philosomia ricini*, popularly known as eri silkworm, is a totally domesticated one and very common in tribal regions of Northeastern Indian states. Silk, the mystic fiber, has been the attire of royals and a symbol of prestige from time immemorial both in Western and Eastern civilizations. The historically famous Silk Road, connecting Asia and Europe, has been the main trade route for many centuries.

The earliest reference of cultivated silk production dates back to roughly 2640 BC, when the Chinese empress Hsi Ling Shi patronized the silk industry. For several hundred years, the art of producing silk was kept as a secret and guarded safely within the country. Hoshomin, the Queen of China, introduced sericulture in China for the first time. For a long time, sericulture was considered to be a national secret by the Chinese government. Later, it was introduced to Europe and Japan. Sericulture was introduced to India about 400 years back, and since then, the silk industry flourished as an agro-industry par excellence. The list of different sericigenous (silk-producing) fauna and their food plants are listed in Table 12.1. According to Mukerji (1899), the Aryans discovered silkworm in sub-Himalayas beyond Kashmir. Indian epics like the Rigveda, Ramayana, and Mahabharata (Nanavati, 1990) also mention the use of *'Pitambara'* (silk clothes of Indian origin) and *chidambara* (silk clothes of Chinese origin). Despite the long history of Indian sericulture, it is only in the independent India that concerted efforts were initiated for characterization and conservation of silkworm breeds in different agro-climatic regions.

As could be seen from the previous table, except *B. mori*, *A. pernyi*, and *A. yamamai*, *the* all species originated in India. This is mainly because of the availability of suitable food plants throughout the year to successfully complete their life cycle and sustain the generation. Hence, to conserve this insect diversity, the conservation of their food plants is also imperative.

TABLE 12.1

Commercially Exploited Sericigenous Insects of the World and Their Food Plants

Common name	Scientific name	Origin	Primary food plants
Mulberry silkworm	*Bombyx mori*	China	*Morusindica alba, M. multicaulis, M. bombycis*
Tropical tasar silkworm	*Antheraea mylitta*	India	*Shorea robusta, Terminalia tomentosa, T. arjuna*
Oak tasar silkworm	*Antheraea proylei*	India	*Quercus incana, Q. serrata, Q. himalayana, Q. leucotricophora, Q. semicarpifolia, Q. grifithi*
Oak tasar silkworm	*Antheraea frithi*	India	*Q. dealdata*
Oak tasar silkworm	*Antheraea compta*	India	*Q. dealdata*
Oak tasar silkworm	*Antheraea pernyi*	China	*Q. dendata*
Oak tasar silkworm	*Antheraea yamamai*	Japan	*Q. acutissima*
Muga silkworm	*Antheraea assama*	India	*Litsea polyantha, L. citrata, Machilus bombycine*
Eri silkworm	*Philosamia ricini*	India	*Ricinus communis, Manihot utilisma, Evodia fragrance*

12.2 Part A: Mulberry Silkworm (*Bombyx Mori*)

The mulberry silkworm (*Bombyx mori*) is totally domesticated for the production of most commercially valuable silk for the last 4,000 years. The mulberry silkworm finds its commercial origin from China, tracing back to more than 5,000 years. It is believed that two wild silkworms, *Bombyx mandarina* and *Theophila religiosae*, are the closest relatives of the domesticated silkworm.

Taxonomic position

Phylum	:	Arthropoda
Class	:	Insecta
Subclass	:	Pterygota
Division	:	Endopterygota
Order	:	Lepidoptera
Family	:	Bombycidae
Genus	:	*Bombyx*
Species	:	(i) *mori* (domestic silkworm) (ii) *mandarina* (wild silkworm)
Genus	:	*Theophila*
Species	:	*religiosae* (wild silkworm)

The wild silkworm (*Theophila religiosae*) is commonly found in the Himalayan tract, especially in northeastern part of the country.

Bombyx, though a single genus, has three variants from different geographical locations—(1) *Japanese* (peanut-shaped cocoons), (2) Chinese (elliptical or spherical cocoons), and (3) *European* (large, with little constriction in the cocoons)—as suggested by Yokoyama (1959), Dandin et al. (2005), Tazima (1958). Due to continuous breeding, few promising silkworm breeds have been bred in all the three types of silkworms. More than 58 countries are rearing this silkworm (Figure 12.4), and the mori silkworm feeds mainly on various species of mulberry (*Morus* spp.). The origin and distribution of different races is shown in the following section.

12.2.1 Evolution and Distribution

Bombyx mori silkworms are classified into Japanese, Chinese, European, Korean, and tropical races. Earlier studies indicate that based on mutual relationship, the races became differentiated over a long period of time from the Chinese univoltine races, as illustrated in the following representation.

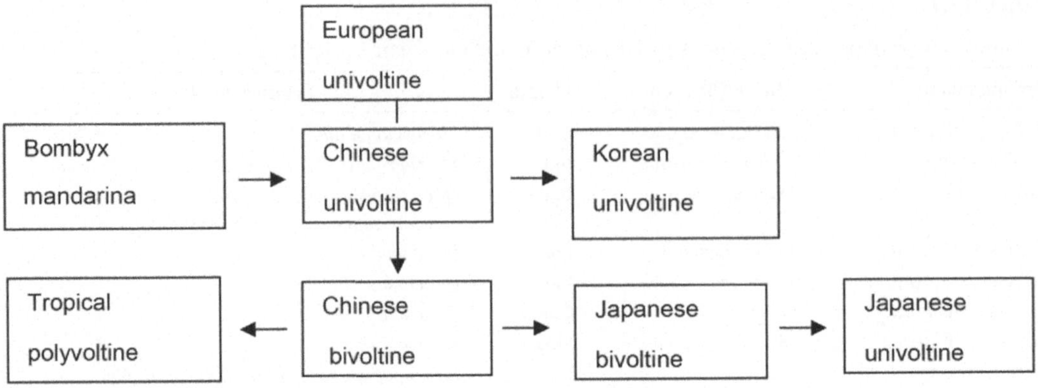

FIGURE 12.1 Evolution and geographical distribution of silkworm races.

(*Source:* Tazima (1964))

For commercial purposes, the Japanese race (constricted shape) and the Chinese race (oval shape) are mainly used in the temperate zone. In the tropical zone, polyvoltine races are used as one of the parents crossing with constricted or oval race. The improvement of traits of interest have been carried out over years, which has resulted in a large number of breeds (Basavaraja et al., 1995; Datta et al., 2000; Mano et al., 1994; Mukherjee 1994). They can be classified as follows:

1. Univoltine: They are suitable for cooler places, the growth period of larvae is longer, and they have big bodies. The cocoon and silk quality are excellent. High temperature and high humidity are not suitable.
2. Bivoltine: They are suitable for warmer places. The larval stage is shorter compared to the univoltine's. Larvae are robust. Cocoon and silk quality are good. They are moderately adapted to high temperature and high humidity.
3. Polyvoltine: Theya are adapted to warmer places. The larval stage is shorter (except *Pure Mysore*). Larvae are robust, cocoons are small, and the silk quality is poor. They lay only non-hibernating eggs.

The evolution and geographical distribution of different geographical races indicates the multicenter origin and bimodal evolution. Univoltine races are mostly temperate in origin and confined to Europe and Russia. They complete only one life cycle in a year and mostly produce white cocoons. Bivoltine races complete their life cycle twice and are exploited commercially. Though the genus had its origin in China, due to continuous selection for various commercial characteristics—namely, survival rate, fecundity, hatching, larval duration, cocoon weight, filament length, resistance to different biotic, and abiotic stresses—numerous races have been evolved mainly in China, Japan, Korea, and India. The silkworm is perhaps the only insect of commercial importance that has been studied extensively both in the field of genetics and biology. The larval and cocoon stages and the whole life cycle of the wild and the domesticated silkworms, as well as the variability in cocoon shape, size, and color, are given in Figures 12.2, 12.3, 12.4, and 12.5. Similarly variability in larval markings and sizes and cocoon shapes and colors are shown in Figure 12.6, 12.7, and 12.8.

12.2.2 Genetic Resources of *Bombyx Mori*

Multivoltine races are mostly tropical in origin and complete many life cycles in a year. Cocoons are small and elongated and have shorter filaments. These races are more resistant to both biotic and abiotic stresses. The list of commercially exploited races of different countries is given in Table 12.2. Similarly, genetic resources of *B. mori* maintained in in different centers of the country is given in Table 12.3. The

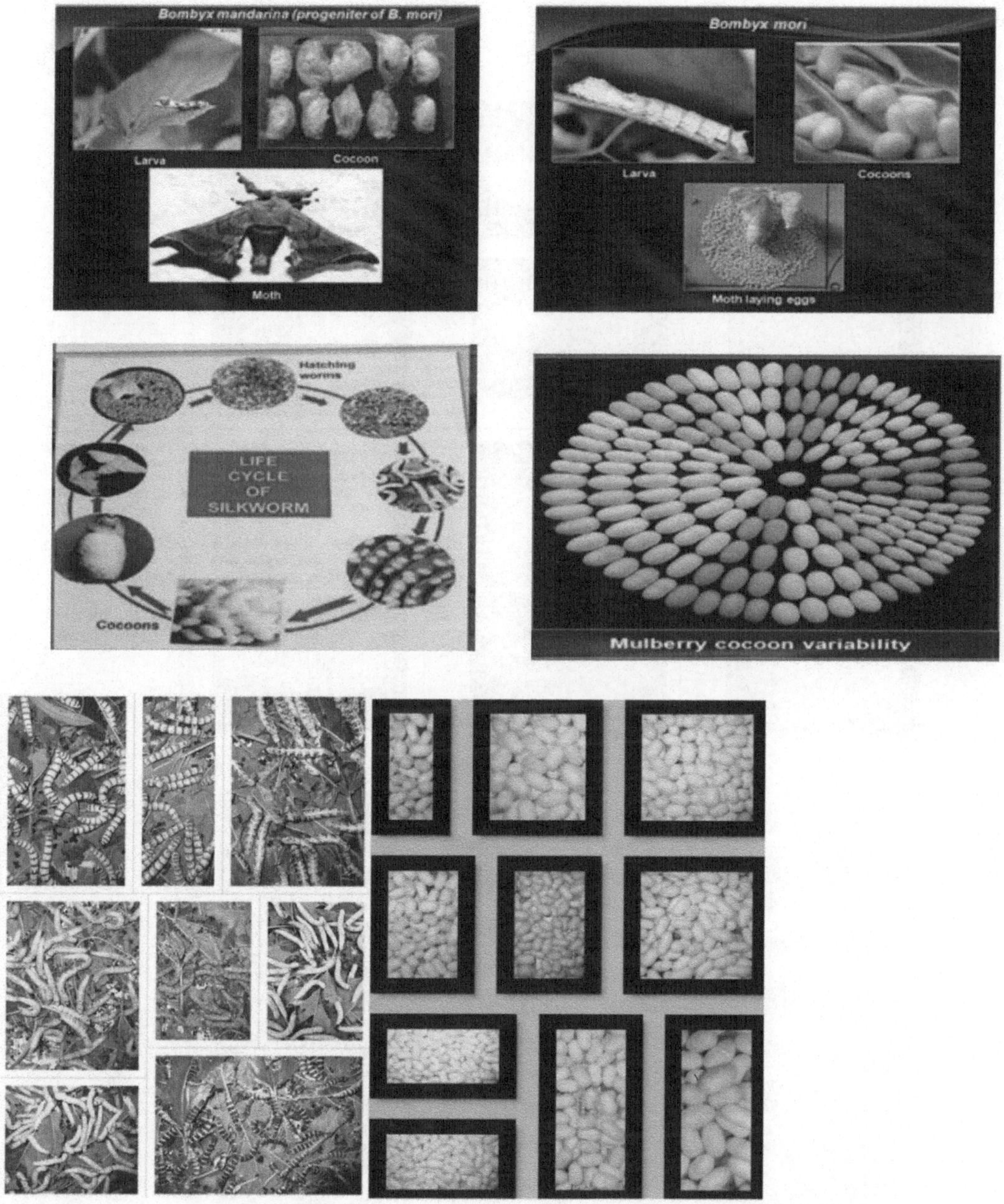

FIGURE 12.2 *Bombyx mandarina* and *B. mori* life cycle and variability in cocoon characteristics of *B. mori*.
(*Source:* CSGRC, CSB, Hosur)

important of silkworm races reared in different zones in the country is given in Table 12.4. Germplasms of all the silkworm breeds maintained in various places is given in Table 12.5.

Silkworm genetic resources consist of geographically isolated races, genetic stocks, and breeds extracted from indigenous and exotic hybrids. It also includes races/strains collected from different institutes of India and other sericulturally advanced countries like Japan, China, Russia, South Korea, and France. The laboratory of silkworm stock culture at Kabuchizawa in Yamanshi Prefecture, Japan, maintains around 500 breeds. MRIS in Russia has a stock culture of around 197 breeds (Basavaraja et al.,

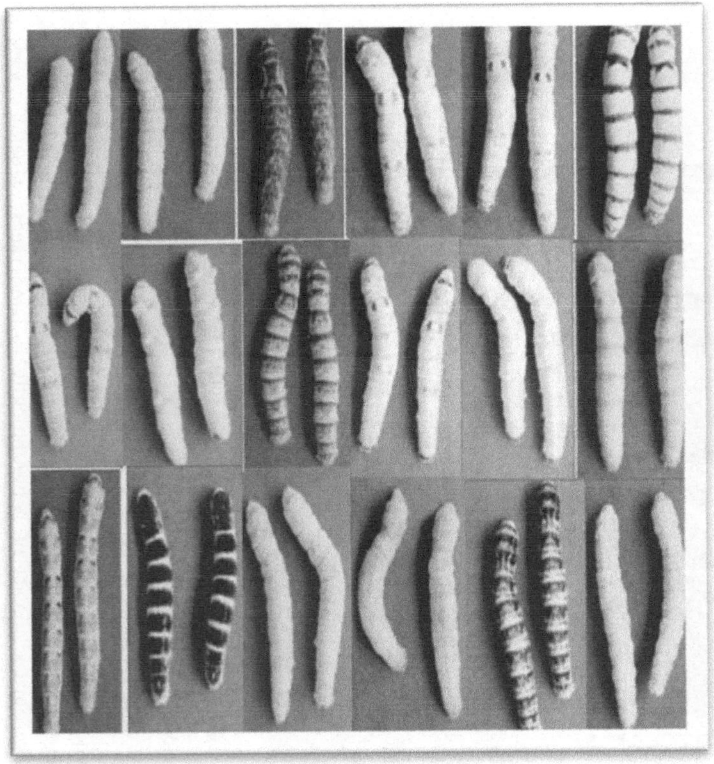

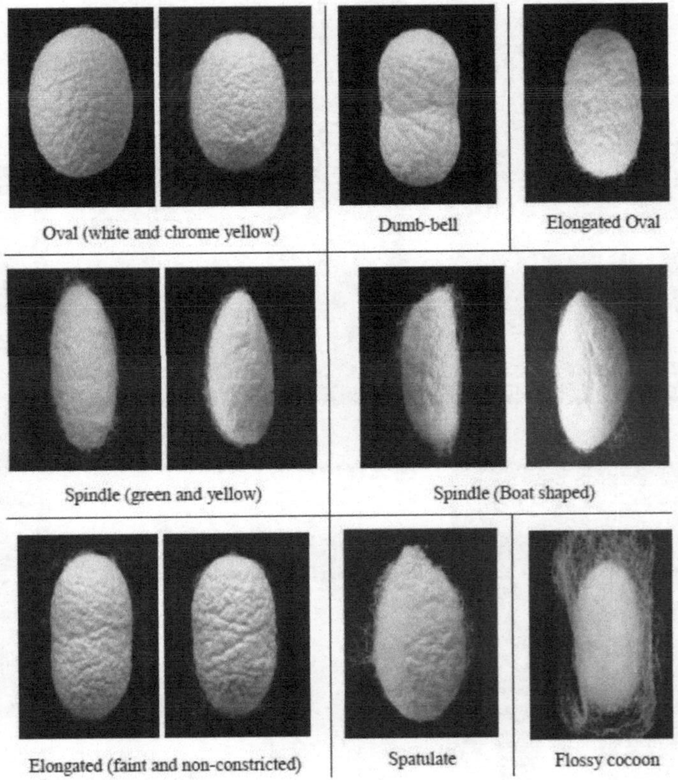

FIGURE 12.3 Variability in larval markings of *B. mori*.

(*Source:* CSGRC, CSB, Hosur)

1995 and Koji, 2020), as shown in Table 12.2. Considering the importance of silkworm biodiversity and conservation, the Central Silk Board, under the National Sericulture Project, established a separate center exclusively for seri-biodiversity conservation in India at Hosur (Tamil Nadu)—the Central Sericultural Germplasm Resource Centre (CSGRC). This center is maintaining 490 bivoltine breeds, and 100 polyvoltine strains have been maintained at different breeding centers/research institutes, with more than 400 silkworm genetic resources representing 14 countries (Table 12.3). The center is also maintaining the genetic resources collected from different centers in India (Table 12.4)

In addition to these mentioned pure silkworm races/parents being maintained at different sites all over the world, silkworm single and double hybrids being maintained in different countries and in India are shown in Tables 12.5 and 12.6.

TABLE 12.2

Country-Wise Collection of Silkworm Genetic Resources

SN	Country	Multivoltine	Bivoltine	Total
1.	Brazil	10	65	75
2.	Bulgaria	–	183	183
3.	China	20	580	600
4.	France	–	53	53
5.	India	150	450	600
6.	Iran	–	50	50
7.	Italy	–	30	30
8.	Japan	30	1,542	1,572
9.	N. Korea	5	281	286
10.	Russia	–	500	500
11.	S. Korea	6	300	306
12.	Thailand	5	25	30
13.	Vietnam	5	20	25
	Total	**231**	**4,079**	**4,310**

(*Source:* Basavaraja, HK)

TABLE 12.3

Country-Wise Collection of Silkworm Genetic Resources Maintained at CSGRC, CSB Hosur (as of September 1, 2021)

		Bivoltine		Multivoltine				
SN	Country	Ind.	Exo.	Ind.	Exo.	BV Total	MV Total	Grand Total
1	India	223	31	73	-	237	71	308
2	Bangladesh	–	0	–	3	–	3	3
3	Brazil	–	3	–	–	3	–	3
4	China	–	40	–	4	40	4	44
5	France	–	11	–	–	11	–	11
6	Indonesia	–	1	–	–	1	–	1
7	Iraq	–	1	–	–	1	–	1
8	Japan	–	64	–	3	64	3	67
9	Poland	–	3	–	–	3	–	3
10	Russia	–	19	–	–	19	–	19
11	South Korea	–	1	–	–	1	–	1
12	Thailand	–	4	–	–	4	–	4
13	Ukraine	–	2	–	–	2	–	2
14	Vietnam	–	3	–	–	3	–	3
TOTAL		**223***	**183**	**73**	**10**	**406**	**83**	**489**

TABLE 12.4
Institute-Wise Available Silkworm Genetic Resource in India (as of September 1, 2021)

		Bivoltine		Multivoltine		Mutant		BV	MV	Mutant	Grand
SN	**Donor Institute**	**Ind.**	**Exo.**	**Ind.**	**Exo.**	**Ind.**	**Exo.**	**Total**	**Total**	**Total**	**Total**
1	APSSRDI, Hindupur	9	–	4	–	–	–	9	4	–	13
2	CSB, Bangalore	2	25	–	–	–	–	27	–	–	27
3	CSRTI, Berhampore	37	1	21	1	–		38	22	–	60
4	CSRTI, Mysore	48	50	40	9	–	22	98	49	22	169
5	CSRTI, Pampore	64	72	–	–	–	–	136	–	–	136
6	DOS, Coonoor	5	–	2	–	–	–	5	2	–	7
7	DOS, J&K	14	–		–	–	–	14	–	–	14
8	KSSRDI, Bangalore	3	–	1	–	–	–	3	1	–	4
9	KSSRDI, Thallagatapura	13	–	1	–	–	–	13	1	–	14
10	NSP, Botate	1	–	–	–	–	–	1	–	–	1
11	NSP, Majra	4		–	–	–	–	4	–	–	4
12	SSBS, Coonoor	11	8	–	–	–	–	19	–	–	19
13	RSRS, Jammu	3	4	–	–	–	–	7	–	–	7
14	RSRS, Jorhat	2	–	–	–	–	–	2	–	–	2
15	RSRS, Sahaspur	7	–	–	–	–	–	7	–	–	7
16	Mysore University	–	–	3	–	–	–	–	3	–	3
17	SSTL, Kodathi	–	–	1		–		–	1	–	1
18	SBRL, Bengaluru	–	–	–	–	–	1	–	–	1	1
	Total	**223**	**160**	**73**	**10**	–	**23**	**383**	**83**	**23**	**489**

(*Source:* CSGRC, CSB, Hosur)

TABLE 12.5
Popular Silkworm Races/Hybrids in Major Sericultural Countries of the World

Country	Season	Silkworm races/hybrids
Japan	Spring	Taihei × Choan; Shungetsu × Hosho; J124 × C122; Shunrei × Shogetsu; Shuko × Ryhaku; Asa.hi × Tokai; Fuyo × To.kai
	Summer	Kinshu × Showa; J124 × C124
	Autumn	(J122.C124) × (J115.C124)
	Special breeds	Akebano (N511.N512) × (C511.C512); Hakugin (C514 × N515); Sakigake (N507.N508) × (C507.C508); Ariake (N509.N5010) × (C509. C5010); Nsd-2 (non-susceptibility to DNV2; Nid-1 (non-infection to DNV)
USSR		Saniish-8 (uni); Saniish-9 (uni); Belakokkonay-1; Belakokkonay-2; Azarbaijan; Azad
China		NanNung-7; Tongsu34 × Shanshu12
	Zhejiang	Tungwe × Waha; Tongfei × Huahe; Giansu16 × Giansu17; Chun3 × Chun4; Giansu12 × No.1; Giansu12 × Tongshan4; Tong34 × Su12 Disease-tolerant strains: Chinese strain (Zhenjiang) Zhnz, 22; Da-09; 7201; Nong-42; Rong TaZhao, Nong-15; 115 Nai; Ye 5; etc. as CPV-resistant and DP Dazhong, 137, etc., as IFV-resistant; Dazhao, 137, resistant to more than one disease, namely Dazhao, Mez, and Dazhong, resistant to both IFV and CPV.
	Spring	Quingsong × Haoyue; Su5 × Su6; Chunlei × Zhenzhu (SR –25%);
	Sumer and Autumn	Furong × Xianghui (Su3 × Qui3) × Su4; Feng1 × 54A; Zhongqiu × Jinling; Xuhua and Quiming (21–23% SR)

Country	Season	Silkworm races/hybrids
	Tropical zone	Quingsong; Haoyue, Sufang; Chunhin, (Chunlei × Zhongzhu), (57A . 57B) × (24 . 46); Zhelei × (Chunxiao)
	Temperate	Sichnun, Zhejang and Jianggsu, Hua; He (Ch) × Tong; Fei (Jap) and Su16 (Jap) × Su17 (Jap)
	Multivoltine	NanNung-7; Guangdong3
	Bivoltine	Tung-34 (Ch) × Su-12 (Jap)
Thailand		Chul Thai No. 1,6; ChulThai No. 7; Thai hybrid silkworm (yellow cocoon); Chinese × Japanese
Bulgaria		B1 × Merefa 2; Svila2 × Hesa1; Vratza35 × Svila2; Super 1 × Svila2; B1 × Hesa2
Brazil (Cocomar)		(M11A.M12B) × (C124A.C25B); (M18.M19) × (C124A.C25B); (HaA. HaB) × (C12A.C122B); (Na. 1. Hi.1) × (Ki.1.Yo.1)

(*Source:* Basavaraja et al.)

TABLE 12.6

Commercially Exploited Mulberry Silkworm Hybrids in India

Region	Bivoltine hybrids	Polyvoltine × bivoltine hybrids
Southern zone	CSR2 × CSR4	PM × CSR2
	FC1 × FC2 (double hybrid)	MH1 × CSR2
	APS9 × APS8	APM1 × APS8
	APS5 × APS4	
	NP2 × KS01	
Northern zone	DUN6 × NB4D2	
	SH6 × NB4D2	
	YS3 × SF19	
	PAM101 × NB4D2	
	SKUAST-1 × SKUAST-6	
Eastern zone	Bcon1 × Bcon4	Nistari × NB4D2
	SK6 × SK7	Nistari × (SK6.SK7)
		Nistari × M12 (w)

(*Source:* Central Silk Board, www.csb.gov.in)

12.2.3 Conservation Strategies

Conservation strategies are the most important aspects and mainly depend on reproductive behavior and life cycle pattern of the individual species/biotypes. *Bombyx mori* is a monophagous insect that solely depends on mulberry foliage for feeding, and food availability of desired quality is of utmost importance to ensure to complete life cycle. As indicated earlier, there are three types of mori silkworms—univoltine having only on life cycle in a year, bivoltine with two, and multivoltine with five to six. The silkworm germplasm maintenance and conservation as recommended by CSGRC, Hosur, is in the flowchart in the following image. Egg disposal/hibernation is the common aspect in the life cycle and needs to be followed meticulously to keep all original characteristics of the genotype intact and also maintain the continuity.

For univoltine silkworms, at least the genotype has to be reared once in a year, especially during spring, which is the most suitable season for rearing. For the bivoltine silkworms, a minimum of two rearings can be conducted, and proper hibernation can be scheduled, followed with ensuring that the storage

Flow Chart on conservation of silkworm genetic resources

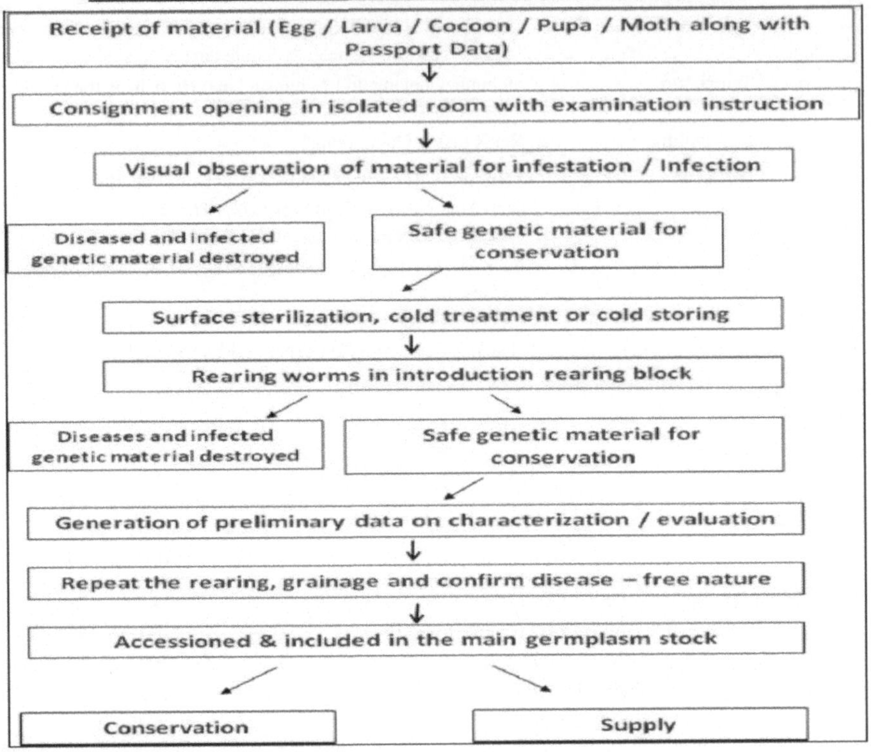

(*Source:* CSGRC, CSB Hosur)

facilities are cold. On the contrary, all the multivoltine races are reared at least five times a year by following all the standard rearing practices and multiplication protocols for all the races (Tazima, 1964). The one-way multiplication system shall be followed for the bivoltine, as suggested by the Japanese experts (Koji, 2020). The Central Silk Board has taken the job very seriously, and an exclusive institute, the Central Sericulture Germplasm Resource Station (CSGRC), Hosur, has been established as early as 1991 with all the required facilities and clear mandate. Well-trained staff have been employed to meticulously maintain the breed characteristics. The center has been recognized by NBPGR and NBAGR of the government of India, extending all the technical support. The center has collected all the available silkworm and mulberry races and varieties all over the world, as shown in the Tables 12.2 and 12.3. The majority of the collections have been characterized and cataloged per the international parameters, and the center is supplying the genetic material to all the sericulture institutes of the country, as well as outside the country, as a resource center. As indicated earlier, enabling the breeds to express maximum genetic potential by supplying the nutritious feed under the defined niche condition is highly essential. In this direction, the center is maintaining the identified mulberry varieties to ensure constant supply of nutritious leaf.

12.3 Part B: Non-Mulberry (Vanya) Silkworm Biodiversity

India has the distinction of being the only country that has domesticated all the four commercially known wild silkmoths—tropical tasar (*Antheraea mylitta*, D.), oak tasar (*A. proylei*, J.), muga (*A. assamensis*, Helfer), and eri (*Samia ricini*, Donovan). Of the five varieties of natural silks, mulberry silk is the chief variety in India. The next in commercial importance is tasar silk, followed by eri and muga silk. Hence, the vanya (non-mulberry) silks are of considerable importance in the Indian context.

Collection of wild silk cocoons and drawing handspun yarn from them had been a profitable occupation of our tribal folks since time immemorial. Besides earning a livelihood for thousands of rural poor, silk cultivation and silk weaving has been the part of our cultural heritage. Hence, it is essential that these traditional cottage industries of native India, along with associated traditional knowledge sustain at any cost.

The wild silkmoths or non-mulberry silkmoths mostly belong to Family Saturniidae are reared out door except the eri silkworms. The Indian sub-Himalayan belt extending from Jammu and Kashmir in the northwest and northeastern regions in general and the northeastern region in particular of India is the natural abode of large variety of sericigenous insects and their food plants, as reported by Jolly et al. (1976), Baruah et al. (2000), Singh et al. (2000), Singh and Maheswari (2003) and Thangavelu (1991). The wild silkmoths play an important role in the conservation and utilization of biodiversity as reported by Frankel (1982), Peigler (1993) and Kioka (1998).

Many indigenous species and eco-races of silkworms along with their food plants are still naturally available in different parts of India (Tables 12.7, 12.8, and 12.9 and Figures 12.9 and 12.10). In addition, hundreds of newly evolved breeds that are commercially exploited in the hybrid forms have also been introduced in the habitat. Some of the traditional silkworm races are lost due to habitat destruction as well as rapid replacement by high yielding silkworm breeds. The wild silkworms have several ecotypes feeding on different food plants and distributed in different eco-climatic conditions of the country. Due to large-scale disturbances of habitat, wild silkworm species are on the verge of extinction. Thus, for sustained utilization of silkworm bio-resources, it is necessary to adopt appropriate strategies for *in situ* and *ex situ* conservation of existing seri-bio-resources.

TABLE 12.7

Available Species of Wild Silkmoths and Their Wild Relatives in India

SN	Scientific name	SN	Scientific name
Family Saturniidae			
1	*Actias selene*, Hübner, 1806	23	*Rhodinia newara*, Moore, 1872
2	*Antheraea assamensis*, Helfer, 1837	24	*Salasa tonkiniana*, Le Moult, 1933
3	*Antheraea compta*, Rothschild, 1899	25	*Samia canningii*, Hutton, 1860
4	*Antheraea frithi* Moore, 1858	26	*Samia kohlli*
5	*Antheraea mylitta* Drury, 1773	27	*Samia ricini*, Boisduval, 1854
6	*Antheraea pernyi*, Guérin-Meneville, 1855	28	*Satrunia pyretorum*, Westwood, 1847
7	*Antheraea roylei*, Moore, 1858	29	*Saturnia simlaensis*, Westwood, 1847
8	*Antheraea helferi*, Moore, 1858	30	*Saturnia thibeta*, Westwood
9	*Antehraea andamana*, Moore, 1877	31	*Saturnia* sp.
10	*Archaeoattacus edwardsii*, White, 1859	32	*Rhodinia newara*, Moore, 1872
11	*Archaeoattacus staudingeri*, Rothschild, 1895		**Family Bombycidae**
12	*Argem amaenas*, Doubleday, 1847	33	*Andraca* sp.
13	*Argema sinensis*, Walker, 1855	34	*Bombyx mori*, Linnaeus, 1758
14	*Attacus atlas*, Linnaeus, 1758	35	*Bomby xincomposita*, van Eecke, 1929
15	*Cricula Andrei*, Jordan, 1909	36	*Norasuma javanica*, Moore, 1872
16	*Cricula trifenestrata*, Helfer, 1837	37	*Gunda ochracea*, Walker, 1862
17	*Loepa katinka*, Westwood, 1848	38	*Gunda* sp.
18	*Loepa megacore*, Jordan, 1911	39	*Ocinara bifurcula*, Dierl, 1978
19	*Loepa sikkima*, Moore, 1865	40	*Triuncina religiosae*, Helfer, 1837
20	*Loepa* sp. 1	41	*Triuncina* sp.
21	*Loepa* sp. 2	42	*Trilochavarians*, Walk, 1855
22	*Loepa miranda*, Moore, 1865		

(*Source:* Kumar et al., 2017)

TABLE 12.8

Host Plants of Wild Silkmoths

Vanya silk	Scientific name	Host plants	Distribution in India
Tasar silk	*Antherea mylitta* *A. proylei* *A. pernyi* *A. yamamai* *A. frithi*	*Terminalia arjuna* *T. tomentosa* *Shorea robusta* *Quercus incana* *Q. serrata* *Q. semicarpifolia* *Q. himalayana* *Q. citrate* *Q. serrata* *Q. acutissima*	Jharkhand, Orissa, Chattisgarh, Manipur, Andhra Pradesh, Nagaland, Assam
	Attacus atlas, L.	*Maesa indica* (Roxb.), Wall *Evodia flaxinifolia*, Hook *Litsea polyantha*, Juss	Assam, Nagaland, Mizoram
Muga silk	*Antheraea assamensis*, Helfer	*Litsea polyantha*, Juss *L. citrate* *Persea bombycina*, Kost	Assam, Meghalaya, Nagaland
Eri silk	*Samia ricini*, Donovan	*Ricinus communis*, L. *Heteropanax fragrans* (Roxb.), Seem *Evodia flaxinifolia*, Hook *Manihot esculanta*, Crantz *Ailanthus grandis* *A. excelsa* *A. malabarica* *A. altissima*	NE Region, Orissa, Andhra Pradesh, Uttar Pradesh
Wild eri silk	*Samia canningi*, Hutton	*Ricinus communis*, L. *Heteropanax fragrans* (Roxb.), Seem *Evodia flaxinifolia*, Hook *Manihot esculanta*, Crantz *Anthocephalus cadamba*, Lamk *Ailanthus grandis* *Litsea salicifolia*, Roxb. *Litsea cubeba*, Pers *Psidium guajava*, L.	Assam, Manipur, Nagaland, Mizoram
Wild Muga	*Cricula trifestrata*, Helfer	*Persea bombycina*, Kost *Litsea cubeba*, Pers *Mangifera indica* *Psidium guajava*, L.	Assam, Nagaland, Manipur, Meghalaya

TABLE 12.9

Host Plants of Wild Silkmoths

Vanya silk	Scientific name	Host plants	Distribution in India
Tasar silk	*Antherea mylitta* *A. proylei* *A. pernyi* *A. yamamai* *A. frithi*	*Terminalia arjuna* *T. tomentosa* *Shorea robusta* *Quercus incana,* *Q. serrata* *Q. semicarpifolia,* *Q. himalayana* *Q. citrate* *Q. serrata* *Q. acutissima*	Jharkhand, Orissa, Chattisgarh, Manipur, Andhra Pradesh, Nagaland, Assam
	Attacus atlas, L.	*Maesa indica* (Roxb.), Wall *Evodia flaxinifolia*, Hook *Litsea polyantha*, Juss	Assam, Nagaland, Mizoram
Muga silk	*Antheraea assamensis*, Helfer	*Litsea polyantha*, Juss *L. citrate* *Persea bombycina* Kost	Assam, Meghalaya, Nagaland

Vanya silk	Scientific name	Host plants	Distribution in India
Eri silk	*Samia ricini*, Donovan	*Ricinus communis*, L. *Heteropanax fragrans* (Roxb.), Seem *Evodia flaxinifolia*, Hook *Manihot esculanta*, Crantz *Ailanthus grandis* *A. excelsa* *A. malabarica* *A. altissima*	NE Region, Orissa, Andhra Pradesh, Uttar Pradesh
Wild eri silk	*Samia canningi*, Hutton	*Ricinus communis*, L. *Heteropanax fragrans* (Roxb.), Seem *Evodia flaxinifolia*, Hook *Manihot esculanta*, Crantz *Anthocephalus cadamba*, Lamk *Ailanthus grandis* *Litsea salicifolia*, Roxb. *Litsea cubeba*, Pers *Psidium guajava*, L.	Assam, Manipur, Nagaland, Mizoram
Wild muga	*Cricula trifestrata*, Helfer	*Persea bombycina*, Kost *Litsea cubeba*, Pers *Mangifera indica* *Psidium guajava*, L.	Assam, Nagaland, Manipur, Meghalaya

(*Source:* Ahmed and Rajan, 2011)

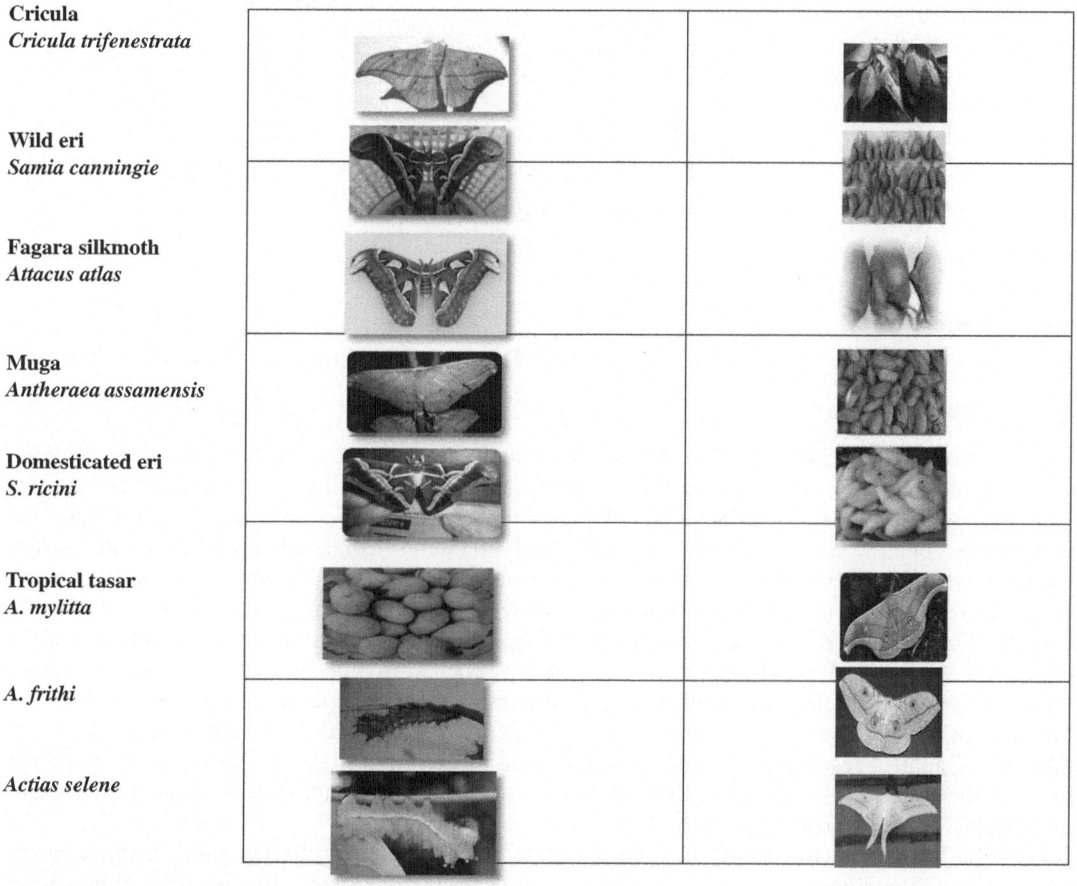

FIGURE 12.4 Wild sericigenous insects of NE India.

(*Source:* Ahmed and Rajan, 2011)

FIGURE 12.5 Geographical distribution of non-mulberry (vanya) silks.

12.4 Status of Vanya Sericigenous Insect Diversity in India

12.4.1 Tropical Tasar

The rich biological diversity of *A. mylitta* is largely due to its wide range of distribution and foraging of silkworms on variety of food plants. Owing to its wide range of distribution, the species has encountered diverse geo-climatic situations, like varying annual precipitation, temperature, day length, and plant succession. Factors like latitude, longitude, and altitude of distinct areas lead to marked differences, expressing wide variations in phenotypic, physiological, behavioral, commercial, and technological traits. Various populations, thus isolated geographically over centuries, have adapted to a particular ecological niches and are referred to as eco-races. These eco-races constitute unique genetic resources of tropical tasar for today and tomorrow. The existing genetic resources of *A. mylitta* can be utilized effectively only if sufficient information on the intra and interspecific genetic diversity available. The inventory of tropical tasar genetic diversity will serve as ready record of the silkworm *Antheraea mylitta* (Drury) for applied researchers, taxonomists, geneticists, environmentalists, entomologists, physiologists, textile technologists, foresters, fashion designers, farmers, dietitians, cosmeticians, pharmacologists, and silk users for reasons of their own.

Available information reveals that *A. mylitta* was bred in Europe during the beginning of 19th century, while typical *A. mylitta* occurs only in the Asian continent. It is represented in the continent by a large number of similar forms, many of which were treated as distinct species, though they are local races: *A. fasciata*, *A. versicolor*, *A. pulchra*, and *A. ochripicata* (Jolly et al., 1974) stated that the cocoons of *A. paphia* and *A. mylitta* imported to Europe often yield a series of aberrations and transitions. Thus, he expressed doubt about the value of forms and the justification of the names. Due to its variable appearance, it was described under a number of synonyms.

12.4.1.1 Natural Habitats

Natural habitat of majority of the eco-races is restricted mainly to tropical moist deciduous forest where the average rainfall varies between 1,200 and 2,000 mm and the deciduous zone of the tropical dry forest where the average rainfall is up to 1,000 mm. A discrete population of *A. mylitta* is also found in subtropical mountains and tropical wet evergreen and thorny forests. Details of natural habitats of economically important eco-races are presented in Table 12.10.

Different stages of life cycle of tasar silkworm is shown in Figure 12.11. Due to the wide range of distribution over different altitudes and other climate determents and also different food plants on which silk tasar worm feeds, an enormous amount of variation is observed in larvae, cocoons, and moths, as shown in Figures 12.12 and 12.13.

The tropical tasar silkworm, *Antheraea mylitta*, is distributed in wide forest regions of the country, and there is a report of availability of 44 eco-races of the tasar silkworm (Suryanarayana and Srivastava, 2005). Among them, *Shorea robusta* (sal) eco-races available in Jharkhand (Laria), Chhattisgarh (Raily), and Odisha (Modal) are promising because of the richness in silk content and the availability of the sal forest. The geographical distribution of these eco-races is shown in Table 12.11.

12.4.1.2 Source of Silkworm Germplasm

The continuous collection of naturally grown cocoons without caring for the population dynamics and survival often results in overexploitation of natural resources for commercial/economic gain by mankind, pushing the species to the threshold of extinction. The important minor forests' production is something that thousands of tribal people depend on for their livelihood. Wild/natural populations are supposed to harbor many beneficial alleles developed through the evolutionary process over centuries. They need to be conserved for posterity. So there is a need to take measures not only to increase their population in nature but also to make sustainable use for commercial purpose. India is the only country in the world where the germplasm of the silkworm *Antheraea mylitta* (Drury) is being maintained and evaluated. Under the guidance of the Central Tasar Research and Training Institute, Nagri, Ranchi-835303, Jharkhand, India (CTR&TI), and its different regional tasar research stations (RTRSs), the following eco-races (Tanle,12), in the form of germplasm material for tropical tasar crop improvement, are kept. This is basically an on-farm conservation coupled with *ex situ* maintenance of the silkworm gene pool in artificially created cold-room facilities to preserve pupae and eggs, as the case may be, following different hibernation schedules/cold-storage methods. This method will help in the timely release of new stock for both commercial and conservation purposes.

12.4.1.3 Host Plant

Though the tasar silkworm is known for its polyphagous nature, it mainly feeds on *Terminalia arjuna, T. tomentosa*, and *Shorea robusta* and sometimes on *Lagerstroemia*, *Anogeissus*, and *Zizyphus* spp. As the tasar silkworm has the widest geographical distribution, the host plants' range is also quite large, as shown in Table 12.14.

12.4.1.4 Conservation Efforts Made So Far for the Wild Tasar Silkworm

A few decades back, naturally grown tasar cocoons contributed substantially to the total tasar silk production in India. But in later years, it declined alarmingly due to rampant collection by local tribals, rapid deforestation and pollution due to industrialization in areas where wild tasar is grown. Raily is a wild eco-race found naturally in dense tropical moist deciduous forest, on *Shorea robusta*, in the forests of Bastar in Chhattisgarh. It is not amenable to handling, loses its original characteristics on interference, and exhibits a low survival rate. Hence, an attempt was made to create a suitable *in situ* conservation

TABLE 12.10

Natural Habitats of Economically Important Eco-Races of *Antheraea mylitta* (Drury)

Eco-race	Area of habitat	Geographical coordinates			Forest type	Soil type	Avg. max. temp. (°C)	Avg. min. temp. (°C)	Annual precipitation	Primary food plant
		°N	°E	MASL						
Andhra local	Adilabad (Andhra Pradesh)	19.40	79.18	170	Tropical dry deciduous	Black clayey	31.52	21.80	925.25	*Terminalia* spp.
Bhandara	Bhandara (Maharashtra)	21.09	79.42	311	Tropical dry deciduous	Black clayey	34.50	20.80	939.40	*Terminalia* spp.
Daba	Singhbhum (Jharkhand)	22.12	86.23	209	Tropical moist deciduous	Red loamy	28.90	18.40	1,095.00	*Terminalia* spp.
Japla	Palamau (Jharkhand)	23.52	84.17	253	Tropical moist deciduous	Red loamy	29.30	19.50	1,075.00	*Zizyphus jujuba*
Laria	Peterbar (Jharkhand)	23.04	85.40	613	Tropical moist deciduous	Red loamy	28.40	17.66	1,100.10	*Shorea robusta*
Raily	Bastar (Chattisgarh)	19.05	82.05	670	Tropical moist deciduous	Sandy red	31.50	18.62	1,275.62	*Shorea robusta*
Modal	Keonjhar (Orissa)	21.40	86.40	423	Tropical moist deciduous	Red loamy	30.40	20.30	1,218.20	*Shorea robusta*
Sarihan	Santhal Pargana (Jharkhand)	24.30	87.00	140	Tropical moist deciduous	Red loamy	31.54	20.52	1,200.20	*Terminalia* spp.
Sukinda	Sukindagarh (Orissa)	21.00	86.00	209	Tropical moist deciduous	Red loamy	31.45	20.60	1,096.00	*Terminalia* spp.
Tira	Purulia (West Bengal)	23.14	87.07	210	Tropical moist deciduous	Red loamy	29.92	19.42	1,010.52	*Lagerstroemia* spp.

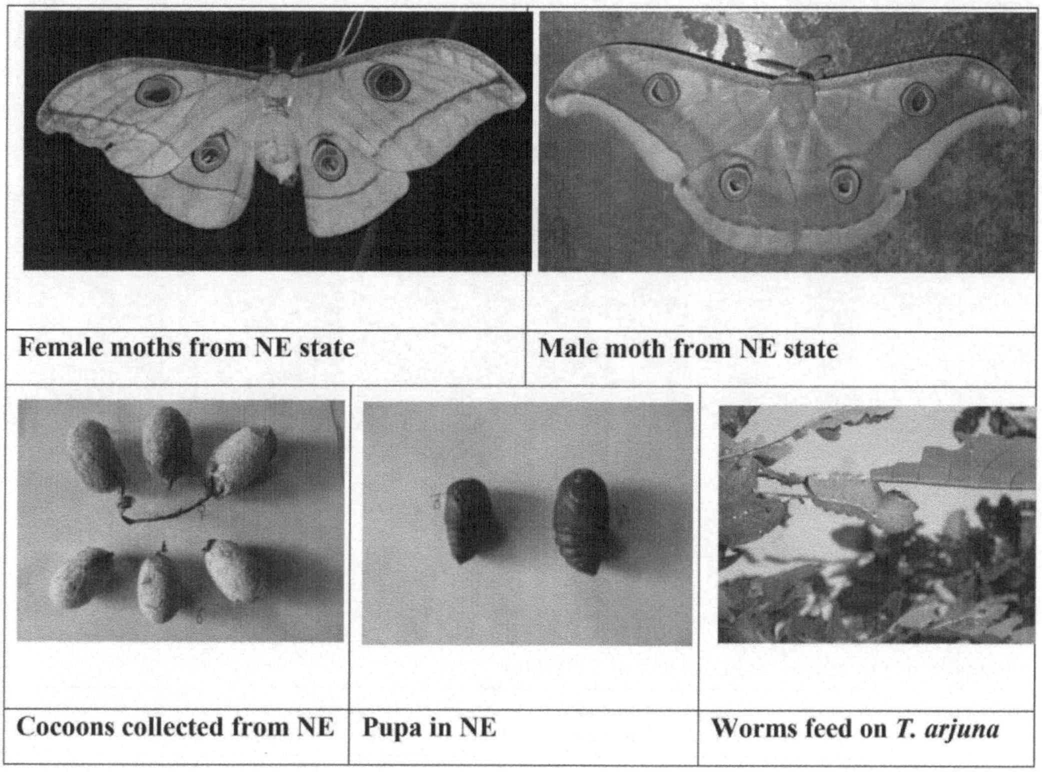

FIGURE 12.6 Stages of the tasar silkworm, *Antheraea mylitta*.

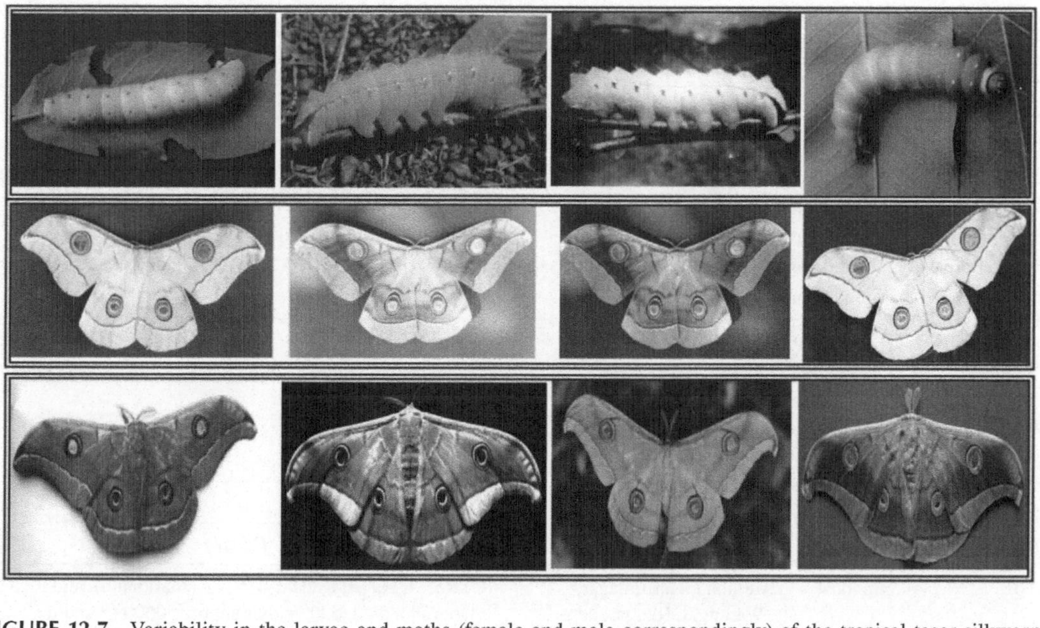

FIGURE 12.7 Variability in the larvae and moths (female and male correspondingly) of the tropical tasar silkworm, *Antheraea mylitta*.

(*Source:* CSRTI, Ranchi)

FIGURE 12.8 Variability in the cocoons of various eco-races of the tropical tasar silkworm, *Antheraea mylitta*; a number of eco-races/varieties recognized.

(*Source:* CTRTI, Ranchi, CSB)

TABLE 12.11

Distribution and Occurrence of Eco-Races of *A. Mylitta*

No.	Name of eco-race	State	No.	Name of eco-race	State	No.	Name of eco-race	State
1.	Sivalica	J&K	16.	Sarihan	Jharkhand	31.	Mandalla	Chattisgarh
2.	Sultanpur	UP	17.	Palma	Jharkhand	32.	Kurudh	Chattisgarh
3.	Monga	UP	18.	Daba	Jharkhand	33.	Raily	Chattisgarh
4.	Mizapur	UP	19.	Lodhma	Jharkhand	34.	Bhopalpatnam	Chattisgarh
5.	Nowgong	Assam	20.	Tira	WB	35.	Janghbhir	Chattisgarh
6.	NE_1-95	Assam	21.	Bankura	WB	36.	Jhabua	MP
7.	NG-94	Nagaland	22.	Tesera	Rajasthan	37.	Piprai	MP
8.	NE_2-95	Meghalaya	23.	Simlipal	Orissa	38.	Seoni	MP
9.	Jiribam	Manipur	24.	Modal	Orissa	39.	Multai	MP
10.	Japla	Jharkhand	25.	Nalia	Orissa	40.	Dadar & Nagar Haveli	Maharashtra
11.	Modia	Jharkhand	26.	Sukinda	Orissa	41.	Bhandara	Maharashtra
12.	Kowa	Jharkhand	27.	Boadh	Orissa	42.	Andhra local	AP
13.	Laria	Jharkhand	28.	Omarkote	Orissa	43.	Belgaum	Karnataka
14.	Barharwa	Jharkhand	29.	Sukly	Chattisgarh	44.	KE-02	Kerala
15.	Munga	Jharkhand	30.	Korbi	Chattisgarh			

(*Source:* CTRTI, CSB, Ranchi)

TABLE 12.12

Genetic Stocks of Tropical Tasar, *Antheraea Mylitta*, and Their Sources

Resource number	Germplasm	Source of availability: *In situ*	Source of availability: *Ex situ*
TT01	Andhra local	Adilabad, Karimnagar, Warangal, Rangareddy (Andhra Pradesh)	CTR&TI, Nagri, Ranchi-835303 (Jharkhand), and RTRS, Door No. 5-11-529, Vidyarnapuri-506009 (Andhra Pradesh)
TT02	Bhandara	Bhandara Chandrapur (Maharashtra)	CTR&TI, Nagri, Ranchi-835303 (Jharkhand), and RTRS, Swami Vivekananda Road, Opp: Sai Mandir, Bhandara-441001 (Maharashtra)

Resource number	Germplasm	Source of availability	
		In situ	*Ex situ*
TT03	Daba	Singhbhum (Jharkhand)	CTR&TI, Nagri, Ranchi-835303 (Jharkhand)
TT04	Japla	Palamau (Jharkhand)	CTR&TI, Nagri, Ranchi-835303 (Jharkhand)
TT05	Laria	Peterbar, Hazaribag (Jharkhand)	CTR&TI, Nagri, Ranchi-835303 (Jharkhand)
TT06	Modal	Keonjhar, Mayurbhanj (Orissa)	CTR&TI, Nagri, Ranchi-835303 (Jharkhand) and RTRS, Baripada-757001, Dist: Mayurbhanj (Orissa)
TT07	Raily	Bastar (Chhattisgarh)	CTR&TI, Nagri, Ranchi-835303 (Jharkhand), and RTRS, Kalipur Road, Dharampura, Jagdalpur-494005 (Chhattisgarh)
TT08	Sarihan	Santhal Pargana (Jharkhand)	CTR&TI, Nagri, Ranchi-835303 (Jharkhand), and RTRS, Shiva Pahar Chowk, Dumka-814101 (Jharkhand)
TT09	Sukinda	Sukindagarh (Orissa)	CTR&TI, Nagri, Ranchi-835303 (Jharkhand)
TT10	Tira	Purulia, Bankura (West Bengal)	CTR&TI, Nagri, Ranchi-835303 (Jharkhand)

(*Source:* CTRTI, CSB, Ranchi)

TABLE 12.13

Different Secondary Host Plants of *Antheraea Mylitta* (Drury)

Order	Family	Host plant
Sapindales	Anacardiaceae	*Anacardium occidentale* *Semecarpus anacardium*
Gentiales	Apocynaceae	*Carissa carandus*
Celastrales	Celastraceae	*Celastrus paniculatus*
Myrtales	Combretaceae	*Terminalia arjuna* *Terminalia tomentosa* *Terminalia alata* *Terminalia belerica* *Terminalia catappa* *Terminalia citrina* *Terminalia chebula* *Terminalia crenulata* *Terminalia myriocarpa* *Termilalia paniculata* *Anogeissus latifolia*
Rosales	Leguminoceae	*Cassia lanceolata* *Bauhinia variegata* *Hardwickia binate*
Parietales	Dipterocarpaceae	*Shorea robusta* *Shorea talura*
Myrtales	Lythraceae	*Lagerstroemia indica* *Lagerstroemia parviflora* *Lagerstroemia speciosa*
Malvales	Malvaceae	*Bombax malabaricum*
Malvales	Moraceae	*Ficus religiosa*
Myrtales	Myrtaceae	*Careya arborea*
Rhammales	Rhamnaceae	*Zizyphus jujuba*
Rosales	Rosaceae	*Prunus domestica*
Rubiales	Rubiaceae	*Canthium diecocum*
Sapindales	Sapindaceae	*Dodonaea viscose*
Ebenales	Sapolaceae	*Bassia longifolia*
Lamiales	Verbenaceae	*Tectona grandis*
Euphorbiales	Euphorbiaceae	*Ricinus communis*

(*Source:* CTRTI, Ranchi)

FIGURE 12.9 Primary tasar host plants.

(*Source:* CTRTI, CSB, Ranchi)

FIGURE 12.10 Secondary food plants of tasar.

(*Source:* CTRTI, CSB, Ranchi)

model. Five methods were studied by the Regional Tasar Research Station, Jagdalpur, in a forest patch that has no continuity with the adjacent forest and is devoid of a natural Raily population. The five methods were (1) release of seed cocoons, (2) release of male and female moths, (3) release of gravid moths, (4) release of eggs in leaf cups, and (5) release of chawki worms. The results obtained indicated that all the methods helped in genetic conservation of the eco-races. Of the five methods, release of eggs in leaf cups has been found to be economically viable and maintain the original commercial characteristics. The Chhattisgarh State Sericulture Department has adopted this method and succeeded in arresting the decline of wild cocoon production and preserving the inherent genetic variations on the Raily eco-race. Similar was the case of the Modal eco-race in Odisha, where the conservation of this eco-race in the Similipal Biosphere Reserve was developed by RTRS, Baripada (2001–2005). Modal is being conserved through the Modal Eco-Race Conservation Program implemented by the Department of Textiles, Handlooms, and Handicrafts of Odisha since 2006.

In all the conservation methods adopted for Raily and Modal eco-races, the following observations were found in common:

1. Unlike the economic plantation conditions, the yield per disease-free laying was much lower, ranging from 3 to 15 cocoons per tree, and this also varies with generation and year. But this productivity in natural conditions is also sufficient for the perpetuation and proliferation in natural sal forests.
2. Wild populations are subjected to pest and predator attack, which is a natural phenomenon of food web, which is balanced in due course. However, the wild cocoons are liable to being collected by local inhabitants, which can be checked during breeding seasons to facilitate natural perpetuation.
3. Disease incidence is either nil or negligible in the wild cocoons, which is an advantage for the silkworm itself.
4. Outbreeding is the natural phenomenon in the wild tasar populations, where male moths from distant areas mate with females. This leads to increasing and maintaining vigor, in addition to increasing immunity toward disease.
5. The habitat of wild tasar populations is not fixed. Although there is no published information, it has been observed that wild population shifts their habitat from generation to generation and year to year. Exploration of new suitable habitat by the insect itself indicates the evolutionary feature of coevolution of host plant and the silkworm species with exerting minimum parasitic effect on host plant.
6. Wild tasar eco-races are not amenable to human handling; as such, they need to be conserved in their own habitat or *in situ*.

12.4.2 Oak Tasar

Oak tasar silkworms are distributed in sub-Himalayan oak belt of India (Assam, Arunachal Pradesh, Nagaland, Manipur, Meghalaya, Mizoram, Himachal Pradesh, Jammu and Kashmir, and Uttaranchal) and also subtropical and temperate rain forest of sub-Himalayan oak belt. In India four species of oak tasar silkworm—*Antheraea frithii*, *Antheraea proylei*, *Antheraea roylei*, and *Antheraea pernyii*—are reported. Depending on the life cycle and availability of foliage on host plants, the season of different oak species silkworms varies, as shown in Table 12.14.

The different stages of the life cycle of all the four species is shown in Figures 12.16–12.18.

TABLE 12.14

Seasons of Natural Occurrence

SN	Name of species/breed	Period of natural occurrence
1	*Antheraea proylei*	February to May and July to October
2	*Antheraea frithii*	March to May and July to October
3	*Antheraea roylei*	Spring (February to May) and autumn (August to October)
4	*Antheraea pernyii*	Spring (February to May) and autumn (August to October)
5	PRP_{12}	Spring (February to May) and autumn (August to October)
6	PRP_5	Spring (February to May) and autumn (August to October)

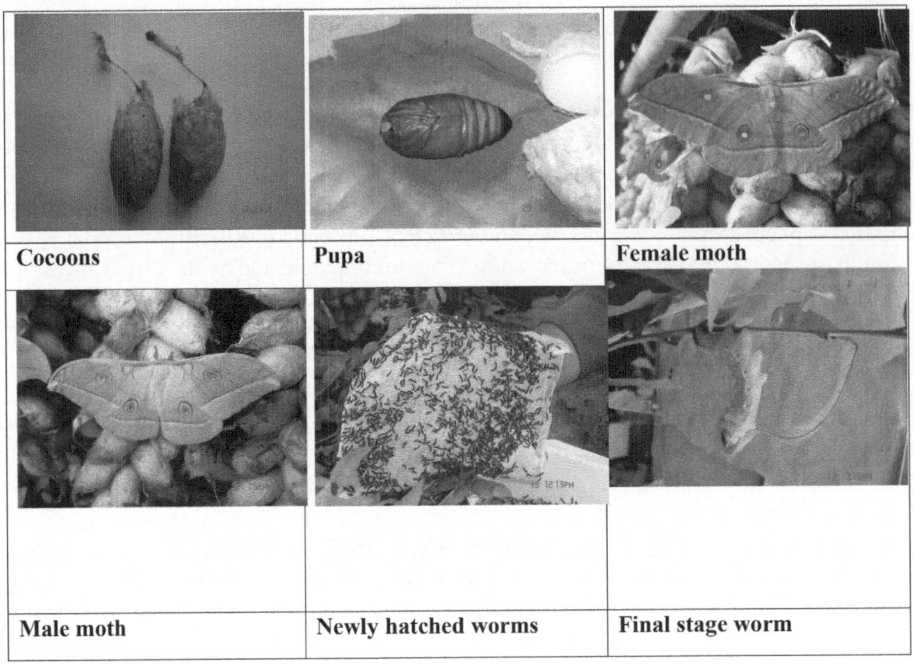

FIGURE 12.11 *Antheraea proylei* (Jolly).

(*Source:* CTRTI, Ranchi, CSB)

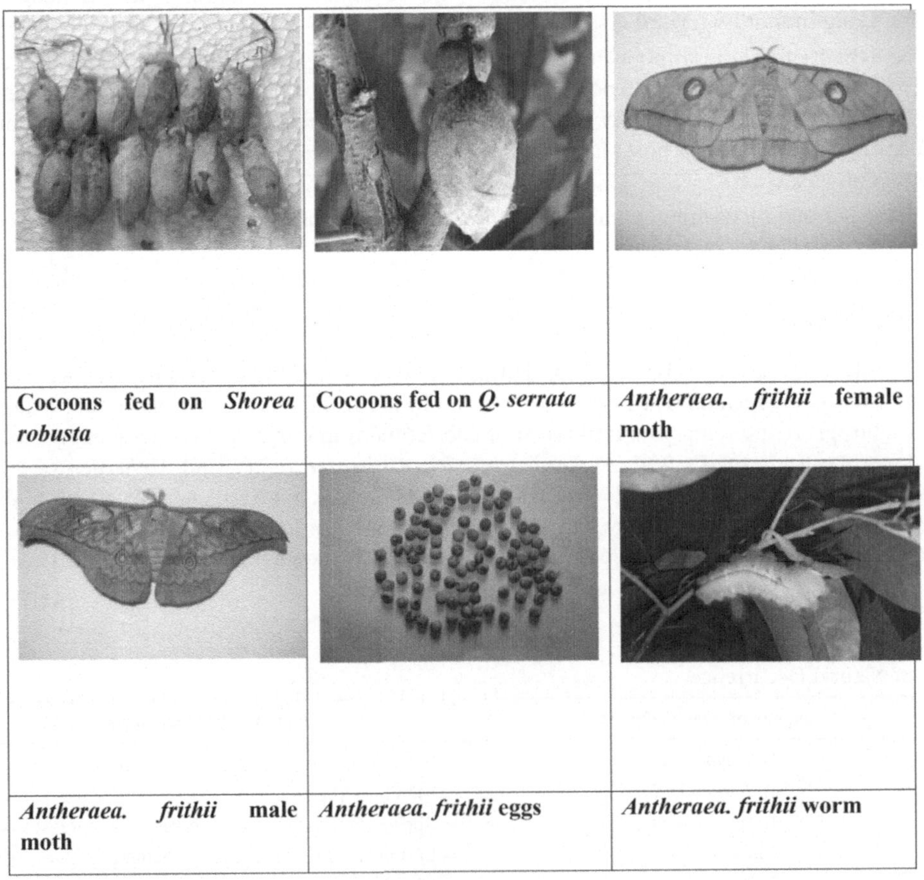

FIGURE 12.12 *Antheraea frithii.*

(*Source:* CTRTI, Ranchi, CSB)

	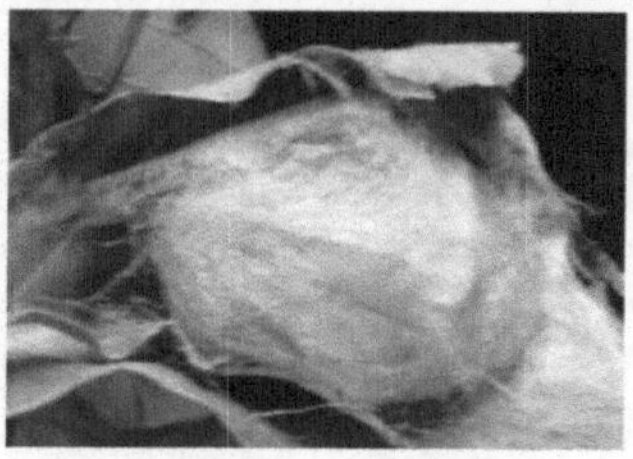	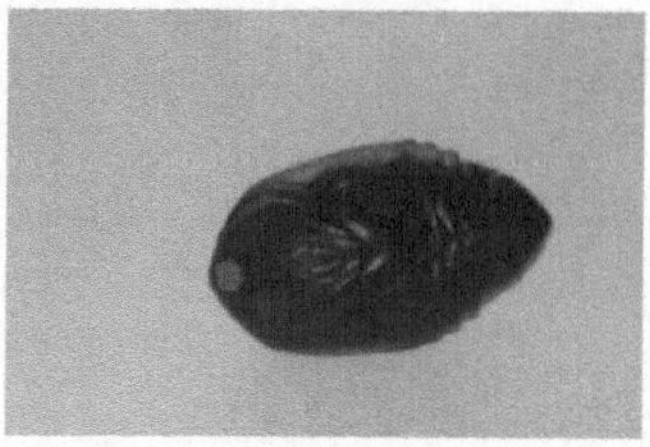
***Antheraea roylei* cocoons**	***Antheraea roylei* cocoon**	***Antheraea roylei* pupa**
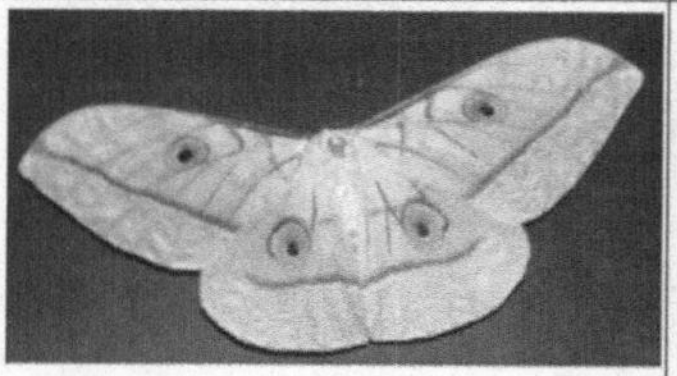		
***Antheraea roylei* female moth**	***Antheraea roylei* male moth**	***Antheraea roylei* worms**

FIGURE 12.13 *Antheraea roylei.*

(*Source:* CTRTI, Ranch, CSB)

Natural variability, as observed in the morphology of male and female moths of all the four species and their wild relatives, is shown in Figure 12.18.

12.4.2.1 Host Plant

A. proylei (J.) is extremely polyphagous and differs in its adaptation to different food plants. The species recognized with preferred host plants are given in Table 12.15.

12.4.2.2 Scenario of Oak Tasar and Other Wild Sericigenous Insects

Sericigenous are wild or semi-domesticated insects, which produce lustrous silk and exhibit a great range of variation in life history from eggs to adults, with characteristically different physiological, morphological, and feeding parameters. Some wild silkmoths not only yield valuable silk but also play an important role in the ecosystem. Their diversity varies radically in regions and types of biological environments. However, the majority of the wild silkmoths belong to Family Saturniidae, the largest group in Lepidoptera, comprising 2,010 described species in 176 genera and nine subfamilies, of which, 47 species of wild silkmoths have already been reported in India (Singh and Suryanarayana, 2003). From Northeast India alone, 24 species are listed (Singh and Chakravorty, 2006). Being an

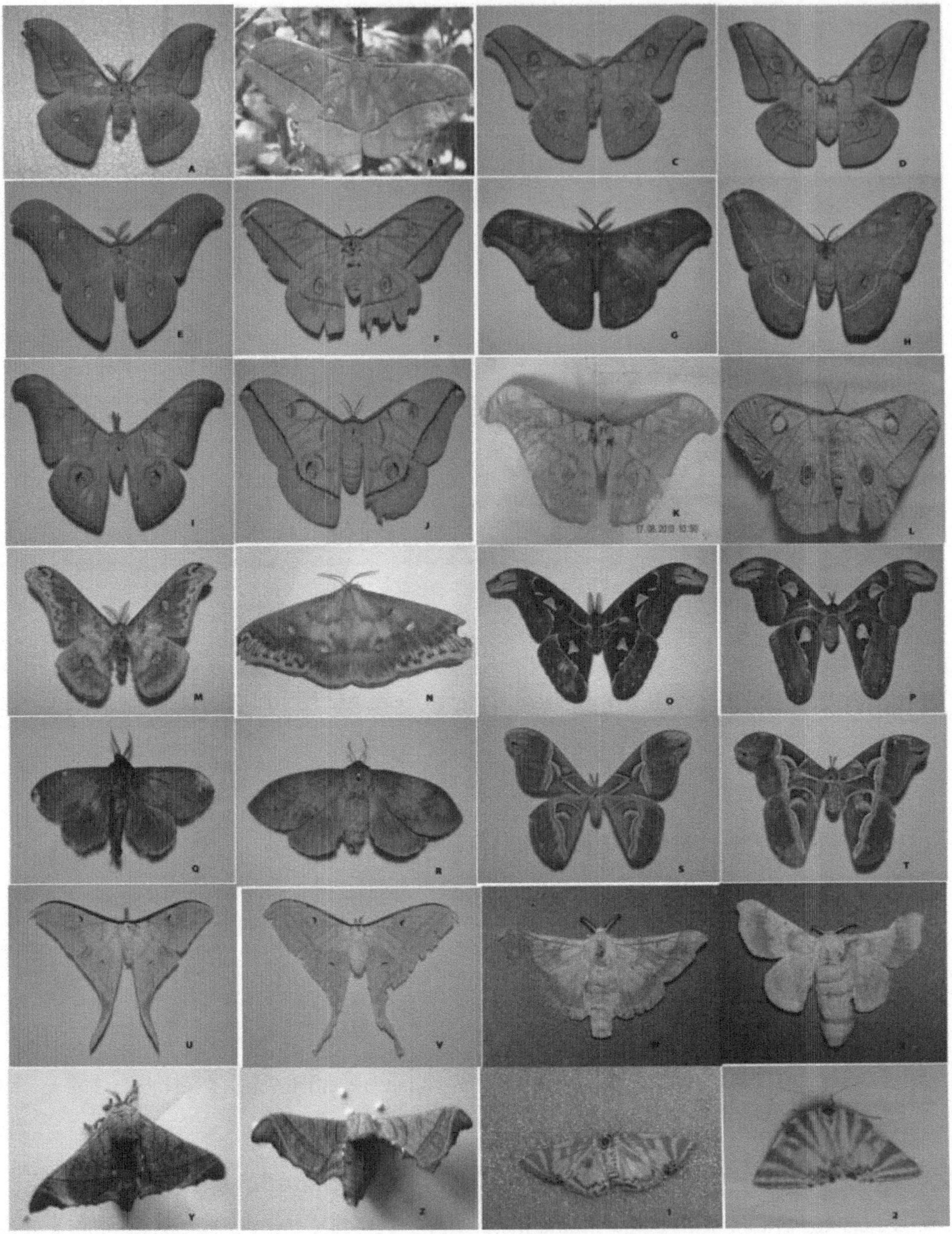

FIGURE 12.14 A & B (♂ & ♀) = *Antheraea proylei* (Jolly), C & D (♂ & ♀) = *Antheraea frithi* (Moore), E & Fig. 19 (♂ & ♀) = *Antheraea roylei* (Moore), G & H (♂ & ♀) = *Antheraea assamensis* (Helfer), I & J (♂ & ♀) = **Antheraea helferi* (Moore), K & L (♂ & ♀) = *Antheraea compta* (Roth), M & N (♂ & ♀) = **Rhodinia newara* (Moore), O & P (♂ & ♀) = *Attacus atlas* (Linnaeus), Q & R (♂ & ♀) = **Dendrolimus grisea* (Moore), S & T (♂ & ♀) = *Samia canningi* (Hutton), U & V (♂ & ♀) = *Actias selene* (Hubner), W & X (♂ & ♀) = *Bombyx* sp. (local name "leimaren"), Y & Z (♂ & ♀) = *Bombyx huttoni* 1 & 2 (♂ & ♀) = *Lebada nobilis* (Walker).

(*Source:* CTRTI, Ranchi, CSB)

TABLE 12.15

Different Species of Oak Tasar and Their Host Plants

SN	Name of species/breed	Host plant
1	*Antheraea proylei*	*Quercus serrata*, *Q. dealbata* (synonym to *Lithocarpus dealbata*), *Q. grifithii* and *Q. incana* (synonym to *Q. leucotrichophora*), *Q. himalayana* (synonym to *Q. floribunda*), *Q. glauca*, *Q. semecarpifolia*, etc.
2	*Antheraea frithii*	*Lithocarpus dealbata*, *Quercus serrata*, *Q. accutissima*, *Q. grifithii*, *Q. incana*, *Q. semecarpifolia*
3	*Antheraea roylei*	*Quercus serrata*, *Q. accutissima*, *Q. grifithii*, *Q. himalayana* (synonym to *Q. dilatata* and *Q. floribunda*), apple leaves, etc.
4	*Antheraea pernyi*	*Quercus serrata*, *Q. accutissima*, *Q. grifithii*, *Q. himalayana* (synonym to *Q. dilatate* and *Q. floribunda*)
5	PRP_{12}	*Quercus serrata*, *Q. dealbata* (synonym to *Lithocarpus dealbata*), *Q. grifithii* and *Q. incana* (synonym to *Q. leucotrichophora*), *Q. himalayana* (synonym to *Q. floribunda*), *Q. glauca*, *Q. semecarpifolia*, etc.
6	PRP_5	*Quercus serrata*, *Q. dealbata* (synonym to *Lithocarpus dealbata*), *Q. grifithii* and *Q. incana* (synonym to *Q. leucotrichophora*), *Q. himalayana* (synonym to *Q. floribunda*), *Q. glauca*, *Q. semecarpifolia*, etc.

Indo-Burma biodiversity hotspot region, NorthEastern India makes an ideal home for a number of wild sericigenous insects and is a center of wild silk culture, including muga, eri, tropical tasar, temperate tasar, and fagara silks (Peigler and Naumann, 2003). The forest fauna in the sub-Himalayan belt of India is the natural abode of large variety of sericigenous insects, as reported by Jolly et al. (1976), Baruah et al. (2000), Singh et al. (2000), Singh and Maheswari, (2003), and Thangavelu (1991), which plays an important role in the conservation and utilization of biodiversity (Frankel, 1982; Peigler, 1993).

The exploration of wild sericigenous insects from Northeastern India is reported by several workers (Thangavelu and Borah, 1986; Singh et al., 2008). However, descriptive and detailed information of these sericigenous insects is meager as their distribution is restricted to highly inaccessible areas—Manipur, in particular. In support of the previous researchers and considering the importance of sericigenous genetic resources that are facing major threats due to rapid transformation of original vegetation and intensive land-use practices, it is pertinent to characterize, classify, and document the status of these precious faunal species in the region. Conservation links genetic diversity to utilization, protecting diverse gene pools, habitats, or ecosystems for human socio-economic needs. In the present context, it is important to evaluate, identify, and characterize the economic characteristic features of the wild silkmoth species in-depth for commercial and conservation importance. Hence, an attempt has been made to survey, collect, characterize, and identify the sericigenous insects, along with their host plants, in Manipur to link conservation with socio-economic enhancement. Survey and collection of the wild saturniid species in Manipur, in the Indo-Burma region. Surveys, exploration, and collection of wild sericigenous insects along with their food plants were conducted at the hill (Chandel, Churachandpur, Senapati, Tamenglong, and Ukhrul) and valley (Bishnupur, Imphal East, Imphal West, and Thoubal) districts of Manipur in different months/seasons from 2011 to 2015, showing the sites of sample collection (Figure 12.19). The natural habitats and undisturbed forests areas were surveyed extensively to record the existence of wild silkmoths, worms, cocoons, and their host plants. During the survey, tribal folk, farmers, and private rearers were interviewed for gathering information on presence/absence of wild sericigenous insects, their preferred host plants and attempts to rearing the wild silkmoths. The incidences patterned of all the wild silk species were counted and recorded throughout the year. Pooled data of three years were analyzed by following standard statistical methods for a conclusive interpretation.

TABLE 12.16
Number of Eco-Races/Varieties Recognized with the Preferred Host Plants

SN	Name of the eco-race	Host plant
1	Borduar	Castor, Kesseru, Payam
2	Titabar	Castor, Kesseru
3	Khanapara	Castor, Kesseru
4	Dhanubhanga	Castor, Kesseru
5	Mendipathar	Castor, Kesseru
6	Sille	Castor, Kesseru
7	Nongpoh	Castor, Kesseru
8	Kokrajhar (red)	Castor

12.4.3 Eri Silkworm

12.4.3.1 Geographical Distribution

The tribal inhabited areas of Northeastern India, particularly the Brahmaputra Valley of Assam and its neighboring states, are considered as the natural home of raising eri silkworms. Extensive survey was conducted in different areas of Northeast India during the early ears for collection and characterization of different eri silkworm genetic resources. Twelve eri silkmoth types were collected and characterized to make an inventory. Studies related to the collection, characterization, and evaluation of different stocks of eri silkmoths were made and attempted for establishment of a germplasm bank and to evolve productive hybrids since establishment of Central Muga Eri Research Station (CMERS) at Titabar, Assam, during 1974.

Ample information is available on the characterization of eri stocks collected from different parts of Northeast India. Though it was obscure for a long time, the structure of the genitalia, wing pattern, and chromosome number demonstrated that *Samia ricini* (Donovan) is derived from its wild form, *Samia canningi* (Hutton) (Peigler and Naumann, 2003). *Samia ricini* does not occur in the wild. Domestication for centuries has resulted in striking physical and behavioral changes between the wild and cultivated populations (Tazima, 1964). *S. ricini* is multivoltine and has eight eco-races—Nongpoh, Kokrajhar Red, Borduar Local, Titabar Local, Sille, Dhanubhanga, Mendipathar, and Khanapara, based on the location of collection (Sharma et al., 2002 and Singh et al., 2003). Further, six homozygous strains—Greenish Blue Plain, Greenish Blue Spotted, Greenish Blue Zebra, Yellow Plain, Yellow Spotted, and Yellow Zebra, based on external body color and markings—have been isolated from the Borduar and Titabar eco-races. These eco-races and homozygous strains are being maintained at the Regional Eri Research Station, Mendipathar, East Garo Hills, Meghalaya. The eri silkworms feed mainly on castor (*Ricinus communis*) (Thangavelu and Phukan, 1983). NBR-1 is the most prominent variety for eri silkworm rearing (Sharma et al., 2002). The other food plants are kesseru (*Heteropanax fragrans*), payam (*Euodia fraxinifolia*), tapioca (*Manihot utilissima*), borpat (*Ailanthus grandis*), and borkesseru (*Ailanthus excelsa*) (Table 12.16).

12.4.3.2 Range of Variation Recorded

Larva:

Color: Head yellowish green, integument light blue and covered with white powdery wax secretion. Six strains were isolated on the basis of larval color and markings as follows: Yellow Plain (YP), Yellow Spotted (YS), Yellow Zebra (YZ), Greenish Blue Plain (GBP), Greenish Blue Spotted (GBS), and Greenish Blue Zebra (GBZ) (Figure 12.20).

Size: The length of fifth instar larva is between 4.85 and 4.94 cm and the breadth ranges from 1.60 to 1.63 cm.

Weight: Ranges from 5.67 to 8.7 g.

Cocoon

Colou: White, pale white, and brick red

Characteristics: Cocoons are oval in all cases. Cocoon weight ranges from 2.4 to 4.9 g, shell weight 0.41 g to 0.7 g, and shell ratio from 12.41% to 17.89%.

Size: Maximum length of cocoon recorded was 5.6 cm (Figure 12.21)

Moth:

Color: Variable, but always solidly white dorsal surface of the abdomen. Ground color is usually gray or grayish brown, rarely reddish, but occasionally olive gray. Wing pattern is heavily marked by white on antemedian and postmedian lines.

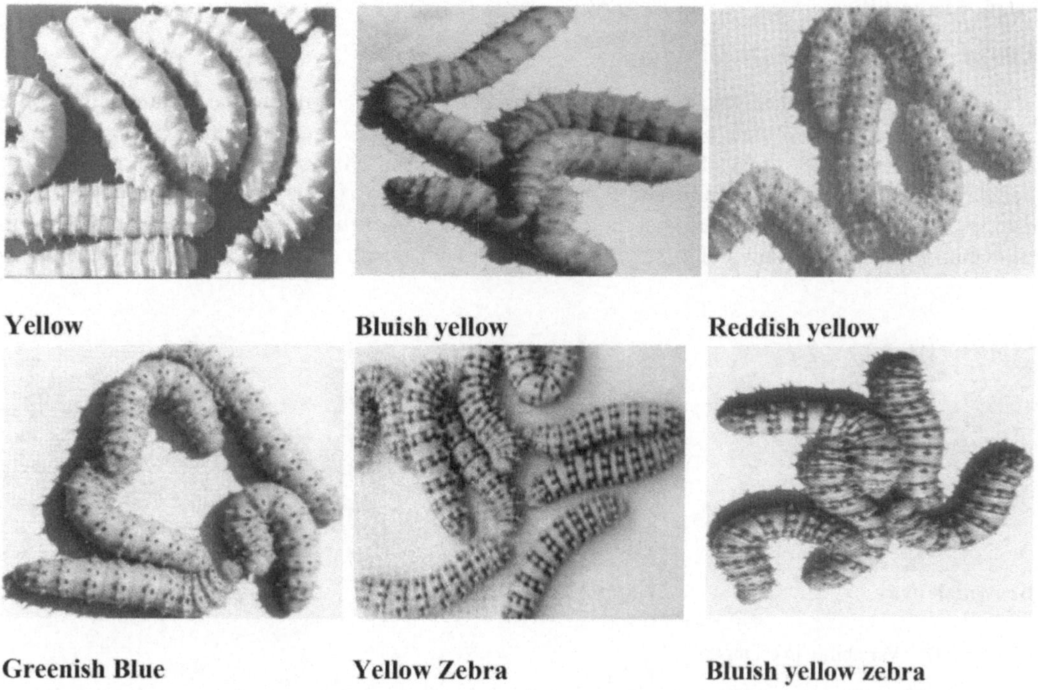

FIGURE 12.15 Variability in eri larvae.

(*Source:* CMERTI, CSB, Lahdoigarh

FIGURE 12.16 Variability in eri cocoons.

(*Source:* CMERTI, CSB, Lahdoigarh)

Size: Body length of the male moth ranges from 2.9 to 3.21 cm, while the female moth ranges from 4.1 to 4.6 cm. Forewing expansion size of the male moth ranges from 13.2 to 14.1 cm and that of the female moth ranges from 13.2 to 14.46 cm. Hindwing expansion size of the male moth ranges from 7.6 to 8 cm and that of the female moth ranges from 7.5 to 8.3 cm (Figures 12.22 and 12.23).

12.4.3.3 Host Plant

S. ricini, being a polyphagous insect, has a wide range of food plants. In the forest of NE India, different food plants of the eri silkworm are found abundantly as listed in the following table

12.4.4 Muga Silk

12.4.4.1 Geographical Distribution

The famous muga silk is endemic to India and is found widely occurring in the Brahmaputra Valley of Assam of the country. *Persea bombycina* (Kost and Soalu) and *Litsaea polyantha* (Juss) are the primary food plants of the muga silkworm, *Antherea assamensis*. Himachal Pradesh, Sikkim, Assam, Meghalaya, and Pondicherry in India and Bangladesh (in Sylhet), Sri Lanka, and Indonesia (Arora and Gupta, 1979). Later studies assumed that the natural distribution of this insect is mainly confined to the Brahmaputra Valley of Assam and its surrounding hills (Thangavalu et al., 1988 and Chaudhuri, 1981).

FIGURE 12.17 Variability in eri moths.

(*Source:* CMERTI, CSB, Lahdoigarh)

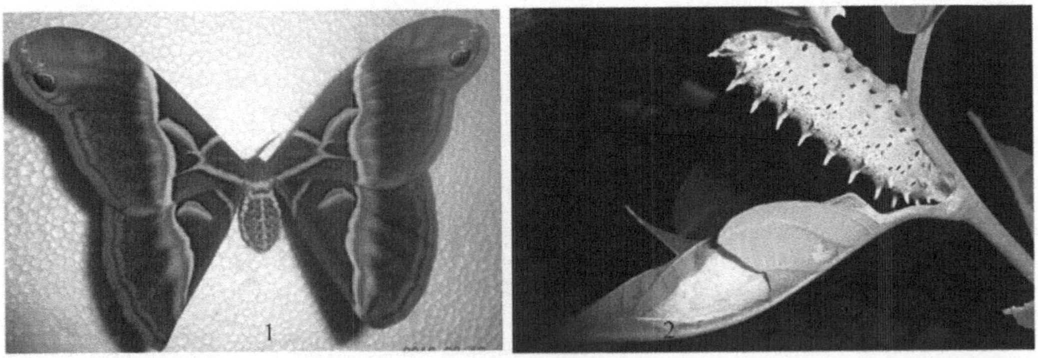

FIGURE 12.18 1–5, *Samia canningi*: (1) adult moth, (2) larva and cocoon on *Litsea monopetala*, (3) adult moth, (4) cocoons, (5) larvae.

(*Source:* CMERTI, CSB, Lahdoigarh)

TABLE 12.17
Food Plants of the Eri Silkworm

SN	Scientific name	Vernacular name	Family	Distribution
1	*Ricinus communis* (Linn.)	Castor	*Euphorbiaceae*	Throughout India
2	*Manihot esculenta* (Crantz)	Tapioca	*Euphorbiaceae*	Eastern, Southern, and Northeastern India
3	*Sapium eugeniifolium* (Buch-Ham)	Korha	*Euphorbiaceae*	Northeastern India
4	*Heteropanax fragrans* (Roxb.), Seem	Kesseru	*Araliaceae*	Nagaland and Arunachal Pradesh
5	*Evodia fraxinifolia* (Hook)	Payam	*Rutaceae*	Northeastern, Western, and Eastern India
6	*Ailanthus grandis* (Prain)	Barpat	*Simaroubaceae*	Northeastern, Western, and Eastern India
7	*Ailanthus excelsa* (Roxb.)	Barkesseru	*Simaroubaceae*	Throughout India
8	*Plumeria acutifolia* (Poir)	Gulanch	*Apocynaceae*	Throughout India
9	*Gmelina arborea* (Roxb.)	Gamari	*Verbanaceae*	Norheastern and Eastern India

(*Source:* Chowdhury, 2000, and Peigler, 2003)

FIGURE 12.19 Food plants of the eri silkworm.

(*Source:* CMERTI, CSB, Lahdoigarh)

12.4.4.2 Number of Eco-Races/Varieties Recognized with Preferred Food Plants

Rearers refer to three types of eco-races available in the wild environment. These are *Harujukia* (Harubhagia), *Barjukia* (Barbhagia), and *Lebang* (Lebong) (Chowdhury, 1981). But these are not recorded during the current study (Figure 12.25).

12.4.4.3 Secondary Host Plant

In addition, this silkworm also feeds on the leaves of numbers of secondary food plants as listed in Table 12.18.

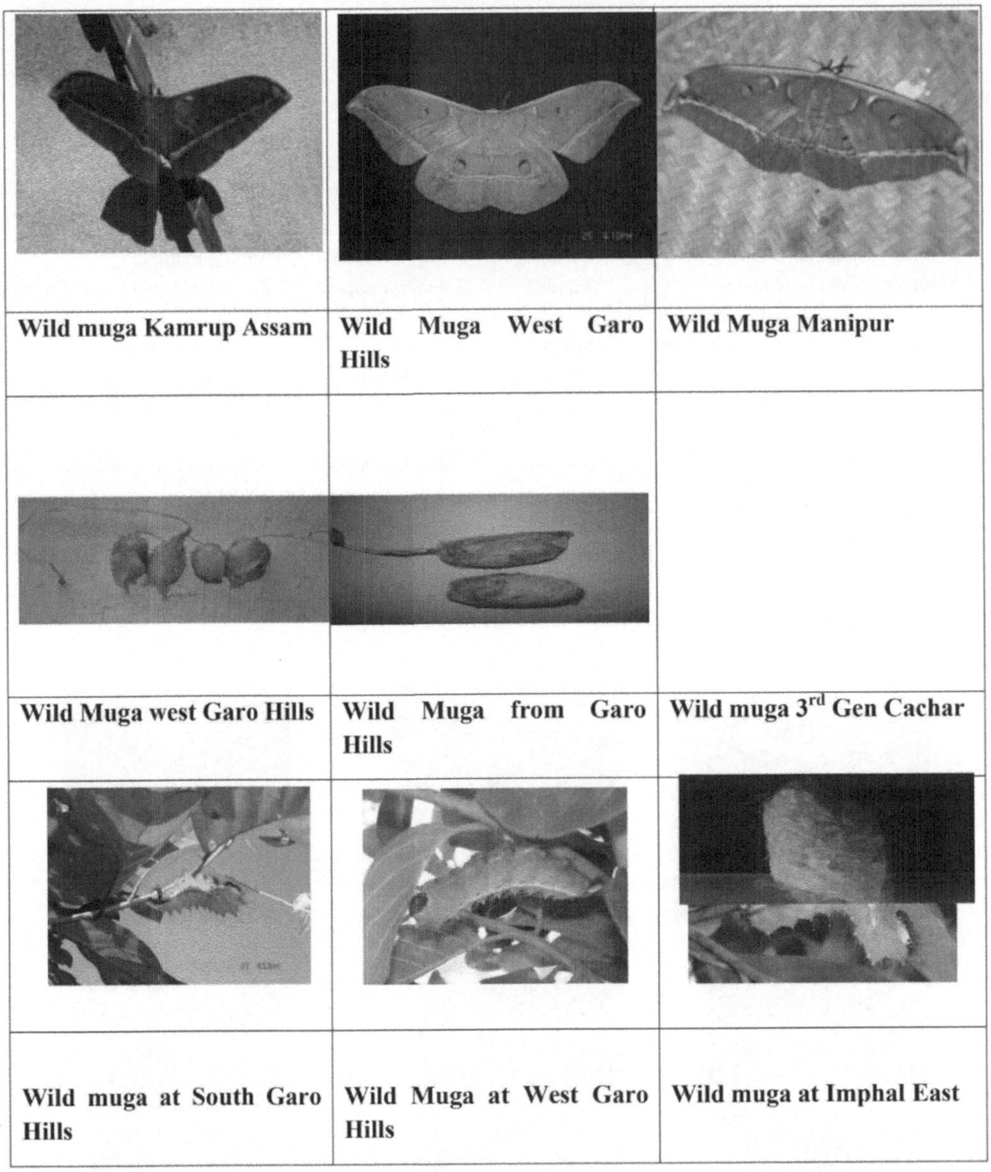

FIGURE 12.20 Wild muga silkmoths and cocoons.

Persea bombycina *Litsaea salicifolia* *Litsaea polyantha* *Litsaea citrate*

FIGURE 12.21 Primary food plants of the muga silkworm.

TABLE 12.18

List of Secondary Food Plants of Muga

SN	Scientific name	Vernacular name	Family	Distribution
1	*Persea bombycine* [Kost] (=*Machilus bombycina* [King])	Som	Lauraceae	Prevalent throughout Northeast India up to an altitude of 1,500 feet AMSL. Lower Himalayas extending from Garowal to Nepal.
2	*Litsaea polyantha* (Juss) (=*L. monopetala*)	Sualu	Lauraceae	Extending northward from Punjab up to the salt range along the foothills of Himalayas ascending to 3,000 feet AMSL, eastward to Northeast India and southward to Satpura hill range. Prevalent throughout Assam, Meghalaya, Mizoram, and Nagaland. Common in plains of NE India.
3	*Litsaea salicifolia*	Dighloti	Lauraceae	Distributed throughout Northeast India, Sikkim, and Nepal, ascending up to an altitude of 6,000 feet AMSL.
4	*Litsaea citrate*	Mejankari	Lauraceae	Eastern Himalayas, covering Sikkim to Arunachal Pradesh, Assam, Manipur, Meghalaya, and Naga Hills up to an altitude of 6,000 feet AMSL.
5	*Cinnamomum glaucescans*	Gandhsarai	Lauraceae	Distributed throughout Northeast India.
6	*Actinodaphnae obovate*	Patihonda	Lauraceae	Common tree in Northeast India. Found in Sikkim and Bangladesh.
7	*Zizyphus zuzuba*	Bogori	Rhamnaceae	Common throughout India, Afghanistan, Malaya, China, and Australia.
8	*Michelia champaca,*	Champa	Magnoliaceae	Tropical Himalayan forests from Nepal to Assam, Meghalaya, Bangladesh, and Burma.
9	*Magnolia sphenocarpa*	Panchapa	Magnoliaceae	Tropical Himalayan forests from Nepal to Assam, Meghalaya, Bangladesh, and Burma.
10	*Xanthozylum rehsta*	Bajramani	Rutaceae	Prevalent throughout Northeast India.
11	*Gmelia arborea*	Gamari	Verbanaceae	Prevalent throughout Northeast India in hills and plains.
12	*Celastrus momosperma*	Bhomloti	Celastraceae	Common in Arunachal Pradesh, Assam, Nagaland, and Meghalaya.
13	*Actinodaphnae anquistifolia*	Baghnala	Lauraceae	Prevalent throughout Northeast India.
14	*Cinnamomum glanduliferum*	Gansarai	Lauraceae	Distributed throughout Northeast India.

(*Source:* Thangavelu et al., 1988, and *Handbook of Muga Culture*, published by CSB, Bangalore)

12.5 Conservation Strategies

In case of conservation of domesticated animals and cultivated crops, the conservation strategies, methodologies, and utilization aspects are well defined and being followed. Specific institutions have been given the exclusive responsibility for the collection, conservation, characterization, and utilization of genetic resources of the said species. However, in case of wild and semi-domesticated fauna and flora, the approaches and strategies are different and needs to be determined based on geographical distribution, extent of variability, reproduction cycle and influencing factors, nature and product of economic exploitation, and so on. The strategies shall be multidirectional, involving all the stakeholders right from the beneficiaries to policymakers on participatory mode. The sericigenous flora and fauna of tasar, oak tasar, and muga are basically a forest produce and reared outdoors, directly on food plants, whereas eri is indoor and reared on harvested leaves. Since there are two components—namely, insects and their food plants—that need to be considered as a single entity rather than individually, conservation of sericigenous insects presupposes the conservation of food plants automatically. In view of this, the conservation of the ecosystem where the dominant diversity of food plants and insects coexist appears to be imperative.

12.6 Conservation Methods

As explained earlier, conservation methods differ from species to species; type of reproduction; nature and duration of life cycle; stage of dormancy etc, so a multipronged approach appears to be imperative. The wild silkmoths and their food plants need conservation both under *in situ* and *ex situ* condition for the sustainable utility. The economic values, employment potential and global demand in general and bio-diversity conservation in particular have motivated the conservation of wild silkworm genetic resources, which has been rapidly depleting and getting endangered. The conservation of wild silkworm's strategies fauna and flora are as follows.

1. ***In situ* conservation:** Conservation of germplasm under natural habitat is referred to as *in situ* conservation and is the process of protecting and augmenting plant or animal species in its natural habitat, along with the evolutionary processes and the ecosystems that host the given genetic diversity. The *in situ* method of conservation is the cheapest and easiest method of conservation. In this approach, the process of natural evolution remains operative. It requires establishment of natural or biosphere reserved national parks/reserves for the protection of existing natural genetic resources, including habitats and ecosystems. In this method of conservation, the existing genetic diversity of both insects and host plants are preserved under complete natural or semi-natural ecosystems. Except for the domesticated eri silkworm's genetic resources, the other silkworms can be conserved following the *in situ* methods. Sometimes the in situ method is also referred to as an on farm conservation system wherein domesticated and cultivated species are conserved where they have developed their distinctive properties (CBD Article 2). It is the sustainable management of genetic diversity of locally developed crop animal varieties (landraces and ecotypes), along with their wild associated relatives. As the habitat for domesticated and cultivated species is on the farm, *in situ* conservation of agriculture/sericulture/silviculture biodiversity is also on farm conservation. Based on the geographical locations, the extent of the area, and other legal prerequisites, the *in situ* sites can be designated as (1) national parks, (2) sanctuaries, (3) biosphere reserves, (4) nature/conservation reserves, and (5) projects for endangered species. The only fear in this approach is devastation due to natural calamities, such as floods, fires, prolonged droughts, and hailstorms, and pests and epidemics, such as pebrine, tiger band disease, and uzi flies, which needs to be monitored constantly.

Proposed areas for *in situ* conservations: Based on the abundant occurrence of silkworm fauna and host plants and also the extent of existing genetic variability, geographical regions of naturally occurring suitable sites need to be identified and demarcated to each of these species. However, the following sites have been suggested for *in situ* conservations and to be notified as biospheres/natural genetic diversity parks

2. ***Ex situ* conservation:** The *ex situ* conservation approach involves conservation of genetic resources outside their original habitat in the form of seed/eggs, embryos, tissues, or parts of the plants. Methods of conservation include cold storage of seeds/eggs, *in vitro* storage, or field gene banks, depending on the propagules or reproductive unit used. This is a supplementary method of conservation of desired biodiversity away from the natural habitat of the given fauna and flora, which complement the *in situ* activities. Besides being a safety backup, it serves as source material for conducting applied and basic research related to characterization, multiplication, and yield analysis, including quality assessment. Apart from establishing a plot as genetic diversity park, silkworm genetic material needs to be conserved by rearing silkworms on food plant blocks, followed by egg production and preservation. Based on the life cycle and hibernation stage, cold-storage schedules need to be developed. Crop colanders for different species and their eco-races have to be developed for regular multiplication and maintenance. Besides, conservation through regular rearing/cultivation schedule, genetic material can also be preserved by cryopreservation of eggs/pupae (cocoons) in silkworms and by cryopreservation of seed/vegetative material (buds, tissue, pollen, etc.) for host plants for medium- and long-term periods. In case of cryopreservation of material, the standardization of regeneration protocol is a must.

12.7 Partners of the Conservation Strategies

The biodiversity of sericigenous insects and large number of their host plants spread over different states of India. The conservation of these genetic resources is massive and uphill task that requires multi stakeholder approach. As suggested by Sinha and Sinha (1994), Shreevatsava et al. (2009), Das et al. (2000), Sarkar (1980), and Ibotombi et al. (2020), the following organizations could be the major players of the program;

1. The Central Silk Board, a statutory body, is involved in promotion of all types of silks in India has established five research institutes to work exclusively on different types of silks. The Central Tasar Research and Training Institute (CTRTI); the Ranchi and Central Muga and Eri Research Institute (CMERTI) in Lahdoigarh, Assam; and the Central Silkworm and Mulberry and Germplasm Resource Center (CSGRC), with a network of seed production and extension units, are conducting research in all aspects of both insects and host plants of the four major types of vanya silks—tasar, oak tasar, eri, and muga.
2. State departments invoved in sericulture and textiles are supporting egg production, transfer of technology, marketing, and reeling activities.
3. The Ministry of Environment and Forestry (MoEF) has the responsibility of protection and promotion of forests and forest produce, both timber and non-timber. Major policies toward conservation and utilization are developed and monitored by MoEF and plays a vital role in this important task.
4. National Biodiversity Authority (NBA), in addition to the three previously mentioned organizations, is supported by the state biodiversity boards and is also involved in documentation, conservation, and monitoring of biodiversity resources of respective states.
5. The Zoological Survey of India (ZSI) and the Botanical Survey of India (BSI), the two national organization interested with survey, identification, and documentation of the faunal and floral wealth of the country.
6. The National Bureau of Agriculturally Important Insects (NBAII) is an ICAR Institute vested with the responsibility of collecting maintenance, management, registration, and regulating the import and export of all agriculturally important insects and related microorganisms.
7. The most important partners of the program are the tribal population, who are the local custodians of this important natural resource, deriving their livelihood from nature since time immemorial. Directly or indirectly, they are playing important role the conservation and utilization of this genetic wealth.

For effective conservation and sustainable exploitation of the valuable wealth of genetic diversity, all these seven organizations need to work in a coordinated fashion with specific roles to be played in this mega program.

12.8 Basic Guidelines and Stepwise Action Plan

- Identification of natural habitats of all the wild silkworms and their food plants
- Delimitation of the geographical areas of naturally occurring silkworms and host plants, along with documentation of physiographic, edaphic, and climatological information
- Revisiting for the exploration, identification, and documentation of silkworms and larval host plants
- Survey and documentation of interspecific and intraspecific genetic variability, besides the identification specific types

- Conducting in-depth studies on population density, reproduction cycle, dormancy, pests and predators, and so on of both host plants and silkworms
- Developing catalogs following DUS guidelines/standard descriptors and registration of the material with an appropriate authority and proper accession numbers
- Identification and describing of the qualifying material for registration as geographical indicators (GIs)
- Declaration of the identified area as (1) national parks (2), sanctuaries, (3) biosphere reserves, (4) nature/conservation reserves, and (5) projects for endangered species, per the set guidelines through an appropriate act followed by notification
- Developing sound long-term action plans with clear roles for all partners involved
- Awareness generation among local community and involving them in conservation process
- Bringing out publications on the guidelines for maintenance, precautions, and dos and don'ts, through pictorial guides, posters, newsletters, and websites for nationwide and worldwide publicity
- Periodical replenishment of genetic resources by releasing of pupae/eggs/larvae into the *in situ* site and allowing them to multiply in nature to replace losses occurring due to natural calamities, disease epidemics, and climatic aberrations and also to maintain genetic heterozygosity to defend against decline
- Formation of local biodiversity clubs and communities and empowering them with the required know-how and do-how on the subject
- Creating proper mechanisms for implementing, monitoring, and reviewing, involving all the stakeholders to oversee the entire program, besides having an apex body at the national level, such as a national coordination committee, for a successful coordination of the project
- Creating a corpus for proper maintenance and augmentation of the genetic resources
- Capacity building at the grassroots level in order to form a conservation working group and a network to study the muga ecosystem

12.9 Way Forward

Conservation and sustainable use of sericigenous biodiversity of the county is imperative for present and future too as vanya (non-mulberry) silks are the backbone of the livelihood and socio-economic sustainability of large number of tribal communities in the country. Because of the greed of the rich and need of the poor, destructive and uneconomical harvesting of these natural resources unabatedly have posed severe threats to very existence and continuity of the genetic diversity of both insects and their food plants. Conservation of seri-biodiversity involves two biological entity which are closely linked needs to consider in tot rather than in isolation. Hence, an integrated approach for conservation involving all the stakeholders directly or indirectly concerned with the industry is the need of the hour. Most of the vanya silkworms and their food plants are naturally grown, and rearing is done outdoors, *in situ*, and on-farm conservation, which appears to be the better approach in the long run. Further, as the end products are by and large non-timber forest products collected by tribal people and local inhabitants, the conservation policy needs to be multi-stakeholder participatory approach. Keeping all these in mind, a comprehensive and nationally coordinated multi-stakeholder approach needs to be developed with a strong institutional framework for proper monitoring and sufficient funds for long-term sustainability.

12.10 Acknowledgments

The authors wish to acknowledge the member secretary of the Central Silk Board for according the permission to use the photographs and other statistical information published by the Central Silk Board and its nested units. References and copies of abstracts and lead papers of the Eighth International Conference on Wild Silkworms, provided by Shri Dileep of ISC, CSB, was of great use, and the same is gratefully acknowledged.

REFERENCES

Ahmed, S.A. and R.K. Rajan. 2011. Exploration of vanya silk biodiversity in north eastern region of India: Sustainable livelihood and poverty alleviation. *International Conference on Management, Economics and Social Sciences (ICMESS'2011)*. Bangkok, pp. 485–489.

Arora, G.S. and I.J. Gupta. 1979. Taxonomic studies on some of the Indian non-mulberry silk moths. *Memoirs of Zoological Survey of India*, 16: 1–63.

Baruah, B., D. Duarah, and D. Chakroborty. 2000. Performance of wild silkmoths. *International Journal of Wild Silkmoth & Silk*, 5: 376–378.

Basavaraja, H.K., S. Nirmal Kumar, N. Suresh Kumar, N. Mal Reddy, Kshama Giridhar, M.M. Ahsan, and R.K. Datta. 1995. New productive bivoltine hybrids. *Indian Silk*, 34: 5–9.

Chaudhuri, S.N. 1981. *Muga Silk Industry: Directorate of Sericulture & Weaving*. Govt. of Assam, p. 178.

Dandin, S.B., H.K. Basavaraja, and N. Sureshkumar. 2005. *Silkworm Breeds & Hybrids at Glore 2005*. Central Sericultural Research and Training Institute, Mysore, India, p. 51.

Das, P.K., A.K. Sahu, and Babulal. 2000. Management of muga silkworm and host plant germplasm. In *Proceedings of the National Workshop on Management of Sericultural Germplasm for Posterity Held on 26–27 July 2000 at Central Sericultural Germplasm Resources Centre*, Hosur, India, pp. 133–139.

Datta, R.K., H.K. Basavaraja, N. Mal Reddy, S. Nirmal Kumar, M.M. Ahsan, Suresh N. Kumar, and Ramesh Babu. 1996. *Manual on Bivoltine Rearing, Race Maintenance and Multiplication*, Central Sericultural Research and Training Institute, Mysore, India, p. 28.

Everlyn, K., Ngauku, and Nigoka Boniface. 2018. Wild Silk moths in Kenya, Anaphe Pand. *8th International Conference on Wild Silkworms*, www.csb.gov.in 22–12 January 2018, p. 59.

Frankel, O.H. 1982. Can genetic diversity survive? In: R.B. Singh, R.M. Singh, and B.D. Singh (eds.) *Advances in Cytogenetics and Crop Improvement*. New Delhi: Kalyani Publications.

Ibotombi Singh, N., L. Somen Singh, Y. Debraj, and A.K. Snha. 2020. Wild Oak tasar silk moth, Antheraea frithii Moor: Need for conservation and utilization. *Indian Silk*, 11(1): 20–23.

Jolly, M.S. 1974. New dimension of tasar industry in India. *Souvenir. Central Silk Board, Bombay*, XII: 83–86.

Jolly, M.S., S.N. Chowdhury, and S.K. Sen. 1975. *Non-Mulberry Sericulture in India*. Central Silk Board, Bombay, 25p.

Jolly, M.S., M.N. Narasimhana, S.S. Sinha, and S.K. Sen. 1969. Interspecific hybridization in Antheraea. *Indian Journal of Heredity*, 1: 45–48.

Jolly, M.S., T.N. Sonwalkar, S.K. Sen, and G.K. Prasad. 1976. *Non-mulberry Silks: Sericultural Manual*. FAO United Nations, Rome.

Kioka, E.N. 1998. *Biodiversity of Wild Silk Moths (Lepidoptera) and Their Potential for Silk Production in East Africa*. Ph.D. Thesis, Knyatta University, Nairobi, Kenya.

Koji, Yamada. 2020. *16 Years of JICA Technical Cooperation for Bivoltine Sericulture*, JICA, Japan, p. 111.

Kumar, R., D. Goswami, and B.K. Singh. 2017. *North East Indian Wild Silk Moths—an Illustrated Male Genitalic Identification Guide*. Published by Director, CMERTI, Lahdoigarh, p. 100.

Mano, Y., R.K. Datta, N. Malreddy, H.K. Basavaraja, S. Nirmal Kumar, and N. Sureshkumar. 1994. Potentiality of Indian and Japanese Hybrids. *Indian Silk*, June, pp. 5–7.

Mukerji, N.G. 1899. *Handbook of Sericulture*. Culcutta: Bengal Sericulture Press.

Mukherjee, P. 1994. Hybrid vigour in different crossing systems in silkworm Bombyx mori. In: Silkworm Breeding and G. Srirama Reddy (ed.) *Oxford and IBH*. New Delhi, Calcutta, India: Publishing Co. Pvt. Ltd., pp. 197–208.

Nagesh, Prabhu H., and R.S. Teotia. 2018. Emerging needs to establish sanctuaries for conservation of genetic biodiversity of Vanya silkworms. *8th International Conference on Wild Silkworms*, www.csb.gov.in 22–12 January 2018, p. 111.

Nanavati, M. Mahesh. 1990. *Silk Production, Processing and Marketing Central Silk Board*, Central, Silk Board, Bombay.

Pachuau, L., B.N. Choudhary, and S.N. Gogai. 2018. Cnservation of wild Muga in Mizoram. *8th International Conference on Wild Silkworms*, www.csb.gov.in, 22–12 January, p. 120.

Peigler, R.S. 1993. Wild silks of the world. *American Entomologist*, 39(3): 151–161.

Sarkar, D.C. 1980. *Eri Culture in India*. Central Silk Board, Bombay, pp. 1–51.

Sharma, M.C., Y. Debaraj, B.K. Singh, P.K. Das and N. Suryanarayana. 2002. Status paper on Eri host plant germplasm. *Proc. National Workshop on Sericultural Germplasm Management and Utilization*, Hosur, Tamil Nadu, 6–7th February 2002, pp. 59–61.

Singh, A.K. 2018. Present status of non-mulberry sericulture in India: Colossal growth and employment potential: Abstract book. *8th International Conference on Wild Silkworms*, www.csb.gov.in, 22–12 January 2018, pp. 1–4.

Singh, B.K. 2018. Recent advances in Muga and Eri culture, its prospectus and strategies for future developments. *8th International Conference on Wild Silkworms*, www.csb.gov.in 22–12 January 2018, pp. 28–31.

Singh, K.C. and N. Suryanarayana. 2003. Wild silk moth—wealth of India. *Proceeding of National Conference on Tropical Sericulture for Global Competitiveness*, NASSI, Bangalore, p. 113.

Singh, N.I., N.I. Singh, L.S. Singh, and N. Suryanarayana. 2005. An overview of the salient features of *Antheraea frithii* Moore—A promising wild silk moth of India. *20th Congress of International Sericultural Commission on Non-Mulberry Silkworm*, www.csb.gov.in Vol. II, pp. 74–77.

Singh, N.I., T.J. Singh, Y.R. Keisa Singh, and K.C. Singh. 2000. Conservation and utilization of Indian oak fed *Antheraea* fauna. *International Journal of Wild Silk Moth & Silk, Tsukuba*, Japan, 5: 330–331.

Singh, R.N. and M. Maheswari. 2003. Conservation and utilization of sericigenous insects in North East Region of India. *Sericologia, Bangalore*, 43: 1–15.

Sinha, S.S. and A.K. Sinha. 1994. Conservation strategies for important wild silk moth population of *Antheraea mylitta* Drury (Lepidoptera: Saturniidae). *International Journal of Wild Silkmoth*, 2: 159–162.

Srivastava, A.K., P.K. Kar, R.B. Sinha, and N.B.V. Prakash. 2009. Present status of biodiversity of wild sericigenous insects in India and strategy for conservation. *Proceedings of National Workshop on Seri-Biodiversity Conservation*, March 7–8, CSGRC, Central Silk Board, Hosur, India, pp. 167–172.

Suryanarayana, N. and A.K. Srivastava. 2005. *Monograph on Tropical Tasar Silkworm: Central Tasar Research and Training Institute*, Ranchi, India, p. 134.

Tazima, Y. 1958. *Report on Sericulture Industry in India*, Published by Central Silk Board, India, pp. 1–56.

Tazima, Y. 1964. *The Genetics of the Silkworm*. Logos Press, London, 421p.

Thangavelu, K. 1991. *Wild Sericigenous Insects of India: A Need for Conservation*. Wild Silkmoths'91, Tsukuba, pp. 71–77.

Thangavelu, K., A.K. Bhagowati, and A.K. Chakroborty. 1987. Studies on some wild sericigenous insects of North-eastern India. *Sericologia*, 27: 91–99.

Thangavelu, K. and A. Borah. 1986. Occurrence of Antheraea mylitta Drury (Lepidoptera: Saturniidae) in North eastern India—distributional significance. *Current Science*, 55: 940.

Thangavelu, K., A.K. Chakravorty, A.K. Bhagowati, and M.D. Isa. 1988. *Handbook of Muga Culture*. Central Silk Board, Bangalore, pp. 9–14.

Thangavelu, K. and A.K. Sahu. 1983: Domestication of muga silkworm (*Antheraea assaminensis*): A major breakthrough in muga culture. *Sericologia*, 23: 153–158.

Yokoyama. 1959. *Silkworm Genetics Illustrated*. Japan Society for the Promotion of Science, Tokyo.

13

Status, Conservation, and Economic Importance of Oak Tasar Silkworm, Antheraea Proylei *(Lepidoptera: Saturniidae) in the Western Himalayan Region of India*

P. C. Joshi, Pashupati Nath, Namita Joshi, and Induja Mishra

CONTENTS

13.1 Introduction

India ranks second in world in the production of tasar silk next to China. Among the four varieties of silk produced in 2015–2016, mulberry accounted for 71.79%; tasar, 9.88%; eri, 17.74%; and muga, 0.58% of the total raw silk production. However, this was 20,582 tonnes during in 2016–2017, in which mulberry and vanya raw silk output amounts to 13,865 tonnes and 6,717 tonnes, respectively (CSB, GOI), the share of non-mulberry silk was only 2.3%. From non-mulberry production, eri, tasar, and muga silks

DOI: 10.1201/9781003220398-13

account for 69.78%, 24.85%, and 5.37%, respectively (Shetty and Samson, 1998). The availability of host food plants for rearing of wild silkworms is crucial factor. There is practically no systematic plantation of food plants for rearing of non-mulberry silkworms in India. In India, about 35 well-known species of oak occur all along the northwestern sub-Himalayan region at 1,200–3,500 m above sea level and northeastern region at 600–1,800 m above sea level, out of which 15 species belong to genus *Quercus* distributed in the sub-Himalayan belt of India (Negi and Naithani, 1995). A total number of 14 species of *Quercus* have been also described by Jolly, out of which five species—*Q. serrata, Q. semecarpifolia, Q. dealbata, Q. dilatata, and Q. incana*—have been identified and tested for temperate tasar silkworms in different states of India (Goel and Rao, 2004). Most insects, including lepidopteron, are phytophagous in nature, feeding upon green leaves; they feed on a wide range of food plants for their minimum necessity of essential elements—carbon, nitrogen, phosphorus, sulfur, and so on. Oak tasar silkworms (*Antheraea* spp.) are also polyphytophagous insects completing their life cycle on oak plants, though species specificity varies from region to region and altitude to altitude because of the availability of specific food plants at a particular altitudinal range. Among all the species of oak tasar silkworms, *Antheraea proylei* (Jolly) is considered as the most prominent, viable, and commercially exploited species in the temperate regions of the Indian subcontinent. Goel and Rao (2004) were successful in raising three crops of this species on different food plants and at different altitudes.

Oak tasar is found on a number of oak species of the genus *Quercus*. It is a finer variety of tasar generated by the silkworm *Antheraea proylei* (Jolly) in India, which feeds on natural food plants of oak, found in abundance in the sub-Himalayan belt of India, covering the states of Manipur, Himachal Pradesh, Uttarakhand, Assam, Meghalaya, and Jammu and Kashmir. In the world the major producer of oak tasar is China, and this comes from another silkworm, which is known as *Antheraea pernyi*. Oak tasar has been till recently the monopoly of China, which produces nearly 4,000 tonnes of raw silk per annum and accounts for 90% of tasar production in the world. India had, until recently, been depending on only *Antheraea mylitta* for the production of tasar silk. A general survey conducted by the Central Tasar Research and Training Institute of the Central Silk Board in 1965–1966 revealed the existence of rich flora of oak plantation in the sub-Himalayan belt of the country (Jolly, 1970). The survey also resulted in the discovery of a wild silkworm variety known as *Antheraea roylei* (Moore), which was incidentally an oak-feeding species. This species had not been of commercial value because of the fact that its larva spins a double-layered cocoon, which was found unreelable. Jolly *et al.* (1973) were successful in synthesizing fertile interspecific hybrid when the Chinese oak tasar silkworm, *Antheraea pernyi*, was crossed with its Indian sister species, *Antheraea roylei*. This hybrid was named as *Antheraea proylei*.

Sericulture is the main source of livelihood, either directly or indirectly, to a large number of people, the majorities of whom belongs to economically or socially deprived strata (SC/ST) and are dependent upon this industry. This industry provides employment at various levels, from food plant cultivation, rearing, reeling or spinning, and weaving to garment production, thus sustaining the livelihood of local communities. Wild sericulture remained obscure as an exclusive craft of tribal and hill folk inhabiting the forests of central India, sub-the Himalayan region, and northeastern India for a long time. However, in recent years, this traditional tribal craft has gained tremendous importance. Due to its rich production potential, eco-friendly nature of the activities and steady demand for handmade textile products within and outside the country, wild sericulture is commercially exploited from traditional craft into an industry of high potential. A large number of families in the western Himalayan region of Garhwal are practicing the sericulture and earning their livelihood. In the recent past, with the changing eco-climatic conditions due to various natural and man-made phenomena, the wild population of silk moths declined to the extent that now no more wild silkmoths are being reared by the local communities. Keeping this fact in view, the present study has been carried out in the Garhwal region of the state of Uttarakhand in India.

13.2 Review of Literature

In mountainous regions, there employment is limited, and the forest is the main source of income for the rural people, particularly in the western Himalayan region. (Singh, 1981; Singh and Rawat, 2012a, 2012b; Malik *et al.*, 2014). Silk insects, bees, and lac insects are used by local communities to earn their livelihood, and silkworms are very important to India from an economic point of view (Reddy, 1993; Nayak *et al.*, 2000; Suryanarayana and Srivastava, 2005; Hansda *et al.*, 2008; Gupta *et al.*, 2008; Zamal *et al.*, 2010; Manohar *et al.*, 2009; Ojha *et al.*, 2009; Gangopadhyay, 2009; Reddy *et al.*, 2010; Dewangan *et al.*, 2011; Singh *et al.*, 2014; Prasad *et al.*, 2016). Goswami and Singh (2012) have carried out studies on ecology of Indian golden silkmoth, *Antheraea assamensis* (Helfer; Lepidoptera: Saturniidae) and focused on the natural habitat, morphological characteristics, and developmental stages of *Antheraea assamensis.* Another worker (Choudhury, 1970) has also studied the polyphagous behavior and feeding behavior of different food plants of *Antheraea assamensis.* Few scientists have carried out work on ecological aspects of wild silkmoths (*Antheraea* spp.) in the northeastern region of India (Jolly *et al.*, 1976; Thangavelu, 1991; Singh and Singh, 1998; Singh *et al.*, 2000; Singh and Maheswari, 2003). Kakati and Chutia (2009) have carried out studies on the diversity and bioecology of wild sericigenous insect (i.e., silk-producing moths in Nagaland) and have reported 14 species of sericigenous insects and a large number of their food plants. Many workers have also studied the seri-biodiversity and its potential as a source of natural silk in the Indian subcontinent (Arora and Gupta, 1979; Thangavelu and Borah, 1986; Thangavelu *et al.*, 1987; Thangavelu, 1991; Nassig *et al.*, 1996; Bhattacharya and Teotia, 2000; Bhattacharya *et al.*, 2004; Srivastava and Thangavelu, 2005). Lefory and Ghosh (1912) reported the commercial importance of saturniids, the diverse wild silkmoths as silk-producing individuals of the world; they also reported that the wild saturniid moon moth, *A. selene*, was geographically distributed in tropical and deciduous forests. Many workers also studied the wild silkmoth *Antheraea mylitta* on *Terminalia catappa* under the laboratory conditions (Jolly, 1972; Sathe and Jadhav, 2001; Shamitha, 2007; Kavane and Sathe, 2008 and Jadhav *et al.*, 2014).

In oak tasar culture, the climatic conditions and man-made activities also play a major role because tasar worm has a long life cycle, a unique feeding behavior, and a body weight larger than other silkworms (e.g., *Bombyx mori* and *Phylasomia ricinia*). In the Indian Himalayan region, oaks are the dominant climax tree species of the moist temperate forest (Troup, 1921). According to Singh and Singh (1992), the five species of evergreen oak—*Quercus gluca* (phaliyant/hrinj), *Quercus leucotricophora* (banj), *Quercus lanuginosa* (rianj), *Quercus floribunda* (tilonj/moru), and *Quercus semecarpifolia* (brown/kharsu)—grow naturally in the western Himalayan region of Uttarakhand. Many scientists have carried out quantitative analysis of diversity in different oak-dominated (*Quercus* spp.) forests in Garhwal Himalaya (Singh and Rawat, 2012a, 2012b). Oak forests are providers of numerous ecosystem services and intricately associated with agro-ecosystem, as source of wood fuel and timber and also correlated with natural springs and wildlife (Singh, 1981; Ralhan *et al.*, 1982; Upreti *et al.*, 1985; Singh and Singh, 1986, 1987).

A considerable amount of work has been carried out on changing climate trends of Uttarakhand, India (Singh *et al.*, 1995; Pant *et al.*, 1999; Pant, 2003; IINC, 2004; Kumar *et al.*, 2005; IPCC, 2007; Bhutiyani *et al.*, 2007; Basistha *et al.*, 2008; Basistha *et al.*, 2009, Mishra, 2014). Due to climate change, global monsoon circulation has also been affected, leading to the reduction of forest cover, agriculture, and habitat destruction (Meher-homji, 1991; Kothyari *et al.*, 1997; Pielke, 2001; Vedwan *et al.*, 2001; Chase *et al.*, 2003; Ramankutty *et al.*, 2006; Douglas *et al.*, 2006; Basistha *et al.*, 2009; Chiu *et al.*, 2009; Chaudhary and Bawa, 2011; Nandargi and Dhar, 2011; Negi *et al.*, 2012; Kamboj and Belwal, 2013; Pranuthi *et al.*, 2014; Kotal *et al.*, 2014). Climate change has widespread impacts on human and natural systems (IPCC, 2014), and it is a major and growing challenge in mountain areas, where a minor shifts in any meteorological condition can disturb the fragile balance of the natural environment (Kotru *et al.*, 2014). It has been reported that climate change in the Himalayan region will impact the people and their livelihoods, health, and natural resources (Sharma *et al.*, 2009; Dhanai *et al.*, 2014). Insects are highly sensitive to slightest changes in environmental factors, such as temperature, humidity, air, light, and water, which affect growth, reproduction, and survival of different stages of insects, and the change in population size of some insects is directly related to any disturbance in climate.

Several scientific studies have been carried out on alternate assessment and mitigation of various anthropogenic disturbances, which affect the Himalayan biodiversity and also affect the survival of human being (Malik *et al.*, 2014; Malik *et al.*, 2016). Huge forest degradation in this region has taken place due to excess of increment of human population, tourism, overexploitation, widespread logging, overgrazing, and removal of leaf and wood litter from the forest cover (Gorrie, 1936; Chaturvedi, 1985; Singh and Rawat, 2012a, 2012b). Many villagers are involved in forest-based industry, particularly tasar culture, which is highly affected by parasites, predators, and climatic unevenness. Mohanty *et al.* (2003) revealed that the population of *Anthereaea mylitta* (Drury) is affected by various parasites and predators. Several studies have reported a large number of predators, such as ichneumon flies, sting bugs, and praying mantises, which are their natural enemies and affect the field crop of silkworms (Singh *et al.*, 1992; Mathur *et al.*, 1998). Shivakumar and Shamitha (2013) have made studies on the rearing of the tasar silkworm, *Antheraea mylitta* (Drury), outdoors, with the help of tribal farmers, and there was 80–90% crop loss due to pests, predators, natural calamities, and diseases. Some have also reported that *Canthecona furcellata* feed on the larvae of tropical and temperate tasar silkmoth species, *Antheraea myllita, Antheraea proylei*, and *Antheraea roylei*, (Sen *et al.*, 1971; Jolly *et al.*, 1979 and Singh *et al.*, 1992). Singh and Thangavelu (1991), in their studies have focused on reasons for the low production of silk due to the pest problem associated with food plants and silkworms. They reported that there are 100 insect pest species that attack silkworms at different developmental stages. They have also reported seven to ten species of parasites and predators of silkworm larvae and found that diseases are also a major factor of crop loss (70–80%), which includes sporozoasis, bacteriosis, virosis, and mycosis. An extensive report has been published on fungal diseases, such as sooty mold, rust disease, and powdery mildew, on host plants (*Quercus* spp.) of the oak tasar silkmoth (Ng and Chhetry, 2006). Another study has also been conducted on *Antheraea mylitta*, concluding that starvation is also a factor that leads to bacterial and viral infection in silkworms (Thangavelu et al., 1995). Various scientists have advocated that various crops of *Atheraea proylei* are mostly affected by microbial diseases (viral, bacterial, fungal, and microsporidian) (Rao and Nigam, 1986)). Bidyapati and Naomani (1995) have reported the infectious diseases in *Antheraea proylei*. Varatharajan et al. (1998) have also reported that CPV and gregarine infection in *Antheraea proylei* led to the loss in the different life stages during spring crop in Nagaland.

13.3 Materials and Methods

Study Area

The present study was carried out in the Rudraprayag and Uttarkashi districts of Uttarakhand, India (Figure 13.1). Rudraprayag is situated at 30°16′N to 30°28′N latitude and 78°58′E to 78°98′E longitude, while Uttarkashi is situated at 30°43′N to 30°73′N latitude and 78°27′E to 78°45′E longitude. On the basis of occurrence of host plants (*Quercus* spp.) of *Antharaea* spp. in abundance and suitable climatic conditions, the following four sites were selected for survey in these two districts.

1. Site I (Badeth)

It has an average elevation of 1,400–2,400 m. The area is rich in flora and fauna and has relatively fewer human habitations. The area provides a good habitat for finding and rearing oak tasar silkworms. Besides the oak plantation (*Quercus serrata, Quercus leucotricophora*), the area possesses a rich diversity of mixed vegetation, such as *Ficus roxburghii, Prunus cerasoides, Mangifera indica, and Pinus roxbughii.*

2. Site II (Ghangasubangar)

It has an average elevation of 1,800–3,000 m. The area is rich in oak plantation and provides a suitable habitat for wild oak tasar silkworm. The forest is mostly dominated by mixed vegetation, such as *Quercus lucotricophora, Quercus glauca, Fiscus roxburghii, Abies pindrow, Mangifera indica*, and *Euphobia lathyris.*

3. Site III (Radi Top)

This study area is situated at an altitude of 2,496 m. This is densely covered with oak trees and provides a good habitat for oak tasar silkworm rearing. Besides oak trees, the forest cover includes other common trees, such as *Rhododendronarboreum, Lyonia ovalifolia, Pyrus pashia, Symplocos paniculata, Myrica esculenta, Neolitsea pallens*, and *Ilex dipyrena*.

4. Site IV (Chaurangi Khal)

It is situated at an altitude of 2,452 m. The forest at this site is mostly dominated by *Quercus floribunda, Quercus leucotrichophora, Rhododendron arboreum, Lyonia ovalifolia, Pyrus pashia, Symplocos paniculata, Myrica esculenta, Neolitsea pallens*, and *Ilex dipyrena*, while the dominant shrubs were *Sarcococca salinga, Arundinaria falcata, Berberis asiatica, Indigofera* sp., and *Daphne papyracea*.

13.4 Abiotic Factors

Record of Various Meteorological Data

The abiotic factors of an area play an important role in the distribution of oak tasar silkmoth and its food plants. Keeping this fact in view, the temperature, humidity, and rainfall of each study site were recorded on a monthly basis. Temperature and humidity were measured with the help of thermo-hydrometer. Data was also collected from State Meteorological Station, Indian Metrological Department, Dehradun, and Uttarakhand Jal Sansthan, Haridwar.

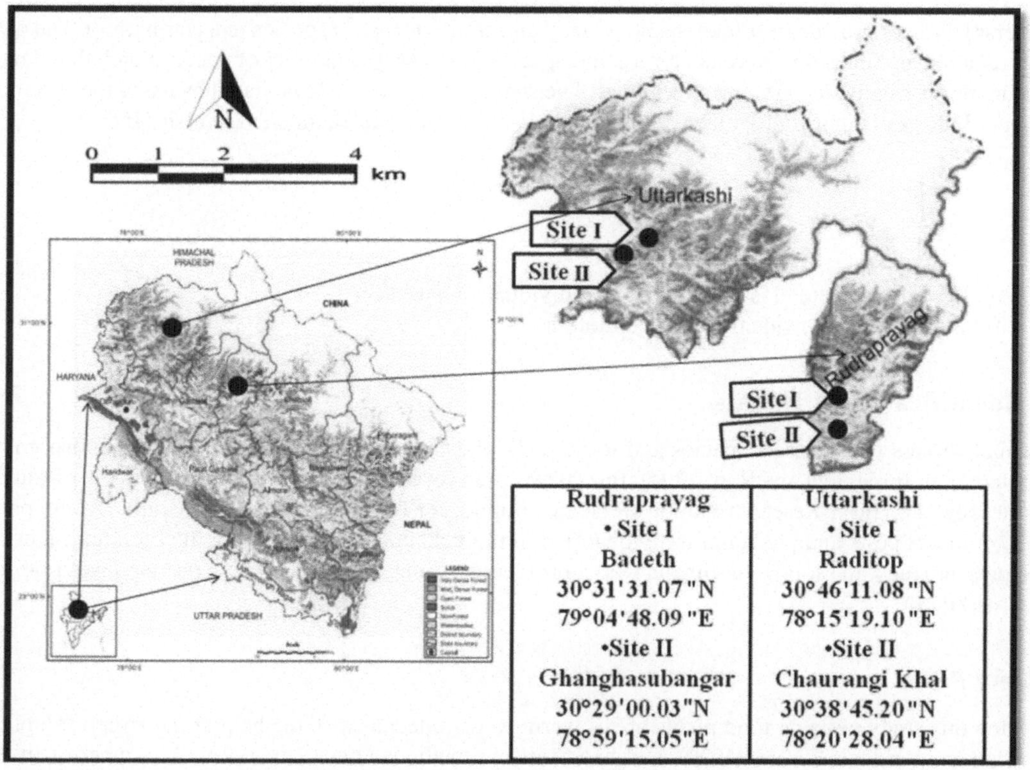

FIGURE 13.1 Locations of different study sites in Rudraprayag and Uttarkashi in Uttarakhand.

13.5 Biotic Components

1. Population Dynamics of Insects

Extensive surveys were conducted at different selected sites located at different altitudes of two districts of Uttarakhand, covering various seasons and forest types, from January 2014 to December 2015. Natural habitats of tasar worms, located at Badeth, Sangur, Mathya, Ghanghasubangar, Biroon, Basti, Patiun, Bakshir, Uchaoula, Radi Top, Sarnol, Badkot, and Chaurangi Khal, were surveyed for wild populations of silkworms/cocoons/moths and their host plants. During the survey, officials of the state and central sericulture departments (Regional Tasar Research Station, Bhimtal), as well as other private rearers, including local villagers, were interviewed for information on the presence or absence of wild populations of silkworms, their preferred host plants, and the present status on the rearing of the insects. A questionnaire was also developed and discussed with the private rearers, including local villagers. The fecal pellet method (Moi and Veenakumari, 2002) was also used to report the presence or absence of larvae/cocoons/oak tasar worms. The samples of adult moths and their larval forms were collected by following standard practices, using simple hanging of light traps, and they were visually observed at the study site and identified with the help of literature (Arora and Gupta, 1979; Nassig *et al.*, 1996; Peigler and Naumann, 2003; Goel and Rao, 2004) and scientists specializing on the subject from the Regional Tasar Research Station, Bhimtal, Nainital. The population density (individuals per hectare) of oak tasar silkmoths was determined by dividing the total number of insects collected/observed on each sampling month from each study site with five, as the area of each selected site was five hectares.

2. Species Diversity

Diversity indices provide information about rarity and commonness of species in a community. The data recorded during this study were analyzed, and species richness (the number of species) and abundance (the number of individuals) were determined. The species diversity was calculated by using the Shannon Wiener Index (Shannon and Weiner, 1963), which is also termed as Shannon's Diversity Index:

$$H' = -\sum_{i-1}^{S} pi \ln pi$$

Where, Pi = ni/N, where ni is the number of individuals in species i
N = Total number of individuals in the community

3. Identification of Threats

Various threats to the insect species and to the host plants were identified in the field and also going through literature (Goel and Rao, 2004). Insects were surveyed for disease, which were further brought to the Regional Tasar Research Station (RTRS), Bhimtal, for further confirmation. Similarly, the presence of insect pests and predators were identified in the field and confirmed with the literature. During the present study, the following threats were considered as major threats to the wild population of oak tasar silkmoth.

Insect Pests

During this study pests on food plants of silkworm were collected with the help of the sweep sampling method, per Gadagkar *et al.* (1990), and hand-picking methods (Jonathan, 1990). A number of insect species have been reported as pest on food plants of silkworms and silkmoths. Many species of insects were observed as damaging agents in oak forests; hence, it was necessary to identify them and to prepare inventory of harmful insects in oak forests. The inventory of pest species was prepared on the basis of feeding nature and visual surveys during the study period. It was also confirmed with the help of published literature (Goel and Rao, 2004).

1. Predators: During the present study, several predators, including birds, insects, and mammals, were identified, which predate the silkworms during various stages of their life cycle.
2. Pathogens: For detection of pathogens in infected samples collected during the rearing season, the help of the Regional Tasar Research Station, Bhimtal, was taken.

13.6 Major Observations

1. Abiotic Components

Temperature and humidity play a significant role in the development of different insect stages. Keeping this fact in mind, these two parameters were recorded on each sampling date.

The maximum and minimum temperature in Rudraprayag ranged between 28.33 (June) and 3.81°C (January) in the first year of study, while it ranged between 34.03 (June) and 4.87°C (January) in Uttarkashi. During the second year of study, the monthly mean value of temperature varied from 28.80 (June) to 3.94°C (January) and 31.43 (September) to 5.06°C (January) in Rudraprayag and Uttarkashi, respectively. Similarly, the humidity in Rudraprayag ranged between 93.06 (July) and 34.33% (June), while in Uttarkashi, it ranged between 93.03 (August) and 34.90% (June).

2. Biotic Components

Population dynamics of different species of *Antheraea* : A total of 330 individuals of *Antheraea* spp. Were collected from the study sites of Rudraprayag during the study period in 2014–2015. The maximum numbers of individuals (100) were reported from Ghangasubangar, followed by 71 individuals from Badeth. A total of 171 individuals were present in both the study sites during the first year of study. During the second year of study, a total of 159 individuals were recorded, which included 80 individuals from Badeth and 79 individuals from Ghangasubangar.

Similarly, a total 315 individuals of *Antheraea* spp. Were collected from the Uttarkashi during the study period. The maximum numbers of individuals (87) were collected from Chaurangi Khal, followed by 60 individuals from the Radi Top site of Uttarkashi. A total of 147 individuals were collected from both the study sites during the first year of study. Similar trends were followed in the second year of study, wherein 92 individuals were collected from Chaurangi Khal and 76 from Radi Top.

Taxonomic composition of *Antheraea* spp.: During our study two species of *Antheraea* were recorded from the study site that included *Antheraea roylei* and *Antheraea proylei.*

Number of different developmental stages of *Antheraea* spp. Recorded in Rudraprayag: During the first year of study, a total of 19 cocoons of *Anthearea roylei* were collected. The highest numbers (14) of cocoons were recorded from Ghangasubangar, while only five were recorded from Badeth. However, only one adult moth was collected from these study sites.

During this period, 55 larvae, 42 cocoons, 54 adults of the *Antheraea proylei* were collected. The highest number of larvae (30), cocoons (24), and adults (31) were collected from Ghangasubangar, while 25 larvae, 18 cocoons, and 23 adults were collected from Badeth.

During the second year of study, 20 cocoons of the *Anthearea roylei* were collected, which included er of 12 cocoons from Ghangasubangar and 8 cocoons from Badeth. Four adult moths were also collected from these study sites. During this period, 58 larvae, 42 cocoons, and 35 adults of *Antheraea proylei* were collected. The highest number (31) of larvae, 21 cocoons and 17 adults were collected from Badeth, while 27 larvae, 21 cocoons, and 18 adults were collected from Ghangasubangar.

Number of different developmental stages of *Antheraea* spp. Recorded in Uttarkashi: During our regular surveys in the area, eight cocoons of the *Anthearea roylei* were collected from study sites in the first year of study, which included five cocoons from Chaurangi Khal and three from Radi Top (Uttarkashi). Only one adult moth was collected from the study site. During this period, 37 larvae,

48 cocoons, and 53 adults of the *Antheraea proylei* were collected. The highest number (22) larvae, 27 cocoons, and 32 adults were collected from Chaurangi Khal, while 15 larvae, 21 cocoons, and 21 adults were collected from Radi Top. During the second year of study, 13 cocoons of the wild silkmoth *Anthearea roylei* were collected, which included seven cocoons from Radi Top and six cocoons from Chaurangi Khal. A total of four adult moths were also collected from the study sites. During this period, 54 larvae, 52 cocoons, and 45 adults of the *Antheraea proylei* were collected. A total of 27 larvae, 27 cocoons, and 31 adults were collected from Chaurangi Khal, while 27 larvae, 25 cocoons, and 14 adults were collected from Radi Top.

3. Species Diversity (Shannon-Weiner Index)

Species diversity is a measure of the number and occurrence of species in a community. It has two different components—number of species present (species richness) and relative abundance (termed dominance or evenness). Species diversity is the effective number of different species that are represented in a collection of individuals.

During the first year of study, the species diversity was maximum (0.380) at Ghangasubangar and minimum (0.254) at Badeth in Rudraprayag, while during second year the maximum and minimum values were 0.440 and 0.4004, respectively.

The species diversity was recorded as 0.25 and 0.198 at Chaurangi Khal and Radi Top, respectively, in Uttarkashi, during the first year of study, while during second year it was 0.389 at Radi Top and 0.269 at Chaurangi Khal. The species diversity (H) and abundance of *Antheraea spp.* In different study sites have been presented in Tables 13.1 and 13.2.

Oak Tasar Rearing Practices in the Area

To collect the data related with oak tasar practices in the area, 12 private rearers were selected randomly and interviewed. Rearing practices of *Anthearea proylei* (Jolly) by different farmers in different villages of the study area were also taken into consideration with the help of a questionnaire.

During the first year of survey, it was found that the maximum number of oak tasar worm (50,000) were reared by Mr. Satte Singh in the spring crop, while minimum number of worms (15,000) were reared by Mr. Pancham Singh in Rudraprayag. In Uttarkashi, the maximum (30,000) and minimum (20,000) numbers of worm were reared by Mr. Virendra Singh Rana and Mr. Ravindra Singh, respectively. In autumn crop of 2014, the maximum number of oak tasar silkworms (30,000) were reared by Mr. Kuwar Singh, while Ms. Gayatri Devi reared 10,000 number of oak tasar silkworms. A total of

TABLE 13.1

The Shannon Winner Index of Two Species of *Antheraea* in Rudraprayag in 2014 and 2015

		Study period 2014		Study period 2015	
SN	Study site	Abundance	Diversity (H)	Abundance	Diversity (H)
1	Site I (Badeth)	71	0.254	80	0.4004
2	Site II (Ghangasubangar)	100	0.380	77	0.440

TABLE 13.2

Shannon Winner Index of Two Species of *Antheraea* in Uttarkashi in 2014 and 2015

		Study period 2014		Study period 2015	
SN	Study Site	Abundance	Diversity (H)	Abundance	Diversity (H)
1	Site I (Radi Top)	60	0.198	76	0.398
2	Site II (Chaurangi Khal)	87	0.250	92	0.269

255,000 of worms in spring season and 160,000 of worms in autumn season were reared by different farmers in both the districts.

During the second year (2015), it was found that the maximum number (60,000) of oak tasar silkworms were reared by Mr. Satte Singh in the spring crop, and the minimum number of worms (15,000) were reared by Mr. Pancham Singh again in Rudraprayag, whereas in Uttarkashi, the maximum numbers of worms (35,000) were reared by Mr. Virendra Singh Rana. In autumn crop of 2015, the maximum number of worms (20,000) were reared by Mr. Satte Singh, and the minimum number of worms (10,000) were reared by Ms. Gayatri Devi in Rudraprayag, whereas in Uttarkashi, the maximum number of worms (30,000) were reared by Mr. Virendra Singh Rana and the minimum number of worms (20,000) reared by Mr. Ravindra Singh. A total of 275,000 tasar worms in spring and 160,000 worms in autumn were reared by different farmers in both districts.

4. Threats to Oak Tasar Worm *Antheraea* spp.

During this study (2014 and 2015), an extensive survey was carried out at different areas and after discussion with farmers and local people of the village and tehsil, some important facts have been explored about the threats to oak tasar worms. These factors compelled the people to leave this practice. A large part of the western Himalayan region of Uttarakhand is covered by oak forests (*Quercus* spp.; 1,000–2,000 m above sea level), it forms the dominant forest vegetation (Singh and Singh, 1992). In this region, oak covers about 12.3% of the total geographical area of Uttarakhand. Local people and tribes of this state depend on these forests for a variety of goods and facilities for their livelihood. The oak forest is significant for the people in terms of provisioning of goods, such as quality fuel wood, year-round green fodder, nutrient-rich leaf litter used as organic manure, nutrient-rich soil, and a variety of minor forest products, including medicinal plants. (*Glimpses of Forestry Research in the Indian Himalayan Region Special Issue on International Year of Forests, 2012*). In the present study, the following threats to the oak tasar worm were identified.

a. *Habitat Fragmentation*

Habitat fragmentation is one major cause of concern in the Garhwal Himalayas. This phenomenon is affecting to the loss of biodiversity of oak tasar worm because of damage to food plants. This natural vegetation or oak covering forest is being removed due to increasing human activities. The villages are escalating day by day. Natural disaster is also a major reason of habitat fragmentation in Garhwal Himalayas.

Overgrazing is also a major cause of habitat loss and decline in the insect fauna, including the silkmoth in the oak forest of Garhwal Himalaya. Because the larvae of the wild silkworm *Antheraea spp.* feed on *Quercus spp.* After feeding they start spinning of cocoon on the leaves of *Quercus* plant and complete their cocoon formation. The collection of fodder by the villagers causes damage to the cocoons.

b. *Disease of Silkmoths*

The following diseases have been reported in the wild silkmoth by various authors:

i. The bacteriosis characterized by three symptoms—rectal protrusion, chain-type excreta, and sealing of anal lips. Singh *et al.* (2011) have been reported that anal lips were closed by sticky semi-fluid excreta and shrinking in length of larvae. In chain-type excreta, larvae excreted out a jellylike substance in the form of chain. The hemolymph filled in the transparent bag (rectum protrudes) and opens out of the anus (Sahay *et al.* 2000).

ii. The larva of silkworm is affected by polyhedrosis virus, which turned it into a pale and sluggish form and made it lose its legs' hold from the host twig, except the claspers, with help of which it hangs its head downward after death. As a result of infection, there was complete fragmentation of tissue. It is caused the polyhedral inclusion bodies (PIBs). The integument turns dirty and brownish, and the skin becomes fragile and blackish. After the degeneration of tissues, the larva secretes from its mouth a dark brownish fluid that gives off an obnoxious odor.

iii. Microsporidiosis is caused by the protozoan parasite *Nosema milittensis* (Chakrabarti and Manna, 2006). Black spots appear all over the body of larvae. The infected larvae show irregular molting, and their growth is threatened. Larvae infected with *Nosema* spp. show a discontinued development period and reduced size and larval weight in comparison to uninfected ones (Rath *et al.*, 2003).

iv. Mycosis is caused by the fungus *Penicillium citriinum*. The common symptoms of mycosis are body hardness and dorsal bending. The body is covered with white powdery spores of fungus and gets laterally compressed, dry, hard, and wrinkled (Kumar *et al.*, 2011).

c. Pest and Predators

The main predators of silkworm insects are birds; they play a key role in maintaining insect abundance of any area. Generally, insect abundance and sometime richness is too low where these insectivorous birds are present. Some insectivorous birds have been observed during the rearing seasons from the different study sites. Some insectivorous mammals were also reported to damage the worms, which include *Rhesus macaque* and *Rattus rattus*. It has been reported that approximately 1,000 insect larvae are destroyed at a time during the rearing seasons at particular rearing stations. However, *Rattus rattus* was reported as the predator of the large number of cocoons of this insect during rearing season.

A total 19 species of pests and predators were recorded, which were detrimental to food plants of silkworms and silkworm larvae and their cocoons at the rearing station, which include *Alcaeoorhynchus grandis*, *Dalpada* spp., *Phymata* spp., *Hypomeces squomosus, froghoppers, and Lilioceris* spp. As major pests. *Alcaeoorhynchus grandis*, *Dalpada* spp., *Phymata* spp., *Hypomeces squomosus*, and *Lilioceris* spp. Damage the leaves of food plants of silkworms, while froghoppers suck the sap of food plants. The predators included three species of insects (*Podisus spp.*, *Vespa affinis*, and *Praying mantis*), eight species of birds (spotted dove, house sparrow, white-throated kingfisher, *Verditer flycatchers*, jungle crow, Himalayan bulbul, red-billed blue, magpie, *Urocissa erythrorycha*), and two species of mammals (*Rhesus macaque, Rattus rattus*) .

13.7 Summary

From these observations, we may conclude that the studied area has only two species of tasar moths—*Anthearea roylei* and *Antheraea proylei*. The population/abundance of the tasar silkmoth (*Antheraea proylei* and *Anthearea roylei*) has decreased drastically during last two decades, and no silk is being produced by the local communities using the population of moth collected from wild. However, rearing practices are being adopted by only some families after procuring eggs of *Antheraea proylei* from the regional tasar research center. It has also been revealed that local communities were largely practicing the sericulture using the wild variety of oak tasar in the area. Adopting sericulture practices by some of the rearers is proving to be a good source of alternative income generation. There are various factors that are responsible for this decline in the number of moths, which include anthropogenic disturbances like forest fires, grazing by livestock, and collection of fodder by the villagers, thus affecting the food plants of the moths, and change in some of the climatic variables. Though the food resources of the worms are still available in the study areas, it appears that the interest of the local community has deviated from sericulture. Thus, making people aware of these resources and using them for rearing of silkworms as an alternate source of increasing their economy seems to be the current need.

REFERENCES

Arora, G. S. and Gupta, I. J. (1979). Taxonomic studies of some of the Indian non-mulberry silk moths (Lepidoptera: Saturniidae). *Memoirs of Zoological Survey of India*, 16, 1–63.

Basistha, A., Arya, D. S. and Goel, N. K. (2008). Spatial distribution of rainfall in Indian Himalayas–a case study of Uttarakhand region. *Water Resources Management*, 22(10), 1325–1346.

Basistha, A., Arya, D. S. and Goel, N. K. (2009). Analysis of historical changes in rainfall in the Indian Himalayas. *International Journal of Climatology*, 29(4), 555–572.

Bhattacharya, A., Singh, B. K. and Das, P. K. (2004). Biodiversity of wild silkmoths in Assam (North East India). *Annals of Forestry*, 12, 208–216.

Bhattacharya, A. and Teotia, R. S. (2000). Conservation strategies of wild silk moths in the North-Eastern Region of India. *International Journal of Wild Silk Moths Silk* (Speical issue), 5, 311–313.

Bhutiyani, M. R., Kale, V. S. and Pawar, N. J. (2007). Long-term trends in maximum, minimum and mean annual air temperatures across the Northwestern Himalaya during the twentieth century. *Climatic Change*, 85(1–2), 159–177.

Bidyapati Devi, L. and Naomani, M. K. R (1995). Pests and diseases of oak tasar silkworm and its food plants. *Workshop in Oak Tasar Culture; Regional Tasar Research Station, Imphal. Base Paper*, 11, 1–6.

Chakrabarti, S. and Manna, B. (2006). Three new species of Nosema from non-mulberry silkworms in Assam: Light and electron microscopy studies. *Parasitic Diseases*, 30(2), 125–133.

Chase, T. N., Knaff, J. A., Pielke, R. A. and Kalnay, E. (2003). Changes in global monsoon circulations since 1950. *Natural Hazards*, 29(2), 229–254.

Chaturvedi, A. N. (1985). Fuel and fodder trees for man-made forests in hills. In *Environmental Regeneration in Himalaya: Concepts and Strategies*, edited by J. S. Singh. Central Himalayan Environment Association and Gyanodaya Prakashan, Dehradun.

Chaudhary, P. and Bawa, K. S. (2011). Local perceptions of climate change validated by scientific evidence in the Himalayas. *Biology Letters*, rsbl20110269.

Chiu, C. A., Lin, P. H. and Lu, K. C. (2009). GIS-based tests for quality control of meteorological data and spatial interpolation of climate data: A case study in mountainous Taiwan. *Mountain Research and Development*, 29(4), 339–349.

Choudhury, S. N. 1970. *Muga Silk Industry*. Directorate of Sericulture, Govt. of Assam.

Dewangan, S. K., Sahu, K. R. and Achari, K. V. (2011). Sericulture-a tool of eco-system checking through tribal. *Journal of Environmental Research and Development*, 6(1), 165–173.

Dhanai, R., Negi, R. S., Singh, S. and Parmar, M. K. (2014). The effects of climate change on natural resources and socio-economic condition of Himalayan communities of Uttarakhand, India. *International Journal of Modern Communication Technologies & Research (IJMCTR)*, 2(11), 21–25.

Douglas, E. M., Niyogi, D., Frolking, S., Yeluripati, J. B., Pielke, R. A., Niyogi, N. and Mohanty, U. C. (2006). Changes in moisture and energy fluxes due to agricultural land use and irrigation in the Indian Monsoon Belt. *Geophysical Research Letters*, 33(14).

Gadagkar, R., Chandrasekara, K. and Nair, P. (1990). Insect species diversity in tropics: Sampling methods and a case study. *Journal of the Bombay Natural History Society*, 87, 337–353.

Gangopadhyay, D. (2009). *Sericulture Industry in India: A Review*, www.nistads.res.in/indiasnt2008/t6rural/t6rur16.htm.

Goel, R. K. and Rao, J. V. K. (2004). *Oak Tasar Culture*. A.P.H. Publishing Corporation, Delhi.

Goswami, D. and Singh, H. I. (2012). Bioecological studies of Indian golden silk moth, Antheraea assamensis Helfer (Lepidoptera: Saturniidae). *Munis Entomology & Zoology*, 7, 274–283.

Gorrie, R. M. (1936). Tree lopping on a permanent basis. *Indian Forester*, 63(1), 20–31.

Gupta, R., Singh, A. K., Chatterjee, K. K., Chandra, H. and Chakraorty, D. (2008). Tasar culture showed them the way. *Indian Silk*, 46(10), 20–21.

Hansda, G., Reddy, R. M., Sinha, M. K., Ojha, N. G. and Prakash, N. B. (2008). Ex-situ stabilization and utility prospects of 'Jata' ecorace of tropical tasar silkworm *Antheraea mylitta* Drury. *International Journal of Industrial Entomology*, 17(2), 169–172.

IINC. (2004). India's Initial National communication to the United Nations framework convention on climate change. *Ministry of Environment and Forests*, Govt. of India, 265.

IPCC. (2007). Climate change 2007: The physical science basis. In *Contribution of Working Group I to the Fourth Assessment Report of the Intergovernmental Panel on Climate Change*, edited by S. Solomon, D. Qin, M. Manning, Z. Chen, M. Marquis, K. B. Averyt, M. Tignor and H. L. Miller. Cambridge University Press, Cambridge and New York.

IPCC. (2014). *Climate change 2014: Synthesis Report-Headline Statement from Summary for Policymakers, Impacts, Adaptation, and Vulnerability, Part A: Global and Sectoral Aspects, Contribution of Working*

Group II to the Fifth Assessment Report of Intergovernmental Panel on Climate Change, edited by C. B. Field, V. R. Barros, D. J. Dokken, K. J. Mach, M. D. Mastrandrea, T. E. Bilir, M. Chatterjee, K. L. Ebi, Y. O. Estrda, R. C. Genova, B. Girma, E. S. Kissel, A. N. Levy, S. MacCracken, P. R. Mastrandrea and L. L. White. Cambridge University Press, Cambridge and New York, NY, pp. 1–32.

Jadhav, A. D., Bhusnar, A. R., Sathe, T. V., Yankanchi, S. R. and Kirwale, K. (2014). Rearing performance of Tasar silkworm *Antheraea mylitta* Drury (Lepidoptera: Saturniidae) on different food plants from Kolhapur district of Western Maharashtra. *Biospectra*, 9(1), 141–146.

Jolly, M. S. (1970). Oak oriented tasar culture in the offing. *Indian Silk*, 8, 3–5.

Jolly, M. S. (1972). New technique of tasar silkworm rearing. *Indian Silk*.

Jolly, M. S. and Sen, S. K. (1972). Infection of *Antheraea mylitta* drury (Lepidoptera: Saturniidae). *Indian Journal of Sericulture*, 11, 52–57.

Jolly, M. S., Sen, S. K., Sonawalker, T. N., and Prasad, G. K. (1979) Non mulberry silks. FAO United Nation. *Service Bulletin*, 29, 1–178.

Jolly, M. S., Sinha, B. R. R. P. and Sinha, S. S. (1973). Inter specific hybridization in *Antheraea* with special reference to *Antheraea roylei* and *Antheraea pernyi*. Document, 12th Int. Silk Cong., Barcelona, p. 6.

Jolly, M. S., Sonwalker, T., Sen, S. K. and Prasad, G. K. (1976). *Non-mulberry Silks. Sericultural Manual.* FAO. United Nations, Rome.

Jonathan, J. K. (1990). *Collection and Preservation of Animals (Hymenoptera)*. Zoological Survey of India, Calcutta, pp. 147–150.

Kakati, L. N. and Chutia, B. C. (2009). Diversity and ecology of wild sericigenous insects in Nagaland, India. *Tropical Ecology*, 50(1), 137–146.

Kamboj, N. K. and Belwal, O. K. (2013). Effect of climate change on out-migration in Uttarakhand: A statistical analysis based on cox-proportional hazard model for Kimsararea and Rajaji National Park. *International Journal of IT, Engineering and Applied Sciences Research*, 2(11).

Kavane, R. P. and Sathe, T. V. (2008). Indoor rearing of tasar silk worm *Antheraea mylitta* Drury. on *Terminalia catappa* Cooke. *Biotechnological Approaches in Entomology*, 178–183.

Kotal, S. D., Roy, S. S. and Bhowmik, S. R. (2014). Catastrophic heavy rainfall episode over Uttarakhand during 16–18 June 2013–observational aspects. *Current Science*, 234–245.

Kothyari, U. C., Singh, V. P. and Aravamuthan, V. (1997). An investigation of changes in rainfall and temperature regimes of the Ganga Basin in India. *Water Resources Management*, 11(1), 17–34.

Kotru, R., Choudhary, D., Fleiner, R., Khadka, M., Pradhan, N. and Dhakal, M. (2014). Adapting to climate change for sustainable agribusiness in high mountain watersheds: a case study from Nepal. *ICIMOD Working Paper* (2014/1). https://www.cabdirect.org/cabdirect/abstract/20163181483

Kumar, K. K., Sinha, A. K., Singh, G. P. and Madhusudhan, K. N. (2011). Efficacy of systemic fungicides for control of white muscardine in tasar silkworm, *Antheraea mylitta* D. *Research Journal of Microbiology*, 6(11), 805–812.

Kumar, V., Singh, P. and Jain, S. K. (2005). Rainfall trends over Himachal Pradesh, Western Himalaya, India. In *Conference on Development of Hydro Power Projects—A Prospective Challenge*, Shimla, pp. 20–22.

Lefory, H. M. and Ghosh, C. C. (1912). Eri silk. *Memoirs of the Department of Agriculture in India*. Vol. 4, No. 1. Thacker, Spink & Company, 130 pp.

Malik, Z. A., Bhat, J. A. and Bhatt, A. B. (2014). Forest resource use pattern in Kedarnath wildlife sanctuary and its fringe areas (a case study from Western Himalaya, India). *Energy Policy*, 67, 138–145.

Malik, Z. A., Hussain, A., Iqbal, K. and Bhatt, A. B. (2014). Species richness and diversity along the disturbance gradient in Kedarnath Wildlife Sanctuary and its adjoining areas in Garhwal Himalaya, India. *International Journal of Current Research*, 6(12), 10918–10926.

Malik, Z. A., Pandey, R. and Bhatt, A. B. (2016). Anthropogenic disturbances and their impact on vegetation in Western Himalaya, India. *Journal of Mountain Science*, 13(1), 69–82.

Manohar, R. R., Hansda, G., Ojha, N. G. and Suryanarayana, N. (2009). Heterobeltiosis in Fl hybrids of wild and domesticated ecoraces of tropical tasar silkworm *Antheraeamylitta Drury Sericologia*, 49, 189–200.

Mathur, S. K., Singh, B. M., Sinha, A. K. and Sinha, B. R. R. P. (1998). Integrated package for rearing of *A mylitta* D. *Indian Silk*, 37(2), 15–18.

Meher-Homji, V. M. (1991). Probable impact of deforestation on hydrological processes. In *Tropical Forests and Climate*. Springer, Dordrecht, pp. 163–173.

Mishra, A. (2014). Changing climate of Uttarakhand, India. *Journal of Geology and Geosciences*, 3(4), 163.

Mohanty, P. K. (2003). *Tropical Wild Silk Cocoons of India, Book*. Daya Publishing House, New Delhi.

Moi, P. and Veenakumari, K. (2002). Biology of *Antheraea andamana* (Saturniidae) on the Andaman Islands, Indian Ocean. *Journal of the Lepidopterists' Society*, 56(3), 123–128.

Nandargi, S. and Dhar, O. N. (2011). Extreme rainfall events over the Himalayas between 1871 and 2007. *Hydrological Sciences Journal*, 56(6), 930–945.

Nassig, W. A., Lempe, R. E. J. and Kger, S. (1996). The Saturniidae of Sumatra (Lepidoptera). *Heterocera Sumatrana*, 10, 3–10.

Nayak, B. K., Dash, A. K. and Patro, K. B. G. (2000). Biodiversity conservation of wild tasar silk moth Antheraea paphia L. of Simlipal Biosphere reserve and strategy for its economic utilization. *International Journal of Wild Silk Moth*, 5, 367–370.

Negi, G. C. S., Samal, P. K., Kuniyal, J. C., Kothyari, B. P., Sharma, R. K. and Dhyani, P. P. (2012). Impact of climate change on the western Himalayan Mountain ecosystems: An overview. *Tropical Ecology*, 53(3), 345–356.

Negi, S. S. and Naithani, H. B. (1995). *Oaks of India, Nepal and Bhutan*. International Distribution, Dehradun, India, pp. 1–275.

Ng, T. C. and Chhetry, G. K. N. (2006). Fungal diseases of Oak Tasar host plants (*Quercus* spp.) in Manipur. *Journal of Mycopathological Research*, 44(2), 211–217.

Ojha, N. G., Reddy, R. M., Hansda, G., Sinha, M. K., Suryanarayana, N. and Prakash, N. V. (2009). Status and potential of Jata, a new race of Indian tropical tasar silkworm (*Antheraea mylitta* Drury). *Academic Journal of Entomology*, 2(2), 80–84.

Pant, G. B. (2003). Long-term climate variability and change over monsoon. *Asia Journal of Indian Geophysical Union*, 7(3), 125–134.

Pant, G. B., Rupa Kumar, K. and Borgaonkar, H. P. (1999). Climate and its long-term variability over the western Himalaya during the past two centuries. In *The Himalayan Environment*. New Age International (P) Limited, New Delhi, pp. 172–184.

Peigler, R. S. and Naumann, S. (2003). *A Revision of the Silkmoth Genus Samia*. University of the Incarnate Word, San Antonio, Texas.

Pielke, R. A. (2001). Influence of the spatial distribution of vegetation and soils on the prediction of cumulus convective rainfall. *Reviews of Geophysics*, 39(2), 151–177.

Pranuthi, G., Dubey, S. K., Tripathi, S. K. and Chandniha, S. K. (2014). Trend and change point detection of precipitation in urbanizing Districts of Uttarakhand in India. *Indian Journal of Science and Technology*, 7(10), 1573–1582.

Prasad, R., Seshagiri, S. V., Das, D., Nandi, S. and Mishra, R. K. (2016). Evaluation of Tasar elite seed Rearing performance under Replenishment program. *International Journal for Research in Agricultural and Food Science*, 2(2), 63–73.

Ralhan, P. K., Saxena, A. K. and Singh, J. S. (1982). Analysis of forest vegetation at and around Nainital in Kumaun Himalaya. *Proceedings of Indian National Science Academy*, 48, 121–137.

Ramankutty, N., Delire, C. and Snyder, P. (2006). Feedbacks between agriculture and climate: An illustration of the potential unintended consequences of human land use activities. *Global and Planetary Change*, 54(1–2), 79–93.

Rao, S. and Nigam, M. P. (1986). Studies on bacterial infection of *Antheraea proylei* Jolly. *Proc. of the Seminars on Prospects and Problems of Sericulture in India Sponsored by U.G.C.*, 23, 207–209.

Rath, S. S., Prasad, B. C. and Sinha, B. R. (2003). Food utilization efficiency in fifth instar larvae of *Antheraea mylitta* (Lepidoptera: Saturniidae) infected with *Nosema* sp. and its effect on reproductive potential and silk production. *Journal of Invertebrate Pathology*, 83, 1–9.

Reddy, K. S. S. (1993). Women the strength of sericulture. *Indian Silk*, December. www.silkwormmori.blogspot.com/2011/05/Indian-sericulture-industry-over-view.html.

Reddy, R. M., Suresh, R., Srivastava, A. K., Kar, P. K., Sinha, M. K. and Prasad, B. C. (2010). Heterosis pattern and commercial prospective of assorted F1 hybrids of Indian tropical tasar silkworm, Antheraea mylitta Drury. *Journal of Entomology*, 7(3), 160–167.

Sahay, D. N., Roy, D. K. and Sahay, A. (2000). Diesease of tropical tasar silkworm *Antheraea mylitta* D., Symptoms and control measures in: Lessons on tropical tasar. In *Central Tasar Research and Training Institute*, edited by K. Thagavelu. Piska Nagari, Ranchi, p. 104.

Sathe, T. V. and Jadhav, A. D. (2001). *Sericulture and Pest Management*. Daya Publ. House, New Delhi, pp. 1–197.

Sen, S. K., Jolly, M. S. and Jammy, T. R. (1971). Biology and life cycle of *Canthecona furcellata* wolff (Hemipetra: pentatomidae) predator of tasar silkworm *Antheraea myllita* Drury. *Indian Journal of Sericulture*, 10, 53–56.

Shamitha, G. (2007). Total indoor rearing of the tasar silkworm. *Every' Man's Science*, XIII(4), 198–202.

Shannon, C. E. and Weiner, W. (1963). *The Mathematical Theory of Communication Urban*. University Illinois Press, Champaign, Illinois, 125p.

Sharma, E., Chettri, N., Tse-ring, K., Shrestha, A. B., Jing, F., Mool, P. and Eriksson, M. (2009). *Climate Change Impacts and Vulnerability in the Eastern Himalayas*. International Centre for Integrated Mountain Development, Kathmandu.

Shetty, K. K. and Samson, M. V. (1998). Non—Mulberry sericulture in India. *Indian Silk*, 37(6 & 7), 21–25.

Shivakumar, G. and Shamitha, G. (2013). Studies on Larval mortality: Disease, pest and predator menace in outdoor and indoor reared Tasar silkworm. *Antheraea mylitta* drury (Daba TV). *Research Journal of Animal, Veterinary and Fishery Sciences*, 1(4), 1–7.

Singh, G. and Rawat, G. S. (2012a). Depletion of oak (*Quercus* spp.) forests in the Western Himalaya: Grazing, fuelwood and fodder collection. In *Global Perspectives on Sustainable Forest Management*, edited by Clement A. Okia. InTech, UK. ISBN: 978-953-51-0569-5.

Singh, G. and Rawat, G. S. (2012b). Quantitative analysis of tree species diversity in different oak (*Quercus* spp.) dominated forests in Garhwal Himalaya, India. *Notulae Scientia Biologicae*, 4(4), 132–140.

Singh, G. P., Sinha, A. K., Kumar, P. K. and Prasad, B. C. (2011). Characterization and identification of bacterial infection Indian tropical tasar silkworm, *Antheraea mylitta* D. *Research Journal of Microbiology*, 6, 891–897.

Singh, J. S. (1981). Rural ecosystems and development in the Himalaya. In *Science and Rural Development in Mountains*, edited by J. S. Singh, S. P. Singh, and C. Shastri. Gyanodaya Prakashan, Nainital, India, pp. 74–87.

Singh, J. S. and Singh, S. P. (1987). Forest vegetation of the Himalaya. *The Botanical Review*, 53(1), 80–192.

Singh, J. S. and Singh, S. P. (1992). *Forests of Himalaya: Structure, Functioning and Impact of Man*. Gyanodaya Prakashan, Nainital.

Singh, K. C. and Singh, N. I. (1998). Biology and ecology of temperate tasar silkmoths, *Antheraea proylei* Jolly and *Antheraea pernyi* Guerin-Meneville (Saturniidae): a review. *Indian Journal of Sericulture*, 37(2), 89–100.

Singh, N. I., Singh, N. I., Kesia, T. J., Singh, Y. R. and Singh, K. C. (2000) Conservation and utilization of Indian oak fed *Antheraea* Fauna. *International Journal of Wild Silkmoths and Silk*, 5, 330–331.

Singh, P., Ramasastri, K. S. and Kumar, N. (1995). Topographical influence on precipitation distribution in different ranges of western Himalayas. *Hydrology Research*, 26(4–5), 259–284.

Singh, R. N. and Maheswari, M. (2003). Conservation and utilization of sericigenious insects in North East region of India. *Sericologia*, 43, 1–15.

Singh, R. N., Sinha, M. K., Bajpeyi, C. M., Sinha, A. K. and Tikader, A. (2014). *Tasar Culture*. A.P.H. Publishing Corporation, New Delhi.

Singh, R. N. and Thangavelu, K. (1991). Parasites and predators of Tasar silkworm, *Antheraea mylitta* and their control measures. *Indian Silk*, 29(12), 33–36.

Singh, S. P. and Singh, J. S. (1986). Structure and function of the Central Himalayan oak forests. *Proceedings: Plant Sciences*, 96(3), 159–189.

Srivastava, P. K. and Thangavelu, K. (2005). *Sericulture and Seri-biodiversity*. Associated Publishing Company, New Delhi.

Suryanarayana, N. and Srivastava, A. K. (2005). Monograph on tropical tasar silkworm. In *Central Tasar Research and Training Institute*, Central Silk Board, Government of India Ranchi, Jharkhand, India, pp. 1–87.

Thangavelu, K. (1991). Recent studies in tasar and other wild silkmoths. *Wild Silkmoth*, 20–29.

Thangavelu, K., Bansal, A. K., Shukla, R. M., Saxena, N. N., Jayaswal, J. and Goel, A. K. (1995). Evolution of some chemicals for prevention of some diseases in tasar silkworm, *Antheraea mylitta* Drury (Lepidoptera: Saturniidae). *Indian Journal Sericulture*, 34(2), 142–144.

Thangavelu, K., Bhagawati, A. K. and Chakraborty, A. K. (1987). Studies on some wild sericigenous insects of North-Eastern India. *Sericologia*, 27, 91–98.

Thangavelu, K. and Borah, A. (1986). Occurrence of *Antheraea mylitta* Drury (Lepidoptera: Saturniidae) in North eastern India-distributional significance. *Current Science*, 55, 940.

Troup, R. S. (1921). *Silviculture of Indian Trees*. Vol. I–III. Clarendon Press, Oxford.
Upreti, N., Tewari, J. C. and Singh, S. P. (1985). The oak forests of the Kumaun Himalaya (India): Composition, Diversity, and Regeneration. *Mountain Research and Development*, 5(2), 163–174.
Varatharajan, R., Sudhakar, S., Reeta, L., Ibohal Singh, N. and Mathavan, S. (1998). Occurrence of reovirus in oak tasar silkworm *Antheraea Proylei*. *Current Science*, 75(7), 724–726.
Vedwan, N. and Rhoades, R. E. (2001). Climate change in the Western Himalayas of India: A study of local perception and response. *Climate Research*, 19(2), 109–117.
Zamal, T., Sarmah, B., Hemchandra, O. and Kalita, J. (2010). Global warming and its impact on the productivity of muga silkworm (*Antheraea assamensis* Helfer). *The Bioscan*, 1, 199–209.

14

The Cnidarian Journey of Myxozoa

Harpreet Kaur, Aditya Gupta, and Prabhjot Kaur

CONTENTS

14.1 Introduction

Myxozoans are diverse, microscopic obligate endoparasites representing almost one-fifth of presently known cnidarian species (Zhang et al., 2010). These are parasites of lower vertebrates, such as fish, and invertebrates, such as oligochaetes. Recent studies have shown that myxozoans evolved as an endoparasitic radiation of cnidarians, exploiting invertebrate and vertebrate hosts (primarily fish) in freshwater and marine environments. They are present in the form of pseudocysts/plasmodia in the infected organ of the fish, as shown in Figure 14.1. Currently, there are approximately 2,600 nominal species, classified to 67 genera and 17 families (Fiala, 2010). The typical vertebrate host is the teleost fish, but several myxosporean species have been found in the cartilaginous fish. Myxozoans were discovered for the first time by Jurine (1825) and were detected as cysts in the muscles of a whitefish, *Coregonus fera*. In the light of recent discoveries, Myxozoa is now part of Phylum Cnidaria, having contribution of about 19% of the species, with others like Anthozoa contributing 52%, Hydrozoa 27%, and jellyfishes 2% (Atkinson et al., 2015). Before 2015, they were considered as a separate phylum under Myxozoa, but on the basis of their molecular phylogeny, protein-encoding cnidarians' specific genes, and the similarity of their polar capsules to the nematocyst of Phylum Cnidaria support their place within Cnidaria. Hence, myxozoans are now placed under Phylum Cnidaria and Class Myxozoa (unranked).

Myxozoan spores have proteinaceous shell valves, nematocyst-like polar capsules with coiled extrusible polar filaments and an amoeboid infective sporoplasm. They have complex plasmodium, with some of nuclei differentiating with the surrounding cytoplasm into generative cells, which later undergo sporogenesis. Myxozoan spores differ markedly from microsporean spores as later are parasitic principally in insects, fish and occasionally in mammals, including humans. Spores are uniformly small sized, 10–25 μm long, but valvular extensions attain lengths of 30–60 μm and have two to three shell valves while actinospores may have size up to 300 μm. The polar filaments are for anchoring function. Microsporean spores have simpler plasmodium with all nuclei equipotential and all become nuclei of sporoplasm. Sporoplasm produces protective shell and organelles for extrusion. The microsporean spores are ovoid and ellipsoid and have a smooth, chitinous outer wall, and unlike myxozoan spores, there are no valves or sutures. The polar tube is used to inject the sporoplasm into the potential host. The microsporean genera most often seen in fish are *Glugea*, *Pleisophora*, and *Nosema*.

DOI: 10.1201/9781003220398-14

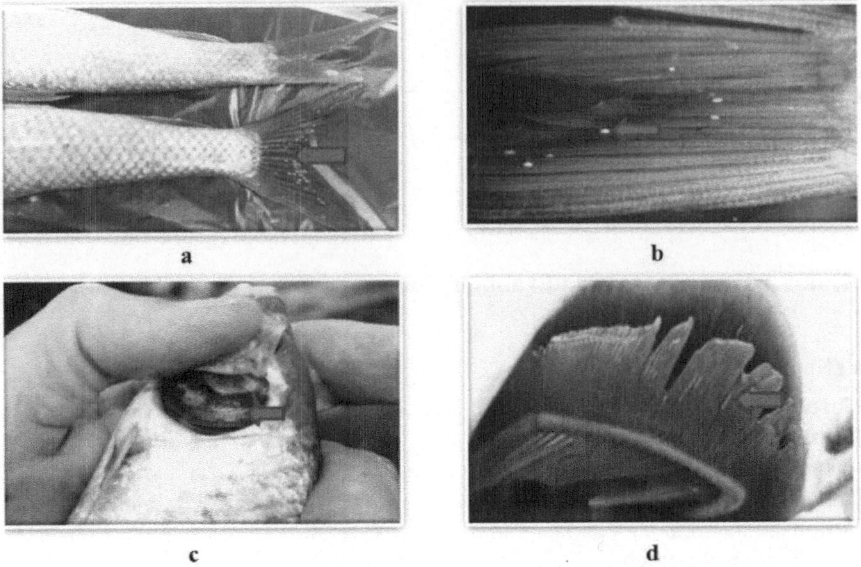

FIGURE 14.1 (a) Infected fish scales. (b) Plasmodia on the fins of *Puntius sophore*. (c) Plasmodia on the gills of *Cirrhinus reba*. (d) Plasmodia on the gills of *Labeo rohita*. Picture designed during the present study.

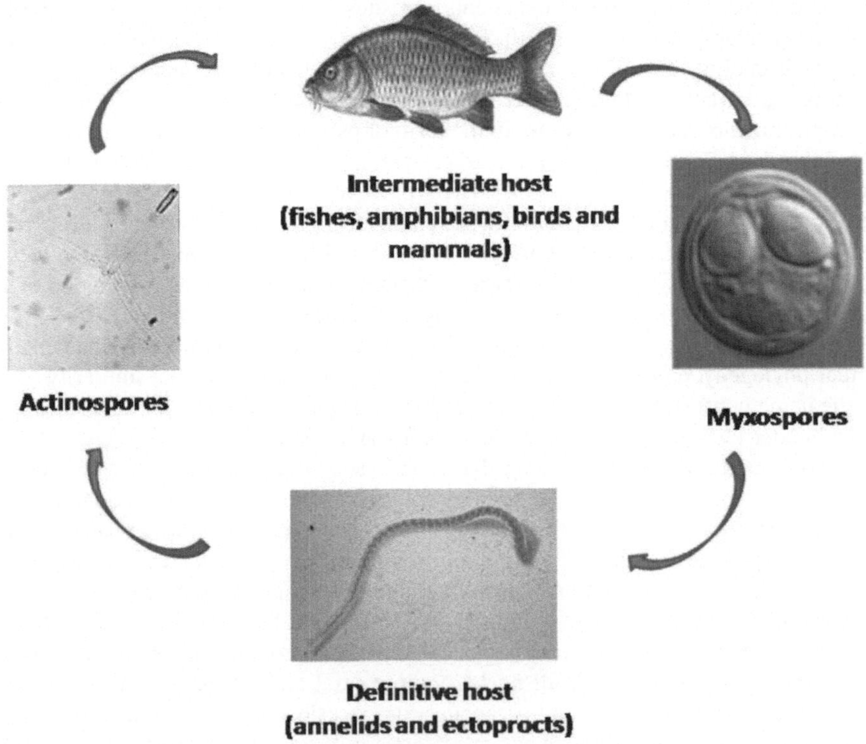

FIGURE 14.2 Life cycle of Myxozoa.

(*Source*: Based on Lom and Dykovà, 1992).

Recent developments in myxozoan biology has resurrected interest in this group, as it has now been well documented that parasites that were once thought to comprise two separate classes, Myxosporea and Actinosporea in fact represented the alternating developmental stage (i.e. Actinosporea) in the life history of a single myxozoan species (Levine et al., 1980). Wolf and Markiw (1984) established that the actinosporean genus *Triactinomyxon* as an alternating stage in the life history of species of *Myxobolus*. The repeatability of these findings led to a complete reconsideration of the systematics of the group, heralded by Kent et al. (1994) as the "demise of a class of protists," Class Actinosporea.

The myxosporean life cycle includes the alternation of two hosts. The myxospore phase takes place in the vertebrate, whereas the stages of the actinospore phase undergo their development in the invertebrate, as seen in Figure 14.2. The typical vertebrate host includes the teleost fish however, but several myxosporean species have been found in the cartilaginous fish (Lom and Dykovà, 1992). Other hosts from the aquatic environment include amphibians, reptiles, and ducks (Bartholomew et al., 2008). The recent findings of a myxosporean species in shrews demonstrated that myxozoans can also infect terrestrial homeothermic vertebrates (Prunescu et al., 2007). The invertebrate hosts of the actinosporean phase of the freshwater myxosporeans include annelids (mainly oligochaetes and rarely polychaetes). As for the marine myxosporeans, polychaetes and rarely sipunculids act as hosts of the actinosporean phase. Vertebrates are considered as the intermediate hosts since the sexual process (gametogony) has been observed during the actinospore phase in the invertebrates (definitive host).

14.2 Phylogenetics and Cladistics

Myxozoan classification had been based exclusively on spore morphology, host taxon, and organ and tissue specificity, but this system was criticized as these morphological features showed high structural variability among sister taxa and cryptic species, resulting in problems with species nomenclature.

Myxozoa has been reviewed and compiled in the form of synopses and checklists by many workers. Some major synopses on myxozoan genera include Eiras (2002) on *Henneguya* (146 species), Eiras et al. (2005) on *Myxobolus* (744 species), and *Eiras* (2006) on *Ceratomyxa* (147 species). Furthermore, Eiras (2005) contributed and revised 19 myxosporean species infecting amphibians and reptiles belonging to genera *Myxobolus*, *Myxidium*, *Hoferellus*, *Chloromyxum*, *Caudomyxum*, and *Sphaerospora* and gave detailed account of characteristics, pathology, host, geographical range, host specificity, and life cycle. Basu et al. (2006) provided a synopsis of 32 Indian species belonging to the genus *Thelohanellus*, including one new species, *T. disporomorphus*, infecting Indian major carp, *Cirrhinus mrigala*. Lom and Dykovà (2006) reviewed myxozoan genera and assigned 790 valid species to the genus *Myxobolus*, including seven in amphibians. More recent synopses include Eiras et al. (2011) on *Myxidium* (232 species), Kaur and Singh (2012) on *Myxobolus* (131 species), Eiras et al. (2012) on *Chloromyxum* (140 species) and *Henneguya* (43 species), Zhang et al. (2013) on *Thelohanellus* (108 species), Eiras et al. (2014b) on *Kudoa* (95 species), Rangel et al. (2014) on *Myxobilatus* (39 species), and Kaur et al. (2016) on *Thelohanellus* (52 species). Pazooki and Masoumian (2012) provided a synopsis of 247 species of parasites from Iranian freshwater fish.

Many workers have contributed in the taxonomy and molecular systematic of myxozoan parasites (e.g., (Smothers et al., 1994; Anderson et al., 1998; Okamura et al., 2002; Eszterbauer, 2006; Fiala, 2010; Kaur and Gupta, 2017a–c; Gupta and Kaur, 2017a–b; 2018a–f, 2020; Gupta et al., 2017; Kaur et al., 2020; Nissa and Kaur, 2021). There are various approaches for constructing phylogenetic trees, but the one that provides an explicit and testable hypothesis of interspecific and intraspecific relationships among various individuals has been chosen for solving the current research problem. Phylogeny was initially predominantly based on morphological data (measuring and quantifying the phenotypic properties) of different type species. Molecular systematics holds a cladistic perspective pivoting on two principles. The prime assumption is that the taxa chosen must be a monophyletic, and second, the classification must correspond to phylogenetic descent.

Recently, many research findings have been known for identifying whether they make a place in Phylum Cnidaria or not as earlier it was not so clear. For this, researchers done both molecular and proteomics studies. Feng et al. (2014) and Chang et al. (2015) concluded that myxozoa is a sister to Medusozoa within Cnidaria by studying its phylogenomic analyses, as shown in Figure 14.3. They also studied the cnidarian-specific minicollagen genes and detected 11 of the 19 proto-mesodermal genes in the *Thelohanellus kituae* genome. During evolution, they lost most of the cnidarian's characteristics, except for the polar capsules; thus, this structure is the only prominent structure that unites all members of Phylum Cnidaria. They illustrated that the evolutionary adaptation of cnidarian nematocysts into polar capsule could be a part of an organismal shift from free living to parasitic mode of life. Their data showed that the basic function such as protein content and osmotic machinery needed for everting the polar tubule retained, but they lost toxins and tubule injection capabilities. Holzer et al. (2018) examined the history of adaptive radiations in myxozoans and their hosts by determining the degree of congruence between their phylogenies and by timing the emergence of myxozoan lineages in relation to their hosts. They provided evidence that, through invertebrate invasion, fish hosts were acquired multiple times, which led to the co-speciation of major phylogenetic lineages. They also found that acquisition of vertebrate hosts facilitates alternative transmission and also dispersion strategies of myxozoans, which act as one of the reasons for their distinct success. They concluded that myxozoans are the oldest metazoan parasites on Earth, and these parasites shows high flexibility regarding exploring host of different animal groups. They demonstrated that myxozoans emerged long before fish populated Earth and also, by their phylogenetic congruence with the invertebrate hosts, indicated that bryozoans and annelids were their original hosts.

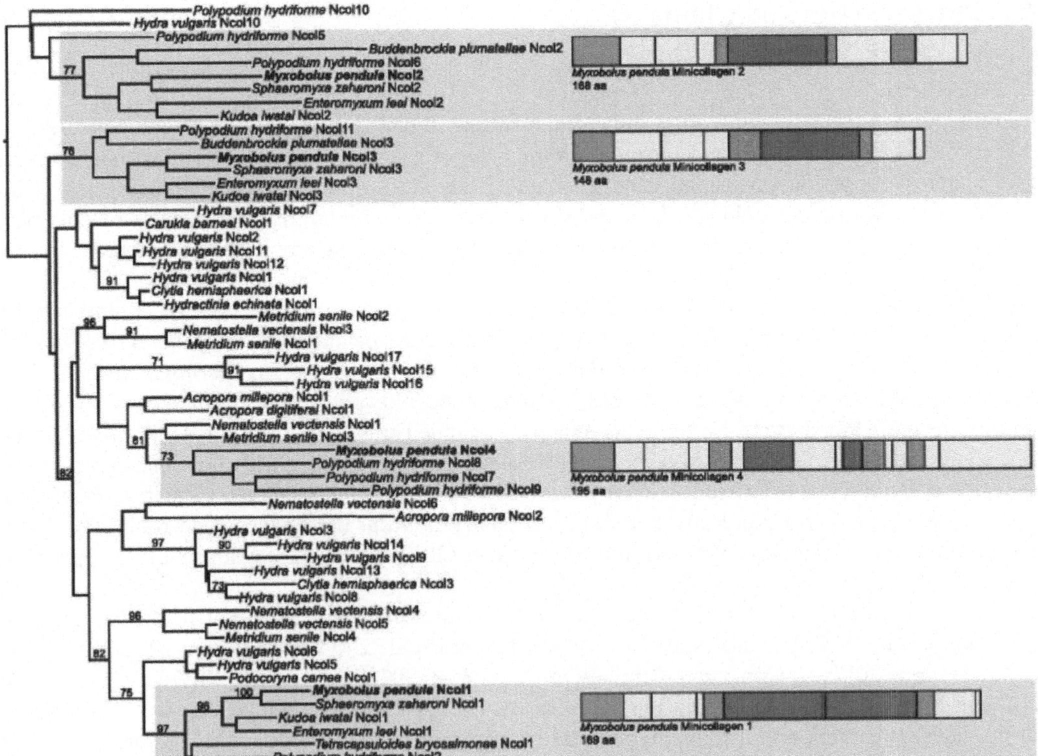

FIGURE 14.3 Phylogenetic tree generated from a matrix of 51,940 amino-acid positions and 77 taxa using Bayesian inference under the CAT model.

(*Source*: Adapted from Chang et al., 2015).

14.3 SSU Gene Markers in the Study of Myxozoa

Phylogenetic studies, in particular, have propagated a large number of rRNA gene sequences on public genetic databases, as the organismal universality and typically high gene copy number per cell facilitates gene amplification and sequencing. Forro and Eszterbauer (2016) observed the correlation between host specificity and genetic diversity of the parasite *Myxobolus pseudodispar.* They suggested that these parasites undergo host-shifting that is a gradual shift from primary host to secondary host. Its primary host being roach and secondary host is common bream that was susceptible to a very lower extent. Moreover, phylogenetic and experimental findings suggested that *M. pseudodispar* is a cryptic species complex. Their experimental findings conformed by molecular analysis suggested that *M. pseudodispar* from roach have the capacity to infect other cyprinids like *Abramis brama* although the prevalence and intensity was not that high.

Figure 14.4 concluded that both polar capsules and nematocysts were homologous structures based on the minicollagens, NOWA and other cnidarian-specific proteins (Shpirer et al., 2014). This was further demonstrated in the malacosporean *Tetracapsuloides bryosalmonae* by Holland et al. (2010). During the study based on medusozoan nematocysts, Reft and Daly (2012) found some differences in the morphology of polar capsules and medusozoan nematocysts but were found to be similar to each other on the basis of SSU genes, minicollagens, and cnidarians-specific proteins.

Okamura and Gruhl (2015) emphasized to resolve the discrepancies between the myxosporean phylogeny and taxonomy. New light has been shed by the mapping of morphological and bionomical character evolution on the SSU-based tree. The performed analyses have shown that the only synapomorphic character corresponding to both myxozoan taxonomy and phylogeny was the character of polar filament. This structure was flat and zigzag folded only in the *Sphaeromyxa* species contrary to other myxosporeans with spirally wound polar filament. Other morphological characteristics mainly relating to the spore characteristics were homoplastic.

According to Okamura and Gruhl (2015), recent phylogenetic trees showed that all of the myxosporeans group according to their site specificity rather than to the type of host. Despite many evolutionary trends in myxosporeans have been revealed by the analyses of the SSU rDNA data, certain relationships among myxosporean groups have still not been fully resolved. Under question were the interrelationships of the main clades within the marine lineage, the position of the Sphaerosporid clade to other myxosporeans. The insufficient resolution of the single-gene analyses has been ascribed to the limited number of alignable nucleotides. Moreover, the discrepancies between the myxozoan phylogeny and taxonomy have introduced a question if the phylogeny based on the single SSU rRNA gene corresponds to the real organismal evolution.

14.4 Minicollagens in Myxozoa and Cnidaria

Holland et al. (2010) reported the first inference on the phylogenetic placement of Myxozoa on the basis of phylum-specific genes within Metazoa. The much-debated phylogeny of Myxozoa within Metazoa was addressed by the sequencing and characterization of homologues of cnidarian-specific genes called minicollagens from the myxozoan *Tetracapsuloides bryosalmonae.* Minicollagens are taxonomically restricted cnidarian specific genes encoding for important proteins of the nematocysts wall and its functionality comprising a central collagen triple helix domain characterized by 12–16 Gly-X-Y repeats, which provide flexibility to the wall followed by variable length proline sequences with terminating cysteine-rich domains (CRD). Homologue of group-1 minicollagens named as Tb-Ncol-1 was sequenced from the full-length library of cDNA from the myxozoan *T. bryosalmonae* spore sacs and was verified of its gene origin by PCR analysis.

Maximum-likelihood (ML) employing JTT + I + G model and neighbor-joining methods were used to perform phylogenetic analysis. The BLAST results demonstrated that Tb-Ncol-1 is homologous to the cnidarian minicollagen genes consisting of specific central collagen-like domain with repeated GLY-X-Y sequences with adjacent proline sequences on both N-terminal and C-terminal, flanked by six cysteine residues in the unique CRDs, characteristic of the Tb-Ncol-1. Further, several features of the Tb-Ncol-1

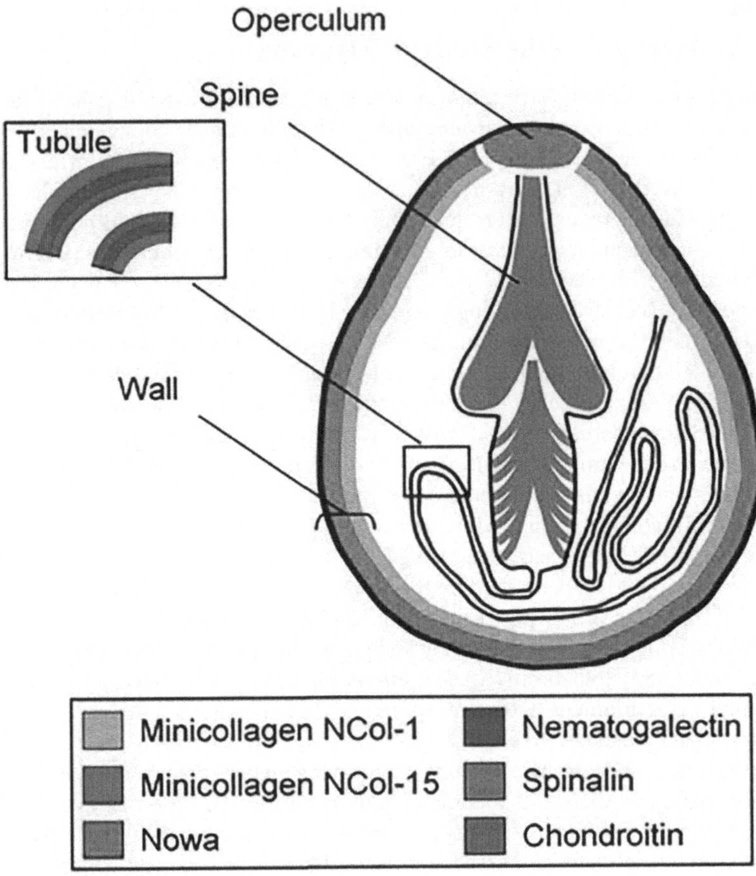

FIGURE 14.4 Similarities between myxozoan polar capsules and nematocysts of cnidarians.

(*Source*: Adapted from Shpirer et al. (2014).

gene are related to the group-1 minicollagens of medusozoan and anthozoans including the conservative 14 + 13 disposition of GLY-X-Y repeats and similar cysteine sequence pattern in the CRDs, which was delineated using multiple sequence alignments. In addition to this, the genomic organization of the *T. bryosalmonae* minicollagen resembled the genomic pattern of available cnidarian group-1 minicollagens. Homology modeling of the CRD domain foresees a similarity with the *Hydra* CRD with similar cysteine and side chain positions. All these facts provide robust evidence for the placement of Myxozoa within Cnidaria and sister to Medusozoa. It was also suggested that a novel protein sequence variation elucidates the differences between the rather less complex polar capsules of myxozoans and complex cnidarian nematocysts, as shown in Figure 14.5.

Foox et al. (2015) independently discovered the presence of cnidarian specific minicollagen genes from different myxozoan species both supporting its inclusion within Cnidaria. Shpirer et al. (2014) investigated genome and transcriptomic assemblies of four cnidarian-specific genes from distinct myxozoan species, as well as sister taxon, Polypodium, out of which three minicollagens were recovered, while Foox et al. (2015) discovered the fourth copy of minicollagen genes from the transcriptome of myxozoan *Myxobolus pendula*. Out of the four well-defined families of cnidarian specific genes, which are minicollagens, nematogalectins, spinalin, and NOWA, only two of the gene families that is minicollagen and nematogalectin were identified from three myxozoan species—*Kudoa iwatai*, *Enteromyxum leei*, and *Sphaeromyxa zaharoni*—while from the fourth species, *Buddenbrockia plumatellae*, only minicollagens were identified. Three different minicollagen sequences were recovered from the species *Kudoa iwatai*, two from *Buddenbrockia*

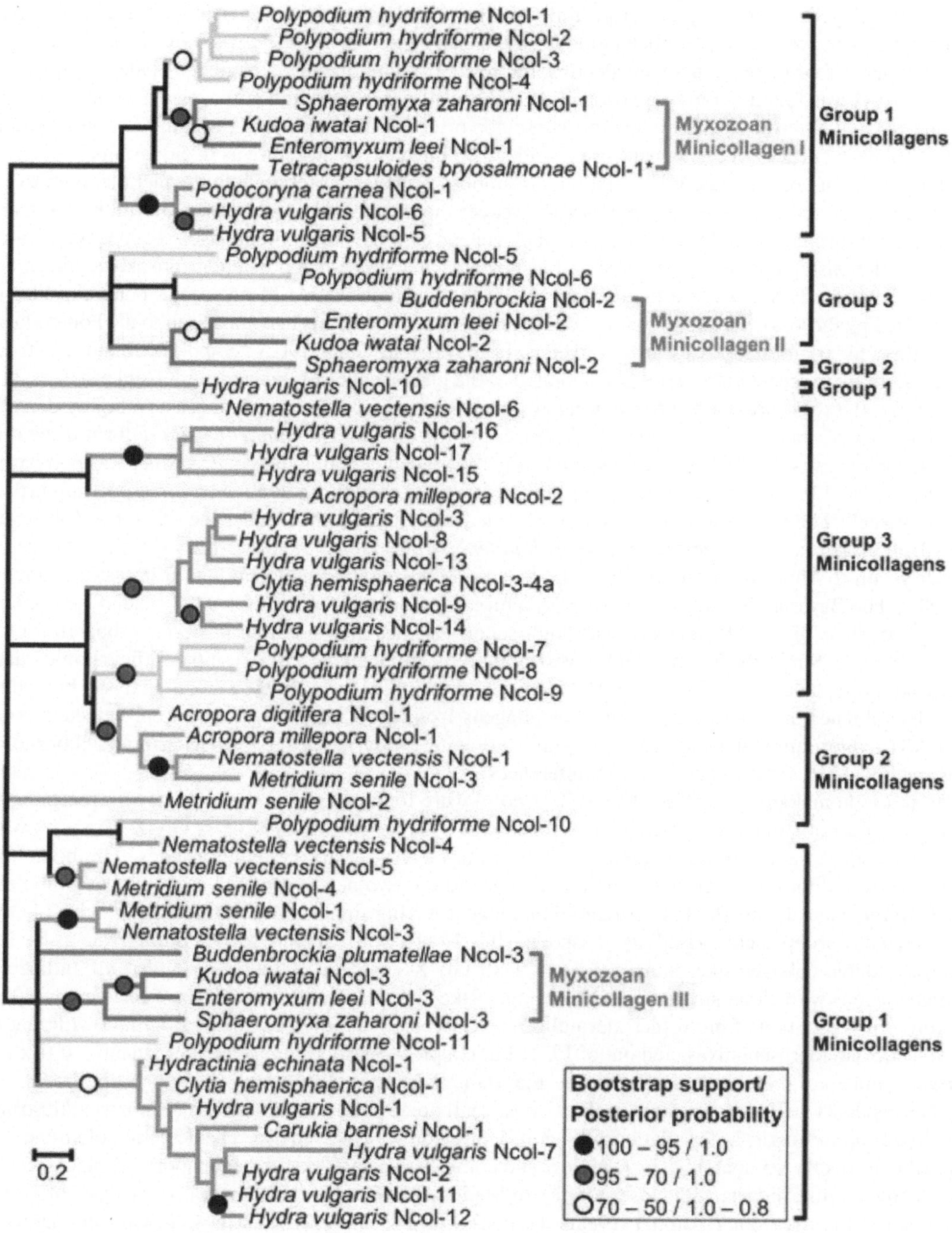

FIGURE 14.5 ML phylogenetic tree reconstructed based on the minicollagen dataset. Circles indicate node supports. Black circles: 100 ≥ BP ≥ P ≥ 95, PP = 1.0. Anthozoa, Medusozoa, Polypodioza, and Myxozoa are depicted by blue, green, orange, and red branches, respectively.

(*Source*: Shpirer et al. (2014).

and several from *Polypodium*. These were distinguished into clades I, II, and III on the basis of CRD organization. The minicollagens of clades I and II clustered with those of *Polypodium*.

In the study of the fourth minicollagen, similarities between the adjoining polytripeptide and cysteine-rich domains were reported with that of Ncol7–9 of *Polypodium*, which was not found in any other minicollagen gene, suggesting a strong affinity for the homology of the genes. Among the nematogalectins, three genes were identified belonging to myxozoans and four from *Polypodium*, while cnidarians typically have four different nematogalectins. Interestingly, a unique nematogalectin clade called nematogalectin-C was found in only myxozoans, *Polypodium*, mesozoans, and anthozoans characterized by a long varying N-terminal end, indicating a common origin of the sequences. Also, the nematogalectin-related gene sequences of myxozoans possess identical intron/exon structure as in cnidarians.

The results of genomic and transcriptomic assemblies of genes corroborate to the fact that myxozoan polar capsules and cnidarian nematocysts are indeed homologous and, therefore, the myxozoans are grouped within cnidarians. Furthermore, the phylogenetic assessment of minicollagen sequences indicates a sister relationship among *Polypodium* and myxozoans and suggests a possible medusozoan placement within Cnidaria. It was also postulated that the dramatic simplification of myxozoan polar capsules is delineated by the loss of diversity of minicollagens and nematogalectins in comparison to the Polypodium and other cnidarians that possess a vast diversity of these genes and thus have complex nematocyst structures. Both these recent discoveries corroborated to the placement of Myxozoa within Cnidaria

Feng et al. (2014) provided corroborating evidence for the placement of Myxozoa as a highly derived cnidarian taxon based on the genome and transcriptomic data of *Thelohanellus kitaue* form three different perspectives. In the first analysis, the newly sequenced genomic data of *T. kitauei* was compared to each of the 128 protein-encoding genes of 57 different taxa with production of corresponding orthologs for each. The results were obtained by ML analysis using RaxML under the LG + F + C4 model in which myxozoans were nested in a clade (*Buddenbrockia*, *Thelohanellus*, *Myxobolus*) and formed a sister group to Medusozoa within Cnidaria, leaving Myxozoa monophyletic with a bootstrap support of 98%. The Bayesian inference analysis with a mixture model of CAT + POI + C4 yielded concordant results as those of the ML analysis with high support values and maximal posterior probability of 1 were obtained. Similarly, ML tree and Bayesian analysis obtained the same results for the trimmed data set with monophyletic Myxozoa sister to Medusozoa within Cnidaria. In the second analysis, two putative homologues of cnidarian-specific minicollagens from the *T. kitauei* genome were detected using tBLASTn algorithm and evaluated for its similarity with Cnidarian minicollagens using neighbor-joining methods, ML analysis and Bayesian inferences.

Both the homologues Tk_Ncol-1 and Tk_Ncol-2 were found to possess similarity with the group-1 minicollagens of Cnidaria with characteristic central collagen-like domain of 14 Gly-X-Y repeats with two adjacent variable proline sequences and terminal CRDs of six cysteine residues. The homologs were recovered into three groups: Group A comprised the two newly detected minicollagen homologs of *T. kitauei* placed with all medusozoan minicollagens having single collagen-like domain with 14 Gly-X-Y repeats. Group B comprised malacosporean Tb_Ncol-1 with all medusozoan minicollagens having similar double collagen-like domain with 14 + 13 Gly-X-Y repeats. Group C included all anthozoan minicollagens with either single or double collagen-like domain. The third analysis of Feng et al. (2014) involved investigation of proto-mesodermal genes recovered from the genome of *T. kitauei* with that of five metazoan representatives, and out of 19, 11 human proto-mesodermal genes were identified from *T. kitauei*, and all of these were similar to the cnidarian *Hydra magnipapillata*.

The results of new phylogenomic analysis consistently recovered a monophyletic Myxozoa with strong evidence supporting its sister relationship with Medusozoa within Cnidaria. The detection of cnidarian-specific minicollagen homologs in *T. kitauei* and comparative analysis further support the placement of Myxozoa within Cnidaria with closer affinity to Medusozoa. Finally, the comparative analysis of proto-mesodermal genes from *T. kitauei* reveals that their number is higher than those in Choanoflagellata, Ctenophora, Porifera, and Placozoa but lesser than that in Bilateria and the cnidarian *N. vectensi*, but the noteworthy result was the presence of all components of *T. kitauei* proto-mesodermal genes in *H. magnipapillata* corroborated toward an affinity to Cnidaria.

14.5 Conclusion

A meticulous study of the history of the work attempted to resolve the controversial phylogeny of Myxozoa has led to the inference that myxozoans are indeed metazoan in origin based on the morphological evidence, such as multicellularity of its spores, presence of cellular junctions, terminal cell differentiation, and phylogenetic data based on 18S rDNA. Within Metazoa, the parasites group within Cnidaria confirmed by a series of ultrastructural and phylogenomic studies outweighing the alternative hypothesis that Myxozoa is sister to Bilateria. The results of molecular data also corroborated the sister relationship between Polypodium and Myxozoa, both forming a sister clade to Medusozoa, which in turn refutes the alternative hypothesis supporting the bilaterian origin since *Polypodium* is a cnidarian parasite. The unambiguous similarity between the polar capsules and nematocysts of Cnidaria was held as indisputable evidence for this inclusion, which was further reinforced by the ultrastructural studies on the enigmatic malacosporean *Buddenbrockia* and molecular evidence presented by 18S rDNA data sets, broad sampling of protein-coding genes, and taxonomically restricted minicollagen genes. The current study supports the placement of Myxozoa within Phylum Cnidaria.

14.6 Scope for Further Research

The present study gives an insight on the cnidarian journey of Myxozoa. Various genetic tools, along with computational biology, enabled the scientific community to shift the highly evolving Myxozoa to Phylum Cnidaria. But it is noteworthy to work on new approaches and technologies that will continue to facilitate reassessment and re-description of existing taxa, discovery of new taxa, and identification of species complexes, if any. It is also predicted that DNA bar-coding and whole genome sequencing will continue to reveal that many myxozoans, especially those presently regarded as having restricted geographic and host ranges as well as anaerobic respiration as found in *Henneguya salminicola*.

14.7 Acknowledgments

The authors acknowledge financial support by University Grants Commission (UGC), Government of India, under UGC-CAS-II and DST-Purse grant (no. F4-28/2015/CAS-II), to the Department of Zoology, Panjab University, Chandigarh.

REFERENCES

Anderson, C.L., Canning, E.U. and Okamura, B. 1998. A triploblast origin for Myxozoa. *Parasitology* 124:215–223.

Atkinson, S.D., Bartošová-Sojkova, P., Whipps, C.M. and Bartholomew, J.L. 2015. Approaches for characterizing myxozoan species. In *Myxozoan Evolution, Ecology and Development*, eds. B. Okamura et al., 111–123. Springer International Publishing, Switzerland.

Bartholomew, J.L., Atkinson, S.D., Hallett, S.L., Lowenstine, L.J., Garner, M.M., Gardiner, C.H., Rideout, B.A., Keel, M.K. and Brown, J.D. 2008. Myxozoan parasitism in waterfowl. *International Journal of Parasitology* 38(10):1199–1207.

Basu, S., Modak, B.K. and Haldar, D.P. 2006. Synopsis of the Indian species of the genus *Thelohanellus* Kudo, 1933 along with description of Thelohanellus disporomorphus sp. n. *Journal of Parasitology and Applied Animal Biology* 15(1–2):81–94.

Chang, E.S., Moran, N., Rubinsteind, N.D., Diamant, A., Philippe, H., Huchon, D. and Cartwright, P. 2015. Genomic insights into the evolutionary origin of Myxozoa within Cnidaria. *Proceedings of National Academics of Science USA* 112(48):14912–14917.

Eiras, J.C. 2002. Synopsis of the species of the genus *Henneguya* Thélohan, 1892 (Myxozoa: Myxosporea: Myxobolidae). *Systematic Parasitology* 52:43–54.

Eiras, J.C. 2005. An overview on the myxosporean parasites in amphibians and reptiles. *Acta Parasitologica* 50:267–275.

Eiras, J.C. 2006. Synopsis of the species of *Ceratomyxa* Thélohan, 1892 (Myxozoa: Myxosporea: Ceratomyxidae). *Systematic Parasitology* 65:49–71.

Eiras, J.C. and Adriano, E.A. 2012. A checklist of new species of *Henneguya* Thélohan, 1892 (Myxozoa: Myxosporea, Myxobolidae) described between 2002 and 2012. *Systematic Parasitology* 83:95–104.

Eiras, J.C., Molnár, K. and Lu, Y.S. 2005. Synopsis of the species of *Myxobolus* Bütschli, 1882 (Myxozoa: Myxosporea: Myxobolidae). *Systematic Parasitology* 61:1–46.

Eiras, J.C., Saraiva, A., Cruz, C.F., Santos, M.J. and Fiala, I. 2011. Synopsis of the species of *Myxidium* Bütschli, 1882 (Myxozoa: Myxosporea: Bivalvulida). *Systematic Parasitology* 80:81–116.

Eiras, J.C., Saraiva, A. and Cruz, C. 2014b. Synopsis of the species of *Kudoa* Meglitsch, 1947 (Myxozoa: Myxosporea: Multivalvulida). *Systematic Parasitology* 87:153–180.

Eszterbauer, E. 2006. Genetic relationship among gill-infecting *Myxobolus* species (Myxosporea) of cyprinids: Molecular evidence of importance of tissue-specificity. *Diseases of Aquatic Organisms* 58:35–40.

Feng, J.M., Xiong, J., Zhang, J.Y., Yan, Y.L., Yao, B., Zhou, Z.G. and Miao, W. 2014. New phylogenomic and comparative analyses provide corroborating evidence that Myxozoa is Cnidaria. *Molecular Phylogenetics & Evolution* 81:10–18.

Fiala, I. 2010. Searching for the origin of Myxozoa and everything changes again. *Conference 38*. Jírovçovy Protozoologické dny, Jáchymov, Czech Republic.

Foox, J. and Siddall, M.E. 2015. The road to Cnidaria: History of phylogeny of the Myxozoa. *Journal of Parasitology* 101:269–274.

Forro, B. and Eszterbauer, E. 2016. Correlation between host specificity and genetic diversity for the muscle-dwelling fish parasite *Myxobolus pseudodispar*: Examples of myxozoan host-shift? *Folia Parasitologica* 63:019.

Gupta, A. and Kaur, H. 2017a. A new pathogen, *Myxobolus holzerae* (Myxosporea: Myxozoa) causing severe gill disease in an Indian major carp *Labeo rohita* in a cold-water wetland, Punjab (India). *Microbial Pathogenesis* 111:244–251.

Gupta, A. and Kaur, H. 2017b. *Myxobolus okamurae* sp. nov. (Myxosporea: Myxozoa) causing severe gill myxoboliosis in the cyprinid Labeo bata in a cold-water wetland, Punjab (India). *Microbial Pathogenesis* 115:86–92.

Gupta, A. and Kaur, H. 2018a. A report on two Myxobolids (Cnidaria: Myxozoa) infecting freshwater fishes in Ranjit Sagar Wetland of Punjab (India). *Invertis Journal of Science & Technology* 11(1):12–17.

Gupta, A. and Kaur, H. 2018b. Prevalence of Myxozoan parasites in freshwater fishes of Ranjit Sagar Wetland, Punjab (India). *Invertis Journal of Science & Technology* 11(1):18–23.

Gupta, A. and Kaur, H. 2018c. 18S and 28S rDNA identity and phylogeny of two novel myxosporeans infecting gills of cyprinid carps inhabiting a cold-water wetland in northern India. *Microbial Pathogenesis* 120:97–108.

Gupta, A. and Kaur, H. 2018d. *Thelohanellus gabori* sp. nov. (Myxosporea: Myxozoa) infecting gill filaments of *Crossocheilus latius* (Hamilton, 1822) inhabiting a cold-water wetland in Punjab (India). *Parasitology Research* 117(9):2715–2723.

Gupta, A. and Kaur, H. 2020. Two new species of Myxobolus (Cnidaria: Myxosporea) infecting freshwater fishes of Ranjit Sagar Wetland, Punjab, India. *Microbial Pathogenesis* 147:104421.

Gupta, A., Kaur, H. and Katoch, A. 2017. Record of new host, site of infection and locality of two Myxobolids infecting freshwater fishes in Ranjit Sagar Wetland of Punjab (India). *International Journal of Biology Research* 2(4):06–11.

Holland, J.W., Okamura, B., Hartikainen, H. and Secombes, C.J. 2010. A novel minicollagen gene links cnidarians and myxozoans. *Proceedings of Royal Society London Biological Sciences* 278:546–553.

Holzer, A.S., Hartigan, A., Patra S., Pecková, H. and Eszterbauer, E. 2018. Molecular fingerprinting of the myxozoan community in common carp suffering swim bladder inflammation (SBI) identifies multiple etiological agents. *Parasites & Vectors* 7:398.

Jurine, L. 1825. Histoire abrégée des poissons du lac Léman. *Mémoires de la Société de Physique et d'Histoire Naturelle de Genève* 3:133–235.

Kaur, H. and Gupta, A. 2017a. Molecular and phylogenetic characterization of *Thelohanellus boggoti* Qadri, 1962 (Cnidaria, Myxosporea, Bivalvulida) infecting the fin of Indian minor carp *Labeo dero* (Hamilton, 1822). *Molecular Biology and Research Communications* 6(1):13–21.

Kaur, H. and Gupta, A. 2017b. Morphological, histopathological and molecular characterization of *Thelohanellus muscularis* n. sp. (Cnidaria: Myxosporea) infecting head muscles of *Labeo rohita* from Ranjit sagar wetland, Punjab. (India). *Journal of Applied Biology & Biotechnology* 5(1):021–028.

Kaur, H. and Gupta, A. 2017c. Morphological, histopathological and molecular characterization of *Thelohanellus pathankotensis* n. sp. (Cnidaria: Myxosporea: Myxozoa) infecting an Indian minor carp, *Labeo dero* Hamilton, 1822 from a cold-water wetland in Punjab (India). *Zootaxa* 4353(1):161–173.

Kaur, H., Gupta, A. and Attri, R. 2020. Highlights on the molecular identification of closely related members of genus *Thelohanellus* (Cnidaria: Myxosporea) forming species complex from the Indian subcontinent. *Microbial Pathogenesis* 150:104690. http://doi.org/10.1016/j.micpath.2020.104690.

Kaur, H. and Singh, R. 2012. A synopsis of the species of *Myxobolus* Bütschli, 1882 (Myxozoa: Bivalvulida) parasitizing Indian fishes and a revised dichotomous key to myxosporean genera. *Systematic Parasitology* 81:17–37.

Kaur, H., Singh, R., Katoch, A., Attri, R., Dar, S.A. and Gupta, A. 2016. Species diversity of the genus *Thelohanellus* Kudo, 1933 (Myxozoa: Bivalvulida) parasitizing fishes in Indian subcontinent. *Journal of Parasitic Diseases* 41(2):305–312.

Kent, M.L., Margolis, L. and Corliss, J.O. 1994. The demise of a class of protists: Taxonomic and nomenclatural revisions proposed for the protist phylum Myxozoa Grasse, 1970. *Canadian Journal of Zoology* 72:932–993.

Levine, N.D., Corliss, J.O., Cox, F.E.G., Deroux, G., Grain, J., Honigberg, B.M., Leedale, G.F., Loeblich, A.R., Lom, J., Lynn, D., Merinfeld, E.G., Page, F.C., Poljansky, G., Sprague, V., Vavra, J. and Wallace, F.G. 1980. A newly revised classification of the Protozoa. *Journal of Protozoology* 27:37–58.

Lom, J. and Dykovà, I. 1992. Protozoan parasites of fishes. In *Developments in Aquaculture and Fisheries*, eds. J. Lom and I. Dykovà, 159–235. Elsevier Publishers, Amsterdam.

Lom, J. and Dyková, I. 2006. Myxozoan genera: definition and notes on taxonomy, life-cycle terminology and pathogenic species. *Folia Parasitologica* 53:1–36.

Nissa, K. and Kaur, H. 2021. First record of the genus *Hennegoides* Lom, Tonguthai and Dyková, 1991 from Punjab (India) infecting the catfish, *Sperata seenghala* (Sykes, 1839). *International Journal of Parasitology: Parasites & Wildlife* 14:7–12.

Okamura, B., Curry, A., Wood, T.S. and Canning, E.U. 2002. Ultrastructure identifies *Buddenbrockia* as a myxozoan and reveals/identifies the bilaterian origin of the Myxozoa. *Parasitology* 124:215–223.

Okamura, B. and Gruhl, A. 2015. Myxozoan affinities and route to endoparasitism. In *Myxozoan Evolution, Ecology and Development*, eds. B. Okamura, A. Gruhl and J.L. Bartholomew, 23–44. Springer, Berlin.

Pazooki, J. and Masoumian, M. 2012. Synopsis of the parasites in Iranian freshwater fishes. *Iranian Journal of Fisheries Science* 11(3):570–589.

Prunescu, C.C., Prunescu, P., Pucek, Z. and Lom, J. 2007. The first finding of myxosporean development from plasmodia to spores in terrestrial mammals: *Soricimyxum fegati* gen. et sp. n. (Myxozoa) from *Sorex araneus* (Soricomorpha). *Folia Parasitologica* 54:159–164.

Rangel, L.F., Rocha, S., Borkhanuddin, M.H., Cech, G., Castro, R., Casal, G., Azevedo, C., Severino, R., Székely, C. and Santos, M.J. 2014. *Ortholinea auratae* n. sp. (Myxozoa: Ortholineidae) infecting the urinary bladder of the gilthead seabream *Sparus aurata* (Teleostei: Sparidae), in a Portuguese fish farm. *Parasitology Research* 113(9):3427–3437.

Reft, A.J. and Daly, M. 2012. Morphology, distribution, and evolution of apical structure of nematocysts in *Hexacorallia*. *Journal of Morphology* 273(2):121–136.

Shpirer, E., Chang, E.S., Diamant, A., Rubinstein, N., Cartwright, P. and Huchon, D. 2014. Diversity and evolution of myxozoan minicollagens and nematogalectins. *BMC Evolutionary Biology* 14(1):205.

Smothers, J.F., Von Dohlen, C.D., Smith, L.H. and Spall, R.D. 1994. Molecular evidence that the myxozoan protist are metazoan. *Science* 265:1719–1721.

Wolf, K. and Markiw, M.E. 1984. Biology contravenes taxonomy in the Myxozoa: New discoveries show alternation of invertebrates and vertebrates hosts. *Science* 225:1449–1452.

Zhang, J.Y., Gu, Z.M., Kalavati, C., Eiras, J.C., Liu, Y., Guo, Q.Y. and Molnár, K. 2013. A synopsis of the species of *Thelohanellus* Kudo, 1933 (Myxozoa: Myxosporea: Bivalvulida). *Systematic Parasitology* 86:235–256.

Zhang, J.Y., Yokoyama, H., Wang, J.G., Li, A.H., Gong, X.N., Ryu-Hasegawa, A., Iwashita, M. and Ogawa, K. 2010. Utilization of tissue habitats by *Myxobolus wulii* Landsberg & Lom, 1991 in different carp hosts and disease resistance in allogynogenetic gibel carp: redescription of *M. wulii* from China and Japan. *Journal of Fish Diseases* 33:57–68.

15

Evolutionary Journey of Experimental Animal Models: Contribution for Understanding Life Phenomenon

Gurfateh Singh, Ramica Sharma, and Parvinder Singh

CONTENTS

15.1 Introduction

Animal models are regarded as non-human models that are used in the field of research for the prevention, diagnosis, and treatment of various disease/disorders (Ferreira et al., 2019). Medical research remains dependent on models and systems for establishing efficacy prior to clinical trials (Fedele et al., 2016). The use of animal in research was established a long back that about in fourth century BC. These models are employed for experimental purpose in order to prove the hypothesis generated during experimentation (Khorramizadeh and Sadat, 2020). New approaches are inculcated now in order to validate cellular mechanisms and animal models of disease that harmonize their behavior with human disease (Breyer et al., 2015). Animal models impart better understanding of mechanism behind the progression of disease and management of the same. While selecting animal models various factors are to be kept into consideration that includes natural history of the disease and susceptible population, exposure, frequency of exposure, and virulent pathogen strain (NRC, 2006). Many species are used in biomedical research, such as insects (*Drosophila*), nematodes (*Caenorhabditis elegans*), fish (*Danio rerio*, or zebrafish), frogs (*Xenopus*), and many mammals, such as mice, rats, dogs, cats, pigs, and monkeys, due to their phylogenetic proximity to human beings (ICMR, 1991). Modification is sometimes required per the specific features and requirement of the study. Nowadays, with the progress and advent of new genetic

DOI: 10.1201/9781003220398-15

and genomic tools, genetic engineering techniques can easily be applied in order to develop knockout or transgenic animals that are used in research such as genetic model was developed by Gurumurthy et al. in 2020 to combat the deleterious effects of COVID-19 (Gurumurthy et al., 2020). It is important to note that beyond adequate technical knowledge, animal experimentation requires that ethical concerns speak louder than scientific interests. Each researcher must have a complete understanding of the animal model being used and of the biology and behavior of that species, and they must be aware of the importance of the work being conducted and consider all the premises that justify each specific project based on a solid scientific background. With this regard, the book chapter has been designed that depicted journey of animal models from higher animals to smaller animals and then finally use of cell line study, in silico methods, and cell docking.

15.2 Historical Journey of Animals

Animal experimentation has played a central role in biomedical research throughout history and has also been an issue of heated public and philosophical discussion. Aristotle and Erasistratus in the ancient Greeks, in fourth and third centuries BCE, began animal testing (Maehle and Tröhler, 1987; Franco, 2013) (Table 15.1).

15.3 Recent Milestones in the Development of Experimental Animal Models

By the end of the 19th and beginning of the 20th century, the pharmacopeia was adopted effectively that marks a landmark and increased awareness among common people to understand the importance and validity of scientifically sound medical knowledge and, with it, the relevance of animal-based research (Davis, 1903; Pearce, 1912; Ericsson et al., 2013). The 20th century witnessed surprising advancements in the field of medical knowledge and the management of disease (Franco, 2013; Blake et al., 2021).

15.4 Rodent Models in Biomedical Research

To understand the fundamental mechanisms and to discover safer and innovative methods to prevent, diagnose, and treat diseases, animal experiments are always essential (Vandamme, 2014). Rodent models are the most preferred due to the easiness for the experiments and the possibility to modify genetically these model animals. Phylogenic trees estimated that both humans and rodents shared a common ancestor (Bejerano et al., 2004).

TABLE 15.1
Early Milestones of Experimental Animals

Years	Milestones
6th century BCE	Alcmaeon of Croton used dogs to determine the function of the brain as "seat of intelligence" and sensory integration based on studies using dogs.
4th century BCE	Embryogenesis and ontogeny in chicks were studied by Aristotle.
3rd century BCE	The cardiovascular system in live animals was studied by Erasistratus, and he deduced that the heart functions as a pump, whereas Galen of Pergamum, in the 2nd century BCE, studied the cardiovascular system and neuroanatomy extensively, using live animals.
12th century	Avenzoar practiced surgical techniques on animals.
17th century	William Harvey evaluated anatomy of several species of live animals and provided accurate and detailed physiology of cardiovascular and all other systems.

TABLE 15.2

Recent Milestones in Experimental Animal Modeling

Researchers	Experimental work
William Castle (1902)	Initiated breeding of mice for genetic studies
Clarence Little (1909)	Began inbreeding mice to eliminate variation
Frederick Banting (1920)	Isolated canine insulin and effectively treated diabetic dogs
Little and MacDowell (1930)	First fully inbred mouse achieved by mating 20 pairs of brothers and sisters
John Cade (1940)	Studied the use of lithium salts as an anticonvulsant in guinea pigs and translated his findings to treatments of depression
Rudolf Jaenisch et al. (1976)	Developed the first transgenic mouse
Year 1980	Extensively testing of drug safety and dosing regimens for HIV, performed on rhesus macaques
Capecchi et al. (1987)	Developed the first knockout mouse
Wilmut and Campbellin (1997)	Produced Dolly the sheep, the first animal cloned from an adult somatic cell
2002 and 2004	First genome sequencing of mouse and rat developed, respectively
Aron Geurts et al. (2009)	Developed the first knockout experimental rats

15.4.1 Mouse (*Mus Musculus*)

The mouse, a small mammal scientifically known as *Mus musculus,* belongs to Order Rodentia, is most commonly used for biomedical research. Geneticists, biologists, and other scientists are rapidly finding new uses for the domestic mouse in research (Perlman, 2016). Different strains of mice (transgenic), such as BALB/Cj, C55BL/6 DBA/2J, and NOD.CB17-Prkdscscid/j Apoe-ko are employed in the field of biomedical research (Nishimura et al., 1995). Mice has been used as research tool for the evaluation of various disorders, such as hypertension, diabetes, cataracts, obesity, seizures, respiratory problems, deafness, Parkinson's disease, Alzheimer's disease, various cancers, cystic fibrosis, and acquired immunodeficiency syndrome (AIDS), heart disease, muscular dystrophy, and spinal cord injuries. Further, they are also employed for screening of behavioral, sensory, aging, nutrition, and toxicological screening, along with genetic studies (Hickman et al., 2017).

15.4.2 Rat (*Rattus Norvegicus*)

Rats and humans have a long history of coexistence. Experimental rats were used in the biomedical research for the first time in the year 1828 (Hedrich, 2000). Rats had been documented to possess a number of qualities, which make them a highly suitable and much preferred animal model in the field of drug discovery and preclinical testing. Evidence indicates that rats have been used as experimental animal models in numerous areas of research starting from space exploration to finding appropriate key of more basic scientific questions regarding nutrition, genetics, cardiology, immunology, neurology, infectious disease, metabolic disease, and behavioral disorders (Hickman et al., 2017). Further, it has been evident that different strains of rats, such as Sprague Dawley (SD), Fischer F344, and Hannover Wistar, were employed in the toxicological screening (Weber et al., 2011).

15.4.3 Rabbits (*Oryctolagus Cuniculus*)

Rabbits are very social, nocturnal creatures that possess a specific scent-marking and chinning behaviors (Sohn and Couto, 2012). In the field of research, rabbits have been used as a model of human pregnancy and for the production of polyclonal antibodies for use in immunology research (Hanly et al., 1995; Ema et al., 2010; Ito et al., 2011; Fischer et al., 2012). In addition, rabbits are routinely used in the evaluation of various disorders, such as atherosclerosis, osteoporosis, ocular disease, and immunological disorders (Southard et al., 2000; Arslan et al., 2003; McMahon et al., 2005; Castaneda et al., 2008; Habjanec et al., 2008; Manabe et al., 2008; Xiangdong et al., 2011; Panda et al., 2014; Sriram et al., 2014; Wei et al., 2014; Zhou et al., 2014).

15.5 Development of the 3 Rs Model

Many rules and acts are followed at the international level in order to protect the animals against the cruelty and misuse. The organizations like ICH (International Conference on Harmonization of Technical Requirements for Registration of Pharmaceuticals for Human Use), CPCSEA (Committee for Purpose of Control and Supervision on Experiments on Animal), NIH (National Institutes of Health), and OECD (Organization for Economic Cooperation and Development) provide the guidelines for animal housekeeping, breeding, feeding, and transportation and mainly for their use in scientific experiments. Hence, Russell and Burch, 1959 proposed the 3 Rs model—reduction, replacement, and refinement—which are widely acceptable principles in the governance of experimentation that involves humane animal research (Russell and Burch, 1959; Ormandy et al., 2009) that provide alternatives to animal testing. Various pieces of evidence emphasized that the concept of 3 Rs is economical and ethical and possesses academic basis behind sharing experimental animal resources, making biomedical research data scientifically and easily available. Hence, the 3 Rs animal model involves the reduction in the number of animals used and the refinement of testing to reduce suffering, which should be important goals for the industries involved (Doke and Dhawale, 2015). Further, different methods and alternative organisms are applied to implement this strategy.

15.6 An Approach Toward Fulfillment of the 4 Rs and 5 Rs

The 4 Rs model was developed in 1995 (Bank, 1995). The fourth R of research is *responsibility*, added to the original 3 Rs model of Russell and Burch. It has grown into a new era of performance-based outcomes, which reflects integrity, honesty, and scientific correctness in appropriate and reasonable use of laboratory animals. This model ensures that for the advancement of biomedical science and research, animal life is required (Arora et al., 2011). Nowadays, the 5 Rs model—replace, reduce, refine, reuse, and rehabilitate—is effectively communicated in biomedical research field (Kumar, 2012).

15.6.1 Zebrafish Model

The zebrafish is a well-established model organism that is used extensively in the field of developmental biology. In the year 1970, George Streisinger selected larvae of zebrafish for the development of the first vertebrate assay, enabling forward genetic screening (Streisinger et al., 1981). The zebrafish has been used to identify potential safety liabilities before the selection of any candidate and lead the process of drug discovery (Barros et al., 2008). Nowadays, zebrafish models has also been used to evaluate various behavioral assays that are used as into screening tool for the assessment of locomotor activity (Padilla et al., 2011), convulsions (Sharma et al., 2021; Kumari et al., 2019; Mazumder et al., 2019; Kumari et al., 2020), and cognitive impairment (Yendapalli et al., 2019). Further, the zebrafish also offers interesting possibilities for evaluating effects on bone density and gastrointestinal function.

15.7 New Trends *In Vitro* Animal Experimental Models

Industrialization and advancement have paved the way to explore using emerging technologies, such as organs-on-chips to replace animals in research. Organoids and organs-on-chips are emerging technologies nowadays that is replacing animal models (Silva-Almeida et al., 2020). *In vitro* models are designed to allow us to test toxicity and efficacy of drug compounds—for example, to check the irritancy of chemicals used in earlier Draize test, which requires rabbits—and to use a new animal every time is unethical and painful. So various scientists and researchers fashioned organs-on-chips that use human cells that mimic the anatomy and physiological functions of human organ system. Further, organs-on-chips are used now in research process and testing of drugs in various disorders such as colon cancer, testicular cancer, and gastrointestinal problems (Alves-Lopes et al., 2018; Silva-Almeida et al., 2020; Lannagan et al., 2021).

Further, stem cell research is another complementary alternative to animals used in *in vitro* models of disease and for toxicological testing. Stem cells have been found to possess specific properties, such as self-renewal and infinite proliferation. Thus, human-stem-cell-based *in vitro* systems have been proven to be valuable in increasing the predictive power of toxicology (Liu et al., 2017; Kim et al., 2019; Bremer and Hartung, 2004).

15.7.1 Computer (In Silico) Modeling

An in silico experiment is one performed on computer or via computer simulation in experimental field. Multiple biochemical and computational methods are introduced as an alternate to animal models in order to study the pathogenesis of various disorders, such as CNS disorders (e.g., Parkinson's) and cardiovascular disorders (Muhammad and Fatima, 2015; Jean-Quartier et al., 2017; Jaiteh et al., 2018). It has been proposed that the computer-based approach in the long run will reduce laboratory work in the first place and effectively succeed in 3 Rs. Computational and bioinformatics studies are both studied to store and process large-scale experimental data, extract and provide information, and develop integrative tools to support analysis tasks and produce biological insights (Villaverde et al., 2015; Badano, 2021).

Further, computer-aided drug design (CADD) software is used in biomedical sciences to predict the binding site of lead compounds for receptors (Vedani, 1991). Structure activity relationship (SARs) offers another tool in computer programs that aids in predicting biological activity of a drug based on the presence of chemical moieties attached to the parent compound (Doke and Dhawale, 2015). In addition, another mathematical description between physiochemical properties of a drug molecule and biological activity was determined with the help of a well-known phenomenon known as quantitative structure activity relationship (QSAR) and is used to predict activities like *carcinogenicity* and *mutagenicity* (Toma et al., 2021).

15.7.2 Cell Docking

Cell docking is the method that has its ability to predict the binding-conformation of small molecule ligand to the appropriate target binding site (Kitchen et al., 2004). Docking software comprises a particular algorithm that is used to identify the active site of the protein by allowing binding of the ligand in different parts of the protein and thus exploring the best possible binding position of the ligands with the protein. Characterization of the binding behavior plays an important role in rational design of drugs and elucidates fundamental biochemical processes (French et al., 2017). Literature review documented that network pharmacology and molecular docking technique was used recently to explore the various pathways and pharmacological potential of various drugs (Huang, 2011; Yuan et al., 2021). Moreover, structure-based multitargeted molecular docking analysis was used in the detection of various types of carcinomas (Acharya et al., 2019). The Flex-X technique is used as possess an excellent ligand flexibility that can be attained by changing the conformations of the ligand in active site making the protein rigid (Schellhammer and Rarey, 2004).

During the pandemic, various bioengineering tools emerged to speed up the discovery and preclinical testing of vaccines for SARS-CoV-2 and therapeutic agents for COVID-19 (Raimondi et al., 2020). Further, CRISPR/Cas9 allowed the researchers to knock out specific genes quickly from human cell cultures in order to screen for desired properties. Chemo-genomics also emerged as new research area that focused on the systematic examination of the biological impact of a broad series of minute molecular-weighing ligands of various macromolecular target spots (Ezzat et al., 2019a). Advancement and complexity in the algorithm paved the way for the development of big data technology like Spark (Kaushik et al., 2020). Ensemble learning technique is another computational method that is used to study various drug interactions and properties (Ezzat et al., 2019b). Recently, the AutoDock Vina method was used to perform the docking in the identification of SARS-CoV-2 spike receptor-binding domain attached to ACE2 (Lehrer and Rheinstein, 2020).

FIGURE 15.1 Principles for experimental animal care.

International Guidelines for Animal Use in Biomedical Research

In 1985, the International Guidelines for Animal Use in Biomedical Research were put forth by Council of International Organization of Medical Science (CIOMS) and approved by both the European Medical Research Council and the World Health Organization (WHO). (Figure 15.1)

15.8 Conclusion

Indubitably, we are using new and innovative technology/methods for interpretation of data, but still, nothing can totally replace experimental animal in field of biomedical research. So keeping all these things in mind, various models are proposed that emphasized using and handling the animal with care and minimum pain. Initially, we are employing larger animals to elucidate the pathogenesis of various disorders. Then the evolutionary started and scenario shifts toward smaller animals and then to tiny animals like zebrafish in biomedical research and preclinical evaluations. The shift is due to the prevalence of the 4 Rs and 5 Rs models, which give more importance to animal-related ethical issues. More efforts need to be undertaken for the effective implementation of these models during laboratory use of animals. Various alternative approaches, such as computer models, docking systems, bioinformatics tools, *in vitro* cell cultures, enzymatic screens and, model organisms, provide alternatives to animal use, which need to be implemented in an effective manner and are preferential in the process of drug discovery. Hence, this chapter enlightens the evolution of animal models and their contribution in understanding the life phenomenon in a better manner.

15.9 Acknowledgment

I wish to express our gratefulness to Prof. Parvinder Singh (Worthy Vice Chancellor) and S. Gurvinder Singh Bahra (Honorable Chancellor), Rayat and Bahra University Mohali (Punjab), for their guidance, praiseworthy inspiration, platform, and constant support for the completion of this study.

REFERENCES

Acharya R, Chacko S, Bose P, Lapenna A, Pattanayak SP (2019). Structure based multitargeted molecular docking analysis of selected furanocoumarins against breast cancer. *Sci Rep.* 9:15743. https://doi.org/10.1038/s41598-019-52162-0.

Alves-Lopes JP, Stukenborg JB (2018). Testicular organoids: A new model to study the testicular microenvironment in vitro? *Hum Reprod Update.* 24(2):176–191. https://doi.org/10.1093/humupd/dmx036.

Andersen ML, Winter LMF (2017). Animal models in biological and biomedical research—experimental and ethical concerns. *Anais da Academia Brasileira de Ciências.* 91(Suppl. 1):e20170238, 1678–2690. http://doi.org/10.1590/0001-3765201720170238.

Arora T, Mehta AK, Joshi V, Mehta KD, Rathor N, Mediratta PK, Sharma KK (2011). Substitute of animals in drug research: An approach towards fulfillment of 4R's. *Indian J Pharm Sci.* 73(1):1–6. https://doi.org/10.4103/0250-474X.89750

Arslan H, Ketani A, Gezici A, Kapukaya A, Necmioglu S, Kesemenli C, Subasi M (2003). The effects of osteoporosis on distraction osteogenesis: an experimental study in an ovariectomised rabbit model. *Acta Orthop Belg.* 69(1):67–73.

Badano A (2021). In silico imaging clinical trials: Cheaper, faster, better, safer, and more scalable. *Trials.* 22:64. https://doi.org/10.1186/s13063-020-05002-w.

Banks RE (1995). The 4th R of research. *Contemp Top Lab Anim Sci.* 34:43.

Barros TP, Alderton WK, Reynolds HM, Roach AG, Berghmans S (2008). Zebrafish: An emerging technology for in vivo pharmacological assessment to identify potential safety liabilities in early drug discovery. *Br J Pharmacol.* 154(7):1400–1413. http://doi.org/10.1038/bjp.2008.249.

Bejerano G, Pheasant M, Makunin I, Stephen S, Kent WJ, Mattick JS, Haussler D (2004). Ultraconserved elements in the human genome. *Science.* 304(5675):1321–1325. https://doi.org/10.1126/science.1098119.

Blake JA, Baldarelli R, Kadin JA, Richardson JE, Smith CL, Bult CJ (2021). The Mouse genome database group, mouse genome database (MGD): Knowledgebase for mouse–human comparative biology. *Nucleic Acids Res.* 49(8):D981–D987. https://doi.org/10.1093/nar/gkaa1083.

Bremer S, Hartung T (2004). The use of embryonic stem cells for regulatory developmental toxicity testing in vitro—the current status of test development. *Curr Pharm Des.* 10(22):2733–2747. https://doi.org/10.2174/1381612043383700.

Breyer MD, Thomas LA, Cifra A (2015). From bench to patient: Model systems in drug discovery. *Dis Model Mech.* 8:1171–1174. https://doi:org/10.1242/dmm.023036.

Castañeda S, Calvo E, Largo R, González-González R, de la Piedra C, Díaz-Curiel M, Herrero-Beaumont G (2008). Characterization of a new experimental model of osteoporosis in rabbits. *J Bone Miner Metab.* 26(1):53–59. https://doi.org/10.1007/s00774-007-0797-1.

Davis N (1903). *History of Medicine.* Cleveland Press: Chicago, IL: History of Medical Progress During the Last Half of the Nineteenth Century Completed. The Present Status of Medical Investigations and Progress, and the Indications of Further Advancement.

Doke SK, Dhawale SC (2015). Alternatives to animal testing: A review. *Saudi Pharm J.* 23(3):223–229. https://doi.org/10.1016/j.jsps.2013.11.002.

Ema M, Naya M, Yoshida K, Nagaosa R (2010). Reproductive and developmental toxicity of hydrofluorocarbons used as refrigerants. *Reprod Toxicol.* 29:125–131. https://doi.org/10.1016/j.reprotox.2009.11.005

Ericsson AC, Crim MJ, Franklin CL (2013). A brief history of animal modeling. *Mo Med.* 110(3):201–205.

Ezzat A, Wu M, Li XL, Kwoh CK (2019a). Computational prediction of drug-target interactions using chemogenomic approaches: An empirical survey. *Brief Bioinform.* 20(4):1337–1357. http://doi.org/10.1093/bib/bby00.

Ezzat A, Wu M, Li XL, Kwoh CK (2019b). Computational prediction of drug-target interactions via Ensemble learning. *Methods Mol Biol.* 1903:239–254. http://doi.org//10.1007/978-1-4939-8955-3_14.

Fedele M, Gulillo O, Vecchione A (2016). Animal models of human pathology. *BioMed Res Int.* Article id 4023982. https://doi:org/10.1155/2016/4023982.

Ferreira GS, Veening-Griffioen DH, Boon WPC, Moors EHM, Gispen-de Wied CC, Schellekens H, Hvan Meer PJK (2019). Correction: A standardised framework to identify optimal animal models for efficacy assessment in drug development. *PLOS One.* 14(7):e0220325. https://doi.org/10.1371/journal.pone.0220325.

Festing S, Wilkinson R (2007). The ethics of animal research. Talking Point on the use of animals in scientific research. *EMBO Rep.* 8(6):526–530. https://doi.org/10.1038/sj.embor.7400993

Fischer B, Chavatte-Palmer P, Viebahn C, Naverrete Santos A, Duranthon V (2012). Rabbit as a reproductive model for human health. *Reproduction*. 144:1–10. https://doi.org/10.1530/REP-12-0091.

Franco NH (2013). Animal experiments in biomedical research: A historical perspective. *Animals: An Open Access Journal from MDPI*. 3(1):238–273. https://doi.org/10.3390/ani3010238.

French KC, Antonyak MA, Cerione RA (2017). Extracellular vesicle docking at the cellular port: Extracellular vesicle binding and uptake. *Semin Cell Dev Biol*. 67:48–55. https://doi.org/10.1016/j.semcdb.2017.01.002.

Gurumurthy CB, Quadros RM, Richardson GP, Poluektova LY, Mansour SL, Ohtsuka M (2020). Genetically modified mouse models to help fight COVID-19. *Nat Protoc*. 15:3777–3787. https://doi.org/10.1038/s41596-020-00403-2.

Habjanec L, Halassy B, Vdović V, Balija ML, Tomasić J (2008). Comparison of mouse and rabbit model for the assessment of strong PGM-containing oil-based adjuvants. *Vet Immunol Immunopathol*. 15;121(3–4):232–240. https://doi/org/10.1016/j.vetimm.2007.09.014.

Hanly WC, Artwohl JE, Bennett BT (1995). Review of polyclonal antibody production procedures in mammals and poultry. *ILAR J*. 37:93–118. https://doi.org/10.1093/ilar.37.3.93.

Hedrich HJ (2000). History, strains, and models. In: Krinke G, editor. *The Laboratory Rat*. Academic Press: San Diego.

Hickman DL, Johnson J, Vemulapalli TH, Crisler JR, Shepherd R (2017). Commonly used animal models. In: *Principles of Animal Research for Graduate and Undergraduate Students*, 117–175. https://doi.org/10.1016/B978-0-12-802151-4.00007-4.

Huang LS (2011). Cell docking, movement and cell-cell interactions of heterogeneous cell suspensions in a cell manipulation microdevice. *Sensors (Basel)*. 11(10):9613–9627. https://doi.org/10.3390/s111009613.

Ito T, Ando H, Handa H (2011). Teratogenic effects of thalidomide: Molecular mechanisms. *Cell Mol Life Sci*. 68:1569–1579. https://doi.org/10.1007/s00018-010-0619-9.

Jaiteh M, Zeifman A, Saarinen M, Svenningsson P, Bréa J, Loza MI, Carlsson J (2018). Docking screens for dual inhibitors of disparate drug targets for Parkinson's disease. *J Med Chem*. 61(12):5269–5278. https://doi.org/10.1021/acs.jmedchem.8b00204.

Jean-Quartier C, Jeanquartier F, Jurisica I, Holzinger A (2018). In silico cancer research towards 3R. *BMC Cancer*. 18:408. https://doi.org/10.1186/s12885-018-4302-0.

Kaushik AC, Mehmood A, Dai X, Wei DQ (2020). A comparative chemogenic analysis for predicting Drug-Target Pair via machine learning approaches. *Sci Rep*. 2020;10(1):6870. https://doi.org/10.1038/s41598-020-63842-7.

Khorramizadeh MR, Saadat F (2020). Animal models for human disease. *Anim Biotechnol*:153–171. https://doi:org/10.1016/B978-0-12-811710-1.00008-2.

Kim TW, Che JH, Yun JW (2019). Use of stem cells as alternative methods to animal experimentation in predictive toxicology. *Regul Toxicol Pharmacol*. 105:15–29. https://doi.org/10.1016/j.yrtph.2019.03.016.

Kitchen DB, Decornez H, Furr JR, Bajorath J (2004). Docking and scoring in virtual screening for drug discovery: Methods and applications. *Nat Rev Drug Discov*. 3(11):935–949. https://doi.org/10.1038/nrd1549.

Kumar AH (2012). Effectively communicating the 5R's (replace, reduce, refine, reuse, and rehabilitate) of research ethics, biomedical waste, personalized medicines and the rest. *J Nat Sci Biol Med*. 3(1):1–2. https://doi.org/10.4103/0976-9668.95932.

Kumari S, Mazumdera AG, Bhardwaja A, Singh D (2019). Early α-linolenic acid exposure to embryo reduces pentylenetetrazolinduced seizures in zebrafish larva. *Prostaglandins Leukot Essent Fatty Acids*. 143:15–20. https://doi.org/10.1016/j.plefa.2019.02.002.

Kumari S, Sharma P, Mazumder AG, Rana AK, Sharma, S, Singh D (2020). Development and validation of chemical kindling in adult zebrafish: A simple and improved chronic model for screening of antiepileptic agents. *J Neurosci Methods*. https://doi.org/10.1016/j.jneumeth.2020.108916.

Lannagan TR, Jackstadt R, Leedham SJ, Sansom OJ (2021). Advances in colon cancer research: In vitro and animal models. *Curr Opin Genet Dev*. 66:50–56. https://doi.org/10.1016/j.gde.2020.12.003.

Lehrer S, Rheinstein PH (2020). Ivermectin docks to the SARS-CoV-2 spike receptor-binding domain attached to ACE2. *In Vivo*. 34(5):3023–3026. http://doi.org//10.21873/invivo.12134.

Liu S, Yin N, Faiola F (2017). Prospects and frontiers of stem cell toxicology. *Stem Cells Dev*. 26(21):1528–1539. https://doi.org/10.1089/scd.2017.0150.

Maehle AH, Tröhler U (1987). Animal experimentation from antiquity to the end of the eighteenth century: Attitudes and arguments. In: Rupke NA, editor. *Vivisection in Historical Perspective*. Croom Helm: London, UK.

Manabe YC, Kesavan AK, Lopez-Molina J, Hatem CL, Brooks M, Fujiwara R, Hochstein K, Pitt ML, Tufariello J, Chan J, McMurray DN, Bishai WR, Dannenberg AM Jr, Mendez S (2008). The aerosol rabbit model of TB latency, reactivation and immune reconstitution inflammatory syndrome. *Tuberculosis (Edinb)*. 88(3):187–196. https://doi/org/10.1016/j.tube.2007.10.006.

Mazumder AG, Kumari S, Singh D (2019). Anticonvulsant action of a selective phosphatidylinositol-3-kinase inhibitor LY294002 in pentylenetetrazole-mediated convulsions in Zebrafish. *Epilepsy Res*. 157:10620.

McMahon AC, Kritharides L, Lowe HC (2005). Animal models of atherosclerosis progression: Current concepts. *Curr Drug Targets Cardiovasc Haematol Disord*. 5(6):433–440. https://doi.org/10.2174/156800605774962031.

Morrissey B, Blyth K, Carter P, Chelala C, Jones L, Holen I, Speirs V (2017). The sharing experimental animal resources, coordinating holdings (search) framework: Encouraging reduction, replacement, and refinement in animal research. *PloS Biol*. 15(1):2000719. https://doi.org/10.1371/journal.pbio.2000719.

Muhammad SA, Fatima N (2015). In silico analysis and molecular docking studies of potential angiotensin-converting enzyme inhibitor using quercetin glycosides. *Pharmacogn Mag*. 11(Suppl 1):S123–S126. https://doi.org/10.4103/0973-1296.

National Research Council (US) Committee on Animal Models for Testing Interventions Against Aerosolized Bioterrorism Agents. (2006). *Overcoming Challenges to Develop Countermeasures Against Aerosolized Bioterrorism Agents: Appropriate Use of Animal Models*. Washington, DC: National Academies Press, 2, Selection or Development of an Animal Model.

Nishimura M, Hirayama N, Serikawa T, Matsushima Y, Katoh H, Wakana S, Kojima A, Hiai H (1995). The SMXA: A new set of recombinant inbred strain of mice consisting of 26 substrains and their genetic profile. *Mamm Genome*. 6:850–857. https://doi.org/10.1007/BF00292434.

Ormandy EH, Schuppli CA, Weary DM (2009). Worldwide trends in the use of animals in research: The contribution of genetically-modified animal models. *Altern Lab Anim*. 37(1):63–68. https://doi.org/10.1177/026119290903700109.

Padilla S, Hunter DL, Padnos B, Frady BS, MacPhail RC (2011). Assessing locomotor activity in larval zebrafish: Influence of extrinsic and intrinsic variables. *Neurotoxicol Teratol*. 33(6):624–630.

Panda A, Tatarov I, Masek BJ, Hardick J, Crusan A, Wakefield T, Carroll K, Yang S, Hsieh YH, Lipsky MM, McLeod CG, Levine MM, Rothman RE, Gaydos CA, DeTolla LJ (2014). A rabbit model of non-typhoidal Salmonella bacteremia. *Comp Immunol Microbiol Infect Dis*. 37(4):211–220. https://doi/org/10.1016/j.cimid.2014.05.004.

Pearce RM (1912). Research in medicine. II. The development of laboratories for the medical sciences. *Popul Sci Month*. 80:548–562.

Perlman RL (2016). Mouse models of human disease: An evolutionary perspective. *Evol Med Public Health*. 1:170–176. https://doi.org/10.1093/emph/eow014.

Raimondi MT, Donnaloja F, Barzaghini B, Bocconi A, Conci C, Parodi V, Jacchetti E, Carelli S (2020). Bioengineering tools to speed up the discovery and preclinical testing of vaccines for SARS-CoV-2 and therapeutic agents for COVID-19. *Theranostics*. 10(16):7034–7052. https://doi.orf/10.7150/thno.47406

Russell WMS, Burch R (1959). *The Principles of Humane Experimental Technique*. London.

Schellhammer I, Rarey M (2004). FlexX-Scan: Fast, structure-based virtual screening. *Proteins*. 57(3):504–517. https://doi.org/10.1002/prot.20217.

Sharma P, Kumari S, Sharma J, Purohit R, Singh D (2021). Hesperidin Interacts with CREB-BDNF signaling pathway to suppress pentylenetetrazole-induced convulsions in Zebra fish. *Front Pharmacol*. 11:607797. http://doi.org/10.3389/fphar.2020.607797

Silva-Almeida C, Ewart MA, Wilde C (2020). 3D gastrointestinal models and organoids to study metabolism in human colon cancer. *Semin Cell Dev Biol*. 98:98–104. https://doi.org/10.1016/j.semcdb.2019.05.019.

Sohn J, Couto MA (2012). Anatomy, physiology, and behavior. In: Suckow MA, Stevens KA, Wilson RP, editors. *The Laboratory Rabbit, Guinea Pig, Hamster, and Other Rodents*. Academic Press: Boston.

Southard TE, Southard KA, Krizan KE, Hillis SL, Haller JW, Keller J, Vannier MW (2000). Mandibular bone density and fractal dimension in rabbits with induced osteoporosis. *Oral Surg Oral Med Oral Pathol Oral Radiol Endod*. 89(2):244–249. https://doi.org/10.1067/moe.2000.102223.

Sriram S, Gibson DJ, Robinson P, Pi L, Tuli S, Lewin AS, Schultz G (2014). Assessment of anti-scarring therapies in ex vivo organ cultured rabbit corneas. *Exp Eye Res*. 125:173–182. https://doi.org/10.1016/j.exer.2014.06.014.

Streisinger G, Walker C, Dower N, Knauber D, Singer F (1981). Production of clones of homozygous diploid zebra fish (Brachydanio rerio). *Nature*. 291(5813):293–296. https://doi.org/10.1038/291293a0.

Toma C, Manganaro A, Raitano G, Marzo M, Gadaleta D, Baderna D, Roncaglioni A, Kramer N, Benfenati E (2021). QSAR models for human carcinogenicity: An assessment based on oral and inhalation slope factors. *Molecules*. 26(1):127. https://doi.org/10.3390/molecules26010127.

Vandamme TF (2014). Use of rodents as models of human diseases. *J Pharm Bioallied Sci*. 6(1):2–9. https://doi.org/10.4103/0975-7406.124301.

Vedani A (1991). Computer-aided drug design: An alternative to animal testing in the pharmacological screening. *Altex*. 8:39.

Villaverde AF, Henriques D, Smallbone K, Bongard S, Schmid J, Cicin-Sain D, Crombach A, Saez-Rodriguez J, Mauch K, Balsa-Canto E, Mendes P, Jaeger J, Banga JR (2015). Biopredyn-bench: A suite of benchmark problems for dynamic modelling in systems biology. *BMC Syst Biol*. 9:8.

Weber K, Razinger T, Hardisty JF, Mann P, Martel KC, Frische EA, Blumbach K, Hillen S, Song S, Anzai T, Chevalier HJ (2011). Differences in rat models used in routine toxicity studies. *Int J Toxicol*. 30(2):162–173. https://doi.org10.1177/1091581810391818.

Wei C, Zhu M, Petroll WM, Robertson DM (2014). Pseudomonas aeruginosa infectious keratitis in a high oxygen transmissible rigid contact lens rabbit model. *Invest Ophthalmol Vis Sci*. 55(9):5890–5899. https://doi.org/10.1167/iovs.14-14235.

Xiangdong L, Yuanwu L, Hua Z, Liming R, Qiuyan L, Ning L (2011). Animal models for the atherosclerosis research: A review. *Protein Cell*. 2(3):189–201. https://doi.org/10.1007/s13238-011-1016-3.

Yendapalli PR, David DC, Balasundaram A (2019). Evaluating the combined cognitive enhancement effect of *Brassica Juncea* and *Cynadon Dactylon* extract in scopolamine induced amnesia zebrafish model. *Toxicol Environ Health Sci*. 11:190–196. https://doi.org/10.1007/s13530-019-0393-5.

Yuan C, Wang, Meng-Heng, Wang, Fei, Chen, Peng-Yu, Ke, Xin-Ge, Yu, Bing, Yang, Yan-Fang, You, Peng-Tao, Wu, He-Zhen (2021). Network pharmacology and molecular docking reveal the mechanism of Scopoletin against non-small cell lung cancer. *Life Sci*. 270:119105. https://doi.org/10.1016/j.lfs.2021.119105.

Zhou Q, Liu XY, Ruan YX, Wang L, Jiang MM, Wu J, Chen J (2015). Construction of corneal epithelium with human amniotic epithelial cells and repair of limbal deficiency in rabbit models. *Hum Cell*. 28(1):22–36. https://doi.org/10.1007/s13577-014-0099-6.

16

Animals: Pioneers or Victims of Experimental Biology?

Anuradha Sharma, Indu Sharma, and Ranbir Chander Sobti

CONTENTS

16.1 Introduction

The theory of common descent of all living beings and the conservation of genetic and developmental pathways carried over the evolution process prompts the researchers to utilize the animals as the experimental models. The animals are being used in basic science research of human anatomy and physiology since the sixth century, but their use expanded in the 18th and 19th centuries owing to the scientists like Francesco Redi, John Needham, Lazzaro Spallanzani, Antoine Lavoisier, and Louis Pasteur who conducted animal experiments to study the origin of life (Oparin, 1957). The animal use lies as the test subjects of human diseases and testing of drugs, vaccines, and other compounds, such as hormones and antibodies. Animals have also been employed in the surgical trials in medical science. For example, Ibn Zuhr, in 12th century, performed tracheostomy on goats, and Kellings, in the 1900s, practiced laparoscopy on dogs (Litynski, 1997; Haddad, 2004), Thomas and Blalock developed the dog model for tetralogy of Fallot (TOF) in the 1940s (Timmermans, 2003).

16.2 Animal Models

An ideal animal model is the one that accurately mimics the desired disease or pathway, provides extrapolatable data to humans, is easy to be handled, has a moderate survival rate, has a sufficient size to provide multiple samples, and is multiparous to generate multiple offspring. An animal model is a particular

DOI: 10.1201/9781003220398-16

animal species that, when given a specific challenge by a specific route, produces a disease which multi-dimensionally represents human disease or any other condition of interest. There is no denying the fact that animals have helped the medical science to set a milestone in discovering the vaccines, antibiotics, and detailed knowledge of human diseases. Without the use of canines in 1920s and monkeys in 1930s the discovery of polio vaccine by Albert Sabin (Sabin, 1965) and insulin (King, 2012) would not have been possible. Approximately all the antibiotics are tested on the animals before being recommended to the humans.

Animal research has undoubtedly provided invaluable data to promote human healthcare and alleviate human suffering. Nearly 90% of Nobel Prize research in physiology and medicine to date used animal experiments in their discoveries. Most non-human animal species are used as experimental models because of their phylogenetic proximity to humans. These include insects (*Drosophila melanogaster*), nematodes (*Caenorhabditis elegans*), fish (*Danio rerio*), frogs (*Xenopus*), and many mammals, such as mice, rats, dogs, cats, pigs, and monkeys, due to their phylogenetic proximity to humans (National Research Council (US) Committee for the Update of the Guide for the Care and Use of Laboratory Animals, 2011). According to the CPCSEA guidelines amendment in 2006, a special permission from CPCSEA is required to carry out any experiments on the large animals like dogs, cats, and non-human primates. However, the Institutional Animal Ethics Committee (IAEC) of the respective establishment is empowered to permit experiments on small animals like mice, fish, and rats (The recommendations of the subcommittee on rehabilitation of animals after experimentation set up by CPCSEA, 2006).

16.2.1 History

Darwin's theory of evolution in 1859 served as the scientific rationale for using animals as models for the biomedical research. Per the literature concerned, the animal use in research was initiated by Hippocrates (450 BC), who emphasized on the similarity between diseased human organs and those of animals. Later on, the anatomists like Alcmaeon (500 BC), Herophilus (330–250 BC), and Erasistratus (305–240 BC) performed vivisection on animals to understand the anatomy and physiology of their organs (Petroianu, 1996). "Vivisection" is the term used by those who oppose the use of animals in research; this means cutting into or dissecting a living animal. The concerns regarding the ethics of animal-based research came into light when the first society for the protection of animals was formed in 1824 in England, and the first law regulating the use of animals in research was proposed in 1876 in Great Britain, known as the British Cruelty to Animals Act. Later on, UNESCO announced the Universal Declaration of Animal Rights in 1978 in Brussels (Miziara, 2012). The opening phrases of this declaration included the equal rights for the existence of all animals and the right to be respected by humans.

The enlightenment toward unethical animal use in research further flourished when René Descartes's described animals as "machine-like" (Descartes, 1989) in his work and was greatly criticized by many of his equals. This brought attention toward the numerous gruesome experiments that were being carried out on animals at the time when the anesthesia was not available, neither for humans nor for animals (Maehle & Tröhler, 1987; Boden, 2006; Allen, 2011).

16.2.2 Dichotomy in the Use of Animals

It is a common notion that medicine cannot be taught without the exposure to clinics, wards, and surgery; similarly, the basic science of living beings cannot be taught without animal experimentation. On the other hand, it is argued that animals should not be unnecessarily sacrificed just to gain skills and experimental expertise. The animal use in biomedical research has been a hot topic of debate and puts the scientific community in turmoil for a long period of time. This is because of the fact that only a fraction of animal-based studies is able to produce promising results that can be applied to humans, while others fail to demonstrate any similarity between the animal model and the human model and cannot justify the useless killing and pain or suffering imposed on the animals.

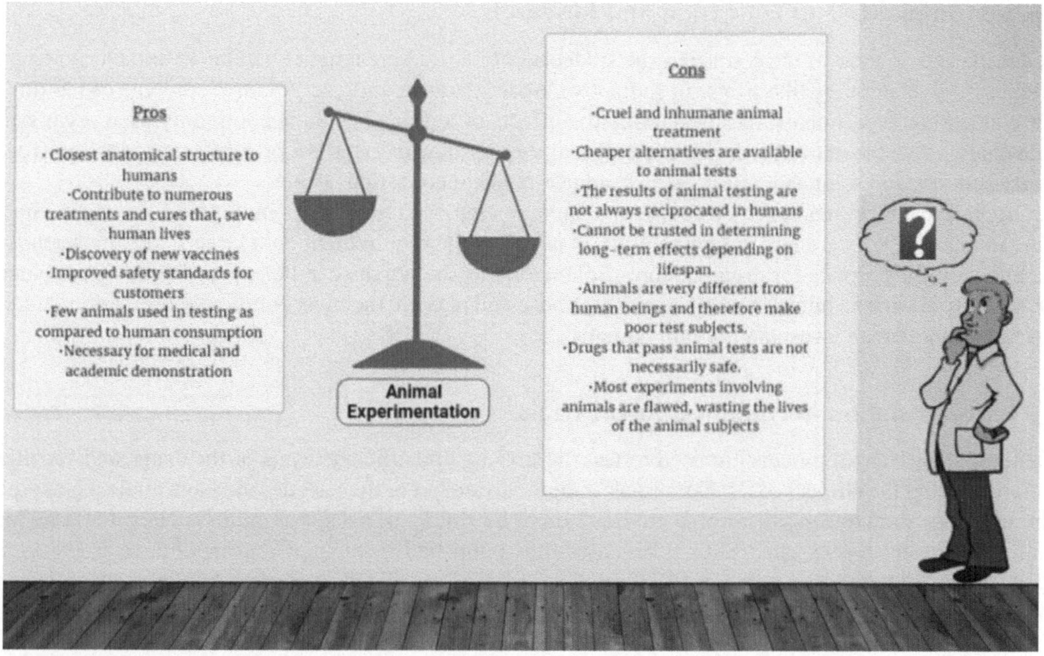

FIGURE 16.1 Animal experimentation a dilemma for researcher: pros and cons.

Most people refute the use of animals in scientific studies, emphasizing the reason that the animals are denied their freedom and subjected to unwanted pain and suffering. Some acknowledge the cruelty imposed on the animals during experiments but consider it inevitable when no other alternative is available. Moreover, the critics of animal use in research imply that there are so many differences between the animals and humans at the genetic and phylogenetic levels that it is impossible to produce similar results in both, thus proving the animal use and euthanasia to be useless. Contrariwise, the proponents of using animals claim that humans are not the only beneficiaries of the animal-based research but the animals also get profited, for there is advancement in veterinary treatments, services, vaccines, and so on. Dogs and cats are living longer and healthier lives owing to vaccines discovered for rabies, parvovirus, tetanus, and feline leukemia, among other breakthroughs made in veterinary medicine, thanks to animal research.

According to a report of US Department of Agriculture (USDA), 7,92,168 animals were used in experiments at USDA-registered facilities in fiscal year 2017. This number does not include amphibians, birds, fish, mice, rats, and reptiles used, which otherwise account for the majority of animals used in scientific studies. Scientists also demand the use of non-human primates, like chimpanzees, for experimental studies as they share close proximity with humans at the genetic and physiological levels. The National Aeronautics and Space Administration (NASA), for example, sent several non-human primates into space before sending human astronaut Alan Shepard in 1961.

The opponents of animal-based research believe in the notion "all lives matter." Hence, it is unethical to use the animals for the sole purpose of benefitting mankind, even when it has been reported in several studies that the results obtained in animal models do not necessarily always translate in humans. A systematic review compared the treatment effects in six separate animal studies with human clinical trials and reported that the concordance between the outcomes of animal and human trials was highly variable. Take for example the drug trial of tirilazad, a drug evaluated for its efficacy in acute ischemic stroke. The results of all the 18 separate animal studies showed improved neurologic outcomes and reduced overall infarct size. On the other hand, in humans, the use of tirilazad led to an increased risk of mortality and dependency (Perel et al., 2007). As a result of heterogeneity like this, the results of animal studies do not always carry over when humans are the trial subjects.

16.2.3 Animal Use in Education and Research

A decade ago, in most of the countries, the students of biology were taught anatomical and physiological studies on live animals, like frogs. In India also, undergraduate and postgraduate students use animals for conducting experiments, like evaluating the effects of a drug. The most commonly used organs are intestines, eyes, the central nervous system, and the reproductive system of rabbit and also the frog's heart, rectum, and so on (Regulations on graduate medical education, 1997).

Organ transplantation is a very common technique employed these days in the medical world. Since human organ donors are few in number and do not meet the requirement on a large scale, the scientists turn toward the animals for organ xenotransplantation as the organs can be exposed to the physiologic complexity similar to human body. Currently, mouse and pigs are the most commonly used animal models to promote organ transplantation in humans.

16.2.4 Translation in Human Clinical Trials

Generally, the animal models are used to test the toxicity and efficacy levels of the drugs and vaccines before starting the clinical trials. Numerous studies carried out in the past decade have made it clear that the inference drawn from the animal studies cannot be simply translated in human trials. Per a review summarizing 60 highly cited animal-based studies published in reputed journals between 1980 and 2000, it was reported that only a third of total studies was translated in random human trials (Hackam & Redelmeier, 2006).

Taking the case of vaccine against HIV-like viruses. More than 100 vaccines have been developed till now, which showed efficacy in animal models but none proved to be efficient for humans (Gamble & Matthews, 2010; Hall et al., 1988). Moreover, the translation rate of animal-based data in humans in cancer research lies at 8% only (Mak et al., 2014). Ferdowsian in her review in 2011, found that in various federally approved trials, no justification was provided for the harm and injury inflicted upon the test animals, and the studies led to questionable outcomes, which could not be applied to humans (Ferdowsian, 2011). In some other studies, animals were subjected to pain and injury without any analgesics or measures taken to alleviate the suffering.

16.3 The Principle of 3 Rs

William Russell, a psychologist, and Rex Burch, a microbiologist, worked together to form the well-known principle of 3 Rs for the ethical aspects in laboratory techniques, which stand for reduction, refinement, and replacement (Russell & Burch, 1959) (Figure 16.2).

Reduction comprises the use of such techniques that reduce the number of animals to be used in an experiment, without comprising statistical significance. This may be guaranteed by using the animals having genetic similarities and by detailed planning of the experiment. Many websites are available these days that calculate the statistically accurate number of animals to be used in a study.

Refinement deals with the proper and ethical use of animals which alleviate their pain and suffering. This includes the use of anesthesia during a surgical procedure and pain relievers during the process of recovery, employing non-invasive techniques wherever possible, comfortable and safe housing facilities, and acclimatization of the animal to the laboratory before initiation of the experiment.

Replacement refers to the use of alternatives for animals in science experiments. These alternatives may include other models like microorganisms or other invertebrates, cell cultures, organs, or computer-simulated models. As an initiative by the National Centre for the Replacement, Refinement, and Reduction of Animals in Research (NC3Rs), the ARRIVE (Animal Research: Reporting of *In Vivo* Experiments) guidelines came into being in 2010 and are currently endorsed by scientific journals and by funding agencies. The ARRIVE guidelines comprise a checklist of 20 items, including information such as specific characteristics of animals used (like species, sex, strain, and genetic background), housing and husbandry, and the experimental, statistical, and analytical methods. The goal is to improve the design, analysis, and reporting of animal-based research, thereby maximizing the information

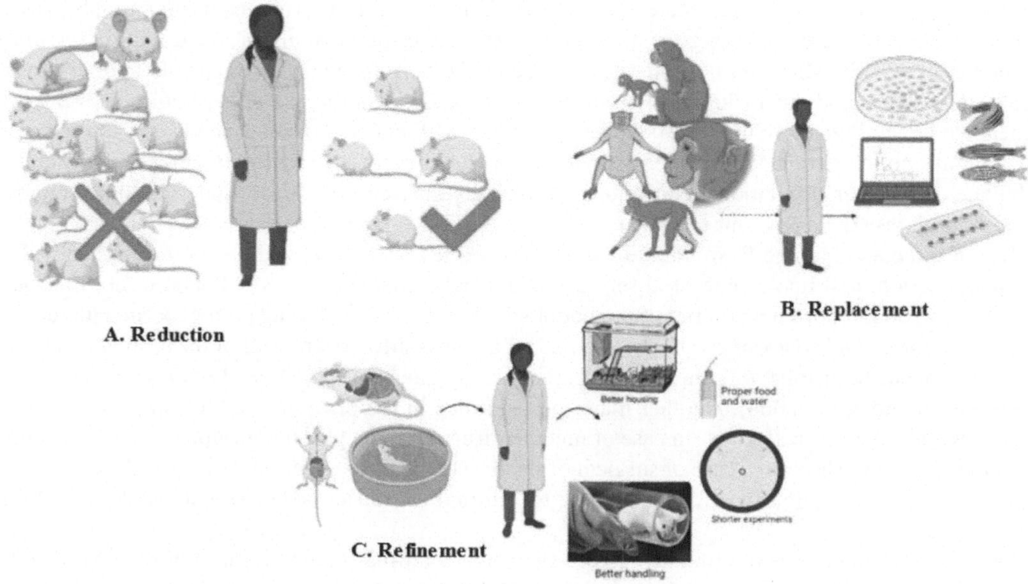

FIGURE 16.2 The principle of 3 Rs for animal experimentation.

obtained at the expense of minimum possible number of animals and eliminating the unnecessary studies (Kilkenny et al., 2010).

In India, the Committee for the Purpose of Control and Supervision of Experiments on Animals (CPCSEA), the University Grants Commission (UGC, 2011), and the Medical Council of India (MCI, 2009) recommend the use of 3 Rs for animal experimentation, with an additional R added, which stands for rehabilitation.

16.4 Arguments Favored by the Opponents

Some intellectuals argue that animal experimentation, in addition to being disadvantageous to the life of animals, is an immoral and non-consensual practice as the animals involved are not capable of providing consent and do not involve themselves in the inhumane and painful procedures voluntarily. Most scientists and a part of the general public agree to the use of animals where it is absolutely necessary and no other alternative is available and that, too, if proper ethical guidelines are followed. They acknowledge that the animal use can be justified in case of understanding the mechanism underlying a disease, testing a vaccine or drug, evaluating drug toxicity, and so on (Kehind, 2013).

Some very cruel experiments conducted on the animals attracted worldwide attention. iBritches, a macaque, was used for an experiment to test sensory substitution devices for blind people in University of California, Riverside. The eyes of the monkeys were sewn shut, and this inhumane activity attracted a lot of criticism in 1985. Later, the laboratory was raided by Animal Liberation Front, and the animals were rescued (PETA, 2013). Similarly, in 2006, a trial of a monoclonal antibody in primates triggered a disastrous immune reaction and widespread organ failure in the six trial animals. This happening in London also generated a lot of media attention and public dispute (BBC news, 2013).

The animal models should be treated with similar amount of respect as given to the human organ or tissue donors as the results obtained from this study will be applicable on both humans and animals. An animal model, if kept alive, even after the completion of an experiment, should be rehomed rather than again subjecting to reuse in another research project. Reusing may not only interfere with the results of the study but also subject the animal to continuous period of insufferable pain and injury. The Animal

Welfare Act, passed in the United States in 1970, stipulates that all experiments using animals provide adequate veterinary care, including the appropriate use of anesthetic, analgesic, or tranquilizing drugs (Cardon et al., 2012). Moreover, scientists have discovered various alternatives for the animal models in research. These methods include *in vitro* tests (tissues and cells), the use of vegetables, non-invasive clinical studies in human volunteers, conducting studies with corpses instead, the use of lower organisms that are not classified as protected animals (e.g., shrimp and water flea larvae), physicochemical techniques, computer simulations, educational software, films, mathematical models, nanotechnology, and test dummies (Balls & Combes, 2017).

Most of the data obtained from animal studies to date has been criticized for its inconsistency, lack of randomization, and inadequate or absent use of formal statistical analysis. Pound et al., in a book published in 2004, reported a number of methodological problems, including poor experimental design, inadequate power, and a lack of established statistical analysis. Moreover, most of the animal trials were run simultaneous to or following the human trials, thereby questioning the need of animal study in the first place (Pound et al., 2004). Another major problem lies in the practices and techniques adopted in medical world, like surgical trials. In case of medical drug trial, the placebo group is given an inert pill that poses no harm, whereas in case of surgical trials, the sham group experiences the surgical procedure without any therapeutic intervention, subjecting the animal to considerable risk without benefit (Louw et al., 2017).

Non-human primates are the first choice of scientists for animal model because of their astonishing similarity at the phylogenetic, genetic, behavioral, and biochemical levels with the humans. They are most commonly used for research in the area of human diseases like AIDS, hepatitis, and Parkinson's disease, as well as psychiatric disorders, transplant testing, and vaccine and drug trials. Chimpanzees are the most commonly employed non-human primates. The United States, in 2015, passed a law declaring the chimpanzees in captivity (even as laboratory animals) an endangered species. Consequently, the main supporter of research in the US, the National Institutes of Health (NIH), decided not to fund any research that used chimpanzees as experimental animal model.

In most of the cases, it is seen that the results found in the animal research cannot be translated in clinical trials. This contradicts the principle of biomedical research to provide useful knowledge that is translationally similar to clinical trials. On the other hand, it generates skepticism in general public, and the value of animal research falls into question.

The British Parliament in 1876 introduced the Cruelty to Animals Act, which emphasized on the key points: for example, animal experiments should only be carried out in case of absolute need of knowledge aimed toward saving or prolonging life or alleviating suffering, the animals must be anesthetized, and the animals must be killed immediately after the experimental procedure if they are injured or in pain as a result of the experiment.

16.5 Alternatives to Animal Experimentation

16.5.1 Cell Culture

The most common alternative to the animal experimentation is the use of human-derived cells, cultured in *in vitro* systems. Evidence suggests that the *in vitro* models provide more accurate, fast, and extrapolatable information as compared to the animal models. In the last decade, there has been a tremendous improvement in the complexity of human cell cultures. This includes co-culture systems, 3D cell constructs, perfusion chambers, and more recent microfluidic-based organ-on-a-chip. For example, the researchers have developed a three-dimensional human-cell-derived model EpiDerm (MatTek Life Sciences), which can be used to replace rabbits in painful experiments conducted for assessing the chemicals that may corrode the skin. Also, EpiAlveolar, a first-of-its-kind three-dimensional model composed of human lung cells, can be used to study the effects of inhaling a variety of chemicals, pathogens, and cigarette smoke. This will replace the current cruel some method where the rats are confined to tiny tubes and forced to inhale toxic substances for hours before being killed. The Organization for Economic Cooperation and Development (OECD) has also approved these tissue culture techniques as a substitute to the corresponding animal-based experiments. The ability of these cell constructs to recapitulate the organ functions in particular has been demonstrated by several researchers in their studies. However, replacing animal models completely with *in vitro* models relies on the fulfilment of an assumption that the numerous indicators of toxicity observed *in vivo* are initiated by relatively fewer number of distinguishable pathways. But this, owing to the complexity of animal or human body, is not true. For example, the pharmacokinetic factors like transformation of the drug *in vivo* to an active metabolite are hardly carried out by the *in vitro* model. Therefore, there is still doubt whether the *in vitro* cell cultures can completely replace the animal studies in case of toxicological studies.

16.5.2 Micro-Dosing

Micro-dosing refers to a technique of evaluating the effects of a drug by directly testing on the human volunteers prior to large-scale human trials. The drug is administered at such sub-therapeutic level (low doses) that it fails to produce the whole-body pharmacological effects but allows the cellular response to be examined. This dose is decided by the analytical methods that detect substances in blood and plasma at pg/ml concentrations. These methods include liquid chromatography coupled with mass spectrometry. The functional magnetic resonance imaging (fMRI) and other advanced brain imaging and recording techniques done with human volunteers can be used to replace the outdated experiments that cause damage to the brains of rats, cats, and monkeys. Moreover, these modern techniques provide the advantage of studying the human brain down to the level of a single neuron, such as intracranial electroencephalography. Micro-dosing is considered to be only the phase 0 of a clinical drug trial. Therefore, animal testing is still needed with the full dose of a drug to determine its safety, efficacy, and approval. However, the practice of micro-dosing forces the scientists to design subject-friendly experiments as they can empathize more easily with someone of their own species.

16.5.3 Mathematical Models and Computer Simulations

The mathematical models employ the use of computer graphics in actively designing drugs and other chemicals. These are based on the quantitative structure activity relationships (QSARs) that work on the correlations between molecular structure and biological activity to predict the potential anticipated and undesired effects of drugs, thus excluding the need of animal testing. This works on the principle that the biological effects of a chemical depend on its size, shape, and other properties of its molecules. QSAR makes sophisticated estimates of a substance's possibility for acting hazardous, based on its similarity to existing pre-tested substances and the knowledge of human biology. Per an article, the database guiding such systems, of course, takes its information from animal experiments. But once the relation between molecular structure of a chemical and its activity is established, the toxicity of any new chemical can be predicted with a computer instead of sacrificing a mouse (Watts, 2007).

Brazilian scientists have also established mathematical models of aerodynamic and viscoelastic equations for the larynx where a computer software simulates the anatomy and physiology of the larynx. These models may be used as teaching tools to avoid the unnecessary killing of animals (Rosa et al., 2003). Later, a Canadian group also developed a virtual station for teaching laryngeal anatomy and surgery based on computer tomography and magnetic resonance imaging sections (Hu et al., 2009). The use of virtual reality (VR) in laparoscopic skills training has already been validated and widely accepted (Seymour et al., 2002; Ahlberg et al., 2007). This technology has the potential to dramatically reduce the number of animals used in the academic teaching setting while simultaneously reducing the long cost in high-fidelity models (Figure 16.2).

16.5.4 Organ-on-a-chip

Another breakthrough in the scientific world is the discovery of the organ-on-a-chip (OOAC), which comprises a miniature form of tissues and organs generated *in vitro* to model the human physiology and disease. The microenvironment of the chip mimics the tissue interfaces and mechanical stimulation of a

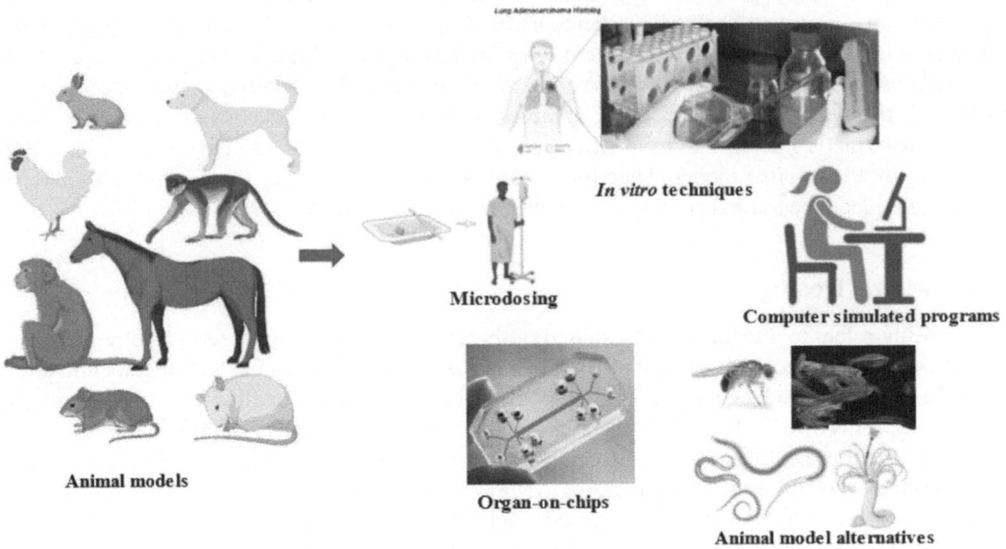

FIGURE 16.3 Alternatives available for animal models of experimentation

particular organ. These chips adequately reflect the structural and functional properties of human tissue and, therefore, can predict response to various stimuli, like drug responses and environmental effects. Multi-organ integration can be brought up by linking individual organ chips via microfluidic channels and keeping volume ratios and flow distribution similar to physiology *in vivo*. The first developed OOAC was the lung-on-a-chip, which mimics the physiological breathing motion of lungs. It consists of vacuum channels on each side of the central channel, which stretch the middle membrane. OOAC have also been employed as cancer models in various studies. For example, Jeon and coworkers designed a microfluidic 3D *in vitro* model to demonstrate the seed-and-soil paradigm of human breast cancer extravasation into muscle- and bone-mimicking microenvironments (Jeon et al., 2015). The OOAC approach has already been successfully employed in modeling various living tissues such as cystic fibrosis (Dekkers et al., 2016), brain cultures (Pasca, 2018), kidney glomerulus (Musah et al., 2017), and lung functions (Huh et al., 2010). Hence, considering these facts, these OOACs may prove to be an efficient alternative to the animal experimentation.

Replacing animal experimentation does not mean putting human lives at risk or hindering medical progress. Rather, it accounts for improving the quality and the humaneness of science. The quest of alternatives for animal models mainly revolves around the 3 Rs principle framed by Russell and Burch. Conducting research should not be permitted in cases where the damage to the animal is greater than the gain in knowledge since no scientific advancement can be justified based on the suffering of other living beings (Singh et al., 2016).

The minimum number of animals that yields statistical results should be used for an experimental study. Suitable anesthesia and analgesics should be used during the study to minimize pain. In medical science teaching programs, multiple procedures can be carried out using the same anesthetized animal and that animal should be euthanized before coming to conscious.

16.5.5 Machine-Learning-Based Computational Models

Machine learning (ML) comprises a part of artificial intelligence and is considered as the study of computer algorithms that improve automatically through experience and learn from data. ML is classified into two broader categories—supervised and unsupervised learning. Supervised learning is when the computer has to approximate the human performance by predicting a known output or target, like identifying handwriting of a person, identifying images of objects or animals, and distinguishing various documents. On the other hand, in unsupervised learning, the computer detects naturally occurring patterns or groupings within the data (Valletta et al., 2017).

Most animal-tested toxicological experiments, like the Draize rabbit eye test and chemical skin sensitization tests, where chemicals are tested numerous times in large number of animals, can be replaced by using artificial-intelligence-based models. These models work on simple one-nearest-neighbor approach, where a chemical is classified with respect to its closest analog requiring a minimum level of similarity between an earlier known chemical and the new tested one (Luechtefeld et al., 2018).

With the help of ML, bio-researchers are trying to develop novel methods to completely replace animal studies for chemical safety-related projects. A recent technique called read-across structure activity relationships (RASAR) developed by Thomas Hartung and colleagues from John Hopkins University combines chemical similarities with supervised learning (Hartung & Hoffmann, 2009). They compiled the information regarding 10,000 substances based on 800,000 animal tests to design their algorithm. This software algorithm proved that some animal tests are inconsistent because a single compound can produce different reactions in different animals.

Performing machine learning involves creating a model, which is accomplished on some training data and can process additional data to make predictions. Various types of models have been used and researched for machine learning systems. An artificial neural network (ANN) is a model based on a collection of connected units or nodes called artificial neurons, which loosely mimic the neurons in a biological brain. This model is designed to mimic the learning processes of the human brain and comprises an input layer, which is connected to a series of "hidden" nodes, which are in turn connected to an output layer. Statistically significant correlations can be developed between the signifiers and an endpoint result by modulating these hidden nodes.

Therefore, it is speculated that artificial intelligence in the form of machine learning could greatly help in establishing chemical similarities between a large number of toxic chemicals and their information stored on a database. This would reduce the countless animal sacrifices for testing each drug individually. But still there is a long way to go to completely replace animal testing with machine learning, due to limited information stored on databases.

16.6 Conclusions

Jeremy Bentham, a British philosopher, in his book rightly mentioned that "the question is not whether animals can reason, nor can they talk. The true question is this: Can they suffer?" The use of animals in experimental studies has long been a controversial topic in the scientific community. Performing animal-based research should not be allowed where the damage or injury caused to the animals is much greater than the knowledge gained. No advancement in information database can justify the cruelty toward animals. Animal models can only provide approximations rather than final answers because no single animal model can completely duplicate a disease in man. Moreover, when the data obtained from animal studies is analyzed, it is most commonly seen that they fail to predict the human responses. Animals are sentient beings and, hence, should not be subjected to unnecessary pain and suffering. However, where the teaching sessions are concerned in basic and medical science, where the animal use is inevitably necessary, only one or two animals of a kind should be used to demonstrate to the entire group rather than giving an animal to each individual. In experimental arena, the preference should be given to the alternatives available, like *in vitro*, microorganisms, plants, and computer-simulated programs. One should carry out all the experiments ethically and not abuse humans' authority over animals. The investigators are responsible for providing the animals with access to proper diet and water, protection, warm and safe housing, and relief from pain and injury. Therefore, it is the need of the hour to design experiments while wisely keeping in mind the compassion for animals so that less and less animals are harmed and more useful knowledge can be generated from a limited number of sacrifices.

REFERENCES

Ahlberg G et al. (2007). Proficiency-based virtual reality training significantly reduces the error rate for residents during their first 10 laparoscopic cholecystectomies. *Am J Surg* 193(6): 797–804.

Allen C (2011). Animal consciousness. In: Zalta EN, editor. *The Stanford Encyclopedia of Philosophy*. Summer 2011 ed. Center for the Study of Language and Information, Stanford University, Stanford, CA.

Balls M, Combes R (2017). Animal experimentation and alternatives: revealed preferences. *Altern Lab Anim* 45(1): 1–3.

BBC News (2013). *Two Drug Trial Men Critically Ill*. http://news.bbc.co.uk/2/hi/uk_news/england/london/4808836.stm

Boden ME (2006). *Mind as Machine: A History of Cognitive Science*. Volume 1. Oxford University Press, Oxford, UK.

Cardon AD, Bailey MR, Bennett BT (2012). The Animal Welfare Act: from enactment to enforcement. *J Am Assoc Lab Anim Sci* 51(3): 301–305.

Dekkers JF, Berkers G, Kruisselbrink E, Vonk A, de Jonge HR, Janssens HM, Bronsveld I, van de Graaf EA, Nieuwenhuis EE, Houwen RH, Vleggaar FP, Escher JC, de Rijke YB, Majoor CJ, Heijerman HG, de Winter-de Groot KM, Clevers H, van der Ent CK, Beekman JM (2016). Characterizing responses to CFTR-modulating drugs using rectal organoids derived from subjects with cystic fibrosis. *Sci Transl Med* 8(344): 344ra84.

Descartes R (1989). Animals are machines. In: Regan T., Singer P., editors. *Animal Rights and Human Obligations*. 2nd ed. Prentice Hall, Englewood Cliffs, NJ, 13–19.

Ferdowsian H (2011). Human and animal research guidelines: aligning ethical constructs with new scientific developments. *Bioethics* 25: 472–478.

Gamble LJ, Matthews QL (2010). Current progress in the development of a prophylactic vaccine for HIV-1 *Drug Des Dev Ther* 5: 9–26.

Hackam DG, Redelmeier DA (2006). Translation of research evidence from animals to humans. *J Am Med Assoc* 296(14): 1731–1732.

Haddad FS (2004). Ibn Zuhr and experimental tracheostomy and tracheotomy. *J Am Coll Surg* 199(4): 665.

Hall ED, Yonkers PA, McCall JM, Braughler JM (1988). Effects of the 21-aminosteroid U74006F on experimental head injury in mice. *J Neurosurg* 68(3): 456–461.

Hu A, Wilson T, Ladak H, Haase P, Fung K (2009). Three-dimensional educational computer model of the larynx: voicing a new direction. *Arch Otolaryngol Head Neck Surg* 135(7): 677–681.

Huh D, Matthews BD, Mammoto A, Montoya-Zavala M, Hsin HY, Ingber DE (2010). Reconstituting organ-level lung functions on a chip. *Science* 328(5986): 1662–1668.

Jeon JS, Bersini S, Gilardi M, Dubini G, Charest JL, Moretti M, Kamm RD (2015). Erratum: Human 3D vascularized organotypic microfluidic assays to study breast cancer cell extravasation. *Proc Natl Acad Sci U. S. A.* 112: 214.

Kehind EO (2013). They see a rat, we seek a cure for diseases: the current status of animal experimentation in medical practice. *Med Princ Pract* 22(1): 52–61.

Kilkenny C, Browne WJ, Cuthill IC, Emerson M, Altman DG (2010). Improving bioscience research reporting: the ARRIVE guidelines for reporting animal research. *PLoS Biol* 8(6): e1000412.

King AJ (2012). The use of animal models in diabetes research. *Br J Pharmacol* 166(3): 877–894.

Litynski GS (1997). Laparoscopy-the early attempts: spotlighting Georg Kelling and Hans Christian Jacobaeus. *J Soc Laparoendosc Surg* 1(1): 83–85.

Louw A, Diener I, Fernández-de-Las-Peñas C, Puentedura EJ (2017). Sham surgery in orthopedics: a systematic review of the literature. *Pain Med* 18(4): 736–750.

Maehle AH, Tröhler U (1987). Animal experimentation from antiquity to the end of the eighteenth century: Attitudes and arguments. In: Rupke NA, editor. *Vivisection in Historical Perspective*. Croom Helm, London, UK.

Mak IW, Evaniew N, Ghert M (2014). Lost in translation: animal models and clinical trials in cancer treatment. *Am J Trans Res* 6(2): 114–118.

Medical Council of India, New Delhi, Amendment Notification of 8 July 2009 to the Minimal Standard Requirements for Medical Colleges with 150 Admissions Annually, Regulations (1999). www.mciindia.org/helpdesk/how_to_start/STANDARD%20FOR%20150.pdf.

Miziara ID, Magalhães ATM, Santos MA, Gomes EF, Oliveira RA (2012). Research ethics in animal models. *Braz J Otorhinolaryngol* 78(2): 128–131.

Musah S, Mammoto A, Ferrante TC, Jeanty SSF, Hirano-Kobayashi M, Mammoto T, Roberts K, Chung S, Novak R, Ingram M, Fatanat-Didar T, Koshy S, Weaver JC, Church GM, Ingber DE (2017). Mature induced-pluripotent-stem-cell-derived human podocytes reconstitute kidney glomerular-capillary-wall function on a chip. *Nat Biomed Eng* 1: 0069.

National Research Council (US) Committee for the Update of the Guide for the Care and Use of Laboratory Animals (2011). *Guide for the Care and Use of Laboratory Animals* [Internet]. Eighth ed. National Academies Press (US), Washington, DC. www.ncbi.nlm.nih.gov/books/NBK54050/

Oparin AI (1957). *The Origin of Life on the Earth*. Academic Press, New York.

Pasca SP (2018). The rise of three-dimensional human brain cultures. *Nature* 553(7689): 437–445.

People for the Ethical Treatment of Animals (PETA) (2013). http://en.wikipedia.org/wiki/People_for_the_Ethical_Treatment_of_Animals.

Perel P, Roberts I, Sena E, Wheble P, Briscoe C, Sandercock P, Macleod M, Mignini LE, Jayaram P, Khan KS (2007). Comparison of treatment effects between animal experiments and clinical trials: systematic review. *BMJ* 334(7586): 197.

Petroianu A (1996). Aspectos éticos na pesquisa em animais. *Acta Cir Bras* 11(3): 157–164.

Pound P, Ebrahim S, Sandercock P, Bracken MB, Roberts I (2004). Where is the evidence that animal research benefits humans? *BMJ* 328(7438): 514–517.

The Recommendations of the Sub-Committee on Rehabilitation of Animals after Experimentation set up by CPCSEA (2006). http://envfor.nic.in/divisions/awd/Rehabiliaion.

Regulations on Graduate Medical Education, 1997 (amended up to 2010). Medical Council of India. www.mciindia.org/Rules-and-Regulation/GME_REGULATIONS.pdf.

Rosa Mde O, Pereira JC, Grellet M, Alwan A (2003). A contribution to simulating a three-dimensional larynx model using the finite element method. *J Acoust Soc Am* 114(5): 2893–2905.

Russell WMS, Burch R L (1959). *The Principles of Humane Experimental Technique*. Reprinted 1992. Universities Federation for Animal Welfare. Methuen & Co. Ltd., London.

Sabin AB (1965). Oral poliovirus vaccine. History of its development and prospects for eradication of poliomyelitis. *J Am Med Assoc* 194(8): 872–876.

Seymour NE et al. (2002). Virtual reality training improves operating room performance. *Ann Surg* 236(4): 458–464.

Singh VP, Pratap K, Sinha J, Desiraju K, Bahal D, Kukreti R (2016). Critical evaluation of challenges and future use of animals in experimentation for biomedical research. *Int J Immunopathol Pharmacol* 29(4): 551–561.

Timmermans S (2003). A black technician and blue babies. *Soc Stud Sci* 33(2): 197–229.

University Grants Commission (2011). *Guidelines for Discontinuation of Dissection and Animal Experimentation in Zoology/Life Sciences in a Phased Manner*. www.gujaratuniversity.org.in/web/NWD/NewsEvents/8999_UGC%20Guidelines%20for%20Animal%20Dissection%20and%20Experimentation.pdf

Watts G (2007). Alternatives to animal experimentation. *BMJ* 334: 182.

17

Understanding of Adaptive Response: A Study from Animal Systems

V. Vasudev and H. P. Gurushankara

CONTENTS

DOI: 10.1201/9781003220398-17

17.1 Introduction

Due to rapid industrialization, modernization, and changing lifestyles, many chemical agents have continuously entered our environment. The use of fertilizers and pesticides in agriculture, drugs, cosmetics, and food additives in humans' day-to-day life releases effluents from industries; uncontrolled burning of fossil fuels by vehicles and mining contribute to the addition of exogenous chemicals into the environment. These, in turn, pollute the air, water, and soil. Most decompose very slowly, and their residues persist in our environment in various forms. Many of these detrimental chemicals find their way into the body of organisms through the food chain. They may lead to multiple harmful biological effects: abortions, congenital malformations, inherited diseases, immunological impairments, inefficient tissue repairs, premature aging, carcinogenesis, and/or cell death (Poplawski, 2019). Above all, all living organisms' DNA is continuously exposed to several exogenous and endogenous agents, eventually resulting in cytological and genetic damages; every cell experiences up to 10^5 spontaneous or induced DNA lesions per day (De Bont and van Larebeke, 2004). Estimated rates of DNA damage per human cell per day is 50,000 single-strand breaks (SSBs), 10,000 depurination, 600 deaminations, 2,000 oxidative base damages, 5,000 alkylated bases, 10 intra-strand cross-links, 10 double-strand breaks (DSBs), and total DNA-damaging events approximate to 70,000 lesions per day (Tubbs and Nussenzweig, 2017). It also said that the entire DNA-damaging events would be 2,500 per cell per hour! Furthermore, the genome of all organisms, including humans, is susceptible to damage caused by DNA-damaging agents. The DNA lesions caused due to interactions of chemical agents include chemical modification of bases or sugars, SSBs, DSBs, and inter- or intra-strand cross-links (Helleday et al., 2007). These lesions express as point mutations, sister chromatid exchanges (SCEs), chromosomal aberrations (CAs), genetic disorders, and/or cancer (Figure 17.1). Thus, there is a direct link between DNA damage and several disease conditions (Chatterjee and Walker, 2017).

Some lesions also cause cell cycle arrest leading to necrosis or apoptosis (da Silva, 2016), and mutations are the practical carcinogenesis mechanisms. If it continues through generations, the questions that come up are as follows: How can living organisms survive? How can genetic information be transmitted with high fidelity from one generation to the next? What are the extent of mutations and their effects? Considering the mutation rate in different organisms, including humans in other generations, one can hypothesize that all living cells must rectify mechanisms in chemically transformed DNA to maintain genetic integrity and high fidelity. Experiments have proven that several mechanisms are involved in repairing the damaged DNA in the living organisms responsible for maintaining genetic material's stability and integrity.

Further, repairing such DNA damages plays a vital role in cellular functions. These are highly conserved among bacterial, yeast, insect, and mammalian cells (Samson et al., 1991). These mechanisms are universal and ubiquitous processes essential for the survival of organisms. They act to prevent the

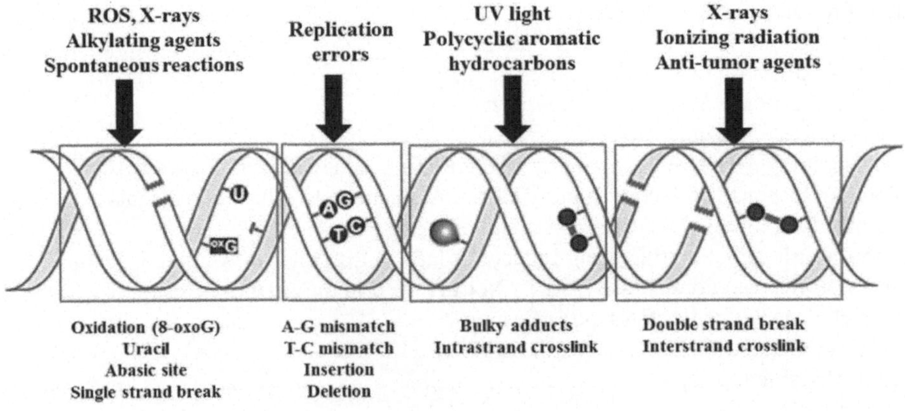

FIGURE 17.1 Structural alterations created by DNA-damaging agents.

consequent high mutation rate called DNA repair mechanisms or genetic repair mechanisms, which are built and are the gift of nature to the living cells that counteract genotoxicity.

17.2 Types of DNA Damages

Tautomerization

The changed pairing qualities of the bases (pairing purine with purine and pyrimidine with pyrimidine) are due to a tautomerism phenomenon. Tautomers are the alternate forms of bases produced by rearrangements of electrons and protons in the molecules, and the process is called tautomerization. These rearrangements allow for hydrogen bonding of mismatched base pairs. Tautomeric shifts, spontaneous rearrangements of the nitrogenous bases, are caused by certain chemical mutagens. In the replication, purines pair with pyrimidines, and the base pair is altered at a particular locus. The distinctive forms are unstable, and the cycle reverts to its typical forms at the subsequent replication. Due to tautomerization, the amino (-NH_2) group of cytosine and adenine is converted into the imino (-NH) group, and likewise, the keto (C=O) group of thymine and guanine is converted into enol (-OH) group. Tautomeric thymine pairs with normal guanine and cytosine with adenine (Figure 17.2). Such pairing of nitrogenous bases is known as forbidden or unusual base pairs (Steven and William, 2018).

Base Substitutions (Replacements)

Base substitutions result from a change of one nucleotide base in the DNA sequence. They are also called point mutations, as there is a change in the DNA sequence at one specific point. It is further divided into three subtypes—transition, transversion, and frameshift. In transition, a purine (adenine or guanine) or a pyrimidine (cytosine or thymine or uracil) in triplet code of DNA or mRNA is replaced by its type, i.e., a purine replaces purine and pyrimidine replace pyrimidine. Transversions are substitution gene mutations in which a purine (adenine or guanine) is replaced by pyrimidine (thymine or cytosine) or vice versa (Figure 17.3). Transitions are more common than transversions because DNA repair enzymes recognize the wrong

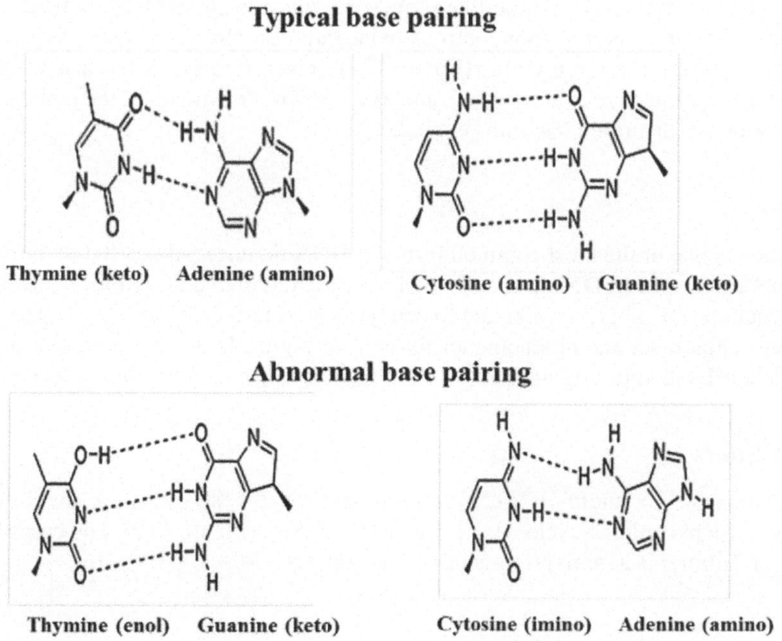

FIGURE 17.2 Normal and abnormal pairing of bases.

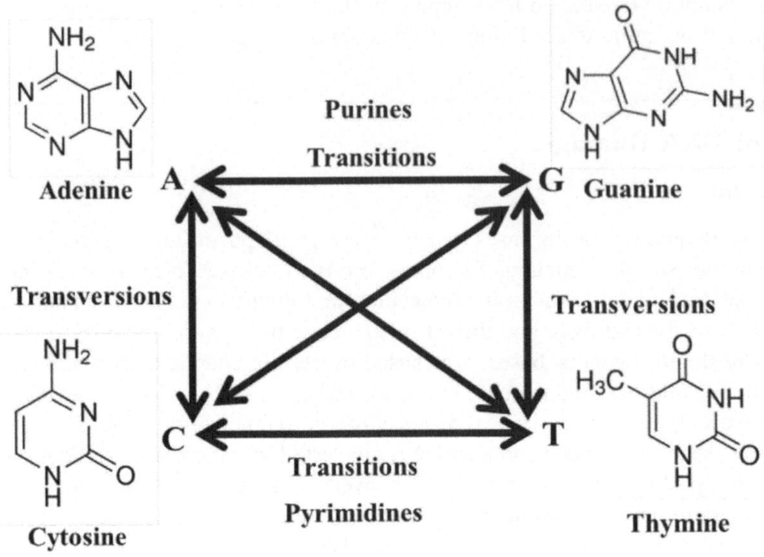

FIGURE 17.3 Transitions and transversions.

insertion representing a transition better than a transversion. In the frameshift type of gene mutation, a single nitrogenous base is added or deleted. None of the codons remains in the same original position, and the reading of the genetic code is shifted laterally either in the forward or backward direction (Maki, 2002).

Deamination

It is a spontaneous base degradation mechanism. Some chemicals, like nitrous oxide, deaminate (removal of NH_2, or amino group) the nitrogen base and change DNA codon. Nitrous oxide deaminates adenine to form hypoxanthine, which has guanine-like properties. A-T pairing in DNA molecule is replaced by G-C pairing, thus causing gene mutation. Nitrous oxide can also change cytosine to uracil, guanine to xanthine, and 5-methylcytosine to thymine (Figure 17.4). It has been estimated that spontaneously 100 uracils per haploid genome per day are formed, and as a result of deamination, the replication, transcription, and translation are disturbed (Griffiths et al., 2000).

Depurination

Base depurination is one of the most common forms of DNA damage. About 10,000 purine bases (adenine or guanine) are lost from DNA per cell per day in mammalian cells. In spontaneously occurring depurination reactions, the N-glycosyl bound to deoxyribose is broken by hydrolysis, leaving the DNA's sugar-phosphate chain intact and producing an abasic site (Figure 17.5). The average rate of loss at normal physiological pH and ionic strength, at 37 °C, is approximately 3×10^{-11}/sec (Clancy, 2008).

Pyrimidine Dimers

Ultraviolet radiations (UV), mainly UV-C (<280 nm) and UV-B (280–320 nm), are absorbed by DNA and commonly cause pyrimidine cyclobutyl T-T and T-C dimers (Figure 17.6). The bases become cross-linked, and the T-T dimer is a more prominent one (Kuzminov, 2013)

Oxidation

Oxidation is a process in which free radicals of oxygen damaged bases. Nuclear and mitochondrial DNA (mtDNA) is attacked by reactive oxygen species (ROS), but mtDNA is affected more because it does not

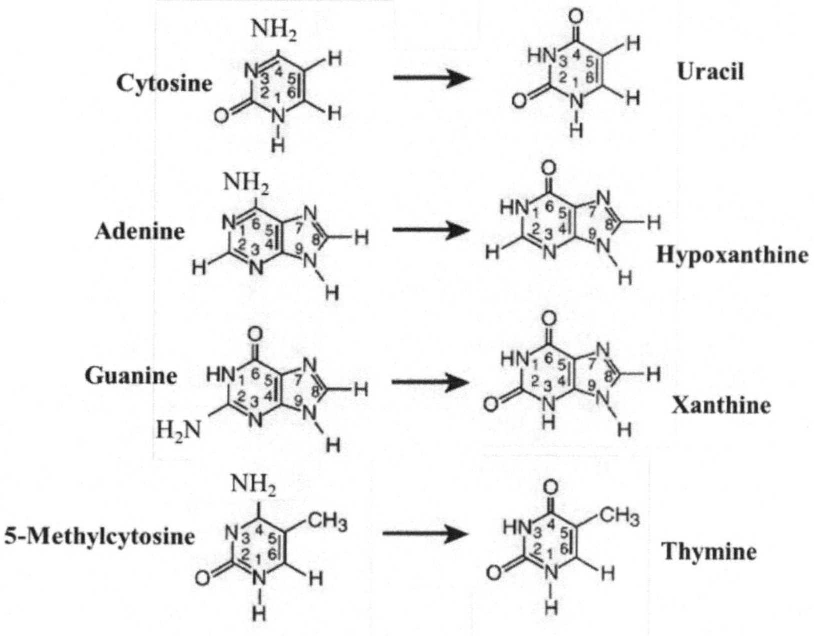

FIGURE 17.4 Deamination of bases.

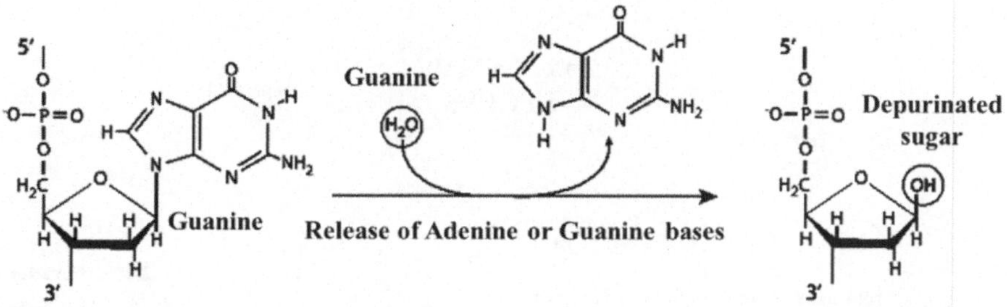

FIGURE 17.5 Depurination of guanine (or adenine).

contain histones; it is in proximity to the ROS generation limited antioxidant repair system. DNA strand breaks, oxidized base adducts, DNA-DNA, and DNA-protein cross-links (Cooke et al., 2003) are formed by ROS (Figure 17.7).

Alkylation

It is the addition of alkyl groups to the bases or backbone of DNA. Some cause mutations by mispairing during replication or stopping polymerase altogether. Alkylating agents are a group of chemicals that induce mutations in most model organisms. They are the oldest class of anticancer drugs still commonly used to treat cancer. These are potent cytotoxic, mutagenic, and carcinogenic agents. Three categories of alkylating agents are grouped depending on functional groups: (1) Monofunctional (e.g., ethyl methanesulfonate [EMS], methyl methanesulfonate [MMS]), (2) bifunctional (e.g., diepoxybutane [DEB], endoxan), and (3) polyfunctional (e.g., trenimone, triethylene melamine [TEM]). Monofunctional agents are more extensively used to induce point mutations than the other agents because the latter agents induce intra- and inter-strand cross-links. Further, the majority of monofunctional agents cause a dozen DNA lesions. They are adenine (N-1, N-3, N-7), guanine (N-3, N-7, O^6), thymine (O^2, N-3, O^4), cytosine

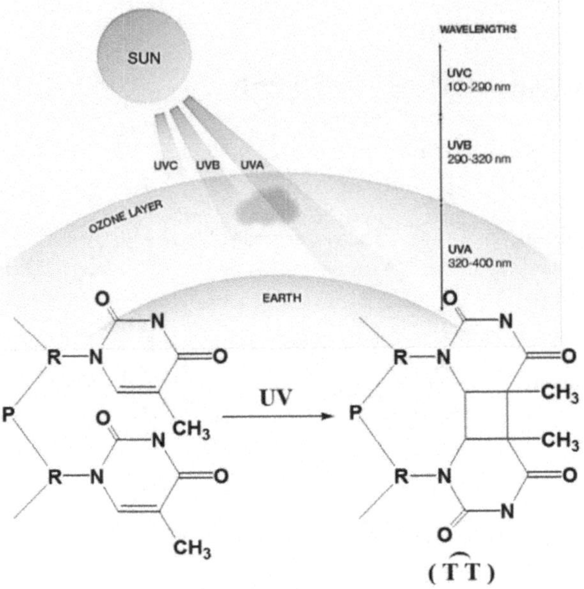

FIGURE 17.6 Dimerization.

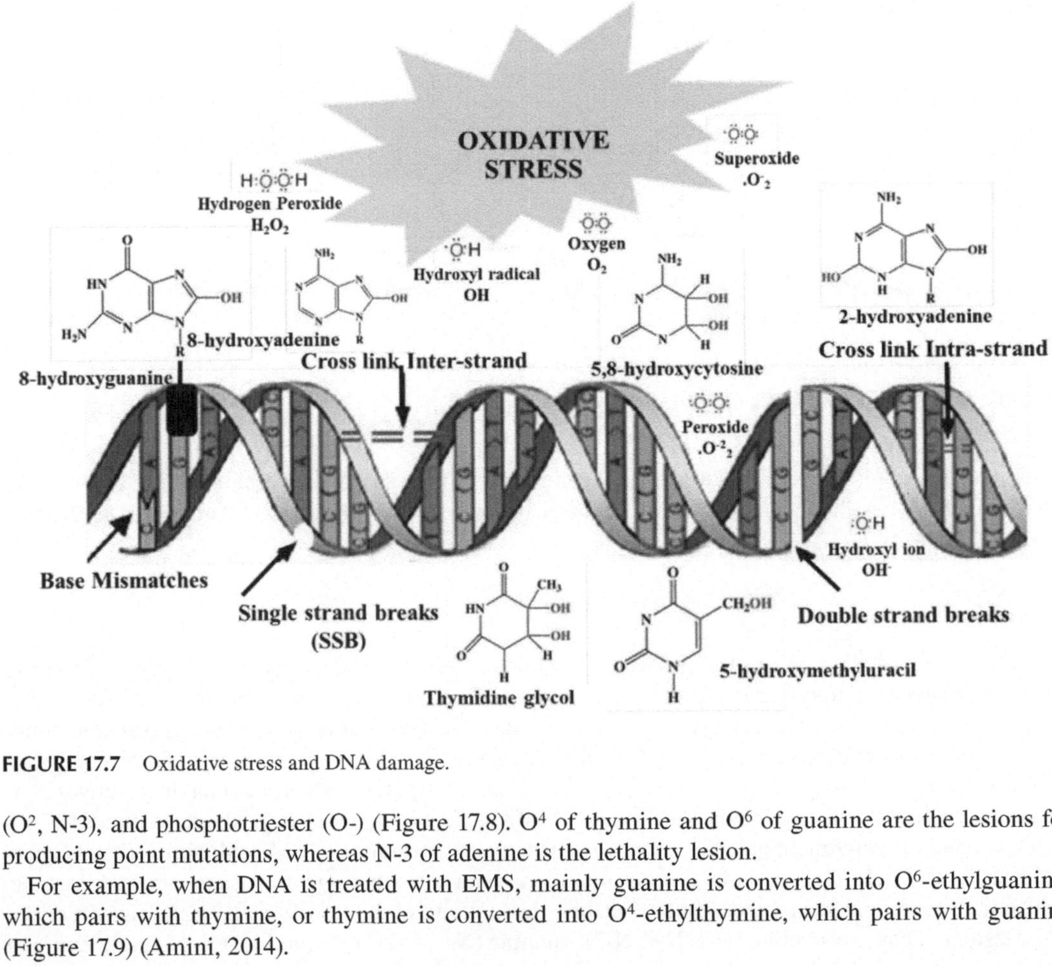

FIGURE 17.7 Oxidative stress and DNA damage.

(O^2, N-3), and phosphotriester (O-) (Figure 17.8). O^4 of thymine and O^6 of guanine are the lesions for producing point mutations, whereas N-3 of adenine is the lethality lesion.

For example, when DNA is treated with EMS, mainly guanine is converted into O^6-ethylguanine, which pairs with thymine, or thymine is converted into O^4-ethylthymine, which pairs with guanine (Figure 17.9) (Amini, 2014).

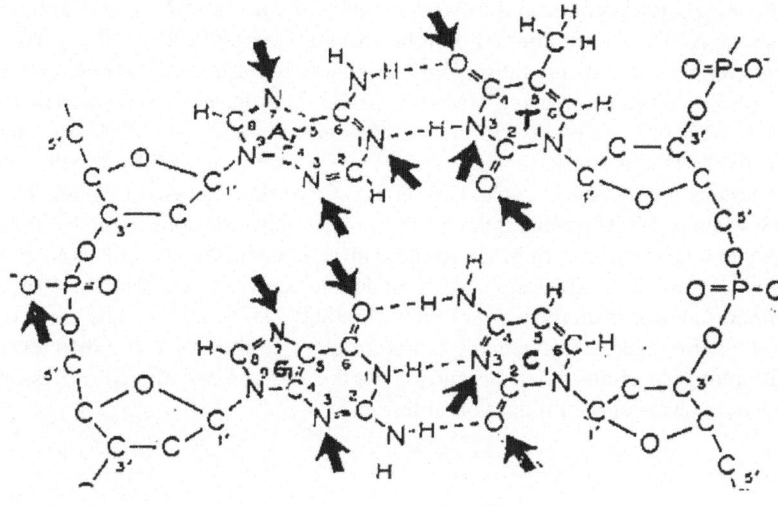

FIGURE 17.8 Sites of alkylation.

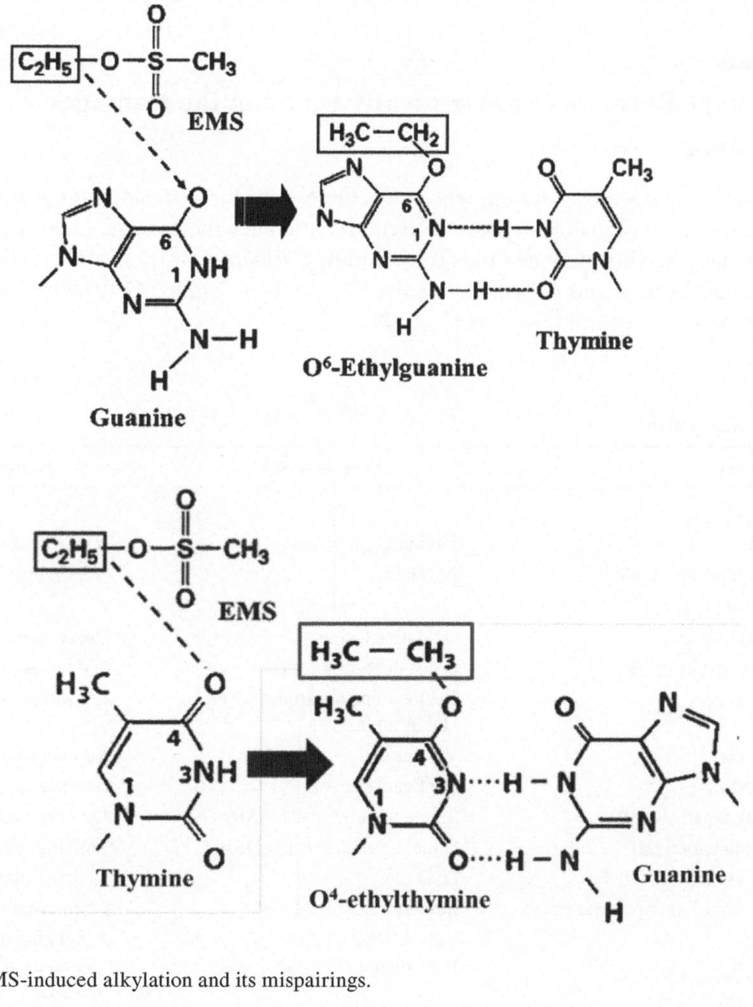

FIGURE 17.9 EMS-induced alkylation and its mispairings.

Singer and Grunberger (1983) conducted *in vitro* experiments using DNA from salmon sperm, calf thymus, salmon testes, rat liver, brain, human fibroblasts, and liver cells. They also performed *in vivo* experiments using mammalian cells in culture (V79, human fibroblasts, and liver cells) to get a picture of relative amounts of DNA alkylation using MMS and MNU. They have demonstrated that N-7 guanine lesions are the highest (83) induced by MMS *in vitro* and *in vivo* systems, and MNU is 67 and 69 *in vitro* and *in vivo*, respectively. Furthermore, 0.3 guanine O^6 lesions are produced by MMS, whereas MNU induces 6.3 and 5 lesions *in vitro* and *in vivo.* Conversely, O^4 of thymine is not detected when DNA is treated with MMS, whereas MNU produces 0.4 lesions *in vitro.* These lesions in the DNA induced point mutations. Contrary to this, treating with MMS resulted in 10.4 and 8.7 lesions of N-3 adenine, and similarly, 9 lesions *in vitro* and 4.2 *in vivo* lesions of N-3 adenine have been demonstrated by using MNU. This lesion is an important one from the point of view of lethality. When the results are pooled together, it is evident that alkylating agents produce alkylation products in the DNA, but different methylating agents produce different proportions of lesions and are not the same. Hence, there is necessary to analyze each alkylating agent in terms of the production of lesions.

17.3 Overview of Genetic Repair Mechanisms

Genetic repair mechanisms are operative in living systems and are divided by the type of repair they perform in the cells. Broadly they are divided into (1) damage reversal and (2) damage removal. The genotoxicants cause lesions, and their repair systems are provided in Table 17.1.

17.4 (A) Damage Reversal (Enzymatically Undoing the Damage)

Photoreactivation

Photo-reactivation is an operating system when cells are exposed to UV radiation, which produces thymine dimers. When an organism is exposed to light, it stimulates the enzyme photolyase production at a particular wavelength. This enzyme binds to pyrimidine dimers, catalyzes a photochemical reaction, breaks the cyclobutane ring, and reforms two adjacent thymines (Figure 17.10). *phr A* and *phr B* genes encode the enzyme's two subunits (Reef et al., 2009).

TABLE 17.1

Genetic Repair Mechanisms

DNA repair Systems	Type of lesion	Genotoxic source
(A) Damage reversal		
Photo-reactivation	Thymine dimmers	UV radiation
Alkyltransferases (removal of alkyl groups)	Alkylation	Alkylating agents
(B) Damage removal		
Base excision repair (BER)	Oxidative lesions	Reactive oxygen species (ROS)
Nucleotide excision repair (NER)	Helix-distorting lesions	UV radiation
uvrABC excision repair	Cyclobutene pyrimidine dimers, DNA adducts	UV radiation
Translesion synthesis	Various lesions	Various sources
Mismatch repair (MMR)	Replication errors	Replication
Single-strand-break repair (SSBR)	Single-strand breaks (SSB)	Ionizing radiation, ROS
Homologous recombination (HR)	Double-strand breaks (DSB)	Ionizing radiation, ROS
Non-homologous end joining (NHEJ)	DSB	Ionizing radiation, ROS
DNA inter-strand cross-link repair pathway	Inter-strand cross-links	Chemotherapy
Adaptive response	SSB or DSB Base damages	Alkylating agents, heat stress, oxidants, radiation, and heavy metals

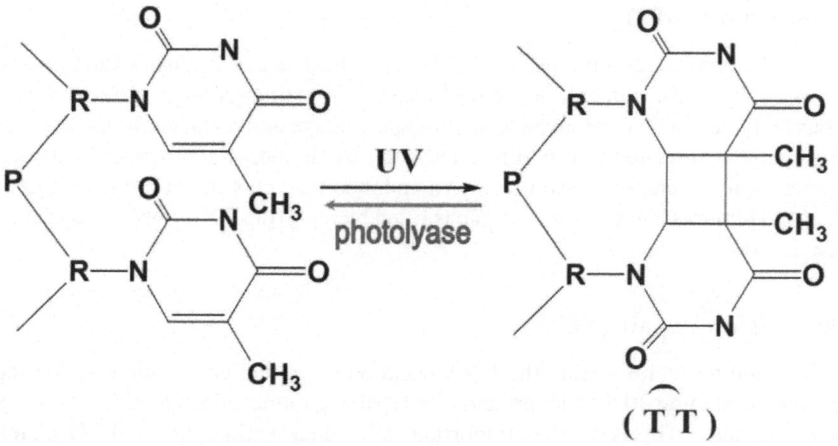

FIGURE 17.10 Photoreactivation of the thymine dimers.

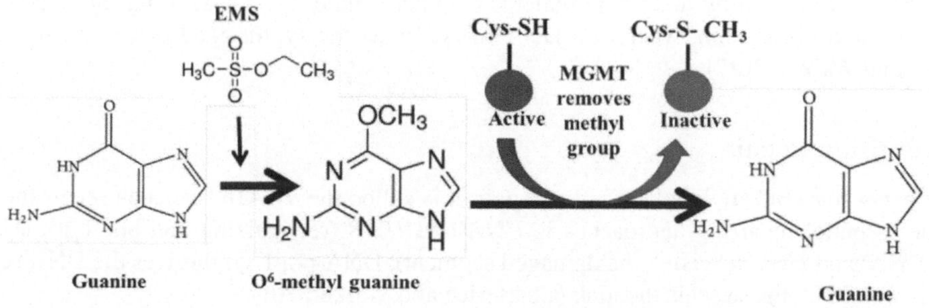

FIGURE 17.11 Methylation of guanine and repairing it by methylguanine-methyltransferase (MGMT) recognizing O6-methylguanine in DNA and removing the methyl group normalcy is maintained.

Alkyltransferases (Removal of Alkyl Groups)

Alkylation products of the DNA induced by the alkylating agents are removed by an enzyme system, maintaining the DNA's normalcy without breaking the DNA backbone. Uracil, a cytosine deamination product, is detected, removed by uracil N-glycosylase, and replaced by cytosine. Hypoxanthine is an adenine deamination product that is recognized by the hypoxanthine N-glycosylase enzyme and then replaced by adenine. Methylguanine produced by methylation is repaired by the methyl group's enzymatic transfer by methylguanine-DNA-methyltransferase (MGMT) and restores the guanine (Figure 17.11). This enzyme also repairs 1-methyl adenine and 3-methyl cytosine (Pegg, 1990).

17.5 (B) Damage Removal

The repair enzymes first act on the damaged DNA by removing the damaged part. The first and foremost explained method in the damage removal mechanism is excision repair.

Excision Repair

As the word indicates, excision is to cut the damaged part of DNA. There are three types of excision repairs: base excision repair (BER), nucleotide excision repair (NER), and mismatch repair (MMR). These repair mechanisms can also work during dark conditions and are called dark repair mechanisms.

Base Excision Repair (BER)

This process is called base excision repair if the damaged base is cut. It repairs small, non-bulky DNA lesions, such as methylated, oxidized, or reduced bases. It is estimated to occur 20,000 times a day in each cell in our body; the same is removed from its sugar linkage and replaced. In BER, a defective DNA in which cytosine is deaminated to uracil is acted upon by the enzyme uracil DNA glycosylase. This results in the removal of defective base uracil. An endonuclease cuts the backbone of the DNA strand near the defect and removes a few bases. The gap is filled by repairing DNA polymerase action and ligase seal (Wallace, 2014).

Nucleotide Excision Repair (NER)

Removal of damaged nucleotides from the DNA backbone is called nucleotide excision repair (NER). The complex unit containing 30 distinct proteins for repairing damaged nucleotides is known as nucleotide excision repairosome. This is the most important DNA repair pathway for bulky DNA lesions, which creates a block to DNA replication and transcription. The steps are (1) double incision of the damaged strand several nucleotides away from the damaged site by an endonuclease on both sides; (2) removal of the short segment (about 24 to 32 nucleotides) containing the damaged region by an exonuclease; (3) filling in the resulting gap by a DNA polymerase, which synthesizes DNA using the opposite strand as a template; and (4) ligation, wherein a DNA ligase binds the synthesized piece into the backbone (Chatterjee and Walker, 2017).

*uvr*ABC Excision Repair

When the genes *uvr*ABC are involved in the repair, it is called the *uvr*ABC excision repair mechanism. One of the essential repairing approaches is *UvrA* and *UvrB* (recognizing and binding the damaged region of DNA) and *UvrC* (excising the damaged segment); DNA-pol 1 synthesizes the DNA segment to fill the gap, and DNA ligase seals the nick (Chatterjee and Walker, 2017).

Mismatch Repair

This process occurs after DNA replication as a last "spell-check" on its accuracy. Although high accuracy in replication, DNA polymerase III actions, including proofreading and exonuclease, result in one misincorporation per 10^8 bases. Mismatch repair reduces this rate to one change in every 10^{10} or 10^{11} bases. In other words, the mismatched bases in DNA, for example, G-T or A-C base pairs, are repaired. In mismatch repair, the coordinated action of three genes—*MUT S, MUT L*, and *MUT H*—are involved. *MUT S* recognizes the mismatched heteroduplex (TG). *MUT L*, a dimer, in ATP's presence, binds to *MUT S* heteroduplex complex to activate *MUT H*. This *MUT H* is an endonuclease that cleaves 5′ to the G in an unmethylated GATC, which leaves a nick. *MUT H* translocates and cuts 3′ to G: T mismatch, then *Uvr D* helicase removes the DNA with the mismatch. DNA pol III fills this gap, and DNA ligase seals the nick (Fishel, 2015).

Recombination Repair

Double-strand break (DSB) repair was researched in the Holliday junction, where homologous recombination was observed in all organisms, such as bacteria, archaea, and eukarya (Haber et al., 2004). The large segment of damaged DNA in bacteria is primarily repaired by employing the RecBCD pathway (Dillingham and Stephen, 2008). In this pathway, the RecBCD complex binds to the blunt end of the DSB. The complex can pause upon recognizing the Chi site 5′-GCTGGTGG-3′. RecBCD loads RecA protein onto the nicked strand 5′ of the Chi site and the complex disassembles. DNA/RecA nucleoprotein performs strand invasion on the identified homologous sequence, forming a displacement loop (D-loop). Conversion to a Holliday structure leads to repair by crossing over. The resolution of the Holliday structure is carried out by the RuvABC complex (Smith, 2012). Eukaryotes typically use either DSBR or synthesis-dependent strand annealing (SDSA). Shortly after a DSB occurs, the MRN (*Mre11, Rad50,* and *Nbs1*) complex initiates repair. It leads to forming 3′ overhangs coated in *Rad51,* the *RecA* homolog. The invading 3′ strand is extended by DNA synthesis, forming the Holliday structure. In DSBR, the non-invading DNA/Rad51 nucleoprotein is base-paired by the newly synthesized strand and restored by DNA polymerase/ligase. Double Holliday structures were produced. In SDSA, only one Holliday structure is formed by the invasion of the nucleoprotein. The top strand is synthesized and ligated, resulting in a non-crossover homologous repair (Savir and Tlusty, 2010).

17.6 Inducible Repair Processes

Error-Prone Repair

It is also known as SOS repair. Two genes control SOS (save our soul): *Rec A* and *Lex A*. Hence, it is also called a *Lex-Rec* repair system. The SOS response is a conserved regulatory mechanism induced when bacterial cells are exposed to stress conditions that cause DNA damage. This process is also known as the inducible repair process. *Rec A* and *Lex A* fall into the category of induction occurring within five minutes of DNA damage. These processes culminate in the mutagenic polymerization of DNA past a lesion. In essence, genetic integrity is sacrificed for cell survival. A network of more than 30 proteins (SOS gene products) is involved in these mutagenic bypass events. The Lex A repressor protein regulates the expression of the corresponding genes. *Rec A* is induced too much higher levels under SOS conditions. Excessive DNA damage leads to complete blockage of replication. *Rec A* binds to these damages and undergoes autocatalytic proteolysis. This stimulates cleavage and inactivation of *Lex A* as a repressor. This results in decreased active *Lex A* and a contrary increase in *Rec A* gene expression. Protein products of *umu C, umu D, and din B* are coordinated to increase processivity and decrease accuracy (Rodgers and McVey, 2016).

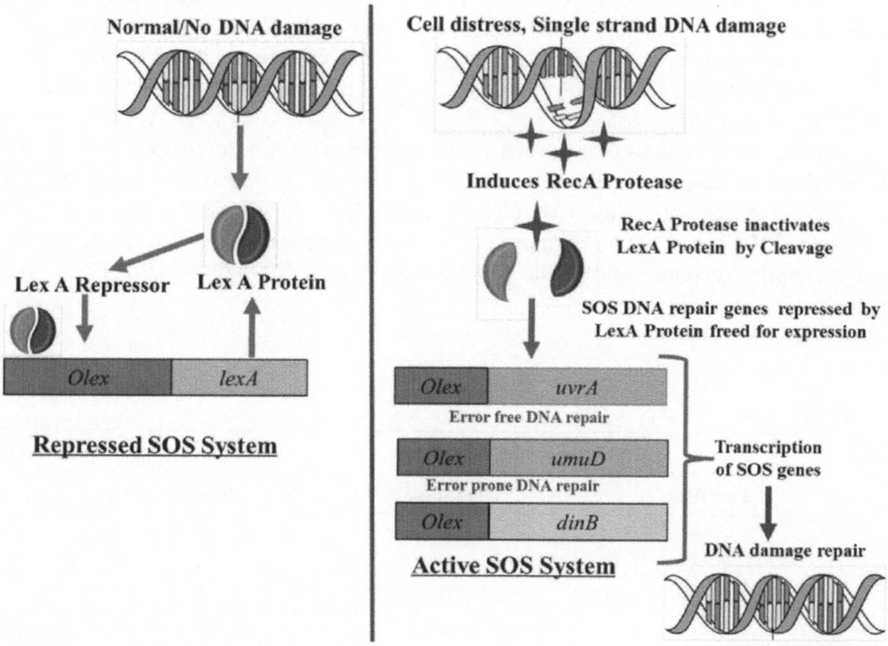

FIGURE 17.12 SOS repair mechanism.

17.7 Perspectives in Adaptive Response

Error-Free Repair Mechanism

It is an inducible repair process because it also requires stimulating a particular set of genes to induce repair. For the first time, Hillova and Drasil (1967), working with the *Chlamydomonas reinhardii* under the stress condition, proposed the name adaptive response. It has not attracted the attention of scientists for another 10 years. In 1977, Samson and Cairns discovered this mechanism when working with a monofunctional and anticancerous alkylating agent, N-methyl-N-nitro-N-nitrosoguanidine (MNNG), in *E. coli.* They named this mechanism once again as an adaptive response. They defined it as "when the cells are continuously exposed to nontoxic levels of simple alkylating agent MNNG, it protects the bacteria against the killing and mutagenic effects of high toxic doses of the same agent." After the continuous experimentations, they hypothesized that a protein responsible for the repair (error-free) of DNA mutagenic insults was induced by exposure to a low MNNG level. Hence, it is also termed an inducible repair process. The survival of pretreated cells was enhanced several folds, and the yield of induced mutations was markedly reduced (Table 17.2). This adaptive response was entirely different from the SOS repair pathway. This repair system makes the cells less mutable, and hence, it is error-free. The adaptive response, however, also protects the cells from lethal damage. Subsequent works confirmed that this explanation is indeed correct.

After Samson and Cairns's (1977) work, several microorganisms were tested for an adaptive response, which led to increased resistance to toxic and mutagenic effects of alkylating agents after treatment with sublethal doses. The adaptive response has been demonstrated in *Caenorhabditis elegans*, *Bacillus subtilis*, *B. thuringiensis*, *B.lichenifonnis*, *B.amyloliquefaciens*, *Vibrio cholera*, etc. (Chankova et al., 2007). On the other hand, the non-presence of adaptive response is displayed by *Haemophilus influenza* (Kimball, 1980), *Salmonella typhimurium* (Ather et al., 1984), and *Saccharomyces cerevisiae* (Chen and Samson, 1991). The adaptive response was also absent in some radiosensitive cancer cell lines and ataxia-telangiectasia patients' cells (Lambinet al., 1994). Furthermore, alkylating agents, heat stress, oxidants, UV irradiation, ionizing radiation (IR), and heavy metals induce an adaptive response. This phenomenon is an evolutionarily conserved cellular defense mechanism observed in various organisms, from prokaryotes to eukaryotes. Reports are also available on adaptive response using alkylating agents *in vitro* mammalian systems, such as Chinese hamster ovary cells, V79 Chinese hamster cells, human lymphocytes, human fibroblasts, human keratinocytes, rat liver cells, and rat lymphocytes (Dimova et al., 2008; Laval and Laval, 1984; Shadley and Wolff, 1987; Bosi and Olivieri, 1989; Sankaranarayanan et al., 1989; Wolff, 1996).

For the first time in the human opportunistic pathogen, *Aspergillus fumigates*, a study has demonstrated that it elicits an adaptive response to sublethal doses of the mono-functional alkylating agent N-methyl-N′-nitro-N-nitrosoguanidine (MNNG) (Figure 17.13). Identified the genes encoding methylphosphotriester (MPT) and O^6-alkylguanine DNA alkyltransferase (AGT), DNA repair proteins abolishing the adaptive response and sensitizing the organism to MNNG. *In vitro* DNA repair assays

TABLE 17.2

Adaptive Response to MNNG in *E. Coli*

	MNNG (μg/ml)		Adaptive response	
Strain	**Low dose**	**High dose**	**Killing**	**Mutation**
WP2	0.75	100	+	+
WP2s	0.75	100	+	+
CM561	0.1	10	+	+
CM611	0.1	10	+	+
CM571	0.1	5	+	+

(*Source*: Data from Samson and Cairns, 1977; Samson et al., 1991)

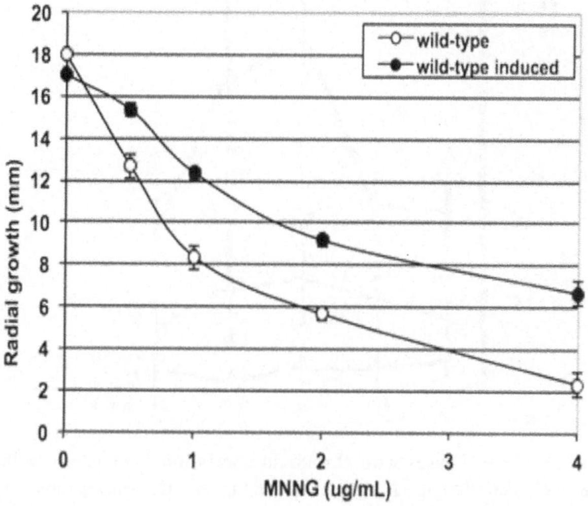

FIGURE 17.13 Adaptive response to the alkylating agent, MNNG in *A. fumigatus.*

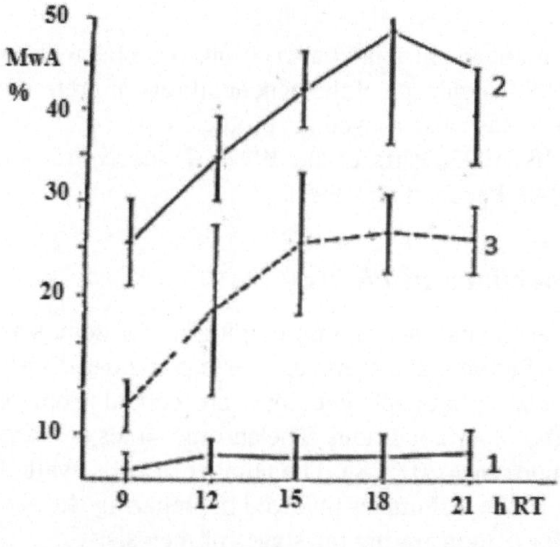

FIGURE 17.14 The frequency of metaphases with chromatid aberrations (%) after conditioning treatment with a low TEM concentration and 2 h later, a challenging treatment with a higher concentration of TEM. Each data point represents the mean percentage of metaphases with chromatid aberrations. The vertical lines represent the standard error of the mean. RT (recovery time [h]). (1) 30 min 10^{-5} M TEM, (2) 30 min 10^{-4} M TEM, and (3) 30 min 10^{-5} M TEM—2 h RT—30 min 10^{-4} M TEM.

confirmed MPT and AGT's ability to repair methylphosphotriester and O^6-methylguanine lesions in *A. fumigates* (O'Hanlon et al., 2012).

Overnight exposure of *A. fumigatus* conidia to MNNG (0.5 mg/ml; adapted) allowed subsequent growth on elevated concentrations of MNNG (0.5–4.0 mg/ml, $p < 0.001$) compared to an unadapted control. The radial growths of colonies were measured after 72 h. The mean ± standard error of three experiments is shown. Key: ***p<0.001 (Reproduced from O'Hanlon et al., 2012).

Rieger and coworkers (1982, 1984, 1993), using alkylating agents in *Vicia faba*, demonstrated an adaptive response phenomenon. They termed this protective process as clastogenic adaptation. They defined it as reduction by pretreatment with low (conditioning) clastogen doses of the yield of chromatid aberrations induced by a subsequent "challenge" treatment separated from conditioning by particular time intervals (Figures 17.14 and 17.15) (Rieger et al., 1984). Reduction of maleic hydrazide (MH) or

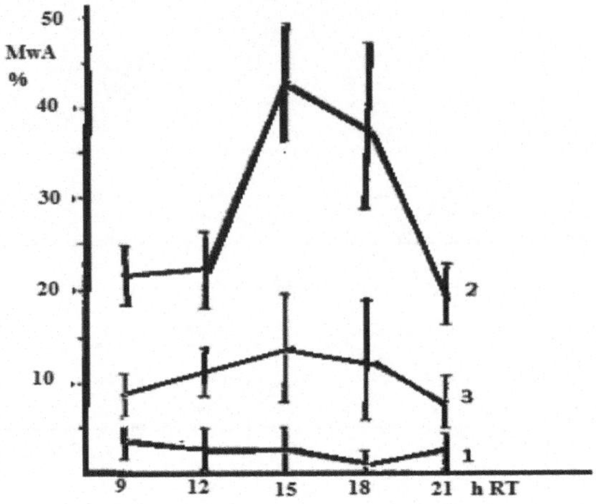

FIGURE 17.15 The frequency of metaphases with chromatid aberrations (%) after conditioning treatment with a low concentration of MH and 2 h later, a challenging treatment with a higher MH concentration. Each data point represents the mean percentage of metaphases with chromatid aberrations. The vertical lines represent the standard error of the mean. RT (recovery time [h]). (1) 30 min 5×10^{-5} M MH, (2) 30 min 4×10^{-4} M MH, and (3) 30 min 5×10^{-5} M MH—2 h RT—30 min 4×10^{-4} M MH.

triethylenemelamine (TEM)-induced chromatid aberrations have been observed in *V. faba* pre-exposed to heat shock (Rieger et al., 1988), evidence of the beneficial role of pretreatment of heat shock in reducing DNA damages. Researchers have also proved the protective process in other plants, such as *Hordeum vulgare* (Rieger and Michaelis, 1993; Patra et al., 1997), *Tradescantia* (Veleminsky et al.,1983), and *Allium cepa* (Cortes et al., 1990; Panda et al., 1997).

Adaptive Response in *Poecilocerus Pictus*

Grasshoppers are classic experimental material for cytologists for demonstrating meiotic chromosome-related research works since Sutton's discovery of spermatological divisions in 1902. Especially the grasshopper *Poecilocerus pictus*, a monophagous pest on the Calotropis plant, is easy to handle and economical to maintain in laboratory conditions. The chromosomes are large and relatively few in number: male: 2n = 19 (X0); female: 2n = 20 (XX). The number of cells available for scoring chromosomal anomalies is high. Nonetheless, the techniques involved in preparing slides of this material are quick and simple; therefore, it is ideal for demonstrating the stages of meiosis.

Furthermore, the abundance of meiotic cells in all stages is an asset of grasshoppers over mammalian test systems. It is also interesting to note that this species' natural populations occur in their habitat as isolated pockets. Hence, inbreeding is very high, resembling an isogenic state, if not exactly achieving it like laboratory breed animals. *P. pictus* has been employed as a sub-mammalian *in vivo* system to analyze the adaptive response in meiotic cells with all the previously mentioned points. This is the first report on this animal and the first on *in vivo* systems' meiotic cells.

For the first time, Riaz Mahmood and Vasudev (1990), using this animal system, have demonstrated adaptive responses in meiotic cells. They employed two monofunctional alkylating agents, EMS and MMS. Meiotic cells were exposed to a low dose of 0.03 M EMS or 0.012 M MMS and found to be resistant to the induction of chromosomal anomalies by a subsequent challenge dose of 0.12 M EMS or 0.048 M MMS as compared to cells that were not exposed or challenged with high dose (Table 17.3). They have also used different time lags such as 2 h and 30 min (Riaz Mahmood and Vasudev, 1992) between conditioning and challenging doses to understand the influence of time lag on adaptive response. From these experiments, they termed this adaptive response phenomenon protective processes. They defined it as "treatment with a low concentration of a clastogen before treatment with a high clastogen concentration may render the genetic system less responsive to the high concentration concerning the yield of

TABLE 17.3

Percentage of Chromosomal Anomalies After Challenging and Combined Treatments of Meiotic Cells of *P. Pictus* at Different Recovery Times (RT) with a Two-Hour Time Lag

Recovery time (h)	Mutagen	Challenge treatment	Combined treatment *
12	EMS	49.78	17.71
	MMS	92.16	79.65
24	EMS	43.86	39.6
	MMS	86.99	75.2
36	EMS	73.37	65.77
	MMS	94.28	84.80
48	EMS	84.82	41.5
	MMS	86.85	64.82

Control anomalies 4–6%; condition dose induced 10–15% anomalies. Percent anomalies of metaphase I (MI) and metaphase II (MII) and anaphase I (AI), and anaphase II (AII) are given for challenging and combined treatments.
* Significantly reduced compared to challenge treatment ($p < 0.05$).

(*Source*: Data from Riaz Mahmood and Vasudev, 1990)

induced genetic effects than the effects observed without conditioning" (Riaz Mahmood and Vasudev, 1990).

Results also demonstrated that EMS induced up to 66% reduction in chromosomal anomalies whereas MMS could reduce to only 15% at 12 h recovery time. The reduction is true for other recovery times. This clearly shows that the ethylating agent is more inducer of adaptive response than the methylating agent in *P. pictus*. This has to be proved with sufficient data by further research.

Adaptive Response in Mouse

The adaptive response phenomenon was shown in the Swiss albino mouse *in vivo* system using mitotic cells (Riaz Mahmood and Vasudev, 1991, 1993; Riaz Mahmood et al., 1996). Two monofunctional alkylating agents, EMS and MMS, were employed to investigate the induction of adaptive response (inducible protective processes) in Swiss albino mice's mitotic cells. Animals were treated with a low (conditioning) dose of EMS 80 mg/kg body weight (b.w.) or MMS 50 mg/kg b.w. and then challenged with a subsequent high (challenging) dose of EMS 240 mg/kg b.w. or MMS 150 mg/kg b.w. After different time lags (2, 5, 8, or 10 h), the yield of chromosomal aberrations in bone marrow cells was significantly reduced compared to the additive effects of both conditioning and challenging doses (Table 17.4). It appears that a low dose of EMS or MMS offered resistance to the mitotic cells against further clastogenic effects of any challenging dose of EMS or MMS employed. It is clear from the results that an adaptive response can also be encountered in mammalian *in vivo* systems. Furthermore, it is also observed that ethylating agent EMS is a poor inducer of adaptive response than its corresponding methylating agent MMS in the mouse's bone marrow cells. It is also very clear from the table that there is a slow repair activity with a peak at 8 h time lag and declining activity at 10 h.

Adaptive Response *in Human Lymphocytes*

Since the discovery of adaptive response in *E. coli*, the same phenomenon has been demonstrated in some mammalian strains or cell lines in culture using low doses of N-nitroso-N-methyl nitroguanidine (NMNG), employing different endpoints (Samson and Schwartz, 1980). The first report of radio-adaptive response was published by Olivieri et al. in 1984, where they exposed human lymphocytes to a low dose of ionizing radiation and subsequently treated them with a high dose of ionizing radiation. Of about 5, 7, 9, and 11 h after post-treatment, the yield of chromatid aberrations was less than the sum of the

TABLE 17.4

Percentage of Reduction of Chromosomal Aberrations in Mouse Bone Marrow Cells After Combined EMS or MMS Treatments with Various Time Lags (Between Conditioning and Challenging Doses) at Different Recovery Times (RTs)

		% Reduction**	
RTs (in h)	**Time lag (in h)**	**EMS**	**MMS**
24	2	−16.99*	27.64
	5	11.13	19.13
	8	35.90	47.72
	10	32.42	34.64
48	2	3.44*	27.1
	5	48.14	15.34
	8	54.18	24.65
	10	51.78	6.68
72	2	2.85*	14.19
	5	7.67	14.59
	8	12.34	24.74
	10	7.40	20.06

* Values are insignificant compared to additive effect ($p > 0.05$).
** Values are significant compared to additive effect ($p < 0.05$).

TABLE 17.5

Frequency of Sister Chromatid Exchanges (SCEs) in G1 Human Lymphocytes After Conditioning, Challenging, and Combined Treatments with EMS or MMS

	No. of SCEs/cells	
Treatments	**EMS**	**MMS**
Control	2.46 ± 0.08	2.46 ± 0.08
Conditioning (L)	8.58 ± 0.63a	20.12 ± 0.63a
Challenging (H)	21.78 ± 0.55a	43.28 ± 0.43a
L + 6h TL + H	15.17 ± 0.56b	31.54 ± 0.51b

a Significant compared to controls ($p < 0.01$).
b Significant compared to challenging concentration ($p < 0.01$).

(*Source:* Data from Harish et al., 1998)

yield of aberrations induced by high dose and low dose. They termed this phenomenon as an adaptive response to ionizing radiation and proposed that the low dose induced a chromosomal repair activity. Many publications in this area have appeared since then by using radiations and chemicals. Here in this report, we present the data of the Indian population concentrating on sister chromatid exchanges (SCEs) to detect the adaptive response.

Human peripheral blood lymphocytes stimulated *in vitro* for 6 h were exposed to a low (conditioning) dose of EMS 1.5×10^{-4} M or MMS 1.5×10^{-5} M. After 6h, the cells were treated with a high (challenging) concentration of the same agent ($1.5x10^{-3}$ M EMS or $1.5x10^{-4}$ M MMS). The cells that received both conditioning and challenging doses became less sensitive to the induction of SCEs than those which did not receive the pretreatment with EMS or MMS (Harish et al., 1998). They responded with lower frequencies of SCEs. This suggests that EMS or MMS's conditioning dose has offered the lymphocytes to have decreased SCEs (Table 17.5). This led to the realization that pre-exposing lymphocytes to low doses can cause the induction of repair activity. This is a clear indication of adaptive response induced by alkylating agents, whether ethylating or methylating in human lymphocytes *in vitro.*

TABLE 17.6

Aberration Frequencies of Cells Treated with Various Concentrations of MNU in Chinese Hamster V-79, with MNU Pretreatment 6 H Before MNU Post-Treatment and Fixation 14 Hours After Post-Treatment

Treatment of MNU		
Pre (mM)	**Post (mM)**	**Breakage events per cell**
–	–	0.010
–	0.3	0.040
–	0.6	0.740
0.01	–	0.067
0.01	0.6	0.471
0.03	–	0.050
0.03	0.3	0.180
0.03	0.6	0.280
0.05	–	0.020
0.05	0.3	0.160
0.05	0.6	0.317
0.08	–	0.020
0.08	0.6	0.360
0.10	–	0.060
0.10	0.3	0.160
0.10	0.6	0.530

(*Source:* Data from Kaina, 1982).

Exposure of V79 Chinese hamster cells to a single very low sub-toxic and a sub-clastogenic dose of N-methyl-N-nitrosourea (MNU) or N-methyl-N'-nitro-N-nitrosoguanidine (MNNG) made these cells resistant to the toxic, mutagenic, and clastogenic activities of the same agents given 6 h later. Cell survival was increased nearly twofold under optimal conditions when compared with the non-pretreated controls. Chromosomal aberration frequencies were reduced to nearly half the control values (cells not pretreated) (Table 17.6). This was observed for a wide range of pretreatment concentrations and different recovery times (Kaina, 1982).

Adaptive Response in *Drosophila*

Since the discovery of mutagenic effects of x-rays in *Drosophila melanogaster* by Muller in 1927, this organism has been widely used in genetic toxicology to identify potential carcinogens and mutagenesis studies of chemicals (Vogel et al., 1999). This is because *Drosophila* has a short generation time of 8 to 12 days, a broad range of suitable reporter genes, and a variety of genetic damage in germ cells. It produces many eggs and is economical and easy to handle. Further, the results of *Drosophila* can be extrapolated to other mammalian systems. After decades of experimentations, *D. melanogaster* is considered one of the best sub-mammalian systems for laboratory works on genotoxicity. Nonetheless, spectrums of protocols can be used to screen varied end effects of environmental pollutants (Vasudev and Gurushankara, 2016). Kaya et al. (2000), using *D. melanogaster* as an *in vivo* test system, have reported the presence of adaptive response in two-day-old larvae by employing ethylating agent EMS, methylating agent MMS, and N-ethyl-N-nitrosourea (ENU). They pretreated the two-day-old larvae with two different adaptive doses for 24 h and then challenged them with a high dose until the end of the development. The wing-spot-test analysis showed that the selected adaptive doses had reduced the frequency of mutant clones induced by the challenge dose. Savina et al. (2003) observed that after low-dose treatment of alkylating agents, offspring production increased, and dominant lethality in F1 and sex-linked recessive lethals (SLRL) F2 were significantly reduced as compared to challenging dose in Drosophila. Dhananjaya (2005), using EMS and MMS, not in wild flies but transheterozygotes of *Drosophila melanogaster*, analyzed adaptive response with different conditioning and challenging doses to understand the intricacies involved in it. The results demonstrated adaptive response irrespective of combinations of conditioning and challenging doses (Tables 17.7 and 17.8). Demir et al. (2010), using UVB radiation in *D. melanogaster*, also demonstrated the presence of adaptive response.

Tejaswi Kumar (2012) used wing spot assay and dominant lethal assay to analyze adaptive response in once again not wild flies but transheterozygotes of mutant flies of *D. melanogaster* using different time lags (TLs). This was a new direction to understanding the adaptive response. Furthermore, to understand this, he selected not one conditioning dose or one challenging dose but three lower and three challenging doses of ENU or MNU. In addition, different time lags were used instead of one, as has been worked out by many investigators. This was to decipher the intricacies involved in the adaptive response if any.

As it was the first time the transheterozygotes were used in the experiments, it was necessary to get the picture of conditioning and challenging doses of ENU and MNU. Pilot toxicity studies fixed this.

TABLE 17.7

Dominant Lethals or Wing Spots Obtained After Treatment with Conditioning, Challenging, and Combined Doses of EMS at 72 ± 4 H Aged *D. melanogaster* Larvae Using Different Time Lags (TLs)

Treatment	Wing spots (Fr)#	% Dominant lethals
Control	0.19	4.02
Conditioning (L: 0.5 mM)	0.5*	8.04
Challenging (H: 15 mM)	4.26*	49.04*
L-2h TL—H	0.71**	14.04**
L-4h TL—H	0.91**	12.04**
L-6h TL—H	1.42**	14.04**
Challenging (H: 20 mM)	5.18*	58.01*
L-2h TL—H	1.12**	19.00**
L-4h TL—H	1.28**	16.01**
L-6h TL—H	1.72**	18.02**
Challenging (H: 25 mM)	6.88*	69.05*
L-2h TL—H	1.31**	28.02**
L-4h TL—H	1.77**	25.00**
L-6h TL—H	2.08**	29.00**

* Significant ($p < 0.05$) compared to control.
** Significant ($p < 0.05$) reduction compared to challenging dose.
# Statistical diagnosis following Frei and Wurgler (1988). Fr = Frequency/wing.

TABLE 17.8

Dominant Lethals or Wing Spots Obtained After Treatment with Conditioning, Challenging, and Combined MMS Doses at 72 ± 4 H Aged *D. melanogaster* Larvae Using Different Time Lags (TLs)

Treatment	Wing spots (Fr)#	% Dominant lethals
Control	0.19	4.02
Conditioning (L: 0.1 mM)	0.32*	5.97
Challenging (H: 3 mM)	1.87*	46.98*
L-2h TL—H	0.3**	10.98**
L-4h TL—H	0.41**	19.02**
L-6h TL—H	0.48**	20.98**
Challenging (H: 4 mM)	3.32*	55.95*
L-2h TL—H	0.48**	16.02**
L-4h TL—H	0.57**	21.02**
L-6h TL—H	0.67**	20.00**
Challenging (H: 5 mM)	4.42*	66.97*
L-2h TL—H	0.66**	19.71**
L-4h TL—H	0.7**	19.00**
L-6h TL—H	0.8**	18.03**

Note: Similar to Table 17.7.

In this direction, the two protocols—the development and viability rate—have been studied. Rate of development and survival value are the best characteristics to observe the somatic effects, which reveal the extent of toxicity of the chemical (Luning, 1966). Different concentrations ranging from 0.1 to 4 mM of ENU and 0.01 to 2 mM of MNU, when analyzed for toxicity, the results have revealed that there was a significant delay in the rate of development; significant reduction in the viability, a significant increase in the percentage of dominant lethals and frequencies of somatic mutations, in a dose-dependent manner in both 48 ± 4 h or 72 ± 4 h aged transheterozygous larvae. The conditioning doses of 0.1, 0.25, and 0.5 mM ENU and 0.01, 0.05, and 0.1 mM MNU were selected. As far as challenging doses were considered, doses, namely, 2.5, 3.0, and 3.5 mM ENU and 1.0, 1.4, and 1.8 mM MNU were opted to understand the adaptive response. The results have revealed a significant reduction in the percentage of dominant lethals and total wing spot frequency in all treatment schedules of combined chemical agents combined with respective challenging doses (Tables 17.9 and 17.10). These data indicate the induction of adaptive response by ENU and MNU in somatic and germinal *Drosophila* cells *in vivo*.

TABLE 17.9

Percentage Reduction of Dominant Lethals or Wing Spots by Combined Treatment of ENU with 2, 4, and 6 H TLs, with Combined Treatment Given to 72 ± 4 H Aged *D. melanogaster* Larvae.

		% Reduction after combined treatment* with a different TLs					
		2 h		4 h		6 h	
Conditioning dose (L) (in mM)	Challenging dose (H) (in mM)	Dominant lethals	Wing spots	Dominant Lethals	Wing spots	Dominant Lethals	Wing spots
0.1	2.5	78.39	69.88	69.11	57.41	67.17	44.47
	3.0	71.19	71.18	78.23	59.78	65.57	42.80
	3.5	71.94	67.40	65.70	51.61	67.67	41.26
0.25	2.5	76.22	74.02	69.89	64.85	62.24	51.75
	3.0	73.93	67.47	62.40	61.45	69.20	50.40
	3.5	72.54	68.65	76.10	58.52	65.07	49.20
0.5	2.5	67.34	74.10	71.99	65.54	67.91	55.78
	3.0	68.47	67.53	65.08	53.32	61.09	41.33
	3.5	72.24	66.37	73.33	58.11	65.96	51.02

* Combined treatment (L—2, 4, or 6 h—H)

TABLE 17.10

Percentage Reduction of Dominant Lethals or Wing Spots by Combined Treatment of MNU with 2, 4, and 6 H TLs, with Combined Treatment Given to 72 ± 4 H Aged *D. melanogaster* Larvae.

		% Reduction after combined treatment with a different TLs					
		2 h		4 h		6 h	
Conditioning dose (L) (in mM)	Challenging dose (H) (in mM)	Dominant lethals	Wing spots	Dominant lethals	Wing spots	Dominant lethals	Wing spots
0.01	1.0	65.83	87.25	60.05	85.54	58.52	84.07
	1.4	67.46	84.58	57.32	83.08	65.75	81.94
	1.8	75.99	83.30	70.20	81.47	66.69	77.43
0.05	1.0	64.07	88.86	69.19	87.34	57.08	85.37
	1.4	70.11	85.12	63.69	83.73	74.08	81.15
	1.8	71.49	82.18	74.26	80.67	67.71	78.49
0.1	1.0	64.85	87.50	62.27	86.48	64.86	84.43
	1.4	73.26	84.46	69.25	82.77	65.61	1.09
	1.8	75.47	83.04	68.35	82.08	71.69	79.84

Note: Per the Table 17.9.

Understanding Adaptive Response by Using Mating Behavior of *D. melanogaster*

Over the last 100 years, since the first discovery of courtship behavior in Drosophila by Sturtevant (1915), there has been a stream of reviews about fly sexual behavior (Spieth, 1952; Billeter et al., 2002). As most of the activity and courtship behaviors are performed by male *D. melanogaster*, the extensively studied species and the early works concentrated on it and descriptive. Indeed, Duncan (1930) reported that one particularly promiscuous male mated 123 times during his 72-day life span. Furthermore, these behavioral approaches culminated in a description of courtship in Drosophila's 101 species (Spieth, 1952). Recent works have shown that genes govern behaviors' repertoires with direct and indirect roles in courtship behaviors (Singh and Singh, 2016). To name a few genes like *cacophony, hite, fruitless* (*fru*), *voila, courtless, desat*, and so on, and variants of *fru* and *dsx* expressed in different regions of the brain involved in male-specific courtship patterns, like wing song (Dauwalder, 2008; Mackay et al., 2005; Massey et al., 2019). Similarly, female-specific behaviors are also under the control of specific genes such as *rofn, spinster* (*spn*), and *chaste* (*chst*) (Juni and Yamamoto, 2009). It is very clear from the works on genes in Drosophila's behavior that genes control behavior patterns and that these genes interact with environmental agents (Beckwith and French, 2019).

Even though many protocols of *D. melanogaster* have been validated for genotoxicity studies, time and again, attempts are being made to introduce inexpensive, short-duration, and efficient parameters to understand the genotoxic effects of environmental pollutants in general and adaptive response in particular. It is pertinent to mention here that *D. melanogaster*, with its series of sequential stereotyped elements of courtship behavior (Spieth and Ringo, 1983) and the genes that control it, is well established. An attempt has been made to use this protocol to understand the genetic effects of environmental pollutants. The works of Vasudev et al. (2013) have demonstrated that Dithane M-45, a carbamate pesticide, has no impact on male qualitative behaviors, such as orientation, tapping, wing vibration, licking, curling the abdomen, and mounting. However, there were quantitative differences between the control and different treated groups. It was undoubtedly proved that various elements of courtship were significantly increased in a dose-dependent manner. Furthermore, higher doses above 100 ppm of different groups were highly significant compared to controls. Thus, the higher doses of Dithane M-45 were effective in altering the courtship behavior of *D. melanogaster*. Nonetheless, Dithane M-45 was also shown to be an inducer of toxicity in the form of an increased development rate and reduced viability. It significantly reduced morphological characteristics, such as body length, wing length, and size of pupae above 50 ppm compared to controls (Vasudev and Krishnamurthy, 1982). Thus, it can be opined that courtship activities of *D. melanogaster* as a parameter can be used in routine genotoxicity studies in general and adaptive response in particular (Vasudev et al., 2013). The courtship behavior of *D. melanogaster* has been employed to understand the presence of adaptive response. Here researcher has used EMS and MMS, the standard alkylating agents, to unravel the adaptive response. Toxicity experiments showed that the development rate had increased from 10.03 ± 0.02 to 16.07 ± 0.24 in the highest dose of MMS tested. Similar results were observed using EMS as an agent. Depending on the doses, higher doses of either MMS or EMS reduced the viability significantly. On the other hand, similar to established toxicity protocols, courtship behavior elements also showed increased values considerably compared to controls in EMS or MMS treated groups. These preliminary results demonstrate that courtship behavior can be used to screen toxicity because (1) *D. melanogaster* is a small but not minute organism (mating behavior can be observed through the naked eye if not possible be observed under stereo zoom without any problem), (2) there is sexual dimorphism, (3) single pair mating can be easily achieved, (4) there is elaborate courtship behavior, (5) courtship behavior can be analyzed qualitatively and quantitatively, and (6) many behavioral mutants are available (Divyashree and Vasudev, 2021; Divyashree et al., 2021).

The dose of 0.1 mM MMS or 0.5 mM EMS as a conditioning dose and 3 mM of MMS or 15 mM of EMS as a challenge dose was employed in adaptive response experiments. When the two were combined with a time lag of 2 h, the preliminary results revealed that the courtship elements are significantly altered compared to the challenge dose. For example, copulation duration, which is the time between initiations of copulation to termination of copulation of each pair, is longer in combined treatment compared to challenge treatment of MMS or EMS. This is true for all elements. Thus, it is clear from these results that there is a repair mechanism to enhance the genetic system's efficacy to overcome the drastic effects of higher doses (unpublished data). Further research is in the pipeline using different mutants, alkylating and non-alkylating agents, enzyme inhibitors, and so on to understand the intricacies of adaptive response in *D. melanogaster.*

Adaptive Response of *Ascites* Tumor Cells to Anticancerous Alkylating Agents

Scrutiny of literature has revealed protection against radiations in many cancerous cells, such as human malignant melanoma, epidermal laryngeal carcinoma, fibrosarcoma, glioblastoma, and epithelial carcinoma (Azizova et al., 2018). On the other hand, only a few reports are available concerning a chemically induced adaptive response. Nonetheless, literature also showed the absence of adaptive response in cancerous cells, such as mouse papilloma and lymphoma (Park et al., 1999), human bladder carcinoma (Schaffer et al., 2004), erythroleukemia, promyelocytic leukemia, solid tumor cell lines (Jiang et al., 2008), colon carcinoma (Sowa et al., 2011), and gastric cancer cells (Wang et al., 2013). When the data were pooled together, it gave the impression that there was a contradictory opinion, and hence, there was a necessity for further experimentations to decipher this.

Studies with ascites tumor mice showed that 80 mg/kg b.w. of EMS was the conditioning dose, and 240 mg/kg b.w. was the challenging dose. In MMS, 50 mg/kg b.w. was used as a conditioning dose, and 150 mg/kg b.w. as a challenging dose. The time lag between conditioning and challenge was found to be 8 h (peak of activity) per literature; hence, the same has been employed. The low dose induced a low frequency of chromosomal aberrations, and the challenging dose induced a high frequency of aberrations. Contrary to this, combined treatment has reduced the frequency of chromosomal aberrations compared to challenge treatment. This reduction was seen in all recovery times (RTs) tested, such as 24, 48, and 72 h (Table 17.11). This proved beyond doubt that mitotic delay and others did not influence damaged cells' action. In other words, the possibility of reduced frequency of chromosomal aberrations was due to some repair mechanism operative in cancerous cells. This is the first report on the induction of adaptive response by chemical agents *in vivo* cancer system using a clastogenic endpoint (Mahadimane et al., 2014). *In vivo* results of ascites cells were consistent with those of Lee et al. (2000). They have demonstrated adaptive response *in vitro* sarcoma 180 cells using the cells' survival as the endpoint.

TABLE 17.11

Alkylating Agents Induced Adaptive Response in Cancerous Cells: Total Number of Chromosomal Aberrations Observed After Conditioning, Challenging, and Combined Treatments of EMS or MMS *In Vivo* Ehrlich Ascites Tumor Cells

	Total no. of chromosomal aberrations at different RTs					
	24 h		48 h		72 h	
Treatment	**EMS**	**MMS**	**EMS**	**MMS**	**EMS**	**MMS**
Control	7.67 ± 0.88[a]	7.67 ± 0.67[a]	7.33 ± 0.67[a]	9.00 ± 1.00[a]	8.00 ± 1.15[a]	8.00 ± 1.53[a]
Conditioning (L)	42.33 ± 1.33[b]	54.00 ± 4.62[b]	35.67 ± 1.45[b]	42.67 ± 2.40[b]	26.67 ± 2.60[b]	37.33 ± 0.67[b]
Challenging (H)	182.00 ± 8.54[c]	250.33 ± 7.88[c]	157.33 ± 7.31[c]	195.00 ± 3.06[c]	130.67 ± 2.96[c]	133.33 ± 2.91[c]
Combined (L-8h-H)	142.33 ± 2.73[d]	173.67 ± 6.39[d]	118.33 ± 2.96[d]	132.00 ± 7.09[d]	91.33 ± 4.98[d]	78.67 ± 3.93[d]
% Reduction*	36.55	42.94	38.68	44.46	41.94	53.91

Note: Mean value of 3 independent experiments; 3 animals per experiment were used; 100 cells per animal were scored, and a total of 900 cells were scored per dose. Values with the same superscripts are not significantly different ($p > 0.05$), whereas values with different superscripts are significantly different ($p < 0.05$) from one another.

* Percentage of reduction was calculated by using formula $C = (B/A \times 100) - 100$. Where A = additive effect of conditioning and challenging dose (%), B = combined treatment of conditioning and challenging dose (%), and C = reduction (%).

Adaptive Response in Diabetic Mice

Of the many diseases of humans, diabetes is one disease from which approximately 463 million people were suffering in 2019, and it is estimated that 700 million will be diabetic worldwide by 2045, per International Diabetic Federation (IDF), 2020. Elevated blood glucose levels characterize this chronic disease. Nonetheless, long-term diabetes leads to damage, dysfunction, and various organs' failure to reduce life expectancy and quality of life. IDF (2020) has reported that each person is to die every seven seconds. The diabetic condition can be duplicated in the mouse system by using streptozotocin. Thus, this model can be used to extrapolate the results to humans. Furthermore, there were no reports of adaptive response in diseased animals and humans in particular. Hence, the experiments were designed to decipher the underlying aspect of adaptive response by using alkylating agents in diabetic mouse.

Toxicity studies were conducted to get the picture of conditioning and challenge doses required to analyze the adaptive response in the diabetic mouse. When diabetic mice were treated with different EMS doses ranging from 50 to 300 mg/kg b.w., there was a dose-dependent production of chromosomal aberrations. Similarly, the same results were obtained when MMS was used at a range of 35 to 175 mg/kg b.w. This was true for all recovery times tested, such as 24, 48, and 72 h. On par with earlier experiments, in diabetic mice, a very low dose of MMS was sufficient to induce significant chromosomal aberrations compared to EMS. To produce the same frequency of aberrations, a higher EMS dose was required. For example, 105 mg/kg b.w. of MMS induced 33.3% chromosomal aberrations, whereas to generate almost the same number of breaks (34.68%), 250 mg/kg b.w. of EMS was required in the diabetic mouse (Table 17.12). Thus, it was opined that MMS is a more potent inducer of chromosomal aberrations in diabetic mice similar to normal mice, cancerous mice, or other mammals.

Based on the results obtained, 100 mg/kg b.w. of EMS or 40 mg/kg b.w. of MMS was selected as the conditioning dose and 300 mg/kg b.w. of EMS or 160 mg/kg b.w. of MMS as the challenging dose. When combined treatment of EMS with an 8 h time lag between conditioning and challenge was made, a significant reduction of aberrations was observed at 24 h recovery time. The same results were noticed at 48 and 72 h recovery times. Even in the case of MMS, a reduction of aberrations was shown when combined treatment was given. These results show that both EMS and MMS induced adaptive responses in the diabetic mouse (Table 17.13). Furthermore, the results also demonstrated that methylating agents are more effective inducers of adaptive response than ethylating agents. These data are on par with earlier works with mammalian systems.

Biochemically, experiments were conducted to understand the adaptive response in diabetic mice. Catalase, an enzyme known to reduce an oxidant H_2O_2 into H_2O and O_2, showed increased activity in EMS or MMS combined treated diabetic mice compared to challenging doses at 24 or 48 recovery time. Similarly, SOD, an enzyme that catalyzes the dismutation of the superoxide (O_2) radical into either ordinary molecular oxygen (O_2) or hydrogen peroxide (H_2O_2), also increased in combined treated individuals compared to challenge-dosed animals. These increased activities of catalase and SOD reflect the repair activity of these enzymes and reduce the effect of challenging doses. On the other hand, combined treatment of either EMS or MMS did not alter the GSH or glycogen level in the diabetic mouse at all recovery times compared to challenge treatment.

Contrary to the said findings, MDA has been reduced in combined EMS, or MMS treated diabetic mice compared to challenging treatments at 24, 48, or 72 h recovery time. These results indicated an adaptive response in diabetic mice (Dada Khalandar and Vasudev, 2015). In these days of increased diabetic conditions in the human population, further research is required in adaptive response because of the use of drugs, the chemical agents in therapy.

17.8 Hyperthermia Ameliorating Drug-Induced Adaptive Response in Animal Systems

Hyperthermia is an excellent therapeutic tool and acts as a synergistic agent and potent adjuvant treatment in cancer therapy. Further, heat treatment of CHO cells and human lymphocytes in the S-phase of the cell cycle leads to chromatid type aberrations and micronuclei production. Pretreatment of hyperthermia

TABLE 17.12

Percentage Frequency of Chromosomal Aberrations (Mean ± SE) Induced by Conditioning, Challenging, and Combined Doses of EMS or MMS in Non-Diabetic and Diabetic Mice Bone Marrow Cells Recovered After 24 H, 48 H, and 72 H

			% Chromosomal aberrations	
RT in h	**Treatments**	**Mouse system**	**EMS**	**MMS**
24	Control	Non-diabetic	2.22 ± 0.22^{a}	2.22 ± 0.22^{a}
		Diabetic	2.44 ± 0.40^{a}	2.45 ± 0.4^{a}
	Conditioning	Non-diabetic	29.74 ± 0.80^{b}	22.00 ± 2.6^{b}
		Diabetic	17.33 ± 0.38^{b}	16.11 ± 0.67^{b}
	Challenging	Non-diabetic	93.78 ± 10.48^{e}	103.89 ± 2.61^{c}
		Diabetic	69.22 ± 5.00^{d}	72.67 ± 2.00^{d}
	Combined	Non-diabetic	63.45 ± 3.28^{cd}	52.89 ± 3.56^{e}
		Diabetic	50.33 ± 3.46^{c}	45.44 ± 1.82^{f}
48	Control	Non-diabetic	2.22 ± 0.31^{a}	2.22 ± 0.31^{a}
		Diabetic	2.44 ± 0.4^{a}	2.44 ± 0.4^{a}
	Conditioning	Non-diabetic	18.89 ± 0.80^{b}	18.00 ± 0.96^{b}
		Diabetic	12.11 ± 1.5^{b}	11.78 ± 0.59^{ab}
	Challenging	Non-diabetic	77.89 ± 5.79^{e}	71.33 ± 9.51^{c}
		Diabetic	53.23 ± 3.37^{c}	52.89 ± 1.46^{d}
	Combined	Non-diabetic	57.78 ± 2.48^{d}	40.33 ± 0.67^{fe}
		Diabetic	41.66 ± 3.18^{f}	37.78 ± 1.22^{e}
72	Control	Non-diabetic	2.22 ± 0.31^{a}	2.22 ± 0.31^{a}
		Diabetic	2.67 ± 0.51^{a}	2.44 ± 0.4^{a}
	Conditioning	Non-diabetic	16.33 ± 0.19^{b}	14.11 ± 1.06^{b}
		Diabetic	11.33 ± 2.60^{b}	10.11 ± 0.78^{b}
	Challenging	Non-diabetic	65.78 ± 4.56^{e}	48.11 ± 1.55^{d}
		Diabetic	46.33 ± 3.06^{c}	35.78 ± 0.59^{e}
	Combined	Non-diabetic	49.11 ± 2.99^{c}	35.67 ± 1.86^{e}
		Diabetic	35.33 ± 2.08^{d}	30.44 ± 0.78^{f}

Note: Pooled data from three independent experiments; 900 cells were analyzed per dose; 3 animals were used for each treatment. Values with the same superscripts are not significant ($p > 0.05$); values with different superscripts are significantly different from one another ($p < 0.05$), according to the Duncan post hoc test.

TABLE 17.13

Percentage Reduction of Chromosomal Breaks (Mean ± SE) After the Combined Treatment of EMS or MMS at Different Recovery Times in Bone Marrow Cells of Non-Diabetic and Diabetic Mice

		Reduction (%)	
RT in h	**System**	**EMS**	**MMS**
24	Non-diabetic	49.04 ± 1.19	59.59 ± 0.63
	Diabetic	40.00 ± 6.07	51.74 ± 1.38
48	Non-diabetic	40.82 ± 4.13	50.92 ± 1.37
	Diabetic	35.01 ± 2.47	45.13 ± 1.30
72	Non-diabetic	41.33 ± 1.28	46.36 ± 2.67
	Diabetic	36.90 ± 1.57	32.03 ± 1.97

(40 °C, 10 min) to MH- or TEM-exposed *Vicia faba* resulted in reduced chromatid aberrations induction. The combination of hyperthermia and drugs produced fewer chromatid and iso-chromatid breaks in human lymphocytes. Cai and Liu (1992) showed enhanced adaptive cytogenetic response by hyperthermia (41°C and 43°C) in human lymphocytes. Even in human melanoma cells, hyperthermia with etoposide (chemotherapeutic drug) showed decreased drug cytotoxicity.

Contrary to these, potentiating the frequency of chromosome aberrations was reported when CHO-K1 cells and human lymphocytes were exposed to hyperthermia with anticancer drugs (Vasudev and Obe, 1987). In *in vivo* system of *P. pictus*, it was used to analyze the response of meiotic cells to hyperthermia. Hyperthermia of 40°C or 45°C and conditioning dose of 0.03 M of EMS or 0.012 M of MMS combined with a challenging dose of 0.12 M EMS or 0.048 M MMS were employed. Pre-, inter-, and post-treatments of hyperthermia were used to test the efficacy.

Similarly, from pilot toxicity studies *in vivo* mouse, hyperthermia of 40°C or 45°C, conditioning dose of 50 mg/kg b.w. of MMS or 80 mg/kg b.w. of EMS, and challenging dose of 150 mg/kg b.w. of MMS and 240 mg/kg b.w. of EMS were tested. The results have revealed that pre-, inter-, and post-treatments of hyperthermia significantly decreased the production of anomalies induced by EMS or MMS in the meiotic cells of the grasshopper *P. pictus* (Tables 17.14 and 17.15). The results also revealed an ethylating agent combined with hyperthermia-induced more cytogenetic adaptive response than a methylating agent (Venu et al., 2016; Gurushankara et al., 2018). Even in mice, when mitotic cells were exposed, the results have shown that hyperthermia ameliorated the adaptive cytogenetic response induced by alkylating agents (Tables 17.16 and 17.17). In the mammalian system, methylating agents were more potent inducers of adaptive response than ethylating agents, even though both agents induced adaptive response. Furthermore, both in *P. pictus* and mice, hyperthermia acted as a cross-adaptor for alkylating agents. These results underline the importance of further research by using more chemotherapeutic

TABLE 17.14

Percentage of Chromosomal Anomalies (Mean ± SE) Observed After Hyperthermia (HT), EMS, or Different Treatment Schedules in Meiotic Cells of *P. Pictus* at Different RTs

	% Chromosomal anomalies at different RTs (in h)			
Treatment groups	**12**	**24**	**36**	**48**
Control	7.09 ±0.249	7.10±0.125	7.12±0.071	7.17±0.096
HT[40°C-30min]	7.67±0.270	7.55±0.097	7.41±0.166	7.35±0.167
HT[45°C-30min]	7.89±0.066	7.70±0.056	7.67±0.212	7.58±0.088
EMS-L	37.45±0.318	32.83±0.306	31.86±0.252	28.16±0.208
EMS-H	84.41±1.271	78.92±0.264	76.14±0.334	68.22±0.497
L-2h-H	67.54±0.963	60.39±0.207	53.62±0.432	51.17±0.463
% Reduction#	44.51±1.001	45.94±0.352	50.34±0.441	46.89±0.544
Pre-treatment of HT				
HT[40°C-30min]-4h-L-2h-H	42.83±0.120 (57.69)	42.01±0.130(43.75)	36.13±0.154 (48.41)	35.03±0.099 (46.07)
HT[45°C-30min]-4h-L-2h-H	41.05±0.093 (64.53)	40.04±0.123 (50.82)	34.68±0.055 (54.61)	34.00±0.082 (50.50)
Inter-treatment of HT				
L-1h-[HT-40°C-30min]-1h-H	47.26±0.068 (42.91)	45.02±0.083 (34.14)	43.18±0.082 (24.18)	42.03±0.097 (21.75)
L-1h-[HT-45°C-30min]-1h-H	45.29±0.125 (49.13)	43.00±0.067 (40.44)	41.02±0.064 (30.72)	40.57±0.066 (26.13)
Post-treatment of HT				
L-2h-H-4h[HT-40°C-30min]	58.50±0.042 (15.45)	55.50±0.058 (8.81)	49.51±0.078 (8.30)	47.95±0.068 (6.72)
L-2h-H-4h[HT-45°C-30min]	56.09±0.057 (20.41)	53.25±0.041 (13.41)	48.06±0.058(11.57)	46.46±0.053 (10.14)

Treatment groups	% Chromosomal anomalies at different RTs (in h)			
	12	24	36	48
##HT challenging with a high dose of EMS				
[HT-40°C-30min]-2h-H	57.26±0.024 (47.42)	55.39±0.067 (42.48)	51.29±0.033 (48.45)	48.13±0.073 (41.74)
[HT-45°C-30min]-2h-H	55.05±0.090 (53.33)	54.26±0.300 (45.45)	48.49±0.079 (57.02)	46.17±0.050 (47.76)

Note: Pooled data from three independent experiments; minimum of 500 cells in each meiotic stage per dose were scored; 4 grasshoppers per experiment were used.

Individual chromosomal anomalies were scored and pooled to make data concise. Thus, the percentage anomalies for different recovery times are given in this table.

#Calculation of percent reduction: (A) additive effect: sum of chromosomal anomalies observed in both conditioning (L) and challenging (H) dose (L + H); (B) combined effect: chromosomal anomalies observed in combined treatment of conditioning and challenging doses with 2 h time lag (L-2h-H); percentage of reduction in chromosomal anomalies (C) was calculated by using the formula: $C = (B/A \times 100) - 100$.

TABLE 17.15

Percentage of Chromosomal Anomalies (Mean ± SE) Observed After HT, MMS, or Different Treatment Schedules in Meiotic Cells of *P. Pictus* at Different RTs

Treatment groups	% Chromosomal anomalies at different RTs (in h)			
	12	24	36	48
Control	7.09±0.249	7.10±0.126	7.12±0.071	7.17±0.096
HT[40°C-30min]	7.02±0.433	7.07±0.394	7.07±0.403	7.15±0.289
HT[45°C-30min]	7.07±0.404	7.12±0.556	7.14±0.510	7.17±0.470
MMS-L	40.77±0.885	39.02±0.985	38.11±0.650	37.17±0.585
MMS-H	94.49±0.646	89.07±0.754	86.10±0.743	83.05±0.722
L-2h-H	76.31±0.729	73.24±0.903	70.07±0.798	67.03±0.823
% Reduction#	43.58±0.980	42.82±0.878	43.59±0.683	44.24±0.652
Pre-treatment of HT				
HT[40°C-30min]-4h-L-2h-H	60.07±0.727 (27.04)	55.49±0.573 (31.99)	53.13±0.326 (31.88)	47.46±0.106 (41.23)
HT[45°C-30min]-4h-L-2h-H	58.09±0.436 (31.36)	53.65±0.415 (36.51)	53.23±0.427 (31.64)	46.83±0.118 (43.13)
Inter-treatment of HT				
L-1h-[HT-40°C-30min]-1h-H	63.11±0.374 (20.90)	58.98±0.504 (24.18)	55.09±0.343 (27.20)	52.33±0.159 (28.10)
L-1h-[HT-45°C-30min]-1h-H	60.07±0.357 (27.04)	54.48±0.365 (34.43)	51.33±0.494 (36.51)	48.92±0.563 (37.02)
Post-treatment of HT				
L-2h-H-4h[HT-40°C-30min]	71.22±0.706 (7.14)	70.09±0.563 (4.49)	68.98±0.456 (1.58)	60.12±0.874 (11.50)
L-2h-H-4h[HT-45°C-30min]	69.04±0.883 (10.53)	68.97±0.784 (6.19)	66.91±0.634 (4.72)	59.01±0.346 (13.59)
##HT challenging with a high dose of MMS				
[HT-40°C-30min]-2h-H	71.14±0.386 (32.82)	68.28±0.614 (30.45)	65.45±0.459 (31.55)	62.02±0.537 (33.09)
[HT-45°C-30min]-2h-H	69.03±0.837 (36.88)	67.13±0. 540 (32.68)	63.61±0.574 (35.36)	61.18±0.615 (35.75)

Note: As Table 17.14.

TABLE 17.16

Percentage of Chromosomal Aberrations (Mean ± SE) Observed After HT, EMS, or Different Treatment Schedules in Mitotic Cells in Mouse Bone Marrow at Different RTs

	% Chromosomal aberrations at different RTs (in h)		
Treatment groups	**24**	**48**	**72**
Control	1.67±0.167	1.56±0.176	1.44±0.176
HT[40°C-15min]	2.33±0.167	2.22±0.222	2.11±0.200
HT[45°C-15min]	2.56±0.176	2.44±0.176	2.33±0.167
EMS-L	13.78±0.324	12.11±0.261	11.11±0.261
EMS-H	66.78±0.434	60.11±0.351	52.00±0.373
L-2h-H	39.11±0.484	36.44±0.294	30.78±0.494
% Reduction#	51.43±0.731	49.27±0.452	50.91±0.830
Pre-treatment of HT			
HT[40°C-15min]-4h-L-8h-H	32.44±0.377 (20.56)	28.44±0.338 (28.13)	24.00±0.333 (28.25)
HT[45°C-15min]-4h-L-8h-H	35.89±0.389 (8.97)	32.11±0.423 (13.48)	25.56±0.503 (20.42)
Inter-treatment of HT			
L-4h-[HT-40°C-15min]-4h-H	27.11±0.539 (44.26)	22.11±0.423 (64.81)	19.67±0.527 (56.48)
L-4h-[HT-45°C-15min]-4h-H	30.33±0.289 (28.95)	26.11±0.512 (39.56)	22.44±0.338 (37.17)
Post-treatment of HT			
L-8h-H-4h[HT-40°C-15min]	36.11±0.261 (8.31)	34.00±0.408 (7.18)	26.33±0.500 (16.90)
L-8h-H-4h[HT-45°C-15min]	37.44±0.503 (4.46)	35.44±0.444 (2.82)	27.56±0.475 (11.68)
##HT challenging with a high dose of EMS			
[HT-40°C-15min]-8h-H	34.11±0.423 (95.78)	31.00±0.471 (93.90)	24.11±0.423 (115.68)
[HT-45°C-15min]-8h-H	53.89±0.389 (23.92)	33.00±0.500 (82.15)	24.89±0.512 (108.92)

Note: Pooled data from three independent experiments; minimum of 900 cells were analyzed per dose; 3 animals were used for each treatment.

Individual chromosomal aberrations were scored and pooled to make data concise. Thus, the percentage aberrations for different recovery times are given in this table.

#Calculation of percent reduction: (A) additive effect: sum of chromosomal aberrations observed in both conditioning (L) and challenging (H) dose (L + H); (B) combined effect: chromosomal aberrations observed in combined treatment of conditioning and challenging doses with 2 h time lag (L-2h-H); percentage of reduction in chromosomal aberrations (C) was calculated by using the formula: $C = (B/A \times 100) - 100$.

TABLE 17.17

Percentage of Chromosomal Aberrations (Mean ± SE) Observed After HT, MMS, or Different Treatment Schedules in Mitotic Cells in Mouse Bone Marrow at Different RTs

	% Chromosomal aberrations at different RTs (in h)		
Treatment groups	**24**	**48**	**72**
Control	1.67±0.167	1.56±0.176	1.44±0.176
HT[40°C-15min]	2.33±0.167	2.22±0.222	2.11±0.200
HT[45°C-15min]	2.56±0.176	2.44±0.176	2.33±0.167

	% Chromosomal aberrations at different RTs (in h)		
Treatment groups	**24**	**48**	**72**
MMS-L	19.22±0.401	18.11±0.200	17.00±0.167
MMS-H	80.11±0.512	72.22±0.324	68.11±0.351
L-2h-H	52.22±0.278	45.11±0.512	39.56±0.412
% Reduction#	47.40±0.423	49.44±0.999	52.78±1.087
Pretreatment of HT			
HT[40°C-15min]-4h-L-8h-H	47.11±0.200 (10.85)	43.00±0.408 (4.91)	34.56±0.503 (14.47)
HT[45°C-15min]-4h-L-8h-H	50.11±0.539 (4.21)	43.44±0.475 (3.84)	35.44±0.444 (11.63)
Inter-treatment of HT			
L-4h-[HT-40°C-15min]-4h-H	34.22±0.641 (52.60)	27.89±0.512 (61.74)	24.22±0.683 (63.34)
L-4h-[HT-45°C-15min]-4h-H	35.11±0.611 (48.73)	31.11±0.423 (45.00)	27.11±0.261 (45.92)
Post-treatment of HT			
L-8h-H-4h[HT-40°C-15min]	48.11±0.455 (8.54)	42.44±0.530 (6.29)	36.22±0.434 (9.22)
L-8h-H-4h[HT-45°C-15min]	50.22±0.465 (3.98)	47.22±0.619 (−4.47)	37.11±0.423 (6.60)
##HT challenging with a high dose of MMS			
[HT-40°C-15min]-8h-H	44.44±0.556 (80.27)	42.22±0.596 (71.06)	35.33±0.441 (92.78)
[HT-45°C-15min]-8h-H	46.11±0.539 (73.74)	43.89±0.423 (64.55)	36.56±0.294 (86.30)

Note: Table 17.16.

agents in different model organisms to draw medical scientists' attention in particular and cytogeneticists in general.

Values in parentheses are percentage reduction of chromosomal anomalies compared to combined treatment (L-2h-H). Calculation of percent reduction: (A) chromosomal anomalies observed in respective HT treatment schedule; (B) chromosomal anomalies observed in combined treatment (L-2h-H) or (## high dose of mutagen); percentage of reduction in chromosomal anomalies (C) was calculated by using the formula: $C = (B/A \times 100) - 100$.

Values in parentheses are percentage reduction of chromosomal aberrations compared to combined treatment (L-2h-H). Calculation of percent reduction: (A) chromosomal aberrations observed in respective HT treatment schedule; (B) chromosomal aberrations observed in combined treatment (L-2h-H) (## high dose of mutagen: H); percentage of reduction in chromosomal aberrations (C) was calculated by using the formula: $C = (B/A \times 100) - 100$.

17.9 Poly(ADP-Ribose) Polymerase (PARP) Not Involved in the Adaptive Response

Although there are considerable data on the adaptive response, the molecular mechanism remains unclear. Different repair enzymes are implicated in this repair pathway. Poly(ADP-ribose) polymerase (PARP) is one such repair enzyme reported to participate in DNA repair processes. PARP is a nuclear enzyme activated by DNA strand breaks induced by alkylating agents or x-rays (Cleaver et al., 1983; Chatterjee and Berger, 1994; Kleczkowska and Althaus, 1996a). Furthermore, the adaptive response was prevented *in vitro*, when PARP inhibitors were administered 2 h after the adaptive treatment (Wiencke et al., 1986; Shadley and Wolff, 1987; Wiencke, 1987). These reports suggest the involvement of PARP in the adaptive response. A more effective method to understand the involvement of PARP is to look for the active proteins expressed and to test for possible functions. An alternative approach is to selectively inactivate individual protein/enzyme (i.e., PARP) in the animal systems. The DNA repair enzyme PARP inhibitors, such as nicotinamide (N) and benzamide (B), were employed to study the role of PARP in EMS- or MMS-induced adaptive response in *P. pictus*, mice, and *in vitro* human lymphocytes. To understand the adaptive response, the conditioning and challenging doses of 0.03 M and 0.12 M of EMS and 0.012 M and 0.048 M of MMS and 80 mg/kg b.w. and 240 mg/kg b.w. of EMS and 50 mg/kg b.w. and 125 mg/kg b.w. of MMS, which were already established in the previous experiments with *P. pictus* and mice, respectively, were employed. Based on the dose-effect relationship studies, 5 mM/kg b.w. of nicotinamide and 2 mM/kg b.w. of benzamide were selected in *P. pictus*. Similarly, in mice, 600 mg/kg b.w. (5 mM) of nicotinamide and 625 mg/kg b.w. (2.5 mM) of benzamide were selected. The inter-treatment of nicotinamide (conditioning-1h-N-1h-challenging) with EMS or MMS yielded significantly less frequency of chromosomal anomalies compared to combined treatment (conditioning-2h-challenging), indicating the enhancement of adaptive response. Treatment of nicotinamide 2 or 4 h earlier than combined treatment also produced a significantly lower frequency of chromosomal anomalies than combined treatment. A similar reduction in the frequencies of chromosomal anomalies was observed in the post-treatment of nicotinamide with EMS or MMS. Inter-, pre-and post-treatment of benzamide with EMS or MMS also yielded reduced frequencies of chromosomal anomalies compared to combined treatment (Table 17.18). Thus, nicotinamide and benzamide enhance the adaptive response in meiotic cells of *P. pictus*. PARP is an NAD+-dependent enzyme. NAD+ acts as a substrate for PARP activity; thus, the NAD+ amount gives indirect evidence of the activity of PARP. The data reveals that NAD+ level is significantly increased in nicotinamide or benzamide inter-treated meiotic cells of *P. pictus*. This increase is observed at all RTs tested.

When similar experiments were done with mice, the inter-treatment of nicotinamide 2 or 4 h after the conditioning dose produced significantly lower frequencies of chromosomal aberrations than the combined treatment (conditioning-8h-challenging). Similarly, the pretreatment of nicotinamide 6 or 4 h earlier than the combined treatment also produced significantly less frequency of chromosomal aberrations compared to the combined treatment of EMS or MMS. Interestingly, in the post-treatment, the treatments of nicotinamide after 12 or 18 h of combined treatment yielded significantly lower frequencies of chromosomal aberrations than combined treatments. On the other hand, the frequency of

TABLE 17.18

Chromosomal Anomalies Observed in EMS and MMS with Nicotinamide or Benzamide (PARP Inhibitors) Treated *P. Pictus* at 24 H Recovery Time

	Chromosomal anomalies (%)			
	Nicotinamide*		**Benzamide***	
Treatments	**EMS**	**MMS**	**EMS**	**MMS**
Combined treatment L-2h-H	53.29±0.81	77.2±2.13	37.81±1.03	74.0±1.12
Inter-treatment L-1h-I-1h-H	26.95±0.29	43.74±1.71	33.8±3.41	49.5±0.59
Pre-treatments L-2h-L-2h-H	35.24±3.52	55.6±3.06	43.8±2.84	62,4±1.25
L-4h-L-2h-H	29.51±1.13	52.5±3.12	40.2±2.60	53.3±2.17
Post-treatments L-2h-H-2h-I	36.43±1.75	59.8±1.90	38.0±1.53	63.5±0.94
L-2h-H-4h-I	39.83±3.03	60.0±1.51	45.3±2.98	65.3±0.04

* Significant compared to combined treatment

TABLE 17.19

Chromosomal Breaks/Cells Observed in EMS and MMS with Nicotinamide or Benzamide (PARP Inhibitors) in the Treated Mouse at 24 H RT

	Chromosomal breaks/cell			
	Nicotinamide*		Benzamide*	
Treatments	EMS	MMS	EMS	MMS
Combined treatment L-8h-H	0.51±0.03	0.59±0.030	0.58±0.005	0.51±0.013
Inter-treatments L-2h-I-6h-H	0.35±0.003	0.36±0.021	0.48±0.011	0.39±0.010
L-4h-I-4h-H	0.30±0.006	0.34±0.010	0.41±0.12	0.34±0.014
Pre-treatments I-6h-L-8h-H	0.35±0.000	0.36±0.021	0.40±0.012	0.34±0.008
I-4h-L-8h-H	0.31±0.003	0.34±0.026	0.45±0.013	0.34±0.008
Post-treatments L-8h-H-6h-I	0.46±0.006	0.57±0.031	0.57±0.021	0.49±0.009
L-8h-H-12h-I	0.33±0.003	0.39±0.010	0.45±0.015	0.32±0.007
L-8h-H-18h-I	0.27±0.003	0.30±0.022	0.38±0.026	0.29±0.003

* Significant compared to combined treatment

chromosomal aberrations after 6 h of nicotinamide post-treatment is insignificant to the combined treatment. Similar results were observed in pre-, inter-, and post-treatment benzamide with EMS or MMS. These results indicate the enhancement of adaptive response in mouse bone marrow cells with PARP inhibitors (Table 17.19). It has been observed that there is a significant increase in the NAD+ level in the liver cells of mice that received nicotinamide or benzamide conditioning and a challenging dose of EMS or MMS at all RTs. This indicates the non-involvement of PARP in adaptive response (Guruprasad and Vasudev, 2001).

In cultured human peripheral lymphocyte system, the conditioning and challenging doses of EMS or MMS were already established in the previous experiments, and the same conditioning dose of 1.5×10^{-4} M and challenging dose of 1.5×10^{-3} M of EMS and conditioning dose 1.5×10^{-5} M and challenging dose 1.5×10^{-4} M of MMS were used. Two mM of nicotinamide and 1 mM of benzamide were selected based on the earlier experiments of nicotinamide and benzamide. Further, they were selected based on the frequency of SCEs. The pre-, inter-, and post-treatments of nicotinamide or benzamide to EMS or MMS-adapted human lymphocytes produce a significantly reduced yield of SCEs compared to combined treatment (conditioning-6h-challenging) (Table 17.20). PARP inhibitors, such as nicotinamide or benzamide, enhance the EMS- or MMS-induced adaptive response *in vivo* systems of *P. pictus* and mice and *in vitro* human lymphocyte test systems. These studies also indicate the non-involvement of PARP in EMS- or MMS-induced adaptive response (Guruprasad et al., 2002).

17.10 Molecular Mechanisms of Adaptive Response

The adaptive response can arise from different stress conditions in various biological systems (Koval et al., 2020). Even though the molecular mechanism is known in prokaryotic systems at the genomic level, it is still in its infancy in eukaryotes. However, it is said that many human genes are known to be involved in DNA repair processing (Wood and Lowry, 2020). The adaptive response is probably involved in the transcription of genes responsible for activation of various signaling pathways that trigger cell defenses, which include more efficient detoxification of free radicals, DNA repair systems, induction of new proteins, enhanced antioxidant production, cell cycle regulation, and induction of apoptosis (Sekiguchi et al., 1986; Rigaud and Moustacchi, 1996; Dimova et al., 2008; Takahashi and Ohnishi, 2009). There have been numerous pieces of evidence regarding the occurrence of adaptive

TABLE 17.20

Sister Chromatid Exchanges (SCEs) Observed in EMS and MMS with Nicotinamide or Benzamide (PARP Inhibitors) Treated Human Peripheral Blood Lymphocytes at 24 H RT

	SCEs			
	Nicotinamide*		Benzamide*	
Treatments	**EMS**	**MMS**	**EMS**	**MMS**
Combined				
L-6h-H	16.9±0.95	31.49±0.41	16.27±0.19	32.37±0.53
Inter-treatments				
L-2h-I-4h-H	11.37±0.33	8.03±0.58	11.91±0.24	23.85±0.028
L-4h-I-2h-H	11.25±0.37	19.47±0.33	11.74±0.21	23.62±0.45
Pre-treatments				
I-2h-L-6h-H	6.77±0.59	16.07±0.40	10.03±0.20	19.28±0.82
I-4h-L-6h-H	8.27±0.49	14.93±0.26	9.36±0.24	18.40±0.28
Post-treatments				
L-6h-H-2h-I	11.77±0.19	20.07±0.23	10.92±0.52	23.75±0.30
L-6h-H-4h-I	11.75±0.41	19.48±0.41	11.23±0.26	24.84±0.31

* Significant compared to combined treatment

response at the molecular level. Fritz-Niggli and Schappi-Buchi (1991) demonstrated adaptive response in *Drosophila* germ cells after exposure to ionizing radiation in flies. However, they showed that the adaptive response mechanism is unrelated to standard post replication and excision repair mechanisms. Induction of alterations in the cell cycle probably by signal transduction mechanism (Meyers et al.,1992) may result in protection. However, Laval (1998) believed that the inducible protective process might be due to superoxide dismutase (SOD) expression. This enzyme can protect cells from the harmful effects of free radicals. It is also implicated that metallothionein's induction protects against oxidative damage and toxicity, resulting in an adaptive response (Sato and Bremmeer, 1993; Lazo and Pitt, 1995).

Molecular studies suggested that alternative dose-specific pathways of radio adaptive response could exist in mammalian cells: one response activated at low doses by the protein kinase C through p38 MAP kinase resulting in p53 activation, and another activated at higher doses resulting in the inactivation of ERK and JNK kinases and WIP phosphatase (Lanza et al., 2005). Low levels of damage could trigger events that signal DNA repair systems (Boreham and Mitchel, 1991; Wolff, 1998; Matsumoto et al., 2004). Experiments with restriction enzymes indicated that DNA double-strand breaks (DSBs) with blunt or cohesive ends could induce an adaptive response (Wolff, 1996, 1998). Some radiosensitive DSB repair-deficient mutants exhibited no induced radioresistance, suggesting the DSB rejoining involvement (Skov et al., 1994). Changes in chromatin conformation could result in less sensitivity to damage by the effect of a dose or increased accessibility of damaged sites to repair enzymes (Belyaev et al., 1996; Hauer and Gasser, 2017; Kleczkowska and Althaus, 1996b). DNA repair mechanisms could be responsible for the adaptive response, which has been strengthened by experiments in which an inhibitor of poly (ADP-ribose) polymerase (PARP), an enzyme implicated in rejoining of the DNA strand break, prevents the adaptive response (Wiencke et al.,1986). This was confirmed that the adaptive response could be interpreted in terms of increased non-homologous end-joining of DSB or increased homologous recombination (Marples et al., 2004; Raaphorst et al., 2006). On the other hand, experiments have demonstrated that PARP's non-involvement in the induction of adaptive response after treatment with alkylating agents in mouse bone marrow cells and *P. pictus*, an insect system (Guruprasad and Vasudev, 2001; Guruprasad et al., 2002).

When alkylating agents interact with DNA, they transfer unsubstituted alkyl groups, such as methyl or ethyl, to DNA nucleophilic sites (Lawley, 1984). Alkylation occurs at N^7, O^6, N^3, and N^2 of guanine; N^1, N^3, and N^7 of adenine; O^2, O^4, and N^3 of thymine; and O^2 and N^3 of cytosine and phosphate groups (Sega, 1984). O^6, N^7-alkylguanine, and N^3-alkyladenine are the most relevant adducts. The majority of the alkylation reaction occurs at the N^7 position of guanine. Still, it is biologically less harmful than a reaction at the O^6 position of guanine and O^4 position of thymine (Margison and O'Connor, 1990), which contribute to the mutagenicity and carcinogenicity (Loveless, 1969; Paaknan et al., 1997). N^3-alkylation contributes to toxicity by blocking DNA synthesis (Klungland et al., 1992). Two defense mechanisms maintain the genetic integrity of an organism to alkylation. The constitutive mechanism depends on the expression of the *ogt* gene encoding O^6-alkylguanine alkyltransferase and the *tag* gene encoding 3-methyl adenine DNA glycosylase. The inducible repair mechanism relies on the induction of a protein/enzyme by low doses of alkylating agents to suppress the mutagenic or clastogenic effect. Defective mutants showed that different gene products are essential for the adaptive response (Jeggo et al., 1978). The *ada, alkA, alkB,* and *aidB* are involved in adaptive response (Shevell et al.,1990) in *E. coli*, and the *ada* gene, which encodes the Ada protein, is a regulator of adaptive response (Lindahl et al., 1988). Among *ada, alkA, alkB,* and *aidB,* defective mutants *alkA* and *alkB* are sensitive to the killing but not to mutagenesis, and *aidB* mutants are resistant to killing. The *ada* mutants are sensitive to mutation and killing (Lindahl et al., 1988). The Ada protein is a 354-amino acid methyltransferase that removes alkyl groups from DNA strands and alkylated nucleotides (Yarosh et al.,1986). This 39 kDa protein, organized in two domains, each with one methyl-acceptor site, plays a dual role. It serves as a DNA repair enzyme to remove O^6-alkylguanine and the adaptive response's positive regular (Teo et al., 1984). During the repair process, this protein dealkylates O^6-alkylguanine and phosphotriester lesions of DNA by self-alkylation of cysteine residues due to the noncatalytic mode of suicidal action of the methyltransferase protein. It is considered a repair protein called "the kamikaze protein" (Wright, 1986). Alkylation is centered at positions cys-69 and cys-321. The C-terminal cys-321 acquires an alkyl group from either O^6-methylguanine or O^4-methylthymine, and the N-terminal cys-69 is required for the dealkylation of alkylphosphotriesters (Lindahl et al., 1988; Paaknan et al., 1997). Alkylation of cys-69th position converts the Ada protein as a potent activator of *ada, aidB, alkA,* and *alkB* genes, whose products are closely linked with multiple forms of alkylation damage. Higher concentrations of unmethylated *ada* can inhibit *ada* transcripts' activation by methylated Ada (Saget and Walker, 1994). However, Ada's same concentration did not deter the activation of *alkA* transcription by methylated Ada. Deleting the carboxyl-terminal 67 amino acids of Ada abolished the protein's unmethylated form to inhibit activation of ada transcription but not the ability of methylated form to activate ada or alkA protein.

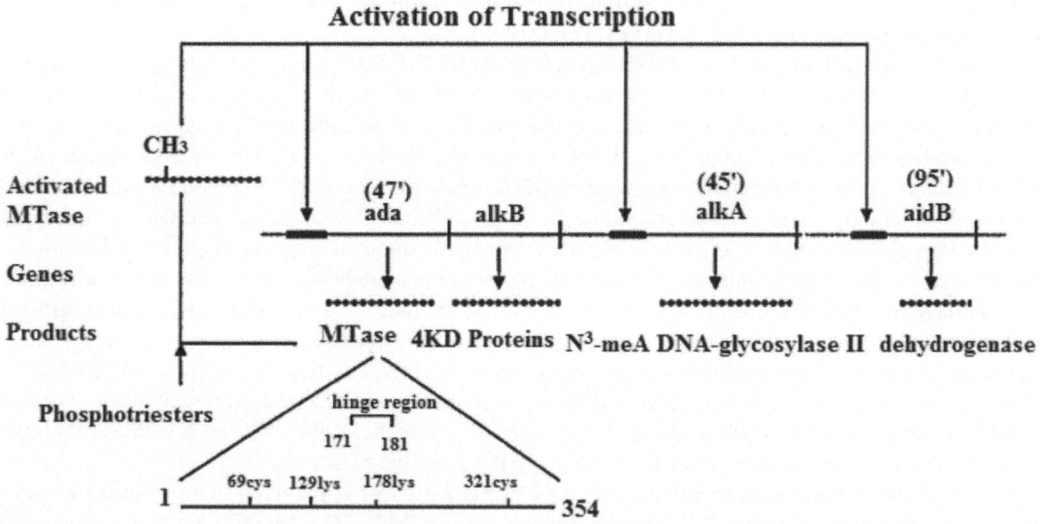

FIGURE 17.16 Genetic structure and regulation of the adaptive response in *E. coli.* Four genes are involved in the adaptive response and their corresponding proteins. Active cysteine residues and two sites adjacent to lysines are sensitive to proteolytic cleavage close to the central hinge region.

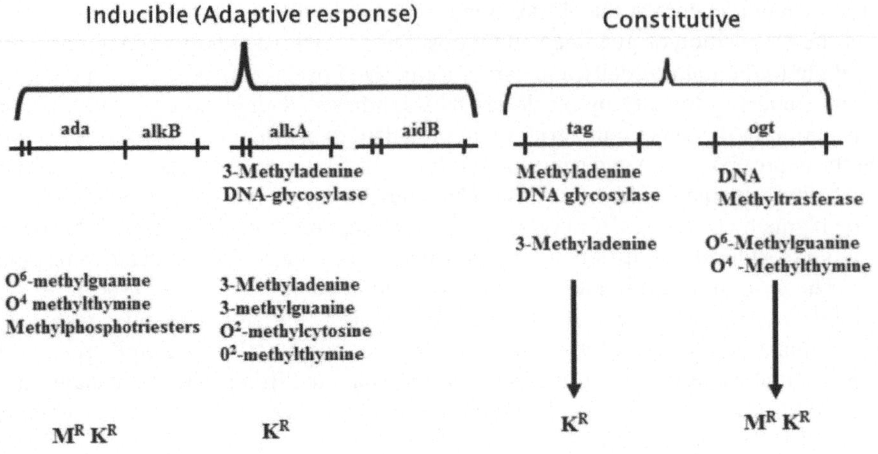

FIGURE 17.17 The genes and proteins are involved in the specific repair of DNA alkylation damage in *E. coli*. Where possible, the alkylated DNA substrates and the biological consequences of repairing these substrates are indicated; MR and KR indicate mutation resistance and killing resistance, respectively.

The second type of inducible repair enzyme in adaptive response is 3-methyladenine DNA-glycosylase II, *alkA* gene product (Evenson and Seeberg, 1982; Karran et al., 1982). A similar gene, the 3-methyladenine-DNA-glycosylase gene (*MAG*), was characterized in *Saccharomyces cerevisiae* (Chen and Samson, 1991) and induced to protect the cells against the alkylation-induced killing but not mutation. This enzyme is induced approximately 20-fold during adaptive response and removes alkylated bases by cleaving the N-glycosylic bond between the target base and deoxyribose. Thus, releasing a freebase and leaving an apurinic/apyrimidinic (AP) site is repaired by AP endonucleases. The DNA alkylpurine-glycosylases are bifunctional enzymes with a lyase activity that cleaves the phosphodiester backbone 3′ of the AP site generated by glycosylase activity (Karran et al., 1979). Also, 3-methyladenine DNA glycosylase excises the methylated bases more efficiently than the ethylated bases in *E. coli* and humans (Tudek et al., 1998). O^6-alkylguanine DNA alkyltransferase activity has been found in many species, including insects, fish, and mammals, similar to one observed in *E. coli* is named as mammalian O^6-methylguanine-methyltransferase (MGMT) (Pegg, 1990).

The molecular mechanism of adaptive response in the eukaryotic system is based on repairing proteins and their protective processes. There is evidence that DNA repair underlies the adaptive response induced by low radiation doses in human and plant cells by increasing the amount and rate of DNA repair (Lambin et al., 1994; Joiner et al., 1999; Patra et al., 2003). Lower levels of micronuclei (MN), higher numbers of apoptotic cells, and enhanced DNA repair were observed in blood mononuclear cells from residents of a high natural background radiation area (Ramsar) when irradiated with a challenging dose of γ rays (Mohammadi et al., 2006). Activation of UVB-induced adaptive response in human skin cells with p53-induced cell cycle arrest and DNA repair (Decraene et al., 2005). The repair kinetics of DNA damage in Chinese hamster V79 cells showed that the radio adaptive response could result from DNA repair mechanisms (Ikushima et al., 1996). An adaptive response was observed through reduced micronuclei formation, and neoplastic transformation in murine cells resulted from an enhanced DSB repair (Azzam et al., 1994). A reduction in deletion-type mutants in adapted cells may also result from DSB DNA repair in various cell systems (Rigaud and Moustacchi, 1996). All these studies mentioned have hypothesized the involvement of DNA repair in the adaptive response.

In the eukaryotic system, as mentioned, the DNA repair process is involved in an adaptive response mediated by O^6-methylguanine DNA methyltransferase (MGMT). The MGMT enzyme transfers the methyl group from the O^6-methylguanine DNA adduct to a cysteine residue in the enzyme and becomes irreversibly inactivated. The methyltransferase is a monomeric protein of molecular weight 18,000–21,000, 207 amino acids, containing four cysteine residues, only one of which serves as an acceptor.

The enzyme can accept alkyl groups from O^6-methylguanine and the analogous lesion O^6-ethylguanine in DNA but not N-alkylated purine residues. The transfer reaction is rapid (completed in 1 sec at 37°C) but occurs by unusual suicide kinetics, reflecting the lack of active enzyme regeneration. *Ada* reacts at a very high rate, and human *MGMT* has a lower but still high rate constant (Mitra, 2007). The number of Ada molecules is estimated to rise from 1–2 molecules in an unadapted state to ~3,000 in a fully adapted cell (Shrivastav et al., 2010). Mammalian MGMT is induced moderately (up to sixfold) (Grombacher et al., 1996). Pieper (1997) stated that in human cells, MGMT seemed to be "neither highly inducible nor repressible," while Shrivastav and his colleagues (2010) said it to be "not inducible." Human tissues and cells have a higher content of MGMT than rodents. The human liver and spleen have the highest activities, whereas the nervous and mammary glands have the lowest (Pegg and Byers, 1992). The question of whether MGMT can be described as a "true" enzyme. MGMT is considered a DNA repair enzyme, and its mechanism is only viewed as such. MGMT is an enzyme of paramount importance in defense of the genome and is widely researched and used in developing therapeutic drugs, once again highlighting its importance. However, its unusual mechanism of action does not appear to support its proposed function. Further refined research is required to get the complete picture of MGMT at the molecular level.

It is pertinent to mention here that the Nobel Prize in Chemistry for 2015 was awarded to the scientists for their remarkable research to map the proofreading and repair mechanisms adopted by proteins in a cell's DNA. DNA is the chemical code for making and sustaining life. Cells divide or replicate billions of times throughout our lifetime. Molecular machines seek to copy the code correctly. Still, random errors in their work can cause the daughter cells to die or malfunction. Proteins monitor the process of replication of DNA. They proofread the code and repaired the damage. From Sweden, Lindahl showed DNA decayed rapidly, which led him to uncover the counteracting repair process, work on the adaptive response, and identify DNA repair mechanisms. Paul Madrich, a professor of biochemistry at Duke University in the USA, discovered the correcting errors during cell division and laid bare a complex DNA-mending process called mismatch repair. Aziz Sancar, a Turkish American professor of biochemistry at the University of North Carolina in the USA, worked out how cells repair UV damage—dissected the molecular mechanism of photoreactivation. The impact of noble research is the targeted attacks on cancer, better ways to treat diabetes, and the development of life-saving treatments. These three scientists have shared the 2015 Nobel Prize for their outstanding works "for mechanistic studies of DNA repair" (The Nobel Prize in Chemistry, 2015).

17.11 Conclusion

Alkylating agents are environmental pollutants extensively used for human health and welfare, especially in cancer therapy. Thus, one must take precautionary measures if the individuals resist chemotherapy. In this direction, understanding the adaptive response may lighten future prospective treatments. The adaptive response is a novel type of repair mechanism, which is an error-free inducible type. It is ubiquitous and controlled by genes.

The adaptive response has been demonstrated in various test systems, such as *E. coli, Bacillus subtilis*, *B. thuringiensis*, *B. lichenifonnis*, *B. amyloliquefaciens*, *Caenorhabditis elegans*, and *Vibrio cholera*. Mammalian *in vitro* systems, in *in vivo* systems, such as *Vicia faba*, *H. vulgare*, *P. pictus*, *Drosophila melanogaster*, and mice.Different conditioning and challenging doses in *D. melanogaster* with different time lags showed an adaptive response in transheterozygotes. With preliminary reports on the utility of mating behavior of *D. melanogaster* to understand toxicity, the same protocol has been used to analyze the adaptive response. Positive results were obtained from an extensive experimental schedule. It has been shown that mating behavior can be used to analyze the adaptive response.

Hyperthermia-mediated amelioration of drug-induced adaptive response has been demonstrated in the meiotic cells of *P. pictus* and mitotic cells of mice. Cancerous cells in the form of ascites tumor cells in mice have been used to screen the adaptive response. The results have revealed that there is a presence of an adaptive response. Diseased conditions in the form of diabetic mice have been employed to test the presence of adaptive response. Induction of adaptive response has been demonstrated in diabetic mice.

These results underline the importance of further experimentations using different alkylating agents on the one hand and non-alkylating chemotherapeutic agents on the other. If the data proves the contention of adaptive response, treatment modalities need to be looked into, and precautionary measures to overcome drug inefficacy need to be taken.

17.12 Acknowledgment

The authors thank the University of Mysore, Kuvempu University, and the Central University of Kerala for the facilities and the University Grants Commission for awarding the project (F.3-58/93SR-II).

17.13 Declaration of Interest

The authors declare that they have no competing interests.

REFERENCES

Amini, M., 2014. Ethyl Methanesulfonate. Module in biomedical sciences. In: *Encyclopedia of Toxicology* (3rd edition), Academic Press, Elsevier Inc. pp. 522–524.

Ather, A., Ahmed, Z., Riazuddin, S., 1984. Adaptive response or *Micrococcus luteus* to alkylating chemicals. *Nucl. Acids Res.*, 12: 2111–2126.

Azizova, T.V., Bannikova, M.V., Grigoryeva, E.S., Rybkina, V.L., 2018. Risk of malignant skin neoplasms in a cohort of workers occupationally exposed to ionizing radiation at low dose rates. *PLoS One*, 13(10): e0205060.

Azzam, E.I., Raaphorst, G.P., Mitchel, R.E., 1994. Radiation-induced adaptive response for protection against micronucleus formation and neoplastic transformation in C3H 10T1/2 mouse embryo cells. *Radiat. Res.*, 138: S28–31.

Beckwith, E.J., French, A.S., 2019. Sleep in Drosophila and its context. *Front. Physiol.*, 10: 1167.

Belyaev, I.Y., Spivak, I.M., Kolman, A., Harms-Ringdahl, M., 1996. Relationship between radiation-induced adaptive response in human fibroblasts and changes in chromatin conformation. *Mutat. Res.*, 358: 223–230.

Billeter, J.C., Goodwin, S.F., O'Dell, K.M., 2002. Genes mediating sex-specific behaviors in Drosophila. *Adv. Genet.* 47: 87–116.

Boreham, D.R., Mitchel, R.E., 1991. DNA lesions that signal the induction of radioresistance and DNA repair in yeast. *Radiat. Res.*, 128: 19–28.

Bosi, A., Olivieri, G., 1989. Variability of the adaptive response to ionizing radiations in humans. *Mutat. Res.*, 211: 13–17.

Cai, L., Liu, S.Z., 1992. Study on the mechanism of cytogenetic adaptive response induced by low dose radiation. *Chin. Med. J. (Engl).*, 105: 277–283.

Chankova, S.G., Dimova, E., Dimitrova, M., Bryant, P.E., 2007. Induction of DNA double-strand breaks by zeocin in *Chlamydomonas reinhardtii,* and the role of increased DNA double-strand breaks rejoining in the formation of an adaptive response. *Radiat. Environ. Biophys.*, 46: 409–416.

Chatterjee, N., Walker, G.C., 2017. Mechanisms of DNA damage, repair, and mutagenesis. *Environ. Mol. Mutagen*, 58(5): 235–263.

Chatterjee, S., Berger, N.A., 1994. Growth phase-dependent response to DNA damage in poly (ADP-ribose) polymerase deficient cell lines: basis for a new hypothesis describing the role of PARP in DNA replication repair. *Mol. Cell. Biochem.*, 138: 61–69.

Chen, J., Samson, L., 1991. Induction of S. cerevisiae MAG 3-methyladenine DNA glycosylase transcript levels in response to DNA damage. *Nucl. Acids Res.*, 19(23): 6427–6432.

Clancy, S., 2008. DNA damage and repair: mechanisms for maintaining DNA integrity. *Nat. Educ.*, 1(1): 103.

Cleaver, J., Bodwell, W.J., Morgan, W.F., Zelle, B., 1983. Differences in the regulation by poly(ADP-ribose) of repair of DNA damage from alkylating agents and ultraviolet light according to cell type. *J. Biol. Chem.*, 258: 9059–9068.

Cooke, M.S., Evans, M.D., Dizdaroglu, M., Lunec, J., 2003. Oxidative DNA damage: mechanisms, mutation, and disease. *FASEB J.*, 17(10): 1195–1214.

Cortes, F., Dominguez, I., Mateos, S., Pinero, J., Mateos, J.C., 1990. Evidence for an adaptive response to radiation damage in plant cells conditioned with X-rays or incorporated tritium. *Int. J. Radiat. Biol.*, 57: 537–541.

Da Silva, J., 2016. DNA damage induced by occupational and environmental exposure to miscellaneous chemicals. *Mutat. Res.*, 770: 170–182.

Dada Khalandar, B.B., Vasudev, V., 2015. Inducible protective processes in animal systems XIV: cytogenetic adaptive response induced by EMS or MMS in bone marrow cells of diabetic mouse. *Egyptian J. Med. Hum. Genet.*, 16(4): 301–312.

Dauwalder, B., 2008. Systems behavior: of male courtship, the nervous system and beyond in *Drosophila. Curr. Genom.*, 9(8): 517–524.

De Bont, R., and van Larebeke, N., 2004. Endogenous DNA damage in humans: a review of quantitative data. *Mutagenesis*, 19: 169–185.

Decraene, D., Smaers, K., Maes, D., Matsui, M., Declercq, L., Garmyn, M., 2005. A low UVB dose, with the potential to trigger a protective p53-dependent gene program, increases the resilience of keratinocytes against future UVB insults. *J. Invest. Dermatol.*, 125: 1026–1031.

Demir, E., Kocaoglu, S., Kaya, B., Marcos, R., 2010. Induction of adaptive response in Drosophila after exposure to low doses of UVB. *Int. J. Radiat. Biol.*, 86: 957–963.

Dhananjaya, S.G., 2005. *Studies on Adaptive Response of Drosophila melanogaster to a Few Alkylating Agents.* Thesis. Kuvempu University, Shivamogga, Karnataka, India.

Dillingham, M.S., Stephen, C.K., 2008. RecBCD enzyme and the repair of double-stranded DNA breaks. *Microbiol. Mol. Biol. Rev.*, 72: 642–671.

Dimova, E.G., Bryant, P.E., Chankova, S.G., 2008. Adaptive response—some underlying mechanisms and open questions. *Genet. Mol. Biol.*, 31: 2.

Divyashree, D.U., Vasudev, V., 2021. Utilization of courtship behavioural elements to understand the toxicity of Methyl methanesulfonate in *Drosophila melanogaster. Toxicol. International*. 28: 245-253.

Divyashree, D.U., Krishna, M.S., Vasudev, V., 2021. Analysis of courtship behaviour as a parameters of toxicity of Ethyl methanesulfonate (EMS) in *Drosophila melanogaster. J. Advanced Zool.* 42(01): 34-42.

Duncan, F.N., 1930. Some observations on the biology of the male *Drosophila melanogaster. Am. Nat.*, G4: 545–551.

Evenson, G., Seeberg, E., 1982. Adaptation to alkylation resistance involves induction of DNA glycosylase. *Nature*, 296: 113–115.

Fishel, R., 2015. Mismatch repair. *J. Biol. Chem.*, 290(44): 26395–26403.

Frei, H., Würgler, F.E., 1988. Statistical methods to decide whether mutagenicity test data from Drosophila assays indicate a positive, negative, or inconclusive result. *Mutat. Res.*, 203: 297–308.

Fritz-Niggli, H., Schappi-Buchi, C., 1991. Adaptive response to dominant lethality of mature (class A) and immature (class B) Oocytes of *D. melanogaster* to low doses of ionizing radiation: effects in repair-proficient (yw) and repair-deficient strains (mei 41D5 and mus 302D1). *Int. J. Radiat.*, 59: 1175–1184.

Griffiths, A.J.F., Miller, J.H., Suzuki, D.T., et al., 2000. Spontaneous mutations. In: *An Introduction to Genetic Analysis* (7th edition). New York: W. H. Freeman.

Grombacher, T., Mitra, S., Kaina, B., 1996. Induction of the alkyltransferase (MGMT) gene by DNA damaging agents and the glucocorticoid dexamethasone and comparison with the response of base excision repair genes. *Carcinogenesis*, 17: 2329–2336.

Guruprasad, K.P., Vasudev, V., 2001. Inducible processes in animal systems. VIII. Enhancement of adaptive response by nicotinamide. *Mutagenesis*, 16: 257–263.

Guruprasad, K.P., Vasudev, V., Anilkumar, M.N., Chethan, S.A., 2002. Inducible protective processes in animal systems. X. Influence of nicotinamide in methyl methanesulphonate-adapted mouse bone marrow cells. *Mutagenesis*, 17: 108.

Gurushankara, H.P., Venu, R., Vasudev, V., 2018. Hyperthermia enhances methyl methanesulfonate-induced adaptive response in meiotic cells of grasshopper *Poecilocerus pictus. Mutagenesis*, 33: 215–224.

Haber, J.E., Gregorz, I., Anna, M., Neal, S., 2004. Repairing a double-stranded chromosome break by homologus recombination: revisiting Robin Holliday's model. *Phil. Trans. R. Soc. Lond.* 359: 79–86.

Harish, S.K., Guruprasad, K.P., Mahmood, R., Vasudev, V., Manjunath, K.R., Chethan, G.K., 1998. Adaptive response to low dose of EMS or MMS in human peripheral blood lymphocytes. *Indian J. Exp. Biol.*, 36: 1147–1150.

Hauer, M.H., Gasser, S.M., 2017. Chromatin and nucleosome dynamics in DNA damage and repair. *Genes Dev.*, 31(22): 2204–2221.

Helleday, T., Lo, J., Vangent, D., Engelward, B., 2007. DNA double-strand break repair: from mechanistic understanding to cancer treatment. *DNA Repair*, 6: 923–935.

Hillova, J., Drasil, V., 1967. The inhibitory effect of iodoacetamide on recovery from sub-lethal damage in *Chlamydomonas reinhardti*. *Int. J. Radiat. Biol. Relat. Stud. Phys. Chem. Med.*, 12: 201–208.

Ikushima, T., Aritomi, H., Morisita, J., 1996. Radioadaptive response: efficient repair of radiation-induced DNA damage in adapted cells. *Mutat. Res.*, 358: 193–198.

International Diabetic Federation (IDF), 2020. Diabetes facts & figures on diabetes worldwide. In *The IDF Diabetes Atlas* (9th edition 2019). www.diabetesatlas.org.

Jeggo, P., Defais, M., Samson, L., Schendel, P., 1978. An adaptive response of *E. coli*, to low level of alkylating agent. In: Molineux, I., Kohiyama, M. (Eds.), *DNA Synthesis, Present, and Future*. New York and London: Plenum, pp. 1011–1024.

Jiang, H., Li, W., Li, X., Cai, L., Wang, G., 2008. Low-dose radiation induces adaptive response in normal cells, but not in tumor cells: *in vitro* and *in vivo* studies. *J. Radiat. Res.*, 49: 219–230.

Joiner, M.C., Lambin, P., Marples, B., 1999. Adaptive response and induced resistance. *CR Acad. Sci. III*, 322: 167–175.

Juni, N., Yamamoto, D., 2009. Genetic analysis of chaste, a new mutation of *Drosophila melanogaster* characterized by extremely low female sexual receptivity. *J. Neurogenet.*, 23(3): 329–340.

Kaina, B., 1982. Enhanced survival and reduced mutation and aberration frequencies induced in V79 Chinese hamster cells pre-exposed to low levels of methylating agents. *Mutat. Res.*, 93: 195–211.

Karran, P., Hjelmgren, T., Lindahl, T., 1982. Induction of a DNA glycosylase for N-methylated purines is part of the adaptive response to alkylating agents. *Nature*, 296: 770–773.

Karran, P., Lindahl, T., Griffm, B., 1979. Adaptive response to alkylating agents involves alteration in situ of O^6-methyl-guanine residues in DNA. *Nature*, 280: 76–77.

Kaya, B., Creus, A., Velazquez, A., Yanikoglu, A., Marcos, R., 2000. Induction of an adaptive response in *Drosophila* imaginal disc cells exposed *in vivo* to low doses of alkylating agents. *Mutagenesis*, 15(4): 337–340.

Kimball, R.F., 1980. Further studies on the induction of mutation in *Haemophilus influenzae* by N-methyl-N'-nitro-N-nitrosoguanidine: lack of an inducible error-free repair system and the effect of exposure medium. *Mutat. Res./Fundam. Mol. Mech. Mutagen.*,72: 361–372.

Kleczkowska, H.A., Althaus, F.R., 1996a. The role of poly(ADP-ribosyl)altion in the adaptive response. *Mutat. Res.*, 358: 215–221.

Kleczkowska, H.E., Althaus, F.R., 1996b. Biochemical changes associated with the adaptive response of human keratinocytes to N-methyl-N-nitro-N-nitrosoguanidine. *Mutat. Res.*, 368: 121–131.

Klungland, A., Fairbaim, L., Watson, A.J., Margison, G.P., Seeberg, E., 1992. Expression of the *E. coli* methyladenine DNA glycosylase I gene in mammalian cells reduces the toxic and mutagenic effects of methylating agents. *EMBO J.*, 11: 4439–4444.

Koval, L., Proshkina, E., Shaposhnikov, M. et al., 2020. The role of DNA repair genes in radiation-induced adaptive response in *Drosophila melanogaster* is differential and conditional. *Biogerontology*, 21: 45–56.

Kumar, T., 2012. *Contribution to the Knowledge of Induction of Adaptive Response in Drosophila melanogaster Exposed to Ethyl Nitrosourea and Methyl Nitrosourea*. Thesis. Submitted to Kuvempu University, Shivamogga, Karnataka, India.

Kuzminov, A., 2013. Pyrimidine dimers. In: Stanley, M., Kelly, H. (Eds.), *Brenner's Encyclopaedia of Genetics* (2nd edition) Vol 1. San Diego: Academic Press.

Lambin, P., Fertil, B., Malaise, E.P., Joiner, M.C., 1994. Multiphasic survival curves for cells of human tumor cell lines: induced repair or hypersensitive subpopulation? *Radiat. Res.*, 138: 32–36.

Lanza, V., Pretazzoli, V., Olivieri, G., Pascarella, G., Panconesi, A., Negri, R., 2005. Transcriptional response of human umbilical vein endothelial cells to low doses of ionizing radiation. *J. Radiat. Res.*, 46: 265–276.

Laval, F., 1998. Pretreatment with oxygen species increases the resistance of mammalian cells to hydrogen peroxide and gamma-rays. *Mutat. Res.*, 201: 73–79.

Laval, F., Laval, J., 1984. Adaptive response in mammalian cells: crossreactivity of different pretreatments on cytotoxicity as contrasted to mutagenicity. *Proc. Natl. Acad. Sci. USA*. 81: 1062–1066.

Lawley, P.D., 1984. Carcinogenesis by alkylating agents. In: Searle, E.C. (Ed.), *Chemical Carcinogens*. Washington, DC: American Chemical Society, pp. 326–484.

Lazo, J.S., Pitt, B.R., 1995. Metallothionein and cell death by anticancer drugs. *Annu. Rev. Pharmacol. Toxicol.*, 35: 635–653.

Lee, J.H., Choi, E.S., Oh, K.S., Lee, D.W., Chang, J.H., Um, K., 2000. Effect of low dose mutagens on adaptive response and plasma membrane glycoconjugates in sarcoma 180 cells. *Korean J. Biol. Sci.*, 4: 293–297.

Lindahl, T., Sedgwick, B., Sekiguchi, M., Nakabeppu, Y., 1988. Regulation and expression of the adaptive response to alkylating agents. *Annu. Rev. Biochem.*, 57: 133–157.

Loveless, A., 1969. Possible relevance of O^6-alkylation of deoxyguanosine to the mutagenicity and carcinogenicity of nitrosamines and nitrosamides. *Nature*, 223: 206–207.

Luning, K.G., 1966. Drosophila test in pharmacology. *Nature*, 209: 84–86.

Mackay, T.F., Heinsohn, S.L., Lyman, R.F., Moehring, A.J., Morgan, T.J., Rollmann, S.M., 2005. Genetics and genomics of Drosophila mating behavior. *Proc. Natl. Acad. Sci. USA.*, 102: 6622–6629.

Mahadimane, P.V., Vasudev, V., 2014. Inducible protective processes in animal systems XIII: comparative analysis of induction of adaptive response by EMS and MMS in *Ehrlich Ascites Carcinoma* cells. *Scientifica*, 2014: 703136.

Maki, H., 2002. Origins of spontaneous mutations: specificity and directionality of base-substitution, frameshift, and sequence-substitution mutageneses. *Ann. Rev. Genet.*, 36: 279–303.

Margison, G.P., O'Connor, P.J., 1990. Biological consequences of reaction with DNA: role of the specific lesion. In: Cooper, C.S., Grover, P.L. (Eds.), *Handbook of Experimental Pharmacology*. Berlin: Springer, pp. 547–571.

Marples, B., Wouters, B.G., Collis, S.J., Chalmers, A.J., Joiner, M.C., 2004. Low-dose hyper-radiosensitivity: a consequence of ineffective cell cycle arrest of radiation-damaged G2-phase cells. *Radiat. Res.*, 161: 247–257.

Massey, J.H., Chung, D., Siwanowicz, I., Stern, D.L., Wittkopp, P.J., 2019. The yellow gene influences Drosophila male mating success through sex comb melanization. *eLife*, 8: e49388.

Matsumoto, H., Takahashi, A., Ohinishi, T., 2004. Radiation-induced adaptive response and bystander effects. *Biol. Sci. Space*, 18: 247–254.

Meyers, M., Schea, R.A., Petrowski, A.E., Seabury, H., McLaughlin, P.W., Lee, I., Lee, S.W., Boothman, D.A., 1992. Role of X-ray inducible genes and proteins in adaptive survival responses. In Sugahara, T., Sagan, L.A., Aoyama, T. (Eds.), *Low Dose Irradiation and Biological Defense Mechanisms*. Amsterdam, The Netherlands: Elsevier, pp. 263–266.

Mitra, S., 2007. MGMT: a personal perspective. *DNA Repair*, 6(8): 1064–1070.

Mohammadi, S., Taghavi-Dehaghani, M., Gharaati, M.R., Masoomi, R., Ghiassinejad, M., 2006. Adaptive response of blood lymphocytes of inhabitants residing in high background radiation areas of Ramsar-micronuclei, apoptosis, and comet assays. *J. Radiat. Res.* 47: 279–285.

The Nobel Prize in Chemistry, 2015. *Nobel Media AB 2020*. Friday, 4 December 2020. NobelPrize.org. www.nobelprize.org/prizes/chemistry/2015/summary/.

O'Hanlon, K.A., Margison, G.P., Hatch, A., Fitzpatrick, D.A., Owens, R.A., Doyle, S., Jones, G.W., 2012. Molecular characterization of an adaptive response to alkylating agents in the opportunistic pathogen *Aspergillus fumigatus*. *Nucl. Acids Res.*, 40(16): 7806–7820. https://doi.org/10.1093/nar/gks522

Olivieri, G., Bodycote, J., Wolff, S., 1984. Adaptive response of human lymphocytes to low concentrations of radioactive thymidine. *Science*, 223: 594–597.

Paaknan, S.R., Sung, C., Clark, N.D., 1997. Specificity of DNA repair methyltransferase determined by competitive inactivation with oligonucleotide substrates: evidence that *Escherichia coli* Ada repairs O-methylguanine and 0^6-methylamine with similar efficiency. *Biochemistry*, 36: 11118–11124.

Panda, K.K., Patra, J., Panda, B.B., 1997. Persistence of cadmium-induced adaptive response to genotixicity of maleic hydrazide and methyl mercuric chloride in root meristem cells of *Allium cepa* L.: differential inhibition by cycloheximide and buthionine sulfoximine. *Mutat. Res./Genet. Toxicol. Environ. Mutagen.*, 389: 129–139.

Park, S.H., Lee, Y., Jeong, K., Yoo, S.Y., Cho, C.K., Lee, Y.S., 1999. Different induction of adaptive response to ionizing radiation in normal and neoplastic cells. *Cell Biol. Toxicol.*, 15: 111–119.

Patra, J., Panda, K.K., Panda, B.B., 1997. Differential induction of adaptive responses by paraquat and hydrogen peroxide against the genotoxicity of methyl mercuric chloride, maleic hydrazide, and ethyl methane sulfonate in plant cells *in vivo*. *Mutat. Res./Genet. Toxicol. Environ. Mutagen.*, 393: 215–222.

Patra, J., Sahoo, M.K., Panda, B.B., 2003. Persistence and prevention of aluminium- and paraquat-induced adaptive response to methyl mercuric chloride in plant cells *in vivo*. *Mutat. Res.*, 538: 51–61.

Pegg, A.E., 1990. Properties of mammalian O^6-alkylguanine DNA transferases. *Mutat. Res.*, 233: 165–175.

Pegg, A.E., Byers, T.L., 1992. Repair of DNA containing O^6-alkylguanine. *FASEB J.*, 6(6): 2302–2310.

Pieper, R.O., 1997. Understanding and manipulating O^6-methylguanine-DNA methyltransferase expression. *Pharmacol. Ther.*, 74(3): 285–297.

Poplawski, T., 2019. DNA Damage Response as a pharmacological target for cancer and infectious diseases. *Curr. Med. Chem.*, 26: 1423–1424.

Raaphorst, G.P., Li, L.F., Yang, D.P., 2006. Evaluation of adaptive responses to cisplatin in normal and mutant cell lines with mutations in recombination repair pathways. *Anticancer Res.*, 26: 1183–1187.

Reef, R., Dunn, S., Levy, O., Dove, S., Shemesh, E., Brickner, I., Leggat, W., Hoegh-Guldberg, O. 2009. Photoreactivation is the main repair pathway for UV-induced DNA damage in coral planulae. *J. Exp. Biol.*, 212(17): 2760–2766.

Riaz Mahmood, Vasudev, V., 1990. Inducible protective processes in animal systems: I. Clastogenic adaptation triggered by Ethyl methanesulfonate (EMS) in *Poecilocerus pictus*. *Biol. Zent. Bi.*, 109: 41–43.

Riaz Mahmood, Vasudev, V., 1991. Inducible protective processes in animal systems: II. Absence of adaptive response when mitotic cells of mouse are exposed to low dose of EMS and challenged after short time-lag. *Cell. Chromosome. Res.*, 14: 15–16.

Riaz Mahmood, Vasudev, V., 1992. Inducible protective processes in animal systems: III. Adaptive response after meiotic cells of grasshopper *Poecilocerus pictus* to the low dose of Ethyl methanesulfonate (EMS). *Mutat. Res.*, 283: 243–247.

Riaz Mahmood, Vasudev, V., 1993. Inducible protective processes in animal systems: IV. Adaptation of mouse bone marrow cells to a low dose of Ethyl methanesulfonate. *Mutagenesis*, 8: 83–86.

Riaz Mahmood, Vasudev, V., Harish, S.K., Guruprasad, K.P., 1996. Inducible protective processes in animal systems: adaptive response to the low dose of methyl methanesulfonate in mouse bone marrow cells. *IJEB*, 34: 502–507.

Rieger, R., Michaelis, A., 1993. Combined application of MH and TEM for conditioning treatment simultaneously triggers two protective functions against induction of chromatid aberrations in *Vicia faba* root tip meristem cells. *Mutat. Res.*, 302: 161–164.

Rieger, R., Michaelis, A., Nicoloff, H., 1984. "Clastogenic adaptation" of the *Vicia faba* root tip meristem as affected by various treatment parameters. *Mutat. Res.*, 140: 99–102.

Rieger, R., Michaelis, A., Nicolofif, H., 1982. Inducible repair processes in plant root tip meristem? "Below additivity effects" of unequally fractionated clastogen concentrations. *Biol. Lent. BL.* 101: 125–138.

Rieger, R., Michaelis, A., Schubert, I., Veleminsky, J., Gichner, T., Angelis, K.J., 1988. Induction of chromatid aberrations by TEM and maleic hydrazide is differently affected by pretreatment of *Vicia faba* root tip meristems with methyl iodide. *Mutat. Res.*, 208: 101–104.

Rigaud, O., Moustacchi, E., 1996. Radioadaptation for gene mutation and the possible molecular mechanisms of the adaptive response. *Mutat. Res.*, 358: 127–134.

Rodgers, K., McVey, M., 2016. Error-prone repair of DNA double-strand breaks. *J. Cell. Physiol.*, 231(1): 15–24.

Saget, B.M., Walker, G.C., 1994. The Ada protein acts as both a positive and a negative modulator of *Escherichia coli* response to methylating agents. *Proc. Natl. Acad. Sci., U.S.A.*, 91: 9730–9734.

Samson, L., Cairns, J., 1977. A new pathway for DNA repairing in *E.coli. Nature*, 267: 2SI–2S3.

Samson, L., Derfler, B., Bossalis, M., Call, K., 1991. Cloning and characterization of a 3-methhyladenine DNA glycosylase cDNA from human cells whose gene maps to chromosome 16. *Proc. Natl. Acad. Sci. U.S.A.*, 88: 9127–9131.

Samson, L., Schwartz, J.L., 1980. Evidence for an adaptive DNA repair pathway in CHO and human skin fibroblast cell lines. *Nature*, 287: 861–863.

Sankaranarayanan, K., Duyn, A.V., Loos, M.J., Natarajan, A.T., 1989. Adaptive response of human lymphocytes to low-level radiation from radioisotopes or X-rays. *Mutat. Res.*, 211: 7–12.

Sato, M., Bremmer, I., 1993. Oxygen-free radicals and metallothionein. Free Radio. Biol. Med., 14: 325–337.

Savina, N., Dalivelya, O., Kuzhir, T., 2003. Adaptive response to alkylating agents in the Drosophila sex-linked recessive lethal assay. Mutat. Res., 535: 195–204.

Savir, Y., Tlusty, T., 2010. RecA-mediated homology search as a nearly optimal signal detection system. *Mol. Cell.*, 40: 388–396.

Schaffer, M., Schwarz, S.B., Kulka, U., Busch, M., Dahmke, E., 2004. Adaptive doses of irradiation-an approach to a new therapy concept for bladder cancer? *Radiat. Environ. Biophys.*, 43: 271–276.

Sega, A.G., 1984. A review of the genetic effects of ethyl methanesulfonate. *Mutat. Res.*, 134: 113–142.

Sekiguchi, M., Kondo, H., Sakumi, K., Nakabeppu, Y., 1986. Molecular mechanism of adaptive response to alkylating agents. In: Shankel, D.M., Hartman, P.E., Kada, T., Hollaender, A., Wilson, C.M., Kuny, G. (Eds.), *Antimutagenesis, and Anticarcinogenesis Mechanisms*. Basic Life Sciences, 39. Boston, MA: Springer.

Shadley, J.D., Wolff, S., 1987. Very low doses of X-rays can cause human lymphocytes to become less susceptible to ionizing radiation. *Mutagenesis*, 2: 95–96.

Shevell, D.E., Friedman, B.M., Walker, G.C., 1990. Resistance to alkylation damage in *Escherichia coli*: role of the ADA protein in the induction of the adaptive response. *Mutat. Res.*, 233: 53–72.

Shrivastav, N., Li, D., Essigmann, J.M., 2010. Chemical biology of mutagenesis and DNA repair: cellular responses to DNA alkylation. *Carcinogenesis*, 31, 59–70.

Singer, B., Grunberger, D., 1983. *Molecular Biology of Mutagens and Carcinogens*. New York: Plenum Press.

Singh, B.N., Singh, A., 2016. The genetics of sexual behavior in *Drosophila*. *Adv. Genom. Genet.*, 6: 1–9.

Skov, K., Marples, B., Matthews, J.B., Joiner, M.C., Zhou, H., 1994. A preliminary investigation into the extent of increased radioresistance or hyper-radiosensitivity in cells of hamster cell lines known to be deficient in DNA repair. *Radiat. Res.*, 138: S126–129.

Smith, G.R., 2012. How RecBCD enzyme and Chi promote DNA break repair and recombination: a molecular biologist's view. *Microbiol. Mol. Biol. Rev.*, 76: 217–228.

Sowa, M.B., Goetz, W., Baulch, J.E., Lewis, A.J., Morgan, W.F., 2011. No evidence for a low linear energy transfer adaptive response in irradiated RKO cells. *Radiat. Protect. Dosim.*, 143: 311–314.

Spieth, H.T., 1952. Mating behavior within the genus Drosophila (Diptera). *Bull. Am. Mus. Nat. Hist.*, 99: 395–474.

Spieth, H.T., Ringo, J.M., 1983. *Mating Behavior and Sexual Isolation in Drosophila* (Vol. 3c). New York: Academic Press.

Steven, A., William, H., 2018. Effects of environmental factors on DNA: damage and mutations. In: Török, B., Dransfield, T. (Eds.), *Green Chemistry*. Elsevier, pp. 109–128.

Sturtevant, A.H., 1915. Experiments on sex recognition and the problem of sexual selection in Drosophila. *J. Anim. Behav.* 5: 351–366.

Takahashi, A., Ohnishi, A., 2009. Molecular mechanisms involved in adaptive responses to radiation, UV light, and heat. *J. Radiat. Res.*, 50(5): 385–393.

Teo, I.B., Sedgwick, B., Demple, B., Li, B., Lindahl, T., 1984. Induction of resistance to alkylating agents in *E. coli*: the ada+ gene product serves both as a regulatory protein and as an enzyme for the repair of mutagenic damage. *EMBO J.*, 3: 2151–2157.

Tubbs, A., Nussenzweig, A., 2017. Endogenous DNA damage as a source of genomic instability in cancer. *Cell*, 168(4): 644–656.

Tudek, B., VanZeeland, A.A., Kusmierek, J.T., Laval, J., 1998. Activity of *Escherichia coli* DNA-glycosylases on DNA damaged by methylating and ethylating agents and influence of 3-substituted adenine derivatives. *Mutat. Res.*, 407: 169–176.

Vasudev, V., Gurushankara, H.P., 2016. *Understandings in Cytogenetic Techniques*. New Delhi: Today and Tomorrow Publishers.

Vasudev, V., Gurushankara, H.P., Mahadimane, V.P., Khalandar, D., 2013. Effects of fungicide Dithane M 45 in *Drosophila melanogaster* on courtship behavior. *Drosophila Inform. Serv.*, 96: 94–98.

Vasudev, V., Krishnamurthy, N.B., 1982. Effect of Dithane M-45 on the rate of development, viability, morphology and fecundity in *Drosophila melanogaster*. *J. Mys. Univ.*, 29: 79–86.

Vasudev, V., Obe, G., 1987. Effect of heat treatment on chromosomal aberrations induced by the alkylating agent trenimon or the restriction endonuclease Alu I in Chinese hamster ovary (CHO) cells. *Mutat. Res.* 178: 81–90.

Veleminsky, J., Gichner, T., Satava, J., 1983. Reduction in the frequency of N-methyl-N-nitrosourea-induced somatic mutations in Tradescantia by pretreatment with low doses of alkylating agents. *Mutat. Res. Lett.*, 122: 229–234.

Venu, R., Gurushankara, H.P., Dada Khalandar, Vasudev, V., 2016. Inducible protective processes in animal systems XV: hyperthermia enhances the Ethyl methanesulfonate induced adaptive response in meiotic cells of grasshopper *Poecilocerus pictus*. *Egyptian J. Med. Hum. Genet.*, 17: 173–184.

Vogel, E.W., Graf, U., Frei, H.J., Nivard, M.M., 1999. The results of assays in Drosophila as indicators of exposure to carcinogens. *IARC Sci. Publ.*, 146: 427–470.

Wallace, S.S., 2014. Base excision repair: a critical player in many games. *DNA Repair*, 19: 14–26.

Wang, S., Jiang, G., Yu, H., Liu, X., Xu, C., 2013. Effect of low-dose X-ray radiation on adaptive response in gastric cancer cell. *Chinese-German J. Clin. Oncol.*, 12: 171–174.

Wiencke, J.K., 1987. Nicotinamide deficiency in human lymphocytes prevents (3H) thymidine-induced adaptive response for the repair of X-ray induced Chromosomal damage. *Exp. Cell Res.*, 171: 518–523.

Wiencke, J.K., Afzal, V., Olivieri, G., Wolff, S., 1986. Evidence that the [H] thymidine-induced adaptive response of human lymphocytes to subsequent doses of X-rays involves the induction of a chromosomal repair mechanism. *Mutagenesis*, 1: 375–380.

Wolff, S., 1996. Aspects of the adaptive response to very low doses of radiation and other agents. *Mutat. Res.*, 358: 135–142.

Wolff, S., 1998. The adaptive response in radiobiology: evolving insights and implications. *Environ. Health Persp.* 106: S277–283.

Wood, R., Lowry, M., 2020. Human DNA repair genes. An update of the table cited in Wood RD, Mitchell M, & Lindahl T, Mutation Research, 2005, in Science, 2001, in the reference book DNA Repair and Mutagenesis, 2nd edition, 2006 Nature Reviews Cancer, 2011.

Wright, P., 1986. New move by cancer scientists. *The Times London*, 31 July.

Yarosh, D.B., Hurst-Calderone, Babich, M.A., Day, R.S., 1986. Inactivation of 0^6-methylguanine-DNA methyltransferase and sensitization of human tumour cells of killing by chloroethyl nitrosourea by O6-methylguanine as a freebase. *Cancer Res.* 46: 1663–1668.

18

Seasonal Abundance, Status, and Diversity of Hymenopterans (Insecta: Hymenoptera) in a Protected Landscape

Manoj Kumar Arya, Hem Chandra, Surabhi Bisht, and Aman Verma

CONTENTS

18.1 Introduction

The global diversity of hymenopteran fauna is about 1,53,088 species (Zhang, 2013) of which approximately 10,605 species are known from India. Hymenopterans not only play crucial role in maintenance of ecosystem but also boots up human economy by, for example, providing honey and wax (Hussain et al., 2019). These insects originated in the Triassic Period and are adapted to specific habitats or host, predators, and parasitoids. Hymenopteran insects, due to their economic and biological importance, are considered to be one of the most beneficial animal groups. Majority of these insects are valuable to both man and environment, as they make food and pollinate flowers, and small number of them are pests of fruits, crops, and forests (Gupta, 1995, 1997). Hymenopteran insects are also well recognized for providing valuable ecosystem services in the form of pollination for the successful fertilization of flowers, the formation of fruits and seeds, and the sustainable production in the forest and agro-ecosystems (Kumar et al., 2019). Several species of bees and other associated pollinators are responsible for the successful fertilization and production of approximately 87 species of the world's leading food crops (e.g., fruits, vegetables, and seeds), comprising 35% of global food production (Losey and Vaughan, 2006; Klein et al., 2007; Gallai et al., 2009). Various environmental factors, such as habitat structure, abiotic and biotic factors, altitude, seasonality, vegetation types, flowering resources, and microclimatic conditions also influence the abundance, richness, and diversity of hymenopteran insects (Wardell et al., 1998; Klein et al., 2002; Tylianakis et al., 2006; Kumar et al., 2019). In recent past, studies on hymenopteran species' composition, richness, abundance, and diversity in the different ecosystems of the world have been carried out by various workers (Singh and Singh, 1993; Jonathan, 1995; Gupta, 1995, 1997; Uniyal and Singh, 1996; Joshi, 1997; Ent and Shaw, 1998; Singh et al., 2010; Arya and Joshi, 2011; Kannagi et al., 2013; Arya et al., 2013; Gupta, 2013; Azim et al., 2014; Sharma et al., 2016; Sheeja and Jobiraj, 2017; Rajamohana et al., 2018; Hussain et al., 2019; Kumar et al., 2019). Despite the fact of their large

DOI: 10.1201/9781003220398-18

ecological functions in a forest ecosystem, the study of insects is still not very exhaustive, and no information is available regarding hymenopteran insects of the Nandhaur Wildlife Sanctuary, Uttarakhand, which is the most suitable habitat for the Bengal Tiger (*Panthera tigris tigris*) after the Corbett Tiger Reserve in Uttarakhand. The present study, therefore, aims to (1) assess the taxonomic composition of hymenopterans and their relative abundance, (2) analyze their seasonal and species diversity, and (3) outline the status of hymenopterans in and around the Nandhaur Wildlife Sanctuary.

18.2 Materials and Methods

General Description of Study Area

The present study was carried out from March 2018 to February 2020 in the landscape of Nandhaur Wildlife Sanctuary (NWLS). The NWLS was established in 2012 is a newly carved wildlife sanctuary lying in the Champawat and Nainital Districts of the State of Uttarakhand. It is the seventh wildlife sanctuary in the state and lies in the two districts Nainital and Champawat. It comprises 269.95 km^2 areas of reserve forests of the Haldwani and Champawat forest divisions. Its geographical location lying between 29°09′ N latitude and 79°80′ E longitude is bounded by the Himalayas and Tarai regions, rewarding the NWLS with a remarkable variety of landscapes (Figure 18.1). Due to its varied topography, geology, and climate, it harbors a huge diversity of flora and fauna. The vegetation is generally dominated by subtropical broad leaved plant species. The canopy cover mainly comprises evergreen sal (*Shorea robusta*) and its associated species, like shisham (*Dalbergia sisso*), sagon (*Tectona grandis*), khair (*Acacia catechu*), and chir pine, as well as riverine forests. Among the animals, the Bengal tiger, leopard, jungle cat, Asiatic elephant, sambar deer, spotted deer, and barking deer are the dominant residents of the sanctuary. In addition to these, the sanctuary is well known for its huge avian diversity, and more than 325 species of birds are found here. To conduct the present study, eight different sampling stations were selected in and around the NWLS.

Collection, Preservation, and Identification of Hymenopterans

The sampling of insects was conducted at an interval of 30 days from March 2018 to February 2020. The insects were collected by the sweep sampling method, per Gadagakar et al. (1990). The net sweeps were carried to collect the insects. The nets used in systematic sweeping were made of thick cotton cloth with a diameter of 30 cm at mouth and a beg length of 60 cm. The collected hymenopterans were transferred into jars containing ethyl-acetate-soaked cotton. These jars were brought to the laboratory, and the hymenopterans were stretched and pinned. The species, which could not be identified in the laboratory, were sent to Northern Regional Station of Zoological Survey of India, Dehradun, and the Entomological Section of Forest Research Institute, Dehradun, India. Photography using Nikon D3500 was also used for the sampling of hymenopterans in the landscape of the NWLS. Each sampling year was divided into three seasons, winter (November–February), summer (March–June), and rainy (July–October), respectively.

Quantitative Analysis

Status of species of hymenopterans was assigned as very common (VC) when counted in large numbers of individuals, common (C) when observed regularly, uncommon (UC) when recorded occasionally, and rare (R) when recorded rarely. Lastly, the collected data on hymenopterans for each parameter of diversity indices were analyzed using PAST 3.4 (Hammer et al., 2014). Various measures of diversity (Simpson, Shannon-Wiener, evenness, Margalef) were computed by using the program PAST 3.4. The Shannon-Wiener information index helps to determine the overall species diversity (H) for each family of insects.

Temperature and Humidity

To determine the effects of abiotic parameters on the abundance of hymenopteran insects, the temperature and humidity of study area were also recorded during the sampling of insects by using thermo-hygrometer. The annual variation pattern of temperature and humidity recorded during the study period (2018–2020) are given in Figure 18.2.

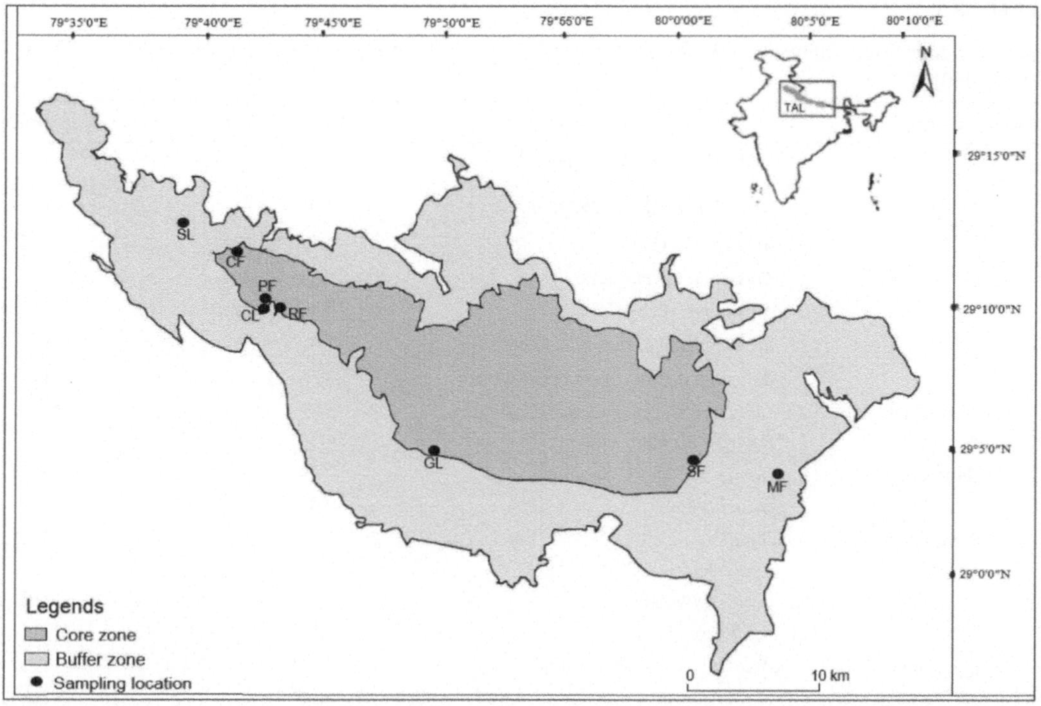

FIGURE 18.1 The geographical position of the protected landscape of Nandhaur in TAL (Terai Arc Landscape), India, showing the eight sampling locations (land-use types) selected for the present study (black circles).

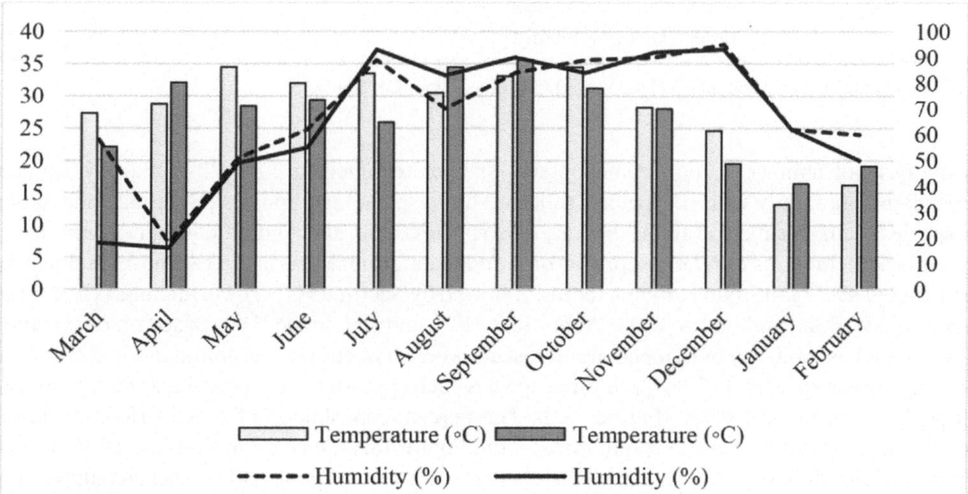

FIGURE 18.2 Monthly variation in temperature and humidity of the study area.

18.3 Results and Discussion

A total of 1,014 individuals of 23 species belonging to eight families under Order Hymenoptera were recorded in 2018–2020 from eight study sites selected in the Nandhaur landscape. The species composition of different species of hymenopteran insects recorded from the NWLS is shown in Table 18.1.

TABLE 18.1

Species Composition, Status, and Relative Abundance of Hymenoptera Recorded During the Study Period (2018–2020)

S. No.	Family	Species composition	Relative abundance	Status
1	Andrenidae	*Andrena cineraria* (Linnaeus)	2.36	C
2	Apidae	*Amegilla cingulata* (Fabricius)	2.66	C
3		*Apis cerana* (Fabricius)	9.27	VC
4		*Apis dorsata* (Fabricius)	25.83	VC
5		*Apis florea* (Fabricius)	5.52	C
6		*Bombus haemorrhoidalis* (Smith)	1.97	C
7		*Xylocopa auripennis* (Lepeletier)	3.35	C
8	Formicidae	*Camponotus compressus* (Fabricius)	5.91	C
9		*Polyrhachis simplex* (Mayr)	4.43	C
10	Halictidae	*Nomia curvipes* (Fabricius)	0.29	R
11		*Halictus* sp.	0.88	R
12	Pompilidae	*Pepsis* sp.	1.28	UC
13	Scoliidae	*Phalerimeris* sp.	3.74	C
14		*Scolia affinis* (Guerin)	15.97	VC
15		*Scolia* sp.	2.07	C
16	Sphecidae	*Ammophila atripes* (Smith)	2.26	C
17		*Sceliphron* sp.	2.66	C
18		*Sphex* sp.	1.47	UC
19	Vespidae	*Labus* sp.	0.39	R
20		*Polistes dorsalis* (Fabricius)	2.56	C
21		*Polistes stigma* (Fabricius)	2.07	C
22		*Vespa velutina* (Lepeletier)	1.57	UC
23		*Vespula flaviceps* (Smith)	1.38	UC

Abbreviations used: VC (very common), C (common), UC (uncommon), R (rare)

On the basis of number of total hymenopteran species recorded in 2018–2020, Family Apidae was the most dominant family with six species, followed by Vespidae (five species), Scoliidae and Sphecidae (three species each), Formicidae and Halictidae (two species each), and Andrenidae and Pompilidae (a single species each). In terms of relative number of individuals, Apidae was again the most abundant family and constituted 49% of the total hymenopterans, followed by Scoliidae (22%), Formicidae (11%), Vespidae (8%), Sphecidae (6%), Andrenidae (2%), Pompilidae (1%), and Halictidae (1%). Based on observations 13 species (the highest number) of hymenopteran insects were characterized as common (56.52%), followed by four uncommon speices (17.39%), and three species each (13.04%) were recorded as very common and rare in and around the sanctuary (Figure 18.3). The percent contribution of species richness and abundance of different families have been provided in the Figure 18.4. During the study period (2018–2020), Shannon's species diversity (H_S) was 2.589, Margalef's richness (H_M) was 3.178, Fisher's alpha (A_F) was 4.186, Simpson's dominance (D_S) was 0.882, and Pielou's equitability (J_P) was 0.825. Various diversity measures calculated for hymenopteran fauna in the study area have been provided in Table 18.2.

Across the total number of species of hymenopteran insects recorded in the NWLS, *Apis dorsata* (Fabricius) was the most abundant species and constituted 25.83% of total Hymenoptera. The maximum number of individuals of this species was recorded during the rainy season. *Scolia affinis* (Guerin) was the second most abundant species and constituted 15.97% of Hymenoptera. *Apis cerana* (Fabricius) was the third most abundant species and constituted 9.27% of total Hymenoptera. On the other hand, *Nomia curvipes* (Fabricius), *Labus* sp., and *Halictus* sp. were the less abundant species and constituted 1.56% of total individuals of Hymenoptera during the entire study period. Variations in the number of species and individuals of Hymenoptera recorded across the different families in NWLS have been provided in Figure 18.5.

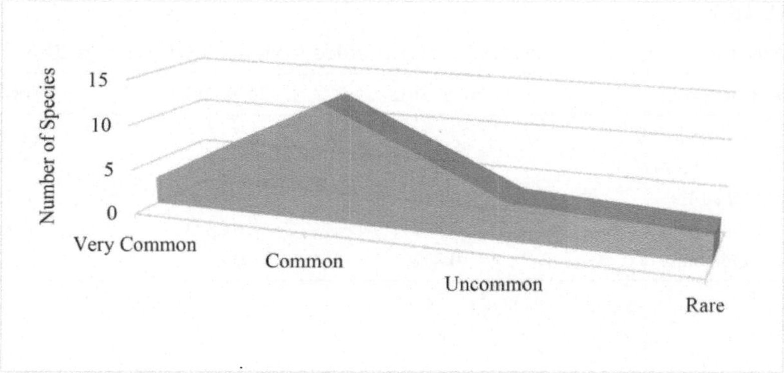

FIGURE 18.3 Status of hymenopteran species recorded from the NWLS.

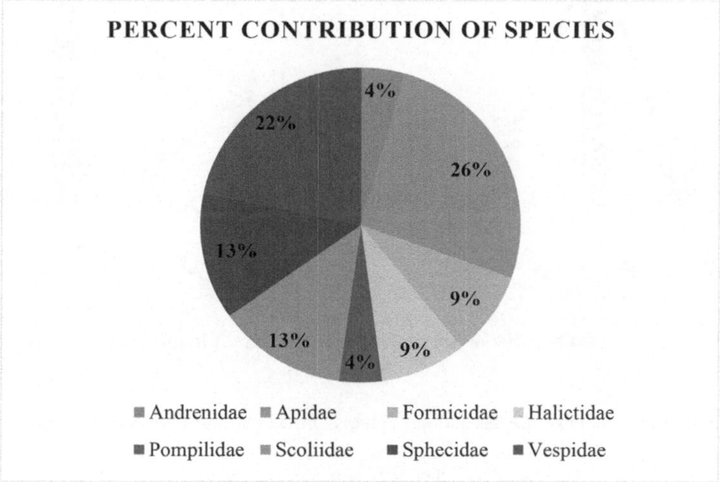

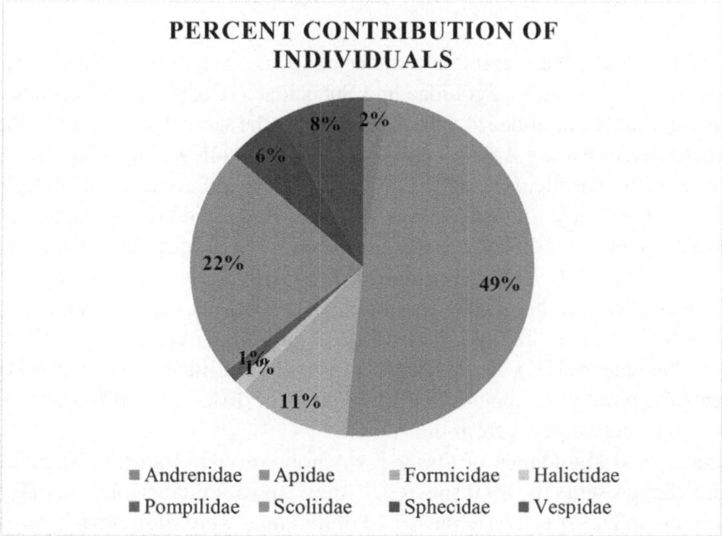

FIGURE 18.4 Percent contribution of relative number of species and individuals of different families of Order Hymenoptera recorded from the NWLS.

TABLE 18.2

Diversity Indices for Hymenopteran Fauna Recorded from the Study Area in 2018–2020.

Diversity measures	2018–2019	2019–2020	2018–2020
Shannon's diversity (H_S)	2.574	2.589	2.589
Margalef's richness (H_M)	3.385	3.359	3.178
Fisher's alpha (A_F)	4.718	4.659	4.186
Simpson's dominance (D_S)	0.880	0.884	0.882
Pielou's equitability (J_P)	0.832	0.837	0.825

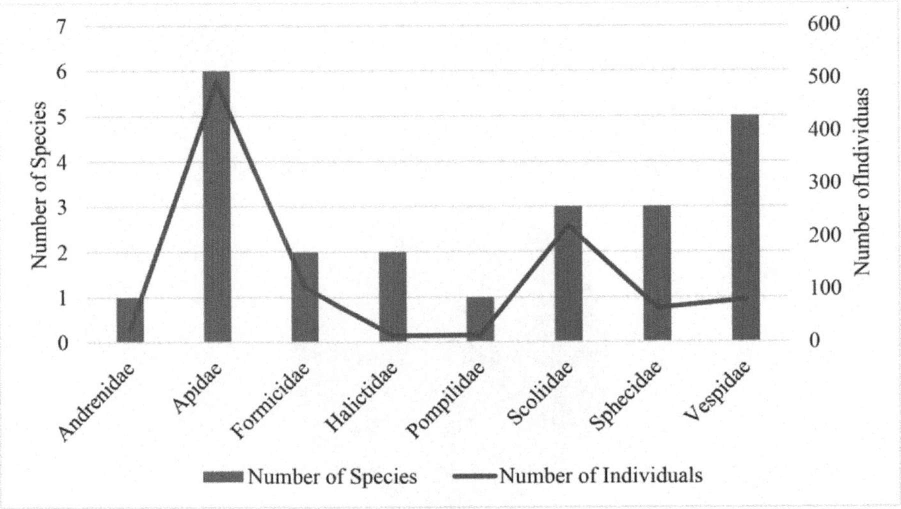

FIGURE 18.5 Variations in number of species and individuals of different families of Hymenoptera recorded during the study period.

During the first year of study, the maximum number of species belonged to Family Apidae, constituting 27.27%, followed by Vespidae (18.18%), Scoliidae and Sphecidae (13.63% each), Formicidae and Halictidae (9.09% each), Andrenidae and Pompilidae (4.54% each). During the second year of study, the maximum number of species also belonged to Family Apidae constituting 27.27%, followed by Vespidae (22.72%), Scoliidae and Sphecidae (13.63% each), Formicidae (9.09%), and Halictidae, Andrenidae, and Pompilidae (4.54% each).

During first year of the study, the maximum number of individuals belonged to Family Apidae, which constituted 49.09% of the total individuals, followed by Scoliidae (21.61%), Formicidae (10.30%), Vespidae (7.67%), Sphecidae (6.46%), Andrenidae (2.02%), Halictidae (1.61%), and Pompilidae (1.21%). During the second year of study, the maximum number of individuals also belonged to Family Apidae, which constituted 48.16% of the total individuals, followed by Scoliidae (21.96%), Formicidae (10.40%), Vespidae (8.28%), Sphecidae (6.35%), Andrenidae (2.69%), Pompilidae (1.34%), and Halictidae (0.77%). Table 18.3 represents the number of species and individuals of different families of Order Hymenoptera recorded in the two consecutive years from the NWLS.

The species richness and abundance of Order Hymenoptera were found to be maximum during the rainy season. Figure 18.6 presents the total species richness and abundance of Order Hymenoptera across seasons in the study area. The species richness and abundance were high during rainy season (21 species with 58.48% abundance), followed by summer (17 species with 31.55% abundance) and winter (11 species with 9.96% abundance). During first year of the study, the maximum number of species and

TABLE 18.3

Species Richness and Abundance of Different Families Under Order Hymenoptera Recorded from the Study Area in 2018–2020

	2018–2019		2019–2020		2018–2020	
Families	**Richness**	**Abundance**	**Richness**	**Abundance**	**Richness**	**Abundance**
Andrenidae	1	10	1	14	1	24
Apidae	6	243	6	250	6	493
Formicidae	2	51	2	54	2	105
Halictidae	2	8	1	4	2	12
Pompilidae	1	6	1	7	1	13
Scoliidae	3	107	3	114	3	221
Sphecidae	3	32	3	33	3	65
Vespidae	4	38	5	43	5	81
Total	22	495	22	519	23	1,014

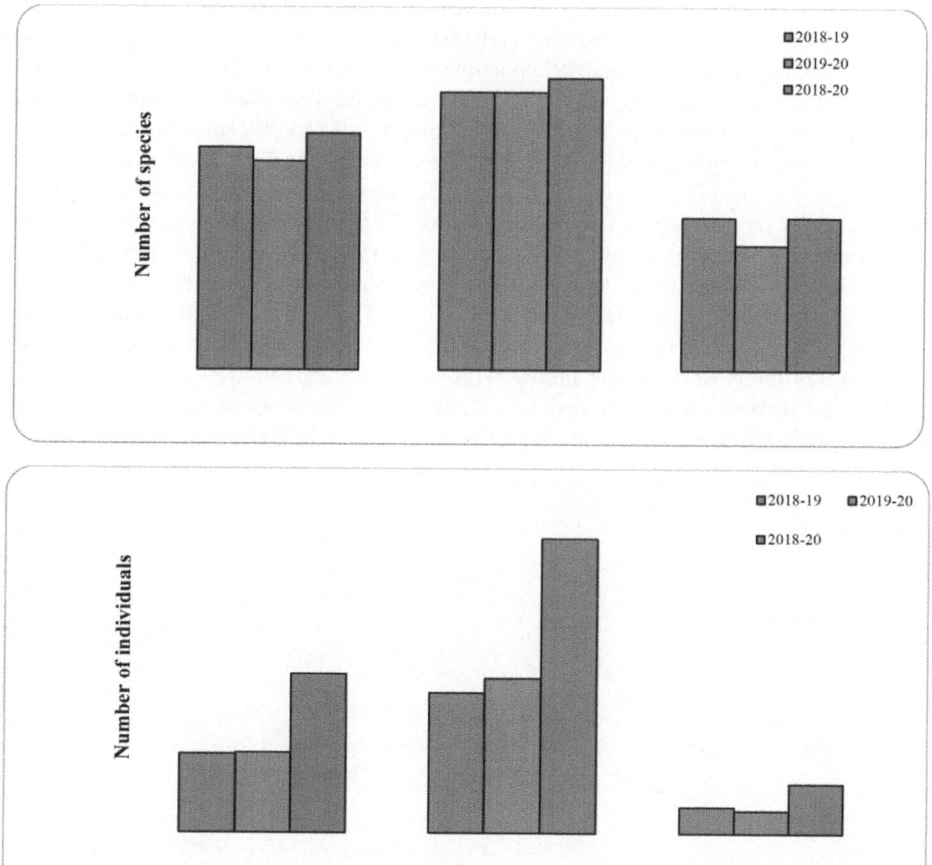

FIGURE 18.6 Species richness and abundance of hymenopteran insects across seasons during 2018–2020.

individuals was recorded during rainy season (20 species with 56.96% abundance), followed by summer (16 species with 32.12% abundance) and winter (11 species with 10.9% abundance). Similarly, during second year of the study, the maximum number of species and individuals was recorded during rainy season (20 species with 59.92% abundance), followed by summer (15 species with 31.02% abundance) and winter (9 species with 9.05% abundance). The annual abundance pattern of hymenopteran insects during the study period (2018–2020) are given in Figure 18.7. During the study period, the generally increasing trend of abundance values started from the summer season and reached maximum during the rainy season within the study period. Across the years, the lowest values of abundance were recorded during the month of winter season. This may be attributed to the maximum availability of host plant and food resources during the summer and rainy season. During both the years, abundance of hymenopteran insects significantly increased with the increase in temperature and relatively decreased during the winter months (i.e., December to February). Our findings were also supported by similar findings by other workers regarding the effects of several other factors, such as floral resources, availability of food, and microclimatic conditions, on the abundance of hymenopteran insects (Arya and Joshi, 2011; Arya et al., 2013; Hussain et al., 2019; Kumar et al., 2019). The maximum insect activities in the present study area were also observed during rainy season, which is the flowering season and has optimal temperature and rainfall required for the survival of the insects and co-incidence with availability of large number of host plants in the landscape of the NWLS. Minimum values were obtained in the winter season due to the minimum number of host plants along their productivity and less temperature.

Different ecosystem of the world harbors different species composition and species richness of hymenopterans. In comparisons, Gupta (1995) documented 12 families of hymenopterans from Western Himalaya, Uttar Pradesh, India. Gupta (1997) gave a detailed systematic account of 24 species of hymenopterans belonging to 19 genera and 10 families from Nanda Devi Biosphere Reserve, India. Uniyal and Singh (1996) reported eight species of hymenopterans belonging to four families from Tirthan Valley of Great Himalayan National Park, Himachal Pradesh. During the investigation, Apidae was found as the most dominant family represented by four species, followed by Vespidae (three species) and Formicidae and Chalcidae (one species each). Joshi (1997) reported nine species belonging to four families of Order Hymenoptera. Apidae was the most dominant family with four species, followed by Vespidae (three species) and Formicidae and Chalcidae (one species each). Singh et al. (2010) reported 13 species of hymenopterans under three families from the Kane Wildlife Sanctuary, Arunachal Pradesh, Northeastern India. Vespidae was the most dominant family with seven species, followed by Apidae (four species) and Anthophoridae (two species). Arya and Joshi (2011) recorded 20 species of hymenopteran insects that belonged to eight families from Nanda Devi Biosphere Reserve, Western Himalayas, India. Apidae

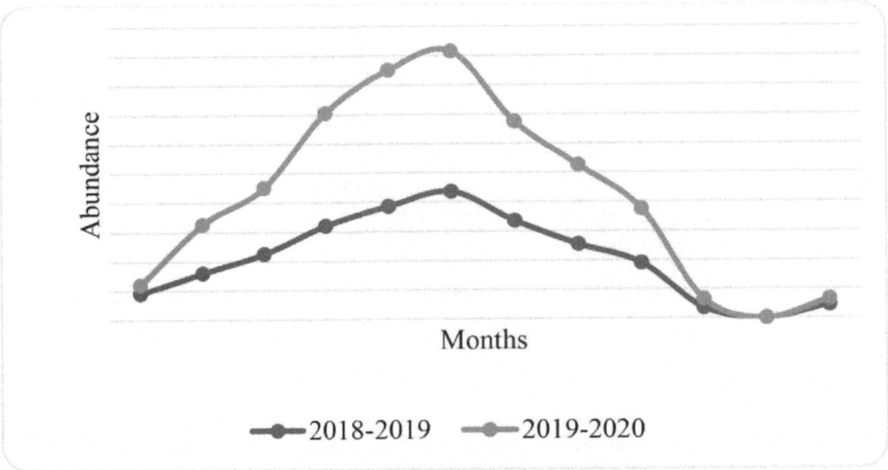

FIGURE 18.7 Monthly variation in abundance of hymenopteran insects during the study period.

was the most dominant family with seven species followed by Vespidae (four species); Ichneumonidae, Pompilidae, and Scoliidae (two species each); and Ropalidae, Specidae, and Xylocopidae (one species each). Kannagi et al. (2013) recorded a total of 38 collected species in a deciduous forest from South India, of which 36 species were identified belonging to 21 genera and distributed over nine families. Vespidae was found to be the most dominant family during the study period, followed by Apidae, Formicidae, Sphecidae, and Megachilidae. Carpenter bees, honeybees, and leaf-cutter bees and ants were seen most abundant during the study. The highest hymenopteran diversity was found in the month of May, where the Shannon index was calculated to be 1.95. Arya et al. (2013) reported 17 species of hymenopteran insects belonging to eight families from the Shyampur forest range in the Shivalik foothills, Haridwar, India. Scoliidae represented by five species and was found to be the most dominant family, followed by Vespidae (four species); Apidae and Sphegidae (two species each); and Roalidae, Eumenidae, Tiphidae, and Ichneumonidae (one species each). Gupta (2013) listed 180 species of bees under 33 genera and distributed over five families from Uttarakhand, India. Apidae was the most dominant family with 72 species, followed by Megachilidae (57 species), Halictidae (36 species), Andrenidae (10 species), and Collectidae (5 species). Azim et al. (2014) reported nine species of hymenopteran pollinators belonging to five genera and three families from different districts of Kashmir. Apidae was the most dominant family, with six species, followed by Vespidae (two species) and Megachillidae (one species). Sharma et al. (2016) documented twenty-two species of bees belonging to fourteen genera, three subfamilies, and three families from the Sundarbans Biosphere Reserve, West Bengal, India. Apidae was the most dominant family with ten species, followed by Halictidae (nine species) and Megachilidae (three species). Sheeja and Jobiraj (2017) studied the bee fauna of the Vanaparvam Biodiversity Park, Kerala, India, and reported the total of 21 species, of which 18 species were identified, which belonged to nine genera under three families—Apidae, Halictidae, and Megachilidae. Apidae was the most dominant family and among the species identified. *Apis cerana* (Fabricius), *Ceratina heiroglyphica* (Smith), and *Apis dorsata* (Fabricius) were the most observed species during the study. In a more recent study, Hussain et al. (2019) reported 41 species of Hymenoptera under 19 genera and 8 families from District Bagh, Pakistan. Vespidae was the most dominant family, with twelve species, followed by Formicidae (nine species), Apidae (seven species), Pompilidae (four species), Ichneumonidae and Bracoinidae (three species each), Braconidae (two species), and Chysidae (one species). Similarly, Kumar et al. (2019) reported twenty species of Hymenoptera from seven families under seven genera from the Nanda Devi Biosphere Reserve, India. Apidae was the most dominant family with seven species, followed by Scoliidae (five species); Ichneumonidae, Pompilidae, and Vespidae (two species each); and Eumenidae and Sphecidae (one species each).

The NWLS, established in 2012, is a newly carved wildlife sanctuary lying in Champawat and Nainital in Uttarakhand. Few studies conducted so far on flora and fauna and conservation practices viewed the NWLS as a low-profile protected area in comparison to the other 16 protected areas located in Uttarakhand. Many studies in the NWLS have been conducted on diversity of mammals and other wildlife fauna by different naturalists, conservationists, and scientists of different institutions, but there is little known on insect biodiversity, or it is unfamiliar so far. The present study provides an addition of different group of insect fauna (Order Hymenoptera) from the landscape of the NWLS. In the recent years, few studies have been carried out related to insects in this sanctuary, regarding the ecology and behavior of the lepidopteran group of insects (butterflies and moths) by Arya and Dayakrishna (2017), Sanwal et al. (2017), Arya et al. (2020), Arya et al. (2021). However, the significant contribution of other major group of insects has remained untapped. Therefore, the present study explores the unexplored biodiversity of hymenopteran insects and their crucial role in the sustainability of the landscape of the NWLS.

18.4 Acknowledgments

The authors are thankful to the SDO, the Nandhaur Wildlife Sanctuary, the DFO, and the Haldwani Forest Division and Champawat Forest Division of Uttarakhand Forest Department for their support during the study. We also thank the scientists of the Northern Regional Station of Zoological Survey of India, Dehradun, and the Forest Research Institute, Dehradun, for their cooperation in identifying coleopteran insects.

REFERENCES

Arya, M.K., Chandra, H., Bisht, S. and Farooq, F. (2021). Preliminary report on moth diversity in the landscape of Nandhaur Wildlife Sanctuary: adding subset to Lepidopteran insects of Protected Area Network. *Uttar Pradesh Journal of Zoology*, 42(15): 12–22.

Arya, M.K. and Dayakrishna (2017). Species richness and diversity of butterflies in the landscape of Nandhaur Wildlife Sanctuary, Nainital, Uttarakhand. *Journal of Environment and Bio-Sciences*, 31(2): 307–315.

Arya, M.K. and Joshi, P.C. (2011). Species composition, abundance and density of Hymenopteran insects in Nanda Devi Biosphere Reserve, Western Himalayas, India. *Journal of Environment and Bio-Sciences*, 25(2): 175–179.

Arya, M.K., Joshi, P.C. and Agarwal, A. (2013). Hymenopteran community dynamics of Shyampur forest range in Shivalik foot hills, Haridwar, India. *Journal of Environment and Bio-Sciences*, 27(1): 5–9.

Arya, M.K., Verma, A. and Chandra, H. (2020). Sex-specific extra floral feeding preferences among subtropical butterfly fauna of Terai Arc Landscape, Southern Himalaya, India. *Indian Journal of Ecology*, 47(1): 196–204.

Azim, M.N., Ahmad, S.T. and Akhter, F. (2014). Diversity of Hymenopteran pollinators in Kashmir region. *New York Science Journal*, 7(7): 84–88.

Ent, L.J.V.D. and Shaw, S.R. (1998). Species richness of Costa Rican Cenocoeliini (Hymenoptera: Braconidae): a latitudinal and altitudinal search for Anomalous diversity. *Journal of Hymenoptera Research*, 7(1): 15–24.

Gadagakar, R., Chandrashaekara, K. and Nair, P. (1990). Insect species diversity in the tropics: sampling method and case study. *Journal of Bombay Natural History Society*, 87(3): 328–353.

Gallai, N., Salles, J.M., Settele, J. and Vaissiere, B.E. (2009). Economic valuation of the vulnerability of world agriculture confronted with pollinator decline. *Ecological Economics*, 68(3): 810–821.

Gupta, R.K. (2013). Apoidea (Insecta: Hymenoptera) of Uttarakhand state (India): a checklist with Synonymies and distribution records. *Journal of Environment and Bio-Sciences*, 27(1): 57–70.

Gupta, S.K. (1995). Hymenoptera. In *Fauna of Western Himalaya (U. P.)*. Himalayan Ecosystem Series: Part 1. Ed. Director, Zoological Survey of India, Kolkata, India, pp. 81–89.

Gupta, S.K. (1997). Hymenoptera: In *Fauna of Conservation Area 9: Fauna of Nanda Devi Biosphere Reserve*. Ed. Director, Zoological Survey of India, Kolkata, India, pp. 97–104.

Hammer, O., Harper, D.A.T. and Ryan, P.D. (2014). PAST—Palaeontological Statistics Version 3.04.

Hussain, R., Nazir, N., Qureshi, S.A., Ahmed, S., Hamza, A.M. and Ali, S. (2019). Diversity and distribution pattern of Hymenoptera in District Bag, Pakistan. *Indian Journal of Entomology*, 81(3): 403–405.

Jonathan, J.K. (1995). Hymenoptera: Ichneumonidae. In *Himalayan Ecosystem Series: Fauna of Western Himalaya Part 1 (Uttar Pradesh)*, Ed. Director, Zoological Survey of India, Kolkata, India. pp. 91–110.

Joshi, P.C. (1997). Population dynamics and biomass of Hymenoptera in a temperate grassland. *The Indian Forestry*, 123(3): 245–248.

Kannagi, A., Sivakumar, V., Santhi, V. and Borgia, J.F. (2013). Hymenopteran diversity in a deciduous forest from South India. *International Journal of Biodiversity and Conservation*, 5(10): 666–670.

Klein, A.M., Steffan-Dewenter, I., Buchori, D. and Tscharntke, T. (2002). Effects of land-use intensity in tropical agroforestry systems on flower-visiting and trap-nesting bees and wasp. *Conservation Biology*, 16: 1003–1014.

Klein, A.M., Vaissiere, B.E., Cane, J.H., Steffan-Dewenter, I., Cunningham, S.A., Kremen, C. and Tscharntke, T. (2007). Importance of pollinators in changing landscapes for world crops. *Proceedings of Royal Society of London*, 274: 303–313.

Kumar, K., Joshi, P.C., Arya, M.K. and Nath, P. (2019). Altitude, Seasonality of the Hymenopteran insects associated with high altitude forest of Nanda Devi Biosphere Reserve, Western Himalaya, India. *Journal of Environment and Bio-Sciences*, 33(1): 19–29.

Losey, J.E. and Vaughan, M. (2006). The economic value of ecological services provided by insects. *Bioscience*, 56: 311–323.

Rajmohana, K., Saini, J., Kumar, G.P. and Patra, S. (2018). Insecta: Hymenoptera. In *Faunal Diversity of Indian Himalaya*. Ed. Director, Zoological Survey of India, Kolkata, pp. 353–398.

Sanwal, C.S., Dhakate, P.M., Joshi, K.K. and Neethu Laxmi, M. (2017). Habitat conservation through butterfly zone: a new approach to *in-situ* conservation of butterfly diversity. *Journal of Entomology and Zoology Studies*, 5(6): 2195–2199.

Sharma, G., Kumar, P.G. and Gupta, R.K. (2016). Studies on diversity of Bees (Apoidea: Hymenoptera: Insecta) from Sunderbans Biosphere Reserve including fourteen new distributional records of Bees species from West Bengal, India. *Bio Bulletin*, 2(1): 43–51.

Sheeja, K. and Jobiraj, T. (2017). The bee fauna of Vanaparvam Biodiversity Park, Kerala, India (Hymenoptera: Apoidea). *International Journal of Agricultural Sciences*, 7(7): 1338–1341.

Singh, O.T., Chakravorty, J. and Varatharajan, R. (2010). Entomofauna of Kane Wildlife Sanctuary, Arunachal Pradesh, Northeastern India. *Journal of Threatened Taxa*, 2(13): 1392–1400.

Singh, S. and Singh, S.P. (1993). Altitudinal distribution of Apoidea (Hymenoptera) and their association with high altitude Angiosperms in the Kullu and Lahaul Spiti region of N.W. Himalaya. *Annals of Entomology*, 11(2): 33–47.

Tylianakis, J.M., Klein, A.M., Lozada, T. and Tscharntke, T. (2006). Spatial scale of observation affects alpha, beta and gamma diversity of cavity-nesting bees and wasps across a tropical land-use gradient. *Journal of Biogeography*, 33: 1295–1304.

Uniyal, V.P. and Singh, S.K. (1996). A preliminary study of the ecological relationship between high altitude flowering plants and insects in the Tirthan Valley of the Great Himalayan National Park, Himachal Pradesh. *Indian Bee Journal*, 58(5): 137–139.

Wardell, G.A., Berhardt, P., Binter, R., Burquez, A., Burchmann, S., Cane, J., Cox, P.A., Dalton, V., Feinsinger, P., Ingram, M., Inouye, D., Jones, C.E., Kennedey, K., Kevan, P., Koopowitz, Medellin, R. and Morakesm, S.M. (1998). The potential consequences of pollinator declines on the conservation of biodiversity and stability of food crop yield, *Conservation Biology*, 12(1): 8–17.

Zhang, Z.Q. (2013). Phylum Arthropoda. Animal biodiversity: an outline of higher-level classification and survey of taxonomic richness. *Zootaxa*, 3703: 17–26.

19

Application of Molecular Phylogeny for the Identification of Species

Archana Chauhan, Phuntsog Dolma, Ahmad Ali, Anuragini Kadwalia, and Ranbir Chander Sobti

CONTENTS

19.1 Introduction

Although over 260 years have elapsed since the publication of the tenth edition of *Systema Naturae*, by Carl Linnaeus (1758), which marked the starting point of zoological taxonomy, the original task of cataloging animal species is still far from completion. At present, 1,600,000 species of animals have been formally described (Zhang, 2013), but this number seems to comprise quite a small percentage of the actual species diversity, the greatest part of which, at least among the multicellular animals, undoubtedly belongs to insects and reptiles (Whittaker et al., 2001; Stork et al., 2015; Xiao et al., 2010; Larsen et al., 2017; Singhal et al., 2018; Singhal et al., 2021). Revealing and describing this diversity will remain the primary challenge and one of the main goals of taxonomists long into the future.

The diagnostic version of the phylogenetic concept defines a species as a set of populations distinguished by one or more unique (diagnostic) characteristics (Cracraft, 1989a; Rosser et al., 2015). The biological species concept deals with quite different entities, such as reproductively or morphologically isolated communities. Their isolation may be incomplete, but it should be sufficient to prevent two communities from merging into a single gene pool (Mayr, 1968; Coyne & Orr, 2004; Rosser et al., 2015).

19.1.1 Species Complex

Species boundaries are frequently hard to delimit based on morphology only, a fact that has called for integrative taxonomy, including additional sources of information such as molecular data, biogeography, behavior, and ecology (Dayrat, 2005; Padial et al., 2010). Founding a species description on a variety of characteristics from different, independent datasets is generally regarded as best practice (Padial & Riva, 2010). Taxonomists are urged to discriminate characteristics according to their quality and suitability for species delineation rather than to just add more and more data (Valdecasas et al., 2008). For instance,

DOI: 10.1201/9781003220398-19

TABLE 19.1
Depicting the Various Species Complexes with Examples

	Cryptic species	Polymorphic species	Haplotype species
Definition	Cryptic species are defined as a two or more species which are morphologically similar and hidden under one species name. Such species are morphologically similar but genetically distinct (Bickford et al., 2007).	Polymorphism is the occurrence of two or more clearly different morphs or forms, also referred to as alternative phenotypes, in the population of a which are genetically identical but a few can interbreed and few cannot interbreed (Dobzhansky, 1970).	While haplogroup is a group of similar haplotypes in an individual that share a common ancestor with a single-nucleotide polymorphism (SNP) mutation (Arora et al., 2015).
Examples	• Cryptic species shown by skipper butterflies (Burns et al., 2008)	• Light and dark forms of Peppered moth (Van't Hof et al., 2016)	• *Microhyla sholigari* (Seshadri et al., 2016)
Picture with characteristic phenomenon			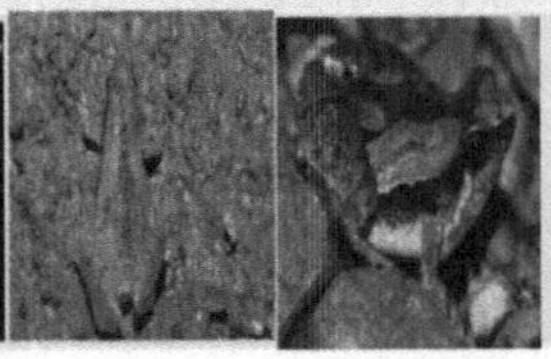

the steelhead and rainbows fish pose a conundrum. They are not the exactly same thing, but they are not entirely different either. They are two forms of salmon diversity that include a bewildering collection of shapeshifting individuals, alternate life cycle pathways, erratic migratory runs, linked breeding populations, and closely related species. Biologists refer to this as a species complex (Alagona, 2016). Different forms of species complex, including cryptic species, polymorphic species and haplotype species, are illustrated in Table 19.1.

19.2 Approaches to Identify Species Complex

19.2.1 Traditional Methods

Traditional methods basically depend on morphological taxonomy—that is, differentiating organisms on the basis of morphological parameters, like shape, size, color, beak, wings, dewlap, and claws, which depend on the type of organism being studied. For example, in Diptera, an order of insects, species identification is based on male genitalia (Valentini et al., 2009), and in case of fish larvae, morphological identification is focused on characteristics like body size ratio, eye size and shape, position of fins throughout life stages, gut morphology, and pigmentation, all of which is done with the help of morphological keys (Friedheim, 2016). For the past 250 years, taxonomy has been defined on the basis of morphological structures and the phenotypic characteristics of individual organisms (Hebert & Gregory, 2005). But for the identification of cryptic species, polymorphic species and haplotype species' morphological taxonomy is not much help as all these species are morphologically identical (Figure 19.1). For example, in case of fig wasp, *Pleistodontes imperialis*, the morphological analysis revealed no clear differences, except for colour, and they are considered to be the same species, but the mitochondrial DNA divergence suggested the existence of four cryptic species (Haine et al., 2006). In addition to this, there is a lack of taxonomic keys to identify immature specimens of many species, and also, traditional taxonomy requires high levels of skillfulness in any given group and is, therefore, confined to specialists (Pires & Marinoni, 2010; Hebert et al., 2003). Haplogroups are mostly related to a single line of descent. Therefore, membership of

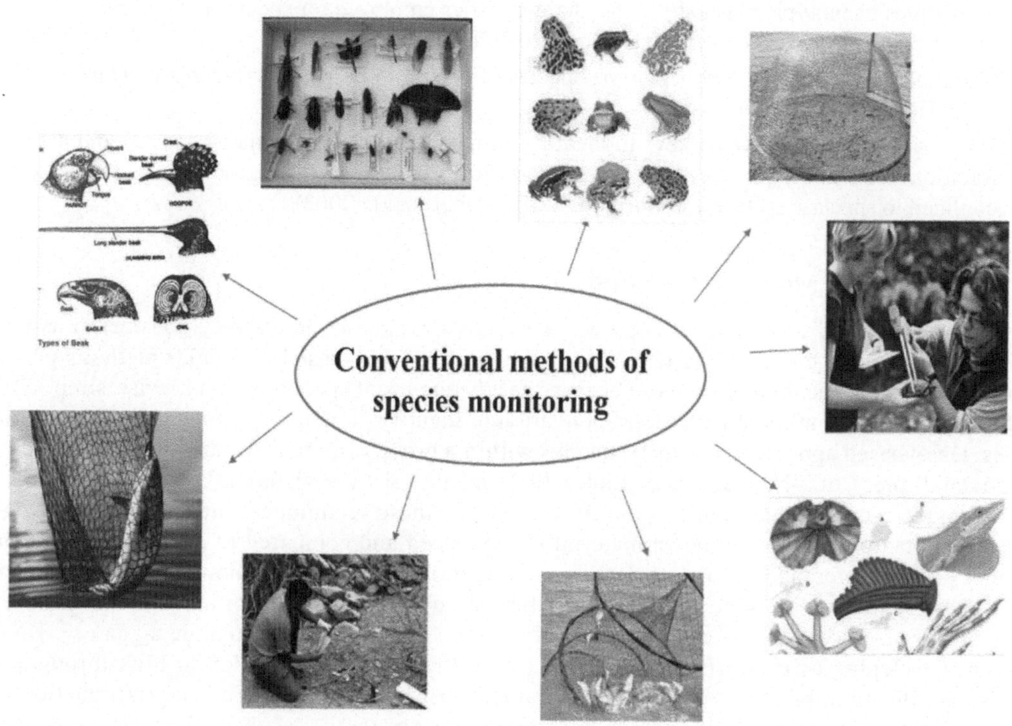

FIGURE 19.1 Traditional approaches to identify species.

a haplogroup, by any individual, relies on a relatively small proportion of the genetic material possessed by that individual (Tishkoff & Williams, 2002). Understanding haplotype provides the basis for the phylogenetic relationship between the species of a group and also tracks their evolutionary pattern across the distribution range (Roff 1986; Lloyd et al., 2016). Haplotypes give information about the recombinational, mutational, selective, and demographic dynamics of genomic regions in order to understand the population genetic variations (Tishkoff et al., 2000; Mateu et al., 2001). But due to inappropriate or non-availability of advanced techniques, these species were not identified and were considered under polymorphic species till late 20th century. Therefore, traditional approaches to identify species complex possessed certain limitations.

Limitations

- Few physical characteristics are common among major groups of organisms, such as eubacteria and eukaryotes, which is one of the issues with morphological systematics (Hillis, 1987).
- The majority of the species including microbial species are difficult, or nearly impossible to see with the naked eye; thus, this method does not help to identify them (Savolainen et al., 2005).
- Cryptic species have morphologically similar cryptic, yet they are not same (Buth, 1984), while with polymorphic species, even though they are morphologically different, they are genetically similar. Hence, this method can sometimes be deceiving in identifying species within the population.
- Many species go through different life stages, for example, caterpillars to butterflies (Park, 2008), and their larvae look very similar to one another and very different compared to what the species looks like at its final stage of life (Park, 2008)
- Many species use camouflage or mimicry as a defense mechanism (Park, 2008). This also makes it difficult and often confusing when using morphological features to identify one species from another.

- Sometimes phenotypic plasticity in the characteristics employed for species recognition led to incorrect identifications (Pires & Marinoni, 2010; Hebert et al., 2003).
- Morphologically cryptic and haplotype species often fail to be identified (Pires & Marinoni, 2010; Hebert et al., 2003).
- There is a lack of taxonomic keys to identify immature specimens of many species and also traditional taxonomy requires high levels of skillfulness in any given group and is therefore confined to specialists (Pires & Marinoni, 2010; Hebert et al., 2003).

19.2.2 DNA-Based Molecular Methods

Molecular methods of species identification are rapidly developing and widely considered as fast and efficient methods for resolving taxonomic ambiguities. DNA-based molecular methods play a vital role in the identification of species because the advantages of specificity, sensitivity, simplicity, and low cost can be combined with their application to identify complex, cryptic, and polymorphic species. DNA-based approaches identify species within a group of organisms and unknown species using a small piece of DNA sequence. Tables 19.2 and 19.3 list the different DNA-based molecular techniques used for the identification of species. In these techniques, little amount of DNA taken from a taxonomically unknown material is sequenced and compared to previously available sequences in the reference library or databases to identify a specific/unknown species (Wilson, 2012). These sequences are species-specific, so they serve as a "nucleotide or molecular signature" for the species from which it is easiest to identify. Each species has its nucleotide signature. DNA chips are developing as promising instruments for multiple-species detection in high-throughput species identification, but sensitivity and application must be improved. With the introduction of next-generation sequencing (NGS), which has quickly become a crucial technology in fundamental science and has just lately been utilized in species identification and with recent technological advancements genomics has been transformed.

19.2.3 Integrative Molecular Approach

High gene flow among the population is thought to limit the opportunities for genetic divergence or convergence, which results in species formation or extinction. Therefore, to deal with various complexes in an ecosystem more integrative approaches is needed. To some extent the combination of ethno-classification and morphological character have identified most of the organism, but in case of complex species, they failed to do so. Because of this limitation and advancement in science, molecular techniques come into consideration that can give the exact phylogenetic classification and nomenclature. The combination of molecular techniques with the traditional techniques has opened a new window for taxonomists and biologists to understand the species complex present in various ecosystem (Figure 19.2). By using the techniques of molecular biology, we have begun to understand clearly and completely the mechanisms by which a group of organism adjusts to environmental changes.

19.3 Conclusion

The introduction of different molecular techniques has helped scientists to overcome the limitations, leading to the description of many different new species across most taxonomic groups. Molecular tools can provide crucial information for an integrative taxonomy (Padial et al., 2010). Moreover, DNA-based molecular data allow evolutionary biologists to undergo deeper exploration underlying population-level processes behind speciation events (Tosi et al., 2003; Carstens et al., 2013). Nevertheless, the use of molecular tools to define species boundaries is not straightforward, and many different molecular approaches have been developed to test hypotheses about species delimitation over the last decade.

TABLE 19.2

DNA-Based Molecular Techniques Used by Ecologists for Species Complex Identification

Technique	Method	Application	Species complex	Reference
Restriction fragment length polymorphism (RFLP)	Variation in length of restriction fragments between two species differs due to point mutation in the restriction enzyme site. Restriction enzymes are originally found in bacteria where they cleave foreign DNA of bacteriophages by detecting specific sequences of DNA and cleave it.	Identification of two cryptic species of the temperate bumblebee using partial CO1 (700 bp) from 39 individuals' species that live in sympatry.	Cryptic species	(Murray et al., 2008)
		Identification of game species with the help of the mt genome where 464-bp products were cut with different restriction endonucleases (RE) developing species-specific RFLP. Phylogenetically related deer species were characterized by the application of one or two RE.	Polymorphic species	(Wolf et al., 1999)
		RFLP analysis of a 535 bp segment of the mt 16S rRNA gene with nine RE in 240 brine shrimp shows clustering of haplotypes in two definite groups corresponding to the two species.	Haplotype species	(Gajardo et al., 2004)
Random amplified polymorphic DNA (RAPD)	RAPD amplifies random sections of the target DNA from PCR with the help of arbitrary, short primers that are 8–12 nucleotides long. These primers will then randomly bind to the target DNA and amplify fragments of DNA that are approximate 0.5 to 5 kb in length.	D-loop region of mitochondrial DNA by RFLP, and this region was amplified by PCR, and the product was digested with nine REs, which clearly distinguishes the three species of *Calomys* co-inhabiting in central Argentina.	Cryptic species	(Ittig et al., 2002)
		Identification of four different varieties of lamprey larvae species. Sixty-five RAPD loci were discovered, out of which seven loci possessing species-specific fragments were obtained.	Polymorphic species	(Yamazaki et al., 2005)
		The amplification of 19 species using 18S and 18S plus ITS from larval DNA. Restriction analysis of the 18S + ITS PCR product generated 13 haplotypes in fruit fly.	Haplotype species	(Armstrong et al., 1997)
PCR—single-strand conformational polymorphism (SSCP)	Change in the conformation of a single-stranded DNA is determined depending on its nucleotide sequence and was detected using nondenaturing polyacrylamide gel.	Allowed the differentiation of *Aplysina* sponges within a population.	Cryptic species	(Lamarão et al., 2010)
		Identification of nematode species variation patterns within 14 species inhabiting different geographical isolates.	Polymorphic species	(Gasser & Monti, 1997)
		Identification of bivalve larvae species using the 18S region of the 538 fragment gives at least 20 species among the 130 larvae.	Haplotype species	(Livi et al., 2006)

(Continued)

TABLE 19.2
(Continued)

Technique	Method	Application	Species complex	Reference
Amplified fragment length polymorphism (AFLP)	From the digested genomic DNA, the double-stranded nucleotide adapters were ligated to the restriction DNA fragments to serve as primer binding sites for PCR amplification, causing selective amplification of the fragments.	Hawaiian populations of *Laupala* showed the aptness of AFLP for a wide investigation of cryptic species population in the island of Hawaii.	Cryptic species	(Parsons & Shaw, 2001)
		MtDNA sequences of 524 base pairs indicated huge divergence between barred and spotted owls (13.9%). These genetic markers make it possible to understand hybrids species, which can be used for research, conservation, and law enforcement purpose.	Haplotype species	(Haig et al., 2004)
Single-nucleotide polymorphism (SNP)	The variation of a single base pair change occurs in DNA sequences due to substitution, deletion, or insertion. The dissimilarity between alleles is less in SNP, whereas the length of the DNA will be similar.	Identification of four cryptic species of voles (genus *Microtus*) from the sampled 38 voles.	Cryptic species	(Belfiore et al., 2003)
		Investigated SNPs of two passerine bird species (pied and collared flycatchers along with human SNPs), and it was revealed that high nucleotide diversity shown by the avian genome	Polymorphic species	(Primmer et al., 2002; Backström et al., 2008)
		Investigation on the genetic diversity of sheep breeds from three countries of the Mediterranean basin were conducted. D-loop sequence analysis detected 154 unique mtDNA haplotypes 93 polymorphic sites.	Haplotype species	(Pariset et al., 2011)
Microarrays	Amplification of a mtDNA gene fragment and hybridization on a microarray containing diagnostic oligonucleotides. It also analyzes the expression of genes at once.	Identified 40 mycotoxigenic fungi that were characterized from different food supplies, as well as the genes that are involved in the mycotoxin synthetic pathways, were analyzed.	Polymorphic species	(Lezar & Barros, 2010)
		Identification of bat species from guano mixed samples using 16S mtDNA sequences.	Haplotype species	(Pariset et al., 2011)
DNA barcoding	The precise ordering of individual nucleotides of DNA that are composed and well-curated at a time. It involves the production of PCR amplicons from particular regions to sequence them, and these sequence data are used to identify or barcode that organism to make a distinction from other species.	Investigated 89 *Thrips* morphospecies using DNA barcoding, and two important pest and vector species, *F. schultzei* and *T. palmi*, were detected as involving cryptic species.	Cryptic species	(Belfiore et al., 2003)
		Analysis of Irish solitary bees were done using standard barcoding sequence where 51 of 55 species identified shows polymorphic speciation among its population.	Polymorphic species	(Magnacca & Brown, 2012)
		Obtained 21 of the 49 sampled Hawaiian *Hylaeus* species showed amplified multiple haplotypes from within individual specimens. And out of which 18 of these species, all were tested heteroplasmic, whereas remaining three species showed limited heteroplasmy.	Haplotype species	(Magnacca & Brown, 2010)

DNA minibarcoding	DNA minibarcodes target comparatively smaller regions of the gene.	New set of mt COI gene amplified all 11 snake species with a 175 bp fragment, while the 245 bp amplicon amplified 10 species.	Cryptic species	(Dubey et al., 2011)
		A total of 25 samples from across four batoid orders were amplified. With is technique, 81% of minibarcodes have matched back to a single species and 100% correctly back to all other taxonomic ranks.	Polymorphic species	(Wannell et al., 2020)
		The new set of primers effectively amplified genomic DNA from 81 specimens from the Japan lake. Among them, 18 haplotypes species were detected.	Haplotype species	(Mabuchi & Nishida, 2020)
Next-generation sequencing (NGS)	NGS method uses array-based sequencing, which combines the techniques developed in Sanger sequencing to process millions of reactions in parallel, resulting in a very high speed and throughput at a reduced cost.	Discovered three sets of PCR primers to amplify the mtCOI gene, which will differentiate 131 *B. tabaci* specimens collected from nine countries. Four main different haplotype species were reported.	Cryptic species	(Elfekih et al., 2018)
		Mustelus mustelus from Africa and Mediterranean Sea was explored using NGS and microsatellite. Two microsatellite multiplex panels were assembled from 11 polymorphic loci and categorized in two populations of *Mustelus mustelus*.	Polymorphic species	(Maduna et al., 2017)
		Investigation on *Ngirhaphium caeruleum* from South China Sea was carried out using NGS barcodes (313 bp), which were compatible with traditional taxonomic delimitation and hence identified 32 haplotype species among population.	Haplotype species	(Samoh et al., 2019)

TABLE 19.3

List of Specific Molecular Techniques Used by Ecologists for Species Complex Identification

Polytene chromosome method	These are very large and hence display characteristic band-interband morphology, and a library of DNA clones was constructed from the region and used to study phylogenetic relationships among species and to distinguish among members of species complexes.	Analyzed the cryptic diversity shown by 272 larvae belonging to black flies in Vietnam where 88 different chromosomal rearrangements, mostly paracentric inversions, were seen. Fifteen cytoforms and nine morphoforms were discovered from Vietnam; far more than any other country.	Cryptic species	(Adler et al., 2016)
Small-insert genomic libraries	Ligation of small-insert genomic libraries labeled with whole genomic DNA isolated from the different species of interest and other sibling species. This type of screen usually identifies highly repeated sequences that differ substantially in either copy number or sequence between the species being compared.	Identification of species within the *Anopheles gambiae* andgenomic libraries were prepared with number of clones and were differentiated into two groups based on the presence of homologous sequence. These two groups are not sex-specific in the species.	Cryptic species	(Adler et al., 2016)
Single-cell amplification	Analyzing a single cell to uncover the mechanisms in genome without the consideration of bulk population of cell.	48 single-cell genomes data were derived from a polyclonal infection collected from Chikhwawa, Malawi, which allowed for unambiguous determination of haplotype diversity	Haplotype species	(Trevino et al., 2017)
Cloning of a genomic region	A small piece or bp of DNA taken from the genome of a cell or animal and spliced into a bacteriophage or other cloning vector.	Analyzed a total of 27 inbred mouse strains of BCG-R and BCG-S phenotypes for the presence of nucleotide sequence variations within the coding portion of *Nramp* and have carried out haplotype typing.	Haplotype species	(Malo et al., 1994; Martínez-Arias et al., 2001)
Allele-specific amplification (AS-PCR)	Detection of the single-nucleotide sequences using the allele-specific primers that have 3′ end base, which is complementary to the SNP site.	10 SNPs were identified from partial (i.e., 570 bp) DNA sequence analysis of the mt 16s rRNA gene from 96 Korean and Chinese individual *Acetes chinensis*. Origin of *Acetes chinensis* with 100% accuracy using two external primers and C220F primers were authenticated.	Haplotype species	(Kang et al., 2015)
Somatic hybrid cell construction	To segregate a chromosome of interest from the full chromosome complement so as to acquire a permanent source of the chromosome.	Probe generated from two somatic hybrid cell lines by Alu- and LI-PCR can be used for chromosome painting in normal human lymphocyte metaphase spreads by chromosomal *in situ* suppression (CISS) hybridization.	Haplotype species	(Lengauer et al., 1990)
Insertion/deletion polymorphisms (indels)	Genetic variation in which a specific nucleotide sequence is present (insertion) or absent (deletion), known as indels and widely spread across the genome.	One hundred candidate markers (4-bp indels) were selected and genotyped in unrelated dogs (n = 7) and wolves (n = 18). Eighty-one and 76 out of 94 were confirmed as polymorphic loci.	Polymorphic species	(Väli et al., 2008)

Polymorphic repetitive sequences	Tandem repetitive DNA elements, including microsatellites and simple sequence repeats, are extremely common throughout the genomes. It may include transposons or pseudogenes.	Comparisons of 16 bacterial species and the orthopoxviruses, including the variola virus and three of its close neighbors.	Polymorphic species	(Denœud & Vergnaud, 2004)
Structural and copy number variations	Source of genomic variability where region of DNA can be 1 kb or large in size, including balanced translocation, inversions, and genomic imbalances due to insertion or deletion, allude to CNVs (copy number variations).	Comparative intraspecific genomic hybridizations were carried out to identify loci sheltering CNV in four great ape species. Differences in CNV location and frequency were seen and identified a a number regions that represents high-frequency CNVs in all great apes.	Polymorphic species	(Gazave et al., 2017)
MLEE (multi-locus enzyme electrophoresis)	Constructed on the electrophoretic mobility pattern of enzymes acquired from mutations in genes. Up to 15–30 independent enzyme loci will be studied over each and every sample and were compared to a single gene.	*Trypanosoma cruzi* variants have two distinct phylogenetic lineages that were further partitioned into five sharply delineated phylogenetic clusters. The first two clusters were found in sylvatic environments, whereas the three latter were restricted to domestic transmission cycles.	Polymorphic species	(Brisse et al., 2000)
ISSR-PCR (inter simple sequence repeat–PCR)	Recommended for intrapopulation genetic variations where DNA segments located in microsatellite regions were amplified using PCR followed by gel electrophoresis for separation.	Genetic analysis of the 13 silkworms (*Bombyx mori*) were carried out with the help of RAPD, ISSR-PCR, and SSRs, where the highest diversity index was observed for ISSR-PCR (0.957) and the lowest for RAPDs (0.744).	Polymorphic species	(Nagaraju et al., 2001)
IGS-PCR (intergenic spacer region–PCR)	IGS structure is prominent for rDNA transcription and for growth rate.	Identification of two *Giardia intestinalis* genotypes from humans using the intergenic spacer (IGS) region of rDNA. Thirteen *G. intestinalis* isolates were classified to assemblages A and B.	Polymorphic species	(Lee et al., 2006)

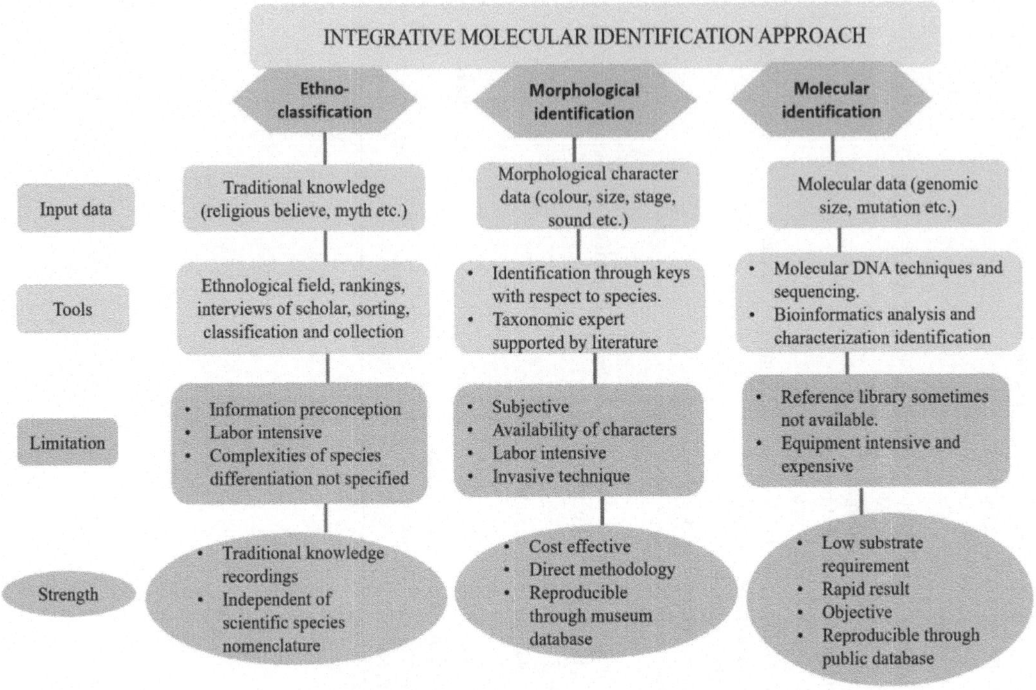

FIGURE 19.2 Flowchart of integrative molecular approa0ch.

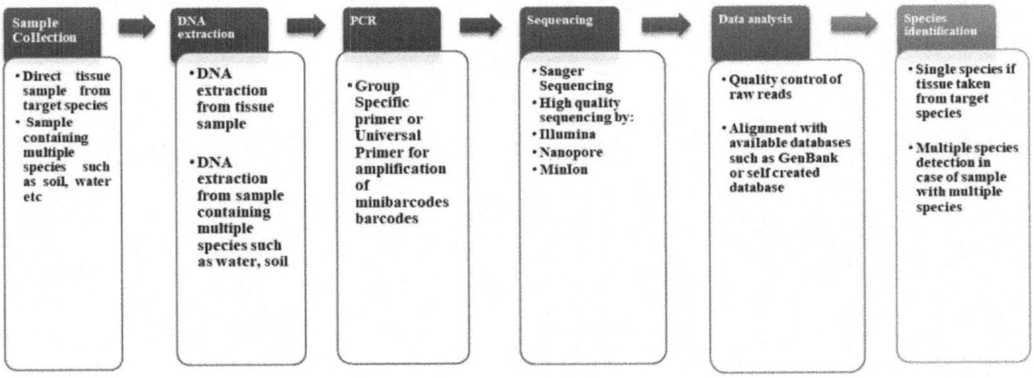

FIGURE 19.3 A diagrammatic process of DNA barcoding for identifying animal species.

DNA barcoding uses molecular information to classify species, whereas morphological taxonomy uses morphology and phenotypic characteristics to distinguish between one species and another. Every method has their own pros and cons, and over time, new techniques overcome the old one. It is always a challenge to conclude which method supersedes another. Both these methods complement each other and have proven to identify species complexes. Both methods should be required and used in combination to identify and compare species process as accurate as possible. There are so many other molecular methods to identify species, such as SNP, MLEE, and RAPD. Effective identification of species using short DNA fragments (DNA barcoding, minibarcoding, and metabarcoding) requires reliable sequence reference libraries of known taxa. Both taxonomically comprehensive coverage and content quality are important for sufficient accuracy (Weigand et al., 2019). Future scope of species complex identification is going to be mostly on the next-generation sequencing technologies along with the morphological tools. In an imagined future world, DNA identification could provide cost-effective tools for some unknown species and a hint to additional species level units, hypothesis-testing, phylogeny, or classification. So in conclusion, DNA-based molecular techniques, especially DNA barcoding and sequencing technology, play a very significant role in identifying the phylogenetic assessment of species and the conservation of biodiversity in the ecosystem.

REFERENCES

Adler, P. H., Takaoka, H., Sofian-Azirun, M., Low, V. L., Ya'cob, Z., Chen, C. D., . . . & Pham, X. D. (2016). Vietnam, a hotspot for chromosomal diversity and cryptic species in black flies (Diptera: Simuliidae). *PLoS ONE*, 11(10), e0163881.

Alagona, P. S. (2016). Species complex: classification and conservation in American environmental history. *ISIS*, 107(4), 738–761.

Armstrong, K. F., Cameron, C. M., & Frampton, E. R. (1997). Fruit fly (Diptera: Tephritidae) species identification: a rapid molecular diagnostic technique for quarantine application. *Bulletin of Entomological Research*, 87(2), 111–118.

Arora, D., Singh, A., Sharma, V., Bhaduria, H. S., & Patel, R. B. (2015). HgsDb: Haplogroups Database to understand migration and molecular risk assessment. *Bioinformation,* 11(6), 272.

Backström, N., Karaiskou, N., Leder, E. H., Gustafsson, L., Primmer, C. R., Qvarnström, A., & Ellegren, H. (2008). A gene-based genetic linkage map of the collared flycatcher (Ficedula albicollis) reveals extensive synteny and gene-order conservation during 100 million years of avian evolution. *Genetics,* 179(3), 1479–1495.

Bickford, D., Lohman, D. J., Sodhi, N. S., Ng, P. K., Meier, R., Winker, K., & Das, I. (2007). Cryptic species as a window on diversity and conservation. *Trends in Ecology & Evolution*, 22(3), 148–155.

Brisse, S., Barnabé, C., & Tibayrenc, M. (2000). Identification of six Trypanosoma cruzi phylogenetic lineages by random amplified polymorphic DNA and multilocus enzyme electrophoresis. *International Journal for Parasitology*, 30(1), 35–44.

Burns, J. M., Janzen, D. H., Hajibabaei, M., Hallwachs, W., & Hebert, P. D. (2008). DNA barcodes and cryptic species of skipper butterflies in the genus Perichares in Area de Conservacion Guanacaste, Costa Rica. *Proceedings of the National Academy of Sciences,* 105(17), 6350–6355.

Buth, D. G. (1984). The application of electrophoretic data in systematic studies. *Annual Review of Ecology and Systematics*, 15(1), 501–522.

Carstens, B. C., Pelletier, T. A., Reid, N. M., & Satler, J. D. (2013). How to fail at species delimitation. *Molecular ecology,* 22(17), 4369–4383.

Coyne, J. A., & Orr, H. A. (2004). *Speciation* (Vol. 37). Sunderland, MA: Sinauer Associates.

Cracraft, J. (1989a). Speciation and its ontology: the empirical consequences of alternative species concepts for understanding patterns and processes of differentiation. *Speciation and Its Consequences*, 28, 59.

Cracraft, J. (1989b). Species as entities of biological theory. In *What the Philosophy of Biology Is* (pp. 31–52). Dordrecht: Springer.

Dayrat, B. (2005). Towards integrative taxonomy. *Biological Journal of the Linnean Society*, 85(3), 407–417.

Denœud, F., & Vergnaud, G. (2004). Identification of polymorphic tandem repeats by direct comparison of genome sequence from different bacterial strains: a web-based resource. *BMC Bioinformatics*, 5(1), 1–12.

Dobzhansky, T., & Dobzhansky, T. G. (1970). Genetics of the evolutionary process (Vol. 139). Columbia University Press.

Dubey, B., Meganathan, P. R., & Haque, I. (2011). DNA mini-barcoding: an approach for forensic identification of some endangered Indian snake species. *Forensic Science International: Genetics*, 5(3), 181–184.

Elfekih, S., Tay, W. T., Gordon, K., Court, L. N., & De Barro, P. J. (2018). Standardized molecular diagnostic tool for the identification of cryptic species within the Bemisia tabaci complex. *Pest Management Science*, 74(1), 170–173.

Friedheim, S., 2016. Comparison of species identification methods: DNA barcoding versus morphological taxonomy. *Horizons*, 1, 13.

Gajardo, G., Crespo, J., Triantafyllidis, A., Tzika, A., Baxevanis, A. D., Kappas, I., & Abatzopoulos, T. J. (2004). Species identification of Chilean Artemia populations based on mitochondrial DNA RFLP analysis. *Journal of Biogeography*, 31(4), 547–555.

Gasser, R. B., & Monti, J. R. (1997). Identification of parasitic nematodes by PCR-SSCP of ITS-2 rDNA. *Molecular and Cellular Probes*, 11(3), 201–209.

Gazave, E., Darré, F., Morcillo-Suarez, C., Petit-Marty, N., Carreño, A., Marigorta, U. M., . . . & Navarro, A. (2011). Copy number variation analysis in the great apes reveals species-specific patterns of structural variation. *Genome Research*, 21(10), 1626–1639

Haig, S. M., Mullins, T. D., Forsman, E. D., Trail, P. W., & Wennerberg, L. I. V. (2004). Genetic identification of spotted owls, barred owls, and their hybrids: legal implications of hybrid identity. *Conservation Biology*, 18(5), 1347–1357.

Haine, E. R., Martin, J., & Cook, J. M. (2006). Deep mtDNA divergences indicate cryptic species in a fig-pollinating wasp. *BMC Evolutionary Biology*, 6(1), 1–11.

Hebert, P. D., Cywinska, A., Ball, S. L., & DeWaard, J. R. (2003). Biological identifications through DNA barcodes. *Proceedings of the Royal Society of London. Series B: Biological Sciences*, 270(1512), 313–321.

Hebert, P. D., & Gregory, T. R. (2005). The promise of DNA barcoding for taxonomy. *Systematic Biology*, 54(5), 852–859.

Hillis, D. M. (1987). Molecular versus morphological approaches to systematics. *Annual Review of Ecology and Systematics*, 18(1), 23–42.

Hof, A. E., Campagne, P., Rigden, D. J., Yung, C. J., Lingley, J., Quail, M. A., Hall, N., Darby, A. C. & Saccheri, I. J. (2016). The industrial melanism mutation in British peppered moths is a transposable element. *Nature,* 534(7605), 102–105.

Ittig, R. G., Chiappero, M. B., Blanco, A., Provensal, C., & Gardenal, C. N. (2002). Accurate identification of three cryptic species of rodents of the genus Calomys using RAPD-PCR and mitDNA RFLP markers. *Biochemical Systematics and Ecology*, 30(5), 425–432.

Kang, J. H., Noh, E. S., Park, J. Y., An, C. M., Choi, J. H., & Kim, J. K. (2015). Rapid origin determination of the Northern Mauxia Shrimp (Acetes chinensis) based on allele specific polymerase chain reaction of partial mitochondrial 16S rRNA gene. *Asian-Australasian Journal of Animal Sciences*, 28(4), 568.

Lamarão, F. R. M., Reis, E. C., Simão, T. A., Albano, R. M., & Lôbo-Hajdu, G. (2010). Aplysina (Porifera: Demospongiae) species identification through SSCP-ITS patterns. *Journal of the Marine Biological Association of the United Kingdom*, 90(4), 845–850.

Larsen, B., Toldam-Andersen, T. B., Pedersen, C., & Ørgaard, M. (2017). Unravelling genetic diversity and cultivar parentage in the Danish apple gene bank collection. *Tree Genetics & Genomes,* 13(1), 1–12.

Lee, J. H., Lee, J., Park, S. J., Yong, T. S., & Hwang, U. W. (2006). Detection and genotyping of Giardia intestinalis isolates using intergenic spacer (IGS)-based PCR. *The Korean Journal of Parasitology*, 44(4), 343.

Lengauer, C., Riethman, H., & Cremer, T. (1990). Painting of human chromosomes with probes generated from hybrid cell lines by PCR with Alu and L1 primers. *Human Genetics*, 86(1), 1–6.

Lezar, S., & Barros, E. (2010). Oligonucleotide microarray for the identification of potential mycotoxigenic fungi. *BMC Microbiology*, 10(1), 1–14.

Livi, S., Cordisco, C., Damiani, C., Romanelli, M., & Crosetti, D. (2006). Identification of bivalve species at an early developmental stage through PCR-SSCP and sequence analysis of partial 18S rDNA. *Marine Biology*, 149(5), 1149–1161.

Lloyd, S. S., Steele, E. J., & Dawkins, R. L. (2016). Analysis of haplotype sequences. Next Generation Sequencing-Advances, Applications and Challenges. *InTechOpen,* 345–368.

Mabuchi, K., & Nishida, K. (2020). A DNA mini-barcoding system for endangered unionid mussels in the Lake Biwa system in Japan. *Conservation Genetics Resources*, 12(4), 581–584.

Maduna, S. N., Rossouw, C., Da Silva, C., Soekoe, M., & Bester-van der Merwe, A. E. (2017). Species identification and comparative population genetics of four coastal houndsharks based on novel NGS-mined microsatellites. *Ecology and Evolution*, 7(5), 1462–1486.

Magnacca, K. N., & Brown, M. J. (2010). Mitochondrial heteroplasmy and DNA barcoding in Hawaiian Hylaeus (Nesoprosopis) bees (Hymenoptera: Colletidae). *BMC Evolutionary Biology*, 10(1), 1–16.

Magnacca, K. N., & Brown, M. J. (2012). DNA barcoding a regional fauna: Irish solitary bees. *Molecular Ecology Resources*, 12(6), 990–998.

Malo, D., Vogan, K., Vidal, S., Hu, J., Cellier, M., Schurr, E., . . . & Gros, P. (1994). Haplotype mapping and sequence analysis of the mouse Nramp gene predict susceptibility to infection with intracellular parasites. *Genomics*, 23(1), 51–61.

Martínez-Arias, R., Calafell, F., Mateu, E., Comas, D., Andrés, A., & Bertranpetit, J. (2001). Sequence variability of a human pseudogene. *Genome Research*, 11(6), 1071–1085.

Mateu, E., Calafell, F., Lao, O., Bonné-Tamir, B., Kidd, J. R., Pakstis, A., Kidd, K. K. & Bertranpetit, J. (2001). Worldwide genetic analysis of the CFTR region. *The American Journal of Human Genetics,* 68(1), 103–117.

Mayr, E. (1968). The role of systematics in biology. *Science*, 159(3815), 595–599.

Murray, T. E., Fitzpatrick, U., Brown, M. J., & Paxton, R. J. (2008). Cryptic species diversity in a widespread bumble bee complex revealed using mitochondrial DNA RFLPs. *Conservation Genetics*, 9(3), 653–666.

Nagaraju, J., Reddy, K. D., Nagaraja, G. M., & Sethuraman, B. N. (2001). Comparison of multilocus RFLPs and PCR-based marker systems for genetic analysis of the silkworm, Bombyx mori. *Heredity*, 86(5), 588–597.

Padial, J. M. & De La Riva, I. (2010). A response to recent proposals for integrative taxonomy. *Biological Journal of the Linnean Society*, 101(3), 747–756.

Padial, J. M., Miralles, A., De la Riva, I. & Vences, M. (2010). The integrative future of taxonomy. *Frontiers in Zoology*, 7(1), 1–14.

Pariset, L., Mariotti, M., Gargani, M., Joost, S., Negrini, R., Perez, T., . . . & Valentini, A. (2011). Genetic diversity of sheep breeds from Albania, Greece, and Italy assessed by mitochondrial DNA and nuclear polymorphisms (SNPs). *The Scientific World Journal*, 11, 1641–1659.

Park, K. T. (2008). Four new species of Torodora Meyrick and a new species of Antiochtha Meyrick from the Philippines (Lepidoptera, Lecithoceridae). *Journal of Asia-Pacific Entomology*, 11(2), 89–95.

Parsons, Y. M., & Shaw, K. L. (2001). Species boundaries and genetic diversity among Hawaiian crickets of the genus Laupala identified using amplified fragment length polymorphism. *Molecular Ecology*, 10(7), 1765–1772.

Pires, A. C., & Marinoni, L. (2010). DNA barcoding and traditional taxonomy unified through integrative taxonomy: a view that challenges the debate questioning both methodologies. *Biota Neotropica*, 10, 339–346.

Primmer, C. R., Borge, T., Lindell, J., & Sætre, G. P. (2002). Single-nucleotide polymorphism characterization in species with limited available sequence information: high nucleotide diversity revealed in the avian genome. *Molecular Ecology*, 11(3), 603–612.

Roff, D. A. (1986). The evolution of wing dimorphism in insects. *Evolution*, *40*(5), 1009–1020.

Rosser, N., Kozak, K. M., Phillimore, A. B., & Mallet, J. (2015). Extensive range overlap between heliconiine sister species: evidence for sympatric speciation in butterflies? *BMC Evolutionary Biology*, 15(1), 1–13.

Samoh, A., Satasook, C., & Grootaert, P. (2019). NGS-barcodes, haplotype networks combined to external morphology help to identify new species in the mangrove genus Ngirhaphium Evenhuis & Grootaert, 2002 (Diptera: Dolichopodidae: Rhaphiinae) in Southeast Asia. *Raffles Bulletin of Zoology*, 67, 640–659.

Savolainen, V., Cowan, R. S., Vogler, A. P., Roderick, G. K. & Lane, R. (2005). Towards writing the encyclopaedia of life: an introduction to DNA barcoding. *Philosophical Transactions of the Royal Society B: Biological Sciences*, 360(1462), 1805–1811.

Seshadri, K. S., Singal, R., Priti, H., Ravikanth, G., Vidisha, M. K., Saurabh, S., Pratik, M. & Gururaja, K. V. (2016). Microhyla laterite sp. nov., a new species of Microhyla Tschudi, 1838 (Amphibia: Anura: Microhylidae) from a laterite rock formation in South West India. *PLoS One*, 11(3), e0149727.

Singhal, S., Colston, T. J., Grundler, M. R., Smith, S. A., Costa, G. C., Colli, G. R., Moritz, C., Pyron, R. A. & Rabosky, D. L. (2021). Congruence and conflict in the higher-level phylogenetics of squamate reptiles: an expanded phylogenomic perspective. *Systematic Biology*, 70(3), 542–557.

Singhal, S., Hoskin, C. J., Couper, P., Potter, S., & Moritz, C. (2018). A framework for resolving cryptic species: a case study from the lizards of the Australian wet tropics. *Systematic Biology*, 67(6), 1061–1075.

Singhal, S., Huang, H., Grundler, M. R., Marchán-Rivadeneira, M. R., Holmes, I., Title, P. O., Donnellan, S. C. & Rabosky, D. L. (2018). Does population structure predict the rate of speciation? A comparative test across Australia's most diverse vertebrate radiation. *The American Naturalist*, 192(4), 432–447.

Stork, N. E., McBroom, J., Gely, C., & Hamilton, A. J. (2015). New approaches narrow global species estimates for beetles, insects, and terrestrial arthropods. *Proceedings of the National Academy of Sciences*, 112(24), 7519–7523.

Tishkoff, S. A., Pakstis, A. J., Ruano, G., & Kidd, K. K. (2000). The accuracy of statistical methods for estimation of haplotype frequencies: an example from the CD4 locus. *The American Journal of Human Genetics*, 67(2), 518–522.

Tishkoff, S. A., & Williams, S. M. (2002). Genetic analysis of African populations: human evolution and complex disease. *Nature Reviews Genetics,* 3(8), 611–621.

Tosi, A. J., Morales, J. C., & Melnick, D. J. (2003). Paternal, maternal, and biparental molecular markers provide unique windows onto the evolutionary history of macaque monkeys. *Evolution*, 57(6), 1419–1435.

Trevino, S. G., Nkhoma, S. C., Nair, S., Daniel, B. J., Moncada, K., Khoswe, S., Banda, R. L., Nosten, F. & Cheeseman, I. H. (2017). High-resolution single-cell sequencing of malaria parasites. *Genome Biology and Evolution*, 9(12), 3373–3383.

Valdecasas, A. G., Williams, D. & Wheeler, Q. D. (2008). "Integrative taxonomy" then and now: a response to Dayrat (2005). *Biological Journal of the Linnean Society*, 93(1), 211–216.

Valentini, A., Pompanon, F., Taberlet, P., 2009. DNA barcoding for ecologists. *Trends in Ecology & Evolution*, 24, 110–117.

Väli, Ü., Brandström, M., Johansson, M., & Ellegren, H. (2008). Insertion-deletion polymorphisms (indels) as genetic markers in natural populations. *BMC Genetics*, 9(1), 1–8.

Van't Hof, A. E., Campagne, P., Rigden, D. J., Yung, C. J., Lingley, J., Quail, M. A., . . . & Saccheri, I. J. (2016). The industrial melanism mutation in British peppered moths is a transposable element. *Nature*, 534(7605), 102–105.

Wannell, G. J., Griffiths, A. M., Spinou, A., Batista, R., Mendonça, M. B., Wosiacki, W. B., . . . & Gubili, C. (2020). A new minibarcode assay to facilitate species identification from processed, degraded or historic ray (batoidea) samples. *Conservation Genetics Resources*, 12(4), 659–668.

Weigand, H., Beermann, A. J., Čiampor, F., Costa, F. O., Csabai, Z., Duarte, S., Geiger, M.F., Grabowski, M., Rimet, F., Rulik, B. & Ekrem, T. (2019). DNA barcode reference libraries for the monitoring of aquatic biota in Europe: Gap-analysis and recommendations for future work. *Science of the Total Environment*, 678, 499–524.

Whittaker, R. J., Willis, K. J., & Field, R. (2001). Scale and species richness: towards a general, hierarchical theory of species diversity. *Journal of Biogeography*, 28(4), 453–470.

Wilson, J. J. (2012). DNA barcodes for insects. In DNA Barcodes (pp. 17–46). *Humana Press,* Totowa, NJ.

Wolf, C., Rentsch, J., & Hübner, P. (1999). PCR–RFLP analysis of mitochondrial DNA: A reliable method for species identification. *Journal of Agricultural and Food Chemistry*, *47*(4), 1350–1355.

Xiao, J. H., Wang, N. X., Li, Y. W., Murphy, R. W., Wan, D. G., Niu, L. M., Hu, H. Y., Fu, Y. G., & Huang, D. W. (2010). Molecular approaches to identify cryptic species and polymorphic species within a complex community of fig wasps. *PLoS One*, 5(11), e15067. https://doi.org/10.1371/journal.pone.0015067

Yamazaki, Y., Fukutomi, N., Oda, N., Shibukawa, K., Niimura, Y., & Iwata, A. (2005). Occurrence of larval Pacific lamprey Entosphenus tridentatus from Japan, detected by random amplified polymorphic DNA (RAPD) analysis. *Ichthyological Research*, 52(3), 297–301.

Zhang, Z. Q. (2013). Animal biodiversity: An update of classification and diversity in 2013. In: Zhang, Z.-Q. (ed.) Animal biodiversity: an outline of higher-level classification and survey of taxonomic richness (addenda 2013). *Zootaxa*, 3703(1), 5–11.

20

In Vitro Fertilization and Embryo Transfer

Mani Chopra, Muskan Budhwar, Madhu Sharma, and Ranbir Chandra Sobti

CONTENTS

20.1 Introduction

Infertility is a phenomenon that has increasingly become a medical condition among the human population. Most couples, roughly about 85%, achieve conception within a year of trying and approximately 7% of the remaining couples have the greatest likelihood of conception in the second year. Therefore, in a biomedical context, infertility is defined as the inability to conceive a pregnancy after 12 months of unprotected intercourse. Infertility can have various causes, including age, reproductive health of

DOI: 10.1201/9781003220398-20

both sexes, stress, smoking, and drinking (Griel et al. 2010; Gurunath et al. 2011). The medicalization of infertility has proceeded at an astounding pace with the development of new assisted reproductive technologies like *in vitro* fertilization, embryo transfer, intra-cytoplasmic sperm injection (ICSI), gamete intrafallopian transfer, and cryopreservation. Due to one or several other reasons, *in vitro* fertilization and embryo transfer technique facilitate conception within a female, which was naturally not possible under normal conditions. According to the Centers for Disease Control and Prevention (CDC), 1% of babies born in the United States are through IVF, whereas in Sweden and Australia, 1 in 50 and 60 births is through IVF (Sunderam et al. 2009). Hence, around eight million babies have been born worldwide through IVF and similar assisted reproductive technology.

20.2 *In Vitro* Fertilization and Embryo Transfer (IVF-ET)

IVF is a well-established method of culture and fertilization of gametes in an *in vitro* laboratory environment and has vast implementations in basic and applied sciences. This technology facilitates the production of embryos for experimental research investigations and the preservation of endangered mammals. It is more likely advantageous to infertile human couples, homosexual couples, and single parents. The standard IVF-ET protocol involves several steps: retrieval of gametes from parents, fertilization of gametes in a medium in a petri dish, *in vitro* maintenance of embryos, and at last, transfer and implantation of embryos into the uterus (Bavister 2002; Choe et al. 2020).

Retrieval of gametes→ Fertilization in a petri dish→ *In vitro* maintenance of the embryo → Embryo transfer to the uterus

Although IVF technology has various applications, it took almost a century to reach its total potential usage. Hence, the world should not overlook its origin, as many experiments that started in the year 1878 successfully contributed to the first convincing reports of IVF. However, the discovery of sperm capacitation in 1951 had enormous significance in developing IVF technology. A well-recognized study by Chang (1959) gave the idea that eggs of rabbits fertilized *in vitro* could grow into a normal young ones. In IVF, two groundbreaking articles in 1963 and 1964 by Yanagimachi and Chang reported *in vitro* fertilization of hamster gametes within a suitable medium that formed a strong base of human IVF (Johnson 2019). In 1969, the first definitive achievement of IVF made by three enormous scientists named Edwards, Bavister, and Steptoe, which first demonstrated human *in vitro* oocyte fertilization, and a decade later, the first human embryo transfer following IVF resulted in the live birth of a child named Louise Brown in July 1978 (Steptoe and Edwards 1978). Earlier, this technology was popularly known as test-tube baby, and for the development of revolutionary IVF technology, Robert Edwards received a Nobel Prize in Physiology and Medicine in the year 2010 (Wang and Sauer 2006).

IVF-ET is not directly carried out or never the first step of couples who cannot normally conceive until they lie under specific IVF eligibility criteria (National Collaborating Centre for Women's and Children's Health, UK 2004; Van Loendersloot 2010). IVF-ET is preferred choice in cases where there is (1) absence or blocked fallopian tubes due to surgery, tubal infections, or tuberculosis; (2) endometriosis, as endometrial function and its receptivity is the limiting factor for determining pregnancy rate, and the adherence of endometrium to the ectopic pelvic region, outside endometrial cavity, creates hindrances in the movement of spermatozoa and eggs; (3) reduced sperm count or motility, a condition in which sperm count is less than five million per milliliter, there are abnormalities in the sperm shape, or the sperm is unable to survive within cervical mucous; (4) problems with ovulation and poor oocyte quality generated from hormonal imbalance; (5) hereditary disease in which there is probability or risk of transferring the genetic disease from either parent to offspring; or (6) unexplained immunological infertility originating from certain antibiotics in males, despite of usual sperm motility and function. IVF-ET can also be the last option for people who have regular checkups and routine treatment but fail to conceive conventionally. Therefore, couples suffering from these reasons will face difficulties in conception that can be resolved using various ART.

20.3 Procedure of IVF-ET

IVF-ET treatment aims to mimic nature within an artificial laboratory system as much as possible. Its protocol breaks down into few essential components, including ovulation induction, oocyte fertilization, embryo development and transfer, and maintenance into the uterus. The treatment of IVF-ET is a multistep and complex procedure requiring high-quality types of equipment and personnel. The elective process designed for IVF involves the following principal steps.

20.3.1 Pre-IVF Evaluation

The success rate of IVF-ET is influenced by many factors, including the patients' age, medical conditions, and treatment approaches. The examination of patients before IVF-ET is the first part of the process and a mandatory step consisting of various tests and evaluations:

- Ovarian reserve testing: This testing involves anti-Mullerian hormone (AMH) and antral follicle count (AFC) standard tests, which predict risks of poor ovarian response, oocyte size, and quality of oocytes or eggs present inside the ovary (La Marca and Sunkara 2014).
- Response of women to fertility drugs: The females with high FSH are more resistant to ovarian stimulation, which is affected by age.
- Presence of endometriosis: The females undergo hysteroscopy to see inside the uterus to determine if there is no fibroid produced by endometriosis.
- Uterine lining evaluation: The examination of the uterine lining before ovulation is vital for implantation, which is done by performing a sonogram to generate an image of the uterus. Specific patterns of uterine lining development are associated with a risk of poor pregnancy rate. Estradiol treatments are given to take care of any uterus-associated problem (Alshehre et al. 2021).
- Semen analysis: This test involves men's semen sample analysis for sperm quality parameters, such as number, shape, and size. A procedure known as intracytoplasmic sperm injection (ICSI) is performed in IVF if the sperm cells are weak, damaged, or low in number (Back Louis et al. 2014).
- Infectious or hereditary disease screening: Both parents are to be screened for genetic diseases and infectious diseases, such as HIV and hepatitis.

20.3.2 Hormonal Medication for IVF-ET

The IVF-ET process consists of various stages, which are carefully regulated via the administration of hormonal medications. The human reproductive cycle of females produces only one oocyte per month, decreasing the chances of a successful IVF, as there will be a possibility of low quality or immature egg retrieval and less fertilization. However, to achieve multiple oocytes, females undergo various hormonal medication to mature several oocytes per cycle. The significant role of hormonal therapy is involved in downregulation of the pituitary, controlled ovarian hyperstimulation (COH), ovulation induction and supporting luteal phase. The mechanism of action of various drugs is different; however, some suppress the production of body's hormones that stimulate oocyte development and ovulation, thereby downregulate ovarian physiological functions, whereas to get super-ovulation, controlled hyperstimulation of ovaries is carried out by GnRH analog, clomiphene, urofollitropin, and menotropins (Shrestha et al. 2015; Siristatidis et al. 2015; Panesar 2016; Quaas and Legro 2019). Some of the significant medications are briefly explained in the following list.

- GnRH antagonists (Antagon [ganirelix acetate] and Cetrotide [cetrorelix acetate]): These antagonists bind to the GnRH receptors and block their ability to receive any signal, thereby immediately shutting down the production of gonadotropins that are follicle-stimulating hormone (FSH) and luteinizing hormone (LH). They are also used in later stages of ovarian stimulation for a short period to prevent premature ovulation (Yildiz et al. 2019).

- GnRH agonists (Lepron [leuprolide acetate], Synarel [nafarelin], and Zoladex [goserelin]): GnRH agonists mimic the naturally occurring gonadotropin-releasing hormone (GnRH), and by flooding the body with GnRH-a, the ovaries become increasingly submerge and, eventually after around three weeks, shut down the production of LH and FSH. This medication is of two types: short-acting medication, including daily injections and a long-acting medication that lasts for one to three months. This medication prevents premature LH surge; otherwise, it could result in ovulation of oocytes before retrieval (Georgiou et al. 2019; Yilmaz et al. 2020).
- Gonadotropins: Gonadotropins or injectable fertility drugs are natural hormones that stimulate the ovary to induce several oocytes' simultaneous growth (eggs). All fertility drugs have FSH, which stimulates the growth of ovarian follicles containing eggs. Some of them also include LH, which works with FSH to enhance estrogen and follicular development (Ezra et al. 2019).
- Human chorionic gonadotropin (hCG): This natural hormone induces eggs or oocytes to become mature and fertilizable. The administration timing of hCG approximately 34–36 hours before retrieval is critically essential to retrieve mature oocytes. It is also continued for the production of progesterone after oocyte retrieval (Yilmaz et al. 2020).
- Human menopausal gonadotropin (hMG) or menotropin: It is a mixture of LH and FSH purified from the urine of menopausal female donors, and in conjugation with hCG, it supports multiple follicular developments and ovulation induction in females who have earlier received pituitary suppression.
- Clomiphene: It binds to estrogen receptors and induces ovulation by enhancing the output of pituitary gonadotropins. It has estrogen receptor binding for a much longer time than estrogen. Consequently, on GnRH release, estrogen's negative feedback is inhibited, thereby promoting a rise in circulating levels of LH and FSH, which further stimulate the ovaries and enhance ovarian follicle maturation, eventually leading to ovulation and development of the corpus luteum (Moffat et al. 2021).
- GnRH analogs: GnRH analogs include hMG or menotropins, urofollitropin, highly purified FSH, and rh-FSH (recombinant human-FSH). These GnRH analogs stimulate ovarian follicular production in females without primary ovarian failure, therefore leading to follicular growth and maturation.
- Progesterone and estradiol: Ovaries usually produce these hormones after ovulation, but some female ovaries do not produce enough of these hormones for a long time to fully support a pregnancy; hence, these ensure adequate hormonal support for the maintenance of uterine lining (Bu et al. 2019).

TABLE 20.1

Hormonal Medications for IVF

Medication	Market name
GnRH antagonists	Ganirelix, Cetrotide
GnRH agonists	Lupron, Synarel, Zoladex
Gonadotropins	Follistim, Gonal-F, Bravelle, Menopur
hCG	Profasi, Novarel, Pregnyl, Ovidrel
Menotropin or hMG	Repronex
Clomiphene	Clomid or Serophene
Progesterone	Endometrin, Crinone, Prochieve, Prometrium
Estradiol	Estrace

The patient's response to these drugs is monitored by sonography, and progesterone administration is continued even after oocyte retrieval until the follicles achieve a diameter of 15 cm.

20.3.3 Retrieval of Gametes

i. Collection and Processing of Oocytes

The IVF procedure's essential requirement is the gametes (i.e., sperm and ovum) produced by male and female parents. A million sperms using the masturbation process are collected from males, but females release only one egg or ovum per month and are collected through the aspiration process. However, there are three types of oocytes residing in the ovary of females. The first type is the growing oocytes, which have not acquired their full size, and when released from follicles and cultured *in vitro*, they are unable to resume maturation. The second type is medium-sized oocytes, which can continue their development in *in vitro* culture as oocytes are arrested in the metaphase I (MI) stage. A mature oocyte is the third type; it is full-sized and, in response to gonadotrophins, is released from follicles and cultured *in vitro*, and it resumes its maturation. This process is regarded complete as oocytes have reached the metaphase II (MII) stage, and it is the only stage when oocytes can be successfully fertilized (Mrazek and Fulka 2003).

The oocyte retrieval process is a quick, painless procedure with no cuts and stitches. It is done before ovulation and 34–36 hours after the final injection of medication. Intravenous sedation medication is used on patients to avoid discomfort or pain during the egg retrieval process. The usual retrieval method is transvaginal ultrasound follicular aspiration (as shown in Figure 20.1). A thin hollow needle and ultrasound probe guide are inserted through the vagina into the ovary to identify and retrieve mature follicles.

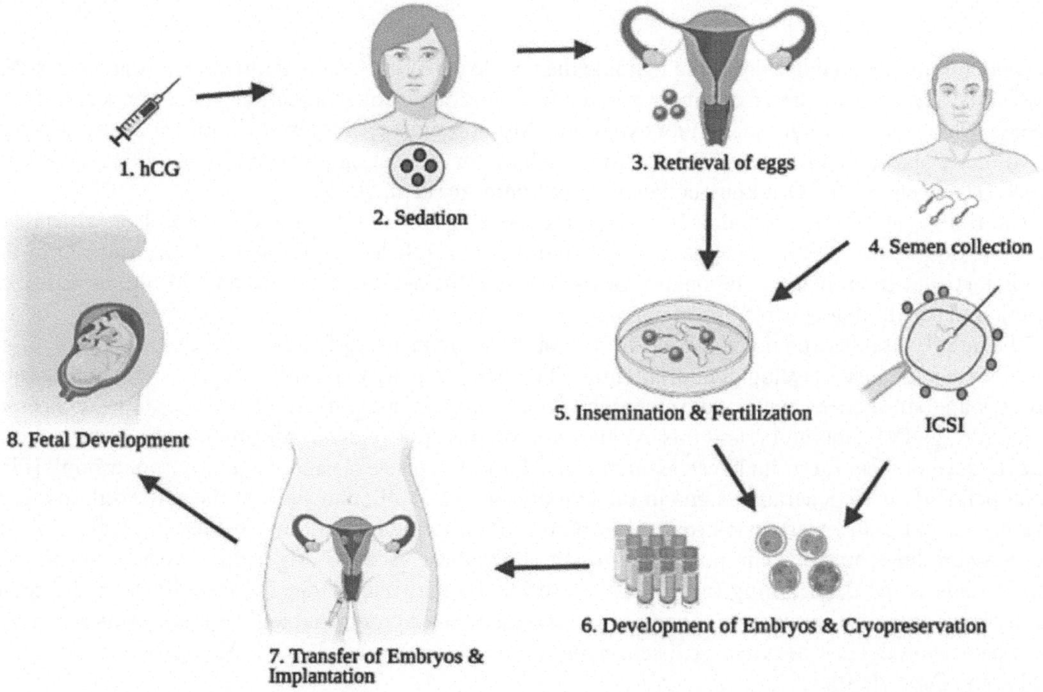

FIGURE 20.1 A schematic flow diagram of *in vitro* fertilization. (1) & (2) Stimulation of ovaries with hormones (fertility drugs) to produce more than one egg per cycle. (3) Egg retrieval: eggs are collected at the time of ovulation. (4) Semen collection. (5) Insemination and fertilization: eggs and semen are placed in an environmentally controlled chamber, and fertilization occurs a few hours after insemination, or a single sperm is directly injected into an oocyte cytoplasm through a glass micropipette (ICSI). (6) Embryo culture: fertilized eggs become embryos and may be cryopreserved for transfer at later stages. (7) Embryo transfer and implantation: one to three embryos are transferred to women's womb three to five days after fertilization. (8) Fetal development.

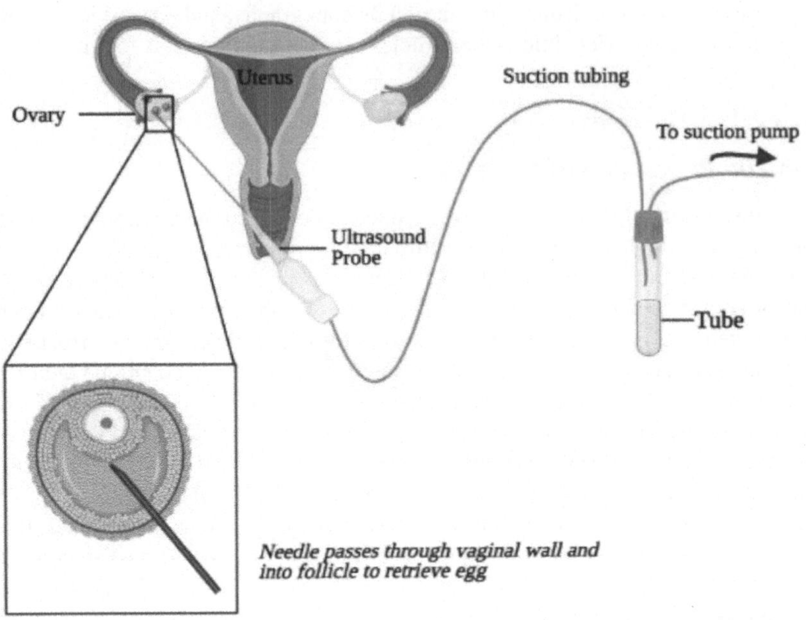

FIGURE 20.2 Ultrasound-guided transvaginal needle aspiration of follicles for *in vitro* fertilization.

An abdominal ultrasound is preferred to guide the needle when transvaginal ultrasound is not accessible to ovaries. The eggs are drawn out and separated from follicles through a suction device connected to the needle, which creates a light suction for extraction. Multiple eggs are removed within 15–20 minutes and are directly placed in a nutritive liquid culture medium for incubation (Wikland et al. 1987; Howe et al. 1988; Osemwenkha and Osaikhuwuomwan 2016; Sonmezer et al. 2020).

The timing of oocyte extraction is crucial because eggs retrieved too early or too late will fail to develop. To prevent damage to eggs, it is ensured that the follicles are immediately transferred to the laboratory and examined for the quality of eggs. The follicles may be reflushed with the heparinized solution to certify that oocytes have come out.

The quality of retrieved oocytes is an important limiting factor of female fertility in IVF, reflecting the oocyte's intrinsic developmental potential. The oocyte quality has a vital role in the fertilization process and subsequent embryo development. Morphological assessment of the oocyte is superficial in an average IVF laboratory, and this evaluation provides approximate information about the stage of oocyte development (germinal vesicle, metaphase I, or metaphase II) and its quality (polar body [PB], zona pellucida or degenerative signs in the cytoplasm). The follicular fluid of the retrieved oocyte is examined under an inverted microscope, which is then graded according to the appearance of the corona-cumulus complex. The polar body of the metaphase II stage or germinal vesicle of the prophase stage is the determining factor for the pre-incubation time before the insemination. To determine the timing of insemination, oocytes are classified based on morphological features as immature oocytes, revolutionary oocytes, or inseminable oocytes (Rienzi et al. 2011; Kenigsberg et al. 2017; Elder and Dale 2020).

Immature oocytes have morphological characteristics, like compact corona layer, little amount of expanded cumulus, and granulosa cells enveloping the oocyte, showing minor intercellular expansions and compactly aggregated. The insemination of these eggs is carried out only after excluding polar bodies, possibly the next day.

Revolutionary oocytes display expanded corona radiata and expansion of extra coronal cumulus.

Inseminable oocytes have markedly expanded corona layer and cellular expression, and these granular cells are with marked acing (Khosravi et al. 2019; De Santis et al. 2005).

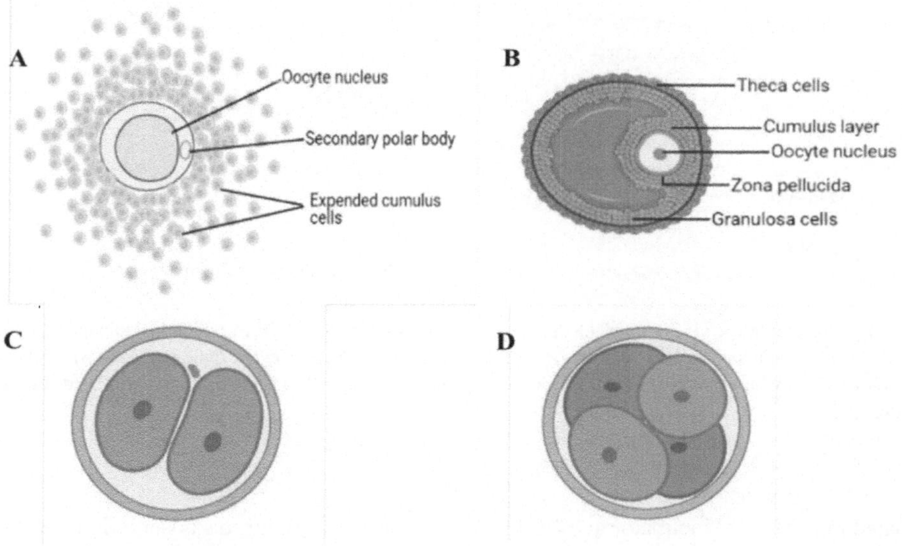

FIGURE 20.3 *In vitro* oocyte maturation and cleavage of human germinal vesicle oocytes. (A) Human germinal vesicle with attached corona cumulus and expansion corona cumulus cells. (B) Mature oocyte follicle with attached cumulus layer. (C) A fertilized two-celled pronuclear zygote. (D) A 42 h cleaved embryo.

ii. Collection and Processing of Sperms

Collection of sperms is done through the masturbation process on the day of egg retrieval. Sometimes other methods are required, such as testicular aspiration that uses a needle or surgical procedures to extract sperms directly from testicles. The sperm is processed before the insemination, which involves the liquefaction of semen at room temperature for 30 minutes; washing of sperms in a modified culture medium; centrifugation of sperms at 20 G for 50 minutes; washing of intact pallet with 1 ml of sperm-washing medium, following supernatant decant; and incubation of sperms in 5% CO2 for one hour, and the sperm will swim out of the pallet. This process will capacitate the sperm, and it becomes ready for fertilization (Balaban et al. 2012; Esteves et al. 2015; Iketubosin 2018).

20.3.4 Fertilization

The fusion of the two parental gametes, the spermatozoon and the oocyte, is known as fertilization (Neri et al. 2014). In IVF, fertilization is performed through several steps: at first, in a separate test tube, the separation of the uppermost layer of motile spermatozoa is done and counting for sperm is conducted under a microscope insemination process that can be carried out using two standard methods:

- Intracytoplasmic sperm injection (ICSI): In this method only, a single healthy spermatozoon is injected directly into each of the mature oocytes by using a microneedle or micropipette under the microscope.
- Conventional insemination: In this method, 500,000 healthy motile sperms are mixed with mature eggs in a petri dish.

Incubation of eggs after insemination is carried out in a humid atmosphere with 5% CO2 at 37°C temperature for 18–21 hours, and oocytes are observed under an inverted dissection microscope for the presence of pronuclei and polar bodies. The dispersal of cumulus mass and loosening of the coronal layer around oocytes makes the fertilization process more convenient. The fertilized healthy embryos

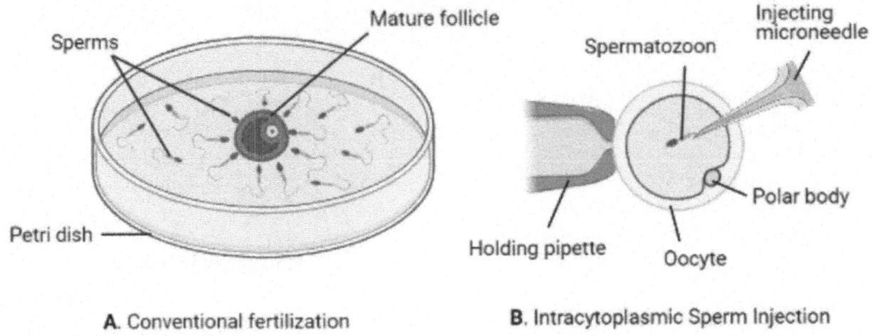

FIGURE 20.4 Representative diagrams of (A) conventional insemination in IVF in which an oocyte is surrounded by thousands of sperm cells in a petri dish, and the oocyte gets naturally fertilized by the male gamete, (B) intracytoplasmic sperm injection (ICSI), which involves the injection of a single sperm directly into an oocyte cytoplasm through a glass micropipette.

are incubated in a 5% CO2 incubator at 37°C for 22–24 hours. The embryo's observation following one-day incubation confirms an even-sized two to eight blastomere cells whose culturing is carried out in a suitable medium, and examination of pre-embryos is performed based on the number of blastomeres and thickness of the corona pellicle layer. Finally, for the implantation into the uterus, the embryos are allowed to hatch for about five to six days from their surrounding layer, zona pellucida. In certain situations, other procedures are recommended before embryo transfer:

- Assisted hatching: It is a technique for the embryos having thick corona pellucid or when the female is aged or/and had multiple failed IVF attempts. In this process, a dent or hole is created in the zona pellucida by dissolving a portion of the zona pellucida with an acid solution or using a laser just before the transfer to help the embryos hatching process and implantation. It is also helpful for the previously frozen eggs or embryos, as the process can harden the zona pellucida. However, this procedure increases the chances of pregnancy; still, it may cause damage to the embryo as well as increase the cost of IVF (Lu et al. 2019).
- Pre-implantation genetic testing: Approximately after five to six days of embryo development, a small sample of the embryo is removed and tested for specific genetic diseases or the correct number of chromosomes. Only those embryos without any defective genes or chromosomes will be transferred to the uterus. This preimplantation genetic testing reduces or eliminates the likelihood of passing a genetic problem to the child from parents (Practice Committee of Society for Assisted Reproductive Technology and Practice Committee of American Society for Reproductive Medicine 2008; Harper and Sengupta 2012).

20.3.5 Embryo Transfer

Transferring a fertilized embryo into a female's uterus is an integral part of the *in-vitro* fertilization procedure. Embryo transfer occurs after two to five days after the follicular aspiration process, and it usually takes very few minutes. The process is painless, and hence anesthesia or sedative is avoided. In this process, a speculum is placed into the female's vagina to keep the walls of the vagina open. A long, thin, flexible plastic tube called a catheter is inserted into the vagina, through the cervix and into the womb or uterus, using ultrasound guidance. One or more 4 to 10 staged embryos are passed through the tube into the womb by a syringe containing embryos suspended in a small amount of fluid attached to the end of the catheter. Some females experience bloating, cramps, and vaginal discharge after the procedure. Once embryos are placed in the uterus, females are advised to lie on the bed for some time and are examined to confirm pregnancy (Schoolcraft et al. 1999; Csokmay et al. 2011). Depending on the patient's age, embryo quality and choice, there are a few different options available and opted for the transfer of embryo that includes:

- **Fresh embryo transfer**: In this embryo transfer, eggs fertilized *in vitro* are cultured for one to two days, and the best quality embryos are screened and transferred directly to the female's uterus. For females under the age of 40, two healthy embryos are considered ideal for the transfer.
- **Blastocyst embryo transfer**: If many healthy viable embryos develop after the fertilization, it is recommended to stand by for few days to ensure embryo development into blastocysts. The chances of a successful pregnancy are higher in the case of blastocyst transfer than standard embryo transfer. However, recent studies suggest that blastocyst transfer poses some risks later in the pregnancy; hence, it should not always be recommended (Maheshwari et al. 2016).
- **Frozen embryo/blastocyst transfer**: The healthy supernumerary embryos are frozen and stored for future use if not utilized in the first transfer attempt. When the first attempt fails to develop a successful pregnancy and to avoid repeating IVF's previous steps, these frozen embryos or blastocysts are thawed and directly transferred to the uterus for further attempts (Pan et al. 2020).

To curtail the problem of multiple pregnancies, national guidelines published in 2013 recommended limits on the number of embryos to transfer. These limits differ depending on the embryo's quality and developmental stage and the patient's personal history (Kamath et al. 2020).

• *Luteal Phase Management*

The luteal phase after oocyte retrieval is majorly supported by exogenous progesterone, and from the egg retrieval process to the embryo transfer process, females are injected with the progesterone hormone to maintain the uterine lines and pregnancy. Therefore, the administration of exogenous progesterone is continued up to ten weeks of pregnancy. Several natural progesterone preparations in oil base are available, which are administered intramuscularly, vaginal progesterone suppositories and capsules of micronized progesterone (Fanchin et al. 2001; Yanushpolsky 2015; Bu et al. 2019).

20.3.6 Refinements in IVF-ET Techniques

Recent advance modifications have been made in the current IVF-ET technology, and these enhancements effectively increased the conception rates.

i. Cryopreservation (CP)

CP is a technique introduced into human-assisted reproduction and has become a routine *in vitro* fertilization procedure. It includes freezing of gametes or embryos for future use/purpose. Embryos can be cryofreeze any time after fertilization, and embryos cultured for a long time lower the pregnancy rates. Embryo CP is an essential tool in case of failed or canceled embryo transfer to avoid or reduce the risks of ovarian hyperstimulation, hormonal imbalances, or any other unplanned events and allow an additional attempt of pregnancy. The number of fresh embryo transfers to the uterus has been decreased due to the embryo CP, and it has increased the effectiveness of the IVF process by avoiding multiple pregnancies. However, there are more advantages of oocyte CP compared to embryo freezing, including egg donation, fertility preservation in females who are at risk of becoming infertile due to chronic disease or oncological treatments, and postponing childbirth, and CP also eliminates the moral, ethical, and legal concerns of embryo freezing (Konc et al. 2014). Generally, two CP methods include a traditional slow cooling method known as equilibrium cooling and the rapid/ultrarapid nonequilibrium cooling procedure, also known as vitrification. There are still debates or arguments on the best stage, procedure, and cryoprotective additives to use in CP (Fuller et al. 2004).

ii. Blastocyst Culture

The blastocyst stage is cultured up to the fifth or sixth day in a sequential culture medium. Its implantation rate after the transfer is significantly higher than that of embryos transferred before the blastocyst stage. However, fewer blastocyst embryos are required for transfer than a three-day embryo. As a result, blastocyst transfers reduce the risk of multiple gestations without affecting pregnancy rate and improving IVF outcomes (Karaki et al. 2002).

iii. Assisted Hatching

The eggs or embryos surrounded by the zona pellucida layer prevent the attachment or implantation to the endometrial linings of the uterus. In natural conditions, when the embryo reaches the uterus, the zona pellucida layer gets dissolved automatically, leading to the embryo hatching. However, sometimes, zona pellucida becomes thick and hard, which is broken down using assisted hatching techniques explained in the earlier section.

iv. Microsurgical Epididymal Sperm Aspiration (MESA) or Testicular Sperm Extraction (TESE)

Sometimes, the patient's semen samples do not contain any spermatozoa, which may occur due to several problems such as vasectomy, failed vasectomy reversal, congenital obstruction of sperm duct, or primary testicular failure. Therefore, in these conditions or circumstances, sperms are surgically obtained either from the epididymis or testis; hence, these procedures are known as microsurgical epididymal sperm aspiration and testicular sperm extraction, respectively. Also, sperms extracted from these procedures can be cryopreserved or frozen (Morin et al. 2020).

20.3.7 Variation in Assisted Reproduction Procedure

i. Gamete Intrafallopian Transfer (GIFT)

It is a similar technique to IVF in which retrieved eggs and sperms are placed directly into the fimbriated end of the fallopian tube using a catheter and are allowed to fertilize. As fertilization occurs within female's fallopian tube, there is no need for embryo culture. This technique mimics the way of a naturally fertilized egg, which begins its journey to the uterus for its implantation. However, this procedure is used rarely against IVF (Da Silva Junior et al. 2020).

ii. Zygote Intrafallopian Transfer (ZIFT)

This procedure involves fertilization of gametes in an *in vitro* environment, and after fertilization, the single-celled zygote stage is transferred directly either into the fallopian tube or uterus for implantation. Zygote intrafallopian transfer is well accepted as it combines attributes of both GIFT and IVF procedures (Schwarz et al. 2020).

iii. Pronuclear Stage Tubal Transfer (PROST)

It is a technique that includes IVF of oocytes, which is then followed by the transfer of pronuclear oocytes into the fallopian tubes. It is developed for its predictive value of confirming fertilization in couples with oligospermia or asthenospermia. Also, it enables fertilization in cases with anti-spermatozoal antibodies (ASAB). PROST has provided useful diagnostic information in managing couples who have failed to conceive in other treatment programs and has particular advantages over IVF for those receiving fresh donated oocytes for ovarian failure (Salian et al. 2018).

iv. Tubal Embryo Stage Transfer (TEST)

When different embryogenesis stages, such as zygote, embryo, and blastocyst, are transferred into the fallopian tube, this technique is known as tubal embryo stage transfer. Transfer of early-stage embryos into an ampullary portion of the fallopian tube allows further development of the embryo before entering into the uterine cavity, as various growth factors and cytokines within fluid contribute to the early embryo development and also enhance embryo implantation. Therefore, the embryo enters the uterus at a more advanced stage or in high synchronization with the uterus. However, this procedure requires general anesthesia and laparoscopy and a longer post-operation recovery time (Habana and Palter 2001).

v. Intracytoplasmic Sperm Injection (ICSI)

Intracytoplasmic sperm injection (ICSI) has been a revolutionary technique in the case of male infertility. In this, only a single sperm is directly injected into the oocyte's cytoplasm with the help of a microneedle. The techniques are advantageous to males with meager to no sperm in the semen or irregular sperm motility, shape, and size. For this procedure, a limited number of sperms are required, and it also takes care of people who have undergone vasectomy by using direct sperm aspiration from the testis or epididymis to fertilize the eggs (Sustar et al. 2019).

20.4 Outcomes of IVF-ET Conceptions

Various factors are affecting the outcomes of the IVF-ET conceptions, such as the patient's age, sequential culture medium, embryo transfer stage, and maternal or embryonic factors, which include uterine anatomic abnormalities, endometrium receptivity or characteristics, immunological factors, and embryonic causes from genetic abnormalities and other embryonic intrinsic factors that hamper its hatching and implantation (Simon and Laufer 2012). The outcomes of IVF-ET may result in some of the common situations listed:

- Miscarriages: The rate of miscarriages after IVF-ET is almost the same as that of natural conception, which is up to 15–25%, and mainly occurs within the first eight weeks of pregnancy. Also, the rate of miscarriages increases with maternal age.
- Multiple births: IVF-ET usually has more than single embryo transfer (SET) and results in multiple births, increasing the risk of early or preterm labor and low birth weight of the babies compared to the singletons (Halliday 2007).
- Premature delivery and low birth weight: There are slightly high chances of early birth or premature delivery in IVF-ET, and the baby will be born with low weight (Fauser et al. 2014).
- Ectopic pregnancies: About 1% of all IVF-ET-treated female patients have high chances of ectopic pregnancy, which may cause fatal problems to the females and the embryo. Hence, these ectopic pregnancies are avoided (Winter et al. 2002).
- Several reports have shown the association of increased incidence of gestational hypertension, gestational diabetes mellitus, intrahepatic cholestasis of pregnancy, preeclampsia, placenta previa, placental abruption, placental adherence, postpartum hemorrhage, and polyhydramnios with the IVF-ET conceptions than to naturally conceived pregnancies. But the children born with the advancement of IVF-ET have no greater risk of developmental or learning defects (Zhu et al. 2016).

20.5 Risks of IVF and Infertility Therapies

Despite several advantages of infertility therapies and IVF-ET, there are also some risks associated with these treatments, explained below:

20.5.1 Ovarian Hyperstimulation Syndrome (OHSS)

The significant impact of ovarian stimulation injections involves a severe complication known as ovarian hyperstimulation syndrome (OHSS). It is characterized by ovarian cystic enlargement, fluid leakage from follicles, and increased capillary permeability that cause leakage of fluid through porous blood vessels within and outside the ovary. This massive fluid shift from the intravascular bed to the third space leads to intravascular hypovolemia, along with the development of edema, ascites, hydropericardium, and hydrothorax. This syndrome occurs due to human chorionic gonadotrophin (hCG) administration and its analogs for final oocyte maturation trigger (Kumar et al. 2011). Some studies have notably demonstrated that the final oocyte maturation triggered by GnRH-a has eliminated the risk of severe OHSS. However, the administration of GnRH-a along with hCG (dual triggering) still has a chance of developing OHSS. Symptoms of mild OHSS typically include mild to moderate abdominal pain, nausea, abdominal bloating, vomiting, and diarrhea, which usually last a week, whereas if pregnancy occurs, a severe form of OHSS lasts from days to weeks, with symptoms of rapid weight gain, shortness of breath, blood clots, severe abdominal pain, vomiting, and persistent nausea (Ioannidou et al. 2021).

20.5.2 Multiple Gestations

With the rapid development of IVF-ET, the chances of multiple pregnancies have significantly increased. Initially, when couples discover twin or even triplet conception, they feel very happy. However, it is associated with lots of discomfort to the mother, as there are some complications involved during the pregnancy and are reported to have four to five times more risk than single conceptions. Sometimes they may end up having a premature birth or, even worse, miscarriages. The twin pregnancy rate is more than 40% even after embryo transfer is restricted to two. An obvious way to reduce the high multiple pregnancy rate is to transfer only one embryo; however, it will affect embryo implantation rates, but the use of cryopreserved embryos for further cycles may enhance the overall success of IVF (Catt et al. 2003).

In some cases, when there is low implantation potential, it requires multiple embryo transfer, resulting in multiple pregnancies. Thus, to improve the chances of a healthy pregnancy, a surgery called selective pregnancy reduction or multi-fetal reduction lowers the number of fetuses growing within the uterus. The procedure occurs before the first trimester and usually involves injecting particular drugs within fluid-filled pouches containing the fetus through the vagina or belly, which quickly stops the fetus's heartbeat. This procedure often reduces two fetuses. Similarly, a technique called radiofrequency ablation is used to terminate one or more fetuses, with the help of a small needle device that uses electric currents to cut off the blood flow from the umbilical cords attached to the fetuses. Fetal pregnancy reduction is a safe procedure with fewer complications, and it takes only a few minutes (Ouyang et al. 2020).

20.5.3 Complications in Pregnancies

Pregnancies conceived by IVF are significantly associated with a higher prevalence of perinatal complications than spontaneously conceived pregnancies. Even after correction of confounding factors, IVF pregnancies in the third stage of labor are proven to be an independent risk factor for a higher rate of postpartum hemorrhage (PPH), manual removal of retained placenta, and blood transfusion (Wertheimer et al. 2020). Unlike pregnancies conceived naturally, conceptions resulting from IVF treatments are initially linked with an enhanced chance of ectopic implantations. The major controversial factors related to ectopic pregnancy risk are the day of embryo transfer, fresh or frozen-thawed embryo cycle, and single or double embryo transfer. Day 3 of embryo transfer, compared to day 5 or 6 of blastocyst transfer, increases the ectopic pregnancy rate; similarly, fresh embryo transfer has elevated risk of ectopic conceptions relative to frozen-thawed embryo transfer, whereas single embryo transfer decreases the chances of ectopic pregnancies compared to that of double embryo transfers. Moreover, ectopic implantation remains a significant cause of maternal morbidity and mortality in pregnancy's first trimester (Karadag and Caliskan 2020).

Ectopic tubal pregnancies sometimes require tubal surgery, causing further impairment to fertility. The surgery may also cause excessive blood loss, thereby creating more complications in the pregnancy, and in rare instances, shock from blood loss requires transfusion or other treatments. Statistical data shows that approximately 2–5% of all IVF pregnancies become ectopic, and 15–23% of IVF pregnancies end up in miscarriage. Sometimes, for oocyte collection, bleeding has been reported in 0.03–0.5% cases and infections in 0.02–0.3% during or after embryo transfers. In rare cases, damage to the bladder or bowel may occur (Andersen et al. 2009).

20.5.4 Long-Term Risks

The increased duration of up to one year of infertility and IVF treatments and the high doses of medications, followed by defecation and treatment failure, are linked with an increased risk of ovarian cancer (Farhud et al. 2019).

20.6 Future Perspectives

The significant drawbacks of IVF involve its high cost and limitation of IVF centers in the sub-Saharan region. Due to its high treatment price, IVF has poor government supports. In most cases, health insurance companies or health management organizations do not include IVF treatment funds, which leave the patients with huge pocket expenditure and money debts for their treatment. It is, therefore, essential to ensure appropriate patient selection before the beginning of IVF treatment as sometimes, it may benefit patients from other cheaper fertility treatments. Recent technology developments should be made in the whole IVF-ET procedure to be conducted more efficiently and cheaply, making it feasible for low-income society members. Also, several parts of the world have been misled under religion and have ethical issues against IVF, which can be controlled using proper educational awareness programs. Also, IVF is complex and time-consuming and often requires multiple cycles, which causes psychological, emotional, and physical stress. Hence, shortening the treatment's period and increasing its success rates may aid patients in many indirect ways. In conclusion, several new challenges are always ahead of researchers, but our scientific community has the intellectual potential to take on and overcome these challenges.

REFERENCES

Alshehre, S. M., Narice, B. F., Fenwick, M. A., Metwally, M. 2021. The impact of endometrioma on in vitro fertilisation/intra-cytoplasmic injection IVF/ICSI reproductive outcomes: a systematic review and meta-analysis. *Arch Gynecol Obstet* 303(1):3–16. https://doi.org/10.1007/s00404-020-05796-9

Andersen, A. N., Goossens V., Bhattacharya, S., Ferraretti, A. P., Kupka, M. S., de Mouzon J., Nygren K.G. 2009. Assisted reproductive technology and intrauterine inseminations in Europe, 2005: results generated from European registers by ESHRE: ESHRE. The European IVF Monitoring Programme (EIM), for the European Society of Human Reproduction and Embryology (ESHRE). *Hum Reprod* 24(6):1267–1287. https://doi.org/10.1093/humrep/dep035.

Back Louis, G. M., Sundaram, R., Schisterman, E. F., Sweeney, A., Lynch, C. D., Ki M. S., Chen, Z. 2014. Semen quality and time to pregnancy: the longitudinal Investigation of Fertility and the Enivronment Study. *Fertil Steril* 101(2):453–462.https://doi.org/10.1016/j.fertnstert.2013.10.022.

Balaban, B., Barut, T., Urman, B. 2012. Assessment of oocyte quality. In: *Practical Manual of In Vitro Fertilization*, 1st ed. New York: Springer, pp 105–119. https://doi.org/10.1007/978-1-4419-1780-5_13.

Bavister, B. D. 2002. Early history of in vitro fertilization. *Reproduction* 124(2):181–196. https://doi.org/10.1530/rep.0.1240181.

Bu, Z., Yang, X., Song, L., Kang, B., Sun, Y. 2019. The impact of endometrial thickness change after progesterone administration on pregnancy outcome in patients transferred with single frozen-thawed blastocyst. *Reprod Biol Endocrinol* 17(1):99. https://doi.org/10.1186/s12958-019-0545-0.

Catt, J., Wood, T., Henman, M., Jansen, R. 2003. Single embryo transfer in IVF to prevent multiple pregnancies. *Twin Res* 6(6):536–539. https://doi.org/10.1375/136905203322686554.

Chang, M. C. 1959. Fertilization of rabbit ova in vitro. *Nature* 184:466–467. https://doi.org/10.1038/184466a0.

Choe, J., Archer, J. S., Shanks, A. L. 2020. *In Vitro Fertilization*. Treasure Island: StatPearls. www.ncbi.nlm.nih.gov/books/NBK562266/

Csokmay, J. M., Hill, M. J., Chason, R. J., Hennessy, S., James, A. N., Cohen. J., Decherney, A. H., Segars, J. H., Payson, M. D. 2011. Experience with a patient-friendly, mandatory, single-blastocyst transfer policy: the power of one. *Fertil Steril* 96(3):580–584. https://doi.org/10.1016/j.fertnstert.2011.06.043.

Da Silva Junior, J. G., Da silva, D. J., De oliveria, J. R. S., Nascimento, W. M. D., Callou, M. A. M., Do monte, T. V. S. 2020. Investigation of in vivo and in vitro fertilization methods: a review of the literature. *Hum Reprod Arch* 34:e002919. http://doi.org/10.4322/hra.002919.

De Santis, L., Cino, I., Rabellotti, E., Calzi, F., Persico, P., Borini, A., Coticchio, G. 2005. Polar body morphology and spindle imaging as predictors of oocyte quality. *Reproduct BioMed Online* 11(1):36–42. https://doi.org/10.1016/s1472-6483(10)61296-5.

Elder, K. and Dale, B. 2020. Oocyte retrieval and embryo culture. In *In-Vitro Fertilization*. Cambridge: Cambridge University Press, pp. 213–253. https://doi.org/10.1017/9781108611633.012.

Esteves, S. C., Roque, M., Bedoschi, G., Haahr, T., Humaidan, P. 2015. Intracytoplasmic sperm injection for male infertility and consequences for offspring. *Nat Rev Urol* 15(9):535–562. https://doi.org/10.1038/s41585-018-0051-8.

Ezra, O., Haas, J., Nahum, R., Maman, E., Cohen, Y., Segev-Zahav, A., Orvieto, R. 2019. Do poor-responder patients undergoing IVF benefit from splitting and increasing the daily gonadotropin dose? *Gynecol Endocrinol* 35(10):890–893. https://doi.org/10.1080/09513590.2019.1608520.

Fanchin, R., Righini, C., de Ziegler, D., Olivennes, F., Ledee, N., Frydman, R. 2001. Effects of vaginal progesterone administration on uterine contractility at the time of embryo transfer. *Fertil Steril* 75(6):1136–1140. https://doi.org/10.1016/s0015-0282(01)01787-3.

Farhud, D. D., Zokaei, S., Keykhaei, M., Zarif Yeganeh, M. 2019. Strong evidences of the ovarian carcinoma risk in women after IVF treatment: a review article. *Iran J Public Health* 48(12):2124–2132.

Fauser, B. C., Devroey, P., Diedrich, K., Balaban, B., Bonduelle, M., Delemarre-van de Waal, H. A., Estella, C., Ezcurra, D., Geraedts, J. P., Howles, C. M., Lerner-Geva, L., Serna, J., Wells, D. 2014. Health outcomes of children born after IVF/ICSI: a review of current expert opinion and literature. *Reprod Biomed Online* 28(2):162–182. https://doi.org/10.1016/j.rbmo.2013.10.013.

Fuller, B. J., Lane, N., Benson, E. 2004. *Life in Frozen State*, 1st ed. New York: CRC Press, pp. 3–67.

Georgiou, E. X., Melo, P., Baker, P. E., Sallam, H. N., Arici, A., Garcia-Velasco, J.A., Granne, I. E. 2019. Long-term GnRH agonist therapy before in vitro fertilisation (IVF) for improving fertility outcomes in women with endometriosis. *Cochrane Database Syst Rev* 2019(11):CD013240. https://doi.org/10.1002/14651858.cd013240.pub2.

Griel, A.L., Slauson-Blevins, K., Mcquillian, J. 2010. The experience of infertility: a review of recent literature. *Social Health Illn* 32(1):140–162. https://doi.org/10.1111/j.1467-9566.2009.01213.x

Gurunath, S., Pandian, Z., Anderson, R. A., Bhattacharya, S. 2011. Defining infertility-a systematic review of prevalence studies. *Hum Reprod Update* 17(5):575–588. https://doi.org/10.1093/humupd/dmr015.

Habana, A.E., Palter, S. F. 2001. Is tubal embryo transfer of any value? A meta-analysis and comparison with the Society for Assisted Reproductive Technology database. *Fertil Steril* 76(2):286–293. https://doi.org/10.1016/s0015-0282(01)01879-9.

Halliday, J. 2007. Outcomes of IVF conceptions: are they different? *Best Pract Res Clin Obstet Gynaecol* 21(1):67–81. https://doi.org/10.1016/j.bpobgyn.2006.08.004.

Harper, J. C., Sengupta, S. B. 2012. Preimplantation genetic diagnosis: state of the art 2011. *Hum Genet* 131(2):175–186. https://doi.org/10.1007/s00439-011-1056-z.

Howe, R. S., Wheeler, C., Mastroianni, L. Jr., Blasco, L., Tureck, R. 1988. Pelvic infection after transvaginal ultrasound-guided ovum retrieval. *Fertil Steril* 49(4):726–728. https://doi.org/10.1016/s0015-0282(16)59849-5.

Iketubosin, F. 2018. In vitro fertilization embryo transfer processes and pathway: a review from practice perspective. *Trop J Obstet Gynaecol* 35(3):227–232. https://doi.org/10.4103/TJOG.TJOG_83_18.

Ioannidou, P. G., Bosdou, J. K., Lainas, G. T., Lainas, T. G., Grimbizis, G. F., Kolibianakis, E. M. 2021. How frequent is severe ovarian hyperstimulation syndrome after GnRH agonist triggering in high-risk women? A systematic review and meta-analysis. *Reprod Biomed Online* 42(3):635–650. https://doi.org/10.1016/j.rbmo.2020.11.008.

Johnson, M. H. 2019. A short history of in vitro fertilization (IVF). *Int J Dev Biol* 63(3–5):83–92. https://doi.org/10.1387/ijdb.180364mj.

Kamath, M. S., Mascarenhas, M., Kirubakaran, R., Bhattacharya, S. 2020. Number of embryos for transfer following in vitro fertilisation or intra-cytoplasmic sperm injection. *Cochrane Database Syst Rev* 8:CD003416.https://doi.org/10.1002/14651858.CD003416.pub5.

Karadag, C. and Caliskan, E. 2020. Ectopic pregnancy risk with assisted reproductive technology. *Curr Obstet Gynecol Rep* 9:153–157. https://doi.org/10.1007/s13669-020-00292-y.

Karaki, R. Z., Samarraie, S. S., Younis, N. A., Lahloub, T. M., Ibrahim, M. H. 2002. Blastocyst culture and transfer: a step toward improved in vitro fertilization outcome. *Fertil Steril* 77(1):114–118. https://doi.org/10.1016/s0015-0282(01)02939-9.

Kenigsberg, S., Wyse, B. A., Librach, C. L., da Silveira, J. C. 2017. Protocol for exosome isolation from small volume of ovarian follicular fluid: evaluation of ultracentrifugation and commercial kits. *Methods Mol Biol* 1660:321–341. https://doi.org/10.1007/978-1-4939-7253-1_26.

Khosravi, P., Kazemi, E., Zhan, Q., Malmsten, J. E., Toschi, M., Zisimopoulos, P., Sigaras, A., Lavery, S., Cooper, L. A. D., Hickman, C., Meseguer, M., Rosenwaks, Z., Elemento, O., Zaninovic, N., Hajirasouliha, I. 2019. Deep learning enables robust assessment and selection of human blastocysts after in vitro fertilization. *NPJ Digit Med* 2:21. https://doi.org/10.1038/s41746-019-0096-y.

Konc, J., Kanyo, K., Kriston, R., Somoskoi, B., Cseh, S. 2014. Cryopreservation of embryos and oocytes in human assisted reproduction. *Biomed Res Int* 2014:307268. https://doi.org/10.1155/2014/307268.

Kumar, P., Sait, S. F., Sharma, A., Kumar, M. 2011. Ovarian hyperstimulation syndrome. *J Hum Reprod Sci* 4(2):70–75. https://doi.org/10.4103/0974-1208.86080.

La Marca, A., Sunkara, S. K. 2014. Individualization of controlled ovarian stimulation in IVF using ovarian reserve markers: from theory to practice. *Hum Reprod Update* 20(1):124–140. https://doi.org/10.1093/humupd/dmt037.

Lu, X., Liu, Y., Cao, X., Lui, S. Y., Dong, X. 2019. Laser-assisted hatching and clinical outcomes in frozen-thawed cleavage-embryo transfers of patients with previous repeated failure. *Lasers Med Sci* 34(6):1137–1145. https://doi.org/10.1007/s10103-018-02702-3.

Maheshwari, A., Hamilton, M., Bhattacharya, S. 2016. Should we be promoting embryo transfer at blastocyst stage? *Reprod Biomed Online* 32(2):142–146. https://doi.org/10.1016/j.rbmo.2015.09.016.

Moffat, R., Hansali, C., Schoetzau, A., Ahler, A., Gobrecht, U., Beutler, S., Raggi, A., Sartorius, G., De Geyter, C. 2021. Randomised controlled trial on the effect of clomiphene citrate and gonadotropin dose on ovarian response markers and IVF outcomes in poor responders. *Human Reprod* 36(4):987–997. https://doi.org/10.1093/humrep/deaa336.

Morin, S. J., Hanson, B. M., Juneau, C. R., Neal, S. A., Landis, J. N., Scott, R. T., Hotaling, J. M. 2020. A comparison of the relative efficiency of ICSI and extended culture with epididymal sperm versus testicular sperm in patients with obstructive azoospermia. *Asian J Androl* 22(2):222–226. https://doi.org/10.4103/aja.aja_58_19.

Mrazek, M. and Fulka, J. Jr. 2003. Failure of oocyte maturation: possible mechanisms for oocyte maturation arrest. *Hum Reprod* 18(11):2249–2252. https://doi.org/10.1093/humrep/deg434.

National Collaborating Centre for Women's and Children's Health (UK). 2004. *Fertility: Assessment and Treatment for People with Fertility Problems*. London, UK: RCOG Press.

Neri, Q. V., Lee, B., Rosenwaks, Z., Machaca, K., Palermo, G. D. 2014. Understanding fertilization through intracytoplasmic sperm injection (ICSI). *Cell Calcium* 55(1):24–37. https://doi.org/10.1016/j.ceca.2013.10.006.

Osemwenkha, A. and Osaikhuwuomwan, J. 2016. Transabdominal follicular aspiration in an in vitro fertilization cycle: experiences with an unusual but necessary intervention in a resource-limited setting. *Clin Exp Reprod Med* 43(1):54–57. https://doi.org/10.5653/cerm.2016.43.1.54.

Ouyang, Y., Cai, P., Gong, F., Lin, G., Qiun, J., Li, X. 2020. The risk of twin pregnancies should be minimized in patients with a unicornuate uterus undergoing IVF-ET. *Sci Rep* 10(1):5571. https://doi.org/10.1038/s41598-020-62311-5.

Pan, Y., Hao, G., Wang, Q., Liu, H., Wang, Z., Jiang, Q., Shi, Y., Chen, Z. J. 2020. Major factors affecting the live birth rate after frozen embryo transfer among young women. *Front Med* (Lausanne) 7:94. https://doi.org/10.3389/fmed.2020.00094.

Panesar, K. (2016) Medications used for in vitro fertilization. *US Pharm* 41(9):HS8–HS11.

Practice Committee of Society for Assisted Reproductive Technology, Practice Committee of American Society for Reproductive Medicine. 2008. The role of assisted hatching in in vitro fertilization: a review of the literature. A Committee opinion. *Fertil Steril* 82:S164–165. https://doi.org/10.1016/j.fertnstert.2008.08.043.

Quaas, A. M., Legro, R. S. 2019. Pharmacology of medications used for ovarian stimulation. *Best Pract Res Clin Endocrinol Metab* 33(1):21–33. https://doi.org/10.1016/j.beem.2018.10.002.

Rienzi, L., Vajta, G., Ubaldi, F. 2011. Predictive value of oocyte morphology in human IVF: a systematic review of the literature. *Hum Reprod Update* 17(1):34–45. https://doi.org/10.1093/humupd/dmq029.

Salian, S. R., Nayak, G., Kumari, S., Patel, S., Gowda, S., Shenoy, Y., Sugunan, S. Rajanikant G. K., Managuli, R. S., Mutalik S, Dahiya V, Pal S, Adiga SK, Kalthur G. 2018. Supplementation of biotin to sperm preparation medium enhances fertilizing ability of spermatozoa and improves preimplantation embryo development. *J Assist Reprod Genet* 36(2):255–266. https://doi.org/10.1007/s10815-018-1323-1.

Schoolcraft, W. B., Gardner, D. K., Lane, M., Schlenker, T., Hamilton, F., Meldrum, D. R. 1999. Blastocyst culture and transfer: analysis of results and parameters affecting outcome in two in vitro fertilization programs. *Fertil Steril* 72(4):604–609. https://doi.org/10.1016/s0015-0282(99)00311-8.

Schwarz, L., Karnaushenko, D. D., Hebenstreit, F., Naumann, R., Schmidt, O. G., Medina-Sanchez, M. 2020. A rotating spiral micromotor for noninvasive zygote transfer. *Adv Sci (Weinh)* 7(18):2000843. https://doi.org/10.1002/advs.202000843.

Shrestha, D., La, X., Feng, H. L. 2015. Comparison of different stimulation protocols used in in vitro fertilization: a review. *Ann Transl Med* 3(10):137. https://doi.org/10.3978/j.issn.2305-5839.2015.04.09.

Simon, A. and Laufer, N. 2012. Assessment and treatment of repeated implantation failure (RIF). *J Assist Reprod Genet* 29(11):1227–1239. https://doi.org/10.1007/s10815-012-9861-4.

Siristatidis, C. S., Gibreel, A., Basios, G., Maheshwari, A., Bhattacharya, S. 2015. Gonadotrophin-releasing hormone agonist protocols for pituitary suppression in assisted reproduction. *Cochrane Database Syst Rev* (11):CD006919. https://doi.org/10.1002/14651858.CD006919.pub4.

Sonmezer, M., Gulumser, C., Sonmezer, M., Ukur, E. S., Atabekog, C. 2020. Transabdominal ultrasound guided oocyte retrieval using vaginal ultrasound probe: definition of the technique. *J Obstet Gynaecol Res* 47(2):800–806. https://doi.org/10.1111/jog.14618.

Steptoe, P. C. and Edwards, R. G. 1978. Birth after the reimplantation of a human embryo. *Lancet* 2(8085):366. https://doi.org/10.1016/s0140-6736(78)92957-4.

Sunderam, S., Chang, J., Flowers, L., Kulkarni, A., Sentelle, G., Jeng, G., Macaluso, M. 2006. Centers for Disease Control and Prevention (CDC) (2009) Assisted reproductive technology surveillance—United States. *MMWR Surveill Summ* 58(5):1–25.

Sustar, K., Rozen, G., Agresta, F., Polyakov, A. 2019. Use of intracytoplasmic sperm injection (ICSI) in normospermic men may result in lower clinical pregnancy and live birth rates. *Aust N Z J Obstet Gynaecol* 59(5):706–711. https://doi.org/10.1111/ajo.13004.

Van Loendersloot, L. L., Van Wely, M., Limpens, J., Bossuyt, P. M., Repping, S., Van der Veen, F. 2010. Predictive factors in in vitro fertilization (IVF): a systematic review and meta-analysis. *Hum Reprod Update* 16(6):577–589. https://doi.org/10.1093/humupd/dmq015.

Wang, J. and Sauer, M. V. 2006. In vitro fertilization (IVF): a review of 3 decades of clinical innovation and technological advancement. *Ther Clin Risk Manag* 2(4):355–364. https://doi.org/10.2147/tcrm.2006.2.4.355.

Wertheimer, A., Melamed, S., Ashwal, E., Sapir, O., Oron, G., Ben-Haroush, A., Shochat, T., Wiznitzer, A., Shufaro, Y. 2020. Complications of the third stage of labor are more prevalent in IVF pregnancies. *J Matern Fetal Neonatal Med* 23:1–5. https://doi.org/10.1080/14767058.2020.1730797.

Wikland, M., Enk, L., Hammarberg. K., Nilsson, L. 1987. Use of a vaginal transducer for oocyte retrieval in an IVF/ET program. *J Clin Ultrasound* 15(4):245–251. https://doi.org/10.1002/jcu.1870150405.

Winter, E., Wang, J., Davies, M. J., Norman, R. 2002. Early pregnancy loss following assisted reproductive technology treatment. *Hum Reprod* 17(12):3220–3223. https://doi.org/10.1093/humrep/17.12.3220.

Yanushpolsky, E. H. 2015. Luteal phase support in in vitro fertilization. *Semin Reprod Med* 33(2):118–127. https://doi.org/10.1055/s-0035-1545363.

Yildiz, S., Turkgeldi, E., Angun, B., Eraslan, A., Urman, B., Ata, B. 2019. Comparison of a novel flexible progestin primed ovarian stimulation protocol and the flexible gonadotropin-releasing hormone antagonist protocol for assisted reproductive technology. *Fertil Steril* 112(4):677–683. https://doi.org/10.1016/j.fertnstert.2019.06.009.

Yılmaz, N., Ceran, M. U., Ugurlu, E. N., Gulerman, H. C., Ustun, Y. E. 2020. GnRH agonist versus HCG triggering in different IVF/ICSI cycles of same patients: a retrospective study. *J Obstet Gynaecol* 40(6):837–842. https://doi.org/10.1080/01443615.2019.1674262.

Zhu, L., Zhang, Y., Liu, Y., Zhang, R., Wu, Y., Huang, Y., Liu, F., Li, M., Sun, S., Xing, L., Zhu, Y., Chen, Y., Xu, L., Zhou, L., Huang, H., Zhang, D. 2016. Maternal and live-birth outcomes of pregnancies following assisted reproductive technology: a retrospective cohort study. *Sci Rep* 20(6):35141. https://doi.org/10.1038/srep35141.

21

Erythrocytes as an Indicator of Diseases in a Freshwater Catfish Heteropnuetes Fossilis *(Bloch.)*

B. D. Joshi, Induja Mishra, Namita Joshi, and Pashupati Nath

CONTENTS

21.1 Introduction

Blood, as a liquid tissue working as a transportation medium, also plays vital role in the defense mechanism and related immunological aspects. Therefore, hematological studies give us few of the simplest but most valuable tools to monitor not only the status of fish health under normal conditions but also reflects the status and impacts of ambient environment and the physio-pathological conditions of the vertebrate concerned. Fish are no exception to this generalization. Indeed, fish are most sensitive than the majority of the other higher vertebrates to any impact on its body, physiology, and environment, being very specifically aquatic animals with most primitive vertebrate (Joshi, 1980a, 1987; Abowei et al., 2012).

The cells of erythrocytic and leucocytic series arise from the hemocytoblast. Fish erythrocytes are ellipsoidal nucleated cells of different sizes (Joshi, 1987). The size of erythrocytes is species-specific and impacted by the nutritional status, activity, seasons, and habitat conditions, like water temperature and dissolved oxygen, as well as other physiological and environmental factors to some extent but temporarily. There is presence of immature cells in the circulation of fish's blood because the barrier connecting hemopoietic tissue and blood circulation is not very strong. Mature erythrocytes are most abundant types of cells in the blood. Joshi (1987) has described a variety of erythrocytic cells in circulation—namely, erythroblasts, normocytes, microcytes, macrocytes, and crenated erythroblasts, besides anisocytes and poikilocytes. Erythrocytes show structural and behavioral changes quite conspicuously and, therefore, can be trusted in diagnostic observations and to monitor fish health and their environment too (Joshi et al., 2015; Jeronimo et al., 2011; Mishra et al., 2021).

This paper briefly describes structural deformities and aberrations among large population of erythrocytes of the catfish *Heteropneustes fossilis* on account of diseases in a freshwater Indian subcontinent species, *H. fossilis*, found abundantly all over India.

DOI: 10.1201/9781003220398-21

21.2 Materials and Methods

Collection of Samples

The freshwater catfish species *H. fossilis* was obtained from Bharat Heavy Electricals Limited, Haridwar, fish market of Uttarakhand and brought alive in plastic bags and transferred to the aquarium and maintained under laboratory conditions for further investigation. The transportation time was 15–20 minutes from the sampling site to the laboratory. The fish were rested for acclimation of at least for three to four days. Blood was collected from the live fish from the caudal peduncle with the help of sterilized syringe described by Hickey (1976). It was immediately pooled into sterilized vials, coated with the anticoagulant dipotassium ethylenediaminetetraacetic acid (EDTA) Adeyemo (2007) and immediately used for the hematological analyses, while the remaining portion of blood was stored without anticoagulant for further biochemical analyses, including slide preparations following the methods as used by Tandon and Joshi (1973)

Morphological Analysis

Thin blood smears were air-dried and fixed in absolute methanol, and slides were stained with Giemsa stain and examined under a microscope by means of 100× oil immersion objective. Images were taken with the help of digital camera (Lobovision) and a scanning electron microscope (Model No.: JEOL-JSM-6490 LV). The first 100 red and white blood corpuscles examined on each slide were counted and listed. In the present study, only variations in the differential blood cell counts of erythrocytic series have been made along with comparative cytomorphological characteristics for healthy and diseased fish. The first 100 red corpuscles seen on each slide were counted and listed as either erythroblast or erythrocytes.

Results and Discussion

The stained preparations exhibited four types of infections in as many as 12 specimens of *H. fossilis* during the course of this study. Out of a group of 40 specimens, 4 specimens were found infected with hemoflagellate *Trypanosoma* sp. (A, F, G; see Figure 12.1), two specimens with *Haemogregarina* sp. (I), two with *Coccidia* sp. (D), and two with *Haemohormedium* sp. (B, E). A number of hematozoans have a tendency to be intra-erythrocytic (A, B, E, F), while few others are only extra-erythrocytic, living in the blood plasma (A, D, E, F). The fish's mature erythrocyte contains a centrally placed nucleus. It is densely packed with nuclear chromatin, in clusters or in loosely woven form. The vascular system is one of the first to expose to any extrinsic components (e.g., chemicals or pathogens), so therefore, it also immediately expresses any exogenous or indigenous impact on the physio-biochemical metabolic activities or pathological mirroring of homeostasis, indicating general homeostasis. Erythrocytes show structural and behavioral changes quite conspicuously; therefore, they can be used and trusted to monitor fish health and their environment too (Joshi 1980a, 1986). The erythrocytes were seldom normo-chromic, in the diseased fish. These showed a high degree of poly-chromasia, hypo-chromasis (J–S), or baso-philia and appear mostly impacted by the status and tenure of a disease and parasitemia. Anisocytosis (B), poikilocytosis (H, P, Q, R), dislocation of nuclei (K, L, M), denucleated conditions (F, L), nuclear aberrations (M), vacuolized conditions (P), echinus conditions (J), micro nuclear appearances (S) and degenerating forms of RBCs at different stages of erythrolysis degeneration (B) were quite often found. Such diverse variations among freshwater teleost have been widely described by Joshi (1982b, 1986). A large number of various stages of hematozoans were found within the erythrocytes (A, B, E, G). One specimen of *H. fossilis* also had PEN (pycnotic, erythrocytic necrosis), a viral infection (C, F).

This showed a very heavy infestation of intraerythrocytic condition. On the contrary some specimens infected with any type of hematozoan did not show any aberration of any type. There were also two specimens of *H. fossilis*, which had multiple infections (A, G). Erythrocytes showing hypertrophy were also found. The hypertrophy were of two kinds—first, where intra-erythrocytic parasite caused it, according to the size and position in side its erythrocytes (E, B, J, P) and the host cell nucleus pushed by the parasite, and second, just due to parasitic presence even at extra-erythrocytic (D, F) condition.

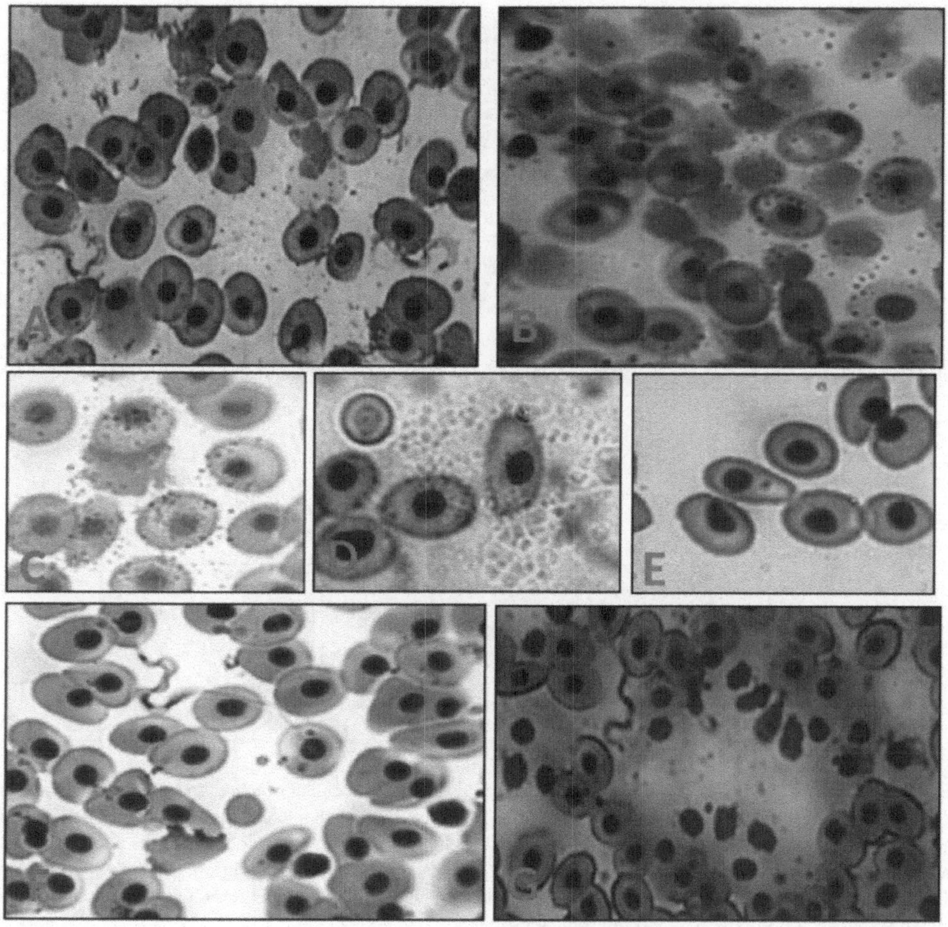

FIGURE 21.1 (A) Case of the multiple hematozoan infection of H. fossilis: Trypanosomes, sporozoans, and necrotic viroids. (B) Showing an intra-erythrocytic stage of Haemohormodium and degenerating forms of RBC. (C) Showing VEN (viral erythrocytic necrosis). (D) Showing extracellular and intracellular stages of coccidian infection. (E) Showing an intra-erythrocytic stage of Haemohormodium. (F) Showing Trypanosoma and Cryptobia, VEN (viral erythrocytic necrosis), intra-erythrocyticsporozoans, and their parasitophores, besides polychromasia, hypertrophy, and host cell nuclear aberrations. (G) Again, showing Trypanosoma and sporozoans besides hypochromasia and aberrations of the erythrocytes.

The infected erythrocytes showing the phenomenon of hypertrophy appeared infected with any type of sporozoan or hemoflagellate like *Trypanosomes*, *Hemogregarine*, and *Haemohormodium*.

Parasitophorous vacuoles, which separate the parasites from the host cell nucleus (H, E), making an isolated niche, were also found among the diseased fish. Among most erythrocytes, the nucleus of the host cell erythrocytes did not show any changes in their shape and structure due to parasite and only the position of the nucleus is displaced to give space to parasite. Joshi (2005) and Hassan et al. (2007) suggested that there is also potential compensatory release of immature erythrocytes due to stress in the organism excited by the effect of such infections. The nuclei of a number of erythrocytes that were deformed in shape and indented in the infected fishes show the phenomenon of karyorrhexis; it is conspicuous in the mildly infected fishes. Crenation, also referred to as echinus forms, of the erythrocytes were also observed conspicuously in mild and strong conditions of infection during the studies (J, N, S).

The erythrocyte numbers of the fish infected with parasite have been conspicuously found lower in comparison to those of the non-infected ones among many fishes, as reported by earlier workers (Tandon

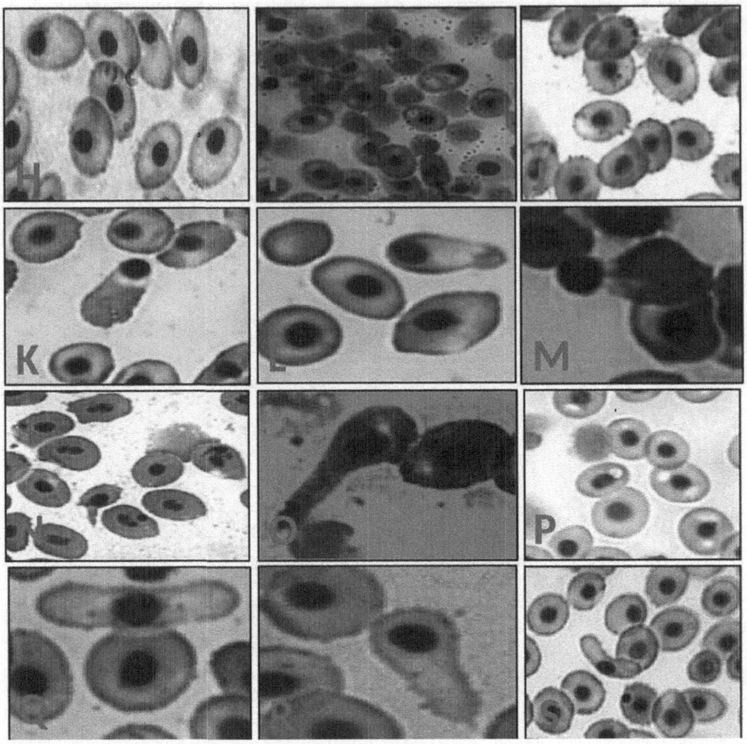

FIGURE 21.2 (H) Showing hypochromasia, cell hypertrophy, and displacement of nucleus (I) Showing intra-erythrocyticsporozoan their parasitophores, besides hypochromasia, hypertrophy, and displacement of host cell nuclei. (J) Showing *Haemogregarina* sp. (K) Showing echinus conditions of erythrocytes. (L) Showing dislocation of the nucleus. (M) Showing dislocation of nucleus and an erythroid. (N) Dislocation of nucleus and basophilic condition. (O) Showing nuclear aberrations. (P) Showing vacuolized conditions. (Q) Showing anisocytosis. (R) Showing echinus condition and anisocytosis. (S) Showing anisocytosis, poikilocytosis, and micronucleus.

and Joshi, 1973a, 1973b; Genç et al., 2005; Boon et al., 1989). Most species of *Trypanosoma*-infected fish show pathogenic symptoms of clinical interest and importance, where trypanosomiasis' impacts range from mild anemia associated with low levels of parasitemia to severe pathological changes due to heavy parasite burden Islam and Woo (1991); leukocytosis, hypoglycemia, and erythropenia are the frequent outcomes of trypanosomiasis (Joshi, 1973, 2005; Joshi and Dabral, 1981; Tandon and Joshi, 1973; Joshi 1982b; Gupta and Jairajpuri, 1983; Joshi 1980a, 1980b). Our present observations are also in the same line, ascertaining the effects of parasites on variable blood cell structure. The present study demonstrated that there were alterations in the fish erythrocytes, and the intensity of these changes depends on the prevalence of pathogens and conditions of the infected fish.

21.3 Conclusion

The cytomorphological evaluation of fish erythrocytes can be clinically used as bioindicator for evaluating environmental stresses in fish but also for clinical pathological observations to ascertain the effect of diseases, infections, infestations, and other types of physiological stresses. The morphological distortion of red blood cells is a relatively steady index as the erythrocytes change according to the internal and external factors and stresses of aquatic organisms. It is the most important and trustworthy indicator of amalgamation of aquatic environment in general and a disease as observed in the present case of *Heteropneustes fossilis* for identifying stress- and disease-caused effects. The present study is significant for indicating simpler microscopic but very trustworthy changes in morphological alternations of fish's erythrocytes during diseases; these changes will also be helpful for fishes in further required direction to maintain health parameters in fish health management and aquaculture.

21.4 Acknowledgments

The authors are very grateful to the University Grant Commission for providing the financial assistance via the UGC BSR scheme for their research work. The authors are also thankful to the Department of Zoology and Environmental Science, GurukulaKangari University, Haridwar, for providing the necessary experimental facilities during the research work.

REFERENCES

Abowei, J. F. N., Briyai, O. F. and Bassey, S. E. (2012) A review of some basic parasite diseases in culture fisheries flagellids, dinoflagellides and ichthyophthriasis, ichtyobodiasis, coccidiosistrichodiniasis, heminthiasis, hirudinea infestation, crustacean parasite and ciliates. *British Journal of Pharmacology and Toxicology*, 2(5), 213–226.

Adeyemo, O. K. (2007) Haematological profile of *Clariasgariepinus* (Burchell, 1822) exposed to lead. *Turkish Journal of Fisheries and Aquatic Sciences*, 7(2), 163–169.

Boon, J. H., Lokin, C. J. A., Ceusters, R. and Ollevier, F. (1989) Some properties of the blood of European eel (Anguilla anguilla) and the possible relationship with *Anguillicolacrassus* infestations. *Aquaculture*, 76(3–4), 203–208.

Genç, E., Şahan, A., Altun, T., Cengizler, İ. and Nevşat, E. (2005) Occurrence of the swim bladder parasite Anguilli colacrassus (Nematoda, Dracunculoidea) in European eels (Anguilla anguilla) in Ceyhan River, Turkey. *Turkish Journal of Veterinary and Animal Sciences*, 29(3), 661–663.

Gupta, N. and Jairajpuri, D. S. (1983) Trypanosoma batrachi Qadri, 1962 and its effect on the biochemical composition of the blood of "*Clarias batrachus*." *ActaProtozoologica*, 22, 79–85.

Hassan, L. G., Umar, K. J. and Atiku, I. (2007) Nutritional evaluation of Albizialebbeck (L.) pods as. *American Journal of Food Technology*, 2(5), 435–439.

Hickey, C. R. (1976) Fish hematology, its uses and significance. *New York Fish and Game Journal*, 23(2), 170–175.

Islam, A. K. and Woo, P. T. K. (1991) Trypanosomadanilowskyi in Carassiusauratus: the nature of the protective immunity in recovered gold fish. *Journal of Parasitology*, 77, 258–262.

Jeronimo, D., Cruz, C. and Saraiva, A. (2011) Polycystic liver in farmed turbot, *Scopthalmusmaximus* (L.). *Journal of Fish Diseases*, 34(12), 937–938.

Joshi, B. D. (1973) *Studies on the Blood of Freshwater Fishes of India*. Ph.D. Thesis, University of Lucknow, Lucknow, India.

Joshi, B. D. (1980a) On some normal heme topological values of some fresh-water teleosts. *National Academy Science Letters-India*, 3(8), 251–254.

Joshi, B. D. (1980b) Sex-related variation in some haeamtological values of some fresh water teleosts. *Japanese Journal of Ichthyology*, 27(2), 139–143.

Joshi, B. D. (1982a) Changes in the blood glucose and liver glycogen contents of healthy and trypanosome infected fish, *Clarias batrachus*, following intra muscular injection of glucose solution. *Angewandte Parasitologie*, 23(3), 121–124.

Joshi, B. D. (1982b) Studies on some haematological changes in a fresh water cat fish *Clarias batrachus* infected fish. *The Indian Journal of Zootomy*, 23(2), 99–103.

Joshi, B. D. (1986) Differential, red cell counts and relative abundance of erythroblasts in normal circulating blood of certain fresh water teleosts. *Matsya*, 12, 76–79.

Joshi, B. D. (1987) Cyto-morphological classification and key to the identification of normal circulating blood corpuscles of freshwater teleosts. *Himalayan Journal of Zoology*, 1, 98–113.

Joshi, B. D. (2005) Relevance of fish haematology in monitoring their health & the aquatic environment. In: *Presidential Address delivered at 92nd Indian Science Congress, January 2005*. Part II. Advanced Abstracts, pp. 1–36.

Joshi, B. D. and Dabral, R. (1981). Some haematological changes in a freshwater catfish Heteropneustes fossilis, infected with the trypanosome, Trypanosomamaguri. *Proceedings: Animal Sciences*, 90(3), 295–301.

Joshi, B. D., Joshi, N., Mishra, I. and Rawat, S. (2015) First report on the occurrence of acute viral erythrocytic necrosis in the freshwater Catfish *H. fossilis* (Bloch), from India. *Journal of Environmental Biology*, 29, 85–86.

Mishra, I., Joshi, N., Nath, P. and Joshi, B. D. (2021) The influences of haematozoans on different haematological and morphological parameters in a freshwater catfish *Heteropneustes fossilis* (Bloch.) *Aquaculture International*, 29(4), 1669–1681. https://doi.org/10.1007/s10499-021-00706-5

Tandon, R. S. and Joshi, B. D. (1973a) Blood glucose and lactic acid levels in the fresh water fish, Heteropneustes Fossilis, following stress. *Zeitschrift für Tierphysiologie Tierernährung und Futtermittelkunde*, 31(1–5): 210–216.

Tandon, R. S. and Joshi, B. D. (1973b) Studies on the physiopathology of blood of freshwater fishes infected with two new forms of trypanosomes. *Zeitschrift für wissenschaftliche Zoologie Leipzig*, 185: 207–221.

22

Insects as Forensic Indicators and Their Utility in Crime Investigation

Madhu Bala and Anika Sharma

CONTENTS

22.1 Introduction

The utilization of insects and other arthropods as forensic evidence in both civil and criminal investigation is known as forensic entomology (Williams and Villet, 2006). The most important approach of forensic entomology is the determination of time of colonization (TOC), which is also known as the postmortem interval (PMI). When a corpse is discovered, the time since death, otherwise known as PMI, needs to be determined. A medical pathologist can give an accurate PMI up to 72 hours, or 3 days, after death depending on the state of the body and the climate (the milder the conditions the faster the decomposition process) (Gennard, 2007). However, the longer the body is discovered after this window of time, the more inaccurate the PMI will be, which is where insect evidence can be utilized, often with a great deal of success. Methods currently used to estimate the minimum PMI (PMI_{min}) includes two main processes:

1. For the early PMI estimations, the oldest insect specimens that have developed on the carrion must be identified and aged.
2. For late PMI estimations, the succession patterns of arthropods present on the carrion must be examined (Amendt et al., 2007).

DOI: 10.1201/9781003220398-22

22.2 Retrospective

The first forensic entomology case was reported by Sung Tźu, a Chinese lawyer and death investigator in the 13th century in the medico-legal textbook *The Washing Away of Wrongs* (Tzu, 1981). This book has more than 15 cases describing PMI estimation and colonization of human remains as well. Most notable was about a farmer who found murdered in a field with a sharp weapon; all the suspects were told to place their sickle on the ground. Only one sickle attracted blowflies to the trace amount of blood hidden to the naked eye, which resulted in the confession by the murderer. Precedent to Tzu, forensic entomology was used by Yan, a prefectural governor to investigate a case, in which women found guilty to murder her husband. Yan's retinue investigated the corpse and found flies in the head. Investigators separated the hair and observed an iron nail into the corpse head. Without the flies surrounded the wounds, Yan would not have found the nail. Until the 17th century, it was believed that maggots spontaneously arise from the rotten meat. In 1668, Francesco Redi demolished the hypothesis of the spontaneous generation of life and proved by his experiments that maggots come from fly's eggs deposited on rotten meat or putrefying carcasses (Cruz, 2006).

At the beginning of the 19th century, it was noted that flies are attracted to corpses at a very early stage of decomposition. Mende (1829) compiled a list of necrophagous insects, including flies and beetles, as well as other taxa. Orfila and Lesueur (1831) provided a more precise account but did not link flies to the time of death. Krahmer (1857) described the opportunities and problems associated with using insects for the estimation of the PMI, many of which are still relevant today.

In the West, the first forensic entomology case was investigated by a French doctor Bergeret in 1855 (Bergeret, 1855). This case can be contemplated as a quantum leap in forensic entomology. Skeletonized remains of a child were found behind a chimney, and insect evidence was accepted as proof. In estimating PMI, he assumed that metamorphosis occurred within one year and concluded that the current occupants of the building could not have been the murderers. In reappraisal, it must be understood that Bergeret did not focus on forensic entomology in his report for the legal procedure in court but used it as one of the forensic tools (Benecke, 2001). At that time, forensic examiners had only a poor understanding of insect biology (Amendt et al., 2004). In 1878, another case of a newborn was reported by Brouardel. The mummified body was occupied by several arthropods including butterfly larvae and acari (i.e., mites), which led to a request to Monsieur Perier, a professor at the Museum of Natural History in Paris, and Pierre Megnin, an army veterinarian, for further assistance (Brouardel, 1879). He counted numbers of live and dead mites that developed every 15 days, and comparing this with the initial count of mites on the infant, he was able to estimate how long that infant was dead. Megnin was considered the accredited founder of forensic entomology and published many articles and books on various subjects including the books *Faune des Tombeaux* and *La Faune des Cadavres*, which are considered the most important forensic entomology books in history (Klotzbach, 2004).

It is only in the last few decades that forensic entomology true potential is being utilized. Professional organizations like the North American Forensic Entomology Association (NAFEA) and the European Association of Forensic Entomology (EAFE) were established. NAFEA was founded by Dr. John Wallace and Dr. Jeffery Tomberlin to promote recognition of insects as an essential component of the crime investigation. In November 2002, while attending the Entomological Society of America Conference in Ft. Lauderdale, Florida, Dr. John Wallace and Dr. Jeffery Tomberlin had a very long discussion about the problems that forensic entomology is facing on its way to becoming an important forensic tool in medicolegal investigations. During this discussion, both set a goal to organize a conference that could be attended by national and international forensic entomologist, especially from developing countries. This resulted the first North American forensic entomology conference in Las Vegas, Nevada, and the professional organization was named NAFEA. EAFE, on the other hand, was founded in 2002, and Jens Amendt played a major role in its establishment. EAFE has almost 100 members and 41 associate members. Their aim is to encourage the forensic entomology throughout the Europe and establish the collaboration with related international bodies.

22.3 Forensic Entomology Types

Insects are attracted to a body immediately after death, often within minutes (Erzinclioglu, 1983; Smith, 1986; Anderson and VanLaerhoven, 1996; Haskell et al., 1997; Anderson, 2001). Whenever human remains are found, the most important question usually is how, when, where, and why the person died. Historically, various body conditions, such as core body temperature (Nelson, 2000), muscular flaccidity, rigor mortis, lividity, pallor of the skin, and others (Smith, 1986), have been utilized to determine the PMI. However, the entomological specimens have now been proved to be reliable indicators for estimating the PMI in both early and advanced stages of carcass decomposition (Smith, 1986; Kashyap and Pillay, 1989b; Greenberg, 1991; Byrd and Butler, 1998).

According to Lord and Stevenson (1986), forensic entomology has three components:

- Urban entomology: Legal proceedings involve insects and other animals that affect man-made structure and another aspect of the human environment.
- Stored product entomology: It involves disputes over arthropods infesting stored commodities, such as cereals and other kitchen products. Many arthropods damage the food resources, and the intake of infested food can cause allergy and other health issues.
- Medicolegal entomology: Sometimes termed "forensic medical entomology" and in reality "medico-criminal entomology" because of its focus on violent crime, relates primarily to the following:
    1. Determination of the time of death (PMI)
    2. Site or place of death
    3. Mode of death (Anderson, 2001)
    4. Traffic accidents with no immediately apparent cause

Insects are known to have been used in the detection of crimes for a long time, and many researchers contributed to the field of forensic entomology (Benecke, 2001; Greenberg and Kunich, 2002). To interpret a crime scene, it is essential to know which insect species has colonized the body and something about their habits and environmental requirements. Although identification keys are available, the identification of the larvae of forensically relevant insects and proper differentiation of the larval stage using morphological criteria is quite a difficult job. Accurate identification and age determination of an insect specimen are crucial steps in forensic entomological analysis.

Age determination is done by comparing the stage of the oldest immature specimens sampled from the cadaver and the environmental conditions, to which it is presumed that they were exposed, with known growth rate data recorded from baseline rearing of Diptera from the same zoogeographical area at a known temperature. The crucial element in the determination of larval age is the progressive increase in total body size (length and weight) with time. The larval instar, and hence approximate age, can be determined by internal morphological analyses. To determine larvae age, it is essential to do the following:

- Measure the larvae's size (length and weight) and their stage of development (first, second, or third instar), and determine whether they are at feeding or post-feeding stage.
- Accurately determine the temperature to which the larvae were exposed during their development.
- Relate the size and feeding stage of the larvae to their age using suitable experimental references or a database of size and stage against age at different temperatures.

The age of immature stages of insects found on a dead body can provide evidence for the estimation of a minimum PMI ranging from one day to more than one month, depending on the insect species involved and the climatic conditions at the death scene. Exact species identification of insect samples is the first essential step in estimating the age of the larvae found. Calliphorid flies are known to travel over long distances to find their food source and start the postmortem clock as they visit the dead remains both for oviposition and feeding. Studies have shown that due to their early arrival at a corpse, they are the major sources used by forensic practitioners to provide an accurate estimation of the time since death (Archer and Elgar, 2003; James and Harwood, 1969; Borror and White, 1970; Bland and Jacques, 1978; Liu and Greenberg, 1989; Hall and Doisy, 1993; Hogue, 1993).

22.4 Mammalian Decomposition

Decomposition is the gradual breakdown of dead organic matter (Spitz, 1993), which starts within few minutes (depending on various environmental factors, surroundings, and situation of corpse) after death and continues over a period (Vass et al., 2002). During decomposition, the carcass goes through various physical, chemical, and biological changes and releases volatile organic compounds (VOCs) to which insects like blowflies, flesh flies, and other carrion feeders are attracted. According to Greenberg, Diptera and Coleoptera represent most of the carrion-eating insects (Greenberg, 1991). The insects invade the carrion in a predictable sequence or succession, and each group of insects is attracted to dead body at certain stage of decomposition. Some species are attracted to the corpse very shortly after death, while others are attracted during active decay, and still others visit to the dry skin and bones. Thus, the type and composition of insect fauna found on a corpse are indicative of its stage of decomposition (Anderson, 2009). It is estimated that Diptera and Coleoptera account for 60% of the fauna associated with decomposing carcasses, and because of that, they are indeed very important in forensic science (Moretti et al., 2008).

22.4.1 Stages of Decomposition

Various stages of decomposition have been proposed by many forensic entomologists. Bornemissza (1957) worked with dead guinea pigs in Australia and recognized five stages of decomposition—initial decay (0 to 2 days), putrefaction (2 to 12 days), black putrefaction (12 to 20 days), butyric fermentation (20 to 40 days), and dry decay (40 to 50 days). In studies conducted in Hawaii, USA, the five stages of the decomposition have been recognized, and these appear to be easily applied to studies conducted in temperate areas (Lord and Goff, 2003). These stages are fresh, bloated, active decay, advance decay, and skeletal/remains (Figure 22.1). Tomberlin et al. (2011) divided the changes into five stages incorporated

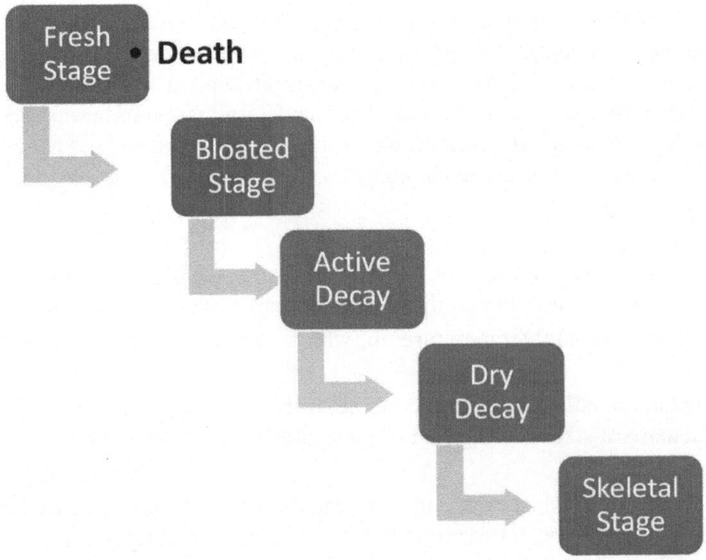

FIGURE 22.1 Decomposition stages.

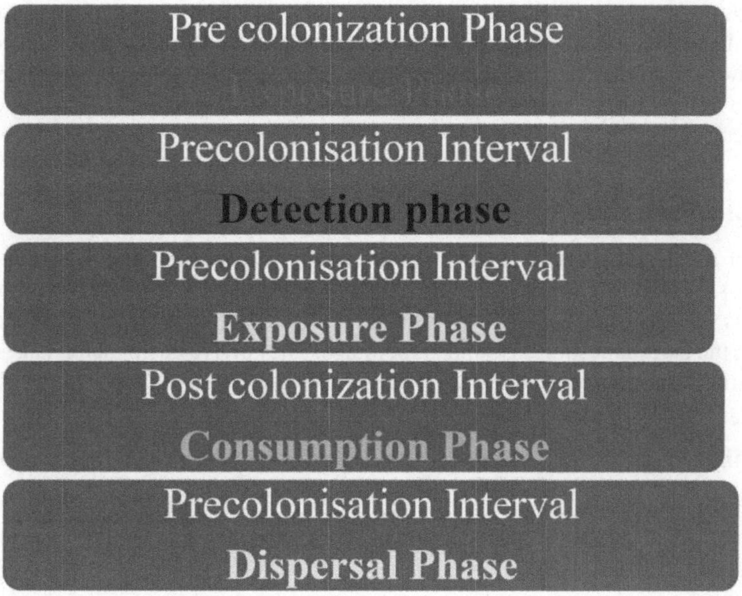

FIGURE 22.2 Entomological decomposition process.

into pre-colonization and post-colonization, as indicated in Figure 22.2. The five stages of decomposition and insects associated with them are briefly outlined in the following sections.

Fresh Stage

The fresh stage starts after death until bloating of the body begins. This stage has only few noticeable changes associated with corpse, like green discoloration of the abdomen, livor mortis, skin cracking, and tache noire, may be observed (Goff, 2010). Discoloration of the body is evident due to decomposition of internal organs and accumulation of bacteria. The insect species associated with this stage and the first to arrive on the body are the Calliphoridae, which are then followed by the Sarcophagidae. These insect species feed and lay eggs or larvae on the dead body, especially in the natural openings of the head (eyes, nose, ear, and mouth), anus, genitals, and wounds, if any, present on the body. During this stage, eggs laid by flies starts hatching and feed internally. Depending on weather conditions, it represents 1–3 days of PMI.

Bloated Stage

The second stage of decomposition is the bloated stage, where putrefaction starts. During this stage, there is breakdown of the soft tissues of the body. Putrefaction is characterized by the green or purple discoloration of the skin, occurring usually between 36 and 72 hours after death (Dent et al., 2004). Putrefaction results in breakdown of carbohydrates, proteins, and lipids into gases, liquids, and smaller molecules (Vass et al., 2002). These gases are responsible for the bloating of carcass. This stage is mostly associated with blowfly and maggot mass activity. Maggots' metabolic activities results an increase in the internal body temperature. A bloated body's temperature can be significantly above ambient temperature (>50°C), and the body itself becomes a distinct habitat in many ways, independent of the surrounding environment (Goff, 2010). Depending on weather conditions, it represents 2–6 days of PMI.

Active Decay/Decay Stage

Following the release of the trapped gases, the active decay stage commences. This is when the combined activities of the maggot feeding and bacterial putrefaction result in the breakdown of the outer layer of the skin and the release of the gasses from the abdomen. Additional compounds, including volatile fatty acids, are produced from the continued breakdown of muscle and protein. During this stage body deflates and step into decay stage. Early insect colonizers could be at the post-feeding stage and start migrating away from the resource searching for a suitable place to pupate. By the end of the active decay stage, most of the flesh get removed from the body by Diptera larvae. Depending on weather conditions, it represents 5–11 days of PMI.

Advanced/Post-Decay Stage

The next stage of the decomposition stage is the dry stage, where colonization by Diptera cease. Any remaining moist skin and tissue becomes attached to the bone with a leather-like appearance (Vass et al., 2002). This stage is usually composed of bone, hair, cartilage, and skin. Coleoptera particularly Dermestidae, are the dominant species found at this stage. They arrive during later stages of active decay but become dominant during post-decay stage of decomposition.

Skeletal/Remains Stage

The final stage is the skeletal stage, and at this stage, only bones and hairs are present. Carrion-feeding insect activity is mostly absent at this stage. There is little or no smell associated with carcass. This stage may represent PMI of month to years.

22.5 Life Cycle of Blowfly

Blowflies (Diptera: Calliphoridae) are the first group of the insects that invades a dead body due to the odor of fresh blood. Thus, it is the most important insect when it comes to estimate the time since colonization. Colonization of resource by blowflies can be within minutes of exposure based on factors, such as moisture, pheromones, VOC, bacterial assemblage (Tomberlin et al., 2012), other insects, and other abiotic factors, such as shade. After finding an appropriate place for oviposition, the female blowfly deposits eggs directly on the dead body (Smith and Wall, 1997) in a position where eggs are protected and in an idle environment (moist environment). This ensures nutrient supply for the hatching process. The life cycle of blowflies includes four stages: egg, larvae, pupae, and imago stage (Tao, 1927). During the larval stage, three instars can be separated: first, second, and third. The third instar stage is much longer than the first and second and is divided to feeding and post-feeding larvae because of behavioral change (Figure 22.3). Post-feeding is the stage where larvae stop feeding and migrate away from the food substrate, seeking cool, dark, and damp conditions to pupate. Adult flies emerge after the pupae have undergone metamorphic changes and the cycle begins.

22.5.1 Biology of *Chrysomya rufifacies* (Macquart)

The hairy maggot blowfly, *Chrysomya rufifacies* is an indigenous species to the Australasian region of Old World tropics (Norris, 1959; Baumgartner, 1993) and is widely distributed throughout the world. This species was first described in the neotropics in 1978, from a human carcass in Costa Rica (Jirón, 1979). *C. rufifacies* was introduced to Central America in 1978 and is still increasing its geographic range northward into the southern states of America (Baumgartner, 1993). *C. rufifacies* is one of the predominant necrophagous species in central Texas, USA, and always arrive spontaneously on a corpse. It is more adapted to tropical condition and found throughout the year (Greenberg and Povolny, 1971). *C. rufifacies* larvae (second and third instar) are facultative predators and cannibalistic (Baumgartner, 1993). *C. rufifacies* is easily identified based on greenish metallic appearance and pale genal dilation (Figure 22.4). Trials in the laboratory showed that *C. rufifacies* can switch from necrophagy to cannibalism and predating on other blowfly larvae in situations of limited resource (Goodbrod, 1990). *C. rufifacies* has a unique life history trait, with females being able to produce offspring of the same sex (Ullerich, 1977). Thus, this makes it difficult to keep colonies of this fly and require greater population size to exclude colonies of all being the same sex.

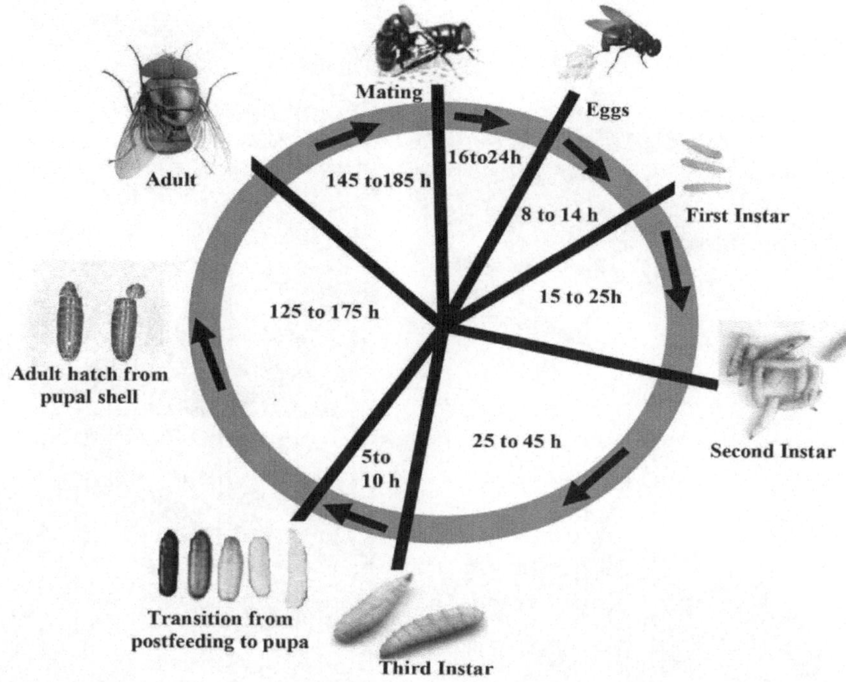

FIGURE 22.3 Life cycle of blowfly.

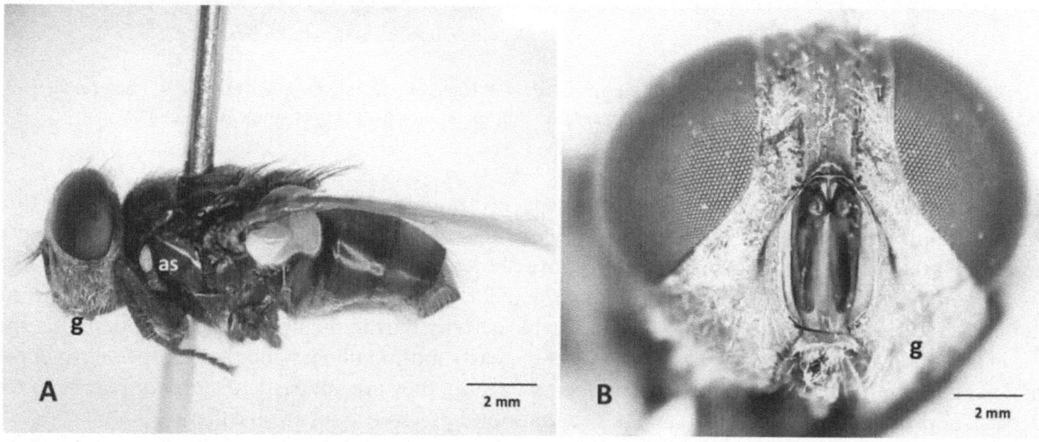

FIGURE 22.4 (A) Lateral view of a female *C. rufifacies.* (B) Head of a female *C. rufifacies.* Abbreviations: as (anterior spiracle); g (gena).

22.5.2 Biology of *Chrysomya megacephala* (Fabricius)

Chrysomya megacephala (Fabricius), the Oriental latrine fly, is widely distributed in the Oriental and Australian regions and adjacent parts (Afghanistan, China, and Japan) of the Palearctic region (Zumpt, 1965; Greenberg and Povolny, 1971; Smith, 1986; Essar, 1991). Its relationship to man varies from hemisynanthropic to eusynanthropic category (Greenberg and Povolny, 1971). The adults of this species are greenish blue in color and can easily be separated from other calliphorids by the presence of a blackish-brown anterior thoracic spiracle (Figure 22.5), while in others it is white (Zumpt, 1965; Prins, 1979, 1982).

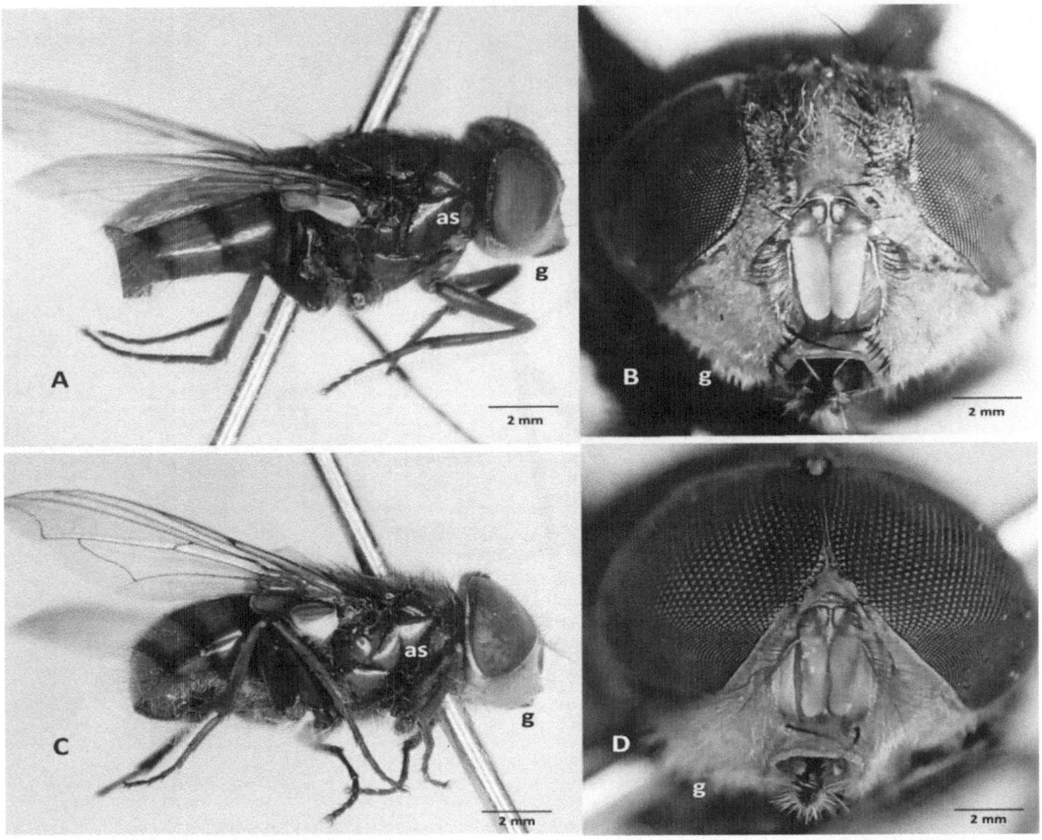

FIGURE 22.5 (A) Lateral view of a female *C. megacephala.* (B) Head of a female *C. megacephala.* (C) Lateral view of a male *C. megacephala.* (D) Head of a male *C. megacephala.* Abbreviations: as (anterior spiracle); g (gena).

Another vital character that distinguishes this species from others is that in males the upper two-third of each eye orbit has large facets, sharply distinct from smaller facets of lower one-third. However, in females, all the facets are uniform. The adults have short stout body similar in appearance to *C. rufifacies*, but the head is larger. The adult flies are attracted to carrion and sweet foods, as well as urine and excrement. Although *C. megacephala* has noticeable activity during the heat of the afternoon, this species is one of the first species to become active in the early morning hours and the last species to depart carrion at nightfall. Once the adults have settled on carrion, they are not easily disturbed. Larvae of this species are primarily carrion feeder, and the adults shows preference for fresh cadaver.

22.6 New Trends and Development in Forensic Entomology

The use of insects as evidence in forensic investigations requires, first, the identification of the insect species present at the crime scene. This is crucial because closely related insect species that colonize vertebrate carrion can have considerably different developmental rates and hence false identification can lead to inaccurate estimation of PMI. In investigations where immature stages are collected from the crime scene, the forensic entomologist will have to rear the larvae to adult flies. This is only possible if specimens are alive when they reach to the analyst. Due to certain limitations, DNA-based analysis has been suggested by Sperling and coworkers to aid the current identification method as it can be performed at all developmental stages (Sperling et al., 1994). DNA-based techniques have also been studied as a potential for aging insect species. Tarone and Foran (2011) investigated the gene expression of the blowfly *Lucilia sericata* during its development to improve age estimation. Zehner and colleagues investigated the potential of using gene expression analysis as an age estimation tool for *Calliphora vicina* puparia. Their results gave age specific expression patterns, allowing for the puparia to be aged in three stages: the beginning, the middle, and the end of their metamorphosis. Hence, a method that could be used for identification to the species level at the immature stage and aging without having to rear the insects would be essential and shorten the turnaround time in the estimation of PMI.

22.6.1 Age Grading and Species Identification Techniques

To establish the PMI, the forensic entomologist aims to determine the age of the oldest colonizing insects' species. However, for the age to be determined, the insect species first needs to be established. Species identification is challenging in the immature stage than in the adult form (Smith, 1986). For species identification and PMI estimation, forensic entomologists rear the immature insect's stage till the adult flies. Insect rearing at the same temperature and humidity of the surroundings from where the insects were collected is important as these factors have been shown to affect the rate of the development (Marchenko, 2001; Tarone et al., 2011). Also, previous studies have shown that the type of tissue on which the insect species were fed during the developmental process could impact on the size (length and weight) and developmental rate (Clark et al., 2006; Kaneshrajah and Turner, 2004). The muscle attachment site (MAS) is found to be a good tool for species identification in case of larval stage.

During this research work, attempts have been made to determine the age of immature stages of forensically important insect species (*Chrysomya megacephala* and *Chrysomya rufifacies*) with the help of some recent age estimation techniques. Cuticular hydrocarbons (CHCs) have also been utilized in the field of entomology for aging, gender identification, and species identification (Tregenza et al., 2000; Savarit and Freveur, 2002; Brown et al., 2000; Desena et al., 1999; Hugo et al., 2006; Chen et al., 1990; Mpuru et al., 2001; Jackson and Bartelt, 1986; Steiner et al., 2006; Marican et al., 2004). CHCs may give the same accuracy as DNA techniques for aging the insects. Composition of CHC changes with age in some of the insect's species. These compounds are very useful to age the post-feeding larvae, which is quite long and difficult to differentiate morphologically. Analysis of volatile organic compounds (VOCs) released by insects' stages (immature and mature) can enable the entomologist to distinguish the feeding third instar from the post-feeding one and younger pupae from older ones. According to Frederickx and coworkers, the volatile profile has shown a variation in both composition and quantity in the larvae and pupae of *Calliphora vicina* (Robineau-Desvoidy) (Diptera: Calliphoridae) (Frederickx et al., 2012).

Gene expression is another alternative approach that is a cost-effective and quantitative measurement to age immature insects. Research has shown that temporal patterns of gene expression exist throughout immature insect's development. Dipteran gene expression has been of recent interest in forensic entomology, and colleagues have shown that quantitative studies with gene expression can be used to age immature blowflies (Tarone et al., 2007; Tarone and Foran, 2011; Boehme et al., 2013, 2014). Such biological information can be used to break lengthier stages (e.g., third instar or pupa) into smaller temporal pieces and improve the precision of insect age estimates. Internal morphological analysis of blowfly pupae by CT scan techniques provides additional internal development information to that of external morphological analysis, allowing a more accurate age and thus PMI estimation. The development of the internal organs inside the pupal shell have the potential of age determination.

22.7 Conclusions

If a forensic entomologist knows the blowfly species present, the respective development times for that species, and how it is affected by temperature and other environmental conditions, a minimum PMI can be obtained. The estimation depends on correct identification, a high degree of accuracy of the environmental conditions the insects are thought to have been exposed to, and the reliability and credibility of experimental data used when performing the calculations. Once the identity of the insects is established, the next step to estimate the PMI based on the age of the insects collected from the carcass. When presented with a crime scene, a forensic entomologist must be able to establish the age of the oldest insect present to acquire a time since colonization, which in turn enables PMI estimation. These are the initial vital factors a forensic entomologist must undertake when presented with a crime scene. The field of forensic entomology could greatly benefit from a technique that facilitates identification information and age determination. These techniques have been proven to offer a significant advantage over traditional morphological and aging techniques because it can be accomplished at any instar stage of the larval life cycle. In contrast, older techniques require larvae to be in the third instar stage due to their size and still require the time-consuming step of rearing to adult flies for species confirmation.

Rumination of larvae length and weight (i.e., body size) can ameliorate the precision of age estimates, but only for the first, second, and early third instars when the larvae are actively growing. Once the third instars leave the food source, they begin to shrink and exhibit increased variance in body size (Grassberger and Reiter, 2001; Greenberg, 1991; Anderson, 2000), adding to the difficulty of using size to estimate age (Wells and Kurahashi, 1994; Wells and Lamotte, 1995; Tarone and Foran, 2008). Likewise, size measurements do not strengthen pupal age estimates as they do not change during development. New approaches to predicting the age of immature stages seem warranted.

REFERENCES

Amendt, J., Campobasso, C. P., Gaudry, E., Reiter, C., LeBlanc, H. N. and Hall, M. J. R. 2007. Best practice in forensic entomology-standards and guidelines. *International Journal of Legal Medicine*, 121:90–104.

Amendt, J., Krettek, R. and Zehner, R. 2004. Forensic entomology. *Naturwissenschaften*, 91:51–65.

Anderson, G. S. 2000. Minimum and maximum development rates of some forensically important Calliphoridae (Diptera). *Journal of Forensic Science*, 45:824–832.

Anderson, G. S. 2001. Succession on carrion and its relationship to determining time of death. In: Byrd, J. H. and Castner, J. L. (eds.), *Forensic Entomology: The Utility of Arthropods in Legal Investigations*. CRC Press, Boca Raton, FL, pp. 143–175.

Anderson, G. S. 2009. Factors that influence insect succession on carrion. In: Byrd, J. H. and Castner, J. L. (eds.), *Forensic Entomology: The Utility of Arthropods in Legal Investigations*, 2nd ed. CRC Press, Boca Raton, FL, pp. 201–250.

Anderson, G. S. and VanLaerhoven, S. L. 1996. Initial studies on insect succession on carrion in Southwestern British Columbia. *Journal of Forensic Science*, 41:617–625.

Archer, M. S. and Elgar, M. A. 2003. Effect of decomposition on carcass attendance in a guild of carrion-breeding flies. *Medical and Veterinary Entomology*, 17(3):263–271.

Baumgartner, D. L. 1993. Review of *Chrysomya rufifacies* (Diptera: Calliphoridae). *Journal of Medical Entomology*, 30:338–352.

Benecke, M. 2001. A brief history of forensic entomology. *Forensic Science International*, 120:2–14.

Bergeret, M. 1855. Infanticide: Momification naturelle du cadaver. *Annales d'Hygiéne Publique et de* Médecine Légale, 4:442–452.

Bland, R. G. and Jacques, H. E. 1978. *How to Know the Insects*. William C. Brown Co., Dubuque, IA, pp. 1–414.

Boehme, P., Spahn, P., Amendt, J. and Zehner, R. 2013. Differential gene expression during metamorphosis: a promising approach for age estimation of forensically important *Calliphora vicina* pupae (Diptera: Calliphoridae). *International Journal of Legal Medicine*, 127:243–249.

Boehme, P., Spahn, P., Amendt, J. and Zehner, R. 2014. The analysis of temporal gene expression to estimate the age of forensically important blow fly pupae: results from three blind studies. *International Journal of Legal Medicine*, 128:565–573.

Bornemissza, G. F. 1957. An analysis of arthropods succession in carrion and the effect of its decomposition on the soil fauna. *Australian Journal of Zoology*, 5:1–12.

Borror, D. J. and White, R. E. 1970. *A Field Guide to Insects of America North of Mexico*. Houghton Mifflin, Boston, pp. 1–404.

Brouardel, P. 1879. De la determination de I' epoque de la naissance et de la mort d' un nouveau-né, faite a I'aide de la presence des acares et des chenilles d' aglosses dans un cadaver momífié (Determination of the postmortem interval/time of death of a new-born child by use of larvae from the genus Aglossa on a mummified corpse). *Annales d'hygiènepublique et de médecinelégale*, 2:153–158.

Brown, W. V., Rose, H. A., Lacey, M. J. and Wright, K. 2000. The cuticular hydrocarbons of the giant soil-burrowing cockroach *Macropanesthia rhinoceros* Saussure (Blattodea: Blaberidae: Geoscapheinae): analysis with respect to age, sex and location. *Comparative Biochemistry and Physiology Part B, Biochemistry and Molecular Biology*, 127:261–277.

Byrd, J. H. and Butler, J. F. 1998. Effects of temperature on *Sarcophaga haemorrhoidalis* (Diptera: Sarcophagidae) development. *Journal of Medical Entomology*, 35(5):694–698.

Chen, C. S., Mulla, M. S., March, R. B. and Chanley, J. D. 1990. Cuticular hydrocarbon patterns in *Culex quinquefasciatus* as influenced by age and sex and geography. *Bulletin of the Society for Vector Ecology*, 15(2):129–139.

Clark, K., Evans, L. and Wall, R. 2006. Growth rates of the blowfly, *Lucilia sericata*, on different body tissues. *Forensic Science International*, 156(2–3):145–149.

Cruz, A. M. 2006. *Crime Scene Intelligence: An Experiment in Forensic Entomology*. The National Defense Intelligence College, Washington, DC.

Dent, B. B., Forbes, S. L. and Stuart, B. H. 2004. Review of human decomposition processes in soil. *Environmental Geology*, 45:576–585.

Desena, M. L., Clark, J. M., Edman, J. D., Symington, S. B., Scott, T. W., Clark, G. C. and Peters, T. M. 1999. Potential for aging female *Aedes aegypti* (Diptera: Culicidae) by gas chromatographic analysis of cuticular hydrocarbons, including a field evaluation. *Journal of Medical Entomology*, 36:811–823.

Erzinclioglu, Y. Z. 1983. The application of entomology to forensic medicine. *Medicine, Science and Law*, 23:57–63.

Essar, J. R. 1991. Biology of *Chrysomya megacephala* (Diptera: Calliphoridae) and reduction of losses caused to the salted-dried fish industry in South-East Asia. *Bulletin of Entomo*logical *Research*, 81:33–41.

Frederickx, C., Dekeirsschieter, J., Brostaux, Y., Wathelet, J. P., Verheggen, F. J. and Haubruge, E. 2012. Volatile organic compounds released by blowfly larvae and pupae: new perspectives in forensic entomology. *Forensic Science International*, 219(1–3):215–220.

Gennard, D. E. 2007. *Forensic Entomology: An Introduction*. John Wiley & Sons Ltd, West Sussex, UK, pp. 1–224.

Goff, M. L. 2010. Early postmortem changes and stages of competition. In *Current concepts of forensic entomology,* Springer Dordrecht.

Goodbrod, J. R. 1990. Effects of larval population-density on rates of development and interactions between two species of Chrysomya (Diptera, Calliphoridae) in laboratory culture. *Journal of Medical Entomology*, 27(3):338–343.

Grassberger, M. and Reiter C. 2001. Effect of temperature on *Lucilia sericata* (Diptera: Calliphoridae) development with special reference to the isomegalen- and isomorphen-diagram. *Forensic Science International*, 120:32–36.

Greenberg, B. 1991. Flies as forensic indicators. *Journal of Medical Entomology*, 28(5):565–577.

Greenberg, B. and Kunich, J. C. 2002. *Entomology and the Law: Flies as Forensic Indicators*. Cambridge University Press, Cambridge, UK, pp. 332.

Greenberg, B. and Povolny, D. 1971. Bionomics of flies. In: Greenberg, B. (ed.), *Flies and Disease*, Vol. 1. Princeton University Press, Princeton, pp. 57–83.

Hall, R. D. and Doisy, K. E. 1993. Length of time after death: Effect on attraction and oviposition or larviposition of midsummer blowflies (Diptera: Calliphoridae) and flesh flies (Diptera: Sarcophagidae) of medicolegal importance in Missouri. *Annals of the Entomological Society of America*, 86:589–593.

Haskell, N. H., Hall, R. D., Cervenka, V. J. and Clark, M. A. 1997. On the body: Insect's life stage presence, their postmortem artifacts. In: Haglund, W. D. and Sorg, M. H. (eds.), *Forensic Taphonomy: The Postmortem Fate of Human Remains*. CRC, Boca Raton, FL, pp. 415–448.

Hogue, C. L. 1993. *Latin American Insects and Entomology*. University of California Press, Berkeley, CA, pp. 1–536.

Hugo, L. E., Kay, B. H., Eaglesham, G. K., Holling, N. and Ryan, P. A. 2006. Investigation of cuticular hydrocarbons for determining the age and survivorship of Australasian mosquitoes. *The American Journal of Tropical Medicine and Hygiene*, 74(3):462–474.

Jackson, L. L. and Bartelt, R. J. 1986. Cuticular Hydrocarbons of Sex. *Insect Biochemistry*, 16:433–439.

James, M. T. and Harwood, R. F. 1969. *Herm's Medical Entomology*, 6th ed. The Macmillan Company, Toronto, ON, pp. 1–114.

Jirón, L. F. 1979. Sobremoscascalifóridas de Costa Rica (Diptera: Cyclorrhapha). *Brenesia*, 16:221–222.

Kaneshrajah, G. and Turner, B. 2004. *Calliphora vicina* larvae grow at different rates on different body tissues. *International Journal of Legal Medicine*, 118(4):242–244.

Kashyap, V. K. and Pillay, V. V. 1989a. Efficiency of entomological method in estimation of postmortem interval: a comparative analysis. *Forensic Science International*, 40(3):245–250.

Kashyap, V. K. and Pillay, V. V. 1989b. Insects and crime investigations, what is forensic entomology? *Journal of Indian Council of Philosophical Research*, 40:12–17.

Klotzbach, H., Krettek, R., Bratzkec, H., Puschela, K., Zehnerc, R. and Amendt, J. 2004. The history of forensic entomology in German-speaking countries. *Forensic Science International*, 144(2–3):259–263.

Krahmer, F. L. 1857. *Handbuch der gerichtlichenMedizin*, 2. Aufl. Berlin.

Liu, D. and Greenberg, B. 1989. Immature stages of some flies of forensic importance. *Annals of the Entomological Society of America*, 82(1):80–93.

Lord, W. D. and Goff, M. L. 2003. Forensic entomology: application of entomological methods to the investigation of death. In: Froede, R. C. (ed.) *Handbook of Forensic Pathology*, 2nd ed. College of American Pathologists, IL, Boca Raton, Florida, USA, pp. 423–436.

Lord, W. D. and Stevenson, J. R. 1986. *Directory of Forensic Entomologist*, 2nd ed. Defense Pest Information Analysis Center; Walter Reed Army Medical Center, Washington, DC, p. 42.

Marchenko, M. I. 2001. Medicolegal relevance of cadaver entomofauna for the determination of the time of death. *Forensic Science International*, 120:89–109.

Marican, C., Duportets, L., Birman, S. and Jallon, J. M. 2004. Female-specific regulation of cuticular hydrocarbon biosynthesis by dopamine in *Drosophila melanogaster. Insect Biochemistry and Molecular Biology*, 34:823–830.

Mende, L. J. K. 1829. *AusfuhrlichesHandbuch der gerichtlichenMedizin fur Gesetzgeber‖, Rechtsgelehrte, Aerzte und Wundaerzte*, Vol. 5. Dyk, pp. 385.

Moretti, T. C., Ribeiro, O. B., Thyssen P. J. & Solis D. R. 2008. Insects on decomposing carcasses of small rodents in a secondary forest in Southeastern Brazil. *European Journal of Entomology*, 105:691–696.

Mpuru, S., Blomquist, G. J., Schal, C., Roux, M., Kuenzli, M., Dusticier, G., Clément, J. L. and Bagnères, A. G. 2001. Effect of age and sex on the production of internal and external hydrocarbons and pheromones in the housefly, *Musca domestica. Insect Biochemistry and Molecular Biology*, 31:139–155.

Nelson, E. L. 2000. Estimation of short term postmortem interval utilizing core body temperature: a new algorithm. *Forensic Science International*, 109(1):31–38.

Norris, K. R. 1959. The ecology of sheep blowflies in Australia. In: Keast, A., Crocker, R. L. and Christian, C. S. (eds.), *Biogeography and Ecology in Australia*. Monographiae Biologicae Book Series, Springer, Dordrecht, Netherlands, pp. 514–544.

Orfila, M. J. B. and Lesueur, C. A. 1831. *Traite des exhumations juridiques‖*, Paris, pp. 387.

Prins, A. J. 1979. Discovery of the oriental latrine fly *Chrysomya megacephala* (Fabricius) along the southwestern coast of South Africa. *Annals of South African Museum*, 78:39–47.

Prins, A. J. 1982. Morphological and biological notes on six South African blowflies and their immature stages (Diptera: Calliphoridae). *Annals of South African Museum*, 90(4):201–217.

Savarit, F. and Ferveur, J. F. 2002. Temperature affects the ontogeny of sexually dimorphic cuticular hydrocarbons in *Drosophila melanogaster. Journal of Experimental Biology*, 205:3241–3249.

Smith, K. E. and Wall, R. 1997. The use of carrion as breeding sites by the blowfly *Lucilia sericata* and other Calliphoridae. *Medical and Veterinary Entomology*, 11(1):38–44.

Smith, K. G. 1986. *A Manual of Forensic Entomology*. The Trustees of British Museum, London, pp. 1–195.

Sperling, F. A. H., Anderson, G. S. and Hickey, D. A. 1994. A DNA-based approach to the identification of Insect species used for postmortem interval estimation. *Journal of Forensic Sciences*, 39(2):418–427.

Spitz, W. U. 1993. *Medicolegal Investigation of Death: Guidelines for the Application of Pathology to Crime Investigations*, 3rd ed. Charles C Thomas Pub Ltd, Springfield, IL.

Steiner, S., Hermann, N. and Ruther, J. 2006. Characterization of a female-produced courtship pheromone in the parasitoid *Nasoniavitripennis. Journal of Chemical Ecology*, 32(8):1687–1702.

Tao, S. 1927. A Comparative Study of the early larval stages of some common flies. *American Journal of Epidemiology*, 7:735–761.

Tarone, A. M. and Foran, D. R. 2011. Gene expression during blow fly development: improving the precision of age estimates in forensic entomology. *Journal of Forensic Sciences*, 56(1):112–122.

Tarone, A. M., and Foran, D. R. 2008. Generalized additive models and *Lucilia sericata* growth: assessing confidence intervals and error rates in forensic entomology. *Journal of Forensic Sciences*, 53(4):942–948.

Tarone, A. M., Jennings, K. C. and Foran, D. R. 2007. Aging blow fly eggs using gene expression: a feasibility study. *Journal of Forensic Sciences*, 52(6):1350–1354.

Tarone, A. M., Picard, C. J., Spiegelman, C. and Foran, D. R. 2011. Population and temperature effects on *Lucilia sericata* (Diptera: Calliphoridae) body size and minimum development time. *Journal Medical Entomology*, 48(5):1062–1068.

Tomberlin, J. K., Crippen, T. L., Tarone, A. M., Singh, B., Adams, K., Rezenom, Y. H., Benbow, M. E., Flores, M., Longnecker, M., Pechal, J. L., Russell, D. H., Beier, R. C. and Wood, T. K. 2012. Interkingdom responses of flies to bacteria mediated by fly physiology and bacterial quorum sensing. *Animal Behaviour*, 84(6):1149–1456.

Tomberlin, J. K., Mohr, R., Benbow, M. E., Tarone, A. M. and VanLaerhoven, S. 2011. A roadmap for bridging basic and applied research in forensic entomology. *Annual Review of Entomology*, 56:401–421.

Tregenza, T., Buckley, S. H., Pritchard, V. L. and Butlin, R. K. 2000. Inter- and intra-population effects of sex and age on epicuticular composition of meadow grasshopper, *Chorthippusparallelus*. *Journal of Chemical Ecology*, 26:257–278.

Tzu, S. 1981. *The Washing Away of Wrongs* (*Hsi yuan chi lu*), 2nd book, 5th chapter, translated by McKnight, B. E. Center for Chinese Studies, University of Michigan, Ann Arbor, pp. 69–70.

Ullerich, F. H. 1977. Production of male and female offspring in the strictly monogenic fly *Chrysomya rufifacies* after ovary transplantation. *Naturwissenschaften*, 64(5):277–278.

Vass, A. A., Barshick, S. A., Sega, G., Caton, J., Skeen, J. T., Love, J. C. and Synstelien, J. A. 2002. Decomposition chemistry of human remains: a new methodology for determining the postmortem interval. *Journal of Forensic Sciences*, 47(3):542–553.

Wells, J. D. and Kurahashi, H. 1994. *Chrysomya megacephala* (Fabricius) (Diptera: Calliphoridae) development: rate, variation, and the implications for forensic entomology. *Japanese Journal of Sanitary Zoology*, 45:303–309.

Wells, J. D. and Lamotte, L. R. 1995. Estimating maggot age from weight using inverse prediction. *Journal of Forensic Science*, 40:585–590.

Williams, K. A. and Villet, M. H. 2006. A history of southern Africa research relevant to forensic entomology. *South African Journal of Science*, 102:59–65.

Zumpt, F. 1965. Calliphoridae (Diptera: Cyclorrhapa) Part I: Calliphorini and Chrysomyiini. *Exploration du Parc National Albert, Mission G.F. de Witte*, 87:1–200.

23

Potential Applications of Animal-Friendly Affinity Reagents in Forensics

Jagdish Rai

CONTENTS

23.1 Introduction

Several 3R research organizations were established for finding alternatives to reduce using animals in scientific research and industry (Rai & Kaushik 2018). Antibody production is one of the major contributors to using animals in scientific research and industries related to diagnosis, therapeutics, and analyte detection. The regulations in Europe are becoming stricter on the use of animals in research and industry (Adler et al. 2011). In the future, they are expected to even more strict given the activism of animal rights groups, and on the other hand, technologically, it is becoming feasible to avoid using animal due to research by 3R organizations like the National Centre for the Replacement Refinement and Reduction of Animals in Research (NC3Rs), European Centre for the Validation of Alternative Methods (ECVAM) (Marafante et al. 1994), Center for Alternatives to Animal Testing (CAAT) (https://caat.jhsph.edu/), Universities Federation for Animal Welfare (UFAW) (www.ufaw.org.uk), and Interagency Coordinating Committee on the Validation of Alternative Methods (ICCVAM) (Stokes et al. 2002). The publicly funded organization like Recombinant Antibody Network (RAN) is developing recombinant antibody clones for the complete human proteome so that the animal-derived antibodies used in research can be phased out with these antibodies expressed in bacteria. RAN has robotic setups for screening clones for binding to a target and they supply plasmid clones free of cost for the monoclonal recombinant antibodies they have developed. They have three expert centers at the University of Chicago, the University of Toronto, and the University of California at San Francisco (UCSF) and keep a list of antibodies developed so far in their web portal (Hornsby et al. 2015) Many contract research organizations (CROs), like Yumab and Biobench, also provide custom-made recombinant antibodies (Schafer et al. 2017). Other than immunoglobulin protein scaffold, small proteins that are bacterial homologues of immunoglobulin or otherwise suitable for developing a binding protein for a given chemical entity are being developed as a substitute for antibodies. Table 23.1 is the list of some important alternatives of antibodies, which are called affibodies, or affirmers, or sometimes the name is based on the original protein scaffold, such as anticalins, darpins, fynomers, alphabodies, monobodies, nanoCLAMPs, Kunitz domain peptides, and avimers (Gebauer & Skerra 2009; Arora et al. 2004). Polynucleotides are also used as an alternative for molecular recognition of target molecules, which is called aptamers (Lakhin et al. 2013). Although there has been a lot of research activity around aptamers for diagnostics and research applications, the

DOI: 10.1201/9781003220398-23

TABLE 23.1
Various Alternatives to Antibodies That Can Be Used in the Lateral Flow Assay (LFA) Format of Immunochromatography

Antibody alternative	Derived from natural protein	Reference (year-wise)
Avimers	Various membrane receptor	(Silverman et al. 2005)
Kunitz domain peptides	Kunitz domains of protease inhibitor	(Nixon & Wood 2006)
Affilins	Gamma crystallin B	(Ebersbach et al. 2007)
Fynomers	SH3 domain of Fyn	(Grabulovski et al. 2007)
Monobodies	Fibronectin	(Koide & Koide 2007)
Affibodies	Protein A	(Nygren 2008)
Anticalins	Lipocalins	(Nygren 2008)
DARPins	Ankyrin	(Stumpp et al. 2008)
Affimers	Cystatin	(Johnson et al. 2012)
Alphabodies	Triple-helix coiled coil	(Desmet et al. 2014)
nanoCLAMPs	Carbohydrate-binding module 32–2	(Suderman et al. 2017)

secondary structure and tertiary structure that nucleotides can adopt are very limited in diversity in comparison to polypeptides. Therefore, the focus again is on the protein-based binding molecules for molecular recognition of target biomolecules and other chemicals. Despite the aptamer binding to hCG being available, the antibody-based method is commercially successful for detection of hCG in urine to detect pregnancy (Chang et al. 2014). In nature, the polynucleotides have the primary function of carrying information from generation to generation and within a cell. Although polynucleotides are very stable compared to polypeptides, the polypeptides excel in the diversity of structure. Nature has used these molecules for molecular recognition work—for example, protein-protein interactions in cell signaling networks and identification of small molecules by enzymes in metabolic networks. The proteins can present surfaces of very diverse stereochemical properties for molecular recognition and pockets of very extreme microenvironments for biological catalysis. Therefore, the choice of heteropolymer for molecular recognition type of application is polypeptide only unless other synthetic heteropolymers come up in the future (Feringa 2017). These animal-friendly affinity (AFA) reagents are alternatives to antibodies, produced without immunization of animals. The AFAs for a target are generally designed computationally and then refined by random changes and screening through phage, ribosome, or yeast display technique. In a comparative analysis between animal-derived antibodies and AFAs, AFAs are found to be less time-consuming, superior in quality, and more reliable, reproducible, and cost-effective except that initial investment and expertise is required to move from animal antibodies to AFAs (Gray et al. 2016b). At present, an 80 billion USD industry exists that creates millions of animal-derived monoclonal and polyclonal antibodies for use in routine diagnostics and detection. This data does not include the undetermined amount of custom-made antibodies produced by companies and research institutes. In principle, all these animal-immunization-based antibodies can be replaced by alternatives available, such as naive B lymphocytes or recombinant antibodies expressed by phage display. Therefore, regulatory measures are soon expected to ban animal immunization for mere production of antibodies (Gray et al. 2016a). The antibody-based detection is already in use in forensic sciences, which can be replaced with the mentioned alternatives of antibodies. The body fluids at crime sites are detected using rapid stain identification (RSID) kits, which utilize antibodies against various tissue specific protein markers (table 23.2). The species of sample is also detected in forensic labs using the popular serological ring test. Recently the antibody-based home test kits are becoming available for detecting narcotic drugs in body fluids or trace amounts on the surface of articles. These tests are mainly based on polyclonal antibodies against popular narcotic drugs, such as heroin, barbiturates, benzodiazepines, cocaine, amphetamines, and cannabinoids. In a recent evidence-based study, the commonly misused drugs are ranked by their physical harm and risk of dependence in following descending order: heroin, cocaine, barbiturates, street methadone, alcohol, ketamine, benzodiazepines, amphetamine, tobacco, buprenorphine, and cannabis. This study puts alcohol and tobacco at higher risk than cannabis (Nutt et al. 2007). The United Nations

Office of Drugs and Crime has prepared a report in 2019 on drug use in Asia and India. According to the report (UNODC Report 2019), the use of opiates is increasing faster than expected in India. Law enforcement agencies are grappling with menace of drugs in society despite having many other priority areas of their duties. Parents can rather monitor their children for drug use if low-cost kits are available to easily test the presence of narcotic substances or its metabolites in body fluids like sweat and urine. It will help in saving children from a dangerous habit, and it can be a national resource on tackling the crime arising from drug use and narcotics trade. The suspected material in possession of a narcotic dealer at the airport, streets, or a house can be detected on the spot using these kits. The fake drugs traded in the black market can also be checked by buyers. The animal-immunization-based antibody generation against these molecules have several drawbacks; for example, the body may contain natural analogs of these molecules, and it is less reproducible and more time-consuming. Antibodies can be raised for immunogens of specific size and non-self molecules, whereas affibodies can be screened *in vitro* against a wider range of molecules. The affibody generated by phase display can also be attached with dye and used to detect all these drugs in one kit/strip. Unlike antibodies, they can be raised against toxins, metabolites, and a wide range of other molecules. They can have higher specificity and affinity and are more cost-effective. The same as antibodies, there are various formats of the assay in which affibodies can be used to detect and measure the concentration of the target molecule. ELISA (enzyme-linked immunosorbent assay), Ouchterlony assay, crossed immunoelectrophoresis (CRIE), immunochromatographic assay (lateral flow test), and so on are formats of an assay suitable for various situation demanding sensitivity, quantitation or convenience. The immunochromatographic method called lateral flow assay (LFA) is becoming a widely used method of detection of biomacromolecules as it is cost-effective, requires no instrument, and is quick. For the detection of narcotic drugs in forensic investigation, the lateral flow technique is useful as it can be used on the spot at crime site investigation, and therefore, there is no need to send the sample at the forensic lab that demands additional security for preventing manipulation of evidence. The efforts by such organizations as the Netherlands Forensic Institute (NFI) will make most of forensic investigation shift to mobile labs (Kloosterman et al. 2015). LFA is emerging as a cost-effective point-of-care tool for diagnosis and detection (Syedmoradi et al. 2017). It is expanding the scope of application in environment monitoring and quality check. It uses a protein-like antibody for specific identification of a target molecule in a sample. The molecular recognition between the target molecule and the antibody for it is because of shape and charge complementarity between them. The binding between target molecule and antibody is then used to produce detection signals.

In the immunochromatography method LFA, there is a plastic case containing a sample well, as well as a test zone, as shown in Figure 23.1. The sample containing the target molecule is dissolved in aqueous medium and applied in the sample well. In the sample well, it binds to its antibody. Then due to capillary force, this antigen antibody complex moves in an aqueous medium toward the test zone. In the test zone there are immobilized antibodies separately on the two lines. First is the test line, which has an immobilized antibody for the target molecule, whereas the second line has an immobilized antibody against an antibody that was in the sample well itself. The antibody that is in the sample well is color labeled covalently with dye or a high-quantum-yield tag like a quantum dot. Binding of these tagged antibodies on the second line (control line) makes it a visibly colored line. If the target molecule is there in a sample that will bind at the test line to its antibody, but it is also bound to another antibody against it that was in well. Therefore, the binding of the target molecule on the test line turns it visibly colored as the target molecule is also bound to the color-labeled antibody in the sample well. Therefore, the appearance of two colored lines indicates the presence of the target molecule in the sample, whereas only one line (control line) is indicative of absence of target molecule. No color on the control line indicates failure of the test due to the presence of some denaturant in the sample or fluid not reaching the test zone. Sometimes too much of the target molecule concentration in the sample can also give false negative results due to the hook effect (Tate & Ward 2004). Too much of the target molecule will saturate the antibodies in the sample well, and the remaining unbound target molecules will bind at the test line. These target molecules that are not bound to any color tagged antibody from sample well, are unable to produce colored test lines. Due to less amount of material required and small size, this is an inexpensive and portable device. It does not require any additional instrument for testing therefore it is a point of care and on-site detection method.

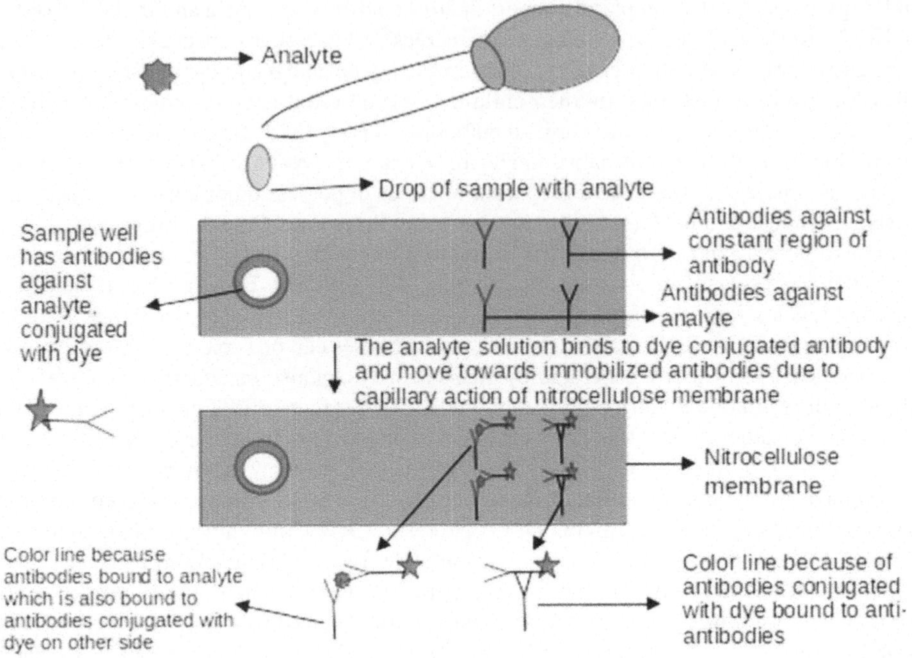

FIGURE 23.1 The concept of immunochromatography method, lateral flow assay (LFA).

23.2 Historical Perspective

The discovery of antibodies, and specifically monoclonal antibodies, revolutionized the research in biological sciences. It made detection of biomacromolecules very easy. It has been taken granted that a specific antibody can be generated for any molecule of size above the 500 dalton. The antibodies can be generated against molecules smaller than 500 daltons called haptens, too, by conjugating them to carrier proteins such as albumin and polylysine. Before the advent of MALDI (matrix-assisted laser desorption/ionization) mass spectroscopy, antibodies were the main tool to discriminate between similar proteins and other macromolecules. Owing to the very specific biomolecular recognition by antibodies, they soon became very popular for diagnosis of various infectious diseases. They are also used now for detecting various biomarkers of cancer, hormonal imbalances, and pregnancy. Many formats of detection of molecules using affinity reagents are emerging, ranging from simple lateral flow to optoelectronic devices for precise quantification of the analyte (Fodey et al. 2011). At present in forensics, sniffer dogs are used to follow body odor of criminals and traces of explosives. At molecular level, the sense of smell in dogs is through the binding of analyte molecules to the olfactory receptors. Artificial affinity molecules are also being developed for explosives, which in the future may be integrated in microfluidic optoelectronic devices to replace the sniffer dogs (Ramoni et al. 2007). There has been a long history of various formats of using antibody for detection of target molecules, starting from ring assay, western blot, Ouchterlony ELISA (enzyme-linked immunosorbent assay), and so on and finally lateral flow immunochromatography being very popular for point of care diagnosis. Detection of pregnancy using antibodies against hCG (human chorionic gonadotropin) is the commercially most successful product in immunochromatography. The antibody-based immunochromatography kit has been developed against the very toxic mustard gas, but they have not been very successful because the small size of analyte limits the specificity (Sathe et al. 2014). Similarly, the antibody-based detection kits for toxic metal ions like mercury (Jin et al. 2019) and chromium (Liu et al. 2012) have also been developed, but other traditional methods are more reliable instead. Due to animal ethics concerns and high cost and limitations of antibody development against some molecules, other protein folds are being explored for applications in chemical detection.

23.3 Major Techniques and Important Observations

A novel protein for binding to the target molecule of forensic significance can be either (1) designed de novo using computational methods or (2) developed through high-throughput screening in *in vitro* evolution. Generally, the approach used is a hybrid of these two, mixing in proportion depending on the type of resources at disposal. A completely automated protein design was demonstrated in 1997 (Dahiyat & Mayo 1997). Since then, a lot of progress has been made in simulation of protein folding at the molecular mechanical (MM) and quantum mechanical (QM) levels. The Nobel Prize of the year 2013 was awarded to Michael Levitt, Arieh Warshel, and Martin Karplus for developing various methods for molecular dynamics and specifically for using multiscale modeling methods like MM and QM in the same complex chemical systems (Karplus 2014). CASP (critical assessment of protein structure prediction) is a biennial competition to compare the performance of various methods of protein design and structure prediction (Moult et al. 1995). Rosetta and *I-TASSER* web servers have shown excellence continuously for many years in algorithms for the design and prediction of protein structure (Ovchinnikov et al. 2018; Roy et al. 2010). But the computational resources required for such simulations is prohibitive for designing novel proteins for industrial application. So far, the state of art in computational design of protein is capable of designing only overall architecture of protein and secondary structures, but not the finer structural features like the stereochemical properties of a binding surface or the precise positioning of side-chain functional groups in the active site for catalysis. Rather, it is more practical to use *in vitro* evolution to screen random sequences of protein to search for a target functional protein if there is a good screening assay. For example, a novel beta-lactamase enzyme was evolved by putting a selection pressure of increasing ampicillin concentration on the E. coli population transformed with a plasmid carrying random protein sequences generated by binary patterning (Chopra & Ranganathan 2003). Often, after the initial computational design of the protein, the function or stability or binding affinity of protein is refined by *in vitro* evolution (Lee 2005). The bacteriophage display method is a very efficient screening method for optimizing the specificity and affinity of a binding protein for the target molecule. Even commercial kits are available for developing artificial antibodies using bacteriophage display and biopanning. These artificial antibodies and other similar small binding protein scaffolds called affibodies are suitable for mass production through protein expression in bacteria, whereas gene-cloning of antibodies raised in animals can have many problems in expression, folding, and post-translational glycosylation, ultimately resulting in no binding to the target analyte (Fodey et al. 2011). In an algorithm similar to "humanizing antibodies," the binding sites of affibodies can be grafted onto a scaffold of human antibodies for their use in therapeutics in humans (Queen et al. 1989). Unique peptide-binding sites have been successfully grafted within the Fab domain by Donaldson, J. M., et al. (Donaldson et al. 2013). A computational method for transplanting the binding site from one protein to another has been described by Liang, S., et al. (Liang, Liu et al. 2000). A detailed strategy for grafting a binding site on a protein has been described by Shide et al. (Liang, Xiao et al. 2000). Bio-inspired design based on natural protein-ligand complexes can be a good start for the rational protein design for binding to narcotic drugs like morphine. In Figure 23.2 a, the cocaine bound to dopamine transporter in PDB structure 4XPB is shown. The molecular recognition of cocaine by this transporter is achieved by its binding to a domain, which has the complementary shape and charge distribution with respect to the shape and charge distribution of dopamine. The structural requirements for such a complementary surface can give important insights for designing a small and stable protein, having the surface/pocket for the binding of cocaine. The biomolecular recognition of metabolites by various signaling proteins, receptors, and enzymes can give a starting point for designing AFA proteins for structurally similar molecules of forensic importance. Most of the narcotic drugs have a natural counterpart in the body, having affinity to receptors and the enzymes involved in its synthesis. The natural analog of heroin is met-enkephalin (Vuong et al. 2010). The binding site of met enkephalin with DPP3 available in pdb structure is 5E33 (Kumar et al. 2016). The shape and interacting residues in a known binding site can be very useful for creating a binding site for target molecules on affibodies. The synthesis pathway of enkephalin has the last enzyme carboxypeptidase E, which can be engineered to work as affinity proteins for heroin (Fricker 1988). The last enzyme in the enkephalin synthesis pathway can be mutated to make it catalytically inactive so that it does bind to enkephalin but does not modify it further. Proteins in the pathway of tetrahydrocannabinol (THC) can be modified into specific affinity molecules for THC class of narcotics. The artificial antibody/affibody so created will be used for making a lateral flow test kit for specific, easy, and sensitive tests for narcotic drugs. In a bio-inspired design, the minimalist binding domain of these proteins can be stabilized and refined by computational design. The binding and stability can be further fine-tuned using *in vitro* evolution for the designed protein (Figure 23.2). There had been attempts to replace antibody with polynucleotide aptamers, but since the structural diversity of polynucleotides is very limited compared to the proteins, it does not have desired specificity and affinity (Walsh & DeRosa 2009). The *in vitro* evolution, based on a phage display system, can be used to select an artificial antibody from a pool of phages with diverse sequence at the binding site of an antibody displayed on its surface (Miersch et al. 2015). Here, the phenotype and genotype of a phage are linked, and the DNA inside the phage is coding for the protein on the surface. By randomly varying the DNA sequence, we can change the protein sequence at the binding site region of the protein displayed on the surface. After that, these phages are added to a plate where analyte of interest like cocaine is immobilized to the surface of the plate. Sometimes, the polymerization of a small molecule antigen can be beneficial because it will make a large enough epitope for binding to affinity protein. Eventually, such affinity proteins will recognize two or more free target molecules together. The epitope size can be also increased by covalently binding these target molecules to another protein or adjuvant like albumin. The phage having the right antibody will bind to analyte and then during washing all other non-binders will be washed away. These binder phages can be further mutated to improve affinity or specificity and this cycle is repeated several times. Then the binder phage

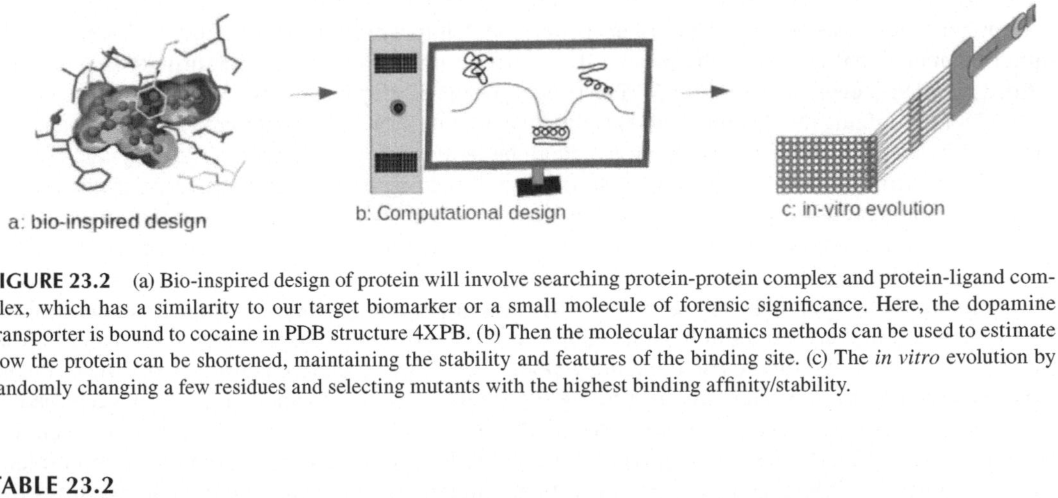

FIGURE 23.2 (a) Bio-inspired design of protein will involve searching protein-protein complex and protein-ligand complex, which has a similarity to our target biomarker or a small molecule of forensic significance. Here, the dopamine transporter is bound to cocaine in PDB structure 4XPB. (b) Then the molecular dynamics methods can be used to estimate how the protein can be shortened, maintaining the stability and features of the binding site. (c) The *in vitro* evolution by randomly changing a few residues and selecting mutants with the highest binding affinity/stability.

TABLE 23.2

Tissue Identification at Crime Sites Using Immunochromatographic Assays in Which Antibodies Can Be Substituted with AFAs

Target molecules	Developed method	Commercially popular kit
Saliva (amylase)	RSID™-saliva (Sinelnikov et al. 2013)	Independent Forensics https://ifi-test.com/
Semen (prostate-specific antigen)	ABAcard-PSA (p30), PSA-check-1(VED-LAB), (Kearsey et al. 2001)	Abacus Diagnostics www.abacusdiagnostics.com
Semen (seminogelin)	RSID™-semen Nanotrap Sg (Old et al. 2012)	Independent Forensics https://ifi-test.com/
Blood (human hemoglobin)	ABAcard® HemaTrace (Horjan et al. 2016)	Abacus Diagnostics www.abacusdiagnostics.com
Species origin of blood (human glycophorin A)	RSID™-Blood (Howard et al. 2019)	Independent Forensics https://ifi-test.com/

can be amplified as a single colony, and their DNA sequence can be determined to find the protein sequence of the binding antibody. This DNA sequence of antibodies can be then also cloned into bacteria or other expression systems to make monoclonal artificial antibodies. Similarly, yeast display system and ribosome display system can be used for artificial evolution of a protein binding to an analyte (Van Deventer et al. 2015). The recombinant antibody against the whole proteome of the human body is also being pursued (Dübel et al. 2010). These artificial antibodies against human proteins can be utilized for forensic detection of various tissues and body fluids (Table 23.2). Lectins, which are proteins of plant origin binding to carbohydrates, can be used to identify blood group antigens (Gorakshakar & Ghosh 2016). But despite these antibody alternatives being available, mostly the antibody-based kits are used for blood group identification in medical and forensic applications. AFAs can also be developed to detect a wide range of other molecules in forensics, like a new bioterror agent, pathogen, explosive, and new narcotic drug, and a person can be followed by using a peculiar biomolecular profile (Table 23.3).

TABLE 23.3

Examples of Target Analytes with Potential for Applications in Forensics and Their Immunochromatographic Assays That Have Scope for Replacement by AFAs

Type of target molecules	Target molecules	Developed methods of detection		Commercially popular kit based on antibodies from corporation (website)
		Antibody-based	AFA-based	
Narcotic drugs	Morphine-3-glucuronide	(Dillon et al. 2003)	(Brennan et al. 2003)	www.alere.com
	Benzimidazoles	(Keegan et al. 2009)		www.creative-diagnostics.com
	Cocaine	(Marckel et al. 2019)	(Du et al. 2010)	www.alere.com
Biotoxins (chemical warfare and terrorism)	Aflatoxin	(Daly et al. 2000)	(Dunne et al. 2005)	www.creative-diagnostics.com
	Warfarin	(Fitzpatrick & O'Kennedy, 2004)		www.creative-diagnostics.com
	Ricin	(Wiley & Oeltmann 1989)	(Anderson et al. 2008)	www.creative-diagnostics.com
Forensic toxicology—pesticides	Endosulfan	(Zhang et al. 2006)	(Chronopoulou et al. 2019)	www.indifoss.com
	Carbofuran	(Guo et al. 2009)	(Wang et al. 2009)	www.indifoss.com
	Chlorpyrifos-methyl	(Hua et al. 2010)	(Alcocer et al. 2000)	www.indifoss.com
Food safety threats	Zearalenone	(Pei et al. 2013)	(Choi et al. 2009)	www.indifoss.com
	Domoic acid	(Tsao et al. 2007)	(Lotierzo et al. 2004)	www.biooscientific.com
Biological warfare or bioterrorism agents category A (highest risk) (per the classification of the CDC, USA)	Ebola virus	Patent US20180180614	(Maruyama et al. 1999)	www.orasure.com
	Anthrax	(Pelat et al. 2007)	(Mazumdar 2009)	www.oxfordimmunotec.com
	Clostridium botulinum	(Chahboun et al. 2011)	(Amersdorfer et al. 2002)	www.alexeter.com
Bioterrorism agents category B (per the classification of the CDC, USA)	*Staphylococcus aureus*	(Singh et al. 2010)	(Larkin et al. 2010)	https://hardydiagnostics.com
Other potential bioweapons	West Nile virus	(Gould et al. 2005)	(Rizzo et al. 2020)	www.rapidtest.com
	SARS-CoV	(H.-K. Lee et al. 2008)	(Sui et al. 2004)	https://hardydiagnostics.com

23.4 Patents, Regulatory Concerns, and Entrepreneurial Scope

Compared to the development of drugs and therapeutic antibodies/affibodies, the development of affibodies for diagnostics and other detection applications, has far less regulatory hassles. Therefore, small entrepreneurs can afford to complete the development cycle of these products, whereas products that require clinical human trials are the domain of only big corporations. The component technologies required to develop these kits are not controlled by big corporations, which is otherwise a big issue in most other technological products. There were several design patents around lateral flow immunochromatography—(EP) 0810436A1, (US) 5654162, (WO) 95/16207, (US) 5075078, (US) 5028535—but they all could be easily circumvented so far. The design patent of LFA is even challenged in court by Grassroot Webgroup, and at present, anybody can use it. The latest patent, US6485982B1, held by Alere (now owned by Abbott), titled "Test device and method for colored particle immunoassay," has even expired on 26-11-2019. A patent can be taken for a protein sequence that has affinity to a target molecule, but it is quite easy to develop another affinity molecule with claims of higher affinity, specificity, stability, and ease of use. Polyclonal antibody-based kits for detection of narcotic substances are there in the market, which has low reproducibility and specificity. Available kits internationally are based on polyclonal antibodies like "Six Panel Drug Abuse Test Kits" (https://drugtestkitusa.com/) and CLIA Waived Drug Tests (www.confirmbiosciences.com/products/clia-waived-drug-tests). The market for such products is expected to increase with increasing awareness. But at the same time, regulatory agencies are expected to restrict the use of animals for antibody production, which will open new opportunities for startup entrepreneurs. The artificial antibody for these detection kits will decrease the cost and increase specificity, reproducibility, and the shelf life of kits. This will lead to development of indigenous technology for manufacturing these highly valuable products. All point-of-care detection and diagnosis based on antibodies can be replaced with affibodies, which are AFA reagents. These kits can detect many drugs in one test as affinity reagents for various narcotic drugs can be immobilized on strip in form of separate dots on strip. The endogenously developed technology can be used to mass-produce detection kits in India. These kits for narcotic substance detection can be used by forensic laboratories across the country and family members for diagnosing drug use in teenagers. Forensic labs in India are already using antibody-based RSID kits for body fluid detection, which is also a huge market for such kits developed using affibodies. The presently used methods of narcotic drugs in forensic labs are colorimetry or chromatography based which are far more expensive and time-consuming compared to LFA. LFA kits can also be developed for new applications in forensics, like the detection of explosives, pathogens, and toxin bioterror agents, or trace volatile molecules for identification and tracking of the culprit (Table 23.3). The use of LFA is also expected to expand into authentication of other consumer products and their purity.

24.5 Concluding Remarks

The binder proteins expressed in bacteria (1) are cost-effective, (2) can be developed to bind toxins and small molecules that fail to generate an immune response, (3) have more affinity, and most importantly, (4) are animal-friendly. The great potential of designed proteins for industrial applications will be realized starting from the use of affinity proteins for the detection of various biomolecules and chemicals. The proteins were envisioned as molecular machines of the future and wonder molecules for therapeutics, but these dreams could not be realized so far, respectively, due to the scale of complexity and long development cycle. But developing proteins for forensic applications like the identification of prohibited substances/chemicals and the detection of biomolecular markers for body fluid identification are easy for the present state of the art in protein science. Therefore, a big commercial potential for the startup biotech industry is there in the development of AFAs and manufacturing on-site detection kits for forensic applications.

REFERENCES

Adler, S., Basketter, D., Creton, S., Pelkonen, O., van Benthem, J., Zuang, V., Andersen, K. E., Angers-Loustau, A., Aptula, A., Bal-Price, A., Benfenati, E., Bernauer, U., Bessems, J., Bois, F. Y., Boobis, A., Brandon, E., Bremer, S., Broschard, T., Casati, S., . . . Zaldivar, J.-M. (2011). Alternative (non-animal) methods for cosmetics testing: Current status and future prospects—2010. *Archives of Toxicology*, *85*(5), 367–485. https://doi.org/10.1007/s00204-011-0693-2

Alcocer, M. J. C., Doyen, C., Lee, H. A., & Morgan, M. R. A. (2000). Properties of polyclonal, monoclonal, and recombinant antibodies recognizing the organophosphorus pesticide chlorpyrifos-ethyl. *Journal of Agricultural and Food Chemistry*, *48*(9), 4053–4059. https://doi.org/10.1021/jf990917l

Amersdorfer, P., Wong, C., Smith, T., Chen, S., Deshpande, S., Sheridan, R., & Marks, J. D. (2002). Genetic and immunological comparison of anti-botulinum type A antibodies from immune and non-immune human phage libraries. *Vaccine*, *20*(11–12), 1640–1648. https://doi.org/10.1016/s0264-410x(01)00482-0

Anderson, G. P., Liu, J. L., Hale, M. L., Bernstein, R. D., Moore, M., Swain, M. D., & Goldman, E. R. (2008). Development of antiricin single domain antibodies toward detection and therapeutic reagents. *Analytical Chemistry*, *80*(24), 9604–9611. https://doi.org/10.1021/ac8019398

Arora, P., Oas, T. G., & Myers, J. K. (2004). Fast and faster: A designed variant of the B-domain of protein A folds in 3 microsec. *Protein Science: A Publication of the Protein Society*, *13*(4), 847–853. https://doi.org/10.1110/ps.03541304

Brennan, J., Dillon, P., & O'Kennedy, R. (2003). Production, purification and characterisation of genetically derived scFv and bifunctional antibody fragments capable of detecting illicit drug residues. *Journal of Chromatography B*, *786*(1–2), 327–342. https://doi.org/10.1016/S1570-0232(02)00807-3

Chahboun, S., Hust, M., Liu, Y., Pelat, T., Miethe, S., Helmsing, S., Jones, R. G., Sesardic, D., & Thullier, P. (2011). Isolation of a nanomolar scFv inhibiting the endopeptidase activity of botulinum toxin A, by single-round panning of an immune phage-displayed library of macaque origin. *BMC Biotechnology*, *11*, 113. https://doi.org/10.1186/1472-6750-11-113

Chang, C., Chen, C., Lee, C., Chen, C., & Lin, C. (2014). Colorimetric detection of human chorionic gonadotropin using catalytic gold nanoparticles and a peptide aptamer. *Chemical Communications*, *50*(92), 14443–14446 . https://doi.org/10.1039/c4cc06366j.

Choi, S.-W., Chang, H.-J., Lee, N., Kim, J.-H., & Chun, H. S. (2009). Detection of mycoestrogen zearalenone by a molecularly imprinted polypyrrole-based surface plasmon resonance (SPR) sensor. *Journal of Agricultural and Food Chemistry*, *57*(4), 1113–1118. https://doi.org/10.1021/jf804022p

Chopra, S., & Ranganathan, A. (2003). Protein evolution by "codon shuffling." *Chemistry & Biology*, *10*(10), 917–926. https://doi.org/10.1016/j.chembiol.2003.09.007

Chronopoulou, E. G., Vlachakis, D., Papageorgiou, A. C., Ataya, F. S., & Labrou, N. E. (2019). Structure-based design and application of an engineered glutathione transferase for the development of an optical biosensor for pesticides determination. *Biochimica Et Biophysica Acta. General Subjects*, *1863*(3), 565–576. https://doi.org/10.1016/j.bbagen.2018.12.004

Dahiyat, B. I., & Mayo, S. L. (1997). De Novo protein design: fully automated sequence selection. *Science*, *278*(5335), 82–87. https://doi.org/10.1126/science.278.5335.82

Daly, S. J., Keating, G. J., Dillon, P. P., Manning, B. M., O'Kennedy, R., Lee, H. A., & Morgan, M. R. A. (2000). Development of surface plasmon resonance-based immunoassay for aflatoxin B1. *Journal of Agricultural and Food Chemistry*, *48*(11), 5097–5104. https://doi.org/10.1021/jf9911693

Desmet, J., Verstraete, K., Bloch, Y., Lorent, E., Wen, Y., Devreese, B., Vandenbroucke, K., Loverix, S., Hettmann, T., Deroo, S., Somers, K., Henderikx, P., Lasters, I., & Savvides, S. N. (2014). Structural basis of IL-23 antagonism by an Alphabody protein scaffold. *Nature Communications*, *5*, 5237. https://doi.org/10.1038/ncomms6237

Dillon, P. P., Daly, S. J., Manning, B. M., & O'Kennedy, R. (2003). Immunoassay for the determination of morphine-3-glucuronide using a surface plasmon resonance-based biosensor. *Biosensors and Bioelectronics*, *18*(2–3), 217–227. https://doi.org/10.1016/S0956-5663(02)00182-3

Donaldson, J. M., Zer, C., Avery, K. N., Bzymek, K. P., Horne, D. A., & Williams, J. C. (2013). Identification and grafting of a unique peptide-binding site in the Fab framework of monoclonal antibodies. *Proceedings of the National Academy of Sciences*, *110*(43), 17456–17461. https://doi.org/10.1073/pnas.1307309110

Du, Y., Chen, C., Yin, J., Li, B., Zhou, M., Dong, S., & Wang, E. (2010). Solid-state probe based electrochemical aptasensor for cocaine: A potentially convenient, sensitive, repeatable, and integrated sensing platform for drugs. *Analytical Chemistry*, *82*(4), 1556–1563. https://doi.org/10.1021/ac902566u

Dübel, S., Stoevesandt, O., Taussig, M. J., & Hust, M. (2010). Generating recombinant antibodies to the complete human proteome. *Trends in Biotechnology*, *28*(7), 333–339. https://doi.org/10.1016/j.tibtech.2010.05.001

Dunne, L., Daly, S., Baxter, A., Haughey, S., & O'Kennedy, R. (2005). Surface plasmon resonance-based immunoassay for the detection of aflatoxin B 1 using single-chain antibody fragments. *Spectroscopy Letters*, *38*(3), 229–245. https://doi.org/10.1081/SL-200058689

Ebersbach, H., Fiedler, E., Scheuermann, T., Fiedler, M., Stubbs, M. T., Reimann, C., Proetzel, G., Rudolph, R., & Fiedler, U. (2007). Affilin-novel binding molecules based on human gamma-B-crystallin, an all beta-sheet protein. *Journal of Molecular Biology*, *372*(1), 172–185. https://doi.org/10.1016/j.jmb.2007.06.045

Feringa, B. L. (2017). The art of building small: From molecular switches to motors (nobel lecture). *Angewandte Chemie International Edition*, *56*(37), 11060–11078. https://doi.org/10.1002/anie.201702979

Fitzpatrick, B., & O'Kennedy, R. (2004). The development and application of a surface plasmon resonance-based inhibition immunoassay for the determination of warfarin in plasma ultrafiltrate. *Journal of Immunological Methods*, *291*(1–2), 11–25. https://doi.org/10.1016/j.jim.2004.03.015

Fodey, T., Leonard, P., O'Mahony, J., O'Kennedy, R., & Danaher, M. (2011). Developments in the production of biological and synthetic binders for immunoassay and sensor-based detection of small molecules. *TrAC Trends in Analytical Chemistry*, *30*(2), 254–269. https://doi.org/10.1016/j.trac.2010.10.011

Fricker, L. D. (1988). Carboxypeptidase E. *Annual Review of Physiology*, *50*, 309–321. https://doi.org/10.1146/annurev.ph.50.030188.001521

Gebauer, M., & Skerra, A. (2009). Engineered protein scaffolds as next-generation antibody therapeutics. *Current Opinion in Chemical Biology*, *13*(3), 245–255. https://doi.org/10.1016/j.cbpa.2009.04.627

Gorakshakar, A. C., & Ghosh, K. (2016). Use of lectins in immunohematology. *Asian Journal of Transfusion Science*, *10*(1), 12–21. https://doi.org/10.4103/0973-6247.172180

Gould, L. H., Sui, J., Foellmer, H., Oliphant, T., Wang, T., Ledizet, M., Murakami, A., Noonan, K., Lambeth, C., Kar, K., Anderson, J. F., de Silva, A. M., Diamond, M. S., Koski, R. A., Marasco, W. A., & Fikrig, E. (2005). Protective and therapeutic capacity of human single-chain Fv-Fc fusion proteins against West Nile Virus. *Journal of Virology*, *79*(23), 14606–14613. https://doi.org/10.1128/JVI.79.23.14606-14613.2005

Grabulovski, D., Kaspar, M., & Neri, D. (2007). A novel, non-immunogenic Fyn SH3-derived binding protein with tumor vascular targeting properties. *The Journal of Biological Chemistry*, *282*(5), 3196–3204. https://doi.org/10.1074/jbc.M609211200

Gray, A. C., Sidhu, S. S., Chandrasekera, P. C., Hendriksen, C. F. M., & Borrebaeck, C. A. K. (2016a). Animal-based antibodies: Obsolete. *Science*, *353*(6298), 452.2–453. https://doi.org/10.1126/science.aag3305

Gray, A. C., Sidhu, S. S., Chandrasekera, P. C., Hendriksen, C. F. M., & Borrebaeck, C. A. K. (2016b). Animal-friendly affinity reagents: Replacing the needless in the haystack. *Trends in Biotechnology*, *34*(12), 960–969. https://doi.org/10.1016/j.tibtech.2016.05.017

Guo, Y.-R., Liu, S.-Y., Gui, W.-J., & Zhu, G.-N. (2009). Gold immunochromatographic assay for simultaneous detection of carbofuran and triazophos in water samples. *Analytical Biochemistry*, *389*(1), 32–39. https://doi.org/10.1016/j.ab.2009.03.020

Horjan, I., Barbaric, L., & Mrsic, G. (2016). Applicability of three commercially available kits for forensic identification of blood stains. *Journal of Forensic and Legal Medicine*, *38*, 101–105. https://doi.org/10.1016/j.jflm.2015.11.021

Hornsby, M., Paduch, M., Miersch, S., Sääf, A., Matsuguchi, T., Lee, B., Wypisniak, K., Doak, A., King, D., Usatyuk, S., Perry, K., Lu, V., Thomas, W., Luke, J., Goodman, J., Hoey, R. J., Lai, D., Griffin, C., Li, Z., . . . Wells, J. (2015). A high through-put platform for recombinant antibodies to folded proteins. *Molecular & Cellular Proteomics*, *14*(10), 2833–2847. https://doi.org/10.1074/mcp.O115.052209

Howard, D., Chaseling, J., & Wright, K. (2019). Detection of blood on clothing laundered with sodium percarbonate. *Forensic Science International*, *302*, 109885. https://doi.org/10.1016/j.forsciint.2019.109885

Hua, X., Qian, G., Yang, J., Hu, B., Fan, J., Qin, N., Li, G., Wang, Y., & Liu, F. (2010). Development of an immunochromatographic assay for the rapid detection of chlorpyrifos-methyl in water samples. *Biosensors and Bioelectronics*, *26*(1), 189–194. https://doi.org/10.1016/j.bios.2010.06.005

Jin, X., He, R., Ju, X., Zhang, J., Wang, M., Xing, C., & Yuan, J. (2019). Development and optimization of an immunoassay for the detection of Hg(II) in lake water. *Food Science & Nutrition*, *7*(5), 1615–1622. https://doi.org/10.1002/fsn3.991

Johnson, A., Song, Q., Ko Ferrigno, P., Bueno, P. R., & Davis, J. J. (2012). Sensitive affimer and antibody based impedimetric label-free assays for C-reactive protein. *Analytical Chemistry*, *84*(15), 6553–6560. https://doi.org/10.1021/ac300835b

Karplus, M. (2014). Development of multiscale models for complex chemical systems: From H+H 2 to biomolecules (nobel lecture). *Angewandte Chemie International Edition*, *53*(38), 9992–10005. https://doi.org/10.1002/anie.201403924

Kearsey, J., Louie, H., & Poon, H. (2001). Validation study of the "Onestep ABAcard® PSA Test" kit for RCMP casework. *Canadian Society of Forensic Science Journal*, *34*(2), 63–72. https://doi.org/10.1080/00085030.2001.10757518

Keegan, J., Whelan, M., Danaher, M., Crooks, S., Sayers, R., Anastasio, A., Elliott, C., Brandon, D., Furey, A., & O'Kennedy, R. (2009). Benzimidazole carbamate residues in milk: Detection by surface plasmon resonance-biosensor, using a modified QuEChERS (Quick, Easy, Cheap, Effective, Rugged and Safe) method for extraction. *Analytica Chimica Acta*, *654*(2), 111–119. https://doi.org/10.1016/j.aca.2009.09.028

Kloosterman, A., Mapes, A., Geradts, Z., van Eijk, E., Koper, C., van den Berg, J., Verheij, S., van der Steen, M., & van Asten, A. (2015). The interface between forensic science and technology: how technology could cause a paradigm shift in the role of forensic institutes in the criminal justice system. *Philosophical Transactions of the Royal Society B: Biological Sciences*, *370*(1674), 20140264. https://doi.org/10.1098/rstb.2014.0264

Koide, A., & Koide, S. (2007). Monobodies: Antibody mimics based on the scaffold of the fibronectin type III domain. *Methods in Molecular Biology (Clifton, N.J.)*, *352*, 95–109. https://doi.org/10.1385/1-59745-187-8:95

Kumar, P., Reithofer, V., Reisinger, M., Wallner, S., Pavkov-Keller, T., Macheroux, P., & Gruber, K. (2016). Substrate complexes of human dipeptidyl peptidase III reveal the mechanism of enzyme inhibition. *Scientific Reports*, *6*(1), 23787. https://doi.org/10.1038/srep23787

Lakhin, A. V., Tarantul, V. Z., & Gening, L. V. (2013). Aptamers: problems, solutions and prospects. *Acta Naturae*, *5*(4), 34–43.

Larkin, E. A., Stiles, B. G., & Ulrich, R. G. (2010). Inhibition of toxic shock by human monoclonal antibodies against staphylococcal enterotoxin B. *PLoS ONE*, *5*(10), e13253. https://doi.org/10.1371/journal.pone.0013253

Lee, H.-K., Lee, B.-H., Dutta, N. K., Seok, S.-H., Baek, M.-W., Lee, H.-Y., Kim, D.-J., Na, Y.-R., Noh, K.-J., Park, S.-H., Kariwa, H., Nakauchi, M., Mai, L. Q., Heo, S.-J., & Park, J.-H. (2008). Detection of antibodies against SARS-Coronavirus using recombinant truncated nucleocapsid proteins by ELISA. *Journal of Microbiology and Biotechnology*, *18*(10), 1717–1721.

Lee, S. (2005). *Encyclopedia of chemical processing (Online)*. CRC Press.

Liang, S., Liu, Z., Li, W., Ni, L., & Lai, L. (2000). Construction of protein binding sites in scaffold structures. *Biopolymers*, *54*(7), 515–523. https://doi.org/10.1002/1097-0282(200012)54:7<515::AID-BIP40>3.0.CO;2-0

Liang, S., Xiao, L., Mao, F., Jiang, L., Han, Y., & Lai, L. (2000). Grafting of protein-protein binding sites. *Chinese Science Bulletin*, *45*(18), 1707–1712. https://doi.org/10.1007/BF02898992

Liu, X., Xiang, J. J., Tang, Y., Zhang, X. L., Fu, Q. Q., Zou, J. H., & Lin, Y. (2012). Colloidal gold nanoparticle probe-based immunochromatographic assay for the rapid detection of chromium ions in water and serum samples. *Analytica Chimica Acta*, *745*, 99–105. https://doi.org/10.1016/j.aca.2012.06.029

Lotierzo, M., Henry, O. Y. F., Piletsky, S., Tothill, I., Cullen, D., Kania, M., Hock, B., & Turner, A. P. F. (2004). Surface plasmon resonance sensor for domoic acid based on grafted imprinted polymer. *Biosensors and Bioelectronics*, *20*(2), 145–152. https://doi.org/10.1016/j.bios.2004.01.032

Marafante, E., Smyrniotis, T., & Balls, M. (1994). ECVAM: The European centre for the validation of alternative methods. *Toxicology in Vitro*, *8*(4), 803–805. https://doi.org/10.1016/0887-2333(94)90072-8

Marckel, J. A., Wetzel, H. N., Amlal, S., Amlal, H., & Norman, A. B. (2019). A recombinant humanized anti-cocaine monoclonal antibody alters the urinary clearance of cocaine and its metabolites in rats. *Drug Metabolism and Disposition*, *47*(3), 184–188. https://doi.org/10.1124/dmd.118.083857

Maruyama, T., Parren, P. W. H. I., Sanchez, A., Rensink, I., Rodriguez, L. L., Khan, A. S., Peters, C. J., & Burton, D. R. (1999). Recombinant human monoclonal antibodies to Ebola virus. *The Journal of Infectious Diseases*, *179*(s1), S235–S239. https://doi.org/10.1086/514280

Mazumdar, S. (2009). Raxibacumab. *MAbs*, *1*(6), 531–538. https://doi.org/10.4161/mabs.1.6.10195

Miersch, S., Li, Z., Hanna, R., McLaughlin, M. E., Hornsby, M., Matsuguchi, T., Paduch, M., Sääf, A., Wells, J., Koide, S., Kossiakoff, A., & Sidhu, S. S. (2015). Scalable high throughput selection from phage-displayed synthetic antibody libraries. *Journal of Visualized Experiments: JoVE*, *95*, 51492. https://doi.org/10.3791/51492

Moult, J., Pedersen, J. T., Judson, R., & Fidelis, K. (1995). A large-scale experiment to assess protein structure prediction methods. *Proteins*, *23*(3), ii–v. https://doi.org/10.1002/prot.340230303

Nixon, A. E., & Wood, C. R. (2006). Engineered protein inhibitors of proteases. *Current Opinion in Drug Discovery & Development*, *9*(2), 261–268.

Nutt, D., King, L. A., Saulsbury, W., & Blakemore, C. (2007). Development of a rational scale to assess the harm of drugs of potential misuse. *Lancet (London, England)*, *369*(9566), 1047–1053. https://doi.org/10.1016/S0140-6736(07)60464-4

Nygren, P.-A. (2008). Alternative binding proteins: affibody binding proteins developed from a small three-helix bundle scaffold. *The FEBS Journal*, *275*(11), 2668–2676. https://doi.org/10.1111/j.1742-4658.2008.06438.x

Old, J., Schweers, B. A., Boonlayangoor, P. W., Fischer, B., Miller, K. W. P., & Reich, K. (2012). Developmental validation of RSIDTM-Semen: A lateral flow immunochromatographic strip test for the forensic detection of human semen. *Journal of Forensic Sciences*, *57*(2), 489–499. https://doi.org/10.1111/j.1556-4029.2011.01968.x

Ovchinnikov, S., Park, H., Kim, D. E., DiMaio, F., & Baker, D. (2018). Protein structure prediction using Rosetta in CASP12. *Proteins: Structure, Function, and Bioinformatics*, *86*, 113–121. https://doi.org/10.1002/prot.25390

Pei, S.-C., Lee, W.-J., Zhang, G.-P., Hu, X.-F., Eremin, S. A., & Zhang, L.-J. (2013). Development of anti-zearalenone monoclonal antibody and detection of zearalenone in corn products from China by ELISA. *Food Control*, *31*(1), 65–70. https://doi.org/10.1016/j.foodcont.2012.09.006

Pelat, T., Hust, M., Laffly, E., Condemine, F., Bottex, C., Vidal, D., Lefranc, M.-P., Dübel, S., & Thullier, P. (2007). High-affinity, human antibody-like antibody fragment (single-chain variable fragment) neutralizing the lethal factor (LF) of Bacillus anthracis by inhibiting protective antigen-LF complex formation. *Antimicrobial Agents and Chemotherapy*, *51*(8), 2758–2764. https://doi.org/10.1128/AAC.01528-06

Queen, C., Schneider, W. P., Selick, H. E., Payne, P. W., Landolfi, N. F., Duncan, J. F., Avdalovic, N. M., Levitt, M., Junghans, R. P., & Waldmann, T. A. (1989). A humanized antibody that binds to the interleukin 2 receptor. *Proceedings of the National Academy of Sciences*, *86*(24), 10029–10033. https://doi.org/10.1073/pnas.86.24.10029

Rai, J., & Kaushik, K. (2018). Reduction of animal sacrifice in biomedical science & research through alternative design of animal experiments. *Saudi Pharmaceutical Journal*, *26*(6), 896–902. https://doi.org/10.1016/j.jsps.2018.03.006

Ramoni, R., Bellucci, S., Grycznyski, I., Grycznyski, Z., Grolli, S., Staiano, M., De Bellis, G., Micciulla, F., Pastore, R., Tiberia, A., Conti, V., Merli, E., Varriale, A., Rossi, M., & D'Auria, S. (2007). The protein scaffold of the lipocalin odorant-binding protein is suitable for the design of new biosensors for the detection of explosive components. *Journal of Physics: Condensed Matter*, *19*(39), 395012. https://doi.org/10.1088/0953-8984/19/39/395012

Rizzo, S., Imperato, P., Mora-Cárdenas, E., Konstantinidou, S., Marcello, A., & Sblattero, D. (2020). Selection and characterization of highly specific recombinant antibodies against West Nile Virus E protein. *Journal of Biotechnology*, *311*, 35–43. https://doi.org/10.1016/j.jbiotec.2020.02.004

Roy, A., Kucukural, A., & Zhang, Y. (2010). I-TASSER: A unified platform for automated protein structure and function prediction. *Nature Protocols*, *5*(4), 725–738. https://doi.org/10.1038/nprot.2010.5

Sathe, M., Srivastava, S., Merwyn, S., Agarwal, G. S., & Kaushik, M. P. (2014). Competitive immunochromatographic assay for the detection of thiodiglycol sulfoxide, a degradation product of sulfur mustard. *The Analyst*, *139*(20), 5118–5126. https://doi.org/10.1039/c4an00720d

Schafer, S., Viswanathan, S., Widjaja, A. A., Lim, W.-W., Moreno-Moral, A., DeLaughter, D. M., Ng, B., Patone, G., Chow, K., Khin, E., Tan, J., Chothani, S. P., Ye, L., Rackham, O. J. L., Ko, N. S. J., Sahib, N. E., Pua, C. J., Zhen, N. T. G., Xie, C., . . . Cook, S. A. (2017). IL-11 is a crucial determinant of cardiovascular fibrosis. *Nature*, *552*(7683), 110–115. https://doi.org/10.1038/nature24676

Silverman, J., Liu, Q., Lu, Q., Bakker, A., To, W., Duguay, A., Alba, B. M., Smith, R., Rivas, A., Li, P., Le, H., Whitehorn, E., Moore, K. W., Swimmer, C., Perlroth, V., Vogt, M., Kolkman, J., & Stemmer, W. P. C.

(2005). Multivalent Avimer proteins evolved by exon shuffling of a family of human receptor domains. *Nature Biotechnology*, *23*(12), 1556–1561. https://doi.org/10.1038/nbt1166

Sinelnikov, A., Kalinina, A., Old, J., Boonlayangoor, P., & Reich, K. (2013). Evaluation of Rapid Stain IDentification (RSIDTM) reader system for analysis and documentation of RSIDTM tests. *Applied Sciences*, *3*(3), 624–635. https://doi.org/10.3390/app3030624

Singh, P. K., Agrawal, R., Kamboj, D. V., Gupta, G., Boopathi, M., Goel, A. K., & Singh, L. (2010). Construction of a single-chain variable-fragment antibody against the superantigen Staphylococcal enterotoxin B. *Applied and Environmental Microbiology*, *76*(24), 8184–8191. https://doi.org/10.1128/AEM.01441-10

Stokes, W. S., Schechtman, L. M., & Hill, R. N. (2002). The Interagency Coordinating Committee on the Validation of Alternative Methods (ICCVAM): A review of the ICCVAM test method evaluation process and current international collaborations with the European Centre for the Validation of Alternative Methods (ECVAM). *Alternatives to Laboratory Animals: ATLA*, *30*(Suppl 2), 23–32. https://doi.org/10.1177/026119290203002S04

Stumpp, M. T., Binz, H. K., & Amstutz, P. (2008). DARPins: a new generation of protein therapeutics. *Drug Discovery Today*, *13*(15–16), 695–701. https://doi.org/10.1016/j.drudis.2008.04.013

Suderman, R. J., Rice, D. A., Gibson, S. D., Strick, E. J., & Chao, D. M. (2017). Development of polyol-responsive antibody mimetics for single-step protein purification. *Protein Expression and Purification*, *134*, 114–124. https://doi.org/10.1016/j.pep.2017.04.008

Sui, J., Li, W., Murakami, A., Tamin, A., Matthews, L. J., Wong, S. K., Moore, M. J., Tallarico, A. St. C., Olurinde, M., Choe, H., Anderson, L. J., Bellini, W. J., Farzan, M., & Marasco, W. A. (2004). Potent neutralization of severe acute respiratory syndrome (SARS) coronavirus by a human mAb to S1 protein that blocks receptor association. *Proceedings of the National Academy of Sciences*, *101*(8), 2536–2541. https://doi.org/10.1073/pnas.0307140101

Syedmoradi, L., Daneshpour, M., Alvandipour, M., Gomez, F. A., Hajghassem, H., & Omidfar, K. (2017). Point of care testing: The impact of nanotechnology. *Biosensors and Bioelectronics*, *87*, 373–387. https://doi.org/10.1016/j.bios.2016.08.084

Tate, J., & Ward, G. (2004). Interferences in immunoassay. *The Clinical Biochemist. Reviews*, *25*(2), 105–120.

Tsao, Z.-J., Liao, Y.-C., Liu, B.-H., Su, C.-C., & Yu, F.-Y. (2007). Development of a monoclonal antibody against domoic acid and its application in enzyme-linked immunosorbent assay and colloidal gold immunostrip. *Journal of Agricultural and Food Chemistry*, *55*(13), 4921–4927. https://doi.org/10.1021/jf0708140

UNODC Report, U. Report 2019. (2019). *World drug report 2019*. United Nations Publication, Sales No. E.19.XI.8.

Van Deventer, J. A., Kelly, R. L., Rajan, S., Wittrup, K. D., & Sidhu, S. S. (2015). A switchable yeast display/secretion system. *Protein Engineering, Design & Selection: PEDS*, *28*(10), 317–325. https://doi.org/10.1093/protein/gzv043

Vuong, C., Van Uum, S. H. M., O'Dell, L. E., Lutfy, K., & Friedman, T. C. (2010). The effects of opioids and opioid analogs on animal and human endocrine systems. *Endocrine Reviews*, *31*(1), 98–132. https://doi.org/10.1210/er.2009-0009

Walsh, R., & DeRosa, M. C. (2009). Retention of function in the DNA homolog of the RNA dopamine aptamer. *Biochemical and Biophysical Research Communications*, *388*(4), 732–735. https://doi.org/10.1016/j.bbrc.2009.08.084

Wang, H., Yang, J., Liu, X., Liang, Y., Lei, H., Shen, Y., Xu, X., Sun, Y., Xu, Z., & He, Y. (2009). Cloning, expression, and identification of anti-carbofuran single chain Fv gene. *Biotechnology Progress*, *25*(4), 1018–1024. https://doi.org/10.1002/btpr.297

Wiley, R. G., & Oeltmann, T. N. (1989). Anti-ricin antibody protects against systemic toxicity without affecting suicide transport. *Journal of Neuroscience Methods*, *27*(3), 203–209. https://doi.org/10.1016/0165-0270(89)90081-2

Zhang, C., Zhang, Y., & Wang, S. (2006). Development of multianalyte flow-through and lateral-flow assays using gold particles and horseradish peroxidase as tracers for the rapid determination of carbaryl and endosulfan in agricultural products. *Journal of Agricultural and Food Chemistry*, 54(7), 2502–2507. https://doi.org/10.1021/jf0531407

24

Biomimetics Approach of Biodiversity through Interventions of Bio-Designing and Technology

Rashmi Srivastava, Anita Yadav, Neeshma Jaiswal, Anshika Yadav, Sandeep K. Malhotra, Suman Mishra, and Ranbir Chander Sobti

CONTENTS

24.1 Introduction

Human characteristics that are best suited for survival become part of the human population, and they adapt to change as the populations change. These are, in the long run, chosen for perpetuation in the forthcoming human populations through the process of genetic continuity. Such adaptations are transformed in nature into the principles of human engineering. These comprise biomimicry, which drives the rapid changes in species diversity to trigger adaptations under natural conditions and thus regulate variations, as well as natural selection. Biomimicry incorporates technology-based inputs for the transformation of processes in nature into practice. The concept of biomimicry strikes a balance from the standpoint of conservation of nature's resources mainly to function in accordance with the global framework encompassing responsible and sustainable development. The appropriate way to optimize the conservation of useful natural resources and resist their wastage would be the replacement by an adapted form of wild life. The development and upgrading of technology-driven resources could thus also be achieved to facilitate the upliftment of the economy and the society as a whole. The solution to a whole

DOI: 10.1201/9781003220398-24

lot of environmental problems is by using an adaptive system, with nature as a prototype, to reinvent sustainability and surpass the hurdles that are current challenges to humanity (Mead and Jeanrenaud, 2017; Nkandu, 2018).

The word "biomimicry" is often used synonymously with the word "biomimetics" owing to their similarities in their objective and functionality, and they go hand in hand with modern technology. Actually, in the opinion of Vincent et al. (2006), von Gleich et al. (2009), Lepora et al. (2012) and Goldstein and Johnson (2015), biologically produced products (e.g., enzymes and silk) are studied under biomimetics for their synthesis and configuration so that a biological framework can be molded artificially to generate identical chemicals that imitate the naturally functioning chemicals and/or processes. Biomimetics can be applied in several fields (Morris et al., 2020). The number of qualities that can be replicated is huge due to the diversity and complexity of biological systems. Biomimetic applications are at various stages of development, from prototypes to technologies that might become commercially usable (Bhushan, 2009; Stroble et al., 2009). Murray's law, which, in conventional form, determines the optimum diameter of blood vessels, has been re-derived to provide simple equations for the pipe or tube diameter that gives a minimum mass engineering system (Williams et al., 2008).

Living beings have adapted to constantly changing environments through evolution—that is, through mutation, recombination, and selection. The main idea of the biomimetic philosophy is that nature's inhabitants, including animals, plants, and microbes, have the most experience in solving problems and have already found the most appropriate ways to last on planet Earth (Collins, 1963). Mathews (2011), Jacobs (2014), and Dicks (2016) have emphasized that the notion of biomimicry imperatively propounds that the skills acquired by the constituents of nature, which also encompass microbes, vegetation, and animals, are equipped to resolve issues to sustain an appropriate manner in life. Correspondingly, biomimetic architecture seeks solutions for building sustainability present in nature. The 21st century has seen an omnipresent waste of energy due to inefficient building designs, in addition to the overutilization of energy during the operational phase of their life cycle (Gebeshuber et al., 2009; Gehan and Nouran, 2016). Resultantly, substitutes of operational procedures, computerized imagery, and refabricated implements have been generated, which have opened up newer avenues to establish functioning alternatives in nature. This has strengthened human approach to expedite developmental activities to achieve sustenance in energy efficacy. The framework of biomimicry is structured to employ nature's implements for the purpose of power conservation and operative infrastructure management (Gebeshuber and Majlis 2010; Gamage and Hyde, 2012; Shahda et al., 2014; Dicks, 2018).

Human engineering's inspiration from the nature's operating principles and natural selection is the main aspect of biomimicry. This phenomenon intends to adopt the naturally tested phenomena that could conceptually be applied to operative principles that have emerged over a period of time or failed in times of resolute difficulties. Biomimicry has emanated from the considered analyses of nature's formulations that could be emulated to refine designs and processes for the resolution of adversities in human life. According to the biomimicry website and emerging movement, founded by science writer Jeanine Benyus (1998), drawing ideas from Mother Nature to create technologies and products mimicking nature is known as biomimicry. Her Biomimicry 3.8 joined hands with the American industry's Nike, General Electric (GE), and Boeing for the fabrication of refined products and polished amenities. The multifaceted approaches have been adopted to fabricate innumerable refurbished products and energized technologies, several of which remain undescribed till today.

Biomimicry represents the new focus on mimicking natural processes to find innovative solutions to complex problems; instead of focusing on what can be extracted from nature, biomimicry pays attention to what they can learn from nature (Benyus, 2002; Mathews, 2019). Those inspired by biomimicry study nature and then imitate or take inspiration from the designs and processes inherent in nature to solve human problems. Biomimicry occurs at the juncture where ecology meets agriculture, medicine, manufacturing materials science, energy, computing, and commerce. Biomimicry is expanding in areas such as architecture, engineering, and product development (Benyus, 2020). With appropriate investment, it offers major potential for new markets. Biodiversity is integral to economic growth and poverty reduction (UNEP, 2010). According to a report, about 40% of the world's economy is contingent on biodiversity. More than 80% of the world's poor directly or indirectly rely on biodiversity for their survival (Hirsch and Secretariat of the Convention on Biological Diversity, 2010). The versatility of a field like biomimicry generates best natural alternatives that are perceived to help resolve human problems. This approach has been formulated in the foregoing text. The approach of quickly imitating natural functions to achieve vital performance emerged from recent research analysis. However, the premise was hypothetically conceptualized outlining fundamental principles.

It is implied, in the end, that the substitute of ecosystem services could easily be found in time-tested parallel approaches with assessed ecological performance standards, particularly in context of the available research data from nature, over 3.8 billion years, and have also been applied in biomimicry (Dicks, 2007; Pedersen and Storey, 2007). For example, carbon dioxide dissolved in seawater is used by corals, which can be mimicked in the production of cement. Velcro fasteners and straps are other widely known examples of biomimicry. It was invented by a Swiss engineer, George de Mestral, in 1941 after he removed some plant material adhering to his dog's fur.

24.2 Classification and Applied Aspects of Biomimicry

Fieldwork, research, and technical demand analysis are required for the development of continuous biomimicry technology and knowledge. The functional classification of biomimicry was developed in this study for the efficient linkage between the biological and ecological characteristics and technology utilization (Dicks, 2019; Zwart, 2019). Although there is a well-known classification system called biomimicry taxonomy, developed by the Biomimicry Institute (Ask Nature, 2018), it is necessary to develop a unique system for each field of research and technology. Each aspect of biology of an organism, like ecology, developmental biology, and other aspects of biology, should form the basis of such classificatory schemes, once advanced research in the field of biomimicry is conducted. The operability of structural elements of human engineering to imitate a living entity and their interaction with biotransformed entities that participate in the phenomenon of bio-allocation could constitute biomimicry (Figure 24.1).

24.2.1 Bionics

The term "bionics" can be defined as the biologically inspired engineering that is largely referred to as the application of biological methods and systems found in nature utilized for the study and design of engineering systems and modern technology (Iouguina, et al., 2014). Bionics opens the doors of imitated operability (Wahl, 2006).

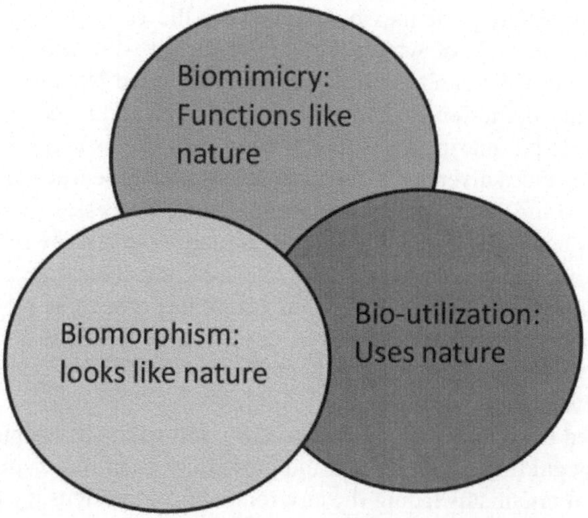

FIGURE 24.1 The structural and functional elements of biomimicry.

24.2.2 Biotransformed Entity

Biotransformation is a metabolic process that takes place mainly in the liver and helps to facilitate the excretion of both exogenous and endogenous substances. A series of reactions alter the chemical structures of these substances; however, the biotransformed entity is identical to the natural (Uchenna, Amadi et al., 2018; Amadi et al., 2018; Phang-Lyn and Llerena, 2020).

24.2.3 Bio-Allocation

Bio-allocation can be accomplished by biodistribution, which is a method of tracking where compounds of interest travel in an experimental animal or human subject. A radioactive isotope is chemically joined with a peptide while fabricating newer compounds for PET (positron emission tomography). This isotope releases antiparticles, positrons, which have an equivalent mass to electrons and have a positive charge. Two gamma rays bearing converse directional properties are emitted on the ejection of positrons from the nucleus, which are decimated on contact with electrons. These gamma rays can be measured and quantified when compared to a standard, and this has a lot of biomedical applications (Raty et al., 2006; Zhang, 2012).

Several innovative means have been proposed through the significant emergence of biomimicry (Kennedy et al., 2015b). Its recent examples include altering bathing suits by adopting sharkskin patterns, using metal components of airplanes to achieve bacterial resistance, and using nano-fabrication to reduce friction in water. Biomimicry has made contributions to this world as many sustainable solutions have emerged in connection to this field (Mead and Jeanrenaud, 2017). (Mead and Jeanrenaud, 2017). Current applications, innovations and examples related to biomimicry include the following:

1. Porcupine quills used as a model to develop medical needles and adhesives. Currently, with the intention to cut down the use of fossil fuels for the production of materials and energy, biological processes are being mimicked as well.
2. Artificial photosynthesis system (also known as photo-electrochemical cells) is being developed to replicate plant photosynthesis. In the future, when exposed to sunlight, it will produce electricity and also supply pure oxygen into the atmosphere.
3. Camouflaging sources of inspiration are chameleon, octopus, and the like.

24.3 Insights into Biodiversity

The Canadian Biodiversity Strategy defines biodiversity as the variety of species and ecosystems on Earth and the ecological processes of which they are a part of—including ecosystems, species, and genetic diversity components (Vold and Buffett, 2008). The United Nations Convention on Biological Diversity provides a similar definition for biodiversity: "the variability among living organisms from all sources including terrestrial, marine and other aquatic ecosystems and the ecological complexes of which they are part; this includes diversity within species, between species and of ecosystems." In short, the term is used to refer to life in all its forms and the natural processes that support and connect all life-forms. Biodiversity is not easily defined because it is more than just the sum of its parts, as all of its elements, regardless of whether we understand their roles or know their status, are integral to maintaining functioning, evolving, resilient ecosystems. Complex concepts such as biodiversity are often easier to grasp if reduced to their component parts (Prati et al., 2014). While this approach does not give a complete picture of how these pieces interact and combine to create biodiversity, it helps us understand different aspects of biodiversity (Vold and Buffett, 2008).

Koshoo (1994) proposed that biodiversity is one of the assets of the developing countries. Nature acts as an important source by guiding the architects and providing structural inspiration through its natural formations. The biological organisms facing the environmental conditions try to endure or withstand to the best of their capabilities; thus, survival of the fittest is favored by the environment through coping and sustaining within their limits of their capability and resource availability and even trying to develop and continue to survive while facing the challenges of changing environmental conditions. The mimicking of the ecosystem is an integral part of biomimicry. The term "eco-mimicry" has also been used to describe the mimicking of the ecosystem in design, which also uses the term in the visionary approach of sustainable development, thereby formulating the design that can be used in conjunction with other levels of biomimicry (organism and behavior), besides applying the principles of sustainability.

As defined by the United Nations Convention on Biological Diversity (as part of the United Nations Environmental Program, or UNEP), "Biotechnology means any technological application that uses biological systems, living organisms, or derivatives thereof, to make or modify product or processes for specific use" (UNEP, 2011). In simple words, biotechnology is the use of life (*bios* in Greek) to develop products or processes, which perform tasks for mankind. Living organisms can include plants, animals, and microorganisms. The human use of biological systems or processes varies widely from the use of simple technologies based in natural processes, such as using microorganisms to produce yoghurt or the composting of organic material, to advanced genetic modifications, which allow the cloning of animals and the replication of desired characteristics in different species. Biotechnology is today a widely explored field of research, and applications in industry also cover a broad scope, from food and medicine production to the development of new materials and biomedical applications, among others. All these applications are often referred to as industrial biotechnology. In relation to design, and specifically industrial design, some of the most relevant applications of industrial biotechnology are bioenergy, biofuels, biorefining, biomass/feedstocks, agricultural sciences, biomaterials (bioplastics, biofilms), biobased chemicals and enzymes (bulk, fine, specialty), food and beverages, feed processing, cosmeceuticals and personal care, fibers (pulp, paper, and textiles), lubricants, surfactants, detergents, automotive, biodefense, bioremediation, bioprospecting, marine biotechnology, and nanobiotechnology.

24.4 Use of Biomimicry in the Conservation of Biodiversity

It is very apparent in today's world that human development must coexist with nature (Paul, 2010). The world is progressing toward modernization each and every day, and every field needs to understand the global trends to remain workable and survive in the future, and biomimicry is the important solution to this problem. According to Paul (2010), the ecosystem is dependent on the population explosion, aging population, health challenge, scarcity of resources, rising costs, migration of population, extinction of species, transparency, and awareness. The efficacious strengthening and perfection of skills to the mentioned business ecosystems with lower expenditures, under the camouflaged biomimicry, would trigger upgrading of human societal living standards, to enhance conservation status of the environment, and a cost-benefit evaluation points to the merits of biomimicry as reducing three major sources of costs—the economic cost of pollution, the economic cost of waste disposal, the economic cost of natural resource depletion (Paul, 2010).

Bioinspiration refers to the biological phenomena that stimulate nonbiological science and technology research (Kennedy et al., 2015a; Whitesides, 2015). Finally, the term "nature-based solutions" was adopted at the 2016 World Conservation Congress and was developed by International Union for Conservation of Nature (IUCN, 2019). It is defined as a series of actions to protect, continually manage, and restore nature and the changing ecosystem. It provides for human welfare and the benefits of biodiversity and solves societal challenges. By mimicking optimized natural systems, it is possible to adapt to nature and develop efficient and sustainable environmental solutions for technologies, such as robots, electronics, energy, textiles, chemicals, and architecture.

24.5 Applications of Biomimicry

Biomimicry has made contributions to this world as many sustainable solutions have emerged in connection to this field. Current applications, innovations, and examples related to biomimicry are discussed in the following sections.

24.6 Imitation of Shark skin

The imitation of shark skin patterns onto fabrics for swim wears and metallic structures of airplanes and ships to make them bacteria resistant and decrease friction in water with the help of nano-fabrication.

Viscous secretion (mucus), which could withstand even flowing water, used as a model to develop strong glue to reseal incised organs which can also be useful for septal defects. Materials within the abalone shells are twice the strength of high-tech ceramics. Porcupine quills used as representation to develop medical needles and adhesives. Currently, with the intention to cut down the use of fossil fuels for production of materials and energy, biological processes are being mimicked as well. Artificial photosynthesis system (also known as photo-electrochemical cells) is being developed to replicate plant photosynthesis. In the future, when exposed to sunlight, it will produce electricity and also supply pure oxygen into the atmosphere. Camouflaging & its source of inspiration are chameleon, octopus, etc. Plants (mushroom)—building styles, Morph butterfly-light reflection, Amazon lily pads-the crystal palace, a solar cell is inspired by a leaf.

24.7 Biomimicry of Natural Structural Colors

One of the most studied species, in terms of structural colour production and biomimetics (Parry and Savin, 2016), is the Morpho butterfly, whose wing-scales exhibit a unique metallic-blue colour. Morpho is an excellent example of the combined diversity of physical mechanisms in colour production, hence it is all the more challenging to mimic. The iridescent blue colour with high reflectivity is a result of coherent scattering in the periodic arrays of scales

24.8 Future Innovations in the Field of Biomimicry

From the time when Leonardo Da Vinci created the "flying machine," biomimicry was prevalent, which has further led to more innovative inventions presently. The design and engineering aspect of biomimicry enables products to be not only appealing to the human eye but also more energy-efficient. Some of the natural resources that are used as a source of innovative bio-designing ideas with the help of biomimicry are mentioned in Table 24.1.

Biomimicry is not simply a transition from biology and ecology to technology; rather, it involves systematically finding out the functional characteristics that already exist in the natural world—that is, in biology and ecosystems. Biomimicry, from the point of view of a solution-driven approach, or biology push (ISO, 2015), applies biological knowledge obtained by examining principles in biological functions and focuses on the activities of biological and ecological resources that promote the value of biodiversity. Some of the natural resources and their respective modules used as innovative bio-designing ideas with the help of biomimicry are mentioned in Table 24.1.

TABLE 24.1

Innovations in Biomimicry with Respect to Their Natural Sources

S. No.	Natural sources	Innovations
1	Ant colony network	Repair system
2	Chimpanzees	Medicinal research
3	Sharks	Antibiotics, cancer treatment
4	Bat	Ultra-cane (walking cane)
5	Fish sensitivity	Fish robots to search for pollution in water
6	Spider silk	Elastic and tensile materials

24.9 Biomimicry Approaches and Their Impact on Society

Biomimicry has not only made us close to the nature with our innovative and sustainable ideologies but also tried to solve human problems by linking the built-in environment with the model designed as a standard of measure and mentor, which becomes more evident from the rationale of Janine Benyus: "the more our world function like the natural world, the more likely we are to endure on this home that is ours, but not ours alone." Thus, comparative analysis between nature as a model, measure, and mentor clearly establishes the approach of biomimicry. There are several models of biomimicry:

1. Nature as model: Biomimicry is a new science that studies nature's models and then imitates or takes inspiration from these designs and processes to solve human problems (e.g., a solar cell inspired by a leaf).
2. Nature as measure: Biomimicry uses an ecological standard to judge the "rightness" of our innovations. After 3.8 billion years of evolution, nature has learned what works, what is appropriate, and what lasts.
3. Nature as mentor: Biomimicry is a new way of viewing and valuing nature and introduces an era based not on what we can extract from the natural world but on what we can learn from it.

The designing approaches that have been utilized in several applications after incorporation of biomimicking ideas are as follows:

1. Designs looking to biology
   a. Design spiral by the Biomimicry Institute
   b. Steel cable technology in muscles
2. Biology influencing design
3. Biomimicry to support architectural sustainability
4. Biomimicry to support economic sustainability

24.10 Social Impact of Biomimicry

Biomimicry has been evolving in the areas where there is still a lack of scientific knowledge. For the time being, humans have found that the nature acts as an influential aspect to develop engagement and empowerment among communities. As the involvement of psychology is considered along with ecology, eco-psychology was developed to help companies and entrepreneurs to face complex situations. Thus, biomimicry is aiding humans to build resilience to face any issues.

24.11 Environmental and Economic Impact of Biomimicry

Businesses have started looking at the pros and cons of their products. Rather than focusing on short-term profit, they are keen on achieving more sustainable outcomes. For instance, bacteria are used to make cement. Manufacturing cement needs a very long time, and it also contributes to an increased carbon footprint, leading to loss of money as well. The bacterium *Sporosarcina pasteurii* has been used to produce bricks as they secrete calcium carbonate under ambient temperatures. This reduces the overproduction of wastes rather than useful materials.

24.12 Levels of Biomimicry

24.12.1 The First Level: Nature Being the Inspiration for the Formation

Nature acts as an important source by guiding the architects and providing structural inspiration through its natural formations.

24.12.2 The Second Level: Mimicry of How an Organism Behaves

The biological organisms facing the environmental conditions try to endure or withstand to the best of their capabilities; thus, survival of the fittest is favored by the environment through coping and sustaining within their limits of their capability and resource availability and even trying to develop and continue to survive while facing the challenges of changing environmental conditions.

24.12.3 Third Level: Ecosystem Level

The mimicking of the ecosystem is an integral part of biomimicry. The term "eco-mimicry" has also been used to describe the mimicking of the ecosystem in design, which also uses the term in the visionary approach of sustainable development, thereby formulating the design that can be used in conjunction with other levels of biomimicry (organism and behavior), besides applying the principles of sustainability.

It has been emphasized that the incorporation of an understanding of the living world into architectural design could be a significant step toward the creation of a built environment that allows for positive integration with and restoration of natural carbon cycles. Increased positive integration with ecosystems and a restorative rather than damaging effect on ecosystems may have significant benefits on maintaining biodiversity and the ecosystem of goods and services that humans are dependent on for survival (Daily et al., 2000). If the built environment is to become truly regenerative, it will increase human and community well-being and that of ecosystems.

One of the most studied species in terms of structural color production and biomimetics is the morpho butterfly, whose wing scales exhibit a unique metallic blue color. It is an excellent example of the combined diversity of physical mechanisms in color production; hence, it is all the more challenging to mimic. The iridescent blue color with high reflectivity is a result of coherent scattering in the periodic arrays of scales (Kinoshita et al., 2002; Prum et al., 2006).

24.13 Conclusions

There may be a growing need for biomimicry because it is often a drive for sustained social innovation and growth to resolve environmental problems supported ecological systems. Although most successful cases support technological engineering demands, it is time to get the groundwork from a biological and ecological perspective for the application of the structure and functioning principles of biological traits to technology. The results of this study lead to a plan to create a biomimicry system that categorizes the biological and ecological characteristics, then link each functional characteristic to practical areas where technology is often utilized. The advancements and innovative investigations pertaining to biomimicry could be instrumental to achieve targeted developmental objectives and would further restore confidence of future generations in the role of biodiversity in economic upliftment.

24.14 Acknowledgments

N. J. is thankful to the University Grants Commission, New Delhi, for its financial support (UGC-BSR Start-Up Grant No. F. 30–460/2019 [BSR]). A. Y. is grateful to UGC, and to the faculty and administration of CMP (PG) College (a constituent college of the University of Allahabad), Prayagraj, UP, for its laboratory and other forms of support.

REFERENCES

Amadi, P., Ogunka-Nnoka, C., Bene, A. (2018). Biotransformation of plantain pseudostem fibres using local enzyme sources: Analysis of their potential as commercial poultry feed. *Biocatal. Biotransform.*, 36: 1–9.

Ask Nature (2018). *Biomimicry Taxonomy*. https://asknature.org/resource/biomimicry-taxonomy/

Benyus, J. (2020). Biomimicry: Innovation Inspired by Nature; Harper Collins Publishers: New York City, NY, USA, 1997. *Biomimetics*, 5(33): 13–14.

Benyus, J.M. (1998). *Biomimicry: Innovation Inspired by Nature*. New York: Perennial (Harper Collins).

Benyus, J.M. (2002). *Biomimicry: Innovation Inspired by Nature* (3rd ed.). New York: Williams and Morrow & Co.

Bhushan, B. (2009). Biomimetics: Lessons from nature—An overview. *Philos. Trans. R. Soc. A: Math. Phys. Eng. Sci.*, 367: 1445–1486.

Collins, G.R. (1963). Antonio Gaudi: Structure and form. *Perspecta*, 8: 63–90.

Daily G.C., et al. (2000). The value of nature and the nature of value. *Science*, 289: 395–396.

Dicks, H. (2007). A new way of valuing nature: Articulating biomimicry and ecosystem services. *Environ. Ethics*, 30: 307–323.

Dicks, H. (2016). The philosophy of biomimicry. *Philos. Technol.*, 29: 223–243.

Dicks, H. (2018). Nature as mentor: Foundations of biomimetic epistemology. *J. Biomim. Eng.*, [Pr.: Not required because it's a Book.] Available online: https://www.researchgate.net/publication/328191528_Nature_as_Mentor_Foundations_of_Biomimetic_Epistemology

Dicks, H. (2019). Being like Gaia: Biomimicry and ecological ethics. *Environ. Values*, 28: 1–28.

Gamage, A., Hyde, R. (2012). A model based on biomimicry to enhance ecologically sustainable design. *Architect. Sci. Rev.*, 55: 224–235.

Gebeshuber, I.C., Gruber, P., Drack, M. (2009). A gaze into the crystal ball: Biomimetics in the year 2059. *Proc. Inst. Mech. Eng. Part C J. Mech. Eng. Sci.*, 223: 2899–2918.

Gebeshuber, I.C., Majlis, B.Y. (2010). Biomimetic nanotechnology: A powerful means to address global challenges. *arXiv:1001.3319*. Retrieved from arXiv e-reprint services Cornell University. http://arxiv.org/ftp/arxiv/papers/1001/1001.3319.pdf, accessed November 10, 2012.

Gehan R.A.N., Nouran, O. (2016). Biomimicry, an approach, for energy efficient building skin design. *Procedia Environ. Sci.*, 34: 178–189.

Goldstein, J., Johnson, E.R. (2015). Biomimicry: New natures, new enclosures. *Theory Cult. Soc.*, 32: 61–81.

Hirsch, T., Secretariat of the Convention on Biological Diversity, eds. (2010). *Global Biodiversity Outlook 3*. Montreal, Quebec, Canada: Secretariat of the Convention on Biological Diversity.

Iouguina, A., Dawson, J.W., Hallgrimsson, B., Smart, G. (2014). Biologically informed disciplines: A comparative analysis of bionics, biomimetics, biomimicry, and bio-inspiration among others. *Int. J. Des. Nat. Ecodynamics*, 9: 197–205.

ISO (2015). *Biomimetics—Terminology, Concepts and Methodology*. International Organization for Standardization 18458:2015. Berlin: Beuth Verlag.

IUCN (2019). www.iucn.org/commissions/com mission-ecosystem-management/our-work/naturebased-solutions

Jacobs, S. (2014). Biomimetics: A simple foundation will lead to new insight about process. *Int. J. Des. Nat. Ecodyn.*, 9: 83–94.

Kennedy, B., Buikema A., Nagel, J.K. (2015a). Integrating biology, design, and engineering for sustainable innovation. In: *2015 IEEE Integrated STEM Education Conference*. The Institute of Electrical and Electronics Engineers, Princeton, New Jersey, USA, pp. 88–93.

Kennedy, E., Fecheyr-Lippens D., Hsiung B. K., Niewiarowski P.H., Kolodziej M. (2015b). Biomimicry: a path to sustainable innovation. *Des. Issues*, 31: 66–73.

Kinoshita, S., Yoshioka, S., Kawagoe, K. (2002). Mechanisms of structural color in the Morpho butterfly: Cooperation of regularity and irregularity in an iridescent scale. *Proc. Biol. Sci.*, 269(1499): 1417–1421.

Koshoo, T.N. (1994). India's biodiversity: Tasks ahead. *Curr. Sci.*, 67(25): 577–582.

Lepora, N.F., Verschure, P.F.M.J., Prescott, T.J. (2012). The state-of-the-art in biomimetics. In: *Proceedings of the Lecture Notes Computer Science*. (Including Subser. Lecture Notes in Artificial Intelligence Lecture Notes Bioinformatics) 7375 LNAI, Barcelona, Spain, 9–12 July 2012, pp. 367–368.

Mathews, F. (2011). Towards a deeper philosophy of biomimicry. *Organ. Environ.*, 24: 364–387.

Mathews, F. (2019). Biomimicry and the problem of praxis. *Environ. Values*, 28, 1–47.

Mead, T., Jeanrenaud, S. (2017). The elephant in the room: Biomimetics and sustainability? *Bioinspired Biomim. Nanobiomater.*, 6: 113–121.

Morris, E.K., Caruso, T., Buscot, F., Fischer, M., Hancock, C., Maier, T.S., Meiners, T., Müller, C., Obermaier, E. (2020). Choosing and using diversity indices: Insights for ecological applications from the German biodiversity exploratories. *Biomimetics*, 5(33): 14.

Nkandu, M.I. (2018). Biomimicry as an alternative approach to sustainability. *Archit. Res.*, 8: 1–11.

Parry, A., Savin T. (2016). Recent advances in the biomimicry of structural colors. *Chem. Soc. Rev.* 45: 6698–6724.

Paul, R. (2010). *Design Inspiration from Nature: Biomimicry for a Better Planet.* http://inhabitat.com/finding-design-inspiration-innature-biomimicry-for-a-better-planet (accessed May 29, 2013).

Pedersen Zari, M., Storey, J.B. (2007). *An Ecosystem Based Biomimetic Theory for a Regenerative Built Environment.* Paper read at Lisbon Sustainable Building Conference 07, at Lisbon, Portugal.

Phang-Lyn S., Llerena, V.A. (2020). Biochemistry, biotransformation. In: *StatPearls.* Treasure Island, FL: StatPearls Publishing.

Prati, D., et al. (2014). Choosing and using diversity indices: Insights for ecological applications from the German Biodiversity Exploratories. *Ecol. Evol.*, 4: 3514–3524.

Prum, R.O., Quinn, T., Torres, R.H. (2006). Anatomically diverse butterfly scales all produce structural colours by coherent scattering. *J. Exp. Biol.* 209(4): 748–765.

Raty, J.K., Liimatainen, T., Wirth, T., et al. (2006). Magnetic resonance imaging of viral particle bio distribution in vivo. *Gene Therapy*, 13(20): 1440–1446.

Shahda, M., Elmokadem, A., Elhafeez, M.A. (2014). Biomimicry levels as an approach to the architectural sustainability. *Port-Said Eng. Res. J.*, 18(2), 117–125. doi:10.21608/pserj.2014.45298

Stroble, J.K., Stone, R.B., Watkins, S.E. (2009). An overview of biomimetic sensor technology. *Sensor Rev.*, 29: 112–119.

Uchenna, A., Uchechi, O.-N., Bene, A. (2018). Properties of oils from plantain pseudostem biotransformed using crude local enzyme sources: a comparison of poultry feed oil. *Recent Pat. Food Nutr. Agric.*, 10(2), 140–151. doi:10.2174/2212798410666181217141311

UNEP (2010). *Linking Biodiversity Conservation and Poverty Alleviation: A State of Knowledge Review.* www.jstor.++org/stable/2709362.

UNEP (2011). *United Nations Environmental Program and United Nations Convention on Biological Diversity.* Definition of Terms. www.cbd.int/convention/articles/?a=cbd-02.

Vincent Julian, F.V., et al. (2006). Biomimetics: Its practice and theory. *J. R. Soc. Interf.*, 3(9): 471–482.

Vold, T., Buffett, D.A. (eds.) (2008). *Ecological Concepts, Principles and Applications to Conservation.* BC. 36 pp. Library and Archives Canada Cataloguing in Publication Data. Available at: www.biodiversitybc.org

von Gleich, A., Pade, C., Petschow, U., Pissarskoi, E. (2009). *Potentials and Trends in Biometics.* Berlin/Heidelberg: Springer. ISBN 9783642052453.

Wahl, D.C. (2006). Bionics vs. biomimicry: From control of nature to sustainable participation in nature. *WIT Trans. Ecol. Environ.*, 87: 289–298.

Whitesides, G.M. (2015). Bioinspiration: Something for everyone. *Interf. Focus*, 5: 20150031

Williams, H.R., Trask, R.S., Weaver, P.M., Bond, I.P. (2008). Minimum mass vascular networks in multifunctional materials. *J. R. Soc. Interf.*, 5(18): 55–65.

Zhang, G. (2012). Biomimicry in biomedical research. *Organogenesis*, 8: 101–102.

Zwart, H. (2019). What is mimicked by biomimicry? Synthetic cells as exemplifications of the three fold biomimicry paradox. *Environ. Values*, 28: 1–23.

25

Biodiversity for Sustainable Development

Mehreen Manzoor Gora, M. R. Mir, Shaista Mehraj, R. K. Sharma, Aadil Hussain, Naveena Nazim, and I. L. Khan

CONTENTS

25.1 Introduction

The term *biodiversity* is used to describe the enormous variety of life on Earth. It can be used more specifically to refer to all of the species in one region or ecosystem. Biodiversity refers to every living thing, including plants, bacteria, animals, and humans. In other words, it is the biological variety and

DOI: 10.1201/9781003220398-25

variability of life on Earth and is a measure of variation at the genetic, species, and ecosystem levels. Biodiversity is the basis of all *ecosystem services* that provide supporting, provisioning, regulating, and cultural services and to which human *well-being* is intimately linked. No feature of Earth is more complex, dynamic, and varied than the biosphere that occupies its surfaces and its seas, but no feature is experiencing more dramatic change due to anthropogenic activities.

Biodiversity plays a very important role in maintaining natural cycle and ecological balance, besides forming the basis of existence, human wealth, and preservation of nature on the Earth. It actually describes the biological capital held within an area and refers particularly to the differences between living organisms at different levels of biological organization—gene, individual, species, and ecosystem. Biodiversity encompasses multiple values and is vital for the production of food and conservation of the ecological foundations needed to sustain people's livelihood.

Biological diversity is often understood at three levels:

1. Species diversity refers to the variety of different species.
2. Genetic diversity corresponds to the variety of genes contained in plants, animals, fungi, and microorganisms.
3. Ecosystem diversity refers to all the different habitats that exist—tropical and temperate forests, hot and cold deserts, wetlands, rivers, mountains, coral reefs, and so on (Yucel, 2015).

25.1.1 Species Diversity

Species diversity is the variety of species within a habitat or a region. Species are the basic units of biological classification and thus the normal measure of biological diversity. Species richness is the term that describes the number of different species in a given area (Naughton-Treves et al., 2005). The world is estimated to have 5 to 10 million species, though only 1.75 million have been named scientifically so far and hence remain unexploited. Some habitats, such as rainforests and coral reefs, have many species. For example, tropical North and South America has about 85,000 flowering plant species, and tropical and subtropical Africa has about 35,000. In contrast, all of Europe has 11,300 vascular plants. Yet other areas such as salt flats or polluted streams have fewer species. Species are grouped together into families according to shared characteristics.

25.1.2 Genetic Diversity

Every species on Earth is related to some other species through genetic connections. The more closely related any two species are, the more genetic information they will share, and the more similar they will appear. An organism's closest relatives are members of its own species, or organisms with which it has the potential to mate and produce offspring. Members of a species share genes, the bits of biochemical information that determine, in part, how the animals look, behave, and live. One eastern gray squirrel, for example, shares the vast majority of its genes with other eastern gray squirrels, whether they live in the same area or are separated by thousands of miles. Members of a species also share complex mating behaviors that enable them to recognize each other as potential mates. Virtually for every species there is a similar and closely related species in an adjacent habitat. Each species also has other, more remotely related species with which it shares a more general set of characteristics. Gray squirrels, chipmunks, marmots, and prairie dogs all belong to the squirrel family because they share a number of features, such as tooth number and shape and details of their skull and muscle anatomy. All of these animals are rodents, a large group of more distantly related animals who share similar, chisel-like incisor teeth that grow continuously. All rodents are related to a broader group, mammals. Mammals have hair, raise their young on milk, and have three bones in the middle ear. All mammals, in turn, are more distantly related to other animals with backbones or vertebrates. All these organisms are animals but share a common cell structure with plants, fungi, and microbes. Finally, all living organisms share a common molecule, ribonucleic acid (RNA) and almost all have deoxyribonucleic acid (DNA). While all species have descended from a single, common ancestor, species diverge and develop their own peculiar attributes with time, thus making their own contribution to biodiversity.

25.1.3 Ecological Diversity

Ecological diversity is the intricate network of different species present in local ecosystems and the dynamic interplay between them. An ecosystem consists of organisms from different species living together in a region and their connections through the flow of energy, nutrients, and matter or the biogeochemical cycling. Those connections occur as the organisms of different species interact with one another. The ultimate source of energy in almost every ecosystem is the sun. The sun's radiant energy is converted to chemical energy by plants. That energy flows through the system when animals eat the plants and then are eaten, in turn, by other animals. Fungi derive energy by decomposing organisms, which releases nutrient back into the soil. Thus, an ecosystem is a collection of living components (microbes, plants, animals, and fungi) and non-living components (climate and chemicals) that are connected by energy flow. Measuring ecological diversity is difficult because each of Earth's ecosystems merges into the ecosystems around it.

25.2 Global Biodiversity with Special Reference to India

Global biodiversity is the measure of biodiversity on the planet Earth and comprises the total variability of living forms. The Earth's current species are estimated to range from two million to one trillion of which more than 80% have still not been described. The total amount of DNA base pairs is estimated at 5×10^{37} and weighs 50 billion tons, whereas the total mass of the biosphere has been estimated to be as much as four trillion tons of carbon (TtC) (Kenneth et al., 2016). The countries with higher biodiversity than India are located in South America, such as Brazil, and Southeast Asian countries, such as Malaysia and Indonesia. The species found in these countries are different from what we have in India.

India is one of the most diverse nations in the world. It ranks ninth in terms of plant species richness. It is the origin of important crop species, such as pigeon pea, eggplant, cucumber, cotton, and sesame. India is also center of various domesticated species, such as millets, cereals, legumes, vegetables, and medicinal and aromatic crops. India is equally diverse in its faunal wealth. There are about 91,000 animal species found here. However, diversity is depleting at a drastic rate, and various programs on biodiversity conservation are being launched to conserve nature. India has 59,353 insect species, 2,546 fish species, 240 amphibian species, 460 reptile species, 1,232 bird species, and 397 mammal species, of which 18.4% are endemic and 10.8% are threatened. India is the home to at least 18,664 species of vascular plants, of which 26.8% are endemic. With only 2.4% of the total land area of the world, the known biological diversity of India contributes to about 8% to the known global biodiversity (Balasubramanian, 2017). In terms of species richness, India ranks seventh in mammals, ninth in birds, and fifth in reptiles. In terms of endemism of vertebrate groups, India's position is tenth in birds with 69 species, fifth in reptiles with 156 species, and seventh in amphibians with 110 species. India's share of crops is 44% as compared to the world average of 11%. India also has 23.39% of its geographical area underforest and tree cover. Of the 34 globally identified hotspots, India harbors four hotspots—Himalaya, Indo-Burma, Western Ghats and Sri Lanka, and Sundaland. Sundaland is a biogeographical region of Southeast Asia that encompasses the Sunda shelf, the part of the Asian continental shelf that was exposed during last ice age.

25.3 Emerging Issues

Loss of biodiversity is a growing trend in virtually all Earth ecosystems. It is under threat, and the main culprit is the one who paradoxically depends on it—the human being. Some of the emerging issues are listed in the following sections.

25.3.1 Rapid Urbanization and Infrastructure Development

The global urban population is expected to reach five billion by 2030, and more than 60% of the area projected to become urban by 2030 is yet to be developed. In addition to the impact of expansion of urban land use, changing consumption patterns lead to an increased ecological footprint—that is, demand for food, water, fiber, energy, and other goods and services, as well as waste and wastewater discharge, impacting ecosystems near and far due to associated land use changes. Such degradation of the ecosystem and its services potentially affect the well-being of both urban and rural populations that rely on them. Efforts to achieve some of the Sustainable Development Goals closely linked to urbanization—such as Goal 6 on water and sanitation, Goal 7 on energy, Goal 9 on industry, innovation, and infrastructure, and Goal 11 on cities and human settlements—will require substantial investment for construction of various infrastructures. In particular, the energy and transportation sectors are likely to receive large investments, followed by water and sanitation and then communications. Some estimates suggest that, in order to deliver on the Sustainable Development Goals, investments in infrastructure as high as USD 6 trillion annually are needed. Infrastructure and urbanization are among the major drivers of the fragmentation, degradation, or loss of habitats, leading to significant biodiversity loss. Ensuring careful management of such investments in a way that does not undermine ecosystem services or biodiversity, especially those ecosystem services that are of particular importance for the poor, is an urgent task. Thus, in order to be "sustainable," infrastructure investments must not only be low-carbon and resilient but must also support the conservation and sustainable use of biodiversity.

25.3.2 Agriculture and Fisheries

Injudicious practices in agriculture and forestry, such as pollution by fertilizers, chemicals, and pesticides, conversion of habitats, and excessive water withdrawal, cause substantial environmental degradation and biodiversity loss. Agricultural expansion is said to account for 70% of the projected loss of terrestrial biodiversity. Demand for fertile land is projected to increase substantially by 2050 due to increasing population and urbanization. The combination of expanded agriculture and bioenergy could result in a global land squeeze in which there would not be sufficient room to conserve natural terrestrial habitats, leading to large declines in biodiversity. Under prevailing production and consumption patterns, biodiversity loss and natural resource degradation will continue unabated or accelerate without additional policies, with the poor being disproportionally affected. The provision of food, water, and energy to the poor becomes more difficult when available natural resources are not managed sustainably or are degraded. Fisheries account for 17% of global intake of animal protein, with small-scale fishers directly dependent on coastal and marine biodiversity for their livelihoods. However, the world's oceans and coasts are highly threatened and subject to rapid environmental change. Coral reefs continue to be degraded, and nearly 90% of fisheries stocks are said to be fully fished or overfished. Significant challenges exist in controlling major threats to coastal marine ecosystems, including unsustainable fishing practices, impact of climate change, land-based pollution, and eutrophication. This requires the enforcement of effective regulations through cooperation among different global, national, and subnational governing bodies and the private sector.

25.3.3 Climate Change

Global temperature increases of 0.4 to 2.6°C by 2055 and 0.3 to 4.8°C by 2090 would be accompanied by rising sea levels, changes in precipitation patterns, substantial loss of summer Arctic sea ice and increasing ocean acidification. These changes would have a broad range of impacts on biodiversity at the genetic, species, and ecosystem levels, including shifts in the distribution of species and ecosystems, changes in species abundance, and increased risk of extinctions. This, in turn, will affect vital ecosystem services, such as air and water purification, pollination, food production, and global nutrient and carbon cycles. Efforts to mitigate climate change could also have very large impacts, both positive and negative, on biodiversity. Poor populations are at higher risk from climate-related shocks. For instance, agriculture is one of the most important economic sectors in many poor countries, but also one of the most sensitive to climate change due to its dependence on weather conditions and other climate-related stressors, such as pests, diseases, or sea level rise. One scenario estimates that an additional 100 million people will fall in poverty by 2030 due to climate change, mostly due to rising food prices. In addition to addressing the issue of food security, there is a need to increase the resilience of communities through restoration of ecosystems, which contributes to the protection of people and infrastructure from the negative impacts of extreme climatic events.

25.3.4 Judicious Use of Resources

Halting deforestation and appropriately implementing reforestation could make important contributions to climate mitigation and protection of biodiversity. Nevertheless, a substantial degree of climate change by 2050 and beyond is already committed due to long lags in the Earth's climate system; therefore, measures for adaptation of communities, including the poor, are needed. Such measures include conservation and restoration of coastal habitats, such as mangroves, and enhancing diversity of crops and their wild relatives to help farmers adapt to climate change by switching to drought- or flood-resistant varieties. From the perspective of biodiversity conservation, adaptation will require, for instance, anticipating climate change in the design of protected area systems. Major transformations to food systems are among the key areas of actions for achieving sustainability. There is a need for improved management of agriculture, aquaculture, and wild capture fisheries. Realistic changes in management of crops and livestock could substantially reduce both water consumption and pollution. Significant reductions in fishing pressure and changes in fishing techniques in most marine fisheries would lead to rebuilding of fisheries over the next one to two decades. It is also essential to restore land and water resources by shifting to more sustainable agricultural practices. Food waste needs to be reduced roughly a third of harvested food is lost either in the food transport and transformation chain (primarily in developing countries) or in the home. Diverse diets, combined with global convergence to moderate levels of calorie and meat consumption, would improve health and food security in many areas and also substantially reduce impacts on biodiversity.

25.4 Threats to Biodiversity

There are a number of issues threatening our planet's biodiversity, from climate change to invasive species. In the following sections are some of the biggest threats facing biodiversity.

25.4.1 Climate Change

The rapid anthropogenic climate change that is being experienced in the early 21st century is intimately entwined with the health and functioning of the biosphere. Climate change is impacting ecosystem very much through changes in mean condition and in climate variability, coupled with other associated changes, such as increased ocean acidification and atmospheric carbon dioxide concentrations (Lenton et al., 2019). Changes in climate has altered life on Earth in the long run, ecosystems have come and gone, and species routinely go extinct. But rapid, man-made climate change speeds up the process, without affording ecosystems and species the time to adapt. For example, rising ocean temperatures and diminishing Arctic sea ice affects marine biodiversity and can shift vegetation zones, having global implications. Overall climate is a major factor in the distribution of species across the globe, and climate change forces them to adjust. But many are not able to cope, causing them to die out.

25.4.2 Deforestation and Habitat Loss

Deforestation is a direct cause of extinction and loss of biodiversity. An estimated 18 million acres of forest are lost each year, due in part to logging and other human practices, destroying the ecosystems on which many species depend. The dramatic alteration of habitats directly threatens biodiversity. When such habitats are lost due to deforestation and other anthropogenic activities, such as mining, the respective environments are unable to provide shelter, food, water, or breeding grounds for the living organisms. It also leads to unhealthy and unbalanced ecosystems that result in the loss of biodiversity and extinction.

25.4.3 Overexploitation

Overexploitation is one of the main threats to global diversity. Overhunting, overfishing, and overharvesting contribute greatly to the loss of biodiversity, killing off numerous species over the past several hundred years.

25.4.4 Invasive Species

The introduction of non-native species into an ecosystem can threaten wildlife (either as predators or as competitors for resources), affect human health, and upset economies. Currently, about a third of the world's endangered vertebrates are threatened by overexploitation.

25.4.5 Pollution

From the burning of fossil fuels (releasing dangerous chemicals into the atmosphere and in some cases depleting ozone levels) to dumping 19 billion pounds to plastic into ocean every year, pollution completely disrupts the Earth's ecosystems. Atmospheric and hydrologic pollution have far-reaching negative effects on biodiversity. While it may not necessarily cause extinction, pollutants do have the potential to influents species habits. For example, acid rain, which is typically caused by burning fossil fuels, can acidify smaller bodies of water and soil, negatively affecting the species that live there by changing breeding and feeding habits. In addition, chemical pollutants such as pesticides and herbicides leach into soils and watersheds. Some fish species, such as salmonids, require small freshwater streams to spawn. Polluted streams results in the abandonment of traditional spawning areas and ultimately in the loss of salmon populations. Species sensitivity to pollution is variable. However, many species are vulnerable to the indirect effects of pollution through the concentration of toxic chemicals in top predators of food chains and disruption of predator-prey interactions.

25.4.6 Global Climate Change

Carbon dioxide released from burning fossil fuels and biomass, deforestation, and agricultural practices contributes to greenhouse gases, which prevent heat from escaping the Earth's surface. With the increase in temperature expected from increasing greenhouse gases, there will be higher levels of air pollution, greater variability in weather patterns, and changes in the distribution of vegetation in the landscape. Some species will not be able to adapt to these changes in the environment and will become extinct. However, it is expected that many plant and animal species will attempt to disperse to higher latitudes and altitudes as the temperature increases. Therefore, any barriers in the landscape, such as highways and urban areas that prevent movement to more hospitable environments, will result in loss of biodiversity.

25.4.7 Urbanization and Developmental Projects

Development of colonies, roads, railway tracts, power plants, dams, and so on cause a huge loss of biodiversity. While construction and execution of different construction projects, a huge number of species is lost besides making conditions unfavorable for the surviving species through water, air, soil, and noise pollution (Almond et al., 2020).

25.5 Benefits of Biodiversity

Biodiversity is important to humans for many reasons. It plays significant role in the lives of humans (Adom et al., 2019). It provides functioning ecosystems that supply oxygen, clean air and water, pollution of plants, pest control, wastewater treatments, and ecosystem services. Some of the benefits are listed in the following sections.

1. Soils

Biodiversity includes not only the large plants and animals we see but also microscopic bacteria, fungi, algae, and a host of tiny insects and invertebrates. These smaller organisms are the ones responsible for creating soil and maintaining the quality of soil. Earthworms are well known for conditioning soil by digging through and aerating it and providing nutrients from their castings or waste. Bacteria and fungi degrade organic material, which then further breaks down in the soil, where plants can use the nutrients.

2. Water Quality

The benefits of biodiversity include keeping quality of water pure. In wetlands, plants take up contaminants in water and process and purify the water. Shellfish such as mollusks take in nutrients from the water, thereby preventing a condition called eutrophication, which can cause a huge increase in organisms in the water that leads to oxygen depletion and mass die-offs. If any species is removed from the food webs of an ecosystems, the ecosystem has the possibility to crash, and in the case of water quality, contaminants can remain and cause immense additional problems.

3. Waste Removal

The biodiversity of microbes, fungi, and other small organisms is important in decomposing waste matter. Organic material in nature, such as leaves, logs, and twigs and dead animals and insects, is all degraded and decomposed by the biodiversity of organisms in the ecosystem. It is a delicate balance, in which certain insects or microbes perform a vital functioning in removing waste from the environment, making it cleaner and less ripe for the spread of disease.

4. Pollination and Seed Dispersal

Bees are not the only organisms for pollinating plants and crops. In fact, birds, bats, and butterflies play a vital role in spreading pollen and in dispersing seeds. Butterflies and hummingbirds, for instance, feed on nectar and can transport pollen from flower to flower. Birds and fruit bats eat the fruit and nuts from plants, then disperse the seeds.

5. Medicine

Many modern medicines incorporate natural ingredients from plants that are found in jungles or forests, notably the Amazon rainforest. There is a huge number of plant species used in modern medicine, somewhere around 50,000 to 70,000. Should any of these plants become extinct, medicinal cures would suffer. Examples of plants used in modern medicine include quinine from the cinchona tree in South America, used to treat malaria.

6. Ecotourism

Awareness of biodiversity has recently led to a boom in ecotourism. Travel companies promote tours into off-the-beaten-path locations, such as forests and jungles, where tourists can go on river cruises or hikes and observe and enjoy nature without disturbing the ecosystem. Money from ecotourism goes back into the region and helps to preserve the diversity of plants and animals (Gemma, 2018).

25.6 Sustainable Development

Sustainable development, according to the Brundtland Report of 1987, is development that meets the needs of the present generation without compromising the ability of future generations to meet their own needs. Sustainable development is the integration of economic, social, and environmental development considered as the interdependent and mutually reinforcing pillars that operate at the local, national, regional, and global levels. Poverty eradication, the change in unsustainable patterns of production and consumption, and the protection and management of natural resources base of economic and social development are regarded as the prime objectives and requirements for sustainable development. Biodiversity is essential for sustainable development and human well-being. As it has provisions for food, fiber, and water, it mitigates and provides resilience to climate change; it supports human health and provides livelihood through agriculture, fisheries, forestry, and many other sectors.

25.7 Sustainable Development Goals

The 2030 Agenda for Sustainable Development sets out an ambitious framework of universal and indivisible goals and targets to address a range of global societal challenges. It has been reported by Ministry of External Affairs in June 7, 2021, that India's overall score across SDGs (Sustainable Development Goals) improved by 6 points—from 60 in 2019 to 66 in 2020–2021. Biodiversity and ecosystems feature prominently across many of the SDGs and associated targets. They contribute directly to human well-being and development priorities. Biodiversity is at the center of many economic activities, particularly those related to crop and livestock agriculture, forestry, and fisheries. Globally, nearly half of the human population is directly dependent on natural resources for its livelihood, and many of the most vulnerable people depend directly on biodiversity to fulfill their daily subsistence needs.

The Strategic Plan for Biodiversity 2011–2020 and its Aichi Biodiversity Targets adopted under the Convention on Biological Diversity has been recognized as setting the global framework for priority actions on biodiversity. The SDGs and the Strategic Plan are mutually supportive and reinforcing, as implementation of one contributes to the achievement of the other. Here is how biodiversity plays an essential role in achieving the SDGs:

1. SDG 1: End poverty in all its forms everywhere. Biodiversity provides resources and income, particularly for the rural poor. Ecosystem services and other non-marketed goods make up between 50% and 90% of the total source of livelihood among poor rural and forest-dwelling households.
2. SDG 2: End hunger, achieve food security, and improve nutrition; promote sustainable agriculture. Biodiversity is a key element of food security and a means of improving nutrition. Many of the most vulnerable people depend on food gathered from natural ecosystems, such as forests, grasslands, oceans, and rivers. Biodiversity also underpins ecosystem functions, such as pollination, maintenance of soil fertility, and water quality, central to agricultural productivity.
3. SDG 3: Ensure healthy lives and promote well-being for all at all ages. Nearly one in four deaths globally is attributed to environmental risk factors. Healthy ecosystems help mitigate the spread and impact of pollution by both sequestering and eliminating certain types of air, water, and soil pollution. Agricultural biodiversity contributes to increased sustainable production, reducing the need for pesticides and other chemical inputs, resulting in benefits to human health. Further, a substantial proportion of the world's population depends on traditional medicines derived from biodiversity for their healthcare needs.
4. SDG 5: Achieve gender equality and empower all women and girls. Women play a vital role in agriculture, nutrition, and the well-being of families and communities. Recognizing women's roles as key land and natural resource managers is central to sustainable development. In addition, loss of biodiversity and associated ecosystem services can perpetuate gender inequalities by increasing the time spent by women and children in performing certain tasks, such as collecting valuable resources, including fuel, food, and water.
5. SDG 6: Ensure the availability and sustainable management of water and sanitation for all. Ecosystems help maintain water supply and quality and guard against water-related hazards and disasters. For example, wetlands play a role in surface, subsurface, and groundwater storage and reduce the risk of flooding. They also help capture, process, and dilute pollutants. Similarly, vegetation, such as grasslands and forests, supports the healthy functioning of watersheds. Managing ecosystems to maintain these types of services is generally more cost-effective than employing built technologies.
6. SDG 8: Promote sustained, inclusive, and sustainable economic growth, full productive employment, and decent work for all. Biodiversity and ecosystems underpin many national and global

economic activities, including those related to agriculture, forestry, fisheries and aquaculture, energy, tourism, transport, and trade. Biodiversity conservation and sustainable use can lead to higher productivity, more efficient resource use, and long-term viability of resources.

7. SDG 9: Build resilient infrastructure; promote inclusive and sustainable industrialization and faster innovation. Biodiversity and healthy ecosystems can provide reliable and cost-effective natural infrastructure. For example, coral reefs and mangrove forests protect coasts against flooding that are expected to increase with climate change. Natural infrastructure, such as vegetation in cities, can reduce the runoff of pollution into water bodies. Such green infrastructure can offer multiple benefits and are often more effective than built infrastructure in terms of cost, longevity, and effectiveness.
8. SDG 11: Make cities and human settlements inclusive, safe, resilient, and sustainable. Ecosystems and biodiversity underpin the day-to-day functioning of human settlements by delivering the basic services and conditions that enable, support, and protect human production, consumption, and habitation. Biological resources provide many of the foods, building materials, energy, and medicines consumed in urban centers. Urban planning that integrates biodiversity considerations can contribute to more sustainable, cost-effective, and healthy human settlements.
9. SDG 12: Ensure sustainable consumption, and production patterns. Consumption and production of all goods and services require the transformation of many natural resources, which in turn impacts biodiversity. Current unsustainable consumption and production patterns can undermine the ability of ecosystems to provide services for industries and communities that rely upon them. Using cleaner and more resource-efficient approaches that minimize wastes and pollutants can bring about economic opportunities and better quality of life for consumers and producers alike and at the same time benefit biodiversity.
10. SDG 13: Take urgent action to combat climate change and its impacts. Forests, peatlands, wetlands, and ocean and coastal ecosystems represent globally significant carbon stores, and their conservation and sustainable use is a critical element for avoiding dangerous changes to the Earth's atmospheric temperature and climate system. Efforts to protect and restore habitats offer cost-effective and proven ways to mitigate climate change. Such ecosystems can also serve as natural buffers against climate extremes and other disasters and strengthen adaptation to climate change.
11. SDG 14: Conserve and sustainably use the oceans, seas, and marine resources for sustainable development. The conservation and sustainable use of biodiversity in marine and coastal ecosystems is a key aspect of sustainable development. Biodiversity underpins all fishing and aquaculture activities, as well as other species harvested for foods and medicines. Conservation and sustainable use of marine and coastal biodiversity is essential to ensure that the world's oceans, seas, and marine resources remain vital.
12. SDG 15: Protect, restore, and promote sustainable use of terrestrial ecosystems, sustainably manage forests, combat desertification and halt and reverse land degradation, and halt biodiversity loss. The conservation, restoration, and sustainable use of terrestrial ecosystems is essential for sustainable development. Targets under this goal include a call to integrate ecosystem and biodiversity values into national and local development planning, poverty reduction strategies and accounts.
13. SDG 16: Promote peaceful and inclusive societies for sustainable development, provide access to justice for all, and build effective, accountable, and inclusive institutions at all levels conflicts over natural resources. Environmental degradation and contamination can be one of the factors leading to social insecurity and violence. Vulnerable people are often disproportionately affected. Strengthening community rights over natural resources management, combating illegal exploitation and corruption, and ensuring transparent decision-making on social and environmental issues constitute an important process toward building an inclusive society based on justice.
14. SDG 17: Strengthen the means of implementation and revitalize the global partnership for sustainable development. The Strategic Plan for Biodiversity provides opportunities for

strengthening global partnership on science, technology, and innovation, for dissemination of environmentally sound technologies, and for building national capacity for monitoring the progress of the 2030 Agenda.

25.8 How Does Biodiversity Impact Sustainable Development

Without biological diversity, there is no other life on Earth, including our own. Even though we are often oblivious to it, this diversity of life is what provides clean water, oxygen, and all other things that end up being part of our diet, as well as clothing and shelter. It provides a lot of psychological benefits too, which are not much appreciated. Biodiversity boosts ecosystem productivity where each species, no matter how small, all have important role to play:

- A larger number of plant species means a greater variety of crops.
- Greater species diversity ensures natural sustainability for all life-forms.
- Healthy ecosystems can better withstand and recover from a variety of disasters.

Healthy and rich biodiversity provides a number of natural services for everyone:

1. Ecosystem services, such as protection of water resources, soil formation and protection, nutrient storage and recycling, pollution breakdown and absorption, contribution to climate stability, maintenance of ecosystems, and recovery from unpredictable events
2. Biological resources, such as food, medicinal resources and pharmaceutical drugs, wood products, ornamental plants, breeding stocks, population reservoirs, future resources, diversity in genes, species, and ecosystems
3. Social benefits, such as research, education and monitoring, recreation and tourism, and cultural values

The cost of replacing these, if possible, would be extremely expensive. Biodiversity, therefore, makes economic and development sense to move toward sustainability (Shah, 2021).

25.9 Policy Recommendations for Biodiversity Protection

- Most of the legal provisions pertain mainly to use/exploitation of biological resources rather than their conservation. Even the Wildlife Protection Act, 1972, focuses on protection rather than conservation. Protection under the Wildlife Protection Act is largely directed toward large animal species (charismatic terrestrial species) rather than the large spectrum of fauna and flora also found in the marine realm.
- Examination of existing laws relating to biodiversity makes them in tune with the provisions of convention to reflect current understanding of biodiversity conservation.
- Formulation of policies for protection of wetlands, grasslands, sacred groves, marine flora and fauna, and other areas is significant from the point of view of biodiversity.
- Documentation of biodiversity is needed.
- There mus be an increase in the allocation of financial resources for the conservation of biodiversity.
- Conservation and development must go hand in hand.
- Continuous monitoring of biodiversity use for review of results of implementation of policies and programs.

25.10 Conclusion

Biodiversity is getting eroded at a very alarming rate. While there are myriad of reasons, human activities are the principal reason for biodiversity loss. Development, because of its lack of concern for nature and its associated importance, has led to uncertainty of a sustainable future. Clearing land for agriculture, introducing invasive alien species, constructing infrastructure without specific knowledge, and overexploiting resources have all contributed to the cause. Lately, climate change and global warming are also posing as threats to the ecosystem. This is creating a vicious cycle of environmental degradation. This is a critical moment in Earth's history, a time when humanity must choose its future. Our planet Earth is perhaps the only human habitat in the vast universe, and we owe it to posterity to preserve the divine heritage of our biosphere without pollution, degradation, and destruction. While progress toward sustainable development has been made through meetings, agreements, and changes in environmental governance, real change has been slow. To effectively address environmental problems, policymakers should design policies that tackle both pressures and the drivers behind them.

REFERENCES

Adom, D., Umachandran, K., Ziarati, P., Sawicka, B. and Sekyere, P. (2019). The concept of biodiversity and its relevance to mankind. *Journal of Agriculture and Sustainability*, 12:219–231.

Almond, R.E.A., Grooten, M. and Peterson, T. (2020). *Bending the Curve of Biodiversity Loss.* Living Planet Report, pp. 978–999.

Balasubramanian, A (2017). *Biodiversity Profile of India.* Mysuru: Centre for Advanced Studies in Earth Science, University of Mysore, pp. 1–11.

Gemma, A (2018). Benefits of biodiversity. *American Association for the Advancement of Science*, 529:300–500.

Kenneth, L., Lennon, J. and Jay, T. (2016). Scaling laws predict global microbial diversity. *Proceedings of the National Academy of Sciences*, 113(21):5970–5975.

Lenton, T.M., Rockstrom, J., Gaffney, O., Rahmstorf, S., Richardson, K., Steffen, W., and Schellnhuber, H.J. (2019). Climate tipping points. *Nature*, 575:592–595.

Naughton-Treves, L., Holland, M.B., and Brandon, K. (2005). The role of protected areas in conserving biodiversity and sustaining local livelihood. *Annual Review of Environmental and Resources*, 30:1–219.

Shah, A. (2021). Why is biodiversity important. *Global Issues.*

Yucel, E. (2015). Biological diversity. *A paper presented at the 2015 Wood Day Symposium Booklet pdf*, vol. 12:219–231.

26

Health and Helminths: Revisiting the Paradigm of Host-Parasite Relationship

Abdur Rehman and S. M. A. Abidi

CONTENTS

26.1 Introduction

"Health" is a very broad term that comprises a number of factors, and being disease-free is actually a part of being healthy. According to WHO, health is a state of complete physical, mental, and social well-being and not merely the absence of disease or infirmity. The immune system is a complex network of intricate pathways in the body that protect us against a wide variety of pathogens and a number of diseases. Immunity in general could thus be considered as a biological system well endowed with the capability to recognize and tolerate whatever belongs to the self and to recognize and reject what is foreign (non-self). A healthy immune system defends the body against both communicable and noncommunicable diseases, but if it malfunctions, it causes autoimmune disorders—that is, it attacks its own healthy cells, tissues, and organs, resulting in impaired body functions and even death. Therefore, a functional and efficient immune system is the real key to good health. There are over 100 autoimmune diseases reported among humans. A few are well known, such as type 1 diabetes, multiple sclerosis, Crohn's disease, ulcerative colitis, and rheumatoid arthritis, whereas others are quite rare and difficult to diagnose that may include various syndromes of unknown etiology. Most of the autoimmune diseases still have no permanent cure and require lifelong treatment to ease the suffering/symptoms of the patients. In United States of America alone, these diseases are estimated to affect more than 24 million people and appear to be rapidly increasing (Dinse et al., 2020). Although our understanding of the causes of these autoimmune diseases is still incomplete, the current consensus points to the interactions between genetic and environmental factors. The reasons for the increase in autoimmune diseases are not clear, but there is a big concern as they are likely to increase in near future (Dinse et al., 2020).

Parasites for long have been associated with a large number of pathological and adverse effects on the host's body. Beyond their sanitary or economic importance, parasites are an integral part of the biosphere. They are so diverse and pervasive that they virtually infect every free-living organism, potentially influencing the host's health, behavior, and population size (Raga et al., 2009). A very close association between helminth parasites and humans has been reported from time immemorial. With the improvement in the healthcare facilities and better hygiene practices, helminths, to a very large extent, have been eradicated from developed countries, which interestingly coincides with the sharp increase in

DOI: 10.1201/9781003220398-26

the cases of autoimmune disorders. A number of epidemiological studies have revealed that the absence of helminth infections inversely correlates with a global rise in chronic inflammation-associated diseases (Sobotková et al., 2019). Further, mounting evidence supports the view that the complete loss of these organisms has dire consequences for the human health. In one of the studies carried out more than four decades ago it was demonstrated that seasonal allergies could be successfully reversed by exposure to the worms, thus reflecting the beneficial effects of helminths on human health and well-being (Turton, 1976). Such studies resulted in the formulation of the hygiene hypothesis in late 1980s, which was later refined into the old friends hypothesis or the biome depletion theory (Parker, 2014).

26.2 Helminths and Autoimmune Diseases

The autoimmune diseases are one of the leading causes of death in women less than 65 years of age. The increase in the rate of emergence of these diseases during past few decades indicates that there might be some environmental factors also that might be contributing to the immune dysfunction leading to autoimmunity problems, allergies, and numerous other inflammatory diseases (Elliott and Weinstock, 2012). Many of the experts in the scientific community believe that although many striking environmental changes that have occurred since industrialization began in the 20th century, one that stands out is the loss of exposure to parasitic worms, especially the helminths. Human beings and helminths share a very close association that dates back to the era of our earliest recorded history, and the evidence of this association has been established through the recovery of eggs of intestinal helminths from the mummies, coprolites, and skeletons, dating back to thousands of years (Camacho et al., 2018). Also, there is a written evidence about helminth infections from the ancient literature of Hippocrates, Egyptian medical papyri, and the Bible (Hotez et al., 2008). Breaking this long-standing relationship between the helminth and humans could have immediate effects on our immune system. Following adaption to the micro-environment within their definitive host, the helminth parasites establish a successful host-parasite relationship by overcoming the host-mounted rejection responses. Helminths are known to synthesize and excrete/secrete a number of immunomodulatory molecules that help in their successful establishment by altering both the innate and adaptive immune responses (Ryan et al., 2020). Some of the immunomodulatory molecules derived from the liver flukes, *Fasciola* spp., are summarized in Figure 26.1.

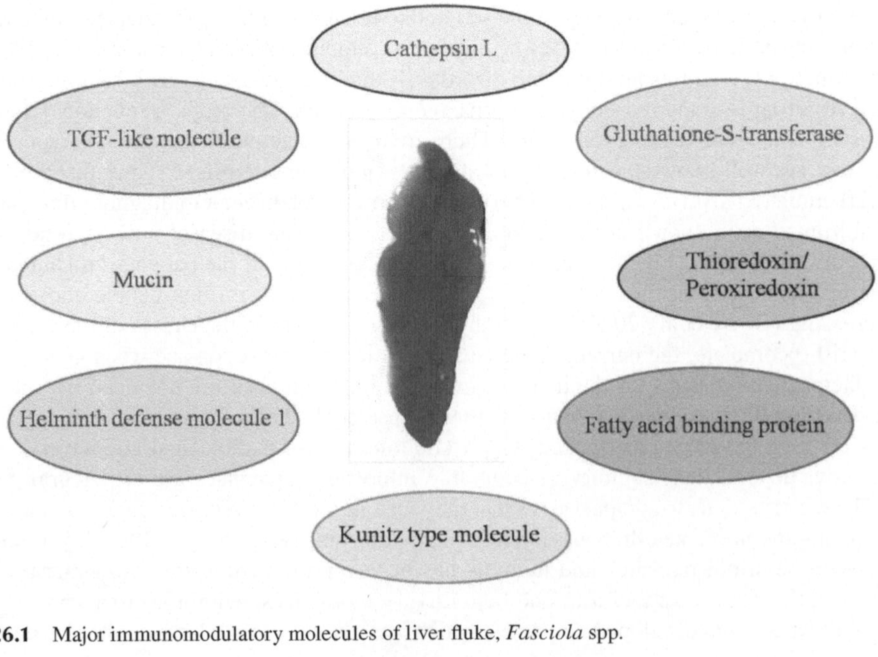

FIGURE 26.1 Major immunomodulatory molecules of liver fluke, *Fasciola* spp.

It is assumed that a complete eradication of the helminth infection might remove an important immune regulator that could give rise to an autoimmune disease and exaggerated inflammatory responses. In the present communication, we discuss about the increasing evidence for helminth-host interactions that suggest the possible roles of helminth and their derived products for the prevention and control of immune-mediated disorders.

26.3 Helminths and Inflammatory Bowel Disease

Inflammatory bowel disease (IBD) is a collective term used to describe disorders like ulcerative colitis and Crohn's disease, involving chronic inflammation of the digestive tract. Although the actual reasons for these inflammatory diseases still remain a mystery, it is assumed that they arise from inappropriately aggressive mucosal immune responses to the luminal substances. Recent studies have highlighted the impact of different genetic alterations in imparting risk for or protection from acquiring these conditions. Although the number of potential IBD susceptibility genes continue to increase, the evidence suggests that the genetic factors contribute only a very small part toward the disease pathogenesis indicating the importance of epigenetic mechanisms and microbial and environmental factors in the development and progression of IBD (Loddo and Romano, 2015). Earlier, it was believed that IBD is restricted to the developed nations, but now it is emerging rapidly in many developing and underdeveloped countries (M'Koma, 2013). The poorly defined environmental factors could be responsible for a rapid increase in disease frequency on a global level. Also, better hygiene practices and alteration in gut microbiota could be postulated as a major risk factor in the development of these chronic diseases.

A number of epidemiological and clinical studies suggest that the loss of exposure to helminth infections as a result of improved health measures and better hygienic conditions could prove significant in the development of IBD in humans as these worms are known to be excellent regulator of host's immune system (Elliott and Weinstock, 2012; Weinstock and Elliott, 2009). The extent to which these infections are helpful in the prevention and/or treatment of IBD is still being investigated. The ongoing research does suggest that helminths are responsible for mitigating the adverse effects of both ulcerative colitis and Crohn's disease.

A number of animal models of IBD are currently being utilized to have a better understanding of the mechanism through which IBD is developed in the first place and how these mechanisms are influenced by the helminth infections. There are three major types of animal models based on the method of induction of disease that are commonly used for the study of IBD under laboratory conditions.

1. Chemically induced model of IBD: It is one of the most widely used method for the induction of IBD in animals. Experimental colitis could be induced in healthy mice either by the intra-rectal instillation of trinitrobenzene sulfonic acid (TNBS) or dinitrobenzene sulfonic acid (DNBS) or through the oral administration of dextran sulfate sodium (DSS). Both these routes trigger a strong Th-1 type pro-inflammatory response within the intestine (Smallwood et al., 2017).
2. Knockout model of IBD: Another important model for studying IBD is the knockout (KO) mice model. A number of KO mice like TGF-1 -/-, IL-10 -/- and STAT -/- are currently being used to understand the mechanisms of disease establishment and progression during the acute and chronic phases (Mizoguchi et al., 2016).
3. Defective intestinal epithelial model of IBD: The compromised immunity of the intestinal epithelial layer could lead to the development of spontaneous colitis. Mouse models, such IKK-γ (NEMO), IKK-β, and mdr1a-/- are few examples where spontaneous colitis is developed via impaired intestinal epithelial responses (Smallwood et al., 2017).

A number of regulatory pathways might be involved in the control of IBD by the helminth parasites. The components of both innate and adaptive immunity along with a number of cytokines are known to mediate the protective responses by the helminths against the IBD.

26.4 Helminths and Crohn's Disease

In recent years, there is growing experimental and clinical evidence that infection with parasitic helminths protects the host from IBD. A case control study carried out to examine the association between self-reported childhood helminth infection and the development of IBD in South Africa revealed an inverse association between the childhood helminth infection and the development of IBD at a later stage in life. The results are in agreement with the hygiene hypothesis. The improvement in the quality of life through better hygienic conditions and non-exposure to the parasitic infections may be contributing to an increase in the incidence of IBD, which might be low in individuals pre-exposed to immunomodulatory molecules of helminths (Chu et al., 2013).

The TNBS-induced inflammation in the colon of mice was greatly reduced following treatment with the eggs of *Schistosoma japonicum* through the suppression of Th-1 cytokine level, whereas a significant elevation in the level of Th-2 cytokines was observed along with an increase in the number of regulatory T cells (Tregs) as compared to the control animals, which did not receive helminth eggs (Mo et al., 2007). In case of another study the prior exposure to *Schistosoma mansoni* eggs not only attenuated TNBS-induced colitis in mice but also provided protection from lethal inflammatory responses by decreasing the IFN-gamma level and enhancing the IL-4 production (Elliott et al., 2003). In case of *Toxoplasma gondii*–induced ileitis in mice co-infected with *S. mansoni*, there is a marked reduction in the inflammation of ileum by preserving the mucosal integrity and downregulation of the local inflammatory responses through the activation of NF-kB and mitogen-activated protein kinases compared to the mice infected alone with *T. gondi* (Pêgo et al., 2019). The prophylactic therapy with *S. japonicum* ova has been shown to maintain epithelial barrier function in mice with TNBS-induced colitis through increased tight junction proteins like zonula occludens protein-1 and occludin, hence causing less exposure of nucleotide-binding oligomerization domain 2 (NOD2) to the luminal antigens, which may initiate a series of inflammatory responses and induce colitis (Xia et al., 2011). Despite the promising results obtained in these *in vivo* studies, there is a concern about the active helminth infection associated with the live worm therapy. Therefore, it is better to identify and utilize the biomolecules that could bring about the immunomodulatory effects leading to the suppression of pathological condition in diseased individuals.

P28 glutathione-S-transferase (P28GST), a protein derived from *Schistosomes* spp. was found to be effective in reducing intestinal inflammation in experimental colitis by downregulating the Th1/Th17 response. The evaluation of safety of P28GST administered to patients with mild Crohn's disease revealed a significant decrease in the Crohn's disease activity index (CDAI) scores and blood calprotectin levels (Capron et al., 2019). The administration of P28GST in mice slightly modulated the diversity and composition of fecal microbiota along with significant reduction in the TNBS-induced colitis (Foligné et al., 2019). However, further studies are required with large sample size to ensure the utility/efficacy of P28GST in the treatment of Crohn's disease considering the potential of P28GST as a safe therapeutic option.

Experimental infection with the intestinal nematode *Trichinella spiralis* significantly reduced the severity of DNBS induced colitis along with decreased mortality in the mouse model. The protective effect was accompanied with a significant downregulation in the myeloperoxidase (MPO) activity and Th1-type immune responses, while the Th2-type immune responses were found up regulated in the colon tissue (Khan et al., 2002). In an another study the intestinal inflammation in the TNBS-induced experimental colitis in mice was ameliorated following the establishment of *Trichinella spiralis* infection through the induction of Th2 immune responses and balanced Th1 responses (Xu et al., 2018). The rectal submucosal administration of *Trichinella spiralis* antigens in mice prior to the induction of DNBS-induced colitis significantly reduced the severity of the disease which was accompanied with a decline in the MPO activity, IL-1 beta production, and iNOS expression along with an increase in IL-13 and TGF-beta production in colon, thereby establishing a protective role of the helminth-derived antigens, an alternative to the live-parasite therapy, for the treatment of experimental colitis (Motomura et al., 2009). *Trichinella spiralis*–derived cystatin, a cysteine protease inhibitor, as well as *Trichinella spiralis* serine protease inhibitors, induced Th2-type immune responses in the mouse Crohn's disease model, which balanced the harmful Th1-type immune responses induced by TNBS administration, thereby relieving the colitis symptoms (Xu et al., 2018; 2019). The bone marrow-dendritic cells exposed to the excretory/secretory products of *Trichinella spiralis* muscle larvae when injected intravenously prior to the TNBS challenge ameliorated the severity of TNBS-induced colitis by skewing the Th1 immune responses toward the Th2-type and regulatory T cell responses (Jin et al., 2019). The administration of a 49-kDa recombinant newborn-larvae-specific serine protease from *Trichinella spiralis* (rTsSp) prior to the establishment of DSS-induced colitis significantly reduced the severity of intestinal inflammation and the disease index. It also suppressed the macrophage infiltration along with the reduction in TNF-α and increased IL-10 levels (Long et al., 2021). Therefore, there is an emergent need to identify the immunomodulatory proteins not only from the adult worms but also from various larval stages of helminth parasites because the virulent larval forms are much better equipped to evade the host's immune responses in order to successfully establish themselves within their hosts. It would not be wrong to assume that using these larval-specific proteins would help in the better management of IBDs.

The excreted/secreted (E/S) products of *Trichuris suis*, the porcine whipworm, were also found to reduce the barrier function and the expression of the tight junction proteins like epithelial membrane protein-1 and claudin-4 in a glycan-dependent manner in the murine colorectal epithelial cells (CMT93/69). Such an effect would result in an increased passage of soluble compounds to the basolateral side affecting the dendritic cell function and modulating the immune responses (Hiemstra et al., 2014). Also, the E/S products of *T. suis* suppressed the LPS-induced pro-inflammatory cytokine production while inducing the Th2 responses, suggesting that *T. suis* E/S glycans could reduce the symptoms of autoimmune and allergic diseases without the need for the administration of live worms in the diseased individuals (Hiemstra et al., 2014).

A number of clinical trials have been undertaken to establish the safe dosage and the benefits of *Trichuris suis* ova (TSO) therapy in the treatment and prevention of Crohn's disease (Summers et al., 2003, 2005). The patients with active Crohn's disease, when given 2,500 live *T. suis* ova at an interval of three weeks for a period of 24 weeks, responded to the treatment as indicated by a decrease in the Crohn's disease activity index (CDAI), and there was no adverse effect on the health of the patients receiving ova therapy. These findings support the premise that natural exposure to helminths such as *T. suis* provides protection from immunological diseases and offers a safe and efficacious alternative for Crohn's disease treatment (Summers et al., 2005). In subsequent studies a single dose of 7,500 ova of *Trichuris suis* was well tolerated and did not result in a short-term or long-term treatment-related side effects (Sandborn et al., 2013). A clinical study carried out to establish the efficacy of TSO therapy in patients with Crohn's disease revealed that administration of 250–7,500 TSO fortnightly over a period of 12 weeks was safe but did not show any clinically relevant effects over the placebo group, with respect to the clinical remission or response in mild to moderately active cases of ileocolonic Crohn's disease (Schölmerich et al., 2017). Later, it was shown that the spontaneous inflammatory changes in the small intestine of NOD2-/- mice could be significantly ameliorated following the infection with *Trichuris muris*. Therefore, it was speculated that helminth therapy could be beneficial in patients carrying defective NOD2 variants. However, the clinical TSO trial with six functional NOD2 genetic variants showed no significant correlation between the NOD2 variants or their haplotypes with clinical outcome after TSO treatment (Jaeger et al., 2018). An inverse association between Crohn's disease and prior hookworm infection has also been observed in patients with Crohn's disease, a reflection of the application of hygiene hypothesis (Kabeerdoss et al., 2011).

The identification of immunomodulatory effects of extracellular vesicles (EVs) released in the excretory/secretory products of a number of helminth parasites has given investigators an opportunity to explore these EVs for various purposes. Immunization of mice with EVs obtained from *Fasciola hepatica* have been shown to ameliorate the pathological symptoms in DSS-induced colitis by reducing the levels of pro-inflammatory cytokines and interfering with both mitogen activated protein kinase and NF-kB pathways. Interestingly, these protective effects were not mediated by the T cells (Roig et al., 2018).

Pre-infection with *Taenia crassiceps*, a cestode parasite of rodents and canines, significantly reduced the pathological manifestations of DSS-induced colitis in mice by decreasing the systemic levels of pro-inflammatory cytokines while increasing the levels of anti-inflammatory cytokines. Further, adoptive transfer of alternatively activated macrophages from *T. crassiceps*–infected mice into the mice with DSS-induced colitis suppressed the severity of inflammation in the colon (Ledesma-Soto et al., 2015).

Intraperitoneal or intravenous injection of alternatively activated macrophages from the mice infected with *Hymenolepis diminuta*, a rat tapeworm, significantly reduced the severity of DNBS-induced colitis in mice. Studies on the biopsies of the patients with active Crohn's disease revealed a lesser number of alternatively activated macrophages as compared to the patients with latent Crohn's disease. This indicates the essential role of alternatively activated macrophages in mitigating the pathological manifestations in Crohn's disease, which might open up the possibility for adoptive transfer of alternatively activated macrophages in order to subside the inflammatory responses in IBD patients (Hunter et al., 2010).

26.5 Helminths and Ulcerative Colitis

In a DSS-induced colitis model the exposure of mice to *Schistosoma japonicum* attenuated the inflammatory responses through endoplasmic reticulum stress beside the Th1/Th2/Th17 and NF-κB pathways and reducing apoptosis. Thus, endoplasmic reticulum stress could serve as an important aspect in unraveling the relationship between the helminth infection and IBD that could be utilized for the development of novel therapeutic interventions (Liu et al., 2017). Pre-infection with the intestinal trematode *Metagonimus miyatai* ameliorated DSS-induced colitis in mice as reflected by the reduction in the clinical severity of DSS-induced colitis and a decline in the expression of TNF-α, IL-1 band cyclooxygenase-2 was observed as compared to the DSS-treated mice that did not receive helminth infection (Lee et al., 2020).

Intraperitoneal injection of *Schistosoma japonicum*–derived cysteine protease inhibitor (SjCystatin) significantly inhibited the inflammatory responses in DSS-induced acute ulcerative colitis in mice (Chu et al., 2018). A 16-kDa recombinant protein of *Schistosoma japonicum* (rSj16) also attenuated the production of pro-inflammatory cytokines along with an increase in the production of immunoregulatory cytokines and an increase in the population of the Treg cells in the experimental mice with DSS-induced colitis. Moreover, treatment of such mice with rSj16 also downregulated the genes of the peroxisome proliferator activated receptor-α (PPAR-α) signaling pathway, which might be playing an important role in the induction of experimental colitis, and therefore, it could be explored for further therapeutic studies on IBD (Wang et al., 2017a).

A study carried out to gain insight about time-related interactions between the intestinal nematode *Trichinella spiralis* infection and experimentally induced colitis in mice revealed that *T. spiralis* infection ameliorated the severity of inflammatory responses generated due to acetic-acid-induced colitis. The remission of the symptoms was even more pronounced when the *T. spiralis* infection preceded the induction of colitis. *T. spiralis* infection–induced recruitment of Foxp3-expressing Treg cells to the sites of inflammation that could limit the inflammatory responses in colon through the inflammatory-regulatory axis (Ashour et al., 2014). The amelioration of intestinal inflammation in mice with TNBS-induced colitis was observed when the animals were infected with *T. spiralis* primarily through the induction of protective Th2 responses and the suppression of Th1 responses (Xu et al., 2019).

The prophylactic and therapeutic effects of *Trichinella spiralis* serine protease inhibitors, *T. spiralis* Kazal-type serine protease inhibitors (TsKaSPI) and *Trichinella spiralis* adult serine protease-like protein (TsAdSPI) on the CD mouse models revealed an intervention effect, possibly because of the induction of Th-2 type of immune response and suppression of the TNBS-induced Th1-type immune response (Xu et al., 2018). In the follow-up study, the percentage of classically activated macrophages decreased, while the proportion of alternatively activated macrophages was significantly increased in the mouse model of CD treated with *Trichinella spiralis* serine protease inhibitors. Also, the key factors of IL-6/JAK2/STAT3 signaling pathway were found to be significantly elevated, while the expression of IL-33 and ST2 were found to be significantly reduced (Xu et al., 2020). Moreover, the administration of recombinant *Trichinella spiralis* adult serine protease-like protein (rTs-ADSp-7) significantly reduced the clinical disease in mice with TNBS-induced colitis. There was a significant decline in the production of Th1- and Th17-related cytokines, while there was an increase in the expression of Th2- and Treg-related cytokines in the rTs-ADSp-7-treated animals as compared to controls. There was also an increase in the population of Th2 and Treg cells in mice treated with rTs-ADSp-7 (Pang et al., 2020).

The therapeutic potential of *Brugia malayi* derived recombinant cystatin (rBmCys) in attenuating clinical symptoms of experimental colitis has also been established following treatment of DSS-induced colitis where the intraperitoneal injection of rBmCys reduced the severity of colitis in mice along with down regulation of IFN-γ, TNF-α, IL-5, IL-6 and IL-17 while the expression of IL-10 was found to be significantly upregulated, thus suggesting anti-inflammatory potential of rBmCys that should be explored as a promising therapeutic agent in ulcerative colitis (Khatri et al., 2015). Further studies have revealed that the rBmCys induced suppression of inflammation in the colon of mice with DSS-induced colitis is mediated through the Treg-cells as well as alternatively activated macrophages along with suppression of pro-inflammatory responses (Bisht et al., 2019). Moreover, the combination of filarial *Brugia malayai* proteins, *Brugia malayi* abundant larval transcript 2 (BmALT2) and BmCys, resulted in improvement in the clinico-pathological status of chronic colitis through suppression of pro-inflammatory immune response preferably in NF-κB dependent manner (Togre et al., 2018b). The recombinant antigens of another filarial nematode, *Wuchereria bancrofti,* have also been used to investigate their utility in the amelioration of ulcerative colitis. The prophylactic and therapeutic potential of the recombinant protein of L2 stage of *Wuchereria bancrofti* (rWbL2) has also been shown in DSS induced ulcerative colitis in a mouse model. The underlying immunological mechanism may involve the upregulation of Th2 immune responses along with downregulation of Th1 cytokines (Togre et al., 2018a,c).

The colonization of intestine with human whipworm, *Trichuris trichiura,* has been shown to reduce symptomatic colitis by inducing the goblet cell hyperplasia and stimulating mucus production through the involvement of Th-2 cytokines and also IL-22 which is involved in the mucosal healing (Broadhurst et al., 2010). The efficacy and safety of the TSO therapy for ulcerative colitis was investigated in a randomized, double-blind, placebo-controlled study that revealed that an oral dose of 2500 TSO at two-week intervals for 12 weeks was well tolerated and led to the improvement of the disease activity index, an indicator of the promising application of ova therapy in the treatment of patients with UC (Summers et al., 2005). In view of the attention garnered by TSO therapy in the treatment and prevention of IBD, a meta-analysis was performed to assess the effectiveness of TSO therapy in IBD. Although, the meta analysis of the clinical trials revealed no statistical benefit of TSO therapy for IBD patients, more studies with larger sample size are required to ascertain the actual impact of TSO therapy in IDB (Huang et al., 2018).

The acute infection of mice with *Strongyloides venezuelensis* has been shown to cause clinical improvement in DSS-induced colitis through decreased production of pro-inflammatory cytokines and induction of mucosal repair mechanisms (Rodrigues et al., 2018). The changes in the protein expression in the lamina propria and the intestinal epithelial cells of mice with DSS-induced colitis following treatment with *Ancylostoma caninum* excretory/secretory products revealed significant alteration in the expression of proteins involved in various physiological processes like antioxidant activity, cell death, and immunological responses. The expression of different mucins like MUC13 which is involved in the gastrointestinal homeostasis, was found to be upregulated in the lamina propria, while MUC2, a major component of mucus, was upregulated in the intestinal epithelial cells of the mice with DSS-induced colitis and treated with *A. caninum* proteins (Sotillo et al., 2017). A better understanding of the proteins involved in disease manifestation and targeting them using helminth-derived products can go a long way in developing novel treatment regimes for auto-immune disorders.

The infection of humans with a number of helminth parasites has been explored as biological and curative interventions for the control immune disorders. Although the complete mechanism by which the protection is generated in such diseases is still not known, the identification and characterization of the biomolecules of parasite origin which could be exploited for their therapeutic potential against the autoimmune disorders, is essentially required for the development of novel therapies to manage the IBD prevalence. Recent advancement in the next-generation sequencing techniques as well as bioinformatics tools, the transcriptomic analysis of various helminth parasites have been performed. The available data provides a deeper insight into the biomolecules of the parasite origin that could possibly be involved in the parasite-host interactions, and immunomodulatory or immuno-regulatory functions (Cantacessi et al., 2011).

26.6 Helminths and Rheumatoid Arthritis

Rheumatoid arthritis (RA) is a chronic inflammatory disorder with a predilection for specific diarthrodial (freely movable) joints with unknown etiology (Plenge, 2013). RA is mainly known to affect the joints in the body but in some patients damage to a variety of extra-articular organs, including eyes, heart, lungs, and skin, as well as the renal and nervous systems, have also been reported (Cojocaru et al., 2010). It is an age-related disorder, with females being at greater risk for developing the disease as compared to males (Intriago et al., 2019).

Although there is no permanent cure for RA, anti-rheumatic medications and surgery can reduce the symptoms and provide temporary relief. However, there are distinct side effects and risks associated with each treatment option. Therefore, alternative measures are currently being explored for the prevention and better management of RA. Helminths with their exemplary capability to modulate the host's immune system and successful establishment inside their host are being considered as available option for the treatment of various autoimmune diseases, including RA.

Intra-articular injection of the N-terminal modified tetradecapeptide derived from *Trichinella spiralis* paramyosin (Ts-pmy) has been shown to significantly ameliorate swelling and joint damage in knees of the antigen-induced arthritis (AIA) in mice. The Ts-pmy-derived peptide modified with a membrane-targeting signal could improve the therapeutic effects of the Ts-pmy peptide, which has been shown to bind to human complement component C9 to inhibit its polymerization and, therefore, could be an important candidate for therapeutic interventions for membrane attack complex related disorders including, RA (Chen et al., 2020).

The role of PD-1 in CD4+ T cells in providing protection against the collagen-induced arthritis (CIA) in *Trichinella spiralis* infection was investigated in mice model using anti PD-1 antibody and PD-1 knockout mice. The animals infected with *T. spiralis* showed higher expression of PD-1 in the CD4+ splenocytes, which was accompanied with lower arthritis incidence and significantly attenuated pathology when compared with the uninfected group. The anti-PD-1 antibody reversed the therapeutic effects of *T. spiralis* infection on CIA which has been supported by similar results obtained when *T. spiralis* infection was established in the PD-1 knockout mice, thus establishing the critical role of PD-1 in helminth infection-attenuated CIA in a mouse model and providing new insights into the mechanisms involved in helminth-induced immunomodulatory responses in the host (Cheng et al., 2018).

Schistosoma mansoni infection reduced the severity of collagen induced arthritis (CIA) in mice model. *S. mansoni* infection suppressed the production of IL-17A and TNF-α. The study revealed that signaling via STAT6 (presumably IL-4 and/or IL-13) and IL-10 is required for the suppression of CIA by *S. mansoni* infection, but through separate mechanisms. STAT6 was found to be essential for *S. mansoni*–induced reduction of IL-17A on the other hand arthritis severity was reduced through IL-10 production (Osada et al., 2019).

Prophylactic injection of recombinant protein of *Schistosoma japonicum* cystatin (rSjCystatin) attenuated paw clinical scores, incidence, and histopathology scores of joints in CIA mice. The anti-arthritic effects were closely associated with an increased IL-4, IL-10, and collagen-specific IgG1 levels, along with reduction in the IFN-γ, collagen-specific IgG2a, and pro-inflammatory cytokines IL-6, IL-17, and TNF-α and RANKL. This clearly suggest that rSjCystatin may prevent cartilage damage and inflammation of joints in CIA mice through the upregulation of Tregs and Th2 responses and downregulation of Th1 and Th17 responses (Liu et al., 2017).

The *S. japonicum* infection in mice prior to administration of type II collagen for the induction of arthritis significantly reduced the severity of CIA. The *S. japonicum* infection evoked a Th2 response and suppressed the Th1 response and pro-inflammatory mediators. The infection with both the sexes of the worms upregulated the Treg response and downregulated the Th17 response, resulting in an amelioration of arthritis (Song et al., 2011).

The administration of a recombinant 16-kDa protein of *Schistosoma japonicum* (rSj16), significantly reduced the paw swelling in a dose-dependent manner along with the suppression in the production of TNF-α, IL-1β, and NO, while the level of IL-10 in the serum was found to be significantly elevated in mice model of adjuvant-induced arthritis. Further, the bone-marrow-derived dendritic cells primed with rSj16 exhibited increased production of anti-inflammatory IL-10 and IL-4 cytokines and decreased production of IL-12p70 and IFN-γ. The rSj16 alleviated CFA-induced arthritis possibly through the interruption of maturation and function of DCs (Sun et al., 2010).

In another study the protective effect of *Schistosoma japonicum* infection against collagen-induced arthritis was shown to be stage-dependent. The pre-established egg-producing stage at seven weeks post-infection (PI) provided protection against CIA but not the pre-established early phase (two weeks PI) of the infection. In mice with acute *S. japonicum* infection, the level of IL-4 was significantly higher than the uninfected controls and remained high during the entire course of arthritis development. The protection against CIA correlated with reduced levels of anti-collagen II IgG, particularly the subclass of IgG2a (He et al., 2010).

The *Schistosoma japonicum* HSP60-derived peptide, SJMHE1, was shown to protect against CIA by suppressing the clinical signs and blocking the progression of joint erosion in mice. The therapeutic effect was accompanied with downregulation of IFN-γ, TNF-α, IL-6, IL-17, and IL-22, which are the key cytokines involved in the pathogenesis of CIA along with an up-egulation of IL-10, an inhibitory cytokine. There was also an increase in the expression of TGF-β1 and the number of Foxp3$^+$ regulatory T cells in the SJMHE1 treated mice compared to the controls (Wang et al., 2017b), thus providing a safe alternative to the worm therapy, which the patients are sometimes reluctant to undertake given the risk associated with the active helminth infections. This highlights the utility of helminth-derived product(s) as an anti-inflammatory and therapeutic molecule(s), which could be further explored for the prevention and treatment of autoimmune diseases like RA.

Infection of rats with *Toxocara canis*, the roundworm of dog, significantly modulated the immune response in Freund's complete adjuvant induced arthritis by alleviating inflammatory cytokines and increasing the TGF-β level. The severity of swelling was markedly reduced along with faster recuperation of the diseased individuals (Shayesteh et al., 2020).

The infection of the rat hookworm, *Nippostrongylus brasiliensis*, has been reported to induce Th2 response and accumulation of eosinophil in the joints associated with robust inhibition of arthritis and protection from bone loss in two independent models of arthritis. Mechanistically, IL-4/IL-13-STAT6 signaling pathway polarizes the macrophages into the anti-inflammatory types in the joints. In addition, eosinophils are activated and further contribute to the resolution of disease. The identification of Th-2 cells and eosinophils in the synovium of RA patients reflects the presence of cellular machinery to counteract arthritis in human cases that could be activated to treat RA. Hence, activation of Th-2 responses and eosinophils in both early and established disease may emerge as a new strategy to treat arthritis (Chen et al., 2016).

The therapeutic efficacy of *Brugia malayi* recombinant cystatin (rBmCys) in methylated BSA (mBSA) induced arthritis was investigated using animal model. The administration of rBmCys was shown to suppress the severity of mBSA-arthritis in rats by reducing the swelling of paws and significantly lowered arthritic index. The histopathological examination revealed a decreased synovitis score along with lesser bone erosion and fibrosis in paws in the rBmCys treated rats. Further, the anti-arthritic effect of cystatin was associated with a significant reduction in the production of pro-inflammatory cytokines, IFN-γ and TNF-α, while the levels of anti-inflammatory cytokines, IL-4 and IL-10, was significantly elevated, thereby indicating that rBmCys treatment alleviates mBSA-induced arthritis (Yadav et al., 2016).

ES-62 is a phosphorylcholine-containing glycoprotein secreted by filarial nematode *Acanthocheilonema viteae*, which has been reported to possess immunomodulatory properties. When ES-62 was administered during collagen priming, it significantly reduced initiation of inflammatory processes in mice model of CIA. The ES-62 was also found to suppress the severity and progression of CIA even when the administration of the same was delayed until the onset of disease (McInnes et al., 2003). The ES-62 treatment was found to restore the number of IL-10-producing B cells, which decreased in the spleens of mice with CIA. The IL-10-producing regulatory B cells have been proposed to suppress pathogenic Th1/Th17 responses in CIA. The exposure of mice to ES-62 also decreased the infiltration of effector B cells, mainly the plasma cells in the joints where these cells showed significantly lower levels of Toll-like receptor-4 and the activation markers CD80 and CD86 that would suppress B cell activation and plasma cell generation, in turn reducing the production of pathogenic collagen-specific IgG2a antibodies (Rodgers et al., 2014).

The ES-62 treatment in CIA mice inhibited the priming and polarization of IL-17 responses by targeting a complex IL-17-producing network, involving signaling between dendritic cells and γ/δ or $CD4^+$ T cells. In addition, ES-62 directly targeted Th17 cells by downregulating the expression of myeloid differentiation factor 88 which suppresses the response mediated by IL-1 and TLR ligands. Finally, there was a strong suppression of IL-17 production by the osteoclasts within the bone areas. Taken together, ES-62 targeted the initiation and effector phases of the IL-17 inflammatory network (Pineda et al., 2012). The IL-22, a Th17-associated cytokine has been reported to play a critical role, being pathogenic during the early phase and anti-inflammatory during the established phase of CIA. Therefore, IL-22-mediated protection against bone damage could be exploited for therapeutic intervention in the advanced stage of RA (Pineda et al., 2014). The ES-62 has also been shown to selectively stimulate γδ T cells, which express TLR4 to induce IL-22 production but not IL-17. Further, an active phosphorylcholine (PC) moiety of ES-62, conjugated to BSA, mimics the ability of ES-62 to upregulate IL-22 production to provide protection against CIA, thus establishing the utility of small drug-like PC-based compounds with therapeutic properties as potential candidates against RA (Harnett et al., 2014). Further, a number of synthetic drug-like small molecule analogue (SMA) of ES-62 like SMA-11a and SMA-12b have been shown to prevent development of pathology in CIA, a model of the musculoskeletal disease. The ES-62 and its analogues could modify bone remodeling by altering the bone marrow progenitors and thus impacting the osteoclastogenesis. Moreover, this effect seems to be achieved by induction of antioxidant response genes, which resulted in the reduction of ROS production (Rzepecka et al., 2015; Doonan et al., 2018). The increased ROS production has been associated with osteoclastogenesis during RA, and it also serves as important intracellular signaling molecule that amplify the synovial inflammatory–proliferative responses (Hitchon and El-Gabalawy, 2004; Doonan et al., 2018); therefore, the antioxidant-mediated responses of ES-62 and its analogues could also serve as an effective strategy to combat the increased ROS level, which is detrimental in RA.

Following oral or subcutaneous administration of tuftsin-phosphorylcholine (TPC) prior to induction of CIA in mice, a significant reduction in the arthritis score was observed as compared to the control group. The joints architecture was found to be maintained in the TPC-treated mice. Further, TPC treatment decreased the levels of circulating collagen type II antibodies and the pro-inflammatory cytokines TNF-α, IL-17, and IL-1β, along with an increase in the production of anti-inflammatory cytokine IL-10. TPC was also found to significantly expand the Treg cells and Breg cells and thus could be used as a natural therapy during the early phase of RA (Bashi et al., 2016). In the follow-up study by the same group, the interrelationship between TPC immunomodulatory activity and the gut microbiota in CIA mice was also analyzed. The prophylactic treatment with TPC maintained the physiological arthritis score and a steady gut microbial environment, which was comparable to the healthy controls (Ben-Amram et al., 2017). The ES-62 has also been shown to modulate the gut microbiome and could be a better substitute of the live worm therapy. The subcutaneous administration of ES-62 not only provided protection against CIA in a mouse model of RA but also lead to the normalization of gut microbiota and prevention of loss of intestinal barrier integrity (Doonan et al., 2019). The immunomodulatory effect of helminth-derived products and their influence in the maintenance of healthy gut microbiota could be further explored for the prevention and treatment of RA.

The liver flukes *Fasciola hepatica* and *Fasciola gigantica* are common parasites of both cattle and humans. The antigen of both species has been shown to possess immunomodulatory properties. The immunization of mice with dendritic cells pre-exposed to type II collagen pulsed with somatic extract of the temperate liver fluke, *F. hepatica*, and synthetic oligonucleotides containing unmethylated CpG sequences (CpG-ODN), a TLR9 ligand led to a strong immunostimulatory effect by modulating the inflammatory responses in CIA. The intraperitoneal injection of the treated dendritic cells reduced both the disease severity and incidence of symptoms associated with CIA through the production of anti-inflammatory cytokine, IL-10, and increased Treg population (Carranza et al., 2012).

The therapeutic efficacy of somatic antigens of *Fasciola gigantica* and *Gigantocotyle explanatum*, the two common digenetic trematodes of Indian water buffalo was investigated in rat model of CIA. The treatment of rats with the somatic antigens ameliorated the disease through the down regulation of TNF-α and IFN-γ and upregulation of IL-4 and IL-10 cytokines, and the inhibition of matrix metalloproteinases and nitric oxide production in knee joints and an improved patellar morphology with interspecific differences were observed (Khan et al., 2015). In another study, a synthetic peptide corresponding to first 34 C-terminal amino acids of *Fasciola hepatica* helminth defense molecule-1 (C-FhHDM-1) was shown to protect against the cartilage loss and destruction of joint's architecture in CIA mice model. Further, C-FhHDM-1 significantly reduced the CIA-induced expression of TNF, IL-17, and IFN-γ in joints, but their levels were not altered in the serum of the treated animals (Khan et al., 2020). Such localized effects are now being considered as vital in the treatment of autoimmune diseases rather than the suppression of systemic immune responses caused by the immunosuppressant drugs, which render the individuals susceptible to various infections and increasing the risk of cardiovascular diseases and malignancies (Cangemi et al., 2019; Russel, 2017).

One important feature that is common in a wide variety of the autoimmune disorders is the polarization of the immune response toward the pro-inflammatory Th1/Th17 type. The helminths and their products have a very uncanny ability to switch these potentially harmful responses in favor of the Th2/Treg type (Figure 26.2). It can be concluded from the foregoing information that the helminth parasites could

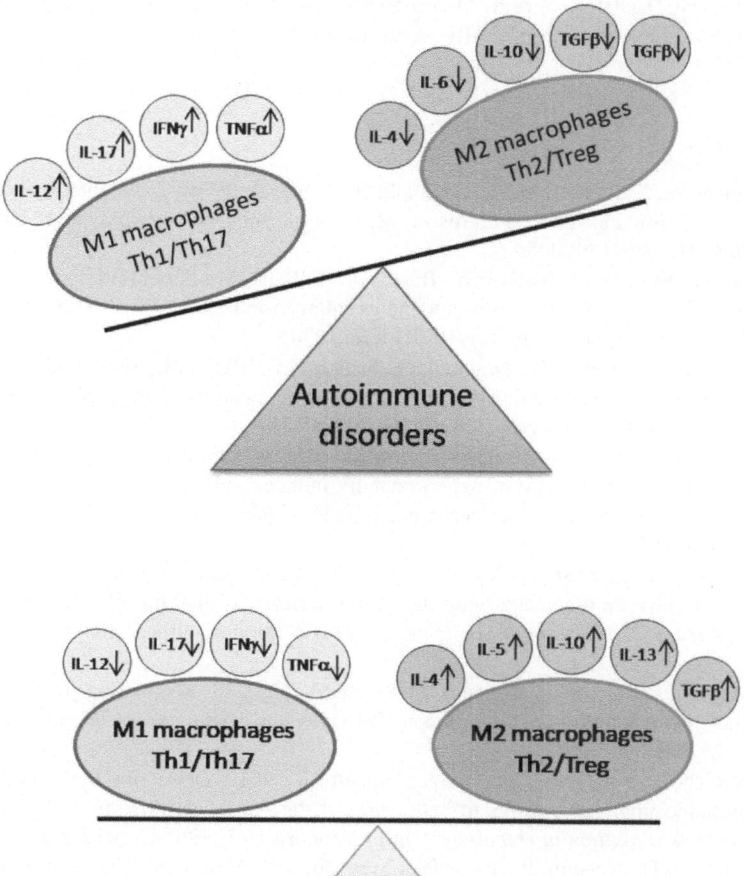

FIGURE 26.2 The autoimmune disorders are typically characterized by the immune responses skewed in favor of Th1/Th17 type with an increased production of the pro-inflammatory cytokines. The parasitic helminths and helminth-derived products, such as eggs/ova, excretory/secretory products, purified antigens, and extracellular vesicles, shift the balance toward the anti-inflammatory Th2 type.

serve as a vast reservoir of biomolecules with great immunomodulatory properties for the development of novel drugs for both prophylaxis and the treatment of autoimmune diseases, which are responsible for debilitating a huge human population across the globe. Despite the promising results obtained in various studies regarding the efficacy and safety of helminth-based therapy, there are few important questions—like the identification of the best source of immunomodulatory molecules, the optimization of dosage, frequency of treatment, and mode of administration of the prophylactic/therapeutic agents—that need to be answered before we can utilize this pharmacopeia of anti-inflammatory molecules for the eradication of autoimmune disorders.

26.7 Acknowledgments

The authors would like to thank the chairman of the Department of Zoology, Aligarh Muslim University, Aligarh, India, for providing necessary laboratory facilities established under the BBSRC-UK-CIDLID project and the DBT-BUILDER program. The financial assistance provided by CSIR, New Delhi, to AR under the CSIR-RA scheme is also gratefully acknowledged.

REFERENCES

Ashour, D.S., Othman, A.A., Shareef, M.M., Gaballah, H.H., Mayah, W.W., 2014. Interactions between trichinella spiralis infection and induced colitis in mice. *Journal of Helminthology* 88, 210–218. https://doi.org/10.1017/S0022149X13000059

Bashi, T., Shovman, O., Fridkin, M., Volkov, A., Barshack, I., Blank, M., Shoenfeld, Y., 2016. Novel therapeutic compound tuftsin-phosphorylcholine attenuates collagen-induced arthritis. *Clinical and Experimental Immunology* 184, 19–28. https://doi.org/10.1111/cei.12745

Ben-Amram, H., Bashi, T., Werbner, N., Neuman, H., Fridkin, M., Blank, M., Shoenfeld, Y., Koren, O., 2017. Tuftsin-phosphorylcholine maintains normal gut microbiota in collagen induced arthritic mice. *Frontiers in Microbiology* 8. https://doi.org/10.3389/fmicb.2017.01222

Bisht, N., Khatri, V., Chauhan, N., Kalyanasundaram, R., 2019. Cystatin from filarial parasites suppress the clinical symptoms and pathology of experimentally induced colitis in mice by inducing T-Regulatory cells, B1-cells, and alternatively activated macrophages. *Biomedicines* 7, 85. https://doi.org/10.3390/biomedicines7040085

Broadhurst, M.J., Leung, J.M., Kashyap, V., McCune, J.M., Mahadevan, U., McKerrow, J.H., Loke, P., 2010. IL-22+ CD4+ T cells are associated with therapeutic Trichuris trichiura infection in an ulcerative colitis patient. *Science Translational Medicine* 2, 60ra88–60ra88. https://doi.org/10.1126/scitranslmed.3001500

Camacho, M., Araújo, A., Morrow, J., Buikstra, J., Reinhard, K., 2018. Recovering parasites from mummies and coprolites: An epidemiological approach. *Parasites and Vectors* 11, 1–17. https://doi.org/10.1186/s13071-018-2729-4

Cangemi, M., Montico, B., Faè, D.A., Steffan, A., Dolcetti, R., 2019. Dissecting the multiplicity of immune effects of immunosuppressive drugs to better predict the risk of de novo malignancies in solid organ transplant patients. *Frontiers in Oncology* 9. https://doi.org/10.3389/fonc.2019.00160

Cantacessi, C., Young, N.D., Nejsum, P., Jex, A.R., Campbell, B.E., Hall, R.S., Thamsborg, S.M., Scheerlinck, J.P., Gasser, R.B., 2011. The transcriptome of Trichuris suis—First molecular insights into a parasite with curative properties for key immune diseases of humans. *PLoS ONE* 6, e23590. https://doi.org/10.1371/journal.pone.0023590

Capron, M., Béghin, L., Leclercq, C., Labreuche, J., Dendooven, A., Standaert, A., Delbeke, M., Porcherie, A., Nachury, M., Boruchowicz, A., Dupas, J.-L., Fumery, M., Paupard, T., Catteau, S., Deplanque, D., Colombel, J.-F., Desreumaux, P., 2019. Safety of P28GST, a Protein derived from a schistosome helminth parasite, in patients with Crohn's disease: A Pilot Study (ACROHNEM). *Journal of Clinical Medicine* 9, 41. https://doi.org/10.3390/jcm9010041

Carranza, F., Falcón, C.R., Nuñez, N., Knubel, C., Correa, S.G., Bianco, I., Maccioni, M., Fretes, R., Triquell, M.F., Motrán, C.C., Cervi, L., 2012. Helminth antigens enable CpG-activated dendritic cells to inhibit the symptoms of collagen-induced arthritis through Foxp3+ regulatory T cells. *PLoS ONE* 7, e40356. https://doi.org/10.1371/journal.pone.0040356

Chen, Y., 2020. Therapeutic efficacy of a Trichinella spiralis paramyosin-derived peptide modified with a membrane-targeting signal in mice with antigen-induced arthritis. *Frontiers in Microbiology* 11, 608380. https://doi.org/10.3389/fmicb.2020.608380

Chen, Z., Andreev, D., Oeser, K., Krljanac, B., Hueber, A., Kleyer, A., Voehringer, D., Schett, G., Bozec, A., 2016. Th2 and eosinophil responses suppress inflammatory arthritis. *Nature Communications* 7, 1–12. https://doi.org/10.1038/ncomms11596

Cheng, Y., Zhu, Xing, Wang, X., Zhuang, Q., Huyan, X., Sun, X., Huang, J., Zhan, B., Zhu, Xinping, 2018. Trichinella spiralis infection mitigates collagen-induced arthritis via programmed death 1-mediated immunomodulation. *Frontiers in Immunology* 9, 26. https://doi.org/10.3389/fimmu.2018.01566

Chu, K.M., Watermeyer, G., Shelly, L., Janssen, J., May, T.D., Brink, K., Benefeld, G., Li, X., 2013. Childhood helminth exposure is protective against inflammatory bowel disease: A case control study in South Africa. *Inflammatory Bowel Diseases* 19, 614–620. https://doi.org/10.1097/MIB.0b013e31827f27f4

Chu, L., Li, H.H., Wang, S.S., Yuan, Y., Jiang, H., Xu, L.S., He, W.X., Wang, S.X., Zhao, M.Z., Bai, Y.S., Wei, M., Liu, T., Sheng, J., Chen, X.Z., Fang, Q., Yang, X. di, 2018. Effect of cystatin from Schistosoma Japonicum on DSS-induced ulcerative colitis in mice. *Chinese Journal of Schistosomiasis Control* 30. https://doi.org/10.16250/j.32.1374.2017193

Cojocaru, M., Cojocaru, I., Silosi, I., Maedica, C.V.-, 2010, undefined, 2010. Extra-articular manifestations in rheumatoid arthritis. *Maedica* 5, 286–291.

Dinse, G.E., Parks, C.G., Weinberg, C.R., Co, C.A., Wilkerson, J., Zeldin, D.C., Chan, E.K.L., Miller, F.W., 2020. Increasing prevalence of antinuclear antibodies in the United States. *Arthritis and Rheumatology* 72, 1026–1035. https://doi.org/10.1002/art.41214

Doonan, J., Lumb, F.E., Pineda, M.A., Tarafdar, A., Crowe, J., Khan, A.M., Suckling, C.J., Harnett, M.M., Harnett, W., 2018. Protection against arthritis by the parasitic worm product ES-62, and its drug-like small molecule analogues, is associated with inhibition of osteoclastogenesis. *Frontiers in Immunology* 9. https://doi.org/10.3389/fimmu.2018.01016

Doonan, J., Tarafdar, A., Pineda, M.A., Lumb, F.E., Crowe, J., Khan, A.M., Hoskisson, P.A., Harnett, M.M., Harnett, W., 2019. The parasitic worm product ES-62 normalises the gut microbiota bone marrow axis in inflammatory arthritis. *Nature Communications* 10. https://doi.org/10.1038/s41467-019-09361-0

Elliott, D.E., Li, J., Blum, A., Metwali, A., Qadir, K., Urban, J.F., Weinstock, J. v., 2003. Exposure to schistosome eggs protects mice from TNBS-induced colitis. *American Journal of Physiology—Gastrointestinal and Liver Physiology* 284. https://doi.org/10.1152/ajpgi.00049.2002

Elliott, D.E., Weinstock, J. v., 2012. Helminth-host immunological interactions: Prevention and control of immune-mediated diseases. *Annals of the New York Academy of Sciences* 1247, 83–96. https://doi.org/10.1111/j.1749-6632.2011.06292.x

Foligné, B., Plé, C., Titécat, M., Dendooven, A., Pagny, A., Daniel, C., Singer, E., Pottier, M., Bertin, B., Neut, C., Deplanque, D., Dubuquoy, L., Desreumaux, P., Capron, M., Standaert, A., 2019. Contribution of the gut microbiota in P28GST-mediated anti-inflammatory effects: Experimental and clinical insights. *Cells* 8, 577. https://doi.org/10.3390/cells8060577

Harnett, M., Harnett, W., Pineda, M., 2014. The parasitic worm product ES-62 up-regulates IL-22 production by γδ T cells in the murine model of collagen-induced arthritis. *Inflammation and Cell Signaling* 1. https://doi.org/10.14800/ics.308

He, Y.K., Li, Jia, Zhuang, W.J., Yin, L., Chen, C.X., Li, Jun, Chi, F.L., Bai, Y.S., Chen, X.P., 2010. The inhibitory effect against collagen-induced arthritis by Schistosoma japonicum infection is infection stage-dependent. *BMC Immunology* 11, 28. https://doi.org/10.1186/1471-2172-11-28

Hiemstra, I.H., Klaver, E.J., Vrijland, K., Kringel, H., Andreasen, A., Bouma, G., Kraal, G., van Die, I., den Haan, J.M.M., 2014. Excreted/secreted Trichuris suis products reduce barrier function and suppress inflammatory cytokine production of intestinal epithelial cells. *Molecular Immunology* 60, 1–7. https://doi.org/10.1016/j.molimm.2014.03.003

Hitchon, C.A., El-Gabalawy, H.S., 2004. Oxidation in rheumatoid arthritis. *Arthritis Research and Therapy* 6. https://doi.org/10.1186/ar1447

Hotez, P.J., Brindley, P.J., Bethony, J.M., King, C.H., Pearce, E.J., Jacobson, J., 2008. Helminth infections: The great neglected tropical diseases. *Journal of Clinical Investigation* 118, 1311–1321. https://doi.org/10.1172/JCI34261

Huang, X., Zeng, L.R., Chen, F.S., Zhu, J.P., Zhu, M.H., 2018. Trichuris suis ova therapy in inflammatory bowel disease: A meta-analysis. *Medicine (Baltimore)* 97. https://doi.org/10.1097/MD.0000000000012087

Hunter, M.M., Wang, A., Parhar, K.S., Johnston, M.J.G., van Rooijen, N., Beck, P.L., McKay, D.M., 2010. In vitro-derived alternatively activated macrophages reduce colonic inflammation in mice. *Gastroenterology* 138, 1395–1405. https://doi.org/10.1053/j.gastro.2009.12.041

Intriago, M., Maldonado, G., Cárdenas, J., Ríos, C., 2019. Clinical characteristics in patients with rheumatoid arthritis: Differences between genders. *Scientific World Journal*. https://doi.org/10.1155/2019/8103812

Jaeger, S.U., Schaeffeler, E., Winter, S., Tremmel, R., Schölmerich, J., Malek, N., Stange, E.F., Schwab, M., Wehkamp, J., 2018. Influence of NOD2 variants on Trichuris suis ova treatment outcome in Crohn's disease. *Frontiers in Pharmacology* 9, 764. https://doi.org/10.3389/fphar.2018.00764

Jin, X., Yang, Y., Bai, X., Shi, H., Zhang, W., Zhang, Z., Jia, W., Lin, J., Liu, M., Liu, X., 2019. Dendritic cells treated by Trichinella spiralis muscle larval excretory/secretory products alleviate TNBS-induced colitis in mice. *International Immunopharmacology* 70, 378–386. https://doi.org/10.1016/j.intimp.2019.02.028

Kabeerdoss, J., Pugazhendhi, S., Subramanian, V., Binder, H.J., Ramakrishna, B.S., 2011. Exposure to hookworms in patients with Crohn's disease: A case-control study. *Alimentary Pharmacology and Therapeutics* 34, 923–930. https://doi.org/10.1111/j.1365-2036.2011.04824.x

Khan, W.I., Blennerhasset, P.A., Varghese, A.K., Chowdhury, S.K., Omsted, P., Deng, Y., Collins, S.M., 2002. Intestinal nematode infection ameliorates experimental colitis in mice. *Infection and Immunity* 70, 5931–5937. https://doi.org/10.1128/IAI.70.11.5931-5937.2002

Khan, Y.A., Maurya, S.K., Kulkarni, C., Tiwari, M.C., Nagar, G.K., Chattopadhyay, N., 2020. Fasciola helminth defense molecule-1 protects against experimental arthritis by inhibiting osteoclast formation and function without modulating the systemic immune response. *FASEB Journal* 34, 1091–1106. https://doi.org/10.1096/fj.201901480RR

Khan, Y.A., Umar, S., Abidi, S.M.A., 2015. Somatic antigens of tropical liver flukes ameliorate collagen-induced arthritis in Wistar rats. *PLoS ONE* 10. https://doi.org/10.1371/journal.pone.0126429

Khatri, V., Amdare, N., Tarnekar, A., Goswami, K., Reddy, M.V.R., 2015. Brugia malayi cystatin therapeutically ameliorates dextran sulfate sodium-induced colitis in mice. *Journal of Digestive Diseases* 16, 585–594. https://doi.org/10.1111/1751-2980.12290

Ledesma-Soto, Y., Callejas, B.E., Terrazas, C.A., Reyes, J.L., Espinoza-Jiménez, A., González, M.I., León-Cabrera, S., Morales, R., Olguín, J.E., Saavedra, R., Oghumu, S., Satoskar, A.R., Terrazas, L.I., 2015. Extraintestinal helminth infection limits pathology and proinflammatory cytokine expression during DSS-induced ulcerative colitis: A role for alternatively activated macrophages and prostaglandins. *BioMed Research International*. https://doi.org/10.1155/2015/563425

Lee, M.R., Jeong, Y. il, Kim, H.J., Shin, H.E., Lee, S.E., Ju, J.W., Sohn, W.M., Cho, S.H., 2020. Metagonimus miyatai ameliorates dextran sodium sulfate-induced colitis in mice. *Parasitology International* 74. https://doi.org/10.1016/j.parint.2019.05.002

Liu, F., Cheng, W., Pappoe, F., Hu, X., Wen, H., Luo, Q., Wang, S., Deng, F., Xie, Y., Xu, Y., Shen, J., 2016. Schistosoma japonicum cystatin attenuates murine collagen-induced arthritis. *Parasitology Research* 115, 3795–3806. https://doi.org/10.1007/s00436-016-5140-0

Liu, Y., Ye, Q., Liu, Y.L., Kang, J., Chen, Y., Dong, W.G., 2017. Schistosoma japonicum attenuates dextran sodium sulfate-induced colitis in mice via reduction of endoplasmic reticulum stress. *World Journal of Gastroenterology* 23, 5700–5712. https://doi.org/10.3748/wjg.v23.i31.5700

Loddo, I., Romano, C., 2015. Inflammatory bowel disease: Genetics, epigenetics, and pathogenesis. *Frontiers in Immunology* 6. https://doi.org/10.3389/fimmu.2015.00551

Long, S.R., Liu, R.D., Kumar, D.V., Wang, Z.Q., Su, C.-W., 2021. Immune protection of a helminth protein in the DSS-induced colitis model in mice. *Frontiers in Immunology* 12, 1438. https://doi.org/10.3389/fimmu.2021.664998

M'Koma, A.E., 2013. Inflammatory bowel disease: An expanding global health problem. *Clinical Medicine Insights: Gastroenterology* 6, 33–47. https://doi.org/10.4137/CGast.S12731

McInnes, I.B., Leung, B.P., Harnett, M., Gracie, J.A., Liew, F.Y., Harnett, W., 2003. A novel therapeutic approach targeting articular inflammation using the filarial nematode-derived phosphorylcholine-containing glycoprotein ES-62. *The Journal of Immunology* 171, 2127–2133. https://doi.org/10.4049/jimmunol.171.4.2127

Mizoguchi, A., Takeuchi, T., Himuro, H., Okada, T., Mizoguchi, E., 2016. Genetically engineered mouse models for studying inflammatory bowel disease. *Journal of Pathology* 238, 205–219. https://doi.org/10.1002/path.4640

Mo, H.M., Liu, W.Q., Lei, J.H., Cheng, Y.L., Wang, C.Z., Li, Y.L., 2007. Schistosoma japonicum eggs modulate the activity of CD4+ CD25+ Tregs and prevent development of colitis in mice. *Experimental Parasitology* 116, 385–389. https://doi.org/10.1016/j.exppara.2007.02.009

Motomura, Y., Wang, H., Deng, Y., El-Sharkawy, R.T., Verdu, E.F., Khan, W.I., 2009. Helminth antigen-based strategy to ameliorate inflammation in an experimental model of colitis. *Clinical and Experimental Immunology* 155, 88–95. https://doi.org/10.1111/j.1365-2249.2008.03805.x

Osada, Y., Horie, Y., Nakae, S., Sudo, K., Kanazawa, T., 2019. STAT6 and IL-10 are required for the anti-arthritic effects of Schistosoma mansoni via different mechanisms. *Clinical and Experimental Immunology* 195, 109–120. https://doi.org/10.1111/cei.13214

Pang, J., Ding, J., Zhang, L., Zhang, Y., Yang, Y., Bai, X., Liu, X., Jin, X., Guo, H., Yang, Y., Liu, M., 2020. Effect of recombinant serine protease from adult stage of Trichinella spiralis on TNBS-induced experimental colitis in mice. *International Immunopharmacology* 86. https://doi.org/10.1016/j.intimp.2020.106699

Parker, W., 2014. The "hygiene hypothesis" for allergic disease is a misnomer. *BMJ* (Online). https://doi.org/10.1136/bmj.g5267

Pêgo, B., Martinusso, C.A., Bernardazzi, C., Ribeiro, B.E., de Araujo Cunha, A.F., de Souza Mesquita, J., Nanini, H.F., Machado, M.P., Castelo-Branco, M.T.L., Cavalcanti, M.G., de Souza, H.S.P., 2019. Schistosoma mansoni coinfection attenuates murine toxoplasma gondii-induced Crohn's-like ileitis by preserving the epithelial barrier and downregulating the inflammatory response. *Frontiers in Immunology* 10, 442. https://doi.org/10.3389/fimmu.2019.00442

Pineda, M.A., McGrath, M.A., Smith, P.C., Al-Riyami, L., Rzepecka, J., Gracie, J.A., Harnett, W., Harnett, M.M., 2012. The parasitic helminth product ES-62 suppresses pathogenesis in collagen-induced arthritis by targeting the interleukin-17-producing cellular network at multiple sites. *Arthritis and Rheumatism* 64, 3168–3178. https://doi.org/10.1002/art.34581

Pineda, M.A., Rodgers, D.T., Al-Riyami, L., Harnett, W., Harnett, M.M., 2014. ES-62 protects against collagen-induced arthritis by resetting interleukin-22 toward resolution of inflammation in the joints. *Arthritis and Rheumatology* 66, 1492–1503. https://doi.org/10.1002/art.38392

Plenge, R.M., 2013. Rheumatoid arthritis, in: *Genomic and Personalized Medicine*. Elsevier Inc., pp. 839–852. https://doi.org/10.1016/B978-0-12-382227-7.00071-9

Raga, J.A., Fernández, M., Balbuena, J.A., Aznar, F.J., 2009. Parasites, in: *Encyclopedia of Marine Mammals*. Elsevier, pp. 821–830. https://doi.org/10.1016/B978-0-12-373553-9.00193-0

Rodgers, D.T., Pineda, M.A., Mcgrath, M.A., Al-Riyami, L., Harnett, W., Harnett, M.M., 2014. Protection against collagen-induced arthritis in mice afforded by the parasitic worm product, ES-62, is associated with restoration of the levels of interleukin-10-producing B cells and reduced plasma cell infiltration of the joints. *Immunology* 141, 457–466. https://doi.org/10.1111/imm.12208

Rodrigues, V.F., Bahia, M.P.S., Cândido, N.R., Moreira, J.M.P., Oliveira, V.G., Araújo, E.S., Rodrigues Oliveira, J.L., Rezende, M. de C., Correa, A., Negrão-Corrêa, D., 2018. Acute infection with Strongyloides venezuelensis increases intestine production IL-10, reduces Th1/Th2/Th17 induction in colon and attenuates dextran sulfate sodium-induced colitis in BALB/c mice. *Cytokine* 111, 72–83. https://doi.org/10.1016/j.cyto.2018.08.003

Roig, J., Saiz, M.L., Galiano, A., Trelis, M., Cantalapiedra, F., Monteagudo, C., Giner, E., Giner, R.M., Recio, M.C., Bernal, D., Sánchez-Madrid, F., Marcilla, A., 2018. Extracellular vesicles from the helminth Fasciola hepatica Prevent DSS-induced acute ulcerative colitis in a T-lymphocyte independent mode. *Frontiers in Microbiology* 9, 1036. https://doi.org/10.3389/fmicb.2018.01036

Russell, S.D., 2017. The effects of immunosuppressive and cytotoxic drugs on the heart, in: The *Heart in Rheumatic, Autoimmune and Inflammatory Diseases: Pathophysiology, Clinical Aspects and Therapeutic Approaches*. Elsevier Inc., pp. 645–657. https://doi.org/10.1016/B978-0-12-803267-1.00026-0

Ryan, S., Shiels, J., Taggart, C. C., Dalton, J. P., Weldon, S. (2020). Fasciola hepatica-derived molecules as regulators of the host immune response. *Frontiers in Immunology*, 11, 2182. https://doi.org/10.3389/fimmu.2020.02182

Rzepecka, J., Pineda, M.A., Al-Riyami, L., Rodgers, D.T., Huggan, J.K., Lumb, F.E., Khalaf, A.I., Meakin, P.J., Corbet, M., Ashford, M.L., Suckling, C.J., Harnett, M.M., Harnett, W., 2015. Prophylactic and therapeutic treatment with a synthetic analogue of a parasitic worm product prevents experimental arthritis and inhibits IL-1β production via NRF2-mediated counter-regulation of the inflammasome. *Journal of Autoimmunity* 60, 59–73. https://doi.org/10.1016/j.jaut.2015.04.005

Sandborn, W.J., Elliott, D.E., Weinstock, J., Summers, R.W., Landry-Wheeler, A., Silver, N., Harnett, M.D., Hanauer, S.B., 2013. Randomised clinical trial: The safety and tolerability of Trichuris suis ova in patients with Crohn's disease. *Alimentary Pharmacology and Therapeutics* 38, 255–263. https://doi.org/10.1111/apt.12366

Schölmerich, J., Fellermann, K., Seibold, F.W., Rogler, G., Langhorst, J., Howaldt, S., Novacek, G., Petersen, A.M., Bachmann, O., Matthes, H., Hesselbarth, N., Teich, N., Wehkamp, J., Klaus, J., Ott, C., Dilger, K., Greinwald, R., Muellerp, R., 2017. A randomised, double-blind, placebo-controlled trial of Trichuris suis ova in active Crohn's disease. *Journal of Crohn's and Colitis* 11, 390–399. https://doi.org/10.1093/ecco-jcc/jjw184

Shayesteh, Z., Hosseini, H., Nasiri, V., Haddadi, Z., Moradi, N., Beikzadeh, L., Sezavar, M., Heidari, A., Zibaei, M., 2020. Evaluating the preventive and curative effects of Toxocara canis larva in Freund's complete adjuvant-induced arthritis. *Parasite Immunology* 42, e12760. https://doi.org/10.1111/pim.12760

Smallwood, T.B., Giacomin, P.R., Loukas, A., Mulvenna, J.P., Clark, R.J., Miles, J.J., 2017. Helminth immunomodulation in autoimmune disease. *Frontiers in Immunology* 8, 1. https://doi.org/10.3389/fimmu.2017.00453

Sobotková, K., Parker, W., Levá, J., Růžková, J., Lukeš, J., Jirků Pomajbíková, K., 2019. Helminth therapy—From the parasite perspective. *Trends in Parasitology*. https://doi.org/10.1016/j.pt.2019.04.009

Song, X., Shen, J., Wen, H., Zhong, Z., Luo, Q., Chu, D., Qi, Y., Xu, Y., Wei, W., 2011. Impact of Schistosoma japonicum infection on collagen-induced arthritis in DBA/1 mice: A murine model of human rheumatoid arthritis. *PLoS ONE* 6, e23453. https://doi.org/10.1371/journal.pone.0023453

Sotillo, J., Ferreira, I., Potriquet, J., Laha, T., Navarro, S., Loukas, A., Mulvenna, J., 2017. Changes in protein expression after treatment with Ancylostoma caninum excretory/secretory products in a mouse model of colitis. *Scientific Reports* 7, 1–11. https://doi.org/10.1038/srep41883

Summers, R.W., Elliott, D.E., Qadir, K., Urban, J.F., Thompson, R., Weinstock, J. v., 2003. Trichuris suis seems to be safe and possibly effective in the treatment of inflammatory bowel disease. *American Journal of Gastroenterology* 98, 2034–2041. https://doi.org/10.1111/j.1572-0241.2003.07660.x

Summers, R.W., Elliott, D.E., Urban, J.F., Thompson, R.A., Weinstock, J. v., 2005. Trichuris suis therapy for active ulcerative colitis: A randomized controlled trial. *Gastroenterology* 128, 825–832. https://doi.org/10.1053/j.gastro.2005.01.005

Sun, X., Liu, Y.H., Lv, Z.Y., Yang, L.L., Hu, S.M., Zheng, H.Q., Hu, W., Cao, J.P., Fung, M.Q., Wu, Z.D., 2010. rSj16, a recombinant protein of Schistosoma japonicum-derived molecule, reduces severity of the complete Freund's adjuvant-induced adjuvant arthritis in rats' model. *Parasite Immunology* 32, 739–748. https://doi.org/10.1111/j.1365-3024.2010.01240.x

Togre, N., Bhoj, P., Amdare, N., Goswami, K., Tarnekar, A., Shende, M., 2018a. Immunomodulatory potential of recombinant filarial protein, rWbL2, and its therapeutic implication in experimental ulcerative colitis in mouse. *Immunopharmacology and Immunotoxicology* 40, 483–490. https://doi.org/10.1080/08923973.2018.1431925

Togre, N., Bhoj, P., Goswami, K., Tarnekar, A., Patil, M., Shende, M., 2018b. Human filarial proteins attenuate chronic colitis in an experimental mouse model. *Parasite Immunology* 40, e12511. https://doi.org/10.1111/pim.12511

Togre, N., Bhoj, P.S., Khatri, V.K., Tarnekar, A., Goswami, K., Shende, M.R., Reddy, M.V.R., 2018c. SXP–RAL family filarial protein, rWbL2, prevents development of DSS-induced acute ulcerative colitis. *Indian Journal of Clinical Biochemistry* 33, 282–289. https://doi.org/10.1007/s12291-017-0671-4

Turton, J. A., 1976. IgE, parasites, and allergy. *The Lancet*. Vol. 2, 7987: 686. https://doi.org/10.1016/S0140-6736(76)92492-2

Wang, L., Xie, H., Xu, L., Liao, Q., Wan, S., Yu, Z., Lin, D., Zhang, B., Lv, Z., Wu, Z., Sun, X., 2017a. rSj16 protects against DSS-induced colitis by inhibiting the PPAR-α signaling pathway. *Theranostics* 7, 3446–3460. https://doi.org/10.7150/thno.20359

Wang, X., Li, L., Wang, J., Dong, L., Shu, Y., Liang, Y., Shi, L., Xu, C., Zhou, Y., Wang, Y., Chen, D., Mao, C., 2017b. Inhibition of cytokine response to TLR stimulation and alleviation of collagen-induced arthritis in mice by Schistosoma japonicum peptide SJMHE1. *Journal of Cellular and Molecular Medicine* 21, 475–486. https://doi.org/10.1111/jcmm.12991

Weinstock, J. v., Elliott, D.E., 2009. Helminths and the IBD hygiene hypothesis. *Inflammatory Bowel Diseases* 15, 128–133. https://doi.org/10.1002/ibd.20633

Xia, C.M., Zhao, Y., Jiang, L., Jiang, J., Zhang, S.C., 2011. Schistosoma japonicum ova maintains epithelial barrier function during experimental colitis. *World Journal of Gastroenterology* 17, 4810–4816. https://doi.org/10.3748/wjg.v17.i43.4810

Xu, J., Liu, M., Yu, P., Wu, L., Lu, Y., 2019. Effect of recombinant Trichinella spiralis cysteine proteinase inhibitor on TNBS-induced experimental inflammatory bowel disease in mice. *International Immunopharmacology* 66, 28–40. https://doi.org/10.1016/j.intimp.2018.10.043

Xu, J., Wu, L., Yu, P., Liu, M., Lu, Y., 2018. Effect of two recombinant Trichinella spiralis serine protease inhibitors on TNBS-induced experimental colitis of mice. *Clinical and Experimental Immunology* 194, 400–413. https://doi.org/10.1111/cei.13199

Xu, J., Wu, L., Yu, P., Sun, Y., Lu, Y., 2020. Effect of T. spiralis Serine protease inhibitors on TNBS-induced experimental colitis mediated by Macrophages. *Scientific Reports* 10, 3147. https://doi.org/10.1038/s41598-020-60155-7

Yadav, R.S.P., Khatri, V., Amdare, N., Goswami, K., Shivkumar, V.B., Gangane, N., Reddy, M.V.R., 2016. Immuno-modulatory effect and therapeutic potential of Brugia malayi cystatin in experimentally induced arthritis. *Indian Journal of Clinical Biochemistry* 31, 203–208. https://doi.org/10.1007/s12291-015-0515-z

27

Physical Growth of Cephalo-Facial Region in Adolescents—a Study from Himachal Pradesh, India

Kewal Krishan, Bahadur Singh, Kawaljit Kaur, Raj Kumar, and Tanuj Kanchan

CONTENTS

27.1 Introduction

All human beings belong to a single species, *Homo sapiens*, although they differ in their physical structure, shape, size, and so on. This variation is more reflected while making a comparison of the different populations, countries, regions, and endogamous groups of the world. The head and face are the most variable part of the human body, and they display the most conspicuous variation of human species. They permit distinction between the major races of the world, various ethnic groups, sexes, and even the members of the same family. Consequently, there is a possibility to recognize various endogamous groups by a set of measurements and indices of the cephalo-facial region.

The human cephalo-facial region is unique also because of their pattern of growth. It is well established that alterations in head shape are the result of unusual brain growth. Moss and Young (1960) explained that it is the growth of the brain that determines the cranial shape and expansion. The brain attains 90% of its adult weight by six years of age and 95% by ten. Accordingly, the cephalo-facial dimensions change while the body as a whole has attained only 40% of its adult value at six years and 50% at ten (Scammon 1930). The end of the childhood development stage is marked by the cessation of brain growth. However, the majority of somatic growth takes place in the late adolescent growth stage (Bogin 1999).

The growth of the head and face is quite different from the rest of the body. It is well known that at a young age, the growth of the head is well advanced in its development as compared with height (Singal & Sidhu 1986).

The head develops earlier than any other part of the fetal skeleton. At birth, the brain averages about 25% of its adult weight; at six months, nearly 50%; at two years, about 75%; at five years, 90%; and at ten years, 95%. This is in contrast with the weight of the whole body, which is at birth is about 5% of young adult weight and at ten years about 50%. The brain, together with skull covering it, and the eyes and ear develop earlier than any other part of the body and thus have a characteristic postnatal curve that does not follow the height curve (Tanner 1978)

DOI: 10.1201/9781003220398-27

The development of the shape of the head and face depends upon a variety of interactive genetic and environmental forces. Intrinsic and extrinsic pressure and the neuromuscular function contribute to the overall shape. Both genetic and environmental components contribute to the cephalo-facial structure (Richardson 1980). The change in the shape and size of the human head and face is the most important aspect of recent human evolution. Modern human evolution attained their highly encephalized (large brain relative to body mass) condition due to a pattern of gradual increase in brain size from *Australopithecus africanus* through *Homo habilis* and *Homo erectus* to modern humans (Hoppa & Fitzgerald 1999). The enlargement of the brain occurs due to the development of a high degree of visual equity, and bipedalism or upright posture also caused changes in the head and face.

The knowledge of normal age changes in cephalo-facial dimensions is important for clinical research and ergonomic purposes. Cephalo-facial measurements also tell us about certain diseases and disorders as these diseases cause a deformity in head and facial morphology—for example, in Down's syndrome (Fink, Madaus, & Walker 1975; Shukla et al. 2014), cleft lip (Young, Wat, Diewert, Browder, & Hallgrímsson 2007), cleft palate (Kurisu, Niswander, Johnston, & Mazaheri 1974), thalassemia major (Cooley's anemia) (Başsimitçi, Yücel-Eroğlu, & Akalar 1996), and dentofacial abnormalities (Magalhaes, Stella, & Epker 1995). Cephalometry also has various applications in aeronautics and space research especially in designing various headgears and masks for civilians and for defense personals. Cephalo-facial anthropometry has been considered as an important and useful tool in personal identification (Mane, Kale, Bhai, & Hallikerimath 2010), and recently, its applications in the reconstruction of skull and face and making sketches of suspects are well documented. Therefore, the present study is an attempt to evaluate the pattern of age changes in cephalo-facial measurements of a North Indian population.

27.2 Materials and Methods

27.2.1 Subjects

A random sample of 152 school going adolescent *Koli* boys ranging in age from 12 to 18 years was taken for the present study. The data were collected from five schools in Sirmaur District of Himachal Pradesh State in North India. The *Kolis* of North India are an endogamous caste group (marrying within the same caste group). They are considered as one of the most backward, poor, and remote people of North India. *Kolis* are professional weavers, having no agricultural land, and some of them work as laborers (Krishan & Kumar 2007). This investigation is a part of a large study (Krishan & Kumar 2007) conducted on the Kolis of North India.

27.2.2 Methodology

Sixteen cephalo-facial anthropometric measurements were taken on each subject. The measurements include maximum head length, maximum head breadth, head vault, minimum frontal diameter, horizontal circumference of head, bizygomatic breadth, bigonial diameter, nasal height, nasal breadth, morphological facial length, morphological superior facial length, physiognomic facial length, physiognomic superior facial length, physiognomic ear length, and physiognomic ear breadth. All the subjects included in the study were living under more or less similar environmental, nutritional, and socio-economic conditions. The subjects were healthy and free from any symptomatic cephalo-facial or any other physical deformity.

As the assessment of correct age is an important factor in any physical growth study, the date of birth of each subject was noted from school admission registers and converted to decimal age with the help of decimal age calendar devised by Tanner, Whitehouse, and Takaishi (1966).

Cephalo-facial measurements were taken using standard anthropometric instruments in centimeters to the nearest millimeter using the procedures and techniques devised by Lohman, Roche, and Martorell (1988). The anatomical landmarks used in taking measurements are identified keeping the head in

Frankfort horizontal plane. All the measurements were taken in a well-lighted room. The cephalo-facial measurements are defined as follows:

Maximum head length. It measures straight distance between the glabella (the most prominent point on the frontal bone above the root of the nose, between the eyebrows) and the opisthocranion (the most posterior point on the posterior protuberance of occipital bone in the midsagittal plane).

Maximum head breadth. It is the maximum biparietal diameter and is the distance between the most lateral points on the parietal bones (euryon) on each side of the head.

Head vault. It is the horizontal arc of the head measured from tragion to tragion (the point in the notch just above the tragus of the ear) with the measuring tape passing over vertex.

Minimum frontal diameter. It is the least breadth of the forehead between the two frontotemporal points (the most medial point on the temporal crest just above and lateral to orbits identified by palpation) on the temporal ridges.

Horizontal circumference of head. It is measured from glabella to glabella with the measuring tape passing over the opisthocranion.

Bizygomatic breadth. It is the maximum distance between the most lateral points on the zygomatic arches (zygion to zygion).

Bigonial diameter. It is the maximum breadth of the lower jaw between two gonion points (it is the most posterior, inferior, and laterally situated point on the external angles of the mandible) on the angles of mandible.

Nasal height. It is the distance from the nasal root (nasion) to the nasal base (subnasale).

Nasal breadth. It is the distance between the lateral most aspect of one ala nasi to the lateral most aspect of other ala nasi.

Morphological facial length. It is the straight distance from the nasal root (nasion) to the lowest point on the lower border of mandible in the midsagittal plane (gnathion).

Morphological superior facial length. It is the straight distance from the nasion to the point on the gum between the upper central teeth (prosthion) in the midsagittal plane.

Physiognomic facial length. It is the straight distance between the landmarks trichion (it is the anterior limit of the hair on the midline of the forehead; this limit is considered to be the place where hair really begin—in other words, it is the point where the hairline meets the midpoint of the forehead) and gnathion.

Physiognomic superior facial length. It is the length of the upper face between nasion and to a point where the oral fissure (slit of the mouth) is cut by a medial line in between the two lips (stomion).

Physiognomic ear length. It is the maximum distance from the superior aspect (supra-aural) to the inferior aspect of the external ear (sub-aural).

Physiognomic ear breadth. It the maximum width of the external ear (pinna) taken from pre-aural and post-aural.

The data so obtained were then grouped into seven age groups having interval of one year. For example, in age groups, 12–13, boys from 11.5 to 12.499 years of age are included, and in age group 13–14, the boys from 12.5 to 13.499 years of age are included, and so on. Table 27.1 gives the age-wise distribution of the subject in different groups. The data were subjects to the elementary statistics like mean, standard deviation, standard error of mean, range, coefficient of variation, and so on using statistical package for social science (SPSS) for Windows Professional.

27.3 Results

The descriptive statistics of the cephalo-facial anthropometric variables and indices of the growing *Koli* boys of Himachal Pradesh are presented in this paper. To comprehensively elucidate the growth pattern within the cephalo-facial region, the age group of 12 to 18 years has been split to seven groups with one year difference. Each age group is having at least 30 or more subjects, which make the group reasonably enough to bring out reliable inferences. The descriptive statistics of the maximum head length and maximum head breadth is presented in Table 27.1 and 27.2, respectively. The mean head length of the first age group (12–13 years) and last age group (18–19 years) came out to be 15.453 cm and 16.584 cm, respectively. The rest of the age group showed gradual increment in the mean head length. The mean head breadth of the first (12–13 years) and the last age group (18–19 years) came out to be 13 cm and 13.13 cm, respectively. The highest mean head breadth has been shown by the 14–15 years age group. The rest of the groups showed variable mean head breadths.

Descriptive statistics of the maximum head vault, minimum frontal diameter, and horizontal circumference of the head are shown in the Tables 27.3, 27.4, and 27.5, respectively. The mean maximum head vault, minimum frontal diameter, and horizontal circumference of the first age group (12–13 years) came out to be 33.21 cm, 9.49 cm, and 51.03 cm, respectively. The mean maximum head vault, minimum frontal diameter and horizontal circumference of the last age group (18–19 years) came out to be 34.39 cm, 10.23 cm, and 53.98 cm, respectively. In all the other age groups, variability in mean values has been observed.

Tables 27.6 and 27.7 represent the descriptive statistics of the bizygomatic breadth and bigonial diameter. The mean bizygomatic breadth and bigonial breadth of the first age group (12–13 years) came out to be 9.28 cm and 7.87 cm, respectively. The mean bizygomatic breadth and bigonial breadth of the last age group (18–19 years) came out to be 10.52 cm and 8.86 cm, respectively. Other age groups in both bizygomatic and bigonial breadth showed increment in mean values from the first to the seventh age group.

TABLE 27.1

Descriptive Statistics of Maximum Head Length (cm) in *Koli* Boys of Himachal Pradesh from 12 to 18 Years of Age

							Observed range	
S. no.	**Age group**	**N**	**Mean**	**SD**	**SEM**	**CV**	**Min.**	**Max.**
1	12–13	36	15.453	1.1841	0.1973	7.6625	10.1	17.4
2	13–14	41	16.032	0.6366	0.0994	3.9708	14.6	17.3
3	14–15	33	16.233	0.8451	0.1471	5.2065	14.1	18.3
4	15–16	34	16.353	0.7580	0.1301	4.6407	14.8	18.4
5	16–17	30	16.400	0.7887	0.1400	4.8091	14.9	18.9
6	17–18	35	16.491	0.7249	0.1225	4.3950	14.9	17.9
7	18–19	43	16.584	0.6821	0.1040	4.1130	15.2	18.1

TABLE 27.2

Descriptive Statistics of Maximum Head Breadth (cm) in *Koli* Boys of Himachal Pradesh from 12 to 18 Years of Age

							Observed range	
S. no.	**Age group**	**N**	**Mean**	**SD**	**SEM**	**CV**	**Min.**	**Max.**
1	12–13	36	13.003	0.9729	0.1622	7.4821	11.4	16.1
2	13–14	41	12.837	0.9940	0.1128	5.6270	11.4	14.0
3	14–15	33	13.312	1.3186	0.2295	9.9053	11.6	18.4
4	15–16	34	12.800	0.8794	0.1508	6.8703	10.3	14.1
5	16–17	30	12.850	0.5637	0.1029	4.3867	11.3	14.1
6	17–18	35	13.194	0.7231	0.1220	5.4805	12.1	15.1
7	18–19	43	13.137	0.8764	0.1337	6.6710	11.3	15.1

TABLE 27.3

Descriptive Statistics of Maximum Head Vault (cm) in *Koli* Boys of Himachal Pradesh from 12 to 18 Years of Age

							Observed range	
S. no.	**Age group**	**N**	**Mean**	**SD**	**SEM**	**CV**	**Min.**	**Max.**
1	12–13	36	33.211	1.493	0.2489	4.497	29.1	36.5
2	13–14	41	33.376	1.1262	0.1759	3.373	30.5	37.5
3	14–15	33	34.055	1.4750	0.2568	4.331	30.5	37.5
4	15–16	34	33.991	1.2962	0.2220	3.813	31.5	37.5
5	16–17	30	34.117	0.974	0.1784	2.864	32.0	36.5
6	17–18	35	34.506	1.5039	0.2542	4.358	31.2	39.1
7	18–19	43	34.398	1.5974	0.2436	4.642	31.3	38.0

TABLE 27.4

Descriptive Statistics of Minimum Frontal Diameter (cm) in *Koli* Boys of Himachal Pradesh from 12 to 18 Years of Age

							Observed range	
S. no.	**Age group**	**N**	**Mean**	**SD**	**SEM**	**CV**	**Min.**	**Max.**
1	12–13	36	9.494	0.8318	0.1386	8.761	8.2	12.1
2	13–14	41	9.700	0.5874	0.0917	6.055	8.5	11.3
3	14–15	33	9.885	0.5896	0.1026	5.964	8.4	11.0
4	15–16	34	9.821	0.6295	0.1080	6.409	8.7	11.4
5	16–17	30	9.980	0.5792	0.1057	5.803	8.5	11.1
6	17–18	35	10.349	0.8790	0.1486	8.493	9.0	13.4
7	18–19	43	10.233	0.7147	0.1090	6.984	8.7	12.3

TABLE 27.5

Descriptive Statistics of Horizontal Head Circumference (cm) in *Koli* Boys of Himachal Pradesh from 12 to 18 Years of Age

							Observed range	
S. no.	**Age group**	**N**	**Mean**	**SD**	**SEM**	**CV**	**Min.**	**Max.**
1	12–13	36	51.033	1.802	8.300	3.531	47.1	55.1
2	13–14	41	51.912	1.772	0.276	3.414	49.0	56.1
3	14–15	33	52.606	2.382	0.405	4.425	50.2	63.5
4	15–16	34	52.706	2.132	0.365	4.046	49.1	60.0
5	16–17	30	52.720	1.610	0.294	3.043	49.5	56.5
6	17–18	35	53.780	2.249	0.380	4.181	50.0	60.8
7	18–19	43	53.981	1.845	0.281	3.419	49.9	57.0

Descriptive statistics of the nasal height and nasal breadth has been presented in table 27.8 and 27.9 respectively. The mean values of nasal height and nasal breadth for the first age group (12–13 years) came out to be 4.38 cm and 1.73 cm respectively. The mean values of nasal height and nasal breadth for the last age group (18–19 years) came out to be 5.03 cm and 2.08 cm respectively. All other age groups showed variable mean values.

TABLE 27.6

Descriptive Statistics of Bizygomatic Breadth (cm) in Koli Boys of Himachal Pradesh from 12 to 18 Years of Age

							Observed range	
S. no.	**Age group**	**N**	**Mean**	**SD**	**SEM**	**CV**	**Min.**	**Max.**
1	12–13	36	9.289	0.698	0.115	7.484	8.2	11.2
2	13–14	41	9.593	0.644	0.100	6.720	8.1	19.6
3	14–15	33	9.842	0.657	0.114	6.670	8.5	11.2
4	15–16	34	9.641	0.536	0.091	5.559	8.6	10.7
5	16–17	30	10.217	0.799	0.145	7.820	9.2	12.8
6	17–18	35	10.509	1.026	0.173	9.765	8.4	13.8
7	18–19	43	10.521	0.996	0.151	9.469	7.1	13.9

TABLE 27.7

Descriptive Statistics of Bigonial Diameter (cm) in *Koli* Boys of Himachal Pradesh from 12 to 18 Years of Age

							Observed range	
S. no.	**Age group**	**N**	**Mean**	**SD**	**SEM**	**CV**	**Min.**	**Max.**
1	12–13	36	7.878	0.691	0.115	8.757	6.3	9.6
2	13–14	41	7.988	0.745	0.116	9.326	4.9	9.7
3	14–15	33	8.239	0.087	0.501	6.090	7.3	9.3
4	15–16	34	8.238	0.505	0.086	6.145	7.1	9.1
5	16–17	30	8.473	0.545	0.099	6.441	7.4	9.7
6	17–18	35	8.751	0.614	0.103	7.018	7.5	10.2
7	18–19	43	8.863	0.718	0.109	8.106	6.5	9.9

TABLE 27.8

Descriptive Statistics of Nasal Height (cm) in Koli Boys of Himachal Pradesh from 12 to 18 Years of Age

							Observed range	
S. no.	**Age group**	**N**	**Mean**	**SD**	**SEM**	**CV**	**Min.**	**Max.**
1	12–13	36	4.386	0.613	0.102	13.983	3.6	7.4
2	13–14	41	4.541	0.367	0.057	8.090	3.9	5.5
3	14–15	33	4.761	0.538	0.093	11.134	3.9	6.8
4	15–16	34	4.932	0.882	0.151	17.891	4.2	9.5
5	16–17	30	4.820	0.452	0.082	9.396	3.9	5.6
6	17–18	35	4.974	0.448	0.075	8.926	3.9	5.8
7	18–19	43	5.035	0.438	0.066	8.709	4.2	6.1

TABLE 27.9

Descriptive Statistics of Nasal Breadth (cm) in Koli Boys of Himachal Pradesh from 12 to 18 Years of Age

							Observed range	
S. no.	**Age group**	**N**	**Mean**	**SD**	**SEM**	**CV**	**Min.**	**Max.**
1	12–13	36	1.739	0.450	0.075	25.900	1.2	3.2
2	13–14	41	1.710	0.286	0.044	16.725	1.2	2.5
3	14–15	33	1.930	0.272	0.042	14.093	1.3	2.4
4	15–16	34	1.897	0.290	0.051	15.820	1.2	2.4
5	16–17	30	2.080	0.378	0.069	17.788	1.3	2.8
6	17–18	35	2.026	0.345	0.058	17.673	1.3	2.6
7	18–19	43	2.084	0.329	0.050	15.816	1.4	2.7

Tables 27.10, 27.11, 27.12, and 27.13 describe the descriptive statistics of morphological facial length, morphological superior facial length, physiognomic facial length, and physiognomic superior facial length. The mean value of the morphological facial length, morphological superior facial length, physiognomic facial length, and physiognomic superior facial length for the first age group (12–13 years) came out to be 9.51 cm, 5.91 cm, 15.43 cm, and 6.83 cm, respectively. The mean value of the morphological facial length, morphological superior facial length, physiognomic facial length, and physiognomic superior facial length for the last age group (18–19 years) came out to be 10.72 cm, 6.56 cm, 17.08 cm, and

TABLE 27.10

Descriptive Statistics of Morphological Facial Length (cm) in *Koli* Boys of Himachal Pradesh from 12 to 18 Years of Age

							Observed range	
S. no.	Age group	N	Mean	SD	SEM	CV	Min.	Max.
1	12–13	36	9.511	0.638	0.106	6.710	8.3	11.6
2	13–14	41	9.622	1.180	1.840	12.271	8.5	11.2
3	14–15	33	10.127	0.589	0.102	6.817	8.9	11.3
4	15–16	34	10.471	0.551	0.094	5.265	9.4	11.8
5	16–17	30	10.548	0.926	0.172	8.779	9.2	14.3
6	17–18	35	10.714	0.623	0.105	5.820	9.6	12.1
7	18–19	43	10.726	0.704	0.107	6.570	9.3	12.5

TABLE 27.11

Descriptive Statistics of Morphological Superior Facial Length (cm) in *Koli* Boys of Himachal Pradesh from 12 to 18 Years of Age

							Observed range	
S. no.	Age group	N	Mean	SD	SEM	CV	Min.	Max.
1	12–13	36	5.919	0.4927	0.082	8.320	5.1	6.7
2	13–14	41	6.093	0.3663	0.057	6.011	5.4	6.8
3	14–15	33	6.358	0.3623	0.063	5.693	5.6	6.9
4	15–16	34	6.388	0.4277	0.073	6.684	5.1	7.2
5	16–17	30	6.280	0.3498	0.063	5.537	5.8	7.2
6	17–18	35	6.540	0.7445	0.125	11.373	5.6	8.9
7	18–19	43	6.560	0.0810	0.531	8.102	5.0	7.7

TABLE 27.12

Descriptive Statistics of Physiognomic Facial Length (cm) in *Koli* Boys of Himachal Pradesh from 12 to 18 Years of Age

							Observed range	
S. no.	Age group	N	Mean	SD	SEM	CV	Min.	Max.
1	12–13	36	15.436	0.983	0.163	6.368	13.5	18.3
2	13–14	41	16.015	0.855	0.133	5.338	14.2	17.6
3	14–15	33	16.006	1.288	0.224	7.997	10.1	18.1
4	15–16	34	16.497	1.319	0.226	7.998	10.7	18.3
5	16–17	30	16.737	0.818	0.149	4.890	14.2	18.2
6	17–18	35	16.949	0.977	0.165	5.768	14.8	18.7
7	18–19	43	17.088	0.170	0.114	6.519	14.0	18.07

7.26 cm, respectively. The rest of age groups showed variable mean values. In general, the mean values are increasing from lower age groups to higher age groups. However, the age group (16–17 years) for physiognomic superior facial length exhibited decrement.

The descriptive statistics for the physiognomic ear length and physiognomic ear breadth are provided in Tables 27.14 and 27.15, respectively. The mean values of physiognomic ear length and breadth for the first age group (12–13 years) came out to be 5.15 cm and 3.02 cm, respectively. Similarly, the mean values of physiognomic ear length and breadth for the last age group (18–19 years) came out to be 5.59 cm and

TABLE 27.13

Descriptive Statistics of Physiognomic Superior Facial Length (cm) in *Koli* Boys of Himachal Pradesh from 12 to 18 Years of Age

							Observed range	
S. no.	**Age group**	**N**	**Mean**	**SD**	**SEM**	**CV**	**Min.**	**Max.**
1	12–13	36	6.833	0.468	0.078	6.849	5.9	7.6
2	13–14	41	6.871	0.343	0.053	4.992	6.1	7.6
3	14–15	33	7.027	0.414	0.072	5.908	6.1	7.8
4	15–16	34	7.138	0.540	0.092	7.565	5.4	7.8
5	16–17	30	6.997	0.069	0.380	5.436	6.4	7.9
6	17–18	35	7.277	0.073	0.437	5.979	6.4	8.1
7	18–19	43	7.263	0.456	0.725	6.549	6.4	7.9

TABLE 27.14

Descriptive Statistics of Physiognomic Ear Length (cm) in Koli Boys of Himachal Pradesh from 12 to 18 Years of Age

							Observed range	
S. no.	**Age group**	**N**	**Mean**	**SD**	**SEM**	**CV**	**Min.**	**Max.**
1	12–13	36	5.158	0.496	0.082	9.625	4.2	6.0
2	13–14	41	5.180	0.464	0.072	8.963	4.2	5.9
3	14–15	33	5.415	0.312	0.544	5.761	4.7	6.1
4	15–16	34	5.453	0.640	0.079	8.448	4.4	6.7
5	16–17	30	5.447	0.564	0.103	10.370	3.6	6.7
6	17–18	35	5.446	0.437	0.073	8.033	4.3	6.2
7	18–19	43	5.593	0.424	0.064	7.589	4.8	6.4

TABLE 27.15

Descriptive Statistics of Physiognomic Ear Breadth (cm) in *Koli* Boys of Himachal Pradesh from 12 to 18 Years of Age

							Observed range	
S. no.	**Age group**	**N**	**Mean**	**SD**	**SEM**	**CV**	**Min.**	**Max.**
1	12–13	36	3.025	0.389	0.064	12.859	2.1	4.0
2	13–14	41	3.027	0.347	0.054	11.460	2.1	3.6
3	14–15	33	3.103	0.325	3.103	10.425	2.3	3.7
4	15–16	34	3.129	0.287	0.049	9.172	2.5	3.9
5	16–17	30	3.147	0.293	0.053	9.228	2.5	3.8
6	17–18	35	3.094	0.336	0.569	10.888	2.1	3.7
7	18–19	43	3.114	0.361	0.055	11.612	2.2	4.0

3.11 cm, respectively. The in-between age groups showed variable increment in mean values, moving from lower age group to higher age group.

Height and weight have been considered being the most important parameters to study the overall growth in young people. In the present study, height and weight, along with other facial variables, have also been recorded and the descriptive statistics of these has been given in Table 27.16 and 27.17, respectively. The average height and weight for the first age group (12–13 years) came out to be 134.85 cm and 28.51 kg, respectively. Similarly, the mean values of height and weight for the last age group (18–19 years) came out to be 162.68 cm and 51 kg, respectively. In general, increment in mean values has been observed from lower age group to higher age group in height and in weight.

TABLE 27.16

Descriptive Statistics of Height Vertex (cm) in *Koli* Boys of Himachal Pradesh from 12 to 18 Years of Age

							Observed range	
S. no.	**Age group**	**N**	**Mean**	**SD**	**SEM**	**CV**	**Min.**	**Max.**
1	12–13	36	134.857	6.439	1.073	4.746	122.6	154.36
2	13–14	41	143.422	9.816	1.533	6.840	123.5	160.5
3	14–15	33	150.215	8.838	1.538	5.883	130.3	164.9
4	15–16	34	153.532	9.655	1.655	6.288	131.8	168.8
5	16–17	30	159.520	7.820	1.427	4.902	140.8	178.9
6	17–18	35	164.280	8.880	1.501	5.405	132.9	178.9
7	18–19	43	162.686	7.182	1.095	4.415	139.1	174.4

TABLE 27.17

Descriptive Statistics of Body Weight (kg) in *Koli* Boys of Himachal Pradesh from 12 to 18 Years of Age

							Observed range	
S. no.	**Age group**	**N**	**Mean**	**SD**	**SEM**	**CV**	**Min.**	**Max.**
1	12–13	36	28.51	4.485	0.747	15.728	22.5	48.5
2	13–14	41	33.56	6.224	0.972	18.544	23.0	48.5
3	14–15	33	39.21	7.263	1.264	18.519	25.0	51.0
4	15–16	34	39.37	6.565	1.125	16.671	29.9	55.0
5	16–17	30	40.17	9.591	1.751	21.78	31.5	58.0
6	17–18	35	48.67	7.198	1.216	14.853	31.0	62.0
7	18–19	43	51.00	6.234	0.95	12.438	30.0	60.0

TABLE 27.18

Descriptive Statistics of Total Face Index in *Koli* Boys of Himachal Pradesh from 12 to 18 Years of Age

							Observed range	
S. no.	**Age group**	**N**	**Mean**	**SD**	**SEM**	**CV**	**Min.**	**Max.**
1	12–13	36	73.13	5.32	0.887	7.282	60.19	83.33
2	13–14	41	75.66	4.32	0.656	5.717	68.0	86.21
3	14–15	33	74.73	6.53	1.137	8.746	54.65	83.87
4	15–16	34	77.00	6.44	1.105	8.366	69.06	97.87
5	16–17	30	77.38	4.52	0.825	5.815	68.94	86.78
6	17–18	35	78.59	7.31	7.319	9.312	67.38	101.65
7	18–19	43	78.05	5.36	0.817	6.868	68.87	92.00

In the present study, five cephalo-facial indices have been computed for each age category and their respective descriptive statistics has been provided in Tables 27.18, 27.19, 27.20, 27.21 and 27.22. The mean values of total face index, upper face index, frontozygomatic index, nasal index, and ear index for the first age group (12–13 years) came out to be 73.13, 63.98, 102.49, 70.69, and 58.79, respectively. Similarly, the mean values of total face index, upper face index, frontozygomatic index, nasal index, and ear index for the last age group (18–19 years) came out to be 78.05, 62.79, 97.81, 67.76, and 55.81, respectively.

TABLE 27.19

Descriptive Statistics of Upper Face Index in *Koli* Boys of Himachal Pradesh from 12 to 18 Years of Age

							Observed range	
S. no.	Age group	N	Mean	SD	SEM	CV	Min.	Max.
1	12–13	36	63.986	6.391	1.065	9.988	52.04	76.19
2	13–14	41	63.841	6.301	0.984	9.869	52.38	79.01
3	14–15	33	64.861	5.539	0.946	8.539	54.29	79.29
4	15–16	34	66.389	4.928	0.845	1.506	55.43	76.34
5	16–17	30	61.745	4.950	0.903	8.016	49.15	70.65
6	17–18	35	62.797	8.900	1.579	14.317	35.64	88.12
7	18–19	43	62.798	6.847	1.04	10.904	46.04	81.69

TABLE 27.20

Descriptive Statistics of Frontozygomatic Index in *Koli* Boys of Himachal Pradesh from 12 to 18 Years of Age

							Observed range	
S. no.	Age group	N	Mean	SD	SEM	CV	Min.	Max.
1	12–13	36	102.499	8.674	1.45	8.476	91.30	135.96
2	13–14	41	101.441	7.624	1.190	7.542	89.47	120.99
3	14–15	33	100.689	6.739	1.173	6.693	87.85	117.4
4	15–16	34	102.06	7.290	1.250	7.142	89.72	121.43
5	16–17	30	98.022	6.76	1.237	6.912	83.59	110.87
6	17–18	35	9.987	9.336	1.578	9.428	75.0	124.07
7	18–19	43	97.811	8.326	1.269	8.513	77.70	126.78

TABLE 27.21

Descriptive Statistics of Nasal Index in *Koli* Boys of Himachal Pradesh from 12 to 18 Years of Age

							Observed range	
S. no.	Age group	N	Mean	SD	SEM	CV	Min.	Max.
1	12–13	36	70.69	8.67	1.445	12.63	53.06	91.67
2	13–14	41	65.22	7.62	1.190	14.23	28.21	80.49
3	14–15	33	68.90	6.73	1.173	13.57	45.59	91.30
4	15–16	34	65.32	7.29	1.250	17.26	25.0	83.33
5	16–17	30	69.26	9.93	1.813	14.34	50.98	89.74
6	17–18	35	66.82	6.20	1.049	9.20	53.45	84.62
7	18–19	43	67.76	7.29	1.113	10.77	50.82	84.44

In the present study, a total of 18 cephalo-facial anthropometric variables have been studied to access the pattern of facial growth in growing *Koli* boys of Himachal Pradesh. The calculation of growth velocities is an important step to understand the trends and pattern in any given society. In the present study, growth velocities for all the 18 cephalo-facial anthropometric variables in each age category have been computed and are given in Table 27.23. In general, the age group of 13–14 years showed the highest growth velocity but for all the variables. Some cephalo-facial measurements showed variable trends with age.

The results of the present study have been compared with the earlier published study (Singh 1964) to investigate the ethnic and regional variability in cephalo-facial growth. The comparison of the cephalo-facial anthropometric variables along with cephalo-facial indices with Singh's (1964) is presented in Tables 27.24 and 27.25, respectively. In terms of cephalo-facial anthropometric variables, the present study exhibited smaller values as compared to Singh's (1964) in each age category, except for the morphological superior facial length. Morphological superior facial length in each age category showed

TABLE 27.22

Descriptive Statistics of Ear Index in *Koli* Boys of Himachal Pradesh from 12 to 18 Years of Age

							Observed range	
S. no.	**Age group**	**N**	**Mean**	**SD**	**SEM**	**CV**	**Min.**	**Max.**
1	12–13	36	58.79	6.70	1.128	11.51	44.64	76.0
2	13–14	41	58.76	7.64	1.193	13.00	39.62	73.0
3	14–15	33	57.39	5.84	1.017	10.17	43.40	66.67
4	15–16	34	47.67	5.97	1.025	103.64	45.45	67.31
5	16–17	30	59.07	10.05	1.835	17.02	46.43	102.86
6	17–18	35	57.10	7.16	1.211	12.54	48.89	69.81
7	18–19	43	55.81	6.39	0.725	11.46	40.68	76.92

TABLE 27.23

Growth Velocity (cm) per Year in Different Age Groups of *Koli* Boys of Himachal Pradesh from 12 to 18 Years of Age

S. no.	Measurement	12–13	13–14	14–15	15–16	16–17	17–18	18–19
1	Max. head length	–	0.579	0.201	0.12	0.047	0.091	0.093
2	Max. head breadth	–	−0.166	0.475	−0.512	0.05	0.344	−0.057
3	Head vault	–	0.165	0.679	−0.064	0.126	0.389	−0.108
4	Min. frontal diameter	–	0.206	0.185	−0.064	0.159	0.369	−0.116
5	Horizontal head circumference	–	0.879	1.573	0.1	0.214	0.86	0.201
6	Bizygomatic breadth	–	0.304	0.249	−0.201	0.576	0.292	0.012
7	Bigonial diameter	–	0.11	0.251	−0.001	0.235	0.276	0.112
8	Nasal height	–	0.155	0.22	0.171	−0.112	0.154	0.061
9	Nasal breadth	–	−0.155	0.298	−0.081	0.132	0.003	0.082
10	Nasal depth	–	0.029	0.22	−0.033	0.183	−0.054	0.058
11	Morphological facial length	–	0.111	0.5	0.344	0.077	0.166	0.012
12	Morphological superior facial length	–	0.174	0.265	0.03	−0.108	0.26	0.02
13	Physiognomic facial length	–	0.579	0.009	0.491	0.24	0.212	0.139
14	Physiognomic superior facial length	–	0.038	0.156	0.111	−0.141	0.28	−0.014
15	Physiognomic ear length	–	0.022	0.235	−0.038	−0.006	−0.001	0.147
16	Physiognomic ear breadth	–	0.002	0.076	0.026	0.018	−0.053	0.02
17	Height vertex	–	8.547	6.793	3.317	5.988	4.76	−1.594
18	Weight	–	5.044	5.655	0.161	4.638	4.653	2.332

TABLE 27.24

Comparison of Cephalo-Facial Anthropometric Variables of the Present Study with Singh (1964)

S. no.	Measurements	12–13	13–14	14–15	15–16	16–17	17–18	18–19	
1	Max. head length	18.38	18.41	18.52	19.01	18.68	18.93	–	Singh (1964)
		15.45	16.03	16.23	16.35	16.40	16.49	16.58	Present
2	Max. head breadth	13.62	13.85	13.82	13.86	14.06	13.93	–	Singh (1964)
		13.00	12.83	13.31	12.80	12.85	13.19	13.13	Present
3	Horizontal head circumference	52.85	52.65	53.50	53.54	53.88	–	–	Singh (1964)
		51.03	51.91	52.60	52.70	52.92	53.78	53.98	Present
4	Bizygomatic breadth	12.16	12.36	12.38	12.51	13.16	12.99	–	Singh (1964)
		9.28	9.59	9.84	9.64	10.21	10.50	10.52	Present
5	Bigonial diameter	9.00	9.01	9.35	9.70	9.76	10.00	–	Singh (1964)
		7.87	7.98	8.23	8.23	8.47	8.75	8.86	Present
6	Morphological facial length	10.25	10.14	10.16	10.70	10.78	11.12	–	Singh (1964)
		9.51	9.62	10.12	10.47	10.54	10.71	10.76	Present
7	Morphological superior facial length	5.45	5.55	5.74	5.83	5.82	–	–	Singh (1964)
		5.91	6.93	6.35	6.38	6.28	6.54	6.56	Present

TABLE 27.25

Comparison of the Indices of the Present Study with Singh (1964)

S. no.	Indices	12–13	13–14	14–15	15–16	16–17	17–18	18–19	
1	Cephalic index	74.10	74.96	76.20	72.92	76.14	74.55	–	Singh (1964)
		84.69	80.20	82.25	78.50	78.54	80.16	79.37	Present
2	Total face index	84.26	81.94	82.31	85.39	82.80	85.01	–	Singh (1964)
		73.13	75.66	74.73	77.00	77.73	78.59	78.00	Present
3	Upper face index	44.86	44.62	45.11	46.13	44.74	44.38	–	Singh (1964)
		63.98	63.84	64.86	66.38	61.75	62.79	62.79	Present
4	Frontozygomatic index	81.68	78.93	79.61	79.68	79.07	79.85	–	Singh (1964)
		102.44	101.44	100.68	102.06	98.02	99.98	97.81	Present

larger values as compared to Singh's (1964) study. In case of indices, the present study showed higher index values for cephalic index, upper face index, and frontozygomatic index as compared to the values presented in Singh's (1964), except for the total face index in which index values for each age category has lower value for the present study than Singh's (1964) study.

27.4 Discussion

Human variability has been the toughest and the longest academic discourse in human biology. *Homo erectus* was the first hominid to leave Africa and started peopling of the world (Bar-Yosef 2003). Over time, they adapted to the regional climates, and considerable anatomic variability is evident from the fossils (Baab 2008). The present global population of *Homo sapiens* exhibited a great deal of biological variability in terms of skin color, head shape, stature, eye color, hair texture, and more and has been the core of anthropological research (Ousley, Jantz, & Freid 2009). Human variation can be classified into two levels: the individual level and the ethnic level. At the individual level, everyone is different from the others; even monozygotic twins grow differently after interacting with the environment (Fraga et al. 2005). On the other hand, at the ethnic level, different ethnic communities show different morphological features and growth patterns as compared to other ethnic communities (Hiernaux 1968). Different factors, such as genetics (Jorgenson et al. 2005), environment (Kumanyika 2008), nutrition

(Wang & Chen 2011), culture (McKenzie & Crowcroft 1994), migration (Kobyliansky 1983), disease (Hegde, Praveen, & Shetty 2013), and geographical area (Denevan 1983), play important role in human variation. Populations distributed spatially over larger geographical areas show larger variation than the locally constricted population in a particular geographic zone. In a particular geographic zone, social and cultural factors, such as caste, religion, and norms, become the driving factors to produce local variation in the population (McKenzie & Crowcroft 1994). Racial differences in head shapes are well documented in the literature; however, within the particular community, changes in cephalo-facial growth with respect to age are scarce. The present study is an attempt to understand the growth of cephalo-facial structure with age and to compare similar study done in the same Indian state by Singh (1964).

In the present study, gradual enhancement in cephalo-facial structure with age has been observed. Maximum head length showed a regular increase from 12–13 years to 18–19 years. In the age group of 13–14 years, head length exhibited the highest growth velocity while, the slowest being in the age group of 16–17 years. The maximum head breadth showed a maximum increase from 13–14 years to 15–16 years; this is followed by a decrease up to 15–16 years. Highly variable growth velocities have been observed in the case of head breadth. For the age group 13–14 years, 15–16 years, and 18–19 years, head breadth showed negative growth velocity. The head vault showed a small increment with age as compared to the other parameters. Minimum frontal diameter, too, exhibited a gradual increase in the mean values with age. Like head vault, head circumference, too, showed a little increment with age. Growth velocities for parameters such as head vault, head circumference, and minimum frontal diameter have been found to be highly variable. All of the parameters taken on the head in the present study are less than the previous study conducted by Singh (1964) on Kullu Valley Rajputs. The lesser mean values of the present study as compared to Singh's (1964) study provides a ground for meaningful discourse about the socio-economic and genetic differences in both the communities. *Rajputs* of Himachal Pradesh enjoy a higher social and economic status than *Kolis*. The better access to nutrition may have caused a greater mean value of the *Rajputs* in each age category. Both the communities are highly endogamous, so genetic differences may have contributed to the differences in the anthropometric values. More in-depth studies on growth variability, emphasizing on social and economic disparity, should be done. No other study has published data age-wise, so direct comparison with other studies is not possible.

Similar trends have also been observed in facial variables. Bizygomatic breadth showed an increment in mean values from 12–13 years to 18–19 years, except for 15–16 years. A decrease in the age group of 15–16 years has been observed. A gradual increment in bigonial diameter with age has been observed from 12–13 years up to 18–19 years. Mean values for each age category for bizygomatic breadth and bigonial diameter in the present study are larger compared to Singh's (1964) study. Both bizygomatic breadth and bigonial breadth showed negative growth velocity in 15–16 years of age, while for the rest of the age groups, positive trends have been observed. Nasal height, breadth, and depth showed an increment in mean values with respect to growing age. However, the increment at certain ages is not consistent, such as nasal height decreases at 16–17 years and nasal breadth at 15–16 years. Morphological facial length and morphological superior facial length showed steady enhancement from 12 to 19 years. For superior morphological facial length, a slightly lesser mean value for the age group of 16–17 years has been observed. In terms of growth velocities, morphological facial length and morphological superior facial length expressed positive growth, except at the age of 16–17 years, for morphological superior facial length. The mean values for each age category for morphological facial length are lower than provided by Singh's 1964 study. However, morphological superior facial length showed larger values than in Singh's 1964 study for each age category. Physiognomic facial length and physiognomic superior facial length exhibited uniform increment in mean values in each age group from 12 to 19 years. In terms of growth velocities, these two facial lengths showed positive growth velocities for each age group, except at 16–17 years, for physiognomic superior facial length. Physiognomic ear length and physiognomic ear breadth, too, showed a gradual increase in mean values for increasing age groups from 12 to 19 years. A little lesser mean value has been observed at 16–17 years for ear length and at 17–18 years for ear breadth. Ear length and breadth both showed variable growth velocities. Koli boys gained both height and weight with age from 12–19 years. The highest range and mean stature for ages 18–19 years is less compared to those ages 17–18 years, which is unusual as boys tend to gain stature at this age.

In the present study, cephalo-facial indices were also computed and for classification in each index; ranges given by Martin and Saller (1957) were used. Cephalic index computed for each age category revealed that only two age groups, 12–13 years and 14–15 years, fall under the classification of brachycephalic; other age groups fall in the category of mesocephalic. The total face index for each age category was computed, and it was observed that all the indices were less than the lower limit ($X > 79.9$). Therefore, all of the individuals in the present study have hypereuryprosopic face. For the face, upper face index was also computed for each age category. Mean values of upper face index in each age category are larger than the upper range ($60 < X$). Therefore, hyperleptene is the dominant facial type in the present population.

Variability in different cephalo-facial measurements for the present study as compared to the Kullu Valley *Rajputs* (Singh 1964) provides interesting information on how different ethnic groups can take different growth trajectories. These values can be of great utility for the medical coroners and facial surgeons for reconstructing any face. Ethnic dimorphism in cephalo-facial growth confirms that each ethnic group needed to be studied separately, and conclusions made from one ethnic group should not be generalized to other ethnic groups. In a study conducted by Wallace and colleagues on American children with mean age 7.8 ± 4.7 months showed ethnic difference in craniosynostosis surgeries. This study showed that Hispanic/non-white children tend to undergo craniosynostosis repair at older ages and lengthier operations than non-Hispanic/white children (Wallace et al.2019).

In any community, the change in the anthropometric variables with age is the reflection of the overall health of the community. Anthropometry provides an objective understanding on how well the health of the community is. In terms of cephalo-facial growth, *Koli* boys lag behind the Rajputs even if both the population are residing in the same Indian state and share the same ecology. The uneven growth at certain age groups in the present study brings an interesting observation that most of the irregular trend in increment has been observed at 15–16 and 16–17 years of the age. There is a possibility that the children of these age groups could have suffered malnutrition in their childhood. In those two years when the respective children have been born, the community might have undergone a serious economic crisis due to famine, landslide, earthquake, or flood, which is reflected in facial growth. This observation emphasizes that historical and economic data along with biological data are important to comprehensively understand the ethnic growth patterns.

27.5 Conclusion

Assessment of physical growth is important to understand the health status of any ethnic group. In the present study, it can be concluded that cephalo-facial dimensions increase with age in Koli boys of Himachal Pradesh. However, the trends are not always consistent with age in certain age groups. It can also be concluded that the growth velocity of cephalo-facial dimensions is not uniform in all the age groups. The lower socio-economic status of Koli community could be responsible for the lower mean values of cephalo-facial dimensions in each age group. Better opportunities for education and nutrition can improve the growth and health status of Koli community. Moreover, the present study was only conducted in boys, so it cannot be generalized for girls. Multi-ethnic studies with larger samples, including females, along with historical data, socio-economic data, and nutritional data, should be conducted to comprehensively elucidate the growth patterns in different communities.

27.6 Acknowledgements

This study is a part of a master's degree dissertation submitted to the Department of Anthropology, Panjab University, Chandigarh, India, by one of the authors (RK). Kewal Krishan is supported by a DST PURSE grant and UGC Centre of Advanced Study (CAS-II) awarded to the Department of Anthropology, Panjab University, Chandigarh, India. Thanks are also due to the subjects, who voluntarily participated in the study.

REFERENCES

Baab KL (2008) The taxonomic implications of cranial shape variation in Homo erectus. *J Hum Evol* 54:827–847. https://doi.org/10.1016/j.jhevol.2007.11.003

Bar-Yosef O (2003) Early colonizations and cultural continuities in the Lower Palaeolithic of western Asia. In: *Early human behaviour in global context*. Routledge. p 234–291.

Başsimitçi Ş, Yücel-Eroğlu E, Akalar M (1996) Effects of thalassaemia major on components of the craniofacial complex. *Br J Orthod* 23:157–162. https://doi.org/10.1179/bjo.23.2.157

Bogin B (1999) *Patterns of human growth*. 2nd ed. Cambridge University Press.

Denevan WM. 1983. Adaptation, variation, and cultural geography. *Prof Geogr* 35:399–407. https://doi.org/10.1111/j.0033-0124.1983.00399.x

Fink GB, Madaus WK, Walker GF (1975) A quantitative study of the face in Down's syndrome. *Am J Orthod* 67:540–553. https://doi.org/10.1016/0002-9416(75)90299-7

Fraga MF, Ballestar E, Paz MF, Ropero S, Setien F, Ballestar ML, Heine-Suñer D, Cigudosa JC, Urioste M, Benitez J (2005) Epigenetic differences arise during the lifetime of monozygotic twins. *Proc Natl Acad Sci* 102:10604–10609. https://doi.org/10.1073/pnas.0500398102

Hegde S, Praveen BN, Shetty SR (2013) Morphological and radiological variations of mandibular condyles in health and diseases: a systematic review. *Dentistry* 3:154. http://dx.doi.org/10.4172/2161-1122.1000154

Hiernaux J (1968) Ethnic differences in growth and development. *Eugen Q* 15:12–21. https://doi.org/10.1080/19485565.1968.9987748

Hoppa RD, Fitzgerald CM (1999) *Human growth in the past: studies from bones and teeth*. Cambridge University Press.

Jorgenson E, Tang H, Gadde M, Province M, Leppert M, Kardia S, Schork N, Cooper R, Rao DC, Boerwinkle E (2005) Ethnicity and human genetic linkage maps. *Am J Hum Genet* 76:276–290. https://doi.org/10.1086/427926

Kobyliansky E (1983) Changes in cephalic morphology of Israelis due to migration. *J Hum Evol* 12:779–786. https://doi.org/10.1016/S0047-2484(83)80135-3

Krishan K, Kumar R (2007) Determination of stature from cephalo-facial dimensions in a North Indian population. *Leg Med* 9:128–133. https://doi.org/10.1016/j.legalmed.2006.12.001

Kumanyika SK (2008) Environmental influences on childhood obesity: ethnic and cultural influences in context. *Physiol Behav* 94:61–70. https://doi.org/10.1016/j.physbeh.2007.11.019

Kurisu K, Niswander JD, Johnston MC, Mazaheri M (1974) Facial morphology as an indicator of genetic predisposition to cleft lip and palate. *Am J Hum Genet* 26:702.

Lohman TG, Roche AF, Martorell R (1988) *Anthropometric standardization reference manual*. Human Kinetics Books.

Magalhaes AE, Stella JP, Epker BN (1995) Facial anthropometrics versus cephalometry as predictors for surgical treatment in patients with Class III dentofacial deformities. *Int J Adult Orthodon Orthognath Surg* 10:295–302.

Mane DR, Kale AD, Bhai MB, Hallikerimath S (2010) Anthropometric and anthroposcopic analysis of different shapes of faces in group of Indian population: a pilot study. *J Forensic Leg Med* 17:421–425. https://doi.org/10.1016/j.jflm.2010.09.001

Martin R, Saller K (1957) *Lehrbuch der anthropologie*. 3rd ed. (Saller K, editor). G. Fischer. www.worldcat.org/title/lehrbuch-der-anthropologie-in-systematischer-darstellung/oclc/14619557

McKenzie KJ, Crowcroft NS (1994) Race, ethnicity, culture, and science. *BMJ* 309:286–287.

Moss ML, Young RW (1960) A functional approach to craniology. *Am J Phys Anthropol* 18:281–292. https://doi.org/10.1002/ajpa.1330180406

Ousley S, Jantz R, Freid D (2009) Understanding race and human variation: why forensic anthropologists are good at identifying race. *Am J Phys Anthropol* 139:68–76. https://doi.org/10.1002/ajpa.21006

Richardson ER (1980) Racial differences in dimensional traits of the human face. *Angle Orthod* 50:301–311. 10.1043/0003-3219(1980)050%3C0301:RDIDTO%3E2.0.CO;2

Scammon RE (1930) The measurement of body in childhood. In: Harris JA, editor. *The measurement of man*. University of Minnesota Press.

Shukla D, Bablani D, Chowdhry A, Thapar R, Gupta P, Mishra S (2014) Dentofacial and cranial changes in Down syndrome. *Osong public Heal Res Perspect* 5:339–344. https://doi.org/10.1016/j.phrp.2014.09.004

Singal P, Sidhu LS (1986) A study of cephalo-facial measurements from age 20 to 80 in Jat-Sikh and Bania females of Punjab (India). *Anthropol Anz* 44(4):361–371.

Singh L (1964) *Growth changes in body proportions of Kullu valley Rajputs (H.P.).*, PhD Dissertation, Department of Anthropology, Panjab University, Chandigarh, India

Tanner JM (1978) *Foetus into man: Physical growth from conception to maturity.* Harvard University Press.

Tanner JM, Whitehouse RH, Takaishi M (1966) Standards from birth to maturity for height, weight, height velocity, and weight velocity: British children, 1965. I. *Arch Dis Child* 41:454.

Wallace ER, Birgfeld C, Speltz ML, Starr JR, Collett BR (2019) Surgical approach and periprocedural outcomes by race and ethnicity of children undergoing craniosynostosis surgery. *Plast Reconstr Surg* 144. https://journals.lww.com/plasreconsurg/Fulltext/2019/12000/Surgical_Approach_and_Periprocedural_Outcomes_by.31.aspx

Wang Y, Chen X (2011) How much of racial/ethnic disparities in dietary intakes, exercise, and weight status can be explained by nutrition-and health-related psychosocial factors and socioeconomic status among US adults? *J Am Diet Assoc* 111:1904–1911. https://doi.org/10.1016/j.jada.2011.09.036

Young NM, Wat S, Diewert VM, Browder LW, Hallgrímsson B (2007) Comparative morphometrics of embryonic facial morphogenesis: implications for cleft-lip etiology. *Anat Rec* 290:123–139. https://doi.org/10.1002/ar.20415

28

Human-Animal Coexistence and Conflicts—a Gateway to the Zoonotic Infections

Tejinder Kaur, Manish Thakur, and Ranbir Chander Sobti

CONTENTS

28.1 Introduction

Humans are inextricably linked to animals. For millennia, we have relied on animals for food, security, entertainment, transportation, and scientific experimentation. Indeed, many people maintain animals only for the sake of company. Despite the numerous benefits of human-animal partnerships, these defenseless animals have the potential to transmit a variety of illnesses to humans, some of which are quite serious. The term "zoonotic illnesses" refers to infections contracted by humans from animals. Zoonotic diseases are extremely common. According to experts, animals transmit more than 60% of infectious illnesses known to humans. Additionally, around 75% of novel or emerging infectious illnesses in humans are spread by animals (Centers for Disease Control and Prevention, 2021). Each year, around one billion people suffer and millions die as a result of endemic zoonoses (Karesh et al., 2012). Animals, such as mammals, birds, reptiles, and amphibians, may serve as reservoirs for these zoonotic diseases. The term "zoonoses" originates in the Greek words "zoon" (animal) and "nosos" (disease) (Rahman et al., 2020). The World Health Organization defines zoonotic diseases as infectious diseases caused by an infectious agent, such as a bacterium, virus, protozoan, parasite, or prion, and transmitted from a vertebrate (non-human animal) to a human individual. Generally, zoonotic diseases are enzootic, meaning they occur solely in animal populations, but the pathogens that cause them can be accidentally transmitted to humans. This is the point at which the actual difficulty arises. Animal diseases that infect humans may be limited to a single host or may begin a chain of human-human transmission capable of causing localized or worldwide epidemics, as was the case with the recent COVID-19 pandemic, which turned the globe upside down in less than a year and a half. How do zoonotic illnesses develop or reemerge? Is it just cohabitation between humans and animals, or is there anything more to it? The

DOI: 10.1201/9781003220398-28

following section discusses the variables that have had a role in the evolution and reemergence of zoonotic diseases.

28.2 Emergence and Reemergence of Zoonotic Infections

Man himself is a significant factor in the development and spread of these zoonotic illnesses. In recent decades, a continuous stream of novel viral diseases has developed as a result of various threats to wild animal populations and increasing human contact, which increases the probability of pathogen transmission from wild animals to people (Woolhouse et al., 2012). Some of the factors that lead to the emergence and reemergence of zoonotic infections are discussed in the following sections.

28.2.1 Human-Animal Coexistence

Animals have been revered in many civilizations across the globe as a result of human dependence on natural resources. Animals play a vital role in our lives, fulfilling a range of human needs. For instance, dogs have traditionally been employed to deter theft from houses and personal possessions. These canines are beneficial for detecting and locating evidence in criminal investigations, such as bombs and explosives. Nowadays, dogs are also bred for companionship. Regrettably, dogs and other domesticated animals may act as rabies virus reservoirs (RABV). Each year, about 15.69 million people worldwide are exposed to the rabies virus, and roughly 59,000 human fatalities are recorded (Hampson et al., 2011; Depani et al., 2012). Rabies is a public health threat in poor nations such as Africa and Asia, which account for more than 95% of all human rabies cases (Fooks et al., 2014). Domestic animals may also contribute significantly to the amplification of diseases imported from the wild, thus changing the cycle of human infectious disease transmission (Morand et al., 2014). Numerous significant diseases in human civilization are said to have arisen concurrently with the advent of agriculture. This is because of the domesticated animals used in agricultural activities. Zoonotic diseases, such as anthrax, tuberculosis, brucellosis, campylobacteriosis, leptospirosis, toxoplasmosis, balantidiasis, ancylostomiasis, toxocariasis, listeriosis, bovine pustular stomatitis, rotavirus infection, and Q fever, are all transmissible to people through domestic animals (Rahman et al., 2020).

28.2.2 Human-Animal Conflicts

A conflict between people and wildlife arises when the demands and behaviors of animals have a negative impact on humans, or when the desires and activities of humans have an adverse effect on the peaceful existence of animals. Humans have traditionally seen the Earth as a storehouse, a repository of materials that could be mined and used arbitrarily, as if they were limitless and there were no repercussions to material extraction or usage. Anthropogenic disturbances to high biodiversity areas increase the risk of zoonotic emerging infectious disease (Allen et al., 2017). Because of the escalating human population and consequent encroachment on wildlife habitats, pathogen transfers into humans have increased (Schmeller et al., 2020).

28.2.2.1 Climate Change

The emergence of many zoonotic diseases can be attributed to climate change caused by anthropogenic activities (Pearce-Duvet, 2006). Global surface air temperature is expected to rise over the course of the 21st century, owing largely to the fact that human-caused greenhouse gas concentrations (most notably those emitted by fossil fuel combustion) will increase, as will the frequency of severe weather events, such as heat waves, floods, and droughts. Every element of human existence will be impacted, from drinking water to our natural environment (de La Rocque et al., 2008). Thousands of species extinctions are widely recognized as a significant danger to world biodiversity over the next century as a result of human-caused climate change (Cahill et al., 2013). Local extinctions of spiny lizards (*Sceloporus*) are already occurring due to rising temperatures (Sinervo et al., 2010). Climate change has the greatest impact on arthropod vectors. Temperature is critical for the life cycles of mosquitoes, ticks, and sandflies. Temperature fluctuations may contribute to the transmission of disease. Sandflies are more likely to transmit leishmaniasis in a warmer world. This is because in warm temperatures, sandflies are more active and eat more often. Similarly, an area's mosquito population is proportional to its temperature. Because mosquitoes act as vectors for a variety of infections, a rise in mosquito population would almost certainly result in an increase in all mosquito-borne illnesses, such as dengue fever, chikungunya, and others. Zika, like dengue and chikungunya, is spread via mosquito vectors and causes illnesses in non-human primates. Global warming also has an effect on increasing rodent activity, putting the whole Scandinavian population at danger of acquiring hantavirus (Evander and Ahlm, 2009). Worse still, rising water temperatures may increase aquatic species' oxygen needs while decreasing oxygen levels. All of these factors are cause for concern: rising sea levels, melting polar ice caps, and eroding glaciers. With rising sea levels, coastal flooding and water-borne zoonotic diseases will become more prevalent.

28.2.2.2 Urbanization

To begin, urbanization has a tendency to decrease habitat quantity while also increasing habitat fragmentation over a period of time (Liu et al., 2016). The worldwide expansion of people into natural ecosystems causes habitat loss and fragmentation, which in turn leads to a reduction in the diversity of species. Because habitat destruction has an impact on the illnesses that animals carry and the probability of them being transmitted from animals to people, it may jeopardize the survival of humans and their livelihoods. The prevalence of zoonotic infections among animal orders rises with species diversity. Thus, although habitat loss and fragmentation may result in a loss of species variety, human encroachment on species-rich regions may enhance human exposure to new infectious pathogens transmitted by animals (Wilkinson et al., 2018). Until recently, it was believed that biodiverse environments resulted in the development of new zoonotic diseases that posed a threat to humans. This anticipation was founded on the premise that pathogen diversity is a risk factor for the development of zoonotic diseases. However, we have known for decades that biological diversity can help slow the spread of zoonotic diseases by diversifying the hosts within the animal community, thereby sparing humans from becoming hosts to the pathogens due to limited direct contact with infected animals (Keesing and Ostfeld, 2021). Thus, while habitat loss and fragmentation may result in the extinction of species, human expansion into species-rich regions may result in increased human exposure to novel infectious pathogens transmitted by animals (Wilkinson et al., 2018).

28.2.2.3 Agricultural Practices

Agricultural production is a significant contribution to human land usage, with an estimated 37% of the Earth's land area being converted to agricultural production by 2015 (García-Vega and Newbold, 2020). The encroachment of human settlements and agriculture on natural ecosystems leads in the development of ecotones (transition zones between neighboring ecological systems), which are areas where species assemblages from various habitats come together. Pathogen spillover, genetic diversity, and adaptability are all enhanced as a result of these developments. For a number of illnesses, including yellow fever, Lyme disease, hantavirus pulmonary syndrome, Nipah virus encephalitis, influenza, rabies, cholera, leptospirosis, malaria, and human African trypanosomiasis, associations between disease onset and ecotones have been hypothesized (Despommier et al., 2006). The majority of them are zoonoses, and the epidemiology of many of them involves both animals and cattle. It has been suggested that the geographic expansion of the Japanese encephalitis virus (JEV) in Southeast Asia is related to greater irrigated rice farming and pig raising, both of which have risen as a consequence of an increasing human population. Irrigated fields, which enhance the density of vectors and water birds, combined with pig farming, result in an increase in pig production and the risk of virus spillover into humans (Jones et al., 2013). The Kyasanur forest disease originally limited itself to India's 2,000-square-mile endemic zone, infecting just 300 square miles. The tickborne disease, characterized by paralysis, often spreads in evergreen rain forests, even those with deciduous regions and human settlements. Loggers are at a greater risk of death, with a mortality rate of 10%. The lethal epidemic that ravaged monkeys, infected 1,555 people, and killed 150 people in 1983 was especially gruesome (Varma, 2001).

28.2.2.4 Consumption of Animals as Food

Another potential factor for the development of zoonotic illnesses transmitted by animals has been the significant rise in bushmeat eating in many areas of the globe. Six million tons of bushmeat are harvested each year in the Amazon and Congo basins alone, depleting the forest and bringing millions of people into contact with wildlife (Nasi et al. 2011). Numerous examples exist of direct disease transmission from animals to people. HIV is thought to have evolved as a result of non-human primates being hunted for food in central African forests, while Ebola hemorrhagic fever outbreaks have been linked to hunting in Gabon and the Republic of Congo (Jones et al., 2013). Trichinellosis has long been linked to the eating of raw meat from wild animals, such as bears, and the consumption of raw deer and wild boar meat has been linked to the development of severe cases of hepatitis E among Japanese hunters (Takahashi et al., 2004). Wet animal markets also play a crucial role in emergence or reemergence of zoonotic infections. Avian influenza is the widest spread type that infects humans and other animals like pigs and may spread through wild aquatic birds or domestic poultry. The year 2009 witnessed an influenza pandemic, also known as swine flu pandemic, caused by the H1N1 influenza virus. Diseases like severe acute respiratory syndrome (SARS) and the very recent COVID-19 are all considered to be a spillover from the wet markets selling exotic animals like bats and pangolins. The developed world's growing desire for unusual foods has also been associated with a variety of zoonotic diseases or parasites, including protozoans and helminthic parasites.

28.3 Observations Made During the COVID-19 Pandemic Shutdown in Different Nations

The lockdown period during COVID-19 has been called a "ventilator" for global natural resource restoration (Khan et al., 2021). Some of the changes observed during the shutdown were the following.

28.3.1 Decrease in Air Pollution

One significant effect of the lockdown that was visible to everyone was a decrease in air pollution levels. Numerous nations, including China, the United States of America, and Italy, which were severely affected by the virus and emerged as epicenters for COVID-19, were forced to implement stringent lockdowns to prevent the sickness from spreading. NASA and the European Space Agency announced in 2020 that these nations saw an almost 30% reduction in air pollution during lockdown (Zambrano-Monserrate, 2020). Reduced particulate matter, reduced nitrogen dioxide (NO_2) concentrations, and lower greenhouse gas concentrations in the atmosphere were found, which might be a result of companies being closed and vehicle mobility restricted during lockdown (Khan et al., 2021). Some experts even recorded a rise in ozone levels in the atmosphere during the lockdown period in various nations, ranging from 2.4% in Valencia to 115% in Wuhan (Mousazadeh et al., 2021). On the other hand, a clear link between air pollution, COVID-19 infections, and death on a global scale has also been reported (Ali and Islam, 2020). According to current evidence, air pollution may potentially have an effect on COVID-19 transmission. Air pollution exposure has been demonstrated to have a detrimental effect on the prognosis of COVID-19-infected individuals. Additionally, air pollution has been associated with the spread of infectious illnesses. Poor air quality, for example, has been related to an increase in SARS fatalities and influenza flu cases (Ali and Islam, 2020).

28.3.2 Decrease in Water Pollution

The water quality of several Indian rivers, which are typically contaminated by industry and human effluents, has also improved as a consequence of the countrywide shutdown (Lokhandwala and Gautam,

2020). Garg et al. (2020) analyzed the turbidity level along the Ganga River and its waterways, which include Haridwar, Kanpur, Prayagraj, and Varanasi. The COVID-19 lockdown has enabled the pre-polluted Ganga River to meet drinking water standards. According to a CPCB assessment, the Yamuna River's quality increased more than fivefold during the lockdown and industrial closure in India's capital (Paital et al., 2020). Within two months of the lockout, dissolved oxygen levels improved, biological oxygen demand reduced, and nitrate content dropped, leading in an overall improvement in water quality (Paital et al., 2020).

28.3.3 Impact on Animal Life

Since the pandemic began, sightings of wild animals in hitherto unknown locations have shown up on social media, with many individuals posting photographs and videos. During the shutdown, hedgehog roadkill was more than 50% lower than in the pre-pandemic era (Łopucki et al., 2021). According to Derryberry et al. (2020), when traffic noise in California's San Francisco Bay Area was reduced to levels not seen in half a century, the singing frequency of white-crowned sparrows altered. Amphibians and reptiles are less likely to be killed on roadways, and a bird that feeds on flying insects has a better nesting success rate as a result of enhanced species richness in temporarily less disturbed environments. This crisis has had a mixed bag of consequences for animals, with some being beneficial and others being detrimental. The reduced human disturbance induced by lockdown benefitted invasive alien species. The researchers found that the COVID-19 shutdown jeopardized efforts to contain the spread of invasive alien species and to conserve endangered taxonomic species. Additionally, a decrease in enforcement may result in an increase in animal poaching. Along with its socio-economic consequences, the COVID-19 problem may have long-term repercussions for wildlife conservation (Manenti et al., 2020). In a recent study on 20 protected areas, Banke National Park and Bardia National Park demonstrated substantial increases in animal mortality during the lockdown period (Koju et al., 2021). This might be attributed to the increased animal poaching because of lesser checks due to the shutdown.

28.4 Concluding Remarks

Human health is inextricably connected to animal health and the environment in which we live. One Health is a philosophy that takes this into account. All creatures, including humans, are subject to the basic principles of biology, welfare, and health. Domestic animals need care and should be taken to vets on a regular basis for medical examinations. To the greatest degree feasible, we should avoid invading the habitats of wild animals, and if this is not possible, then the animals should be relocated with utmost care to new befitting habitats. We should transfer the creatures with the utmost care to a more suitable environment. Consumption of exotic animals and bushmeat hunting should be totally prohibited. This would not only benefit animals but also protect us from the frequently emerging zoonotic infections. All findings obtained during the COVID-19 shutdown point to the importance of the environment in raising healthy animals and people. Human activities have far-reaching consequences, and we must alter our interactions with other living creatures.

REFERENCES

Ali, N., Islam, F. (2020). The effects of air pollution on COVID-19 infection and mortality: A review on recent evidence. *Front. Public. Health* 8: 580057. https://doi.org/10.3389/fpubh.2020.580057

Allen, T., Murray, K.A., Zambrana-Torrelio, C., et al. (2017). Global hotspots and correlates of emerging zoonotic diseases. *Nat. Commun.* 8: 1124. https://doi.org/10.1038/s41467-017-00923-8

Cahill, A.E., Aiello-Lammens, M.E., Fisher-Reid M.C., Hua X., Karanewsky, C.J., Yeong R.H., Sbeglia, G.C., Spagnolo, F., Waldron, J.B., Warsi, O., Wiens, J.J. (2013). How does climate change cause extinction? *Proc. R. Soc. B.* 280: 20121890. http://doi.org/10.1098/rspb.2012.1890

Centers for Disease Control and Prevention. (2021) www.cdc.gov/onehealth/basics/zoonotic-diseases.html. Accessed 15 September 2021.

de La Rocque, S., Rioux, J.A., Slingenbergh, J. (2008). Climate change: Effects on animal disease systems and implications for surveillance and control. *Rev. Sci. Tech.* (International Office of Epizootics). 27: 339–354.

Depani, S.J., Kennedy, N., Mallewa, M., Molyneux, E.M. (2012). Case report: Evidence of rise in rabies cases in southern Malawi—better preventative measures are urgently required. *Malawi. Med. J.* 24(3): 61–64.

Derryberry, E.P., Phillips, J.N., Derryberry, G.E., Blum, M.J., Luther, D. (2020). Singing in a silent spring: Birds respond to a half-century soundscape reversion during the COVID-19 shutdown. *Science* (New York, N.Y.), 370(6516): 575–579. https://doi.org/10.1126/science.abd5777

Despommier, D., Ellis, B., Wilcox, B. (2006). The role of ecotones in emerging infectious diseases. *EcoHealth* 3(4): 281–289.

Evander, M., Ahlm, C. (2009). Milder winters in northern Scandinavia may contribute to larger outbreaks of haemorrhagic fever virus. *Global Health Action* 2. https://doi.org/10.3402/gha.v2i0.2020. PMID: 20052429; PMCID: PMC2799289.

Fooks, A.R, Banyard, A.C., Horton, D.L., Johnson, N., McElhinney, L.M., Jackson, A.C. (2014). Current status of rabies and prospects for elimination. *Lancet* 384(9951): 1389–1399. https://doi.org/10.1016/S0140-6736(13)62707-5 .

García-Vega, D., Newbold, T. (2020). Assessing the effects of land use on biodiversity in the world's drylands and Mediterranean environments. *Biodivers. Conserv.* 29: 393–408. https://doi.org/10.1007/s10531-019-01888-4

Garg, V., Aggarwal, S.P., Chauhan, P. (2020). Changes in turbidity along Ganga River using Sentinel-2 satellite data during lockdown associated with COVID-19. *Nat. Hazards Risk* 11(1): 1175–1195.

Hampson, K., Cleaveland, S., Briggs, D. (2011). Evaluation of cost-effective strategies for rabies post-exposure vaccination in low-income countries. *PLoS. Negl. Trop. Dis.* 5(3): e982. https://doi.org/10.1371/journal.pntd.0000982

Jones, M.R., Singels, A., Ruane, A. (2013). Simulated impacts of climate change on water use and yield of irrigated sugarcane in South Africa. *Proc. S. African Sugar Technol. Assoc.* 86: 184–189.

Karesh, W.B., Dobson, A., Lloyd-Smith, J.O., Lubrothm, J., Dixon, M.A., Bennett, M., Aldrich, S., Harrington, T., Formenty, P., Loh, E.H., Machalaba, C.C., Thomas, M.J., Heymann, D.L. (2012). Ecology of zoonoses: Natural and unnatural histories. *Lancet* 380(9857): 1936–1945. https://doi.org/10.1016/S0140-6736(12)61678-X

Keesing, F., Ostfeld, R.S. (2021). Impacts of biodiversity and biodiversity loss on zoonotic diseases. *Proc. Nat. Acad. Sci. USA*. 118(17): e2023540118. https://doi.org/10.1073/pnas.2023540118

Khan, I., Shah, D., Shah, S.S. (2021). COVID-19 pandemic and its positive impacts on environment: An updated review. *Int. J. Environ. Sci. Technol.* 18: 521–530. https://doi.org/10.1007/s13762-020-03021-3

Khan, R., Saxena, A., Shukla, S., et al. (2021). Effect of COVID-19 lockdown on the water quality index of River Gomti, India, with potential hazard of faecal-oral transmission. *Environ. Sci. Pollut. Res.* 28: 33021–33029. https://doi.org/10.1007/s11356-021-13096-1.

Koju, N.P., Kandel, R.C., Acharya, H.B., Dhakal, B.K., Bhuju, D.R. (2021). COVID-19 lockdown frees wildlife to roam but increases poaching threats in Nepal. *Ecol. Evol.* 11(14): 9198–9205. https://doi.org/10.1002/ece3.7778.

Liu, Z., He, C. Wu, J. (2016). The relationship between habitat loss and fragmentation during urbanization: An empirical evaluation from 16 world cities. *PLoS ONE*. 11(4): e0154613. https://doi.org/10.1371/journal.pone.0154613.

Lokhandwala, S., Gautam, P. (2020). Indirect impact of COVID-19 on environment: A brief study in Indian context. *Environ. Res.*188: 109807. https://doi.org/10.1016/j.envres.2020.109807.

Łopucki, R., Kitowski, I., Perlińska-Teresiak, M., Klich, D. (2021). How is wildlife affected by the COVID-19 pandemic? Lockdown effect on the road mortality of Hedgehogs. *Animals* 11(3): 868. https://doi.org/10.3390/ani11030868

Manenti, R., Mori, E., Di Canio, V., et al. (2020). The good, the bad and the ugly of COVID-19 lockdown effects on wildlife conservation: Insights from the first European locked down country. *Biol. Conserv.* 249: 108728. https://doi.org/10.1016/j.biocon.2020.108728

Morand, S., McIntyre, K.M., Baylis, M. (2014). Domesticated animals and human infectious diseases of zoonotic origins: Domestication time matters. *Infect. Genet. Evol.* 24:76–81. https://doi.org/10.1016/j.meegid.2014.02.013.

Mousazadeh, M., et al. (2021). Positive environmental effects of the coronavirus 2020 episode: A review. *Environ. Dev. Sustain.* 23(9):12738–12760. https://doi.org/10.1007/s10668-021-01240-3

Nasi, R., Taber, A., Van, V.N. (2011) Empty forests, empty stomachs? Bushmeat and livelihoods in the Congo and Amazon Basins. *Int. For. Rev*. 13: 355–368.

Paital, B., Das, K., Parida, S.K. (2020). Inter nation social lockdown versus medical care against COVID-19, a mild environmental insight with special reference to India. *Sci. Total Environ*. 728: 138914. https://doi.org/10.1016/j.scitotenv.2020.138914.

Pearce-Duvet, J.M. (2006). The origin of human pathogens: Evaluating the role of agriculture and domestic animals in the evolution of human disease. *Biol. Rev. Camb. Philos. Soc*. 81(3): 369–382. https://doi.org/10.1017/S1464793106007020

Rahman, M.T., Sobur, M.A., Islam, M.S., Ievy, S., Hossain, M.J., El Zowalaty, M.E., Rahman, A.T., Ashour, H.M. (2020). Zoonotic diseases: Etiology, impact, and control. *Microorganisms* 8(9): 1405. https://doi.org/10.3390/microorganisms8091405

Schmeller, D.S., Courchamp, F., Killeen, G, (2020). Biodiversity loss, emerging pathogens and human health risks. *Biodiv. Conserv*. 29: 3095–3102. https://doi.org/10.1007/s10531-020-02021-6

Sinervo, B., et al. (2010). Erosion of lizard diversity by climate change and altered thermal niches. *Science* 328: 894–899. https://doi.org/0.1126/science.1184695

Takahashi, K., Kitajima, N., Abe, N., Mishiro, S. (2004). Complete or near-complete nucleotide sequences of hepatitis E virus genome recovered from a wild boar, a deer, and four patients who ate the deer. *Virology* 330: 501–505. https://doi.org/10.1016/j.virol.2004.10.006

Varma, M.G.R. (2001). Kyasanur Forest disease. In: Service MW, editor. *The Encyclopedia of Arthropod-Transmitted Infections*. New York: CABI Publishing, pp. 254–260.

Wilkinson, D.A., Marshall, J.C., French, N.P., Hayman, D.T.S. (2018). Habitat fragmentation, biodiversity loss and the risk of novel infectious disease emergence. *J. R. Soc. Interface*. 5: 20180403. http://doi.org/10.1098/rsif.2018.0403

Woolhouse, M., Scott, F., Hudson, Z., Howey, R., Chase-Topping, M. (2012). Human viruses: Discovery and emergence. *Philos. Trans. R. Soc*. 367: 2864–2871.

Zambrano-Monserrate, M.A., Ruano, M.A., Sanchez-Alcalde, L. (2020). Indirect effects of COVID-19 on the environment. *Sci. Total Environ*. 728: 138813. https://doi.org/10.1016/j.scitotenv.2020.138813

29

A Solution to Healthy Living in Relation to Vector-Borne Encephalitis

Goutam Chandra and Sunanda Burman
Mosquito, Microbiology and Nanotechnology Research Units, Parasitology Laboratory, Department of Zoology, The University of Burdwan, Golapbag, Burdwan, Purba Bardhaman, West Bengal, India, PIN-713104

CONTENTS

29.1 Introduction

Different forms of encephalitis are dreadful diseases involving human beings and some animal of economic importance. Japanese encephalitis virus (JEV), an arthropod-borne virus, has a wide distribution, including Asia, Western Pacific, and northern Australia. The World Health Organization (WHO) has ascribed Japanese encephalitis (JE) to be the most important cause of mosquito-borne pediatric viral encephalitis in endemic Asian countries, mostly affecting pediatric age group of 0–14 years and accounting about 75% of JE cases. Globally, it accounts for about 68,000 clinical cases per year (Campbell et al., 2011).

In India, in the year 2005, an outbreak of JE resulted accounting a majority of pediatric death among 1,700 total casualties (Parida et al., 2006; Halstead and Thomas, 2011). The virus belongs to Family *Flaviviridae*, contains a positive sense RNA genome within a host-derived membrane. They are distributed

DOI: 10.1201/9781003220398-29

throughout the world crossing the national boundaries, which needs global surveillance. JE is considered to be an emerging zoonotic disease and a major public health problem in India (Lindenbach and Rice, 2001). Symptomatic cases are rare and only 1 in 250 subclinical cases were reported. In case of symptomatic cases about 30% fatality can be encountered in infected patients (Kulkarni et al., 2018). A spectrum of clinical features is found, including flu-like symptoms, stiffness in the neck, seizures, spastic paralysis, disorientation, coma, and even death of the infected patients. JEV is considered to be one of the major public health problems owing to its neuro-psychiatric effects that lead to morbidity and mortality in large number of infected patients, high epidemic potential, high fatality rate, and sequelae among the survivors, and this also imparts socio-economic burden due to the increase in the number of disabilities in affected patients (Campbell et al., 2011; Solomon, 2006). However, JE may affect adults, particularly the people who come to JE endemic areas and lacking childhood immunization against JEV. JE mainly occurs in rural areas mainly in association with rice cultivation and irrigated rice fields and ditches associated with rice fields where vector mosquitoes breed profusely in the close proximity with domestic pigs, wading birds, and ducks. Like most of the flaviviruses, its transmission occurs through mosquitoes, especially by genus *Culex*, in a natural cycle between birds and some species of livestock (Kulkarni et al., 2018). Reservoir hosts include pigs, birds, rodents, squirrels, and also bats. While humans and horses (equines) are the dead-end hosts. Ardeid birds play considerable role in the long-distance propagation of JEV and also act as a reservoir of this viral disease (Endy and Nisalak, 2002). Domestic pigs, having long-lasting viremia, are the key amplifiers of the virus, making them mainly responsible for transferring the disease to humans (Van den Hurk et al., 2008). Insufficient viremia in horses and other vertebrates other than birds make them incidental dead-end hosts (Solomon and Vaughn, 2002).

Pigs are the biological amplifiers of JE and act as reservoirs of the virus (Rojanosuphot and Tsai, 1995). Interestingly, in countries like Pakistan, where piggeries can be rarely found, JE outbreaks are reported only very rarely (Igarashi et al., 1994) . Disease manifestation in the livestock varies from asymptomatic infection in some species to acute neurological disorder in others. Pigs are among the animals that are mainly affected by this viral infection, which leads to abortion, still birth, and birth defects, while the adult swine remains asymptomatic. Thus, this virus is responsible for reproductive problems in pigs, which in turn causes great economic loss for stablemen. Infected piglets face neurological disease, ultimately leading to significant mortality in pigs as well (Shimizu et al., 1954; Yamada et al., 2004; Straw et al., 2006).

Besides pigs, the domestic animals that can be infected by JEV are equines. Infected horses show nonspecific signs, including pyrexia, but this can be sporadically leading to major neurological disorders. Syndromic manifestations in horses include hyper-excitable type syndrome, which results in high fevers (41°C or higher), profuse sweating, and muscle tremors. Behavioral changes include aimless wandering, aggression, vision problems, collapse, coma, and finally death. Neurologic sequelae may result in 5% mortality in horses, which may increase to 30% (Animal_Health_in_the_World/docs/pdf). In extreme cases, where disease manifestation is severe, deaths of infected horses may occur. Though veterinary vaccination is available for both pigs and horses, JEV in livestock exists as a great problem, resulting in great economic loss for caretakers, as detection and treatment of the disease in these domestic animals are not always available or the cost of the treatment are not affordable for the cultivators. Control and prevention of JEV in livestock (i.e., within the veterinary sector) is not only important for the economic benefit of the cultivators but also for the prevention of transmission of viral infection from livestock to humans. Such measures are essential in breaking the zoonotic transmission cycle—that is, association of JEV with domestic animals and consequently entering into the human population where humans act as dead-end hosts (Mansfield et al., 2017).

JEV outbreaks significantly affect agricultural economy and public health, particularly in areas that have low vaccine coverage and lack of proper diagnostic facilities.

As viral outbreaks cannot completely be avoidable, preventative strategies should be taken to control JE incidence. Management strategies of JE eradication do not work fully as pathogens from a population cannot be omitted due to its zoonotic potential. Thus, reevaluation of encephalitis infection control in humans, pigs, and horses is much needed. The present discussion focuses on the topics including types of encephalitis, reservoirs, vector-host relationship, geographical distribution, zoophilic and zoo-anthropophagic cycle in relation to transmission of JEV, and most importantly, the solution to control vector-borne encephalitis.

29.2 Different Forms of Vector-Borne Encephalitis

Symptomatic human arboviral diseases are divided into three different syndromes: (1) febrile systemic illness (e.g., dengue fever), (2) hemorrhagic fever (e.g., dengue hemorrhagic fever and yellow fever), and (3) encephalitis, (e.g., Japanese encephalitis, eastern equine encephalitis, western equine encephalitis, Venezuelan equine encephalitis, and La Crosse encephalitis, which are vector-borne) (Morens et al., 2020).

29.2.1 Japanese Encephalitis (JE)

JE is endemic to the areas of South Asia, East Asia, Southeast Asia, and the Pacific region. JE is caused by JEV, a zoonotic flavivirus. Infection of this virus occurs in frogs, snakes, egrets, bats, and most domestic animals, like cattle, poultry, and ducks. Pigs and ardeid birds are considered as the reservoir and amplifier hosts and are unaffected by the infection of this virus (Halstead, 2003; Animal_Health_in_the_World/docs/pdf; Kuno et al., 2017; Longbottom et al., 2017). Horses and donkeys also suffer from JE other than humans. JE does not spread from human to human because of the low and insufficient viremia in man. Humans are accidental, dead-end hosts.

29.2.2 Eastern Equine Encephalitis (EEE)

Eastern equine encephalitis virus (EEEV) belongs to the genus Alphavirus of Family Togaviridae. In the United States, about six to eight affected persons are reported annually, occurring mainly in May through October, mostly in Florida, Georgia, Wisconsin, New Jersey, and Maryland (Curren et al., 2017; Shah and Cherabuddi, 2016; Garlick et al., 2016). Though passerine birds are considered as principal reservoir hosts of EEEV in North America, rodents can also amplify the virus's natural cycle. The primary reservoir hosts for Madariaga virus (South American EEEV) are still uncertain, but small mammals may play a prominent role. Some experiments have suggested that reptiles (especially snakes) may help maintaining EEEV over the winter. EEEV mainly affects horses and other equid species. However, disease manifestations have been seen in sheep, cattle, dogs, llamas, alpacas, pigs, deer, and non-human primates. Per reports, other susceptible animals include goats, moose, certain rodents, bats, reptiles, and even amphibians.

29.2.3 Western Equine Encephalitis (WEE)

Western equine encephalitis virus also belongs to the genus *Alphavirus* of Family Togaviridae. It was first discovered from a horse in 1930, and it is mainly a disease of equines (https://en.wikipedia.org/wiki/Western_equine_encephalitis_virus). WEE is reported from the states in the Mississippi River of the USA and also in some western Canadian states (Simon et al., 2020). The virus, like other encephalitis viruses, is maintained by a bird-mosquito-bird life cycle. Birds and humans are incidental hosts. Passerine birds are main amplifying reservoir hosts of the virus, and often the natural viral cycle involves blacktail jackrabbit, *Lepus californicus* (Arrigo et al., 2010). Mules, horses, and pheasants more often precede epidemics in human. Infections lead to fever, chills, and malaise, and a high viral load often results in translocation of the virus into central nervous system across the blood-brain barrier, causing cerebral and meningeal inflammation and necrosis (www.ncbi.nlm.nih.gov/books/NBK470228). Asymptomatic infections in cattles, squirrels, other rodents, snowshoe hares (Lepus americanus), opossums, snakes, tortoises, and frogs may occur (www.cfsph.iastate.edu/Factsheets/pdfs/easter_wester_venezuelan_equine_encephalomyelitis.pdf).

29.2.4 Tick-Borne Encephalitis (TBE)

Tick-borne encephalitis, caused by the tick-borne encephalitis virus (TBEV), is also an arbovirus belongs to Family Flaviviridae. It mainly occurs in middle and eastern European countries and some parts in Asia and is considered to be a major health menace for humans. Like any other arbovirus, TBEV is associated with complex interactions between vectors, reservoir hosts, and favorable environments for successful viral transmission. Hard ticks are considered as the vectors and reservoir hosts for TBEV, which are responsible for transmission of this virus to a variety of animals. Small rodents are the main hosts, and humans are accidental hosts of TBEV. TBEV causes meningitis, meningoencephalitis, and also meningoencephalomyelitis in infected patients. The severity of the disease increases with age and mainly affects children and adolescents. The long-term sequelae for months to years present in patients affected with TBEV (Kaiser, 2008).

29.2.5 La Crosse (LAC) Encephalitis

La Crosse (LAC) encephalitis is also a mosquito-borne arbovirus, also termed as bunyavirus (CDC, 2009), that causes inflammation of the brain (Encephalitis), which was first found in La Crosse, Wisconsin, in the year 1963 (www.vdh.virginia.gov/Epidemiology/DEE/Vectorborne/factsheets/lacrosse.htm).

This disease is endemic to several Midwestern and Mid-Atlantic states. This disease mainly occurs in children less than 16 years of age. Other than encephalitis symptoms include fever, headache, vomiting, and lethargy. Encephalitis may be accompanied with seizures, and in severe cases, coma and paralysis occurs. As no specific treatment for LAC encephalitis is present, supportive drug therapy is used in severe cases (www.cdc.gov/lac/gen/qa.html).

29.2.6 Venezuelan Equine Encephalitis (VEE)

The causative viral pathogen of Venezuelan equine encephalitis (VEE) is the Venezuelan equine encephalitis virus (VEEV). Outbreaks commonly occur in horses, donkeys, zebras, mules, and humans. VEEV is a single-stranded positive-sense RNA alphavirus belonging to Family *Togaviridae* (Weaver et al., 2004). VEEV was first documented discovered in Venezuela, Colombia, and Trinidad in 1935 after equine outbreaks (Lundberg et al., 2017).

In 1950 it was discovered in humans during an outbreak in Colombia (Aguilar et al., 2011). There are six subtypes of VEEV (I–VI), among which the epizootic strains IAB and IC are responsible for transmission of encephalitis in equines and humans. Equines are the key reservoir hosts and also amplification hosts for epizootic viral strains. Reservoir hosts for enzootic VEE viruses are wild rodents, while birds may be involved in a few cycles. Enzootic VEEVs often infect opossums (*Didelphis marsupialis*), bats, and dogs. The Mexican IE variant causes VEE in equids (Kumar et al., 2018).

Symptoms of VEEV infection varies from flu-like fever to severe encephalitis in infected persons. Patients may develop symptoms, such as leukopenia, tachycardia, and interstitial pneumonia. It is likely to develop into encephalitis more easily in children than in adults and have a higher rate of morbidity and mortality (Kumar et al., 2018). Symptoms of more severe cases may result into tremors, seizures, changes in behavior, hemiparesis, hemichorea, cranial nerve palsies, ataxia, myoclonus, confusion, somnolence, and eventually in to coma (Weaver et al., 2004; Go et al., 2014; Sharma and Knollmann-Ritschel, 2019).

29.3 Role of Vectors in the Transmission Cycle of Encephalitis

JEV has been isolated from over 30 mosquito species of the genera *Aedes*, *Anopheles*, *Armigeres*, *Culex*, and *Mansonia* (Pearce et al., 2018). These vectors can act as primary or secondary vectors in different geographical areas across the world. However, the presence of JEV does not indicate their capability to be vector of JEV. These vectors can act as primary or secondary vectors in different geographical areas across the world. Among different vectors of JEV, *Cx. annulirostris* is primary vector species in Australia. However *Cx. vishnui*, *Cx. pseudovishnui*, *Cx. gelidus*, *Cx. citiens*, and *Cx. fuscocephela* are among secondary vectors in this region. The main vector of JE in Southern Asia, Southeast Asia, and Eastern Asia is *Cx. tritaeniorhynchus* (Pearce et al., 2018). In India, *Cx. vishnui*, *Cx. pseudovishnui*, and *Cx. tritaeniorhynchus* are considered as primary vectors, while *Ma. indiana*, *Cx. whitmorei*, *Cx. epidesmus*, and *An. subpictus* are among the secondary vectors of JE (Chakravarty et al., 1975). In China, the main vector is *Cx. tritaeniorhynchus* (Tao et al., 2014). Different species of mosquitoes (e.g., *Cx. tritaeniorhynchus*, *Cx. tarsalis*, *Cx. annulirostris*, *Cx. gelidus*, *Cx. fuscocephala*, *Cx. vishnui*, *Cx. sitiens*, *Cx. bitaeniorhynchus*, *Cx. pseudovishnui*, *Cx. whitmorei*, *Culiseta inornata*, *Cx. pipiens*, *Cx. salinarius*, *Cx. pipiens pipiens*, *Cx. quinquefasciatus*, *Cx. nigripalpus*, *Ae. vexans*, *Ae. albopictus*, *Ae. japonicas*, and *Ma. uniformis*) are associated with JEV infection in United States (Oliveira et al., 2018).

In case of EEE, transmission cycle of host-virus-reservoir involves birds and its predominant vector *Cs. melanura* mosquitoes found mainly in the freshwater hardwood swamps in America (Lindsey et al., 2018). Other vectors that can transmit the disease are *Coquillettidia perturbans*, *Ae. cinereus*, and *Ae. canadensis* (Armstrong et al., 2010; Oliver et al., 2018). Infected mosquitoes transmit the virus to tree-perching passerine birds and to the opportunistic mammals, reptiles, or amphibians during bloodmeal consumption from these species. Dead-end hosts of these virus are humans, swine, equids, and pheasants. Cross-transmission may occur due to changes in environment and global warming.

WEE is one of the mosquito-borne viral infections. The vectors for humans are *Cx. tarsalis*, *Culiseta* sp., and some species of *Aedes* mosquitoes. Human epidemics lead to outbreaks in mules, horses, pheasants, and other birds. Birds serve as amplifying hosts, which make them reservoirs of the disease (Simon et al., 2021).

Arthropod vectors of TBEV transmission are primarily the hard ticks. In Europe, the major tick vector is *Ixodes ricinus*. In Russia and Asia, *I. persulcatus* is the principal vector species. In Asia, *Haemaphysalis concinna* also serves as vector of this virus. Around other 22 tick species have been reported to carry the virus (Michelitsch et al., 2019).

The principal vector of La Crosse encephalitis is the eastern tree-hole mosquito, *Ae. triseriatus*. Other tree-hole mosquito species, *Ae. albopictus*, is also an efficient vector of this virus (Westby et al., 2015). This zoonotic pathogen is cycled between the daytime-biting tree-hole mosquito, *Ae. triseriatus*, and vertebrate amplifier hosts like chipmunks and tree squirrels in deciduous forest habitats of the Appalachian and Midwestern regions of the United States and in the southeast of the United States (https://en.wikipedia.org/wiki/La_Crosse_encephalitis; accessed on 19-02-2021).

The VEE virus is usually transmitted by mosquito vectors, although certain types of ticks and mites can transmit the virus as well. The mosquito vector *Culex* (Melanoconion) is normally responsible for transmission of enzootic strain of the virus. Other notable vectors are *Ma. titillans*, *Ochlerotatus taeniorhynchus*, *Psorophora columbiae*, *Ps. confinnis*, *Ae.* sollicitans, and *An. aquasalis*, which mainly transmit the epizootic strains of the vee virus (www.msdsonline.com/resources/sds-resources/free-safety-data-sheet).

29.4 Epidemiology and Outbreaks

Epidemiological pattern of JE generally is of two types. Epidemics in northern temperate areas mainly cover countries like Nepal, China, Korea, Japan, Taiwan, Northern Vietnam, Northern Thailand, and Northern India and occur during the summer months, roughly from May to October. In southern tropical areas, such as Sri Lanka, Southern India, Indonesia, Malaysia, Southern Vietnam, Southern Thailand, and the Philippines, JE tends to be endemic; cases occur sporadically throughout the year, with a peak just after starting of the rainy season during the months of July to September (Tiwari et al., 2012; Erlanger et al., 2009, Koskiniemi et al., 1997).

In India, two epidemics generally occur in Karnataka. Severe cases are generally encountered during April to July and a mild form occurs in September to December (Kulkarni et al., 2018). According to WHO, outbreaks of JE occur every 2–15 years. Rainy season is the time when the peak of transmission occurs as the vector population multiples fast. Agricultural development and rice cultivation with irrigation system support the spread in new areas.

About three billion people live in China and India, where JE virus is endemic (Solomon, 2006). However, statistics of JE occurrence should be interpreted with care because the transmission of JE is exceedingly dynamic in terms of its occurrence in endemic areas and occurs typically in epidemics. Thus, the rise and fall in JE cases can be seen globally. Background determinants triggering epidemics mainly include environmental factors and the spillover effects into the human population.

Per the Centers for Disease Control and Prevention (CDC) records, an average of 8 cases of EEEV per year was recorded for the last 15 years in the USA. In 2019, from June to September, 34 cases were recorded. The death rate of patients with EEEV infection was 35% (Lindsey, 2020).

Western equine encephalitis was reported from western United States and Canada. Epidemics are related with warm weather and heavy rainfall in endemic areas. The most noteworthy outbreak occurred in the year of 1941, during which 3,000 positive WEE human cases were reported. Since 1964, fewer than 700 confirmed cases of WEE were reported from United States (Simon et al., 2021).

VEE, first discovered in horses in 1930s in South America, is a highly infectious vector-borne disease that produces illness in 100% of exposed cases, and it is endemic to Central and South America. A VEE outbreak occurred in 1971 when 88 human infections in southern Texas counties, including Cameron, Hidalgo, and Corpus Christi, were reported (Debboun et al., 2019).

According to the CDC reports, annual cases of La Crosse (LAC) encephalitis ranged between 30 and 130 cases per year in the USA (www.cdc.gov/lac/tech/epi.html).

First confirmed case of TBE was diagnosed in 1993 at Hokkaido Prefecture in Japan. About 10,000–12,000 clinical cases of tick-borne encephalitis are reported worldwide each year (www.who.int/health-topics/tick-borne-encephalitis#tab=tab_1).

29.5 Control Strategies of Vector-Borne Encephalitis

The eradication of vector-borne encephalitis is difficult due to its complex eco-epidemiology, which involves transmission of the virus through natural reservoirs like pigs and birds. Effective preventive measures are the foremost strategy to reduce JE burden, which eventually reduces mortality, morbidity, and disability of a large number of people. The strategies include JE vaccination in affected areas, strengthening of surveillance programs and control of vectors.

29.5.1 Vaccination

Elimination of JE is not possible due to its enzootic cycle involving mammals and birds. Hence, vaccination is the most effective strategy and long-term protection to control JE infection. Effective vaccines for JE are accessible, and vaccination in endemic regions significantly decreases the occurrence of JE.

Four types of vaccines are available for JE—inactivated mouse-brain-derived vaccines, inactivated Vero cell vaccines, live attenuated vaccines, and chimeric vaccines. According to WHO, JE immunization is to be the most important evidence-based guideline for reducing burden of JE (Rustagi et al., 2019).

Though vaccination is available in endemic areas, universal JE vaccination is sacrificed due to some reasons: First, herd immunity cannot be developed in humans as they are dead-end hosts of the virus. Second, due to ineffective pig immunization, as piglets acquire antibodies from their immunized mothers, there is low immunization coverage in rural areas, where outbreaks mainly occur, and also there are availability of favorable conditions for JE transmission in endemic areas.

No human vaccine is available in case of EEEV; however, a veterinary vaccine is available for use in equids.

29.5.2 Vector Control Measures

Vector control measures should be taken against its main mosquito vectors *Cx. tritaeniorhynchus*, *Cx. vishnui*, *Cx. pseudovishnui*, and other mosquito species, which prefer to breed in rural agricultural areas in association with rice production and flooding irrigation and can fly up to five kilometers, infecting human with the virus (Rao, 2001). These vector mosquitoes are zoophilic and considered as primary vectors of JEV (Tuno et al., 2017). The use of nitrogenous fertilizer and employment of irrigation schemes has a positive effect on the vector mosquitoes' breeding. Thus, monitoring and proper irrigation strategies in rice fields and also cultivation of mosquito larvae-eating fish in the logged water in ditches and water passages associated with paddy fields may reduce the vector burden by reducing larvae. Use of biofertilizers in the paddy in place of nitrogenous fertilizer may reduce breeding of mosquitoes by creating unfavorable conditions for breeding in rice fields. Controlling adult mosquitoes in human habitations and piggeries with ultra-low-volume fogging with malathion is effective against JE vectors (Srivastava, 2008). Using novel eco-friendly mosquitocides in breeding areas and controlling adult mosquitoes are also an effective strategy to control JE vector mosquitoes (Adhikari and Chandra, 2012; Singha et al., 2012; Bhattacharya and Chandra, 2013; Haldar et al., 2014; Mallick et al., 2014; Rawani et al., 2017; Barik et al., 2019; Singh and Chandra, 2022). Safety from bite of respective vectors has been reported when mixture of Neem oil and Mustard oil was applied on human body surface (Chatterjee et al., 1996). Thus, controlling the vectors can be an effective solution in high-risk JE endemic zones.

In case of TBE, vaccination is recommended in TBE endemic areas or who will travel to these areas. A booster dose should be given after three to five years thereafter. Prevention of tick bites is the best protection against TBE infection.

29.5.3 Surveillance

Surveillance of JE in endemic regions is suggested throughout the year in the endemic areas to understand the early warning signals of JE outbreaks in a particular area, viral infection in human and animal reservoirs, the seasonal variation in JE cases, age groups affected, socio-economic loss measurement, and also the rate of transmission in presence of favorable conditions. Surveillance data is useful in assessment of effectiveness of vaccination programs by the government and JE control programs. Surveillance strategies in the endemic regions may face many challenges. Sentinel surveillance sites may often lack laboratory facilities for diagnosis of JE, and actual incidence of JE cannot be monitored correctly by the surveillance report.

29.5.4 Behavior Change Communication

Like a malaria prevention strategy, the effective use of behavior change communication (BCC) can also be used in vector-borne JE (Rustagi et al., 2019). This strategy includes avoidance of mosquito bites, shifting of pigsties, maintain good hygiene, and keeping houses sanitized. Personal prophylactic measures, like the application of repellents and using chemically impregnated mosquito nets, decrease human contact to reservoir hosts during dusk and dawn, and wearing full-sleeved clothing may be maintained to avoid JE transmission by mosquitoes. Proper immunization with vaccines in childhood is also essential.

29.5.5 Long-Term Therapeutic Measures

JE survivors with disability should have drug therapy, physiotherapy, and other rehabilitation measures to get rid of sequelae, like seizures, paralysis of body parts, psychiatric problems, difficulties in movement, disorientation, incontinence, and even Parkinsonism (Rao, 2001).

29.5.6 Long-Term Preventive Measures

Measures to control mosquitoes, shifting pigsties, environmental sanitation (prevents many other diseases also), avoiding mosquito bites (prevents many other diseases also), and vaccination of the susceptible population (best of all measures) are useful (Srivastava et al., 2008).

29.6 The Scenario of Vector-Borne Encephalitis in India

The first occurrence of vector-borne JEV dates back to 1952 while first case was reported in the year 1955. JEV is widespread in India and outbreaks have been reported in different states of India (Dhillon and Raina, 2008).

Though JE was already a threat to a large population of India since 2005, outbreak due to development in agriculture, mostly the irrigated paddy fields being the breeding place for vector mosquitoes that carry

JEV, made it the risky endemic zone of JE (Banerjee and Chandra, 2004). Because of India's increasing JE cases National Vector-Borne Disease Control Program (NVBDCP) started to monitor the JE scenario around the nation.

The government of India introduced vector control programs to reduce mosquito load by the application of insecticide with little impact on the huge burden of the disease.

Children in Andhra Pradesh and other areas was immunized by giving an India-made mouse-brain-derived vaccine, but the supply of the vaccine fell short due to India's large number of children across the whole country (Center for Vaccine Innovation and Access. Case Study: India's Leadership in the Fight against Japanese Encephalitis). On the other hand, this vaccine has a complicated dosing schedule and is also costly. However, another vaccine, CD-JEV, made in China, solved India's huge requirements of vaccines, to some extent (Wijesinghe et al., 2016). It was effective and cost-effective. India introduced catch-up campaigns in JE endemic areas. Campaigns were made in high-burden and high-priority districts first, followed by routine JE immunization of children. WHO highly recommended India's strategy to control JE in other countries (www.path.org/programs/center-for-vaccine-innovation-and-access/case-study-indias-leadership-in-the-fight-against-japanese-encephalitis/).

29.7 Solution to Vector-Borne Encephalitis Outbreaks

Due to insufficient antiviral treatment against encephalitis, prevention is the first priority to control this brain disease. It is more difficult to control this disease due to high vector abundance, and also it involves high expenses. However, systematic monitoring by surveillance on vector species in endemic areas and detecting the sero-prevalence of viruses in pigs in particular areas should be done routinely for encephalitis outbreak in those areas. Rehabilitation for surviving patients with disabilities needs to be included in controlling programs. To reduce pediatric mortality rate in JE endemic zones, vaccination is a better option. Thus, solving the vaccine deficit in affected areas must be done. Another method to control the disease is the vaccination of horses and pigs. As human infections are linked to the mosquito-swine

TABLE 29.1

List of Vaccines Against Japanese Encephalitis (Kulkarni et al., 2018; Rustagi et al., 2019)

Type of vaccine	Strain	Route of administration	Number of doses given	Vaccine efficacy/ side effects	Manufacturing/ licensing country
Inactivated mouse-brain-derived vaccines	Nakayama/Beijing-1 (P1)	Subcutaneous injection	3	Cases of acute disseminated encephalomyelitis associated with the vaccine.	Manufactured in Japan (1930s)
Inactivated Vero cell vaccines	Beijing-1 (P1)	Intramuscular injection	2		Manufactured in China, recently licensed in South Korea, Nepal, Sri Lanka, and India
Live attenuated vaccines	SA 14–14–2	Subcutaneous injection	1	Vaccine efficacy is about 100%.	Manufactured in China
Chimeric vaccines	Structural genes (pre-membrane and E) from SA 14–14–2 strain incorporated into the backbone of attenuated strain of yellow fever (YF) virus YF 17D	Subcutaneous injection	1	Vaccine efficacy is about 100%.	Licensed in China, Thailand, and India

transmission cycle, pig vaccination or pig slaughter is also an option to eradicate the disease. Vector control always remains a reliable method for the control of any vector-borne disease.

29.8 Conclusion

Factors responsible for encephalitis outbreaks are climatic conditions that influence the typical focal distribution of the virus in endemic regions and occurrence of specific vector species in these regions. The other important factor is the virus reservoir host that is amplifying the virus and subsequently infects the mosquitoes, which transmit the viruses to human, cattle, and equine hosts.

To fight against all types of encephalitis, the most important remedy is to educate people about mosquito and the relationship between mosquito bites and incidence of encephalitis. As the disease is maintained by the reservoir host or amplifier host, taking care about their habitats is also an important step regarding eradication of the disease. If we take necessary steps to stop mosquito bites to pigs, horses, and other domestic hosts, twofold safety measures to encephalitis can be established. As zoophilic female vector mosquitoes of encephalitis imbibe the blood of their hosts for development of their eggs (to propagate their numbers), reduction of mosquito bites to reservoir and amplifier hosts reduces their chances to produce viable eggs. By this method, not only can the mosquito population be controlled to a level, but the transmission of the disease to dead-end hosts can be diminished. If the reservoir host does not get infected with the encephalitis virus, the transmission of the virus to humans and equines could be ended, thus providing the second shield of protection to encephalitis toward healthy living.

REFERENCES

Adhikari, U. and Chandra, G. 2012. Laboratory evaluation of ethyl acetate and chloroform: Methanol (1: 1 v/v) extract of *Swietenia mahagoni* leaf against Japanese encephalitis vector *Culex vishnui* group. *Asian Pacific Journal of Tropical Disease* 2(6):451–455.

Aguilar, P.V., Estrada-Franco, J.G., Navarro-Lopez, R., Ferro C, Haddow, A.D. and Weaver, S.C. 2011. Endemic Venezuelan equine encephalitis in the Americas: Hidden under the dengue umbrella. *Future Virology* 6(6):721–740.

Animal_Health_in_the_World/docs/pdf. www.oie.int/fileadmin/Home/eng/Animal_Health_in_the_World/docs/pdf/Disease_cards/JAPANESE_ENCEPHALITIS.pdf

Armstrong, P.M. and Andreadis, T.G. 2010 Eastern equine encephalitis virus in mosquitoes and their role as bridge vectors. *Emerging Infectious Diseases* 16(12):1869.

Arrigo, N.C., Adams, A.P., Watts, D.M., Newman, P.C. and Weaver, S.C. 2010. Cotton rats and house sparrows as hosts for North and South American strains of eastern equine encephalitis virus. *Emerging Infectious Diseases* 16(9):1373–1380.

Banerjee, A. and Chandra, G. 2004. Role of some factors on the breeding of JE vector *Culex vishnui* group. *The Journal of Communicable Diseases* 36(4):260–263.

Barik, M., Rawani, A., Laskar, S. and Chandra, G. 2019. Evaluation of mosquito larvicidal activity of fruit extracts of *Acacia auriculiformis* against the Japanese encephalitis vector *Culex vishnui*. *Natural Product Research* 33(11):1682–1686.

Bhattacharya, K. and Chandra, G. 2013. Bioactivity of *Acyranthes aspera* (Amaranthaceae) foliage against the Japanese encephalitis vector *Culex vishnui* group. *Journal of Mosquito Research* 3(13): 89–96.

Campbell, G.L., Hills, S.L., Fischer, M., et al. 2011. Estimated global incidence of Japanese encephalitis: A systematic review. *Bulletin of the World Health Organization* 89:766–774.

Center for Vaccine Innovation and Access. Case Study: India's Leadership in the Fight against Japanese Encephalitis. 2021. https://www.path.org/programs/center-for-vaccine-innovation-and-access/case-study-indias-leadership-in-the-fight-against-japanese-encephalitis/ Accessed on 24.02.2021.

Chakravarty, S.K., Sarkar, J.K., Chakravarty, M.S., Mukerjee, K.K., Das, B.C. and Hati, A.K. 1975. The first epidemic of Japanese encephalitis in India-virological studies. *Indian Journal of Medical Research* 63:77–88.

Chatterjee, S.N., Banerjee, B.D. and Chandra, G.1996. Effect of Neem (*Azadirachta indica*) oil on the biting population of *Culex vishnui* (complex). *Journal of Basic and Applied Biomedicine* 4:39–41.

Curren, E.J., Lehman, J., Kolsin, J., et al. 2017. West Nile virus and other nationally notifiable arboviral diseases—United States. *The Morbidity and Mortality Weekly Report* 67:1137–1142.

Debboun, M., Nava, M.R. and Rueda, L., eds. 2019. *Mosquitoes, Communities, and Public Health in Texas*. USA, Academic Press, Elsevier.

Dhillon, G.P. and Raina, V.K. 2008. Epidemiology of Japanese encephalitis in context with Indian scenario. *Journal of the Indian Medical Association* 106(10):660–663.

Disease Control and Prevention (CDC). January 2009. Possible congenital infection with La Crosse encephalitis virus—West Virginia, 2006–2007. *Morbidity and Mortality Weekly Report* 58(1):4–7.

Endy, T.P. and Nisalak, A. 2002. Japanese encephalitis virus: Ecology and epidemiology. In J.S. Mackenzie, A.D.T. Barrett and V. Deubel (Eds.), *Japanese Encephalitis and West Nile Viruses. Current Topics in Microbiology and Immunology*, vol. 267, pp. 11–48. Berlin, Heidelberg: Springer.

Erlanger, T.E., Weiss, S., Keiser, J., Utzinger, J. and Wiedenmayer, K. 2009. Past, present, and future of Japanese encephalitis. *Emerging Infectious Diseases* 15(1):1.

Garlick, J., Lee, T.J. and Shepherd, P. 2016. Locally acquired eastern equine encephalitis virus disease, Arkansas, USA. *Emerging Infectious Diseases* 22(12):2216.

Go, Y.Y., Balasuriya, U.B. and Lee, C.K. 2014. Zoonotic encephalitides caused by arboviruses: Transmission and epidemiology of alphaviruses and flaviviruses. *Clinical and Experimental Vaccine Research* 3(1):58.

Haldar, K.M., Ghosh, P. and Chandra, G. 2014. Larvicidal, adulticidal, repellency and smoke toxic efficacy of *Ficus krishnae* against *Anopheles stephensi* Liston and *Culex vishnui* group mosquitoes. *Asian Pacific Journal of Tropical Disease* 4:S214–S220.

Halstead, S.B. and Jacobson, J. 2003. Japanese encephalitis. *Advances in Virus Research* 61:103–138.

Halstead, S.B. and Thomas, S.J. 2011. New Japanese encephalitis vaccines: Alternatives to production in mouse brain. *Expert Review of Vaccines* 10(3):355–364.

https://en.wikipedia.org/wiki/La_Crosse_encephalitis. Accessed on 19.02.2021

https://en.wikipedia.org/wiki/Western_equine_encephalitis_virus. Accessed on 20.02.2021.

Igarashi, A., Tanaka, M., Morita, K., et al. 1994. Detection of West Nile and Japanese encephalitis viral genome sequences in cerebrospinal fluid from acute encephalitis cases in Karachi, Pakistan. *Microbiology and Immunology* 38(10):827–830.

Kaiser, R. 2008. Tick-borne encephalitis. *Infectious Disease Clinics of North America* 1;22(3):561–575.

Koskiniemi, M., Korppi, M., Mustonen, K., et al. 1997. Epidemiology of encephalitis in children. A prospective multicentre study. *European Journal of Paediatrics* 156(7):541–545.

Kulkarni, R., Sapkal, G.N., Kaushal, H. and Mourya, D.T. 2018. Suppl-2, M8: Japanese encephalitis: A brief review on Indian perspectives. *The Open Virology Journal* 12:121.

Kumar, B., Manuja, A., Gulati, B.R., Virmani, N. and Tripathi BN. 2018. Suppl-2, M5: Zoonotic viral diseases of equines and their impact on human and animal health. *The Open Virology Journal* 12:80.

Kuno, G., Mackenzie, J.S., Junglen, S., Hubálek, Z., Plyusnin, A. and Gubler, D.J. 2017. Vertebrate reservoirs of arboviruses: Myth, synonym of amplifier, or reality? *Viruses* 9(7):185.

La Crosse Encephalitis. Centers for Disease Control and Prevention (CDC). October 16, 2009. www.cdc.gov/lac/gen/qa.html. Accessed on 21.02.2021.

La Crosse Encephalitis. Epidemiology and Geographic Distribution. www.cdc.gov/lac/tech/epi.html. Accessed on 20.01.2021.

La Crosse Encephalitis. Virginia Department of Health. March 29, 2013. www.vdh.virginia.gov/Epidemiology/DEE/Vectorborne/factsheets/lacrosse.htm. Accessed on 23.02.2021.

Lindenbach, B.D. and Rice, C.M. 2001. Flaviviridae: The viruses and their replication. In D.M. Knipe and P.M. Howley (Eds.), *Field's Virology*, 4th edition, vol. 1, pp. 991–1041. Philadelphia, PA: Lippincott Williams & Wilkins.

Lindsey, N.P., Martin, S.W., Staples, J.E. and Fischer, M. 2020. Notes from the field: Multistate outbreak of eastern equine encephalitis virus—United States, 2019. *Morbidity and Mortality Weekly Report* 69(2):50. https://doi.org/10.15585/mmwr.mm6902a4.

Lindsey, N.P., Staples, J.E. and Fischer, M. 2018. Eastern equine encephalitis virus in the United States, 2003–2016. *The American Journal of Tropical Medicine and Hygiene* 98(5):1472–1477.

Longbottom, J., Browne, A.J., Pigott, D.M. et al. 2017. Mapping the spatial distribution of the Japanese encephalitis vector, *Culex tritaeniorhynchus* Giles, 1901 (Diptera: Culicidae) within areas of Japanese encephalitis risk. *Parasites & Vectors* 10(1):1–2.

Lundberg, L., Carey, B. and Kehn-Hall, K. 2017. Venezuelan equine encephalitis virus capsid—the clever caper. *Viruses* 9(10):279.
Mallick, S., Bhattacharya, K. and Chandra, G. 2014. Mosquito larvicidal potentiality of wild turmeric, *Curcuma aromatica* rhizome, extracts against Japanese Encephalitis vector *Culex vishnui* group. *Journal of Mosquito Research* 4(19):1–6.
Mansfield, K.L., Hernández-Triana, L.M., Banyard, A.C., Fooks, A.R. and Johnson, N. 2017. Japanese encephalitis virus infection, diagnosis and control in domestic animals. *Veterinary Microbiology* 201:85–92.
Michelitsch, A., Wernike, K., Klaus, C., Dobler, G. and Beer, M. 2019. Exploring the reservoir hosts of tick-borne encephalitis virus. *Viruses* 11(7):669.
Morens, D.M., Folkers, G.K. and Fauci, A.S. 2020. Eastern equine encephalitis virus—another emergent arbovirus in the United States. *New England Journal of Medicine* 381(21):1989–1992.
Oliveira, A.R., Cohnstaedt, L.W. and Cernicchiaro, N. 2018. Japanese encephalitis virus: Placing disease vectors in the epidemiologic triad. *Annals of the Entomological Society of America* 111(6):295–303.
Oliver, J., Lukacik, G. and Kokas, J. 2018. Twenty years of surveillance for Eastern equine encephalitis virus in mosquitoes in New York State from 1993 to 2012. *Parasites & Vectors* 11(1):1–2.
Parida, M., Dash, P.K. and Tripathi, N.K. 2006. Japanese encephalitis outbreak, India, 2005. *Emerging Infectious Diseases* 12(9):1427.
Pearce, J.C., Learoyd, T.P., Langendorf, B.J. and Logan, J.G. 2018. Japanese encephalitis: The vectors, ecology and potential for expansion. *Journal of Travel Medicine*. 25(Suppl_1):S16–26.
Rao, P.N. 2001. Japanese encephalitis. *Indian Paediatrics* 38(11):1252–1264.
Rawani, A., Ray, A.S., Ghosh, A., Sakar, M. and Chandra, G. 2017. Larvicidal activity of phytosteroid compounds from leaf extract of Solanum nigrum against *Culex vishnui* group and *Anopheles subpictus*. *BMC Research Notes* 10(1):1–8.
Rojanosuphot, S. and Tsai, T.F. 1995. Regional workshop on control strategies for Japanese encephalitis. *Southeast Asia Journal of Tropical Medicine and Public Health* 26(S3):1–59.
Rustagi, R., Basu, S. and Garg, S., 2019. Japanese encephalitis: Strategies for prevention and control in India. *Indian Journal of Medical Specialities* 10(1):12.
Shah, K.J. and Cherabuddi, K. 2016. Case of eastern equine encephalitis presenting in winter. *Case Reports* 10(2016):bcr2016215270.
Sharma, A. and Knollmann-Ritschel, B. 2019. Current understanding of the molecular basis of Venezuelan equine encephalitis virus pathogenesis and vaccine development. *Viruses* 11(2):164.
Shimizu, T., Kawakami, Y., Fukuhara, S. and Matumoto, M. 1954. Experimental stillbirth in pregnant swine infected with Japanese encephalitis virus. *The Japanese Journal of Experimental Medicine* 24(6): 363–375.
Simon, L.V., Coffey, R. and Fischer, M.A. 2021. Western equine encephalitis. [Updated 2020 August 22]. In: *StatPearls* [Internet]. Treasure Island, FL: StatPearls Publishing. www.ncbi.nlm.nih.gov/books/NBK470228/
Simon, L.V., Sandhu, D.S., Goyal, A. and Kruse, B. 2020. Encephalitis, Japanese. [Updated 2020 February 10]. In: *StatPearls* [Internet]. Treasure Island, FL: StatPearls Publishing.
Singh, A. and Chandra, G. 2022. Larvicidal, pupicidal, repellence and smoke toxic efficacies of *Nicotiana plumbaginifolia* leaves against *Culex vishnui* mosquito. *International Journal of Tropical Insect Science* 42(2):1187–1195.
Singha, S., Adhikari, U., Ghosh, A. and Chandra, G. 2012. Mosquito larvicidal potentiality of *Holoptelea integrifolia* leaf extract against Japanese encephalitis vector, *Culex vishnui* group. *Journal of Mosquito Research* 2(4):25–31.
Solomon, T. 2006. Control of Japanese encephalitis—Within our grasp? *New England Journal of Medicine* 355(9):869–871.
Solomon, T. and Vaughn, D.W. 2002. Pathogenesis and clinical features of Japanese encephalitis and West Nile virus infections. *Current Topics in Microbiology and Immunology* 267:171–194.
Srivastava, V.K., Singh, A. and Thapar, B.R., 2008. Field evaluation of malathion fogging against Japanese encephalitis vector, *Culex tritaeniorhynchus*. *Journal of Vector Borne Diseases* 45:249–250.
Straw, B.E., 2006. *Diseases of Swine*, 9th edition, pp. 359–364. Ames, IA: Blackwell Publishing.
Tao, Z., Liu, G., Wang, M., et al. 2014. Molecular epidemiology of Japanese encephalitis virus in mosquitoes during an outbreak in China, 2013. *Scientific Reports* 4:4908. https://doi.org/10.1038/srep04908
Tiwari, S., Singh, R.K., Tiwari, R. and Dhole, T.N. 2012. Japanese encephalitis: A review of the Indian perspective. *The Brazilian Journal of Infectious Diseases* 16(6):564–573.

Tuno, N., Tsuda, Y. and Takagi, M. 2017. How zoophilic Japanese encephalitis vector mosquitoes feed on humans. *Journal of Medical Entomology* 54(1):8–13.

Van den Hurk, A.F, Ritchie, S.A., Johansen, C.A., Mackenzie, J.S. and Smith, G.A. 2008. Domestic pigs and Japanese encephalitis virus infection, Australia. *Emerging Infectious Diseases* 14(11):1736.

Weaver, S.C., Ferro, C., Barrera, R., Boshell, J. and Navarro, J.C. 2004. Venezuelan equine encephalitis. *Annual Reviews in Entomology* 49(1):141–174.

Westby, K.M., Fritzen, C., Paulsen, D., Poindexter, S. and Moncayo, A.C. 2015. La Crosse encephalitis virus infection in field-collected *Aedes albopictus*, *Aedes japonicus*, and *Aedes triseriatus* in Tennessee. *Journal of the American Mosquito Control Association* 31(3):233–241.

Wijesinghe, P.R., Abeysinghe, M.N., Yoksan, S, et al. 2016. Immunogenicity of live attenuated Japanese encephalitis SA 14-14-2 vaccine among Sri Lankan children with previous receipt of inactivated JE vaccine. *Vaccine* 34(48):5923–5928.

www.cfsph.iastate.edu/Factsheets/pdfs/easter_wester_venezuelan_equine_encephalomyelitis.pdf

www.msdsonline.com/resources/sds-resources/free-safety-data-sheet index/venezuelan-equine-encephalitis/

www.ncbi.nlm.nih.gov/books/NBK470228/. Accesses on 23.02.2021.

www.path.org/programs/center-for-vaccine-innovation-and-access/case-study-indias-leadership-in-the-fight-against-japanese-encephalitis/. Accessed on 25.02.2021.

www.who.int/health-topics/tick-borne-encephalitis#tab=tab_1

Yamada, M., Nakamura, K., Yoshii, M. and Kaku, Y. 2004. Nonsuppurative encephalitis in piglets after experimental inoculation of Japanese encephalitis flavivirus isolated from pigs. *Veterinary Pathology* 41(1):62–67.

30

Helminthiasis in a Tropical Environment in Immunomodulatory Role Regulating Dynamics of COVID-19 Infections

Anita Yadav, Neerja Kapoor, Shivji Malviya, and Sandeep K. Malhotra

CONTENTS

30.1 Introduction

The basal low immune reactivity, as noticed in the persons with chronic helminth infestation, desisted the outbreak of the cytokine storm in the COVID-19-infected people, which was the cause of ARDS (acute respiratory distress syndrome), which was principally responsible for death of a severely COVID-19-infected person. This would explain the ferocious prevalence of COVID-19 with higher mortalities in the modern world, which was well equipped with the amenities, as compared to the appreciably minimized infections in the helminth-infested parts of the world. In a similar fashion, with the addition of elements of immunity in the persons afflicted by asthma and atopia, the theory of advancements of protective factors against COVID-19 was proposed by Rodriguez and Veciana (2020). The inverse association of co-occurring helminth infections and COVID-19 has long been gaining attention (Bradbury et al., 2021), on the basis of which conclusions are being drawn to emphasize reduction in the severity of COVID-19 manifestations in the developing countries where helminth infections are prevalent to an appreciable extent (Figures 30.1 and 30.2).

30.2 Transmission of Helminths and Human Health

Interactions of Macroparasites and Immunomodulatory Effect of Helminths

Helminths, or worms, are invertebrate animals that are involved in a whole range of different pathogens and are able to affect human health. These parasites are included under two major phyla: the

DOI: 10.1201/9781003220398-30

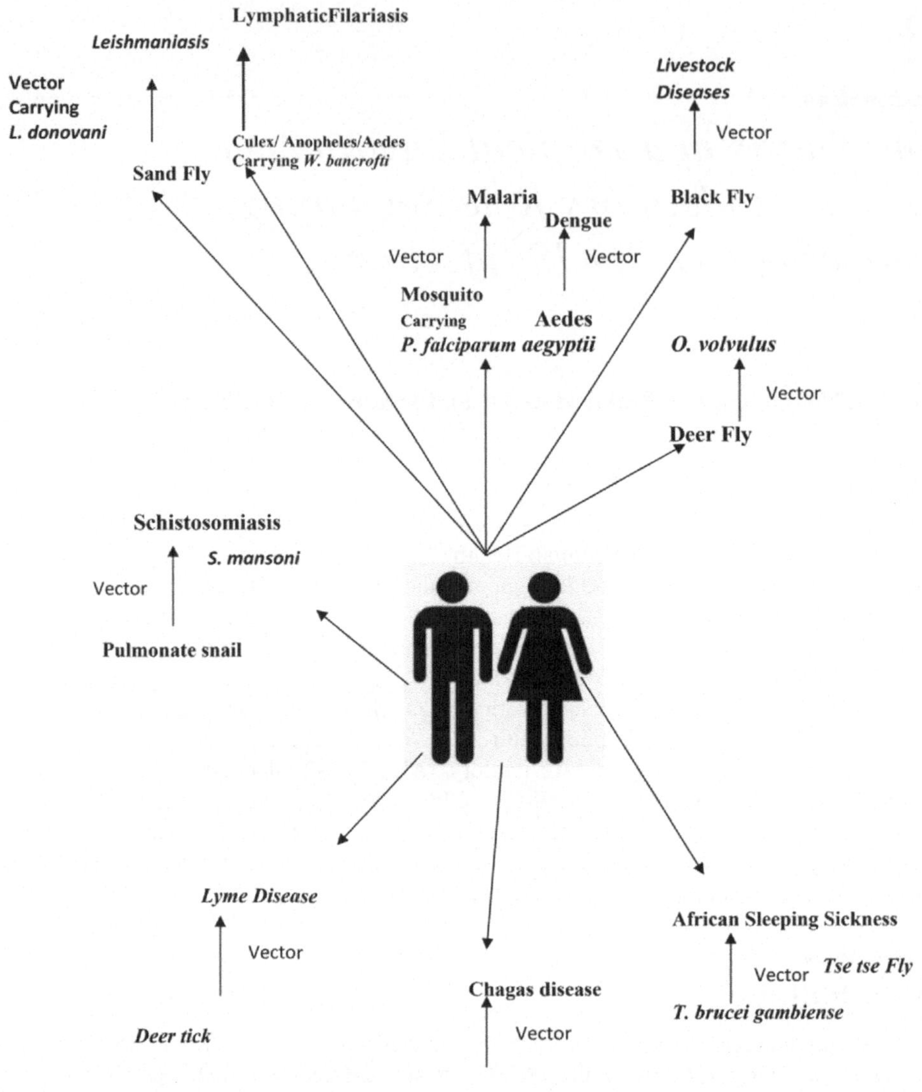

FIGURE 30.1 The pathogenic organisms whose life cycles involve a gastropod or an arthropod vector.

nematodes (or roundworms) and the platyhelminths (or flatworms), the latter in turn being subdivided into trematodes (flukes) and cestodes (tapeworms). Some of them, like hookworms, *Ascaris lumbricoides*, or *Trichiura trichuris* have an oral-fecal mechanism of transmission, while others like hookworms or *Ancylostoma* find their way into the host's blood vessels through the exposed skin. Others, like *Wurchereria bancrofti*, which causes lymphatic filariasis, or *Onchocerca volvulus*, which causes river-blindness disease, are transmitted through permissive dipterans. Finally, some helminths, like *Trichinella spiralis*, are acquired by ingestion of contaminated meat (Johnston et al., 1996; Khetsuriani et al., 2007) (Figures 30.1 and 30.2).

Helminths and Human Health

The prevalence of pathogens comprised spread by worms affect the health of human beings in a significant manner. This is especially the case of developing countries, some of whom claim that the eradication of vectors of certain helminth diseases and the improvement of conditions of health and hygiene will

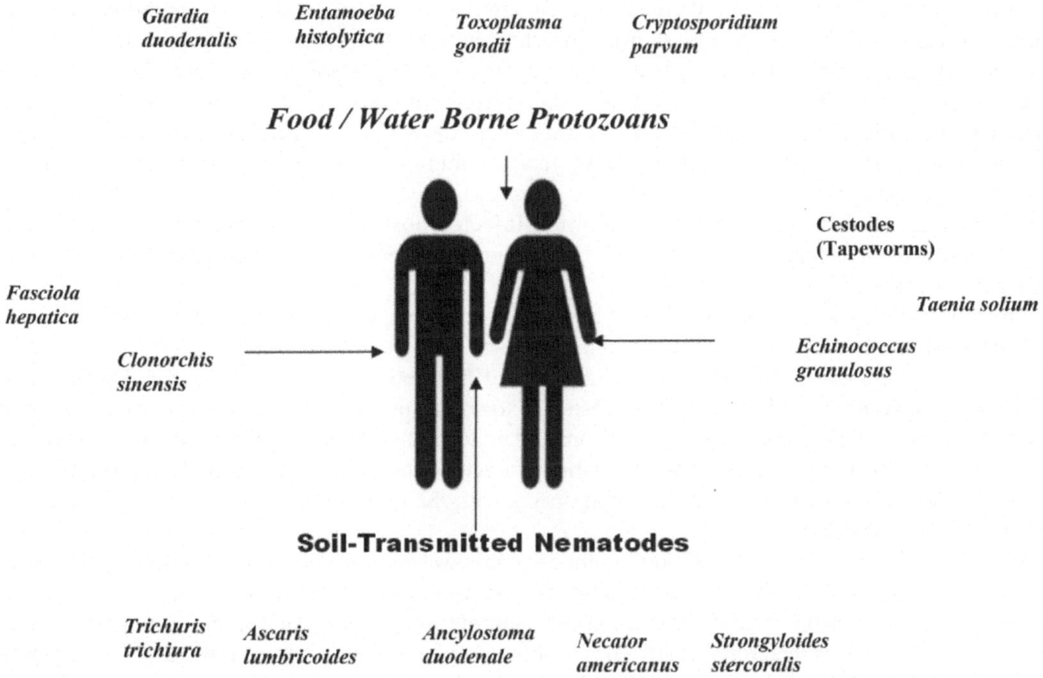

FIGURE 30.2 Protozoans, trematodes, cestodes, and nematodes that are transmitted through the soil or contaminated food or water.

severely restrict the prevalence of vectors of dreadful diseases. At the same time, due to the development of immunity due to long-term intensive helminth infestations in several tropical countries, the immunosuppressive role therein prompted the decline in coronavirus infections. these parasitic helminths are classified into nematodes (comprising roundworms) and platyhelminths (including flatworms), comprising trematodes (flukes) and cestodes (tapeworms). The peculiar ones among roundworms are the hookworms (*Ancylostoma duodenale*) and related species, which possess a specialized teeth mechanism that can penetrate through the skin into human blood vessels and thus reach every part of human body. On the other hand, the oral-fecal route employed by others like *Ascaris lumbricoides* and *Trichuris trichiura* enable them adopt a gastro-parasitic interaction to the detriment of human health. Others like *Trichinella spiralis* find their way into the human body through contaminated meat, while the intermediate hosts, like the permissive dipterans, carry *Wuchereria bancrofti*, the causative agent of lymphatic filariasis, and *Onchocerca volvulus*, the causative agent of onchocerciasis, or the river blindness disease (Wright et al., 2018; Muller, 2002). Diseased persons who harbor *A. lumbricoides* or *T. trichiura* exhibit counteracting interactions between interleukin-5 (probably mediated by eosinophils) and interleukin-13 secretions in case of reinfection (Turner et al., 2003; Jackson et al., 2004, 2006).

The cycle of transmission of helminths is necessarily maintained by the application of a mechanism to avoid acute morbidity, though the facilitation of such regulatory mechanism would certainly require an appropriate environment. This is maintained by the production of interleukin-10 in the role of a regulatory cytokine in the life cycle of soil-transmitted helminth infections, though its influence on the severity of the infection is yet to be determined. However, the crude *Ascaris* antigen from the patients who already harbor *Ascaris* or *Trichuris* evoke high intensity response of interleukin-10 (Geiger et al., 2002). The heavier *Ascaris* infection in older people have higher concentrations of interleukin-10 in the investigations conducted by Jackson et al. (2004).

The survival of soil-transmitted helminths suggests that they succeed by achieving some form of balanced parasitism, in which transmission is maintained and acute morbidity avoided. This ideal homoeostatic state almost certainly needs an environment rich in regulatory mechanisms.

Interleukin-10 is the most abundantly produced regulatory cytokine in soil-transmitted helminth infection. However, its role in maintaining the chronicity of soil-transmitted helminth infection is unclear (Maizels et al., 2004). Geiger and colleagues (2002) reported that interleukin-10 responses to crude ascaris antigen were high in individuals infected with *Ascaris* or *Trichuris*. Whereas Turner and coworkers (Turner et al., 2003) reported that interleukin-10 concentrations declined with intensity of ascaris infection in older individuals, Jackson and colleagues showed that higher interleukin-10 concentrations correlated with heavier *Ascaris* infection in older people. It has been opined by Yazdanbakhsh et al. (2001) that the helminth-infected hosts might be benefited by these down-regulatory immune mechanisms that facilitated to block the predisposition toward developing certain allergic hypersensitivity reactions.

The immune response to these infections has long been known to share key features with the allergic response, especially the enhanced Th2 response. In view of these immunological features and the complementary geographic distribution of soil-transmitted helminth infection and allergic disease, many studies have investigated whether the Th2 response to soil-transmitted helminths protects or preempts the host from developing allergic manifestations linked to Th2, a theory known as the hygiene hypothesis (Yazdanbakhsh et al., 2001). The interactions of secretomes, the set of proteins expressed by the nematodes and secreted into the extracellular space, with the tissues of the host ensure the success of soil-transmitted helminths.

The great economic disparity in middle-, upper-, and lower-income groups is common in most of the tropical and subtropical areas of the world. The variability of environment thus displays a disproportionate infection status of the masses by a variety of helminth infections. The immunity during the host-helminth interactions is therefore greatly compromised, with a skewing impact toward type 2 responses (Harris and Loke, 2017). T helper 1 (TH1) cells are suppressed by the former responses, which thereby skews cytokine response profiles toward IL-4, IL-5, IL-9, and IL-13, triggered initially by TH2 cells in the system, along with alternatively activated macrophages (AAMs) (Harris and Loke, 2017; Brosschot and Reynolds, 2018). Simultaneously, the host's type 1 responses are consequentially hindered by the wider extension of the responses by the regulatory T (Treg) cells and B cells (Harris and Loke, 2017). The pro-inflammatory chemokines' (Rolot, 2018) discharge is impeded, and IL-10 and Treg cell production is induced by the helminth-secreted immunomodulatory proteins. In addition, the systemic immunomodulatory effects were also the outcome of the transitioned state of the microbiome within the gut of humans because of helminth invasion (Harris and Loke, 2017).

The coevolved immunoregulatory effects of a variety of commonly occurring soil-transmitted helminths (STHs) have a special significance. Usually, poverty-stricken areas in different parts of the world are known to harbor (Du et al., 2007) specific bacterial, viral, and other parasitic diseases caused by the pathogens of zoonotic significance. The global public health regulatory bodies enumerate these organisms under the umbrella of neglected tropical diseases. Therefore, population density significant impacts the severity of the STH–SARS-CoV-2 complex. This also necessitated the identification of land areas of various countries all over the globe where STH constituents were harbored by human hosts in great abundance, and more so the segregation on the basis of different income groups need be researched in greater detail.

It is intriguing, however, that the influence of environmental forces is critical to the prevalence of helminth infections in humans, and simultaneously, the combined impact of higher climatic temperature on the pandemic in Uttar Pradesh, which was moderated by low relative humidity in the environment, was recently concluded by Yadav, Kapoor, Malviya and Malhotra (2020). The principal advantage of applying component analysis was to provide an insight to explain the variability that was an outcome of a gamut of interactions between the distribution of total persons infected by novel coronavirus *vis-à-vis* climatic temperature and humidity in Prayagraj, Uttarakhand, and Uttar Pradesh. Therefore, the climatic temperature appeared to be a significant player in the spread of novel coronavirus in the northern states of the country (Yadav et al., 2020).

There are many types of macroparasites that infect humans. STHs, due to their global prevalence and coevolved immunoregulatory effects, are highlighted (Figures 30.1 and 30.2). STH infections are the most prevalent neglected tropical diseases (i.e., a group of bacterial, viral, and parasitic diseases found in areas of extreme poverty) worldwide (Yasui et al., 2008). Notably, STHs infect more than a quarter of the global population, with millions of individuals infected with a few key species: ascarids (800 million cases globally), whipworms (400 million cases globally), and hookworms (400 million cases globally) (Bolles et al., 2011). While often asymptomatic or mild, severe STH infection can result in nutritional deficiencies, mental and developmental delays, organ failure, and death (Tseng et al., 2012).

Fewer of these species at the centerstage encompass ascarids (800 million), whipworms (400 million), and hookworms (400 million) (Jourdan et al., 2018). Ironically, a lot many asymptomatic or mild to severe STH infections co-occurred with COVID-19 outbreak, and both of these, together, resulted in nutritional deficiencies, embryological deformities, organ system failures, and death (Bethony et al., 2006).

30.3 SARS-CoV-2

The prevalence of the COVID-19 virus and the formulation of *the* exit strategy from the pandemic during its second wave by vaccine interventions have recently been published by Yadav, Malviya, and Malhotra (2020). The identification of key mutations, with the listing of variant strains, were succinctly presented therein. Mystery surrounds the international COVID-19 outbreak. The mechanism of entry of the virus into the human body, its evolutionary lineage, and classification of variants were reviewed by Yadav, Malviya and Malhotra (2021). The prevalence of SARS-CoV-2 was recorded world over and at various rates in different countries. The peak of second wave started receding in India at 2.18 lakh of corona-infected people during the last 24 h of May 27, 2021, and 4,039 deaths were reported. The high infection rate, possibility under conditions of comorbidity and mortality due to COVID-19, is related to the demographic, especially affecting older adults and those with chronic infections (CSSE, John Hopkins University, 2020). In a recent assessment by Yadav et al. (2020), the marked influence of climatic temperature with the undercurrent of low humidity having moderating effect on the fluctuating prevalence of SARS-CoV-2 determined the seasonal fluctuations of this virus in areas of Uttar Pradesh in India. Several disease determinants and the involvement of current living styles that could address the prevalence and severity of the disease have been the focus of the study in recent times. The present review deliberates on the immunoregulatory effects of the coexisting helminth parasites that are possibly coevolutionary partners, as they emerged in human beings and contributed to immunoregulatory effects (Bloomfifield et al., 2016; Helmby, 2015 and Yazdanbakhsh, 2002) after dampening harmful inflammatory responses that were typical of COVID-19 (Maizels, 2016).

The most frequently affected body organs during the second wave of COVID-19 was also the gastrointestinal tract, which was not that severely infected during the first wave in 2020, in addition to blood vessels and the central nervous system, under the influence of respiratory distress (Huang et al., 2020; Li et al., 2020; Pan et al., 2020). However, various manifestations to no symptoms whatsoever, along with high variability in the rates of mortality in the 20 most affected countries—1.28% in India, 2% in the Philippines and South Africa, 3% in Brazil and the United States, 6% in Spain, and 11% in the Mexico (Center for Systems Science and Engineering [CSSE], 2020)—reflected inherent mechanisms of influence that were not apparent.

30.4 Zoonotic Transmission of the Novel Coronavirus

The zoonotic transmission of the novel coronavirus was understood to have been the main mode of existence of the virus since its outbreak in 2020. But now, the opinion that SARS-CoV-2 has been artificially created by the scientists in China has been gaining ground, instead of having the view that it naturally emerged from the seafood market in Wuhan City, spreading throughout Hubei Province at China. The virus was thought to have originated from an animal like a pangolin (Figure 30.3) after its transmission from bats, before reaching its final destination, humans. Recently, its widespread human transmission (Figure 30.4) made it into a pandemic (Shereen et al., 2020).

The Vulnerability Dilemma Associated with the COVID-19 Outbreak

COVID-19 affects multiple organ systems, resulting in major respiratory distress, potentially impacting the blood vessels, gastrointestinal tract, and central nervous system (Huang et al., 2020; Li et al., 2020; Pan et al., 2020). Current evidence suggests that up to 81% of infected individuals may experience few to no symptoms (Rodriguez and Vecian, 2020; Yoon et al., 2020; Ariyaratne and Finney, 2019), while mortality rates in the 20 most affected countries are highly variable depending on country (e.g., 2% in the Philippines and South Africa, 3% in Brazil and the United States, 6% in Spain, and 11% in the Mexico) (Yang et al., 2012). The range and variation in disease outcomes suggests that there are individual- and population-level factors at play that shape COVID-19 prognosis, including pre-existing health conditions. Recent evidence has clearly linked obesity, cardiovascular disease, autoimmune diseases, and other chronic diseases with more severe COVID-19 symptoms and an increased risk of hospitalization and death from infection (Jordan et al., 220; Sarzi-Puttini and Anitvalle, 2020; Popkin et al., 2020).

The indication of factors associated with public health at the individual and community levels, bearing the potential to alter the pattern of infection, was obvious. That would mean the severity of disease was a factor in an overreactive inflammatory response, owing to the linkage of comorbidities that induce pro-inflammatory immune activity (Jose and Manuel, 2020; Wu and Yang, 2020).

Socio-Ecological Avenues of Public Health

The growing number of individuals in a community that have been exposed to poor hygienic conditions (Figure 30.3), with an inadequate public health infrastructure, led to malnutrition-oriented survival opportunities in the low-income group countries. Thus, the co-occurring disease outbreaks by

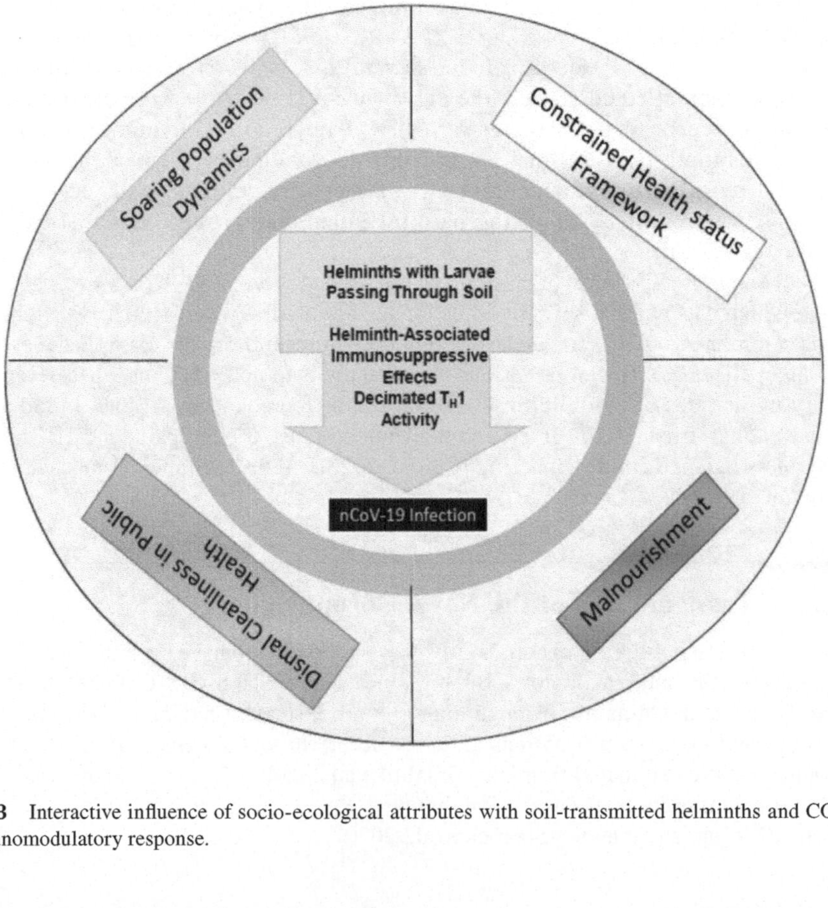

FIGURE 30.3 Interactive influence of socio-ecological attributes with soil-transmitted helminths and COVID-19, generating immunomodulatory response.

soil-transmitted helminth species exposure, combined with higher possibility of air-borne SARS-CoV-2 infectivity, were significantly impacted by the socio-ecological factors. Hence, helminth-induced immunosuppression regulated the severity of COVID-19, propounded through further detailed analysis of susceptibility to lower SARS-CoV-2 intensity, as suggested in several publications.

Nevertheless, an outburst of cytokines (cytokine storm) within a short interval of time, due to derailed immune reaction under sudden deterioration in health of patients, results into hypercoagulation as well as tissue destruction.

The mortality rate variations were also influenced by the age of diseased persons. Besides persons with comorbidities, elderly patients showed greater susceptibility to the disease that resulted into frequent deaths of people under older adult category (Jordan et al., 2020). On contrary to this, the symptoms were not all that severe in children, though they may well be infected. The mortality among the oldest was the highest during the first wave and, therefore, was emphasized in such studies. Children, on the other hand, though exposed to infections for long time, did not show severe manifestations (Bi et al., 2020) owing to the modifications of cell signaling and remodeling of the immune response linked to the inflammation process (Ludvigsson, 2020). In fact, certain recent investigations suggested that the children affected earlier by COVID-19 went into the associated hyperinflammatory responses and developed the potentially lethal multisystem inflammatory syndrome (World Health Organization (WHO), 2020a, 2020b).

There was a clear negative relationship between hookworm burden and IFN-γ responses to the hookworm antigen. Such a negative relationship could indicate hookworm-associated immunosuppressive effects or could be evidence of protective immune responses (Ottesen et al., 1977, Nutman et al., 1987, Sartono et al., 1995; de Jesus et al., 1993). However, that hookworms suppress IFN-γ responses suggests that these responses damage adult worms, and some schistosome studies show a protective role for proliferative (Roberts et al., 1993) and IFN-γ (Acosta et al., 2002) responses. The immunosuppressive effect is apparently Th1-specific because there was no evidence of suppression of IL-4 or IL-5 responses, and the total IgE was positively correlated with hookworm burden and decreased significantly after treatment. Changes in IgE may reflect the level of antigenic stimulation and/or downregulatory IFN-γ responses. Infection with tissue-dwelling filarial nematodes and schistosomes is known to result in the downregulation of parasite-specific immune responses (Ottesen et al., 1977). Immunosuppression has been associated with production of IL-10 and transforming growth factor-β (Doetze et al., 2000; Araujo et al., 1996; King et al., 1996).

Eosinophil Responses During Viral Respiratory/SARS-CoV-2 Infections

The very existence of specialized cells in the form of eosinophils within the body of the parasitized host, whose cytoplasm harbors eosinophilic granules (Rodriguez and Veciana, 2020; Yoon et al., 2020), which are equipped with the twin molecules, the RNAses EDN and ECP, has the inherent potential to overpower RNA viruses (Ariyaratne and Finney, 2019; Yang et al., 2012; Orecchioni et al., 2019), such as SARS-CoV2 (Yoon et al., 2020; Ariyaratne and Finney, 2019), for such eosinophil cells assume front-line defense to fight against such viruses (Honda-Okubo et al., 2015; Chen et al., 2020; Flores-Torres et al., 2019). The NO produced via inducible NO synthetase, by human and murine eosinophils, can directly exhibit antiviral impact on parainfluenza viruses and RSVs (Drake et al., 2016; Phipps et al., 2007; Su et al., 2015).

It has been ascertained that a ribonuclease-dependent mechanism strengthens human eosinophilic constituents to disable RSVs by the activity of the eosinophil-derived neurotoxin (EDN/RNAse2) and eosinophil cationic protein (ECP/RNAse3) (Domachowske et al., 1998a, b). The pro-inflammatory role of eosinophils in the lining of respiratory tract to modulate immune responses illustrates that the inflammatory responses were detrimentally influenced by these cells. These chemically sophisticated arsenals equip eosinophils with molecular memory to determine retaliatory responses against viruses in the respiratory tract (Flores-Torres et al., 2019), as revealed by pre-clinical trials conducted on albino mice.

Scientists have been intrigued with the possibility of (1) any alternative passage of SARS-CoV-2 in a patient who is susceptible to eosinophilia-related diseases; (2) any specifically detected features of disease in patients with declined eosinophilic cells, which could vary from the commonly occurring acute respiratory deterioration during infection with severe acute respiratory syndrome (SARS) coronavirus

(CoV) 2 (SARS-CoV-2), the causative agent of COVID-19 (Zhang et al., 2020); and (3) if the deteriorating health conditions of eosinophil-deficient patients contributed to the altered lung pathology that occur under the influence of COVID-19 in the vaccination studies that were conducted to counter COVID-19 during earlier years (Deming et al., 2006; Du et al., 2007; Yasui et al., 2008; Bolles et al., 2011; Tseng et al., 2012; Iwata-Yoshikawa et al., 2014; Honda-Okubo et al., 2015). The invasive eosinophilic response, accompanied with infective severity, might raise an alarm to apply careful safety evaluations for the enhancement of the immune response by increasing the speed and extent of its development and by prolonging its duration, based on the experience with COVID-19 vaccines (Chen et al., 2020; Flores-Torres et al., 2019).

Specialized cells in the form of eosinophils within the body of the parasitized host, whose cytoplasm harbors eosinophilic granules (Rodriguez and Vecian, 2020; Yoon et al., 2020), which are equipped with the twin molecules, the RNAses EDN and ECP, can overpower RNA viruses (Ariyaratne and Finney, 2019; Yang et al., 2012; Orecchioni et al., 2019) such as the coronavirus (Yoon et al., 2020; Ariyaratne and Finney, 2019), for such eosinophil cells act as front-line defense.

Other Comorbid Associations of SARS-CoV-2 Infections

Very low prevalence of asthma or comorbid atopic disease associated with SARS-CoV-2 infections have been recorded (Zhang et al., 2020). However, the earlier prevalent SARS-CoV-1 infections vehemently triggered viral-induced asthma aggravations (Edwards et al., 2017; Arden et al., 2010; Johnston et al., 2005; Johnston et al., 1996; Khetsuriani et al., 2007). Therefore, no major role of asthma has been assigned to cause the SARS-CoV-2 infections (Yin et al., 2018).

The scientists have been intrigued with the possibility of- i) any alternative passage of SARS-CoV-2 in a patient that was susceptible to eosinophilia-related diseases, ii) any specifically detected features of disease in patients with declined eosinophilic cells, that could vary from the commonly occurring acute respiratory deterioration during infection with severe acute respiratory syndrome (SARS) coronavirus (CoV) 2 (SARS-CoV-2), the causative agent of COVID-19 (Zhang et al., 2020) and iii) if the deteriorating health conditions of eosinophilic deficiency marked patients contributed to the altered lung pathology that occurred under the influence of COVID-19 disease. The vaccination studies were conducted to counter severe acute respiratory syndrome coronavirus 1 (SARS-CoV-1) during earlier years (Deming et al., 2006; Du et al., 2007; Yasui et al., 2008; Bolles et al., 2011; Tseng et al., 2012; Iwata-Yoshikawa et al., 2014; Honda-Okubo et al., 2015). The invasive eosinophilic response accompanied with infective severity might raise an alarm to apply careful safety evaluations for the enhancement of the immune response by increasing the speed and extent of its development and by prolonging its duration, based on the experience with SARS-CoV vaccines (Chen et al., 2020; Flores-Torres et al., 2019).

30.5 Conclusions

1. COVID-19 is affected by rapidly mutating proteins after every strike against the human host.
2. The immunomodulating influence of co-infecting parasitic helminths virtually negated the effects of the coronavirus.
3. The immunomodulatory events triggered by the helminths harbored by the human host prior to COVID-19 infection result into characteristically reduced levels of pro-inflammatory cytokines and enhanced regulatory immune cells that were the predictors to counter clinical deterioration, followed by a decline in COVID-19 severity.

30.6 Acknowledgments

NK and AY are thankful to the principal and head of the Department of Zoology, CMP Post-Graduate Constituent College (A Constituent College of the University of Allahabad), Prayagraj, 211 001, UP, India, for the facilities. SJM acknowledges the cooperation and facilities of UPHESC, Prayagraj, UP.

REFERENCES

Acosta LP, Waine G, Aligui GDL, Tiu WU, Olveda RM, McManus DP (2002). Immune correlate study on human *Schistosoma japonicum* in a well-defined population in Leyte, Philippines. II. Cellular immune responses to *S. japonicum* recombinant and native antigens. *Acta Trop.* 84: 137–149.

Araujo MI, Ribeiro de Jesus A, Bacellar O, Sabin E, Pearce E, Carvalho EM (1996). Evidence of a T helper 2 type activation in human schistosomiasis. *Eur J Immunol.* 26: 1399–1403.

Arden KE, Chang AB, Lambert SB, Nissen MD, Sloots TP, Mackay IM (2010). Newly identified respiratory viruses in children with asthma exacerbation not requiring admission to hospital. *J Med Virol.* 82: 1458–1461.

Ariyaratne A, Finney CAM (2019). Eosinophils and macrophages within the Th2-induced granuloma: Balancing killing and healing in a tight space. *Infect Immun.* 87(10). doi:10.1128/iai.00127-19.

Bethony J, Brooker S, Albonico M, et al. (2006). Soil-transmitted Helminth infections: Ascariasis, trichuriasis, and hookworm. *Lancet*, 367: 1521–1532.

Bi Q, Wu Y, Mei S, et al. (2020). Epidemiology and transmission of COVID-19 in 391 cases and 1286 of their close contacts in Shenzhen, China: A retrospective cohort study. *Lancet Infect Dis.* 20: 911–919.

Blish CA, Sangare´ L, Herrin BR, et al. (2010). Changes in plasma cytokines after treatment of Ascaris lumbricoides infection in individuals with HIV-1 infection. *J Infect Dis.* 201: 1816–1821.

Bloomfifield SF, Rook GAW, Scott EA, et al. (2016). Time to abandon the hygiene hypothesis: New perspectives on allergic disease, the human microbiome, infectious disease prevention and the role of targeted hygiene. *Perspect. Public Health.* 136: 213–224.

Bolles M, Deming D, Long K, Agnihothram S, Whitmore A, Ferris M (2011). A double-inactivated severe acute respiratory syndrome coronavirus vaccine provides incomplete protection in mice and induces increased eosinophilic proinflammatory pulmonary response upon challenge. *J Virol.* 85: 12201–12215.

Bradbury RS, Piedrafita D, Greenhill A Mahanty S (2021). Will Helminth coinfection modulate COVID-19 severity in endemic regions? doi:10.1038/s41577-020-0330-5.

Brosschot TP, Reynolds LA (2018). The impact of a helminth-modified microbiome on host immunity *Mucosal Immunol.* 11: 1039–1046.

Center for Systems Science and Engineering (CSSE) at Johns Hopkins University (2020a). *COVID-19 Dashboard.* https://coronavirus.jhu.edu/map. html (last accessed 6 September 2020).

Center for Systems Science and Engineering (CSSE) at Johns Hopkins University (2020b). *Maps & Trends: Mortality Analyses.* https://coronavirus.jhu.edu/data/mortality (last accessed 6 September).

Channappanavar R, Perlman S (2017). Pathogenic human coronavirus infections: Causes and consequences of cytokine storm and immunopathology. *Semin Immunopathol.* 39: 529–539.

Chen WH, Strych U, Hotez PJ, Bottazzi ME (2020). The SARS-CoV-2 vaccine pipeline: An overview. *Curr Trop Med Rep.* 1–4.

Cooper PJ, Chico ME, Sandoval C, et al. (2000). Human infection with *Ascaris lumbricoides* is associated with a polarized cytokine response. *J Infect Dis.* 182: 1207–1213.

Cui J, Li F, Shi Z-L (2019). Origin and evolution of pathogenic coronaviruses. *Nat Rev Microbiol.* 17(3): 181–192.

de Jesus AMR, Almeida RP, Bacellar O, et al. (1993). Correlation between cell mediated immunity and degree of infection in subjects living in an endemic area of schistosomiasis. *Eur J Immunol.* 23: 152–158.

Deming D, Sheahan T, Heise M, Yount B, Davis N, Sims A (2006). Vaccine efficacy in senescent mice challenged with recombinant SARS-CoV bearing epidemic and zoonotic spike variants. *PLoS Med.* 3: e525.

Doetze A, Satoguina J, Burchard G, et al. (2000). Antigen-specific cellular hyporesponsiveness in a chronic human helminth infection is mediated by Th3/Tr1-type cytokines IL-10 and transforming growth factor-β but not by a Th1 to Th2 shift. *Int Immunol.* 12: 623–630.

Domachowske JB, Dyer KD, Adams AG, Leto TL, Rosenberg HF (1998a). Eosinophil cationic protein/RNase 3 is another RNase A-family ribonuclease with direct antiviral activity. *Nucleic Acids Res.* 26: 3358–3363.

Domachowske JB, Dyer KD, Bonville CA, Rosenberg HF (1998b). Recombinant human eosinophil-derived neurotoxin/RNase 2 functions as an effective antiviral agent against respiratory syncytial virus. *J Infect Dis.* 177: 1458–1464.

Drake MG, Bivins-Smith ER, Proskocil BJ, Nie Z, Scott GD, Lee JJ (2016). Human and mouse eosinophils have antiviral activity against parainfluenza virus. *Am J Respir Cell Mol Biol.* 55: 387–394.

Du L, Zhao G, He Y, Guo Y, Zheng BJ, Jiang S (2007). Receptor-binding domain of SARS-CoV spike protein induces long-term protective immunity in an animal model. *Vaccine* 25: 2832–2838.

Edwards MR, Strong K, Cameron A, Walton RP, Jackson DJ, Johnston SL (2017) Viral infections in allergy and immunology: How allergic inflammation influences viral infections and illness. *J Allergy Clin Immunol.* 140: 909–920.

Flores-Torres AS, Salinas-Carmona MC, Salinas E, Rosas-Taraco AG (2019) Eosinophils and respiratory viruses. *Viral Immunol.* 32: 198–207.

Gaze S, McSorley HJ, Daveson J, et al. (2012). Characterising the mucosal and systemic immune reponses to experimental human hookworm infection. *PLoS Pathog.* 8: e1002520.

Geiger SM, Massara CL, Bethony J., et al. (2002). Cellular responses and cytokine profiles in *Ascaris lumbricoides* and *Trichuris trichiura* infected patients. *Parasite Immunol.* 24: 499–509.

Harris NL, Loke P (2017). Recent advances in type-2-cell-mediated immunity: Insights from helminth infection. *Immunity.* 47: 1024–1036.

Hays R, Pierce D, Giacomin P, Loukas A, Bourke P, McDermott R (2021). Helminth coinfection and COVID-19: An alternate hypothesis. *PLOS Neglect Trop Dis.* doi:10.1371/journal.pntd.0008628.

Helcio RB, Colaco OM, Dulikravich GS, Ferreira LFS (2020). Estimation of state variables and model parameters for the evolution of COVID-19 in the city of Rio de Janeiro. doi:10.1101/2020.05.21.20108407.

Helmby H (2015). Human helminth therapy to treat inflammatory disorders—Where do we stand? *BMC Immunol.* 16: 1471–2172.

Honda-Okubo Y, Barnard D, Ong CH, Peng BH, Tseng CT, Petrovsky N (2015). Severe acute respiratory syndrome-associated coronavirus vaccines formulated with delta inulin adjuvants provide enhanced protection while ameliorating lung eosinophilic immunopathology. *J Virol.* 89: 2995–3007.

Huang C, Wang Y, Li X, et al. (2020). Clinical features of patients infected with 2019 novel coronavirus in Wuhan, China. *Lancet.* 395: 497–506.

IdalíSaboyá M, Catalá L, Nicholls, RS, Ault SK (2013). Update on the mapping of prevalence and intensity of infection for soil-transmitted helminth infections in Latin America and the Caribbean: A call for action. *PLoS Negl Trop Dis.* 7(9): e2419. doi:10.1371/journal.pntd.0002419.

Ing AJ, Cocks C, Green JP. (2020) COVID-19: In the footsteps of Ernest Shackleton. *Thorax.* 75: 693–694.

Iwata-Yoshikawa N, Uda A, Suzuki T, Tsunetsugu-Yokota Y, Sato Y, Morikawa S (2014) Effects of Toll-like receptor stimulation on eosinophilic infiltration in lungs of BALB/c mice immunized with UV-inactivated severe acute respiratory syndrome-related coronavirus vaccine. *J Virol.* 88: 8597–8614.

Jackson JA, Turner JD, Rentoul L, et al. (2004). Cytokine response profiles predict species-specific infection patterns in human GI nematodes. *Int J Parasitol.* 34: 1237–1244.

Jackson JA, Turner JD, Rentoul L, et al. (2006). T helper cell type 2 responsiveness predicts future susceptibility to gastrointestinal nematodes in humans. *J Infect Dis.* 190: 1804–1811. Seminar, www.thelancet.com: 367 May 6, 1531.

Johnston NW, Johnston SL, Duncan JM, Greene JM, Kebadze T, Keith PK (2005). The September epidemic of asthma exacerbations in children: A search for etiology. *J Allergy Clin Immunol.* 115: 132–138.

Johnston SL, Pattemore PK, Sanderson G, Smith S, Campbell MJ, Josephs LK (1996). The relationship between upper respiratory infections and hospital admissions for asthma: A time-trend analysis. *Am J Respir Crit Care Med.* 154: 654–660.

Jordan RE, Adab P, Cheng KK (2020). COVID-19: Risk factors for severed disease and death. *BMJ.* 368. doi:10.1136/bmj.m1198.

Jose RJ, Manuel A (2020). COVID-19 cytokine storm: The interplay between in flammation and coagulation. *Lancet Respir Med.* doi:10.1016/S2213-2600(20)30216-30222.

Jourdan PM, Lamberton PH, Fenwick A, et al. (2018). Soil-transmitted helminth infections. *Lancet.* 391: 252–265.

Khetsuriani N, Kazerouni NN, Erdman DD, Lu X, Redd SC, Anderson LJ (2007). Prevalence of viral respiratory tract infections in children with asthma. *J Allergy Clin Immunol.* 119: 314–321.

King CL, Medhat A, Malhotra I, et al. (1996). Cytokine control of parasite specific energy in human urinary schistosomiasis: IL-10 modulates lymphocyte reactivity. *J Immunol.* 156: 4715–4721.

Lai C-C, Shih T-P, Ko W-C, Tang H-J, Hsueh P-R (2020). Severe acute respiratory syndrome coronavirus 2 (SARS-CoV-2) and corona virus disease-2019 (COVID-19): The epidemic and the challenges. *Int J Antimicrob Agents.* 105924.

Li Z, Liu T, Yang N, et al. (2020). Neurological manifestations of patients with COVID-19: Potential routes of SARS-CoV-2 neuro invasion from the periphery to the brain. *Front Med.* doi:10.1007/s11684-020-0786-5.

Ludvigsson JF (2020). Systematic review of COVID-19 in children shows milder cases and a better prognosis than adults. *Acta Pediatr.* 109: 1088–1095.

Maizels RM, Balic A (2004). Resistance to helminth infection: The case for interleukin-5-dependent mechanisms. *J Infect Dis.* 190: 427–429.

Maizels RM, Balic A, Gomez-Escobar N, Nair M, Taylor MD, Allen JE (2004). Helminth parasites: Masters of regulation. *Immunol Rev.* 201: 89–116.

Maizels RM, McSorley HJ (2016). Regulation of the host immune system by helminth parasites. *J Allergy Clin Immunol.* 138: 666–675.

Orecchioni M, Ghosheh Y, Pramod AB, Ley K (2019). Macrophage polarization: Different gene signatures in M1(LPS+) vs. classically and M2(LPS—) vs. alternatively activated macrophages. *Front Immunol.* 10. doi:10.3389/fimmu.2019.01084.

Mitra AK and Mawson AR (2017). Neglected tropical diseases: Epidemiology and global burden. *Trop Med Infect Dis.* 2: 36.

Mizumoto K, Kagaya K, Zarebski A, Chowell G. (2019) Estimating the asymptomatic proportion of coronavirus disease (COVID-19) cases on board the Diamond Princess cruise ship, Yokohama, Japan. *Euro Surveill.* 25. doi:10.2807/1560–7917.

Muller R (2002). Worms and human disease. *CABI.* doi:10.1079/9780851995168.0252.

Ned S (2020). Covid-19: Why Germany's case fatality rate seems so low. *BMJ.* doi:10.1136/bmj.m1395.

Nishiura H, Kobayashi T., Miyama T, et al. (2020) Estimation of the asymptomatic ratio of novel coronavirus infections (COVID-19). *Int J Infect Dis.* 94: 154–155.

Nutman TB, Kumaraswami V, Ottesen EA (1987). Parasite-specific energy in human filariasis: Insights after analysis of parasite antigen-driven lymphokine production. *J Clin Invest.* 79: 1516–1523.

Ottesen EA, Weller PF, Heck L (1977). Specific cellular immune unresponsiveness in human filariasis. *Immunology.* 33: 413–421.

Pan L, Mu M, Yang P, et al. (2020). Clinical characteristics of COVID-19 patients with digestive symptoms in Hubei, China: A descriptive, cross-sectional multicenter study. *Am J Gastroenterol.* 115: 766–773.

Phipps S, Lam CE, Mahalingam S, Newhouse M, Ramirez R, and Rosenberg HF (2007). Eosinophils contribute to innate antiviral immunity and promote clearance of respiratory syncytial virus. *Blood.* 110: 1578–1586.

Popkin BM, Du S, Green WD, et al. (2020). Individuals with obesity and COVID-19: A global perspective on the epidemiology and biological relationships. *Obes Rev.* 21: e13128.

Quinnell RJ, Pritchard DI, Raiko A, Brown AP, Shaw MA (2004). Immune responses in human necatoriasis: Association between interleukin-5 responses and resistance to reinfection. *J Infect Dis.* 190: 430–438.

Roberts M, Butterworth AE, Kimani G, et al. (1993). Immunity after treatment of human schistosomiasis: Association between cellular responses and resistance to reinfection. *Infect Immun.* 61: 4984–4993.

Rodriguez C, Veciana C (2020). Asthma and Covid-19: The eosinophilic link. *Qeios.* doi:10.32388/5IY4IF.

Rolot M, *et al.* (2018). Helminth-induced IL-4 expands bystander memoryCD8 + T cells for early control of viral infection. *Nat Commun.* 9: 4516.

Santos JHA, Bührer-Sékula S, Melo GC, et al. (2019). *Ascaris lumbricoides* coinfection reduces tissue damage by decreasing IL-6 levels without altering clinical evolution of pulmonary tuberculosis or Th1/Th2/Th17 cytokine profile. *Rev Soc Bras Med Trop.* 52: e20190315.

Sartono E, Kruize YCM, Kurniawan A, et al. (1995). Elevated cellular immune responses and interferon-γ release after long-term diethylcarbamazine treatment of patients with human lymphatic filariasis. *J Infect Dis.* 171: 1683–1687.

Sarzi-Puttini P, Anitvalle MD (2020). How to handle patients with autoimmune rheumatic and inflammatory bowel diseases in the COVID-19 era: An expert opinion. *Autoimmun Rev.* doi:10.1016/j.autrev/2020.102574.

Shereen MA, Khan S, Kazmi A, Bashir N, Siddique R (2020). Covid-19 infection: Origin, transmission and characteristics of human coronaviruses. *J Adv Res.* 24: 91–98. doi:10.1016/jare.2020.03.005.

Simon AK, Hollander GA, McMichael A (2015). Evolution of the immune systemin humans from infancy to old age. *Proc R Soc B Biol Sci.* 282: 2014–3085.

Su YC, Townsend D, Herrero LJ, Zaid A, Rolph MS, Gahan ME (2015). Dual proinflammatory and antiviral properties of pulmonary eosinophils in respiratory syncytial virus vaccine-enhanced disease. *J Virol.* 89: 1564–1578.

Tseng CT, Sbrana E, Iwata-Yoshikawa N, Newman PC, Garron T, Atmar RL (2012). Immunization with SARS coronavirus vaccines leads to pulmonary immunopathology on challenge with the SARS virus. *PLoS One*. 7.

Turner JD, Faulkner H, Kamgno J, et al. (2003) Th2 cytokines are associated with reduced worm burdens in a human intestinal helminth infection. *J Infect Dis*. 188: 1768–1775.

World Health Organization (2020a). *Laboratory Testing for Coronavirus Disease 2019 (COVID-19) in Suspected Human Cases: Interim Guidance*. World Health Organization.

World Health Organization (2020b). *Multisystem Inflammatory Syndrome in Children and Adolescents Temporally Related to COVID-19*. www.who.int/news-room/commentaries/detail/multisystem-inflammatory-syndrome-in-children-and-adolescents-with-covid-19 (last accessed 6 September 2020).

Wrapp D, Wang N, Corbett KS, et al. (2020) Cryo-EM structure of the 2019-nCoV spike in the prefusion conformation. *Science*. 367: 1260–1263.

Wright JE, Werkman M, Dunn JC Anderson RM (2018) Current epidemiological evidence for predisposition to high or low intensity human helminth infection: A systematic review. *Parasites Vect.* 11(1). doi:10.1079/9780851995168.0252.

Wu D, Yang XO (2020). TH17 responses in cytokine storm of COVID-19: An emerging target of JAK2 inhibitor Fedratinib. *J Microbiol Immunol Infect*. 53: 368–370.

Yadav A, Kapoor N, Malhotra, SK (2020). Multivariate analysis to illustrate nCoV-19—Climatic temperature interactions in Northern India. In: *Proceedings of the International Conference on Contemporary Issues of Climate Change and Conservation of Biodiversity and Natural Resources in Himalayan Environments*, 18pp.

Yadav A, Kapoor N, Malviya, SJ, Malhotra, SK (2020). COVID-19 pandemic and the global vaccine strategy. *Sci Temper*. 11(1–2): 23–35. https://connectjournals.com/03960.2020.11.23.

Yadav A, Malviya SJ, Malhotra, SK (2021). The mystery virus, SARS-CoV-2 and its dynamics. In: *Microbased Revolution in Macrosystem: Transformation in Life Systems*. Springer Nature, 22pp.

Yang M, Kumar RK, Hansbro PM, Foster PS (2012). Emerging roles of pulmonary macrophages in driving the development of severe asthma. *J Leukocyte Biol*. 91(4): 557–569. doi:10.1189/jlb.0711357.

Yasui F, Kai C, Kitabatake M, Inoue S, Yoneda M, Yokochi S (2008). Prior immunization with severe acute respiratory syndrome (SARS)-associated coronavirus (SARS-CoV) nucleocapsid protein causes severe pneumonia in mice infected with SARS-CoV. *J Immunol*. 181: 6337–6348.

Yazdanbakhsh M, Kremsner P, van Ree R (2002). Allergy, parasites, and the hygiene hypothesis. *Science*. 296: 490–494.

Yazdanbakhsh M, van den Biggelaar A, Maizels RM (2001). Th2 responses without atopy: Immunoregulation in chronic helminth infections and reduced allergic disease. *Trends Immunol*. 22: 372–377.

Yin Y, Wunderink RG (2018). MERS, SARS and other coronaviruses as causes of pneumonia. *Respirology*. 23: 130–137.

Yoon J, Um H-N, Jang J, Bae Y-A, Park W-J, Kim HJ (2020). Corrigendum: Eosinophil activation by toll-like receptor 4 ligands regulates macrophage polarization. *Front Cell Dev Biol*. 8. doi:10.3389/fcell.2020.00003.

Zhang JJ, Dong X, Cao YY, Yuan YD, Yang YB, Yan YQ, et al. (2020). Clinical characteristics of 140 patients infected with SARS-CoV-2 in Wuhan, China. *Allergy*. doi:10.1111/all.14238.

31

Women as an Edifice of Nation-Building: A Futuristic Approach

Neeshma Jaiswal, Abhishek K. Gupta, Reshu, Shivani Singh, and Rashmi Srivastava

CONTENTS

31.1 Introduction

Before exploring how discoveries, research, and exploration in animal science can help in national development genuine question that pops in mind is, why is animal science important? It is the boon of animal science that keeps us clothed, helps us in rearing and culture of a number of species, and helps us in understanding humans better by a comparative analysis of humans with animals. Biotechnology and animal science, together, have been proven to be useful in the fields of medical testing, cloning, gene therapies, and so on. The spectrum of animal science, including how nutrition, physiology, animal behavior (ethology), genetics, and breeding synchronously play a significant role in sustainable development, is a popular topic of discussion. There are many avenues in the field of animal science that open doors for new futuristic approaches that engage all aspects of biology. It is very integral that determine how we harness the wide range of natural resources and with which manpower. Or is it manpower or *woman*-power? This is an issue that is highly debated though.

Equality or non-discrimination is a state where every individual gets equal opportunities and rights. Every individual in society yearns for equal status, opportunity, and rights. However, it is a general observation that discrimination exists among humans. Discrimination exists because of cultural differences, geographical differences, and gender. Inequality based on gender is a concern that is prevalent in the entire world. Even in the 21st century, across the globe, men and women do not enjoy equal

DOI: 10.1201/9781003220398-31

privileges. Gender equality means providing equal opportunities to both men and women in political, economic, educational, and health aspects.

31.2 Importance of Gender Equality

A nation can progress and attain greater development only when both men and women are entitled to equal opportunities. Women in society are often cornered and prevented from getting equal rights as men in terms of health, education, decision-making, and economic independence or wages. The current social structure that prevails does not allow women to get equal opportunities as men. Women are generally the caregivers in the family. Because of this, women are mostly involved in household activities. There is lesser participation of women in higher education, decision-making, and leadership roles. This gender disparity is a hindrance to the growth of a country. When women participate in the workforce, it increases the economic growth of the country. Gender equality increases the overall well-being of the nation, along with economic prosperity. Gender equality is an important factor in determining a country's overall growth. There are several indexes to measure gender equality:

1. The gender-related development index (GDI) is a gender-centric measure of the human development index and considers parameters like life expectancy, education, and income in assessing the gender equality of a country.
2. The gender empowerment measure (GEM) includes many detailed aspects like the proportion of seats that women candidates hold in a national parliament, the percentage of women in economic decision-making roles, and the income share of female employees.
3. The gender equity index (GEI) ranks countries on three parameters of gender inequality: education, economic participation, and empowerment. However, GEI ignores the health parameter.
4. The global gender gap index, introduced by the World Economic Forum in 2006, focuses more on identifying the level of female disadvantage. The four important areas that the index considers are economic participation and opportunity, educational attainment, political empowerment, health, and survival rate.

31.2.1 Gender Inequality in India

Per the World Economic Forum's gender gap ranking, India stands at rank 108 out of 149 countries. This rank is a major concern as it highlights the immense gap in opportunities in women with comparison to men. In Indian society, for a long time, social structure has been such that women are neglected in many areas like education, health, decision-making, and finance. Another major reason that contributes to discriminatory behavior toward women in India is the dowry system in marriage. Because of this dowry system, most Indian families consider girls as a burden. Preference for sons still prevails. Girls are barred from higher education. Women are not entitled to equal job opportunities and wages. In the 21st century, women are still the preferred gender for homemaking activities. Many women quit their jobs and opt out from leadership roles because of family commitments. However, such actions are very uncommon for men.

31.3 Veterinary Research and National Development by Human Welfare

> *Between animal and human medicine there is no dividing line—nor should there be. The object is different but the experience obtained constitutes the basis of all medicine.*
>
> —Rudolf Virchow (1821–1902)

Virchow's statement is as appropriate and wise today as it was centuries ago. The closeness of animals and humans leads to a common wide platform of research and findings wherein knowledge about one helps in findings of other. There have been considerable and numerous contributions of veterinary research, but its crucial role in public health and food safety has been brought into focus in past 20 years or so after the recognition that many emerging infectious diseases of humans are zoonotic. Such diseases as *Escherichia coli* O157:H7 infection, bovine spongiform encephalopathy (BSE), and avian influenza and the most recent coronavirus emphasized the importance of explorations and research in veterinary science in order to improve public health and food safety.

According to National Research Council (US) Committee on the National Needs for Research in Veterinary Science, veterinary research has evolved to address societal changes. Companion animals play a central role in the quality of life of an increasing proportion of the public; the beneficial psychosocial effects of the human-animal bond are widely accepted. Companion animals are also important sentinels for human disease and toxicant exposure, and companion-animal research improves our understanding of zoonotic diseases and how to address them; diagnostic and therapeutic data from companion animals can often be translated to human medicine (Science, Medicine, and Animals, 1991).). Because the health, well-being, and longevity of companion animals are a growing concern for a substantial portion of society, the demand for research on companion animal health and disease has increased; indeed, it is crucial for improving the health and welfare of these animals, which serve not only as companions but as aides, detectives, and soldiers. The three fields of veterinary science encompass research with domestic animals (such as livestock and poultry), wild animals (such as deer and exotic species), companion animals (such as dogs, cats, and horses), service animals (such as horses and dogs), and laboratory animals. Specifically, the future needs of research were assessed in the three fields of veterinary science, which are not limited to science conducted by veterinarians; they include science performed by professionals who have veterinary degrees and professionals who have various other degrees (National Research Council (US), 2005).

Surgical procedures, pain relievers, psychoactive drugs, medications for blood pressure, insulin, pacemakers, nutrition supplements, organ transplants, treatments for shock trauma, and blood diseases—all these have been developed and tested in animals before being used in humans. According to the American Medical Association, "Virtually every advance in medical science in the 20th century, from antibiotics and vaccines to antidepressant drugs and organ transplants, has been achieved either directly or indirectly through the use of animals in laboratory experiments."

31.3.1 Animal Science and Biotechnology

Animal biotechnology involves the use of science and engineering to modify or transform living organisms. It commonly involves creating transgenic animals, cloning, using gene knockout technology, and so on to increase growth rates, enhance lean muscle mass, enhance resistance to disease, or improve the use of dietary phosphorous to lessen the environmental impacts of animal manure. Transgenic poultry, swine, goats, and cattle that generate large quantities of human proteins in eggs, milk, blood, or urine also have been produced, with the goal of using these products as human pharmaceuticals. Human pharmaceutical proteins include enzymes, clotting factors, albumin, and antibodies. The major factor limiting the widespread use of transgenic animals in agricultural production systems is their relatively inefficient production rate—a success rate of less than 10% (aboutbioscience.org).

A specific example is the transfer of the growth hormone gene of rainbow trout directly into carp eggs, resulting in transgenic carp producing both carp and rainbow trout growth hormones and growing to be one-third larger than normal carp. Another example is the use of transgenic animals to clone large quantities of the gene responsible for a cattle growth hormone. The hormone extracted from the bacterium is purified and injected into dairy cows, increasing their milk production by 10 to 15%. That growth hormone is called bovine somatotropin, or BST. Another major application is the use of animal organs in humans. Pigs are currently used to supply heart valves for insertion into humans, but they also are being considered as a potential solution to the severe shortage of human organs available for transplant procedures.

31.3.2 Animal Science and Sustainability

Rejuvenating animal research is necessary to sustainably tackle and address the global challenge of food security. It has been anticipated that the global food demand from animal agriculture will nearly double in the next 30 years. This is mainly credited to the predicted increase in population to nine to ten billion people in 2050. The United Nations Food and Agriculture Organization (FAO) estimates that there will be a 73% increase in meat and egg consumption and a 58% increase in dairy consumption over 2011 levels worldwide by the year 2050 (Committee on Considerations for the Future of Animal Science Research; Science and Technology for Sustainability Program; Policy and Global Affairs; Board on Agriculture and Natural Resources; Division on Earth and Life Sciences; National Research Council). This will naturally put additional pressure on the availability of land, water, and energy for animal culture and crop production. Hence, animal research enterprise will significantly help to rectify the unequal distribution of animal food. Advances in animal agriculture have been a result of R&D and new technologies, particularly in areas such as food safety, genetics and breeding, reproductive efficiencies, nutrition, disease control, biotechnology, and the environment (Roberts, 2008). Major productivity gains in various species are also attributed to R&D in this field. Developments in reproductive technologies will continue to allow acceleration of genetic selection

31.3.3 Animal Science and Gender Balance

India is facing a major crisis due to such a high population growth rate. According to the livestock census released by the Department of Animal Husbandry and Dairying, the total livestock population is 535.78 million. Livestock species play a very important economic and social cultural role for the well-being of a large population, so there is a need for following conservation strategies to conserve livestock. Women play an important role in sustainable preservation of livestock populations in different ways.

Animal science is a field of science that offers the study of biology and management of animals that are under the control of humankind. Animal science provides multifunctional and future career opportunities, such as business, industry, government, education, and research work. In addition, exploration of animal science strikes out some important nation building steps and that can help to nurture skills useful for a career as an animal breeding program manager, molecular geneticist, national park and wildlife service officer, reproductive technologist, sustainable agriculturist, veterinarian, animal health and quarantine manager, animal nutritionist, production manager, biosecurity scientist, biotechnologist, medical researcher, microbiologist, federal meat inspector, sales and service personnel, government agency official, livestock buyer, marketing and public relation director, or quality control manager.

Animal science offers us with a good career opportunities but also reflects gender imbalance. The gender gap in animal science is still very wide, with females lagging behind; the representation of women in the disciplines of science is generally very low. Gender disparity imposes a devastating effect on the development of the nation and is affecting the economy of India. The development of a nation is a product of equal participation of both genders in every field. Gender parity is the only condition that will help us in achieving the goal of a developed nation with an optimum GDP growth rate. The government of India has made several efforts to bridge the gender gap in the field of applied sciences and many other fields of sciences by formulating a comprehensive national strategies for mainstreaming the female gender.

In many regions, the current use of biomass fuels requires women and children to spend many hours per week collecting and carrying traditional biomass that is burned in highly inefficient and polluting stoves. Historically, fertility and mortality rates have fallen with higher levels of education and economic opportunity for women and girls. On the gender equality front, although the number of women in science and engineering is growing at the global level, men still outnumber women, especially at the upper levels of those professions. Even in countries where girls and boys take math and science courses in roughly equal numbers, and about as many girls as boys graduate secondary school to pursue careers in science and engineering, fewer women than men actually do so. Actively promoting gender equality in the sciences has the potential to lead to substantial knowledge and social and economic gains.

31.4 Women's Involvement in Science

The undeterred importance of gender equality has been proclaimed, but still the world hasn't been the same for men and women. It is surprising that although females constitute nearly half a graduates, still there are less than 4% of women scientists who have managed to reach the leading research facilities. This meagerness of women in top positions creates a vicious cycle that starts from absence of women in policymaking debates and women's paucity in parliament, and the cascade follows. According to UNESCO's Institute of Statistics, less than one in three researchers in the world is a woman. This low representation can be observed from the start of higher education: things are balanced until the end of high school, but after that, only 32% of undergraduates are women. The higher the diploma, the lower this percentage. The gap gets even wider in the professional world: the higher you climb the academic ladder, the fewer the women. Thus, only 11% of the highest academic positions are held by women. But the most striking statistics come from the field of scientific awards: only 3% of Nobel Prize winners are women! And it is no better with the prestigious mathematics award, the Fields Medal: since its creation in 1936, the award was given to a woman for the first time in the year 2014. Also, most of the top positions in science and technology see very few and selective numbers of women candidates, despite the introduction of a lot of women-oriented programs from DST and other governing bodies.

31.5 Women Who Raised the Bar High

31.5.1 Jane Goodall

She is a British primatologist and ethologist widely considered to be the world's foremost expert on chimpanzees. And Goodall wasn't just working in a lab; she climbed trees and mimicked the behavior of chimps in Tanzania to gain their trust and study them in their natural habitat.

31.5.2 Annie Jump Cannon

Annie Jump Cannon was the first woman to earn a scientific doctorate awarded at Oxford University. An astronomer, she worked on classifying and cataloging stars, discovering five novae.

31.5.3 Barbara McClintock

Barbara McClintock made a number of groundbreaking discoveries in genetics. She demonstrated the phenomenon of chromosomal crossover, which increases genetic variation in species. She also discovered transposition—genes moving about within chromosomes, often described as jumping genes—and showed that genes are responsible for switching the physical traits of an organism on or off. Her discovery of transposition was rejected for years by other scientists, but eventually they realized she was right all along. At age 81, she was awarded the Nobel Prize in Physiology or Medicine.

31.5.4 Maria Merian

Before German-born naturalist and scientific illustrator Maria Merian began to study the life cycle of butterflies. She was also one of the first naturalists to observe insects directly, giving her remarkable insights into the way they really lived. Although she emerged as one of the leading entomologists of her day, since she wrote in German and not in Latin, the official language of science at the time, her remarkable discoveries about the metamorphosis of insects were ignored by many scientists. She also raised eyebrows by funding her own, unofficial expedition to Suriname where she described many new insects and plants; a highly unusual venture for a woman of the period to undertake. Even so, her impact on science is undeniable: many of her classifications are still valid today, and her exquisite paintings of plants, animals, and insects have been widely admired throughout the centuries.

31.5.5 Jennifer Doudna

Jennifer Doudna is one of the most culturally significant scientists studying today. She helped developed CRISPR, the genetic-engineering method that could allow for "designer babies" but also for the eradication or treatment of sickle cell anemia, cystic fibrosis, Huntington's disease, and HIV.

31.5.6 Francoise Barre-Sinoussi

French biologist Francoise Barre-Sinoussi helped identify HIV as the cause of AIDS. She shared the Nobel Prize in 2008 with her mentor, Luc Montagnier, for their discovery of the human immunodeficiency virus (HIV).

These respectful and much talented women have revolutionized explorations in science and prove it well that women can bring wonders to science, UNESCO, in its article "Women needs science and science needs women," says, "It is inconceivable that we should deprive ourselves of half humanity's talents," which clearly means women, if deprived of the field of science and research, will unfortunately lead to deprivation of half of the talent, brains, and hard work to science. Not only globally but equally at home, women scientists have always paved their way toward excellence in the field of science.

31.6 Technology and Women Empowerment

The urgent need for sustainable transformations requires strengthening the directionality of science on behalf of a mutually beneficial "moon landing" for humanity and the Earth. Researchers, science policymakers, and funding agencies can use the 2030 Agenda as a shared compass to increase the relevance and benefits of science and technology for the global community. Women's empowerment is essential for supporting transformations to sustainable development.

Technological innovation has long been recognized as crucial to achieving development. In another example, advances in gene-editing technologies, notably CRISPR (clustered regularly interspaced short palindromic repeats), can improve the prospects for gene therapy at the individual level with gains in productivity and control vector-borne diseases, such as malaria, and facilitate the precision breeding of plants and animals. Deploying advanced technologies like artificial intelligence could also play a major role in achieving the Sustainable Development Goals. Many such applications are under development but need careful assessment of potential broader consequences before deployment.

Furthermore, empowering women in science and technology represents another important measure. Thus, women should be supported and encouraged through education and career opportunities for sustainability science but also by building strong networks such as the International Network of Women Engineers and Scientists, a global network of organizations of women in science, technology, engineering, and mathematics. Enhancing opportunities for women and girls in animal science education can have enormous impacts on human well-being and across all the Sustainable Development Goals. Teaching institutions and learning technologies, as well as teachers, must have equal expectations of girls and use gender-balanced curricula that take into account girls' interests and provide hands-on learning opportunities and scholarship funds with public and private contributions.

Gender equality and the empowerment of all women and girls will make a crucial contribution to progress across all the nation-building goals and targets. Women and girls should enjoy equal access to quality education at all levels, healthcare services, economic and natural resources, and civil and political participation, as well as equal opportunities with men and boys for employment, leadership, and decision-making at all levels. Several schemes run by the government of India, like the Women Scientist Scheme-A (WOS-A), has completed a decade of support and encouragement through science and technology to women having break in their career and provide them opportunity to come back in mainstream of science by pursuing research in science and engineering. The scheme Science and Technology for Women is being implemented with the aim of promoting gender equality and empowering women with inputs of S&T to promote the development and adaptation of appropriate technology, transfer proven technologies, and demonstrate live technology models. Apart from the ongoing All India Coordinated Research Programs on Manufacturing of Low-Cost Sanitary Napkins, Biomass Tray Dryers, and so on, several projects in technology sectors, such as women's health and nutrition, agriculture, animal husbandry—including post-harvest technologies, modern agricultural practices, nursery techniques for cultivation of aromatic and medicinal plants, watershed-based soil and water conservation, tissue culture technology, alternate livelihoods, natural resource management, and occupational hazards—are also supported by government organizations to make a distinct impact in the socio-economic status of women.

Realizing the importance of ensuring quality and safety of the products in fisheries sector a project called Food Safety Interventions for Women in Fishery-Based Microenterprises has been taken up in coastal Kerala. This ensures safety of the products developed by women microenterprises, through skill development and capacity building, which lays importance on food safety and its actual requirement, to produce a quality and safe product under hygienic and sanitary conditions without bacterial contamination, to enable better income with quality produce. Capacity-building programs for women entrepreneurs, in coordination with Central Institute of Fisheries Technology, Kochi, and Matysa Fed, have been organized at various coastal locations of Kerala. In a project, "Extension of quality seed production techniques and fast grow-out of Indian and exotic carp in Terai region of West Bengal," being implemented in rural areas of Cooch Behar District, emphasis is being laid on dissemination of quality seed production techniques of carp through the creation of higher numbers of quality fish seed production centers and their proper grow-out culture, utilizing potential water bodies. There is a sustainable income generation for women through good quality seed production and grow-out by improved techniques of induced breeding of Indian and exotic carp, rearing of fish seed, using local feed components, including dried muga silkworm pupae, along with other source of income like muga silk thread reeling and muga farming activities. The spawn production and revenue status for a few villages in the target areas is given in the following table. A project, "Empowering rural women of Himachal Pradesh sustainable dairy entrepreneurship," was taken up to systematically ascertain the reproductive, productive, managerial, and physiological status of rural cattle in respect to the available resources of women farmers in different regions of Himachal Pradesh. The project aims to propagate and analyze the effect of scientific rearing, including breeding, housing, health, and management, in consonance with certain sustainable nutritional interventions in rural cattle involving farm women of Himachal Pradesh. This also generates geographical information system mapping and spatial information for development of region-specific packages and practices on scientific rearing of rural cattle in Himachal Pradesh.

Empowerment of Tribal Women Toward Livelihood Improvement Through Sustainable Sericulture and Associated Integrated Farming System (IFS) Practices envisaged an approach of the Lab to Land Program for capacity building and technical empowerment through various training programs in the state of Karnataka. This project, being carried out jointly by Karnataka State Sericulture Research and Development Institute and BAIF Institute for Rural Development, Tiptur, has bridged the technological gaps between lab and land toward the improvement in quality and enhancement of productivity and crop stability involving the tribal community.

31.7 Conclusion

For overall well-being and growth of a nation, scoring high on gender equality is the most crucial aspect. Countries with less disparity in gender equality have progressed a lot. The government of India has also started taking steps to ensure gender equality. Several laws and policies are prepared to encourage girls. The Beti Bachao, Beti Padhao Yojana (save girl, and make girls educated) campaign is created to spread awareness of the importance of the girl child. Several laws to protect girls are also there. However, we need more awareness to spread the knowledge of women rights. In addition, the government should take initiatives to check the correct and proper implementation of policies.

REFERENCES

Committee on Considerations for the Future of Animal Science Research (2015). *Science and Technology for Sustainability Program: Policy and Global Affairs*. Board on Agriculture and Natural Resources,

Division on Earth and Life Sciences. Washington, DC: National Research Council. National Academies Press (US)

National Research Council (US) (2005). *Committee on the National Needs for Research in Veterinary Science.* Washington, DC: National Academies Press, 2005.

Science, Medicine, and Animals (1991). *Prepared for the Councils of the National Academy of Sciences and the Institute of Medicine.* Committee on the Use of Animals in Research. Washington, DC: National Academy Press.

Roberts TL (2008). Improving nutrient use efficiency. *Turk J Agric For*, *32*(3): 177–182.

www.aboutbioscience.org/topics/animal-biotechnology/

www.amightygirl.com/blog?p=11511

32

New Dimensions of Self-Employment for Sustainability

Ranbir Chander Sobti, Ram Kumar, and Anudeep

CONTENTS

32.1 Introduction

Sustainable development is a global agenda that envisages action to protect the planet, improve the lives and prospects of everyone, everywhere, and end poverty. The issue is so important that all UN member states in 2015 set out a 15-year plan to achieve this by 2030. With a decade left, there is an urgent need today to usher in an ambitious plan so that the target can be achieved.

A three-level action plan has been proposed by the UN involving (1) *global action* to include steps to be taken by international agencies, (2) *local action* based on the specific needs of communities and countries, and (3) *people action* involving youth, society, institutes, and other clients. It is an accepted fact that the central dogma of sustainability is tackling growing poverty, empowering women and girls, and resolving the climate emergency by rebalancing the relationship between people and nature.

Agriculture is the mainstay of Indian economy. If sustainability is the goal, then the way lies through poverty eradication, employment generation, and human resource development through agri-business. As technology base is poor, the small and marginal farmers are engaged in low-productive activities, turning out primary products without adding value to them. Every year, the produce of the small farmers is sold at very low prices due to lack of post-harvest technologies and value-added techniques. The introduction and application of appropriate technologies can get a higher value for produce like fruits, vegetables, honey, milk, and meat by enhancing the market value. The development and application of such agro-based technologies have the potential to generate employment by creating value-added jobs and creating and maintaining infrastructure and organized marketing.

DOI: 10.1201/9781003220398-32

Agri-business contributes significantly to national employment through self-employment opportunities. Some of the self-employment avenues are in the form of micro and small enterprises. These can help in bringing science and technology knowledge inputs to the average illiterate people and low-skilled farmers and provide the benefits of scale to the small and marginal farmers. Some of the activities that hold promise are as follows:

- Micro-propagation through plant tissue culture labs
- Setting up vermiculture, biofertilizers, biopesticides, and biocontrol units
- Setting up beekeeping units
- Providing extension consultancy services
- Facilitating agricultural insurance services
- Setting up aquaculture units
- Setting up value-addition units
- Setting up retail marketing outlets for agri-products

32.2 Major Technologies' Present Scenario and Scope of Apiculture in Combating Employment Issues

The world is facing something unreal and terrifying. COVID-19 has made the mighty kneel, halting the world's progress. The ongoing lockdown has brought economic status to a standstill. Our minds are racing back and forth, longing for a return to normal lives. In the midst of this terrible despair, the mind wanders to explore such employment modes, which can provide economic stability. In the absence of white-collar jobs and skill-based employment, the best option is to adopt available self-employment avenues involving little inputs and providing direct financial support. The pandemic is a perfect opportunity to channel the energy and abilities of the unemployed people. Beekeeping is one such opportunity that can lead to enhancement of income, along with employment generation. The need of the hour is to focus on the practical aspects in favor of the unemployed migrant labor and rural youth in order to make them successful entrepreneurs in the science, art, and skill of beekeeping.

Beekeeping is an ideal agro-based industry providing income to a group of people from a wide sector, including horticulturists, agriculturists, tenant farmers, landless people, and farm laborers who have no access to land for farming for themselves. Besides direct benefits of honey and other bee products, the indirect benefits through pollination also emphasize the importance of apiculture industry. The significance of beekeeping lies in its potential for contributing toward the country's foreign exchange earnings and reducing poverty through job creation (SNV, 2010). Besides providing beneficial products of the honeybee to society, the sector provides self-employment and income to the people involved in the occupation chain (Kumar, 1999; Hilmiet al.,2012; Teferi et al., 2011; Singh et al., 2016; Mickels-Kokwe, 2006; Moniruzzaman and Rahman, 2009), keeping them out of poverty (Figure 32.1). There are four species of honeybees available in India. The giant honeybee (*Apis dorsata*), the dwarf honeybee (*A. florea*), the Asian honeybee (*A. cerana*), and the Italian honeybee (*A. mellifera*). Though *A. dorsata* provides the bulk of market honey, it cannot be used in beekeeping (Anudeep and Kumar, 2012a, b)

Beekeeping as a self-employment opportunity has some unique benefits to offer. It does not require ownership of land. It can be taken up at any level of operation: small, medium, or large scale. The cost of infrastructure is low and can be further reduced by the adoption of eco-friendly low-cost technologies. The know-how and procedure are simple and can be carried out by men, women, or children, by the educated or the illiterate. It opens avenues of income generation for local craftsmen, such as carpenters, blacksmiths, salesmen and, painters, also creating more jobs in these areas. Beekeeping can serve to diversify the economy and earn respectable living thus contributing to elevating the social status of the needy (Figures 32.2 and 32.3).

FIGURE 32.1 A flourishing hill apiary.

FIGURE 32.2 Women taking training in beekeeping.

FIGURE 32.3 Women engaged in beekeeping.

FIGURE 32.4 Langstroth movable frame hive.

The systematic, scientific keeping and rearing of bees for obtaining the benefits of harvesting bee products and the important service of pollination is known as apiculture. It is the science, art, and skill of keeping honeybees for the benefit of man. Presently, two species of honeybees are used in apiculture in India: the native honeybee, *Apis cerana* (Fabr.), and the exotic, European honeybee, *A. mellifera*

(Linn.). Traditional beekeeping with both these species has been practiced in several parts of the world, using structures such as the skeps, basket hives, wall hives, and log hives. Modern beekeeping has been made possible by the discovery of bee space by L. L. Langstroth, who designed the commercially used Langstroth movable frame hive (Figure 32.4). This is now the basis of a flourishing bee industry the world over (Kumar, 1999).

The native honeybee, *Apis cerana*, is an important genetic resource of the Asian continent. It is an efficient pollinator of our agriculture, horticulture, and forest ecosystems and resource base of local bee industry. In order to conserve this species a technology of low-cost hives, which are ecologically viable, locally acceptable, and eco-friendly has been developed (Verma et al., 1996). Eco-friendly material,

FIGURE 32.5 Beekeeper observing a hive.

FIGURE 32.6 Demonstrating a movable bee frame.

available free on the farmers field, such as rice straw, dried wild grass, and mud, has been used in the development of modern hives for the indigenous honeybee, *Apis cerana*. The technology has found direct field application. Beekeepers have accepted the alternative hiving structures with low-cost inputs, giving them increased economic return from the market.

One such structure is the modernized wall hive, which hybridizes the modern movable frame concept with the traditional low-cost hiving structure (Kumar and Kumar, 1997). This is described here as it is easy to construct in rural areas, requires no additional space or infrastructure, can be maintained close to home, and is ideal for migrant laborers returning to their villages because of the pandemic.

Traditional Wall Hive

The honeybee's habit of nesting in the dark enabled man to keep it in indigenously designed hives, such as clay pots, skeps, boxes, logs, logs inserted in wall, wall recesses, and ultimately wooden hives. One successful design for culturing *Apis cerana*, which has been carried through generations and has become an ancestral heritage of villagers in almost all parts of Asia, is the wall hive (Figures 32.7 and 32.8)

Description of a Wall Hive

A wall hive is a cavity as deep as the thickness of the wall. There is considerable variability in the size of wall hive. Data recorded on 1,200 wall hives located in farmhouses in Hindu Kush Himalayas in Himachal Pradesh, comprising the famous hill ranges of Kasauli, Kandaghat, Chail, Shimla, and Kullu, with altitudes ranging from 1,450 to 2,800 m above mean sea level, revealed the average size of wall hive as 37 cm deep × 33 cm wide × 40 cm high, giving the cavity volume of 48,840 cm^3 (L.R. Verma et al., 1997).

This cavity is left deliberately in the wall when the house is being constructed, and it is made at a height of about 150 cm from the floor in kitchens, living rooms, stores, and sheds. It has an entrance hole for honeybees on the outside of the wall and is covered from the inside usually with a plank of wood plastered with mud. If the position of the wall hive is suitable, a swarm of *A. cerana* can come

FIGURE 32.7 A traditional wall hive (outside).

FIGURE 32.8 A traditional wall hive (inside).

and settle, making its hive of parallel combs. Villagers, farmers, and orchardists in many Himalayan villages practice this rich tradition of maintaining this indigenous honeybee in wall hives mainly for honey production.

Drawbacks of a Traditional Wall Hive

With respect to modern scientific methods of beekeeping and management, the traditional wall hive, described earlier, suffers from some drawbacks:

- There is no provision for maintaining cleanliness inside the cavity. Consequently, there is high incidence of scavengers and robbers, such as pseudoscorpions, beetles, ants, of pollen, nectar, and brood, hindering normal colony activity.
- The hive cannot be manipulated for checking the presence and performance of the queen, and the symptoms of queen replacement cannot be determined.
- The traditional wall hive does not provide situations for observing the precursors of supersedure, swarming, and absconding.
- There is a total lack of a facility for providing appropriate colony management concomitant with the changing environmental conditions.
- The traditional method of honey extraction damages the comb, brood, and bees, resulting in the total destruction of the colony.
- The honey extracted is unripe and unhygienic.

In order to be able to scientifically and economically manage the honeybees in the wall hive and to obtain their services for the pollination of our desired crop at a particular time and at a particular place, there is increasing need to improve the traditional wall hive by providing movable comb frames. The introduction of movable comb frames into the wall hive will facilitate colony management, manipulation for pollination, and better and hygienic honey production. No experiments and attempts have earlier been made toward improving or modernizing the wall hive. Singh (1962) and Ahmad (1992), however, mentioned the scope of exploring such possibility.

Improvement of a Traditional Wall Hive

The improved technology provides a workable strategy for easy adoption by the farmer. It overcomes the drawbacks of the traditional hive, simultaneously providing ease of operations for appropriate colony management (Figure 32.9)

The technical manipulations involved in improving the wall hive are outlined in the following:

1. The wall hive is opened from the rear side by removing mud-plastered board covering the back of the hive.
2. The dimensions of the wall hive recess are recorded. Frames of requisite size with proper inter-comb distance and the bee space are made on the spot.
3. Bees are smoked. Combs are cut one by one with the help of a sharp-edged knife, like the one used for uncapping the honeycombs.
4. Honeycombs are separated. Combs with brood, pollen, and honey are cut in such a way as to separate the upper sealed honey from the lower brood area. The brood comb is immediately mounted on to the frame and is supported with wires. The wall recess is thoroughly cleaned to remove pseudoscorpions, mites, wax moths, beetles, ants, and other scavengers. Then two supports are inserted and fixed one each in the upper two opposing sides of the wall recess for resting comb frames. The comb frames are introduced onto the supports one by one and the bees are settled on the combs. Care is taken to complete all the process in as short duration as possible to prevent chilling and loss of brood.
5. Thick sugar syrup (one kilogram of sugar dissolved in one liter of water) is fed to bees to help them to recover from the shock.
6. The wooden cover is replaced, and the wall hive closed by plastering with mud.

Advantages of an Improved Wall Hive

The use of improved wall hives renders the following advantages:

- The modernization process causes very little burden on the beekeeper's pocket. The entire improved structure involves an expenditure of Rs. 100–200 for wooden frames, nails, wire, and labor per wall hive.
- It offers situations for scientifically managing the honeybee colony.
- The movable comb frames can be taken out and observed for colony health, bee strength, and food storage.
- The hive interior can be cleaned to keep off the wax moth, scavengers, and robbers.
- The colony can be checked for disease and thus managed appropriately.
- In times of dearth of forage and during winters, the colony can be managed by providing sugar feeding.
- It facilitates checking for the presence of the queen, replacement of nonproliferating queen, presence of additional queen cells, and swarming and absconding conditions.
- It provides additional combs and comb foundation sheets for increasing space during strength built up of colony and honey flow.
- There is segregation of honeycombs and efficient extraction of hygienic honey without damaging the comb, brood, and bees.

FIGURE 32.9 A traditional and a modernized wall hive.

- It provides for division and multiplication of colonies.
- It promotes easy management for migration for pollination.

Beekeeping in a modernized wall hive can give additional benefits to the amateur beekeeper by engaging in selective breeding of honeybees and royal jelly production. This involves selection of better-performing honeybee varieties on the basis of characteristics, such as gentle temperament, efficient nectar/pollen gathering, frugality, brood rearing, and disease resistance parameters, which will help in stock improvement (Kumar and Kumar, 1993; Kumar and Kumar, 1998; J.S. Verma et al., 1997; Verma et al., 2000)

32.3 Industrial Linkage

- Honey refining, processing, and packaging units directly contact progressive beekeepers using the aforementioned techniques for procurement of honey.
- Other products, such as royal jelly, bee pollen, and propolis are finding novel applications in medicine (Kaur et al., 2020) and are purchased by pharmaceutical industries. Beeswax is used in polishes, crayon making, and candle making.
- Artisans find job opportunities through making of low-cost hives, modern hives, and beekeeping equipment for beekeepers.

32.4 Government Initiative: Apiculture Scaling-Up Program for Income and Rural Employment (ASPIRE)

It is a nationwide program that started in 2013. Its objectives are as follows:

- Provide training to beekeepers, focusing on honey-harvesting methods, to ensure quality (separating ripe honey from combs with unripe honey, pollen, and brood; using honey extractors and clean containers)
- Hold awareness raising campaigns with beekeepers, honey traders, and marketers, informing and educating about the demands of the changing honey and beeswax markets
- Provide basic harvesting, processing, and storage equipment to beekeepers and honey traders

32.5 Apiculture and Beekeeping Organizations in India

Institutions contributing to beekeeping are as follows:

- All India Beekeepers Association
- Agricultural Products Export Development Authority (APEDA), Ministry of Commerce and Industry
- Tribal Cooperative Marketing Development Federation of India (TRIFED)
- Central Bee Research and Training Institute (CBRTI)
- Khadi and Village Industries Commission (KVIC)
- Export Inspection Council (EIC), Ministry of Commerce
- National Bee Board (NBB)

Success Story

A recent report by a correspondent from Lakhimpur gives a living and encouraging example of the prospects of apiculture for rural poor in the times of pandemic and is cited here to validate the claims made in the previous text.

The mass exodus of migrant labor back to their homes necessitated by the pandemic has led to increased unemployment and has become a problem in the state. Under such circumstances, some youths living at Panigaon, under the Dhakuakhana subdivision of Lakhimpur District, have attempted to engage themselves through apiculture. Lakhyajit Koch, Papu Dutta, and Parash Koch started their venture during the initial days of the nationwide lockdown to check the spread of COVID-19. The youths said that they caught a total of 25 swarms of bees from the neighboring jungles in the first week of the lockdown. By adopting special skills, they started rearing the bees in the residential complex of Lakhyajit Koch. For this purpose, they managed to get some wooden boxes so that the bees could build their hives in them. However, though the youths could tame 20 swarms of bees, they could not tame five swarms of bees that left their hives for unknown reasons. According to the youths, lack of adequate food provision might be the main reason why those bees left.

The 20 swarms of the bees have built their hives in the boxes and have also started to produce honey. The enterprising youths expect to get four to five liters of honey from each hive twice in a month. This is the best success story of the importance of small-scale agri-business of apiculture in income generation (by a correspondent, Lakhimpur, Sentinel Digital Desk, July 5, 2020, 9:41 a.m.).

32.6 Summary

Apiculture holds definite promise in creating job opportunities for the rural poor and the unemployed. The present pandemic situation serves as an eye opener for the innumerable migrant laborers returning to their homes to meager resources and several mouths to feed. The technology is a low-cost, very simple, effective, convenient method of harvesting natural resources that would otherwise go to waste, saving energy, and rebuilding the environment, with gainful employment for intending honeybee breeders. In India, the availability of sufficient floral resources, due to its diverse seasons, provides enriching experience for beekeepers. Value-added organic bee farming opens new avenues of income generation for farmers so that apiculture can very well be included as a component of a sustainable lifestyle.

REFERENCES

Ahmad, R. (1992). Present status of beekeeping in Pakistan. In: L.R. Verma (ed.), Honeybees in Mountain Agriculture. New Delhi: Oxford IBH Publishing, pp. 274–279.

Anudeep and Kumar, N.R. (2012a). Biodiversity in *Apis dorsata* F.: SEM comparison of antenna from two different populations. *Geobios*, 39(1): 75–79.

Anudeep and Kumar, N.R. (2012b). Biodiversity of *Apis dorsata* F.: Ultrastructure of the tongue. *Geobios*, 39(4): 251–253.

Hilmi, M., Bradbear, N. and Mejia, D. (2012). *Bee Keeping and Sustainable Livelihoods: Diversification*. Booklet number 1, 2nd edition. Rome: Rural infrastructure and Agro-Industries Division Food and Agriculture Organization of the United Nations. ISBN: 978-92-5-107062-8.

Kaur, R., Kumar, N.R. and Harjai, K. (2020). Therapeutic potential of honey bee collected pollen and bee bread of *Brassica campestris:* A target against oxidative stress induced by *Salmonella typhimurium* in mice. *Indian Journal of Entomology*, 82(2): 217–222.

Kumar, N.R. and Kumar, R. (1993). Successful management of a laying worker colony. *American Bee Journal*, 137(9): 647–649.

Kumar, N.R. and Kumar, R. (1998). Evaluation of honey hoarding quality in the Asian honey bee populations. *Pest Management & Economic Zoology*, 6(2): 101–103.

Kumar, R. (1999). Bees and beekeeping. In: R.C. Sobti and J.S. Yadav (eds.), Some Aspects on the Insight of Insect Biology. New Delhi: Narendra Publishing House.

Kumar, R. and Kumar, N.R. (1997). Traditional hives of *Api scerana* Fabr. *Insect Environment*, 3(1): 9–11.

Mickels-Kokwe, G. (2006). *Small-scale Woodland-Based Enterprises with Outstanding Potential: The Case of Honey in Zambia*. Center for International Forestry Research (CIFOR). www.cifor.org/publications/pdf_files/books/bkokwe0601.pdf. Accessed 15 January 2018.

Moniruzzaman, M. and Rahman, M.S. (2009). Prospects of beekeeping in Bangladesh. *Journal of Bangladesh Agriculture University*, 7(1): 109–116.

Sentinel Digital Desk. (2020, 5 July, 9:41 AM). www.sentinelassam.com/north-east-india-news/assam-news/apiculture-as-a-means-of-self-employment-dhakuakhana-youths-show-the-way-486754

Singh, G., Tiwari, D. and Yadav, S.P. (2016). Income enhancement and employment generation through apiculture enterprise for rural youth in Punjab. *Indian Research Journal of Extension Education*, 16: 112–115.

Singh, S. (1962). *Beekeeping in India*. New Delhi: Indian Council of Agricultural Research. 214pp.

SNV (The Netherlands Development Organisation). (2010). *The Competitiveness of the Zambia Apiculture Industry*. Honey Value Chain Development, SNV Capacity Development Services for Apiculture in Zambia, SNV.

Teferi, M., Yirga, G., Tewelde, T. and Amare, S. (2011). Prospects of beekeeping in the Northern Ethiopian highlands. *Scientific Research and Essays*, 6(29): 6039–6043.

Verma, J.S., Kumar, N.R. and Kumar, R. (1997). Brood rearing activities of *Apis cerana* Fabr populations: Effect of seasonal variation. *Pest Management & Economic Zoology*, 5(1): 21–24.

Verma, J.S., Kumar, N.R. and Kumar, R. (2000). Studies on the correlation between economic and biological performance of *Apis cerana* Fabr. populations. *Indian Bee Journal*, 62(3–4): 13–17.

Verma, L.R., Kumar, N.R. and Kumar, R. (1996). *Development of Ecofriendly, Low Cost Hives for Native Honey Bee ApisceranaFabr*. Technical Bulletin. Solan: Dr. YSP University of Horticulture and Forestry, 7pp.

Verma, L.R., Kumar, R. and Kumar, N.R. (1997). *Modernization of Traditional Wall Hive of Native Honey Bee Apis cerana*. Technical Bulletin. Solan: Dr. YSP University of Horticulture and Forestry, 11pp.

Index

Note: Numbers in **bold** indicate a table. Numbers in *italics* indicate a figure.

B

C

E

F

H

J

M

O

Q

R

T

U

V

Z